UNDERSTANDING
GENETICS

UNDERSTANDING GENETICS

Fourth Edition

NORMAN V. ROTHWELL

Long Island University, The Brooklyn Center

New York   Oxford
OXFORD UNIVERSITY PRESS
1988

Oxford University Press

Oxford   New York   Toronto
Delhi   Bombay   Calcutta   Madras   Karachi
Petaling Jaya   Singapore   Hong Kong   Tokyo
Nairobi   Dar es Salaam   Cape Town
Melbourne   Auckland

and associated companies in
Beirut   Berlin   Ibadan   Nicosia

Published by Oxford University Press, Inc.,
200 Madison Avenue, New York, New York 10016

Library of Congress Cataloging-in-Publication Data
Rothwell, Norman V.
  Understanding genetics.
  Includes bibliographies and index.
  1. Genetics.   I. Title.   [DNLM: 1. Genetics. QH 430 R848u]
QH430.R67 1988   575.1   87-7737
ISBN 0-19-505108-4

9 8 7 6 5 4 3 2 1

Printed in the United States of America
on acid-free paper

To Florence Miller

PREFACE

This fourth edition of *Understanding Genetics,* like the previous ones, presupposes only a passing familiarity with basic genetic principles and little contact with molecular biology. During thirty-one years of teaching, I have tried several approaches in my undergraduate genetics course, and I remain convinced that the best is the one which assumes little more than a nodding acquaintance with the language of genetics. Biology majors often admit a complete lack of understanding of various elementary concepts such as linkage and crossing over. Many critical topics can be slighted as the student is plunged into the details of sophisticated genetic research. Even graduate students who are familiar with details of the chemistry of genetics sometimes have no inkling of the biological significance of such information. It seems to me that a clear, concise approach encompassing the details of classical genetics best enables students to appreciate the excitement of today's research. Moreover, the strides in human genetics that present hope and challenge to today's society demand a firm grasp of basic genetic principles along with the elements of probability and statistics.

Therefore, the sequence of topics in this book starts from the "beginning" and leads to some of the latest advances reported in the literature. In the development of basic topics, elementary or historical points are associated wherever possible with modern ideas and current problems.

The author of any text in a science as dynamic as genetics is faced with the problem of selecting those topics which present a broad picture of the field. Inclusion or expansion of certain subjects at the expense of others becomes necessary in order to prevent the work from growing to a cumbersome size. I have tried to select and balance those topics which I believe can present to the average undergraduate a good picture of genetics from today's perspective. Many of the chapters in the previous editions have been rewritten, reorganized, or expanded to include discussions of advances taking place at the time of writing. Most of the essential material on allelism has now been incorporated into other chapters to provide an additional chapter for introduction and discussion of such topics as gene transfer, oncogenes, RFLP's, and others. The two chapters on population genetics in the third edition have now been consolidated into one without eliminating any essential material. The questions at the end of each chapter, many new to the fourth edition, are intended as a study guide to allow students to evaluate their grasp of the contents of each chapter.

Understanding Genetics can be used in a one-semester course or as a guide for two semesters, depending on the specific goals of the instructor and the preparation of the students. The following would appear to be essential to a one-semester presentation: Chapters 1 through 8, selected parts of Chapter 10, and Chapters 11 through 14. For students with more preparation, the course could begin with Chapter 10, although I recommend a review of Chapter 8, since the fundamentals of linkage and crossing over are so critical to genetic analysis.

While this book has been written to serve students in their undergraduate course in genetics, I feel that the treatment of most topics also can be of use to more advanced students. Although I do not consider this work to be a technical reference, I am confident that it can aid both the student and teacher who want to clarify many genetic points.

It has been a pleasure to work with the professionals at Oxford University Press whose efforts were essential to completion of this edition. Special gratitude is due my editor, William Curtis, whose constant encouragement and suggestions improved the original manuscript, hastened completion of the project, and made the entire undertaking a gratifying experience.

Brooklyn, N. Y.   N. V. R.
July, 1987

CONTENTS

6   PROBABILITY AND ITS APPLICATION   143

7   CONTINUOUS VARIATION AND ITS ANALYSIS   160

14 GENE MUTATION   377

15 BACTERIAL GENETIC SYSTEMS   406

16 VIRUSES AND THEIR GENETIC SYSTEMS   440

17 CONTROL MECHANISMS AND DIFFERENTIATION 474

18 THE IMPACT OF GENETIC MANIPULATION 519

UNDERSTANDING GENETICS

1

FOUNDATIONS OF GENETICS

Achievements before 1900

Before 1900, the history of the science of heredity was closely interrelated with that of cytology, the discipline concerned with cellular structure and function. The origin of both sciences can be traced to the discovery of the cell. This landmark is credited to Robert Hooke, who in 1665 described cellular entities in sections of cork. In the following decade, Leeuwenhoek described many interesting cell types, recognizing many free-living forms in addition to those cells which are associated to form tissues. However, more than a hundred years were to pass before any other significant observations would be made. We can appreciate the main reason for this standstill when we consider that no notable advances in the development of optical tools were made before the nineteenth century. Very little could actually be seen inside the cells. As a result, cell walls and cell boundaries were stressed rather than contents of the cell. Distinguishing the subcellular components was essential to establishing the location of the genetic material.

In 1831, Brown called attention to a body within the cell that he recognized as a regular, constant cellular element. This structure was the nucleus. A few years later (1838–1840), another important contribution was made, the proposal of the "cell theory" by Schleiden and Schwann. This concept is considered by many to be the most significant, all-encompassing generalization in biology: all living things are composed of one cell or more and their products. Schleiden and Schwann were not the first to recognize the cellular nature of organisms; however, they were able to present the idea very convincingly by compiling their own observations with those of others. Once the cell was accepted as the unit of life, several biological disciplines emerged. In addition to cytology, such fields as physiology, embryology, and pathology underwent rapid advances.

Even though cytologists were now concentrating on cellular components, not too much was seen before 1850. This resulted from the still limited optical instruments and the lack of appropriate stains and fixatives. Most of the important improvements in this area occurred in Germany after a government subsidy of the dye industry. Development of the aniline dyes in the 1870s led to better staining methods. About this time, Abbé developed the condenser and the oil immersion lens. Aided by these advances, microscopists were able to see and describe in some detail the nucleus and changes associated with it.

At this time a large number of highly significant descriptions of the chromosome and its behavior suddenly appeared. Several investigators noticed the presence of threads in the nucleus of the cell and reported that these threads became split. Flemming, a zoologist, reported that the halves of the split threads separate, and in 1882, he named the process *mitosis*. The botanist Strasburger was the first to give a complete description of mitotic events. During this period, some biologists were paying special attention to the germ cells. By 1879, fertilization had been recognized in plants by Strasburger and in animals by Hertwig and Fol.

The observations made on mitosis eventually led to the establishment of the "chromosome theory of heredity." In 1884–1885, four different investigators (Hertwig, Kolliker, Weismann (all zoologists), and the botanist Strasburger) concluded independently that the physical basis of inheritance is in the nucleus and the hereditary material is in the substance that makes up the chromosomes. To these workers as well as others, it was apparent that the separation of nuclear threads at mitosis is a very accurate procedure. In contrast to this, other portions of the cell are not so precisely distributed at cell division. Moreover, the cells uniting at fertilization, sperm and egg, differ immensely in size, primarily because of nonnuclear material. So, although cytoplasm can vary greatly in amount, the nuclei of gametes are quite similar. The larger female gamete does not contribute more nuclear material than the male. Inasmuch as heredity appeared to be equal from both parents, it seemed logical to assume that the hereditary material was in the nucleus, carried in the chromosomes. Although the chromosomes seemed to disappear in the "resting phase," they were still somehow carried along from fertilization on and distributed accurately from one cell to the next when the nucleus divided.

In the 1880s, therefore, suggestions were strong that the chromosomes are the carriers of hereditary material, because only they are quantitatively divided. (The word *chromosome* was coined in 1888 by Waldeyer.) However, there was still no real proof of the chromosome theory. Chromosomes can be seen only when the cell is dividing; therefore, it was essential to obtain evidence that they persist throughout all stages of the cell cycle, even when they are not visible. Also needed was some evidence to correlate chromosome behavior with the inheritance pattern of specific traits. Such confirmation was to be presented in the next few decades through the work of several outstanding scientists.

Of the four original proponents of the chromosome

theory, Weismann was the one who recognized its additional implications. He reasoned that if the chromosome theory were correct, then the germ cells of one generation must supply the chromosomes for both the body cells and the germ cells of the next generation. These germ cells, in turn, give rise to still other body cells and germ cells (Fig. 1-1A). From such reasoning, Weismann formulated the idea of the "continuity of the germ plasm." According to this concept, it is the germ cells that connect one generation to the next. Although they provide the body cells as well as other germ cells with chromosomes, the body cells (or somatoplasm) do not form a continuum from one generation to the next; they are a "dead end," so to speak. Any hereditary effect, therefore, must come through the germ plasm, *not* the body cells. Such an idea argued against acquired characteristics, a theory proposed in 1809 by Lamarck that has been popular even in the twentieth century. According to this concept, a change in the body, resulting from an environmental influence during an individual's lifetime, can be passed on to the next generation. But according to the theory of the continuity of the germ plasm, this would not be so, because body cells of one generation do not bridge the gap from one generation to the next. Any inherited change affecting the body must come through the germ cells of the previous generation. To support his idea, Weismann cut off the tails of mice for over 20 generations and demonstrated that the practice had no effect in altering the tail length at the end of the study, even though gametes came from tailless mice (Fig. 1-1B).

Furthermore, Weismann realized that if there is continuity of the germ plasm, some mechanism must exist to prevent the doubling of the chromosome number in each new generation over that in the previous one. The body cells cannot have the same chromosome number as the germ cells because each new generation would have twice the chromosome number of the parents. Weismann proposed that a reduction division must occur, a device that would reduce the number of chromosomes in the germ cells to half that of the body cells (Fig. 1-1C). Although Weismann never found cytological evidence to support his idea of a reduction division, his suggestions directed the attention of other workers to this problem. Cytologists then began to observe the process known today as meiosis. The details of this division, which is so critical to all sexually reproducing life forms, were not worked out until after 1900.

The turn of the century marks a very important

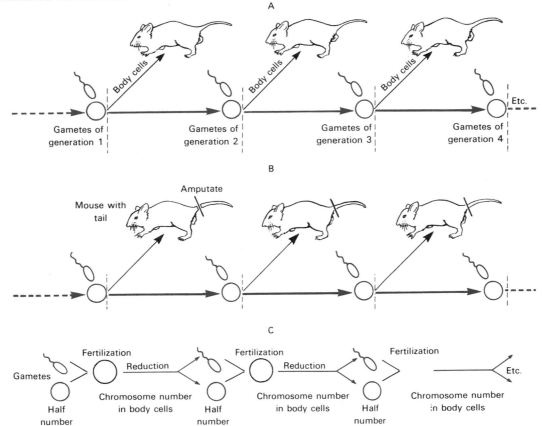

FIG. 1-1. Weismann's concepts. (*A*) The germ plasm (gametes) forms a continuous line connecting one generation to the next; the somatoplasm (body cells) does not link one generation to the other; thus, changes in it cannot influence the next generation. (*B*) Amputation of the tails of mice *before* mating, generation after generation, does not influence the length of the tails of the offspring. The tails will not disappear if it is the germ cells and not the body cells that supply the genetic information for the next generation. (*C*) If the germ line is continuous, the gametes must have half the number of chromosomes present in the body cells. A reduction in number must occur in each generation before gametes are formed; otherwise, the chromosome number would increase with each generation.

date for the science of heredity, for it was in 1900 that the work of George Mendel was discovered. Mendel had actually read his paper at Brunn, Czechoslovakia, 35 years earlier, before any detailed observations had been made on the cell nucleus and the chromosomes. The lack of cytological information accounts in part for Mendel's report lying unappreciated and poorly disseminated among scientists until its significance was pointed out independently by three biologists: De Vries, Correns, and von Tschermak. All three worked with plant material, and their results confirmed those of Mendel. By 1900, a receptive climate prevailed for the recognition of Mendel's principles. For, at that point, some hereditary concept was needed to tie in with the observations made on cell structure and cell division by pioneer cytologists. Moreover, the recognition of Darwin's theory of natural selection called for a solid explanation of the basis of genetic variation.

Before 1900, some breeding work had been performed, but it contributed little to an understanding of the mechanism of inheritance. Several nineteenth-century biologists had actually obtained the same results as Mendel. However, none was able to arrive at a correct interpretation of his findings. One of them actually described the phenomenon of segregation, the basis of Mendel's first law. But, unlike many of these earlier investigators, Mendel most likely had formulated a working hypothesis that enabled him to present a valid interpretation of his results. The nineteenth-century biologists knew that the embryo developed from the fusion of sperm and egg. They realized that all hereditary potential must be transmitted in this way and that the nuclear contributions from each parent to offspring are equal. It was noted that hybrids might be intermediate or at times resemble one parent and not the other. The predominant concept of inheritance in the nineteenth-century was a blending theory. According to this view, the hereditary material was pictured as some sort of fluid, perhaps even blood. The fluids of two parents would come together and blend in some fashion. It would be impossible to separate these hereditary fluids, just as one would fail in attempting to separate a mixture of red and white paints.

A contrasting viewpoint conceived of the hereditary material as particulate in nature. Francis Galton pointed out that particles must be handed down. He based this on the perceptive observation that one parental trait in a pedigree may seem to disappear, only to show up again in a later generation. Galton approached heredity in a statistical way by measuring individuals for a variable characteristic, such as height. He compared the measurements he made on populations from one generation to the next. Galton must be credited with the establishment of biometry, the valuable science of statistics as applied to biological problems. However, he reached erroneous conclusions at times, and his investigations, like those of many others of his day, did not reveal what was really happening when contrasting traits were followed in a family history or lineage.

Mendel and the choice of a genetic tool

The great significance of Mendel's work stems from the fact that it established the particulate concept of heredity and replaced the blending theory. Although many before him had failed, Mendel was able to obtain clear-cut genetic results through his recognition of several requirements that are critical to genetic analysis. The garden pea plant with which he worked enabled him to meet these needs. First of all, it provided several pairs of contrasting traits for study. The plants varied in such characteristics as height, seed texture, seed color, and flower color. Some sort of variation is essential if anything at all is to be learned about the inheritance of any character. Suppose, for example, that all pea plants were the same height and had the same flower color generation after generation. Obviously, no information could be gained from following plant height and flower color in genetic studies, as every individual would continue to look like all the rest in these respects. A characteristic must have alternative traits or variant forms that can

be followed if insight is to be gained into its inheritance.

In all, Mendel was able to study seven distinct characteristics, each with two well-defined traits (Fig. 1-2). Moreover, if a characteristic is to be followed clearly from one generation to the next, its traits must be easy to score. The characteristic height in the pea plant illustrates this point: the varieties Mendel used exhibited contrasting traits that were easily recognized as either tall or dwarf; no continual grades of height existed between the two extremes to obscure the picture. This contrasts with height in humans or some other animal, in which the range varies gradually from very short to very tall, with no distinct categories. The inheritance of height in these cases is more complex than in Mendel's peas. The use of a highly variable characteristic with ill-defined traits in a pioneer genetic investigation is almost certain to lead to confusion.

Mendel also recognized the need to work with a species that could provide a large number of offspring and several generations in a short period of time. Observations made on a few individuals in just one or two generations yield insufficient data for formulating valid conclusions. The elephant would be a very poor animal to select for basic genetic studies. Not only is its gestation period 20 months, but the reward is just a single offspring. At the end of a lifetime of crossing elephants, a worker would accumulate too few data to be meaningful. As we will see more clearly in Chapters 6 and 7, large numbers are usually essential for scientific significance. Conclusions based on small samples are unreliable and increase the probability that the observations may be distorted by chance factors. The garden pea completely satisfied Mendel's need to follow more than one characteristic through many generations composed of large numbers of individuals.

An additional advantage is that the investigator can quite easily manipulate the pea plant to obtain the exact type of cross or mating he desires. Peas are normally self-pollinated; therefore two genetically identical individuals can be crossed. This is so because self-pollination is equivalent to mating one individual with another one exactly like itself. This feature can be put to even further advantage; one may allow plants with a certain set of traits to self-pollinate through one generation or more. If the alternative traits fail to appear among hundreds of offspring, one can feel certain that the strain is pure breeding. In addition, the pea plant can be handled

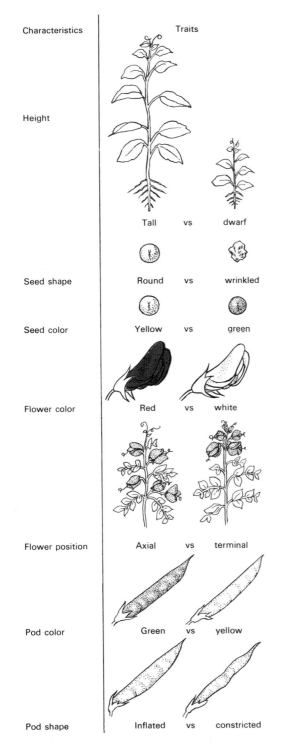

Characteristics		Traits
Height	Tall vs	dwarf
Seed shape	Round vs	wrinkled
Seed color	Yellow vs	green
Flower color	Red vs	white
Flower position	Axial vs	terminal
Pod color	Green vs	yellow
Pod shape	Inflated vs	constricted

FIG. 1-2. Seven characteristics in the pea plant followed by Mendel. Each characteristic has two well-defined traits that are easily recognized. This feature permits ready identification of the traits as they are followed from one generation to the next.

to permit cross-pollination. Consequently, a plant from a particular variety can be used either as a female parent (by removing its stamens and placing pollen from another variety on the stigmas of its flowers) or a male parent (by placing its pollen on the stigmas of a different plant, whose stamens have been removed). Therefore, with the common pea plant, the geneticist can perform crosses in the way he elects, and can easily establish a great deal about the lineage or pedigree of each parent in a particular cross. This is a far cry from the situation in human populations, in which an investigator cannot dictate the mating between two different types and may know little or nothing about the family histories of the individuals in question. The results obtained by many geneticists before 1900 were ambiguous because the species they employed lacked one or more of the features found in the garden pea. Other organisms (the fruit fly, corn, bacteria, etc.) that have also contributed a wealth of information to the science of genetics possess this same combination of desirable attributes: characteristics that show variations but whose traits are well defined and easily scored, the production of appreciable numbers of offspring in a relatively short generation time, and a situation that allows crosses to be controlled with ease.

Mendel's monohybrid crosses

Mendel's genius lay partly in his ability to formulate a scientific problem. The aim of his experiments was to study the numbers and kinds of offspring produced by hybrid individuals and to determine from the observations whether or not any statistical relationship existed among these offspring. Mendel undertook the study of the seven characteristics and their variant forms one at a time: tall versus dwarf, red flower color versus white and yellow seeds versus green, and so on. He performed monohybrid crosses, crosses in which only one pair of contrasting or alternative traits is being followed. In his study of the inheritance of height (Fig. 1-3), Mendel knew that his tall plants and his dwarf ones were all pure breeding, because he had allowed tall and dwarf types to self-pollinate for two generations. When he crossed the pure breeding tall forms with the pure breeding dwarf (the P_1 or first parental generation), the offspring (the F_1 or first filial generation) were all tall. The same result was obtained whether he used the tall plants as male parents or female parents. Mendel realized, however, that these F_1 tall individuals were not the same as

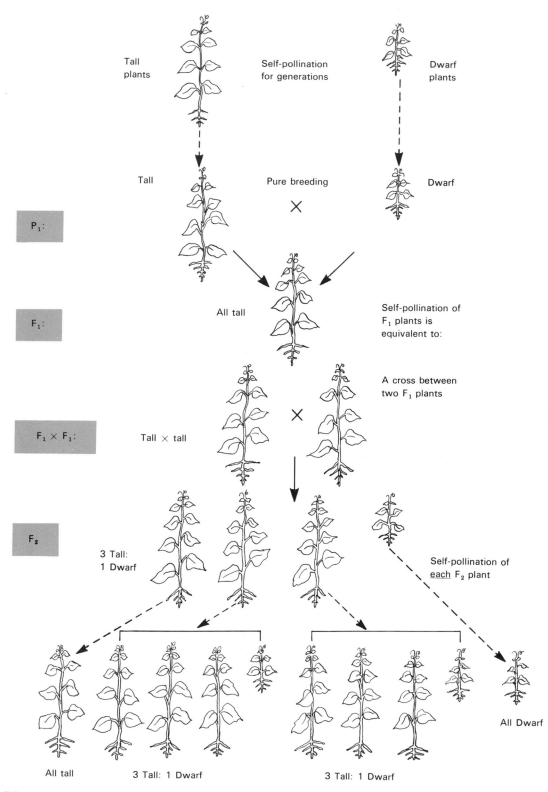

FIG. 1-3. Monohybrid cross in the pea plant followed through three generations. The results of the reciprocal cross, dwarf × tall, are identical.

8

the P_1 tall parents because they had a dwarf parent. He followed these F_1 hybrids to the next generation (the F_2 or second filial generation) by allowing them to self-pollinate. In one cross, he actually obtained 787 tall offspring and 277 short, a ratio that is approximately 3:1. He followed these second-generation plants still further by allowing them to self-pollinate. He found out that the short plants continued to produce only short ones. On the other hand, the tall plants were of two types: approximately one third of them produced only tall offspring upon self-fertilization; the remaining two thirds gave rise to both tall and short, again in a ratio of approximately 3:1 (Fig. 1-3).

Comparable results were obtained when monohybrid crosses were performed for each of the other six characteristics. For example, in the case of red versus white flower color, a cross between the two types of pure breeding varieties yielded all red-flowered plants in the first generation. When these were allowed to self, the second generation gave 705 red offspring and 224 white, again close to a ratio of 3:1. The white-flowered plants produced only white upon self-pollination. The red-flowered ones, as in the case of tallness, were shown to consist of two types: one-third pure breeding for red flowers; the other two thirds producing red and white in a ratio of 3:1. The red trait thus behaved similarly to tall in the first example and white similarly to dwarf.

Mendel's law of segregation

Remarkably, Mendel knew nothing of mitosis or meiosis; he never heard of chromosomes or the units of heredity, the genes, which they carry. Yet his explanation of his results reads almost as if he sensed such phenomena indeed existed. His knowledge of probability and his understanding of the goal of his investigations undoubtedly account in large measure for the perception that is evident in his conclusions. From the data of his monohybrid crosses, he was able to recognize a pattern and formulate what is called "Mendel's first law" or the "law of segregation." In essence, it states that the hereditary characteristics are determined by particulate units or factors. These unit factors occur in pairs in an individual, but in the formation of germ cells, these entities are segregated so that only one member of the pair is transmitted through any one gamete. When the male and female gametes unite, the double number of factors is restored in the offspring. Substituting the word *gene* for

factor, the statements read almost as if the physical basis of heredity had been known to Mendel at the time the experiments were performed.

Mendel coined the term dominant for traits such as tallness or red flower color because they expressed themselves when present with the factors for the contrasting traits and thus seemed to dominate them. He used the word recessive for traits like dwarfness and white flower color; these traits are not expressed when present in the hybrid with factors for the dominant traits. Although Mendel coined the term dominant and recessive to designate traits, throughout the years these expressions have been widely used to refer to the genetic factors associated with the traits as well as to the traits themselves. Throughout our discussions, dominant and recessive will be used both ways.

We can now reexamine the cross of the tall and dwarf plants in light of Mendel's concepts (Fig. 1-4). It should be noted here that when letters are assigned to a pair of contrasting factors (such as those for tallness and dwarfness), the same letter of the alphabet is used. The capital is reserved for the dominant factor (T for tallness) and the small letter for the recessive (t for dwarfness). This serves to avoid confusion when two or more pairs of factors are being followed at the same time in a cross. Still another accepted way of designating genetic factors is presented at length at the end of Chapter 2.

In Fig. 1-4, the Punnett square method, also known as the checkerboard method, is used to bring together the gametes from both parents in all possible combinations. We will see later in this chapter that other devices may be used to achieve the same thing. As the figure shows, each parent in the F_1 generation is a hybrid and can contribute to the F_2 offspring either a factor for tallness (T) or a contrasting factor for dwarfness (t). A sperm with the factor for tallness (T) can unite with an egg bearing the same factor or it can just as well unite with an egg carrying the contrasting factor for dwarfness (t). Similarly, a sperm with the recessive factor for dwarfness (t) has an equal chance of combining with the "T" egg or the "t" egg. Inasmuch as large numbers of gametes would actually be involved, the resulting combinations in the F_2 occur in a ratio of 1 TT: 2 Tt: 1 tt. Because tallness is dominant, the ratio of different physical types is 3 tall: 1 dwarf. It can be seen, however, that of these F_2 tall plants, only one third of them (the TT types) will breed true when self-pollinated to give all tall offspring. The other two thirds are hybrids (Tt). When these are selfed, it is the same as crossing

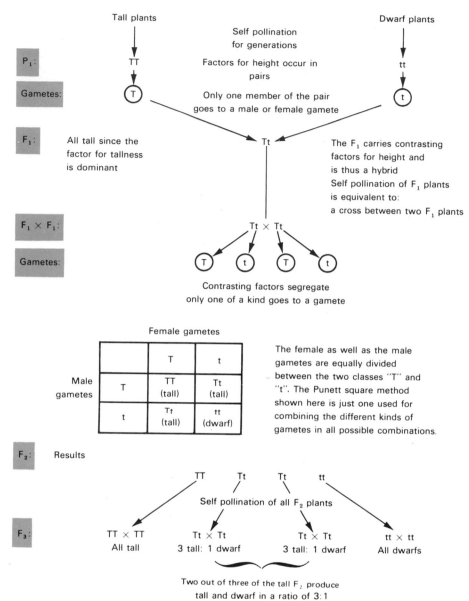

FIG. 1-4. Monohybrid cross of tall and dwarf pea plants. By convention, the female parent is written on the left. The results of the reciprocal cross are the same. Compare with Fig. 1-3.

Tt × Tt, and tall and dwarf offspring will be produced in a ratio of 3:1. The plants that are pure breeding for tallness (TT) have identical genetic factors for that trait. We say they are homozygous. The dwarf plants are also homozygotes—individuals with identical genetic factors for a trait. Indeed, the dwarf plants must be homozygous (tt) because the presence of the dominant factor for tallness would cause a plant to be tall. We see, therefore, that a tall plant may be one of two types: TT or Tt. The latter are not pure breeding and

are said to be heterozygous, meaning that they have contrasting genetic factors (T vs. t) for a particular characteristic. As we will see more clearly later (Chap. 4), contrasting factors such as these are really different forms of the same gene that result in contrasting detectable effects. Contrasting forms of a gene (T vs. t; R vs. r) are called alleles. So the homozygous individual—that is, TT or tt—has only one form of a gene. The same allele is represented twice in each homozygote. On the other hand, the heterozygote possesses both forms—contrasting forms of the same gene. The heterozygote (Tt) therefore contains a pair of contrasting alleles for the characteristic height.

The term allele, therefore, is used to indicate a specific form of a given gene. In this case, we are discussing the gene for the characteristic of "height" in the pea plant, and we are following the transmission of two specific forms of that gene, the allele "T" for tallness and the allele "t" for dwarfness.

We cannot tell simply by looking at a tall plant whether it is homozygous or heterozygous. All we can say is that it is tall in contrast to a plant that is dwarf. We refer to the physical appearance or the detectable attributes of an individual as its phenotype. Knowledge of the phenotype does not necessarily tell us anything conclusive about the genetic endowment of the individual. The term genotype is used to describe the genetic composition or the kinds of genetic factors carried by an individual. If we know that an **allele is** recessive, such as the ones for dwarfness or **white** flowers, then simply by observing a dwarf or white-flowered phenotype, we know the genotype in regard to the particular trait. However, we cannot tell by looking at a tall or a red phenotype whether the genotype is homozygous or heterozygous. We may gain this information only from further knowledge about the ancestry or about the offspring of the individuals expressing a dominant allele.

Blending inheritance disproved

An important point is that regardless of the dominant or recessive nature of an allele, blending inheritance does not occur. In the case of flower color in Mendel's peas, the alleles for redness and whiteness were in no way contaminated or altered by being present

FIG. 1-5. Monohybrid cross: red × white-flowered four-o'clocks. The F_1 is intermediate in color. A cross of two of these pink-flowered plants yields red, pink, and white in a ratio of 1:2:1. This same pattern of inheritance is seen in other plant species, for example, flower color in snapdragons.

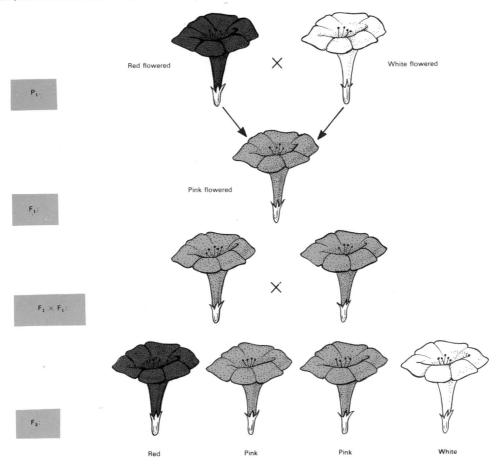

P₁.

Red flowered   ×   White flowered

F₁.

Pink flowered

F₁ × F₁.

×

F₂.

Red   Pink   Pink   White

together in the F_1 hybrids, the heterozygous Rr individuals. Although white seemed to disappear, it *did* show again in the F_2, and these white-flowered plants were identical phenotypically to the original whites in the P_1, the first parental generation. Likewise, the red factor was in no way diluted by the presence of the contrasting allele, the factor for white. This is certainly not to be expected if the fluids from two organisms are mixing, because we would then expect the original qualities to be lost or at least permanently altered in some way. The concept of blending is still entertained by those who are naive about genetic principles.

Let us turn our attention for a moment to another plant species that was not studied by Mendel, but in which the expression of flower color is somewhat different from that in the garden pea. The common horticultural plant known as the four-o'clock may also have red or white flowers. However, if pure breeding white plants are crossed with pure breeding red ones, the F_1 offspring are intermediate in color, pink (Fig. 1-5). When two pink-flowered plants are crossed, the F_2 does not fall into a phenotypic ratio of 3:1 but rather into a ratio of 1 red: 2 pink: 1 white. Does the appearance of the intermediate pink shade mean that blending inheritance has occurred in this case? Far from contradicting Mendelian principles, such an example lends strong support to the concept of particulate inheritance. Neither the factor for redness (R) nor that for whiteness (r) has in any way been altered by being present together in the cells of an individual. The F_2 reds and whites are the same as the original red and white P_1s. The cross illustrates Mendel's first law beautifully and differs from that of the pea plant only in the question of dominance. In this case, neither form of the gene is dominant over the other. Each expresses itself in the presence of its allele to produce an intermediate effect. We say that alleles such as these show *incomplete dominance*. As Fig. 1-6 shows, we may diagram the cross in the four-o'clock in exactly the same way as for the pea plant. It does not matter if we choose to represent the allele for red as R or r, as long as we remember that the heterozygote is an intermediate. An important point that should be evident from this example is that the 3:1 phenotypic ratio, which typically results from a cross of two monohybrids, has been modified to a phenotypic ratio of 1:2:1. This is not surprising. In

FIG. 1-6. Incomplete dominance in the four-o'clock. Neither of the two alleles, R or r, is dominant. Consequently the F_2 phenotypic ratio is modified from 3:1 to 1:2:1, which is identical to the ratio of genotypes.

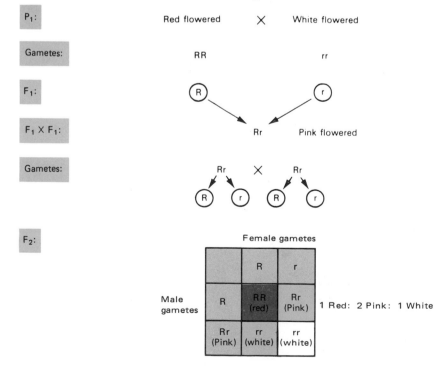

the case of the pea plant, where red (R) is dominant to white (r), the heterozygotes (Rr) are classed as red because they cannot be distinguished from the homozygotes (RR). But in the case of the four-o'clocks, the heterozygotes can be easily recognized phenotypically. Inasmuch as we *can* distinguish them from the homozygous red plants, the phenotypic ratio becomes 1:2:1. It should be noted that this is the same as the *genotypic ratio* resulting when any two monohybrids are crossed. When dominance is incomplete, therefore, a cross of two monohybrids (e.g., Rr × Rr pink plants) gives a phenotypic ratio (1 red: 2 pink: 1 white) identical to the genotypic ratio (1 homozygous RR: 2 heterozygous Rr: 1 homozygous rr). Where complete dominance is operating in a monohybrid cross, the phenotypic ratio (3:1) and the genotypic ratio (1:2:1) are different.

Dominance is not always absolute

Although Mendel noted dominance and recessiveness among the seven pairs of alleles he followed, he did not say that dominance must always pertain. *Dominant* and *recessive* are often used as terms of convenience, but strictly speaking, they are not very precise. One example from the human illustrates this point very well. A certain allele, D, when present in a single dose can produce a serious skeletal disorder, achondroplasia. Persons heterozygous for this allele (Dd) have a dwarfed stature, short limbs, and a large head with bulging forehead. The allele is generally referred to as a dominant. However, persons homozygous for the allele (DD) have a much more abnormal phenotype, and they usually do not survive infancy. Obviously, the terms dominant and recessive are not 100% distinct in this case, because a double dose of the so-called dominant allele produces a far greater effect than a single dose in the heterozygote.

Although heterozygotes often appear superficially to be identical to homozygotes, the so-called recessive allele may actually be producing some effect. The expression *incomplete dominance* refers to such situations in which both alleles express themselves in the hybrid but the effect of one allele appears greater than that of the other. In the absence of a refined examination, one allele may even appear to be completely dominant over the other. Another frequently encountered term is *codominance;* this is also applied in cases in which both alleles produce an effect in the heterozygote. However, in its strict sense, codominance implies that a definite product or substance

controlled by each allele can be identified. As we will see in Chapter 4, the genetic factors that govern the A and B blood types in man are alleles. Each controls the formation of a different red blood cell protein or antigen, antigen A in one case, antigen B in the other. Neither allele is dominant over the other. The individual who is blood type AB is a heterozygote and has both the allele for antigen A and that for antigen B. Both of these proteins, A and B, are easily detected in equal amounts in the red cells. This is a clear-cut case of codominance between a pair of alleles.

In other situations it may not be so easy to distinguish codominance from incomplete dominance because use of the proper term depends on our ability to detect in the heterozygote distinct substances controlled by each member of the allelic pair. An example from the human illustrates the difficulty often involved in a precise distinction between codominance and incomplete dominance. Many of us are familiar with the blood disorder known as sickle cell anemia, a fatal condition in which abnormal hemoglobin is present. It is so named because the red blood cells tend to assume a sickle shape in the blood vessels (Fig. 1-7), resulting in clogging of the capillaries and an assortment of dire symptoms. The disorder stems from the presence in the homozygote of a genetic factor, which governs the formation of the abnormal hemoglobin. Those who are homozygous for the normal genetic factor possess only the typical hemoglobin in their red blood cells. Heterozygous persons appear identical to the homozygotes under usual conditions.

For convenience, let us represent the alleles responsible for the normal and the sickle cell hemoglobin as "S" and "s," respectively. Superficially it would appear that SS and Ss individuals are the same, in contrast to the homozygotes, ss, who present an abnormal phenotypic picture. We might conclude that absolute dominance pertains. However, the heterozygotes, Ss, *are* actually somewhat different from the SS homozygotes. They may be identified when their blood is examined microscopically, because a certain percentage of their red blood cells can be made to sickle by a test that subjects the cells to certain reducing agents capable of reducing the oxygen tension in the immediate environment. Such an effect in the body of the heterozygote would be very unusual. We see, however, that the allele for sickle cell hemoglobin is not a complete recessive, because the heterozygote can be detected.

Is the normal allele S incompletely dominant over the abnormal one, or are the alleles codominant? In

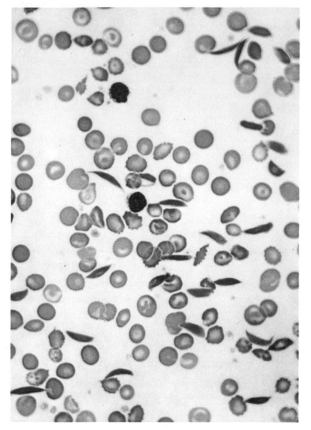

FIG. 1-7. Blood smear from a person with sickle-cell anemia showing typical sickle-shaped cells. A person homozygous for the allele for normal hemoglobin (SS) will have disk-shaped cells under all levels of oxygen concentration. The homozygote for sickle cell anemia (ss) carries red blood cells that assume an abnormal sickling shape under the lower oxygen concentrations in the blood vessels. The heterozygotes (Ss) are revealed by a test that subjects a sample of red blood cells to a reduced oxygen level, causing a large number to sickle. Such sickling does not occur in the body of a heterozygote who typically shows no clinical symptoms and may remain undetected. (Courtesy of Carolina Biological Supply Company.)

the sense that the normal S results in a heterozygote who has normal red blood cells except for those that sickle under test conditions, the allele would appear to be an incomplete dominant. However, refined techniques demonstrate that two different proteins—normal hemoglobin controlled by gene form S and sickle cell hemoglobin controlled by its allele s—are present in the blood. Based on this criterion, therefore, the alleles are codominants. It is evident that the choice of words may depend on the level at which we describe the phenotype of the heterozygote; it also depends on the techniques available for detecting chem-

ical differences. However, it is important that we avoid getting lost in a maze of terminology. The expression we use may depend on our frame of reference. The alleles for the A and B blood groups are clearly codominants. We may say that the alleles R and r in the four-o'clock are incompletely dominant because the heterozygote is intermediate and no distinct substances specific to each has been recognized. The normal and the sickle cell allelic may be described as incompletely dominant at one level of reference (a small percentage of sickling) or as codominants at another (presence of two specific protein products). For convenience, at times the allele for normal hemoglobin is referred to as a dominant.

Misconceptions to be avoided

To avoid erroneous interpretations of Mendelian principles that have been discussed so far, it is essential to pause and reexamine the implications of certain basic ideas. Several important points concern the concept of dominance, which unfortunately is often misunderstood. Although Mendel showed that the factor for red flower color in the pea plant is dominant over its allele for white, the genetic factors for red and white flower in the four-o'clock are incompletely dominant. This example was given to illustrate that a genetic factor which produces a certain phenotypic effect (red, in this example) in one species is not the same as one in a different species, even though the two genetic factors appear to produce identical phenotypes (red flower color in both the pea and the four-o'clock). Simply because a particular factor in one group of animals or plants expresses itself as a dominant is no reason to assume that a different factor with a similar or even identical expression must also be dominant in another group. In animals, a factor for black fur may be dominant in one species (as in guinea pigs) but recessive in another (as in sheep). This presents no contradiction to Mendelian principles, which nowhere imply that similar phenotypes must be due to the presence of similar genes.

Another very common misunderstanding of dominance concerns the frequency of dominant alleles in a group or population. Too often, the student assumes that simply because a particular factor is dominant it must be more abundant in the population than its recessive allele. Many wonder why there are not more curly haired people in the population, because the gene form for that phenotype is inherited as a dominant or incomplete dominant. Implicit in this erroneous idea is the belief that a dominant factor *must* increase in frequency over its recessive allele. A moment's reflection will tell us in the simple case of eye color in humans that this is not so. Although the factor for brown eye color (B) behaves as a dominant to its allele for blue (b), there may be more blue-eyed than brown-eyed people in a population. In Scandinavia this would be so, whereas in Mediterranean countries, we would expect more people to be brown eyed. As we will learn in Chapter 21, the frequency or abundance of an allele is not the same as its degree of dominance. The allele for brown eye color is dominant to the one for blue, which means simply that the factor for brown eyes expresses itself in the presence of its allele, whereas the factor for blue does not. Which allele—the one for brown or the one for blue eye color—will be the more common is a very different matter. The frequency of any allele, dominant or recessive, is related to the survival value it imparts to a population and not just to its dominance or recessiveness.

Nor does the fact that an allele is dominant mean that it is necessarily better or will somehow produce a stronger or more desirable effect. A long list of undesirable dominant alleles in various species could be presented; in humans the following seem to depend on dominants: extra fingers or toes, a form of muscular dystrophy, a type of dwarfism, a form of progressive nervous deterioration resulting in death, and several others. Such conditions are far from desirable and certainly do not make the afflicted person stronger. It is also apparent that the conditions, and hence their associated dominant alleles, are fortunately not as common in the population as the normal conditions and the normal but recessive gene forms. Moreover, these undesirable dominants are not increasing to the disadvantage of the whole population. If their dominance made them increase, everyone would eventually become afflicted.

Interpretations of crosses and ratios

It is also essential at the outset of a study of genetics to appreciate the correct use of the Punnett square (Figs. 1-4 and 1-6) and the interpretation of genetic ratios. We have so far become acquainted with two important ratios: 3:1 and 1:2:1. You will recall that one reason for Mendel's achievement was that he conducted his studies with large numbers of plants.

His F_1 and F_2 generations did not consist of just a few individuals. Few of us would expect a cross between two heterozygous tall plants (Tt × Tt) to yield exactly three tall and one dwarf if only four offspring were obtained. What does the 3:1 ratio (or any other) tell us? It tells us the probability of getting a certain phenotype or genotype when a particular cross is made.

In his experiments, Mendel obtained numbers very close to 3:1 because he worked with many plants; but even then, the ratio was not exactly 3:1. We instinctively know that if we toss a coin many times heads will come up in approximately 50% of the tosses. We also realize that there is a 50% chance for a boy to be born at any birth and a 50% chance for a girl. We do not, however, consider it amazing to find a family in which all five children are girls, nor are we very surprised if five tosses of a coin yield all heads. Only after hundreds of coin tosses or after counting the children in hundreds of families do we come close to realizing 50% heads and 50% tails or equal numbers of boys and girls.

Knowing this, we see how illogical it is to assume that any four offspring from a pair of monohybrid parents must be three of one kind and one of another. This would be the same as expecting a girl baby to be followed by a boy and then by a girl, and so on, just because the chances are equal for the arrival of a boy or a girl. With this in mind, we can avoid absurd replies in situations illustrated by the following. In humans, normal skin pigmentation depends on the presence of a dominant factor, "A." Its recessive allele, "a," results in albinism, a deficiency of melanin pigment. Suppose that two people are known to be monohybrids for this character, Aa. They already have three normal children and inquire about the appearance of their next child. Often, the novice will reply that the child will be an albino. Much confusion and anxiety can result from misuse of genetic concepts. The answer in this case can be given only in terms of probability. *No matter* how many offspring the couple has, the chance is 3:1 *at each birth* in favor of a normal child. In a similar family, it would not be unusual to find three albinos and one normal, even though there is only one chance out of four for an albino at each conception. It must be kept in mind when ratios are discussed that they do not tell us the exact results from a cross; instead, they give the chance, *at any one time*, that a certain event will occur.

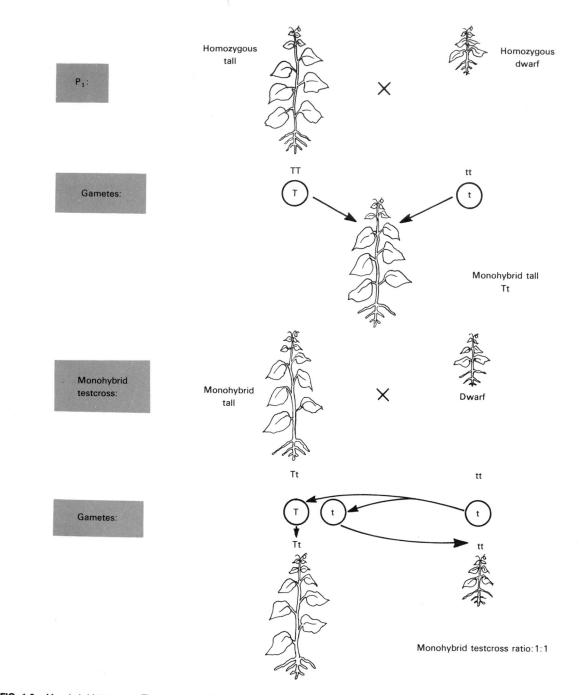

FIG. 1-8. Monohybrid testcross. The testcross ratio, 1 tall: 1 dwarf, is exactly the same as the ratio of the two types of gametes produced by the monohybrid, one kind with allele T, the other with allele t. This is so because the testcross parent (tt) carries only the recessive allele for the character, height, and cannot mask any factors segregating in the hybrid parent. The testcross shown here is also a backcross, because "dwarf" is one of the P₁ types.

16

The testcross and its value

With these points clarified, we can proceed to further experiments of Mendel. Once he formulated the law of segregation, Mendel performed other kinds of crosses to see if the rules in his hypothesis were followed. One of the first types of crosses he made is what we know today as the *testcross*. To perform this, Mendel took his F_1 hybrids and mated them with recessives, such as a cross between hybrid tall plants and dwarf ones (Fig. 1-8). Mendel realized that if the law of segregation were correct, the monohybrid tall plants should produce two classes of gametes in equal amounts, half with the factor for tall (T) and half with the factor for dwarf (t). A dwarf plant, carrying only the recessive allele for height, can produce only one class of gametes for this characteristic, those carrying the allele, t. Consequently, genetic factors for height coming from the dwarf parent will not mask any gene forms for height contributed by the hybrid parent. As a result, it becomes possible to determine the kinds of gametes formed by the hybrid as well as the proportion in which they are formed. To do this it is only necessary to observe the kinds of offspring in the testcross and the frequency with which they occur. As Fig. 1-8 shows, half of the offspring from the testcross should be tall and the other half dwarf—a ratio of 1:1. This is so because the two alleles (T and t) are segregating in the hybrid. They are produced in equal amounts, and it is a matter of chance whether a gamete carrying T or one carrying t combines with the gamete from the recessive parent. Mendel performed many testcrosses of monohybrids and obtained results very close to a 1:1 ratio in each case, thus supporting the ideas expressed in his first law. The testcross is such a valuable procedure in genetic analysis that we refer to it frequently throughout the text.

The term *testcross* is often used interchangeably with the expression *backcross*. However, the two are not necessarily the same. We have just seen that a testcross is a cross of any individual to one that is recessive for a trait being followed in a cross. A backcross, on the other hand, implies a mating of an individual to a type like one of the parents. In our example (Fig. 1-8) we have both a testcross and a backcross. The monohybrid is crossed to the recessive, but the recessive phenotype is also one of the parental types. If the hybrid had been crossed instead to a homozygous tall (TT), we would have an example

of a backcross but not a testcross. Because a backcross may have special value in some genetic analyses, its distinction from the testcross should be recognized.

Dihybrid testcrosses and independent assortment

Mendel's work with monohybrids indicated clearly that a 3:1 ratio is to be expected from a cross of two monohybrids and that a 1:1 ratio will follow from a testcross of a monohybrid. Mendel's further studies entail the behavior of two or more pairs of alleles and led to the formulation of Mendel's second law. A question that occurred to Mendel was how two or more pairs of factors might behave in relationship to one another when followed at the same time in a cross. From his monohybrid crosses, Mendel knew that the color of the pea seed can be either yellow or green and that the factor for yellow (G) is dominant over green (g). When plants homozygous for yellow seeds are crossed with those having green seeds, the F_2 generation contains the yellow and green phenotypes in a ratio of 3:1. Likewise, the shape of the seed, round (W) versus wrinkled (w), depends on a pair of alleles with the factor for the former being dominant. The F_2 ratio is thus three round to one wrinkled. But if both characteristics, seed color and seed shape, are followed at the same time, how will the factors for the two of them behave?

Suppose a plant that is pure breeding for yellow, round seeds is crossed to one from a green, wrinkled variety (Fig. 1-9A). From Mendel's first law, we would expect two factors for each character to be present in each parent. Knowing which genes are dominant, we may represent the genotypes of the parents as GGWW (yellow, round) and ggww (green, wrinkled). The F_1 dihybrid between them will receive one factor for each character from each of the parents and will have the genotype GgWw. The F_1s are all found to have yellow, round seeds. This is not surprising because yellow and round have been shown to be dominant. The question that now arises is, "What will happen when the F_1 dihybrids are followed further?" Will the factors for yellow and round stay together through later generations because they were together in one of the original parents? The same question arises for green and wrinkled. It is also possible that the original gene combinations are not required to remain together but are free to enter into new combinations. The most direct way to settle the matter is to take the F_1 dihy-

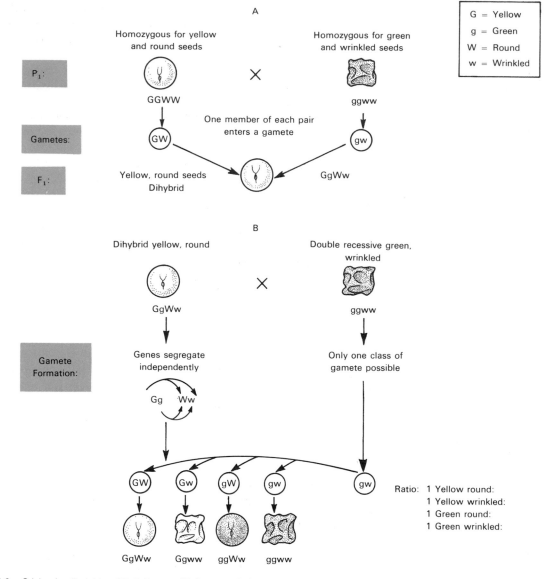

FIG. 1-9. Origin of a dihybrid and its testcross. (*A*) Cross producing a dihybrid. The yellow, round phenotype is to be expected because the factor for yellow (G) is dominant over its allele (g), and the one for round (W) is dominant over wrinkled (w). (*B*) The dihybrid is testcrossed. The phenotypic ratio 1:1:1:1 is exactly the same as that in which the four classes of gametes are produced in the dihybrid: 1GW:1Gw:1gW:1gw.

brid and perform a testcross, because a testcross can tell us the kinds of gametes produced by an individual and the frequency of the different kinds. As Figure 1-9B shows, the offspring resulting from this testcross are of four types, and they occur with equal frequencies, a ratio of 1:1:1:1. This tells us directly that the dihybrid is forming four different classes of gametes. This in turn means that the factors for seed color and those for seed shape are behaving independently. Just

because the factors for yellow and round (G and W) and those for green and wrinkled (g and w) were together in the original parents did not mean they had to stay together. In addition to the old combinations, GW and gw, the F_1 dihybrid also produces new ones, Gw and gW. Nothing seems to tie the factors together; they are free to form new combinations. We say that they "assort independently." As a result, the expected dihybrid testcross ratio is 1:1:1:1, a reflec-

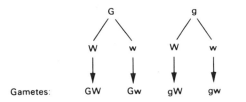

Gg Ww Dihybrid for yellow, round seeds

The alleles for color segregate from each other.

The alleles for seed shape also segregate, but they do so independently of those for seed color. Therefore—

the allele for yellow has an equal chance of entering a gamete with either the allele for round or the one for wrinkled. The same is true for the allele for green seed.

FIG. 1-10. Branching method. One way to determine the different types of gametes that can be formed by a dihybrid.

tion of the types of gametes and their proportions. We see here, as in the case of the monohybrid, that the testcross parent (double recessive ggww) produces only one kind of gamete in relation to the characteristics we are following. Because all the gametes carry both recessives, g and w, they cover up nothing contributed by the other parent. Thus, we can estimate the kinds of gametes and their proportions produced by this other parent simply by counting the kinds of offspring. Figure 1-10 shows one other way to envision the independent behavior of two pairs of alleles in a dihybrid and to determine the kinds of gametes formed.

Independent assortment and the F_2 dihybrid ratio

Knowing that any dihybrid individual forms four different kinds of gametes in equal proportions, we can depict diagrammatically what to expect in a dihybrid cross (one in which two pairs of contrasting factors are being followed) carried through the F_2 generation (Fig. 1-11). We see from the figure that the F_2 will fall into four different phenotypic categories in a ratio of 9:3:3:1. This is exactly what is to be expected from the independent behavior of the two pairs of alleles. When Mendel performed the cross, the actual frequencies he obtained were 315 round yellow, 108 round green, 101 wrinkled yellow, and 32 wrinkled green. Statistical analysis (Chap. 6) shows that these figures are compatible with a 9:3:3:1 ratio. This classic dihybrid ratio is so fundamental to genetic studies and so frequently encountered that we must be able both to recognize it and to predict the phenotypes that are to be expected from a cross of any two dihybrids. Note (Fig. 1-11) that although only four different phenotypes result when dominance is operating in both pairs of factors, there are actually nine different genotypes that arise from the independent assortment of the genes. We can think of a cross

between any two dihybrids in the following general way: AaBb × AaBb. When dominance is involved, the expected results are 9A__B__ (those showing both dominant traits; the dashes indicate that either the dominant or the recessive allele may be present); 3A__bb (those expressing the dominant A and the recessive b); 3aaB__ (individuals expressing the recessive a and the dominant B); and 1aabb (double recessives).

Other crosses of pea plants in which the remaining five pairs of alleles were followed produced comparable results and established the 9:3:3:1 ratio as the one to be expected when two dihybrids are crossed and 1:1:1:1 as the dihybrid testcross ratio. From such data, Mendel formulated his second law, *the law of independent assortment,* which states what we have just seen: the members of one pair of factors (alleles) segregate independently of members of other pairs at the time of gamete formation (pair Gg segregates independently of pair Ww). This rule holds for two, three, or more pairs, but we will see that the law pertains only to those genes located on different pairs of chromosomes (Chap. 3). The complication of linkage (genes found together on the same chromosome) was not involved in Mendel's crosses to cloud the picture of independent assortment.

Separate chromosomes and independent assortment

Knowledge of independent assortment enables us to predict the kinds of offspring in any crosses involving two or more pairs of factors on separate chromosomes. The Punnett square method can aid us in visualizing this; it also provides us with a picture of the different genotypes. However, the method may be cumbersome at times, particularly when more than two pairs of alleles are involved. We can avoid the Punnett square by following an easier way to diagram crosses. This

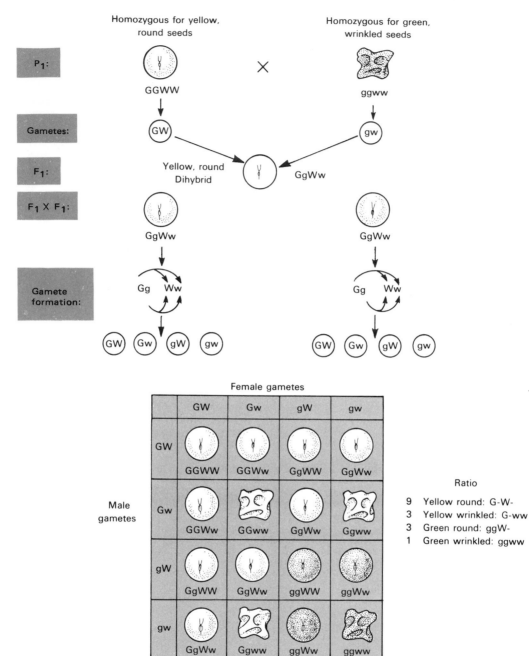

FIG. 1-11. Dihybrid cross followed through the second generation. Inspection of the Punnett square shows that nine different genotypes are found among the F₂ offspring. Because of dominance in one member of each pair of alleles, only four phenotypes are expressed.

requires nothing more than a grasp of a simple mathematical principle, which can be stated as follows: if we know the chances of two independent events happening separately, the chance of the two occurring together is the product of the separate probabilities.

A dihybrid cross can be treated as if it were two separate crosses.

Returning to the dihybrid pea plants whose seeds were round and yellow (GgWw × GgWw), we may think of the cross as two monohybrid crosses. We can consider it to be a cross between the one kind of monohybrid ($Gg \times Gg$) and also a cross between the other kind ($Ww \times Ww$). Our knowledge of monohybrid crosses tells us that in the first case we would expect ¾ yellow and ¼ green and in the second cross ¾ round and ¼ wrinkled. We know that the 3:1 ratio is an expression of probability, the chance of a certain event happening at any one time. In this dihybrid cross, we therefore know the separate chances for round versus wrinkled at any one time, and similarly for the independent occurrences of yellow versus green. To determine the different combinations possible in a dihybrid cross that involves the traits for seed color and for seed texture, we simply multiply the separate probabilities:

3/4 round: 1/4 wrinkled
3/4 yellow: 1/4 green

9/16 round yellow: 3/16 wrinkled yellow:
3/16 round green: 1/16 wrinkled green

This is a much more direct and time-saving method than the Punnett square and accomplishes the same thing in telling us the different phenotypes to be expected and their frequencies. Suppose we were also following a third characteristic at the same time, height, and wanted to know the results of a cross between two trihybrids: $GgWwTt \times GgWwTt$. We obtain the answer quickly by proceeding as if we were dealing with three separate monohybrid crosses. We simply multiply the results of the preceding dihybrid cross by the chances for tall versus dwarf:

9/16 round yellow: 3/16 wrinkled yellow: 3/16 round green: 1/16 wrinkled green
3/4 tall: 1/4 dwarf

27/64 tall round yellow: 9/64 tall wrinkled yellow: 9/64 tall round green: 3/64 tall wrinkled green: 9/64 dwarf round yellow: 3/64 dwarf wrinkled yellow: 3/64 dwarf round green: 1/64 dwarf wrinkled green.

Although the Punnett square will confirm this, it is evident that an appreciable number of squares is required—64—because each trihybrid forms eight different classes of gametes as a result of independent assortment of three pairs of factors. The testcross of the trihybrid ($GgWwTt \times ggwwtt$) shows us this. The trihybrid testcross itself may be treated as if it were three separate monohybrid testcrosses: $Gg \times gg$, $Ww \times ww$, and $Tt \times tt$. Inasmuch as the expected ratio for each is 1:1, the trihybrid testcross becomes

1/2 yellow: 1/2 green
1/2 smooth: 1/2 wrinkled

1/4 yellow smooth: 1/4 green smooth: 1/4 yellow wrinkled: 1/4 green wrinkled
1/2 tall: 1/2 dwarf

1/8 tall yellow smooth: 1/8 tall green smooth: 1/8 tall yellow wrinkled: 1/8 tall green wrinkled: 1/8 dwarf yellow smooth: 1/8 dwarf green smooth: 1/8 dwarf yellow wrinkled: 1/8 dwarf green wrinkled.

This trihybrid testcross ratio of 1:1:1:1:1:1:1:1 is obtained because eight different types of gametes are possible from a trihybrid. Figure 1-12 depicts how the

FIG. 1-12. Trihybrid testcross. The branching method is used here to derive the different kinds of gametes formed by the trihybrid.

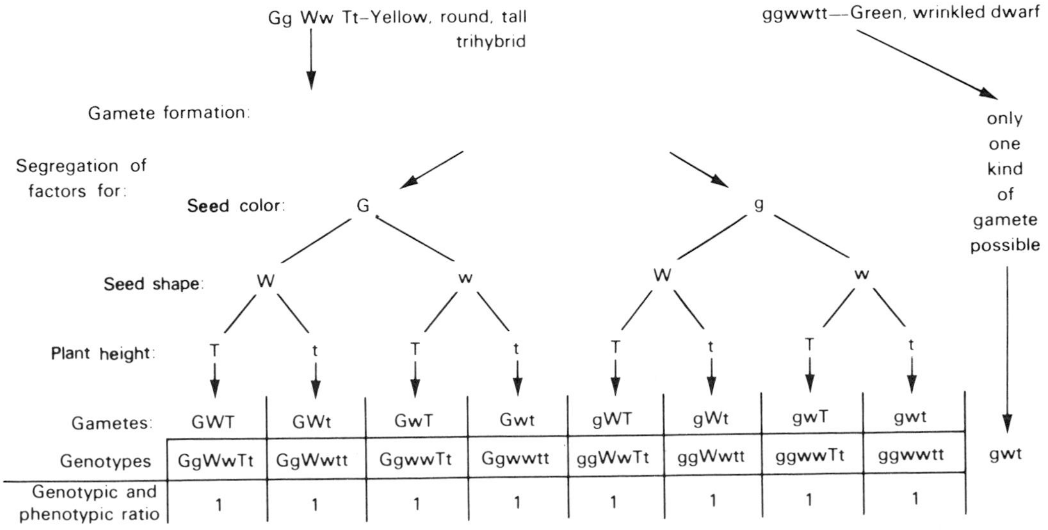

Gametes:	GWT	GWt	GwT	Gwt	gWT	gWt	gwT	gwt		gwt
Genotypes	GgWwTt	GgWwtt	GgwwTt	Ggwwtt	ggWwTt	ggWwtt	ggwwTt	ggwwtt		
Genotypic and phenotypic ratio	1	1	1	1	1	1	1	1		

branching method may also enable us to derive the different combinations and to reach the same results. The independent behavior of the pairs of alleles permits us to estimate the number of different kinds of gametes that will be produced by any hybrid. From the information presented on testcrosses, we have seen that a monohybrid (Gg) forms two different kinds of gametes, the dihybrid and trihybrid (GgWw and GgWwTt), four and eight different types, respectively. A testcross of a tetrahybrid can result in the formation of 16 gene combinations. The number of kinds of possible combinations that can be formed by individuals of different degrees of hybridity forms a geometric progression. This fact can be summarized by the very helpful expression: 2^n. This tells us the number of different kinds of gametes that can be formed by any hybrid, allowing n to stand for the number of hybrid gene pairs. Table 1-1 summarizes a great deal of valuable information that can be derived from the basic genetic principles discussed so far.

We can easily illustrate some of the points contained in Table 1-1. A monohybrid tall plant (Tt) forms two classes of gametes: T and t. If two such monohybrids are crossed, the number of phenotypic classes is again two: tall and dwarf. The number of phenotypic classes in the monohybrid testcross also equals two: tall (Tt) and dwarf (tt). However, the number of different genotypes that can arise when two monohybrids are crossed is three: TT, Tt, and tt. If there is no dominance, as in the case of the four-o'clocks, the number of phenotypic classes will be equal to the number of genotypic classes, because the heterozygotes can be distinguished from the homozygotes (RR, red; Rr, pink; rr, white). If dihybrid, trihybrid, and other crosses are considered in this context, it is

apparent that the simple expression 2^n is indeed a valuable one to keep in mind. It also tells us something about our own species. Humans possess 23 pairs of chromosomes. The human is also highly heterozygous. For any one pair of chromosomes, the human is hybrid for many, many allelic pairs. But to keep the example as simple as possible, let us assume that the human is hybrid for only one pair of alleles on each pair of chromosomes.

We can now consider the possibilities that might arise from independent assortment alone. Table 1-1 tells us that any one parent would form 2^n or 2^{23} different kinds of gametes. From two such parents, at least that many phenotypic classes can appear. However, many more kinds of genotypic classes are possible, 3^{23}. In actuality, this illustration is simplified to the point of absurdity, because countless other allelic pairs on the same chromosome would be involved. We will learn that reassortment among genes on the same chromosome may take place, and this increases the possibilities many times more (Chap. 8). We are also ignoring lack of dominance, which adds to the number of phenotypic classes. At any rate, although 2^{23} is the minimal figure for the kinds of gametes and phenotypic classes, it is so immense that it defies the imagination. We should appreciate why no two people on earth, except for identical twins, are genetically identical. Each of us is unique. The chance for another to be exactly like us, even from our own parents, is so improbable that the idea can be dismissed. Each of us owes his or her distinct nature, in a large degree, to independent assortment. In the two chapters that follow, we will see that this is a consequence of chromosome behavior and the nature of the sexual process.

TABLE 1-1   Summary of important consequences of segregation and independent assortment

NO. OF ALLELIC PAIRS IN HYBRID	NO. OF KINDS OF GAMETES THAT HYBRIDS CAN FORM	NO. OF PHENOTYPIC CLASSES WHEN TWO SUCH HYBRIDS ARE CROSSED (ASSUMING DOMINANCE)	NO. OF PHENOTYPES IN TESTCROSS OF THE HYBRID	NO. OF GENOTYPES WHEN TWO SUCH HYBRIDS ARE CROSSED (ALSO THE NO. OF PHENOTYPES WHEN NO DOMINANCE PERTAINS)
1	2	2	2	3
2	4	4	4	9
3	8	8	8	27
4	16	16	16	81
5	32	32	32	243
6	64	64	64	729
n	2^n	2^n	2^n	3^n

Recognition of the significance of Mendelian principles

Before the discovery of Mendel's paper in 1900, several investigators had almost deduced Mendelian principles from their own experiments. De Vries, one of the three biologists who found Mendel's paper, had made many of the same observations in the course of his studies. In 1900, he reported that Mendelian principles apply to approximately a dozen very different groups of seed plants. De Vries also made an important contribution to genetics as a result of his investigations with the evening primrose, a species that seemed to violate Mendelian rules. Actually, the odd results derive from an unusual chromosome situation in this plant and present no real exception to the chromosome theory (Chap. 10). From the unusual results obtained from work with the evening primrose, De Vries formulated the mutation theory, now a fundamental concept in genetics. He thought that the unexpected appearance of certain phenotypes among his plants resulted from mutations, or sudden inheritable changes. Although the unusual plants were later shown to arise for other reasons, the mutation theory of De Vries focused the attention of geneticists on the fact that sudden inheritable changes may occur and that these may provide the source of the genetic variation required for evolutionary progress.

Correns, another of Mendel's discoverers, realized the full significance of the data on the garden pea. In 1902, he connected the behavior of Mendel's factors with the chromosomes and was thus the first to attribute the operation of Mendelian laws to chromosome behavior. After 1900, Mendel's principles were confirmed and extended to include many other species—animals as well as plants: Cuenot in France with mice, Castle in the United States studying rodents, and Davenport (U.S.) and Bateson (England) working with poultry.

Bateson was another geneticist who was very close to deducing Mendelian principles from his own breeding work. Among his many contributions to genetics is the very name of the science itself, which he coined along with *heterozygous*, *homozygous*, and *allelomorph* (which we shorten to allele).

We owe the expressions *gene*, *genotype*, and *phenotype* to the Danish botanist, W. Johannsen. In one of his classic studies, Johannsen devised a method to distinguish the variation in beans resulting from environment from the variation contributed by genetic factors. Many of his ideas greatly influenced genetic thought and are discussed in later chapters. Many of the investigators, including Mendel, appreciated the need to consider the environment when undertaking genetic analysis.

At the turn of the century, the features of meiosis were being unraveled. Montgomery postulated that the chromosomes occur in corresponding pairs, one member of each pair contributed by each parent. Other workers besides Correns then began to point out that the chromosomes might provide the physical basis for Mendel's factors. In 1902, Sutton gave perhaps the most convincing arguments, building on the ideas of earlier biologists and his own observations. He reasoned that since the gametes are the only connection between the generations, all the hereditary material must be carried in them. The egg contributes a great deal of cytoplasm at fertilization, whereas the sperm sheds most of its cytoplasm as it matures. Since genetic experiments indicated that the hereditary contributions from parents are equal, the factors must reside in the nucleus. The only visible parts of the nucleus that are accurately divided when the cell divides are the chromosomes; therefore, the genes must be carried on them. The chromosomes of an individual occur in pairs; so do Mendelian factors. In the formation of gametes, the chromosomes segregate; Mendelian factors also segregate at some time prior to gamete formation. Mendel's factors segregate independently. The chromosomes seemed to do this also, but there was still no definite proof of this until the studies of Carothers in 1913. The observations of cytologists before this date nevertheless gave no support to the idea that chromosomes are tied together in some way that prevents their independent assortment.

Therefore, before 1910 there was no conclusive proof that the hereditary factors, the genes, were on the chromosomes, but all the evidence favored this interpretation. Perhaps the strongest argument came from the parallel between chromosome behavior and the behavior of the genetic factors, as expressed in Mendel's laws. We examine these points in some detail in the two following chapters.

REFERENCES

Dunn, L. C. *A Short History of Genetics.* McGraw-Hill, New York, 1965.

Jordanova, L. J. *Lamarck.* Past Masters Series. Oxford University Press, New York, 1984.

Mendel, G. *Experiments in Plant Hybridization.* Translated

in C. I. Davern, ed. *Genetics, A Scientific American Reader*. W. H. Freeman, New York, 1981.

Orel, V. *Mendel*. Past Masters Series. Oxford University Press, New York, 1984.

REVIEW QUESTIONS

1. In tomatoes, round fruit (O) is dominant over oblong fruit (o). Write as much as possible of the genotypes of:

   A. A plant from a homozygous round-fruited stock.
   B. A plant from an oblong strain.
   C. A round-fruited plant that resulted from a testcross.
   D. An oblong-fruited plant that resulted from the cross of two round-fruited ones.

2. Using information from Question 1, give the genotypic and phenotypic ratios to be expected regarding fruit shape from each of the following crosses:

   A. Homozygous round × heterozygous round.
   B. Heterozygous round × heterozygous round.
   C. Oblong × oblong.
   D. Heterozygous round × oblong.

3. Two short-haired female cats are mated to the same long-haired male. Several litters are produced. Female 1 produced eight short-haired and six long-haired kittens. Female 2 produced 24 short-haired and no long-haired kittens. From these observations, what deductions can be made concerning hair length inheritance in these animals? Assuming the allelic pair S and s, give the likely genotypes of the two female cats and the male.

4. In cattle, the hornless condition (H) is dominant over the horned condition (h).

   A. A horned bull is mated to a hornless cow that is heterozygous for the condition. What kinds of offspring are to be expected and in what ratio?
   B. If the cow is next mated to a hornless bull that is a heterozygote, what is the chance that the first calf will be horned?
   C. Assuming that the first calf born has horns, what is the chance that the second calf will be hornless?

5. In cattle, the alleles for red coat (R) and white coat (r) behave as codominants. Both red and white hair are produced, so that the heterozygote is intermediate, or roan colored.

   A. Give the phenotypic and genotypic ratios to be expected among the offspring from a cross of two roan animals.
   B. What are the expected genotypic and phenotypic ratios from a cross of a roan animal and a white one?

6. Write as much as possible of the following genotypes:

   A. A horned white bull.
   B. A hornless roan cow.

   C. A hornless red cow whose male parent was horned and roan.

7. In guinea pigs, short hair (L) is dominant over long (l). Black fur (B) is dominant over albino (b). A female from a strain that is pure breeding for black fur and short hair is mated to a male from a strain pure breeding for the albino condition and long hair.

   A. What will be the phenotypes of the F_1?
   B. If members of the F_1 are mated among themselves, what percentage of offspring can be expected to be homozygous for both traits? Give the genotypes and phenotypes of these homozygotes.

8. In certain breeds of chickens, the factor (B) is responsible for the production of black feathers and its allele (b) produces feathers that are basically white (except for flecks of black). The heterozygote is blue-feathered. Another factor (F) produces straight feathers, whereas its allele (f) results in the frizzled condition, giving brittle feathers. The heterozygote is mildly frizzled. What is to be expected from the following crosses? Give the phenotypic ratios.

   A. A black, frizzled hen × a white, straight rooster.
   B. A white, frizzled hen × a blue, straight rooster.
   C. A blue, mildly frizzled hen × a white frizzled rooster.
   D. A blue, mildly frizzled hen × a blue, mildly frizzled rooster.

9. In the human, the allele for normal skin pigmentation (A) is dominant to (a) for the absence of pigment, the albino trait. A certain type of migraine headache (M) behaves as a dominant to the normal condition (m), absence of headache. Write as much as possible of the genotypes of the following three persons:

   A. A phenotypically normal individual whose mother was an albino with migraine.
   B. A person suffering from migraine who has normal skin pigmentation. Both parents have normal skin pigmentation, and one of them suffers from migraine.
   C. An albino person who does not have headaches, whose normally pigmented parents both suffer from migraine.

10. A person who is dihybrid with respect to both the skin pigmentation and headache characteristics marries a person who is heterozygous at the skin pigmentation locus and who does not have headaches.

    A. Give the genotype of each and the kinds of gametes each will produce.
    B. What kinds of offspring are to be expected and in what phenotypic ratio?

11. The factors for normal hemoglobin (S) and for sickle cell hemoglobin (s) may be considered codominants. A blood examination can detect the heterozygote, since a small percentage of the cells will show sickling. A man whose cells show no sickling upon examination marries

a healthy woman who is found to have a certain percentage of sickle cells.

A. What is the chance of their having a baby with severe anemia?
B. The woman later marries a healthy man whose blood examination reveals sickle cells. They have three children free of sickle cell anemia. What is the chance that the next child will have the severe disorder?
C. What is the chance that any one of the healthy children is a carrier of the sickle cell allele?

12. Assume that medical science finds a treatment for sickle cell anemia. When instituted in infancy, it permits an otherwise doomed person to live a normal life span without any serious effects of the anemia. If two such persons, saved from the fatal effects of the disease, become parents, what will be the genotypes of their offspring? Would they require treatment? Explain. Also give an explanation that would be expected from a supporter of the Lamarckian concept.

13. In the human, free ear lobes (F) are dominant over attached lobes (f). Using this information and that on sickle cell hemoglobin in Question 11, answer the following:

A. Give the genotypes of the following persons:
   (1) A healthy woman with attached ear lobes whose red blood cells show some sickling.
   (2) A man with free ear lobes whose mother also had free lobes but whose father had attached ones. This man's red blood cells show some sickling.
B. What kinds of offspring are to be expected if the man and woman marry, and in what ratio?

14. The fatal disorder, Huntington's disease, results in progressive nervous deterioration and death. It usually strikes well after the age of puberty so that afflicted persons may have produced offspring. The genetic factor responsible for the disorder is a dominant one (H), which

is very rare in the population and does not seem to be on the increase.

A. Explain why more and more persons do not seem to be suffering from the disorder resulting from this dominant.
B. If a young person had one parent who died of the disorder and another who is apparently free of it, what does he or she know about the chances of developing the disorder?

15. Give the number of phenotypic classes expected from each of the following:

A. Testcross of AaBbCC.
B. Testcross of AaBbCcDd.
C. Testcross of AaBbCcDdEE.

16. Give the number of phenotypic classes to be expected when the following crosses are made and when there is dominance.

A. Cross of two individuals who are AaBbCcDd.
B. Cross of two individuals who are AaBbCcDDEE.
C. Cross of two individuals who are AaBbCcDdEe.

17. Give the number of genotypic classes to be expected in A, B, and C of Question 16.

In the pea plant, tallness (T) is dominant over shortness (t). Yellow seed color is dominant (G) over green (g), and round seed shape (W) is dominant over wrinkled (w).

18. Show the different kinds of gametes that can be formed by individuals of the following genotypes: (1) TtGG, (2) TtGGWw, (3) TtGgWw.

19. What would be the expected phenotypic ratio if each of the three plants described in Question 18 were testcrossed?

20. Suppose plants 2 and 3 in Question 18 were crossed. What would be the expected phenotypic ratio among the offspring?

2

CHROMOSOMES AND THE DISTRIBUTION OF THE GENETIC MATERIAL

All cells must solve certain problems

Early in the evolution of life, long before the origin of the first cell, primitive systems composed of aggregates of macromolecules and simpler substances probably settled out of the primordial seas. The only systems able to survive were those that by chance came to incorporate several features that guaranteed their persistence and continuity. Such systems were able to maintain themselves and eventually gave rise to the first primitive cells. From these, the familiar cell types of today have evolved. We find in all of them, from the bacterial cell to the most specialized ones in higher organisms, those features that the ancestral cell types acquired as they solved the problems of survival and reproduction.

All cells have some way of separating themselves from the external environment. This is accomplished almost universally by the presence of a cell membrane, which in turn may be surrounded by a wall (typical of plants) or a pellicle. However, this separation from the external environment must not be one of complete isolation because the continuation of internal cellular activities depends on communications with the outside. Substances must enter the cell when new building blocks are required; useless or toxic substances from cellular metabolism must pass out of

the cell. Therefore, the barrier between internal and external environment must not be complete, nor may it be haphazard.

Remarkable features of the cell membrane solve this dilemma and enable it to *selectively* control the entry and exit of substances to and from the internal cellular environment. This property of selective permeability possessed by the cell membrane is not just a passive process but depends on living cellular activities, most of which require an energy source. And thus we find that all cells have some means to release and transfer energy. This energy is needed for the cell to grow, but it is also required just to maintain stable metabolic processes. In most plant and animal cells, the release of energy and its transfer are mediated largely through the activities of certain intracellular bodies (organelles), the mitochondria, which can be seen with the light microscope but whose morphology is resolved only by the electron microscope (Fig. 2-1). These cytoplasmic organelles, which may be very abundant, contain all the enzymes required for aerobic respiration and the manufacture of adenosine triphosphate (ATP), the molecule in which high energy is stored in most cells. Energy release through anaerobic processes takes place outside the mitochondria in the cytosol, the fluid portion of the cytoplasm, but the mitochondria are by far the most

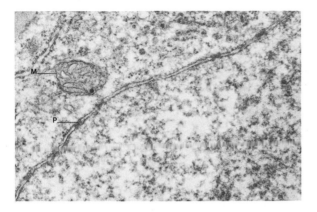

FIG. 2-1. Electron micrograph of portion of a cell of the grass Panicum, showing mitochondrion (*M*) in the cytoplasm. Note the pores (*P*) in nuclear membrane. (Courtesy of R. F. Lewis.)

important and efficient cell components in the manufacture of high-energy bonds. In addition to mitochondria, certain plant cells may also contain chloroplasts, chlorophyll-containing organelles in which energy of the sun may be trapped and converted to ATP. This trapped energy may then be used to manufacture food in the chloroplast. As we will learn in Chapter 19, both the mitochondria and the chloroplasts contain genetic information that plays a role in extrachromosomal inheritance.

However, not all cells possess mitochondria for the formation and transfer of energy. Cells may be classified into two general categories. Those that compose the most familiar plants and animals are termed *eukaryotic*—cells that contain an assortment of intracellular organelles in which certain specific processes are packaged and carried out. Eukaryotic cells also possess a distinct nucleus with a membrane that surrounds the genetic material. On the other hand, bacterial cells and blue-green algae lack true membrane-bound nuclei and never contain certain organelles, such as the mitochondria and chloroplasts. Cells of this type are called *prokaryotic*. Although prokaryotes are faced with the same problems as eukaryotes, they solve them in a somewhat different fashion, as we will see in a moment.

For a cell to persist from one generation to the next, it must have some mechanism to store and transfer information. Otherwise, newly formed cells would lose the ability to build essential cellular products and to carry on metabolic activities in a coherent manner. Moreover, cellular identity would be lost. Such chaotic systems would probably collapse and cease to exist.

All cells have solved this problem through the coding of information in certain large molecules. These are the nucleic acids, deoxyribonucleic acid (DNA) and ribonucleic acid (RNA). The former is the chemical substance of the gene and in eukaryotes is confined primarily to the nucleus. RNA, however, occurs as several molecular types, which are widely distributed throughout the nucleus and cytoplasm. All cells contain ribosomes, tiny organelles, each of which is composed of two subunits made up of proteins and RNA. Ribosomes are quite complex in their organization, and it is through their activities that the messages encoded in DNA and RNA are translated into the diversity of proteins associated with a given kind of cell (see Chap. 13).

General features of the prokaryotic cell

Our description of the cell so far has focused on eukaryotes. Acquaintance with some general features of the prokaryotic cell is now in order. Most prokaryotes consist of smaller cells than those of eukaryotes, in the size range of 1 to 10 μm. The prokaryotic cell possesses ribosomes but contains no membranous organelles such as Golgi, endoplasmic reticulum, nuclear envelope, or plastids. Although some possess flagella, these are not equivalent in structure to eukaryotic flagella. The prokaryotic cell has a plasma membrane, but no other permanent membrane system. We can say that the prokaryotic cell, in contrast to the eukaryotic, is not compartmentalized. Its ribosomes are smaller than those in the cytoplasm of eukaryotes, and the two classes of ribosomes, prokaryotic and eukaryotic, differ in their specific molecular constitutions (see Chap. 13).

Lacking a nuclear envelope, the genetic material of the prokaryotic cell is confined to a region called the *nucleoid*, which can be distinguished from the surrounding cytoplasm (Fig. 2-2). The DNA molecule composing the genetic material of the cell is naked and not complexed with proteins to form chromosomes like those typical of eukaryotes. With the electron microscope, fibrils of naked DNA can be seen in the irregularly shaped nucleoid, which is less dense in appearance than the cytoplasm. More than one nucleoid may be present in a given prokaryotic cell. Those prokaryotes that carry on photosynthesis have their pigments and enzymes bound to folds of the plasma membrane but not separated into compartments or any organelle equivalent to a plastid. The cell wall in prokaryotes is very different in composition

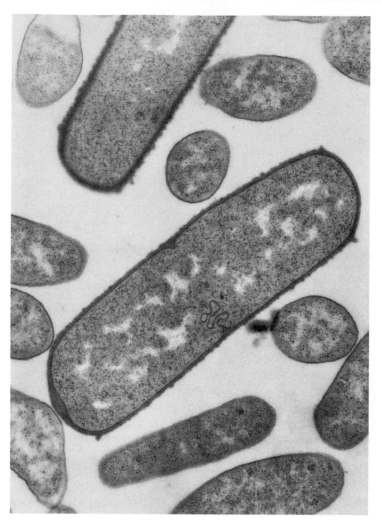

FIG. 2-2. Electron micrograph of bacterial cell (Erwinia sp.) undergoing division. The cell is packed with ribosomes that appear as deep-staining bodies. An invagination of the cell membrane is obvious. This is the mesosome to which the DNA is attached, which plays a role in DNA distribution as the cell divides. Irregularly shaped nucleoid regions appear as light areas in which the DNA can be seen as thin threads. (Courtesy R. F. Lewis.)

from the cellulose wall of eukaryotic plant cells. Still other distinctions exist between prokaryotes and eukaryotes; these are discussed in later chapters.

Communication between nucleus and cytoplasm

The organisms used in early classic genetics studies were mainly eukaryotic. We confine our attention to these in the next several chapters. The nuclear region of eukaryotes is bounded by an envelope, a system composed of a double membrane. The nuclear envelope does not completely seal the nuclear contents off from the cytoplasm. As in the case of the cell membrane, a selective process, which depends on living activities, regulates the passage of substances across the envelope. In addition, the electron microscope reveals that the nuclear envelope of most eukaryotic cells is perforated by pores, which may provide an additional passageway for certain molecules. Communication between nucleus and cytoplasm can also be visualized by electron micrographs, which show that the nuclear envelope is continuous with membranes of the endoplasmic reticulum (ER). The latter

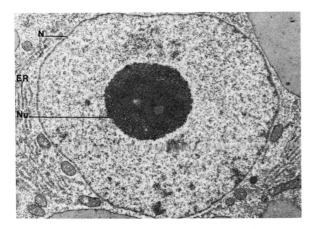

FIG. 2-3. Electron micrograph of a portion of a Panicum cell showing endoplasmic reticulum (*ER*) covered with ribosomes and the nucleus (*N*). Note the conspicuous nucleolus (*Nu*). (Courtesy of R. F. Lewis.)

(Fig. 2-3) is an intracellular system that performs several functions. Not only does it supply channels through which nuclear and cytoplasmic products can pass, it also allows communication with the outside environment. Technically the channels of the ER are outside the cell. This membrane-bound network furnishes a great deal of added surface area for interactions between the cytoplasm and external environment. Of utmost importance is the fact that these membranes themselves contain enzymes and provide surfaces on which many of the cell products can be constructed. In many active cells, the membranes of the ER are typically covered with ribosomes, which can be resolved easily with the electron microscope (see Fig. 2-3). In eukaryotes, they may be seen lying free in the cytoplasm or attached to the ER, depending on the particular cell. They may also be found on the outer membrane of the nuclear envelope.

Very conspicuous even with the light microscope is the nucleolus, one or several of which may be present inside the nucleus. The nucleolus is essential for the assembly of the ribosomes, which in turn are critical to the life and economy of the cell. The ribosomes depend on the DNA of the nucleus, which contains the coded information needed for both ribosomal construction and the formation of cellular proteins; nevertheless, the DNA can accomplish nothing without the ribosomes and other components of the cytoplasm. Although it is the chromosomes with their DNA that contain the information required to perpetuate the cell, the genetic information by itself becomes mean-

ingless if there is no way for coded messages to be interpreted and put to constructive use. In addition, the DNA depends on several devices for its orderly distribution to new cells at the time of cell division.

Distribution of the genetic material

Mitosis is the nuclear division which ensures that two newly formed cells will contain genetic information identical to that in the parent cell. Mitosis is a feature of eukaryotes. In prokaryotes, which contain but a single naked "chromosome," the genetic material is distributed in another, much simpler fashion. Several mechanisms were perfected during cellular evolution to make mitosis accurate and precise. While we examine some of these, we will also review Mendelian concepts in relationship to their physical basis in the cell. A good understanding of genetic phenomena depends on comprehension of the details of chromosome structure and behavior.

When nondividing cells are stained with certain dyes, the nuclei become quite distinct. This is due to a reaction of the coloring agent with the nuclear material, which then assumes a granular or netlike appearance. The name *chromatin* was given to this material in the nucleus that has an affinity for dyes, especially those basic in nature (Fig. 2-4A). As a nucleus starts to divide, one of the first detectable changes is seen in the chromatin, which gradually becomes more and more distinct. This is a consequence of the condensation of the chromatin into the bodies we call the *chromosomes*. It is important to realize that the term *chromatin* was not coined to define a specific kind of chemical material. Analysis of the chromatin has shown it to consist of several diverse substances: DNA, basic histone proteins, and the acidic or neutral nonhistone proteins. Therefore, chromatin refers to an assortment of macromolecules, whereas chromosomes are the bodies evident at nuclear divisions that are composed of chromatin.

Very characteristic changes in the chromosome and other parts of the cell accompany the process of mitotic division. These are so typical of cells in general that it has been possible to assign names to representative stages: prophase, metaphase, anaphase, and telophase (Fig. 2-4B–E). However, nuclear division is an uninterrupted process, in which the various activities proceed continuously without abrupt breaks between the successive stages.

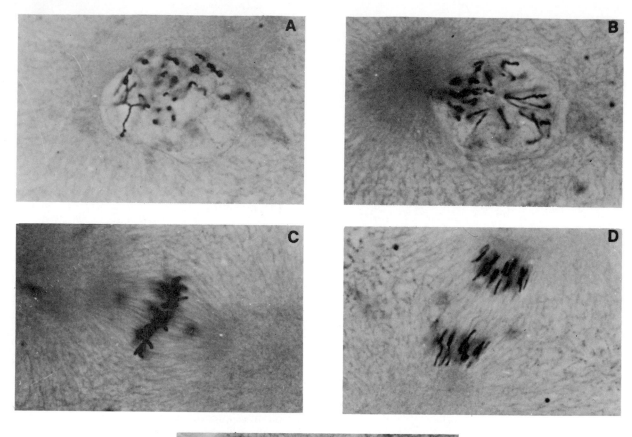

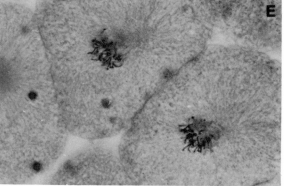

FIG. 2-4. Cells of whitefish embryo showing changes at mitotic division. (*A*) Nucleus of cell in which the chromatin has started to condense as the cell enters prophase. (*B*) Throughout prophase, chromosomes become more and more evident as a spindle composed of microtubules is assembled. By the end of prophase, the nuclear membrane has broken down and nucleoli have disappeared. The dense region seen here at the left of the nucleus is the centrosome, a clear zone containing rays emanating from the center. The rays plus the centrosome form part of the aster. In the typical animal cell, one aster marks each of the two poles of the spindle at metaphase (*C*). The metaphase chromosomes are found at the midplane of the spindle, the equatorial plate. Anaphase (*D*) is recognized by the movement of chromosomes to opposite poles of the spindle. At telophase (*E*) chromosomes become less and less evident as nuclear membranes and nucleoli reform. Meanwhile, the spindle gradually disassembles. Constriction of the cytoplasm that may begin in anaphase achieves the division into two cells, each with one nucleus.

Chromosome morphology

When describing chromosome morphology, cytologists refer to the "mature" or metaphase chromosome. This is customary because the chromosome at metaphase is in a compact state that can be studied easily with the light microscope. We can see clearly that the chromosome is composed of two halves or two sister chromatids. The expression *sister* implies that both structures are identical and have resulted from the replication of an original chromosome strand (Fig. 2-5). This double nature of the chromosome is apparent even at early prophase, the first stage of mitosis. It is a direct consequence of the chromosome material's duplication during *interphase,* the portion of the cell cycle between divisions and before the onset of prophase. The light microscope shows that this double nature extends along the length of the chromosome except at one particular region, the *centromere region,* which appears constricted. This region, one of which is found in every normal chromosome, is also known as the primary constriction, and it is here that the two

FIG. 2-5. Human chromosomes in white blood cell from a human male. The cell was actually in metaphase but was swollen by a treatment that separates the chromosomes so that they can be counted and their morphology observed. Note that each chromosome is composed of two chromatids and is double throughout its length. The chromatids are held together at the centromere region, which is evident here as a constricted area. Human chromosomes range in size from large to small and differ in the position of the centromeric region.

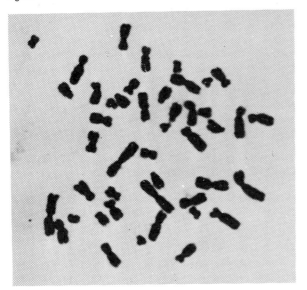

sister chromatids are held together (Fig. 2-5 and 2-6A). The centromere is the dynamic center of the chromosome and is responsible for many of the chromosome movements at nuclear divisions.

The centromere region typically stains faintly, but electron micrographs show that a continuous fiber of chromatin runs through it and the rest of the chromosome. The electron microscope also reveals that some kind of structure, called the *kinetochore,* is present at the centromere region. It is into the kinetochore of the centromere that spindle fibers are inserted. The appearance of this structural component of the centromere is constant for a given species of eukaryote, but variation occurs from one group to the next. Each of the two sister chromatids composing the chromosome at prophase and metaphase contains its own centromere, although the region appears single with the light microscope. The factors that hold the two centromeres of the sister chromatids together from interphase through metaphase are unknown. The separation of the sister centromeres, which takes place at anaphase of mitosis, helps to trigger the movement of the sister chromatids of each chromosome to opposite poles of the spindle. Once the separation of sister centromeres occurs, each chromatid then becomes an independent chromosome. A chromosome, therefore, may be a single entity, or it may be composed of two chromatids, depending on the stage of the cell cycle (Fig. 2-6).

When we observe the mature chromosome complements of many species, we can often see that the chromosomes have different appearances. (Note the size differences of human chromosomes shown in Fig. 2-5.) They may range in size from small to large. Another reason for their variation is related to the position of the centromere. It may be located more or less in the center of the chromosome, in which case the centromere is in a "median position." Consequently, the arms of such a metacentric chromosome appear approximately equal, and during its movements, the chromosome may assume a V shape (Fig. 2-7). If the centromere is just off to one side of the center, it is considered "submedian" in position. This kind of chromosome is submetacentric and will have one arm distinctly shorter than the other; it may resemble an L at anaphase. When the centromere is way off to one side, giving the chromosome a large arm and one that is very small or inconspicuous, the chromosome is termed *acrocentric,* and may have a J or I shape (see Fig. 2-7). When the centromere appears to be at an extremity and the chromosome has

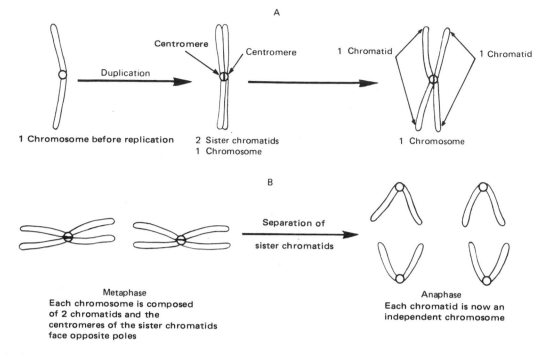

FIG. 2-6. Chromosome and chromatid. (*A*) The chromosome exists as a single structure before it is duplicated. Before the onset of prophase, the chromosome replicates, so that at prophase and metaphase it is composed of two identical sister chromatids, each with its own centromere *(represented as circle for emphasis)*. Both chromatids are held together at the centromere region. (*B*) Separation of the centromeres of the sister chromatids initiates anaphase movement. The sister chromoatids travel to opposite poles, and each is now an independent chromosome.

only one evident arm, the expression *telocentric* is commonly used.

In addition to the narrowing attributed to the centromere, some chromosomes possess still other constrictions, known as *secondary constrictions*. These may be so pronounced that nothing more than a filament appears to connect the two portions of the chromosome. In such a case the smaller portion is called the satellite (Fig. 2-8). Satellites typically are associated with the formation of the nucleolus. The nucleolus does not simply form anywhere in the nucleus, nor does it float about in the nuclear sap. Instead, it arises at a special site on a specific chromosome. In a diploid organism, at least two such chromosomes are present. The special sites are the *nucleolar organizing regions,* and these are generally found in the vicinity of the secondary constrictions or satellites. When nucleoli reform at telophase, it is at these sites. They remain attached to the nucleolar organizing chromosomes until the end of the next prophase. Even in the nondividing cell, the nucleolus remains associated with the chromosomal region on

which it was formed. The number of nucleoli present varies from one species to another, depending on the number of nucleolar organizing chromosomes. Because two or more nucleoli may fuse, the number per cell can vary, even within an organism. Although the nucleolus does not influence chromosome movement and is not involved in the segregation of genetic factors, it is essential for the assembly of the ribosomes, as already noted. Therefore, protein synthesis depends on the presence of nucleoli. On the level of molecular events, the nucleolus is a critical part of the cellular machinery essential for the translation of information stored in the genes.

Homologous chromosomes and genetic loci

When the chromosomes are studied at metaphase, it becomes evident that they exist in pairs. For any one chromosome of a particular length and shape, another can usually be found to match it. We say that the chromosomes exist in *homologous pairs*. This is so, because any sexually reproducing organism receives

Centromere
position:

Shape at
anaphase

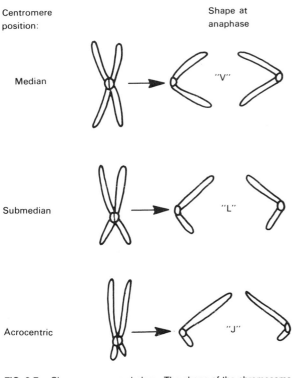

Median

Submedian

Acrocentric

FIG. 2-7. Chromosome morphology. The shape of the chromosome at metaphase depends largely on the position of the centromere. At anaphase, chromosome shape is the result of the arm length and the fact that the centromere leads in the movement to the poles.

one set of chromosomes from each parent. Therefore, in the body cells of sexually reproducing organisms we find the diploid number of chromosomes, two complete chromosome sets. A sex cell contains one complete set of chromosomes, the haploid number. A particular kind of chromosome from the maternal parent should also be found in the chromosome contribution from the paternal parent. An exception to this occurs in the case of those organisms that have sex chromosomes. In such species, the sex chromosomes received from the two parents may be quite different in appearance and may contain extensive regions that are not homologous. In some species the sex chromosome may be completely unpaired, so that one sex has one more chromosome than the other. The sex chromosomes are discussed more fully in Chapter 5.

All our genetic evidence tells us that genes occur in a linear order on the chromosomes. The position of a specific gene on a chromosome is called the *locus* of that gene. A specific locus is found at a characteristic location on a chromosome. Since the chromosomes other than the sex chromosomes exist in homologous pairs, each of these chromosomes in a diploid cell has a mate, which corresponds to it locus for locus. For example, in the fruit fly there is on chromosome III a locus where a gene occurs that affects eye color, which is called the *sepia locus* (the naming of loci is discussed at the end of this chapter). Therefore, any fly would have this locus for eye color rep-

FIG. 2-8. Secondary constrictions. These occur on some chromosomes and may be so pronounced that a portion of the chromosome appears to be set off as a satellite. These secondary constrictions frequently contain nuceolar organizing regions at which nucleoli reform

at telophase. Every typical normal diploid possesses at least two chromosomes with nucleolar organizing regions. All constrictions of the chromosome other than the primary one are known as secondary constrictions, even those not associated with nucleolar organizers.

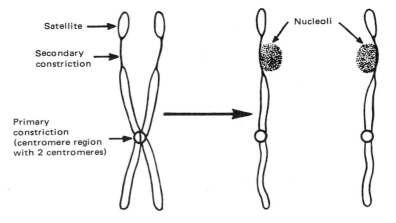

Satellite

Secondary constriction

Nucleoli

Primary constriction (centromere region with 2 centromeres)

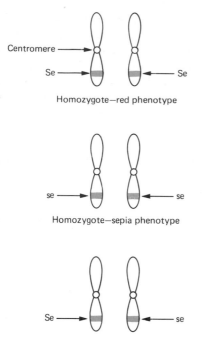

Centromere

Se   Se

Homozygote—red phenotype

se   se

Homozygote—sepia phenotype

Se   se

Heterozygote—red phenotype

FIG. 2-9. Loci on chromosomes. The figure depicts, in a highly schematic way, one pair of chromosomes, Chromosome III of Drosophila. The diploid receives one such chromosome from each parent. The two correspond locus for locus. The sepia locus is therefore present twice in every diploid cell, once on each chromosome of the homologous pair. The individual represented in the top part of the figure has the same gene form on each of the two chromosomes, the allele for red eyes, whose symbol is Se. The fly is homozygous for the allele (Se Se) and would have a red-eyed phenotype. A fly with two recessive alleles for sepia (*middle*) at the sepia locus is also homozygous (se se) but will show sepia eye color. A fly with contrasting genetic factors at the sepia locus (Se se) is a heterozygote with red eyes. The factor for red eye color (Se) is dominant over its allele, or alternative form (se), for sepia eye color.

ticular locus—two forms for red eyes in the one case, and two for sepia in the other.

Because flies with red eyes can be easily distinguished from those with sepia coloration, two distinct phenotypes can be recognized on the basis of eye color. We learned in the last chapter that knowledge of the phenotype indicates the genotype or kinds of genetic factors the individual carries for a particular character. Because the form of the gene for sepia is recessive to the form for red pigmentation, we know that sepia flies are homozygous at the sepia locus; however, we could not infer the genotype of a red fly simply by observing the phenotype. A red fly could receive a factor for red pigmentation from one parent and the alternative gene form for sepia pigmentation from the other parent. This fly would be *heterozygous* for eye color, having a pair of contrasting forms of the eye color gene at the particular locus that affects eye pigmentation (Fig. 2-9). Any locus therefore has contrasting possibilities because it can contain different forms of a gene, which may produce very different phenotypic effects. In the last chapter, we learned that contrasting forms of the same gene are called *alleles*. We now see that allelic forms of a gene are associated with a specific locus on a chromosome. The factor for red eyes and that for sepia are allelic to each other; they produce contrasting effects and are associated with variation at the same chromosome locus.

The chromosome and the cell cycle

DNA is a major constituent of the chromatin and thus of the chromosome. The chromosome, therefore, may be thought of as the vehicle for distribution of the genetic material at nuclear divisions. Since it transports the substance of heredity, its behavior directly determines the genetic composition of newly formed cells. Any abnormalities in chromosome distribution or any damage to the chromosomes through environmental influences can produce serious effects. A normal cell contains at least one complete balanced set of instructions coded in the DNA of its chromosomes. When the cell divides, it is critical for each of the two new cells to receive information identical to that present in the parent cell that gave rise to them. The elimination of information or the addition of extra copies can cause an unbalanced genetic condition resulting in abnormality or cellular death. It is the miotic process, with its characteristic stages, that

resented twice in its body cells, one on the chromosome III received from the maternal parent, the other on the chromosome III from the paternal parent (Fig. 2-9). However, this does not mean that the identical form of the gene must be present twice. Red eye color is normal in the fruit fly, and any fly could have two identical genetic factors for red eye color at this locus: a gene form for red pigmentation on the maternal chromosome and the same one on the paternal. Another fly with sepia eyes could have the gene form for sepia pigmentation at that locus on both members of the homologous pair. We would say that these flies are *homozygous* for eye color, meaning that the individuals have identical forms of the gene at that par-

guarantees the transmission of balanced sets of genetic information from one cell generation to another. The term *genome* refers to the genetic content of a single set of chromosomes.

Portions of a complete set of instructions are packaged in the separate chromosomes. Any one unreplicated chromosome contains a fixed unit of genetic information, which is part of the total amount coded in the DNA of the cell. Upon replication, each of the two sister chromatids composing a prophase or metaphase chromosome contains an identical amount and kind of genetic information. Evidence gathered from electron microscopy, autoradiography, and procedures using refined techniques of physical chemistry indicates that only one DNA molecule is present per chromosome or per chromatid of a replicated chromosome. This molecule, with its associated proteins, can be visualized as a very long fiber running the length of a chromosome (or of a chromatid of a chromosome) and folding back on itself repeatedly. More details on the organization of the chromosome fiber are given in Chapter 17.

Although discrete chromosomes are not ordinarily seen in the nondividing nucleus, it is most important to realize that they are nevertheless present and have not lost their individuality. They are invisible at this time largely because the chromatin fiber of each chromosome is in a highly uncoiled state. The narrow, extended thread is greatly elongated, making visualization difficult and identification of entire separate chromosomes impossible. It is not generally appreciated that the chromosomes in this extended interphase condition are very active metabolically. Indeed, it is in the nondividing cell, when the chromosomes are *not* condensed, that the major molecular activities occur involving DNA, RNA, and protein interactions. On the other hand, a chromosome in the condensed state, typical of division stages, is quite inactive in the dynamics of cellular metabolism. The activities we see at these times are concerned mainly with packing the chromosome material and distributing it; little else is going on. Therefore, it is incorrect to think of the nondividing cell as "resting" and doing nothing. It is wrong to believe that chromosomes are active only when we see them at mitosis or meiosis. A better term than resting for a nondividing nucleus is *metabolic*, because it reflects the true activities of that nuclear stage.

Interphase may be used to denote the state of the metabolic nuclei of cells between divisions. The entire cycle through which dividing cells pass can be separated into two main portions: interphase (the growth period between nuclear divisions) and the actual mitosis or division period, which is referred to as M or D. Early in the 1950s, the technique of autoradiography permitted the recognition of three time periods within interphase itself. It became evident that cells double their DNA content during a certain portion of interphase. This was called the S phase, and all other phases of the cell cycle are defined in terms of it. Immediately preceding the S phase is a period of interphase called G_1; this phase is an interval between the end of the previous nuclear division (M) and the S phase. Following the S phase is another gap period, G_2, before mitosis begins. In various eukaryotic cell types, the following time, in hours, is spent in the various portions of the cell cycle: G_1 (8), S (6), G_2 (4½), M (1) (see Fig. 2-10).

Depending on the cell type and the developmental stage, much variation occurs in the time spent in the phases of the cell cycle, especially G_1. In many animal embryos, cell divisions occur at a very rapid pace. The rate of DNA synthesis accompanying these embryonic divisions is 100 times faster than in adult cells of the same species. Moreover, G_1 may be very brief or even absent. For example, in embryonic cells of the toad *Xenopus* (Chap. 13), the entire cell cycle is only about 25 minutes long, and G_1 is absent. The S phase is

FIG. 2-10. The cell cycle. The cycle shown here is typical for a mammalian cell in culture. Although cell cycles vary from one cell type to the next, that portion of a cycle devoted to mitosis is brief, the major portion being spent in interphase. A cell may enter a G_0 phase in which it no longer undergoes cycling.

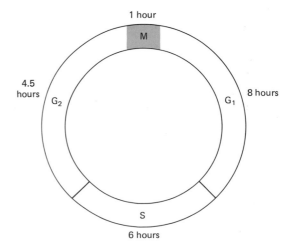

followed by a brief G_2 and then an M. Before the M phase is completed, the following S phase is initiated. In contrast, the adult cells of *Xenopus* typically pass through all the phases of the cell cycle, and the S phase itself lasts about 20 hours! Despite such variations, critical events typically taking place in G_1, S, and G_2 appear to be required for the actual division, M, to take place.

While the length of G_1 varies greatly, certain very important key events must take place during that phase in the typical cycle. These entail preparations for the S phase. When G_1 is absent, such processes must occur at some earlier point, such as immediately after M or even before M is completed. DNA replication at S involves very complex interactions at the molecular level (Chap. 11). It is staggering to realize that, in the human, 46 extended chromosome threads with a total length of about 6 feet must be duplicated while they are packed in a nucleus about 1/2500 of an inch in diameter! In the G_2 portion of interphase, final preparations are made for the mitotic process. No continued cycling of cells occurs without the S, M, and G_2 phases.

We are well aware that cells can live without dividing and that some cells composing an organism rarely, if ever, undergo division. There is very good evidence that the late G_1 period contains a point, called the *restriction point*. We can think of this as a point of no return. If a cell passes this point, it will enter the S phase and then complete the rest of the cell cycle. Cells that remain arrested at this restriction point, and no longer divide, are often considered to have left the cell cycle and to have entered the G_0 phase.

Studies of the cell cycle are extremely important, not only for the insight they provide on the key mechanisms involved in cell division, but also for the information they provide on the cancer cell, for cancer is a problem of the cell cycle. Although a cancer cell's cycle is not unusual, and is similar to that of any other dividing cell, it keeps repeating the cycle over and over again, as if it had been removed from the constraints imposed on a normal cell, which divides only in harmony with its neighbors. The malignant cell is discussed at greater length in Chapter 17.

Changes at mitosis

The interphase chromosomes begin to contract with the onset of prophase and become progressively more condensed and conspicuous as this stage progresses (see Fig. 2-4B). By the end of prophase, we can easily distinguish the two chromatids and the centromere position of each chromosome. Meanwhile, throughout prophase, the spindle apparatus is forming. The spindle is composed of microtubular fibers—that is hollow cylinders assembled from linear arrays of a protein known as tubulin. Actually, each spindle is made up of two *half-spindles*. The most numerous of the spindle fibers are the *polar fibers,* which extend from each end, or pole, of the spindle to the midregion, the equator of the spindle. The midplane of the spindle is often called the *equatorial plate*. The electron microscope reveals that in this area the microtubules from each pole overlap (Fig. 2-11A). Late in prophase, when the nuclear membrane breaks down, another group of spindle fibers arises. These radiate from the kinetochore region of each chromatid of a chromosome. In the human, about 20 to 40 microtubular fibers are inserted into each kinetochore. This number varies greatly from one species to the next. The kinetochore fibers interact with the polar fibers in such a way that the kinetochore fibers of one of the sister chromatids of a chromosome associate with polar fibers of one of the half-spindles, and the kinetochore fibers of the other sister associate with the fibers of the opposite pole (Fig. 2-11B, C). It is the kintechore in the centromere region of each chromatid that provides the only connection between the chromosomes and the mitotic apparatus.

With the disappearance of the nucleoli from the nucleolar organizing region of the satellited chromosomes and the breakdown of the nuclear membrane, interactions between kinetochores and the attached fibers are responsible for movement of the chromosomes to the middle of the spindle. The arrangement of the chromosomes at the equator of the spindle marks the beginning of metaphase (see Figs. 2-4C and 2-11B). Keep in mind that at metaphase it is the centromere of each chromosome that is at the equatorial plate. The arms of the chromosomes project from their centromere regions in various ways, because they are being carried along passively by the dynamic centromere region and therefore do not all lie parallel to the equator as diagrams of metaphase often suggest. At metaphase, the kinetochores of the sister chromatids face opposite poles (Fig. 2-11C).

In the cells of many animals, fungi, algae, and lower plants, the spindle forms in association with *centrioles,* microtubular structures that mark the poles (Fig. 2-12A). Centrioles always occur in pairs. A pair is duplicated in the S phase of the cell cycle, so that two pairs of centrioles are already present at the onset

A

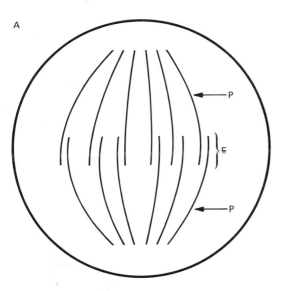

B

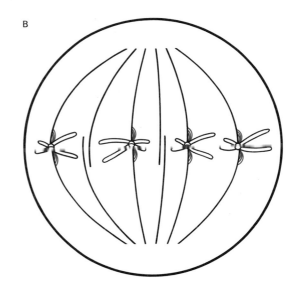

C

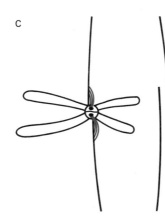

FIG. 2-11. The spindle and chromosome orientation. (A) The spindle is essentially two half-spindles. Polar fibers (P) emanate from opposite poles and overlap in the midregion, the equatorial plate (E). (B) At metaphase, the chromosomes are oriented at the equatorial plate as a result of the interaction of the kinetochore fibers (red) with the spindle fibers. (C) The arrangement of each chromosome at metaphase is such that the kinetochore (black dot) of each chromatid is attached by the kinetochore fibers to a polar microtubule. The kinetochores of sister chromatids are attached to polar fibers emanating from opposite poles.

of prophase. A clear area surrounding the centrioles, the *centrosome,* is particularly conspicuous in some cells (2-4B, C). The two pairs of centrioles gradually move apart throughout prophase as rays of microtubules emanate from their vicinity, passing through the centrosome. This region—composed of a centriole pair, the associated rays, and the centrosome—is known as an *aster.* Because two pairs of centrioles are present at prophase, two asters arise, and these eventually come to mark the poles of the spindle at metaphase (Fig. 2-12A).

An animal cell in which the centrioles have been experimentally destroyed still manages to form spindles. Moreover, the cells of higher plants lack asters and centrioles entirely, yet manage to form spindles. Such observations leave unclear the exact role of the centrioles and asters in spindle fiber assembly. When they are present, they seem to act as a focal point for the assembly of the polar microtubules, though they are not essential to their organization. Such astral spindle fibers look quite different from the less clearly focused anastral spindle of higher plants (Fig. 2-12A, B). The electron microscope has disclosed at the poles of both astral and anastral spindles a region that

A

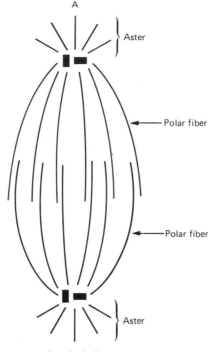

B

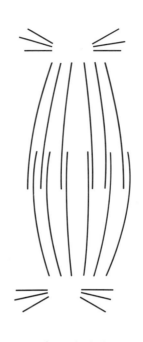

Astral spindle

Anastral spindle

FIG. 2-12. Two types of spindles. (*A*) In the astral spindle, each pole is marked by a pair of centrioles. Each centriole pair, along with the microtubules that radiate from it, constitutes an aster. (*B*) The anastral spindle typical of higher plant cells lacks centrioles and asters. The polar fibers appear much less focused than the fibers of the astral spindle that converge toward the centrioles.

contains lightly staining material from which the polar spindle fibers appear to grow. This rather hazy region may well represent the actual center in the cell that is responsible for the organization of polar microtubules.

One of the most critical events of the mitotic process is the separation of sister chromatids. Their complete separation ends metaphase, and anaphase movement is initiated (Fig. 2-4*D*). The movements of the chromatids at anaphase appear to entail a movement of the kinetochore fibers, which pulls the attached chromatids to the opposite poles (Fig. 2-13). Also, the polar fibers of each half-spindle increase in length. As they elongate, the polar fibers slide past each other, and the poles are pushed farther apart. The generation of the forces involved in the pulling of the kinetochore fibers and the pushing of the polar ones remains obscure, although experiments at the molecular level continue to fill in parts of the puzzle.

As they travel to opposite poles of the spindle, the chromosomes assume various configurations of V, J, and L (Fig. 2-7). This results from the position of the centromere and the fact that it is the leading point of every chromosome. Each of the original sister chromatids is now an independent chromosome. Each chromosome at this part of the cell cycle is single, carrying only one chromosome fiber. If the cell passes through another G_1, S, and G_2, each chromosome will again be composed of two chromatids, each with its own chromatin fiber. It is most important to appreciate the following: Because the two sisters composing a chromosome become independent and move to opposite poles at anaphase, identical genetic material will be found at opposite ends of the cell when this stage is completed. By the end of telophase, the spindle disappears, nuclear membranes and nucleoli are reconstituted, and each new (daughter) nucleus assumes the appearance characteristic of interphase (see Fig. 2-4*E*).

Although mitosis is commonly thought of as a cell division, it is basically a division of the nucleus. *Cytokinesis*, division of the cytoplasm, may or may not take place. The genetically significant point is that the two daughter nuclei resulting from the mitotic division of an original nucleus are identical insofar as their chromosome complements are concerned. Thus,

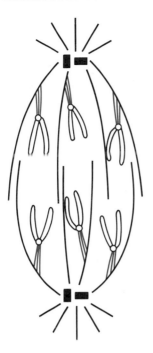

FIG. 2-13. Anaphase movement. At anaphase, the kinetochore fibers appear to pull the chromatids to which they are attached toward the opposite poles.

mitosis is a process that ensures continuity of the identical genetic material from one nuclear division to the next. Barring abnormalities, the same genes will be present in the chromosomes of all those cells that trace back to the same original cell, even though many mitotic divisions may have intervened. No new combinations of genetic factors result from this process, and one kind of gene arrangement is therefore preserved from one cell generation to the next. The importance of such a process in any multicellular organism is obvious. Every normal cell composing the body will contain a full complement of genetic material characteristic of that species. Any departures can bring about abnormalities because the cell may not have a normal or balanced set of genes necessary for the production of a normal phenotype.

Alterations in the normal chromosome picture can result as a consequence of various types of aberrations that arise during a nuclear division. Chromosome aberrations and their consequences, often detrimental to the individual, are the major topic of Chapter 10. Although such accidents occur from time to time in all living things, and may even take place in the course of aging, mitosis is still a remarkably accurate process. A multicellular organism may contain tril-

lions of cells, produced from countless mitotic divisions, which continue throughout the lifetime of the individual in certain tissues. Viewed in this light, mitosis is seen to be an exceptionally precise event, usually free of errors.

The human chromosome complement

The correct number of chromosomes in the human was established in 1956. The great strides that permitted the accurate description of the human chromosome complement were made possible largely through the perfection of a simple procedure in which a small sample of blood is taken and cultured. The cultured cells are exposed to a mitotic stimulator derived from the castor bean plant. This exposure triggers certain white blood cells to undergo mitosis. These cells are then trapped in mitotic metaphase through the application of the drug colchicine, a chemical that disrupts the organization of the mitotic spindle and thus prevents anaphase movement. The cells are than subjected to a chemical fixative (killing agent) and exposed to a hypotonic solution, which swells them and disperses the chromosomes so that they become separated from one another. Finally, the cells are attached to slides and stained for observation with the microscope. The complement of the human cell can then be photographed (see Fig. 2-5). The term *karyotype* is used to refer to the chromosomes of a single cell or of an individual. Each chromosome may be cut out of a photograph and placed in a systematic arrangement (Fig. 2-14). When a karyotype analysis is done, the chromosomes are assembled into homologous pairs and then arranged in order of decreasing size. In any karyotype preparation, the centromeres are placed on the same line or level so that centromere position and length of chromosome arms can be easily compared from one chromosome pair to the next.

When the chromosome complements of most humans are examined, it is seen that 23 pairs of chromosomes are present and that 22 of these pairs are identical when the chromosome complements of males and females are set side by side (Fig. 2-14). There is, however, a difference in one pair, the pair designated the *sex chromosomes*. It is seen that a female karyotype contains two sex chromosomes of appreciable size and identical appearance. These are the X chromosomes. On the other hand, the male karyotype shows one X chromosome and a very small chromosome, the Y. All the chromosomes in the human

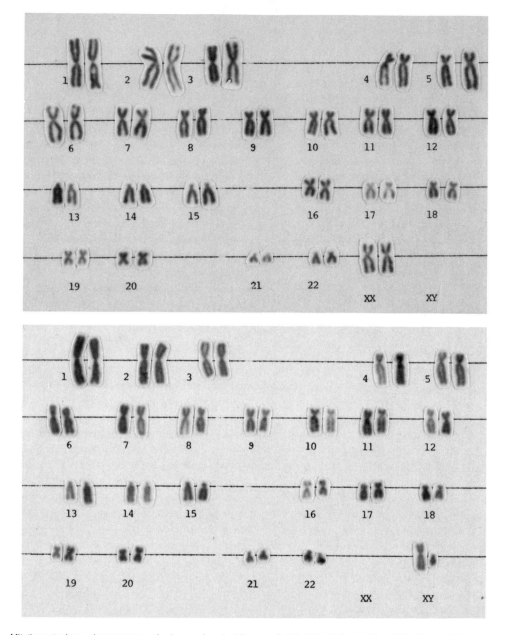

FIG. 2-14. Mitotic metaphase chromosomes of a human female (*A*) and a male (*B*) arranged in homologous pairs. (Courtesy George I. Solish, MD, PhD; Jyoti Roy, MS, Thomas Mathews, MS, Long Island College Hospital, Cytogenetics Lab.)

complement other than the sex chromosomes are called *autosomes*. The human therefore has 22 pairs of autosomes plus one pair of sex chromosomes (2 X chromosomes in the female and 1 X and 1 Y in the male).

Table 2-1 describes the major groups of human chromosomes as recognized in 1960 at a meeting of cytogeneticists in Denver, Colorado. Following the Denver classification, the autosomes are serially numbered 1 to 22 in descending order of length. The sex chromosomes continue to be called X and Y. Following routine staining procedures, the chromosomes can be easily classified into seven groups. Within a group, further identification of individual chromosomes can sometimes be made. However, distinctions are frequently difficult or uncertain, as in Group C,

TABLE 2-1   Classification of human mitotic chromosomes

GROUP	DESCRIPTION
1-3 (A)	Large chromosomes with approximately median centromeres. The three chromosomes are readily distinguished from each other by size and centromere position.
4 and 5 (B)	Large chromosomes with submedian centromeres. The two chromosomes are difficult to distinguish, but Chromosome 4 is slightly longer.
6–12 (C)	Medium-sized chromosomes with submedian centromeres. The X chromosome resembles the longer chromosomes in this group, especially Chromosome 6, from which it is difficult to be distinguished. This large group is the one that presents major difficulty in identification of individual chromosomes.
13–15 (D)	Medium-sized chromosomes with nearly terminal centromeres (acrocentric chromosomes). Chromosome 13 has a prominent satellite on the short arm. Chromosome 14 has a small satellite on the short arm. Chromosome 15 was later found to possess a satellite as well. Satellites may be difficult to detect in preparations.
16–18 (E)	Rather short chromosomes with approximately median (in Chromosome 16) or submedian centromeres.
19 and 20 (F)	Short chromosomes with approximately median centromeres.
21 and 22 (G)	Very short, acrocentric chromsomes with satellites on the short arms. The Y chromosome resembles these chromosomes and is placed in this group. It cannot be distinguished from the other members in many cases.

Source: Reprinted with permission from *JAMA*, 174: 159–162, 1960.

which contains the X chromosome, and in Group G, which includes the smallest chromosomes as well as the Y. Fortunately, since 1969, techniques have been perfected that reveal distinctions among all the chromosomes, even those that appear identical when stained with common dyes. The first of these methods was reported by Caspersson, who exposed chromosome preparations to the fluorescent dye, quinacrine mustard. Following such treatment, characteristic bands, which are designated Q bands, can be recognized on the chromosomes (Fig. 2-15). The banding pattern is distinctive and constant for a given chromosome. The Q bands vary in width and amount of fluorescence along the length of a chromosome, as well as from one chromosome to another. The visualization of Q bands requires a modified form of the light microscope fitted with appropriate lenses. Not only have fluorescent staining procedures permitted precise identification of each human chromosome because of the characteristic banding pattern, they have also revealed that nondividing cells from males contain a fluorescent body that is absent in cells taken from females. This body apparently represents the Y chromosome.

Although fluorescent techniques permit direct viewing and photographing of the chromosomes, they necessitate the use of special optics. Moreover, the fluorescence fades after prolonged illumination and cannot be restored by restaining. These problems were solved in the early 1970s with the announcement of other procedures that use simple methods to bring out bands on the chromosomes and permit permanent staining of the material with the well-known nonfluorescent Giemsa stain.

One type of procedure, which entails a trypsin-digesting pretreatment followed by Giemsa staining, brings out bands that have been designated G bands. These are very similar (although not necessarily identical) to the Q bands; in general, heavily stained G bands correspond to the highly fluorescent Q bands (Fig. 2-16). Another method involving heat denaturation produces a reverse banding pattern that is just the opposite of the Q and G pattern. These so-called R bands stain heavily where the staining is poor in the G band pattern. Those regions that stain poorly in the R band pattern are heavily stained in the G band pattern. R bands can also be brought out when preparations are exposed to certain fluorescent dyes.

Still another banding pattern is seen following a procedure that includes a pretreatment with hydrochloric acid and sodium hydroxide. The C bands that result are very distinct from the other bands we have described and are found only in certain restricted regions of the chromosomes, particularly the centromere region, secondary constrictions, the long arm of the Y chromosome, and the satellites of the G group (Fig. 2-17). These characteristic C bands stain regions

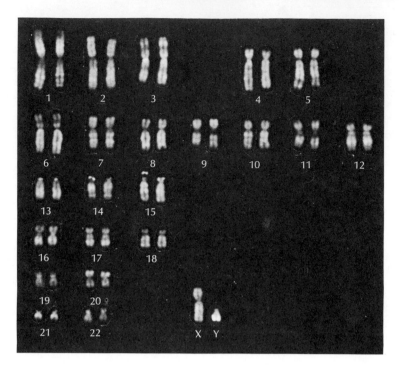

FIG. 2-15. Karyotype of human male showing Q bands. The sex chromosomes are at the bottom of the karyotype. Note the small, highly fluorescent Y chromosome. (Courtesy of T. Caspersson, Karolinska Institute.)

that represent a special type of chromatin found along the length of the chromosome. This kind of chromatin tends to remain condensed in the metabolic or interphase nucleus and in prophase, unlike the rest of the chromosome material. Such chromatin is known as *heterochromatin* as opposed to *euchromatin,* which composes most of the chromosomes and tends to be stretched out at interphase and the early stages of nuclear divisions. In contrast to euchromatin, heterochromatin is late replicating. At the time the genetic material is being replicated at S of interphase, the heterochromatin lags behind the euchromatin. Heterochromatin such as that found at the centromere region and revealed by the C banding procedure is known as *constitutive heterochromatin,* so-called because it is constant in its behavior and always tends to behave differently from the major portion of the chromatin. We will return to this subject in Chapter 5, where we discuss another type of heterochromatin.

The banding techniques have opened up a whole new era in the investigation of human chromosomes. At a conference held in Paris in 1971, various conventions were established to describe the human hap-

loid chromosome complement at mitotic metaphase. The shorter arm of any chromosome was designated the "p" arm and the longer of the two the "q" arm. When a karyotype is prepared, the chromosomes are arranged so that all the p arms are directed upward, and the q arms downward. Based on the G banding pattern, 350 bands were recognized. The dark and light bands of each p and q arm were assigned specific identifying numbers. Improved banding techniques have now increased the resolution of the bands and reveal as many as 2000 bands in a haploid human chromosome set at late mitotic prophase. As the chromosomes continue to condense from prophase to metaphase, the thinner prophase bands coalesce to produce the fewer but thicker bands of metaphase. Because each human chromosome can be readily identified and its two arms characterized by distinct banding patterns, any change in the number of a given chromosome can be recognized as well as any alteration that affects the bands, and hence the structure of a chromosome arm. Such change can then be related to certain specific genetic disorders (Chap. 10).

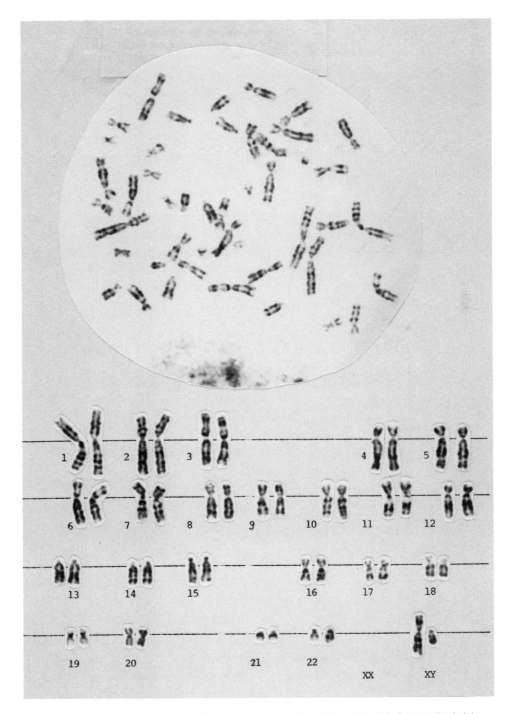

FIG. 2-16. Karyotype of human male showing G bands. (Courtesy George I. Solish, MD, PhD; Jyoti Roy, MS; Thomas Mathews, MS, Long Island College Hospital, Cytogenetics Lab.)

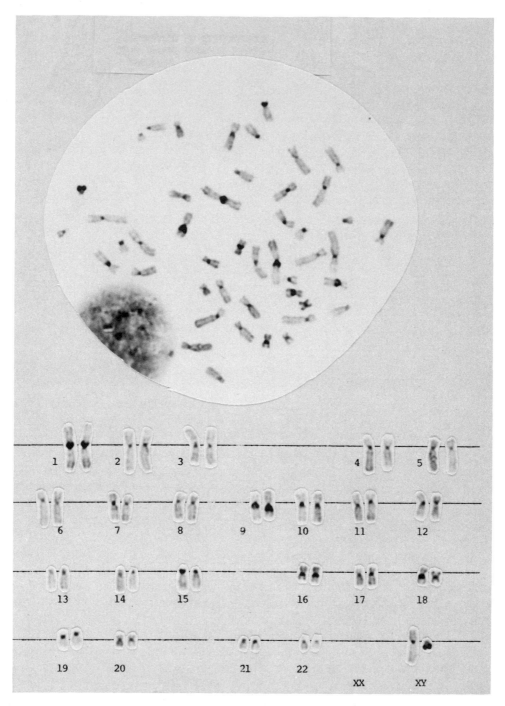

FIG. 2-17. Metaphase chromosomes of human male showing C banding. Note the staining of the centromere regions. In this individual, the two homologous chromosomes composing pair 9 differ in the C banding because of an inverted chromosome segment affecting the region of the centromere. (Courtesy George I. Solish, MD, PhD; Jyoti Roy, MS, Thomas Mathews, MS, Long Island College Hospital, Cytogenetics Lab.)

Giant chromosomes

No discussion of chromosomes would be complete without reference to the unusual chromosomes found in several groups of flies. One of the many features of the fruit fly Drosophila that has contributed to its value as a genetic tool is its giant chromosomes, found in the immense nuclei of the larval salivary gland cells. The larval stage of the fly is concerned primarily with food getting; efficient salivary glands are important to an insect, which literally eats its way through food. These active glands, as well as certain other organs found in some of the dipteran insects (flies), achieve their growth largely through an increase in cell mass and volume rather than an increase in the number of cells. After a certain number of cells is established through mitosis, further cell divisions cease. Growth continues, however, as cells enlarge. As the nuclear size increases, the chromosome threads that represent the chromosomes at metaphase duplicate again and again without an accompanying mitotic division. These extra chromosome sets are probably necessary for the increased activities of cells of greater size and mass. Another unusual feature of these cells is the phenomenon of somatic chromosome pairing. Typically, homologous chromosomes pair only at meiosis (see the next chapter). But in the cells of the salivary glands, not only do homologous chromosomes pair, but the paired chromosome threads duplicate over and over throughout the larval stage. Approximately 11 rounds of duplication take place, producing thousands of individual threads. The very strong forces of pairing that seem to operate in these cells prevent the threads from separating. The end result is that each nucleus appears to contain a haploid number (four in *Drosophila melanogaster*) of giant chromosomes. Each, of course, is actually composed of the many separate chromosome threads of two homologues.

All the threads held together side by side as a large compound unit form a giant structure, a *polytene chromosome*, that is, a chromosome composed of many threads. This growth of the salivary glands through increasing cell size rather than number of cells may be very efficient for the best operation of the enzymatically active cells during the period of active food getting. The glands will finally be resorbed when the larva is transformed into the pupa.

The large salivary cells are not destined to divide again. The large polytene chromosomes are therefore not in any stage of mitosis, but actually in an extended interphase. They are stretched out almost to their maximal length, a condition that makes them superb subjects for cytogenetic studies. They provide an unusual advantage for assigning genes to definite chromosome regions. When the polytene chromosomes are stained, they display deeper staining regions, bands of varying widths alternating with lighter stained regions, the interbands (Fig. 2-18). The width and arrangement of the bands is not random but is so characteristic of every chromosome region that a small chromosome segment by itself may be identified by its banding pattern. The deeply staining bands are also called *chromomeres*. Comparable deep-staining regions are very evident along the length of the chromosomes at early meiosis (see the next chapter). There is reason to believe that the chromosome threads are more tightly folded or packed in the region of the bands than in the interbands. As a result, a greater concentration of hereditary material should occur in the band region, and this seems to be the case, as indicated by stains that are specific for DNA. Since thousands of separate chromosome threads are associated side by side to form a polytene chromosome, the bands, or chromomeres, necessarily stain more deeply than the interbands. Specific genes have been associated with specific bands, although it is still

FIG. 2-18. The salivary gland chromosomes of *Drosophila melanogaster*. (Reprinted with permission from *Hered. J.* 25: 464–476, 1934).

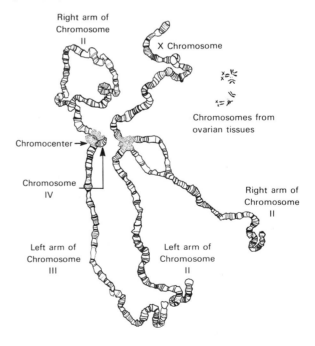

unsettled whether one gene or more are located in a distinct band.

The structure of a chromosome may be altered, and a gene may actually be lost or deleted. Such a loss is often detected by the genetic effects that ensue. Observation of the giant chromosomes may then show that a loss of a certain band accompanies the loss of the gene. Other kinds of changes besides gene deletion also alter chromosome structure, such as inversion, in which the order of the genes becomes reversed, or translocations, in which chromosome arms are shifted to new locations. We will see (Chap. 10) that the genetic effects that accompany these changes can be correlated with corresponding changes in the bands. This enables the cytogeneticist to assign genes in Drosophila to definite chromosome locations. The polytene chromosomes are also of immense value in studies of cell differentiation and gene action (Chap. 17). We will continue to refer to these giant structures throughout our discussions.

Naming of genes and loci

Before proceeding further in our genetic discussions, we must pause to examine the conventions used to assign names to genes and the loci on the chromosomes, as well as to the stocks of organisms used in genetic research. Because Drosophila is such a widely used genetic tool, we refer to it in the following discussion, although the same reasoning applies to other organisms as well.

Work with the fruit fly requires familiarity with different types of pure-breeding stocks, with special attention paid to the "wild" or normal. As commonly used in genetics, the designation "wild" implies the standard form typically found in nature. Any inherited departure from the wild may result from a mutation at a certain gene site, a locus on a chromosome. The name given to any mutant stock usually suggests the kind of phenotypic variation from the wild type attributed to the genetic alteration. For example, red eye color in Drosophila represents wild type. Gene mutation at one of the loci on the X chromosome may result in the absence of eye pigment. Stocks that are homozygous for this change are designated "white," and the position or point on the chromosome where the alteration has taken place is named the "white locus."

The wild gene form for red eye color and its mutant form for white represent a pair of alleles, and either one of the two members may be present at the "white locus" on a particular X chromosome. In elementary genetic discussions such as ours so far, a pair of alleles is usually symbolized by a capital letter for the dominant and a small one for the recessive allele. The symbols W and w are frequently encountered for red eye and white eye, respectively. However, the geneticist does not necessarily follow this system. Instead, the plus sign (+) may be used for *any* wild trait. The factor for red eye is thus simply designated " + " (or w^+ for extra clarification), and its allele for white eye remains w. The name given to any genetic locus—in this example "white"—describes the first mutation or variation that was detected at that site. If there had never been a mutation to alter the color of the eye, we would know nothing about the inheritance of eye color and would be completely unaware of any genes or loci affecting that character. We have noted the fact that the geneticist depends on mutations to learn how a normal character is inherited.

Many recessive mutations in Drosophila are known to affect an assortment of other traits besides eye color, and all are named according to this scheme. The recessive "ebony" can result in a fly with a shiny black body instead of the normal gray. The pair of alleles involved is represented by + (or e^+) for wild and e for ebony. The recessive forked, f, can cause shortened, split bristles in contrast to the longer, unsplit ones determined by the wild gene form, + or f^+. Further examples could be given for a long list of recessive alleles. We will learn (Chap. 4) that very *many* wild alleles exist, not just one. A wild allele corresponds to each mutant one, and every one of the wild alleles is represented as +, no matter what the effect of that particular locus on the phenotype.

In these examples, the first letter of the name of the mutation was used to represent the mutant: w, e, and f for white, ebony, and forked. Since many names begin with the same letter, we often must use more than one letter to represent a locus. As e has been used for ebony, it cannot be selected to describe the recessive allele "eyeless," which results in a reduction of eye size. More than just the first letter is required to stand for eyeless, and the locus is symbolized by ey. The normal eye condition is still + (or ey^+). And similarly for a long list of others such as dp (for "dumpy," a gene form giving reduced wings), ss (for recessive causing reduction in bristle size), and many more

The mutations discussed so far are all recessive to their wild type alleles (+). Gene alterations that produce a dominant effect, although not as numerous as recessive changes, are responsible for several phenotypic departures from the wild. The naming of dominant gene mutations is identical to that for the recessives. The wild allele, although now the recessive form in such cases, is still represented by $+$. The mutant allele, however, is symbolized by a capital letter (or an abbreviation beginning with a capital). A frequently encountered dominant genetic effect in the fruit fly is the "Bar eyed" condition in which eye size is narrowed. It is represented by the capital letter B, the wild condition as $+$ or B^+. There is no reason for confusion if one simply remembers that $+$ always designates *any* wild-type allele, group of alleles, or wild condition; any mutant is represented by a letter or abbreviation. We can tell at a glance that a mutant gene form is recessive if it is designated by a small letter (as w for white eye). A designation beginning with a capital tells us that the gene form is dominant to wild as in B (Bar), Cy (curly wings), and Pm (plum, a brownish eye color). The wild alleles are respectively, B^+, Cy^+, and Pm^+ (or simply $+$ in each case).

It was mentioned that a locus is named after the first detected mutation at a gene site. This implies that more than one kind of alteration is possible in a particular gene, and this is indeed the case as illustrated by any multiple-allelic series (see Chap. 4). The white eye locus is an excellent example. At this particular site on the X chromosome, more than one change has occurred in the genetic material. In addition to the one resulting in complete absence of eye pigment, there are several others that have caused different degrees of reduction in the pigment, producing various shades from red to white. We thus find apricot, blood, coral, and several others. These are all changes at the same locus and are consequently all allelic to one another. Because the change to white was the first one described, the other mutations are all represented by the same base letter, w, indicating that the alteration is at the white locus. To this base letter, superscripts are added, giving w^a for apricot, w^{bl} for blood, w^{co} for coral, and so on. The pair of alleles, w^+ and w^a, implies the wild-type gene for eye color and a recessive allele at the white locus resulting in apricot (pinkish) color.

The system that designates wild by a ($+$) has many benefits. These will become very evident in the discussions related to multiple alleles, linkage, and crossing over. At certain times, we may still prefer to use the older scheme of capital and small letters for dominant and recessive when discussing simple monohybrid or dihybrid crosses.

REFERENCES

Alberts, D. Bray, J. Lewis, M. Raff, K. Roberts, and J. D. Watson. *Molecular Biology of the Cell.* Garland Publishing, New York, 1983.

Cold Spring Harbor Symposium on Quantitative Biology. *Chromosome Structure and Function,* vol. 38. Cold Spring Harbor Laboratory, New York, 1973.

Hartwell, L. H. Cell division from a genetic perspective. *J. Cell Biol.* 77: 627, 1978.

Inoué, S. Cell division and the mitotic spindle. *J. Cell. Biol.* 91: 131s, 1981.

Kavenoff, R., L. C. Klotz, and B. H. Zimm. On the nature of chromosome-sized DNA molecules. *Cold Spring Harbor Sympos. Quant. Biol.,* 38: 1, 1974.

Lewin, B. *Gene Expression, Vol. 2, Eucaryotic Chromosomes,* 2nd ed. Wiley, New York, 1980.

Mazia, D. The cell cycle. *Sci. Am.* (Jan) 54, 1974.

Ruddle, F. H. and R. S. Kucherlapati. Hybrid cells and human genes. *Sci. Am.* (July): 36, 1974.

Yunis, J. J. High resolution of human chromosomes. *Science* 191: 1268, 1976.

Yunis, J. J. (ed.). *Molecular Structure of Human Chromosomes.* Academic Press, New York, 1977.

Yunis, J. J. and O. Sanchez. G-banding and chromosome structure. *Chromosoma* 44: 15, 1973.

REVIEW QUESTIONS

1. Name the stage of mitosis at which:

   A. The nucleolus disappears.
   B. Nuclear membranes reform.
   C. Centromeres are aligned on the equatorial plate.
   D. Microtubules are associating to form the spindle fibers.
   E. The nucleolus reforms.
   F. Chromatids composing a chromosome separate and move to opposite poles.

2. The normal chromosome number in a human body cell is 46.

   A. How many chromatids are present at (1) prophase; (2) interphase following DNA replication?
   B. How many chromosomes are present in a nucleus at (1) prophase; (2) telophase; (3) the interphase following DNA replication?

3. In the human:

   A. How many autosomes are present in body cells of a male?
   B. How many autosomes are present in body cells of a female?
   C. How many sex chromosomes are present in a male?
   D. How many sex chromosomes are present in female?
   E. How many major groups of chromosomes can be recognized on the basis of size and shape?

4. Allow "A" to designate one entire haploid set of human autosomes, and allow "X" and "Y" to represent the sex chromosomes.

   A. How can one represent the chromosome constitution of the body cells of a female in regard to autosomes and sex chromosomes?
   B. Answer Part A for a male.

5. From the column on the right, select the letter of the term that applies best to each of the following statements. A term may be used more than once or not at all.

   1. Does not occur in a cell having two nuclei.
   2. Its position determines the length of the chromosome arms.
   3. Marks the poles of the spindle in some types of cells.
   4. Deeply staining band or region of polytene chromosome.
   5. Holds the chromosome halves together at prophase and metaphase.
   6. Nuclear material composed of DNA and associated proteins.
   7. Chromosome material that remains condensed in the nondividing nucleus.
   8. Associated with the formation of a nucleolus.
   9. Separates from its sister at anaphase.
   10. The genetic content of a single set of chromosomes.

   A. chromomere
   B. cytokinesis
   C. chromatin
   D. chromosome
   E. centromere
   F. centriole
   G. chromatid
   H. genome
   I. heterochromatin
   J. satellite

6. Give at least three features of eukaryotic cells that are absent in prokaryotes.

7. What terms apply to the following definitions:

   A. The specific position on a chromosome that can be occupied by a particular gene.
   B. Chromosomes that correspond in size, shape, and genetic regions.

   C. An individual possessing two identical forms of a gene at a specific genetic region under consideration.
   D. Alternative forms of a gene.
   E. The actual chemical substance composing the gene.
   F. A V-shaped chromosome with its centromere in the middle.
   G. A chromosome composed of many separate chromosome threads that are intimately associated.
   H. The chromosome constitution of a single cell or individual.
   I. That part of some chromosomes that appears as an appendage because of a secondary constriction.
   J. An I-shaped chromosome with one very short arm.
   K. A drug that prevents spindle formation.
   L. The precise period of DNA replication.
   M. Chromosome regions that fluoresce after staining with a fluorescent dye.
   N. Deep-staining chromosome regions occupied by heterochromatin.
   O. The longer arm of a chromosome.

8. Assume that in nine different cells one of the following is experimentally destroyed or eliminated. Indicate whether you would expect mitosis to proceed (+) or to be arrested (−):

   A. Aster
   B. Centriole
   C. Kinetochore
   D. G_2 phase
   E. Polar spindle fibers
   F. Centrosome
   G. Centromere
   H. Cytokinesis
   I. S phase

9. A "new" mutation occurs which changes the flower color in a particular plant species from red to lavender. Breeding experiments show that the lavender flower color depends on a dominant genetic factor. The genetic locus involved is named *lavender,* and the first two letters of the name are used to designate the pair of alleles involved in flower color. Using the " + " system of naming alleles, designate (a) a lavender-flowered plant that is homozygous for flower color; (b) a red-flowered plant; (c) a plant with lavender flowers that is heterozygous for flower color.

3

CHROMOSOMES
AND GAMETE FORMATION

How remarkable that Mendel, totally unaware of any descriptions of cell division, was able to make the deductions that formed the basis of his two laws of inheritance! Perhaps even more incredible is the fact that the intricacies of mitosis and meiosis were being unraveled while Mendel's paper lay ignored. For *there* in the report of Mendel's experiments was the genetic evidence for the existence of factors whose behavior parallels that of the chromosomes. The beginning of the twentieth century brought not only a recognition of Mendel's paper but also independent genetic and cytological evidence to support it.

Although Weismann predicted some sort of reduction division in the life cycle of sexual creatures, the details were slow in coming. Before 1900, the half number of chromosomes had been reported in the gametes of animals (Van Beneden with Ascaris) and in plants (Strasburger). But the many changes taking place in the appearance and behavior of the meiotic chromosomes must have defied interpretation. One problem was the difficulty in recognizing homologous chromosomes. In 1901, Montgomery postulated homologues, one set of chromosomes provided by the female parent, the other by the male. It is these that pair at meiosis; pairing is not a random process. Sutton, a student of Wilson, confirmed Montgomery's findings and said, further, that the paired homologues

must somehow separate. Such behavior, he noted, would provide the physical basis for both of Mendel's laws. Conclusive proof of these points took several more years. Further genetic and cytological studies were needed to place Mendel's factors in the chromosomes and to verify the chromosome theory of inheritance.

At this point, let us examine the main features of the meiotic process to appreciate the problems that confronted cytologists and to enable us to grasp other genetic phenomena that were soon to be discovered.

Meiosis entails two nuclear divisions

Like mitosis, meiosis is fundamentally a nuclear event, and therefore does not occur among prokaryotes. Many features of mitosis and meiosis are identical; however, several distinctions between them produce very different genetic results.

In the last chapter, we saw that mitosis preserves the genetic identity between the parent cell and the two resulting daughter cells. The effect of meiosis is quite the opposite, for it is a process in which the chromosome number of the parent cell is reduced by one half, from diploid to haploid. At meiosis, which entails two nuclear divisions, the chromosome behavior is such that *new* genetic combinations are formed.

Since two nuclear divisions take place, the expected number of nuclei upon completion of meiosis is four. The four nuclei that are typically formed are referred to as the *tetrad,* the immediate products of a meiotic division. Cytoplasmic division usually, but not always, accompanies the meiotic process so that a tetrad of haploid cells is the characteristic end product. Exactly what these four cells will become after meiosis (sperms, eggs, spores) depends on both the species and the sex of the organism. Meiosis and all of its complications are an integral part of the sexual process and appeared early in the evolution of living things.

Morphology of the early meiotic chromosome

The first division of meiosis is preceded by an interphase in which G_1, S, and G_2 phases can be recognized. However, the S of meiotic interphase requires a much longer time than does the S phase of mitosis. In the newt, Triturus, for example, 12 hours is typical for the S of nonembryonic mitotic cells, but the S preceding gamete formation in the male lasts approximately 10 days! Moreover, a very small portion of the DNA may remain unreplicated until the early part of first prophase. It is during first prophase that most of the significant differences between mitosis and meiosis occur. First, its duration is quite prolonged compared with that of mitosis and entails pronounced changes in chromosome appearance and behavior. At least five stages are generally recognized. At the earliest of these, *leptonema* (or the *leptotene stage*), the chromosomes are very stretched out, more so than at any other of the meiotic stages. Individual chromosomes cannot be identified, and the nucleus appears to contain a network of thin, entangled threads. The thin chromosome thread is designated the *chromonema* (Fig. 3-1*A* and *B*). Both ends of each chromonema are attached to the nuclear membrane. Although the chromonemata can be demonstrated at mitosis by chemical treatment of prophase and metaphase chromosomes, they are most apparent during early prophase of meiosis. Indeed, we can think of all the changes in chromosome appearance during nuclear divisions as a result of changes in the coiling of the chromosome thread. In many species, the chromonemata exhibit deeper staining regions along their lengths, giving the threads the appearance of strings of beads. These intensely staining areas are the *chromomeres* and are comparable to deep-staining bands along the lengths of the polytene chromosomes (Chap. 2). Studies with the electron microscope support their

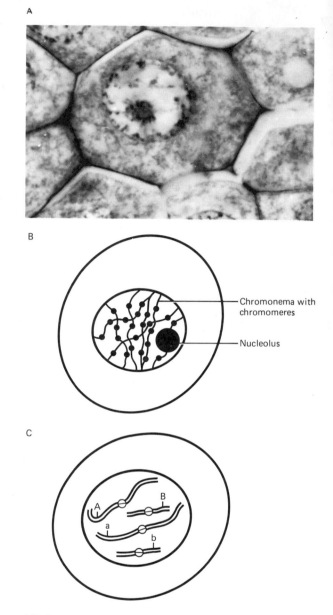

FIG. 3-1. Leptonema. (*A*) microspore mother cell of the lily. (*B*) The chromosomes are in a greatly extended condition and appear as slender threads, the chromonemata. (*C*) The chromosome number is four. Each chromosome thread is actually double, and composed of two chromatids, but this cannot be detected with the light microscope. Two pairs of homologous chromosomes are depicted, one pair with alleles A and a, the other with the alleles B and b. The chromosomes are unassociated and distributed at random in the nucleus.

interpretation as coils in the chromosome thread. Whether or not they represent the sites of single genes is controversial, but nevertheless their positions are constant and appear at characteristic locations along the length of a particular chromosome. Certain specific genes in some organisms have actually been associated with specific chromomeres.

The filament at the leptotene stage is so thin that it appears to be single when viewed with the light microscope. However, autoradiographic procedures that follow the incorporation of radioactive building blocks of DNA clearly tell us that duplication took place in the S phase of premeiotic interphase before the onset of leptonema.

Figure 3-1C is a diagram of an imaginary cell from an organism with a diploid chromosome number of four, in which one pair of chromosomes is larger than the other. At leptonema, these two homologous pairs, although long and extended, are separated and are not associated in any prescribed way. The cell in the figure depicts a heterozygous individual, a dihybrid carrying two pairs of contrasting genetic factors. These are the alleles, A and a, which occupy a locus on the larger of the chromosomes and the alleles B and b, located at a site on the smaller ones.

Homologous chromosomes pair

One of the most critical differences between meiotic and mitotic chromosome behavior involves the pairing or *synapsis* of homologous chromosomes. The act of pairing marks the onset of *zygonema* (the *zygotene stage*), which is defined as the stage of active pairing (Fig. 3-2A). This synapsis is very exact and can be seen to take place chromomere for chromomere. If we recall that genetic loci are distributed along the length of the chromosome, we can envision the homologues pairing locus-for-locus throughout zygonema until all the corresponding loci on all the chromosomes are intimately associated. Therefore, at the zygotene stage the alleles in a heterozygote would be closely paired and no longer apart in the nucleus (Fig. 3-2B). Although the nature of the synaptic force is not understood, it serves to bring the homologous chromosomes together in a close union, even though they may be widely separated in the nucleus before the onset of meiosis. Whatever its exact basis may prove to be, the force is an exact one that operates over large distances on the level of the nucleus. Studies with meiotic cells of the lily show that a very small amount of DNA synthesis (0.3%) takes place at the zygotene stage.

It is very important to appreciate the fact that separate chromosomes are not distributed throughout the nucleus at zygonema, as they are during mitotic pro-

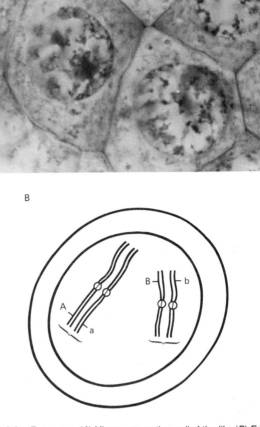

FIG. 3-2. Zygonema. (*A*) Microspore mother cell of the lily. (*B*) Each of the four chromosomes in this imaginary cell is double, giving eight chromatids, although this is not evident with the light microscope. Zygonema is the stage of active pairing of the homologous chromosomes, so that two bivalents (*brackets*) arise in this example. Each of the two bivalents is composed of four threads, and consequently each locus is represented four times: A, A, a, a and B, B, b, b. Each of the two chromatids of a chromosome possesses its own centromere.

phase. What appears to be an individual chromosome is actually an association of two, each of which is double. This association of two homologues is termed a *bivalent*. The expression *tetrad*, referring to the four threads that are present, is also used, but many prefer to restrict usage of the term to the immediate products of meiosis after telophase II. Thus, the nucleus at zygonema contains a number of bivalents, which corresponds to the haploid chromosome number for the species. Figure 3-2B shows that, although four chromosomes are present, they are arranged as bivalents. Since each chromosome thread with all its loci has doubled, two A and two contrasting a genetic factors are present. The same is true for the other pair of alleles, B and b. It is very important to note that, as far as the number of *chromatids* is concerned, the total is the same in both the mitotic and the meiotic prophases. For at prophase of *mitosis,* each chromosome is also composed of two chromatids. The dihybrid represented in Fig. 3-2*B* has eight, the same number that would be present if the cell were undergoing mitosis. Therefore, at both the mitotic and the meiotic prophases, the same situation pertains regarding the number of chromosome threads, because each chromosome in each of the processes is composed of two halves. The main difference is the *distribution* of the chromosomes in the nucleus. At mitotic prophase, the chromosomes are separate and unassociated. At prophase I of meiosis, the homologues are paired as bivalents, and it is these that are found in the haploid amount.

The chromosome thread thickens

After zygonema, the chromosome threads become more conspicuous, and the next portion of meiotic prophase I can be recognized, *pachynema* or the *pachytene stage* (Fig. 3-3*A* and *B*). The thickening that becomes evident is largely due to a folding back of each chromosome fiber upon itself. As a result of the compaction, what appeared to be single threads are now revealed to have a double nature, so that the two chromatids of each chromosome can be recognized. Pachynema is the stage during which the biologically important process of crossing over takes place, but the visible manifestations of this are not evident until the next stage of prophase I and will be described in the following section.

Work with lily meiosis indicates that some digestion of the DNA occurs during pachynema. The digested portion is replaced, however, by a small amount of DNA synthesis, so that the total amount of DNA in the cell does not change. The significance of this small burst of DNA synthesis following premeiotic S will be referred to a bit later in this chapter.

Homologues repel each other

Once intimate pairing is achieved and the chromosomes thicken, a force begins to operate which seems to oppose that of pairing. We recognize *diplonema* (the *diplotene stage*) of prophase I, when the chromosome threads composing each bivalent appear to

A

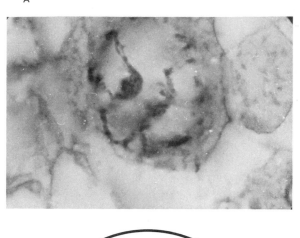

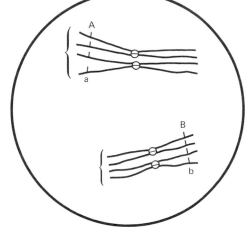

FIG. 3-3. Pachynema. (*A*) Microspore mother cell of the lily. Because thickening of chromosome threads occurs at this stage, the double nature of each thread is evident. (*B*) Chromosome number is four and chromatid number is eight in the nucleus of this imaginary cell. The four chromosomes are arranged as two bivalents (*brackets*).

repel (Fig. 3-4*A*). The diplotene stage is often referred to as the stage in which "opening out" occurs, a recognition of the separation of the chromatids that is evident along the length of the bivalent. As diplonema proceeds, this becomes more pronounced, particularly at the region of the centromere, where the forces of repulsion seem to be strongest (Fig. 3-4*B*). Indeed, the bivalent association would probably fall apart were it not for the presence of certain regions where the homologous chromosomes still remain in intimate contact. These regions are called the *chiasmata* because at each chiasma the threads of the separate chromatids appear to cross. Although their formation has been a matter of debate among cyto-

A

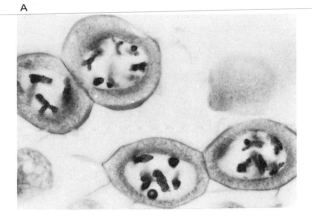

B

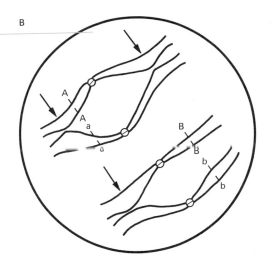

FIG. 3-4. Diplonema. (*A*) Microspore mother cell of the lily. (*B*) Imaginary cell. Chromosome number is four. Chromatid number is eight. The number of bivalents is two. The chromosomes composing a bivalent start to repel each other at the beginning of this stage, and the forces of repulsion progress throughout. The chiasmata (*arrows*) hold the four chromatids together in a bivalent, which would otherwise fall apart. Only the nucleus is represented here.

geneticists, direct evidence has now been presented that the chiasmata are associated with the phenomenon of *crossing over,* the separation of genes linked together on the same chromosome. A full discussion of this topic is presented in Chapters 8 and 9, but a few points about its physical basis are in order here. Genetic analyses have firmly established that homologous chromosomes exchange segments during meiotic prophase. The consensus is that each chiasma reflects a point where a physical exchange of segments between homologues has already taken place. Evidence from studies with microorganisms and certain eukaryotes supports the concept of an actual breakage of chromosome threads. In this process, broken chromatid ends join up reciprocally with chromatid segments of the homologous chromosome (Fig. 3-5). Since this rejoining is reciprocal, the final outcome is an exchange of corresponding blocks of genes between homologous chromosomes. Cytologists have determined that this exchange takes place in pachynema before the onset of diplonema when the consequences of the crossover events are seen as chiasmata.

The diplotene stage may be of very long duration in oocytes, meiotic cells leading down the pathway to the development of eggs. In some species, the chro-

FIG. 3-5. A single crossover event following genes on the same chromosomes: Cc and Dd. In the bivalent, two of the four threads participate in any *one* crossover event (*left*). However, more than one crossover event may take place in a bivalent, and any two homologous threads may be involved. In a *single* crossover event, a reciprocal exchange of chromosome segments occurs between two homologous chromatids. After the event, a new combination of alleles results. On the right, we see that in each chromosome one chromatid shows a new arrangement (C-d in the upper; c-D in the lower). This is the direct result of the reciprocal exchange of homologous chromosome segments past the point of the crossing over. (*Arrow* indicates the point where the reciprocal exchange occurred between the two chromatids.)

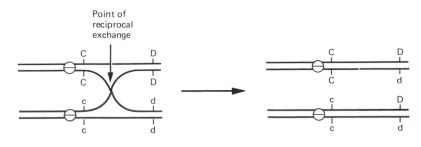

Point of
reciprocal
exchange

mosomes become very indistinct as the chromatin fibers become less folded and appear to decondense. The synthesis of substances important to the egg as storage products takes place during this time. In some species, the diplotene stage lasts for weeks or even many years as it does in the human female.

Crossing over and the synaptonemal complex

The process of crossing over is of utmost importance in the formation of new combinations of genes found together on the same chromosome. It is an integral part of the normal sexual process, and its role in adding to the number of new combinations of the genetic complement of a species cannot be overemphasized. Because of the biological significance of crossing over, the attention of many cytologists has been focused on the early stages of meiosis, and many of their findings are filling in gaps in our understanding of the mechanism of crossing over. One important

discovery has been the recognition of a structure peculiar to early meiotic chromosomes. This structure, first described in 1955 and observable with the electron microscope, is known as the *synaptonemal complex* (SC). The SC is a three-layered proteinaceous structure that develops between the two homologous chromosomes. The development of the SC traces back to the leptotene stage, at which time each chromosome thread forms along its length a longitudinal protein core. At zygonema, when the two homologues pair to form a bivalent, the two protein axes are brought together to form the sides or lateral elements of the SC (Fig. 3-6). The central element then proceeds to develop, and by pachynema, the SC is complete. At pachynema, proteinaceous bodies, often spherical and known as recombination nodules, develop along the length of the central element of the SC. Figure 3-7 summarizes the development of the synaptonemal complex.

The synaptonemal complex does not appear to be

FIG. 3-6. Synaptonemal complex in the tiger lily. (*Top*) An unpaired chromosome at leptonema produces a proteinaceous axial core (×46,000). When two homologous chromosomes are paired (*bottom*) the two juxtaposed cores become the lateral elements of the synap-

tonemal complex. By pachynema, the SC is completed and a central element can be seen between the lateral elements (×80,000). (Courtesy Dr. P. B. Moens.)

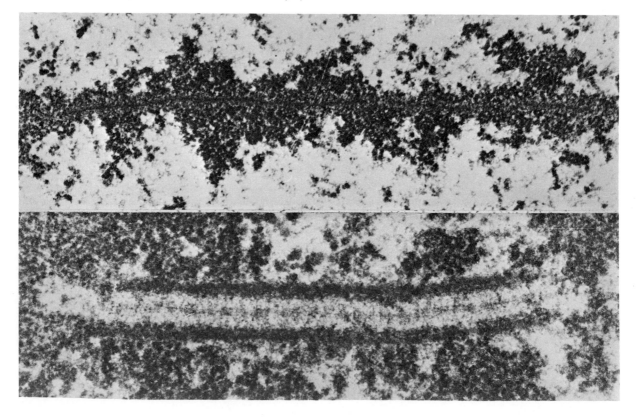

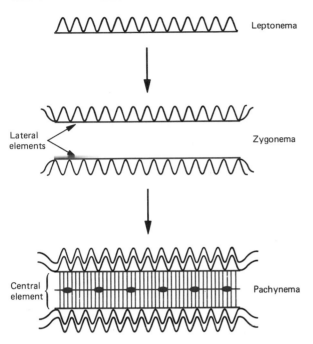

Leptonema

Lateral
elements

Zygonema

Central
element

Pachynema

FIG. 3-7. Development of the synaptonemal complex. At leptonema, each chromosome forms a proteinaceous axis (*red*) along its length. The chromosome thread is actually double, but the two chromatids are so intimately associated that two chromatids are not evident. At zygonema, the homologues pair, and the axial cores of both chromosomes form the lateral element of the complex. By pachynema, the synaptonemal complex is complete with a central element, along which recombination nodules (*red spheres*) develop.

about crossing over. You will recall that a very small amount of DNA synthesis occurs after premeiotic S, during zygonema and pachynema. Experiments using radioactive building units of DNA have shown that during the pachytene stage the DNA precursors are taken up by meiotic cells at the sites of the nodules. DNA synthesis during pachynema may very well be involved in the process of crossing over. This point is explored more fully in Chapter 11.

End of prophase I

The last clearly defined stage of prophase I is *diakinesis,* marked by rather pronounced changes in the appearance and distribution of the chromosomes within the nuclear membrane. The chromosomes, the ends of which have been attached to the nuclear envelope, now become detached. The chromosomes contract markedly, and the chiasmata become increasingly evident as the stage progresses (Fig. 3-8). The bivalents themselves tend to become widely spaced, as if some other force of repulsion were operating to cause them to separate from one another. Consequently, this stage is an excellent one for chromosome counting, because the haploid number is clearly indicated

FIG. 3-8. Diakinesis in the lily. At this stage, the chromosomes are extremely contracted, chiasmata are very evident, and the bivalents tend to be separated from each other in the nucleus. The relationship of the chromatids, however, is basically the same as in the previous stage, diplonema.

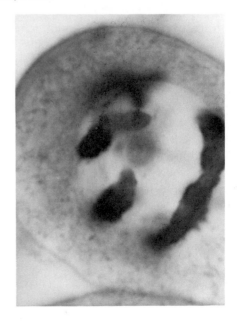

responsible for triggering synapsis, because pairing of the homologous chromosomes starts well before the SC is completely developed, which does not occur until synapsis has been completed. Several lines of evidence, however, indicate that the development of a synaptonemal complex is necessary for chiasma formation and thus for crossing over. If completion of the SC is arrested by protein inhibitors as it starts to form, for example, normal synapsis does not occur and chiasmata do not arise. All the observations indicate that, although the SC does not initiate synapsis, it does maintain the bivalents in pachynema by keeping the homologues together in an intimate and precise state of synapsis. In this way, it increases the probability that crossing over will occur and hence that chiasmata will form. As the homologous regions separate at diplonema, the SC becomes disassembled, in a manner that varies according to the species.

Evidence suggests that the recombination nodules scattered along the central element of the SC contain enzyme assemblies that are instrumental in bringing

by the number of contracted and widely separated bivalents. However, nothing of genetic significance is actually taking place to alter the chromosomes or rearrange the hereditary material. We may consider this stage to be an exaggeration of diplonema.

Chromosome behavior at meiosis and Mendel's laws

As in mitosis, prophase ends with dissolution of the nuclear membrane and the disappearance of the nucleoli. Common to both processes, metaphase movements begin, and the chromosomes arrange themselves at the midplane of the spindle. However, a major distinction exists between the metaphase of mitosis and that of the first meiotic division: in the former, the centromeres of *individual* chromosomes are arranged at the equatorial plate (Chap. 2, Fig. 2-11*B*), whereas at metaphase I, it is the centromere regions of *bivalents* that are found at the equator of the spindle (Fig. 3-9*A* and *B*). Thus, at mitotic metaphase, *single* chromosomes are oriented on the spindle, whereas *pairs* of chromosomes occupy the equator in the meiotic cycle. The centromere regions of the two individual chromosomes composing each bivalent have been repelling each other and are now clearly directed toward opposite poles of the spindle. The fibers coming from the centromeres of the two sister chromatids of each chromosome are oriented toward the *same* pole (Fig. 3-9*C*). Contrast this with mitosis, in which the kinetochores of the sisters at metaphase are oriented toward *opposite* poles (see Fig. 2-11*B*, and *C*).

It is particularly important at this point to note the arrangement of the two pairs of chromosomes in Fig. 3-9*B*. The diagram shows the A allele of one chromosome and the B allele of the other directed to the lower pole and the a and b toward the upper. However, nothing dictates such an orientation requiring factors A and B to go to the same pole. It is equally possible for the chromosome with allele A and the chromosome with allele b to face the same pole and those with a and B to face the opposite one. Assuming that the chromosomes with the A and the B genetic factors came from the maternal parent and those with the a and b factors from the paternal one, we can see that there is no reason to suppose that A and B *must* travel together as well as a and b. The two loci are on completely different chromosomes. How the four chromosomes become arranged in relationship to the poles of the spindle is a matter of chance and forms the foundation of Mendel's law of independent as-

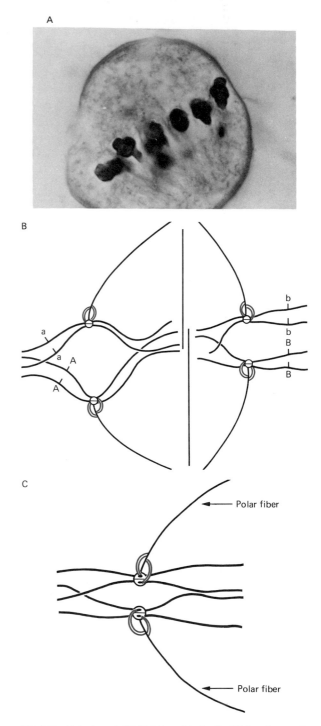

FIG. 3-9. Metaphase I. (*A*) Meiotic cell in the lily. (*B*) Imaginary cell. Chromosome number is four; chromatid number is eight. The two bivalents are arranged at the equator of the spindle. The chromosomes with alleles A and B are directed toward the lower pole, but it is just as possible to find other arrangements, because the separate nonhomologous chromosomes are not tied together in any way. (Kinetochore fibers are in *red*.) (*C*) The kinetochore fibers of the sister chromatids are oriented toward the same pole. (The *black dots* represent the kinetochores in the centromere regions.)

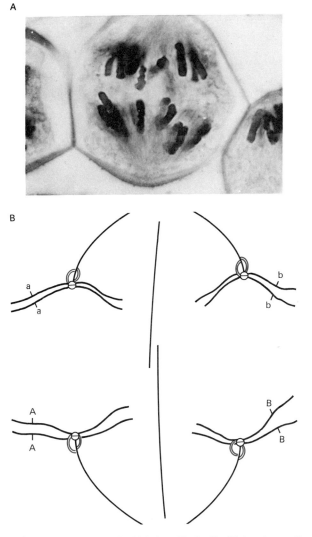

FIG. 3-10. Anaphase I. (*A*) Meiotic cell in the lily. (*B*) Imaginary cell The centromeres of each chromatid in a chromosome remain associated, so that each chromosome moving to a pole is composed of two chromatids.

kinetochore fibers of the sister chromatids are associated with the same pole, the two sisters composing a chromosome remain associated at the centromere region. The centromeres of the homologues in each bivalent repel each other and travel to opposite poles. Consequently, each chromosome is still composed of two chromatids, which remain associated at the region of the centromere (Fig. 3-10*A* and *B*).

Telophase I involves the reformation of nuclear membranes and nucleoli, as in telophase of mitosis (Fig. 3-11*A*). At meiotic telophase, however, the haploid number of chromosomes is present in the newly formed nuclei. In our example (Fig. 3-11*B*), each nucleus contains two chromosomes instead of the diploid number of four. However, each of these chromosomes, unlike those of mitotic telophase, is composed of two chromatids. If we count the chromatid number, in this case four, we see that complete reduction has not yet been achieved. In effect, two genomes are present—two complete sets of genes or

FIG. 3-11. Telophase I. (*A*) Meiotic cells in the lily. At this stage, the nuclear membranes reform. (*B*) Chromosome number per nucleus in this imaginary example is two. Chromatid number per nucleus is four. Each sister nucleus is genetically different; the left one is ab, the right AB. Reduction in chromosome number has occurred, but not in genomic content.

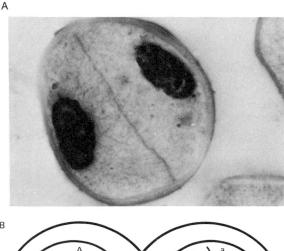

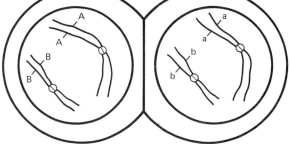

sortment. It is here at metaphase I and the ensuing anaphase I that we find the basis of both of Mendel's primary laws, *segregation* as well as *independent assortment*.

Anaphase I, like anaphase of mitosis, is recognized by the movement of chromosomes to opposite poles of the spindle (Fig. 3-10*A* and *B*). Remember, however, that at mitotic anaphase, the kinetochore fibers of the sister chromatids are associated with opposite poles. At the start of anaphase, the sisters separate, become independent chromosomes, and travel to the opposite poles of the spindle (see Fig. 2-13). This is not the case at first anaphase of meiosis. Since the

genetic information. To bring about the reduction in genome content as well as in chromosome number, the second division of meiosis is required.

Second meiotic division effects true reduction

The onset of the second meiotic division varies with the species involved. Prophase II may begin after a distinct interphase period. In some organisms, the interphase may not occur; thus, an abrupt transition occurs from telophase I to prophase II, and in certain cases even to metaphase II. At any rate, the second division stages of meiosis resemble those of mitosis when viewed with the microscope (Figs. 3-12*A* and 3-13*A*). Unfortunately, this has led to a hasty description of the second meiotic division as a mitotic one. Such a designation is quite incorrect for several reasons. Most obvious is that the double nature of the chromosome at mitotic prophase results from the duplication of the genetic material at S in the preceding interphase. However, no S phase occurs immediately before prophase II of meiosis. Each prophase II chromosome is composed of two chromatids as a result of anaphase I separation (see Fig. 3-10).

It is extremely important to note that the chromatids composing the *mitotic* chromosomes are sisters, meaning they are identical throughout their lengths. Those at prophase II, however, are not truly sisters, because crossing over typically occurs at pachynema to switch chromosome segments between homologues. Only portions of the chromatids of a single chromosome would be identical. We have noted (in Fig. 3-5) that after the point of a crossover a segment of a chromosome is associated with a block of genes from the homologous chromosome. Second division of meiosis is necessary to separate the dissimilar chromatids of each chromosome. This critical division will achieve complete reduction in the amount of genetic material and bring about new combinations of genes from the male and female parent of the preceding generation. The second meiotic division is essential to the sexual process and should not be considered equivalent to a mitosis.

At metaphase II, single chromosomes, each composed of two chromatids held together at the centromere region, are at the equator, and their appearance is similar to that of chromosomes of mitosis (see Fig. 3-12*A* and *B*). The kinetochore fibers of the two chromatids are associated with microtubules emanating from opposite poles, just as in the case of mitotic chromosomes (Fig. 2-11*C*). As in mitosis, separation

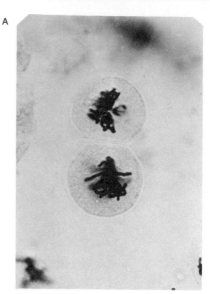

A

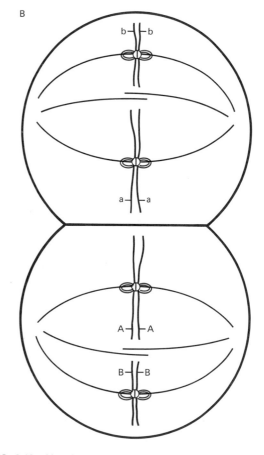

B

FIG. 3-12. Metaphase II. (*A*) Meiotic cells in the lily. (*B*) In this imaginary example the single chromosomes in each sister cell are at the equatorial plate. Contrast this with metaphase I (Fig. 3-9), where bivalents are at the equator.

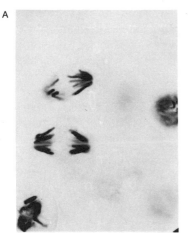

A

of the two chromatids of a chromosome occurs at the centromere region, and anaphase II is initiated (Fig. 3-13 *A* and *B*). *Only* with this separation of the two chromatids is true reduction in the genome content achieved. The nuclei of telophase II (Fig. 3-14 *A* and *B*) contain chromosomes that are now single. Figure 3-14*B* shows the reduction in both chromosome number and genomic content. The figure also shows that the nuclei of the tetrad are genetically different; two of the resulting cells are "AB," having received the chromosome with allele A and the other with allele B, and the other two are "ab." It is essential here to reexamine metaphase I (Fig. 3-9*B*). It was mentioned that the orientation of the bivalents at this stage is not mandatory. The factors on separate chromosomes

FIG. 3-14. Telophase II. (*A*) Tetrad of microspores in the lily. (*B*) The chromosome number per nucleus in our example is two. The chromosome is now single, no longer composed of two chromatids. Each cell of the tetrad has true reduction in both chromosome number and genomic content. The cells are unlike genetically as a result of segregation and independent assortment. In this example, half of the nuclei are AB, and the rest ab.

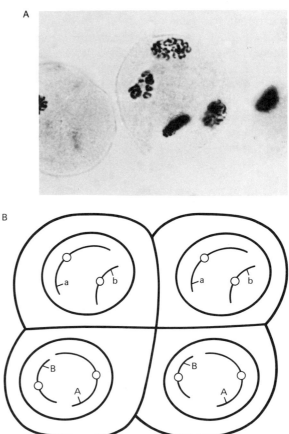

B

FIG. 3-13. Anaphase II. (*A*) Cells in the lily. (*B*) The centromeres of the chromatids now separate. Each chromatid becomes an independent chromosome. Contrast this with anaphase I (Fig. 3-10), where the two chromatids of a chromosome do not separate, and each chromosome therefore remains double.

segregate independently. If the A and the b chromosomes face one pole and the a and B chromosomes face the other, the final products at the completion of meiosis would be nuclei with different genetic arrangements from those shown in Fig. 3-14*B*: Ab and aB, instead of AB and ab. Since both possibilities can arise with equal frequency as the result of chance chromosome orientation on the spindle, meiosis in a dihybrid, AaBb, yields four different kinds of gametes in equal proportions: AB, Ab, aB, and ab. Thus, the origin and frequencies of different kinds of gametes are determined by the physical events of meiosis, which shuffle the maternal and paternal genetic factors into new combinations. Although the details of linkage and crossing over will be left until Chapters 8 and 9, it should be appreciated at this point that even those genes that are linked on the same chromosome are shuffled as a result of the physical exchange of chromosome segments between homologues (see Fig. 3-5). The overall significance of crossing over, like independent assortment, is the formation of new combinations of genetic material in the gametes.

We see that, unlike mitosis, the nuclei formed from the meiotic divisions of a parent nucleus are very different. Just from independent assortment, almost limitless new combinations of allelic pairs are possible (see Table 1-1). The meiotic event, along with independent assortment and recombination through crossing over, is the focal point of the sexual process and has supplied most of the variation for natural selection to work on in the evolution of the diverse forms of life.

Meiosis in the male

All sexually reproducing organisms contain meiotic cells at some portion of their life cycle. Once chromosome behavior during meiosis is fully grasped, it becomes easy to recognize the characteristic stages of the two nuclear divisions of meiosis in specific cells of the animal or plant. In the testis of the mature male animal, a variety of cell types is found in the wall of the seminiferous tubule. Among these is the *spermatagonium*, which is capable of mitotic division. It may enter the prophase stage of first meiosis, however, and once it does so, it is recognized as a primary spermatocyte (Fig. 3-15). Each primary spermatocyte is destined to complete both of the meiotic divisions and yields a tetrad of *spermatids*. Each of these in turn finally undergoes dramatic cytoplasmic transfor-

mations into a sperm (*spermiogenesis*) without further significant changes in the genetic complement. The entire series of events, beginning with meiosis and resulting in the formation of male gametes, is known as *spermatogenesis*. In the human, it has been estimated that approximately 64 days are required for spermatogenesis, starting with a spermatogonium and ending with the development of mature sperms. Any slide of a testis squash or section shows a diversity of cell types because different meiotic stages occur simultaneously along the length of the tubules. The most abundant cell type is the spermatid, which assumes an assortment of shapes as it transforms into a mature sperm.

The grasshopper is used as the species of reference in the following discussion of *spermatogenesis*, because meiotic behavior in the male grasshopper is typical of male animals, and its chromosome number is low, making observations quite easy. Moreover, it has an unpaired X chromosome; no Y is present at all. In this so-called X-O condition, the identification of the X chromosome at meiosis is simpler than in the more common X-Y situation found in man and other mammals.

The characteristic stages of first meiotic division during spermatogenesis all take place in the primary spermatocyte which, as noted previously, is distinguished from a spermatogonium by the onset of meiosis. By the time of pachynema, the nuclei of the primary spermatocytes of the grasshopper show an identifiable X chromosome (Fig. 3-16). At early prophase stages, it is usually more deeply stained and contracted than any other member of the chromosome complement; by diplonema, it is very well defined (Fig. 3-17*A* and *B*). In contrast, the other chromosomes appear fuzzy in outline. The X chromosome in the primary spermatocyte is behaving like heterochromatin, as mentioned in Chapter 2. The cell at diakinesis typically contains a sex chromosome off to one side of the bivalents. When first metaphase is reached, it is often off the equatorial plate (Fig. 3-18). At metaphase I, the unpaired X, which has been the most distinct chromosome of the complement, begins to appear fuzzy and less intensely stained, resembling the early prophase stages of the rest of the chromosomes. The X chromosome of the human and other mammals is paired with the Y at early meiotic prophase, but it too shows this difference in stainability in contrast with the rest of the chromosomes. This differential reaction to staining, often displayed by the sex chromosomes, is called *heteropyknosis*. The phenomenon may also

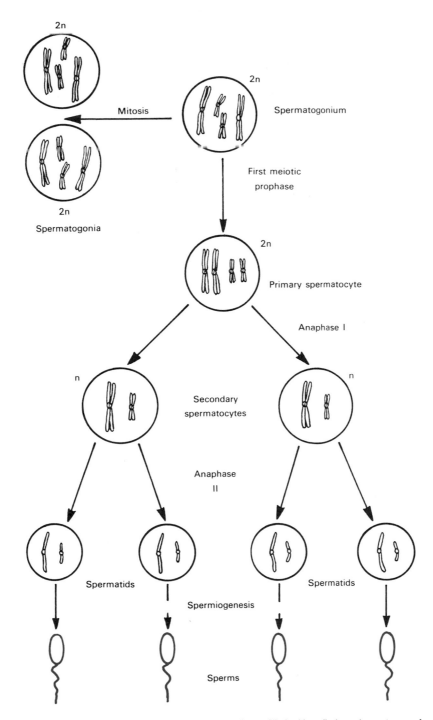

FIG. 3-15. Diagram of spermatogenesis. Among the various types of cells in the seminiferous tubules is the spermatogonium, which may divide mitotically or undergo the meiotic divisions. The tubules are always filled with cells in various stages of gamete formation. The spermatids are the most numerous. Note that the eventual products of one primary spermatocyte are four functional sperms.

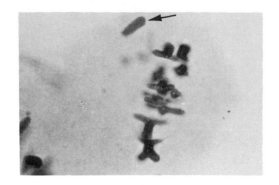

FIG. 3-16. Pachynema in primary spermatocyte of grasshopper. The X chromosome is much more condensed and stains more deeply than the other chromosomes in the complement (*arrow*).

FIG. 3-18. First metaphase in primary spermatocyte of grasshopper. Note that the X chromosome is off by itself and now stains more faintly than the rest of the chromosomes.

FIG. 3-17. Primary spermatocytes of grasshopper. These cells show 11 bivalents and an unpaired X (*arrow*). The X stains more deeply than the other chromosomes at diplonema (*A*). By diakinesis (*B*) all of the chromosomes are highly contracted and react similarly to stains.

A

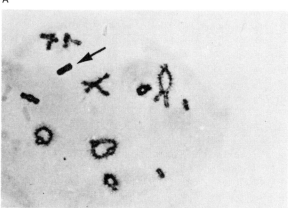

B

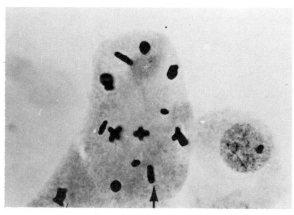

be characteristic of certain regions of other chromosomes. *Positive heteropyknosis* designates the more deeply stained condition of early meiosis and *negative heteropyknosis* the less intensely stained appearance often observed at first metaphase. A difference in staining behavior between the sex chromosomes and the autosomes (all the other chromosomes in the complement) is typical of most animals having the X-O and X-Y mechanisms of sex determination, although details vary among different groups.

At anaphase I, the negatively heteropyknotic X chromosome of the grasshopper passes undivided to one of the poles (Fig. 3-19). After reconstitution of the nuclear membranes at telophase I, two cells are formed, the *secondary* spermatocytes. These are quite different genetically (see Fig. 3-11*B*). A major difference is in the X chromosome. Since it passed undivided to one pole at anaphase I, one of the two

FIG. 3-19. First meiotic anaphase in the male grasshopper. The chromosomes at each pole will become incorporated in nuclei of cells that develop into secondary spermatocytes. Note that 11 chromosomes are at one pole and 12 at the other because the latter contains the X chromosome.

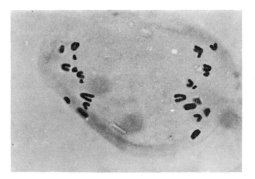

secondary spermatocytes in the grasshopper contains one less chromosome than the other. (In animals where both an X and a Y are present, one of the cells contains the X and the other the Y because the X separates from the Y at first division.) Secondary spermatocytes are mixed with the many other cells of the testis. Their identification is often difficult because they are easily confused with young spermatids; both cell types are undistinguished looking cells with large nuclei.

Each of the two secondary spermatocytes formed from first meiotic division in its turn undergoes the second stages of meiosis. The sex chromosome, displaying negative heteropyknosis, may be recognized at both second metaphase and anaphase. The secondary spermatocyte with the X produces two spermatids, each with an X. The spermatocyte lacking the X gives rise to two spermatids with no X chromosome (but with a Y if the species contains both X and Y). The meiotic end product of each primary spermatocyte in an animal is a tetrad of cells, two of which contain an X and two of which do not. (The latter two contain a Y in those males that are XY.) The significance of this in sex determination is a major subject discussed in Chapter 5.

Meiosis in the female

The nuclear events in both sexes are essentially the same. They differ, however, in some aspects regarding the meiotic process. In females of higher animals, meiosis does not take place continually throughout the age of reproductive maturity as it does in the male. In female mammals, for example, the mitotic divisions of *oogonia*, the counterparts of the spermatogonia, cease during embryological development (Fig. 3-20). Therefore, the entire complement of *primary oocytes* is established by the time of birth. A primary oocyte thus remains in an extended diplonema of first meiotic prophase for many years. The first meiotic division is not resumed until the egg is about to mature in the follicle at the age of sexual maturity. When the primary oocyte does finally divide, it gives rise to two cells, but these are unequal in size. This occurs because the larger one, the *secondary oocyte,* receives most of the cytoplasm. The smaller cell, the *polar body,* may or may not divide again. The secondary oocyte completes the second meiotic division only if a sperm enters the cytoplasm. The result of the division is again two cells of unequal size, a large one, which will become the egg, and another small polar

body. All the polar bodies eventually disintegrate. Throughout oogenesis, the process in the female that produces a gamete from the maturation of an immature germ cell, both X chromosomes stain to the same degree, in contrast to the single X of the male which shows heteropyknosis in the spermatocytes.

Meiosis in the higher plant

The sexual process undoubtedly arose early in the evolution of eukaryotic cells, long before the divergence of the plant and animal kingdoms. We therefore find comparable meiotic stages in plants and animals. Plant life cycles, however, are very diverse and often quite complicated. Meiosis generally takes place in specific plant organs, but the direct result of the process usually is *not* gamete formation as it is in animals. Instead, haploid cells called *spores* are formed. These then undergo a series of mitotic divisions, which eventually leads to the origin of sex cells. In the flowering plant (Fig. 3-21), meiosis occurs in two parts of the flower. On the male side, the anther is the organ involved. Special cells, *microspore mother cells,* arise from the anther wall, and these undergo the meiotic divisions called *microsporogenesis.* (Many of the earlier figures in this chapter show the meiotic divisions and products of microspore mother cells in the lily.) The end product of meiosis of each microspore mother cell is a tetrad of four haploid cells, the *microspores.* Each of the microspores matures into a pollen grain. During this maturation, each haploid nucleus of a microspore divides *mitotically* to produce two haploid nuclei per microspore or pollen grain. Of the two haploid nuclei, one divides again to form two male gametes, one of which fertilizes the egg. Therefore all the stages of meiosis that bring about segregation and independent assortment are completed by the time of microspore formation, well before the production of the male gametes.

Inside the ovary of the flower are found the *ovules,* or immature seeds. In each ovule, *megasporogenesis* occurs—the meiotic divisions that precede the formation of the female gamete. In each ovule, one cell enlarges, the *megaspore mother cell.* It undergoes the second meiotic division, with the formation of a tetrad of four haploid *megaspores.* Of these four cells, three disintegrate, leaving a single cell that then enlarges. The nucleus of this remaining megaspore divides mitotically. Typically, two more mitotic divisions follow, forming a group of eight haploid cells. Only one of these is destined to act as the egg and be fertilized

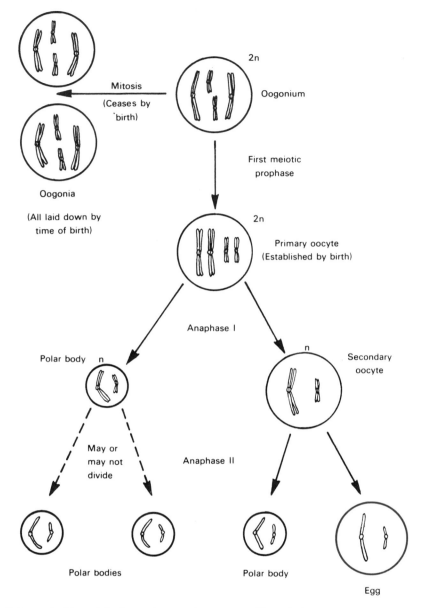

FIG. 3-20. Diagram of oogenesis. The meiotic stages of oogenesis are comparable to those of spermatogenesis (Fig. 3-15). The same chromosome behavior is involved, although cytoplasmic events are very different in the two processes. Note that the eventual product of one primary oocyte is one functional egg.

if pollination is successful. It is easy to recognize a parallel between megasporogenesis in plants and oogenesis in animals. In both processes, a mother cell (megaspore mother cell or primary oocyte) undergoes meiosis, and only one functional cell results. On the male side, in both plants and animals, each tetrad of cells forms four products (microspores or spermatids) from each of which a functional male gamete is derived.

Mitosis and meiosis are common to all eukaryotes

Regardless of differences in detail among species, the meiotic process is remarkably similar in all sexual organisms. This undoubtedly reflects its value and early selection by the forces of evolution, which have refined it since its origin in ancestral sexual cells. By increasing the speed with which a wider variety of

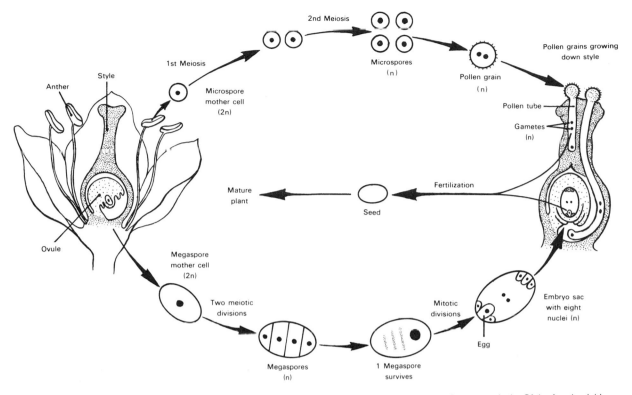

FIG. 3-21. Meiosis in the flowering plant. Meiotic divisions leading to the formation of male gametes occur in the microspore mother cells of the anther. Four microspores, each of which will become a pollen grain, arise from each mother cell. The haploid nucleus of the pollen divides mitotically and yields two haploid nuclei. One of these divides again to produce two male gametes, one of which will fertilize an egg. The ovary contains ovules, and in each of these, one cell—the megaspore mother cell—undergoes meiosis. Of the four haploid megaspores formed, only one survives. It enlarges to give rise to the embryo sac. Mitotic divisions follow in the embryo sac, and one of the nuclei becomes the egg. Growth of the pollen tube down the style delivers one male nucleus in each pollen tube to an egg in the embryo sac of an ovule.

new genetic combinations can be formed, meiosis gives a decided advantage to sexual forms over those that are exclusively asexual. Sexual reproduction with its meiotic mechanism makes possible an increase in the rate of evolutionary progress. It makes available more diverse kinds of living things, which can explore a wider range of environments. Mitosis and meiosis are both fundamental features of eukaryotic cells, and the many genetic phenomena that occur in higher organisms are related directly to both processes. If the essentials of mitosis and meiosis presented in Chapters 2 and 3 have been grasped, the genetic principles encountered in the ensuing discussions should be easily followed.

REFERENCES

Avers, C. J. *Molecular Cell Biology*. Benjamin-Cummings, Menlo Park, CA, 1986.

Hotta, Y., A. C. Chandley, and H. Stern. Meiotic crossing over in lily and mouse. *Nature* 269: 240, 1977.

Moens, P. B. The structure and function of the synaptonemal complex in *Lilium longiflorum* sporocytes. *Chromosoma* 23: 418, 1968.

Moses, M. J., S. J. Counce, and D. F. Paulson. Synaptonemal complex complement of man in spreads of spermatocytes, with details of the sex chromosome pair. *Science* 187: 363, 1975.

Stern, H., and Y. Hotta, Biochemical controls of meiosis. *Ann. Rev. Genet.* 7: 37, 1973.

REVIEW QUESTIONS

1. Name the precise times in the meiotic cell cycle at which the following pertain:

   A. The chromosome threads in the bivalents appear to repel, and chiasmata become evident.

   B. Pairing of homologous chromosomes takes place.

   C. The chromosomes are greatly contracted, and the bivalents are widely spaced in the nucleus.

    D. DNA replication takes place.

    E. The chromonema shows obvious chromomeres and is greatly elongated, appearing as a single filament.

    F. Crossing over takes place.

    G. Chromosomes composed of two chromatids are aligned separately on the equatorial plate.

2. One DNA molecule is believed to be present in an unreplicated chromosome. For the human, with a diploid chromosome number of 46, give the number of DNA molecules, the chromosome number, and the bivalent number for a cell or nucleus at each of the following stages:

    A. Pachynema

    B. Diplonema

    C. Diakinesis

    D. Telophase I

    E. Prophase II

    F. Telophase II

3. How can one distinguish between the following cytologically:

    A. A cell in mitotic metaphase from one at first meiotic metaphase.

    B. A cell at mitotic anaphase from one at first meiotic anaphase.

    C. A mitotic prophase from a first meiotic prophase.

4. A cell is dihybrid for the allelic pairs Aa and Bb. The two loci are found on nonhomologous chromosomes. What possible genetic combinations can arise in the following situations:

    A. When this cell undergoes mitosis.

    B. When a cell of the same genotype undergoes meiosis.

5. A human body cell normally contains 46 chromosomes. Give the number of chromosomes in each of the following:

    A. A cell resulting from the mitotic division of a spermatogonium.

    B. A primary oocyte.

    C. A secondary spermatocyte.

    D. The polar body formed along with a secondary oocyte.

    E. A spermatid.

    F. A cell formed after spermiogenesis.

6. Allow "A" to stand for one set of autosomes in the human, and let "X" and "Y" represent the sex chromosomes.

    A. What will be the chromosome constitution of eggs with regard to autosomes and sex chromosomes?

    B. Answer Question A in regard to sperms.

    C. What will the chromosome constitution in a spermatogonium be?

    D. What will the chromosome constitution in an oogonium be?

7. In which of the cell types in Question 5 are bivalents present?

8. A. How many sperm or male nuclei will arise from each of the following?

    (1) 1000 primary spermatocytes.

    (2) 1000 secondary spermatocytes.

    (3) 1000 spermatids.

    (4) 1000 microspore mother cells.

    B. How many egg cells will arise from each of the following?

    (1) 1000 primary oocytes.

    (2) 1000 secondary oocytes.

    (3) 1000 megaspore mother cells.

    (4) 1000 megaspores that are the immediate products of meiosis?

9. In certain insects such as the grasshopper, which has the X-O condition, the male has only one sex chromosome, an X, whereas the female has two X chromosomes. Assuming the females of such a species have a diploid chromosome number of 12, answer the following questions.

    A. How many autosomes will there be in the wing cells of a female?

    B. Answer Question A for a male.

    C. What will be the total chromosome number in the wing cells of a male?

    D. How many chromosomes will there be in the sperm cells that bear an X chromosome?

    E. How many chromosomes will there be in the non-X bearing sperms?

10. In corn, there are 10 pairs of chromosomes in the somatic cells. What would be the expected chromosome number in the following:

    A. A pollen tube nucleus.

    B. A cell in a petal.

    C. A cell in the embryo of the seed.

    D. A pollen mother cell.

    E. A megaspore.

    F. A cell of the embryo sac.

11. Next to each of the following, place the symbol Bo if the item is associated with both mitosis and meiosis; place the symbol Me if it is much more closely associated with meiosis: (1) chromatids, (2) centromere, (3) bivalent, (4) S phase, (5) synaptonemal complex, (6) independent assortment, (7) kinetochore fibers, (8) cytokinesis, (9) chiasmata, (10) Mendel's first law, (11) DNA digestion, (12) sex chromosomes.

4

GENIC INTERACTIONS

Genotype, phenotype, and characteristic

Throughout the early decades of this century, the particulate nature of inheritance was found to apply to a large number of different plant and animal species. It therefore became essential to define the nature of these factors that are transmitted from one generation to the next. Many of the early twentieth-century geneticists were vague or even confused about the distinction between the unit factors of Mendel and their relationship to an individual's actual visible characteristics. No one person merits more credit for clarifying this critical relationship than the Danish botanist, W. Johannsen. In his first use of the words *phenotype* and *genotype,* he clearly distinguished between the assortment of observable characteristics an individual possesses (the phenotype) and the genetic constitution (the genotype). The genotype is composed of elements (genes) that the individual receives through the gametes. It establishes the foundation from which development of the individual proceeds. The final phenotype, in turn, depends on the interaction of these genetic elements with factors in the environment. Johannsen clearly pointed out that no one gene corresponds directly to any one specific characteristic in the phenotype of an individual. For the phenotype is *not* directly inherited; it results from

a complex interplay of genetic determinants with the various aspects of the environment. Examples of the inheritance of traits in poultry and several other species in the following sections of this chapter clearly distinguish between what *is* inherited (the genotype) and the final phenotype.

Interaction of two or more pairs of alleles

The early work of Bateson with poultry demonstrated that not just one pair of Mendelian factors but that at least two could interact to affect the color of the feathers. It was found, when certain different races of white birds were crossed, that the F_1 offspring were not white but colored! The explanation put forth by Bateson (the correct one) was that two separate pairs of Mendelian units are involved in such cases and these can interact to produce an effect that is distinct from that of either one by itself. We can visualize this by assuming two different dominant genetic factors for pigment formation (C and O) and their recessive alleles for no pigment (c and o). If *any* color at all is to be produced, *both* dominants must be present. In the absence of either dominant allele, no pigment can be formed and the feathers are therefore completely white (Fig. 4-1A). Consequently, two different races may show the same white phenotype but actually

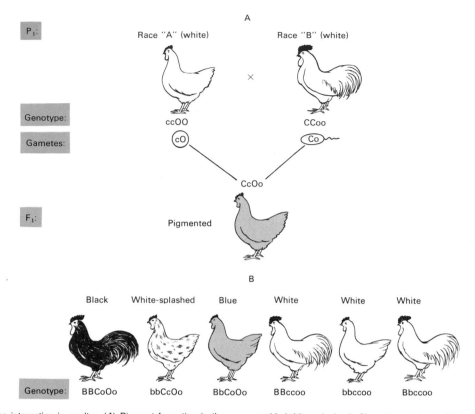

FIG. 4-1. Gene interaction in poultry. (*A*) Pigment formation in the feathers depends on the presence of both dominants, C and O. Races A and B are white because each lacks one of the dominants. The F_1 hybrid between them has colored feathers because it receives a different dominant from each parent. (*B*) The specific color of the feathers depends on several pairs of alleles, among them B (black) and b (white splashed). Since there is lack of dominance, the hybrid (Bb) is blue. However, for the pigment alleles to express themselves, the dominant alleles C and O must be present; otherwise a bird will be white, regardless of the genetic factors present at the pigment locus.

possess different genotypic constitutions. Race A may be white because it is ccOO, whereas race B is white because it is CCoo. A cross between members of the two races brings together the two dominants, one from each race, and the offspring can have colored feathers. *Exactly what* the color will be, however, depends on still other genes.

Another locus is present that can affect the *type* of pigmentation. We may let B represent the determinant that governs the formation of black pigment. Its allele b results in white feathers flaked with pigment (white splashed). Because neither allele is dominant, the heterozygote, Bb, is blue feathered. Any bird may be black (BB), blue (Bb), or white splashed (bb), *but only if both* C and O are present at the other loci (Fig. 4-1*B*). For example, the genotype BbCcOo would produce blue feathers, but Bbccoo can result only in pure white. This example clearly shows that there is not *a gene* for the color of the plumage. At least three

separate loci are involved (really more), the interactions of which may produce different effects. This example should also eliminate the notion that a unit character, such as color, is somehow carried in the cell and transmitted as such to the next generation. Rather, the genetic component in this case includes certain genes that can determine the presence or absence of specific pigments in the feathers. The final color that is realized phenotypically depends on the interaction of these several genetic elements, and, as we will see, environmental factors as well.

Appreciation of genic interaction and the nature of wild type

When we are following a simple monohybrid or dihybrid cross, how do we take into consideration the fact that many genes are actually involved in the expression of a particular character? To answer this,

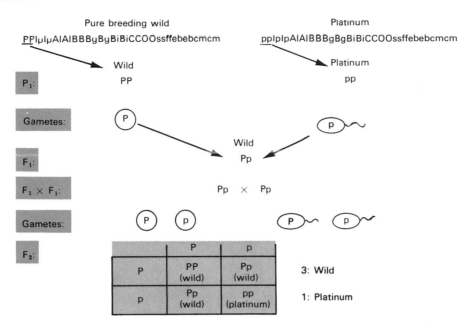

FIG. 4-2. Cross of wild and platinum-coated minks. Several pairs of alleles interact to produce a given coat color. In a simple monohybrid cross, however, only one locus is under consideration. There is no need to represent the others, because the rest of the genotype is identical in the P₁ parents and does not vary in any individual.

let us turn our attention next to the many pairs of loci involved in the production of coat color in the mink. In this animal, at least a dozen genes are known to influence the color of the fur. The standard coat color (wild type) results from a certain combination of genes. Minks that are pure breeding for wild fur color must be homozygotes of the following constitution: PP IpIp AlAl BB BgBg BiBi CC OO ss ff ebeb cmcm.

Substitution at just one of these loci may result in a phenotype quite distinct from the standard. For example, the presence of two recessives, pp, in place of the dominant P causes the fur to be platinum. Replacing the dominant allele B with the recessive condition bb changes the fur to brown, even though two doses of the dominant allele for wild coat color are present at many of the other loci. A substitution of one dominant Eb for just one recessive eb in the genotype can alter the coat color from wild to ebony. How then are we to diagram through the F_2 generation the following monohybrid cross: wild mink × platinum? It would be ridiculous to write more than is shown by the simple Punnett square method in Fig. 4-2. The reason for this is that *only one* locus is under consideration; only one pair of alternatives at a specific locus is being followed. Although we must appreciate that any character such as coat color involves many loci, a particular cross does not necessarily involve more than one or two of them. In any diagram, we symbolize only those loci that vary in the particular cross. In our example, only the variation at the platinum locus concerns us. Although we should appreciate that the genes at the other loci are also exerting an effect, we need not represent these other genes because they are identical in both wild (PP) and the platinum (pp) parents.

Now suppose that a platinum mink is mated with one with brown fur. In this case, two different genes at two different loci are being followed, the p locus and the b locus (Fig. 4-3). We must consider both of these, but we can forget the other 10 because they do not differ. The mating is nothing more than a simple dihybrid cross. It is important at this point to appreciate, fully from Fig. 4-3, that this cross of a platinum and a brown animal involves two loci and that *both loci* and the gene forms present at them must be represented in the genotype of *each parent*. This is essential because the platinum parent is carrying the wild allele (B) in the homozygous condition at the "brown" locus. In a similar way, the brown parent is carrying the wild gene form (P) in the homozygous condition at the "platinum" locus. As a result of this, the F_1 are all wild. It would be a serious error to represent a cross between platinum and brown as pp × bb. This fails to take into consideration that

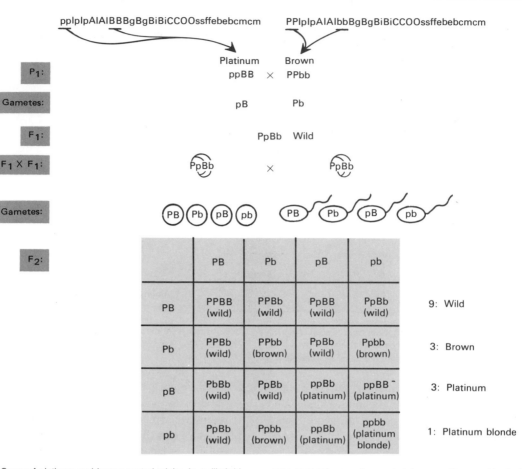

FIG. 4-3. Cross of platinum and brown-coated minks. In a dihybrid cross, variation is being followed at two loci, and these must be represented in each of the individuals. The other loci that also influence coat color are disregarded, because they are identical in every case.

two separate genes and loci concern us in each parent. It results in a meaningless representation of the cross.

Besides showing us the interaction of many genes, the example of coat color in mink also illustrates at least two other important points. The term *wild type* is commonly used to mean "normal" or "standard," the form typically encountered in nature (more details on naming of loci are found at the end of Chapter 2). We see from the case of coat color that *not just one* gene or one form of a gene is responsible for wild. There are as many so-called wild or normal alleles for a specific characteristic as there are separate loci that affect the character.

In Chapter 1, the point was made that we can study a character only if variations in it are available. If everything remained standard, we would never learn the basis of inheritance of any character. But because

a gene can mutate, alternative gene forms arise and produce inheritable changes, which result from alterations that take place in the DNA composing the gene. These may arise spontaneously or be induced by certain environmental factors. All alternative gene forms stem from mutations that have taken place at genetic loci. When a gene mutates and gives rise to an alternative form, we may then recognize the wild form of the gene and its mutant allele. The mutant gene form is a departure from the wild or standard and can result in a phenotypic trait, which varies from the typical or normal form of the characteristic. It is these variations from the standard or wild that provide us with contrasting alternative traits and allow us to study the genetic factors that interact to produce a normal phenotype. Thus, if we have 12 different loci, as we do in the case of mink coat color, we are consequently aware of 12 different kinds of departure from the wild. By studying the inheritance pattern of each pair of traits, we learn that fur color in the mink depends on at least 12 separate allelic pairs, each pair representing a wild or normal gene and its allele, a mutant form.

When the dihybrid cross, platinum × brown is followed through the F_2 generation, another important point comes to light (see Fig. 4-3). Among the offspring of the F_1 wild, dihybrid animals, four different phenotypes appear. The chance for a wild is 9:16, for a platinum 3:16, for a brown 3:16, and 1:16 for a completely different type, which has been designated "platinum blonde" (ppbb). The cross illustrates beautifully one of the important consequences of the sexual process: independent assortment with its production of new gene combinations. Because of the segregation of two pairs of alleles and their independent transfer to the gametes, various genotypic combinations are possible (3^2 or 9; see Chap. 1). Since dominance is involved, they fall into four different phenotypes (2^2): three of them are the familiar wild, platinum, and brown, but a new phenotype—platinum blonde—has been derived. Imagine the many genotypes that could result if all 12 pairs were considered (3^{12}) and the variety of possible phenotypes, more than 2^{12}, because there is no dominance between certain alleles affecting the coat color. We must keep in mind that the physical basis that generates such variation in sexual species is the meiotic cycle of the chromosomes, a foundation on which evolutionary progress has depended.

Genic interaction and modified ratios

Once we understand the concept that many genes interact in the normal development of a character, we can appreciate the basis for various kinds of genetic effects. Let us next take an example from the inheritance of fur color in mice, in which case more than one pair of alleles is known to influence pigmentation. In addition to the wild grayish color, there are other possibilities, such as black or white. Regarding these three types—gray, black, and white—two pairs of alleles are known to be involved. Wild, or gray, depends for its expression on the presence of a dominant, B; black pigmentation depends on the recessive allele, b. However, for any pigmentation at all, gray or black, to develop, the dominant allele, C, must be present at another separate locus. The recessive condition, cc, results in a white animal, regardless of the gene forms present at the first locus. Assume that two

dihybrid animals are crossed: BbCc × BbCc (Fig. 4-4). Both are wild in coat color because each possesses the dominants for gray and for pigment production. Among their offspring, the following can be predicted: 9B − C − (gray): 3 bbC − (black): 3 B − cc (white): and 1 bb cc (white). Notice that these kinds of *genotypes* expected from a dihybrid cross are actually obtained; however, the expression of the genotypes is so altered by genic interaction that the expected phenotypic ratio of 9:3:3:1 is modified to a ratio of 9:3:4. This is so because the genotypes B − cc and bb cc cannot be distinguished from each other. Any pigment in the hair requires the presence of the dominant allele, C. It is as if the c locus can suppress the expression of the pigment factors at the b locus. In one way, this may remind us of dominance, but use of that term is restricted to the interaction between a pair of alleles (B is dominant to b). Another word is needed to describe the suppressive influence of any genetic factor on another that is *not* its allele. Such an effect is called *epistasis*. The example of fur color in mice nicely illustrates *recessive epistasis* because the double recessive condition, cc, is required to mask the expression of genetic factors for pigmentation at the b locus. Epistasis, however, may result from the presence of a dominant allele. Coat color in the dog involves at least two loci. In some varieties, black pigmentation results from the presence of the dominant B; brown pigment depends on the recessive allele, b. But neither color will be expressed if the dominant factor, I (for inhibition of color), is found at another locus on a different chromosome. Figure 4-5 follows the cross of two dihybrid white dogs: Bb Ii × Bb Ii. No pigment is found in their hair because the presence of just one dose of the allele I is sufficient to suppress all pigment formation. Among their offspring, the chances become 12:3:1 for white, black, and brown, respectively.

Epistasis is a term that was coined before 1910 by Bateson, who discovered that this type of genetic interaction in poultry can modify the classic 9:3:3:1 ratio of Mendel. However, once the basis for the modification is understood, we see that there is no exception to Mendelian inheritance; we must simply remember that no one gene or allelic pair acts alone.

We have already noted that feather color in poultry requires the dominants C and O; otherwise, no pigmentation is formed in the feather. We see here an example of two genes (one at the c locus and one at the o locus), either of which can suppress the expression of genetic determinants at a third gene, depend-

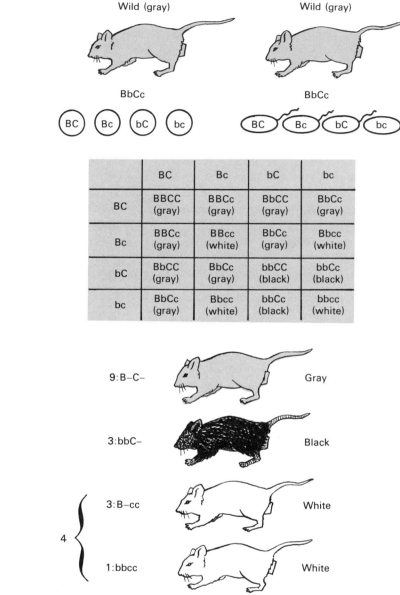

P₁:

Wild (gray)   Wild (gray)

BbCc   BbCc

Gametes:

Offspring:

	BC	Bc	bC	bc
BC	BBCC (gray)	BBCc (gray)	BbCC (gray)	BbCc (gray)
Bc	BBCc (gray)	BBcc (white)	BbCc (gray)	Bbcc (white)
bC	BbCC (gray)	BbCc (gray)	bbCC (black)	bbCc (black)
bc	BbCc (gray)	Bbcc (white)	bbCc (black)	bbcc (white)

9 : B–C–   Gray

3 : bbC–   Black

3 : B–cc   White

4

1 : bbcc   White

FIG. 4-4. Cross of two dihybrid gray mice. The factor B (gray) is dominant to its allele b for black fur. However, no pigment can be produced in the absence of allele C. As a result, two of the four major classes of genotypes cannot be distinguished and the 9:3:3:1 phenotypic ratio is modified to 9:3:4.

ing on which alleles are present. Birds of the genotypes Bb CC oo or Bb cc OO are white, not blue. Only those birds that have at least one of each dominant allele (C__O__) can form feather pigment (see Fig. 4-1B). Consider a cross of two dihybrids: CcOo × CcOo (Fig. 4-6). Color of some type is expressed phenotypically by each parent because both dominants are present. Among their offspring, however, pigmented and white occur in a ratio of 9:7, a modification of the 9:3:3:1 ratio. There are only 9 chances out of 16 for any chick to exhibit some type of color because only 9 out of 16, on the average, will possess both of the required dominant factors. In cases such as this, we say that "duplicate recessive epistasis" is operat-

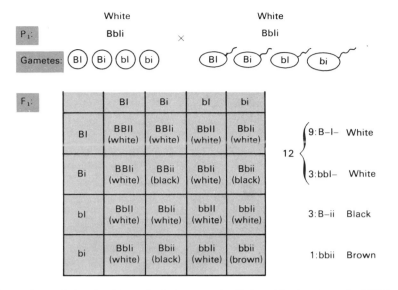

FIG. 4-5. Dominant epistasis. In certain breeds of dogs, the factor for black fur (B) is dominant to its allele for brown (b). However, neither allele can be expressed if the dominant I is present. Because two of the four major classes of genotypes cannot be distinguished and are classified as white phenotypes, the 9:3:3:1 ratio is modified to 12:3:1. Compare this dihybrid cross with that shown in Fig. 4-4, an example of recessive epistasis.

ing, for we have two genes (one at locus "c" and one at locus "o"), *either of which* independently can alter the effects of a second or third one (cc can suppress O— or B—; oo can also suppress C— or B—).

The genetics of poultry provides many illustrations to demonstrate the complexities of genic interaction. Still other genes are involved in feather color. A gene form I at another locus exhibits dominant epistasis and inhibits feather color, just as the allele I in dogs prevented any pigmentation of the hair. (Although the letter "I" is used to symbolize both factors, they are obviously not the same allele, because they occur in very different species.) A pure-breeding variety of black birds must have at least the following genotype; BB CC OO ii. With even more genes involved, we see that the picture approaches that of fur color in the mink. We must remember that these examples of many genes influencing the expression of a character reflect the typical situation; they are not unusual or exceptional in the least.

The basis of genic interaction

Additional examples of epistasis could be presented, but this would simply reemphasize the important point of genic interaction. But an important question should now be answered, "What is the basis of this interaction: how can a factor at one locus affect the expression of one at another locus?" Today we have a great deal of information from the field of molecular biology to supply us with some answers. In later chapters, evidence is presented to show that many genes exert their effects on the phenotype through their control of enzyme production.

A wealth of research, much of it with microorganisms, tells us that the genetic control of metabolism is largely a consequence of gene control of protein formation. Any factor that governs the protein formation of a cell controls its activities. Proteins are the most complex of the chemical compounds in the cell. Not only are they important as units in the structural organizations of various cell parts, but they form the main component of enzymes, the important organic catalysts that enable the cell to carry out most of its chemical steps. We can envision many essential cellular products to be the end result of a series of chemical steps, each of which can take place only if the appropriate enzyme is present in an adequate amount (Fig. 4-7A). Although Fig. 4-7A is a definite oversimplification because it eliminates branches in the sequence, it does enable us to picture the average cellular product as the result of a sequence of steps. No step in the sequence can proceed without the product of a previous step. A breakdown anywhere in the chain prevents the production of the proper end product. Such a breakdown may occur if a step fails

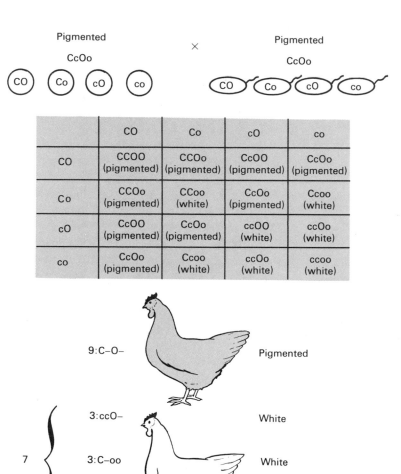

FIG. 4-6. Duplicate recessive epistasis. In poultry, both dominants (C and O) must be present for any pigmentation of the feather. Either double recessive condition (cc or oo) can prevent the expression of genetic factors at other loci that also influence pigment formation. Because three of the four genotypic classes cannot be distinguished phenotypically, the 9:3:3:1 ratio is modified to 9:7.

to progress because of an enzyme that is defective, absent, or present in insufficient quantities.

We may consider pigment production in the feather of the fowl to depend on a developmental sequence that entails several steps, each controlled by a specific enzyme (Fig. 4-7B). We can see that, if a bird lacks the dominant allele C (Fig. 4-7C) or the dominant allele O, no pigment can form. It does not matter what other genetic factors are present for pigment production if a breakdown in a vital step in the chain occurs. If the chain is not interrupted, pigment will form, but the final color depends on still other genes. Inspection of the sequence of steps shows why a ratio

of 9:7 is to be expected after the cross of two dihybrids: CcOo × CcOo. Our knowledge of the classic 9:3:3:1 ratio tells us that, on an average, in 9 cases out of 16, both dominants will be present: C__ O__. Only these genotypes contain the alleles essential to the completion of the developmental sequence through their control of the needed enzymes. The other seven combinations all lack at least one genetic factor that controls one of the enzymes, and hence a step that is essential to a pigment product. If we look at the same chain of reactions (Fig. 4-7C), we see that the presence of allele I blocks completion of the sequence because enzyme i would be lacking. It does not matter

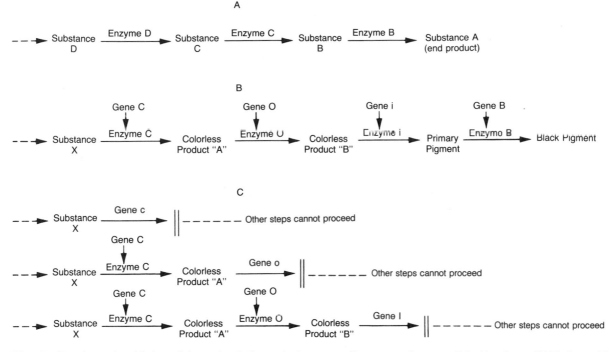

FIG. 4-7. Gene interaction. (*A*) Any cellular product is the result of a series of chemical steps, each catalyzed by a specific enzyme. (*B*) A simplified version of pigment formation in the feathers of the common fowl. A specific gene governs the formation of an enzyme at each step along the pathway leading to the formation of the end product, the pigment. The type of pigment will depend on the enzymes converting the primary pigment. These, in turn, are controlled by specific genes, such as gene B in this scheme. (*C*) Blockage of the pathway leading to pigment formation. The interruption may occur anywhere in the sequence, early (*above*) or later (*below*). Absence of a specific gene form (C, O, or i) and the substitution of its allele (c, o, or I) results in deficiency of the enzyme controlled by the gene. The sequence cannot be completed even though all the other required enzymes are available.

if the factors C and O are both present. The presence of allele I, which causes a lack of enzyme i, interrupts the sequence of steps required for the pigment.

Applying this same reasoning to the case of the dihybrid cross in dogs (see Fig. 4-5), we can see that, on the average, 12 offspring out of 16 will contain I and hence will lack enzyme i:

9 I __B__ white because enzyme i is not present.
3 I __bb white because enzyme i is not present.
3 ii B__ black because enzyme i allows pigment to form and the allele B governs the formation of black pigment.
1 ii bb brown because enzyme i allows pigment to form and the recessive condition bb governs the formation of brown pigment.

An example of duplicate recessive epistasis (9:7 ratio) in humans may be understood in relation to the concepts discussed so far. Two pairs of alleles known to be involved in the development of the hearing apparatus in humans depend on each other for their expression. One pair of alleles, D and d, affects the normal formation of the cochlea. The other pair, E and e, influences the development of the auditory nerve. Absence of either dominant, D or E, thus results in deafness (Fig. 4-8). A normal cochlea without the essential nerve cannot function. Nor can a well-developed auditory nerve operate without all the normal activities of the cochlear canal and its contents. Two dihybrid parents (DdEd) with normal hearing produce gametes that can come together in 16 different ways. Of these, nine will contain both dominant factors (D__E__) needed for a complete, functional hearing apparatus. The other combinations lack one or both of the vital genetic determinants: D__ee (3); dd E__ (3); dd ee (1). Although this example does not explain the genic interaction on the cellular level, it does point out that the normal development of a character (the ear in this case) depends on the orderly development of more than one of its parts. Abnormality in one component resulting from a genetic

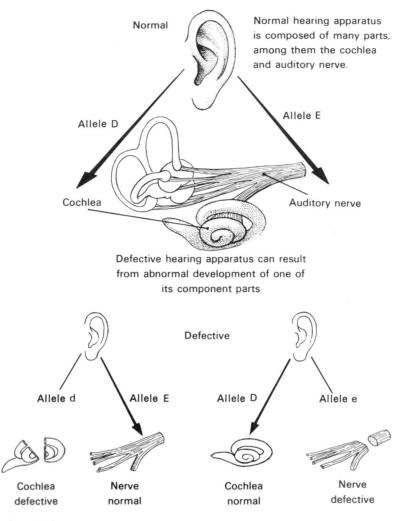

FIG. 4-8. Interaction of genetic factors in development. A normal character may involve the orderly development of many component parts. Abnormality in just one of these because of a genetic block can result in a nonfunctional organ in spite of the presence of the normal components. In the case of the ear, any genotype lacking both dominant alleles (D and E) will be defective as a result of abnormality of the cochlea or auditory nerve.

block in development can prevent the normal expression of another portion, even though the latter may have formed properly under the direction of the required genetic factors.

Interrelated pathways in metabolism

Few synthetic pathways are as simple as suggested by Fig. 4-7A. Most of them are not independent but are linked to others, so that any one product in a sequence may be essential to one additional pathway or more (Fig. 4-9). In the illustration, a lack of C substance leads to defects in two end products, A and Z, because both pathways leading to them require C for their completion. Since metabolic pathways are commonly interrelated in this way, it is not suprising to learn that a gene, when studied carefully, is often found to have more than one phenotypic effect. One of these may be more pronounced than the others, but the latter are real and detectable. For example, in the fruit fly, the recessive allele responsible for white eye color also influences the color of the testes and even the *shape* of the sperm receptacles in the female. In cats, the genetic factor responsible for white fur and blue eyes also results in deafness.

The multiple effects of a single gene or allele are

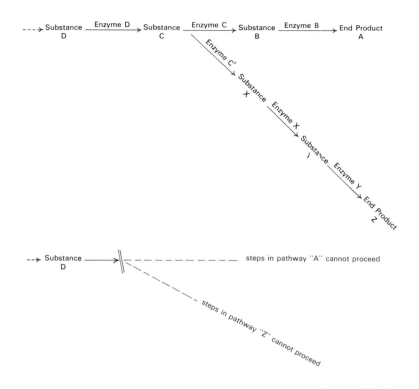

FIG. 4-9. A branched metabolic pathway. Since most pathways are interrelated (*top*), a genetic block at one point may have multiple effects. In this example (*bottom*), a genetic block prevents the formation of substance C, required for the production of two products, A and Z.

termed *pleiotropy*. Although they may seem to be unrelated, study of the chemical or molecular basis of the several effects often reveals a common basis. One excellent example from humans is that of the disorder phenylketonuria (PKU), which is inherited as a simple Mendelian recessive. In this unfortunate condition, severe mental retardation is a typical symptom. Affected children, however, also tend to have light hair and light skin pigmentation. Blood and urine analyses reveal abnormally high levels of phenylalanine, an amino acid concentrated in the protein of milk, cheese, eggs, and several other common foods. The high levels of phenylalanine are associated with high levels of another chemical, phenylpyruvic acid, a substance not found in the body fluids of normal individuals.

Any collection of phenotypic effects, such as these, which defines a clinical condition is commonly called a *syndrome*. The syndrome that describes PKU is now known to result from the pleiotropic effects of a recessive allele that blocks a single step in a metabolic pathway. As Fig. 4-10A indicates, protein from the diet is broken down by the cells into its constituent amino acids. The amino acid phenylalanine is normally converted to another amino acid, tyrosine. This single chemical step involves the substitution of an —OH group for —H and requires a specific enzyme manufactured in the liver. The tyrosine that is normally formed is an amino acid linked to other major pathways—one leading to the formation of melanin pigment, one to the formation of thyroxin, and another to complete oxidation to carbon dioxide and water. Tyrosine may also enter cells as a result of its presence in various dietary proteins; all of it is not formed from the conversion of phenylalanine.

The normal picture of the pathways leading from tyrosine may be altered in the absence of the enzyme that forms tyrosine from phenylalanine (Fig. 4-10B). In this case, the latter substance accumulates in the blood. Appreciable amounts of it become diverted into an alternative pathway, which leads to the formation and excessive accumulation of phenylpyruvic acid. The tyrosine required for the pathways linked to this amino acid must come entirely from preformed tyrosine in the diet. Moreover, excess phenylalanine has been shown to inhibit the activity of the enzyme that

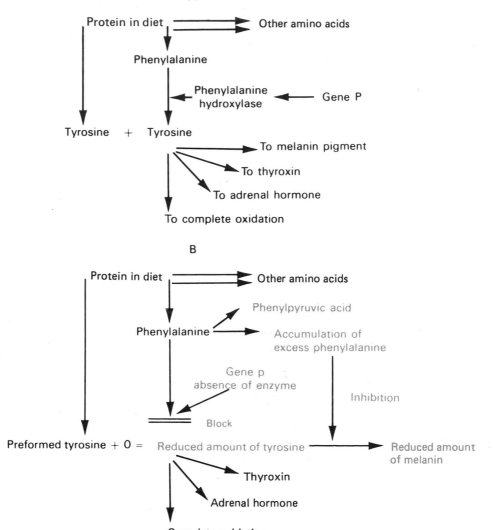

FIG. 4-10. Interrelated pathways in metabolism of phenylalanine and tyrosine. (*A*) Tyrosine normally enters cells through breakdown of dietary protein. In addition, it is formed from phenylalanine by the action of a specific enzyme. The total tyrosine amount then enters other pathways. (*B*) In the absence of the enzyme, because of a recessive allele, phenylalanine accumulates, and some is converted to phenylpyruvic acid. The two substances are toxic. The only available tyrosine is preformed and comes from the dietary protein. The excess phenylalanine inhibits an enzyme activity in the pathway leading to the formation of melanin pigment.

converts tyrosine to another substance needed in the pathway leading to melanin production. We can appreciate why victims of the disorder tend to have less pigment in their skin and hair. The mental retardation is a consequence of the high levels of the accumulated phenylalanine and the phenylpyruvic acid that forms from it. These substances are toxic to the central nervous system and interfere with the metabolism of the brain cells. The unfortunate, irreversible damage

to the nervous system can be prevented if afflicted babies are placed on diets low in phenylalanine. This avoids the accumulation of toxic concentrations of this amino acid and its conversion to phenylpyruvic acid. Fortunately, tests are available for the prompt recognition of affected infants shortly after birth.

The example of PKU demonstrates several ways in which a gene may bring about pleiotropic effects: by the interruption of a chemical step needed for the formation of a substance that is common to more than one metabolic pathway; through the accumulation of a metabolite (phenylalanine here) that may be toxic at high levels and may be converted to another toxic product; and through the inhibitory effect of an accumulated metabolite on enzymes in another metabolic pathway. As in the case of PKU, the assorted phenotypic effects of many genes can be traced to a single chemical block in a sequence of steps. Other examples in humans will be encountered in later discussions concerning genetic phenomena other than pleiotropy.

Molecular basis of certain genetic diseases

Tay-Sachs disease (ganglioside lipidosis) is another serious genetic disorder whose genetic basis has been clarified. This fatal condition is expressed in children homozygous for a certain autosomal recessive. The mutant allele is found with a higher incidence among Jews of Northern European origin than in other groups. The severe symptoms (blindness, growth retardation, insanity, and loss of motor coordination) typically end in death by the age of 4. The severity of the disorder follows from abnormal accumulation of lipids in cells of the central nervous system. The overcrowding produced by the lipid deposits can actually cause death of nerve cells.

We now know that this serious defect is associated with the blockage of a critical step in the breakdown of fats. Victims of Tay-Sachs disease show an abnormality in the enzyme hexosaminidase (Hex). This enzyme is actually composed of two different polypeptide chains, hexosaminidases "A" and "B." Only the latter is found in normal amounts in victims of the disorder. It is a deficiency of Hex A that blocks the normal metabolism of fats in the central nervous system. The parents of affected children are able to handle fats normally, but examination of their blood shows them to have less than the typical amount of Hex A. Prospective parents showing such a deficiency run a risk of one in four of producing a child with Tay-Sachs. It is now possible to detect such a child in utero. Cells shed into the fluid of the amniotic cavity can be obtained through amniocentesis. In this procedure, a sample is obtained of the amniotic fluid that surrounds the fetus. This is accomplished by inserting a hypodermic needle through the woman's abdomen and the wall of the uterus. Those cells that are shed into the fluid are fetal in origin. If cultures of these cells reveal that Hex A is missing, an abortion

may be considered to prevent the birth of a seriously afflicted child.

The molecular bases of other genetic disorders have now been clarified and shown to result from deficient proteins or enzymes that upset steps in biochemical pathways. The knowledge accumulated on the genetic control of metabolic pathways may eventually lead us to ways to treat successfully victims of severe genetic defects, or even to prevent the onset of the condition.

Variation in gene expression

Certain alleles are very constant in their expression. In the human, for example, the genetic factor, S, determines the presence of normal hemoglobin (Hemoglobin A) in the red blood cells, whereas its recessive allele, s, determines the presence of sickle cell hemoglobin (hemoglobin S). A person of genotype SS has only Hemoglobin A in the red blood cells and the ss individual has only Hemoglobin S. The heterozygote has red blood cells with both types of hemoglobin, A and S. The alleles for the two types of hemoglobin can be expected to express themselves whenever they are present in the genotype, and to express themselves in the same way.

In contrast with this, we find a large number of other alleles that do not always express themselves so predictably. A dominant (P) in humans is responsible for the production of extra digits on the hands and feet, an abnormality known as polydactyly. The condition may be present in several members of a family—a parent and one or more children. Occasionally, a phenotypically normal person from such a family marries another normal, unrelated individual but nevertheless produces children with polydactyly. From studying many such cases, it becomes apparent that a person who is heterozygous for the condition (Pp) *may* or *may not* show the trait. The dominant for polydactyly is thus not constant in exerting its phenotypic effect. Because it does not express itself in all the individuals who carry it, we say that the allele has *reduced penetrance*. If an allele always expresses itself, as in the case of Hemoglobins A and S, we say that it is 100% penetrant. If 10 people have the same genotype (such as Pp for polydactyly), but only 9 out of the 10 show the dominant effect of the mutant factor whereas the tenth appears normal, the allele is said to have a penetrance of 90%. The degree of penetrance varies greatly for different genetic factors. For example, another dominant in humans causes the production of bony projections and has a penetrance of only 60%.

Experiments with laboratory animals have enabled us to demonstrate that reduced penetrance is a phenomenon that must be considered in genetic analysis. In the fruit fly, a certain recessive (i) causes an interruption in one of the veins of the wing. When flies with interrupted wing veins are mated, 9 out of 10 of their offspring also show the trait, and the remainder have normal wings. The geneticist has an advantage with laboratory organisms such as flies, because these normal-appearing individuals can be selectively bred together (Fig. 4-11). Among their offspring, 9 out of 10 will show the mutant trait. The remaining normal 10% can be bred further and will again produce 90% mutant progeny and 10% normal. We clearly see from such a study that the 10% that are phenotypically normal individuals are actually homozygous recessives and that the allele for interrupted wing vein has a penetrance reduced to 90%. It is evident that the true genetic constitution of a certain phenotype (e.g., the homozygous recessive genotype of the normal-appearing flies) may remain undetected if selective breeding experiments cannot be performed. Reduced penetrance is therefore a factor that can complicate genetic investigations and also the interpretation of pedigrees, especially those of humans (see Chap. 6).

In addition to the modification of gene expression that can result from reduced penetrance, another type of variation is frequently encountered. In the case of polydactyly, those persons who *do* express the allele when it is present fall into an assortment of phenotypes. Some of them may have an extra digit on each hand and foot, others on only one. Moreover, the extra digit may range in development from complete to just a vestige. Obviously, when the allele for polydactyly *is* penetrant, it varies in the kind of phenotypic effect it produces. We say that an allele whose expression varies in degree is of *variable expressivity*. Note the distinction between *penetrance* and *expressivity*. The former measures the ability of an allele to express itself in any way when it is present in the genotype. Expressivity refers to the kind of phenotype an allele produces when it does express itself (i.e., when it is penetrant). One genetic factor may be 100% penetrant but its expressivity may vary greatly. Another may show reduced penetrance but be very constant

FIG. 4-11. Reduced penetrance. When flies with the mutant trait interrupted wing vein are mated, 10% of their offspring appear normal. When these are mated to each other, 90% of their offspring are mutant and 10% normal. Further breeding of these latter types produces the same results. All of the flies, those normal in phenotype as well as those expressing the mutant trait, are homozygous recessives for interrupted wing vein (ii). Therefore, the recessive genotype is only 90% penetrant.

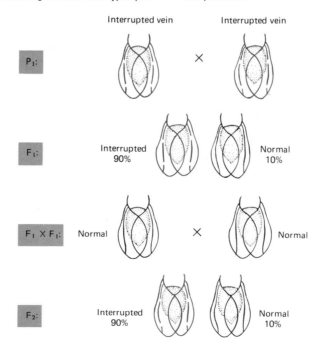

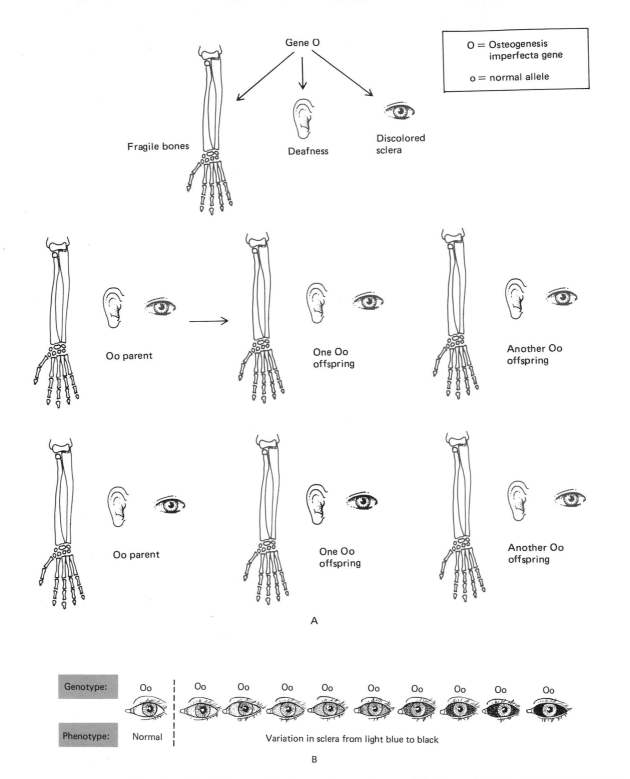

FIG. 4-12. Penetrance, pleiotrophy, and variable expressivity. (A) The dominant allele, O, for osteogenesis imperfecta may affect the bones, inner ear, and eye (top). However, a person who carries the allele and expresses it may show only one abnormality, such as affected eyes. Offspring of such a person who receive the allele may express all three defects, none, or any combination of defects (middle). Since the allele is only 90% penetrant (bottom), a person carrying it may not express it at all in any detectable way. However, offspring of such a person may show none, all, or any combination of abnormalities. (B) A person carrying the allele O may not exhibit any discoloration of the sclera of the eye. Among those who express this trait, there is variation in the shade of the sclera from one person to the next.

in its expression. Besides being 100% penetrant, the factors for Hemoglobins A and S do not vary at all in their expressivity.

A good example from the human illustrates an allele with reduced penetrance that, in addition, has a pleiotropic effect with variable expressivity. The dominant factor (O) is associated with *osteogenesis imperfecta*, a condition in which the bones of the body are quite brittle. The allele is also associated with two other effects. When it is present in the genotype, the sclera (the outermost coat of the eye) may be blue, and the bones of the inner ear may be hardened, producing deafness. However, only 9 persons out of 10 who carry the allele express it in any way at all. Such persons may exhibit only one of the three associated effects, others all three or any combination of two (Fig. 4-12*A*). Moreover, when any one of the traits is expressed, its severity varies markedly from one individual to the next. The effect on the bones may be very pronounced, subjecting such persons to multiple fractures, whereas the condition is so mild in others that it almost escapes detection. The sclera color of those who show any discoloration may range from a tinge of blue to almost black (Fig. 4-12*B*)!

Basis for variation in genic expression

How can we explain the basis of the phenomena of reduced penetrance and variable expressivity, which can cause marked variations in gene expression? Today we have actual experimental evidence to give us an insight into the problem; however, the correct idea goes all the way back to Johannsen, who stressed that characteristics are not inherited but develop from the interaction of many genes and the environment. So far in our discussions we have encountered several examples of genic interaction in which one gene alters the expression of another. In the simple example of complete dominance of the allele for tallness in peas, the recessive allele does not express itself at all in the heterozygote (Tt). In a sense, the recessive (t) is not penetrant when its allele (T) is present. Where epistasis is operating, an allele may not be penetrant because of the presence of another genetic factor at another locus (feather color is not expressed in the genotype (CCII). In these examples, we clearly see that the penetrance of an allele may be the direct result of the influence of some other genetic factor, either its own allele (in the case of dominance) or a nonallelic factor (as in epistasis). Therefore, reduced penetrance should not be regarded as any sort of contradiction to the principles of heredity. It is an expected consequence of the role of the rest of the genotype in the expression of any form of one specific gene.

Similarly, variation in the *kind* of expression an allele has when it *is* penetrant often depends on the presence of certain other genes. Many examples relating to spotting have been studied in cattle and rodents. In mice, white spotting may depend on the presence of a recessive (s). However, although spotting may be expressed in the homozygous animals (ss), the size of the white areas can range from tiny points to large spots to a coat that is entirely white. The amount of whiteness depends on a host of genes, each with a small but definite effect, interacting with the allele for spotting (s). The term *modifier* (or modifying gene) is usually reserved for any such genetic factor whose effect is to alter, in a small quantitative way, the expression of another genetic factor. The cumulative or added influence of modifiers can be very significant. If different modifiers are present in different individuals who possess the same "main allele," the expression of the latter may vary greatly from one individual to the next. The geneticist must consider modifiers when he or she is studying the action of specific genes in experimental animals or plants. The stocks must be kept as similar as possible in regard to modifiers. Failure to do so can lead to erroneous interpretations of inheritance patterns and gene expression.

Some modifiers have such an extreme effect that they completely prevent the expression of some other allele. In the fruit fly, there are several examples of the failure of a mutant genotype to show phenotypically when it is expected. One such case involves the size and shape of the wing. Flies homozygous for the recessive vestigial (vg vg) have reduced and distorted wings. However, the mutant condition may not appear if a certain other allele (called *dimorphos*) is present at another locus. We call the latter a *suppressor*, meaning that it is a genetic factor that can prevent the expression of a mutant allele. Some suppressors behave as dominants, others as recessives. Suppressors as well as modifiers may confuse the genetic picture if they are not considered in crosses. Moreover, their recognition enables us to appreciate part of the basis for penetrance and expressivity, for modifiers and suppressors, along with other genes composing the total genotype, may interact to produce variation in the expression of any one particular gene.

Importance of the environment in gene expression

We must not forget that the activities of all the genes are taking place in the cellular environment, which in turn may be influenced by the external environment. The geneticist must recognize the importance of environmental conditions, because any characteristic depends on a certain environment as well as on a certain genotype for its typical expression. The environment has been shown to be a critical factor in the degree of penetrance and expressivity of many alleles. In the fruit fly alone, temperature changes are known to be able to change the penetrance of many alleles from 0 to 100%. Similarly, the expressivity of an allele may vary greatly with temperature. The genotype that produces blisters on the wings is much more extreme in its expression at 19° than at 25°. Not only temperature, but all aspects of the environment must be considered in relation to their influence on gene expression. Some of the variation in the expression of polydactyly is undoubtedly environmental. The fact that any one person showing the trait may have a complete extra digit on one hand and none at all or just a protuberance on the other argues that environmental influences are operating. We are seeing here an example of variable expressivity in *one* individual. Genetic factors for extra digits are also known in animals in which there is also evidence for an effect of maternal age on the penetrance of the specific factors.

Nutrition or diet is another critical environmental factor to consider. Recall that the dire effects of phenylketonuria are expressed when there is an accumulation of phenylalanine. A diet that eliminates much of this amino acid is a factor in preventing the mental retardation that full expression of the allele for PKU can cause.

The varied responses of different individuals to certain drugs is another good example of the interaction of the genetic component with environmental factors. This is demonstrated in the reaction of some persons to primaquine, a drug used for years to treat malaria. After administration of primaquine, certain patients develop an anemia that follows from a breakdown of red blood cells. These cells in sensitive persons have been shown to be normal until they contact primaquine. Sensitive persons also lack the enzyme glucose-6-phosphate dehydrogenase, although the relationship between this enzyme and the fragility of the red cells is unknown. The condition is inherited as a sex-linked recessive and also predisposes the individual to an increased sensitivity to still other drugs, including sulfanilamide (Chap. 5).

It is significant that those showing this sensitivity are otherwise completely normal. Only an encounter with the specific drugs reveals any difference from the majority of persons. The reaction of the sensitives points out the importance of exercising caution when experimenting with drugs for any reason. Wide differences in response resulting from genetic factors could have disastrous effects for some individuals. This is vividly seen in the violent reaction to barbiturates that can result in the death of susceptible persons. These barbiturate-sensitive individuals are also usually very sensitive to sunlight, a symptom of a condition known as *porphyria*. The basis for these two reactions is definitely genetic, but the responsible genetic factors vary from one pedigree to another. A most significant point is that the genetic disorder is usually not severe unless barbiturates are encountered. Again we see that exposure to a drug may trigger a very unexpected reaction in an apparently normal individual.

A frequently cited example from rabbits concerns the interaction of temperature with the genotypes that influence fur color. Coat color in rabbits, as in other mammals, involves the interaction of many genes. Formation of the normal dark pigment requires a series of steps and enzymatic reactions. Among the several coat patterns is one known as *himalayan,* a phenotype characterized by white body fur and black areas at the tips of the feet, tail, ears, and nose. This pattern results from the presence of an allele (c^h), that controls the production of a certain enzyme capable of catalyzing a reaction needed to form dark pigment. At higher temperatures, however, the enzyme is inactivated. Raised under temperate conditions, rabbits contain this enzyme in its active form *only* in the cooler regions of the body, the extremities such as the feet and tail. That a temperature effect is indeed responsible for the pattern can be shown dramatically by shaving a himalayan rabbit and rearing it under cool conditions. The new fur that grows in is black. If only the white areas of an animal's back are shaved off and kept cool by application of ice packs, the newly emerging fur in these areas will be black. We see here an example of the environmental alternation of gene expression. Indeed, the effect of environment may be such that one cannot tell by mere visual inspection whether the phenotype is the result of the presence of certain alleles or the environment

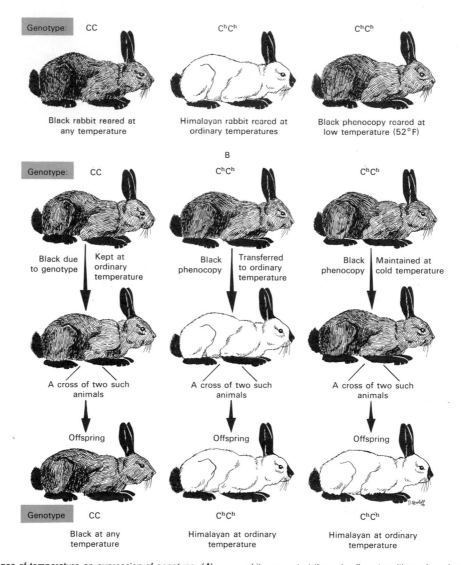

FIG. 4-13. Influence of temperature on expression of genotype. (*A*) If a himalayan rabbit is reared at a low temperature from the time of its birth, its fur will be black, and it will appear identical to the black rabbit whose genetic constitution enables it to form pigment at any temperature. (*B*) The difference between rabbits of the two genotypes becomes apparent when environment is altered. When a black phenocopy is transferred to a warmer temperature, the new hair grows in white, except at the animal's extremities, where body temperature is lower. Even if a black phenocopy is kept under cold conditions so that it remains black, its offspring will still be himalayan at ordinary temperatures. The environment did not change the genetic constitution of the himalayan rabbit that is the phenocopy. However, it did alter the expression of the genotype.

is operating in such a way that it produces a phenotype that mimics the effect of a specific genotype. In the case of the rabbit, for example, there is a variety of animal in which black fur depends on the genic constitution. Such animals are black regardless of the temperature under which they are raised. Rabbits with a himalayan genotype may appear identical to these animals if they are raised under cold conditions, even though their genotype for coat pattern is different (Fig. 4-13*A*). The term *phenocopy* designates the individual whose phenotype has been altered by the environment in such a way that it imitates the phenotype usually associated with a particular genotype. Environmental production of phenocopies can lead to erroneous interpretations unless their true nature is appreciated.

Importance of both environmental and hereditary factors

A very frequently asked question is, "Which is more important, heredity or environment?" We hope it will become clear from the many examples we present that such a question is meaningless on close scrutiny. Does the example of the phenocopy in the rabbit mean that environment is more important? Has it altered the genes? Let us compare a black phenocopy (possessing the allele for himalayan) with an animal that is black as a result of the allele it carries for black pigment formation (Fig. 4-13*B*). The latter will have black fur regardless of the temperature under which it is reared. It does not matter if we cool or heat areas of its skin; it will still produce black fur, and it will pass down to its offspring the genetic factors for the same phenotypic potential. The animal that is the phenocopy, however, will produce white fur if the environment is altered. And no matter how we raise the animal or cool its skin, it cannot pass alleles to its offspring that will permit them to grow a black coat at ordinary body temperature. The environmental influence does not change the genetic factors, but it can alter the manner of their expression.

Concern with the influence of environment on heredity troubled many early biologists and has persisted into the present. Weismann's concept of the continuity of the germ plasm (Chap. 1) focused attention on the importance of the gametes as opposed to the body cells. But the idea persisted that perhaps the environment, acting over long periods of time, could slowly change the genes and direct the course of evolution in the manner suggested by Lamarck. A classic experiment performed in 1909 by Castle and Phillips supported Weismann's theory and demonstrated the fallacy of the Lamarckian argument. These geneticists knew that in guinea pigs black coat color depends on a dominant (A) and albino coat on its recessive allele (a). In their experiment, they removed the ovaries from a white animal and transplanted to this animal ovaries from a young black guinea pig. The albino female, bearing ovaries from a black animal was mated to an albino male (Fig. 4-14). If the whiteness of the female can affect the genes in the gametes of the transplanted ovaries, then we might expect the offspring to be white or some intermediate shade. But his did not occur. In three separate litters resulting from such a cross, the offspring were completely black. Even though the body of the host mother

FIG. 4-14. Ovarian transplant in the guinea pig. An albino animal lacks the dominant required for black pigment formation. When germ tissue from a black animal is transplanted into an albino, the body of the latter in no way alters the germ cells that are being sheltered. A

cross of this animal with an albino will result only in black offspring because the cross is AA × aa. Such experiments demonstrate that the gametes link one generation to the next and that the genetic information they contain is not changed by the body cells.

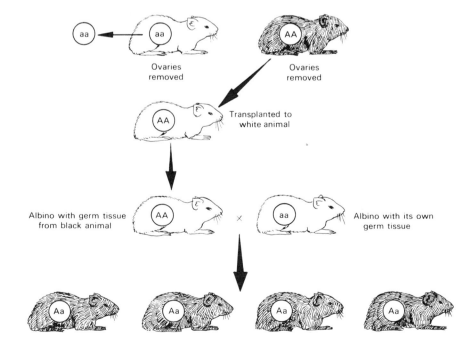

is white, the cross is genetically AA (black female)×aa (white male). This is an elegant demonstration that the body cells in no way affect the transmission of genetic factors to the next generation. The gametes produced in the transplanted ovary carried the factors to be passed to the next generation. Since the ovary contained only genetic factors for black pigment, it was the allele for black pigment production that was handed down. The gametes or their genetic contents were in no way changed just because they were housed in a white body. We will see (Chap. 14) that environmental agents can induce changes in the germ cells, but these mutations are random and are in no way directed by the environmental factor that caused them.

These examples from the rabbit and the guinea pig should help clarify the nature of phenotypes and genotypes. They show us that the phenotype is *not* transmitted. The white phenotype of the host mother was not passed to her offspring. Nor did the black rabbit that was really a phenocopy transmit the potential to produce black fur at higher temperatures. What *is* being transmitted in both cases is the genetic constitution of the gametes. At fertilization, two gametes combine to establish a genotype, a combination of genetic factors from both parents. This genotype sets a potential. What *kind* of phenotype is eventually realized depends on the interaction of all the genes and the environment. And the environmental effect may be the critical one influencing the expression of a certain gene or set of genes and bringing about a particular phenotype.

When we are dealing with very complex characteristics, such as body size or intelligence, the interactions of heredity and environment become so complex they cannot be untangled and measured precisely. But we do know that the expression of the best genotype—let us say one that makes possible the expression of great intelligence—can be suppressed by an unfavorable environment. On the other hand, the most favorable environment will not produce a genius if the hereditary endowment is lacking.

As we learn today of more and more human disorders that have a hereditary basis, we must not despair completely, because the environment can play a major role in alleviating the condition. Recall once more the individuals burdened with the recessive genotype for phenylketonuria. Diet may permit these persons to assume a normal existence so that they appear indistinguishable from other persons. The environmental influence (special diet) has enabled them to escape

the damaging effects of the disorder. Nevertheless, the effect of the environment on their bodies has in no way changed the defective allele they carry. Two normal-appearing PKU persons can produce only offspring who are also metabolic defectives. Their survival and chances for a normal life will again be determined by the environment—proper diet in infancy.

When discussing environment, we must not forget that environment is internal as well as external. The cellular environment provides the background for gene activities, and it may be altered by internal influences, such as hormonal factors inside the body. The expression of those genes that affect the secondary sex characteristics depends on the hormonal background; this is discussed more fully in Chapter 5.

Genes that influence survival of the individual

In this chapter, we have encountered just a few of the numerous ways in which genes interact among themselves as well as with their environment. This interaction often results in modifications of the classic Mendelian ratios, such as 9:3:3:1 to 9:3:4 or 9:7. Expected ratios may be altered by still another factor, which can be illustrated by a familiar example in mice. A cross between any yellow mouse and a gray one always produces yellow and gray offspring in a ratio of 1:1 This makes us suspect that all yellow mice are heterozygotes because the ratio of 1:1 is expected when a monohybrid is testcrossed. The truth of this hypothesis is borne out by following crosses between any two yellow mice (Fig. 4-15). The outcome of such a cross is always two yellow to one gray. And these yellow animals always give the typical 1:1 testcross ratio when crossed to grays.

Evidently some factor is preventing the birth of homozygous yellow animals. Examination of gravid females reveals that these homozygotes are produced but they die as embryos while still in the uterus. The factor for yellow fur (Y) is responsible for killing the individual when it is homozygous, in contrast to its allele for gray. We say that it is a *lethal* allele, meaning that it is responsible for the death of the carrier. Yellow is a *complete lethal*, because it kills the individual before reproductive age. Such lethal alleles are by no means exceptional and must always be considered in populations of plants and animals. Indeed, when a gene mutation arises, it is much more apt to produce an allele that has a harmful or a lethal effect than one that is neutral or conveys a benefit. In the

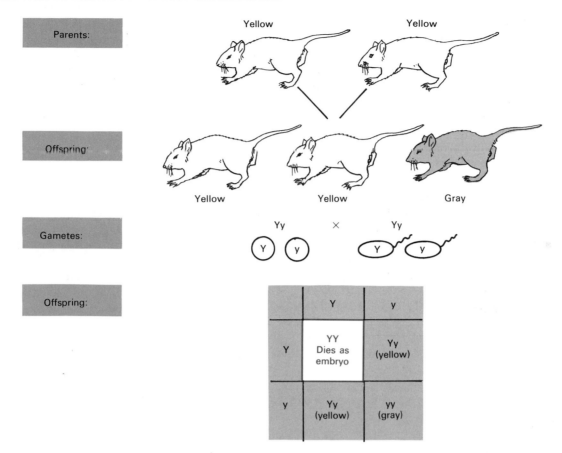

FIG. 4-15. Recessive lethal in mice. When two yellow mice are crossed, the offspring occur in a ratio of 2 yellow: 1 gray (*above*). The explanation is that all yellow animals are heterozygotes. Although the yellow phenotype is inherited as a dominant, the allele that is responsible behaves as a recessive lethal. Therefore, no homozygotes for the allele for yellow fur survive.

fruit fly, invisible lethal mutations outnumber those that produce a detectable phenotypic effect by 10 to 1! The allele for yellow coat in mice is pleiotropic; it has a dominant effect on fur color but a recessive one on viability. Therefore, we can look at the factor for yellow in another way and call it a recessive lethal. Many lethals produce no pronounced effect at all on the phenotype, but they may make their presence known by a decrease in the lifespan or the very early elimination of the carrier. Some geneticists calculate that each human carries, on the average, the equivalent of three lethal alleles.

How can an allele have a killing action? The answer is obvious if we remember that metabolism is the result of many interlocked biochemical pathways (see Fig. 4-9). A defect in just one can upset several others. We can appreciate from what was said about PKU and Tay-Sachs disease that just one defective step can alter the entire chemistry of the body. A lethal, by blocking a critical reaction, can interfere with normal embryological development of an essential organ such as the heart. The death of the embryo may then follow. Lethals can decrease the chances of survival by causing various kinds of abnormalities in development and physiology. Different lethals eliminate individuals at different stages of the life cycle. The complete lethal removes the carrier before reproductive age so that those affected leave no offspring (the allele for yellow fur in mice; in humans the recessive factor for Tay-Sachs disease, which kills in infancy). Some lethals exert their killing action later in the life cycle, after the carrier may have left some offspring. In humans, the dominant factor for Huntington's disease, a fatal deterioration of the nervous system, does not usually express itself before the age of 30. Such genetic determinants, which can result in death but permit the carrier to live to reproductive age, are often grouped as *sublethals*. There are actually no sharp boundaries in the stages of the life cycle at which lethals act to decrease the chances of survival.

Why do most gene mutations tend to be harmful? Keep in mind that mutations are unplanned events. Any unplanned, random change is likely to have dire effects on one or more critical metabolic pathways. Remember that all living things we see around us today, no matter how simple some may seem, are really complex systems on which evolutionary forces have been operating for eons. The force of natural selection has selected those forms that are best fitted to the particular environment in which the species lives. Certain changes may bring about even further advantages in a living system, but unplanned changes take place without consideration of the system they alter. The chances are great, therefore, that a gene mutation will upset the system; thus, the damaging or lethal effect of most gene mutations is not unexpected.

Lethals and the environment

Like any other kind of genetic factor, the lethal can also interact with the environment. In humans, a victim of diabetes, a condition which may involve the complex interaction of many genes, possesses a genotype that may cause death. But today, insulin intake can prevent this lethal effect so that a diabetic may lead an almost normal life and avoid the fatal consequences of the metabolic disorder.

An example from plants illustrates another aspect of environmental influence. In barley, many gene mutations are known that can prevent the development of the chloroplasts. As a consequence, the essential green pigment cannot form, and without it the plant dies. The double recessive white genotype (ww), by preventing chloroplast development, causes the death of the seedling once the reserve food stored in the seed is consumed. A cross of two green heterozygotes (Ww × Ww) produces an average of three green seedlings to one white. The latter type is seen *only* in its seedling stage. If we simply observe the adult plants, we may be unaware of the presence of the lethal. Now suppose we take the offspring from this cross and raise them all in the dark (Fig. 4-16). This prevents the normal development of the plastid, which requires light for its proper formation. In the dark, all the plants will die when the food reserves have been used up. We can keep them all alive, even

FIG. 4-16. Environment and gene expression. The recessive, w, in barley prevents normal development of a green chloroplast. When two monohybrids are crossed and the offspring grown under natural conditions, the white offspring are seen only as seedlings, because they die when food reserves are used up. If the offspring from such a cross are all grown in the dark, all of the plants will be white and must be supplied with nutrients if they are to survive. The plants whose genotypes permit normal chloroplast development cannot be distinguished from the recessives whose genetic factors normally result in their elimination as seedlings.

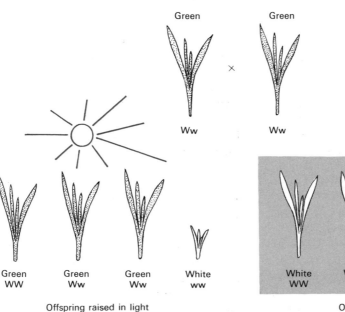

Offspring raised in light

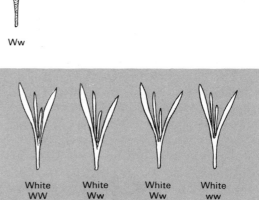

Offspring raised in dark

in the absence of light, by supplying them with organic nutrients. We see in this case that the environment is producing phenocopies; the plants that have the genotypes to permit normal chlorophyll development cannot be distinguished from the mutants. These individuals (WW and Ww) cannot be distinguished from those that possess a genotype (ww) which proves lethal under ordinary environmental conditions. This is just one of numerous illustrations to demonstrate that the best genotype does not ensure normal development; the genetic potential may not be realized in a restrictive environment. However, the most conducive environment will not generate a superior phenotype if the required genetic factors are not there to guide it.

Allelic forms of a gene

In this chapter, we have seen that a genetic locus is a position on a chromosome occupied by a specific gene that has primary effects on a particular characteristic. Therefore, we speak of a locus for eye color or another locus for wing shape, and so on. A gene residing at a given locus may exist in alternative forms, alleles that may produce somewhat different effects on the phenotypic characteristic. For example, a certain locus on Chromosome II of Drosophila influences the size of the wing and its shape. The normal wing is long and smooth in outline. A certain mutation occurred at this locus, producing an alternative form of the normal gene, which causes the wing to shrivel and become nonfunctional. Following the gene naming system discussed in Chapter 2, the locus was designated the "vestigial locus" or vg, because the mutation is recessive to the wild, vg^+. Either of these two alternative forms of the gene may be present at that locus on any one chromosome. Thus, the diploid may be one of three possible genotypes: $vg^+ vg^+$; vg^+ vg; or vg vg.

Now let us turn to a second recessive mutation that

FIG. 4-17. Allelism and complementation. (A) When flies showing vestigial wing are crossed with those showing antlered wing, we might expect the F_1 to be wild if we interpret the cross as one involving separate loci. In such a case, each mutant would be contributing a wild-type allele for which the other is defective. (B) When the cross is actually made, the F_1 flies are mutant, intermediate between vestigial and antlered in wing shape. This indicates that the F_1 did not receive a necessary wild-type allele from each parent. Complementation did not take place. The reason for the results is understandable when the cross is interpreted correctly. Vestigial and antlered are not allelic forms of genes at separate loci, as depicted in A. They are actually variant forms of the same gene that affects the wing. Neither vestigial flies nor the antlered wing flies possess a wild-type form of the gene vg^+, and so the F_1 is mutant. Because the antlered effect is the result of a mutation at the vestigial locus, the correct symbol for antlered wing, vg^a, must be used.

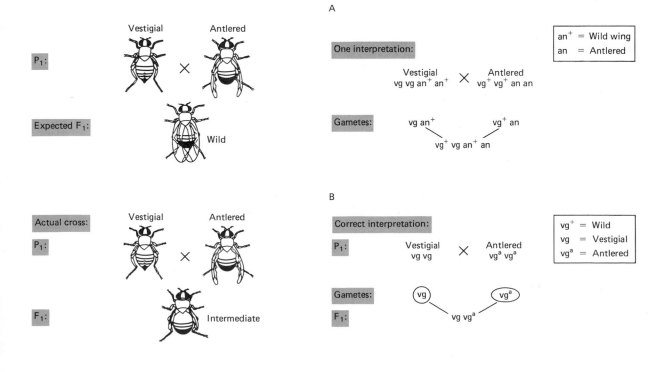

also influences wing shape. This mutation is called *antlered* because it produces a wing form suggestive of an animal's antlers. Suppose we name the recessive mutation "an," and consider it to be a mutation at another locus, the locus "an." We then proceed to mate flies with vestigial wings with those having antlered wings. Believing that we are following two separate loci, vg and an, we would expect our F_1 flies to be wild (Fig. 4-17A). This is the same reasoning we applied to a cross between platinum and brown minks at the beginning of this chapter (see Fig. 4-3). However, the F_1 flies resulting from the cross between vestigial and antlered are not wild. Instead, their wings are distorted and resemble some sort of intermediate between vestigial and antlered. Evidently, vg and an did not complement each other. This means that the F_1 organisms do not have a wild gene form for vestigial and a wild gene form for antlered. As a result, the F_1 flies have a mutant phenotype. It appears, therefore, that the vg and an mutations are two defects in the same gene (Fig. 4-17B). They are not identical defects, and each produces a slightly different phenotype; nevertheless, they are *allelic* to each other, that is, alternative forms of the same gene. This should not surprise us, because the definition of alleles does not imply that only two forms of a gene can exist. Our knowledge of the gene at the molecular level makes this very understandable, for the gene is a stretch of DNA including many sites at which independent changes—gene mutations—may take place (Chaps. 11–14). Each change, though producing a somewhat different effect from the next one, exerts its primary influence on the same characteristic. Any mutation in a gene associated with the wing, for example, has its main effect on the wing, not on some characteristic associated with the eye.

Since antlered (an) and vestigial (vg) are allelic forms of the same gene, we cannot symbolize a cross between antlered and vestigial flies, as shown in Fig. 4-17A, because this implies they are not allelic to each other. As vestigial was the first mutation found at the locus involved in this case, the locus has been designated by the name of that mutation and represented as vg. But now a second mutation has been found in the gene residing at this locus. To designate it, we keep the same base, vg, and add a superscript, giving us vg^a. This tells us we are dealing with another mutation in the gene at the vestigial locus. Figure 4-17B illustrates the cross between an antlered and a vestigial fly in the correct way. We can now see that vestigial and antlered are alleles and that the F_1 will

be mutant. They cannot complement each other because only one locus, vg is involved, and both alleles at that site carry a defect. Many examples are known of genes that exist in three or more alternative forms. We recognize each set as a series of multiple alleles, three or more forms of one gene associated with a specific genetic locus.

Multiple allelic series in animals and plants

Multiple allelic series are known in all groups of organisms, from microorganisms to humans. The number of alleles composing a multiple allelic series has been found in many cases to be quite large, and it is possible that the detection of still additional mutations at a particular locus will increase the number even further. In the fruit fly, as many as a dozen alleles are known for certain loci. On the X chromosome, at the white locus, for example, some 12 alleles have been recognized. These produce a spectrum of pigmentation from white through shades that gradually deepen to the normal red of the wild type. The mutations have been named to suggest their phenotypic effects such as white, ivory, pearl, apricot, and cherry (w, w^i, w^p, w^a, and w^{ch}). A heterozygous female has an eye color intermediate between that of the homozygotes.

The occurrence of numerous mutations at a locus also shows us that, although many changes can occur within a gene and produce other phenotypic effects, the *primary* effect is confined to the one characteristic. Thus, we may find many alterations within the "white" gene that affect eye pigment in different ways, but none of these has a pronounced influence on another characteristic, such as wing shape. This makes perfect sense, because a gene at any locus has a primary effect in a certain chemical pathway as a result of its control of a specific metabolic step (see Fig. 4-7A and B).

In the rabbit, a well-known multiple allelic series influences fur pigmentation (Fig. 4-18). At least four alleles are known in the series: C (wild), c^{ch} (chinchilla), c^h (himalayan), and c (albino). Earlier in this chapter, we discussed the interaction of the himalayan allele, c^h, with the temperature of the surroundings as an example of the influence of environment on gene expression. The expression of the other mutant allele, c^{ch}, is similarly affected by body temperature, because the enzyme controlled by the allele is heat sensitive. The amount of pigment in a chinchilla rabbit is therefore less than in the wild. Any animal

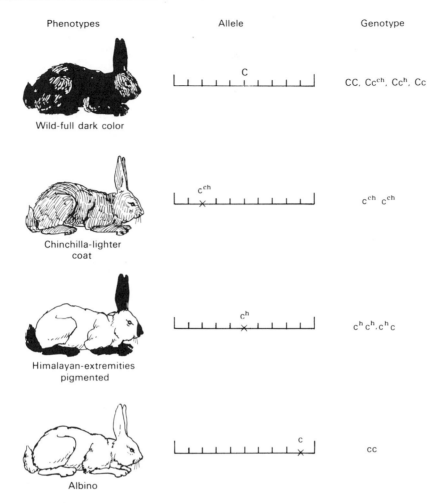

Phenotypes   Allele   Genotype

Wild-full dark color — C — CC, Cc^{ch}, Cc^{h}, Cc

Chinchilla-lighter coat — c^{ch} — $c^{ch} c^{ch}$

Himalayan-extremities pigmented — c^{h} — $c^{h} c^{h}, c^{h} c$

Albino — c — cc

FIG. 4-18. Coat color in rabbits. At least four alleles are known to affect pigment of the fur. In this multiple allelic series, the mutant types are the result of changes at different sites within the pigment gene. The wild-type allele is completely dominant over its variant forms. The himalayan allele is dominant to albino. However, the chinchilla allele is not completely dominant to the himalayan and albino alleles. (The representation of the alleles does not intend to represent actual sites within the gene.)

can have any two of the four alleles. This series illustrates the fact that, in any multiple allelic series, some alleles may be dominant to others; others may show codominance or incomplete dominance. The wild allele, C, is completely dominant to the other three in the rabbit, but different relationships exist between the mutant forms. Himalayan is dominant to albino. Chinchilla animals are homozygous ($c^{ch} c^{ch}$), however, because the chinchilla allele is not completely dominant to albino and himalayan. The heterozygotes are light gray. This series contrasts with the eye pigment alleles in Drosophila, in which the mutant gene forms are all incompletely dominant to each other.

In some plant species, genetic factors that form multiple allelic series have been shown to play very important roles in controlling crossing ability. The situation was first described in tobacco in the 1920s, but comparable series are known to occur in other species. Many plant species are normally only cross-pollinated. For the pollen to function, it must encounter a stigma in a flower on a different plant; otherwise it will cease growing. There is great value to a system that ensures cross-pollination, for this in turn guarantees outbreeding. Crossing with other genetic types increases the chances of hybrid vigor and decreases the probability that harmful recessives will become homozygous and be expressed (see Chap. 7).

The genetic basis for cross-pollination in tobacco resides in a large number of self-sterility alleles, which

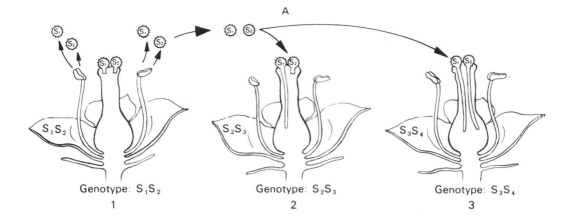

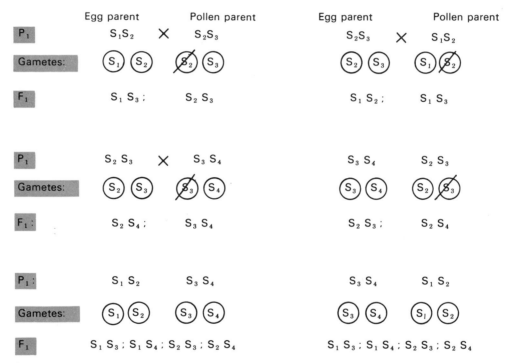

FIG. 4-19. Self-sterility alleles. (*A*) In many plants, the inability to self-pollinate is based on sterility alleles that make it impossible for pollen to grow on the stigma of a plant containing the same allele. All plants must be heterozygous for the sterility alleles. In this example, only half of the pollen from plant 1 can grow on plant 2; the other half that carries the allele in common aborts. Similarly, only half of the pollen from plant 2 will grow on plant 1. However, all of the pollen from plant 1 will grow on plant 3, and vice versa, because they have no sterility alleles in common. Plants 2 and 3 will be able to exchange only half of their pollen successfully. As a result of this system, only heterozygous individuals can develop, as the summary of the reciprocal crosses possible with these three genotypes indicates (*B*).

form a multiple allelic series: S_1, S_2, S_3, S_4, and so on. Any plant can be hybrid for any two of these, but no plant can be homozygous for a particular sterility allele (Fig. 4-19). This is so because pollen carrying a specific sterility allele aborts on the stigma of a flower that contains the same allele. If a plant is S_1S_2, none of its pollen can survive on the stigma of any flower on that same plant. Even if two different plants have the same sterility alleles, they are still incompatible, and so we cannot cross two individuals that are both S_1S_2, for example. However, an S_1S_2 can be crossed with S_2S_3, but only half of the pollen from each is successfully exchanged, because each possesses the S_2 allele in common. On the other hand, the S_1S_2 type and the S_3S_4, having no sterility alleles in common, can freely exchange pollen and genetic material reciprocally. This example of multiple alleles in plants illustrates a genetic mechanism to bring about outbreeding and all the advantages to the species it entails.

Antigens, antibodies, and the ABO blood grouping

In humans, about 15 major blood groups have been recognized, and most of these are inherited on the basis of multiple allelic series. The ability to classify the blood of any individual into the various groupings depends on the presence or absence of specific antigens in the red blood cells. Antigens are large molecules that are often proteinaceous, at least in part. They can elicit the formation of specific antibodies, that is, large protein molecules (immunoglobulins), with which they react to produce clumping or lysis of the cells. The presence of a specific antigen in the red blood cells, A, B, or both A and B, defines the corresponding blood type within this major grouping (Table 4-1). Cells lacking both antigens are classified as type O. Moreover, if an antigen is absent from the cells, the serum has a corresponding antibody. Since antigen A reacts with antibody a (and likewise B with b), knowledge of the ABO type is crucial in blood transfusions. Any transfusion is incompatible if the cells of the donor are clumped by serum of the recipient. A person of type A contains antibody b in his or her serum. The millions of cells from a B donor would be clumped on contact with the b antibody and would clog the smaller blood vessels, resulting in severe illness or death. Only type O can donate blood to the other types with any degree of safety. Although both

types of antibodies, a and b, are present in the serum of type O, and will react with type A and B cells, the serum of the O donor would be diluted by the A, B, or AB recipient so that a severe effect is avoided.

The A and B antigens are actually identical in structure as far as their protein portion is concerned. The difference between them resides in a sugar component to which they are complexed. The A and B antigens are mucopolysaccharides—combinations of a sugar and a protein. We say that their antigenic specificity resides in the sugar component, because it is that portion, not the protein, that reacts specifically with the antibody. The antigenic specificity is under the genetic control of the ABO locus. At its simplest, three different allelic forms may occur at this locus: I^A, I^B, and i. The alleles I^A and I^B determine the presence of antigens A and B, respectively, and each is dominant to the allele, i, for blood grouping O, the absence of both the A and B antigens (Table 4-2). The I^A and I^B alleles provide an excellent example of codominance, because the heterozygote, I^AI^B, produces both the A and B antigens. The number of alleles at the ABO locus is now known to total more than three, as originally described. The A blood types have been shown to include subgroupings: A_1, A_2, and several others. The existence of several kinds of B type is also indicated. The I^{A1} allele appears to act as a dominant to I^{A2}. If we just consider the alleles I^{A1}, I^{A2}, I^B, and i, we see that different subgroupings of type AB are possible, as well as of type A (Table 4-3). Six different types can be recognized, distinguished from one another on the basis of specific antigen–antibody reactions. If we consider additional subgroupings of A and B, the number of possible ABO types increases accordingly. The accurate detection of antigenic differences among the subgroupings has value not only for its medical applications but for its legal aspects as well.

Ignoring the subgroupings for the moment, let us see how knowledge of the inheritance patterns of the

TABLE 4-1 General scheme of antigens and antibodies in the ABO system

BLOOD TYPE	A AND B ANTIGENS IN RED BLOOD CELLS	a AND b ANTIBODIES IN SERUM
A	A	b
B	B	a
AB	A and B	None
O	None	a and b

TABLE 4-2 General scheme of genotypes in the ABO system

BLOOD TYPE	POSSIBLE GENOTYPES	A AND B ANTIGENS IN RED BLOOD CELLS
A	I^AI^A, I^Ai	A
B	I^BI^B, I^Bi	B
AB	I^AI^B	A and B
O	ii	None

TABLE 4-3 Two subgroupings of type A and possible blood groups

	BLOOD GROUP
$I^{A1}I^{A1}$	A_1
$I^{A1}i$	A^1
$I^{A1}I^{A2}$	A^1
$I^{A2}I^{A2}$	A_2
$I^{A2}i$	A_2
$I^{B}I^{B}$	B
$I^{B}i$	B
$I^{A1}I^{B}$	A_1B
$I^{A2}I^{B}$	A_2B
ii	O

blood groups can aid in a case of disputed paternity. Suppose a man is suing his wife for divorce and accuses another man of being the father of the wife's child. Blood typing reveals the following: wife, type B; baby, type AB; husband, type O; and other man, Type A. Our genetic knowledge tells us that the husband's suspicion is well-founded. He cannot be the baby's father because he is blood type O and therefore cannot contribute an allele for the production of either A or B antigen. The baby must have received an allele for antigen A from the male parent. In reaching a decision regarding the correct male parent, however, caution must be exercised. Although the accused man possesses blood type A, an appropriate type for the baby's father, this by itself cannot incriminate him, for many men have blood type A. Moreover, the father could also be type AB, because such a person can contribute either the allele A or the allele B to the offspring. So we see that the genetics of blood types may eliminate as a possible parent one who lacks the allele required for the production of a specific antigen; however, it cannot prove a person *is* the parent in question just because he has the suitable blood grouping.

In our example, suppose that subgroupings of A were identified, and the baby was found to be A_1B and the man A_2. Since an A_2 person (see Table 4-3) can be only genotype $I^{A2}I^{A2}$ or $I^{A2}i$, he cannot give the offspring the I^{A1} allele. The man would thus be exonerated on the basis of this more detailed blood typing. In this way, a person may be eliminated as a possible parent as more detailed blood groupings are examined. Although the probability of parenthood becomes greater as more and more subgroupings between child and suspected parent agree, the evidence by itself is still inconclusive.

A locus that is not linked to the locus for the ABO blood grouping has been found to interact with it in an interesting manner. We have just seen that a person of blood type AB cannot have a parent who is type O. However, very rare cases have been described of individuals carrying the alleles for A or B antigen with neither being expressed. A person of genotype $I^{A}I^{B}$ could thus appear to be type O. The ability to produce A or B antigen at all depends not only on the presence of allele I^{A} or I^{B} at the ABO locus, but also on at least one other locus, the h locus. The presence of the dominant allele H determines the presence of substance H, which can also be detected by antigen–antibody reactions. This H substance or H antigen is essential for the formation of either antigen A or B. It is the substrate from which both antigens are made under the guidance of alleles A and B (Fig. 4-20A). Since most persons have the dominant allele, H, they will produce antigens A and B when the corresponding alleles are present in the genotype. Persons who are blood type O, although they cannot form A or B antigens, nevertheless contain H antigen in its unaltered form. The A, B, and H antigens also occur in cells other than the red blood cells.

Very rare individuals have the genotype hh and therefore cannot produce H substance. Such an individual not only tests negative for H antigen but appears to be blood type O regardless of the actual ABO genotype (Fig. 4-20B). This person illustrates the Bombay phenomenon, named for the city where the first case was reported. Although its occurrence is rare, the Bombay phenomenon is an excellent example in the human of an epistatic effect and underscores the caution that must be exercised before reaching conclusions on the basis of a single pedigree.

Another locus, also not linked to the ABO locus, is the secretor locus. Its importance to genetic counseling is discussed in Chapter 9. About 78% of the

TABLE 4-4 Antigens A, B, and H in secretors and nonsecretors

BLOOD GROUP	ANTIGENS OF SECRETORS		ANTIGENS OF NONSECRETORS	
	Red blood cells	Body fluids	Red blood cells	Body fluids
O	H	H	H	None
A	A	A and H	A	None
B	B	B and H	B	None
AB	A and B	A, B, and H	A and B	None

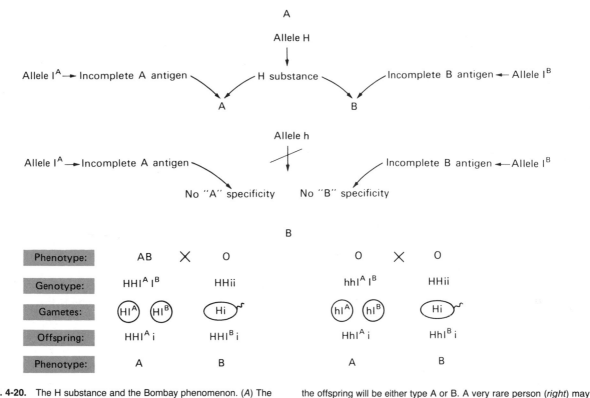

FIG. 4-20. The H substance and the Bombay phenomenon. (*A*) The allele H is present in most persons and is responsible for the production of a substance required for the development of A and B antigens. The recessive allele is unable to produce the substance. A and B antigens do not complete development. A, B, and H antigens cannot be detected in the blood. (*B*) Normally, all persons are homozygous for H. In a typical cross of an AB person with one who is type O (*left*), the offspring will be either type A or B. A very rare person (*right*) may be homozygous for the recessive allele h and manifests the Bombay phenomenon. Although phenotypically type O, the person has the genetic information to produce incomplete A and B antigens. Since the other parent carries the H allele, the offspring will be either type A or B. As seen here, the most unusual situation would arise of two type O parents with offspring who have A or B antigens.

population possesses the dominant secretor allele, Se, which is responsible for the secretion in a water-soluble form of antigens A, B, and H in a number of body fluids, such as saliva. Secretion of the antigens does not occur in persons homozygous for the recessive allele, se, which has no known function. Table 4-4 summarizes the situation regarding antigens A, B, and H in secretors and nonsecretors.

Antigens and the Rh blood grouping

One major blood grouping that has been well publicized is the Rh. Its discovery goes back to 1940. Landsteiner and Wiener injected blood from the Rhesus monkey into the rabbit, which in turn responded by producing antibodies. These antibodies were then able to agglutinate the monkey's blood when mixed with it. The rabbit antiserum containing the antibodies against the monkey's blood was also mixed with blood from humans. It was found that the blood of 85% of those tested was also clumped by the antiserum. Landsteiner and Wiener had removed from the antiserum all those other antibodies known at the time that could possibly react with human antigens. Since clumping still took place, this meant that the rabbit serum contained an antibody against an antigen present in *both* human and monkey blood that had been undetected up to that time. The antigen was designated *Rh* to denote its presence in the Rhesus monkey.

The presence of Rh antigen was shown to be under genetic control. Those persons with the Rh factor were designated Rh⁺. The 15% of the white population whose blood did not react with Rh antibodies were termed Rh⁻ because the antigen or factor was absent from their blood. The inheritance of the Rh type seemed to depend on a single pair of alleles: the dominant R for the presence of the antigen and the recessive r for its absence. Unlike the ABO system, no naturally occurring Rh antibodies of any kind are formed in the blood. The Rh⁻ person (rr) is not born with antibodies against the Rh antigen, nor is the Rh⁺ person born with antibodies against Rh⁻ blood. However, the Rh⁻ individual, it was found, could form antibodies against Rh antigen if he or she received an injection of Rh⁺ blood. These antibodies can clump cells containing Rh antigen once the antibodies are formed by the Rh⁻ person in response to the presence of Rh⁺ blood in the bloodstream. It soon became evident that the Rh factor is just as important as the ABO type in transfusions. An Rh⁻ person may be able to tolerate one transfusion of positive blood, because he or she contains no preformed antibodies. Once sensitized, however, he or she will produce a high level of Rh antibodies on a second transfusion, and their reaction with Rh⁺ antigens in red blood cells can lead to fatal results.

Not only did the Rh type demand consideration in transfusions, it also became implicated in certain cases of blood incompatibility between mother and child. Some babies are born severely jaundiced and seriously anemic. This condition, erythroblastosis fetalis, is the consequence of fetal red blood cell destruction, which not only produces an oxygen defect but also causes clogging of the blood vessels of the liver with damaged red blood cells and the entry of bile pigments into the blood. In very severe cases, the child may suffer permanent damage. Death may even occur shortly before or after delivery. The destruction of the baby's red blood cells results from an antigen–antibody reaction in the child's bloodstream. We have noted that a person may become sensitized to Rh antigen after a transfusion. A high antibody level can be reached in an Rh⁻ woman who has been so sensitized. If her husband is Rh⁺ his genotype may be either RR or Rr. If he is heterozygous there is a probability of ½ that the dominant allele will be transmitted. If the offspring does inherit the dominant allele and is thus Rh⁺, any antibodies against the Rh antigen will react with the blood of the fetus *if* they pass the placental barrier from mother to child because of some defect in the placenta (see later). So an Rh⁻ mother, already sensitized, can possibly pass antibodies to her offspring in utero. Serious consequences can then follow (Fig. 4-21A).

Unfortunately, this is not the only way in which an incompatible Rh reaction can come about (Fig. 4-21B). In an Rh⁻ woman carrying an Rh⁺ child, it is possible for some blood from the fetus to leak into the circulatory system of the mother. Once this happens, the Rh⁻ mother may become sensitized, just as if she had received an injection of Rh⁺ blood. The antibodies formed by the Rh⁻ mother may in turn cross the placental barrier and cause their damaging effects. Although Rh incompatibilities between mother and child are certainly a matter for concern, knowledge of the subject should be used properly. Trouble *may* ensue if the mother is Rh⁻ and the baby Rh⁺. This follows only from the mating between an Rh⁺ man and an Rh⁻ woman. If the woman is positive and the man negative, there is no problem, because an Rh⁻ baby carried by an Rh⁺ woman cannot cause the mother to form antibodies that will react with Rh⁻ blood. But even if the woman is Rh⁻ and her mate is Rh⁺, there still may be no problem. If the male is heterozygous (Rr), there is a 50% chance that the child will be Rh⁺. If the child is Rh⁺, he or she will most probably not be affected if this is the first Rh⁺ born to an unsensitized Rh⁻ mother. This is so because most of the transfer of fetal cells to the maternal circulation occurs late in the pregnancy, at or near the time of birth. And we must keep in mind that an Rh⁺ offspring does not necessarily sensitize the mother. The placenta is normally an effective barrier between mother and child. In most cases, no exchange of blood occurs between mother and child. Erythroblastosis fetalis is actually uncommon. This means that an Rh⁻ mother may have one or more Rh⁺ children and encounter no difficulties. So we must apply the genetic knowledge intelligently. Today, with modern treatment, an Rh⁻ mother who has borne an Rh⁺ child may never become immunized, since immunization to the Rh antigen on the fetal cells can almost always be prevented. This is accomplished by injecting preparations of antibody to the Rh⁺ antigen into the Rh⁻ mother shortly after the birth of an Rh⁺

FIG. 4-21. Rh incompatibilities. (*A*) An Rh⁻ woman and an Rh⁺ man who is heterozygous have a 50:50 chance of having an Rh⁺ child (*above*). If the child is Rh⁺ and the mother is sensitized by a previous transfusion, Rh antibody formation may be evoked, and antibodies may at times pass the placenta (*below*). Consequently, an Rh⁺ child carried by the previously sensitized mother can suffer from *erythroblastosis*. (*B*) A woman not carrying antibodies against Rh⁺ blood may become sensitized by an Rh⁺ offspring, but usually only at the time of birth. The mother will usually not form a sufficient number of antibodies to affect the first Rh⁺ child, even if she becomes sensitized by the pregnancy. An Rh⁻ woman already sensitized by a previous pregnancy with an Rh⁺ child can respond rapidly to antigen from a second Rh⁺ fetus (*right*). Antibody production can be high enough to affect the second Rh⁺ child.

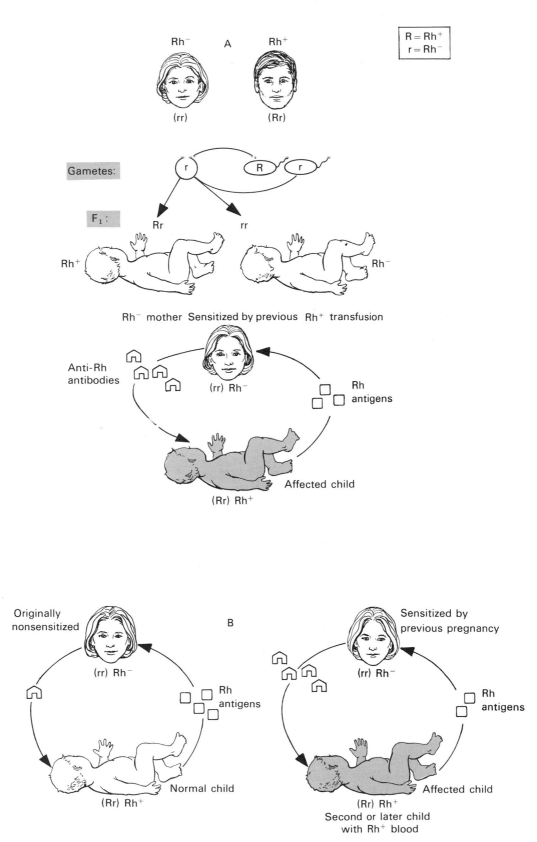

Rh⁻   A   Rh⁺

R = Rh⁺
r = Rh⁻

(rr)   (Rr)

Gametes:

F₁:

Rr   rr

Rh⁺   Rh⁻

Rh⁻ mother Sensitized by previous Rh⁺ transfusion

Anti-Rh
antibodies

(rr) Rh⁻

Rh
antigens

Affected child
(Rr) Rh⁺

Originally
nonsensitized

B

Sensitized by
previous pregnancy

(rr) Rh⁻

Rh
antigens

(rr) Rh⁻

Rh
antigens

Normal child
(Rr) Rh⁺

Affected child

(Rr) Rh⁺
Second or later child
with Rh⁺ blood

97

child. These antibodies remove any Rh$^+$ fetal cells that may have entered the mother's bloodstream. Thus, since these fetal cells cannot act to stimulate antibody formation in the mother, the mother is not sensitized to Rh$^+$ cells. Proper medical counseling can avoid unnecessary alarm and prevent any unfortunate consequences of Rh incompatibility.

Not long after its discovery, the Rh grouping was shown to have a more complex genetic basis than we have just described and to involve more than the one pair of alleles, R and r. With the improvement of techniques for the detection of antigenic differences, it became clear that not all Rh$^+$ individuals are the same. As shown by antibodies produced against Rh$^+$ blood in experimental animals, there are actually several different Rh$^+$ types. These various Rh$^+$ individuals possess different combinations of antigens, which may be detected through antigen–antibody reactions. Indeed, Rh$^-$ people were also found to vary. The many Rh variations stem from the fact that several distinct Rh antigens occur in humans, not just one or two. Even Rh$^-$ persons possess Rh antigens that can bring about antibody production. But the most

commonly occurring antigens in Rh$^-$ people are not responsible for the Rh incompatibilities we have just discussed.

To account for these many Rh types, two ideas have been proposed, which has caused some confusion in terminology. Essentially, one concept, that of Wiener in the United States, pictures the Rh genetic region as a complex locus on the chromosome (Fig. 4-22). We may think of it as a stretch of DNA. Changes at different points within this DNA segment result in the production of different gene forms or alleles, which form part of a multiple allelic series. According to the Wiener theory, at least eight different Rh alleles exist. Each allele controls the formation of a certain *combination* of antigens (Table 4-5). An individual can have any two of the eight alleles and is Rh$^+$ or Rh$^-$, depending on the genes he carries (Table 4-6).

A group of English hematologists, Fisher, Race, and Sanger, interpret the situation somewhat differently. According to their hypothesis, the Rh region on the chromosome includes not just one but three loci (Fig. 4-23). Each locus is the site of a pair of alleles: Cc, Dd, and Ee. Eight different combinations of the six alleles are possible on any one chromosome: CDE, CdE, and so on. Each of the six alleles controls a distinct antigen, and each of these has been detected serologically by antigen–antibody reactions. (Anti-d serum is not easily obtained.) Because eight combinations of alleles are possible, there are eight different combinations of antigens, one for each arrangement of alleles (Table 4-7). The three loci are so closely linked that crossing over in the cde region of the

FIG. 4-22. The Wiener concept of the Rh region. According to this interpretation, a series of multiple alleles exists at one locus. Each allele controls not one but a combination of antigens. For example, the r allele would govern the formation of antigens c, d, and e, which would react with the corresponding antibodies. A total of eight different alleles exists, each controlling a different combination. Only four are depicted here. Actually, the *CDE* designation shown here is employed in the other system (Fig. 4-3) rather than in the Wiener. However, the antigens and antibodies correspond in both, and the one designation is used here for the sake of clarity.

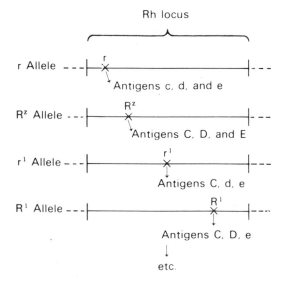

TABLE 4-5   The eight different Rh alleles, according to Wiener[a]

DIFFERENT ALLELES		Anti C	Anti D	Anti E	Anti c	Anti e
Rh$^-$	r				+	+
	r′	+				+
	r″			+	+	
	r^y	+		+		
Rh$^+$	R^0		+		+	+
	R^1	+	+			+
	R^2		+	+	+	
	R^Z	+	+	+		

Header spanning Anti C–Anti e: ANTIBODIES FORMED AGAINST ANTIGENS GOVERNED BY THE ALLELE

[a]Each allele is responsible for the production of a combination of two or more antigens, and these can cause the formation of corresponding specific antibodies (anti-d sera not available).

TABLE 4-6   Some common Rh⁻ and Rh⁺ genotypes[a]

PHENO-TYPE	GENOTYPE		ANTIBODIES
	Weiner	Fisher	
Rh⁻	r/r	cde/cde	c, e
Rh⁻	r″/r	cdE/cde	c, E, e
Rh⁺	R⁰/r	cDe/cde	c, D, e
Rh⁺	R¹/R¹	CDe/CDe	C, D, e
Rh⁺	R²/r	cDE/cde	c, D, E, e
Rh⁺	R¹/R²	CDe/cDE	C, c, D, E, e
Rh⁺	R¹/r	CDe/cde	C, c, D, e

[a]Any person, according to Wiener, can have any combination of the eight alleles, one allele on each chromosome. According to Fisher, any person may have a combination of any three alleles on one chromosome and a combination of any three on the other. No matter how the situation is viewed, the individual forms antibodies against specific antigens. Antigen D is the one most commonly involved in Rh incompatibilities. The Rh positive condition is often defined as "presence of D antigen." (Anti-d serum is not readily available.)

FIG 4-23. The Fisher concept of the Rh region. According to this theory, the Rh region includes three very closely linked loci that rarely are separated by crossing over. At each locus a pair of alleles exists, and each allele controls the formation of a specific antigen. On any one chromosome, any combination of three of the six alleles may occur. A total of eight different combinations of three is possible. Contrast this figure with Fig. 4-22.

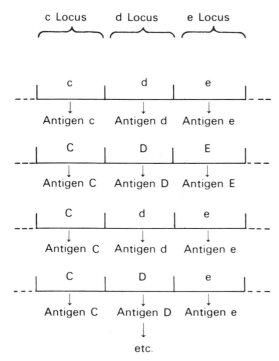

etc.

chromosome is very rare. Therefore, any one of the allelic combinations (CDE, cDe, etc.) would usually stay linked and thus be inherited together as a unit. In essence, each combination appears as if it were one allele, even though three are actually present at each Rh region on the chromosome. Based on this idea, any person possesses six Rh alleles, any two combinations of three each, such as CDE on one chromosome and cdE on the other. The genotype of such a person could be represented as CDE/cdE (see Table 4-6).

No matter how we view the Rh story, either as one locus with a series of multiple alleles or as three tightly linked loci, the various Rh alleles show codominance. As Tables 4-5, 4-6, and 4-7 indicate, each allele when present is expressed and causes the production of its specific antigen. The antigens are in turn detected by reactions with their corresponding antibodies. The maternal–fetal Rh incompatibility is mainly the result of the presence of D antigen (approximately 90% of the cases). The C antigen seems to be involved in the remainder. The Rh⁻ condition is often defined as the absence of D antigen, because that antigen is the most important in the Rh incompatibilities. Following this definition, an Rh⁻ person could contain the C antigen. At any rate, two Rh⁻ persons may have different genotypes. They may contain different combinations of antigens (cde/cde or cde/cdE), but these usually play no significant role in cases of blood type differences. Whether the Wiener or the Fisher concept is correct is still unsettled, and there are good

TABLE 4-7   The eight combinations of the Rh alleles according to Fisher[a]

	Rh alleles COMBINATIONS			ANTIBODIES FORMED AGAINST ANTIGENS GOVERNED BY GENES AT THE LOCI				
				Anti C	Anti D	Anti E	Anti c	Anti e
Rh⁻	c	d	e				+	+
	C	d	e	+				+
	c	d	E			+	+	
	C	d	E	+		+		
Rh⁺	c	D	e		+		+	+
	C	D	e	+	+			+
	c	D	E		+	+	+	
	C	D	E	+	+	+		

[a]Each combination of three alleles is strongly linked and behaves as if it were one allele. Each combination of three, according to this hypothesis, corresponds to one of the alleles, as pictured by the Wiener concept (see Table 4-5).

arguments in favor of each. Nevertheless, we know today that the genetic region associated with the Rh blood grouping is located at a specific region of Chromosome 1. To explain a very complex story in the simplest way, we can think of the Rh grouping based primarily on a pair of alleles, D and d, because it is the D antigen that must be considered primarily when Rh types differ.

It should be evident from the assortment of Rh types that can actually exist among people that the Rh blood groupings go a long way in settling cases of disputed parentage. Added to the information on the ABO, the specific Rh type provides another basis for establishing the possible innocence of an individual accused in a paternity suit.

Our discussion of blood groupings up to this point has by no means exhausted the number identified. Very well known is the MN blood grouping. Each of us falls into one of three classifications: M, N, or MN, due to the possession of M or N antigen in the red blood cells. The average person is unaware of his or her MN type, because the MN antigens play no role in blood transfusions. Antibodies to M or N antigens do not occur naturally in the blood serum. Such antibodies, however, can be elicited by injecting human blood into an experimental animal such as a rabbit. Antibodies to M or N form in the animal's serum and, when mixed with human blood, can reveal the presence of the antigens in the human red blood cells, as they cause the latter to clump. At its simplest, the formation of M or N antigens depends on a pair of alleles, L^M and L^N. The two forms exhibit codominance, so that three possible genotypes and phenotypes are recognizable: $L^M L^M$ (type M), $L^M L^N$ (type MN), and $L^N L^N$ (type N). These react specifically with antibodies M, M and N, and N.

Several other blood groupings are much less familiar to most of us but may be encountered in specific situations. To mention just a few, there are the Kell, the Kidd, the Duffy, Luthcran, and Lewis groupings. The names usually refer to the family in which the grouping was first detected. For example, the Kell factor (K) denotes an antigen found in most persons. Women who are double recessive (kk) may give birth to erythroblastotic Kell positive children. As in the case of the Rh⁻ mothers, the Kell negative women may produce antibodies (anti-Kell), which can react with the Kell antigen in the fetal circulation. These many additional blood groups can also be of aid in legal matters. They also demonstrate how unique each of us is. The number of possible combinations of the many blood group loci, some of which are multiple allelic, defies comprehension.

Antigens, grafting, and transplants

Intimately related to the topic of blood groups is the subject of tissue grafting and organ transplants. We might expect other tissues as well as blood to contain distinct antigens that react with specific antibodies. And indeed, the rejection of an organ such as the kidney or the failure of a skin graft to take are examples of immunological responses. All cells possess antigens, which vary from one individual to the next and may cause the rejection of transplants. Such factors are called histocompatibility antigens and the genes that control them are histocompatibility genes (H genes). The antigens, as well as those associated with the blood types, are present in the cell membrane, where they may play some role in its structure and biological properties.

The most thorough studies on the genetics of grafting have been performed with mice. However, the same principles are believed to apply to other species, including the human. Mice within one family line may be mated for generations to produce strains that are so inbred they are homozygous for an appreciable number of alleles. Valuable genetic information on tissue compatibility factors and their role in tissue transplants may be gained from studying graft tolerance within and between the inbred strains. Such work gives strong evidence for the following concept: a recipient will accept tissue from a donor if the recipient possesses the same kinds of histocompatibility alleles as the donor. Conversely, if the donor, and hence any tissue from the donor, possesses H alleles that are *not* present in the receiver, the graft will be rejected.

Table 4-8 represents a general summary depicting results from two inbred strains, A and B, in which two pairs of histocompatibility alleles are being followed. As was true for the ABO and Rh alleles, the H alleles are codominant in their expression. It is obvious from the table that within an inbred strain, tissue can be successfully grafted and accepted because there is no difference between donor and recipient in the kinds of H alleles they possess. When the two inbred strains are crossed, their F_1 offspring contain H alleles from each parent. These hybrids have all the alleles that are typical of each strain. Thus, they can accept grafts from either parent type. However, the F_1 cannot act as donor to either inbred

TABLE 4-8 Grafting compatibilities between animals of various genotypes

DONOR	DONOR'S GENO-TYPE	RECIP-IENT	RECIP-IENT'S GENO-TYPE	ABIL-ITY TO ACCEPT GRAFTS
Strain A	$A_1A_1B_1B_1$	Strain A	$A_1A_1B_1B_1$	+
Strain B	$A_2A_2B_2B_2$	Strain B	$A_2A_2B_2B_2$	+
F_1 hybrid	$A_1A_2B_1B_2$	Strain A	$A_1A_1B_1B_1$	−
F_1 hybrid	$A_1A_2B_1B_2$	Strain B	$A_2A_2B_2B_2$	−
Strain A	$A_1A_1B_1B_1$	F_1 hybrid	$A_1A_2B_1B_2$	+
Strain B	$A_2A_2B_2B_2$	F_1 hybrid	$A_1A_2B_1B_2$	+
F_2 hybrid	9 Types	F_1 hybrid	$A_1A_2B_1B_2$	+
F_1 hybrid	$A_1A_2B_1B_2$	F_2 hybrid	9 Types	− in most cases

parental strain, because the F_1 has some alleles not found in strain A or B. If the F_1s are interbred to produce F_2 progeny, the latter may have various new combinations (3^2) of the two pairs of alleles, such as $A_1A_1B_1B_2$. But no matter what the new combination in the F_2 animals, the F_1s will be able to accept grafts from them. This is so, because the F_1s will contain *all* the types of alleles present in any F_2. The reverse, however, is not necessarily true, because independent assortment will scramble the H alleles into a variety of new combinations (nine in this case). The only F_2 animals capable of accepting grafts from F_1 animals must be dihybrids, just as the F_1s, and they are in the minority.

This simple example illustrates the principle behind graft tolerance, but it is an oversimplification. In mice, about 30 separate loci are associated with histocompatibility genes. The number of different possible combinations becomes staggering. Moreover, some of these loci exert a greater effect than others. In mice, the genetic region known as the H_2 *complex* is the main one in the sense that the antigens it controls can bring about a very strong immunological response, much more so than other loci. This region has been studied in detail and has been shown to involve several very closely linked loci adjacent to each other.

In addition to circulating antibodies, which form in the body of a mammal in response to foreign antigens, there is another immune response—one that plays the major part in actual rejection of foreign tissue received in a transplant. This is the cellular response, which entails those particular lymphocytes, the so-called T lymphocytes or T cells, that have become differentiated in the thymus gland. The T cells can recognize foreign antigens on the surfaces of cells in tissue grafted to a host. A T cell has antibody molecules embedded in the surface of its cell membrane, and these apparently enable T cells to respond to the presence of antigens on the surfaces of cells in foreign tissue.

As in the mouse, several genetic regions in the human regulate tissue compatibility, and one of them, the *HLA complex,* is of primary importance. This major histocompatibility complex has been assigned to the short arm of Chromosome 6. The HLA complex includes several very closely linked loci that are related to immunity. Three of these (A, B, and C) have been identified as HLA antigen determinants. A multiple allelic series exists at each locus, and each locus determines the structure of a specific histocompatibility antigen. Approximately 20 different alleles are known for each A and B locus, and about 10 for the C locus (Fig. 4-24).

Because the antigen-determining loci in the HLA complex are so tightly linked, crossing over in the region is rare. Therefore, the combination of alleles that occurs together on a chromosome tends to stay together and to be transmitted intact to the next generation. Each given allelic combination (such as $A_7B_{15}C_5$) is termed a *haplotype.* Each person carries two haplotypes, one on each Chromosome 6. Since so many allelic combinations are possible, every person carries two different haplotypes. For the sake of simplicity, we can consider each haplotype (really a combination of alleles) as a genetic unit (say, H_1, H_2, H_3,

FIG. 4-24. The HLA complex on human Chromosome 6. Included in this complex is a D region known to be involved in the production of antibodies to certain foreign antigens. The B, C, and A genes code for histocompatibility antigens. Each gene exists in several allelic forms. (Simplified designations are used here.) Since the genes in the complex are so close together, any one combination of genes, a haplotype, tends to stay together and to be inherited as if it were one genetic unit.

D region		B gene		C gene		A gene	
		↓		↓		↓	
		B^1		C^1		A^1	
		B^2		C^2		A^2	
		B^3		C^3		A^3	
		B^4		C^4		A^4	
		B^5		C^5		A^5	
		↓		↓		↓	
		etc.		etc.		etc.	

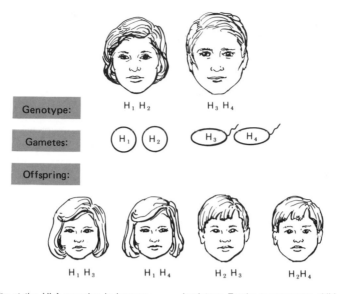

Genotype:

Gametes:

Offspring:

FIG. 4-25. Tissue incompatibility at the HLA complex in humans. Any one person is likely to be heterozygous at the HLA region, carrying two different haplotypes. The parents in this case, as would probably be true of any two unrelated persons, do not possess both haplotypes in common and would be unable to exchange grafts. Among their children, diverse types occur. These cannot give grafts to or accept grafts from either parent, because there will be a difference in one haplotype. For the same reason, children of different genotypes cannot exchange grafts. However, two children can occur in the family who carry the same haplotypes, and these may be able to exchange tissue. However, still other genetic regions exert an influence; differences at these will reduce the chances of free exchange, even if compatibility exists at the HLA complex.

etc.) and think of any individual as being heterozygous at the HLA region for a pair of units, for example, H_1/H_2.

Considering the HLA complex in this simplified way, we can see why most grafts between individuals are not tolerated (Fig. 4-25). A cross between two persons of the genotypes H_1/H_2 and H_3/H_4 results in the following combinations: H_1/H_3, H_1/H_4, H_2/H_3, and H_2/H_4. None of the children will be able to accept a graft from either parent. There is a chance, however, that two offspring will have the same combination of units or haplotypes, such as two who are H_1/H_4, in our example. This means that two siblings could possibly exchange tissue. Fortunately, matching of haplotypes at the HLA region greatly increases the chance that a kidney will be accepted. Although the chance for a successful transplant is greater between ordinary siblings than it is between any other two persons (except for identical twins), we must not forget that several genetic regions are involved in addition to the HLA complex. Even if compatibility should exist for the HLA haplotypes, a graft may eventually be rejected because of incompatibilities at other loci that control tissue antigens. Again, as in the case of blood types, each of us possesses a vast

number of histocompatibility alleles in a combination that is distinct from that of most other persons.

We can thus appreciate why the HLA antigens are much more important in legal matters than the antigens for the blood groupings. Haplotypes for an individual usually can be determined. Any given haplotype is not likely to be widespread in a population sample, which greatly narrows the chances that a given person is the parent in a particular case.

For reasons that are not known, certain HLA haplotypes are associated with specific human disorders—among them, rheumatoid arthritis, chronic liver disease, multiple sclerosis, and Hodgkin's disease.

There is some reason to believe that the evolution of the great complexity in the immune system has resulted largely from the value it imparts in the recognition of those cells in the body that suddenly undergo malignant transformations. Any new cancer call has alterations on its cell surface and is recognized as "foreign." A malignant transformation probably entails a vast assortment of possible alterations, each quite distinct and requiring a specific cellular response if it is to be eliminated.

Striking illustrations of the profound importance of the immune response are seen in those genetic disorders in which there is a decreased efficiency in the ability to respond to foreign antigens. Several types of immune-deficient disorders exist, which are called *agammaglobulinemias*. The types vary in severity, in response to treatment, and in genetic bases. The disorder is a result of a sex-linked recessive in some families, and of an autosomal inheritance pattern in others. In one of the severest forms, both the T lymphocytes and the circulating antibody responses are impaired. Because of the lack of antibodies, such persons are susceptible to every type of infection, viral and bacterial, as well as to malignancies. Such unfortunate persons readily accept grafts, because no antibodies or T cells can participate in a rejection. Still, they are no different from anyone else in the ability to donate tissue, since their unique antigenic determinants are present in cells of their tissues.

REFERENCES

Beadle, G. W. and E. L. Tatum. Genetic control of biochemical reactions in *Neurospora. Proc. Natl. Acad. Sci.* 27: 499, 1941. Reprinted in *Selected Papers on Molecular Genetics.* J. H. Taylor (ed.) Academic Press, New York, 1965.

Cunningham, B. A. The structure and function of histocompatibility antigens. *Sci. Am.* (Oct): 96, 1977.

Dausset, J. The major histocompatibility complex in man. *Science* 213: 1469, 1981.

Eaton, G. J. and M. M. Green. Implantation and lethality of the yellow mouse. *Genetica* 33: 106, 1962.

Hood, L. E., M. Steinmetz, and R. Goodenow. Genes of the major histocompatibility complex. *Cell* 28: 685, 1982.

Mange, A. and E. J. Mange. *Genetics: Human Aspects.* W. B. Saunders, Philadelphia, 1980.

Mulcahy, D. L. and G. B. Mulcahy. Gametophytic self-incompatibility reexamined. *Science* 220: 1247, 1983.

Sutton, H. E. *An Introduction to Human Genetics,* 3rd ed. W. B. Saunders, Philadelphia, 1980.

Tourian, A. Y. and J. B. Sidbury. Phenylketonuria. In *The Metabolic Basis of Inherited Diseases,* 4th ed. J. B. Stanbury, J. B. Wyngaarden, and D. S. Frederickson (eds.). pp. 24–255. McGraw-Hill, New York, 1978.

Wiener, A. S. Blood groups and disease. *Am J. Hum. Genet.* 22: 476, 1970.

REVIEW QUESTIONS

1. A rare allele for extra digits in the human behaves as a dominant (E) to the normal (e). A man with an extra toe on each foot marries an unrelated woman with the normal condition. The first child has an extra finger; the second child is phenotypically normal. In time, the normal offspring marries a normal, unrelated woman and produces a child with an almost complete extra toe on one foot and the normal number of toes on the other.

Offer an explanation and give the probable genotypes of the persons mentioned

2. In the human, study of family histories indicates that 4 out of 10 persons who carry a certain rare dominant allele for a skeletal defect do not express it. What would be the penetrance of this allele?

3. In the human, a certain allele results in severe anemia plus damage to the kidneys and the central nervous system. What genetic phenomenon is illustrated by the expression of this allele?

4. In sweet peas, the two allelic pairs C, c and P, p are known to affect pigment formation in the flowers. The dominants, C and P, are both necessary for colored flowers. Absence of either results in white. A dihybrid plant with colored flowers is crossed to a white one heterozygous at the "c" locus.

   A. What are the genotypes of these two plants?
   B. What kinds of flowers, colored or white, should be expected from the cross, and in what ratio?

5. Assume that another allelic pair in sweet peas also affects pigment formation in addition to the genes mentioned in Question 4. The presence of the dominant, R, is required for red flowers. Its recessive allele, r, produces yellow flowers. What would be the phenotypes of the following plants in relation to flower color?

   A. CcPpRr
   B. CcppRR
   C. CcPPrr
   D. ccPPRR

6. In a certain breed of dogs, the dominant, B, is required for black fur; its recessive allele, b, produces brown fur. However, the dominant, I, is epistatic to the color locus and can inhibit pigment formation. The recessive allele, i, on the other hand, permits pigment deposition in the fur. Describe the phenotypes of the following sets of parents, and the results of their mating.

   A. bbii × BbIi
   B. bbIi × Bbii
   C. bbIi × BBIi

7. In the human, the dominants, D and E, are both required for normal development of the cochlea and the auditory nerve, respectively. The recessives, d and e, can result in deafness because of impairment of these essential parts of the ear. Give the phenotypes of each of the following sets of parents and the chance of a deaf child being born as the first offspring.

   A. DDee × ddEE
   B. DdEE × DDEe
   C. DdEE × DdEe
   D. DdEe × DDEe

8. In poultry, the shape of the comb varies greatly and involves at least two pairs of alleles. The allele, R, can

result in rose comb, and the allele P can result in pea-shaped comb. If both of these dominants are present together, genic interaction produces a walnut comb. When a bird is carrying both recessives, r and p, in the homozygous condition, single comb type results. Give the type of comb of each of the following pairs of birds and the phenotypic ratio to be expected among their offspring as to comb type.

  A. rrPP × RRpp
  B. RrPp × RrPp
  C. RrPp × rrpp

9. Using the information on comb shape given in Question 8, deduce the genotypes of the parents in each of the following crosses.

  A. A walnut bird crossed with a single-comb bird. These parents produce three walnut, four pea, two rose, and three single-comb birds.
  B. A rose-combed bird crossed with a pea. This parental combination gives rise to rose and walnut in a ratio of 1:1 after several matings.
  C. A rose-combed bird crossed with a walnut. These parents produce three walnut, five rose, two pea, and one single-comb birds.

10. In poultry, the allele, B, can result in black feathers. Its allele, b, can produce white splashed feathers. The heterozygote is blue. A bird with black feathers and pure breeding for rose comb is crossed with a bird that is white splashed and pure breeding for pea comb. (Use the information in Question 8 for comb shapes.)

  A. Diagram the cross. What would the F_1 be like?
  B. What kinds of birds are to be expected from crossing F_1s with birds that are black and have single combs?

11. In poultry, the factors, C and O, are both required for any color at all in the feathers. Homozygous recessiveness at either locus will result in white. In each of the following three cases, give the genotypes of the parental birds and their offspring. (Use the information on color in Question 10.)

  A. Two completely white birds which produce an F_1 that is black.
  B. Two black birds which produce 54 black birds and 42 white ones.
  C. Two blue birds which produce 16 black, 37 blue, and 14 white splashed. birds.

12. Two pairs of alleles are involved in cyanide production in white clover. Some strains have a high cyanide content, others a low one. When low Strain A was crossed to low Strain B, the F_1 all had a high cyanide content. When the F_1 plants were crossed among themselves, the following F_2 resulted: high cyanide content 450; low cyanide content 350. Offer an explanation for these observations.

13. Two different highly inbred strains of chickens both have feathers on the upper portion of their legs. When

birds from the two different strains are crossed, the F_1 all have feathered legs. In a typical case, crossing the F_1s to each other gave 323. Of these 301 had feathered legs, and the remainder were unfeathered. When the unfeathered ones were crossed among themselves, only birds with unfeathered legs resulted. Give a likely explanation.

14. Very little hair is found on a Mexican hairless dog. A cross between a Mexican hairless and a dog with a typical coat of hair usually produces litters of pups in which half of the animals are hairless and the other half have hair. On the other hand, a cross between two Mexican hairless dogs tends to produce litters in which two thirds of the pups are hairless and one third have hair. However, in addition to these surviving puppies, usually some are born dead. These appear hairless and occur in about the same frequency as the pups with hair. Offer an explanation and represent the genotypes of the different kinds of animals using any symbols you choose.

15. Several loci affect hair development in dogs. The allele for wire hair (W) is dominant to its allele for straight hair (w). Allow H to represent hairlessness in the Mexican hairless dog and its recessive allele h to represent the gene for typical hair growth in other breeds. The Mexican hairless dog is homozygous for straight hair, but because of the epistatic effect of H, alleles for hair formation cannot be expressed. Write genotypes of the following animals with respect to the two loci mentioned.

  A. A Mexican hairless dog.
  B. A dog with straight hair.
  C. A dog from a strain that is pure breeding for wire hair.

16. Using information in Question 15, give the results of a cross between the following:

  A. A Mexican hairless dog and a dog from a strain that is pure breeding for wire hair.
  B. Two animals of the first generation resulting from the cross given in A. (Cross only the animals which have different genotypes).

17. Assume that in a disease-resistant variety of plant the recessive lethal w occurs with a high frequency, and a considerable proportion of the plants die because they lack the dominant allele W and cannot produce chlorophyll. A heterozygote with one dose of the dominant allele cannot be distinguished from a plant homozygous for the dominant. Suggest a program by which the undesirable recessive might be eliminated from the variety, considering that the plants are exclusively cross-pollinated.

18. In guinea pigs, black fur, A, is dominant over the albino condition, a, which gives white fur. A black female from a pure-breeding strain carries transplanted ovaries from

an albino female. The albino female in turn received the ovaries from the homozygous black animal.

A. What are the expected results among the offspring when the black female is crossed to a black male which had a white parent?

B. What would be expected from this cross if the black female had also had a white parent?

C. What results are to be expected if the albino female is crossed to this same male?

19. The dominant genetic factor, P, is required for the formation of an enzyme needed to convert phenylalanine to tyrosine. The recessive allele, p, results in a lack of the enzyme, and as a consequence PKU can ensue.

A woman who was spared the dire effects of PKU as a result of a special diet during infancy and childhood is married to a man who does not have PKU but does carry the recessive allele. During her pregnancy, the woman is placed on a special diet once more, since excess phenylalanine and other toxic products can cross the placenta in a PKU woman and result in mental retardation in the newborn. What genotypes are to be expected among the offspring in this case, and what would be their phenotypes?

20. Suppose the woman in Question 19 is not placed on the diet during pregnancy. What would be expected now among the offspring?

21. S_1, S_2, S_3, and so on are sterility alleles that form a multiple allelic series in tobacco. A plant cannot be homozygous for any one of them. What will be the constitutions of the F_1 plants regarding the sterility alleles in each of the following crosses?

A. S_1S_2 egg parent $\times S_4S_5$ pollen parent.

B. S_3S_4 egg parent $\times S_4S_5$ pollen parent.

C. S_1S_2 egg parent $\times S_1S_2$ pollen parent.

D. S_4S_5 egg parent $\times S_3S_4$ pollen parent.

22. In the rabbit, a multiple allelic series affects the color of the fur. The wild-type allele (c^+) for dark fur is dominant to all the other alleles in the series. The allele c^h for Himalayan is dominant to albino (c). However, the allele for chinchilla coat color (c^{ch}) gives light gray color when present with either the allele for Himalayan or albino. Give the genotypes and the phenotypes of the offspring from the following crosses.

A. A Himalayan animal whose male parent is albino $\times$ a chinchilla.

B. An animal with wild-type fur that has a chinchilla parent $\times$ an albino.

C. A light gray animal that has an albino parent $\times$ a light gray that has a Himalayan parent.

D. A rabbit with wild-type fur that has an albino parent $\times$ a Himalayan that also has an albino parent.

23. A rabbit has fully pigmented fur that appears to be wild type. However, this animal carries the allele for both the Himalayan pattern and the albino condition. The animal

has been raised under cold, experimental conditions. The animal is mated to a Himalayan animal which also carries the allele for the albino condition.

A. What kinds of offspring will you expect to find and in what proportion if these offspring are raised under normal temperature conditions?

B. Answer the same question for offspring which are reared under cold conditions.

24. Give the probable genotypes of the following pairs of parents with respect to the A, B, O blood type alleles.

A. An A type parent and an O who have one child who is type O.

B. An AB type parent and an A who have one child who is type B and one who is type A.

C. A type A parent and a type B who have eight children who are type AB.

25. Two subgroups of type A blood can be recognized. The allele A_1 is dominant to A_2 and to O. A_2 is also dominant to O. Give the results of the following crosses:

A. An A_1 type person who has an O parent $\times$ an A_2 who has an O parent.

B. An A_2 type person who has an O parent $\times$ an A_1B.

C. $A_1B \times A_2B$.

D. $A_1B \times O$.

26. A woman is blood type B. Man 1 is type O. Man 2 is A_2. The woman's baby is type A_1B. Which man could be the father? Explain.

27. The secretor trait (Se), which enables A and B antigens to be secreted into the body fluids, is dominant to nonsecretor (se). Give the blood type and indicate whether the saliva will contain A or B antigens in the offspring from each of the following parents:

A. I^AI^B Se Se $\times$ ii se se.

B. I^AI^B Se se $\times$ ii Se se.

C. I^Ai Se se $\times I^Bi$ Se se.

28. The allele H is required for the completion of A and B antigens. In its absence, the allele h cannot direct the formation of A and B antigens. The h allele is thus epistatic to the ABO locus and can cause a person to be blood type O regardless of the presence of the A and B alleles. Give the blood types expected among the offspring in each of the following matings, as well as the ratio in which they would occur:

A. I^AI^B Hh $\times I^AI^B$ Hh.

B. ii Hh $\times I^Ai$ Hh.

C. I^Ai Hh $\times I^Bi$ hh.

29. For each of the following four sets of parents, give the ABO blood types that can occur among the offspring and the proportion in which they can be expected. Indicate which, if any, of the antigens involved will be secreted into the body fluids.

A. ii Se Se Hh × ii Se Se Hh.
B. I^Ai Se Se Hh × I^Bi Se Se hh.
C. ii Se se Hh × ii Se se Hh.
D. $I^A I^A$ se se Hh × $I^B I^B$ Se se Hh.

30. A woman is blood type A MN. Her baby is type O, N. Which of the following two men is the father? Explain.

   A. Man 1: type A M.
   B. Man 2: type O N.

31. Considering Rh^+ blood to be associated with the dominant, D, and Rh^- with its recessive allele, d, give the possible offspring from the following matings:

   A. An O, Rh^- woman (who has an Rh^+ parent) and an O Rh^+ man (who has an Rh^- parent).
   B. An A, Rh^+ woman and a B, Rh^+ man, both of whom have one Rh^- parent.
   C. A woman of type O who suffered erythroblastosis fetalis and a man who is AB negative.

32. A woman is blood type O MN Rh^+. Her baby is O MN Rh^-. Man 1 is A N Rh^+. Man 2 is O MN RH^-. Which of the two could be the father?

33. Give the ABO blood grouping of each of the following persons and tell which of the pertinent antigens will be secreted into the body fluids.

   A. ii Se se Hh $L^M L^N$.
   B. $I^A I^A$ se se Hh $L^N L^N$.
   C. $I^A I^B$ Se Se hh $L^M L^N$.
   D. $I^A I^B$ se se HH $L^M L^N$.
   E. $I^A I^B$ Se Se hh DD $L^N L^N$.

34. Of the following matings, which run a risk of producing a baby with erythroblastosis fetalis?

   A. Female $\dfrac{C\,D\,E}{C\,D\,E}$ × Male $\dfrac{c\,d\,e}{c\,d\,e}$.

   B. Female $\dfrac{c\,d\,E}{c\,d\,e}$ × Male $\dfrac{C\,D\,e}{c\,D\,e}$.

   C. Female $\dfrac{C\,d\,e}{c\,d\,e}$ × Male $\dfrac{c\,D\,E}{C\,d\,e}$.

   D. Female $\dfrac{C\,d\,E}{C\,d\,e}$ × Male $\dfrac{C\,d\,E}{C\,d\,e}$.

35. Three genes in the HLA complex, designated B, C, and A, code for histocompatibility antigens. Each gene occurs in many allelic forms. Assume that a woman and her husband have the following genotypes, respectively, for the antigen-determining genes:

$$\frac{B_2 C_1 A_2}{B_1 C_2 A_1} \text{ and } \frac{B_4 C_3 A_4}{B_3 C_4 A_3}$$

   A. Give the genotypes possible among the offspring, ignoring crossing over.
   B. Can the children donate to or accept tissue from either parent?
   C. How would you answer the preceding question if one of the children were a boy with agammaglobulinemia who could not produce T lymphocytes?

36. Following are the phenotypes of members of a family with respect to histocompatibility antigens. Answer the questions regarding this family:

Mother: B_8; B_9; A_1; A_2 Child 1: B_8; B_9; A_2; A_3
Father: B_6; B_8; A_3 Child 2: B_6; B_8; A_1; A_3
                                 Child 3: B_6; B_9; A_2; A_3

   A. Give the genotypes of all these persons.
   B. Suppose another child is born of the following type: B_8; A_2; A_3. Give this child's genotype and explain its occurrence.

37. In the human, there are two different genetic blocks in the pathway leading to melanin pigment formation. One of them, a, interrupts a step that prevents the formation of the immediate precursor of melanin. The second block, b, prevents the conversion of the precursor to melanin pigment. Each of the two types of albinism that result from these blocks is a recessive trait.

   A. What kinds of offspring are to be expected if an albino person of type a has children with an albino person of type b?
   B. What is the chance that two persons of the children's genotype will produce an albino at any birth if two such persons marry?

5

SEX AND INHERITANCE

Sex chromosomes

Cytogenetic studies related to sex differences have played a major role in proving the soundness of the chromosome theory of heredity. Several early geneticists, including Mendel, realized that sex seems to behave as a Mendelian character. Male and female offspring occur in a 1:1 ratio, the classic monohybrid testcross ratio expected from crossing a heterozygote with a homozygous recessive.

In 1902, McClung, working with a species of grasshopper, noted that one chromosome in the male seemed to be unpaired and lacked a corresponding mate. As a consequence, meiosis would produce two kinds of sperms: half with the unpaired element and half lacking it. McClung connected this unpaired chromosome with the determination of sex and inspired other investigations relating chromosomes to sex. Before 1910, Stevens and Wilson had worked out the cytological picture in many insects. They recognized that the chromosomes occur in pairs, but realized that the two sexes often differ in regard to one of the chromosome pairs.

Stevens showed that *Drosophila melanogaster,* the common fruit fly, has eight chromosomes which exist as four pairs (Fig. 5-1A). Cells from females contain one pair of rod-shaped chromosomes (Pair I, which is

acrocentric), two pairs of V-shaped chromosomes (Pairs II and III with median centromeres), and a very small pair designated the *dot chromosomes* (Pair IV with centromere near one end). In the female, the two members of each pair are identical in appearance, but this is not so in the male. The difference is in Pair I. Males show only one rod-shaped chromosome instead of two, but they carry a hook-shaped or J chromosome, which is absent in the female. Similar distinctions between males and females were found in other species as well. The two members of the chromosome complement that differ between the sexes were named the *sex chromosomes*. The one found singly in the male and paired in the female was called the X. The one that is confined to the male was called Y. As noted in Chapter 2, all the chromosomes of the complement other than the sex chromosomes were designated the *autosomes*. Drosophila, then, possesses three pairs of autosomes and one pair of sex chromosomes, the latter represented as two Xs in the female and an X and a Y in the male. Such a chromosome picture is known as an "X-Y" condition and implies that the male has dissimilar sex chromosomes and will produce gametes that differ in their sex chromosome content. For example, male fruit flies form sperms with three autosomes, but half of the gametes carry an X and the other half a Y. Since his

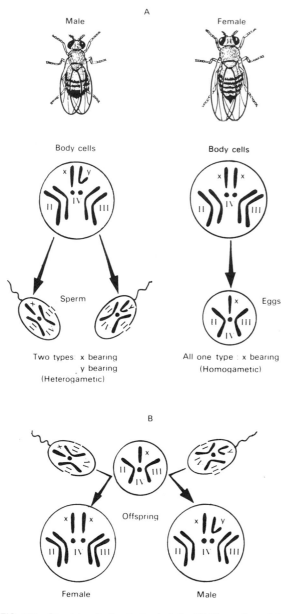

FIG. 5-1. Sex determination in the fruit fly. (*A*) The male and the female both possess two sets of autosomes (Chromosomes II, III, and IV) in their body cells. They differ regarding the sex chromosomes: the male carries 1X + 1Y, whereas the female carries two Xs, also called Chromosome I. All of the eggs produced by a female are alike in that they are X bearing. The male forms two classes of sperms and is designated the heterogametic sex. (*B*) There is an equal chance for any egg to be fertilized by an X-bearing or a Y-bearing sperm. Consequently, male and female offspring occur in equal numbers.

gametes fall into two distinct classes, we say that the male is the *heterogametic sex.* The female is termed *homogametic;* all her eggs are alike in the kinds of sex chromosomes they contain (Fig. 5-1A).

It was realized, however, that this X-Y condition is not universal. Cytological examination had shown that cells of one sex may contain one chromosome more than the other. In such species, the sex chromosome is completely unpaired. Since a Y is not present at all, the X has no mate or pairing partner. We followed the behavior of an X in this so-called X-O condition during our discussion of spermatogenesis in Chapter 3. It was the X-O situation that McClung had observed in his study of the grasshopper. However, he believed that the male had one more chromosome than the female, when actually the reverse is true.

The X-O picture is found in many insects and various other groups as well. In species such as the grasshopper, the male is again the heterogametic sex, because the sperms are of two kinds in relation to their sex chromosome content: half contain one more chromosome than the other. In some organisms with the X-O mechanism, it is the female sex that is heterogametic (XO) and the male that is homogametic (XX). Indeed, the females of certain species may contain an X and a Y, whereas the males are XX. Such an arrangement is found in moths and butterflies and is typical of birds. This is designated "W-Z," to imply that the female (ZW) forms two classes of sex cells, one with the large Z and one with the small W. The female is thus the heterogametic sex. The male is homogametic (ZZ), possessing two large Z chromosomes. This is the reverse of the more common and familiar X-Y arrangement found in Drosophila as well as in the human. In our species, we find 22 pairs of autosomes and one pair of sex chromosomes (two Xs in the female; one X and one Y in the male).

Geneticists recognized that the presence of sex chromosomes could account for the determination of equal numbers of male and female offspring. It could thus explain what appears to be a monohybrid 1:1 testcross ratio. In an animal with the X-Y mechanism, the eggs are all X bearing and are fertilized by either X-bearing or Y-bearing sperms. Since the male produces these two different classes of gametes in equal amounts, it is a matter of chance whether any one egg is fertilized by an X-containing sperm or by one with a Y. The outcome is male and female offspring in equal proportion (Fig. 5-1B).

Genes on the X chromosome: sex linkage

Although such a mechanism seemed to account for sex determination in many organisms, it was not known whether the sex chromosomes contain genes for characteristics *other than* sex or whether the autosomes may also influence sex in some way. The work with the fruit fly done by T. H. Morgan and his associates at Columbia University was begun in 1909 and contributed a wealth of information that is basic to an understanding of sex determination. The investigations of this outstanding group also uncovered other fundamental principles of inheritance that established genetics firmly as a scientific discipline.

Morgan was particularly interested in mutations and their role in evolution. One of the first mutations found under laboratory conditions affected the eye color of Drosophila, changing it from the red found in nature to unpigmented or white. The white-eyed fly that arose was a male, and it was crossed to a normal red-eyed female. All of the offspring were red eyed, suggesting that the genetic determinant for white eyes was recessive to the one for red. A cross between males and females of the first generation produced red-eyed and white-eyed flies in the familiar Mendelian ratio of 3:1. The results indicated that a pair of alleles was involved, W for the dominant red condition and w for the recessive white. However, there was something unusual about this 3:1 ratio. All of the flies with white eyes were males (Fig. 5-2A)! Half of the males had red eyes, but *all* of the females were red eyed. Morgan realized that such results could be explained by assuming that the locus determining red versus white eye color is on the X chromosome (Fig. 5-2B). The females used in these crosses would have two X chromosomes, each bearing the dominant W. They would thus be homozygous for red eye color and can be represented simply as WW. The white-eyed male, however, has but one X chromosome and therefore has the eye color locus represented only once. The Y chromosome would not carry a locus for any eye color trait. The white-eyed male would thus have an X chromosome with the recessive allele w and a Y chromosome with no corresponding eye color locus. The genotype of such a fly can be represented as wY. The Y chromosome must be shown when the genotype is written, to indicate that the male is heterogametic and carries only one factor for white eye color. We cannot say that the male is either homozygous or heterozygous for eye color, because, by definition, both of these terms imply that a particular gene is represented twice for a particular character. Since only one dose of a gene can be present in the case of an X-linked gene in the heterogametic sex, we say that the individual is *hemizygous*. Figure 5-2B shows why all the F_2 white-eyed flies must be

males after a cross between a homozygous red-eyed female and a hemizygous white-eyed male.

Morgan realized that if his hypothesis were correct, it should be possible to obtain females with white eyes. It can be seen from the cross in Figure 5-2B that all of the red-eyed females of the first generation must be heterozygotes, Ww. Mating them with white-eyed males should produce red-eyed and white-eyed flies of both sexes (Fig. 5-3). When matings of this type were performed, white-eyed females were actually obtained. These white-eyed females were then crossed to red-eyed males. This is the reciprocal of the original cross (red-eyed female × white-eyed male). As Figure 5-4A shows, the results of this reciprocal cross were strikingly different from the original mating and strongly supported Morgan's hypothesis. In the first generation, we see that all of the females were red eyed like the male parent, and all of the males were white eyed like their mothers. This is exactly what would be expected if the gene for eye color is on the X chromosome (Fig. 5-4B). We say that a *criss-cross* pattern of inheritance has occurred. This simply means that a phenotype present in the female parent appears among all the sons, whereas the contrasting phenotype of the male parent shows in all the daughters. Note from Fig. 5-4B that the cross of the F_1 red-eyed females with their white-eyed brothers is really the same as the cross diagrammed in Fig. 5-3, which first yielded the white-eyed female offspring.

Thus, a trait other than sex was definitely shown to be associated with a sex chromosome. Morgan then found a short-wing mutant he called *rudimentary*. It gave the same pattern of inheritance as eye color. Such genes were called *sex linked*, genes whose loci are found only on the X and not on the Y. It was soon realized that sex linkage occurs in species other than Drosophila, and a sex-linked recessive pattern was recognized about this time in humans for red-green color blindness and hemophilia.

Actually, the first case of sex linkage had been found in 1906 by Doncaster who was following inheritance of body color in the moth Abraxis. The genetic results suggested that the female is the heterogametic sex and the male is homogametic. This was later verified cytologically. Knowing now that the female is indeed heterogametic in this species, we can diagram sex-linked crosses in the moth, as shown in Fig. 5-5. Notice that the reciprocal crosses are different and a crisscross pattern occurs when dark-bodied females are mated with light-bodied males. Such results strongly indicate sex linkage. The fact

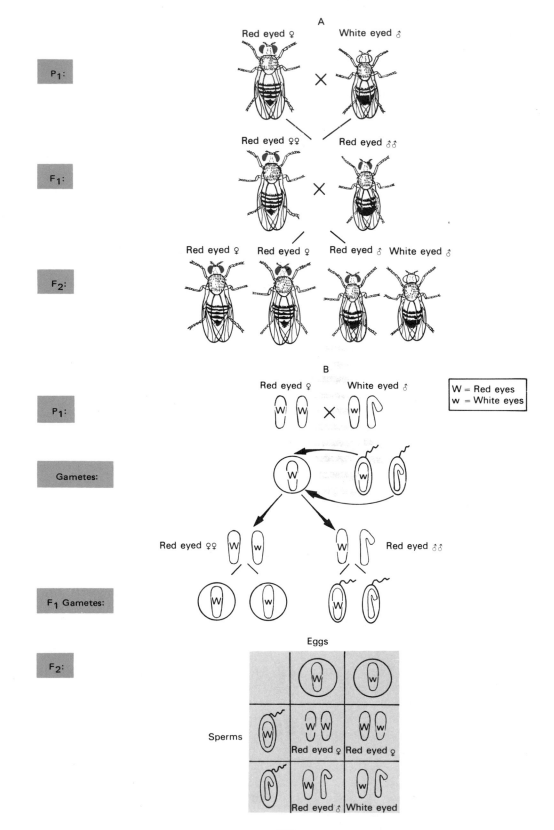

A

P₁: Red eyed ♀   ×   White eyed ♂

F₁: Red eyed ♀♀   ×   Red eyed ♂♂

F₂: Red eyed ♀   Red eyed ♀   Red eyed ♂   White eyed ♂

B

W = Red eyes
w = White eyes

P₁: Red eyed ♀   ×   White eyed ♂

Gametes:

Red eyed ♀♀   Red eyed ♂♂

F₁ Gametes:

Eggs

F₂:

Sperms

Red eyed ♀   Red eyed ♀

Red eyed ♂   White eyed

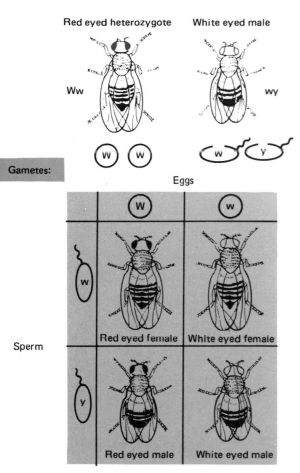

Gametes:

Sperm

FIG. 5-3. Derivation of white-eyed females. Red-eyed heterozygotes are obtained from a cross such as that depicted in Fig. 5-2. When these F$_1$ females are crossed with white-eyed males, half of the female as well as the male offspring will be white eyed.

FIG. 5-2. Inheritance of white eye color in Drosophila. (*A*) A cross of a female from a red-eyed stock with a white-eyed male yields an F$_1$ in which all are red eyed. A mating of the F$_1$ gives an F$_2$ in which red-eyed and white-eyed flies occur in a ratio of 3:1. However, the white-eyed flies are always males. (*B*) The results of the preceding cross are explained by the fact that the locus for white eye color is on the X chromosome. No corresponding locus is found on the Y. All of the F$_2$ females are red eyed because they receive an X from the F$_1$ males, and these all carry the dominant allele. The males of the F$_2$, like any males with an X–Y mechanism, receive the X from their mothers, not their fathers. Half of these carry the recessive, w. The Y each male receives from the male parent carries no locus for eye color. Therefore, the males express any factors on the X chromosome.

that the sex chromosome arrangement is opposite to the one found in Drosophila is no contradiction to the concept of sex linkage. However, this feature of the moth made it difficult at the time to compare it with well-known organisms having the X-Y arrangement and in which the complications of sex linkage were being worked out. As noted, we now know that this type of pattern seen in Abraxis, in which the female is heterogametic, is typical in birds and certain other groups.

Sex linkage in humans

After the discovery of sex linkage in Drosophila, attention was called to similar inheritance patterns for certain human traits. Among the most familiar are red-green color blindness and hemophilia, both of which behave as sex-linked recessives. Approximately 8% of American males have a defect in color recognition, whereas the trait in females is well under 1%. We can easily understand this once we realize that the loci for the major types of color blindness are on the X chromosome. The ability to perceive color depends not on a single genetic locus; at least four are involved, and three of these are X-linked. About three fourths of all color-blind persons carry a sex-linked recessive, which causes the green-sensitive cones in the retina to be defective. Consequently, difficulty is encountered in recognizing green color and distinguishing it from red. This is known as the deutan type of color blindness. Less common is the protan type, which affects about 2% of males in the population. This sex-linked recessive causes a defect in the cones sensitive to red. In this situation, red cannot be distinguished, and again red and green are confused. Since red and green color distinction is poor in both deutan and protan types, it was at first thought that only one locus and one sex-linked recessive was responsible, but more refined observations led to the recognition of two separate X-linked loci that affect red-green color perception.

Very uncommon is the tritan form of color blindness in which recognition is poor for all colors. This recessive is found at still another locus on the X and causes defects in cones sensitive to blue as well as those sensitive to green and red. A very rare autosomal recessive causes total color blindness, since no cones at all develop.

Since the loci for recessives responsible for red-green color blindness are sex linked, we can understand why this condition is much more common in

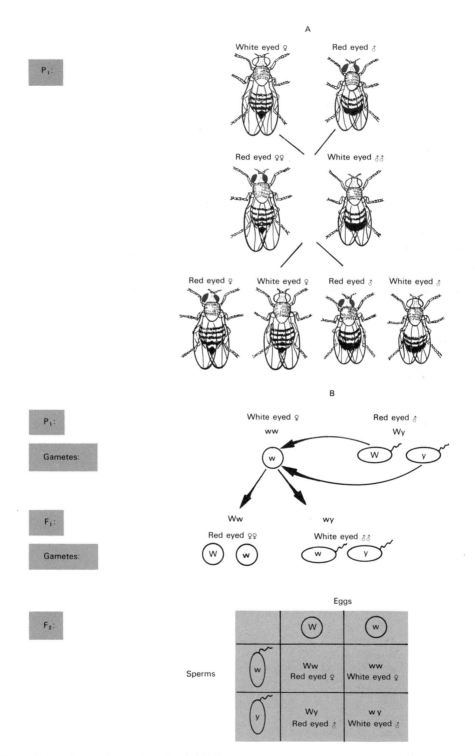

FIG. 5-4. White-eyed female *Drosophila* × red-eyed male (*A*). The results are very different from the reciprocal cross (Fig. 5-2). In the F₁, a crisscross pattern occurs. The trait seen in the female parent occurs in the male offspring; the trait expressed by the male parent is found in the female offspring. (*B*) The diagram of the cross shows that in the case of a sex-linked allele, the males will express the trait that the gene governs, regardless of dominance. This is so because males receive the Y chromosome from their fathers, and thus have no homologous region to mask any factors on the X contributed by their mothers.

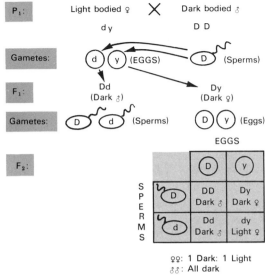
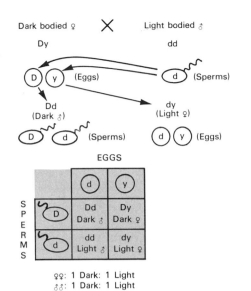

FIG. 5-5. Sex-linked inheritance in the moth Abraxis. A pair of alleles, D (dark body) and d (light body), is associated with the X chromosome. The reciprocal crosses diagrammed here produce very different results. The crisscross pattern (*on the right*) indicates sex linkage. In moths, the female is the heterogametic sex. However, the principles of sex linkage remain exactly the same as in the more familiar situation in which the male is heterogametic. Compare this figure with the crosses in Figs. 5-2, 5-3, and 5-4.

the male. Any female of normal chromosome constitution carries two X chromosomes. For her to express red-green color blindness (let us say of the more common deutan type), her father must show the trait. The mother must show deutan color blindness herself or at least be a heterozygote. This is so because the daughter must receive a deutan color-blind recessive from each parent to express the trait (Fig. 5-6A and B). The same reasoning applies to the protan type of red-green color blindness, as well as to the rare tritan type.

It should be apparent also that a male with the normal XY chromosome constitution will always express any gene that is sex linked because he has only one X chromosome. Being hemizygous, he cannot carry the normal allele and the defective form at the same time. Therefore, any sex-linked allele that expresses itself in a heterogametic male must have come from the female parent. In humans, a son receives the Y from his father; this is a requirement for maleness. Moreover, all the sons of a woman who expresses a sex-linked recessive, such as color blindness, would be affected, even if her mate is normal (Fig. 5-6C). The cross of a color-blind woman and a normal man results in a crisscross pattern and is comparable to a mating in Drosophila between a white-eyed female and a red-eyed male.

The sex-linked recessive responsible for the lethal effect of hemophilia, a condition in which the blood fails to clot normally, is of special interest, for both its consequences for the sufferer and its historical import. The affliction, which has been recognized for hundreds of years, received attention when it appeared among European royalty during the reign of Victoria of England. There is little doubt that Queen Victoria was heterozygous for the recessive, Hh. The probability is great that a mutation from the normal allele (H) to a recessive form (h) was carried in either the sperm from her father or the egg from her mother. At any rate, of the queen's nine children, several definitely received the recessive allele from her. One of her sons was consequently a hemophiliac, and some of her daughters were carriers who were responsible for passage of the hemophilia factor through several royal houses, even down to the present. This is easy to understand from what we know about the inheritance of sex-linked recessives (Fig. 5-7A and B). We would expect from such a mating that half of the daughters would be carriers and half of the sons would express the factor for hemophilia.

Today we know a great deal about the molecular basis of hemophilia. We may think of blood clotting as the end result of a series of reactions requiring various enzymes and cofactors. The end result of this

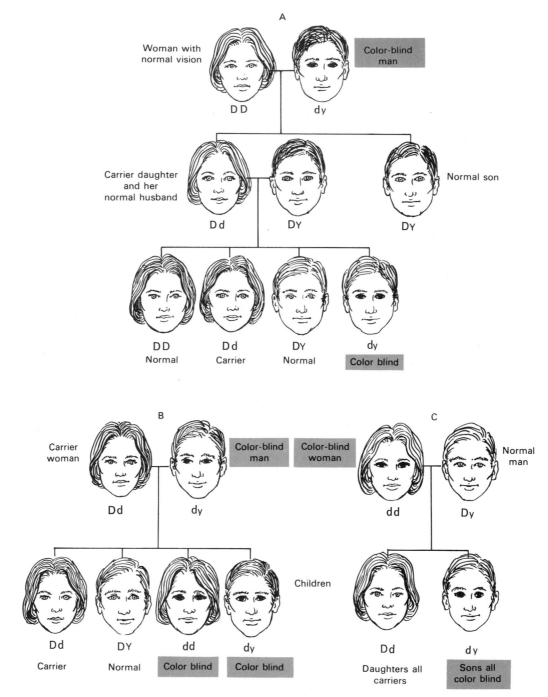

FIG. 5-6. Color-blind inheritance (*A*) The locus for green color perception is on the X chromosome. D represents the dominant allele for normal color distinction and d its recessive allele for deutan color blindness. The trait passes from a color-blind man to half of his grandsons by way of his carrier daughters. Since a male contributes his Y chromosome to his sons and not the X, the sons of a color-blind man cannot receive the recessive from their father. (*B*) For a color-blind female to arise, her mother must at least be a carrier and her father must be color blind. The color-blind female receives from each parent an X that carries the recessive allele (*C*). All of the sons of a color-blind woman must be color blind because a male receives his X from his mother. The daughters of a color-blind woman will be carriers with normal vision if the male parent has normal vision.

114

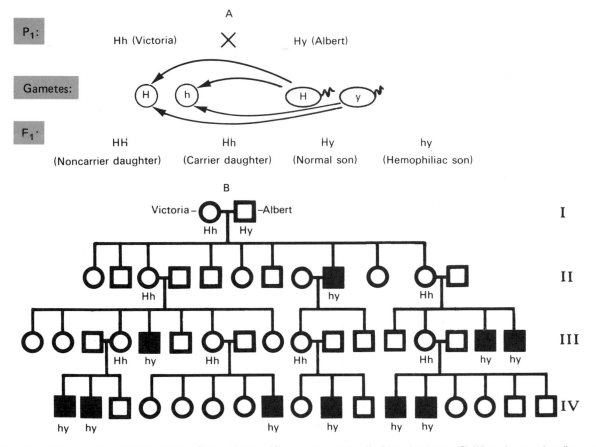

FIG. 5-7. Inheritance of hemophilia through Queen Victoria. (*A*) Hemophilia among European royalty can be traced back to Queen Victoria. The trait was not found among her ancestors, but she was undoubtedly heterozygous for the defective allele. Since hemophilia is sex linked, we would expect her to pass the allele to half of her sons and half of her daughters. (*B*) When the actual pedigree of Victoria and her descendants is studied, the transmission of hemophilia is easily understood on the basis of the expectations summarized in *A*. In the pedigree, circles represent females and squares are males.

series of interactions is the formation of insoluble fibrin from soluble fibrinogen. This last step, which produces the final fibrin clot, requires the action of the enzyme thrombin. The many protein substances (enzymes and cofactors) leading to the activation of thrombin are governed by genes. A defect in any one of these genes leads to defective clotting, since a required enzyme or cofactor would be either defective or absent.

Hereditary disorders have been described in which one or more of these substances is deficient in the blood plasma. It is therefore not surprising that more than one kind of hemophilia is known. About 85% of the cases, designated hemophilia A, result from a deficiency of a protein cofactor known as factor VIII. The gene for factor VIII resides on the X chromosome. Factor VIII is needed for full activity of one of the

enzymes, factor IX, in the series of events leading to the activation of thrombin (Fig. 5-8). Absence of functional factor VIII interrupts the steps leading to the activation of thrombin, and consequently fibrin cannot form.

The remaining 15% of persons with hemophilia suffer from hemophilia B, which results from a defect in another X-linked gene, which governs the formation of factor IX. In its active form, factor IX acts as an enzyme to activate the next enzyme (factor X) in the reaction series. Factor VIII (the substance lacking in hemophilia A) is a cofactor required to work along with active factor IX in activating factor X.

The techniques of molecular biology (Chap. 18) have made it possible to isolate the genes coded for factors VIII and IX and also to characterize factor VIII. This substance has been shown to be a large

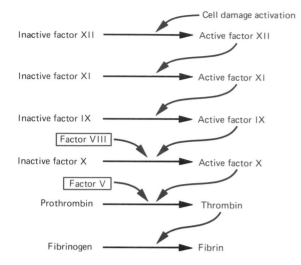

FIG. 5-8. Some steps in blood clotting. Normal coagulation involves a series of reactions leading to the formation of insoluble fibrin. Many of the steps in the chain entail the conversion of a gene product to its active enzymatic form by the enzymatic action of the active product of a previous step. Cofactors (e.g., factors VIII and V) are required to work along with some of the active enzymes. Factor VIII, absent in hemophilia A, is a cofactor required along with factor IX to activate factor X. If factor VIII is absent as a result of a sex-linked recessive, the sequence is interrupted, and the end result is defective clotting. In hemophilia B, factor IX is deficient.

protein, approximately 2332 amino acids long. It is also known that on the molecular level more than one kind of defect can occur in the factor VIII gene, and hence in the factor VIII protein. Different defects are associated with different degrees of severity in cases of hemophilia A. The information and techniques on the molecular level provide hope that factor VIII may soon become available on a wide scale for sufferers of hemophilia A and enable them to avoid the dangers entailed in transfusions of factor VIII from pooled blood.

For years it was thought that hemophilia could not occur in females, because the homozygous condition hh was believed to be lethal to the embryo. We are now aware of a few authentic cases of female bleeders. As we might well expect, they are very rare. As in the case of color blindness, the female parent of an afflicted girl must at least be heterozygous for the condition, and the father must actually suffer from the recessive disorder (see Fig. 5-6). Until the 1960s, most males with hemophilia did not survive to reproductive age. The few that did manage to have offspring would almost always have mated with women who are homozygous normal HH, because the reces-

sive allele involved is not common in the population. When a rare mating between a hemophiliac and a carrier female does take place, an afflicted daughter can be produced.

Well over 100 sex-linked genes have been recognized in humans. Most of them are not as familiar as the one affecting blood clotting and the one for color vision. We should realize that some sex-linked defective alleles have a dominant expression, as does the one responsible for a type of rickets, a skeletal defect. Any trait inherited on the basis of a sex-linked dominant will be more common among females because they possess two Xs and therefore have a greater chance of receiving the X-linked allele. Remember that in cases of color blindness and hemophilia, the dominant, normal condition occurs more frequently among females. For the same reason, an abnormal phenotype will be more common among females if it results from a sex-linked dominant allele.

Sex linkage and the chromosome theory of inheritance

Although the early studies on sex linkage in fruit flies demonstrated that the X carried loci affecting many different characteristics, they did not indicate whether or not chromosomes other than the X and Y were involved in sex determination. The studies of Bridges settled this point and gave insights into several other basic genetic phenomena. His investigations made use of "exceptional" white-eyed females that had been isolated by Morgan. These flies were exceptional because they produced offspring of unexpected phenotype. Remember that when an ordinary white-eyed

FIG. 5-9 Unexpected results in Drosophila. (*A*) When a certain exceptional white-eyed female was crossed with an ordinary red-eyed male, the results yielded the expected red-eyed females (A) and white-eyed males (B). However, a small number of unexpected white-eyed females (C) and red-eyed males (D) occurred. (*B*) Bridges explained the unusual results by assuming that in a certain percentage of oocytes in the exceptional female the two X chromosomes would not separate at meiosis but would move to the same pole. Thus, in addition to the normal eggs, two exceptional types would arise: one with two Xs and the other lacking an X entirely. When fertilized by normal sperms, these exceptional eggs would give rise to exceptional zygotes. (*C*) When Bridges examined the offspring cytologically, he found that the expected flies (A and B) possessed the normal chromosome complement. The unexpected females (C) carried a Y; the red-eyed males (D) lacked a Y and were sterile. No flies were found with three Xs. These were discovered later and are unusual females (superfemales) that are weak and tend to die. Flies lacking an X are never found, as at least one X is required for normal development.

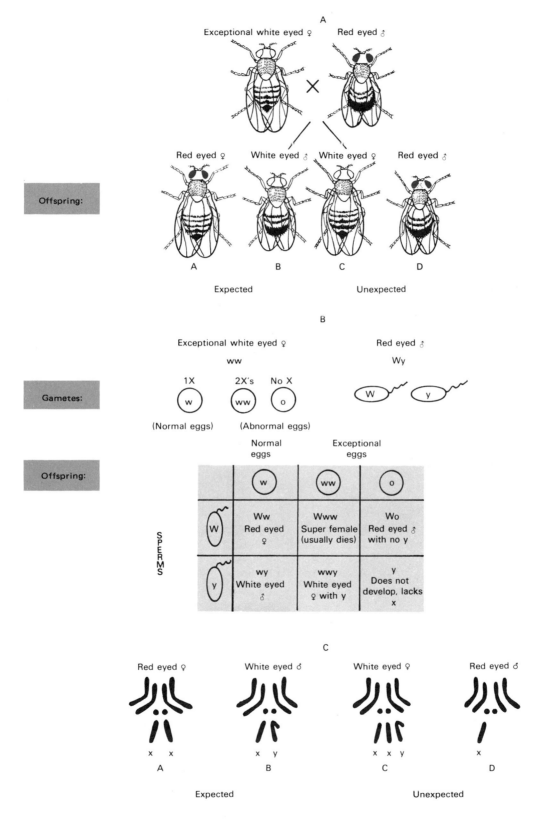

A

Exceptional white eyed ♀   Red eyed ♂

                    ✕

Offspring:

Red eyed ♀   White eyed ♂   White eyed ♀   Red eyed ♂

A   B   C   D

Expected   Unexpected

B

Exceptional white eyed ♀   Red eyed ♂
          ww   Wy

Gametes:

1X   2X's   No X   W   y

w   ww   o

(Normal eggs)   (Abnormal eggs)

Offspring:

	Normal eggs	Exceptional eggs	
	w	ww	o
W	Ww Red eyed ♀	Www Super female (usually dies)	Wo Red eyed ♂ with no y
y	wy White eyed ♂	wwy White eyed ♀ with y	y Does not develop, lacks x

SPERMS

C

Red eyed ♀   White eyed ♂   White eyed ♀   Red eyed ♂

x   x   x   y   x   x   y   x

A   B   C   D

Expected   Unexpected

117

female is crossed to a red-eyed male, a crisscross results (Fig. 5-4). However, although the exceptional white-eyed females, when mated to red-eyed males, produced the expected white-eyed sons and red-eyed daughters, some red-eyed sons and white-eyed daughters always appeared among the offspring (Fig. 5-9A). According to the concept of sex linkage, these phenotypes should not arise from the cross.

To account for these unexpected offspring, Bridges proposed that somehow a mishap was taking place at meiosis. He reasoned that perhaps the two X chromosomes in the exceptional females for some reason had a tendency to stick together during gamete formation. This would mean that abnormal eggs, in addition to those of normal constitution, would be produced (Fig. 5-9B). In a mating of an exceptional female to a red-eyed male, both the normal and the exceptional eggs would be fertilized by the two kinds of sperms. The offspring of the genotypes Ww and wY would represent the expected red-eyed females and white-eyed males. But what would the other four combinations be phenotypically, and how could they be identified? Bridges took the unexpected white-eyed females and red-eyed males arising from the cross and studied their chromosome complements. Cells of the exceptional white-eyed daughters carried the normal number of autosomes as well as two Xs. But in addition, a Y was present! The unexpected red-eyed males showed the normal autosome number and one X, but no Y was present at all (Fig. 5-9C)! These males proved to be sterile because of the immobility

of their sperm. Bridges found no flies with three X chromosomes or any with just the Y and no X. Because one X at least is necessary for life, the latter case never arise. Flies with three Xs were discovered in subsequent work and will be referred to shortly.

The cytological observations completely supported Bridge's hypothesis on the origin of the unexpected offspring. He coined the term *nondisjunction* to describe the failure of the homologous chromosomes to separate at anaphase of meiosis. We now know that nondisjunction may also occur at mitosis when both chromatids composing a chromosome move to the same pole at anaphase. The expression is generally used to describe the failure of chromatid separation at either mitosis or meiosis.

Bridges tested his hypothesis further. He took some of the exceptional white-eyed daughters containing a Y chromosome and crossed them to ordinary red-eyed males (Fig. 5-10). In the unusual females, the two X chromosomes pair in some of the meiotic cells. At anaphase, the Xs therefore pass to opposite poles, whereas the Y chromosome may migrate to either pole. The result is the formation of two classes of gametes: those of the constitution w and those that are wY. In other meiotic cells, the Y chromosome may pair with one of Xs. In such a cell, this X and the Y pass to opposite poles. The unpaired X is free to migrate to either pole. This makes possible the formation of four classes of gametes: ww and Y (the unpaired X moved to the same pole as the X, which paired with the Y) and w and wY (the unpaired X

FIG. 5-10. Exceptional female × ordinary male. When Bridges mated females that carry a Y chromosome, unusual genetic results followed. A small number of white-eyed females and red-eyed males occurred. Cytological examination showed that flies with exceptional chromo-some complements were also being produced at the same time. Extra Ys were found in half of the males and females. Along with the results depicted in Fig. 5-9, we see a perfect correlation between genetics and cytology.

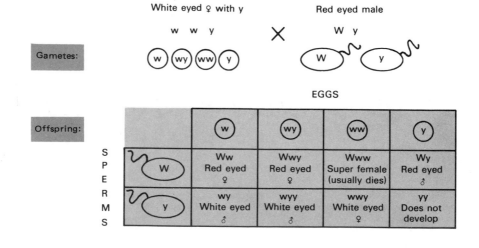

moved to the same pole as the Y chromosome). Therefore, an overall total of four different kinds of gametes can be formed from these pairing possibilities. However, the four types of eggs were produced in very unequal amounts by the wwY females: 92% were w and wY, whereas 8% were ww and Y. This indicates that the two Xs tend to pair more frequently than an X and a Y. Perhaps this should be expected, because the two Xs are completely homologous

In Fig. 5-10, the four kinds of eggs are combined in all possible ways with the normal sperm types, W and Y, to produce an assortment of zygotic combinations. How were these different types represented among the offspring? Cytological examinations were performed, and several of them were identified. No flies were XXX, because this combination usually dies. YY flies were never produced at all. The results clearly showed that when exceptional females with a Y chromosome are bred, exceptional results follow. Their white-eyed daughters have a Y chromosome. Although their red-eyed sons have a normal complement of one X and one Y, the Y chromosomes they carry must come from their exceptional female parents! In addition, extra Ys are present in half of the white males and half of the red females.

This story is an example of perfect correlation between genetic and cytological results. The work is of great significance for many reasons, not the least of which is its demonstration that the genes are in the chromosomes. For when the chromosome constitution is not determined by the normal process, atypical results arise. This evidence was of extreme importance in the early days of genetic experimentation because it gave very strong support to the chromosome theory of inheritance.

Sex and genic balance

A more subtle but equally important concept, that of genic balance, was also demonstrated by this work. The results show that the presence of a Y chromosome in a zygote does not mean a male fly will necessarily develop. In Drosophila, it seems that the Y chromosome is not needed at all for life or even for maleness, because a fly can be male without a Y or can be female with one. Although the Y is essential for fertility, and is thus critical to the survival of the species, it is not required for the life or the sex determination of any one individual. Sex in Drosophila seems to depend not just on the presence or absence of a single chromosome but rather on the balance of

the autosomes with the sex chromosomes. When the balance is two X chromosomes to two sets of autosomes, a female develops. A balance of one X to two sets of autosomes determines a male.

That this concept of genic balance in sex determination is correct was confirmed by the later work of Bridges. Flies with entire extra sets of chromosomes were produced in the laboratory. Individuals with three or four chromosome sets instead of the normal two are *triploids* and *tetraploids,* respectively. Inasmuch as three or four homologous chromosomes are present in the cells of such flies, the meiotic behavior is irregular (see Chap. 10 for details). As a consequence, when these flies are bred, their offspring may be atypical in chromosome number, having different combinations of autosomes and X chromosomes. Normal males and females, being diploid, possess two sets of autosomes. As shown in Fig. 5-1*B,* one set of autosomes in Drosophila includes two different V-shaped chromosomes (Chromosomes II and III) and a small dot chromosome (Chromosome IV). Therefore, in the common fruit fly, one haploid set of autosomes is composed of three chromosomes (Chromosomes II, III, and IV). Normally, two of these sets of autosomes plus two Xs result in a female. Two sets of autosomes plus one X produce a male. Unusual flies that developed from mating the triploids and tetraploids did not all have this normal relationship of autosomes to X chromosomes. Studies of their cytology indicated that genes for maleness and femaleness were distributed throughout all the autosomes and the X. As the summary of results in Fig. 5-11 indicates, the X would contain genes with a strong female-determining tendency. *One set* of autosomes would lean in the male direction as a result of the many genes for maleness scattered throughout. However, the strength of one X toward the female side is greater than the strength of one set of autosomes in the male direction. They are not 1:1; rather, it seemed that the femaleness of an X was about 1.5 to the maleness of *a set* of autosomes (1.5:1). This should not be construed to mean that there is any kind of fluid or substance that has male or female-determining properties. The ratio 1.5:1 only symbolizes the comparative strengths of the two major factors (one X versus one set of autosomes) in sex determination.

The observations indicate that the sex of a fly is determined by a balance between many genes distributed throughout the autosomes and those found on the X. The absolute number of chromosomes is not the sole factor, because *any* fly will be a female when

IX = ♀ Determining
strength of 1.5

1 Set of autosomes = ♂ determining
strength of 1.0

Number of x's	Number of sets of autosomes	Ratio femaleness:maleness	Sex
1 (1.5)	2 (1)	1.5: 2	Male
2 (1.5)	2 (1)	3.0: 2	Female
3 (1.5)	3 (1)	3.0: 2 (4.5:3)	Female
4 (1.5)	4 (1)	3.0: 2 (6.0:4)	Female
2 (1.5)	3 (1)	3.0: 3.0	Intersex
3 (1.5)	2 (1)	4.5: 2.0	Superfemale
1 (1.5)	3 (1)	1.5: 3.0	Supermale

x x   x x y   x x x   x x x x

Diploid   Diploid +y   Triploid   Tetraploid

All ♀ since ratio of x to sets of
autosomes is the same in each
case: 1x to 1 set of autosomes

BUT

x x x   x x

Adding a single x
tips the balance
toward femaleness

Superfemale   Intersex

Adding a set of autosomes
equalizes strength of
male and femaleness

In both of these the ratio of
x to sets of autosomes departs
from the balance of 1x to 1 set
of autosomes

FIG. 5-11. Genetic balance and sex determination in Drosophila. A female results whenever one X chromosome is present for each set of autosomes. A male develops when one X occurs with two sets of autosomes. Any departure from these ratios results in flies that are in some way abnormal in sexual features.

the ratio of femaleness to maleness is 3:2 (which in turn results from a ratio of one X to one set of autosomes). It does not matter if four sets of autosomes and four Xs are present or if there is only one set of autosomes and one X. The ratio is still the same (1:1 between X and autosomes). Any departure from a certain balance, however, produces a fly that is abnormal in some way. The presence of just one extra X tips the balance heavily in the female direction (4.5:2), giving a fly with exaggerated secondary sex characteristics—a superfemale or metafemale. Such a fly is weak and usually dies. It was not found by

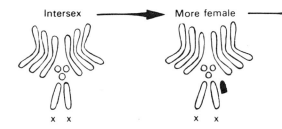

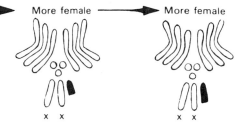

FIG. 5-12. Comparison of intersex types. Flies with two X chromosomes and three sets of autosomes cannot be classified as male or female. Those flies with additional pieces of an X appear more female. The larger the portion of the X, the more female the fly is in appearance. This indicates that genes for femaleness are scattered along the length of the X.

Bridges in his original study of the exceptional white-eyed females, but its possible viability was recognized at that time.

A comparable story holds for the determination of maleness. A ratio of one X chromosome to two sets of autosomes (a female to male ratio of 1.5:2) will always result in a male regardless of the presence or absence of a Y. A supermale or metamale, also generally weak, results from the extra strength of maleness provided by one extra set of autosomes. When the female:male ratio is equal (3:3), a sterile intersex, completely intermediate in secondary sex characteristics, is produced.

That many genes are involved in sex determination was shown by Dobzhansky's study of such intersex flies containing two Xs and three set of autosomes. After crossing normal flies whose chromosomes had been damaged by X-rays, Dobzhansky found offspring that contained three sets of autosomes and two Xs plus additional portions of the X (Fig. 5-12). The larger the piece of the X that was present in a 2X:3A intersex type, the more it resembled a female. Clearly, genes determining femaleness must be scattered along the length of the X.

The involvement of genes on the autosomes is very dramatically shown by a rare recessive autosomal allele in Drosophila known as transformer (tra). This is so named because a double dose of this recessive causes flies that have two X chromosomes and two sets of autosomes (and therefore should be normal females) to be phenotypically male (Fig. 5-13)! Such recessives cannot be distinguished from normal flies by observation; however, the transformed flies are sterile. Because the recessive (tra) is rare in fly populations, most females are Tra Tra XX and most males Tra Tra XY. This is another good example of the interaction of many genes in the determination of a

FIG. 5-13. The transformer allele in Drosophila. The recessive tra can transform XX flies, which would normally be females, into sterile males. Most flies (above) are homozygous for the dominant allele (Tra), so that the ratio of X chromosomes to autosomes usually determines the sex. When the recessive tra is present, however, it may upset the sex ratio (below). In this case, a female carrying the recessive is crossed with a male which is homozygous for it (the recessive does not affect the fertility of X–Y males). Among the offspring, half of the expected females are transformed to sterile males. An autosomal gene can thus clearly influence sex.

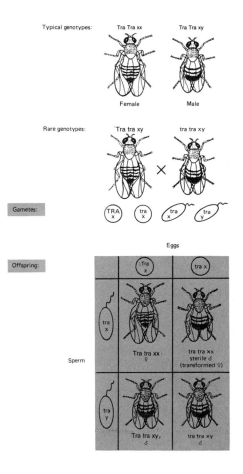

phenotypic characteristic, in this case sex. In the last chapter, we emphasized the fact that any characteristic involves not just one gene or allele but the interaction of many with the environment. Sex is a characteristic whose normal development is no different from that of any other in this respect.

Sex chromosome anomalies in humans

Although the concept of genic balance pertains to all living things, not just fruit flies, we must remember that the genetic picture varies in detail from one species to the next. We have already seen that, in the grasshopper, the male is normally X-O. No Y chromosome is needed even for sperm development, but a balance of X chromosomes to autosomes is also involved here in normal sex determination. Genic balance certainly plays an important role in human sex determination, but there are distinct differences from the situation of the fruit fly. The Y chromosome in Drosophila is apparently devoid of genes for male development. Thus, an XO individual is male and an XXY is female. Such is not the case in humans and other mammals, however, in which the Y chromosome is essential for male attributes. Examples of XO and XXY persons are well known. Those of the former constitution have a chromosome number of 45 and are sterile females who display the various abnormalities of Turner syndrome, a condition that affects the development of both sexual and other bodily characteristics. A Turner female, found in about 1 in 2000 female births, is usually of short stature, possesses a webbed neck, has undeveloped ovaries, and an immature uterus, plus cardiovascular defects and other somatic aberrations (Fig. 5-14). Persons who are of the chromosome constitution XXY are definitely male, but they exhibit an assortment of deviations from normal that describes Klinefelter syndrome. A Klinefelter male arises in about 1 out of 600 male births, a higher figure than that for Turner syndrome. These males typically show some breast development, small testes, sparse body hair, and some mental deficiency (Fig. 5-15). Klinefelter persons are almost always sterile. The Y chromosome in other mammals plays a role in maleness similar to that in humans. In cats, XXY animals occur and have been found to be the sterile equivalents of human Klinefelter males.

To appreciate the cytology of various clinical conditions in humans, it is important to have an acquaintance with the appearance of the normal human chromosome complement or normal *karyotype* (Chap. 2).

Many defects in humans are associated with certain specific chromosome abnormalities, such as variation in number and alteration in chromosome structure. Analysis of the chromosome complement is almost routinely performed today in cases of certain suspected conditions such as Turner and Klinefelter syndromes. Figures 5-14 and 5-15 show karyotypes typical of patients with these syndromes.

However, persons exhibiting Klinefelter syndrome have been identified with chromosome situations more complex than the XXY. Individuals of the following chromosome constitutions have been found: XXXY, XXXXY, XXYY, and XXXYY. On the average, the XXY male has a lower intelligence than the XY male. There is evidence that mental deficiency and body abnormalities are even more pronounced in those Klinefelter males with three and four X chromosomes.

Other kinds of atypical sex chromosome conditions have been found among humans. Surveys among newborns and adults indicate that the incidence of the XYY chromosome constitution in the general population is about 1 per 1000 males. XYY males are taller as a group than XY males, and many of them are below average in intelligence. Cytological studies conducted on inmates of prisons and mental institutions have revealed a significantly high proportion of men with an extra Y chromosome, particularly among those over 6 feet tall. Data from several institutions strongly suggest that the XYY male faces a greater risk of confinement in an institution than the XY male.

These findings have incited much public interest and led to many speculations, some of them rash, regarding the relationship between extra doses of genetic factors on the Y and aggressive or antisocial behavior. One suggestion has been that tallness calls attention to a person and may subject him to greater societal stresses, which in turn could trigger antisocial behavior. However, the equally tall XY male does not appear to be at any greater risk of displaying aggressive behavior than the average male. Moreover, an extra Y chromosome is present in some males who appear to be normally adjusted and who are indistinguishable from other males in the population. The XY brothers of institutionalized XYY men do not exhibit a significantly high tendency for aggression. Proper evaluation has yet to be made of the complex of environmental and biological factors that could trigger aggressive behavior. Until that time, the reasons for the apparently higher risk of confinement for the XYY male will remain unclear. It is clear, however, that

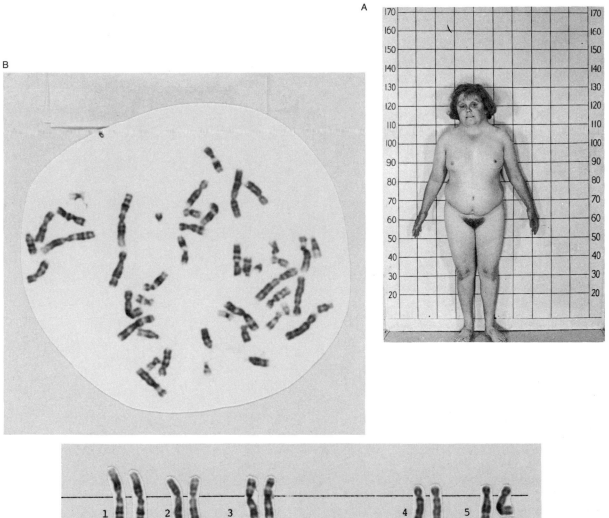

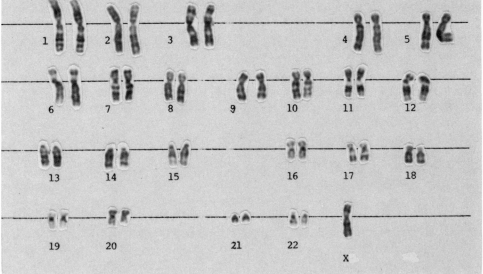

FIG. 5-14. The Turner syndrome. (*A*) Patient. (Reprinted with permission from V. A. McKusick, *Medical Genetics* 1958–1960. C. V. Mosby, St. Louis, 1961.) (*B*) Karyotype of a patient with Turner syndrome. The chromosomes are G banded. Note that only one sex chromosome, the X, is present. (Courtesy George I. Solish, MD, PhD; Jyoti Roy, MS, Thomas Mathews, MS, Long Island College Hospital, Cytogenetics Lab.)

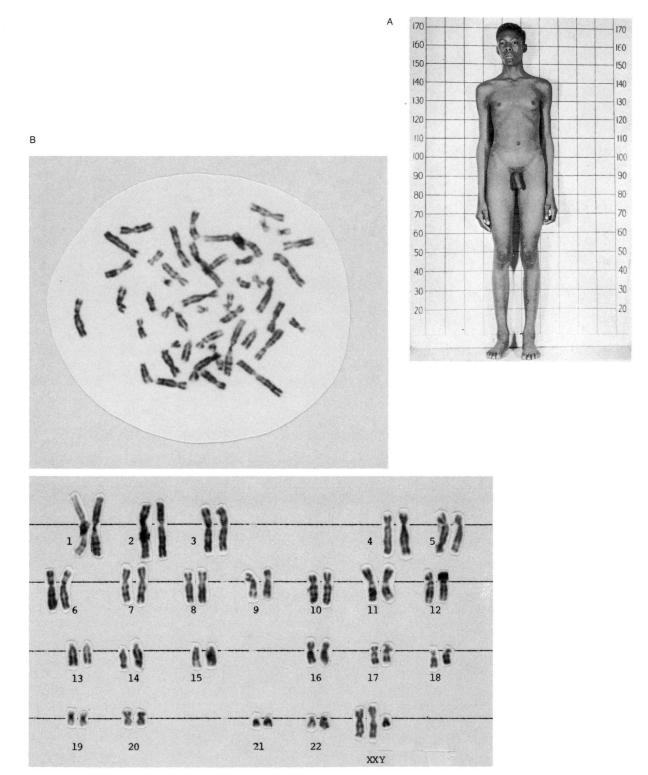

FIG. 5-15. (*A*) Patient showing features of Klinefelter syndrome. (Reprinted with permission from V. A. McKusick, *Medical Genetics* 1958–1960. C. V. Mosby, St. Louis, 1961). (*B*) Karyotype of a patient with the syndrome. The chromosomes are G banded. Note that two X chromosomes are present in addition to the Y chromosome, which is grouped with them. (Courtesy George I. Solish, MD, PhD; Jyoti Roy, MS, Thomas Mathews, MS, Long Island College Hospital, Cytogenetics Lab.)

XYY males are fertile and their offspring are usually typical XX females or XY males in chromosome constitution. Some factor at meiosis that remains unknown seems to prevent an extra Y from becoming part of the chromosome constitution of a functional sperm in the XYY male.

It should be appreciated at this point that the XO, XXY, XYY, and other unusual sex chromosome constitutions can be explained on the basis of nondisjunction at the first or second meiotic divisions in one of the parents. At zygonema of meiosis in the male, the X and Y chromosomes may form a pairing association referred to as the *sex bivalent*. The pairing does not affect the entire length of the chromosomes but appears to be confined to short segments in one arm of each (Fig. 5-16). This is a reflection of some homology in these segments of the X and Y chromosomes. It has now been established that crossing over regularly takes place in these short regions (more later on). The two chromosomes normally separate at first anaphase to produce two cells: one with the X, the other containing the Y (recall spermatogenesis, in Chap. 3). Each of these secondary spermatocytes in turn undergoes the second meiotic division. The final outcome is X and Y containing sperm in equal numbers (Fig. 5-17A).

Failure of separation of the two chromatids of the Y chromosome at the second meiotic division will produce a spermatid with two Y chromosomes and a spermatid with no sex chromosome at all. The former

FIG. 5-16. Meiotic chromosomes of human male at diakinesis. The tiny Y chromosome appears attached terminally to the much larger X. (Reprinted with permission from W. V. Brown, *Textbook of Cytogenetics.* C. V. Mosby; St. Louis, 1972. Courtesy of Dr. J. Melnyk, City of Hope Medical Center, Duarte, Calif.)

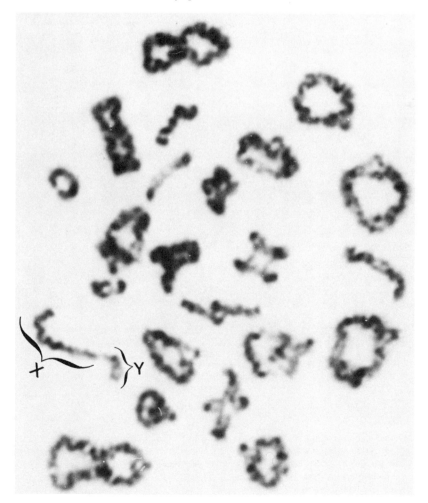

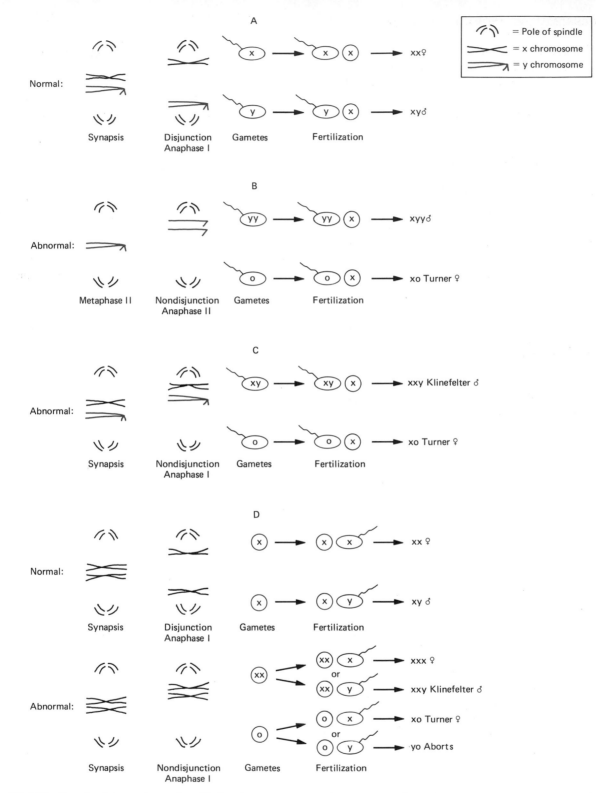

FIG. 5-17. Normal and abnormal separation of the sex chromosomes at meiosis. (A) In the male, normal disjunction of the X and Y chromosomes at first meiotic division, followed by a normal second meiosis, yields normal X-bearing and Y-bearing sperm. (B) Failure of separation of the two chromatids of the Y chromosome at second meiosis yields sperm of exceptional chromosome constitution, one with two Y chromosomes and the other with no sex chromosome at all. (C) Nondisjunction of the X and Y chromosomes at first meiotic division yields sperm of unusual chromosome constitution, those that contain both an X and a Y and those that contain no sex chromosome at all. (D) Nondisjunction of the two X chromosomes in the female may also occur, resulting in eggs of exceptional chromosome constitution.

is responsible for the production of the XYY males just discussed. The latter can be responsible for an XO person if it fertilizes a normal, X-bearing egg (Fig. 5-17B). Consider the effects of nondisjunction of the X and Y at first meiotic division (Fig. 5-17C). Sperm of the constitutions XY and O can be produced. If normal, X-bearing eggs are fertilized by such gametes, the result is XXY (Klinefelter) and XO (Turner). Nondisjunction can also involve a female parent and give rise to abnormal eggs of the constitutions XX and O (Fig. 5-17D). This is reminiscent of the situation first discovered by Bridges in Drosophila. Fertilization of the exceptional eggs by normal sperm can also produce zygotes that are unbalanced, among them XXY and XO types as well as XXX. The latter constitution occurs in 1 out of 1600 female births. Many such females have normal intelligence, but a slightly higher frequency of XXX persons is found in mental institutions. Triple-X women are fertile and usually give birth to offspring normal in chromosome constitution. As in the case of XYY males, some factor appears to operate in them at meiosis to prevent the extra chromosome from entering a functional gamete. Several XXXX women have been found, and these individuals suffer from mental deficiency.

Sex mosaics

The fact that individuals with only one Y chromosome but two, three, and even four Xs are still males, shows the importance of the Y in determining the secondary sex characteristics in humans. This is quite a contrast to the situation in Drosophila and is even more impressively illustrated by comparing certain other chromosome anomalies in the fruit fly and the human. In the fly, certain mishaps that affect the sex chromosomes occasionally take place during embryological development. At rare times during the mitotic divisions in some female embryos, one of the X chromosomes may lag and fail to reach the pole at anaphase (Fig. 5-18A). It may be left out in the cytoplasm where it disintegrates. The result is one cell with two sets of autosomes plus two Xs and another cell with two sets of autosomes plus one X. In Drosophila, this means that some cells will be genetically female and others

FIG. 5-18. Gynandromorph origin in Drosophila. (A) A zygote with two X chromosomes plus two sets of autosomes normally develops into a female. However, at mitotic divisions during embryology, one X chromosome at rare times may lag and be eliminated. The outcome is two cells, one with a nucleus balanced in the female direction and the other with a nucleus balanced in the male direction. If the mishap occurs at the first division of the zygote, a fly that is male on one side of its body and female on the other can arise. (B) A bilateral gynandromorph that is male on the right side of the body and female on the left. The eye color on the male side is white, because only one X is present, and this carries the recessive for white eye color (genotype wO). The X that was lost carried the dominant allele for red color. On the female side, both Xs are present, one with the dominant, the other with the recessive (genotype Ww), and so the eye is red. (From E. J. Gardner, *Principles of Genetics,* 4th ed. Wiley, New York 1972.)

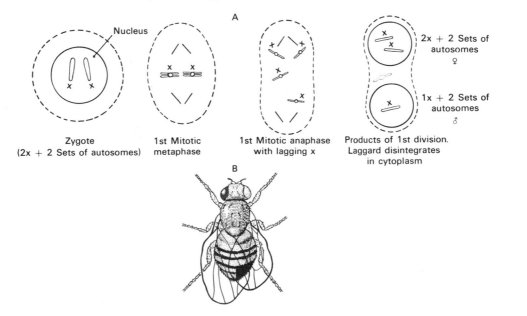

male. Flies such as these actually arise and are called *gynandromorphs,* mosaic individuals with discrete female and male body segments. If the loss of the X takes place at the first division of the zygote, the fly will be a bilateral gynandromorph, male on half of the body, female on the other (Fig. 5-18*B*). If the mishap occurs later, smaller patches of male tissue will be present among the female background.

Mosaic persons, those who are mixtures of different cell lines, are well known. Actually about 15% of persons with Turner syndrome prove to be mosaics, persons with two distinct cell lines. The karyotype of some of these persons can be represented as X/XX, since one population of cells contains only one X chromosome whereas the other has the normal two. Other persons have the karyotype X/XXX, one of the cell lines with three X chromosomes, the other line with only one. The mosaic individuals differ in their phenotypes. This is to be expected, because the mitotic accident resulting in the loss of one X chromosome from one cell line can occur at any time during prenatal development. The earlier the accident, the greater the number of XO cells and the closer the phenotype to that typical of Turner syndrome.

It must be pointed out that these mosaic persons are *not* gynandromorphs. No discrete patches of definite male or definite female tissues are seen in any kind of human mosaic. The mosaics may show a range of abnormal phenotypes, but hormonal activity in mammals will not allow the development in any one of them of two distinct classes of cells, those with typical female features intermingled with those that are obviously male.

Departures from the normal number of chromosomes because of nondisjunction and other mishaps can affect autosomes as well as sex chromosomes. Moreover, any chromosome may become altered in its structure. These and other chromosome anomalies are pursued further in Chapter 10.

Sex chromosomes and sex determination

By no means are all factors involved in the differentiation of a testis or an ovary in the mammal known. However, certain key points have been elucidated. In the human embryo, the gonads are first represented by the gonadal ridges, which appear about 30 days after the formation of the zygote. These are capable of developing into either testes or ovaries. Thickenings appear in the gonadal ridges, which indicate the po-

sitions of the Wolffian and Mullerian duct systems. The former have the potential to develop into the male duct system, the latter into the female duct system. Normal differentiation is determined by the sex chromosome constitution of the cells of the bipotential primitive gonad.

The Y chromosome is essential to the development of a testis. In its absence, no testes will develop, regardless of the number of X chromosomes present. A testicular-organizing substance is believed to exist, and much research is being directed toward its recognition. If a Y *is* present, testes will develop, *even* in persons of unusual karyotypes (XXY, XXXY, XXYY, XXXXY). The testis of the fetus secretes substances critical to normal male development. Among these is one that inhibits the development of the Mullerian system, which otherwise will evolve into Fallopian tubes and uterus. Testosterone, the major male hormone, then guides the Wolffian duct system to form the prostate, seminal vesicles, and tubes of the male system. Testosterone also organizes the development of the penis and scrotum (Fig. 5-19*A*).

In the absence of a Y chromosome, as in the XX condition, no testis differentiation can be triggered, and the Wolffian duct system regresses. Since no Mullerian inhibitor is present, the Mullerian ducts are free to develop their potential in the female pathway (Fig. 5-19*B*). The primitive gonad in the XX individual differentiates into an ovary. Oogonia develop in the cortex of the gonads, and primary oocytes will be present before birth. An ovarian-organizing substance has also been proposed, which presumably is responsible for normal ovarian differentiation.

The X-Y mechanism of sex determination in mammals undoubtedly evolved through the pressure of natural selection, to guarantee equal numbers of males and females. However, the human sex ratio at birth is slightly in favor of males: 106 boys to every 100 girls. The precise explanation for this is still a matter of debate. According to one theory, the Y chromosome is somewhat lighter than the much larger X, and thus a sperm carrying a Y chromosome can travel faster than one carrying an X and has a greater chance of encountering the egg. Another suggestion arrived at from studies of karyotypes of spontaneous abortuses is that the true sex ratio is 1:1, equal numbers of XX and XY zygotes being produced, and that more females are lost in utero. Regardless of the correct explanation for the slight excess of males at birth, the ratio of males to females is about 1:1 at the age of sexual maturity.

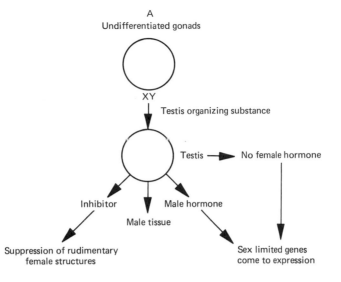

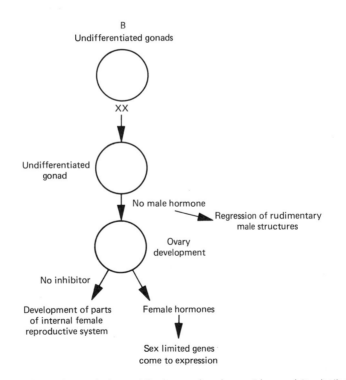

FIG. 5-19. Summary of the interaction of genetic constitution and the hormonal environment in sex determination. (*A*) Development in the male direction. (*B*) Development in the female direction.

The Y chromosome and the H-Y antigen

Any gene that occurs exclusively on the Y chromosome is said to be *holandric*. Such a Y-linked gene normally occurs only in males and is transmitted only from father to son. In Drosophila, we have learned that genes that affect sperm motility are holandric but that the Y chromosome in the fruit fly is not necessary for secondary male characteristics. In the mammal, however, it appears that the only genes on

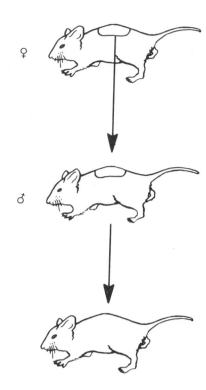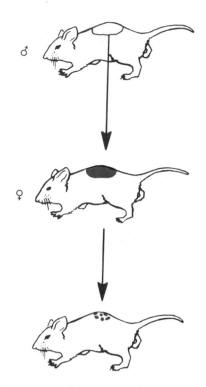

FIG. 5-20. Tissue transplants in isogenic strains of mice. Skin taken from a female (*left*) can be transplanted to a male of the same strain with acceptance of the transplant. Skin from a male, however (*right*), transplanted to a female of the same strain, is eventually rejected.

the Y are those that are necessary for the development of a testis. The testis must be formed early and, in turn, the hormones required for masculinity must be produced by the testes very early in development.

A very interesting discovery made in the mouse has implications for the existence of a testicular organizing substance. Mice can be inbred for so many generations that isogenic strains can be developed. These are populations within which all the individuals have the same genotype. Nevertheless, in an isogenic strain, a genetic difference still exists between males and females, since the two sexes differ in respect to the sex chromosomes. Any male would possess all the alleles found in a female of the strain. However, any genes on the Y would be present only in the male. It was demonstrated in 1955 that skin could be successfully transplanted from female mice to males of the same isogenic strain; the reverse did not hold, however. Male skin transplanted to females from the same highly inbred strain is eventually rejected (Fig. 5-20). It appears that the male tissue must possess certain antigens that cause females to respond by the production of antibodies when male tissue is grafted onto them.

Support for this concept was obtained when it was shown that serum from female mice, which had already rejected male tissue, contained antiserum that could kill male cells! It was concluded that the Y chromosome contained a gene or genes responsible for the antigenic determiners on the male tissue. The Y-linked region or locus was named H-Y. The term *H-Y antigen* was given to the transplantation antigen on the male tissue. This antigen has been found on all normal male tissue of the human and other mature male mammals so far tested. Men who have an XYY genetic constitution were found to possess more H-Y antigen than XY males.

It was this finding that led to the proposal that testis-determining genes are actually on the Y and that the H-Y antigen in mammals is the male-determining substance required for testis organization. Therefore, any genetic factors on the Y chromosome responsible for H-Y antigen would be responsible for the organization of the testes and hence determination of maleness.

However, more recent studies have caused us to reexamine ideas concerning the H-Y antigen. A strain of male mice has been discovered with two X chromosomes into one of which a piece of the Y has been inserted. Such animals develop testes, but they fail to produce both sperm and the H-Y antigen. Moreover, sterile XX men are now known who, as in the case of the animals, carry a portion of the Y in one of the two X chromosomes. In addition, XY females have been found who carry a deletion in a portion of the Y. All the XX men have now been shown to carry in one X chromosome a piece of the Y which is normally located on the *short* arm of the Y. In the XY females, this same small region has been deleted from the Y. These observations argue against the H-Y antigen as the testis-determining substance, because it has now been demonstrated utilizing techniques of molecular biology that the factor or factors responsible for the H-Y antigen normally reside on the *long* arm of the Y, far from the region on the short arm which is associated with testis formation. Whereas the actual role of the H-Y antigen in determination of maleness is still not clear, some investigators believe that it may be required for normal spermatogenesis, because individuals lacking the long arm of the Y chromosome produce no sperm.

Still another factor must be considered in normal sex development. This is illustrated dramatically in rare cases of certain XY individuals who are often reared as females. Such persons typically lack ovaries and the normal female parts derived from the Mullerian duct system. Female breast development occurs at puberty. External female genitalia may be present, or the genitalia may be ambiguous; however, examination shows that testes are present. At puberty, some persons actually virilize in response to hormones after having been raised as females. The defect in these cases, known as testicular feminization, appears to be due to the inability during fetal development of the appropriate tissue to respond to male hormone and thus develop properly into organs typical of the normal male. This is apparently a result of the failure of receptor protein in target cells to bind the male hormone, thus preventing development of organs in the normal male direction. Testicular feminization, which also occurs in the mouse, is believed to be caused by a sex-linked recessive allele, which can be transmitted by normal XX females to their male offspring. Since males with the condition are sterile, the responsible genetic factor cannot be transmitted by them.

Cases are known of XY females who are H-Y antigen positive but lack testis development. These cases have been explained by the proposal that target gonadal cells, which normally respond to H-Y antigen, lack normal receptor protein. Consequently, in the absence of a normal response to the H-Y antigen, a testis cannot form, and the H-Y positive individuals develop in the female direction.

Sex chromatin

For many years, biologists wondered about the fact that in mammals the female carries a double dose of all the genes on the X while males carry a single dose. This might mean that a female homozygous for the dominant allele for the antihemophilia factor has twice as much of the AHF as a normal male. The idea is disturbing when considered in relation to the concept of genic balance. In contrast to the X-linked loci, all autosomal loci are represented twice in both sexes. It seemed to geneticists that something must compensate for the difference in dosage of the sex-linked genes to preserve genic balance.

An answer to this problem dates back to 1949 with the discovery by Barr of a nuclear body found in the neurons of the female cat. This small intranuclear entity was completely absent from neurons of the male (Fig. 5-21). Examination of other tissues and other species showed that a constant distinction can be made between nuclei of cells taken from females and from males. A simple way to demonstrate this in the human is by scraping epithelium from the buccal mucosa and staining it with some common dye. The nuclei of most of the cells from a female show a small rod-shaped structure, more deeply stained than the surrounding chromatin and usually located at the

FIG. 5-21. Barr body in neuron of the female cat (×1050). This body, which is very evident in the nucleus of the female, is often found in the vicinity of the nucleolus (the large intranuclear body). It is entirely absent from the nucleus of the male.

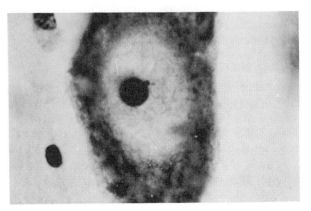

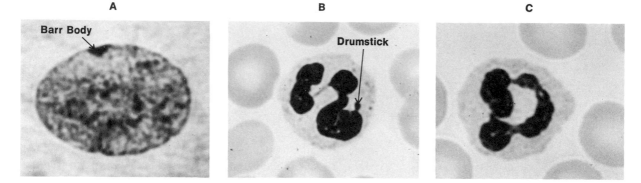

FIG. 5-22. Sex chromatin in humans. (*A*) Barr body in nucleus of a squamous epithelial cell from the buccal mucosa of a female. (*B*) The drumstick in a polymorphonuclear leukocyte of a female. (*C*) Poly-

morphonuclear leucocyte of a male. (Courtesy of Carolina Biological Supply Company.)

periphery of the nucleus (Fig. 5-22*A*). Comparable cells from a normal male show no such body, which has been designated the *Barr body* or the *sex chromatin*

One of the many other tissues that show sex distinctions is the blood, specifically the polymorphonuclear leukocytes. In some of these white blood cells taken from a female, a "drumstick" can be recognized (Fig. 5-22*B*). This is seen as a small appendage attached by a filament to one of the lobes of the nucleus. Although they can be identified in only about 2% of the leukocytes from a female, drumsticks are entirely absent from cells derived from a male (Fig. 5-22*C*).

The discovery was then made that drumsticks and Barr bodies are absent from females with Turner syndrome. On the other hand, cells from XXY males contain one Barr body or one drumstick; males who are XXXY and XXXXY show two and three, respectively. Cases were found of women with multiple X conditions: XXX and XXXX. These persons were found to have two and three sex chromatin bodies, respectively. It became apparent that the number of sex chromatin bodies and drumsticks was one less than the number of Xs present in the cells of the individual.

The X chromosome and the Lyon hypothesis

Such observations led to the formulation of a theory that accounts for the relationships between the number of sex chromatin bodies and the number of Xs. More important, it suggests a mechanism that compensates for the presence of different dosages of X-linked genes in males and females. Dr. Mary Lyon has presented some of the best arguments for the

concept. According to the Lyon hypothesis, all normal females are sex mosaics, composed of different cell lines insofar as the X chromosomes are concerned. Only one X is functioning entirely in a given cell; the other is present largely in an inactive state. Which specific X is active in a cell—the one from the maternal or the one from the paternal parent—is a matter of chance. Any female has the X she received from her mother functioning in some cells and the one from her father in others. Which X will be the active one in a cell is believed to be decided during embryological development.

Several lines of evidence support the Lyon hypothesis. It will be recalled from the discussions of mitosis and meiosis that chromosomes in a compact state, as at the time of nuclear divisions, actually do not participate in cellular activities other than those concerned with the division process itself. Evidence from the molecular level has shown that DNA in its compact state is relatively inactive in directing the synthesis of other molecules. Cells can be taken from human females and maintained for study in tissue culture. Observations have revealed that at early mitosis one of the two X chromosomes in any cell tends to be more compact and to stain more deeply than the other. It also tends to undergo synthesis of its DNA later. All in all, the behavior of one of the Xs at mitosis has been shown to be distinctly different from the other members of the chromosome set, and some of the features it displays are typical of chromatin, which is relatively inactive.

Moreover, sex chromatin is not present in oocytes. During prophase of oogenesis, both X chromosomes appear identical and react the same way to stains.

Both are evidently equally active at this time. During first meiotic prophase, they pair and engage in crossing over. They continue to behave in the same fashion even in the earlier divisions of the embryo. However, at approximately 16 days, in the late blastocyst stage, sex chromatin can be recognized. These observations strongly support the Lyon hypothesis because they indicate that both Xs are equally functional during meiosis and in the cells of the very early embryo. But then at the blastocyst stage, some mechanism becomes operative in the body cells that keeps one of the Xs (or a large part of one) in a permanently inactive state by condensing the chromatin. The Barr body is a visual manifestation of this. However, as indicated by oogenesis, both Xs remain equally active in the germ line.

In Chapter 2, we encountered examples of chromatin which always remains condensed, the constitutive heterochromatin. We see here in the female that while an X chromosome may become condensed and behave as heterochromatin, the condensation occurs only in somatic cells. Moreover, the X that *does* become inactivated is not the same one in every cell; it is the paternally derived in some cells and the maternally derived one in others. This means that the same kind of chromatin may act as heterochromatin in certain cells and as euchromatin in others. The term *facultative heterochromatin* is used to distinguish the kind of heterochromatin exemplified by the X chromosomes from the constitutive heterochromatin, which is constant from cell to cell.

Barr bodies and drumsticks show various parallels that indicate they are both visible manifestations of the same phenomenon, a compacted X, which is thus in an inactive state. The fact that sex chromatin is not seen in every cell of a female (drumsticks in only about 2% of the polymorphonuclear leukocytes; Barr bodies in 90% of the buccal cells) is considered by many to result from its orientation in a given nucleus. The extreme lobing of the polymorphonuclear leukocyte nucleus could affect visualization of the drumstick.

Further evidence for the Lyon hypothesis

Excellent support for the Lyon hypothesis comes from chemical studies of individuals known to be of different genotypes for sex-linked genes, such as the one for hemophilia. For example, normal homozygous women (HH) and normal men (HY) have the *same* amount of antihemophilia protein in the blood, even though the gene dosage is different. The female carrier (Hh) is almost always without symptoms of the disorder, but blood analysis shows the concentration of antihemophilia protein varies from one carrier to the next. A few of them have about the same amount of the clotting factor as does any normal male or homozygous female, but in a few the essential substance is very low. Most carriers have about half that of the average person. This is precisely what one would expect on the basis of the Lyon hypothesis, which predicts that which X chromosome remains inactive in a given cell is determined by chance. Consequently, in a carrier female (Hh), approximately half of the cells, on an average, should have the maternal X with the normal allele operating (Fig. 5-23). In the rest of the cells, the paternal X with the defective allele is being expressed, and so only half the normal amount of the clotting factor is produced. In a small number of cases, however, by chance the majority of cells have the maternal chromosome operating. An almost normal level of the antihemophilia protein is present as a result. Likewise, in a few carriers the paternal X is active in most of the cells, by chance, and consequently a low level of the factor is present and some symptoms of hemophilia may be manifest.

Another product associated with a gene on the X chromosome is glucose-6-phosphate dehydrogenase, an enzyme that occurs in all kinds of tissues, including blood. The effect of a deficiency of this enzyme is not usually of any consequence. However, following the ingestion of certain drugs and foods, a severe anemic response can be triggered. Females who are heterozygous at the enzyme locus may possess an allele for the normal form of the enzyme and a recessive allele for a variant form. Skin cells may be taken from such a heterozygote and grown in tissue culture. It can be demonstrated that the cell population is a mixture of two cell types, one that produces the normal form of the enzyme and one that produces the variant form.

Single cells can be isolated from a cell population in tissue culture, and any cell can then be allowed to divide to produce a *clone,* a population of genetically identical cells. In the case of the heterozygous female, two types of clones can be established: those from which only normal enzyme can be isolated and those from which only the variant form can be retrieved (Fig. 5-24). Cultures of cells taken from any given male, however, show that no such mixture of cell types is present. Only one kind of cell population can

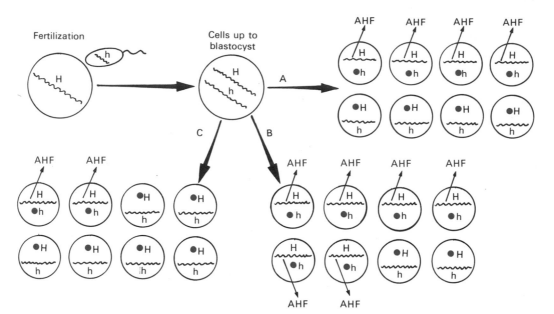

FIG. 5-23. X inactivation and the AHF. A female carrier of hemophilia may arise if a sperm carrying an X with the recessive unites with an egg carrying an X with the dominant allele responsible for the production of AHF. Both X chromosomes are active in each cell up to the time of the blastocyst, when one of them becomes inactive (*arrow A*). The inactive X may be visualized as condensed chromatin, a Barr body, or sex chromatin. It is chance whether the maternally derived X or the paternally derived one remains active. On the average, a carrier will have half of her cells with an active X carrying the dominant, and so half the normal amount of AHF will be present. In a smaller number of cases (*arrow B*), the maternal X will be active in most cells, and the amount of AHF will approach that of a normal homozygote or of a normal male. Conversely (*arrow C*), in a few carriers the paternal X will be active in most cells, and the amount of AHF will be much lower than the normal.

be demonstrated in which all the cells produce only one form of the enzyme.

Many other examples can be given in the human for other X-linked loci, which, in each case, are also associated with a particular product. The story is similar to the one just described, and all therefore support the premises of the Lyon hypothesis. A final example from a natural hybrid gives even more direct evidence for two aspects of the Lyon hypothesis. A female mule is a hybrid resulting from the cross of a horse and a donkey. The X chromosome of the horse has a median centromere and is quite readily distinguished from the X of the donkey, which has its centromere close to one end. Moreover, the enzyme glucose-6-phosphate dehydrogenase (G6PD) from the horse can be distinguished electrophoretically (see Chap. 18) from that of the donkey. Cells can be isolated from the female mule and raised in culture from which two types of clones can be isolated. In one kind it can be demonstrated that the donkey X is late replicating (and thus is the inactivated X). In such a clone, only the horse G6PD is present. In a clone that produces only donkey G6PD, the X chro-

mosome of the horse is late replicating and inactive. We see in this example that not only are two populations of cells present in the female, but the presence or absence of a given product is associated with the inactivation of a specific chromosome.

The precise mechanism that causes inactivation of one X chromosome, and thus compensates for the different dosages of X-linked alleles in males and females, in unknown. However, it has undoubtedly been perfected by natural selection along with the evolution of the X and Y chromosomes, and it ensures genic balance in both sexes. According to Ohno, the X-inactivation mechanism has preserved the ancestral X chromosome almost intact throughout the evolution of mammals. While the Y has become miniaturized, the evolution of the X has been very conservative. Several good arguments support Ohno's law of the conservation of the X chromosome. For example, the X of all mammals is about the same, a good-sized chromosome making up about 5% of the genome. Even the X of the kangaroo is similar to that of placental mammals. The X of the gibbon appears to be very similar to that of the great apes and the

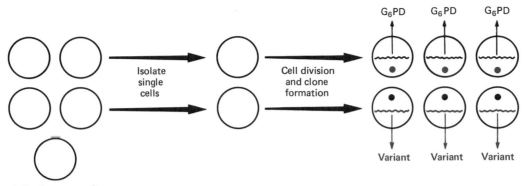

FIG. 5-24. Glucose-6-phosphate dehydrogenase (G6PD) and skin cells. Single cells of carriers of the sex-linked recessive (g) for a variant form of the enzyme can be isolated. Single-cell isolates can then be allowed to divide to form clones. Any one clone of cells either produces the normal enzyme form or lacks the normal enzyme completely. This indicates that the X that carries the allele for the normal enzyme form (G) is active in the cells of clones producing the normal enzyme and that the X carrying the allele for the variant form is inactive. Conversely, a clone in which the variant allele is active contains only the variant form of the enzyme. The X with the standard allele is inactive in these cells.

human, although the autosomes of the gibbon have experienced much evolutionary change. No exception has been found in mammals regarding genes that are sex linked. Demonstration of sex linkage for a given gene in any other mammal is a strong indication that the comparable gene will be sex linked in the human. In spite of the vast amount of speciation and chromosome change that has occurred in mammals in more than 2 million years, changes in the X chromosome appear to have been highly conservative.

As a result of deactivation of one X chrosome in the female, both sexes are, in effect, hemizygous in any one cell for genes on the X. It should be noted, however, that the deactivation of one X in the XX female must be partial and not entire. As an example, a certain X-linked locus (the Xg locus) responsible for the Xg blood types does not undergo deactivation. Another locus, associated with the production of an enzyme required to prevent an abnormal skin condition, is closely linked to the Xg locus and apparently is not deactivated. This region of the X may remain active on both Xs.

Another pertinent fact is that the XX female is quite distinct from the Turner female, who is truly hemizygous for genes on the X. Moreover, it is known that two X chromosomes are essential for female fertility. Infertility results in cases in which just a portion of one X chromosome is missing. The observations indicate that at least some segment of both X chromosomes remains active in XX females, and the devices that keep large parts of one X permanently inactive nevertheless permit some genetic material to remain active in that same X chromosome.

Incomplete sex linkage

In addition to X-linked genes, others must be considered in relation to the sex chromosomes. We have defined sex-linked genes as those found only on the X chromosome. We have also noted that the X and the Y chromosomes pair at first meiotic division, which implies that some homologous segments may be present in each. In the fruit fly, a locus named *bobbed,* which affects the length of the brisltes, is located at the end of the X chromosome near the centromere. A homologous locus, however, also occurs on the Y. This means that a female fly has two bobbed loci present (one locus on each X), and so does the male (one locus on the X, the other on the Y). Thus, a male, like a female, can be heterozygous or homozygous at this locus. The wild allele (+) is needed for long bristles; its recessive allele bb can result in short ones. Alleles such as these are termed *incompletely sex linked,* meaning they occur on both the X and the Y, in a region of homology between the X and the Y chromosomes. This is in contrast with sex-linked alleles, which occur only on the X. The cross diagrammed in Fig. 5-25 brings out some of the unique features of a trait that is inherited on the basis of an incompletely sex-linked gene.

Short regions of homology on the X and Y chromosomes of the human and other mammals are known

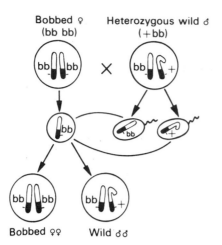

Bobbed ♀   Heterozygous wild ♂
(bb bb)   (+bb)

Bobbed ♀♀   Wild ♂♂

FIG. 5-25. Incomplete sex linkage in Drosophila. The locus "bobbed," which affects bristle length, is found on both the X and the Y chromosomes. In a cross between a heterozygous male, carrying the dominant wild allele on the Y, and a female homozygous for the recessive allele, the normal trait is passed from father to sons. All of the daughters will show the mutant phenotype. This is in contrast with a sex-linked gene. If the locus were on the X chromosome alone and not on the Y, a crisscross pattern would occur.

to exist, and these are located in the p (short) arms of the two sex chromosomes. A gene encoded for a cell surface antigen has now been found on the short arms of both the X and the Y. However, the absence of extensive regions of homology between the X and the Y would appear to be advantageous and to preserve the genes on the X as a single unit as well as those on the Y as a different unit. If the X and Y did possess extensive regions of homology, crossing over could occur freely between the two of them at meiosis. As a result of exchange of genetic segments along the length of the two sex chromosomes in the male, genes for maleness and those for femaleness could become distributed on both the X and the Y, and this would interfere with clear-cut segregation of the genes for sex determination. It seems likely that during the evolution of the mammalian sex-determining mechanism, natural selection favored any changes which made the X and Y chromosomes less homologous. Much of that homology which does remain between the X and the Y may permit the pairing which is seen at first meiotic prophase of spermatogenesis and may be necessary for the orderly behavior of the sex chromosomes at first metaphase and their separation at first anaphase. Crossing over between the X and the Y could be limited further by the condensation of the

X which appears heteropycnotic at the early meiotic stages.

It has now been demonstrated that at meiosis in the human male, the short regions of homology on the X and Y do regularly engage in crossing over. Using techniques of molecular biology (Chaps. 18–19), it has also been shown that most XX males carry on the X chromosome derived from their fathers a portion of the short arm of the Y which bears testis-determining factors. It is suggested that abnormal transfer to the X chromosomes of genetic material typically confined to the Y is related to the normal crossover events which take place between the short regions of homology on the sex chromosomes. The chromosome segment with the factor (or factors) for testis determination does not lie within this homologous region, but it is apparently close to it and may at times be transferred from Y to X during recombination. Such abnormal transfer accompanying a normal crossover event could account for the relatively high frequency of XX males among male births (1 in 20,000).

Sex-influenced alleles

Many species are known in which the manner of expression of certain alleles is affected by the sex of the carrier. Such factors are not necessarily located on the sex chromosomes; most of them are actually autosomal. Among these are various alleles whose expression of dominance depends on the sex of the individual. For example, in certain breeds of spotted cattle, the colored regions on the body may be red or mahogany, a deep reddish brown. When a mahogany female is crossed with a red male (Fig. 5-26A), the results are red-spotted females and mahogany-spotted males. This crisscross pattern suggests a sex-linked recessive in which the female is double recessive and the male hemizygous.

If the red females are mated to their mahogany brothers, however, we see that this is not so. Among the male offspring, mahogany and red occur in a ratio of 3:1. The two phenotypes also show in the females, but the ratio is the reverse: three red to one mahogany. We can recognize the characteristic monohybrid ratio within the females as a group and within the males as a group. The results suggest that the allele for mahogany (M) is dominant in the male and recessive in the female, whereas its allele for red (m) is

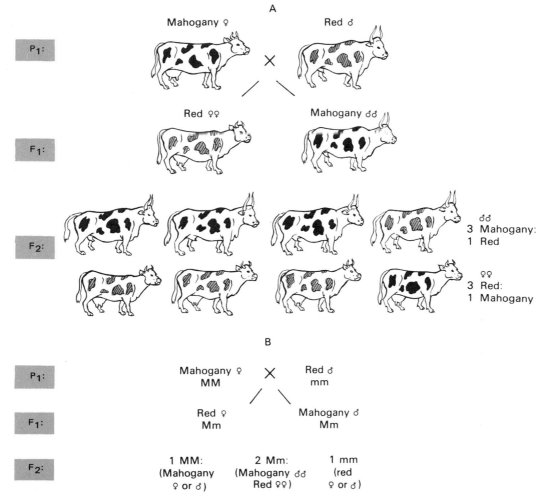

FIG. 5-26. Sex-influenced inheritance in cattle. (*A*) A cross of a mahogany-spotted female and a red-spotted male produces an F_1 of red-spotted females and mahogany-spotted males. The F_2 results indicate the locus involved is not sex linked but that expression of the genes for coat color is influenced by an individual's sex. (*B*) The factor for mahogany (M) behaves as a dominant in males and a recessive in females. Its allele for red color is dominant in females and recessive in males. Therefore, heterozygotes (Mm) will differ in color depending on their sex.

dominant in females and recessive in males (Fig. 5-26*B*). We could just as well have written M for red and m for mahogany. It makes no difference as long as we keep in mind the distinction in dominance between the sexes. Therefore, the heterozygote Mm, will be mahogany if male and red if female.

Alleles like these, whose degree of dominance is determined by the sex of the individual carrying them, are called *sex influenced*. Many other such alleles are known in animals. One is the presence or absence of horns in some breeds of sheep, in which the horned condition behaves as a dominant in males but as a recessive in females; the hornless state is dominant in the female sex but recessive in the male.

A trait that is close to most of us—pattern baldness in humans—seems to be sex influenced. A pair of alleles seems to be involved, B for bald and b for nonbald. Simple monohybrid inheritance occurs, but the allele for baldness (B) behaves as a dominant in males and a recessive in females. Therefore, a person of the genotype Bb is bald if male but nonbald if female. The alleles at the pattern baldness locus interact with other genes that affect the presence of hair on the head. The particular pattern in which the hair is lost, as well as the time of onset, entails the interaction of many genes and the environment. In the last chapter, we encountered several examples of the importance of the environmnent on the expression of the genotype. We might expect the internal environment to exert effects that are as pronounced as those outside the body. We see evidence for this in the case of baldness in women. A woman may become bald if she receives the bald-determining allele, B, from both parents. However, the onset of hair loss in a woman of genotype BB does not occur until later in life because of the effects of the internal hormonal environment.

The dominance difference typical of sex-influenced alleles is mainly the result of hormonal interaction with the genotype. The male hormone is responsible for the expression of the bald-determining allele B when it is present in just a single dose. The involvement of the hormone is clearly shown in cases in which males must take testosterone for some medical reason. This raises the level of the hormone and can result in hair loss in a man who had not yet expressed the trait. Conversely, therapy may prescribe the taking of female hormone by men, with a consequent decrease in the amount of male hormone. In these cases, a balding man may start to resume hair growth. However, the increase in breast size that typically ensues may discourage ingesting of female hormone as a device to grow a good crop of hair.

External environmental factors can also cause baldness: scalp infections and exposure to radiation or certain chemicals. These agents can mimic the effects of genetic baldness to the extent that the individuals are phenocopies, even though the allele for baldness is absent from the genotype. However, the phenocopy may show good hair growth after therapy or removal of the causative agent. On the other hand, the allele B cannot be removed from the man whose baldness stems from his genetic constitution. He will not be able to grow hair on his head as long as his male hormone is there to interact with the genes he has inherited.

Sex-limited genes and the hormonal environment

The effects of the internal environment are perhaps most evident in the case of *sex-limited genes*, so called because their expression is confined or limited to just one sex. Among these factors are those responsible for the obvious secondary sex characteristics. Recall

that the XY condition can trip the development of the rudimentary gonads of the embryo in the male direction, whereas the XX genotype permits ovaries to develop. Once the gonads, male or female, mature they produce their respective hormones. These in turn interact with the genotype, not only to influence the degree of dominance of certain alleles, but also to suppress completely the expression of many others. When we think of secondary sex characteristics, such as breast development in women, it rarely occurs to us that genes for this characteristic have come from both parents, male as well as female. Similarly, the kind of beard a male develops is determined equally by inheritance from his mother and from his father. This means that both sexes carry genes for breast development and for beard growth. That this is indeed the case is illustrated by many clinical conditions. The bearded lady is one of the most familiar examples. The genetic potential for beard development lies dormant in women and requires the presence of the male hormone to trigger it to expression. The adrenal gland normally produces some male hormone in both sexes. An adrenal tumor can cause an excess production of male hormone in a woman and allow the genes she carries for beard growth to express themselves.

We have already noted that the taking of estrogen by men can cause breast development, a phenotype usually limited in expression to females. However, it is not only the presence of a specific hormone that is required for expression of sex-limited genes, the absence of a hormone permits some genes to be expressed. Although the female hormone is needed for breast enlargement, it is not required for female voice. The latter will develop in the absence of the male hormone. If a male is castrated before puberty, thus removing the main source of male hormone, he will develop neither a deep male voice nor a beard. Both of these attributes require the hormone. Today we read of cases of persons who have undergone "sex changes" after surgery. A man who undergoes such a transformation remains genetically male. Having only one X chromosome, such a person can never show a Barr body or a drumstick, because there is no extra X to be deactivated. The individual will be a castrate, but beard growth has been triggered, as has the development of the Adam's apple, also a sex-limited trait. These can be altered only by surgical and cosmetic means. The voice will remain masculine. Breast development will require continual injections of estrogen. The female-to-male transformation

FIG. 5-27. Sex-limited inheritance in the fowl. The dominant H causes the male to have hen feathering. Its recessive allele h is responsible for cock feathering in males. Females, however, will not express the allele h. They will be hen feathered regardless of the presence or absence of allele h.

also requires surgical procedures. Male hormones must be administered to achieve beard growth and the expression of other male secondary sex characteristics. However, the XX chromosome constitution remains unaltered and sex chromatin continues to be present in the body cells.

Some of the most striking examples of sexual dimorphism are found among various groups of birds. The male often appears so different from the female that the novice bird watcher may classify them in separate species. Sex-limited genes are the basis for these striking distinctions. The common fowl is a group that exhibits variation in its plumage pattern. The hen-feathered plumage is quite distinct from the cock-feathered phenotype. Both hen and cock feathering may occur in roosters, but only the hen-feathered phenotype is expressed in females (Fig. 5-27). The allele for hen feathering (H) acts as a dominant in males, whereas its allele for cock feathering (h) behaves as a recessive. But the allele h cannot be expressed at all in females, which are always hen feathered regardless of the genotype.

Experiments with birds have provided a great deal of information on hormonal interaction with the genotype. For example, if the gonads are removed from male or female fowl, so that a young bird is deprived of either sex hormone, the genotypes HH and Hh are expressed phenotypically as cock feathering in both sexes! This indicates that the factor H prevents cock feathering if either the male or the female sex hormone is present. Its recessive allele, h, when homozygous, allows the cock-feathering pattern to develop, but the female hormone does not permit the expression of that phenotype. Therefore, two birds of the genotype hh may show different kinds of plumage. It is expressed phenotypically as cock feathering in the absence of female hormone but gives hen feathering in the presence of the female hormone. Once such interactions between the genotype and the internal environment are appreciated, certain practices in animal breeding no longer seem strange. For example, it is necessary for a cattle breeder to consider the genes for milk production carried by his bulls when he is interested in producing cows with a high milk yield.

We have encountered evidence in this chapter that natural selection has operated for countless years in sexual species to perfect a mechanism that ensures the formation of equal numbers of males and females.

We have also seen that the potential set by a normal genotype for the formation of a fertile individual showing typical secondary sex characteristics can be realized only in the setting of a conducive environment, internal as well as external. Sex is but one of the many phenotypic characteristics of an organism that depends for its normal expression on both the entire genotype and the background in which this genotype is allowed to express itself.

REFERENCES

Anderson, M., D. C. Page, and A. de la Chapelle. Chromosome Y-specific DNA is transferred to the short art of X chromosome in XX males. *Science* 233: 786, 1986.

Cervenka, J., D. E. Jacobson, and R. J. Gorlin. Fluorescing structures of human metaphase chromosomes. Detection of "Y body." *Am. J. Hum. Genet.* 23: 317, 1971.

Epstein, C. J., S. Smith, B. Travis, and G. Tucker. Both X chromosomes function before visible X chromosome inactivation in female mouse embryos. *Nature* 274: 500, 1978.

Goodfellow, P. J., S. M. Darling, N. S. Thomas, and P. N. Goodfellow. A pseudoautomsomal gene in man. *Science* 234: 740, 1986.

Gordon, J. W. and F. H. Ruddle. Mammalian gonadal determination and gametogenesis. *Science* 211:1265, 1981.

Haseltine, F. P. and S. Ohno. Mechanisms of gonadal differentiation. *Science* 211: 1272, 1981.

Kolata, G. Maleness pinpointed on Y chromosome. *Science* 234:1076, 1986.

Koo, G. C., S. S. Wachtel, P. Saenger, M. I. New, H. Dosik, A. D. Amarose, E. Dorus, and V. Ventruto. H-Y antigen: expression in human subjects with testicular feminization syndrome. *Science* 196:655, 1977.

Lawn, R. M. and G. A. Vehar. The molecular genetics of hemophilia. *Sci. Am.* (March): 48, 1986.

Lyon, M. F. Possible mechanism of X chromosome inactivation. *Nature New Biol.* 232: 229, 1971.

Martin, G. R. X-chromosome inactivation in mammals. *Cell* 29: 721, 1982.

McKusick, V. A. The royal hemophilia. *Sci. Am.* (Aug): 88, 1965.

Ohno, S. Evolution of sex chromosomes in mammals. *Ann. Rev. Genet.* 3: 495, 1969.

Rushton, W. A. H. Visual pigments and colorblindness. *Sci. Am.* (March): 64, 1975.

Wachtel, S. S. H-Y antigen and the genetics of sex determination. *Science* 198: 797, 1977.

REVIEW QUESTIONS

1. In the human, the autosomally linked allele A is required for normal skin pigmentation. Its recessive allele is associated with the albino condition a. Several pairs of alleles affect color vision. One of the X-linked recessives responsible for red-green color blindness may be represented as p. Its dominant allele P is required for normal color vision. Write as much as possible of the genotypes of the following three persons:

    A. A woman with normal vision and normal skin pigmentation whose father is a color-blind albino.

    B. A normally pigmented man whose mother is color blind and whose father is albino.

    C. A man with normal vision and normal skin pigmentation whose father is a color-blind albino.

2. Using the information in Question 1, show the different kinds of gametes that can be formed by the following three persons.

    A. An albino woman who has normal vision but carries the allele for color blindness.

    B. A normally pigmented man who carries the allele for albinism and is color blind.

    C. A normally pigmented woman with normal color vision whose father was a color-blind albino.

3. The autosomal recessive (a) described in Question 1 is associated with oculocutaneous albinism, a condition in which pigment is absent from the skin and eyes. A sex-linked recessive (o) produces ocular albinism in which pigment is absent *only* from the eyes. The dominant (O) is required for pigment deposition in the eyes. For each of the following two matings, give the genotypes and the phenotypes to be expected among the children.

    A. A woman with oculocutaneous albinism who is not carrying the sex-linked recessive and a man with ocular albinism who carries no recessive at the autosomal locus.

    B. A woman with no pigment in her eyes but is homozygous for the dominant allele at the autosomal locus and a man who has the oculocutaneous albino condition but has the dominant sex-linked allele.

4. In the human, a certain rare sex-linked recessive, i, can result in a cleft in the iris of the eye. Its allele I is required for normal iris. What would be the results of the following three crosses?

    A. A man with cleft iris and a woman who carries only the normal allele.

    B. A woman with cleft iris and a man with normal iris.

    C. A woman who is heterozygous for the iris character and a man with cleft iris.

5. The dominant allele, M, is associated with a certain form of migraine headache. Its recessive allele, m, is required in homozygous condition for absence of headache. This pair of alleles is autosomal. Using this information, and that given in Question 4 on cleft iris, consider the following. A woman who suffers from no apparent afflictions takes her daughter to a physician because the young girl is suffering from migraine. The doctor notices that the girl has a cleft iris. From just this information, write what the doctor knows about the genotypes of the girl, her mother, and the girl's father.

6. Using the information given in Questions on cleft iris and migraine, give the expected phenotypic ratio among the children from the following cross: mmIi X MmiY.

7. A rare autosomal recessive, t, results in total color blindness. The dominant allele, T, is required for development of rods and cones. Suppose a red-green color-blind

woman (due to the sex-linked recessive, p. noted in Question 1) is homozygous for the dominant autosomal color-vision allele. Her husband is totally color blind but carries the dominant sex-linked allele.

   A. Diagram the cross and give the expected F_1.
   B. Assume a woman and a man with genotypes like the above F_1 have children. What types are to be expected and in what proportion?

8. In the human, a sex-linked dominant allele, R, produces a type of rickets, a skeletal defect. The recessive allele, r, is associated with normal skeletal development. Give the results expected from the following crosses:

   A. A man with rickets and a woman who does not have the condition.
   B. A woman who has rickets but whose father did not, and a man who does not have the condition.

9. In chickens, the Bar feather pattern is due to a sex-linked dominant allele, B, whereas its recessive allele, b, results in non-Bar. Since the female is the heterogametic sex in birds, what would be the results of the following two crosses? (Give the genotypes and phenotypes of the offspring.)

   A. A Bar hen and a non-Bar rooster.
   B. A non-Bar hen and a Bar rooster whose female parent was non-Bar.

10. A. In mice, as in all mammals, the male is the heterogametic sex. Assume that a sex-linked lethal is present in a strain of animals and that this causes the death of the late embryo. How would this affect the sex ratio?
   B. Answer the same question if a sex-linked lethal were present in a strain of chickens.

11. In the human, a certain sex-linked recessive causes a child to exhibit very abnormal behavior in childhood. An assortment of bodily upsets occurs and the affected child dies. Only males suffer from this genetic disorder; it has never been reported in a girl. Explain.

12. In cats, the factor B (for black fur) is codominant to its allele (b) for yellow fur. The heterozygote is tortoise. The pair of alleles is sex linked. In the clover butterfly, the autosomal factor W (white) is dominant over w (yellow), but the expression of the dominant is limited to females. From a distance, you observe a tortoise kitten and a yellow one playing with a white butterfly. What can you say regarding the sex and the genotypes of the cats and the insect?

13. A non-color-blind woman is carrying the sex-linked recessive, p. noted in Question 1, which can result in a type of red-green color blindness. Assume that nondisjunction of the X chromosomes occurred at first meiosis in an oocyte of this woman. Give the kinds of offspring

that could result if one or the other of the possible exceptional eggs is fertilized by sperm from a non-color blind man.

14. Give the possible results if normal eggs resulting from a non-color-blind carrier female (Pp) are fertilized by unusual sperm resulting from nondisjunction of the X and Y at first meiosis in a non-color-blind male.

15. An X-linked dominant allele, H, is required for the production of an enzyme, HGPRT, normally present in human cells. Its allele, h, results in lack of the enzyme and the elevation of uric acid in the blood. Suppose 100 fibroblast cells growing in culture are tested for this enzyme from persons of the following genotypes. In each case, how many cells on the average would be expected to show the presence of the enzyme?

   A. HH   E. hh
   B. HY   F. HhY
   C. Hh   G. HO
   D. hy   H. HYY

16. The multiple alleles w (for white eye), w^a (for apricot eye color), and w^{ch} (for cherry or light red) are found at the w (white) locus on the X chromosome of Drosophila. Heterozygotes for these alleles have intermediate eye colors, since the alleles in the series are incompletely dominant to each other. However, each allele is recessive to the wild-type allele (w^+) for red eye color. Give the results expected from the following crosses:

   A. An apricot female × a white male.
   B. An apricot female × a cherry male.
   C. A cherry female × a red male.
   D. F_1 females from the cross in C × an apricot male.

17. A certain bull is considered a prizewinner on the basis of his musculature and other fine anatomical points. However, when he is mated with cows, the female offspring sired by him produce much less milk than their mothers. Explain.

18. Consider pattern baldness in the human to result from the expression of the autosomal factor (B) as a dominant in males and a recessive in females. Its allele, b (non-bald), behaves as a dominant only in females. What would be the results of the following matings:

   A. A nonbald man and a nonbald woman who is heterozygous for the pair of alleles.
   B. A nonbald man and a woman who became bald later in life.
   C. A bald man whose father never became bald and a nonbald woman who is a homozygote.

19. Consider the alleles for baldness in Question 18 along with the sex-linked alleles that affect color vision, P and p (Question 1). Give the results of a mating between a nonbald color-blind man and a nonbald woman with normal vision whose father was color blind and whose mother became bald.

20. The genetic factor (L) for long index finger behaves as a dominant in females, whereas its allele (1) for short index finger behaves as a dominant in males. This pair of alleles is autosomal. Give the genotypes of (1) a female with short index finger; (2) a male with long index finger. Give the genotypes and phenotypes of (3) their daughter; (4) their son.

21. According to one idea, the autosomal genetic factor, E, for early baldness is dominant to its allele, e, for no early loss of hair. The penetrance of the allele for early baldness is 100% in males but 0 in females. What would be the expected phenotypes among the offspring of a man and a woman, both with genotype Ee?

22. In each of the following cases, assume that the donor and recipient are compatible regarding most antigens that would affect the ability to accept a skin graft. Indicate whether a skin graft could be accepted (+) or rejected (−) on the basis of H-Y antigen or its equivalent. (Assume H-Y is testis determining.)

	Donor	Recipient
A.	XY	XXY
B.	XXY	XX
C.	XXY	XO
D.	XO	XXY
E.	XXX	XYY
F.	XY female	XX female
G.	XX male	XX female
H.	XY male	XY female
I.	XX female	XY female

23. Suppose a 9-year-old girl needs a kidney transplant and that it is not possible to check for tissue antigens. Using information in Chapters 4 and 5, rank the following individuals in order of preference as potential donors: older sister, sister who is fraternal twin to the girl, younger brother, unrelated 9-year-old girl, aunt, mother.

24. Analysis of a buccal smear from a patient reveals the presence of three Barr bodies. Which of the following is indicated? Select the correct answer or answers:

A. The patient is probably suffering from Turner syndrome.
B. The patient is probably a gynandromorph.
C. The patient is probably an XXY Klenefelter male.
D. The patient will probably show three drumsticks in a blood smear.
E. The patient will probably show four X chromosomes in a karyotype analysis.

6

PROBABILITY
AND ITS APPLICATION

Certainty versus probability

When asked his or her opinion on a specific problem, the genetic counselor may be unable to give an exact "yes" or "no" reply. The skilled adviser understands that frequently an unqualified response is impossible from the information at hand. However, he or she can often reply in terms of probability and give the chances between 0 and 100% that a certain event will take place. Starting with extremely simple examples, let us examine the reasoning behind the answers to various questions concerning a few couples and their prospective children.

"When a couple's first child is born, what are the chances that it will be a human being?" Obviously, a clear-cut answer can be given, inasmuch as there is no possibility that the child will be anything other than human. Letting p represent the probability for birth of a human, we can say that the probability here is 1 (representing 100%) or that $p = 1$, because this event is a certainty.

If you had been asked the equally absurd question, "What are the chances that the first offspring of this couple will be a cat?" a precise answer again can be given. The probability of such an occurrence is 0. We can let q symbolize the probability of the alternative event, the arrival of a cat, and say that $q = 0$. Thus, p

and q simply represent alternative probabilities, which in this case have the respective values of 1 (for the occurrence of a human) and 0 (for the birth of a cat).

Very often, the probability of some happening lies between 1 and 0 (i.e., between 100% and 0). If the couple had asked you the chance of their first child's being a boy, you could not answer with certainty but, undoubtedly, you would not hesitate to offer the correct reply, "½." In this instance, $p = 1/2$ (or 50% or 0.5). It is equally obvious that the chance of a girl is also 1/2. This alternative, the birth of a girl, can be represented by q. Therefore, we can say that $q = 1/2$. Note that if there are only two alternatives (p and q) to a particular event, then $p + q$ must equal 1. This is quite obvious in this example, because *either* a male *or* a female must be born at any one birth, and the sum of the separate probabilities, $p + q$ ($1/2 + 1/2$), equals 1. This important point is considered in more detail further on.

Let us next suppose that a husband and wife are concerned about the arrival of an albino child among their offspring because each of them has an albino parent. Allowing A to symbolize the dominant for normal skin pigmentation and a to stand for its recessive allele for albinism, we can represent these two people as Aa, for each one must be a heterozygote carrying the recessive allele. We are now in a position to answer questions about the appearance of albinism among the children of this couple in terms of probability.

"What are the chances that the first child born to these two carriers will be albino?" It is quite obvious that in this case we are considering a simple monohybrid cross: Aa × Aa. The genotypes of the offspring would occur in a frequency of 1AA:2Aa:1aa, giving an expected phenotypic ratio of three normal to one albino. Therefore, the chance of an albino being born at any birth is equal to 1/4, and, consequently, the probability of a normal child is 3/4. We can let $p = 3/4$ and $q = 1/4$. (We could have chosen p to stand for the chance of an albino and q for the normal. It makes no difference as long as we are consistent in any one problem.)

Although these examples may seem trivial, they are presented to emphasize the fact that we deal in probabilities much of the time without being aware of it. There is a tendency to forget even the simplest answers when a basic question is phrased somewhat formally in terms of probability. We might hesitate if asked, "What is the mathematical probability of an albino being born to two heterozygous parents?" Yet we would probably give the quick response, "1/4," if the question were stated in more familiar terms.

Combining probabilities

Also without knowing it, we frequently combine probabilities, a most important and critical operation in situations dealing with chance. It is often essential to contemplate the probability of one event happening in relation to a second or third event. In other words, we rarely consider the chance of one thing happening alone without examining the influence another event may have on it. Combining probabilities is extremely simple, and if the following examples are grasped, little confusion should result in future problems.

Suppose the preceding couple asked the ridiculous question, "What are the chances that our first child will be *either* a boy *or* a girl?" As mentioned previously, it is obvious at any birth that any child *must be* one or the other. If it is not a boy, then it must be a girl and vice versa. That is to say, it is *certain* that the child will be *either* a boy *or* a girl. The probability is 100%, and we say that $p = 1$. Without thinking, we have combined two separate probabilities. In arriving at the answer of one, we have just taken the probability of a boy ($p = 1/2$) and added to it the probability of a girl ($q = 1/2$) to get 1. We are simply acknowledging what was said at the beginning concerning two alternatives to an event. In this illustration, we are dealing with two alternative *mutually exclusive* events. The meaning of this expression is apparent in this case, because the birth of a girl would obviously exclude the birth of a boy at any one time. We call two separate events "mutually exclusive" when the occurrence of one at a particular time prevents or renders impossible the occurrence of the other. Similarly, if we toss a coin and heads appears, the probability of tails showing is now 0. Therefore, when dealing with a set of just two mutually exclusive events, p and q, the sum of their separate probabilities must equal 1. This also holds for more than two probabilities, so that if we have three mutually exclusive events, p, q, and r, then $p + q + r = 1$, and similarly for any number of events.

"What is the probability that the card you pull from a deck of cards will be the ace of clubs, the ace of hearts, the ace of diamonds, or the ace of spades?" Because the chance of pulling any one of them is 1/52, and the events are mutually exclusive, the correct reply is "4/52, or 1/13," the sum of the separate probabilities. The chance of pulling any one of the remaining 48 cards in the deck is 48/52. We see again that the sum of the probabilities of all the mutually exclusive events equals 1 (4/52 + 48/52). To summarize: "The probability that either one or the other of a group of mutually exclusive events will occur at a given time is equal to the sum of the probabilities of the separate events occurring."

Now assume that the couple we are discussing asks about the chance of their having first a boy and then a girl. This is a very different question from the one posed earlier, for now the events are *not* mutually exclusive but are independent. They are independent because the arrival of a child of one sex at a given birth does not affect the sex of the next or of any other children born later. In the same way, the appearance of heads at the toss of a coin does not affect what will show the next time the coin is thrown. So we can say that two (or more) events are independent if the occurrence of one at a given time does not affect the occurrence of the other (others) at any other time. To answer the question put to us, we simply multiply $1/2 \times 1/2$ to give a chance of 1/4 that the first child will be a boy and the second one a girl in the precise order. This is so because the chance that two (or more) independent events will occur together is the product of the chances of each occurring by itself. Since we know that the probability of a boy alone is

$1/2$ ($p = 1/2$) and similarly for a girl ($q = 1/2$), then the probability of their happening together is the product of their separate chances or probabilities. And it should be evident that this rule for combining probabilities can be extended indefinitely for more than two events. "What are the chances of a boy, a girl, a girl, another girl, and then a boy *in that order?*" The answer is quickly calculated by finding the product of each of the single probabilities of the five independent events in this group: $1/2 \times 1/2 \times 1/2 \times 1/2 \times 1/2 = 1/32$.

Many fail to realize that we combine probabilities whenever we use the Punnett square method in a simple genetics problem. For in this procedure, we do no more than multiply diagrammatically the separate chances of the different kinds of gametes uniting in all possible combinations (Fig. 6-1). With just this basic information in mind we can respond to a large number of questions. A quick response can be given to our couple if these two heterozygous individuals ask, "What are the chances that our first child will be a normally pigmented boy?" Since skin pigmentation does not influence the sex of the individual, we are concerned with two independent events and the chance of their occurring together. We know that the chance of a boy at any birth is $1/2$ and that the chance of a child with normal skin is $3/4$. The answer is therefore $1/2 \times 3/4$, or $3/8$. By using the same logic, the chance of an albino girl is $1/4 \times 1/2$, which is $1/8$.

One very important point concerning independent events must not be overlooked—the probabilities remain the same *no matter* how many times any two (or more) may have combined or come together. For example, if this couple has had a boy, then another boy, and a third boy, the chance of yet another boy at the next birth remains $1/2$. And the chance of a normally pigmented boy remains $3/8$, regardless of the number of times this combination between skin pigmentation and sex may have occurred in the family.

What if we were asked the chances of *either* a normal boy *or* an albino girl being the first born in the family? Here we are dealing with mutually exclusive events, so the answer is $3/8 + 1/8 = 1/2$. It should be obvious that the sum of *all* the probabilities in this group would equal 1: $3/8 + 3/8 + 1/8 + 1/8$ (the sum of the separate chances for a normal boy, a normal girl, an albino boy, and an albino girl.)

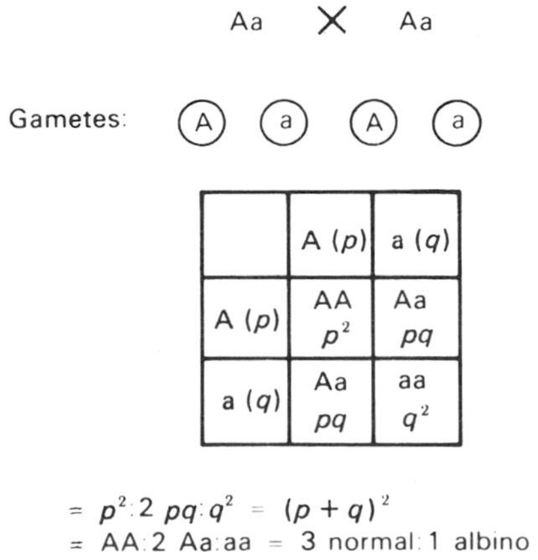

The same combination may arise in different ways

Now let us turn to other simple problems that demand no more than a clear understanding of *what* is being asked. Failure to appreciate the meaning can cause an erroneous reply to the most elementary question. For example, "If a couple has three children, what is the probability that two of them will be boys and one will be a girl?" This is an entirely different question than "What is the chance of having first a boy, then another boy, and finally a girl?" In the latter case, we would simply find the product of the separate probabilities ($1/8$). But in the former question, the *order* in which the three children will be born is *not* prescribed. To obtain the answer, a different procedure must be followed, which makes use of the basic rules of probability that have just been discussed.

In a family of three, the children could arrive in several ways. Remember that the probability for a boy

FIG. 6-1. Monhybrid cross. The Punnett square, which diagrams all the possible combinations of the gametes in a monohybrid cross, is really an expansion of the binomial $(p + q)^2$.

Aa $\times$ Aa

Gametes: Ⓐ ⓐ Ⓐ ⓐ

	A (p)	a (q)
A (p)	AA p^2	Aa pq
a (q)	Aa pq	aa q^2

$= p^2 : 2\ pq : q^2 = (p + q)^2$
$= $ AA : 2 Aa : aa $= $ 3 normal : 1 albino

TABLE 6-1 Different ways in which children may be born in a family of three

	ORDER OF ARRIVAL	
3 boys	$p^a\ p\ p = 1/8$ or p^3	
2 boys and 1 girl	$p\ p\ q^b = 1/8$	
	$p\ q\ p = 1/8$	$= 3/8$ or $3p^2q$
	$q\ p\ p = 1/8$	
2 girls and 1 boy	$q\ q\ p = 1/8$	
	$q\ p\ q = 1/8$	$= 3/8$ or $3pq^2$
	$p\ q\ q = 1/8$	
3 girls	$q\ q\ q = 1/8$ or q^3	

[a] p = probability for a boy = $1/2$.
[b] q = probability for a girl = $1/2$.

will remain 1/2 at each birth ($p = 1/2$), as will the probability for a girl at each birth ($q = 1/2$). The different possibilities are shown in Table 6-1. The question asked the chances of two boys and one girl in any order. A glance at the table shows that there is not one but rather *three* set ways in which this can happen. The probability is not 1/8, as it would be if we were interested in just one of the three possible orders. Instead, it is 3/8, the sum of the three different ways in which two boys and one girl can arrive, *each way* with a probability of 1/8.

Consider the chances of three boys and one girl in any order in a family of four. Again we must acknowledge all the possible manners in which four children can arrive by chance (Table 6-2). We see that four children may be born to a couple in 16 combinations and that there are four separate possibilities for three boys and one girl. These details demonstrate another fact that is often overlooked. Notice that we are simply combining probabilities in all possible manners while at the same time recognizing the various ways in which the same combination can result. There are six different ways of getting the combination 2 boys and 2 girls; there are four different orders for 1 boy and 3 girls, and so on. Looking back at the families of three and four (Tables 6-1 and 6-2), we see that all the possibilities in each example form a binomial

TABLE 6-2 Different ways in which children may be born in a family of four

	ORDER OF ARRIVAL
4 boys	$p^{\mathrm{a}} p\ p\ p = 1/16$ or p^4
3 boys and 1 girl	$p\ p\ p\ q^{\mathrm{b}} = 1/16$ $p\ p\ q\ p = 1/16$ $p\ q\ p\ p = 1/16$ $q\ p\ p\ p = 1/16$ $\left.\right\}= 4/16$ or $4p^3q$
2 boys and 2 girls	$p\ p\ q\ q = 1/16$ $p\ q\ p\ q = 1/16$ $p\ q\ q\ p = 1/16$ $q\ q\ p\ p = 1/16$ $q\ p\ q\ p = 1/16$ $q\ p\ p\ q = 1/16$ $\left.\right\}= 6/16$ or $6p^2q^2$
1 boy and 3 girls	$p\ q\ q\ q = 1/16$ $q\ p\ q\ q = 1/16$ $q\ q\ p\ q = 1/16$ $q\ q\ q\ p = 1/16$ $\left.\right\}= 4/16$ or $4pq^3$
4 girls	$q\ q\ q\ q = 1/16$ or q^4

[a] p = probability for a boy = 1/2.

[b] q = probability for a girl = 1/2.

distribution. Actually, whenever the Punnett square technique is used to illustrate a monohybrid cross, a diagram of a binomial distribution results. This can be seen in the diagram of the cross between the two heterozygous individuals who are carrying the recessive for albinism (see Fig. 6-1).

Expansion of the binomial and its application

By expanding the binomial ($p + q$) to a certain power, n, we can determine the probability of two events occurring in any order without going through the laborious procedure detailed previously. It takes little effort to answer the question, "What are the chances of 3 boys and 1 girl occurring in a family of 4?'" Inasmuch as this case concerns four individuals, the binomial must be expanded to the 4th power. The power to which the expansion is carried is always the total number of trials or occurrences in which we are interested, such as five tosses of a coin, $(p + q)^5$, or the arrival of four children, $(p + q)^4$.

The binomial can be quickly expanded with the use of a few simple rules. Consider first just the exponents of the separate terms in the expansion of $(p + q)^4$ (follow Table 6-3 in the expansion):

1. In the first term, p is simply raised to the power of the expansion (n), in this case 4.

2. In the second term, q enters, and the power of p decreases by 1, producing p^3q. In each of the succeeding terms, the power of p continues to decrease by 1, as the power of q increases in accordance with n, and p no longer appears.

3. The number of separate terms in the expansion always is one more than the power to which the binomial is raised. Because that power here is 4, we have five terms.

4. Note that the sum of the exponents of p and q in each term always equals n and is thus 4 in each of the terms in Table 6-3.

We must next derive the coefficients for each of the terms:

1. The coefficient of the first term is always 1, no matter what the power of the expansion; so here it remains simply p^4.

2. In the second term, the figure is just the same as the power of the expansion, and we have here $4p^3q$.

3. The coefficient of each succeeding term is readily determined by examining the term just preceding it. To obtain the value for the third term, we look at the second, which has just been completed, $4p^3q$. We then multiply the exponent of p (which is 3) by the

TABLE 6-3   Expansion of $(p + q)4$ (follow text for details)

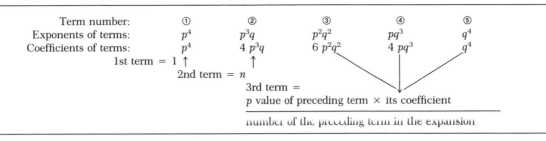

coefficient of the term (which is 4) giving 12. Next we divide this product by the number of the term, which is 2. This results in a coefficient of 6 for the third term, $6p^2q^2$.

4. Continuing in this manner, the fourth term becomes $4pq^3$ and the last one q^4. Actually, the coefficient of the last term, like the first is always 1.

The expanded binomial can tell us the probability of having three boys and one girl in a family of four. We first allow p to represent the chance for the birth of a boy and q for that of a girl. (Again we could have let p represent the chance for a girl and q for a boy. It does not matter as long as we adhere to the same symbols throughout a given problem.) Looking at the expansion $(p+q)^4$ shown in Table 6-3, we select that term in which the exponents are 3 (for three boys) and 1 (for one girl). Since we have decided to let p stand for a boy and q for a girl, it is apparent that three boys and one girl are represented by the term $4p^3q$. Because the probabilities for a boy (p) and for a girl (q) at any one birth are both 1/2, we substitute these for p and q in the term: $4(1/2)^3 \times 1/2 = 4/16 = 1/4$. This figure gives the chance of three boys and one girl occurring in any order in a family of four.

Looking back at the detailed expansion $(p+q)^4$ in Table 6-2, we see that there are indeed 4 ways out of a total of 16 in which this can take place. By expanding the binomial, we have eliminated the need to construct a laborious diagram. Any question involving just two alternatives can be answered by expanding the binomial in this manner as long as we know the separate probabilities. Take the question, "In a family of three in which each parent is genotype Aa, what are the chances of finding two albino boys and one normal girl?" The answer is determined by selecting the term $3p^2q$ from the expansion $(p+q)^3$. Let p stand for the probability at any birth of an albino boy ($1/4 \times 1/2 = 1/8$, which is the chance of two independent events, the birth of an albino and of a boy,

occurring together.) Similarly let q represent the probability of a normal girl ($3/4 \times 1/2 = 3/8$, the product of the separate chances for a normal individual and for a girl). Substituting in the terms $3p^2q$, we have $3(1/8)^2 \times 3/8 = 9/512$.

More than two alternatives

Although we could continue like this for other problems, it would soon become evident that many situations involve more than two alternatives. Suppose the question were asked, "In a family of three, what is the chance for one albino boy, one normal boy, and one normal girl in any order?" To arrive at this answer, it is necessary to expand the multinomial $(p+q+r)^3$. This becomes cumbersome when even more variables are followed. Fortunately, we have a formula that enables us to expand binomials, trinomials, and so on in a general way that permits us to consider any number of events or variables:

$$\frac{n!}{s!\,t!}\,p^s\,q^t$$

Since this formula is merely a general expression allowing us to arrive at any particular term in an expansion, it can be used to answer simpler questions, such as the chances of three boys and one girl in a family of four. In the formula, n stands for the total number of occurrences, in this problem, 4. The familiar symbols p and q represent the respective probabilities for a boy and a girl. The s and the t stand for the possibilities for p and q posed in this question, that is, three boys and one girl. Therefore, $s=3$ and $t=1$. Substituting we have

$$\frac{4!}{3!\,1!}(1/2)^3(1/2)^1 = \frac{4 \times 3 \times 2 \times 1}{3 \times 2 \times 1} \times 1/8 \times 1/2 = 4/16 = 1/4$$

This is the same answer obtained by selecting the term $4p^3q$ after expanding $(p+q)^4$. Similarly, the chance of two albino boys and one normal girl in a family of three, in which the parents are both Aa becomes

$$\frac{3!}{2!\,1!}(1/8)^2(3/8)^1 = \frac{3\times2\times1}{2\times1\times1}\times\frac{3}{512} = \frac{9}{512}$$

Since we can find the same answers by using the other method, why should we remember this formula? It is particularly valuable when we are interested in more than two alternatives, as in the question, "What is the chance of one normal boy, one albino boy, and one normal girl arriving in any order in the family of three?" Here we have three possibilities to consider (p, q, and r), and we represent them in the formula letting p stand for the probability of a normal boy ($3/4\times1/2=3/8$), q for an albino boy ($1/2\times1/4=1/8$), and r for the normal girl ($3/4\times1/2=3/8$). The symbols s, t, and u are their respective possibilities (1, 1, and 1). This gives the following:

$$\frac{n!}{s!\,t!\,u!}p^s\,q^t\,r^u = \frac{3!}{1!\times1!\times1!}(3/8)^1(1/8)^1(3/8)^1$$

$$= 3\times2\times1\times\frac{9}{512} = \frac{54}{512} = \frac{27}{256}$$

The formula can thus be adapted for situations involving many variables. Whether you select it to solve a problem containing only two alternatives or expand the binomial is a matter of choice.

Constructing a pedigree

One important application of probability in genetic studies is its use in the analysis of pedigrees. Examination of family histories is an important approach in the field of genetic counseling, a discipline of increasing significance today. Once the molecular basis of an inherited disorder is understood, it becomes possible, using chemical procedures, to detect persons who are carriers of the defective gene. Genetic counseling of families with histories of genetic impairments can prevent the birth of afflicted children. At the same time, it can avoid needless worry by family members who do not know they are actually free of the mutant gene.

The inheritance patterns of several human defects, such as color blindness, are well known. However, it must be borne in mind that certain families may exhibit their own peculiar pattern of inheritance. Because many genes affect a character, a trait inherited as a dominant in one family may follow a recessive pattern in another. Whenever possible, a pedigree should be assembled for the interpretation of a specific

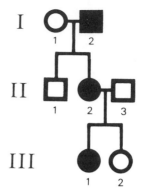

FIG. 6-2. Construction of a pedigree. See text for details.

case history. A complete and accurate analysis of any lineage depends on a thorough understanding of elementary genetic principles and the laws of mathematical probability. In addition to these basic concepts, little more than common sense is needed to avoid certain common pitfalls often overlooked in hasty interpretations.

When constructing a pedigree, circles and squares are commonly used to represent females and males, respectively (Fig. 6-2). A mating between two individuals is represented by a bar to which their offspring are attached. Roman numerals designate successive generations in the pedigree, whereas arabic numbers denote specific persons. Shading indicates the expression of the trait being followed. In the pedigree in Fig. 6-2, persons I-2, II-2, and III-1 exhibit the trait under consideration.

Examining a pedigree

Before examining any pedigree, it is essential to realize that some histories do not provide enough information to permit definitive statements to be made. It is just as important to know that an interpretation is impossible in one case as it is to know that a precise answer can be given in another. These points are illustrated in the following discussion.

When presented with a pedigree, we should attempt to determine whether the trait follows a dominant or recessive pattern of inheritance. At the same time, we should ask if the expression of the trait seems to be influenced by the sex of an individual. Suppose we are just handed the pedigree in Fig. 6-3A with no additional information and are requested to comment on it. We note that the male parent is expressing some trait and that half of the offspring, both sons,

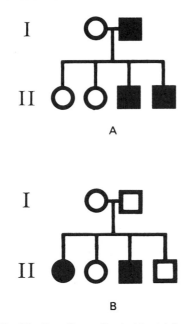

FIG. 6-3. Small family pedigrees. See text for details.

theoretical ratios can become distorted by chance factors when we are working with small samples; the fact that only males in this pedigree show the trait may reflect nothing more than this. Not enough people are involved to indicate anything about the relationship of the trait to sex. Therefore, a valid response to a question about this pedigree is that more information is needed before anything can be said with any degree of certainty.

When studying pedigrees, we are frequently concerned with traits that are not very commonplace in the population. If the preceding pedigree involves a very rare condition, we might suggest that the trait is due to a dominant allele. The chance is greater that a dominant rather than a recessive is involved, for if the trait is rare, genes for it would obviously be rare in the population as a whole. Therefore, the mother, if she is not related by blood to her husband, would probably not be carrying the rare allele. She would probably be homozygous for the normal condition, aa, and the male parent would be the heterozygote, Aa. But unless we know whether or not the parents are blood relatives, we cannot feel secure in our decision. If they are cousins, for example, the probability is greater that they will have more alleles in common than two people taken at random. It has been reported that traits inherited on the basis of rare recessive alleles are expressed with higher frequency in families where first cousin marriages are common than in other families in the general population. Since most of us harbor in the heterozygous condition at least a few defective recessives, it stands to reason that the chances are greater for two of these factors to come together when the parents have obtained a proportion of their genetic material from a common source, a family pool of genes. Thus any recessive in a lineage will have a greater chance of coming to expression if the parents are related than if they come from unrelated stocks that contain their own but different defective, recessive alleles. If the parents are related, therefore, we can make no decision about dominance in a pedigree such as that shown in Fig. 6-3A, in which case half of the offspring expresses a rare trait that is also seen in one of the parents.

But even if we know that the parents are unrelated and that the trait is rare, we still could not be completely certain about a decision made on this pedigree because sex could be exerting an effect in some way. We are therefore forced to conclude that, on the basis of a single pedigree such as this, we cannot answer without some qualifying statements. A definite reply

also show it. Can we say with any degree of certainty how the condition is inherited? The answer is emphatically, "No!" The reason becomes obvious if we allow A to represent any dominant and a its recessive allele. Assuming the trait results from a homozygous recessive condition, then the genotype of the male parent must be aa and that of the female parent Aa. An interpretation is quite possible on this basis. But another conclusion is equally feasible, if we assume the trait is due to a dominant allele and the male parent is the heterozygote, Aa; the female parent would then be homozygous recessive for the normal condition. (Very often the novice assumes that a disorder in a pedigree results from a recessive condition and that the allele for normal expression must be dominant. There is no genetic basis for making such an assumption.)

All three males and none of the females in the pedigree express the trait; therefore, the suggestion might be made that the responsible gene is sex-influenced, sex-limited, or linked to the X or Y chromosomes. Since an argument can also be made for each of these possibilities, it is apparent that nothing definite about the inheritance of the trait can be stated from this pedigree by itself. Those who argue that the sex of the individual influences the expression of the trait should realize that too few people are represented in the pedigree. Elementary probability tells us that

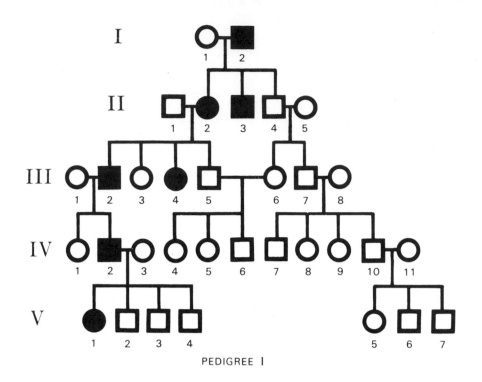

PEDIGREE I

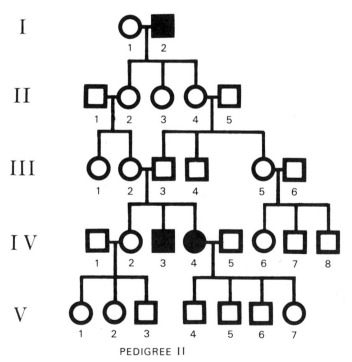

PEDIGREE II

FIG. 6-4. Pedigrees for analysis. See text for details.

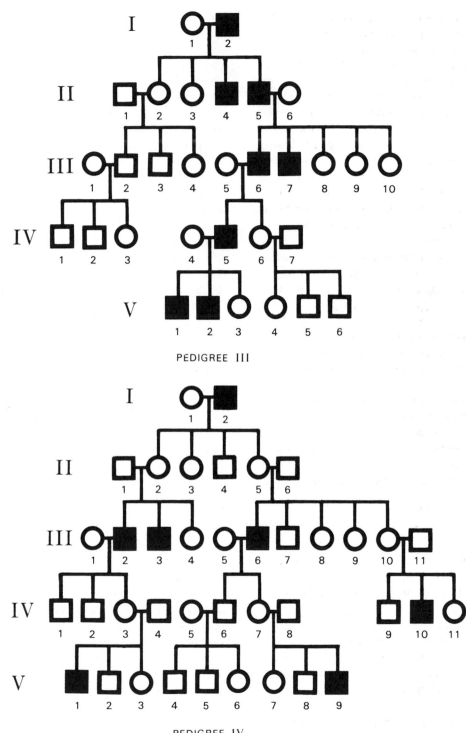

PEDIGREE III

PEDIGREE IV

may require histories of other families in which the trait is expressed or a pedigree encompassing several generations within the one family.

Can anything more definite be said about the small family represented in Fig. 6-3B? As neither parent shows the trait, chances are that the allele inherited here is not a dominant; otherwise, we would expect the trait to show in one of the parents. In the case of this simple pedigree, we can feel fairly sure that an autosomal recessive is involved. If only a son and no daughter had expressed the trait, we could not be certain whether the allele is autosomal or sex linked. It should be apparent that the more persons who are involved in a pedigree, the firmer the basis for reaching a decision becomes.

The phenomenon of penetrance could complicate the picture in Fig. 6-3B. In the case of a dominant with incomplete penetrance, a trait may not show even when the factor for it is present in the genotype. In Chapter 4, we learned that the expression of an allele may be suppressed by the presence of certain modifying genes in the rest of the genotype or even by environmental influences. Although reduced penetrance is certainly a factor to consider in family histories, it will be ignored in our discussion to illustrate characteristic modes of inheritance of alleles with complete penetrance. The student, however, should appreciate the fact that pedigree analysis demands the recognition of many genetic and environmental factors that can alter a classic pattern of inheritance.

Representative pedigrees

Let us examine a few larger pedigrees to see how much information can be derived from them (Fig. 6-4). In all of these, we will assume that the trait is an uncommon one in the general population. In Pedigree I, several individuals, both male and female, are affected. The major thing that should impress us is that the trait always shows in a family line if one of the parents expressed it. It never skips a generation in this sense. If neither parent shows the trait, then it does not appear among the descendants. This is particularly significant in this pedigree, because the lineage involves a mating between two first cousins (persons III-5 and III-6). An autosomal dominant is strongly indicated by the pattern seen in this family.

Pedigree II is quite different from the first one. We see that although the trait appeared in the first-generation male, several generations are free of it.

After skipping two generations, the trait reappears. A dominant allele basis seems unlikely. The pattern suggests an autosomal recessive, an interpretation strengthened by the fact that the parents of those affected in the fourth generation are first cousins. When one of the affected mates outside the family, the trait does not show again. This is to be expected, because the outside members probably would not carry the same rare defective allele.

Pedigrees III and IV should be carefully compared. We note in both that the only affected persons are males. Because many individuals are involved, chances are that the sex of the individual is playing a significant role in the expression of the allele. However, scrutiny of the two pedigrees will reveal important differences between them. In Pedigree III, all the males show the trait if their father expressed it. It is clearly going from father to son, never to daughters. Daughters of affected fathers do not seem to pass it down. The trait cannot be due to a sex-linked recessive, for in that case it would be passed to male offspring by way of a female parent. This is seen in Pedigree IV, which illustrates a sex-linked recessive pattern of inheritance. Here the trait does not go from a father to his sons. When an affected male has offspring, it is not seen among his children but appears among the sons of his daughter. An affected male may have an uncle who showed the trait (persons IV-10 and III-6). Elementary genetics tells us this is exactly the way a sex-linked recessive is inherited.

Pedigrees and probability

Now let us try to answer typical questions that could be asked about family members in the four pedigrees. In Pedigree I, suppose person V-4 is concerned about passing the trait down to his offspring and wants to know the chance of doing so. Since it has been determined that the expression of the trait depends on a dominant allele, we can answer safely that the chance is most unlikely. Obviously, he cannot be carrying the dominant because he would be expressing it, even though it is in the family. (We can ignore the highly unlikely complication of a mutation arising in this person's gametes or those of his mate.) If his affected sister (person V-1) asks the same question, again we can give an answer. We know that her genotype must be Aa. She displays the trait and therefore must be carrying the dominant allele, A. But because her mother did not show it (and is thus genotype aa), she has received a recessive allele for the normal condition from her. If she married a normal man, he too must be aa. The cross is nothing more than a simple testcross: Aa × aa. Basic Mendelian laws tell us that half of her gametes will carry the defective dominant (A) and the other half the recessive (a). The offspring will consequently be Aa and aa in a ratio of 1:1. Therefore, the chance is 1/2 that the trait will show at the birth of any one child. Suppose this woman is married to a normal person and already has one normal child. What is the chance that the next child will be normal? Elementary probability tells us that it is again 1/2 and that the answer will remain 1/2 no matter how many times the question is asked concerning any one birth.

Let us now put the question another way. "If there are two children in the family, what are the chances that both will be normal?" This is the same as asking the chance of first a normal child and then another normal. The answer is the product of the separate probabilities: 1/2 x 1/2, which is 1/4.

Suppose the afflicted person (V-1) is going to marry another similarly afflicted person and wants to know the chance at any birth of a child with the trait. We are now considering the simple monohybrid cross: Aa × Aa. Since the affliction results from a dominant allele, the children will theoretically occur in a ratio of three afflicted to one normal. So at any one birth, the chance will always be 3 to 1 in favor of having a child with the trait.

"What are the chances of three normal and one afflicted if there are four children in the family?" Allowing a to represent the probability for a normal child at any birth and b the probability for an afflicted child, we select the term $4a^3b$ from the expansion $(a+b)^4$. Substituting 1/4 for a and 3/4 for b, the answer is: $4 \times (1/4)^3 \times 3/4 = 12/256 = 3/64$.

The same reasoning applies to similar questions about any of the other pedigrees. We must simply remember the manner in which the trait is inherited and apply the laws of probability. An easily answered question, but one frequently misunderstood, arises in cases involving recessives. Suppose the normal woman in Pedigree II (person IV-2) asks the chance that her first child will be affected if she next marries a man who shows the same recessive trait as her brother and her sister.

To answer such a question, we must first consider the chance that woman IV-2 is a carrier. We do so by referring back to her parents. Neither of them expressed the trait, but each must be a carrier. Letting

A now stand for the normal dominant allele and a for the defective recessive, the parents must both be genotype Aa. A mating between them is again nothing but a Mendelian monohybrid cross. The different genotypes and their frequencies among the offspring would be 1AA:2Aa:1aa. We know that the woman asking the question is not aa because she does not show the trait, but we cannot tell whether she is AA or Aa. However, we do know the probability of her being a heterozygote. Since the normal offspring occur by chance in a ratio of 1AA:2Aa, the chance is thus 2 out of 3 or 2/3 that she carries the recessive. To answer her question, we now simply combine probabilities. Since her prospective husband expresses the trait, he is aa. If she is a carrier, the cross becomes Aa × aa, and the chance of any child being affected in such a cross is 1/2. So we simply multiply this 1/2 by 2/3 (the probability that she is a carrier) and find the answer, 1/3.

In cases like this, we sometimes know the frequency in the general population of normal-appearing people who are carriers of a certain defective gene. Suppose we know that the recessive condition depicted in Pedigree II is carried by 1 out of 100 normal-appearing people. Now we can answer the question, "What are the chances that this woman (IV-2) will have a child showing the trait now that she is married to a normal man who is unrelated to her by blood?" Again we simply combine separate probabilities. We have established that the chance is 2/3 that she is a carrier. If both of them should prove to be carriers, the cross is Aa × Aa, and the chance of an affected child at any birth is 1/4. The answer to the question is easily determined by finding the product of three separate probabilities: $2/3 \times 1/100 \times 1/4 = 2/1200 = 1/600$ (the chance that she is a carrier times the chance that he is a carrier times the chance of a recessive trait in a monohybrid cross). So we see that often a precise answer in terms of probability is possible from pedigree analysis, even when the exact genotype of a concerned family member is in doubt.

The role of sex as a factor in the expression of a gene does not complicate the picture in any way as long as the particular method of inheritance is kept in mind. In the case of a known holandric trait, answers are very obvious. Any affected man will have a chance of 1/2 of passing it to his offspring, because the chance is 1/2 at any birth that he will have a son (Pedigree III). The chance remains 0 that any daughter will show the trait or pass it down, because she has no Y chromosome.

When a sex-linked gene is involved (Pedigree IV), we can tell any male that he has no chance of passing down a recessively inherited trait if he himself does not express it. A sex-linked gene is found on that part of the X with no homologous region on the Y; therefore, the male must express any sex-linked allele he carries, because he is hemizygous for that portion of the X chromosome. What about the normal female (person V-3) in Pedigree IV who has an affected brother? What are the chances that she is a carrier for the sex-linked recessive? Neither of her parents exhibits the trait; so we know that the father carries no allele for it. The mother, however, must be a carrier, because one of her sons expresses it. The genotypes of the mother and father would be, respectively, Aa and AY. The mother will pass the defective allele to half of her children, but none of the daughters will express it. They have received a normal dominant from their father, and this will mask the sex-linked recessive. The chance that the woman with the affected brother is a carrier is 1/2, as a simple illustration shows (Fig. 6-5).

What are the chances the woman under discussion will have a child who shows the trait if she marries a normal, unrelated man? The separate probabilities are again determined and multiplied together. If the man does not show the trait, it does not matter whether or not he is related to his wife as far as any sex-linked recessive is concerned—there is no chance that he is carrying the defective allele. If the woman *is* a carrier, the cross is Aa×AY, exactly as in the case of her parents. Thus, there is again a chance of 1 in 4 that a child would show it. But because there is only a probability of 1/2 that she is a carrier, we must take this into consideration and multiply the two separate probabilities: $1/4 \times 1/2 = 1/8$. If the woman had asked the chance of a son's being affected, the answer is

1/4 (1/2 probability of passing it to a son, combined with the 1/2 chance that she is a carrier).

Keep in mind that women may also express a sex-linked recessive trait. Such a woman must have a father who shows the trait and a mother who either shows it as well or is a carrier for it. All of the sons of any affected woman must also be affected, because every son receives an X from his mother and a Y from his father. If the mother shows the trait, and is thus homozygous for it, both of her X chromosomes will carry the recessive allele.

From the discussion presented so far, it should be possible to answer the additional Questions 1 through 12 on the four pedigrees at the end of the chapter.

Chi-square and its application

Once the elementary rules of probability are grasped, we can proceed to a few other basic but essential concepts of chance. The genetic investigator is almost always concerned with probability, and it must be considered in all crossing experiments performed in the laboratory. For example, assume you are following a cross between brown- and red-eyed fruit flies. You may suspect that the exercise entails a simple monohybrid cross; so you predict for the second generation a phenotypic ratio of three red-eyed flies to one brown, assuming that red is dominant. Suppose 400 F_2 flies emerge. Only the most naive person would expect 300 to be red-eyed and 100 brown-eyed. Instinctively, we realize that ideal figures or ratios are seldom obtained. Many factors, which we lump together under the heading of chance, enter the picture to prevent the realization of the exact hypothetical number.

If we scored 400 flies and found 290 to be red-eyed and 110 brown-eyed, we might not be upset. These figures are close to a true 3:1 ratio, and intuitively we would expect some deviation due to chance. But suppose we had 275:125 or perhaps 250:150. Could these figures also be in agreement with a ratio of 3:1? Have additional chance factors operated to give an even greater distortion from the expected? The answer is uncertain. We would not know where to draw the line, and confidence in our results would be somewhat weakened. We would feel the need to know with some degree of certainty the probability that the results are still compatible with our hypothesis. If the chances are that we are wrong in our assumption that the data agree with a 3:1 ratio, we would certainly want to know this so that we could formulate another hypoth-

FIG. 6-5. Diagram of cross between persons IV-3 and IV-4 in Pedigree IV.

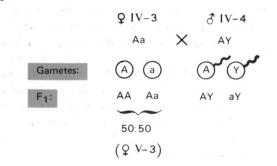

TABLE 6-4   Derivation of a χ^2 value (follow text for details)

CLASS	o[a]	e[b]	d[c]	d^2	d^2/e
Red eyes	290	300	−10	100	$100/300 = 0.33$
Brown eyes	110	100	+10	100	$100/100 = \underline{1.00}$
					$\chi^2 = \Sigma d^2/e = 1.33$

[a] Observed.
[b] Expected.
[c] Deviation.

esis. Or we might reexamine the conditions of our experiment to ascertain what outside factors could be influencing our results. Clearly, it is imperative to know if data, although not ideal, are still in agreement with a hypothesis.

Such a dilemma can be avoided by applying the chi-square (χ_2) formula. This test is a means of determining the probability that departures from the expected values in our data are the result of chance factors. We thus have a stronger basis for accepting our original hypothesis or for rejecting it, whatever the case may warrant.

The χ_2 is easily applied to numerical data; it is indispensable when deciding whether or not actual figures obtained through experimentation agree with those of an ideal ratio. Returning to the 290 red-eyed and 110 brown-eyed flies, how can we determine whether the figures are compatible with a ratio of 3:1? The values, 290 and 110, represent the observed numerical data, those actually obtained from an experimental procedure. As shown in Table 6-4, they are placed in the proper column, labeled o (for observed).

Next to each observed value, we put the expected (e). This value can be determined quickly. We have a total of 400 flies and are applying the χ_2 test to an expected ratio of 3:1. Obviously, four parts are involved (three red and one brown). Dividing the total of 400 by 4 gives us 100, which represents the "1" of the 3:1 ratio. Consequently, we would expect 300:100 in an ideal 3:1 ratio. We next determine how great the deviation (d) is between our actual results and the ideal. It is − 10 and + 10, respectively, in this example. To eliminate the minus sign, each deviation (d) is then squared (d_2). Each of these squared deviations is next divided by the expected value for its particular class (d^2/e). The value of χ^2, as indicated by the formula, is equal to the sum of all these, which gives us 1.33 in our example.

What is the meaning of this figure, and how do we

put it to use? It is now necessary to turn to a table of probabilities for the different values of χ^2 (Table 6-5). Note that the left column in this table is labeled "Degrees of Freedom." The number of degrees of freedom is always equal to 1 less than the number of observed classes. The number of classes here is 2 (red and brown). Hence, the number of degrees of freedom is 1. Some insight into the reason for this can be gained if we regard the situation in this way. We have a total of 400, and this number is distributed between the two classes. If we know how many of the total are included in one of the classes, we automatically know how many are in the other class. Similarly, having four classes and knowing the total as well as the amount in three of the classes, we must know what is in the fourth. So we have 3 degrees of freedom when we have four classes.

Returning to the table (Table 6-5), next to the degrees of freedom we see numbers corresponding to various values of χ^2. Our value is 1.33, and we see that it falls between the values of 0.455 and 1.64. Reading up, we note that this corresponds to a probability value between 50% and 20%, roughly 25%. It is essential to understand what this implies. A probability value of 25% tells us that in the case of an exact 3:1 ratio, chance factors would operate to give a deviation from the expected value at least as large as ours 25% of the time. In other words, when dealing with numbers the size of ours, we should expect a χ^2 as great or greater than the one we calculated in 25 cases out of 100 when we are truly dealing with a ratio of 3:1.

Does this mean we should reject the hypothesis of a 3:1 ratio? On the contrary, a χ^2 value indicating a probability of 25% is not considered significant. Statisticians have agreed that only when a probability of 5% or lower is obtained should a χ^2 be considered significant, that is, large enough to cast doubt on the interpretation of the numerical data. (Return to the χ^2 table and make certain that you see that the higher

TABLE 6-5   Table of χ^2

DEGREES OF FREEDOM	$P = 0.99$	0.95	0.80	0.50	0.20	0.05	0.01
1	0.000157	0.00393	0.0642	0.455	1.642	3.841	6.635
2	0.020	0.103	0.446	1.386	3.219	5.991	9.210
3	0.115	0.352	1.005	2.366	4.642	7.815	11.345
4	0.297	0.711	1.649	3.357	5.989	9.488	13.277
5	0.554	1.145	2.343	4.351	7.289	11.070	15.086
6	0.872	1.635	3.070	5.348	8.558	12.592	16.812
7	1.239	2.167	3.822	6.346	9.803	14.067	18.475
8	1.646	2.733	4.594	7.344	11.030	15.507	20.090
9	2.088	3.325	5.380	8.343	12.242	16.919	21.666
10	2.558	3.940	6.179	9.342	13.422	18.307	23.209
15	5.229	7.261	10.307	14.339	19.311	24.996	30.578
20	8.260	10.851	14.578	19.337	25.038	31.410	37.566
25	11.524	14.611	18.940	24.337	30.675	37.652	44.314
30	14.953	18.493	23.364	29.336	36.250	43.773	50.892

Source: Reprinted with permission of the Macmillan Publishing Co., Inc., from *Statistical Methods for Research Workers*, 14th ed., by R. A. Fisher. Copyright © 1970 by University of Adelaide.

the value of χ^2 is, the lower the value for the probability will be.) Of course, the higher the probability that chance is operating to produce any deviation in the data, the more confident we will feel. However, there is usually no reason to reject an interpretation on the basis of χ^2 alone unless the probability is in the vicinity of 5% or lower.

The importance of the χ^2 cannot be overemphasized. However, it is critical that we use the information it provides correctly and avoid common misconceptions. The following points should be kept in mind. Remember that because χ^2 is a test of numerical data, it must be used only with the actual figures themselves. It is erroneous to apply the method to percentages. Similarly, although we examine ratios by χ^2, we never apply it directly to ratios that have been obtained from the raw data. Wrong concepts may be avoided simply by remembering to use the χ^2 test directly on the actual figures collected and not on anything derived from them.

Another note of caution concerns the sizes of the categories of classes being tested. Anyone would realize the absurdity of working with data that include only two items in one class and one in the other. Obviously, greater numbers are in order, because the larger the totals in each class, the greater the chance of reaching a valid conclusion. To prevent wrong interpretations, one should never apply the χ^2 to any set of data in which the number of individual counts in any class is less than five. Keep in mind that we

do not consider it at all unusual to find human families in which the children are all the same sex. The occurrence of families with a run of four boys or five girls does not shake our belief in the existence of a ratio of male to female that is close to 1:1. For we realize that the numbers involved in a human family are small and therefore chance can distort them from the ideal ratio. Only when we score large populations can we hope to approach the expected ratio. Remember that much of Mendel's success resulted from his choice of the garden pea, a species that provides large numbers of individuals for examination through several generations. The fruit fly, the study of which has uncovered so many genetic principles, is an ideal genetic tool mainly for the same reason. The rapid advance of molecular genetics in the past decade has stemmed largely from extremely fine analyses of bacteria and viruses whose rapid reproductive rates provide numbers that cannot even be approached in higher eukaryotic forms.

An additional point that also entails attention to class size involves the application of a correction factor. The need for this factor may become clearer after the next chapter is read, but it can be appreciated at least in part now. In our genetic experiments, we usually classify flies or other organisms for different traits and assign them to distinct classes, such as red eyes versus brown, black body versus gray, and so forth. In such cases, we are not involved with any continuous gradation of phenotypes from one extreme to the other, such as dark-red eyes through lighter red, pink, light pink, and finally to white. We are concerned primarily with discontinuous distributions containing groups sharply distinguished from one another. However, classification of many characteristics in living things does not lend itself to such precise distinctions. If we were to group 50 people into categories based on their weights, they would not fall into just a few sharp classes but rather into many, ranging from heaviest to lightest, with most of the people falling somewhere between the extremes. We could expect a continuous distribution, that is, the various classes would not be sharply separated, as in red versus white or black versus gray, which are separated by big jumps. The relevant point is that the χ^2 formula has been developed from considerations of continuous distributions. So when the χ^2 test is used for discontinuous distributions, an error is introduced. This is especially so when dealing with classes of fewer than 10 figures, and it can also make a difference when working with just 1 degree of freedom. In borderline cases, failure to apply a correction factor called the *Yates correction* could result in an unwarranted rejection of a hypothesis. The correction is made simply by subtracting a value of 0.5 from each deviation and then proceeding as already described (Table 6-6). We see that the correction in this case made little difference. It can, however, play a role, as stated earlier, and should be applied routinely in all analyses involving only 1 degree of freedom, as well as in those in which any one of the class values is close to 10.

One final point about χ^2 must be appreciated. Many fail to understand that the χ^2 does not tell us that a hypothesis is wrong or correct. Rather, it simply indicates that the numerical data we have assembled either do or do not support the hypothesis. For example, assume we apply χ^2 to simple counts of the number of male and female flies resulting from an ordinary cross in Drosophila. Suppose our hypothesis

is that male and female zygotes are formed in about equal frequency, and we expect roughly 200 males to 200 females. However, we actually obtain 250 females and 150 males. A χ^2 analysis tells us that the data are not in agreement with a ratio of 1:1. The probability is extremely low that this is a distorted 1:1 ratio caused by chance factors. We have good reason to reject the idea of a 1:1 ratio, but the χ^2 does not tell us that the hypothesis about equal numbers of male and female zygotes is wrong. Instead of rejecting it indiscriminately, we should consider several nonchance factors in the conditions of the experiment that could have upset the results. Are we sure that the stocks of flies were free of sex-linked recessive lethals that could kill off more male than female larvae? Were the expected males mainly mutant types whose viability was decreased at certain temperatures? Was the experiment properly controlled? Perhaps we should not be hasty in rejecting the hypothesis but should repeat the experiment with other stocks of flies while controlling temperature and other environmental conditions more closely. We must not reject basic principles because of a misconception of what χ^2 can tell us.

We also see from this that χ^2 can support an incorrect idea. Suppose the hypothesis had been that sex in the fruit fly is not inherited on the basis of the familiar X-Y mechanism but rather is inherited in some fashion that favors an excess of females. The χ^2 would then support the erroneous hypothesis based on the preceding results. Intelligent application of the laws of probability and χ^2 further the advance of genetic research by providing a method that permits discrimination between sound interpretation and one that may mislead the investigator. Misuse of these valuable mathematical tools, on the other hand, can cloud the issue and slow down the progress of valuable efforts. Furthermore, naive ideas about probability can cause unnnecessary anxiety when applied to counseling of human families.

TABLE 6-6   Application of correction factor to χ^2 derivation

CLASS	o[a]	e[b]	d[c]	CORRECTED d	d^2	d^2/e
Red eyes	290	300	−10	−9.5	90.25	0.30
Brown eyes	110	100	+10	+9.5	90.25	0.90
					$\chi^2 = \Sigma d^2/e =$	1.20

[a] Observed.
[b] Expected.
[c] Deviation.

REFERENCES

Chase, W. and F. Brown. *General Statistics*. Wiley, New York, 1986.

Dixon, W. J. and F. J. Massey, Jr. *Introduction to Statistical Analysis*. McGraw-Hill, New York, 1983.

Fleiss, J. L. *Statistical Methods for Rates and Proportions*. Wiley, New York, 1981.

Rostagi, J. S. *Introduction to Statistical Methods,* Vol. 1. Rowman and Allanheld, Totowa, NJ. 1984.

REVIEW QUESTIONS

Questions 1 through 12 constitute an exercise on pedigree analysis based on the pedigrees shown in Fig. 6-4. Allow A to stand for any dominant allele and a for its recessive allele.

Pedigree I:

1. Write as much of the genotype as possible for all persons in the pedigree.

2. If person IV-4 marries a normal man from outside the family, what is the chance that at any birth an abnormal child will be born?

3. Answer Question 2 for person IV-5, assuming she marries her cousin, person IV-7.

Pedigree II:

4. Write as much of the genotype as possible for all persons in the pedigree.

5. What is the chance that person V-1 carries the recessive allele?

6. What is the chance of having an affected child if person V-1 marries an affected person?

Pedigree III:

7. What is the chance that the first child will be affected if person V-1 marries person V-4?

8. Answer Question 7 for person V-3 marrying person V-5.

9. What is the chance that the first boy will be affected if person V-2 marries a normal woman?

Pedigree IV:

10. Write as much of the genotype as possible for each of the following persons: II-2, II-4, II-5, III-2, III-9, III-10, and V-3.

11. What is the chance of having an affected son at any birth if person V-3 marries person V-9?

12. Answer Question 11 for persons V-6 and IV-10.

In the following questions, consider the normal skin pigmentation trait (A) dominant to albino (a). These traits are autosomal. The sex-linked trait, normal iris (I) is dominant to cleft iris (i). Also consider migraine headache to depend on a dominant allele (M) and the normal condition on its recessive allele (m). These alleles are not sex linked.

13. In a family of six children, give the probability that three girls and three boys will occur in this order: girl, boy, girl, boy, girl, boy.

14. In a family of four boys, what is the chance that the next child will be a boy?

15. If six babies are born in a hospital on January 1, what is the chance that the following situations will occur?

   A. They will all be girls.
   B. They will be all boys.
   C. They will all be of the same sex.
   D. Three will be girls and three will be boys.

16. Assume that a man and wife are both known to be heterozygotes, Aa, carriers of the recessive allele for the albino trait.

   A. What is the chance that the first child will be an albino girl?
   B. What is the chance that the first child will be a normal boy?
   C. What is the chance that the first child will be either an albino girl or a normal boy?

17. A. What is the chance that the couple in Question 16 will have children in this order: a normal girl, an albino girl, an albino boy, a normal boy, and finally, a normal girl?
   B. What is the chance that in a family of five children three normal children and two albinos will be born, in any order?
   C. If these two heterozygous persons already have three normally pigmented children, what will the fourth child be like?

18. You are told that an acquaintance of yours has two children. Your informant knows that one of them is a girl but does not know the sex of the other child. What is the chance that the other child is also a girl? Explain.

19. Two parents with normal skin pigmentation carry the recessive autosomal allele (a) for albinism. They have five children, two albinos and three with normal pigmentation.

   A. What is the chance that all the normally pigmented children are homozygous for the dominant allele?
   B. What is the chance that all the normally pigmented children will be carriers?

20. If a woman who is normally pigmented has an albino brother and is married to an albino man, what is the chance that the first child will be albino? (Her parents are not albino.)

21. Suppose that two parents are genotype Mm and thus suffer from migraine headaches. What is the chance that the following will occur?

    A. Their first child will be a girl with migraine and their second a boy without the disorder.
    B. In a family of four these parents will have three children free of migraine and one who suffers from it.

22. Suppose a woman with normal iris whose father had a cleft iris is married to a man with normal iris.

    A. What is the probability that a son will have cleft iris?
    B. What is the chance that the first child will be a boy with cleft iris?
    C. What is the chance that the first child will be a daughter who does not carry the recessive allele?
    D. What is the chance that a daughter will carry the recessive?

23. Assume that two married persons are dihybrids for the skin pigmentation and headache characteristics (AaMm).

    A. What is the chance that they will have children in the following order: normal pigmentation and no migraine, normal pigmentation and migraine, albino with migraine?

    B. What is the chance that the first child will be an albino girl who does not suffer from migraine?

24. Considering the two dihybrid parents in Question 23, what is the chance that if they have four children, two will be completely normal, one will be an albino with migraine, and one will be an albino without migraine?

25. Several pairs of chickens with slightly twisted feathers are crossed. They produce the following kinds of offspring:

    Extremely twisted—33
    Slightly twisted—74
    normal—41

    Submit these data to a χ^2 test for compatibility with a ratio of 1:2:1. What is the χ^2 value? What is the P value? Is this a ratio of 1:2:1?

26. Phenotypically normal, dihybrid fruit flies are test-crossed to flies showing both recessive traits—arc wing and black body. Subject the following data to a χ^2 test on the hypothesis that the genes at the arc and the black loci are undergoing independent assortment. The results are as follows:

    wild—539
    arc—59
    black—712
    black, arc—490

7

CONTINUOUS VARIATION AND ITS ANALYSIS

Continuous versus discontinuous variation

The basic principles of genetics are based on traits that are very well defined. Mendel's pea plants were either definitely tall or short; flowers were either red or white. Morgan's fruit flies also possessed many clear-cut phenotypic features that could be easily followed from one generation to the next (red versus white eye color; short wing versus long wing). Characteristics such as these (height in peas; eye color in the fly, etc.) are said to show discontinuous variation, because contrasting traits can be recognized and easily assigned to distinct categories. In contrast are those characteristics that exhibit continuous variation, in which the phenotypic differences slowly intergrade. Height, weight, or length and breadth of an organ usually describe a continuous distribution from one extreme to the other. No sharp distinctions are found among the phenotypes to permit the recognition of clear-cut classes. Moreover, extra effort must be taken to control environmental effects before the genetic basis of such characteristics can be analyzed. And special methods are needed to describe the variation that is seen, because the expression of the characteristic intergrades from one individual to another. It is no wonder that the nineteenth-century geneticists who selected continuous variation for genetic study encountered great difficulties that obscured the underlying genetic basis. Even after the discovery of Mendel's paper and the demonstration of its relevance to different species, many biologists still believed that Mendelian inheritance did not apply to those characteristics exhibiting continuous variation.

Before 1900, statistical methods were developed to examine complex phenotypic features. The measurement of biological traits and the application of statistics to them form the basis of the discipline of biometry. A mathematical approach is often of great value in the analysis of biological problems, but it may involve many pitfalls. (Recall the discussion on the use of χ^2 in the last chapter.) Using mainly statistical methods and lacking any knowledge of the fundamental principles of genetics, nineteenth-century biometricians often reached erroneous conclusions on the nature of continuous variation. This kind of variation is also referred to as *quantitative variation* and the characteristics that display it are frequently called *quantitative characteristics*. These expressions recognize the fact that the variation shown can be measured and expressed in mathematical terms.

Johannsen was one of the first biologists whose keen insight helped to elucidate the nature of continuous variation. We have already mentioned (Chap. 4) his importance in distinguishing between genotype and phenotype. Johannsen applied statistical methods to study continuous variation in plants. His work showed that mathematics can be very useful when properly applied to quantitative characteristics.

Effects of inbreeding

Johannsen worked with beans and the inheritance of length, breadth, and weight of seeds in this plant. His observations and conclusions held far-reaching implications for the science of genetics. Johannsen was able to demonstrate clearly the distinction between the genetic and the environmental components that influence the expression of quantitative characteristics. Before we discuss his observations, a few words concerning the effects of inbreeding are in order. The bean is normally a self-pollinated plant. Continued selfing over the course of generations renders the offspring more and more homozygous (Fig. 7-1). Self-fertilization, which can take place only in hermaphroditic organisms, is an extreme case of inbreeding, but it nevertheless demonstrates certain important facts. It is evident that the continued mating of individuals within a line or family increases the chances of bringing together hidden recessives. On the other hand, hybrid vigor, also called heterosis, often accompanies heterozygosity; therefore, it will tend to decrease with continued inbreeding.

Heterosis is a phenomenon well known to crop breeders who are familiar with the more robust nature of hybrids compared with individuals from more inbred strains. Each of two commercially inbred strains may possess some desirable trait (greater yield, parasite resistance). When the strains are crossed to produce F_1 plants, it is not at all uncommon to find that members of this generation are larger and more robust in many ways than members of the parental strains.

How can such heterosis be explained? One theory recognizes that when strains are inbred to ensure the establishment of a desirable trait, the strain becomes more and more homozygous at various loci. Consequently, deleterious recessive alleles can come to expression, reducing the vigor of the inbred strains. The F_1 hybrids between the two different inbred strains would contain dominant alleles at many of the loci. This is so because one inbred strain does not necessarily carry the same deleterious recessives as the other, as is seen in the following:

Highly inbred strain A   Highly inbred strain B
AAbbCCddEE × aaBBccDDee
F_1AaBbCcDdEe

In the F_1s, the deleterious effects of the recessives are masked by the dominant alleles, and the individuals are more robust than members of the inbred lines.

Another explanation of hybrid vigor stresses the value of the heterozygous state itself. It does not maintain that dominant alleles will necessarily confer more vigor than recessive ones. Although we cannot yet give definite answers to explain hybrid vigor, knowledge of the relationships between genes and enzymes permits us to formulate some plausible explanations. Sensitive procedures have shown that many so-called normal or wild protein products within populations in nature may actually represent a mixture of two very similar but still detectably different proteins (Chap. 21). Individuals thought to be homozygous wild with regard to a specific locus often prove to be heterozygous. Refined chemical analysis often reveals that the protein governed by the locus in question consists of slightly different molecules, differing somewhat in electric charge as the result of just one or a few amino acid differences. The individ-

FIG. 7-1. Effects of continued inbreeding. With continued self-fertilization, the percentage of homozygosity increases per generation.

Note that the percentage of the heterozygotes decreases by half each generation as the percentage of homozygotes increases.

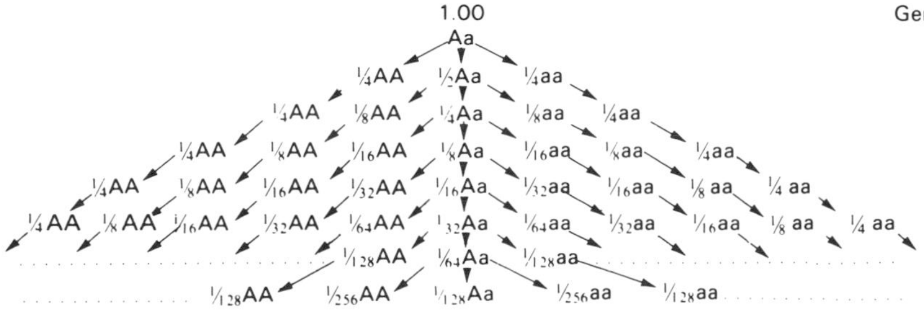

	Generation	Homozygotes AA and aa	Heterozygotes Aa
1.00 Aa	0	0	100%
	1	$\frac{1}{2}$ (50%)	$\frac{1}{2}$ (50%)
	2	$\frac{3}{4}$ (75%)	$\frac{1}{4}$ (25%)
	3	$\frac{7}{8}$ (87.5%)	$\frac{1}{8}$ (12.5%)
	4	$\frac{15}{16}$ (93.75%)	$\frac{1}{16}$ (6.25%)
	5	$\frac{31}{32}$ (96.875%)	$\frac{1}{32}$ (3.125%)
	6	$\frac{63}{64}$ (98.4375%)	$\frac{1}{64}$ (1.5625%)
	7	$\frac{127}{128}$ (99.22%)	$\frac{1}{128}$ (0.78%)

ual appears to be homozygous for the wild allele at the locus, but is actually heterozygous, because two gene forms must be present, each one directing the formation of a slightly different protein. Since heterozygotes are generally more vigorous than highly homozygous individuals, this could mean that being heterozygous for different forms of a gene at a given locus is more advantageous than being homozygous for any one of them. Such studies have revealed that what has often been thought to be a single enzyme turns out to be two closely related but nevertheless slightly different forms. They are concerned with the same catalytic reaction but differ in response to temperature, pH, and other environmental changes. Such similar but slightly different enzymes are called isoenzymes or isozymes. Heterozygosity at a certain locus, that controls a metabolic step would bring about the production of not just one type of enzyme but two forms of one enzyme—two isozymes. The presence of two forms of an enzyme could possibly confer an advantage on an individual. Heterozygosity at many loci would thus result in an overall more vigorous phenotype.

Figure 7-1 follows only one pair of alleles through several generations of self-fertilization. However, it should be evident that intensive inbreeding reduces the amount of variation in a population by bringing about a higher percentage of homozygosity for *all* pairs of alleles. Any organism that has sexual mechanisms but is exclusively self-fertilized, as some plants do, goes through the motions of sex without receiving its main advantage: the production of a large assortment of genotypes and phenotypes on which the force of natural selection can operate.

Exclusive self-fertilization can deprive any plant species of the flexibility needed to produce a range of new allelic combinations, some of which may have a selective advantage in a changed environment. Plant groups that are only self-pollinated are generally at a disadvantage in relation to those in which a certain amount of outcrossing is ensured.

An outstanding example of the consequences of inbreeding and increased homozygosity in an animal is seen in the case of the cheetah, whose fertility has been decreasing at an alarming rate. The male shows a high incidence (71%) of abnormal sperm in a sample of semen. Aberrant sperm is a feature known to accompany inbreeding in laboratory and farm animals. The cheetah also exhibits an increased susceptibility to disease and a high infant mortality rate, all associated with increased inbreeding, an increase in

homozygosity, and a decrease in genetic variation in the population. We will return to this important topic in Chapter 21.

Genetic versus environmental variation

Although continued selfing greatly increases the degree of homozygosity, Johannsen showed that a characteristic such as seed length still displays variation when the offspring from a single self-fertilized plant are studied. If one continually selects seeds of longer length within a highly inbred line, the same variation continues to be shown (Fig. 7-2A); the average value for length of seed does not change. Johannsen continued to select the longest and the shortest seeds within a pure line and found it was ineffective. The same average value persisted from one generation to the next. The variation seen within a pure line is thus environmental, because continued self-fertilization makes all the individuals within an inbred line genetically identical. However, different results were obtained when selection was applied to a mixed population. Such a population is composed of many self-fertilizing individuals, each of which represents a different pure line (Fig. 7-2B). The population thus has genetic as well as environmental diversity. Various pure lines are present that have larger average seed values than other lines in the population. This is an expression of genetic diversity among the lines. Choosing only the largest seeds in the mixed population selects out those pure lines whose genetic factors enable them to produce larger seeds than the average value for the population. This selection can therefore cause a shift in the average size of the seed, but within a line, selection for larger seed remains ineffective. All the variation within the line is due to the environment. Johannsen's work showed the importance of the environment on complex characteristics, but it also demonstrated that these environmental modifications could not be inherited. The concept of Weismann was upheld over the claims of the Lamarckians.

Genes and quantitative inheritance

Not the least of Johannsen's contributions to genetics was his emphasis on the role of the entire genotype in the expression of a characteristic. Too many early geneticists conceived of one unit factor for one character, the erroneous idea that one gene has just one effect, which is in no way influenced by other genes.

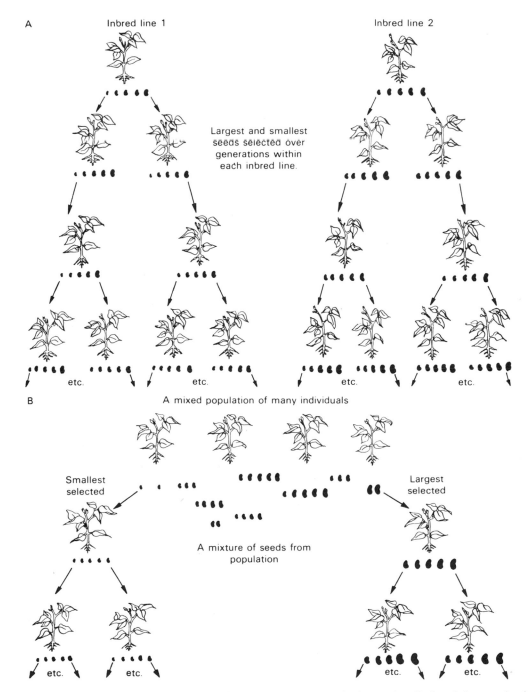

FIG. 7-2. Selection for seed size in beans. (*A*) The seeds from a highly inbred, self-pollinated line show a range in size. When two lines are compared, one may have a larger average size value than the other. Selection of the smallest and the largest seeds within a line generation after generation causes no change in the average value for seed length. The same variation continues to be shown, indicating the variation is due to the environment. (*B*) If the longest and shortest seeds are taken from a mixture obtained from a whole population of

individuals, selection can be effective. A line may be derived that produces smaller seeds on the average than the population as a whole. Another may be selected that bears seed larger than those found in the population as a whole. The variation in seed size results from both genetic and environmental factors. The population is composed of many genetically diverse lines, and these may be selected out. Within each inbred line, a small-seeded or a large-seeded one, environmental variation will continue to be shown.

Johannsen argued that any character is the result of the interaction of many genes and the environment. The demonstration that quantitative characters were dependent on the interaction of several genes was very important in establishing this idea. Perhaps the first clear-cut example was that of color inheritance in wheat grains, in which color intensity varies from white through intermediate shades to red. In 1909, Nilsson-Ehle offered strong evidence that the color characteristic in wheat depends not on one but on several pairs of alleles that undergo random assortment.

When individuals from a pure-breeding red race are crossed with those from a white grained one, the F_1 are intermediates. In the F_2, however, various shades occur among the grains, from full red to white. Among the F_2 offspring, 1 out of 64 is white grained and approximately 1 out of 64 is red, like the red and white of the P_1 parents. Nilsson-Ehle interpreted these results on the basis of three pairs of alleles that assort independently (Fig. 7-3). Each of the factors for redness, R_1, R_2, and R_3, contributes an equal amount of pigment to the phenotype. The effects are cumulative; they can simply be added together. The alleles of the

FIG. 7-3. Quantitative inheritance in wheat. Kernel color depends on three pairs of alleles. Each genetic factor that contributes to pigment formation (R_1, R_2, and R_3) adds an equal dosage. Their alleles (r_1, r_2, and r_3) contribute nothing to pigment formation. The trihybrid carries three alleles for pigment and produces kernels intermediate in color between the parents. The F_1s form eight classes of gametes. When

these combine in all possible combinations, a range in shade is found. The Punnett square shows only the number of effective pigment alleles carried in the offspring. Only 1 out of 64 possesses six pigment factors, and only 1 out of 64 carries none at all. All of the other offspring vary in shade between the original (P_1) parents.

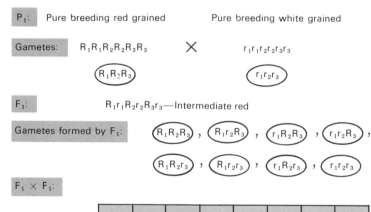

	$R_1R_2R_3$	$R_1r_2R_3$	$r_1R_2R_3$	$r_1r_2R_3$	$R_1R_2r_3$	$R_1r_2r_3$	$r_1R_2r_3$	$r_1r_2r_3$
$R_1R_2R_3$	6	5	5	4	5	4	4	3
$R_1r_2R_3$	5	4	4	3	4	3	3	2
$r_1R_2R_3$	5	4	4	3	4	3	3	2
$r_1r_2R_3$	4	3	3	2	3	2	2	1
$R_1R_2r_3$	5	4	4	3	4	3	3	2
$R_1r_2r_3$	4	3	3	2	3	2	2	1
$r_1R_2r_3$	4	3	3	2	3	2	2	1
$r_1r_2r_3$	3	2	2	1	2	1	1	0

pigment genes, r_1, r_2, and r_3, contribute no pigment and hence no redness to the phenotype; an individual homozygous for all three of these is thus white grained.

Other similar examples of quantitative inheritance were found by others, and these formed the basis for the theory of multiple-factor inheritance. Actually, no new concept is introduced, because the multiple-factor idea explains continuous variation in terms of Mendelian genetics. However, characteristics that exhibit continuous variation typically involve the interaction of more than two pairs of alleles, as well as a pronounced effect by environmental influences. And the effects of the different alleles are quantitative; they may be added up, in contrast to pure dominance or lack of dominance. Multiple factor inheritance, also known as quantitative or polygenic inheritance, implies that several pairs of alleles are interacting and that each has a similar and measurable effect on a characteristic, which can also be influenced by the environment.

The example of grain color in wheat brings out several important points. Many nineteenth-century biologists might have thought that blending inheritance would explain the results. However, although the F_1 has only one phenotypic class (intermediate to the parents), the F_2 results clearly show that blending has not occurred because the extremes (white and red) again appear unchanged. The range of variation seen in the F_2 generation indicates that genes are segregating and that recombination is taking place. It is also evident that there is not one gene for one character. Grain color depends on the interaction of several genes and their cumulative effect. Moreover, in the F_2, environmental factors may operate to eliminate sharp distinctions among the major color groupings. The cross also reveals that two individuals of the same phenotype may have different genotypes. For example (Fig. 7-3), the genotypes $R_1R_1R_2R_2r_3r_3$ and $R_1r_1R_2r_2R_3R_3$ are in the same general color class because each has four pigment alleles, but the two are quite different genetically.

Quantitative inheritance and skin pigmentation

Not long after its formulation, the multiple-factor hypothesis was applied to skin color differences in humans. All normal individuals possess skin pigment, which is found in the melanocytes (pigment cells) of the live layer of the epidermis. The skin pigment is melanin, a dark brown product that combines with protein to form pigment granules in the melanocytes.

Melanin formation depends on the oxidation of the amino acid tyrosine. [Recall the example of phenylketonuria (PKU) in Chap. 4, in which pleiotropy was discussed.] The amount of melanin in the melanocytes is responsible for varying shades of skin color. A certain autosomal recessive gene may block melanin formation and result in albinism, a condition characterized by an abnormally low amount of pigment. This allele is not confined to any one group of humans. Except for those individuals who are albino, the normal range of skin shades varies from the lightest human population to the darkest. Of course, within so-called white or black populations, the degree of pigmentation also varies. The earliest studies of inherited pigment differences in human groups were undertaken by Davenport and Davenport. The method of measuring skin color in black and white parents and their F_1 offspring was very limited, but the observations could be explained on the basis of quantitative inheritance in which approximately two pairs of independent alleles are involved. Recently, more accurate methods using spectrophotometry have been employed to estimate the degree of pigmentation. The equivalents of F_1 and F_2 testcross progeny between blacks and whites have been followed. Although still limited, the results suggest that three or four pairs of alleles are operative in skin color differences. Although the present interpretation may still be an oversimplification, there is no doubt that these color variations have a polygenic basis and that only a relatively small number of genes is directly involved.

For the sake of simplicity, we will assume that two pairs of alleles provide the basis for skin shades in humans (Fig. 7-4). An extreme white person would contain no effective pigment alleles above a certain basic level. The extreme dark parent would have four pigment alleles in our simplified illustration. The principles demonstrated by the red pigment of the wheat grains also apply here to melanin pigmentation of skin. As in the case of grain color, environment also influences gradations of shading in humans, so that no distinct steps exist from shade to shade. Again, individuals of the same phenotype may have very different genotypes. For example (Fig. 7-4), persons of the following three genotypes would have the same intermediate phenotype, ignoring the environment: (1) AaBb, (2) AAbb, and (3) aaBB. Certain consequences ensue from this. A cross of two persons of genotype 1, the dihydrid AaBb, will produce offspring that show a range of variation, indicated in Fig. 7-4. On the other hand, a cross of a person of genotype 2 with one of genotype 3 (Fig. 7-5A) can give rise only to offspring of the same shade as the parents. So we would expect color variation to be great among the offspring in certain families in which the parents are intermediate in degree of pigmentation. In other families, little or no variation in shade would be expected from intermediate parents. All observations suggest that this is the case. It should also be evident that a mating between a person of any shade with a white individual cannot produce any children who are darker than the more pigmented parent (Fig. 7-5B).

Quantitative inheritance and deep-seated characteristics

Although skin and grain pigmentation represent rather superficial characteristics, a moment's reflection tells us that the most deep-seated characters in any species are ones that show continuous variation and are quantitative in nature—such features as height, weight, length, and breadth of limbs or organs. Although the details of the inheritance of such characteristics are unknown for most species, there is little doubt that they have a polygenic basis. Indeed, polygenic differences undoubtedly provide the distinctions between closely related species. An accumulation of many genetic differences that influence basic physiology may affect the ease with which members of separated groups can mate and so lead to genetic isolation between them, even though both arose from one original population (see Chap. 21 for details).

We still know little about the inheritance of such very complex characteristics as behavior, intelligence, and personality. We actually know much more concerning the genetic basis of the more superficial characters of living things than we do about those that are much more fundamental. The reason for this soon becomes obvious. Figure 7-3 shows a Punnett square for the superficial color characteristic in wheat grains in which only three pairs of alleles apparently play a major role. Yet segregation of just these few pairs produces an array of types. Only 1 individual out of 64 has the same genetic combination for pigmentation as does one of the parents, $R_1R_1R_2R_2R_3R_3$ or $r_1r_1r_2r_2r_3r_3$. More complex characteristics must involve many more genes than this. Table 7-1 shows that with just five genes (pairs of alleles) so many new combinations are possible that only a minute number (1/1024) would resemble one of the original parents. This is so because many pairs of alleles are segregating and forming an assortment of new combinations. We would

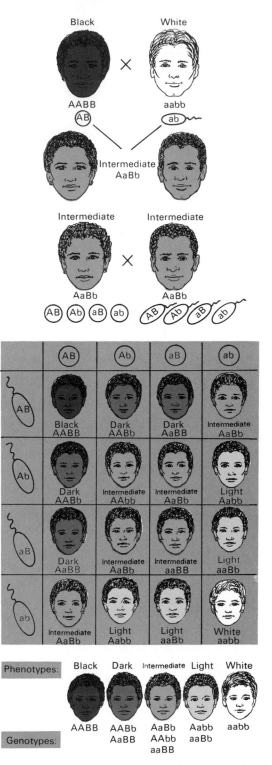

FIG. 7-4. Inheritance of skin color. A mating between a black and a white person produces offspring intermediate in shade. Two parents of this genotype shown here who are intermediates will have offspring who vary in color from black, through various lighter shades, to white. (Probably three or four pairs of alleles are involved, but the principle is the same, as illustrated on the basis of two.)

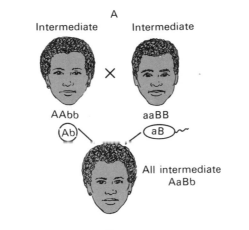

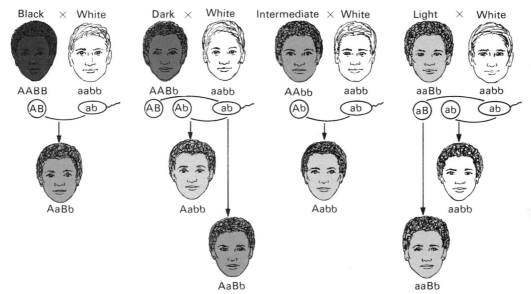

FIG. 7-5. Skin color in certain families. (*A*) Some parents who are intermediate in shade give rise to offspring all of whom are also intermediate. In such cases, each parent is homozygous for pigment alleles at one of the loci. In this example, they happen to be homozygous at different loci, but intermediate offspring would also result if the genotypes of both parents were either AAbb or aaBB. These results should be contrasted with those in Fig. 7-4. (*B*) A cross of a person of any shade with a white individual can never give an offspring darker than the nonwhite parent. This is so because the white contributes no effective pigment alleles that can add to those that may be donated by the darker parent. (Only some of the possible combinations are represented here, but the principle is the same for all of them.)

certainly expect the genetic basis of a very complex character like intelligence to involve more than 20 different allelic pairs. The number of combinations possible becomes staggering. And the expression of the many genotypes is modified further by the influence of the environment. Moreover, for the sake of simplicity, we have assumed that the effects of each pair of alleles are equal and additive. We now know that this is not so in most cases of polygenic inheritance. Some alleles contribute more of an effect than others to a quantitative character. We have also ignored the possible involvement of dominance and epistasis. It is no wonder that the genetic analysis of continuous variation has progressed more slowly than that for discontinuous variation. However, we do have approaches to the study of polygenic inheritance. Although certain assumptions must be made when applying them, they can still give some idea regarding the number of pairs of alleles involved and their quantitative contribution to the characteristic.

TABLE 7-1   Fraction of offspring showing one parental extreme[a]

ALLELIC PAIRS	DIFFERENT KINDS OF GAMETES	FRACTION SHOWING ONE PARENTAL EXTREME
1	$2^1 = 2$	1/4
2	$2^2 = 4$	1/16
3	$2^3 = 8$	1/64
4	$2^4 = 16$	1/256
5	$2^5 = 32$	1/1,024
6	$2^6 = 64$	1/4,096
7	$2^7 = 128$	1/16,384
8	$2^8 = 256$	1/65,536
9	$2^9 = 512$	1/262,144
10	$2^{10} = 1024$	1/1,048,576

[a]This fraction indicates the number of allelic pairs involved in the inheritance of the characteristic under consideration.

Estimation of allelic differences and their quantitative contribution

To approximate the number of pairs of alleles we turn our attention to the fraction of F_2 offspring that expresses a phenotype as extreme as one of the parents (Table 7-1). For example, suppose two races of tobacco plants are crossed (Fig. 7-6A). One race has long flowers in which measurements range from 88 to 100 mm, with a mean or average of 93 mm. The short flowers of the second race range from 34 to 42 mm, with a mean of 39 mm. The F_1 offspring are approximately intermediate in flower length, ranging from 55 to 70 mm, with a mean of 66 mm. When the F_1 are crossed, the 800 F_2 progeny show a much greater range of variation than did the F_1 (from 42 to 91 mm). Of the 800 F_2 offspring, only 12 were in the size range of the longer parent and 10 in the range of the shorter one. The data from the cross indicate several things and suggest polygenic inheritance. On such a basis, we would expect two pure-breeding races to produce an intermediate F_1 when they are crossed. The much greater variation among the F_2 than the F_1 offspring results from the segregation of several allelic pairs and the formation of many new genetic combinations. The 12 largest ones (12 out of 800, or approximately 1/66) and the 10 shortest (10 out of 800, or 1/80) suggest that three pairs of alleles are segregating, because three allelic pairs will produce a fraction of approximately 1/64 (Table 7-1), which is as extreme in phenotype as one of the parents.

We can now consider the small-flower P_1 parent (Fig. 7-6) to have a genotype in which the minimal effect on flower length, an average of 39 mm, is being expressed by an unknown number of genes (Fig. 7-6B). This same minimal effect on length is also being exerted in the long flowered parent. The length in millimeters above the minimum (the smaller average of 39 mm) represents the effective genetic contribution of other genetic factors. It is these that differ between the long- and the short-flowered races to cause a difference in flower length. So the lower average length (representing the minimal or basic length) is subtracted from the larger average. The 54 mm that remains is the effective difference between them. Since six genetic factors (two each of alleles A, B, and C) contributed to that amount, each allele donates an average of 9 mm to petal length. The F_1 received three alleles from the larger parent that contribute an effective difference. The small parent gave no alleles that add to the difference. Therefore, the first-generation offspring would have a mean value approximately midway between the two parental means.

To study continuous variation in this way, environment must be controlled as closely as possible because the length of any organ can certainly be influenced by environmental factors. Two individuals of the identical genotype could have very different phenotypes simply as a result of these factors. To estimate with any degree of accuracy the number of effective alleles that are operating in a specific case, large numbers of individuals must be raised. Where five or more segregating pairs of alleles are involved, it may be impossible in many cases to obtain one of the parental extremes. Moreover, in this example of flower length we assumed equal effects of each contributing allele, and we know this may not be so. Despite these limitations, however, some knowledge may be gained concerning the genetic basis for a character difference between two different stocks or varieties. It must also be emphasized that estimating the number of gene differences in this way does not mean that these are the only genes affecting the character. In our example, the smaller P_1 parent certainly has a flower length, and we noted that an unknown number of genes contribute to it, as it does for that same amount in the longer parent. The same holds in the case of pigment differences in humans. Normal whites, as noted, possess melanin pigment. Considering the average white genotype to be minimal, we try to estimate the number of effective allelic differences between

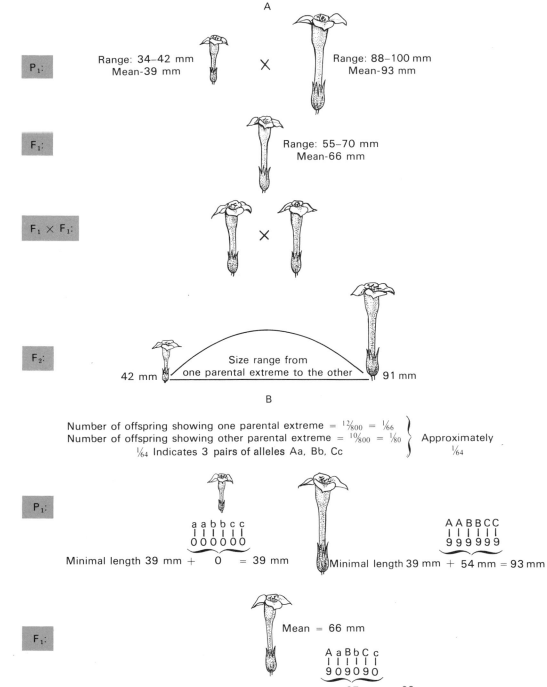

A

P₁: Range: 34–42 mm   ×   Range: 88–100 mm
     Mean-39 mm   Mean-93 mm

F₁: Range: 55–70 mm
                    Mean-66 mm

F₁ × F₁: ×

F₂: Size range from
     42 mm one parental extreme to the other 91 mm

B

Number of offspring showing one parental extreme = ¹²⁄₈₀₀ = ¹⁄₆₆ ⎫
Number of offspring showing other parental extreme = ¹⁰⁄₈₀₀ = ¹⁄₈₀ ⎬ Approximately
¹⁄₆₄ Indicates 3 pairs of alleles Aa, Bb, Cc   ⎭ ¹⁄₆₄

P₁:
        a a b b c c   A A B B C C
        | | | | | | | | | | | |
        0 0 0 0 0 0   9 9 9 9 9 9
Minimal length 39 mm +   0   = 39 mm   Minimal length 39 mm + 54 mm = 93 mm

F₁:
                Mean = 66 mm
                A a B b C c
                | | | | | |
                9 0 9 0 9 0
     Minimal length = 39 mm + 27 mm = 66 mm

FIG. 7-6. Quantitative inheritance in tobacco. (A) When a short-flowered race is crossed with a long-flowered one, the F₁ has a mean about intermediate between the parents. A range in flower size is seen in the P₁ and the F₁. When two F₁ plants are crossed, the F₂ show a range in flower size that greatly exceeds the size range in millimeters shown in the F₁ or P₁. Some of the F₂ plants have flowers as short as those found in the short parental race and some as long as those in the longer parental race. Most of the F₂s have flowers that range in length between these extremes. The observations indi-cate that several pairs of alleles that contribute to flower length segregated in the F₁ to produce the variation seen in the F₂. (B) Calculation of effective allele differences involved in the cross. The genotype of the shorter race can be represented as aabbcc and the longer as AABBCC, because three genes are indicated from the crossing data. The six factors in the longer parent contribute to the difference between the two mean lengths, a difference of 54 mm. Therefore, each allele must add 9 mm.

white and black populations, just as we approximated that three genes are involved in producing the differences between the large and small-flowered races.

Transgressive variation

Let us next examine a cross that produces results that may at first seem bizarre but can easily be explained on the basis of polygenic inheritance. Suppose that each of two inbred varieties of corn has a mean value of 64 in. in height. Suppose that the two varieties are crossed to obtain greater disease resistance and that height is not being considered. As it happens, the F_1 offspring also average approximately 64 in. in height (Fig. 7-7A). When the F_2 progeny arise, however, a spectrum of sizes is found. Of 2000 F_2 plants, 7 are only about 32 in. in height, whereas 9 reach 96 in. The unexpected results can be readily attributed to the segregation of several pairs of alleles in the F_1. Apparently, the P_1 plants, although homozygous for effective alleles that influence height, were not extremes. The possible extremes appeared in the F_2 after segregation and independent assortment in the F_1.

We can now estimate the number of effective pairs of alleles segregating for height, and we can approximate their contribution, again making the assumptions we did previously (Fig. 7-7B). Taking 32 in. as the minimal height resulting from factors common to both races, the effective genetic difference becomes 96 in. $-$ 32 in. $=$ 64 in. (The average largest height minus the smallest.) The number of F_2 plants showing parental extremes (9/2000 and 7/2000) is close to 1/256 and suggests that four pairs of alleles contribute to this difference of 64 in., meaning that four genes are affecting height. This means that each allele in the largest individuals on the average contributes approximately 8 in. to height above the minimal of 32. There is actually no basic difference between this and the preceding examples. The only variation is that we did not start out with either possible extreme phenotype in the original parents. We see from this example that offspring can exceed the measurements of the P_1 parents. Such a distribution, in which the offspring exceed the parental measurements, is called transgressive variation. It can explain cases in which progeny may be much taller or heavier than their parents, both of which may have more modest measurements.

Importance of statistics to genetic problems

In several of our examples, we have noted "average" values and acknowledged that a range of variation exists. Fundamental to the analysis of any characteristic showing continuous variation is the need to describe it. Phenotypes that exhibit continuous variation are much more difficult to describe than those in which the observed variation can be sharply classified. For example, in the case of weight, we could choose to lump measurements into only three classes: light, medium, and heavy. At first glance, this might seem to satisfy our needs, but it would soon become apparent that this approach lacks accuracy. Many of the classifications would be arbitrary, because no sharp boundaries exist between the three categories selected to describe the differences.

Fortunately, a superior approach can be made through the use of elementary statistics. The full meaning of the word statistics may be grasped by a consideration of some very obvious points. Most of us certainly realize that measurements are rarely made on entire populations. In a study designed to compare the weights of adults living in New York and those living in London, it would be absurd to believe that all of the people of each city could be weighed for the purpose. If such an impressive feat could be achieved, we would then have the exact figures to describe weights in the two cities. Such a character value for an entire population is known as a parameter. Although we want parameter values to describe accurately the characteristics of whole populations, their realization is often either impossible or impractical. Instead we must content ourselves with measurements made on samples taken from a population. Such a sample is considered representative of the entire population. Thus, a measurement of any sample character should reflect the true nature of that character for the entire population. Statistics are simply measurements of samples rather than whole populations. We use them to estimate parameters, values that are more difficult or even impossible to obtain.

Various statistics exist to describe any particular characteristic, and it is critical to know the proper one to apply in a certain situation. There are a few basic statistics that have wide, practical application, not only in biological studies but in assorted, unrelated fields in which an analysis of figures is required. We will concentrate here on just a few of the commonly used tools that form a basis for the derivation of more

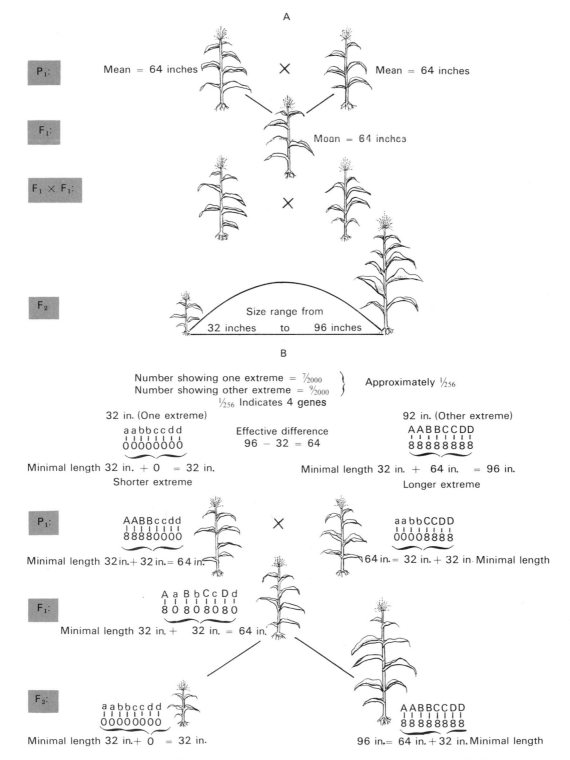

A

P₁: Mean = 64 inches   ×   Mean = 64 inches

F₁: Moon = 64 inches

F₁ × F₁: ×

F₂   Size range from
                   32 inches   to   96 inches

B

Number showing one extreme = 7/2000 }
Number showing other extreme = 9/2000 } Approximately 1/256
        1/256 Indicates 4 genes

32 in. (One extreme)   92 in. (Other extreme)
   a a b b c c d d   Effective difference   A A B B C C D D
   | | | | | | | | 96 − 32 = 64 | | | | | | | |
   0 0 0 0 0 0 0 0   8 8 8 8 8 8 8 8

Minimal length 32 in. + 0 = 32 in.   Minimal length 32 in. + 64 in. = 96 in.
        Shorter extreme   Longer extreme

P₁: A A B B c c d d   a a b b C C D D
     | | | | | | | | × | | | | | | | |
     8 8 8 8 0 0 0 0   0 0 0 0 8 8 8 8

Minimal length 32 in. + 32 in. = 64 in.   64 in. = 32 in. + 32 in. Minimal length

F₁: A a B b C c D d
           | | | | | | | |
           8 0 8 0 8 0 8 0

   Minimal length 32 in. +   32 in. = 64 in.

F₂: a a b b c c d d   A A B B C C D D
     | | | | | | | | | | | | | | | |
     0 0 0 0 0 0 0 0   8 8 8 8 8 8 8 8

Minimal length 32 in.+ 0 = 32 in.   96 in.= 64 in. + 32 in. Minimal length

FIG. 7-7. Transgressive variation in corn. (*A*) In certain crosses, two plants of the same height yield an F₁ that is also similar in height to the parental strains. A cross of these F₁s, however, may yield an F₂ that varies greatly and in which some of the individuals exceed the measurements of the original parents. (*B*) This transgressive variation results when the P₁s are themselves not the extreme types. The number of pairs of alleles involved and the effective contribution of each can be calculated (see the text for details).

171

intricate formulas. Having a full understanding of their application to simple problems, we may be better prepared in later situations to select other appropriate statistics for the most thorough analysis of a complex set of data.

The mean and its calculation

The best known statistic to describe a set of measurements is the numerical average. Most of us have used it at some time to get an idea of a general tendency, whether it concerns a list of test grades or measurements of height or weight, etc. Suppose we are interested in the effects of a certain drug on weight loss when added to the food of experimental animals. After a 2-month period, one group of 20 is weighed, and loss in grams is recorded for each animal. We might first arrange the figures as shown in Table 7-2A. To obtain an average, we could add up all the individual numbers and then divide by the total number of animals, which is 20. This can, however, be simplified a great deal (Table 7-2B). Starting with the lowest value, we can group together those figures that are the same and note their frequency. In this list, we

have five values of 68. We then allow v to represent any individual value (or variable), and f to indicate the frequency of the particular value. By multiplying f by v, we are merely recognizing the individual values and the number of times that each occurs in the sample. This is obviously simpler than adding together five 68s and similarly for the other numbers that appear more than once. All that remains to be done to obtain the mean is to take the total of all the separate values (represented by the symbol Σfv) and divide it by the total number of separate measurements, 20 (n) (Table 7-2C). The choice of symbols to represent different statistical concepts varies from one authority to another, and this often results in confusion for the novice. Here we will try to use those symbols that are the least cumbersome but also frequently encountered.

Once we have the mean for the group of 20 animals, we possess some idea about the tendency toward weight loss in that sample. To obtain additional information from these data, we might reexamine the raw figures to note if any one weight loss value occurs more often than the others. Inspection immediately tells us that "68" is just such a figure. We have now

TABLE 7-2   Recording data and calculation of mean on a set of 20 values[a]

							v	f	fv
(A)	35	28	58	16	35	(B)	16	1	16
					28		28	1	28
	63	63	58	75	58		35	1	35
					16		58	2	116
	68	68	63	75	63		63	3	189
					63		68	5	340
	85	68	78	68	58		69	1	69
					75		75	2	150
	93	69	81	68	68		78	1	78
					68		81	1	81
					63		85	1	85
					75		93	1	93
					85		Totals	$n=20$	$1280=\Sigma fv$
					68				
					78				
					68				
					93	$\dfrac{1280}{20}=64$			
					69				
					81				
					68				
					1280				

$$(C)(\overline{X}=\text{sample means}) \quad \overline{X}=\frac{\Sigma fv}{n}=\frac{1280}{20}=64$$

[a] Key to symbols: A, preliminary recording of data and calculation of average; B, grouping of data and simpler method of summing the values; C, general way to calculate the mean value. v, value or variable; f, frequency; n, total number of values; Σ, sum of.

noted the mode, which is the most frequently repeated value in a list of measurements. The mode may give us an additional way of considering the central tendency.

Standard deviation and its importance

Although it is valuable to have some concept of an average, this knowledge by itself is quite limited and may conceal important aspects of a sample. Such a limitation becomes apparent when we give thought to the following possibility. Suppose we wish to calculate the mean weight loss for another group of 20 animals from a different strain. Conceivably, the two strains could react differently to the presence of the chemical. A tabulation of the weight losses and a calculation of the mean are made just as for the first set of animals (Table 7-3). The mean for this second sample again gives a value of 64. If we were simply given the means for the two strains of animals, we might erroneously conclude that both reacted identically insofar as weight loss was concerned. A glance at the two sets of raw data, however, shows that this is not so (Tables 7-2 and 7-3). The tabulations show that the range of weight loss is much greater in the first group. A few animals lost little (16 and 28 g) in the course of the experiment, whereas some lost an appreciable amount (81, 85, and 93 g).

On the other hand, much less variation is present in the second group, in which the loss in weight of all animals is comparable. With just a knowledge of the means, we would lose sight of the fact that in the first set several individuals were not close to the mean at all, whereas in the second, all clustered around the mean. Fortunately, there is another statistic that provides this information clearly and concisely. This is the standard deviation (SD), also represented by the

TABLE 7-4   Calculation of SD from data in Table 7-3[a]

v	d	d^2	f	fd^2
59	$59 - 64 = -5$	25	1	25
61	$61 - 64 = -3$	9	1	9
62	$62 - 64 = -2$	4	3	12
63	$63 - 64 = -1$	1	5	5
64	$64 - 64 = 0$	0	4	0
65	$65 - 64 = +1$	1	3	3
68	$68 - 64 = +4$	16	1	16
70	$70 - 64 = +6$	36	2	72
	Totals	$n =$	20	$142 = \Sigma fd^2$

$$SD = \sqrt{\frac{\Sigma fd^2}{n-1}} = \sqrt{\frac{142}{19}} = \sqrt{7.47} = 2.73$$

[a]Key to symbols: v, value; d, deviation; f, frequency; n, total number of values; Σ, sum of.

Greek letter σ. It describes for a sample the amount of variation on either side of the mean. Let us calculate the SD for the second group of animals in order to appreciate the valuable information this statistic can give (follow the steps by referring to Table 7-4).

We must first record the deviation (d) of each value (v) from the mean, which is 64. Next, we square each deviation, eliminating minus signs. We must not fail to recognize the number of individuals that deviate from the mean by each set amount. So we multiply the squared deviations by the appropriate frequency (f) for each respective deviation. The SD is then found by applying the formula given in Table 7-4. The reason for dividing by $n - 1$ can be appreciated if we remember that although we are dealing with samples, we are actually interested in a whole population and certain of its parameters. However, in taking a sample, there is the danger of underestimating the variation present in the whole population. To correct

TABLE 7-3   Calculation of mean on a second set of 20 values[a]

v	f	fv	
59	1	59	
61	1	61	
62	3	186	
63	5	315	
64	4	256	
65	3	195	
68	1	68	
70	2	140	
Totals	$n = 20$	$1280 = \Sigma fv$	$\overline{X} = \Sigma = \dfrac{fv}{n} = \dfrac{1280}{20} = 64$

[a]Key to symbols: v, individual value or variable; f, frequency of a value; n, total number of values; Σ, sum of.

for this, 1 is subtracted from the total size of the sample. There may be times when we are interested only in the sample itself and not in the whole population. In such a case, the correction is unnecessary, and we divide by n itself, which in this case is 20.

Now let us proceed to examine the significance of the value, 2.73, which has been calculated as the SD of the second group of animals. To have a basis of comparison, calculate the SD for the first sample, following the illustration given in the table as a guide. When this is completed, a much larger value for the SD will be obtained. This is to be expected, because SD is a measure of the deviation from the mean, and the first group of numbers shows much more variation from the mean than the second.

When a large group of individuals is studied quantitatively for a certain attribute, such as height, weight, or length, most of the values usually cluster around the mean, and fewer are found at the extremes on either side of the mean. If the sample is large enough, a normal or bell-shaped distribution can be expected, with the peak of the curve representing the mean as well as the mode. Typically, any large group of measurements tends to conform to a normal or bell-shaped pattern. The SD is applicable to such distributions, and it indicates how many individuals in a sample are

found with certain values from the mean. One SD indicates that 68% of the individuals measured will fall within ± 1 SD on either side of the mean (Fig. 7-8). For our calculated SD of 2.73, this indicates that approximately 68% of the individuals in the sample would be expected to show a weight loss value in grams from 61.27 to 66.73 (64 − 2.73 to 64 + 2.73). Two SDs tell us that approximately 95% of the individuals will give values between + 2 SD and − 2 SD from the mean. Thus, for our group, knowing the mean (64) and the SD (2.73), we know that 95% of the measurements fall within the range of 58.54 to 69.46 g. As we often want to know where 99% of the individuals in a distribution lie, it is convenient to remember that the mean ± 2.6 SD includes the values for 99% of the individuals.

Clearly, the larger the SD, the greater the departure of the individuals in the sample is from the mean. The SD for the first of the two groups of animals followed here is much larger than the second and turns out to be approximately 18.52. A bit of reflection in the two means and SDs (64 ± 18.52 versus 64 ± 2.73) tells us that 68% includes a much greater departure from the mean in the first group than in the second. In the latter, 68% of the individuals are relatively close to the mean. This can also be seen diagram-

FIG. 7-8. Relationship of standard deviation (SD) to a bell-shaped distribution.

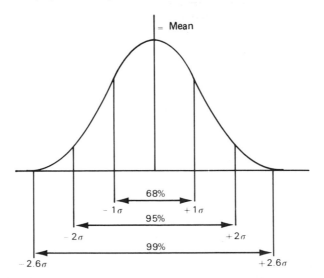

Data from Tables 7–3 and 7–4 give a mean weight loss in grams of 64 and a standard deviation of 2.73. Therefore 99% of the animals in the sample can be expected to show a weight loss value from: 64.0 (mean)   to: 64.0 (mean)
+7.1 (2.6 X 2.73)   −7.1 (2.6 X 2.73)
71.1   56.9

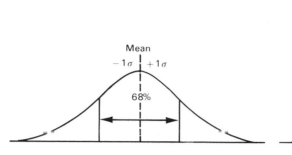

FIG. 7-9. Comparison of curves of different heights. The curve on the left has the larger SD. In the curve on the right, with its smaller SD, 68% of the individual measurements included in the range, + and − 1 SD, are closer to the mean than in the curve on the left.

matically in an examination of bell-shaped curves of different heights (Fig. 7-9). In the curve on the right, 68% of the measurements (from + 1 to − 1 SD) are closer to the mean than in the curve on the left, in which a much larger departure is evident for 68% of the values. A good appreciation of the amount of variation present in a sample can be gained by knowing the SD as well as the mean. Two samples may be compared for their variability by noting their means and SD. The higher the SD, the more variable the sample is. It must be emphasized that SDs can be compared only when they are expressed in the same units. It is erroneous to contrast two characteristics (height and weight, for example) for their relative amounts of variability when they are expressed in different units (centimenters versus grams).

Variance as a measure of variability

Another frequently encountered statistic is the variance, which also measures the amount of variability in a sample. Since this is nothing more than the square of the SD, it is even simpler to derive than the SD. Referring back to the calculations of the SD for the two sets of animals, we see that the variance of the second group is 7.47 (see Table 7-4), the figure we obtained in our calculation before extracting the square root. The variance of the first group is 342.99, the square of 18.52. Immediately, we can tell from the larger value of the variance which sample is the more variable. This statistic is commonly used in this way to compare variabilities, and it also has various other applications in genetic analysis.

Suppose we are studying the variability of a trait, such as length of ears in rabbits. We observe the total variance in a sample and are now interested in how much of that variation can be attributed to the geno-

type of individuals, how much to diet, how much to the temperature under which they were raised, and so on for any number of separate factors that could contribute to the entire amount of variation displayed. After finding the effect of each separate factor on the total, we can look back and see the contribution of each to the entire variance; in other words, we can break up the total variation into different parts. We can add up the amounts of variance produced by each part to estimate the percentage of the total that can be ascribed to each factor. This cannot be done using the SD. Analyses employing variance may have great value in breeding and population studies, so we should be aware of some of its uses and benefits. We have noted that the SD can be related to the normal-shaped curve; however, this cannot be done with the variance. Because the latter is the square of the SD, it must be expressed in square units. This is an awkward feature of the variance that is eliminated by taking its square root, giving us the SD.

Standard error of the mean and its importance

The SD has at least one other very important use in all cases in which means are estimated. We have stressed that in most situations we deal with samples and not with entire populations. Therefore, the mean we calculate in a particular instance is the mean of the sample; it is not the value of the parameter. Thus, if we get a mean on a sample of 20, 50, or 100 animals, we obviously are not examining all the possible individuals composing the population. If we work with second, third, and fourth groups of animals, all from the same population, and calculate the mean for each sample, we would instinctively expect all the individual means to vary from one another. That is, the different sample means will vary among themselves.

Each sample mean is just an estimate of the true mean, which we will be able to derive only in the rare situation. If we estimate means for a large number of samples and plot them, we will find that these separately calculated means also tend to fall into a normal or bell-shaped distribution. We can take advantage of this to estimate how close the mean we have calculated for our sample is to the true mean—a value that might be impossible to determine for the whole population.

What are the chances that the means we calculated for weight loss in the two groups of animals (see Tables 7-2 and 7-3) reflect accurately the true means, which we can learn only from measuring the effects of the chemical on all the animals composing the two strains? The answer is provided through the application of another statistic, the standard error of the mean (SE), which is quickly computed from the following formula:

$$SE = \frac{SD}{\sqrt{n}}$$

A glance at the formula shows that the SE takes into consideration both the SD and the size of the sample (n). The larger the sample, the smaller the SE will tend to be. The SE is really nothing more than the standard deviation of all the individual sample means that could conceivably be taken and that would tend to follow a normal distribution. Consequently, knowing the mean and the SE, we have a way to estimate how close our calculated mean is to the true mean. In the case of the second group of animals, the SE is quickly found to be

$$\frac{SD}{\sqrt{n}} = \frac{2.73}{\sqrt{20}} = \frac{2.73}{4.47} = 0.61$$

A similar calculation for the first sample gives an SE of 4.14. We can now express the two means as 64 ± 4.14 and 64 ± 0.61 for the first and second groups, respectively. This information indicates that in the first case, the chance is 68% that the true mean lies within the range of $64 + 1$ SE and $64 - 1$ SE (Fig. 7-10). The probability is therefore 68% that the true mean lies somewhere between 59.86 ($64 - 4.14$) and

FIG. 7-10. The standard error of the mean (SE). If many separate means ($\bar{X}$) are taken on samples from a poulation, the means will tend to fall into a normal distribution. The SE is the SD of the individual sample means. Approximately 68% of the means will lie within 1 SE on either side of the true mean of the whole population. Approximately 95% of them will lie within 2 SEs on either side of the true mean, and 99% will be 2.6 SEs on either side.

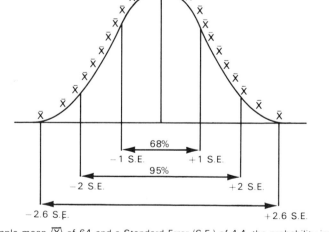

The probability is 95% that the true mean lies between:

68.14 (64 + 4.14). The probability is 95% that it is between 55.72 and 72.28 (64 − 2 SE and 64 + 2 SE). There is a 99% chance that the true mean lies within 64 − (2.6 × 4.14) and 64 + (2.6 × 4.14). Obviously, we are using the identical reasoning applied to the SD because both statistics—the SD and the SE—are based on bell-shaped distributions. It is apparent that the smaller the SE is, the greater the probability is that the sample mean is close to the true mean. In the second sample of animals, there is only a 1% chance that the true mean lies outside the range of 62.42 to 65.58 (± 2.6 × 0.61 from 64). Compare this with the first sample, in which the range is much greater because its SD, and hence its variability, is larger than that of the second sample. Sample means are commonly reported along with SE so that their reliability can be established. Means reported alone without their SEs may arouse skepticism on the part of the reader. If an SE is very high, this suggests the sample may be either too small or very variable, and confidence in the sample mean as a reflection of the true mean may be shaken; additional measurements may have to be taken. Calculation of the SE recognizes both variability and the importance of sample size.

Intelligence and difficulties in its measurement

The human brain is the most complex of all the organs of the body. Its intricacies have enabled the human to overcome environmental challenges to a greater degree than any other animal. The human is far less restricted than other species by environmental problems, and this is largely the result of the superior mental abilities of Homo sapiens. The human not only copes with existing environmental problems, but also changes the environment to suit immediate needs. In so doing, the human influences the course of human evolution as well as that of other species. How the human uses this superior brain to manipulate the environment is a direct reflection of human intelligence. Since human behavior has the potential to influence all life forms, any information that provides an insight into the nature of human intelligence and behavior has a decided value.

It would be impossible to arrive at a satisfactory definition of intelligence that takes into consideration peoples in all cultures and in all the completely different environments of the world. Most students of the problem agree that intelligence involves the ability to handle and interpret abstract symbols in a variety of relationships. The well-publicized intelligence quotient (IQ) is believed by many psychologists to reflect this ability, and hence intelligence. One reason for this is that performance on an IQ test does tend to indicate the degree of both academic success and occupational achievements in later life. The IQ is determined by obtaining the score achieved by an individual on a test that entails handling abstractions. The score is then related to the average score for the age group of the person. The ratio between the former and the latter multiplied by 100 equals the IQ. The peak or average for IQ has been established as 100 points for the Caucasian population. An individual whose score is average for the population age to which he or she belongs has an IQ of 100. If a person scores 120, the intelligence quotient would be 120/100 (the average score) $= 1.2 \times 100 = 120$.

Whether or not the IQ test actually measures intelligence is a matter of debate. Although IQ achievement does tend to indicate accomplishments later in life, it is important to note that the very factors that determine academic achievement are correlated with those for occupational achievements in the society. It is quite possible that intelligence is not the property being measured but, instead, some ability to deal with factors on which a given society places great value. The question of what property is being measured by the IQ test relates to the very construction of the tests. As presented, IQ scores for an age group conform to a normal distribution, such as that shown in Fig. 7-8. As we noted earlier, the peak or average IQ has been established as 100 points. However, the IQ scores from the population do not just happen to fall automatically into a bellshaped curve or normal distribution. The scores are forced to conform to this distribution by the very manner in which the test is constructed. For example, test results indicate that boys perform better than girls on items that deal with mathematics and spatial relationships. Girls, on the other hand, surpass boys in handling test items dealing with language and memory. Therefore, if more items on a test deal with mathematics than with language, the test would favor boys and they would score better as a group. To prevent one sex from continually scoring higher than the other on the tests, items that favor males are balanced by a comparable number of items that favor females. This selection of items eliminates high IQ performance related to sex. It is important to note that items can be selected if so desired to favor a certain group or to blur differences between groups that could be reflected in IQ test performance. Males and females of the same age group and from the same general environmental background can differ in IQ performance, depending on the test items, but this difference can be hidden by selection of items in construction of the test. It would seem, therefore, that if two persons of different sex from the same population differ on test performance, then two persons from very different populations might differ even more in IQ performance. Genetic differences that influence the IQ scores may actually have nothing to do with the intelligence of the two groups.

Although attempts have been made to eliminate as much as possible any environmental effects that could influence IQ performance, the difficulties entailed are overwhelming, and no test has been demonstrated to be entirely culture free. The intangible that we call "normal intelligence" is undoubtedly the result of the interactions of a large number of genes that influence it in a variety of ways. Intelligence itself, moreover, is probably not just a single entity. It must in turn be composed of many separate parts, each influenced by heredity and environment. The IQ score is just a single numerical value that tells us nothing about the separate compartments, how they interact, or the relative importance of heredity and environment in each component. Therefore, although IQ score does have some merit in predicting academic and occupational success in a given society, we still may wonder what it is that is being measured and how hereditary and environmental factors interact in IQ performance.

The exact nature of what is being measured by the IQ test awaits clarification; however, there is no doubt that IQ performance does entail a hereditary component. To estimate that portion of IQ that rests solely on genetic factors, investigations have relied heavily on studies of identical twins, some reared apart, others raised together. These studies have been analyzed in relation to scores achieved by closely related, distantly related, and unrelated persons. From the assemblage of data, an attempt has been made to estimate the relative roles of heredity and environment in IQ performance. The information has been used to obtain a value for heritability. Heritability is a measure of that percentage of the variation shown in the expression of a characteristic that is due to genetic factors. Any complex characteristic, such as height or weight, which involves many genes and their interactions with environmental factors, will exhibit variation. If all that variation can be shown to be due entirely to

genetic factors, then the heritability would equal 1. This would mean that the variation seen from one individual to the next is completely due to genetic influences. If the heritability were 0.5, then half of the variation seen would be due to genetic factors and the other half to environmental influences. A heritability of O tells us that all the variation is environmental.

If heredity alone were involved in IQ performance, the correlation between identical twins should be 1.0, and that between fraternal twins and ordinary siblings 0.5. The correlation between identical twins is significantly higher than for the other groups. The difference between identical twins reared together and identical twins reared apart suggests an environmental difference. From analyses using this approach, some investigators have estimated that environment accounts for approximately 20% of the variance observed in the population on IQ achievement. If this figure is accepted, then 80% of the variance is due to genetic differences among members of the population. Recall that the variance is the square of the standard deviation and that it can be broken up into different components. Since the value of the SD has been estimated to be 15 points for IQ scores, the variance must be 225 (15^2). That portion of the variance that can be attributed to genetic factors would be the heritability. Since the heritability of IQ was estimated to be 80%, this means that 20% of the variance would result from environmental factors. Therefore, of the total variance of 225, a component of 180 would be due to genes and a component of 45 due to environment.

Application of some of these statistical concepts to whites and blacks has led to some questionable conclusions. Few would deny that scores achieved by American blacks on IQ tests tend to be lower than those for American whites. Some studies suggest that the average IQ score for the white population is 100, whereas for the black it is 85. It must be noted that regardless of interpretation, no one can say that any black person must score lower than a white person. Although a difference between an average score of 100 and one of 85 is mathematically significant, the interpretation of this difference has led to many debates. According to many, this difference reflects various disadvantages experienced by American blacks in educational and economic conditions. Little is known about the subtle effects nutritional deficiencies, even those before birth, can have on a character as complex as intelligence. Perhaps all of the difference in IQ performance between the two groups stems from environment alone. Moreover, IQ performance may not be related to intelligence at all.

There are others who hold an opposing view and who adhere rigidly to a heritability value of 0.8, meaning that 80% of the variance has a hereditary basis and only 20% an environmental one. According to Jensen, the mean IQ score for whites is 1 SD above that for blacks, 100 versus 85. Jensen reasons that since 20% of the variance is an environmental component, the IQ achievement could be raised proportionately by eliminating all those factors that prevent higher performance on the tests. Reducing the variance value of 225 by 20% gives a new variance of 180. Since the SD is the square root of the variance, the new SD becomes 13.4. However, this new value is very close to the original one of 15. If whites at present score 1 SD higher than blacks, then even under the best environmental conditions mean IQ scores for blacks could not be raised 15 points but only 1.6 (15, the original SD minus 13.4, the new SD). Thus, the argument has been proposed that the mean IQ value of whites is truly higher than that of blacks and that this IQ value reflects primarily a difference in genetic factors that contribute to intelligence, not mainly unfavorable environmental forces.

Such conclusions are unfortunate from any viewpoint. Besides contributing to naive emotional prejudices and all their ensuing damage, these ill-founded arguments can cast doubt on any conclusions concerning intelligence that are actually based on firm scientific reasoning. Many of the assertions about the inheritance of intelligence as reflected by IQ scores are based on shaky foundations. Not only is there no agreement about what the tests truly measure, but also the heritability value of precisely 0.8 (80%) rests on so many questionable assumptions that it cannot be considered reliable. Heritability values for many characteristics have been shown to vary from one population to another similar population in a different environment. Heritability data on IQ have been obtained largely from studies of white populations in various regions of the United States and Europe. Since heritability for any characteristic is a figure derived from a population in a particular environment at one particular time, we may well question the application of a fixed value for heritability to one population when it has been calculated from data obtained from studies on another. A heritability value applies only to that population in which it is measured. Even if one assumes that the heritability value of 80%

is reliable for the white populations studied, one cannot apply this estimate when comparing two different populations, blacks and whites.

The IQ controversy serves to caution us in our application of statistical concepts to biological problems. These valuable methods, when used naively, may obscure the true basis of a phenotypic feature. We could become more confused about the actual nature of differences in intelligence among individuals than were the nineteenth-century biometricians about the inheritance of the simplest polygenic characters.

Genetic factors and behavior

Perhaps the greatest challenge to humankind will be information yet to be uncovered concerning the genetics of behavior. That behavior which is typical of a species may well prove to be the most complex of all phenotypic characteristics. Like any other aspect of the phenotype, behavior has both genetic and environmental components that interact in complex and very subtle ways. To ask which is the more important is meaningless, because the development of the behavior that is typical or normal for any animal group depends on both. Characteristic behavior patterns of most species depend on many genes whose expression may be altered at various developmental stages by environmental influences. For a few specific behavioral traits in some animals the genetic basis has been recognized. In mice, an aberrant form of behavior occurs which has been well understood for years. This is the waltzing trait characterized by pointless, irregular twirling motions that leave the animal exhausted. When a cross is made between waltzers and wild stock mice, the F_1 exhibit normal behavior, but the F_2 generation produces normals and waltzers in a ratio of 3:1. Clearly, the trait is inherited as an autosomal recessive (Fig. 7-11). The resulting behavior indicates a nervous disorder, and this has been shown to be associated with deafness and various abnormalities of the inner ear. In the homozygous condition, there is undoubtedly a lack of correct genetic information needed for normal development of the inner ear and associated nervous tissue. The result of the block manifests itself in the waltzing behavior.

Another example of simple Mendelian inheritance as the basis of a behavioral trait is seen in honeybees. Bees are particularly interesting in behavior studies, because workers and queens, which are alike genetically, come to display very different types of behavior

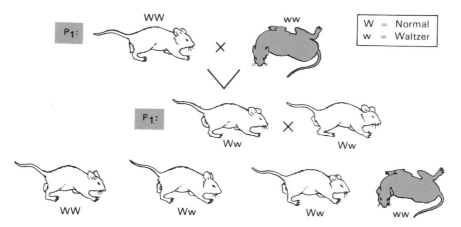

FIG. 7-11. Waltzing behavior in mice. This aberrant waltzing trait is inherited as a simple Mendelian recessive.

FIG. 7-12. Inheritance of hygienic behavior in worker bees. Workers are sterile females, but certain of their behavioral characteristics can be studied genetically by crossing queens and drones of various stocks. The male, or drone, develops from an unfertilized egg of the queen. Thus, males are haploid, and in effect represent the kinds of gametes produced by the queen. The difference between hygienic and nonhygienic behavior in the workers has been shown, as follows, to be based on two pairs of Mendelian alleles. If a queen from a hygienic colony is mated to a drone from a colony that is nonhygienic, all the F₁ workers exhibit nonhygienic behavior. The F₁ drones are all haploid and reflect the genotype of the hygienic queen. The F₁ queen and workers are dihybrids, but only the workers exhibit the behavior characteristic under study. Because the workers are sterile, only the F₁ queen can be crossed. The dihybrid queen will produce four kinds of gametes. Thus four different kinds of drones represent these gametes. If these drones are mated to a queen from a hygienic colony, four different kinds of workers result, in the ratio 1:1:1:1, showing that a two-gene difference is involved. (If the drones from the F₁ queen are mated with a queen from a nonhygienic colony, all the workers will be nonhygienic, because the two dominant alleles, U and R, would be contributed by the nonhygienic queen.)

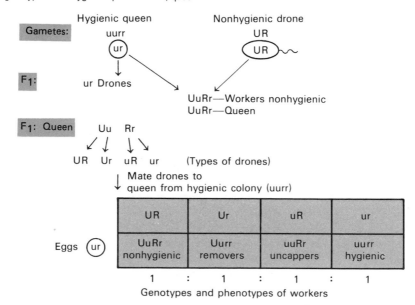

180

as a result of environmental influences operating during larval development. In addition, there are great differences in behavior among strains of bees. Some colonies exhibit "hygienic" behavior, which means that the workers remove dead larvae and pupae from the cells of the hive, an act requiring uncapping the cell and physically removing the dead remains. These hygienic colonies are also resistant to a bacterial infection, American foulbrood, and their resistance stems from the removal of the dead bees, which may harbor the responsible agent. Other bee colonies are "nonhygienic"; no attempt is made by the workers to remove the dead from their cells. Such colonies are thus sensitive to the bacterial disease.

A cross between hygienic and nonhygienic strains produces an F_1 in which the workers display nonhygienic behavior. It will be recalled that worker bees are sterile females; only the queen engages in sexual reproduction. The drones, or males, are all haploid and develop parthenogenetically from unfertilized eggs. Male offspring of the F_1 queen (whose workers are nonhygienic) can be mated to queens from other colonies, which are inbred or pure-breeding for either the hygienic or nonhygienic behavior. When these drones are mated to inbred hygienic queens (Fig. 7-12), the results are a 1:1:1:1 ratio of hygienic bees: nonhygienic bees: bees which just uncap the cell but leave the remains: bees which do not uncap cells but remove remains when the uncapping is performed for them. This typical testcross ratio clearly indicates a two-gene difference between hygienic and nonhygienic behavior.

The example of behavior in bees also illustrates a very significant point. The behavior of the hygienic strain is of definite value to the colony. It is important to bear in mind that the general behavior that is typical of a species is a phenotypic characteristic that has undergone a long history of natural selection. Behavioral patterns have been selected for their survival value, just as any other part of the total phenotype.

Behavior disturbances in the human

Even in humans, the genetic basis is known for certain types of behavior. We have discussed the autosomal recessive responsible for the production of phenylketonuria, a condition that can cause severe mental retardation. However, dietary restrictions can prevent the accumulation of phenylalanine in the blood of afflicted babies and in turn prevent toxic damage to the central nervous system. We see here the interaction of environmental and genetic components in the expression of a phenotype. An unfortunate behavioral outcome may be avoided if the proper environment (low amounts of phenylalanine) is present at a crucial time. This latter point is to be stressed, for the mental retardation cannot be prevented if the dietary regimen is undertaken too late. Nor will higher levels of phenylalanine in the diet impair the mental functions of individuals homozygous for the allele who were given the proper diet in infancy. As in many cases of phenotypic expression, behavioral or otherwise, the environmental influence at a specific time or period of development may be critical.

Other examples of genetic alterations that affect human behavior are discussed in later chapters (cri-du-chat babies, Tay-Sachs infants). There is also the need to investigate more fully the possibility of extra Y chromosomes as genotypic factors predisposing individuals to aggressive and antisocial behavior.

One of the most bizarre behavior disturbances in the human is associated with a single genetic defect, a sex-linked recessive. This is a type of cerebral palsy known as the Lesch-Nyhan syndrome in which afflicted children, all males, develop a compulsion for self-mutilation. While seeming to wish protection against his own self-destructive behavior, a Lesch-Nyhan child will nevertheless bite his tongue, lips, and fingers. Such an unfortunate boy must be fitted with restraints to protect him from inflicting serious damage on his own body. He also demonstrates very aggressive behavior toward others, and when upset will kick and bite those caring for him and, in addition, may vomit over them.

The responsible sex-linked recessive has been shown to produce a molecular defect, the deficiency of a certain enzyme required for normal purine metabolism. Lacking the enzyme, the person accumulates products that produce a high metabolic rate. Lesch-Nyhan victims die before puberty, usually as a result of kidney damage and frequently after developing gout. Persons who are gout sufferers also have a biochemical upset similar to that found in the Lesch-Nyhan syndrome; however, they do not experience nervous derangements.

It is obvious that the Lesch-Nyhan syndrome is extremely relevant to studies of aggressive behavior and compulsions of various kinds. A complete understanding of this syndrome could provide valuable information on behavioral disturbances in general, the biochemical upsets that accompany them, and the genetic and environmental factors that may interact to cause them. Directly related to such investigations is the problem of mental illness. There is good evidence that certain psychoses, the most serious of the mental illnesses, have a genetic as well as an environmental component. Schizophrenia, the most frequently occurring psychosis, is characterized by very unpredictable and unusual forms of behavior associated with the individual's complete withdrawal into a world of his or her own creation. Controversies have been waged up to the present time on the causes of schizophrenia. Many psychiatrists subscribe to the thesis that environmental stresses alone trigger the disturbance. However, present data assembled from studies of identical twins and families in which the illness occurs indicate that genetic as well as environmental factors are involved. As to the genetic component, again there is no consensus. Some favor a single dominant allele hypothesis, others an interpretation based on a recessive. Still others favor an explanation based on polygenes or multiple factors.

All those who recognize a genetic component also implicate the environment in the expression of schizophrenia. There are those who feel that genetic factors may predispose a person to expression of the illness so that he or she has a greater risk than those lacking such factors of developing the disturbance. However, also involved are unknown environmental stresses that may be required before a predisposed person develops the illness. Consequently, such a person could remain healthy. On the other hand, a person lacking the predisposing genetic elements could develop the disorder if subjected to certain stresses.

Data indicate that genetic factors play a role in still other behavioral disturbances. Studies conducted on certain families suggest that the manic-depression disorder is inherited on the basis of a dominant allele with incomplete penetrance. There is hope that further studies will yield information useful for preventing and treating such behavioral upsets.

Behavior as a complex phenotypic characteristic

The basis of most types of behavior is so complex that the genetic and environmental components are only remotely appreciated. We expect animals to behave in certain ways typical of their species, and we sense that such characteristic behavior must somehow be influenced by genetic information that all members

of a species inherit in common. Immediately after hatching, baby ducks will walk, follow their mother, and even swim. Human babies, in contrast, are quite helpless, but at a certain age, they start to speak a language, something no other animal can do, no matter how precocious at birth. Surely children have the potential to speak because the DNA transmitted to them has the encoded information required to equip them with a complex brain, voice box, and other necessary anatomical structures. Animals without the set of genetic instructions needed to guide orderly construction of the features essential to speech will never utter an intelligible word. Does this mean that speaking a language is an inherited trait? Is the precocity of the duckling in recognizing its mother an inherited trait for that species?

An unqualified "yes" to either of these would indicate failure to remember that no phenotypic characteristic is inherited as such. There is no gene for "speaking" any more than there is a special gene for eye color. Despite their structural equipment, children will begin to speak at a certain period in their development only if their environment surrounds them with the proper sound incentives. Whether they will speak English, Chinese, or some other tongue depends on the sounds they hear. Their genetic endowment empowers them with the remarkable ability to assemble separate sound symbols and arrange them in meaningful order. Without ever being taught the rules of grammar, they become able to understand the meaning of sentences and to express themselves by manipulating the symbols of speech.

The magnitude of this feat is appreciated when we note that those children surrounded by the unrelated sound stimuli of more than one language can, without difficulty, sort them out in their proper relationships and so achieve expertise in more than one tongue. However, at about the age of sexual maturity, this talent begins to fade, so that mastery of a language from that time on becomes an effort. It is the rare adult who achieves the perfection and facility with a newly studied tongue that a child acquires so effortlessly. Thus, there is a critical period during development when environmental factors interact with those genetically determined. The former are the proper sounds, the latter the structures needed to receive, interpret, and associate symbols. Should a child be deprived of the proper stimuli at a critical developmental stage, he or she may never acquire the same language facility as one who was surrounded by the sounds that are meaningful in his or her cultural environment. Therefore, group members who are isolated from infancy from appropriate language stimuli would be unable to communicate verbally in any known language. After a series of generations, we might expect descendants of this still isolated group to have developed some means of verbal communication that follows certain rules that give it a structure and grammar, for speaking a language is an aspect of behavior, a potential that is peculiar to the human zygote. Moreover, it is a phenotypic character with survival value. Speech enables the members of the group to transfer an ever-growing amount of cultural information from one generation to the next. This enables the members, building on information from the past, to continue to devise new ways of coping with different environments, which enables their species to survive under a range of diverse conditions. Natural selection has undoubtedly favored this aspect of human behavior, which helps set humans apart from other creatures.

Of course, a spoken language is by no means the only type of communication. An individual born with defective vocal apparatus can still communicate by symbols, as can a person deaf from birth. The latter, although possessing perfect vocal cords, will never speak with the clarity of the more fortunate person who has been able to receive the appropriate sound stimuli during his development. The complex interaction of hereditary and environmental elements is very evident here. Hereditary deafness can prevent the child from reacting with the appropriate environmental factors. However, if he is genetically normal otherwise, such a child will have the kind of nervous system that enables him to associate and manipulate nonverbal abstract symbols that he can substitute for the spoken word. On the other hand, deafness or faulty speech may be environmentally caused. The proper genetic information coded in DNA may be present in all the cells, but suppose a virus crosses the placenta at a critical point in embryonic development. Its interference with the formation of a normal anatomy can prevent the individual from ever speaking clearly, even though he is later given the most conducive environment.

We must remember that speaking is just one form of behavior that permits communication and is certainly not the only kind of language. This is graphically illustrated by the gifted chimpanzee, Sarah, who has achieved a certain ability to read and write by manipulating plastic pieces of various shapes and colors. These plastic bits can be used as symbols to represent parts of speech: nouns, verbs, and adjectives. The chimpanzee was taught to arrange approximately 130 "words" to form meaningful sentences, a feat that has indicated the animal's talent to associate an object with a symbol and also to describe and classify it as well. Sarah also learned to recognize sentences written in plastic symbols and to respond correctly to them, such as following a "written" order to place a banana in a dish. There is even reason to believe that the chimpanzee was eventually able to associate a symbol with an object, even in the absence of the object. If this proves to be correct, it means that the animal can think with language and has the ability to represent mentally, in terms of symbols, objects that are not present. This is perhaps the most important feature of language, as it enables individuals to communicate about things that are not physically there.

Investigations of this type are very important in understanding language and human behavior. A chimpanzee cannot speak a language verbally because its genetic makeup does not endow it with a potential for forming the anatomy required for speech. However, it evidently does possess the genetic information enabling it to grasp certain concepts and to communicate in symbols. The investigators correctly point out that studies such as theirs may clarify the very nature of language. For we can see that language is not unique to humans; human language is just one form of a larger and more general system of communication. Moreover, certain aspects of language that are often considered "human" may yet prove to be part of a larger system humans share with other animals. Further studies may help us recognize some of the specific genetic factors that influence language ability and appreciate their selective value in the evolution of humans and their relatives.

Species-specific behavior

Arguments have raged for years over the inheritance of "innate" behavior, that behavior so typical of a species that it may seem inborn and unalterable. After hatching, the duckling typically follows its mother faithfully; a female rat properly attends to her offspring after their birth. These animals seem to "know" enough to do such things without having in any way been taught. Are there segments of DNA that program this type of behavior? Again, this sounds like the concept of the inheritance of a specific character. From what we have been discussing, we should expect such innate behavior to reflect an interplay of environmental and genetic factors.

The mere fact that we can recognize a behavior that is typical of a species implies there is some genetic basis. That much more is also involved is well illustrated by studies performed with ducks and geese. Years ago, the observation was made that a duckling, which normally forms a rigid attachment for its mother, will form a bond with a human if it sees that person before it sees its mother. Even artificial objects were shown to serve as "parents" if isolated birds were exposed to the inanimate objects before they had any type of social contact. This phenomenon was designated imprinting and has been taken by many students of behavior to represent some form of learning. Lorenz, who called attention to imprinting, considered it a different process, because it happened so quickly and seemed to have a permanent effect. Observations by Hess clearly show that imprinting can be reversed. Laboratory-hatched ducklings exposed to humans formed attachments to them, but these "human-imprinted" birds, when given to a female duck, accepted her and behaved just as other ducklings which had been normally reared by their mother. Hess's observations caution against definitive statements on behavior based solely on laboratory observations. Essential to any interpretation are studies of the animal's behavior in its natural habitat. Rearing an animal in some artificial environment may evoke certain behavioral responses that actually tell little about the normal behavior pattern. Hess's studies of natural imprinting indicate that normal bonding between mother and duckling involves an interplay of sounds between the mother and her offspring, an interaction that starts well in advance of hatching! The unhatched birds emit noises to which the mother reacts by clucking. The stimuli provided by sound, along with tactile stimulation of the progeny (squeezing and scratching), are normal factors in the development of the species' typical bond between ducklings and mother. The process of natural imprinting ensures a very strong filial–maternal attachment. These observations again illustrate that many environmental stimuli may be needed for the development of a normal behavior pattern. And the best place to study this is in nature. The innate behavior is not just encoded in the zygote's DNA so that it will emerge intact at a given moment.

Still other influences must be considered before we can reach conclusions about innate behavior. The female rat apparently does not need to learn to act as a mother. Normally, she cares for her offspring and

will do so with her first litter, even without prior experience or opportunity to observe another mother rat. On the surface, it appears that this maternal behavior is inherited as such. However, it has been demonstrated that a female rat will show hostility toward her offspring, avoid them, or even kill them if she is deprived of certain stimuli. Normally, a rat licks and smells its own body as it develops into adulthood. If a newborn female is fitted with a collar of sufficient size, she will be unable to examine herself in this way. When she bears her young, even after removal of the collar, she will fail to display the species-specific maternal behavior. This example shows that a normal environmental stimulus may be something quite unsuspected. In this case "seeing" maternal behavior is not a necessary prerequisite. Instead, quite a different experience is required, the familiarity of the female rat with her own body.

The lessons learned from animal studies caution us about reaching premature conclusions concerning human behavior. Gaining deeper insights into why we act the way we do may enable us to elevate our species to undreamed of heights. But erroneous conclusions based on failure or reluctance to recognize the subtleties in the interactions of genetics and environment may result in our self-destruction.

REFERENCES

Also see the References listed at the end of Chapter 6.

Baldessarini, R. J. Schizophrenia. *New Engl. J. Med.* 297: 988, 1977.

Bodmer, W. F. and L. L. Cavalli-Sforza. Intelligence and race. *Sci. Am.* (Oct.): 19, 1970.

Feldman, M. W. and R. C. Lewontin. The heritability hang-up. *Science* 190: 1163, 1975.

Hess, E. H. Imprinting in a natural laboratory. *Sci. Am.* (Aug.): 24, 1972.

Jensen, A. R. How much can we boost I. Q. and scholastic achievement? *Harvard Educ. Rev.* 39: 1, 1969.

Jensen, A. R. Race and the genetics of intelligence: a reply to Lewontin. *Bull. Atomic Scientists* (May): 17, 1970.

Kidd, K. K. and L. L. Cavalli-Sforza. Analysis of the genetics of schizophrenia. *Social Biol.* 20: 254, 1973.

Kolata, G. Manic-depression: is it inherited? *Science* 232: 575, 1986.

Lewontin, R. C. Race and intelligence. *Bull. Atomic Scientists* (March): 2, 1970.

McClearn, G. E. Behavioral genetics. *Ann. Rev. Genet.* 4: 437, 1970.

O'Brien, S. J., D. E. Wildt, and M. Bush. The cheetah in genetic peril. *Sci. Am.* (May): 84, 1986.

Premack, A. J. and D. Premack. Teaching language to an ape. *Sci. Am.* (Oct.): 92, 1972.

Rothenbuhler, W. C., J. Kulincevic, and W. E. Kerr. Bee genetics. *Ann. Rev. Genet.* 2: 413, 1968.

Schefler, W. C. *Statistics for Health Professionals*. Addison-Wesley, Reading, MA, 1984.

Smith, J. M. The evolution of behavior. *Sci. Am.* (Sept.): 176, 1978.

Stern, C. Model estimates of the number of gene pairs involved in pigmentation variability of the Negro-American. *Hum. Hered.* 20: 165, 1970.

Thompson, J. N. Quantitative variation and gene number. *Nature* 258, 1975.

Triola, M. F. *Elementary Statistics*, 3rd ed. Benjanin-Cummings, Menlo Park. CA, 1986.

Winston, M. L. and C. D. Michener. Dual origin of highly social behavior among bees. *Proc. Natl. Acad. Sci.* 74: 1135, 1977.

REVIEW QUESTIONS

In Questions 1 through 4, assume that two pairs of alleles, A a and B b, form the basis of skin pigmentation differences in humans and are responsible for the recognition of five classes: black, dark, intermediate, light, and white.

1. Give the possible genotypes of each of the following:

   A. A dark-skinned person.
   B. An intermediate who had a white parent.
   C. A white person who had one intermediate parent.

2. A white person and one who is intermediate in skin color produce nine children, all of whom are light skinned. What are likely genotypes of the two parents?

3. Give the phenotypes and genotypes possible from a cross of an intermediate person who is heterozygous at both loci and a white person.

4. Two persons of intermediate skin color produce seven children, all of whom are about the same in skin color as the parents. However, two other persons of intermediate skin color produce five children who vary greatly from one another, ranging from white through dark. Give the most likely genotypes of the two sets of parents.

5. Assume that in a particular variety of wheat the color of the grains varies from deep red through intermediate shades to white. Six pairs of alleles are involved: R^1r^1, R^2r^2, and so on. Each pigment-contributing allele, R^1, R^2, and so on adds an equal dosage whereas the alleles r^1, r^2, and so on contribute nothing to pigmentation. Plants from a pure-breeding strain that has deep red grains are crossed to a strain with white grains. The F_1s are intermediate in color.

   A. When F_1s are crossed to white-grained plants, how many shades of grain are possible among the offspring?
   B. When the F_1s are crossed among themselves, how many of the offspring can be expected to be white grained, and how many deep red?

6. A sample of plants was measured in inches. The following gives the heights:

Height in inches:	7	8	9	10	11	12
Number of plants:	1	1	4	8	3	3

   Calculate the following:

   A. The mean.
   B. The variance.
   C. The standard deviation.
   D. The standard error of the mean.

7. Suppose height measurements are taken on a random sample of 100 men from a population and that the mean height is found to be 70.50 in. The standard deviation is calculated and found to be 1.50. What range in height would include the following?

   A. 68% of the men in the sample.
   B. 95% of the men in the sample.

8. A. What would be the standard error of the mean in Question 7?
   B. What does this value tell us about the mean?

9. A group of adults is given an IQ test with the following results:

Test score:	90	93	94	97	101	108	111	115	119	125
Number of persons:	1	2	2	1	4	7	1	1	3	3

   Calculate the following:

   A. The mean test score value to the nearest whole number.
   B. The variance.
   C. The standard deviation.

10. A second group gives the following results on an IQ test:

Test score:	99	103	110	111	112	118	119	120	126
Number of persons:	6	2	3	4	3	1	2	2	2

    Calculate the following:

    A. The mean test score to the nearest whole number.
    B. The variance.
    C. The standard deviation.

11. In one variety of rabbit the ear length of the animals averages about 4 in. In a second variety, the average is 2 in. The ear length of hybrid animals obtained after crossing the two varieties averages 3 in. The hybrids, when crossed among themselves, produce offspring whose ear lengths vary much more than that of the F_1 hybrid parents. Of 496 F_2 animals, two have ears that measure 4 in. and two have ears about 2 in. long. How many pairs of alleles would you say seem to be involved? How much does each effective allele contribute to length of ear?

12. In one variety of oats, the weight yield of grain per plant is 10 g, whereas in another variety it is only 4 g. Hybrids between the varieties give an average yield of about 7 g per plant. When the hybrids are crossed to each other, the yield varies greatly. About 4 plants in 253 yield 10 g each and about the same number yield about 4 g each. How many pairs of alleles appear to be involved in weight yield per plant? What contribution does each effective allele make to weight yield?

13. A race of inbred plants has a mean petal length of 12 mm. A second race from another location has the same mean petal length. When the two races are crossed, the F_1s also have a mean petal length of 12 mm. However, the F_2 generation derived from crossing the F_1s to one another shows a very wide spread in petal length. About 3 out of 770 have a length as small as 8 mm, and about 3 out of 770 have a length as long as 16 mm. How many allelic pairs seem to be involved in the difference in petal length? How much does each effective allele contribute to length of the petal?

14. Assuming that height difference in the human depends on four pairs of alleles and that environmental factors are constant, explain why two parents of average height produce children much taller than they.

15. A. Diagram a cross between a queen bee from a non-hygienic hive and a drone from a hygienic colony. Give the genotypes of the F_1 queen and the workers. What is the workers' behavior?
    B. What are the genotypes of the drones produced by the P_1 queen?
    C. What are the genotypes of the drones produced by the F_1 queen?
    D. The drones from the F_1 queen are mated to a queen from a nonhygenic colony. Give the genotypes and the behavior of the resulting workers.

Select the correct answer or answers, if any, in each of the following:

16. Species-specific behavior in higher animals:
    A. Is a phenotypic characteristic.
    B. Has evolved under the force of natural selection.
    C. Depends on environmental stimuli for its normal development.
    D. Requires that the individual be taught by observing the typical behavior.
    E. Is inborn in those cases in which the animal displays the typical behavior pattern without ever having seen it.

17. The following is (are) true concerning IQ:
    A. Performance on an IQ test entails some genetic component.
    B. Achievement on an IQ test may provide some indication of future academic and occupational achievement.
    C. The tests are purposely designed to produce a normal distribution for the population.
    D. A mental age of 5 and a chronological age of 4 indicates an IQ of 125.
    E. The heritability value of 0.8 clearly indicates that 80% of the variability among persons in IQ performance is genetically based.

18. All mammals possess the equivalent of the HLA complex found in humans (see Fig. 4-24). This genetic region includes several very closely linked genes, each of which exists in many allelic forms. The extreme variability at this region accounts for the fact that in mammals graft rejection occurs when tissue is transplanted from one unrelated individual to another. In the case of the cheetah, any animal will reject a skin graft from the domestic cat. However, any cheetah will tolerate a skin graft from any other cheetah, even though the animals may be from different areas. Explain.

8

GENES ON THE SAME CHROMOSOME

Once chromosomes were accepted as the physical basis of Mendelian factors, it became necessary to follow the inheritance patterns of two or more genes located on the same chromosome. The early twentieth-century geneticists realized that certain organisms possess only a small number of chromosome pairs. Does this mean that Drosophila, with just four pairs, contains genes associated into four groups and that the genes in each group must always be inherited together? It was suggested by De Vries and several others that the genes in each chromosome group must undergo some kind of exchange, that is, the genes on homologous chromosomes must somehow be able to switch from one homologue to the other. Mendel did not encounter this problem in his work with peas. All seven pairs of alleles exhibited independent assortment. Therefore, no factor that Mendel followed was tied or linked to any other.

Shortly after Morgan undertook his classic studies with the fruit fly, he showed that two pairs of alleles, and thus two genes, could be associated with the same chromosome. He found this to be so in the case of a gene affecting eye color and one affecting the size of the wing. Recessive alleles of each gene, w (white eye color) and r (rudimentary wing), were found to be sex linked, and hence both located on the X chromosome. When Morgan followed both recessive

alleles of the genes at the same time, he found they did indeed undergo some sort of exchange in relation to each other (Fig. 8-1A). Then an allele of a third gene, the recessive allele for yellow body color, y, was found to be sex linked and it too showed new combinations with respect to the other two, w and r (Fig. 8-1B). Morgan clearly demonstrated from such observations that two or more genes that are linked (whose loci are on the same chromosome) can in some way separate from each other and form new combinations with respect to their allelic forms. To grasp the full significance of this, let us first examine certain crosses in Drosophila that illustrate the nature of linkage and its genetic implications.

Independent assortment versus complete linkage

Suppose we make a cross between two mutant stocks of flies: "eyeless" (having a reduced eye size) and "black" (black body color instead of gray). Both of the responsible alleles are autosomal and are inherited as recessives to the wild. The F_1 offspring are found, as expected, to have the wild phenotype, large eyes, and gray body (Fig. 8-2). A testcross of these F_1 wild flies results in the typical dihybrid ratio, 1:1:1:1. This is exactly what should occur on the basis of independent assortment. The testcross shows the kinds of gametes

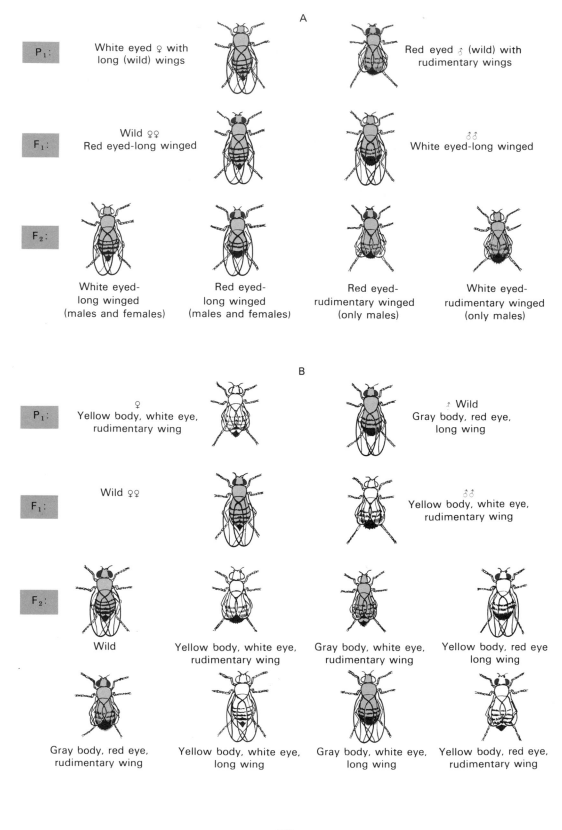

A

P₁: White eyed ♀ with long (wild) wings

Red eyed ♂ (wild) with rudimentary wings

F₁: Wild ♀♀ Red eyed-long winged

♂♂ White eyed-long winged

F₂:

White eyed-long winged (males and females)

Red eyed-long winged (males and females)

Red eyed-rudimentary winged (only males)

White eyed-rudimentary winged (only males)

B

P₁: ♀ Yellow body, white eye, rudimentary wing

♂ Wild Gray body, red eye, long wing

F₁: Wild ♀♀

♂♂ Yellow body, white eye, rudimentary wing

F₂:

Wild

Yellow body, white eye, rudimentary wing

Gray body, white eye, rudimentary wing

Yellow body, red eye long wing

Gray body, red eye, rudimentary wing

Yellow body, white eye, long wing

Gray body, white eye, long wing

Yellow body, red eye, rudimentary wing

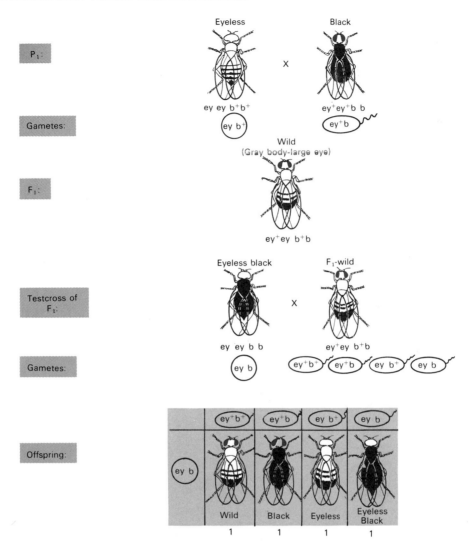

FIG. 8-2. Dihybrid testcross. The eyeless and the black traits are inherited as autosomal recessives. When the F_1 dihybrid is testcrossed to an eyeless black fly, the offspring occur in the expected ratio of 1:1:1:1.

FIG. 8-1. Genes on the same chromosome undergo some kind of exchange. (*A*) White eye color and rudimentary wings are both inherited as sex-linked recessives. When followed through the F_2, a cross of a white-eyed female and a male with rudimentary wings produces flies of four different phenotypes that occur in unequal amounts. The flies with red eyes and long wings and those with white eyes and rudimentary wings represent new combinations that were not found in the P_1 parents. Alleles at the two loci, white and rudimentary, must undergo some sort of exchange to form the new combinations. (*B*) When three sex-linked recessives are followed at the same time, new combinations of traits also appear. In a cross such as this one, eight phenotypes arise among the F_2. The fact that new combinations are formed indicates that somehow genes on the same chromosome (the X here) can form new combinations. The various phenotypes shown here in the F_2 do not occur with the same frequency.

produced by an organism, as well as the proportion in which they are formed. So the results in this case tell us that the eyeless locus and the black are not in any way linked or tied to each other but are free to segregate independently at gamete formation.

Now let us follow the allele "black" (b) along with another recessive, "purple" (p, for dark eye color as opposed to red). Treating the cross between black-bodied and purple-eyed flies exactly as we did in the preceding case, we might again expect all wild flies in the F_1 and a testcross ratio of 1:1:1:1. And we will find after making such a cross that the F_1 flies are all wild. However, the testcross will give quite different results from what we expect. Let us first testcross F$_1$ males with double recessive, purple-eyed, black-bodied females (Fig. 8-3). Instead of four kinds of offspring in equal proportions, we find only two types in a 1:1 ratio. One type is purple and the other is black; they are phenotypically identical to the original P$_1$ parental types. Something different must be involved in this cross because independent assortment has not occurred. The reason for this is that the loci p and b are not found on different chromosomes. They both happen to be located on Chromosome II and thus are linked. As Fig. 8-4 shows, we must diagram a cross a bit differently when linkage is involved. It is essential to indicate clearly which specific allelic combination is present on a particular chromosome. This is necessary, because those alleles

tend to stay together and thus do not assort independently. The cross shows that each P$_1$ parent has, on each chromosome, a recessive in combination with a dominant and that the combination stays together throughout the testcross. The bar or line represents two chromosomes. The combination of alleles above the line tends to remain together, as does the combination below it. We see that p b$^+$ and p$^+$b entered the cross together through the P$_1$ parents. The combination did not come apart but remained together among the testcross progeny. The testcross clearly tells us that the two genes are completely linked, because only the old P$_1$ combinations are formed (purple eyes with wild body color and red eyes with black body).

Now let us make the cross in a slightly different

FIG. 8-3. Unexpected testcross results. An F$_1$ male, dihybrid for purple eyes and black body, is testcrossed. Instead of the expected ratio, only two classes of offspring appear: purple eyed and black bodied. These are exactly like the P$_1$ parents. No new combinations were formed in this case, as seen by the 1:1 ratio instead of the expected 1:1:1:1.

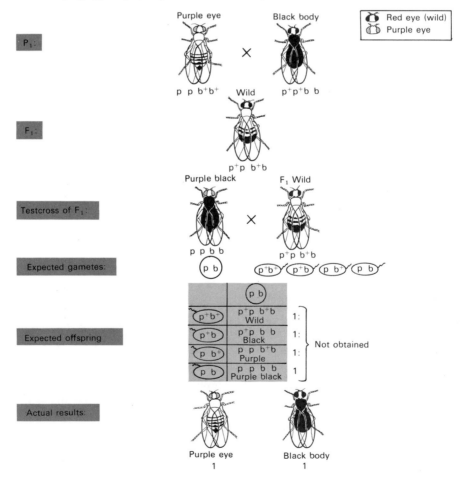

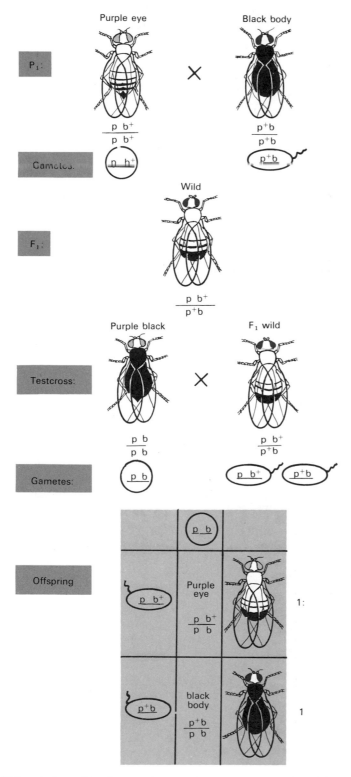

FIG. 8-4. Linkage in Drosophila. The gene for eye color and the gene for body color are on the same autosome. Therefore, the combination of alleles found in the P_1 parents tends to remain together. When a wild, dihybrid male is testcrossed, only two types of offspring result, and these are identical to the P_1 parents. The F_1 male forms only two classes of gametes because the two genes are completely linked.

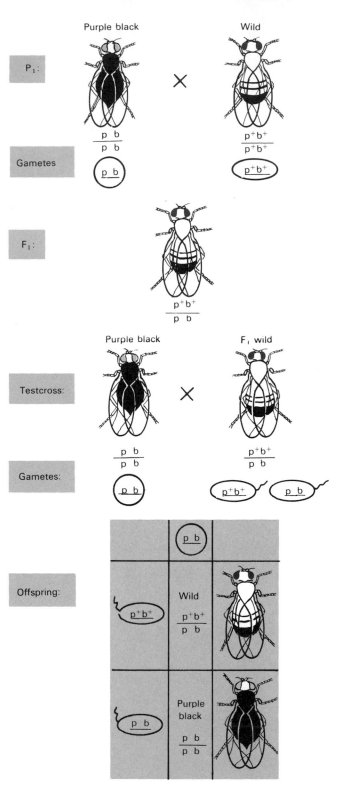

Purple black   Wild

P₁:

$\dfrac{p\ b}{p\ b}$	$\dfrac{p^+b^+}{p^+b^+}$

Red eye (wild)
Purple eye

Gametes

$\boxed{p\ b}$ $\boxed{p^+b^+}$

F₁:

$\dfrac{p^+b^+}{p\ b}$

Purple black   F₁ wild

Testcross:

$\dfrac{p\ b}{p\ b}$ $\dfrac{p^+b^+}{p\ b}$

Gametes:

$\boxed{p\ b}$ $\boxed{p^+b^+}$ $\boxed{p\ b}$

Offspring:

$\boxed{p\ b}$

$\boxed{p^+b^+}$ Wild $\dfrac{p^+b^+}{p\ b}$

$\boxed{p\ b}$ Purple black $\dfrac{p\ b}{p\ b}$

FIG. 8-5. Linkage in Drosophila. The principle in this cross is identical to that shown in Fig. 8-4. However, the F₁ dihybrid males in the two crosses must be compared. They are both wild in phenotype, but their allelic arrangements are different. In this case, the alleles are in the *cis* arrangement as opposed to the *trans* in the other cross. The testcrosses give very different results, but in each case, the testcross offspring are identical phenotypically to the P₁ parents. No new combinations of alleles are formed.

way. In Fig. 8-5, the P_1 parents are purple, black (double recessives), and homozygous red eyed, gray bodied (wild eye, wild body). Again we see that independent assortment does not occur. The combination of alleles in the original P_1 parents stayed together and remained completely linked through the testcross. A very significant difference between this mating and the previous one (Fig. 8-4) is seen in the combination of the alleles. Note that in both cases the F_1 hybrids are wild in phenotype, but the allelic arrangement is very different. In the first case (Fig. 8-4), one chromosome has the combination of a recessive with a wild (p b$^+$). The homologous chromosome has the reciprocal of this, wild with recessive (p$^+$b). This arrangement in the F_1 dihybrid, p$^+$b/pb$^+$, is called the trans or repulsion arrangement and results from a cross between two individuals carrying different recessives on the same chromosome. This trans arrangement is quite different from the cis or coupling arrangement seen in Fig. 8-5. In this case, the dihybrid contains both recessives on one chromosome and both dominants on the other: p $^+$b$^+$/p b. This results from the cross of any individual that is double recessive at two linked loci with an individual homozygous for the corresponding dominant alleles.

Figures 8-4 and 8-5 clearly show that although both dihybrids, the cis and the trans, are identical phenotypically, they represent very different arrangements of alleles and produce disimilar results in the testcross. In each case, however, the testcross offspring are like those of the P_1 parents in phenotype. At times, therefore, it may be very important to know whether two dihybrids showing the same phenotype have the same allelic arrangement, cis or trans, because the two types differ greatly when they are bred. The testcross is the most direct way to determine which of the two arrangements is present in a dihybrid.

Recombination and its calculation

In both of the testcrosses we have been discussing (Figs. 8-4 and 8-5), dihybrid males were selected and crossed with double-recessive females. We now return to these same crosses, but this time, F_1 dihybrid females are mated with double recessive males. As Figs. 8-6 and 8-7 show, testcrosses of the F_1 females give not just two types of offspring, as in the case of the F_1 males, but four kinds! This means that four different kinds of gametes were formed by the F_1 females; however, the numbers show that independent assortment did not occur. It is the old (P_1 or parental) combinations that are the most common. The new combinations, the recombinants, are in the minority. Thus, although there was a definite tendency for the parental allelic combinations to remain intact, a reassortment did occur in a smaller percentage of the gametes. These new combinations of linked genes result from crossing over, an event that entails a reciprocal exchange between two homologous chromosome segments at early meiotic prophase (see Fig. 3-5). In this example, the percentage recombination or the percentage of crossover gametes is 6%, and consequently the strength of linkage is 94%. The recombinant gametes are responsible for the recombinant phenotypes seen in the testcross.

Crossing over does not occur at all in the male fruit fly, in which the strength of linkage is normally 100%. But it must be kept in mind that in most species crossing over is typically characteristic of meiosis in both sexes. The absence of crossing over in the male fruit fly is an exception. This peculiarity, however, is a valuable one when working with Drosophila, for it permits the detection of linkage with relative ease. In most organisms, however, crossing over, and the recombination it generates, is a normal and integral feature of gamete formation in both the male and the female.

Figures 8-6 and 8-7 also contrast the results of testcrossing F_1 females with the alleles in the trans arrangement and F_1 females with the alleles in the cis arrangement. Note that the old or parental combinations among the offspring of the trans females [(p b$^+$/p b) and (p$^+$b/p b)] (Fig. 8-6) are the new combinations among the offspring of the cis females (Fig. 8-7). Likewise, the old combinations occurring among the cis females [(p$^+$b$^+$/p b) and (p b/p b)] are the recombinants or new combinations among the progeny of the trans females. This is exactly what should be expected as a consequence of the gene affecting body color and that affecting eye color being linked together on the same chromosome. It simply means that they tend to remain together in the original P_1 arrangement and thus result in a larger number of parental types among the testcross offspring.

The following example illustrates how the amount of recombination between two linked alleles can be easily calculated from an inspection of raw data obtained in a crossing experiment. A recessive allele of another gene is known to be linked to the one responsible for black body in Drosophila. Its effect is to

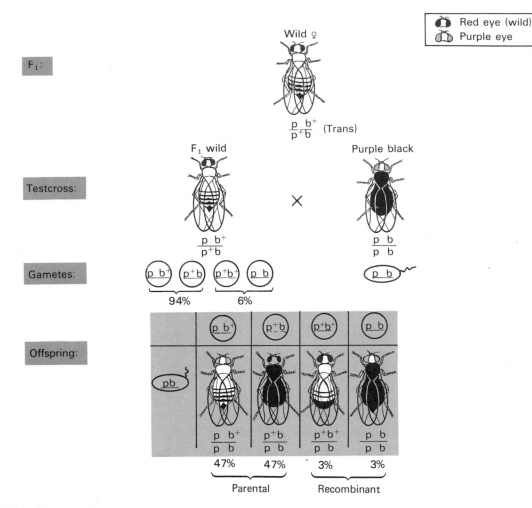

FIG. 8-6. Testcross of F₁ female, *trans* arrangement. When a dihybrid female is testcrossed, the linked genes separate or engage in crossing over. The old combinations are more common than the new ones. Four types of testcross progeny result, but they occur in unequal proportions, which reflect the amount of crossover gametes (6% here). This testcross contrasts with the testcross of the F₁ male shown in Fig. 8-4.

produce a bent wing in the homozygous recessive condition. The allelic pair is "a⁺" (for normal wing) and "a" for arc (bent wing). A cross between arc females and black males produces wild-type progeny (Fig. 8-8). Note that they have the alleles in the trans arrangement. When the F₁ females are testcrossed, four phenotypic classes can be recognized. To determine the amount of recombination, the data are inspected to distinguish the parental types from the recombinants, those with new allelic arrangements. The latter are quickly recognized by their phenotypes (wild-type flies and those that are both black and arc) and are the less numerous types. The amount of recombination is estimated simply by calculating

the percentage of these crossover offspring among the total number of progeny produced ($1020/2392 = 0.426 = 42.6\%$). This example may be used as a guide to estimate the crossover value between any two linked autosomal genes.

In the fruit fly, however, we must remember that the F₁ males will give a very different result in a testcross because of the absence of crossing over (Fig. 8-9). Because linkage strength is 100%, the genes a for wing shape and b for body color will not separate. Therefore half of the flies will show the arc phenotype, and the remaining half will be black. These phenotypes will be about equally distributed among the progeny. It should be apparent that the F₁ males

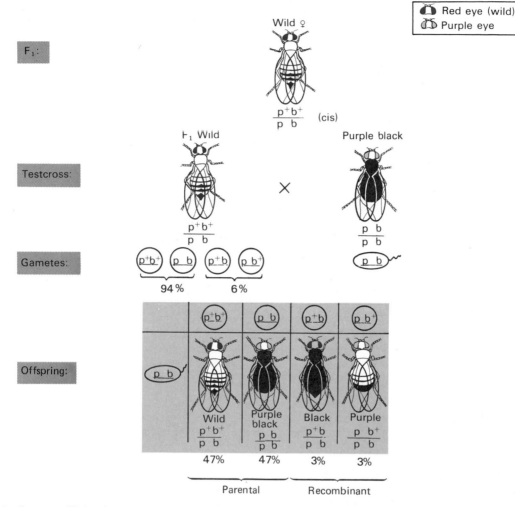

FIG. 8-7. Testcross of F₁ female, *cis* arrangement. Because crossing over occurs, four types of gametes are formed, but the old (parental) combinations are the most common. These results are just the reverse of those shown in Fig. 8-6, but in both cases, it is the parental types that are more common. This cross should be contrasted with the one in Fig. 8-5, in which the corresponding F₁ male is testcrossed.

cannot be used in Drosophila in a testcross to estimate the crossover amount because a ratio of 1:1 will always result regardless of which genes are followed. As mentioned, this absence of crossing over in the male fruit fly is very valuable in showing that two genes are linked and is one of the many advantages of Drosophila as a tool in pioneer genetic studies.

Linkage and the F₂ generation

The discussion so far has stressed the use of the testcross in the detection of linkage and the amount of recombination. Its advantage is clearly seen when we consider a cross that follows two linked genes

through the F₂ generation. Suppose we are working with corn plants and suspect that two particular loci are linked. The alleles of one of the genes affect the leaves (cr⁺, the dominant for normal and cr, its recessive allele for crinkly leaves); the alleles of the other influence height (d⁺ for normal and d for dwarf). Assume that the actual amount of recombination that can take place between these genes is 20% but that we are unaware of this at the outset. Instead of making a testcross, we decide to follow the plants through the F₂ generation and to estimate the crossover percentage from the F₂ data. Figure 8-10A represents the cross of a crinkly plant with a dwarf. Since we are not dealing with Drosophila, each parent (in-

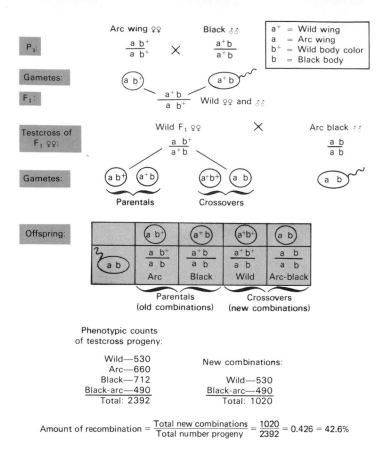

Phenotypic counts
of testcross progeny:

Wild—530
Arc—660
Black—712
Black-arc—490
Total: 2392

New combinations:

Wild—530
Black-arc—490
Total: 1020

$$\text{Amount of recombination} = \frac{\text{Total new combinations}}{\text{Total number progeny}} = \frac{1020}{2392} = 0.426 = 42.6\%$$

FIG. 8-8. Calculation of the amount of recombination. To calculate the amount of recombination between two autosomally linked genes, the F_1 dihybrid is testcrossed. The crossover types and the parentals are easily distinguished phenotypically. Crossover percentage is calculated by dividing the total number of testcross offspring into the number of new combinations.

FIG. 8-9. Testcross of dihybrid males in Drosophila. The testcross of a male that is dihybrid for two autosomally linked genes yields only the parental combinations in a 1:1 ratio. Linkage strength is 100% in the male fruit fly. Thus, crossover percentage cannot be determined using male Drosophila in a testcross. Compare this with the testcross of F_1 females in Fig. 8-8.

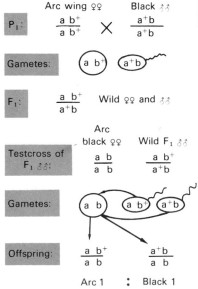

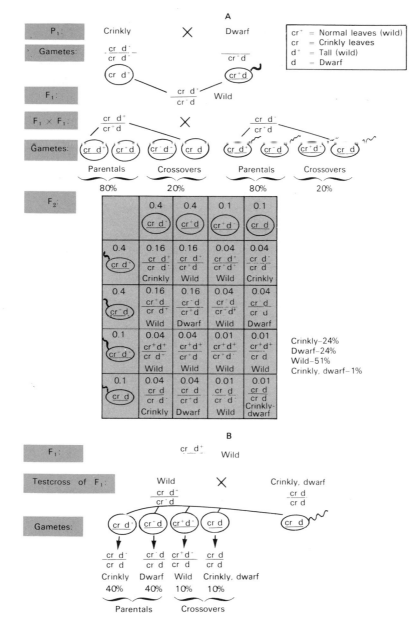

FIG. 8-10. Linkage and the F₂ generation in corn. (*A*) The amount of recombination between cr and d is 20%. Each parent forms crossover as well as parental types of gametes. If the amount of recombination were unknown to begin with, its calculation on the basis of the F₂ data would be difficult. (*B*) A cross of any F₁ hybrid to a double recessive reveals directly the number of new combinations (wild and crinkly, dwarf), and the percentage of recombination is readily determined.

stead of just the female) produces crossover gametes as well as the parental types. When combining the gametes in all possible combinations, we may use the familiar Punnett square method shown in Figure 8-10A, but we must remember that the gametes are not being formed in equal numbers, as is true in independent assortment. Therefore, we must indicate the frequency of each type of gamete. Each frequency, in turn, indicates the probability of the occurrence of that kind of gamete. When we combine two gametes, we must therefore multiply the two separate probabilities to indicate the chance of the two coming together at any one fertilization. This is no different from what we have been doing all along in problems dealing with independent assortment, in which the frequencies of each type of gamete have been equal.

We can see from the diagram of the cross (Fig. 8-10A) that on the basis of 100%, the four different phenotypes would occur in a frequency of 24:24:51:1. If we depended on such an F_2 ratio to tell us the amount of recombination, we could often be misled by fluctuations due to chance or to such factors as lethals and genic interactions. But even at its simplest, the F_2 ratio would usually require intricate calculations to estimate the percentage recombination. Obviously, the testcross is the more direct route (Fig. 8-10B).

Genes linked on the X chromosome

There is a qualification to the statement that using the F_2 is not the best way to estimate the amount of

recombination. This pertains to F_2 ratios obtained from genes linked together on the X chromosome. A cross between two stocks, mutant for different sex-linked alleles, produces an F_2 generation that is particularly valuable because it can be used to determine the amount of recombination. The F_2 generation in a sex-linked cross may be used for this purpose, because at least half of the F_2 in a sex-linked cross are always testcross progeny. This statement can be appreciated if we recall that sex-linked genes are carried on that part of the X that has no homologous region on the Y chromosome. Consequently, all XY males must express the X-linked alleles and thus will be test-cross offspring for any sex-linked genes. This is so, because the Y does not mask any alleles on the X contributed by the female parent. We may, in this sense, consider the Y chromosome devoid of genes being followed on the X.

Figure 8-11 illustrates linkage between two reces-

FIG. 8-11. Genes linked on the X chromosome. When genes on the X are being followed, the F_2 generation may be used to estimate the amount of recombination. This is so because the Y chromosome in the male masks no factors on the maternal X. Therefore, all the F_2 males are testcross offspring, and they can always be used in such crosses to estimate the percentage of recombination. In the cross shown here, *only* the F_2 male offspring can be used because the F_2 females received a dominant allele from the F_1 male parent. This masks recessives on the maternal X chromosome.

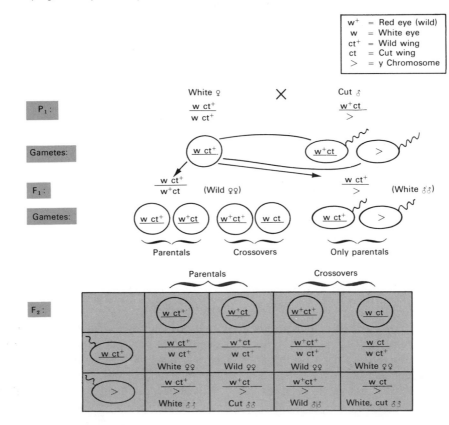

sives on the X of Drosophila: w (white eyes) and ct (cut wings). Note that the alleles in the F_1 female are in the trans arrangement. When allowed to mate with the F_1 males, these females produce crossover as well as parental gametes. The males, on the other hand, contribute only a Y chromosome and an X with the parental combination. The figure shows that the F_2 female progeny fall into only two phenotypic classes, wild and white. However, four different classes can be recognized among the F_2 males, because their fathers contributed the Y chromosome to them. They are thus testcross offspring, having received from their male parents nothing to mask any recessives on the X from their mothers. The F_2 females clearly are not testcross progeny; they received an X chromosome from their fathers that carried a dominant (ct^+), which did mask the presence of the recessive.

We can ignore these female offspring and concentrate only on the different types of males to estimate the crossover frequency. Suppose the mating between

"white" females and "cut" males, which is diagrammed in Fig. 8-11, produced 1000 flies in the F_2 generation, distributed as follows:

Females		Males	
Wild	258	White	191
White	250	Cut	198
	508	White, cut	45
		Wild	38
			472

The F_1 and the F_2 data tell us that the genes are sex linked, because the expression of the genes is seen to depend on the sex of the offspring; only the males are mutant in the first generation; only the males show the cut trait in the F_2. To calculate the crossover frequency, we pay attention only to the F_2 males, the testcross progeny. The parental and crossover combinations among them are recognized and their numbers are recorded, as follows:

FIG. 8-12. F_2 females and males as testcross progeny. In this example, the F_2 females receive only recessives from the F_1 male parent. They are thus testcross offspring in the case of the X-linked genes. All of the offspring in this cross may be used to estimate the amount of recombination. Contrast this with Fig. 8-11.

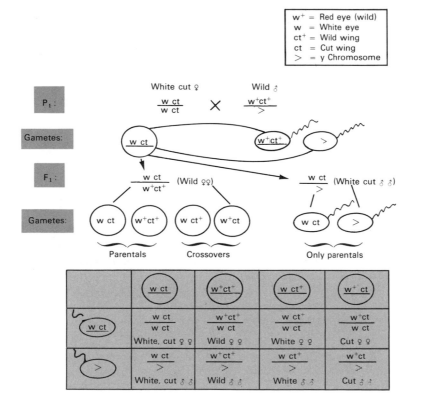

Parentals		Crossovers	
White (w ct^+)	191	Wild (w^+ ct^+)	45
Cut (w^+ ct)	198	White, cut (w ct)	38
	389		83

The percentage of recombination is then obtained by dividing the total number of male crossover types by the total number of male offspring: crossover frequency = $83/472 = 17.6\%$.

We must remember that when experimental crosses are performed, they are done routinely in reciprocal, to reveal any differences that might be related to sex. The student should diagram the reciprocal of the cross followed here: cut females × white males. Although the F_1 male offspring differ phenotypically from those F_1 males diagrammed in Fig. 8-11, the F_2 results among the males are the same. This is so because the F_1 females are identical genotypically in both crosses. In actuality, comparable figures will be obtained if the reciprocal matings are performed under similar conditions.

Although males are always testcross progeny in the case of sex-linked genes, females may also be testcross offspring, depending on the male parent. Consider the cross shown in Fig. 8-12. In the F_2 generation, all of the offspring, female as well as male, are testcross progeny. This is so because the X chromosome of the F_1 males contains both recessives. Consequently, these F_1 males contribute no sex-linked dominants to any of the offspring. Therefore, no factors mask the effects of any sex-linked alleles coming from the F_1 females. The diagram in Fig. 8-12 shows that all of the possible phenotypes are represented equally among the females and the males. Since they are all testcross offspring, every one of the progeny, male and female, can be counted to determine the crossover percentage. In contrast, the reciprocal mating (homozygous wild P_1 females × white, cut males) gives very different results. The student should diagram this cross to convince himself or herself that in this case only the F_2 males can be used in the counts to determine the crossover frequency. This is so because the F_1 males contribute an X to their daughters that contains both of the dominant wild types alleles.

Linkage groups and chromosomes

After Morgan's discovery of linkage of genes on the X chromosome of Drosophila, several autosomal genes in the fruit fly were also found to be associated with linkage groups. It soon became evident that the non-sex-linked genes were associated in one or the other of two large groups. This makes sense, because two other large chromosomes are present in the fruit fly in addition to the X (see Chap. 5). The fourth chromosome is a very small one. Finally, all four groups were established, and the relative sizes of the linkage groups were found to correspond to the sizes of the chromosomes. This is exactly what should be expected if the genes are located on the chromosomes. Linkage was then found in species other than Drosophila, and in these as well, the haploid chromosome number and chromosome sizes corresponded to the number and sizes of the linkage groups. This parallel provided even more support for the chromosome theory of heredity. If the physical basis of heredity is in the chromosomes, there must be many more genes than chromosomes, because the latter are found in relatively low numbers. Therefore, the genes would be associated on specific chromosomes, the linkage groups. The haploid number would thus correspond to the number of linkage groups. Large chromosomes would most probably carry more genes than small ones, and so there should be the parallel size relationship.

Genes that are linked should therefore tend to stay together. But the existence of crossing over shows that linked genes can separate and enter into new combinations. If it were not for crossing over, all allelic combinations on one chromosome would be forced to remain together. The only way a new combination could be formed would be through the slow chance process of gene mutation. Meiotic crossing over, however, brings about the new combination quickly. Although alleles of linked genes may not assort independently, crossing over extends recombination down to the level of the chromosome. Without crossing over much of the value of the sexual process, the formation of new gene combinations, would be lost. Evolution would be greatly slowed down in those species lacking it because fewer variants would be available for the operation of natural selection. Crossing over probably arose long before independent assortment—when only one pair of chromosomes was found in living things. It would have provided the first mechanism to ensure new gene combinations among the earliest cells.

Chromosome mapping

The data on crossing over assembled in Morgan's laboratory from experiments with Drosophila were compatible with the hypothesis that the units of he-

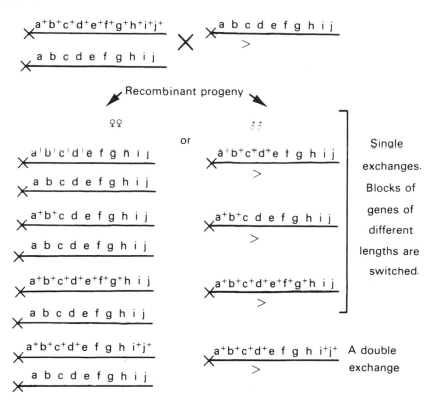

FIG. 8-13. Genes in a linear order. When a group of genes known to be X-linked is followed, the formation of the recombinant types among the offspring is found to be associated with an exchange of blocks of genes. In this schematic representation, females which are multiply heterozygous for 10 X-linked genes are testcrossed. Among the female and male offspring, new combinations may be detected. These recombinant types possess new arrangements of traits associated with blocks of genes. The size of a block that enters into a new combination can vary from one recombinant to the next. Some recombinants (*lower*) possess a block from one of the Xs, then a block from the other, then a block from the first again. These represent double exchanges.

redity are located in linear order on the chromosome. It had been conceivable that the genes are assembled into some complex arrangement with one on top of the other, or even scattered haphazardly in the chromosome. But the linkage and crossing over results showed that this could not be so. For example, in one of his early experiments, H. J. Muller, a student of Morgan, followed 10 mutant alleles linked on the X chromosome. As diagrammed in Fig. 8-13, when multiply heterozygous females were testcrossed, the results showed that the great majority of offspring with new combinations had one whole block of genes from one of the X chromosomes up to a certain point; the rest of the genes represent a block from the remaining part of the homologous X. The blocks might be of any length, indicating that an exchange could occur at any point along the length of the chromosome. Nearly always, there was one block up to a certain point and then a block of genes from the other chromosome.

But in a few cases, arrangements occurred in which a block was present from one chromosome, then a block from the homologue, and then a block from the first one again. These represent chromosomes in which not one but two breaks had occurred between the two chromosomes, giving a double crossover. Results such as these, as well as others that will be discussed shortly, could be explained only on the basis of a genic arrangement on the chromosome in which the genes were side by side, much as numbers are linearly arranged on a ruler.

It was reasoned that if this theory was correct, one should be able to determine the relative positions of genes on the chromosomes by studying strengths of linkage and the amount of recombination. When a particular kind of cross is repeated under the same experimental conditions, the amount of recombination between two genes is found to be constant. The frequency of crossing over between a particular pair

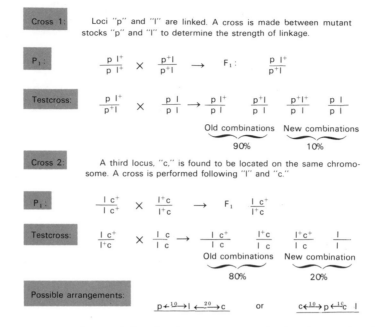

FIG. 8-14. Determining gene arrangement. In cross 1, the value of 10% indicates the amount of recombination, the frequency with which the loci p and l enter into new combinations. Similarly, in cross 2, the value of 20% indicates the amount of crossovers between l and c. The crossover frequencies reflect the distance between the loci. Thus, l and p with a 10% crossover value (90% old combinations) must be closer to each other than l and c with 20% recombination (80% parentals). From the data, two map arrangements are possible. The correct order can be established only by making a cross between stocks p and c to determine the amount of crossovers between them and hence their map distance.

of genetic loci is apparently not haphazard. Crossover values for different pairs of linked genes, however, may be quite dissimilar. Consider the two hypothetical crosses in Fig. 8-14. There would be a crossover value of 10% for p and l and a different constant figure, 20%, for l and c. We can interpret such results to reflect distances among these three genes, assuming a linear order. For if crossing over does entail some kind of breakage and reunion of chromosome threads, it is obvious that the closer together two genes are, the less the chance is that a random break will occur between them. So if l is closer to p than it is to c, there will be fewer chance breaks between p and l. Thus, they will tend to stay together more than will l and c when a single break takes place.

Tentatively, let us arrange the three genetic loci in the following order: p l c. The reasoning here is that the number of new combinations observed in a testcross stems from the amount of crossing over and that this, in turn, reflects the distance between the genes. If these new combinations can be used to indicate distance, what sort of units should be used as a measure? H. J. Muller suggested taking the percentage of new combinations and using them directly as a kind of measure of distance simply by converting the percentage into crossover units. So we would say that genes p and l are 10 units apart, whereas l and c are separated by 20 units. A moment's consideration of our tentative arrangement of the three genes (p l c) tells us that the gene order is not justified from the information at hand so far. It is equally possible that the correct order is c p l. Provisionally, we can only say that l and c are twice as far apart as p and l. However, c could be on either side of p. Either of the two map arrangements shown in Fig. 8-14 is feasible.

Obviously, we must find the number of new combinations between p and c. We cannot place any three genes (1, 2, 3) in the proper relative positions if we find only the crossover values between genes 1 and 2 and between genes 1 and 3. We must also determine the number of new combinations between genes 2 and 3 in order to arrange the loci on the chromosome with any degree of accuracy. Suppose a cross between p and c gives a value of approximately 30%. This would indicate that the first supposition was correct

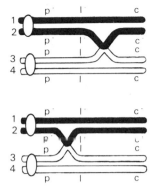 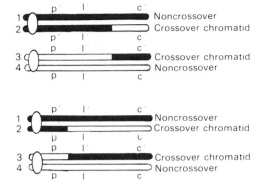

FIG. 8-15.   Physical basis of crossing over. At the time of crossing over, each chromosome is composed of two chromatids. A single crossover event involves only two chromatids of the four that are present in the bivalent. To obtain a crossover between loci p and c, all that is needed is a break and reunion anywhere in the interval between p and c. In the top diagram, the point of crossing over resulted in a new combination between l and c as well as between p and c. To obtain a crossover between p and l, a much shorter distance is involved. Imagine that, for a crossover event to occur within this shorter interval, the threads must bend more sharply. The lower diagram shows a crossover between p and l. Note that in this case and also in the one shown above, there was crossing over between p and c. This is true simply because any one break between p and l must also be a break between p and c. Thus, as a result of their greater distance apart, we would expect more crossing over between p and c than between p and l.

(the left arrangement in Fig. 8-14). A value close to 10%, on the other hand, would indicate that c is to the left of l (the right arrangement, Fig. 8-14).

Assuming that the correct order is found to be p l c, we can appreciate the physical basis for the different crossover values from the illustrations in Fig. 8-15. Notice that in all the illustrations of crossing over, each chromosome is represented as a double structure. Chromosomes are depicted in this way because genetic analyses of such organisms as Neurospora and Drosophila, as well as studies of the cell cycle, indicate that at the time of crossing over, each chromosome thread is already replicated. Each chromosome would thus be composed of two chromatids, and crossing over would take place in the four-strand stage, which simply means that four chromatids are present when it occurs. Any one crossover event, however, involves only two of the four strands, and nothing dictates that it must always be the same two. In the illustrations, strands 2 and 3 were selected as the crossover chromatids only for the sake of diagrammatic clarity. This should not be construed to mean that only these two can participate. Strands 1 and 3, 1 and 4, or 2 and 4 could have engaged equally well.

Although the reasoning discussed in Fig. 8-15 seemed to hold it was soon realized that certain complications could enter the picture. For example, if the gene order is actually p l c, and p and c are truly 30 units apart, it might be possible for crossing over to occur simultaneously at two points between p and c, that is, to get double crossing over between them. Recall that Muller, studying blocks of genes, found evidence for the occurrence of double crossing over.

Let us examine the effects of such an event on the relationship between these two genes and compare them with the results of a single crossover event. If just genes p and c are being followed, any double crossover between them would produce the results in Fig. 8-16A. Compare these with the results of the single crossover event in Fig. 8-16B. We see that the single crossovers produce detectable new combinations between p and c. In the case of the doubles, however, p and c are not placed in a new relationship to each other but still occur together on the same chromatid. Therefore, there would be no way to recognize the fact that two crossovers, not just one, had occurred between them. The doubles would be classified along with the parentals, the types with the old combinations. Because we would be losing sight of two single crossovers for every double that was not detected, p and c would appear to us to remain together in the parental combination more often than they actually do. Since the chromosome map is based on the number of new combinations that we can observe, the genes p and c would appear to be more strongly linked, and hence closer together than they really are.

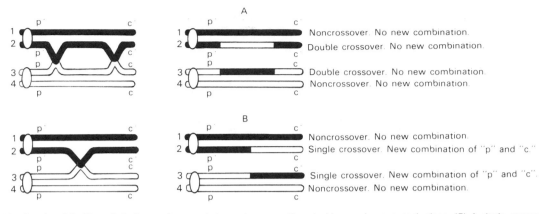

FIG. 8-16. Results of double and single crossing over between two genes. (*A*) A double crossover event between p and c yields no new combinations of alleles. Therefore, if only these two loci are being followed in a cross, we would lose sight of two single crossovers each time double crossing over took place. (*B*) A single crossover event between p and c can be readily detected by the new combination among the offspring.

Trihybrid crosses and mapping

How can we overcome this tendency to underestimate the true number of crossovers? The problem is quite easily solved by making a trihybrid or three-point testcross. Instead of following just two genes at a time in the testcross, we consider three. Trihybrid crosses are especially valuable in chromosome mapping and are routinely performed to ascertain gene locations. The same basic procedures are followed when working with microorganisms or with higher species. Because chromosome mapping is so critical to genetic analysis and is assuming an important role in human genetics, the steps and the reasoning behind them are presented here in considerable detail.

If we have established the gene order as p l c and know that l is between the other two, we can use this knowledge to detect any double crossovers between the more distant genes, p and c. Now consider a double crossover when a trihybrid is being followed

(Fig. 8-17). We see that on either side of gene l, there are two chromosome regions, region 1 and region 2, where single crossover events can occur. If two singles occur together to give a double, then no new combinations between p and c are produced. However, we now have a way to detect these doubles. For when a double crossover event occurs, the gene that is more central, gene l in this case, enters into a new combination with the genes on either side of it. Note the genic parental arrangements in the trihybrid parent. On the one chromosome, the three wild alleles of the three genes are together: p^+ l^+ c^+. On the other, the three mutant alleles are linked: p l c. As a result of the double crossover event, we now have p^+ l c^+ and p l^+ c. It appears to us that l, the gene in the middle, has shifted its position. The apparent shift has resulted from the occurrence of two crossover events, one on either side of l. With the trihybrid testcross, we can now recognize double crossovers that would otherwise be lost, and we add these to the

FIG 8-17. Double crossing over in a trihybrid. A double crossover event involves two single events. There are two regions therefore where single crossing over can take place. A double crossover in a trihybrid can be detected, because the "middle" gene will assume a new relationship with those on either side of it. It will appear to have switched position. The alleles on either side of it will remain in the old combination.

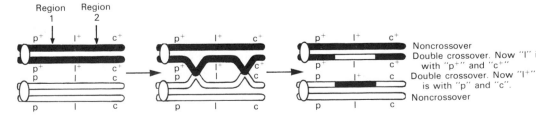
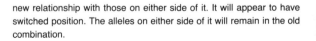

TABLE 8-1 Trihybrid testcross data and construction of a map segment

$$P_1 \; \frac{b^+ \; n^+ \; t^+}{b^+ \; n^+ \; t^+} \times \frac{b \; n \; t}{b \; n \; t} \longrightarrow F_1 \; \frac{b^+ \; n^+ \; t^+}{b \quad n \quad t}$$

Testcross $\dfrac{b^+ \; n^+ \; t^+}{b \quad n \quad t} \times \dfrac{b \; n \; t}{b \; n \; t}$

Testcross progeny			
b^+	n	t	21
b	n^+	t	120
b	n	t	289
b^+	n^+	t	69
b	n	t^+	71
b^+	n^+	t^+	298
b^+	n	t^+	109
b	n^+	t^+	15
Total			992

Construction of Map Segment from Testcross Data

Parentals		
b^+ n^+ t^+	298	
b n t	289	
Total	587	

b n^+ t^+	15	Doubles	
b^+ n t	21		
Total	36		

Alleles at b appear to have switched position. The b locus must be between n and t. One correct way to represent the trihybrid parent is

$$\frac{n^+ \quad b^+ \quad t^+}{n \quad b \quad t}$$

Crossing over may occur in the segment between n and b; it may also occur in the segment between b and t. These regions can be called, respectively, region 1 and region 2, and the trihybrid written as

$$\begin{array}{ccc} & 1 & 2 \\ n^+ & b^+ & t^+ \\ \hline n & b & t \end{array}$$

Single crossover region 1

$$\frac{n^+}{n} \times \frac{b^+}{b} \quad \frac{t^+}{t}$$
$$\downarrow$$

n^+	b	t	120
n	b^+	t^+	109

Region 1 crossovers 229

Total region 1 $\dfrac{36}{265}$

Crossover regions 1 and 2

$$\frac{n^+}{n} \times \frac{b^+}{b} \times \frac{t^+}{t}$$
$$\downarrow \quad \downarrow$$

n^+	b	t^+	15
n	b^+	t	21

Doubles total   36

Single crossover region 2

$$\frac{n^+}{n} \quad \frac{b^+}{b} \times \frac{t^+}{t}$$
$$\downarrow$$

n^+	b^+	t	69
n	b	t^+	71

Region 2 crossovers 140

Total region 2 $\dfrac{36}{176}$

$$\text{Amount of region 1 crossovers} = \frac{265}{992} = 26.7\% \quad \text{Amount of region 2 crossovers} = \frac{176}{992} = 17.7\%$$

$$\text{Map segment} \quad \underline{\text{n} \quad 26.7 \qquad \text{b} \quad 17.7 \quad \text{t}}$$

number of singles to give the true amount of crossing over between p and c. These two genes on either side of l will consequently appear farther apart than they otherwise would. We see from this example that the number of new combinations is not the sole criterion

for an estimation of the amount of crossovers between two distant genes. Whenever possible, three closely linked loci should be followed to arrive at a more accurate estimation of map distance.

Let us now use a trihybrid testcross to map three

hypothetical genes that are autosomally linked in Drosophila. We will call the three genes b, n, and t. Nothing is known about these three sites except that they are linked and the wild type alleles, b^+, n^+, and t^+, are dominant. We do not even know the precise order of the three loci. Since we do not know which gene is the one in between, we cannot say which is the correct way to illustrate the gene arrangement when we diagram the cross. However, if we are consistent, it does not really matter. To start with, we may select the order b n t. This may or may not be accurate, and we could have chosen another arrangement to work with. Data from the actual cross will tell us later which arrangement is actually correct.

Suppose a cross is made between homozygous wild females and triple mutant males. Having decided to use the order b n t, we can depict the trihybrid cross and the testcross as shown in Table 8-1. Eight different categories of offspring can be recognized phenotypically on the basis of the combinations of the wild and mutant traits. Eight phenotypic classes are expected because a trihybrid forms eight classes of gametes ($2^3 = 8$). Remembering the meaning of a testcross, we know that these eight classes actually represent the eight different kinds of gametes produced by the trihybrid and that the data reflect the relative frequency in which they were formed. Therefore, we can use the testcross data to map the three genes, for now we have a way to estimate how often the linked genes separated and entered into new combinations during gamete formation. The steps and reasoning go as follows (follow Table 8-1):

1. Since linked genes tend to stay together, we would expect the noncrossover types to be the most common. The phenotypic classes that are the most numerous among the progeny should represent the parental-type gametes, those in which no crossing over has been involved. The data show us these types, and we record them as the parentals: $b^+n^+t^+$ (298) and b n t (289).

2. Single crossovers would be less common than parentals, but doubles would be even fewer and can be easily identified. So we record the least numerous classes: b n^+t^+ (15) and b^+n t (21).

3. Once both the parental and double crossover classes have been recognized the correct order of the three genes is readily determined. We have seen that when a double crossover event takes place, the gene in the middle appears to have switched its position in relation to the two on either side of it. A comparison of the two classes, the parentals and the doubles,

indicates that b is the gene that is in between. We see that the three wild alleles went into the cross together on one chromosome and the three recessives on the other. In one group of doubles (b n^+t^+), two of the wild types are still together, and in the other group (b^+n t), two of the recessives are still in the parental arrangement. It is alleles of gene b that do not appear in the original combination, and we conclude that gene b must be between the other two.

4. We next rewrite the trihybrid parent to show the proper gene arrangement. It makes no difference whether we write $n^+b^+t^+/n$ b t or $t^+b^+n^+/t$ b n, as long as the proper gene is in the inside position. If this seems confusing, remember that we are primarily concerned with distances. As long as the relationships between the genes are preserved, the sequence selected is of no consequence, just as it would not matter if a road map were turned upside down when it was read. We will choose arrangement $n^+b^+t^+/n$ b t to stand for the trihybrid. To convince yourself that the other possibility is equally valid, you should later redo this cross using the alternative order; you will obtain the same results.

5. Looking at the correctly written trihybrid parent, we can see that there are two places where crossing over can occur to rearrange the three linked genes: one between n and b and the other between b and t. For reference, these regions are designated region 1 and region 2, respectively.

6. A break in region 1 will produce the single crossovers: n^+ b t and n b^+t^+. The data show that they amount to 120 and 109, respectively, giving a total of 229 crossovers in that region. The singles in region 2 are n^+b^+t (69) and n b t^+ (71), totaling 140. It is critical to understand that these single crossovers, that can be recognized from the data do not represent all the crossovers in regions 1 and 2. Double crossing over has occurred; this means that there were additional breaks in both regions. These cannot be detected by considering the singles alone. Since one double masks the effects of two single crossover events, the genes n and t will appear closer together than they should on the final map if we fail to take the doubles into consideration. However, we expressly made a triple testcross to detect these doubles, and we identified (step 2) 36 of them among the progeny. A glance at the trihybrid (Table 8-1) shows that, to get the classes n^+b t^+ and n b^+t, a crossover event must have occurred in *both* regions 1 and 2. Since this happened 36 times in region 1, we must add this figure to the total of singles we have already identified

in this region. Instead of a total of 229 crossovers, there were actually 229 + 36, or 265. Exactly the same reasoning applies to region 2. Again 36 must be added, this time to 140, giving the correct value of 176 for the number of crossovers in this region. The value of the triple testcross should now be quite apparent. It enables us to detect in two regions of the chromosome those single crossovers that otherwise would go unobserved. We must not forget to add the doubles identified in any three point testcross to each set of singles, those in region 1 and those in region 2 as well.

7. All that remains to be done to obtain crossover values among the three genes is to take the total crossover figure for each single region and divide it by the entire number of offspring. The respective percentages, 26.7% and 17.7%, are simply translated into map units.

Significance and use of a chromosome map

Once we have a segment of a chromosome map, it is essential to know how it can be put to use. Although a road map reflects precise distances between points, this is not true of a chromosome map. The number of map units between two genes does indeed depend in part on distance; however, we must not think of the units in terms of accurate spatial measurements. There are several reasons for this. For one thing, the ease with which crossing over occurs along a chromosome is not uniform; there is more crossing over in some parts of the chromosome than in others. For example, the region around the centromere engages in less crossing over than most of the rest of the chromosome. As a result of its influence, genes in the vicinity of the centromere tend to appear more clustered than genes elsewhere. That this is indeed a centromeric effect has been demonstrated through experiments in which the centromere position has been changed as a result of shifts in chromosome segments. Those areas now located away from the centromere undergo more crossing over. When a map is constructed, the genes in such regions appear farther apart than they previously did when they were in the vicinity of the centromere (Fig. 8-18).

In addition to the centromere, there are certain other chromosome regions in which crossing over may be cut down and others in which it seems to take place with more ease. Reasons for these local differences in the chromosomes of living things are

FIG. 8-18. Effect of the centromere. Genes in the vicinity of the centromere generally engage in less crossing over than genes a greater distance away. In this hypothetical map (above), note that loci e through m appear clustered, meaning that they are close together in map units. This is because of the centromere that decreases the frequency of crossing over in its vicinity. The centromere may become shifted (middle) because of a chromosome aberration, such as an inversion in which a segment is turned around. When a map is then constructed (below) on individuals with the new gene order, those genes that had been near the centromere (e.g., e through i) now appear farther apart in map units. They are engaging in more crossing over. Other genes now placed in the vicinity of the centromere (p, q, and r) map closer together as they now engage in less crossing over.

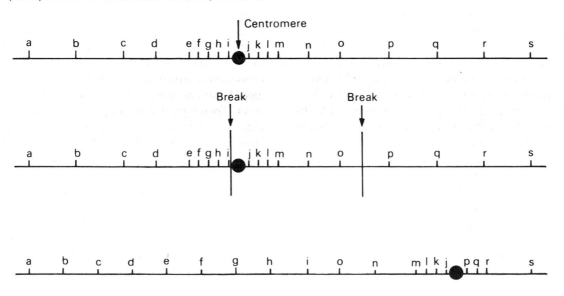

not known. We must also realize that crossing over, like many other genetic phenomena, may be influenced by environment, both external (temperature, radiation, nutrients, etc.) and internal (hormonal). This latter effect is very clearly demonstrated by the absence of crossing over in the male fruit fly. Therefore, to construct accurate maps and compare amounts of crossovers, we must control such factors. Since the amount of crossing over, and thus the number of crossovers, can be influenced in several ways, the map cannot be considered a true reflection of distance. However, the amount of crossing over is constant for a given species under a prescribed set of circumstances. Thus, the map can be used to make predictions when a cross is carried out under a defined set of conditions.

If the map can be used in this way to tell us the expected amount of recombination in different chromosome regions, we should be able to predict from it the number of double crossovers. The map gives us the probabilities of the single events expressed in map units. Since a double crossover is the result of two singles happening simultaneously, we should be able to estimate the probability of a double in a certain chromosome region simply by multiplying together the separate chances for the singles. In our example, the probability is 26.7% that a crossover will occur alone in region 1 and 17.7% that one will take place in region 2. Recalling that the chance for two independent events happening together is the product of their separate probabilities, we simply multiply the two to find the chance of the double event $(0.267 \times 0.177 = 0.047 = 4.7\%)$. The map as a table of probabilities gives the percentage of doubles to be expected in that particular chromosome region. But let us see if the results of the cross bear this out.

We obtained 992 offspring (refer to Table 8-1). Of these 4.7% should be double crossover progeny, by our calculations. Therefore, approximately 47 individuals should be distributed between the types $n^+ b\, t^+$ and $n\, b^+ t$ (992×0.047). The data show that actually only 36 of these types arose from this cross. We might conclude that this lower figure is just the result of chance in this particular instance. However, if we repeated the mating many times, we would continue to find fewer double crossover types than the number predicted from the map. Finding a number of doubles that is smaller than the amount indicated by the map is typical rather than exceptional. This suggests something about the single crossover events themselves. They must not be independent, for if they were, their

product should be a reliable prediction of the actual number of doubles that will be obtained in a cross. We are forced to conclude that the occurrence of one crossover event can influence the occurrence of another in the same chromosome. This is particularly true within short regions of the chromosome, where the influence of one crossover on another may be quite pronounced. The term interference describes the effect crossing over in one chromosome region exerts on the probability of a crossover event in an adjacent region.

This interference varies from one part of the chromosome to another. For a more accurate estimate of the amount of double crossovers that will arise in a cross, interference must be taken into consideration. This factor is measured by the coefficient of coincidence. This is nothing more than an expression of the ratio of double crossovers actually obtained to those predicted from the map (Table 8-2). We expected approximately 47 doubles but found only 36. The coefficient of coincidence equals the actual number of doubles over the expected number, which in this example is $36/47 = 0.76$. This coefficient tells us that we must modify the number of double crossovers predicted from the map. We expected 47, but knowing the coefficient of coincidence, we multiply this figure by 0.76. This in turn indicates that we should expect only 36 doubles, which is what we actually obtained from the cross. The coefficient of coincidence varies between zero for very short distances (meaning that no doubles at all are to be anticipated) and 1 (meaning the expected amount is simply the product of the two singles, and no interference is operating). It is typical for the value of the coefficient to be between the two extremes.

Now let us see how we can use the map to make further predictions. Referring back to the map segment in Table 8-1, we see that we can expect 26.7% crossovers between genes n and b. Similarly, for region 2, we can predict 17.7% crossovers between b and t. Therefore, if 1000 offspring result from a mating, 177 of them $(17.7\% \times 1000)$ should represent crossovers between b and t. However, remember that the crossover amount is not equal to the number of new combinations, since any one double crossover event between two genes will regenerate the old combinations. But knowledge from our map segment can aid us in this regard. We have just calculated that 36 double crossovers are to be expected, taking into consideration the coefficient of coincidence. This figure, 36, was added to each set of singles when the

TABLE 8-2   Calculation of coefficient of coincidence based on data in Table 8-1

Map segment			Number of doubles obtained in actual cross $= 36$
n 26.7 b 17.7 t			

Amount of double crossovers expected from map segment = crossover probability in region 1 × crossover probability in region 2:

$$0.267 \times 0.177 = 0.047 = 4.7\%$$

Number of doubles to be expected on basis of 992 offspring:

$$992 \times 0.047 = 47$$

$$\text{Coefficient of coincidence} = \frac{\text{no. of doubles actually obtained}}{\text{no. of doubles predicted from map}} = \frac{36}{47} = 0.76$$

genes were being mapped, to give a more accurate estimate of the true amount of crossovers. Now, however, we must subtract the 36 from our expected new combinations, since we will get 36 fewer new combinations in each region (1 and 2) than the distance indicates in crossover units. On the basis of 1000 offspring, therefore, we would expect, in region 1, 231 new combinations ($26.7\% \times 1000 - 36$) and, in region 2, 141 new combinations ($17.7\% \times 1000 - 36$). What we are doing is using the map to make predictions in terms of probabilities, and that is basically how a chromosome map serves us.

Mapping three genes on the X chromosome

Once linkage relationships and their major complications are understood, mapping of genes on the sex chromosome should not prove difficult. The procedure is the same as that described earlier for mapping genes on an autosome and requires no more than an appreciation of what has already been discussed for the determination of crossover values on the sex chromosome. Suppose we are following three sex-linked recessive alleles in Drosophila: crossveinless (missing wing vein), echinus (defective eye), and cut (defective wing). At first, all we know is that they are on the X chromosome, and again we know nothing about their arrangement. Suppose we cross females expressing the recessive crossveinless with males showing the other two recessive traits—echinus and cut. We may tentatively summarize the cross as shown in Table 8-3. To construct the map segment, we follow the identical steps just given for linked autosomal genes. The only additional thing to remember is that males are always testcross progeny in the case of

genes on the X chromosome. The F_2 females, on the other hand, may or may not be. Females are testcross offspring only if their fathers were expressing all the recessives. In this example, we see that this is not so. The F_1 male parents are showing the dominant traits wild eye (ec^+) and normal wing (ct^+). Consequently, their daughters will not reveal the results of all the crossing over that occurred in the F_1 females. The F_2 males, however, having received a Y from their fathers, will represent all eight types of gametes (two parental and six crossovers) formed by their mothers. They also indicate the proportion in which those eggs were produced. So, we simply disregard the females and direct our attention solely to the F_2 males for a construction of the map (Table 8-3).

We pick out the parentals that are the most numerous phenotypes: cv^+ ec ct (1099) and cv ec^+ ct^+ (1051). The doubles are clearly cv^+ ec^+ ct^+ (5) and cv ec ct (4). A comparison of the doubles and parentals shows that the cv appears to have shifted its position and thus must be the central gene. So the trihybrid female must be rewritten in the proper way: ec^+ cv ct^+/cv^+ ct. It is most important at this point to note the genic composition of the parental chromosomes. One chromosome has two wild alleles and one mutant (ec^+ ct^+ cv), whereas the other has the two mutants and one wild (ec ct cv^+). The beginner often assumes that in the parents all three wild alleles must occur together on one chromosome and all three mutants on the homologue. Although this was the case in the example given for the autosomal genes, nothing dictates that this must always be so. A moment's thought tells us that a trihybrid individual can arise in various ways, depending on how the cross was made. A trihybrid is any individual heterozygous

TABLE 8-3   Trihybrid cross and F_2 data for three X-linked genes[a]

$$P_1: \quad \frac{cv \;\; ec^+ \;\; ct^+}{cv \;\; ec^+ \;\; ct^+} \times \frac{cv^+ \;\; ec \;\; ct}{Y} \longrightarrow F_1: \frac{cv \;\; ec^+ \;\; ct^+}{cv^+ \;\; ec \;\; ct} \text{ (wild females)}$$

♀ ♀ Crossveinless   ♂ ♂ Echinus, cut

and

$$\frac{cv \;\; ec^+ \;\; ct^+}{Y} \text{ (crossveinless males)}$$

Wild   Crossveinless

$$F_1 \times F_1: \quad \frac{cv \;\; ec^+ \;\; ct^+}{cv^+ \;\; ec \;\; ct} \times \frac{cv \;\; ec^+ \;\; ct^+}{Y}$$

F_2:

Females		Males				
Wild................1340		Echinus, cut	cv^+	ec	ct	1099
Crossveinless..........1308		Wild	cv^+	ec^+	ct^+	5
	2648	Echinus	cv^+	ec	ct^+	111
		Cut	cv^+	ec^+	ct	132
		Crossveinless, cut	cv	ec^+	ct	109
		Crossveinless, echinus	cv	ec	ct^+	135
		Crossveinless	cv	ec^+	ct^+	1051
		Crossveinless, echinus, cut	cv	ec	ct	4
					Total	2646

[a]The order of the loci shown here is only tentative. Consult the text for details.

at three loci that are under consideration in a cross. Thus, it would be incorrect to believe that the F_1 trihybrid *must* be ec^+ cv^+ ct^+/ec cv ct.

This is just one possibility; the F_1 trihybrid could have resulted if the cross had been made differently (by crossing homozygous wild females with triple mutant males or vice versa). When given a list of progeny from a three-point testcross, we must peruse it for the most numerous classes in order to identify the parentals. It is fallacious to assume that the parental types must occur in one set arrangement.

Returning to the correctly written trihybrid, we pick out the crossovers in region 1: ec^+ cv^+ ct (132) and ec cv ct^+ (135)=267. Region 2 crossovers are ec^+ cv ct (109) and ec cv^+ ct^+ (111)=220. The doubles total nine and must be added to each of these sets of singles. The final arithmetic is 276/2646 for the crossover distance between ec and cv and 229/2646 for that between cv and ct. The map for this small chromosome segment is thus ec 10.4 cv 8.7 ct. The coefficient of coincidence is calculated in the usual manner and gives a value of 0.37, as simple arithmetic will show.

Map length, multiple crossovers, and the amount of recombination

Chromosome maps have been constructed in considerable detail for certain organisms, particularly Drosophila, Neurospora, the mouse, and corn. Figure 8-19 shows the location of some of the better known loci in the fruit fly. When detailed chromosome maps are made, they are usually built up by determining the crossover values between genes that are closely linked. Trihybrid testcrosses are used to establish gene order, but if genes are a considerable distance apart, multiple crossovers (doubles, triples, and even higher degrees) may occur between them to obscure the true crossover value. Therefore, more accurate map units are determined by studying recombination values between genes close enough to minimize the effects of these factors.

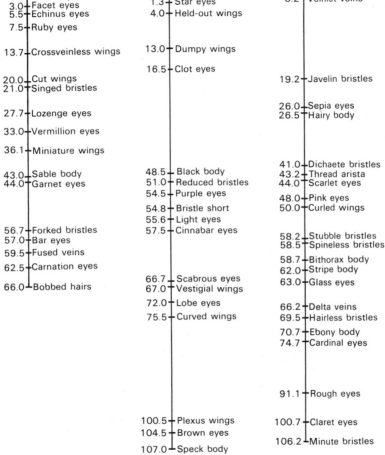

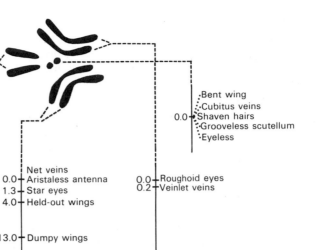

	.Bent wing
	.Cubitus veins
0.0	Shaven hairs
	.Grooveless scutellum
	.Eyeless

	.Yellow body	0.0	Net veins	0.0	Roughoid eyes	
0.0	Scute bristles	1.3	Aristaless antenna	0.2	Veinlet veins	
1.5	White eyes	1.3	Star eyes			
3.0	Facet eyes	4.0	Held-out wings			
5.5	Echinus eyes					
7.5	Ruby eyes					
13.7	Crossveinless wings	13.0	Dumpy wings			
		16.5	Clot eyes	19.2	Javelin bristles	
20.0	Cut wings					
21.0	Singed bristles			26.0	Sepia eyes	
27.7	Lozenge eyes			26.5	Hairy body	
33.0	Vermillion eyes					
36.1	Miniature wings			41.0	Dichaete bristles	
43.0	Sable body	48.5	Black body	43.2	Thread arista	
44.0	Garnet eyes	51.0	Reduced bristles	44.0	Scarlet eyes	
		54.5	Purple eyes	48.0	Pink eyes	
		54.8	Bristle short	50.0	Curled wings	
		55.6	Light eyes			
56.7	Forked bristles	57.5	Cinnabar eyes	58.2	Stubble bristles	
57.0	Bar eyes			58.5	Spineless bristles	
59.5	Fused veins			58.7	Bithorax body	
62.5	Carnation eyes			62.0	Stripe body	
		66.7	Scabrous eyes	63.0	Glass eyes	
66.0	Bobbed hairs	67.0	Vestigial wings			
		72.0	Lobe eyes	66.2	Delta veins	
		75.5	Curved wings	69.5	Hairless bristles	
				70.7	Ebony body	
				74.7	Cardinal eyes	
				91.1	Rough eyes	
		100.5	Plexus wings	100.7	Claret eyes	
		104.5	Brown eyes			
		107.0	Speck body	106.2	Minute bristles	

FIG. 8-19. Genetic or linkage map of the four chromosomes of *Drosophila melanogaster.* (Reprinted with permission from E. W. Sinnott, L. C. Dunn, and T. Dobzhansky, *Principles of Genetics.* McGraw-Hill, New York, 1958.

Suppose a gene order such as the following has been established by trihybrid testcrosses: f g h i j k l. The map would then be more precisely built up in map units by finding the crossover value between h and i, then between h and g, g and f, and so on. Obviously, this would be more accurate than just relying on recombination values between f and l to reflect the number of map units between these two loci, because multiple crossovers could occur in the f–l interval. To have a more reliable estimate of the distance between f and l in map units, we would add up the small crossover values between the genes in the f–l interval.

It soon becomes apparent from glancing at chro-

212   8. GENES ON THE SAME CHROMOSOME

mosome maps that they can include more than 50 map units, even more than 100 (note the two large chromosomes of Drosophila, Fig. 8-19). When estimating map lengths, the relationship between chiasmata and crossover events is taken to be 1:1. Cytogenetic analyses have continued to show a good correspondence between the amount of crossing over detected genetically by observation of the amount of recombination and the number of chiasmata seen with the microscope. A higher chiasma frequency at meiotic prophase is associated with a higher frequency of genetic recombination. Moreover, in those species with large chromosomes, the number of chiasmata present in the bivalents can be determined more accurately than in species with tiny meiotic chromosomes, such as Drosophila. In corn, for example, which has large chromosomes, a very good correlation exists between the amount of genetic recombination and the chiasma frequency. An average of one chiasma in a bivalent would therefore reflect a map length of 50 units. This would be so because genetic recombination between the pairs of genes located at the opposite ends of the chromosome would take place

50% of the time. This follows from the fact that crossing over is the event that takes place in the bivalent to produce recombinant, or crossover, gametes. If a crossover event occurred between two linked genes in 100% of the meiotic cells, half of the resulting gametes would be recombinants and half would be noncrossovers. This is so because only two of the four chromatids in a bivalent participate in any one crossover event (Fig. 8-20A). An average of two, three, or four chiasmata, per bivalent would give, respectively, chromosome map lengths of 100, 150, and 200 units.

What does this mean in terms of the amount of recombination to be expected between very distant genes? Does this mean that 60% or even higher amounts of recombination will occur? The answer is that the number of recombinants cannot exceed 50%. A moment's reflection tells us that 50% recombination is equivalent to independent assortment. If one crossover between two linked genes, a and b (Fig. 8-20A), occurs in every meiotic cell, that means equal numbers of parentals and crossover gametes will form. Indeed, as noted before, if genes are far enough apart

FIG. 8-20. Effects of a high frequency of single crossovers. (*A*) If crossing over occurs between two genes in a meiotic cell, four different kinds of gametes are formed. Only two of the four chromatids participate in any one crossover event. If one crossover event occurs in every meiotic cell between a and b, it will produce recombinants in a frequency of 50%, and the loci will appear to be assorting indepen-

dently. (*B*) The diagram shows independent assortment, assuming a and b are on separate chromosomes. The two possible anaphase separations occur with equal frequency. The result is four kinds of gametes in equal numbers. It can be seen by comparing diagrams A and B that a single crossover event in every cell at meiosis will have the same effect as independent assortment.

A

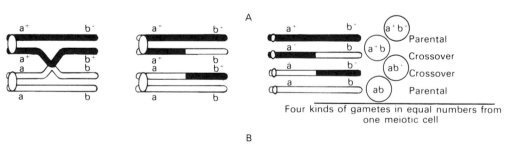

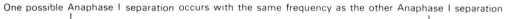

Four kinds of gametes in equal numbers from
one meiotic cell

B

One possible Anaphase I separation occurs with the same frequency as the other Anaphase I separation

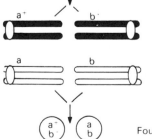   Four kinds of gametes in equal numbers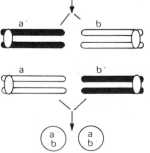

on a chromosome, this tends to occur, and we may not be able to tell at first that the two genes are linked; they would be assorting independently, as if they were on separate chromosomes (Fig. 8-20B).

The same principles of map construction used in lower forms can be applied to higher groups. Well over 200 genetic loci have been assigned to specific chromosomes in the human, and the information is being used in a very practical way (see Chap. 9). An estimate has been made of the map length of the human X chromosome and the total map length of all the autosomes. This has been reasoned from studying the frequency of chiasmata in meiotic cells of testic-

ular tissue. Recall that a single chiasma is equivalent to a crossover value of 50%. Counts of chiasmata average about 552 for all the autosomal bivalents in a meiotic cell. This gives a total map length for all the autosomes of 2600 units. Since the human X chromosome is about 6% the length of one total haploid set of autosomes, its genetic length has been calculated to be at least 160 map units. The figure cannot be considered precise, however, because chiasmata counts were made in male meiotic cells and on autosomes. Apparently less crossing over occurs in the human male, as reflected in less recombination than in the female. Therefore, the true value

FIG. 8-21. Types of double crossovers. Double crossing over can involve any of the four chromatids and may take place in a variety of ways. This figure illustrates just one of these. A three-strand double can always occur in two different manners. Since double crossing over is random, the two-, three-, and four-strand doubles occur with a frequency of 1:2:1, respectively, as indicated here. The number of recombinant types, however, cannot exceed 50%. (The second crossover event usually does not occur at exactly the same position as the first. The diagram here clarifies events between loci a and b.)

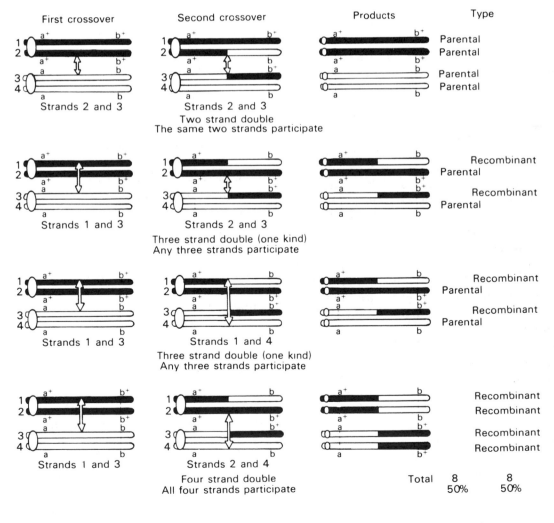

of the X chromosome in map units may be greater, closer to 200 units. Nevertheless, the length of the X in map units is seen to be sufficient to account for the fact that certain genes, known to be X-linked still appear to be undergoing independent assortment. Students of human genetics employ the term *synteny* to describe the situation in which two loci have been assigned to the same chromosome but still may be separated by such a large distance in map units that genetic linkage has not been demonstrated. The failure to show linkage results from the fact that crossing over usually separates the two distant loci.

The complete absence of crossing over in the male fruit fly gives a great advantage in showing that even widely separated genes are located on the same chromosome. Remember that only two of the four chromatids participate at any one crossover. Therefore, even if a crossover event were to occur in every meiotic cell, there would still be only 50% recombination. One might believe that multiple crossovers could increase the recombination value above 50%. A consideration of the different kinds of double crossovers (Fig. 8-21) may clarify this. It was stated in our discussion that a double crossover can involve any of the four strands. We have considered, for the sake of clarity, only two-strand doubles, those double crossovers in which the same two strands participate in each of the two crossover events. Figure 8-21. shows that three- and four-strand doubles may also take place. Because they are independent events, the doubles occur in a random frequency of 1:2:1 (one two-strand double: two three-strand doubles: one four-strand double). If the numbers of parental and recombinants are added up, it is seen that they total 50:50. Again the recombination value does not exceed 50%, even on this basis.

What about even higher numbers of crossover events between two genes? The greater the distance between two genetic loci, the greater is the chance for multiple crossing over to occur between them. Those multiple crossover events with an even number of exchanges between two loci produce only the parental combinations, whereas those with an odd number of exchanges yield recombinants. When two genes are very far apart on the same chromosome, the frequency of even numbered crossovers will be about the same as the frequency of the odd numbered ones. Consequently, the two linked genes will appear to be assorting independently and hence to be unlinked. Nevertheless, recombination frequencies can not exceed 50%. In certain cases, such as that of "black"

and "arc" in Drosophila which are over 50 map units apart, the amount of recombination is actually less than 50%. Therefore, one cannot always look at a map and translate the distance indicated by map units into the amount of recombination to be expected. The number of map units corresponds closely to the amount of recombination only for those genes that are relatively closely linked, usually in the vicinity of 20 map units or fewer.

Linkage and a linear order of genes

Linkage and crossover values with all their complications (interference and multiple crossing over, etc.) can be explained only on the basis of a linear arrangement of genes. No sense can be made of the data on the other assumptions. As Muller showed, exchanges occur in blocks (see Fig. 8-13). There is even cytological evidence for a linear arrangement. When a piece of a chromosome is accidentally lost as a result of two breaks in a chromosome (details in Chap. 10), the missing region at times may be detected with the microscope. Genetic studies of flies with known cytological deletions have revealed that a gene or a group of genes is missing.

We will see in Chapter 11 that the concept of a linear order agrees well with what is known about the molecular biology of the gene. The essentials of linkage and crossing over, which were worked out in much higher species, also have been found to apply to microorganisms. The same concept of a linear order is therefore applicable to these lower groups. Later chapters describe rapid advances in molecular biology and the modern concept of the gene, made possible by detailed linkage analysis in bacteria and viruses. The following chapter discusses the relevance of a knowledge of linkage to humans and also presents additional aspects of the phenomena of linkage and crossing over.

REFERENCES

Allan, G. E. *Thomas Hunt Morgan: The Man and His Science*. Princeton University Press, Princeton, NJ, 1979.

Hotchkiss, R. D. Models of genetic recombination. *Ann. Rev. Microbiol.* 28:445, 1974.

Muller, H. J. The mechanism of crossing over. II. *Am. Naturalist* 50: 284, 1916.

Nicklas, R. B. Chromosome segregation mechanisms. *Genetics* 78: 205, 1974.

Stern, H. and Y. Hotta. Biochemical controls of meiosis. *Ann Rev. Genet* 7: 37, 1973.

Stern, H. and Y. Hotta. DNA metabolism during pachytene in relation to crossing over. *Genetics* 78: 227, 1974.

Sturtevant, A. H. The linear arrangement of six sex-linked factors in Drosophila, as shown by their mode of association. *J. Exp. Zool.* 14: 43, 1913.

von Wettstein, D. The synaptinemal complex and four-strand crossing over. *Proc Natl. Acad. Sci.* 68:851, 1971.

REVIEW QUESTIONS

1. Assume that the two allelic pairs, r^+, r and s^+, s are so closely linked that crossing over can be ignored. What kinds of gametes can be formed by the following?

   A. A double heterozygote with the alleles in the trans arrangement.
   B. A double heterozygote with alleles in the cis arrangement.

2. Assume that the two allelic pairs, t^+, t and u^+, u are linked and that the frequency of crossovers between the two loci is 30%.

   A. What kinds of gametes will be formed by a dihybrid with the genes in the trans arrangement, and what would be the expected frequency of each type considering crossing over?
   B. Answer the same question for a dihybrid with the alleles in the cis arrangement.

3. Assume that a double crossover takes place in a meiotic cell of the dihybrid in Question 2B. What would be the combination of the alleles resulting from the double crossover between t and u?

4. In the fruit fly, the allele for purple eye color (p) is recessive to its allele for red (p^+). The allele for vestigial wings (vg) is recessive to the wild allele for normal wings (vg^+). The two genes are autosomally linked. Females from a purple stock are crossed to males from a vestigial stock. The F_1 flies are all wild (red eyes and normal wings). F_1 females are testcrossed with the following results:

   purple—210
   wild— 40
   vestigial—215
   purple, vestigial— 35

   A. Diagram the testcross.
   B. Calculate the amount of recombination.

5. Give the results to be predicted if an F_1 dihybrid male from the cross described in Question 4 is testcrossed, assuming 500 offspring.

6. In tomatoes, round fruit (o^+) is dominant to long (o). Simple flowering shoot (s^+) is dominant to branching flowering shoot (s). Plants from two different pure-breeding varieties are crossed. One variety bears round fruit and has branched flowering shoots. The other variety has long fruits and simple flowering shoots. F_1 plants were testcrossed, and the following progeny were obtained:

   round, simple—23
   long, simple—83
   round, branched—85
   long, branched—19

   A. Diagram the testcross.
   B. Calculate the amount of recombination.

7. Using information from Question 6, assume that two dihybrids with the alleles in the cis arrangement are crossed to each other.

   A. Give the kinds of gametes and the frequency of each expected from each parent.
   B. Give the phenotypes and the expected frequencies from a cross of two such hybrids with the cis arrangement.
   C. Assuming 1000 plants are obtained in this cross, how many of them would you expect to have the combination long fruits and simple flowering shoots?

8. In chickens, the dominant allele, I, prevents pigment formation so that the feathers are white. The recessive allele (I^+) permits the feathers to be colored. This pair of alleles is autosomally linked to a pair that influences the development of the feather. The allele F^+ results in normal feathers, but the allele, F, causes brittle feathers when in the homozygous condition. However, there is no dominance, so that the heterozygote has mildly brittle feathers.

   A hen with colored, brittle feathers is mated several times to a rooster with white, normal feathers. Over three dozen eggs are produced that yield chicks with white, mildly brittle feathers. Diagram the cross, giving the genotypes of the parents and the F_1.

9. F_1 hens from the cross in Question 8 are crossed to roosters having colored, normal feathers. The following chicks are obtained:

   white, mildly brittle—17
   colored, mildly brittle—64
   white, normal—61
   colored, normal—15

   A. Diagram this cross.
   B. Give the genotypes of all the kinds of birds.
   C. Calculate the amount of recombination.

10. In the fruit fly, the allele w^+ for red eye color is dominant to its allele for white eye, w. This pair of alleles is sex linked, as is another pair that influences body color: the dominant for gray body (y^+) and the recessive for yellow (y). Females from a pure-breeding, white-eyed, gray-bodied stock are crossed to males from a stock pure

breeding for red eye and yellow body. Give the genotypes of the P_1 and the genotypes and phenotypes of the F_1.

11. F_1 females from the cross described in Question 10 are crossed to white-eyed, gray-bodied males. From the results presented below, calculate the amount of recombination between alleles at the white (w) locus and alleles at the yellow (y) locus.

Females: red-eyed, gray-bodied—340
                  white-eyed, gray bodied—350
Males: white-eyed, gray-bodied—346
                  red-eyed, yellow-bodied—324
                  red-eyed, gray-bodied— 9
                  white-eyed, yellow-bodied— 6

12. In the human, two pairs of alleles d^+, d and p^+, p are associated with color perception. The recessives, d and p, can result in color blindness. Both loci, d and p, are found on the X chromosome. Linkage is so close that the amount of crossing over is very low.

   A. Ignore crossing over, and diagram a cross between a color-blind woman who is homozygous for the dominant allele at the p locus and a man who is color blind because of the recessive he carries at the p locus. Show the expected offspring.
   B. Assume a dihybrid female with the color perception alleles in the trans arrangement marries a man who carries neither recessive. Diagram the cross and show the offspring, ignoring crossing over.
   C. In the last case, show the expected results if crossing over occurs.

13. The dominant allele o^+ is required for pigment deposition in the iris of the human eye. Its recessive allele, o, is responsible for ocular albinism. The locus is found on the X chromosome, as is the d locus associated with color blindness that was mentioned in Question 12.

   A. Assuming no crossing over, what are the possible results of a cross between a woman with ocular albinism who is homozygous for the allele for normal color vision and a man who has a pigmented iris but is color blind because of the recessive d.
   B. Assuming no crossing over, give the results of a cross between a woman who is dihybrid at the albinism and color-vision loci (cis arrangement) and a man with both normal traits.

14. The amount of recombination between the linked genes e and s is found to be 11%. The gene r is found to be linked to e and s. Alleles of the r gene undergo recombination with those of the s gene with a frequency of 7% and with those of the e gene with a frequency of 4%. What is the order of these genes on the chromosome and the distance between them in map units?

15. Given the following portion of a chromosome map, answer the following questions:

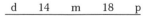

   d   14   m   18   p

   A. The double crossover value in this region is found to be 1.5%. What is the coefficient of coincidence?
   B. Alleles of the s gene are found to be linked to alleles of the genes shown on the map. A crossover value of 13% is found between the s and the p genes. Where should the s gene be placed on the map?

16. In corn, the genes v, b, and l are linked. Using the following data, summarizing the results of a trihybrid testcross, do the following:

   A. Give the correct genotype of the trihybrid parent showing the correct sequence of genes.
   B. Construct a map segment showing the distances among the genes.
   C. Calculate the coefficient of coincidence.

v^+	b^+	l	304
v^+	b	l	119
v	b	l	18
v	b^+	l	70
v^+	b	l^+	64
v^+	b^+	l^+	22
v	b^+	l^+	108
v	b	l^+	295

17. In the Oriental primrose, short style (l^+) is dominant over long style (l). Magenta flower color (r^+) is dominant over red (r), and green stigma (g^+) is dominant over red (g). A trihybrid testcross is performed. From the following data, construct a map giving the positions of the three genes and the distances involved:

l^+	r^+	g	292
l^+	r^+	g^+	153
l	r	g^+	286
l	r	g	139
l^+	r	g^+	36
l^+	r	g	22
l	r^+	g	40
l	r^+	g^+	18

18. In Drosophila, assume the three genes, x, y, and z, are linked and that the alleles x^+, y^+, and z^+ are dominant to x, y, and z. Crosses of trihybrid females and males showing the three dominant traits, x^+, y^+, and z^+, give the following results:

Females: $x^+y^+z^+$ 1012

Males:		
$x^+y^+z^+$	30
x^+y^+z	32
$x^+y\ z^+$	441
$x^+y\ z$	2
x $\ y^+z^+$	3
x $\ y^+z$	430
x $\ y\ z^+$	27
x $\ y\ z$	38

Based on these data, do the following:

   A. Give the genotypes of the P_1 males and females, showing the correct sequence of the genes.

B. Construct a map giving the locations and distances.

C. What can you say about the interference in this map region on the basis of the data?

19. The following is a map segment showing the location of loci, r, s, and t:

$$r \quad 12 \quad s \quad 10 \quad t$$

The coefficient of coincidence in the region is equal to 0.5. The following cross is made in the fruit fly·

$$\frac{r^+s^+t}{rst^+} \times \frac{r^+s^+t^+}{Y \text{ chromosome}}$$

Answer the following based on the expectation of 4000 offspring:

A. How many will be wild-type females (r^+ s^+ t^+)?

B. How many males will be of the constitution r^+ s t?

C. How many males will be r^+ s^+ t and r s t^+?

20. In the fruit fly, the allele for sepia eye color (se) is recessive to the allele for red (se^+). Straight wings (cu^+) are dominant over curled (cu). The two genes, se and cu, are linked on an autosome, Chromosome III. The amount of recombination between them is 24%. A cross is made between a female from a sepia-eyed stock and a male from a curled stock. The F_1s are wild (red-eyed

with straight wings). They are crossed to one another and produce an F_2.

A. What kinds of offspring will appear in the F_2, and in what ratio?

B. The se gene is also linked to the e gene (for ebony body). The allele e is recessive to e^+ for gray body. Diagram a cross between a female from a sepia stock and a male from an ebony stock. Can you tell the amount of recombination between the two genes on the basis of the F_2 results? Explain.

21. Crossing over is the event that produces crossover or recombinant gametes. Suppose a and b are linked and a dihybrid is carrying the combination of alleles in the cis arrangement:

$$\frac{a^+ \ b^+}{a \ \ b}$$

Assume that in every 10 meiotic cells crossing over takes place in 5 cells.

A. What would be the percentage of crossover gametes and the percentage of noncrossovers?

B. Suppose a crossover event occurs between a and b in every meiotic cell. What would be the amount of recombination?

9

FURTHER ASPECTS OF LINKAGE AND CROSSING OVER

Until 1967, the study of family pedigrees was just about the only method available for recognizing human linkage groups. Linage of genes to the X chromosome can be established much more readily than the linkage of two or more genes on an autosome. Consequently, the first advances in assignment of loci were made with the X. Linkage of genes to the X chromosome is more easily detected because pedigree analysis may show more affected males or a crisscross pattern of inheritance. More than 100 human genes have been found to be sex linked. The actual map distances have been calculated for some of these through analyses of many pedigrees and the application of the fundamental principles of sex linkage and genetic recombination. For example, knowing that the locus for deutan color blindness and that for the production of the enzyme glucose-6-phosphate dehydrogenase (G6PD) are sex linked, one can proceed to study family pedigrees that include doubly heterozygous mothers and their sons. The pedigree must be complete enough to establish whether the alleles in any heterozygous woman are in the cis or the trans arrangement. Such information can be obtained if the pedigree also includes information on the woman's father. If the latter, for example, is color blind and cannot produce G6PD, whereas the woman expresses both dominant traits, then we know that her genotype is

$$\frac{gd^+ \ d^+}{gd^- \ d^-}$$

We can deduce this because her father can only give her an X chromosome carrying both recessives.

Similarly, we might reason that another woman carries these alleles in the trans arrangement. The pedigrees are then examined with respect to the sons of such women for whose genotypes the cis or the trans arrangement is known. When this was actually done, it was established that 1 son in 20 of doubly heterozygous mothers showed new combinations of the alleles, meaning that recombination must have occurred in 5% of the cases (Fig. 9-1). The two loci can then be considered 5 map units apart. Studies of this type have established the map distances for several X-linked genes. The loci for hemophilia A and protan color blindness are so closely linked to those for deutan color blindness and G6PD production that the precise order of the loci in the linkage group is still unknown.

Some genes known to be located on the X chromosome, however, seem to be assorting independently from other genes whose loci are also known to

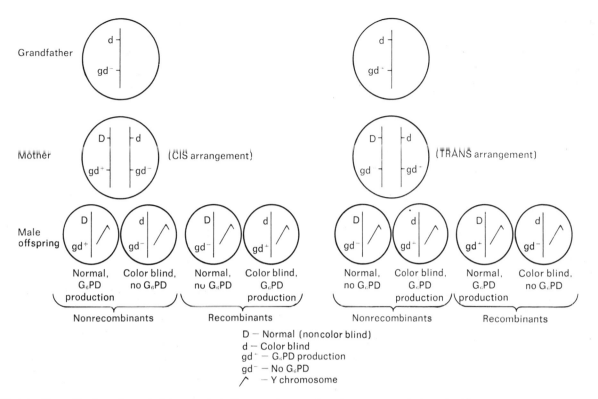

FIG. 9-1. Uses of family pedigrees in linkage analysis. To ascertain the amount of crossovers between the locus for deutan color blindness and the locus for G6PD production, pedigrees must be complete enough to establish whether the doubly heterozygous mothers of sons being studied carry the genes at these loci in the cis or in the trans arrangement. This requires information on the maternal grandfathers. Once the arrangement is determined (only one possible trans arrangement is shown here), a count of the sons is made to determine how many are recombinants and how many nonrecombinants. In this case, 5% of the sons of doubly heterozygous mothers were found to be recombinants, giving a map distance of 5 units between the two loci.

be X-linked. The reason for this is undoubtedly the length of the X (Chap. 8); thus, two loci that are appreciably far apart are frequently separated by crossover events and so appear to be assorting independently. Many students of human genetics would say that two such genes are syntenic (on the same chromosome) but that linkage has not been established between them. In our discussions, we will continue to use the term linkage in the classical sense of two genes being linked if they are on the same chromosome.

In contrast to the X chromosome, it is much more difficult to establish that two or more loci are linked on an autosome, when working with family histories. Even when this is accomplished, pedigree analysis does not allow assignment of the linkage group to a specific autosome. The most useful pedigrees are those in which one parent is doubly heterozygous and the other double recessive (a testcross). Again, it is

necessary to know whether the double heterozygote has the alleles under consideration in the cis or trans arrangement for precision.

Today, family pedigrees continue to be used in the assignment of loci to linkage groups and in the mapping of chromosomes, but complementing this is the important method of somatic cell genetics that has greatly accelerated the assignment of genetic loci in the human.

Somatic cell hybridization

Mapping by somatic cell hybridization entails the fusion of somatic cells of two different species. When grown together in cultures, cells from different sources fuse on rare occasions. It was found that the frequency of cell fusions can be greatly increased by adding to the culture certain chemicals such as polyethylene glycol or by adding Sendai viruses, which have been inactivated by ultraviolet light. Such agents can alter the properties of cell membranes in such a way that fusion between cells is enhanced, even fusions between cells of very different species, such as the mouse and the human. The human cell types commonly employed in such a procedure are the fibroblast, obtained from bits of skin growing in tissue culture, and the white blood cell.

After fusion occurs between cells from different genetic sources, a cell is formed containing two nuclei. Such a hybrid cell containing two genetically different nuclei is called a heterokaryon. When the heterokaryon undergoes mitosis, the chromosomes from the two nuclei align on the single spindle. Two hybrid cells are produced, each with one nucleus containing chromosomes derived from the two different species. However, a problem remains after cell fusion has occurred; this is the need to select out only the hybrid cells that are growing amid all the other kinds of cells in the culture—those that are unfused and those that have resulted from the fusion of two cells of the same origin, such as two human cells and two mouse cells. The problem has been essentially solved by the development of a technique that facilitates the screening out of the hybrid cells.

To appreciate the basis of the procedure, a few essential points must be grasped concerning the growth requirements of cells. Among other things, a cell must be able somehow to obtain the necessary materials to construct its nucleic acids if it is to multiply in a cell culture. The hereditary material, DNA, contains units called purine and pyrimidine nucleotides (see Chap. 11). These nucleotide requirements may be met by two general pathways in a cell (Fig. 9-2). Through the de novo *pathway,* a cell can build up its purine and pyrimidine nucleotides starting only with very simple materials. The other general pathway is referred to as the *salvage pathway,* so called because a cell following this route can use more complex substances, such as breakdown products of nucleic acids, from which it can salvage the required purine and pyrimidine components of nucleic acids.

Certain mutant cells, however, are unable to use this salvage pathway. One such cell type lacks the enzyme hypoxanthine phosphoribosyltransferase (HPRT) and is designed HPRT⁻. (It is this enzyme which is deficient in the Lesch–Nyhan syndrome discussed in Chap. 7). The HPRT⁻ cell cannot form purines from substances in the salvage pathway and must therefore make them de novo. Other mutant cell types are unable to carry out another aspect of

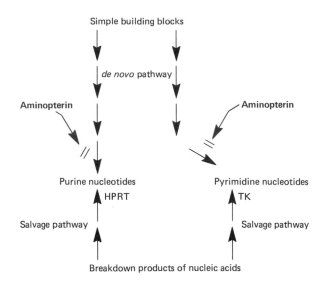

FIG. 9-2. Summary of de novo and salvage pathways in synthesis of nucleotides. A cell requires purine and pyrimidine nucleotides to construct its DNA. A cell can form these in two general ways, by starting with very simple compound (de novo pathway) or by using more complex substances resulting from nucleic acid breakdown (salvage pathway). The enzymes HPRT and TK are needed in the salvage pathway. A cell lacking either of these enzymes (HPRT⁻ or TK⁻) would be unable to perform a necessary step in the salvage pathway and would thus depend on the de novo pathway. The antimetabolite aminopterin can block steps in the de novo pathway, thereby preventing the construction of DNA requirements along this route (see text for details).

the salvage pathway, because they lack the enzyme thymidine kinase (TK). These TK⁻ cells can make the necessary pyrimidine-containing nucleotide (thymidylic acid) along the de novo pathway, but they cannot bring about the essential step (adding a phosphate) required to form thymidylic acid from thymidine (a precursor that lacks the phosphate).

This knowledge of the cell pathways has been put to good use to screen out only hybrid cells on a selection medium known as the HAT medium. The medium is so called because it contains hypoxanthine (which can be used by HPRT⁺ cells as a purine source using the salvage pathway), aminopterin (an antimetabolite that inhibits the de novo pathway), and thymidine (which can be converted to thymidylic acid by TK⁺ cells in the salvage pathway).

Now let us suppose that the human cell line used in a cell hybridization procedure is HPRT⁻. This population cannot grow on the HAT medium, because its cells depend on the de novo pathway to form purines, and the aminopterin in the medium prevents

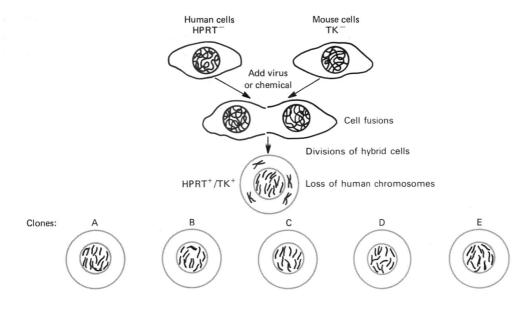

FIG. 9-3. Somatic cell hybridization. Hybrid cells may be formed from the fusion of human and mouse cells. Only hybrid cells in this example can grow on the HAT medium, since they are HPRT⁺/TK⁺. In the subsequent divisions of the hybrid cells, the human chromo- somes tend to become lost. Cells can eventually be isolated that contain just a few human chromosomes. Such cells may give rise to various stable cell lines, each line carrying a different combination of human chromosomes.

this. Only a cell that is HPRT⁺ can use the hypoxan- thine. If the other parental cell line, the mouse line, is TK⁻, it is prevented by the aminopterin in the HAT medium from forming the thymine-containing nu- cleotide, thymidylic acid, along the de novo pathway. Lacking the required enzyme, it cannot use the sup- plied thymidine using the salvage pathway. The only cells that can survive on the HAT medium are those that are HPRT⁺/TK⁺. Only the hybrid cells between the two species will have both of the metabolic re- quirements to grow on the HAT medium.

As the human–mouse hybrid cells divide, there is a pronounced tendency for the human chromosomes to be lost at cell divisions while the full complement of the rodent chromosomes is maintained (Fig. 9-3). Eventually cells are produced with just a small num- ber of human chromosomes. Such single cells can be isolated to establish stable hybrid cell lines from which no further chromosome loss occurs. Each cell line constitutes a clone, since the members of the partic- ular line arose from one cell and must therefore be genetically identical. The human chromosome loss which takes place is a random one so that the estab- lished stable clones differ in regard to the particular combinations of human chromosomes which they

possess. This very fact has been put to good use in associating a particular trait with a specific human chromosome.

Actually the first gene to be assigned to a chromo- some by the somatic cell procedure was the TK gene, the one responsible for the production of thymidine kinase, which we have noted is needed for growth on the HAT medium. The human line used in the hy- bridization was TK⁺, and the mouse line was TK⁻. No clones were established that lacked human Chro- mosome 17. In all stable, established cell lines, human Chromosome 17 was present along with the human enzyme thymidine kinase. Therefore Chromosome 17 must carry the genetic factor responsible for the en- zyme.

The cell hybridization procedure, to be useful in assigning a gene to a chromosome, depends on the ability of the parental genes to be expressed in the hybrid cell. This appears to be the case for many genes in the human–rodent hybrid cells. An added advantage is that a given enzyme found in both the mouse and the human (e.g., G6PD), although it is very similar and performs the same function in both species, exists in different forms—a human form and a rodent form. This allows us to identify a specific enzyme in a cell population. The cell hybridization method also permits assignment of genetic loci in the human that are not associated with the production of a biochemical product. For example, loci affecting the susceptibility of cells to certain viruses and drugs can be assigned by the procedure. Extremely important is the fact that the association of two or more genes with the same chromosome by cell hybridization permits the identification of linkage groups. The genes for HPRT, G6PD, and the enzyme phosphoglycerate kinase were all shown to be linked on the X chromosome. The three enzymes are always present in those clones with an X and are absent when the X is not present in a cell line. The locus for the gene responsible for G6PD was already known to be X-linked from pedigree analysis.

Stable hybrid clones covering all of the human chromosomes are available for use in assigning an unmapped locus to a chromosome. Each clone has its unique combination of human chromosomes. Let us suppose we wish to assign the locus of a gene associated with a certain enzyme (enzyme A) to a given chromosome. We have available the three clones, A, B, and C (Table 9-1). We find that human enzyme A is detectable in clones A and B but absent from C. We can now deduce that the genetic locus for the enzyme is on Chromosome 2. We can easily see this from Table 9-1 by comparing the enzyme pattern (whether it is + or − in a clone) with the chromosome pattern (whether a specific chromosome is present or absent in a clone). It is the vertical columns in each case that we compare. We can quickly see that the vertical pattern + + − for the enzyme is the same as the vertical pattern in column 2 for the chromosomes. None of the other vertical columns are + + −. This is so because clone C lacks Chromosome 2. Chromosome 1 cannot be associated with enzyme A, since that chromosome is present in clone C, which lacks the enzyme. Chromosome 2 is the only other chromosome that clones A and B have in common.

It is important to note that somatic cell hybridization does not give the same kind of linkage information as pedigree analysis. The somatic cell procedure tells us that two or more genes are syntenic, associated with the same chromosome. It tells nothing, however, about the distance between two loci. We rely on classic genetic linkage analysis to give us this information. The two methods actually complement each other. For example, if the family method shows that two or more loci are autosomally linked and a certain map distance apart, the cell hybridization method may establish the exact chromosome to which one of the members of the linkage group belongs because the one gene is associated with a product (say an enzyme) that can be followed in somatic cell studies. The locus for the ABO blood grouping does not produce a detectable product in cell culture and thus cannot be assigned to a chromosome using the cell fusion procedure. However, family studies showed the ABO locus to be linked to a locus that controls the production of an enzyme that is detectable in cell cultures. Since we can show that this locus is associated with Chromosome 9, the ABO locus to which it had been shown to be linked by family studies could be assigned to Chromosome 9.

Somatic cell hybridization has been very useful in assigning genes to specific autosomes, a most difficult task if attempted on the basis of family histories alone. It has also shown that the pictures of synteny in the human and in the chimpanzee are very similar, the same groups of genes being associated together with a specific chromosome and falling into a group. The human and the mouse, however, differ in this way. A very interesting point, however, is that in all mammals the same genes are X-linked (recall Ohno's law, in Chap. 5).

Precise assignment of genetic loci to specific arms or parts of chromosome arms has also been achieved for some genes. This has been made possible by studying human cells in which chromosome breakage and rearrangement of some kind have occurred (see Chap. 10). For example, a piece of one chromosome may be shifted or translocated to another chromosome. The piece may be identified by studying the pattern of chromosome bands (see Chap. 2). If it is found that a band or bands are missing from one chromosome and have been inserted into another, cell hybridization may show that a certain gene product, usually associated with a given chromosome, is now associated with another chromosome. The knowledge that the band is missing from a specific location in

TABLE 9-1   Human chromosome content of three mouse–human hybrid clones

HYBRID CLONE	HUMAN CHROMOSOME CONTENT (+ =present; − =absent)							
	1	2	3	4	5	6	7	8
A	+	+	+	+	−	−	−	−
B	+	+	−	−	+	+	−	−
C	+	−	+	−	+	−	+	−

the first chromosome and is now associated with a new position in another chromosome tells us that the locus associated with the product is normally found in the region of that band on the normal chromosome.

Deletions, chromosome aberrations in which segments of chromosomes are actually missing, have similarly been used to assign genetic loci to specific locations on given chromosomes. A given chromosome band or a portion of a chromosome arm may be missing in some cell lines. If absence of the band can be associated with absence of a product known to be controlled by a locus assigned to that chromosome, and if presence of the band is associated with presence of the product, then the particular genetic locus can be related precisely to that portion of the chromosome. For example, a person who was heterozygous for Rh blood antigen (Rh$^+$ is dominant over rh$^-$) was found to be producing both Rh$^+$ and rh$^-$ red blood cells. The person was then found to have a cell line with a deletion in Chromosome 1 in the short arm, away from the centromere. Thus, the Rh locus, known to be on Chromosome 1, was assigned to that specific region.

Application of linkage information in the human

The close linkage known to exist between the hemophilia A and G6PD loci can be put to use in prenatal diagnosis. Suppose it has been established (Fig. 9-4) that a woman is a carrier for hemophilia. She received from her hemophile father the defective hemophilia allele as well as the linked allele enabling her to produce G6PD. Her mother donated an X carrying the normal allele for blood clotting and the recessive for G6PD production. The alleles are in the trans arrangement. Because the two genes involved are so closely linked, there is just a small chance for crossing over to occur. This woman will, of course, donate one X chromosome to her male offspring, who receive a Y from the male parent. There is, therefore, a 50% chance that a son will receive the defective allele for hemophilia. In this case, the allele is carried along on the same X with the normal allele for enzyme production; therefore, any hemophile son will probably produce G6PD. A nonhemophile son would lack the ability to produce the enzyme. Amniotic cells may be withdrawn during pregnancy. If they indicate a male on the basis of the lack of a Barr body and the

FIG. 9-4. Detection of a defective sex-linked allele. In this pedigree, the P$_1$ female parent is known to have received an X from her father who carries the recessive hemophilia allele and the dominant for the production of G6PD. The X she received from her mother carries the contrasting gene forms. This information, plus the fact that the two loci are tightly linked on the X, can be put to practical use in prenatal diagnosis. Any male offspring receives one X from his mother, making the chance ½ that any son of the P$_1$ woman will be a hemophile. The enzyme locus can be used as a marker to check for the presence in utero of the defective allele. Because the two loci are closely linked, the crossover eggs will be much rarer than the noncrossover types. Therefore, if the amniotic cells indicate a male offspring and if the enzyme is present, the chances are very high that the unborn son has received the parental X carrying the recessive for hemophilia.

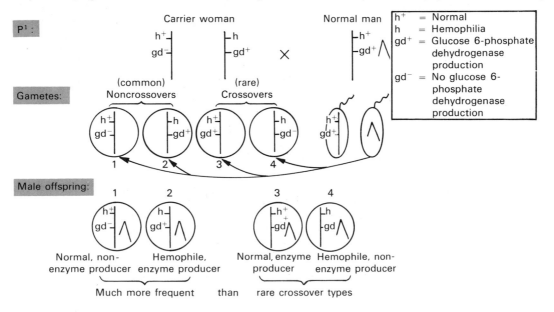

presence of a fluorescent Y body, and if the enzyme is shown to be present, the probability is very high that the woman is carrying a hemophile male embryo. In a case such as this, the enzyme locus is a marker; the presence of the product (the enzyme) associated with the locus, G6PD, tells us that the defective allele is most likely present, too. The mapping of human chromosomes thus is seen to have a decided value; it can increase the number of human disorders susceptible to prenatal diagnosis.

The disorder myotonic dystrophy is the first serious human affliction to be assigned to a gene that is a member of an autosomal linkage group. The dystrophy locus has been found to be linked to two others, one that is responsible for the blood antigens of the Lutheran grouping and the other that determines the secretor trait, the possession of an enzyme that results in the secretion of blood antigens into various body fluids such as the saliva (see Chap. 4). When the secretor allele (Se) is carried by a fetus, it can determine the secretor phenotype of the amniotic fluid (which antigens associated with red blood cells are present in the fluid; see Table 4-4). This fact, plus that of close linkage with the dystrophy locus, has a

very valuable application, because the dystrophy allele has reduced penetrance and delayed onset of expression. Moreover, it produces no product in the body fluids and thus can remain completely undetected until it strikes a victim.

But the knowledge of linkage can guide us in some cases. Suppose, as illustrated in Fig. 9-5, a male parent is known to carry the dominant secretor allele on the same chromosome as the dominant allele for the dystrophy. The man's wife is found to be a nonsecretor. Since she is phenotypically normal, she is not carrying the dominant dystrophy allele. If a couple such as this is concerned about the possibility that their unborn child will be afflicted with dystrophy, amniotic fluid can be withdrawn. If it gives a positive reaction for the presence of blood antigens, this means that the secretor allele from the male parent is present. Since the allele is closely linked to the dystrophy locus, only a small chance exists for crossing over to take place. The probability is therefore, very high (over 90%) that the fetus carries the harmful allele. If the fluid tests negative for the blood antigens, this means the offspring received from the father the chromosome that carries the nonsecretor allele. Since

FIG. 9-5. Detection of a defective autosomal allele. The locus associated with myotonic dystrophy and the secretor locus, which determines the ability to secrete antigens into body fluids, are autosomally linked. In this case, the male parent is known to carry the dominant factors for dystrophy and for the secretor ability on one chromosome; the recessive alleles are on the other. His wife is homozygous recessive. Since the two loci are tightly linked, the crossover gametes produced by the male would be much less common than the noncrossovers. If the amniotic fluid reveals blood antigens, this means the secretor allele is present and the embryo probably carries the dominant for dystrophy.

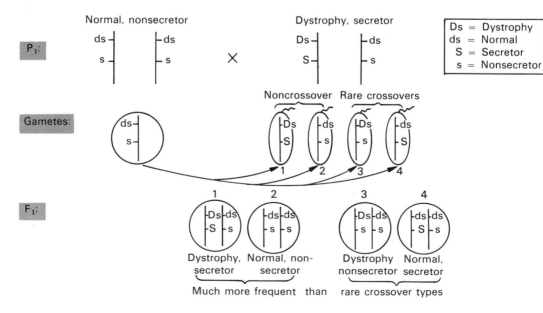

this allele is closely linked to the normal allele, the chances are better than 90% that the dystrophy allele is not present.

The value of the knowledge of linkage is quite obvious in this example. If this technique can be extended to include alleles associated with other serious disorders, prenatal counseling can have a still wider application and provide very accurate predictions, even though the chromosomes on which the alleles are located remain unknown.

Factors influencing crossover frequency

Because crossing over is of such universal importance, biologists have tried to gain a precise understanding of its exact nature and the factors that may influence it. Although we still have no answers to many of the questions concerning the mechanism of crossing over, several important facts have been established. Various factors that influence the frequency of crossing over have been recognized throughout the course of genetic investigations. The effect of the centromere in decreasing the amount of crossing over has already been discussed briefly (see Fig. 8-18). The ability of the internal or hormonal environment to alter the crossover frequency is clearly seen in Drosophila, in which normally no crossing over occurs in the male. This effect of sex is seen in a few other species as well. In the female silkworm, crossing over is normally absent. Although crossing over occurs equally in both sexes as a general rule, it appears that when a difference between the sexes does occur in a species, it is the heterogametic sex that shows the reduced frequency. Female mice show approximately 25% more crossing over than males. Crossover frequencies in humans, although still not firmly established, suggest a 40%–50% excess in females over males.

A definite effect of age on crossover frequency in Drosophila was found by Bridges. More crossing over occurs in young females and tends to decrease in frequency with age of the fly. In humans, the effect of age on crossing over has not been studied sufficiently to give a completely accurate appraisal. However, counts of the number of chiasmata in the meiotic cells of males show an increase with age, whereas in females there is a reduction. The few chiasmata observed in older oocytes seem to be associated with an increase in the number of unpaired chromosomes.

This in turn could reflect the effect of maternal age on the frequency of nondisjunction.

Temperature must be controlled when precise measures of crossing over are desired. Departures from the normal temperature range—that is, extremes of heat and cold—tend to raise the frequency of crossing over. Radiation has a pronounced effect on the amount of crossing over and definitely stimulates it in the fruit fly. Both the temperature and the radiation effects seem to encourage more crossing over in regions where it is normally reduced. For example, both of these environmental factors increase the amount of crossing over in the region of the centromere. They may even induce it in the male fruit fly and stimulate its occurrence in body cells, where it is normally inhibited.

In the 1950s, R. P. Levine showed that certain divalent cations such as Ca^{++} and Mg^{++} could alter the amount of crossing over. Flies raised on a diet with an excess of calcium showed less crossing over after a 4-day period than those on normal food medium. The decrease persisted for more than a week. When the metallic ions were reduced below that of normal (by adding the chelating agent Versene), the crossover frequency increased significantly.

Alterations in chromosome structure (see Chap. 10) may decrease the amount of crossing over. The presence of a deletion, for example, in one chromosome may interfere with crossing over in the chromosome parts that do pair properly (see Fig. 10-16). Inversions, reciprocal translocations, and duplications all seem to cause a decrease in the crossover frequency, as if the alteration in chromosome structure were somehow interfering with the normal forces of attraction between the homologous chromosomes.

Cytological proof of crossing over

In 1909, Janssens pointed out that the paired chromosomes at first meiotic prophase often assume configurations that appear as cross figures. He interpreted each cross, a chiasma, to mean that a breakage and reunion had taken place followed by a reciprocal exchange between two chromatids in the bivalent, one of them paternal and one maternal. This was the physical basis of a crossover event, and the chiasma was its visible manifestation. Janssens's chiasma theory, though offering an explanation for the mechanism of crossing over, could not be tested, and over the years, controversies have raged on the subject. An experimental test of the idea that a definite physical exchange takes place between homologous chromosomes encounters serious difficulties if one considers the following point. If the idea is correct and an exchange of chromosome segments does occur

between maternal and paternal chromosomes, one could not normally detect this exchange cytologically. The reason is that homologous chromosomes are alike in size and shape. Consequently, a reciprocal exchange between them at a particular point will result in two chromosomes identical to the originals. This

FIG. 9-6. Cytological demonstration of crossing over. Stern used females with a pair of X chromosomes that differed in appearance. One of them was broken and carried the recessive "carnation" and the dominant "Bar." The other X chromosome appeared to have a tail because of an attached segment of the Y chromosome. This chromosome carried the wild alleles. Females of this constitution have red eye color (wild) but are Bar eyed. They were crossed with carnation males whose X was normal in appearance. If an exchange of

chromosome segments occurs when crossing over takes place between the carnation and Bar loci, this should alter the appearance of the X in the female parent. Two chromosomes of different morphology should arise in the eggs: a normal-shaped X and a broken one with a tail. When Stern examined the F_1 offspring, he found these chromosomes in the crossover offspring. The noncrossovers carried the original maternal Xs.

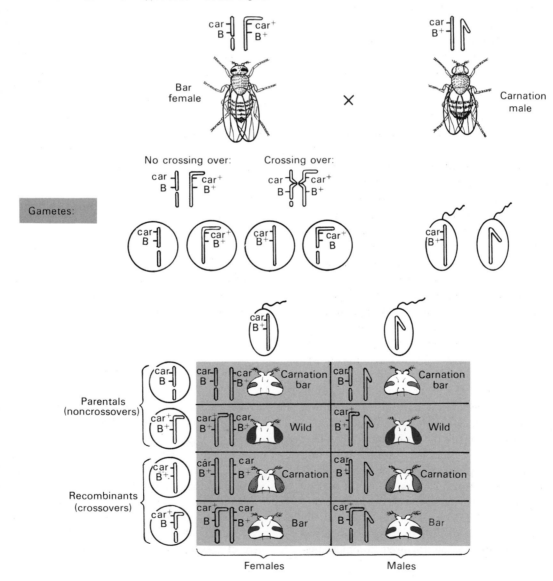

difficulty was surmounted independently and in the same year (1931) by two separate investigations, that of Stern with Drosophila and that of Creighton and McClintock with corn. In both studies, the investigators appreciated the need to obtain for cytogenetic analysis a pair of homologous chromosomes that did not appear identical when observed with the microscope. Both experiments are equally elegant in their design and represent a landmark in the correlation between genetics and cytology. However, only the work of Stern is presented here, because the two procedures have many features in common and rest on the same reasoning.

Making use of accidental chromosome aberrations, Stern was able to develop female flies in which the two X chromosomes differed conspicuously in appearance (Fig. 9-6). One of the Xs was broken into two pieces. The one portion remained attached to the centromere of the X. The second piece became associated with a centromere derived from the small Chromosome IV. The other was also altered in morphology, as a portion of a Y chromosome joined it at one end, giving the impression of a tail. The "broken X" carried two mutant alleles in the part associated with the X centromere. These were the recessive "car" (carnation or pinkish eye color) and the dominant B (bar or reduced eye size). The X with the piece of the Y contained the wild-type alleles.

Stern crossed females with the morphologically different Xs to males which had carnation eye color (refer to Fig. 9-6). He reasoned that if crossing over does entail some kind of physical exchange of chromosome material, then a crossover between the carnation and the Bar loci should produce two chromosomes different in appearance from the two original ones. A chromosome of normal size and shape would result and also a broken one with a tail. Since he knew that car and B were present on the original broken chromosome and the + alleles were on the X with the tail, Stern had the chromosomes marked genetically as well as cytologically. He was actually performing a testcross. Since the male parent had contributed both recessives, car and B^+, Stern was able to recognize recombinant offspring among both the females and the males.

Figure 9-6 shows that the recombinant types would be of two distinct phenotypes: carnation and bar. The two parental types would be carnation–bar and wild. When he examined the flies cytologically, Stern found in the carnation male flies a normal X chromosome. In the bar males, a broken X with a tail could be seen.

Because a male gets his X chromosome from the maternal parent, this means that the new kinds of X chromosomes came from the female parents which did not possess them in their body cells. The X of normal morphology and the one with the tail must have arisen as a result of crossing over in meiotic cells of the female parents. The new chromosomes were also found in the female offspring, as predicted, along with the structurally normal one from the male parents. In the nonrecombinant offspring, only parental kinds of Xs were present. Thus, we see again a perfect correlation between cytology and genetics (recall the work of Bridges on nondisjunction). When something new arose genetically in this example (the new combination of linked alleles), something new was created cytologically (chromosomes of different morphology from the parental ones).

This work of Stern plus that of Creighton and McClintock left no doubt that after crossing over takes place a new physical arrangement of chromosome segments arises. However, it did not show exactly how this arises, and several models other than breakage and reunion were proposed. There is no need to review these, for now recombination studies, particularly with microorganisms, leave no doubt that crossing over entails breakage and reunion (see Chaps. 11 and 15). A great deal of information concerning crossing over on the molecular level has been provided through research on DNA, especially on the manner in which it replicates. We will postpone our treatment of this important subject until Chapter 11, when we have become familiar with certain facts relevant to both DNA replication and crossing over.

Neurospora as a tool in crossover analysis

According to Janssens, only two threads out of four participate in any one crossover event. This idea implies that, at the time of crossing over, each paired chromosome in a bivalent is already double, so that four threads or strands compose the bivalent. The matter cannot be settled by cytological observation, because sufficient resolution of the chromatids is not possible at the earliest meiotic stages. We know now, of course, that DNA replication occurs at the interphase preceding first meiotic prophase. Therefore, each chromosome is, in effect, composed of two chromatids at the onset of meiosis. One could argue, however, that when crossing over occurs, each chromosome behaves as a single entity. The two chromatids would break at exactly the same spot and behave as if they were part of one thread as they enter together into a new combination with the homologue. However, conclusive genetic results with Drosophilia and certain fungi have shown that crossing over takes place when four chromatids are effectively present, that is, in the four-strand stage. Since the red bread mold, Neurospora, has been such a valuable tool in various kinds of genetic analyses, we confine ourselves here to the type of information it provides on crossing over.

Many fungi are composed of a mass of threads known as a mycelium. One of the threads of a mycelium is a hypha. In the sac fungi, or Ascomycetes to which Neurospora belongs, the sexual spores are formed in specialized, sac-shaped hyphae called asci. The life cycles of these molds can be rather complicated. It is sufficient here to understand just a few points. All the nuclei found in the hyphae of a fungus are haploid (Fig. 9-7). Fungi arising from different spores may be of different mating types. If fungi of the proper mating types come into contact, a male nucleus of one fungus may enter the female sex organ of the other. However, fertilization does not take place immediately; instead, the male and female nuclei undergo a series of mitotic divisions without combining. Eventually, asci arise. A pair of nuclei, one from the maternal fungus and one from the paternal fungus, enters each ascus. It is only in the ascus that the two different nuclei finally unite to give true fertilization. Immediately after fertilization, meiosis takes place, producing four haploid nuclei in each ascus.

The crucial point is that each ascus contains a tetrad of nuclei that has resulted from one single meiosis. In other words, when an ascus is formed, the products of one meiotic division are held together in the structure. In Neurospora, each haploid nucleus divides once more in the ascus, but this division is mitotic. The end result is that the nuclei in the ascus become the nuclei of the sexual spores. These spores are arranged in a linear order. Each ascus preserves the products of one meiosis in order and in duplicate. Therefore, Neurospora has an advantage in cytogenetic analysis that most organisms do not. Typically, as in an animal testis for example, the gametes of countless meiotic divisions are all mixed up together. It is impossible to tell which nuclei were formed from the same original cell as the products of one meiotic division. But in Neurospora, we can take an individual ascus, open it, take each spore out in order, and finally test the spore for some trait. We see that Neurospora

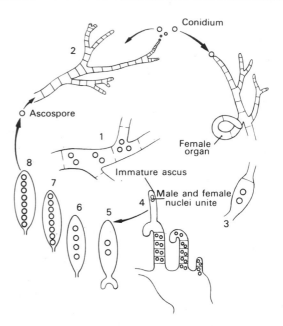

FIG. 9-7. Features of life cycle of Neurospora. (1) A hypha is composed of cells containing haploid nuclei. These may wander from cell to cell through pores in the crosswall. (2) An ascospore, a haploid cell produced after sexual reproduction, can grow into a new mass of hyphae (a mycelium). A new individual may also arise from a conidium, a special cell that can act like an asexual spore. A conidium may also serve as a male gamete. If it contacts receptive hyphae of the opposite mating type, it can act as a male gamete. Its nucleus is able to migrate to the female nucleus contained in the female sex organ. (3) Once it enters the female organ, the male nucleus approaches the female nucleus, but the two do not unite. (4) A number of special hyphae then grow out of the female organ. At the same time, the female–male pair of nuclei undergoes a series of mitotic divisions, still without fusing. Finally, in a cell at the end of each special hypha, male and female nuclei unite. This end cell will become an ascus. Because many special hyphae grow from a female organ, many asci are produced. The fertilization nucleus in each ascus is the same, because each traces back to the original male–female pair. (5) Soon after fertilization, first meiosis takes place. (6) This is followed by second meiosis, producing four haploid nuclei. (7) Each haploid nucleus in the ascus undergoes mitosis to give a duplicate of each of the four products of meiosis. (8) Each one of the eight nuclei becomes the nucleus of an ascospore. Every ascus thus contains the products of one meiotic division in duplicate.

and certain other ascomycetes offer us the opportunity to analyze tetrads, the immediate products of meiosis.

Suppose we are following a single pair of alleles, a^+ and a. A cross of two different molds of the proper mating type will result in a fertilization nucleus that is heterozygous at the a locus we are following. When we later examine the asci that are produced and test the spores from single asci in order, we find that

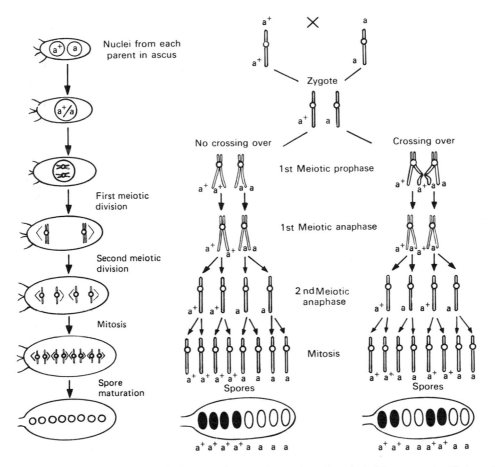

FIG. 9-8. Ordered arrangement and segregation of ascospores in Neurospora. (*Left*) The nuclei in an ascus are held in the same orientation they had after the completion of the meiotic divisions. (*Right*) If no crossing over occurs between a gene and the centromere, the members of a pair of alleles segregate at first anaphase, and the spores are arranged in a 4:4 pattern in the ascus. If crossing over takes place, segregation of a pair of alleles is delayed until the second meiotic division, and the final pattern is thus 2:2:2:2.

different spore arrangements occur. Some show a pattern of 4a$^+$:4a, whereas others have an arrangement that is 2a$^+$:2a:2a$^+$:2a. (Variations of this 2:2:2:2 pattern will be discussed in a moment.) Figure 9-8 gives an explanation for these 4:4 versus 2:2:2:2 arrangements. The 4:4 segregation results from the fact that no crossing over has taken place between the gene and the centromere of the chromosome. Consequently, the alleles segregate at the first meiotic division. In the case of a crossover event, homologous segments are exchanged between two of the four chromatids making up a bivalent. Note from the figure that this results in an association of a pair of alleles with the *same* chromosome. Consequently, segregation of the alleles is now delayed until *second* anaphase of meiosis, at which time the two chromatids of each chromosome are separated. These second

division segregations involving a single crossover event can produce more than one kind of 2:2:2:2 ascospore arrangement, depending on the chromatids involved. Figure 9-9 shows these variations and how they originate.

These second division segregations provided early proof that crossing over in Neurospora occurs when four chromatids are present. If crossing over took place before each chromosome had replicated, then only 4:4 segregations would be possible (Fig. 9-10). We know today that each chromosome is composed of two chromatids during the S portion of interphase, before the onset of meiosis. But one might argue that even though four chromatids are present at the time of crossing over, the two composing each chromosome stay together and behave as if they were one. Again only 4:4 segregations would be possible, for the situ-

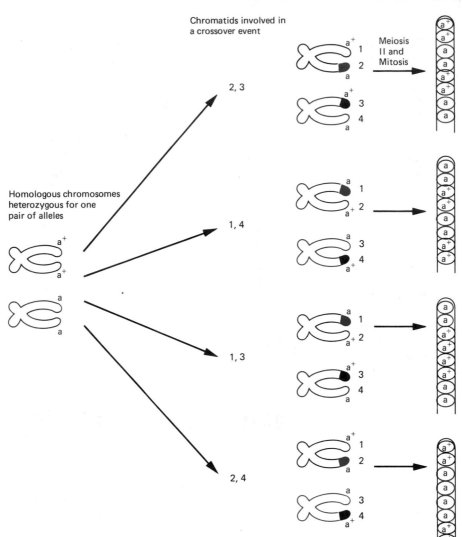

FIG. 9-9. Second division segregations. Different kinds of 2:2:2:2 second division segregation patterns can be produced. A particular pattern depends on the two chromatids involved in the exchange.

ation would be exactly the same as that pictured in Fig. 9-10.

Neurospora and tetrad analysis

The ascus, by keeping intact the segregation pattern of the immediate products of each meiosis, provides us with an excellent opportunity to calculate in map units the distance between a gene and its centromere. Hyphae from two different strains are allowed to grow together on a plate. One strain, let us say, is wild, a⁺, with respect to a given gene. The other carries the mutant allele, a. When asci are produced, they are examined to determine which are hybrid. (Asci in which all the ascopores are a⁺ or in which they are all a would not be hybrid but the result of self-fertilizations.) The numbers of 4:4 and 2:2:2:2 segregations are easily counted. If the characteristic being followed produces a difference in spore color, visual inspection is all that is required. If the gene effect is on a nutrient requirement or some other nonvisible characteristic, the spores can be removed in order

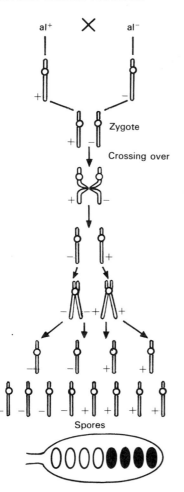

FIG. 9-10. Depiction of crossing over in the two-strand stage. Assume in Neurospora that after zygote formation, crossing over occurs when each chromosome is unreplicated. After the crossover event, the alleles on a crossover segment will simply have shifted to a new centromere attachment. When each chromosome is duplicated, the outcome in each case is a chromosome composed of two identical chromatids. The only segregation possible would be 4:4. Since 2:2:2:2 segregations also occur, this tells us that crossing over must have occurred at the four-strand stage.

from the asci and allowed to grow, and the resulting mycelium can be tested for the particular traits under consideration.

Let us suppose that we count 200 hybrid asci and find that 100 of them show 4:4 segregations. This means that 50% of the asci show second meiosis segregations and that crossing over between the gene and the centromere occured in 50% of the meiotic cells. As explained in Chapter 8, even if crossing over occurs in 100% of the meiotic cells, recombination

cannot exceed 50%, since only two of the four chromatids in a bivalent participate in a single crossover event. Therefore, if crossing over takes place in 50% of the meiotic cells as in our example, only half that amount, 25% are recombinants. The gene to centromere distance is therefore equal to half the percentage of the second division segregations. Gene a in our example would be 25 map units from its centromere.

When two genes instead of one are being followed in a cross, it is a relatively simple task in Neurospora to determine whether or not the two are linked. Assume first that the allelic pairs, a^+, a and b^+, b actually *are* linked. If no crossing over takes place between the two genes in a dihybrid meiotic cell, only parental types will be found in the ascus (Fig. 9-11A). Such tetrads, which show only the original, parental combinations, are referred to as *parental ditypes* (PD). If a single crossover event occurs between the two genes, then all four possible genotypes are represented in the ascus, and the tetrad is classified as a *tetratype* (TT).

Now suppose that double crossing over occurs between the two genes. A reference back to Fig. 8-21 shows what will result from the various possibilities. If a two-strand double occurs (*top*), this means that the same two strands participate in both crossover events and that the spores will represent only the parental types. In Neurospora (since a mitosis follows meiosis), the tetrads would be $4a^+b^+:4ab$. In other words, they would be parental types (Fig. 9-11B), since no new combinations would be generated. Figure 8-22 (*middle*) shows two ways in which three-strand doubles can occur. These would give us tetrads that are $2a^+b$, $2a^+b^+$, $2ab^+$, $2ab$, and all four allelic combinations are found in the asci. These tetrads are tetratypes, which are also generated by a single crossover event (Fig. 9-11A, B). If all four strands participate in the double crossing over (Fig. 8-22, *bottom*), the asci will contain spores of only two genotypes; but unlike the asci with the PD tetrads, the genotypes of these spores will *not* be in the same combination as in the parents. Instead, they will be $4a^+b:4ab^+$. Such tetrads resulting from four-strand doubles are called nonparental ditypes (NPD; see Fig. 9-11B). When two genes are linked, it is apparent that the NPD tetrads will occur in a much smaller percentage than the other types. In summary, three classes of tetrads, and hence three classes of asci, can be recognized when two genes are followed: PD, TT and NPD (Fig. 9-11).

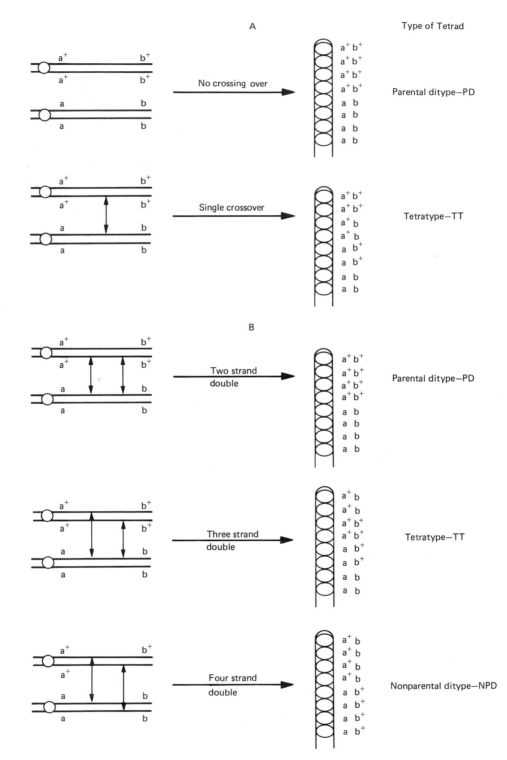

FIG. 9-11. Tetrad patterns and ways in which they may arise. (*A*) A parental ditype includes only the two parental allelic combinations. Such tetrads will arise if no crossing over takes place between the two linked genes. A single crossover between the two genes will generate asci in which all four possible allelic combinations are found. Tetrads of this type are tetratypes. (*B*) Double crossing over between the two genes can also generate tetrads that are PD and TT. A third possibility arises as the result of double crossing over in which all four strands participate. This produces tetrads that are nonparental ditypes. Only two genotypes are present, but these represent allelic combinations that were not found in the parental types.

Now let us assume that two genes are *not* linked but occur on different chromosomes and therefore can assort independently. Figure 9-12A shows that in the absence of any crossing over, parental ditypes and nonparental ditypes will arise in equal frequencies. If a crossover event takes place between either of the genes and its centromere (Fig. 9-12B), tetratypes will arise. However, these will be uncommon if the genes being followed are close to their centromeres. If the unlinked genes are located far from their centromeres, they can be separated from them very often because of a high frequency of crossing over. In such cases, the number of TTs will be high. Nevertheless, the PDs and NPDs will occur in equal frequencies, re-

FIG. 9-12. Tetrad patterns and independent assortment. (A) When no crossing over is involved, independent assortment will bring about the formation of two types of tetrads in equal amounts, the PD and the NPD. (B) A crossover event may occur between a and its centromere or b and its centromere. The frequency of such an event will depend on the distance the gene is from its centromere. We see here a crossover between a and its centromere. Independent assortment followed by meiosis II will produce tetrads that are of the tetratype. If gene a is a great distance from its centromere, crossing over frequency will be high, and the frequency of tetratypes will increase accordingly. Exactly the same would apply if b instead of a had crossed over in relation to its centromere.

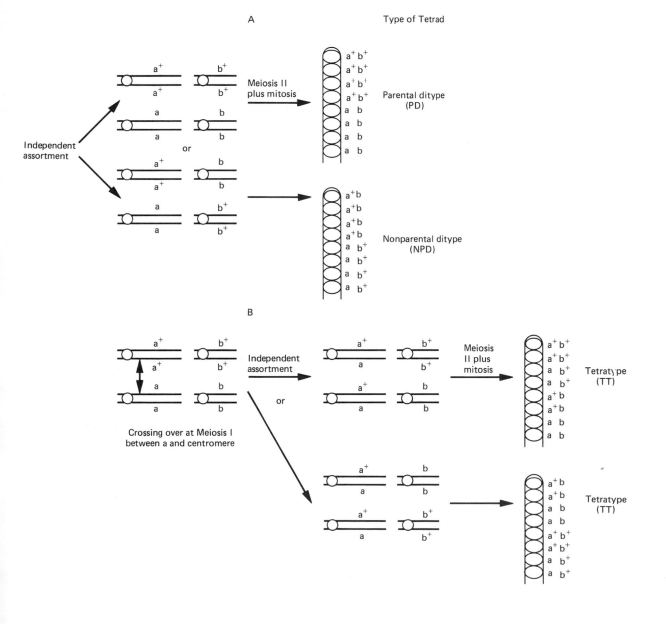

gardless of the number of TTs. If, however, the two genes are linked, as discussed earlier, the frequency of NPDs will be much less common than that of the PDs. An excess number of PDs in a particular cross tells us that the two genes we are following are linked.

Once we know that two genes are linked, we can proceed to calculate the amount of recombination between them. To do this, we must remember that in the NPD asci all the products are recombinant (see Fig. 8-22, *bottom*). In the TT asci, however, only half of the products are recombinant. (They arose from single exchanges and three-strand doubles.) Therefore, the amount of recombination between the genes equals the percentage of NPD asci plus half the percentage of the TT asci among the total number of all the asci examined. If 10% of the asci are NPD and 30% of the tetrads examined are TT, then the map distance between the two genes is 25 units. From these tetrad analyses, we can appreciate that Neurospora has been an ideal tool for studies involving the segregation of genetic factors following meiosis.

Somatic crossing over

We have noted that radiation can stimulate crossing over in somatic cells; this exceptional event has been found to occur in Drosophila and many fungi. At first, whether or not somatic crossing over takes place may seem unimportant. But let us consider the consequences, using a hypothetical example from humans (refer to Fig. 9-13). Suppose a female zygote is formed that is dihybrid for the hemophilia trait and for the ability to produce the enzyme glucose-6-phosphate dehydrogenase. The alleles are carried in the *trans* arrangement. An individual of this genotype normally expresses the normal blood condition and is able to manufacture the enzyme. At birth, every cell of the body should be identical, because normal mitotic divisions guarantee equal distribution of genetic material to daughter cells.

Figure 9-13 shows how somatic crossing over can upset this. In rare cells, the phenomenon of somatic pairing may take place spontaneously or as the result

FIG. 9-13. Effects of somatic crossing over. In this hypothetical example, a zygote is considered to be dihybrid at the hemophilia locus and the locus that governs the production of G6PD. Normally (left) mitotic divisions follow which ensure that every cell of the body will be identical and have the potential to make antihemophilia factor (AHF) and the enzyme G6PD. On very rare occasions, pairing of homologous chromosomes occurs in the body cells. If this is followed by a crossover event, the two resulting chromosomes carry chromatids that are not identical. These chromosomes may arrange themselves

on the spindle in such a way that the products of the ensuing mitotic divisions are different. In this case, a cell results that can make AHF but cannot make the enzyme G6PD. The other cell cannot make AHF but can produce the enzyme. Since most of the mitotic divisions will be normal, the majority of cells in the body in this case will have the potential to manufacture both AHF and G6PD. The individual will be mosaic, composed of three lines of cells. The earlier in development the somatic crossing over takes place, the larger the number of exceptional cells and the extent of the mosaicism.

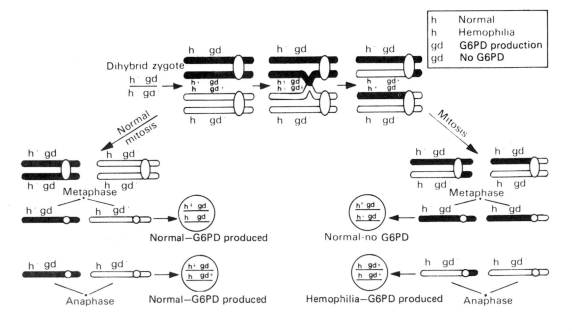

of exposure to radiation. A crossover may then occur between the centromere and the two loci we are following. This crossover event produces two metaphase chromosomes whose chromatids are not identical. Therefore, unlike a normal mitosis, the two chromatids that separate from each other and move to opposite poles at anaphase will be different.

One outcome is that two new cell lines can rise from the one original genotype. The individual would be a mosaic, a mixture of wild-type cells, of cells homozygous recessive for the hemophilia trait, and cells that lack the ability to synthesize G6PD. The extent of the mosaicism depends on how early in development the somatic crossover occurs. The earlier it takes place, the larger will be the number of cells in the two new cell lines, because the recombinant cells will have undergone more mitotic divisions before completion of embryology.

We see from this example that somatic crossing over can uncover recessives in the body, leading to patches of mutant cells scattered among normal cells. Obviously, this can produce undesirable effects. It would therefore be advantageous for crossing over to be suppressed in body cells. It is logical to assume that in the evolution of living things, crossing over provided the first mechanisms of recombination, since the simplest diploid cells must have had only one pair of homologous chromosomes. And so, the most primitive kinds of diploids must have relied on crossing over for recombination. As the diploid stage evolved and the body became differentiated between germ cells and somatic cells, mechanisms must have evolved to allow crossing over in the germ line, where it was needed to provide new combinations of linked alleles, but to prevent it in body cells where the effects could be harmful because of the production of mosaicism. Certainly there are mechanisms that prevent crossing over at meiosis in one sex of certain species, such as the fruit fly and the silkworm. It seems likely that comparable genetic controls must have evolved to prevent its occurrence at mitotic divisions, to ensure a soma composed of cells with the identical genetic information.

Importance of crossing over to evolution

We can aptly conclude this chapter with a consideration of the important role of crossing over in evolutionary progress. An organism that is completely asexual would not have the mechanism of meiosis to bring about new allelic combinations through independent assortment. The only way a new combination can result is through gene mutation. Consider the two diploid members of the same species, with a chromosome number of 4, which are represented in Fig. 9-14. One of them expresses the genotype ab^+, the other the genotype a^+b. The offspring will be identical to each parent: all ab^+ in the one case, all a^+b in the other. If two diploid organisms of these same genotypes were sexual, they would be able to produce haploid gametes that could unite to form dihybrids. These, in turn, by independent assortment can give rise to four kinds of haploid gametes. A cross between the two dihybrids will produce the familiar dihybrid ratio, 9:3:3:1. We now have *four* different phenotypes (2^2) and nine different genotypes (3^2). Without the sexual process, there would be just two phenotypes and two genotypes, one for each of the cell lines. Much more variation is possible through sexual reproduction and the independent assortment that meiosis makes possible.

Now let us consider the importance of crossing over as a source of variation that greatly supplements that contributed by independent assortment. Referring back to Table 8-1, which diagrams a trihybrid testcross, it can be seen that if no crossing over takes place, only two types of gametes will be formed—the old combinations. This means that the main advantage of sexual reproduction, the production of new combinations of alleles, would be lost for genes linked on the same chromosome. However, if all possible crossovers occur among the three genes, then eight different types of gametes are formed, the two parentals plus the six crossover types. It is evident that crossing over, along with independent assortment, contributes greatly to the variation generated by the sexual process. With crossing over, recombination is extended down to the level of the chromosome.

Inasmuch as crossing over typically occurs at every meiosis, new gene combinations are continually being formed. Crossing over, therefore, speeds up the production of new combinations of genetic material. Its contribution to the pool of variations in humans can also be appreciated by referring back to Chapter 1. When discussing new combinations provided through independent assortment, we noted that 2^{23} types of gametes, would be a minimal number for a human. It should now be evident to us that each of the 23 pairs contains thousands of genes and that much heterozygosity is present, because the human is highly

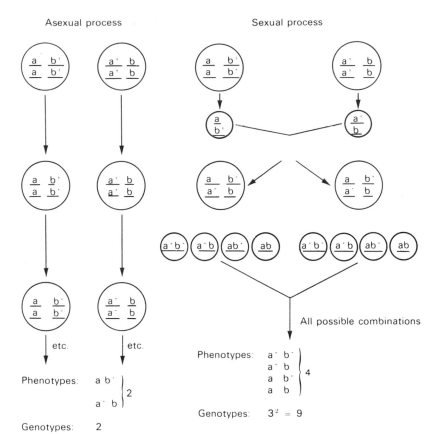

FIG. 9-14. Asexual and sexual reproduction compared. If members of a species are completely asexual, they continue to reproduce their own type indefinitely (left). With two pairs of alleles, and assuming dominance, only two phenotypes and two genotypes are possible in the population. This will remain so until rare spontaneous mutation causes one of the genes to mutate. In contrast (right), the sexual process generates new combinations swiftly. The genetic material of the two types is brought together. With just two pairs of alleles, four phenotypes and nine genotypes are possible. These different combinations may vary in the advantages they give. Natural selection favors those that are more beneficial. Because the sexual process brings about more variation than the asexual in a given length of time, those lines that are sexual have evolved more quickly than those that are exclusively asexual. Crossing over speeds the process up by contributing even more to the variation effected by independent assortment.

hybrid as a result of outbreeding. Since crossing over can occur at any point along a chromosome, the number of new combinations that can be generated is staggering. Each crossover point doubles the number of types of gametes. The new allelic combinations make possible a vast assortment of phenotypes. These are the variations on which the evolutionary force of natural selection can operate. The greater the number of variations, the greater are the chances for types better adapted to different sets of environmental conditions. The larger the number of different types, the greater is the chance for superior forms to occur in a given period of time and, thus, the faster the rate of evolutionary change. Therefore, it is not surprising that sexual organisms have provided the main branches

leading to the evolution of the diverse kinds of living species. The contribution of crossing over to the pool of variation, by extending recombination to linked alleles, has been a major factor in the evolution of higher forms of life and cannot be overestimated.

REFERENCES

Caskey, C. T. and G. D. Kruh. The HPRT locus. Cell 16: 1, 1979.

Creagan, R. P. and F. H. Ruddle. New approaches to human gene mapping by somatic cell hybridization. In Molecular Structure of Human Chromosomes, J. J. Yunis (ed.). Academic Press, New York, 1977.

Creighton, H. S. and B. McClintock. A correlation of cytological and genetical crossing over in Zea mays. Proc. Natl. Acad. Sci. 17: 492, 1931.

Emerson, S. Meiotic recombination in fungi with special reference to tetrad analysis. In Methodology in Basic Genetics. W. J. Burdette (ed.), pp. 167–206. Holden-Day, San Francisco, 1963.

Ephrussi, B. and M. C. Weiss. Hybrid somatic cells. Sci. Am. (April): 26, 1969.

Harper, P. S., M. L. Rivas, and W. B. Bias, et al. Genetic linkage confirmed between the locus for myotonic dystrophy and the ADH-secretion and Lutheran blood group loci. Am. J. Hum. Genet. 24: 310, 1972.

Mayo, O. The use of linkage in genetic counseling. Human Hered. 20: 473, 1970.

McKusick, V. A. The mapping of human chromosomes. Sci. Am. (April): 104, 1971.

McKusick, V. A. and F. H. Ruddle, The status of the gene map of the human chromosomes. Science 196: 390, 1977.

McKusick, V. A. Genetic nosology: three approaches. Am. J. Human Genet. 30: 105, 1978.

O'Brien S. J. and W. G. Nash. Genetic mapping in mammals: chromosome map of the domestic cat. Science 216: 257, 1982.

Ruddle, F. H. Linkage analysis in man by somatic cell genetics. Nature 242: 165, 1973.

Ruddle, F. H. and R. S. Kucherlapati. Hybrid cells and human genes. Sci. Am. (July):36, 1974.

Stern, C. Somatic crossing over and segregation in Drosophila melanogaster. Genetics 21: 625, 1936.

Whittaker, D. L., D. L. Copeland, and J. B. Graham. Linkage of colorblindness to hemophilias A and B. Am. J. Human Genet. 14: 149, 1962.

REVIEW QUESTIONS

1. In the human, the allele for normal blood clotting (h^+) is dominant over the recessive allele (h), which results in hemophilia. The allele for the production of the enzyme glucose-6-phosphate dehydrogenase (gd^+) is dominant over its allele (gd) for the absence of the enzyme. Both allelic pairs are X-linked. A man with hemophilia has a daughter who is concerned about her chance of bearing a child with the disorder. It is found that the hemophiliac man does not produce the enzyme but that his wife, the woman's mother, does. The concerned daughter also produces the enzyme. She is married to a man without hemophilia who is an enzyme producer also. Amniocentesis and chromosome analysis show that the daughter is carrying a male fetus, which is an enzyme producer. What is the chance that the baby has hemophilia, considering that a chromosome map shows the h and the gd loci to be 2 map units apart on the X chromosome?

2. Assume that in the human a dominant allele (A) is responsible for a fatal nervous deterioration, which usually has its onset after the age of 20. The normal condition depends on the recessive allele (a). The presence of the dominant allele is associated with no detectable product before the onset of the disease. Assume that the locus associated with the disorder is on an autosome and is found to be linked to a locus associated with a specific enzyme that can be detected in utero. The allele for enzyme production (E) is dominant over the recessive allele for absence of enzyme (e). Pedigree analyses indicate that the two loci, A and e, are 10 map units apart.

A man with the nervous disorder is also found to be an enzyme producer. Family history reveals he is a dihybrid with the alleles in the *trans* arrangement. The man's normal wife is not an enzyme producer. During her pregnancy, amniocentesis shows that the offspring is not an enzyme producer. What is the chance that the fetus carries the gene for the nervous disorder?

3. The ability to produce the enzyme G6PD depends on a sex-linked dominant allele (see Question 1). Also sex linked is the dominant allele, P, which is necessary for normal color vision. The recessive allele, p, produces protan color blindness. The loci for the enzyme and for color vision are 6 map units apart. A dihybrid woman with alleles in the cis arrangement marries a man with normal vision who is an enzyme producer. Diagram the cross, and show the possible offspring and their expected frequencies.

4. In Neurospora, the loci pro and leu are associated with the production of the amino acids proline and leucine. The two loci are linked. The centromere is to the left of pro, and pro is to the left of leu. A strain of Neurospora that can produce neither amino acid (genotype: pro leu) is crossed to a strain that can produce both of them (genotype: pro^+ leu^+). What will be the sequential arrangement of the ascospores in an ascus derived from a dihybrid nucleus in which each of the following has taken place?

   A. A single crossover event between pro and leu involving the two middle strands.
   B. A double crossover between pro and leu involving the same two strands (the middle ones).
   C. A four-strand double in which each of the crossover events between pro and leu involves a different pair of chromatids.
   D. A single crossover event between the centromere and pro, involving the two middle strands.

5. The ser locus for the production of serine is found to map between pro and leu. Following a cross between a strain that can produce all three amino acids and one that cannot produce any of them, what will be the sequential arrangement of the ascospores in an ascus derived from a trihybrid nucleus in which the following are true?

   A. No crossing over has occurred.
   B. A single crossover event occurred between pro and ser involving the two middle strands.
   C. A single crossover occurred between ser and leu involving the two middle strands.
   D. Double crossing over took place involving the same

two middle strands. One crossover event is to the right of ser and the other is to its left.

6. In Drosophila, the loci for yellow body (y) and distorted or singed bristles (sn) are sex linked. The alleles for gray body (y⁺) and normal bristles (sn⁺) are dominant to their recessive alleles, y and sn. The arrangement on the chromosome is as follows:

A dihybrid zygote is formed with the alleles in the trans arrangement. Assume that somatic synapsis takes place and that it is followed by crossing over between sn and the centromere in one of the cells of the developing fly. What will be the consequences?

7. In the fruit fly, the locus for dumpy wings (dp) is on Chromosome II. The locus for sepia eyes (se) is on Chromosome III, and the locus for the eyeless trait (ey) is on Chromosome IV. The traits dumpy, sepia, and eyeless are recessive to the wild type or normal traits. Suppose a mutation arises in a wild-type laboratory stock of flies and causes the development of extra hairs on the body. The genetic factor causing hairiness (h) is found to be recessive and not sex linked. The following three crosses are made and followed through the second generation: (1) hairy × dumpy; (2) hairy × sepia; (3) hairy × eyeless. The results of the F_2 are as follows:
Cross 1—9 wild; 3 hairy; 3 dumpy; 1 hairy, dumpy.
Cross 2—2 wild; 1 hairy, 1 sepia.
Cross 3—9 wild; 3 hairy; 3 eyeless; 1 hairy, eyeless.

What conclusions can be reached on the assignment of the h locus to a chromosome? Explain.

8. Human cells are fused with those of a mouse, and three clones of hybrid cells are derived with the human chromosome content given in the following panel (+ stands for presence of a chromosome and − indicates the absence). The clones are checked for the presence or absence of three enzymes, X, Y, and Z. Enzyme X is found to be present only in clone A. Enzyme Y is in clones A and B, whereas Enzyme Z occurs in all three. Associate each enzyme with a specific chromosome.

	HUMAN CHROMOSOMES						
	1	2	3	4	5	6	7
Clone A	+	−	+	−	+	−	+
Clone B	+	+	−	−	+	+	−
Clone C	+	+	+	+	−	−	−

9. Assume that production of a certain enzyme in the human has been associated with Chromosome 1 through somatic cell hybridization. However, studies of cells from one person indicate that the enzyme is associated with Chromosome 5. Explain this and how one might get evidence for it.

10. The recessives a, b, and c are sex linked in Drosophila. A cross of two parental strains results in the following F_1:

$$\frac{a^+b^+c^+}{abc} \text{ and } \frac{abc}{Y}$$

If these two F_1s are crossed, which of the following would be true:

A. The amount of recombination cannot be estimated from the resulting F_2.
B. The amount of recombination can be determined from the F_2 from the male offspring alone.
C. The amount of recombination can be determined by considering all of the resulting offspring.
D. The alleles a and b are in the trans arrangement in the female F_1s.
E. Crossing over can be determined only from the F_2 females.

11. Two strains of a mold similar to Neurospora are crossed. Assume that one strain is wild type (al⁺) and can manufacture the amino acid alanine. The other strain (al) is a nutritional deficient and must be supplied with alanine to survive. Following the cross, asci are produced. Spores are removed in order from single asci and tested on medium for their ability to grow in the absence of alanine. A total of 500 asci are studied. Of these, 400 show the following segregation pattern: al⁺ al⁺ al⁺ al⁺ al al al al. The remainder show the following pattern: al⁺ al⁺ al al al⁺ al⁺ al al. Calculate the distance in map units between the centromere and the gene al for alanine synthesis.

12. Suppose that the gene ly for the ability to produce lysine is followed in the same manner as described in Question 11. After a cross between a lysine-forming strain (ly⁺) and one that is deficient in lysine synthesis (ly), all the hybrid asci are found to have the following type of segregation pattern: ly⁺ ly⁺ ly ly ly⁺ ly⁺ ly ly. How far from the centromere in map units would you place the gene ly?

13. Following are the results of two crosses in a mold similar to Neurospora. In each case, classify the types of asci as PD, NPD, or TT. Also determine whether or not the genes are linked, and explain your reasoning. (Spore pairs are listed instead of single spores, since each mitotic division following meiosis results in two identical members, which constitute a pair.)
Cross I: e⁺f⁺ × e f:

	TYPES OF ASCI				
SPORE PAIR	1	2	3	4	5
1–2	e f⁺	e f	e f	e⁺f	e f
3–4	e f⁺	e f	e f⁺	e f⁺	e⁺f
5–6	e⁺f	e⁺f⁺	e⁺f	e f	e f⁺
7–8	e⁺f	e⁺f⁺	e⁺f⁺	e⁺f⁺	e⁺f⁺
Total	150	155	98	6	40

Cross II; $c^+d^+ \times c\ d$:

SPORE PAIR	TYPES OF ASCI			
	1	2	3	4
1–2	c^+d^+	c^+d^+	$c\ d^+$	c^+d^+
3–4	c^+d^+	$c\ d$	$c\ d^+$	c^+d
5–6	$c\ d$	c^+d^+	c^+d	$c\ d^+$
7–8	$c\ d$	$c\ d$	c^+d	$c\ d$
	76	80	4	240

14. From the data given in Question 13 for Cross II (A) give the distance in map units between c and the centromere; (B) give the distance in map units between d and the centromere; (C) give the distance between c and d in map units.

15. The data given in Question 13 for Cross II offer proof that crossing over has taken place in the four-strand stage. Explain.

10

CHANGES IN CHROMOSOME STRUCTURE AND NUMBER

Point mutations versus chromosome anomalies

Throughout the preceding chapters, we have frequently referred to sudden inheritable changes or gene mutations. They are also called *point mutations,* because they represent modifications at specific points along the DNA. These are the alterations that lead to the origin of alleles, alternative forms of a gene. They are not, however, the only kind of modification that can affect the hereditary material and bring about inheritable variation. Some changes alter the actual morphology of the chromosome by modifying its size or shape. Other changes involve the addition or subtraction of one or more entire chromosomes. Although this kind of genetic variation can be technically classified as *mutation,* the term is usually reserved for point mutations. The more gross modifications, which are the subject of this chapter, are generally referred to as *chromosome aberrations* or *anomalies.*

Variations in chromosome structure or number were recognized in early genetic studies (see Chap. 5). Over the years, they have been found in a large number of different plant and animal groups. Their occurrence in humans is a matter of special concern today to the medical practitioner, the genetic counselor, and to certain family groups. The most common cause of congenital defects in humans can be at-tributed to chromosome anomalies. Their alarming frequency is indicated by the fact that more than 90% of abnormal embryos are lost through spontaneous abortion, and among these, chromosome anomalies are very common. An understanding of the nature of chromosome aberrations and their genetic consequences in a variety of plant and animal species can lead us to approach intelligently the problems they present for human society.

Inversion and its consequences

On rare occasions, a chromosome may break spontaneously into one segment or more. Certain environmental agents, such as radiations or certain chemicals, may actually induce breakage and greatly increase the number of resulting aberrations. One of the main types of chromosome alteration is the inversion. As the term itself implies, a stretch of the chromosome is turned around (Fig. 10-1A). This requires two breaks in the chromosome, followed by a healing of the breakage points. It is important to note from Fig. 10-1A that when an inversion arises, the interstitial segment (cdef) does not attach to either of the original chromosome ends (a or g). The reason is that the ends or end genes of a chromosome appear to be polarized and do not attach to other segments. The

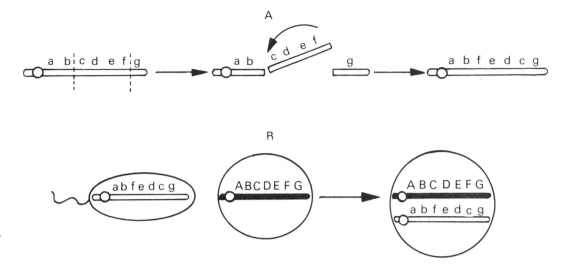

FIG. 10-1. The inversion (*A*) Origin of an inversion after two breaks in a chromosome. (*B*) A sperm carrying an inversion fertilizes an egg in which the homologous chromosome has the normal gene sequence.

The result is a zygote that is an inversion heterozygote. (The centromere is represented by an *open circle*.)

two extremities of each chromosome have been designated the telomeres and are often described as being "nonsticky." Any other gene segments between the telomeres are considered "sticky," meaning that they can attach to one another in various new arrangements, but not to telomeres.

An inversion can arise in a somatic cell or at some time during gamete formation in either sex. Assume that a sperm contains a chromosome with an inverted segment and that it fertilizes an egg in which the corresponding chromosome has the standard sequence of genes (Fig. 10-1B). The zygote would consequently have a normal chromosome and one carrying the inversion. Any individual with a normal chromosome and one in which the structure has been altered in any way at all is called a structural heterozygote. The inversion heterozygote in this example is genically balanced, because no genetic loci are missing or modified in any detectable way. Gene order has been somewhat changed, but there is no change in the kind or number of genes as a result of the inversion. Consequently, we would expect such an individual to be phenotypically normal, and this is generally so. Some position effect is possible, but it can be ignored for the moment.

The inversion heterozygote can usually live to maturity without any ill effects from the anomaly. Why, then, is this condition not more common among living things, since no upset in genic balance occurs in the heterozygote? The answer comes when the meiotic

picture is considered. It will be recalled that homologous chromosome segments undergo synapsis at zygonema of meiotic prophase. Two homologous chromosomes encounter pairing difficulties if one member contains a segment that is reversed. However, the forces of synapsis are powerful enough to ensure pairing of homologous regions if at all possible, even if the chromosomes must undergo contortions. In the inversion heterozygote, the pairing is accomplished through the formation of a loop in one of the chromosomes, either the normal or the inverted one (Fig. 10-2A). The two chromosomes may separate normally at anaphase I, and there would be no complications. However, if the inversion is large enough, a crossover event is certain to occur in some cells. In Fig. 10-2A, a crossover event is indicated in the loop between loci d and e, and its consequences are seen. As a result of crossing over anywhere within the inverted segment, two chromosomes are formed that are very abnormal in structure. One of them possesses two centromeres (a dicentric chromosome); the other lacks a centromere (an acentric). Both are also genically unbalanced; certain loci are represented more than once in each, whereas other loci are missing (in the dicentric, the a and the b loci are present twice but g is absent; the acentric has two doses of the g locus but lacks the a and the b).

If the acentric and the dicentric separate at first anaphase and move to opposite poles, the two nuclei that form will be genically unbalanced. But the af-

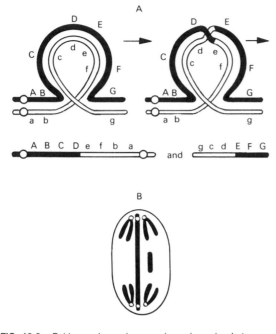

FIG. 10-2. Pairing and crossing over in an inversion heterozygote. (A) Loop formation is shown followed by crossing over. A crossover event between loci d and e (or any other two loci in the inverted segment) results in the origin of a dicentric chromosome and one that is acentric. Both are genically unbalanced. For the sake of clarity, each chromosome is represented as being single. Actually, at the time of crossing over, each is double, but only two chromatids engage in any one crossover event. (B) An inversion may be detected with the microscope, because anaphase I cells will show a bridge and a fragment. These represent the dicentric, being pulled to both poles, and the acentric, which cannot move at all.

fected chromosomes are apt to be involved in still further cytological problems. The dicentric will be attracted to both poles, the acentric to neither. Since the centromere is the dynamic center of the chromosome, the dicentric may be stretched as the two centromeres travel to opposite poles. At the level of the light microscope, this may be visualized as a bridge between the poles (Fig. 10-2B). Next to it may be seen a fragment, actually the acentric chromosome that is unable to move at all and that will disintegrate in the cytoplasm after being left out of a nucleus. The stretched chromosome may break anywhere along its length.

In any case, it is evident that cells that are genetically unbalanced will arise. Since these events take place at meiosis, the consequence is a certain proportion of unbalanced sperms or eggs. If the gametes survive, any offspring resulting from them would be abnormal in some way or would fail to survive at all because of the imbalance. In general, therefore, inversion heterozygotes suffer a certain decrease in fertility as a result of the formation of acentrics and dicentrics formed after crossing over in the inverted segment. Because these structural heterozygotes are at a reproductive disadvantage, leaving fewer offspring in the long run than the normal individuals, natural selection tends to weed the aberration out of the population. If an inversion is tiny, it has a greater chance of persisting in a population, because the probability of crossing over within it is less. Although the small inversions often tend to persist, large ones are not usually found to be widespread in most plant or animal populations.

Inversions in Drosophila

The outstanding exception to a low level of inversions in the wild is found in the genus Drosophila. In many fruit fly populations, inversions of appreciable size are very common and may actually be a normal part of the hereditary composition of the population. How can these persist without causing a reduction in fertility? We have noted that the fruit fly is unusual in that normally no crossing over occurs in the male. Therefore, the complications discussed earlier are not encountered, and no reduction in male fertility ensues from the inversion. In the female, however, crossing over is the rule at meiotic prophase. Still, female fertility is not reduced by the presence of the inversion. Remember that crossing over takes place in the four-strand stage. Only two of the chromatids of the bivalent participate in any one crossover event. In the case of an inversion, two chromatids usually are not involved in a crossover event in the inverted segment and are normal in their genic balance. The abnormal, dicentric chromosome is held back as a result of the chromatid bridge, and the acentric does not move at all and disintegrates in the cytoplasm. Consequently, one of the noncrossover chromatids (either the one with the inversion or the one with the normal order) completes meiosis and enters the egg (Fig. 10-3A). The overall effect of the chromatid bridge in the females of certain animal species, such as those of Drosophila, is to prevent the dicentric chromatid from entering the egg. This preferential inclusion of a normal chromatid in the egg cell of an inversion heterozygote is termed *selective orientation*. The balanced chromatid carrying the inversion has just as

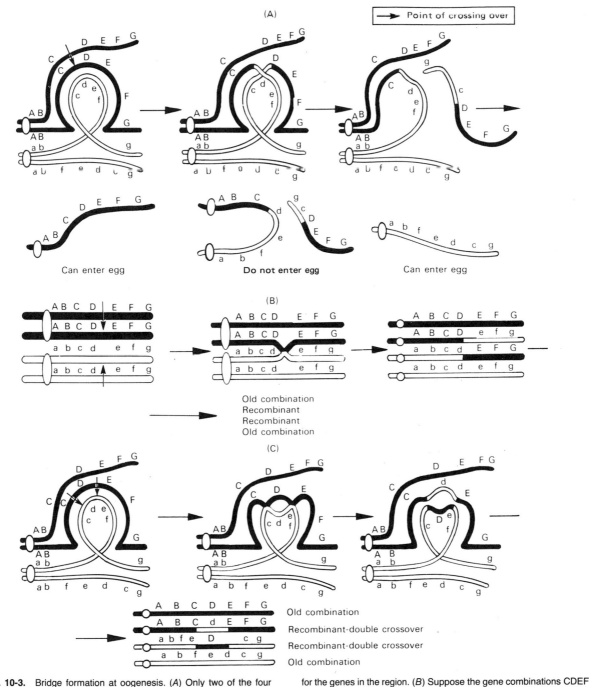

FIG. 10-3. Bridge formation at oogenesis. (*A*) Only two of the four chromosome strands participate in any single crossover event. As a result of the formation of the bridge, the defective dicentric strand is held back at the end of the first meiosis. The strands of normal morphology (those with just one centromere) become directed toward the "outside" of the cell. Upon completion of second meiosis, one of the outside nuclei becomes the nucleus of the egg. The defective dicentric and acentric chromosomes are never oriented toward the outside and do not enter the egg. Note that the egg will therefore contain one of the parental gene combinations, CDEF (the noninverted arrangement) or fedc (the inversion). Since only the parental combinations are present, the chance remains good that many individuals will be heterozygous for the inverted region and also heterozygous for the genes in the region. (*B*) Suppose the gene combinations CDEF and cdef (in any order) bring an advantage when present as a unit on a chromosome. If no inversion is present, any single crossover in the cdef region can break up the combination. The inversion, therefore, as shown in *A*, tends to preserve it. Without the inversion, there is also a greater chance for genes in the region to become homozygous. (*C*) A double crossover within an inverted segment produces chromosomes of normal morphology and also brings about recombination. Since no bridge is present, the recombinant chromosomes can enter an egg nucleus. The advantageous combinations CDEF and cdef will thus be broken up. (For diagrammatic purposes, the top chromatid is shown unpaired with its sister.)

243

much chance as the standard one to enter the egg, and there is no reduction in fertility. So mainly because of a lack of crossing over in the male and selective orientation in the female, inversions can accumulate in Drosophila populations without reducing the fertility.

Because inversions are so common in Drosophila, however, do they carry some sort of advantage? The answer seems to be "Yes." The reason for the value can be appreciated by reconsidering the consequences of a single inversion within the inverted segment (Fig. 10-3A). A single crossover event brings about the recombination of linked alleles, but in the case of the inverted region, the recombination is associated with abnormal chromatids, and these do not enter the egg. Only the old combinations (the chromatids with the standard arrangement and the one with the original inversion) are preserved to enter the egg. Therefore, the effect of the inversion is a decrease in the amount of recombination among genes in the inverted region. We have learned that crossing over and the recombination it generates are highly desirable aspects of the sexual process and have enhanced the variation available for natural selection. The coin, however, has a reverse side (Fig. 10-3B); it is quite possible for a group of certain closely linked alleles to constitute a very desirable combination. The combination and the advantage it gives can be destroyed by recombination. The presence of an inversion ensures that the combination will stay together from one generation to the next. It also cuts down the chances that all the genes in the affected segment will become homozygous for one of their allelic forms in any one individual. Populations of Drosophila have been studied for the possible adaptive value of their inversions. In some natural populations of fruit flies, nearly every individual is heterozygous for one or more inversions on each of its chromosomes. Dobzhansky has followed the distribution and frequency of different inversions and has clearly demonstrated that structural heterozygosity can confer a decided advantage. Studies that followed populations through several years have shown that the frequency of certain specific inversions varies characteristically with the season of the year. Controlled laboratory conditions with flies in population cages have shown that similar variations in the frequency of a particular inversion type occur with temperature changes that duplicate those in nature. It is suspected that small inversions in many animal and plant groups may provide a mechanism for preserving combinations of alleles favorable under a variety of environmental conditions.

If an inversion is sufficiently large, double crossing over may take place within it. Figure 10-3C shows that this will produce two balanced chromatids, each normal in structure and with a new combination of genes. Double crossing over, then, tends to break up old combinations of alleles. However, double crossing over occurs much less frequently than singles and is quite rare within a small inversion.

Different types of inversion

Since chromosome breaks may occur anywhere along the length of a chromosome, countless types of inversions are possible. Two major categories can be recognized. In our discussion so far, both breaks have been in one chromosome arm, to one side of the centromere. Such inversions are called *paracentric*. These are in contrast to pericentric inversions, those in which the breaks occur on either side of the centromere (Fig. 10-4A). The story of synapsis and loop formation is the same for the pericentric heterozygote as for the paracentric. A big difference is seen, however, when the products of crossing over are examined (Fig. 10-4B). No dicentrics or acentrics are formed. The two resulting crossover chromatids are normal morphologically in that each possesses one centromere, but each is genically unbalanced. Since there is no dicentric chromatid, no chromatid bridge is formed at anaphase. The two genically unbalanced chromatids are not held back at meiosis and have as much chance to enter the egg as do the balanced ones. The consequence is some reduction in fertility in the female. Therefore, we find the pericentric inversion a much less common feature of populations, even Drosophila. Again however, small inversions have a chance to accumulate, and there is evidence that they play some role in changing chromosome morphology. From Fig. 10-4C, it can be seen that if the two breaks on either side of the centromere are *not* equidistant, the centromere position will shift. Small shifts of the centromere resulting from pericentric inversions have almost certainly operated in Drosophila during the evolution of the various species to alter chromosome morphology. It accounts for the presence of acrocentric chromosomes in some groups related to those having telocentric or metacentric chromosomes (see Fig. 10-24). Each acrocentric has

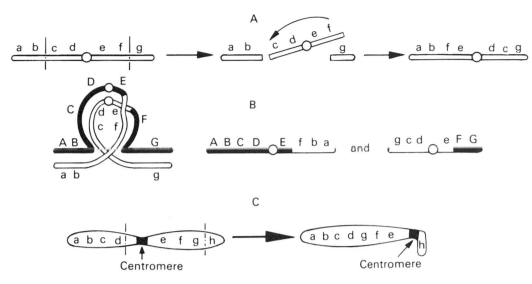

FIG. 10-4. Pericentric inversions. (A) Note that each break is on either side of the centromere. Contrast this with the paracentric inversion (Fig. 10-1A) in which both breaks are on the same side of the centromere. (B) Crossing over anywhere in the inverted region in the case of a pericentric inversion will produce two chromatids, each normal morphologically (with one centromere each) but each genetically unbalanced. Note that in one chromatid loci a and b are represented twice and g is absent. The other has a double dose of g but lacks a and b. Only two chromatids are shown, for the sake of clarity. The other two would not participate in this crossover event and would be balanced parental types. However, because there is no bridge to hold back the unbalanced ones, the egg nucleus may receive a defective chromosome, and so a decrease in fertility ensues. (C) If the two breaks are not equidistant from the centromere, the centromere position shifts. In this case, it is shifted closer to one end, yielding an acrocentric chromosome from a metacentric one.

probably been derived from a telocentric or metacentric chromosome through pericentric inversions that shifted the centromere. In the X chromosome of a stock of Drosophila melanogaster, Muller was able to shift centromere position experimentally over a period of time by using radiations to induce breakage.

Once an inversion has occurred in a chromosome, there is no reason why additional inversions cannot occur later. And each of these would usually involve different points of breakage. However, a second inversion in a chromosome may have one point of breakage in a segment that has already been reversed in a previous inversion. Consequently, one inversion would overlap the other (Fig. 10-5A). If a structural heterozygote possesses the standard chromosome and the one in which the two inversions have taken place, a complex loop, rather than a simple one, is seen at meiosis (Fig. 10-5B). Overlapping inversions have occurred in the evolution of various Drosophila populations. Within a species, certain races of flies differ by two or more inversions. These changes can be detected by studying the complexities of the loops seen in the salivary gland chromosomes, which are in a permanently synapsed state (Fig. 10-5C).

In humans, pericentric inversions have been identified in a few families. Thirteen children have been detected who have inversions in Chromosome 9, and all of them are normal. A pericentric inversion in humans, as well as in other species, may be recognized by a change in the position of one of the centromeres; this is reflected as a change in the length of the two chromosome arms. Crossing over within the inverted segment in a pericentric heterozygote results in the production of unbalanced chromosomes that will differ depending on the point of crossing over. An individual heterozygous for a pericentric inversion could potentially produce offspring with different types of abnormalities. This may explain the case in which several abnormal children were born to a phenotypically normal woman who was found, upon karyotype analysis, to have a pericentric inversion in Chromosome 10. The possibility of inversions in humans as a cause of some complex defects must be considered when karyotypes are examined in families with a history of congenital disorders.

A

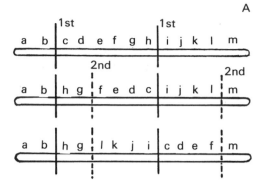

a b | c d e f g h | i j k l m   Standard order
1st   1st

a b | h g | f e d c | i j k l | m   Result of first inversion
         2nd   2nd

a b | h g | l k j i | c d e f | m   Result of second inversion
                                       which overlaps the first

B

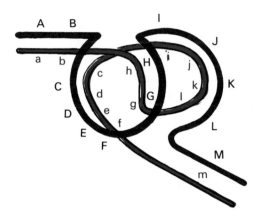

C

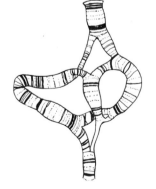

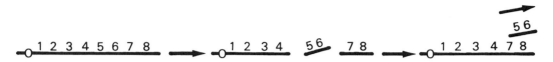

FIG. 10-6. Formation of a deletion. Two breaks may arise in a chromosome, followed by loss of the interstitial piece. The portions with the centromere and telomeres rejoin.

Detection of deletions

Another type of structural aberration, which has now been described in humans as well as in other species, is the deletion. Actually, it was the very first type of structural change to be identified and was recognized in Drosophila in 1915. As its name implies, a deletion entails the actual loss of a portion of the chromosome, so that a block of genes is missing. A deletion can follow a double break in a chromosome, as shown in Fig. 10-6. The affected chromosome is actually smaller as a result of the loss, although a size change is not detectable when the eliminated piece is minute. It should be obvious that the deletion will certainly cause a degree of genic unbalance, because certain loci are completely missing from one of the chromosomes. The possible effects of such unbalance are demonstrated by two early cases in Drosophila.

The first one, described by Bridges, involved a deletion in the Bar region of the X chromosome. The Bar effect, in which the eye is narrowed, is inherited as a sex-linked dominant. As illustrated in Fig. 10-7A, a cross between a normal female and a Bar male should produce Bar females and normal males. However, in a certain cross, Bridges mated a normal female with a Bar male and obtained a normal female among the offspring. At the same time, a lethal seemed to appear. After a series of appropriate crosses, it became clear that a piece of the chromosome, including the Bar locus, had been deleted and that this deletion could be carried in the female but not in the male. Figure 10-7B shows that the deletion causes the elimination of the dominant bar, and so the wild trait, which is recessive, can be expressed. When the female with the deletion is crossed, half of the male offspring die as a result of the missing chromosome region (Fig. 10-7C). Evidently, the deleted segment contains genetic information essential to life, because its complete absence in the hemizygous male prevents development. Further crosses showed that, in addition to Bar, at least one other locus was missing. This proved to be the locus "forked," which has a pronounced effect on the bristles.

A different deletion arose in the X of Drosophila; this affected the development of the wings. A female heterozygous for this deletion has wings that contain a nick or notch at the tip. This notch condition has arisen independently several times in the fruit fly. Each notch fly has been shown to contain a deletion that always includes the locus "facet," a genetic region that affects the texture of the eye. The extent of these different notch deletions varies and may include the "white" locus in addition to facet (Fig. 10-8). All notch flies must be females because the deletion is a complete lethal in the male, for the same reason as described for Bar.

FIG. 10-5. Simple and complex inversions. (*A*) Origin of an overlapping inversion. Assume 1 is the standard order and that the first breakage points result in an inversion. When another inversion occurs, the second points of breakage can involve loci that were inverted in the first rearrangement. (*B*) An individual heterozygous for just one inversion in a pair of chromosomes will show a simple loop at meiosis. However, an individual carrying a chromosome containing two inversions that overlap will show a compound loop. (*C*) Simple loop (*left*) and compound one seen in the salivary gland chromosomes of Drosophila azteca. If two races are crossed and a simple loop forms in a chromosome pair, we know that the races differ by a single inversion. If a compound loop forms, we know there are overlapping inversions. By studying the loops and their complexities, the cytogeneticist can tell by how many inversions two races may differ. Relationships among races may be worked out by making use of this reasoning. (Reprinted with permission from *J. Hered.* 30: 3–19, 1939.)

Deletions and mapping in Drosophila

There have been many cases in the fruit fly in which the presence of suspected deletions has been verified cytologically. If a deletion is present, there will be a region on the normal chromosome that is unable to pair at the time of synapsis because the homologous region has been lost. Consequently, the normal chromosome may form a buckle or loop whose size depends on the extent of the deletion (Fig. 10-9A). The

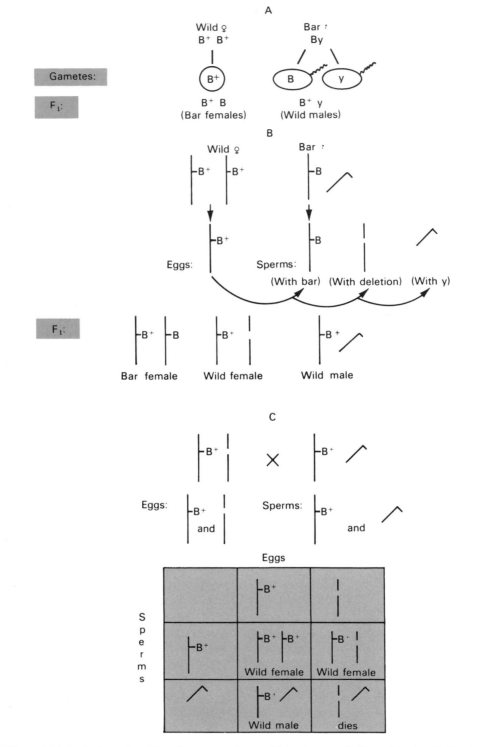

FIG. 10-7. Effects of deletion in Bar region. (A) Ordinary cross of wild with Bar. (B$^+$ = wild; B = bar). (B) A deletion arises in some of the sperm, eliminating the dominant Bar (B). Unexpectedly, some of the females appear to be wild. (C) When the wild female carrying the deletion is crossed with a wild male, all of the female offspring are wild, but half carry the deletion. Since one Bar region is necessary for life, it acts as a sex-linked recessive lethal, and half of the F$_1$ males fail to develop.

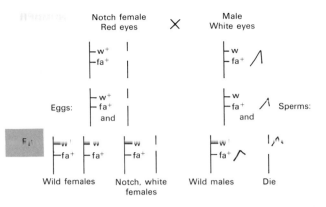

Notch female
Red eyes   ×   Male
White eyes

Eggs:

and

Sperms:

and

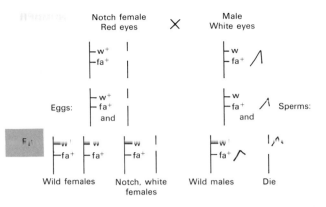

Wild females   Notch, white   Wild males   Die
                females

FIG. 10-8. The notch phenotype always involves a deletion of the locus "facet," which influences eye development. Any deletion may include more than facet, in this case, the white locus. All notch flies must be females because the facet deletion acts as a sex-linked recessive lethal. In the cross shown here, the recessive white expresses itself in half of the females because the white locus is included in the deletion. Only half of the males will develop because of the deleted genes.

loci absent from the deleted chromosome will be present in the loop formed by the normal one.

This fact has enabled the cytogeneticist to ascertain the location of specific genes and frequently to associate them with definite bands in the salivary chromosomes. For example, if the genetic results indicate that a specific locus has been deleted, one may search for a loop in the specific chromosome to which the gene is linked (Fig. 10-9B). One would look at the X in the salivary nuclei in the case of notch or white, because these loci map on the X. If a loop is found, one may compare the bands in the loop with those in the other chromosome. Any bands absent from the latter would thus represent the deletion. In this way, definite loci can be associated with definite chromosome sites. The locations of genes, as determined in this way, correspond well with their positions as determined by genetic mapping.

However, although the linear sequence of the genes is the same, the map may show regions where the loci appear closer together than they do when a map is made cytologically from the salivaries (Fig. 10-10). The reason, of course, is that the genetic map depends on the frequency of crossing over. In some chromosome regions, genes engage in less crossing over than those in other regions. We have noted that loci near the centromere, for example, appear clustered because of its inhibiting effect on crossing over. The genetic map tells us the probability of crossing over, not the exact spatial distances among genes. The latter are more precisely resolved by the cytological map, built up through the study of deletions.

Effects of deletions in various species

The examples of the deletions illustrate several other important points. When a deletion is present, atypical genetic effects usually follow. We see that a recessive

FIG. 10-9. Cytological detection of deletions. (*A*) At the time of meiotic pairing, a buckle may be observed in a bivalent. This results from the fact that the normal chromosome is longer than the homologue with the deletion. The genetic region present in the normal chromosome has no corresponding part with which to pair. The buckling makes possible synapsis of all those homologous regions that are present. (*B*) The giant salivary chromosomes are permanently synapsed and contain conspicuous bands. Any regions in the buckle (*indicated by the arrows*) must be absent from the homologue. Bands in the buckle can be seen to be deleted from the other chromosome. If known loci are absent, as determined genetically, they may be associated with bands in the buckle. In this diagram, the bands indicated are absent from the other chromosome. If certain genes are known to be missing, their chromosome location may thus be associated with these bands.

A

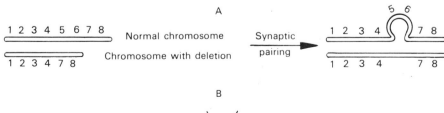

1 2 3 4 5 6 7 8   Normal chromosome

1 2 3 4 7 8   Chromosome with deletion

Synaptic pairing

5 6

1 2 3 4   7 8

1 2 3 4   7 8

B

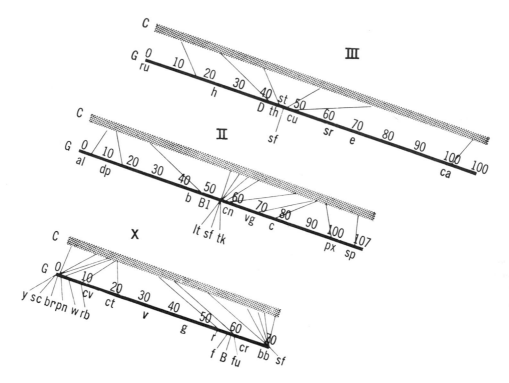

FIG. 10-10. Comparison of the genetic and cytological maps of the third (III), second (II), and X chromosomes (X) of Drosophila melanogaster. C is the cytological map and G the genetic map. Figures indicate the genetic distances in map units. The lines connecting the cytological with the genetic maps indicate the microscopically observed and the genetically determined positions of certain chromosome breakages. (Reprinted with permission from E. W. Sinnott, L. C. Dunn, and T. Dobzhansky, *Principles of Genetics*, 5th ed. McGraw-Hill, New York, 1958.)

may unexpectedly express itself; the recessive wild was expressed when the dominant Bar was assumed to be present. This is often referred to as pseudodominance. In the case of a "notch" deletion, which encompasses the white locus, the recessive white could suddenly appear (see Fig. 10-8). The mutant recessive, white, has superficially acted as if it were dominant, because it was expressed when the wild gene was supposedly present. Actually, it has remained recessive; its pseudodominance is entirely the result of the deletion.

The case of the notch deletion also demonstrates that the deletion itself may produce a characteristic phenotypic effect that acts as a dominant. This is very often true of deletions in a variety of organisms. The notch deletion, moreover, behaves as a recessive lethal (see Fig. 10-8). No males can develop because they have only one X chromosome, and no females homozygous for the notch deletion can arise. Again, this is often the case with small deletions, autosomal as well as sex linked. If their effect on genic balance is not too severe, small deletions may accumulate in a population to some extent. But when they become homozygous, they usually prevent development and act as if a recessive lethal were present. Of course, if a deletion is of sufficient size or includes certain critical genes, it may also act as a dominant lethal and kill the heterozygote.

These observations also tell us something else that reinforces what was said in Chapter 4 and elsewhere concerning the interaction of genes. The notch deletion shows that the elimination of a locus, the facet locus that has a pronounced effect on the eye, may cause a defect in another organ, here the wing. This is another good example that any gene may have more than one effect (pleiotropy) and the normal development of any character depends on the interaction of the entire genotype. The very viability of the organism can also depend on the presence of certain loci, at least in a single dose. Obviously, the gene products of Bar, facet, and others are essential to the development of the embryo in the fruit fly.

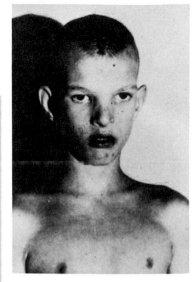

A

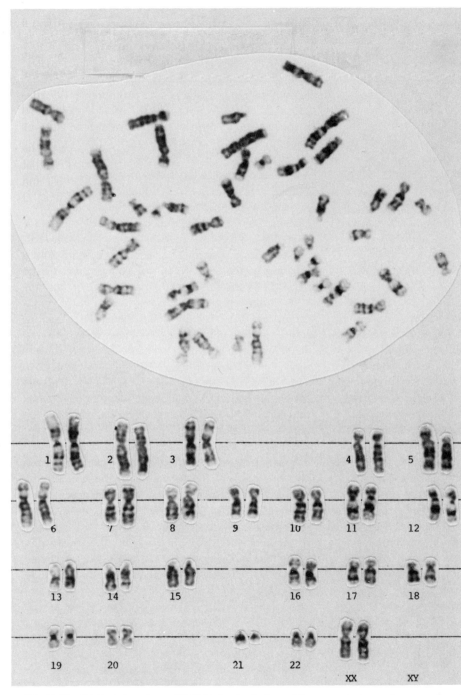

B

FIG. 10-11. Cri du-chat syndrome. (*A*) Child showing features of the syndrome. (Reprinted from *Birth Defects: Atlas and Compendium*, ed. D. Bergsma. The National Foundation—March of Dimes, White Plains, New York, with permission of the editor and contributor.) (*B*) Karyotype of female with cri-du-chat syndrome. Note deletion in the short arm of one member of the Chromosome 5 pair. The chromosomes are G banded. (Courtesy George I. Solish, M.D., Ph.D.; Jyoti Roy, M.S., Thomas Mathews, M.S., Long Island College Hospital, Cytogenetics Lab.)

Work with mice has shown that orderly development may depend on products governed by genes. A certain product may be needed at a critical step or time in the developmental process. If it is absent, a crisis results that can terminate development and prove lethal. Detailed studies of different-size deletions in the X of Drosophila have also illustrated the crucial role of various loci at specific times in differentiation. It is very important to note at this point that a deletion of any size can come to exist in a heterozygous state only if it can be transmitted through haploid cells. Deletions frequently prove lethal to the gametes of an animal—the mouse, for example. Moreover, the haploid (gametophyte) generation of plants is especially vulnerable to deletions. The genes essential to the formation of viable gametes would appear to be widespread throughout the chromosomes in these species, which, unlike Drosophila, cannot tolerate deletions of any substantial size in their sex cells or cells in the haploid generation.

From these observations of the dire effects of deletions in experimental organisms, it comes as no surprise that a deletion in a human can have serious consequences. Unfortunately, this is borne out by the effects associated with a deletion in the shorter arm of Chromosome 5. Afflicted persons exhibit an assortment of effects known as the cri du chat syndrome. The name refers to the catlike cry that such individuals produce in infancy and sometimes even beyond. In addition to various characteristic facial features (Fig. 10-11), the carriers of the deletion suffer growth defects and mental retardation. The products of genes in that particular segment of Chromosome 5 are evidently necessary for normal development.

The importance of deletions in human cells in the assignment of specific genes to definite regions of a given chromosome was noted in Chapter 9. Deletion mapping in the human, which involves techniques of chromosome banding and somatic cell hybridization, along with pedigree analysis, will probably play a very significant role in the localization and ordering of genes on a given chromosome.

Duplications: their origin and their effects

Another kind of structural chromosome alteration that may produce unexpected genetic results is the duplication, a condition in which a portion of a chromosome is present in excess in somatic cells. If a region is present three or more times instead of the normal two, it is said to be duplicated. One way in which a duplication can arise is shown in Fig. 10-12A and B. A simple translocation or shift (details later on) may place a chromosome segment in a new location in the chromosome. The chromosome containing the shift would still be balanced; therefore, there probably would be no phenotypic effect in the individual carrying the altered chromosome along with the one with the normal sequence of genes. The figure shows that pairing problems would again exist at meiosis, and two loops might form to permit synapsis of most of the homologous regions.

Crossing over anywhere between the loops, however, leads to two altered chromosomes, one with a deletion and one with a duplication. The consequences of the former have just been discussed. The chromosome with the duplication can also cause phenotypic effects because of an upset in genic balance. If it enters a gamete, which then combines with a normal gamete, the zygote will contain some loci in three doses instead of two (the loci FG in Fig. 10-12C). The outcome depends on the size of the duplication and the specific loci involved. In some instances, a dose of two recessive alleles along with the dominant allele results in the expression of the recessives. For example, a female Drosophila carrying two doses of the sex-linked recessive v (vermilion eye color) and one of the wild allele for red eye, v^+, has vermilion eyes. The genotypes v^+v v and v v are the same phenotypically.

In contrast with this, many examples could be given in which one dose of the dominant expresses itself over two doses of the recessive. In the fruit fly, the wild allele sp^+, is dominant over two doses of its recessive allele, sp, the gene "speck," which is responsible for a certain body abnormality. Therefore, sp^+sp sp is wild. On the other hand, the recessive pl (for abnormal wing venation), when present in an extra dose with wild pl^+ produces a weak mutant phenotype; the fly that is pl^+ pl pl can be distinguished from both the typical wild (pl^+ pl^+ or pl^+ pl) and the homozygous recessive (pl pl).

In still other cases, intermediate conditions of some type are produced. Obviously, if a duplication is not suspected, its presence in a particular mating can lead to puzzling results. There is evidence that duplications represent a class of structural chromosome changes that has been especially important during the course of evolution (see later in this chapter).

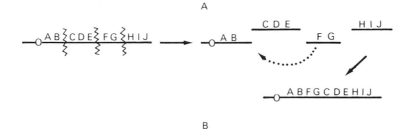

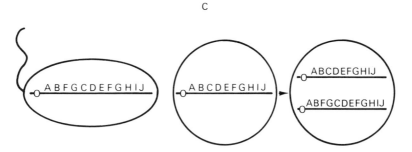

FIG. 10-12. Origin of a duplication. (*A*) A shift of a genetic region takes place after three breaks in one chromosome. (*B*) At meiosis, the chromosome with the shift cannot pair exactly throughout its length with its homologue. A buckle forms in each chromosome to bring together most of the homologous regions. A crossover event anywhere between the loops results in a chromatid with a deletion (FG) and another with a duplication. (*C*) A gamete with the duplication fertilizes a normal gamete. The zygote has one genetic region present three times. The region in excess (FG) is said to be duplicated.

Duplication of the Bar locus and position effect

The unexpected genetic effects of a duplication are well illustrated by the story of the Bar "mutation" in Drosophila. As already noted, the effect of Bar is to decrease the number of facets in the eye so that it resembles a band or bar. The condition is inherited as a sex-linked dominant (actually an incomplete dominant) and was discovered in 1914 (see Fig. 10-7A). It was then found that homozygous Bar stocks, when used in crosses, occasionally give rise to wild flies. This should not be so, because the Bar allele is a dominant. The frequency of the reversal to normal is approximately 1/1500, a figure much too high to

result from mutation of the mutant Bar gene back to normal (B→B$^+$). It was also found that when these unexpected normals arose, another phenotype also appeared, and with the same frequency of 1/1500. This new type was a narrower, more extreme Bar than the typical one, and so it was designated *ultrabar*.

In a series of crosses, Morgan and Sturtevant mated male and female Bar flies separately by crossing them with normals. Their results showed that the normal and the ultrabar arose only when a female was used as the Bar parent, never when a Bar male was crossed with a wild female. For example, the cross of Bar female × wild male (BB × B$^+$Y) can result in some unexpected offspring with wild eyes and some that are ultrabar. The reciprocal cross, wild female × Bar male (B⁺ B⁺ × B Y), gives only the expected results, as diagrammed in Fig. 10-7A.

Sturtevant made a series of crosses in which he followed not only Bar but also genes very closely linked to it. He was able to demonstrate that whenever the normal or the ultrabar appeared, a crossover had taken place in the Bar female in the vicinity of the Bar locus.

The question arose as to why this phenomenon should be happening in this region of the chromosome. Why were similar results not obtained when other loci were studied? Muller suggested that the enigma could be explained on the assumption that the dominant Bar effect is not the result of a point mutation but rather of a duplication. Based on this

theory, a homozygous Bar female should have two Bar loci represented on each X chromosome and would be B B / B B (Fig. 10-13A). The normal female would thus be B/B, with just one dose of Bar on each X.

As a result of the presence of duplicated genetic material in the homozygous Bar female, difficulties could be expected to arise at synapsis. Since four identical regions are present, the pairing need not always be two by two. A certain amount of synaptic confusion might lead to the pairing of the "right-hand Bar" on the one chromosome with the left of the other (or vice versa). Crossing over in the area could then produce unequal products—B and BBB (Fig. 10-13B). The former, according to Muller's hypothesis, would be the normal, nonbar condition; that is, the presence of a single Bar on the X of a male or a single Bar on each X in the female produces the normal-eye phe-

FIG. 10-13. The Bar region. (A) The normal wild fly carries one Bar locus on any X. Because it is sex linked, the wild female carries two X chromosomes, each with one Bar locus. A Bar-eyed fly results when the Bar locus is duplicated on an X, one Bar adjacent to the other. (B) Crossing over may occur within the Bar region. Typically, pairing will be normal (upper diagram) and the two crossover chromatids will be identical to the original. However, forces of pairing may at times lead to some confusion, because four homologous loci are present (lower diagram). Crossing over in the Bar region then leads to production of two different chromatids, a wild type and one with three adjacent Bar loci, which can produce an ultrabar effect. The figure depicts a male that has arisen after fertilization of an egg carrying the X with Bar present three times. (The two chromatids not involved in the crossover event are not represented, for the sake of clarity.)

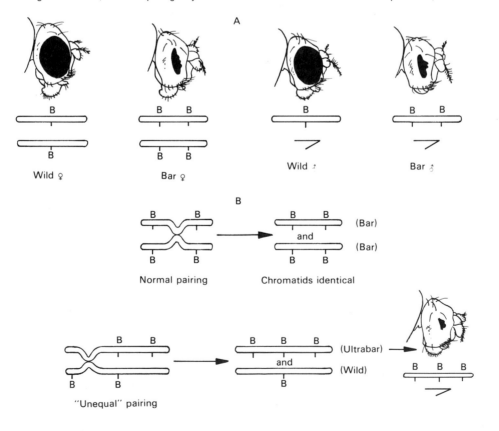

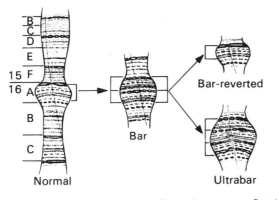

FIG. 10-14. The Bar region in the salivary chromosomes. Certain bands appear duplicated in the X chromosome of flies with the Bar phenotype. In those with the ultrabar condition, the bands are present in triplicate. The normal as well as the ultrabar condition can arise from a Bar female in which "unequal pairing" occurs, followed by crossing over in the Bar region. (Reprinted with permission from C. B. Bridges, Science 83:210–211, 1936.)

son of flies 3 and 4 reveals that they differ phenotypically. Fly 3 has a much wider eye than fly 4. However, when the number of Bar loci is counted, it turns out to be the same; both flies have four Bar genes (BB/BB is Bar; yet BBB/B is ultrabar). This was the first indication that the dosage of a gene is not the only factor that might be involved in the expression of a phenotype. Also to be considered is the arrangement of the genes: three Bars on one chromosome and one Bar on the homologue is different from two Bars on each chromosome. This type of phenotypic effect resulting from the position of one gene with respect to another was named the position effect. Position effects have been studied in detail in the fruit fly and have now been demonstrated in other organisms as well. The possibility of a position effect must be kept in mind when any structural heterozygote is studied.

Heterochromatin and position effect

Not all genes show position effects when they are placed in new locations. Studies with the fruit fly indicate that it depends on the particular gene involved, as well as on the new location. Very interesting position effects are produced when certain loci are shifted to the vicinity of heterochromatin. (Recall our discussions of constitutive and facultative heterochromatin in Chapters 2 and 5.) Most of the genes well known to us are located in euchromatic regions. Relevant to our present discussion is the peculiar type of position effect that can be exerted by heterochromatin on the euchromatin. Suppose the dominant sex-linked allele for red eye color (w^+), normally found in euchromatin, is placed next to heterochromatin as the result of a translocation (Fig. 10-16A). If a female fly contains the dominant w^+ adjacent to heterochromatin on one X chromosome and the recessive allele w in its normal location on the other one, an unex-

notype. The latter chromosome, BBB, with three adjacent Bar loci would result in ultrabar.

Muller suggested a cytological examination of the banding pattern of the X chromosome in the salivaries. In the region of the X, where Bar would be present, as suggested by genetic maps, a certain segment was found to be represented twice in the Bar fly, three times in the ultrabar, and just once in the wild (Fig. 10-14)! Again we see a perfect correlation between genetic results and cytology. Thus, it was shown conclusively that the Bar effect is the result of a tandem duplication (two identical segments next to each other). The more segments that are adjacent, the greater the mutant Bar effect and the narrower the eye.

When flies with the several genotypes shown in Fig. 10-15 are compared phenotypically, a rather unusual fact becomes evident. For example, a compari-

FIG. 10-15. Comparison of female flies with different numbers and arrangements of the Bar locus.

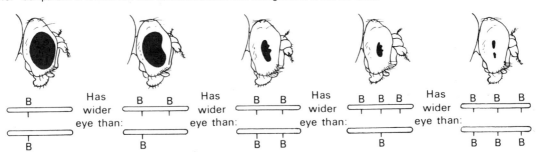

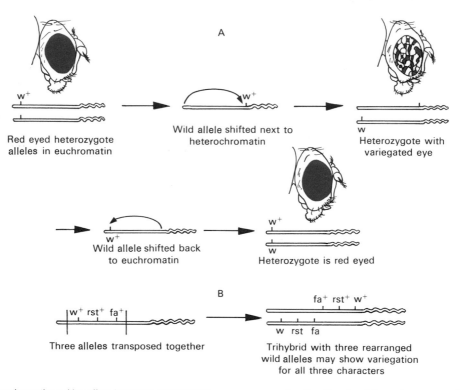

FIG. 10-16. Heterochromatic position effect (*wavy line* indicates heterochromatin). (*A*) Wild-type allele placed next to heterochromatin in the same or in a different chromosome after a chromosome alteration may result in failure of the wild allele to express itself in some cells. When returned to a normal euchromatic region after a second chromosome alteration, the wild allele again expresses itself in every cell. (*B*) Spreading effect is seen when more than one gene is shifted to a region of heterochromatin. If the gene farthest away is influenced (facet, in this case), all those intervening up to the heterochromatin will also show the position effect.

pected phenotype may result. Normally, a fly that is w^+/w should have red eyes. But if such a fly has the w^+ allele next to heterochromatin, it may have a variegated eye, one that has patches of red and patches of white. Many such examples demonstrate that dominant alleles from the euchromatin, placed next to heterochromatin, tend to lose their dominant effect. It is as if the heterochromatin were rendering them inactive in some cells, so that the recessive phenotype is expressed. The degree to which the recessive condition appears depends on both the amount of heterochromatin and the particular kind involved. If exposed to enough heterochromatin of a certain kind, a dominant allele may be unable to express itself at all. Several studies in the fruit fly demonstrate that the dominant allele has not been altered or in any way changed in structure when the variegated effect is produced. In several cases, the rearranged dominant allele has been restored to its normal location in the euchromatin. It then acts again as a normal dominant,

illustrating that it was the neighboring heterochromatin that had been responsible for the variegation and not a mutation in the gene.

Not only may one allele at a time show the variegation effect. If a transposed euchromatic segment contains two or more alleles, more than one of them may be influenced (Fig. 10-16B). The allele closest to the heterochromatin would show the greatest effect. The influence of the heterochromatin decreases with its distance from a particular euchromatic gene. If a more distant allele in a group is affected, those between it and the heterochromatin are certain to be affected also. This illustrates the spreading effect, the tendency of more than one allele to be repressed by the presence of heterochromatin. The closer the allele is to the heterochromatin, the greater its chance of being influenced is.

Position effect has been demonstrated in the mouse. A variegated effect and a spreading effect can result from translocations that place X-derived heterochromatin next to autosomal genes found in the euchromatin. Both facultative and constitutive heterochromatin can apparently bring about a position effect.

The assortment of effects that can arise from duplications and from simple changes in gene position underline the complications that must be considered in cytogenetic analysis. It is important to keep in mind the consequences such alterations in chromosome structure may prove to have for development in humans.

Translocation

A well-known chromosome aberration is the translocation, an alteration in which chromosome segments, or even whole arms, are transported to new locations. A simple type of translocation mentioned earlier, commonly called a shift, involves three breakage points. All three may be in one chromosome (Fig. 10-17A). Other simple shifts may involve two breaks in one chromosome and a single break in a nonhomologue (Fig. 10-17B). The segment from Chromosome 1 is then inserted into Chromosome 2. Some consequences of shifts were discussed in the preceding sections.

A shifting of chromosome segments is termed a reciprocal translocation when a mutual exchange occurs between two nonhomologous chromosomes. Reciprocal translocations have been studied extensively in many plant and animal species. Figure 10-18A illustrates a reciprocal translocation in which whole

arms are exchanged. Assume that the alteration occurs during spermatogenesis, so that a sperm with the chromosome alteration fertilizes a normal egg. The result is a translocation heterozygote that, as in the case of the inversion, would be genically balanced and probably normal phenotypically. Like the inversion heterozygote, the individual with the translocated chromosomes would encounter problems during synapsis at first meiotic prophase. The forces of attraction may achieve pairing by the formation of atypical chromosome associations. When whole arm translocations are present, a characteristic cross figure forms that can be easily seen at pachynema in certain favorable material (Fig. 10-18B). All the homologous segments are paired, and crossing over may occur. The chief problem in the translocation heterozygote does not concern crossing over but rather disjunction, the manner in which the chromosomes will separate at anaphase. The association of four chromosomes at first meiotic division is termed a tetravalent, to distinguish it from the normal meiotic bivalent. At diplonema and diakinesis, the cross figure opens out into a circle as the forces of repulsion become operative. By first metaphase, the circle of four will be at the equatorial plate. The manner in which the chromosomes of the tetravalent disjoin determines the genic balance of the resulting gametes. Inspection of Fig. 10-18C shows that if any adjacent chromosomes in the circle move to the same pole, unbalanced gametes arise. These contain extra doses of some genes and lack others entirely. However, if alternate chromo-

FIG. 10-17. Shifts. (*A*) A shift involving three breaks in one chromosome. The genes c and d are switched to a new location in the same chromosome. (*B*) A shift involving breaks in two nonhomologous chromosomes. The two genes c and d are now switched to a new location in a different chromosome.

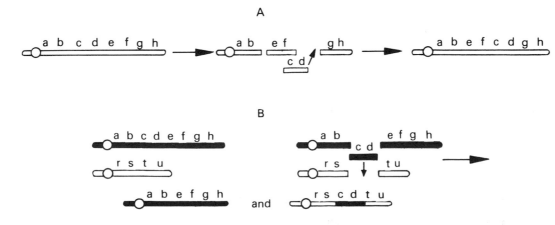

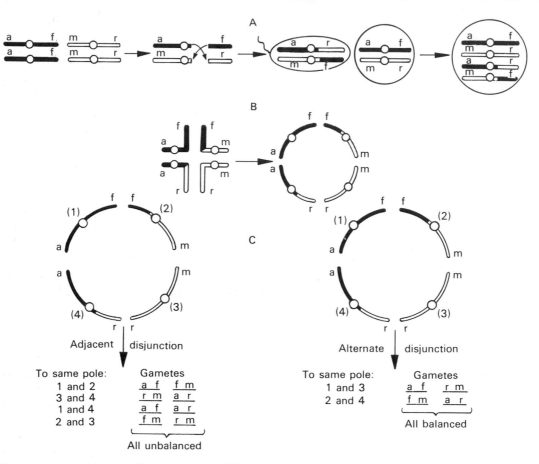

FIG. 10-18. Reciprocal translocation and its consequences. (A) Two pairs of nonhomologous chromosomes engage in a reciprocal translocation. This follows two breaks, one in each of two nonhomologous chromosomes, shown here as occurring next to the centromere. If a sperm carrying the translocation fertilizes a normal egg, the result is a translocation heterozygote. (Only the end genes are labeled in each arm, for sake of clarity.) (B) Since arms have been shifted between nonhomologous chromosomes, the pairing of homologous arms is achieved by formation of an association of four chromosomes, a tetravalent, seen as a cross in early meiosis and then as a circle by late diplonema. (C) Adjacent disjunction (*left*) may occur at anaphase. If Chromosomes 1 and 2, for example, move to the same pole, the gametes will be unbalanced. Some genes will be represented twice (those in arm f) and others will be completely absent. The other adjacent combinations are also seen to be unbalanced. If alternate disjunction occurs (*right*), the gametes will be balanced. Half of them will carry the original arrangement and the other half the translocation.

somes in the circle move to the same pole, balanced gametes are formed. Of those arising from a single meiotic cell, half carry the normal chromosome arrangement, and half have the translocated chromosomes. The end effect, then, of the reciprocal translocation is to produce a reduction in fertility in the structural heterozygote. Usually, the unbalanced gametes will not survive or they will give rise to abnormal offspring.

In some cases, more than two homologous chromosomes may engage in reciprocal translocations. This increases the size of the circle or even the number of circles seen at meiosis, depending on how the arms have been distributed. For example, in the cultivated plant, *Rhoeo discolor*, all the chromosome arms in the complement have been shifted. The result is a multivalent association that includes all 12 chromosomes (Fig. 10-19). At times a chain rather than a circle forms in some meiotic cells. The overall effect of the large number of reciprocal translocations in this plant is to produce a certain number of unbalanced gametes because of irregular chromosome distribution at anaphase.

Decreases in fertility resulting from the presence of reciprocal translocations have been recognized in corn and Drosophila, to mention just two familiar species.

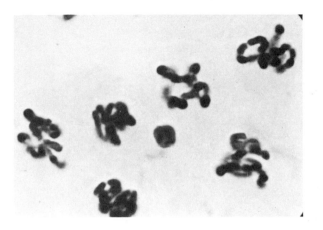

FIG. 10-19. Multivalent associations in *Rhoeo discolor* as a result of translocations.

As in the case of the inversion, reciprocal translocations do not tend to accumulate to any great extent in natural populations, but again there is at least one striking exception to the rule. This is seen in the plant genus Oenothera, the evening primrose.

Importance of translocations in the evening primrose

When meiotic cells of Oenothera are examined, it is found that the 14 chromosomes do not usually form seven pairs, as would be expected. Instead, cytological examination reveals a large circle composed of an end-to-end association of 14 chromosomes, resembling that shown in Fig. 10-19 for the chromosomes of Rhoeo. The work of Cleland clarified the situation and showed that during their evolution, the chromosomes of the evening primroses had undergone a series of translocations. Whole arms of chromosomes were shifted around to such an extent that no one chromosome is completely homologous to any other in the whole chromosome complement. The only way the homologous segments can associate is by forming a large circle (Fig. 10-20).

As we noted earlier, adjacent disjunction leads to unbalanced gametes; so such a large circle might be expected to give rise to an appreciable amount of adjacent disjunction and, consequently, many unbalanced gametes. But this is not the case with Oenothera, in which disjunction is usually alternate. If alternate disjunction always occurs, all of the gametes will be genically balanced. But how can alternate disjunction take place almost exclusively? When the chro-

mosomes of Oenothera are studied, it can be seen that they are all approximately the same size and shape. The arms of any one chromosome are about the same as those of any other. Moreover, the translocations that have taken place have been mainly whole arm translocations in which entire arms, not just small chromosome segments have been exchanged.

The result is that all the chromosomes tend to retain their original size and shape. No matter how the arms are shifted about, the chromosomes all remain quite similar. When such chromosomes separate at anaphase, the forces of segregation are balanced in the circle so that one chromosome is repelled equally by the chromosomes on either side of it. This means that all the chromosomes that alternate with each other stay together, because they always move to the same pole (Fig. 10-20). The alternate chromosomes thus compose a set that tends to remain intact generation after generation. For example, if set 1 comes from an egg and set 2 from the pollen parent, the same sets again are reconstituted after alternate disjunction. Therefore, Oenothera, although it has 14 chromosomes and should consequently have seven linkage groups, behaves as if it has just two chromosomes and one linkage group! A moment's thought tells us that if set 1 is in a sperm nucleus, it could fertilize an egg that also bears set 1 (and similarly for set 2). This would produce plants homozygous for set 1 (or set 2), and we would expect the offspring to have 14 chromosomes that would form seven pairs.

However, Oenothera has another very unusual feature that makes it impossible for any individual to be homozygous for any one kind of chromosome set, because each set carries a specific kind of recessive lethal allele. The lethal of set 1 is different from the lethal of set 2. As Fig. 10-21 shows, this mechanism prevents the development of any individual with two identical sets. Half of the offspring will die because of the homozygous condition of the recessive lethals. Only the hybrids will survive; they contain two different lethals and two different sets. The circle of 14 is thus perpetuated. The situation is just as if two chromosomes instead of 14 are present, because the seven chromosomes of any set tend to remain together as if they were linked.

It might seem that the loss of so many offspring as a result of the lethals would be a disadvantage. However, Oenothera is normally self-pollinated. This ensures the maximal amount of pollination and fertilization that can give rise to a large number of viable

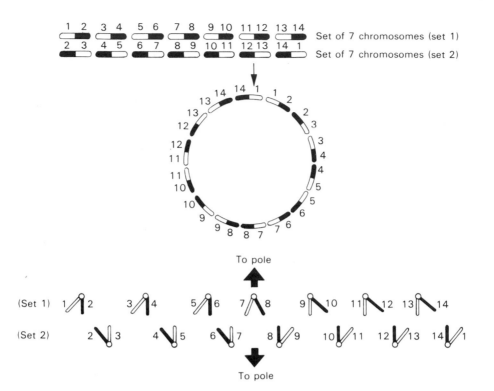

FIG. 10-20. Chromosomes of Oenothera. The arms of all 14 chromosomes have been shifted by translocations so that no one chromosome is completely homologous to any other. Pairing of homologous arms results in a circle of 14 chromosomes. Alternate disjunction always occurs and produces balanced gametes. The same seven chromosomes remain together, constituting a set.

FIG. 10-21. Lethals in Oenothera. Each plant is composed of two sets of seven chromosomes each. The meiotic behavior shown in Fig. 10-20 keeps each set intact. Each carries a different recessive lethal that prevents homozygosity for lethal 1 or 2 and consequently for set 1 or 2. Only heterozygotes survive carrying the two original sets with the lethals.

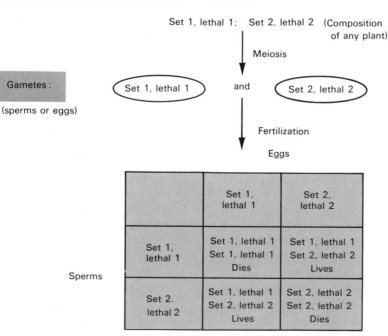

	Eggs	
	Set 1, lethal 1	Set 2, lethal 2
Set 1, lethal 1	Set 1, lethal 1 Set 1, lethal 1 Dies	Set 1, lethal 1 Set 2, lethal 2 Lives
Set 2, lethal 2	Set 1, lethal 1 Set 2, lethal 2 Lives	Set 2, lethal 2 Set 2, lethal 2 Dies

260

seeds. And the plants are highly vigorous because of valuable gene combinations that are held together in each set from one generation to the next. We see here a shining example of two usually disadvantageous features (translocations and lethals) combined in such a way as to produce a highly advantageous mechanism that ensures hybrid vigor. The evening primroses compose a group of plants that is widespread throughout North America, and its success can be attributed mainly to its unusual genetic system.

Oenothera is also of historical interest, because it was this group of plants that led DeVries to formulate his mutation theory. However, the "new" plants that arose suddenly in his gardens were not the consequences of gene mutation as DeVries thought. At the time, DeVries was completely unaware of the presence of both the translocations and the unusual chromosome segregation. He crossed evening primroses from many localities in various combinations. The "new" types or "sports" that arose were actually offspring derived from hybrids in which the circle of 14 did not always form as described previously. And so, these hybrids did not behave as if they had only two chromosomes. The new types were not gene mutants at all but rather recombinants with new gene combinations. The irony is that DeVries focused the attention of geneticists on gene or point mutations which do occur. But he formulated the concept of sudden gene changes on results that were not associated with point mutations at all.

Translocations in humans

Translocations are definitely known to occur in humans, and their consequences are of great concern to the medical geneticist. As will be discussed in more detail later in this chapter, Down syndrome (Fig. 10-22) typically results from the presence of one extra chromosome, Chromosome 21, because of nondisjunction at meiosis. Consequently, an afflicted person has 47 chromosomes. However, this is not the only picture in victims of the disorder, for some Down individuals have the normal chromosome number, 46. Karyotype analyses reveal that a structural chromosome change has taken place and has altered the appearance of one of the chromosomes. The chromosome appears longer than usual as a result of a translocation. Most translocations associated with Down syndrome involve the translocation of 21q (the long arm of chromosome 21) to one of the three members of the D group. Of the D translocations 60% involve

Chromosome 14; the rest are associated with Chromosome 13 or 15. Let us examine a case in which Chromosome 15 happens to be the chromosome involved.

Two normal Number 21 chromosomes are present, but a change is evident in pair 15 in this case. One of the members is normal, but the other is conspicuously larger and does not match its homologue. Therefore, the normal 15 remains unmatched. The change here is explained as the result of a translocation between Chromosomes 21 and 15 (Fig. 10-23A). It can be seen from the figure that the new chromosome at the left contains most of the material from the two original chromosomes, 21 and 15. The tiny chromosome that forms may get lost. This kind of structural chromosome alteration that occurs in Down syndrome is an example of a *Robertsonian translocation*. In this type of translocation, the two nonhomologous chromosomes involved are acrocentrics (as is the case for members of the D and G groups) and the two chromosome breaks occur very close to the centromere regions. The two short arms would be largely heterochromatin, because this type of chromatin surrounds the centromere. The small chromosome that arises would therefore not carry any genes essential to life, so its loss would not prove lethal. The two heterochromatic arms could even be lost without actually joining to form the tiny chromosome at all.

Let us say that an individual inherits a 15/21 translocation through a sperm (Fig. 10-23B). He or she would receive from the female parent a normal 15 and a normal 21. However, the person would have a chromosome total of 45 instead of the normal 46 and can be easily identified by karyotype analysis. The reduced number is present because this individual does not possess a pair of 21 and a pair of 15. Instead, there is a normal 15, a normal 21, and the altered chromosome, 15/21. Since the essential genetic material of Chromosomes 15 and 21 is present, the person would exhibit no ill effects and might be considered completely normal until a karyotype analysis revealed the aberrant chromosome number.

The meiosis in such a person would be quite irregular, because the translocated chromosome, 15/21, can synapse with both the normal 21 and the normal 15 (Fig. 10-23C). A trivalent association of three chromosomes may be seen at meiosis. The segregatation of these three will not be regular and can result in the formation of gametes of different constitutions. Of these, some will be normal, containing one Chromosome 21 and one Chromosome 15. But an appre-

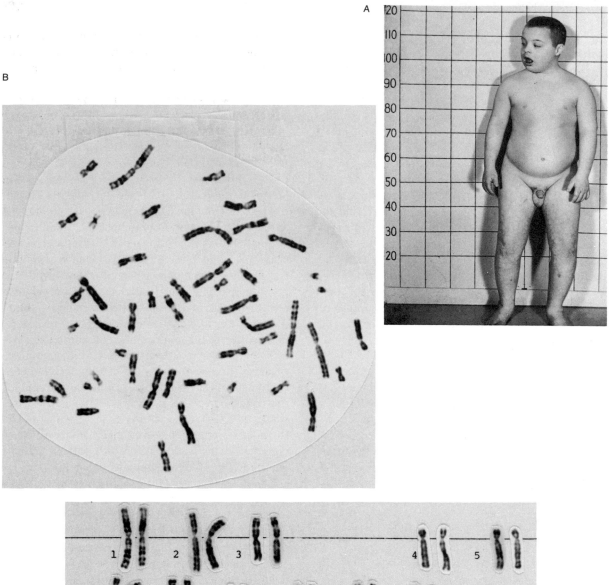

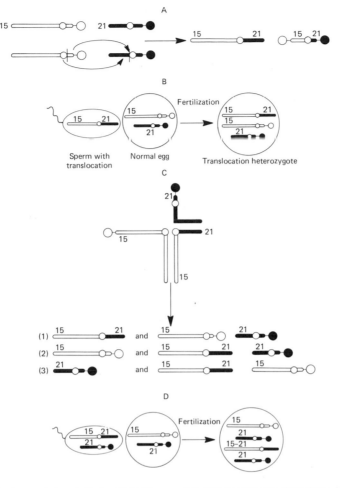

FIG. 10-23. Translocation in the human between Chromosome 21 and a member of the D or G group. In this example. Chromosome 15 of the D group is depicted. (*A*) Arms are exchanged between Chromosomes 15 and 21. The small chromosome that may arise would contain little genetic material. The larger chromosome that arises contains most of the original Chromosomes 15 and 21. (*B*) A sperm carrying the translocated chromosome fertilizes a normal egg. The zygote that results will be essentially balanced, because two Chromosomes 15 and two Chromosomes 21 are, in effect, present. The chromosome number is one fewer than normal because 15 and 21 are joined as a result of the translocation. (*C*) A trivalent forms at meiosis in the translocation heterozygote. This brings about pairing of homologous regions, but it does not guarantee balanced segrega-

tion. Because any one of the three chromosomes may move by itself to one pole, several possibilities follow. Segregation 1 will produce gametes that will pass down the translocation; it will also produce gametes that are normal. Segregation 2 will produce a gamete lacking Chromosome 21 (a lethal condition) and a gamete that can result in Down syndrome, as shown in *D*. The third segregation produces gametes that are very unbalanced. One type lacks Chromosome 15. The other has it present twice. The unbalance proves lethal. (*D*) A sperm formed in a translocation heterozygote as a result of segregation of type 2 (*C*) fertilizes a normal egg. The zygote has three doses of Chromosome 21 and will develop into a victim of Down syndrome. The chromosome number, however, will be normal.

ciable amount will contain number 21 and the translocated chromosome, 15/21. When a gamete of this sort combines with a normal one, Chromosome 21 is in effect present three times in the zygote (Fig. 10-23D). The chromosome number of such an individual would be normal (46), but he or she would be indistinguishable from those sufferers of Down syndrome who have 47 chromosomes. This is so because both

classes of victims have most of Chromosome 21 in three doses. Down syndrome arises with an unfortunate frequency (approximately 1 in 500 newborn); it therefore accounts for a large number of mentally defective children. Most of them possess 47 chromosomes and have arisen as a result of nondisjunction in the female parent, but a percentage arises, as discussed here, through a parent who is a carrier of a translocation. Identification of translocation carriers by karyotype analysis is important where the occurrence of Down syndrome appears to run in a family line, since the heterozygous carrier of the translocation can continue both to transmit gametes that may give rise to Down children and to transmit the carrier condition to other children who may do the same (Fig. 10-23C). Identification and proper genetic counseling of carriers may help reduce the incidence of afflicted children. Those cases of Down syndrome resulting from the presence of one extra Chromosome 21 due to nondisjunction rather than a translocation involving a D group and G group chromosome (Fig. 10-22) are not familial, because the aberration arises in meiotic cells of the immediate parent of the Down individual, who almost always fails to reproduce.

A variety of translocations involving chromosomes from the different human chromosome groups is known to occur. These are all associated with mental retardation and a high probability of death shortly after birth and account for a significant number of spontaneous abortions. A most significant fact is the association of translocations with certain cancers, a topic discussed in the following section. As noted in Chapter 9, studies of cells with translocations are especially valuable in the assignment of a gene to a localized region of a chromosome.

Chromosome aberrations and cancer

For many years, only one cancer, chronic myeloid leukemia, had been found to be associated with a consistent genetic defect. Leukocytes of affected individuals typically contain a Chromosome 22 in which more than half of the long arm is missing. This chromosome with the missing portion was designated the Philadelphia chromosome. Although it was considered a simple deletion at first, the cytological banding technique revealed that the Philadelphia chromosome actually represents the effects of a translocation in which the missing portion of Chromosome 22 has been inserted into Chromosome 9.

About 10% of chronic myeloid leukemia patients lack the translocation, but according to one school of thought, these persons have a different form of the disease. There is an interesting side light to the story of the Philadelphia chromosome. It was uncertain at first whether the Philadelphia chromosome was actually Chromosome 21 or Chromosome 22 with a deletion. The consensus was that the Philadelphia chromosome is the same one as that associated with

Down syndrome, the chromosome that had been designated 21 (see the last section). The reasoning was that a victim of Down syndrome is more likely to develop leukemia than the average individual. However, improvements in straining procedures resolved differences between the two members of the G group of chromosomes and have positively shown that the Philadelphia chromosome is not the one associated with Down syndrome but is the other chromosome in the G group—the chromosome now designated 22.

Since 1980, great improvements have been made in the culturing of tumor cells, especially in methods of chromosome staining and banding. J. J. Yunis has devised chromosome banding procedures that now make it possible to visualize a greater number of bands and to detect precisely chromosome aberrations that previously went unnoticed. This high-resolution staining procedure has revealed the fact that cells composing most malignant tumors have characteristic chromosome alterations. In most leukemias and lymphomas, as well as in some carcinomas, a reciprocal translocation is present in the malignant cells. In chronic myeloid leukemia (CML), Chromosomes 22 and 9 appear to break consistently at specific subbands. Other consistent aberrations have been found to be associated with other malignancies. A few examples include a translocation between Chromosome 8 and 14 in Burkitt's lymphoma, a deletion of certain bands in the short arm of Chromosome 1 in a type of neuroblastoma, and a deletion of certain bands in the small arm of Chromosome 3 in a type of lung carcinoma. In general, those solid tumors (as opposed to leukemias and lymphomas) that have been analyzed cytologically show the loss of a specific chromosome band or segment. In addition to translocations and deletions, some malignant tumors are associated with a given inversion; others have a gain or loss of a specific chromosome.

Although a given malignancy may be associated consistently with a particular chromosome aberration, such as the association of CML with the Chromosome 22 to Chromosome 9 translocation, this does not mean that the alteration is necessarily specific for that one disorder. For example, the defect in CML is also found in acute lymphocytic leukemia and acute myeloid leukemia. The same situation pertains to various other malignancies in which several malignancies may share the same particular chromosome aberration. On the other hand, in certain leukemias the specific chromosome defect so far appears to be unique to that one disorder and is not shared by others.

Characteristic chromosome aberrations are present early in the development of a malignancy; however, further chromosome aberrations appear as the disease progresses. These secondary changes are nonrandom. For example, among the changes in the white blood cells of persons with CML are the gain of extra Philadelphia chromosomes and the gain of extra Chromosomes 8 or 17. This changing cytological picture of the white blood cells is associated with changes in the leukemia from chronic to acute.

Such secondary characteristic chromosome abnormalities associated with many malignancies may confer an advantage on the tumor cells, enabling them to proliferate more rapidly. Many of these additional changes involve duplications of certain chromosomes or certain chromosome segments. Perhaps these duplications contribute to the ability of the malignant cells to spread and invade other tissues with the ultimate death of the individual.

Since the cells of most cancers in the human exhibit some kind of chromosome aberration, it would seem that a rearrangement of the genetic material is a common step that leads to the development of a malignancy. A rearrangement of the chromosome may remove a gene from a particular genetic region that exerts a controlling influence over it, regulating its expression. In Chapter 17, we return to this topic to learn that certain rearrangements place genes that are normally inactive in certain cells next to genetic regions that activate them and bring them to expression. We can well appreciate that chromosome rearrangements as well as gene mutations can bring about profound changes in the expression of the genetic material.

Chromosome alteration and evolution

There is no question that structural changes have altered the size, shape, and number of chromosomes during the evolution of many plant and animal groups. In the genus Drosophila, it is quite evident that paracentric inversions and whole-arm translocations have been involved in the differentiation of species within

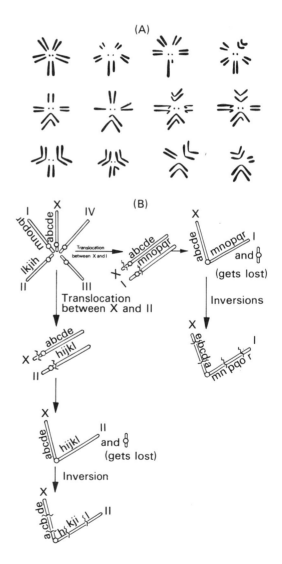

FIG. 10-24. (A) Chromosome complements of some species of Drosophila. The karyotypes are of males, and the X and Y chromosomes are the two lower ones in each case. The different chromosomes (metacentric, acrocentric) appear as arrangements of rods, Vs, and Js. This can be explained on the basis of shifting of chromosome arms and segments during evolution of the species by translocations and inversions. The species shown are (1) *Drosophila virilis*, (2) *Drosophila funebris*, (3) *Drosophila repleta*, (4) *Drosophila monata*, (5) *Drosophila pseudoobscura*, (6) *Drosophila miranda*, (7) *Drosophila azteca*, (8) *Drosophila affinis*, (9) *Drosophila putrida*, (10) *Drosophila melanogaster*, (11) *Drosophila willistona*, and (12) *Drosophila prosaltans*. (B) Alterations in karyotype. In Drosophila, chromosome arms may have been shifted about in a fashion similar to this. From an ancestral type having five acrocentric chromosomes, metacentrics can arise. The X is shown here undergoing a reciprocal translocation with Chromosome I in one case and with Chromosome II in another. The tiny chromosome may be dispensable, because it contains large amounts of heterochromomatin, which is found around the centromere. Eventually two species may be derived in which certain genes are sex linked in one but not in the other. However, a group of genes would be similar and sex linked in both species. These genes would be found in the same arm. This arm would correspond to the single arm in a species having an X with just one prominent arm, because this arm traces back to the same ancestral X. The order of genes in the arms may differ because of different inversions as the species evolved separately.

the genus. When chromosomes of the several fruit fly species are compared, one finds a very striking picture (Fig. 10-24A). The chromosomes are found mainly as metacentrics and acrocentrics with one long arm and one tiny one. The chromosome number may differ from one species to the next, but the number of long arms tends to remain the same. The X chromosome may be acrocentric in some (*D. melanogaster* and *D. virilis*, for example) or it may be metacentric (*D. pseudoobscura, D. miranda*). When chromosome maps are made of the different species, the same types of loci tend to be found linked together in an arm. (Pink and ebony are always in the same arm; so are facet and white.) This implies that the arms in the several species are basically the same and have been derived from an ancestral type. In those species with metacentric chromosomes, the two arms associated to form the large chromosome may be different. For example, the X chromosome of *Drosophila americana* is composed of two long arms, but one of these is very different from one of the arms of the X of *D. pseudoobscura*. The other arm of the X corresponds in both species. It also corresponds to the single long arm that constitutes the X in *D. melanogaster*. The pattern seems to be that genes tend to stay together in an arm but whole arms have been shifted about by translocations. The order of the genes, however, may not be the same in a corresponding arm of different species. The evidence suggests that inversions within corresponding arms alter the gene order (Fig. 10-24B).

Pericentric inversions have undoubtedly played a role in the shifting of the centromere, with a consequent change in the chromosome morphology. Comparisons of banding patterns of the salivary gland chromosomes in closely related Drosophila species indicate that the acrocentric chromosomes with one long and one short arm have been derived from either metacentric chromosomes or acrocentrics with one long arm and one tiny one. Such acrocentrics can be readily explained as the result of inversions in which the two breaks are not equidistant from the centromere (refer to Fig. 10-4).

In Drosophila, it is possible to cross some of the species and to study the association of the arms in the salivary gland chromosomes of the hybrid. The observations support the hypothesis that arms are homologous in the various species but translocations have changed their attachments and inversions have rearranged the gene sequences.

The evolution of human chromosomes can now be studied in a more precise way. The techniques of chromosome banding are exceptionally valuable in revealing the details of chromosome morphology (see Chap. 2), thus permitting comparisons of chromosomes between the human and other species. The number, size, shape, and banding pattern of the chromosomes of the human and some other species may be compared. In this way, karyotype analyses can give some insight into possible relationships among different groups and any chromosome changes that have arisen as they diverged. For example, the chimpanzee probably has the closest common ancestor with the human, as is strongly suggested by similarities in their DNA and proteins. The chromosome number of the chimpanzee is 48; one more pair of autosomes is present than in the human. However, this difference in number can be attributed to a translocation involving two acrocentric chromosomes. In the translocation, the two larger arms of two nonhomologous chromosomes would have become associated, much as seen in Fig. 10-23A. The banding pattern suggests that human Chromosome 2 has arisen in this way from two nonhomologous ancestral chromosomes, each with one very short arm. Many other differences between the chromosome complements of the human and the chimpanzee can be attributed to pericentric and paracentric inversions.

Duplications are considered to have played a very important role in the evolution of living things, since they increase the number of genes in the chromosome complement. Once extra amounts of a genetic region are present, the stage is set for the originally identical genes to diverge. This is true because each of the duplicated regions can undergo its own subsequent course of independent mutation. As different mutations, and hence different gene changes, continue to arise and as the force of natural selection operates, what were once identical gene forms may eventually be recognized as two very different genes, each controlling the formation of a somewhat different product. There is good evidence for this in the fruit fly, mouse, and human. In these species, there are various cases of very close linkage among genes that determine the amino acid composition of similar protein chains. If the two protein chains controlled by two separate genes are much alike in amino acid sequence, this means that the two genes involved must also be quite similar, since their information contents are similar.

Evidence is suggestive that the genes encoded for the various globin chains of hemoglobin are all descended from a common ancestral gene and that a series of duplications, followed by mutation, has resulted in the present families of genes associated with the globin polypeptides (Chap. 19).

Variation in chromosome number

In addition to the various aberrations that change the structure of the individual chromosomes, other anomalies alter the actual chromosome number of an organism. We have seen that normal development depends on genic balance. Therefore, a departure from that chromosome number that is characteristic of a species can be expected to upset this balance and lead to adverse effects. Any variation from the normal number of chromosomes is termed heteroploidy. A heteroploid organism can differ from a typical member of its species by one or more entire sets of chromosomes. This condition is called euploidy. For example, if a diploid number of 10 chromosomes, composed of two haploid sets of five, is the normal number for a species, then the presence of an entire extra set would result in a euploid with 15 chromosomes. Euploidy is discussed more fully in the latter part of this chapter.

In contrast to a euploid, an individual may have one or a few extra chromosomes or perhaps even one or a few chromosomes less than normal. This situation is termed aneuploidy, a condition in which the chromosome number differs from the normal diploid by less than an entire set (2n − 1; 2n + 1; 2n − 2, etc.).

Aneuploidy in humans

We would expect aneuploidy to lead to severe genic unbalance, and we have already encountered several examples of this in humans. Down syndrome (see Fig. 10-22), resulting from an extra Chromosome 21, is a good example that illustrates the serious consequences of the unbalance. Individuals with one extra chromosome are called trisomics, meaning that their body cells contain three doses of a particular chromosome. Trisomics, such as the child with Down syndrome, may result from nondisjunction at meiosis in one of the parents (Fig. 10-25). If two chromosomes fail to separate at anaphase I of meiosis, half of the gametes will contain an extra chromosome and half will have one chromosome fewer than normal. A gamete of the former type, when it combines with a normal gamete, produces a zygote that is trisomic. The zygote receiving only one member of a chromosome pair is designated a monosomic and will not survive.

The extra chromosome present in cases of Down syndrome has been shown by fluorescent staining techniques to be the smaller of the two autosomes in Group G. Since the chromosomes in any group are numbered according to decreasing size, this means that the Down chromosome is technically number 22, not 21 as was thought for so many years. However, inasmuch as Down syndrome has been designated trisomy 21 for many years and the terminology has become well established, geneticists have agreed to

FIG. 10-25. Nondisjunction of an autosome at meiosis. Two homologous chromosomes (assume Chromosome 21 in the human) fail to separate at first anaphase. They move to the same pole and enter the same nucleus. A gamete derived from this will contain two Chromosomes 21, instead of just the normal one. A normal gamete from the other parent will contribute one Chromosome 21. The resulting zygote is thus trisomic, carrying three instead of two. A gamete derived from the other nucleus at meiosis (*bottom* in figure) will lack Chromosome 21 entirely. After fertilization, the zygote will be monosomic, having just a single Chromosome 21, which was contributed by the balanced gamete.

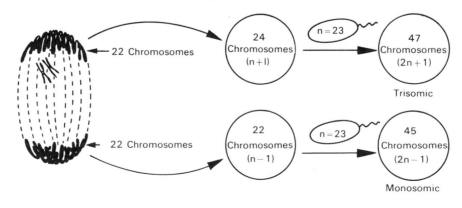

keep the designation 21 for the Down chromosome to avoid confusion. Although the other G group chromosome is actually the larger of the two, it has arbitrarily been designated 22.

The trisomy resulting from nondisjunction, which produces Down syndrome, is definitely correlated with the age of the mother. Women over 40 years of age have a much higher probability of bearing a Down child than younger mothers. The age of the father does not seem to be a factor. It should be noted that the incidence of those cases of Down syndrome that result from translocations is not correlated with maternal age. There is reason to believe that chromosomes of oocytes are much more subject to nondisjunction the longer they remain in first meiotic prophase. It will be recalled that the first meiotic division in the human oocyte is not completed until the follicle matures. This means that an oocyte has been lying in a sustained meiotic prophase for years, perhaps decades, by the time meiosis is resumed. The spermatogonia, on the other hand, provide a supply of spermatocytes that are continually undergoing spermatogenesis throughout adult life.

It is interesting to note that trisomy 21 occurs in the chimpanzee and that the condition results in a syndrome closely resembling Down syndrome in the human.

Trisomy 22 has been definitely recognized in several dozen persons through chromosome banding techniques that permit accurate distinctions to be made between Chromosomes 21 and 22. Trisomy 22 is associated with mental and growth retardation, as well as with cleft palate and an assortment of other disorders. The survival rate of the afflicted individuals appears to be high.

A few other human trisomies are known that also permit survival beyond birth. However, infants trisomic for Chromosome 8, 9, 13, or 18 suffer severe abnormalities of the skeleton and brain in addition to a variety of other disturbances. The unbalance resulting from the excess genetic material is so severe that death usually ensues in early childhood.

The autosomal trisomies we have mentioned are those that permit survival beyond birth. However, tissue from abortuses show that nondisjunction may involve any chromosome. Most trisomies are apparently lethal to the embryo or fetus. Even those mentioned that do manage to survive to birth have an increased risk of being spontaneously aborted during pregnancy. The lethal risk for trisomy 22 is evidently

greater than that for trisomy 21, which is more likely to survive the period of fetal development and is thus more common among trisomic individuals than trisomy 22. A maternal age effect is also associated with the other autosomal trisomies in addition to trisomy 21.

Those aneuploid persons with an XXXY or XXXXY condition (see Chap. 5) are trisomic and tetrasomic, respectively, for the X chromosome. Although they have the abnormalities of Klinefelter syndrome, they nevertheless survive and are in no way as defective as those individuals trisomic for one of the autosomes. This ability to tolerate extra doses of the X may be possible as a consequence of some deactivation of those Xs in excess of one. Remember that one, two, and three Barr bodies are present, respectively, in persons who are XXY, XXXY, and XXXXY. A maternal age effect is also associated with Klinefelter syndrome.

Monosomic cells lacking an autosome are produced as a result of nondisjunction. However, such cells in the human do not survive. There is no clear-cut example in the human of an individual who is monosomic for an autosome. In humans, the ability to tolerate the addition of an extra autosome is far greater than the ability to endure the effects of a missing one.

However, individuals monosomic with respect to the sex chromosomes are well known. These persons with one X chromosome and no other sex chromosome (XO chromosome constitution) show the assortment of characteristics that describe Turner syndrome (see Fig. 5-14). Again, this could arise from nondisjunction during parental gamete formation. If the two X chromosomes of the female move to the same pole during oogenesis, two sorts of eggs could result: XX + 1 set of autosomes and 0 + 1 set of autosomes. Fertilized by a Y-bearing sperm, the former could be responsible for an XXY Klinefelter male. The latter, fertilized by an X-bearing sperm, could produce a victim of Turner syndrome. Turner births are less common than Klinefelter.

We have noted that monosomy in humans causes a greater degree of unbalance than does the presence of an extra chromosome. Only those monosomics who have a missing sex chromosome (Turner females) can apparently survive to birth, but a very large proportion of such XO embryos evidently do not reach full term. The XO chromosome abnormality is actually the most common chromosome aberration among abortuses. The number of XO zygotes that manage to develop and survive to birth is low. Apparently, 99%

abort as a result of severe prenatal disturbances. No age effect of the mother has been associated with Turner syndrome.

Aneuploidy in plant species

Aneuploidy has been widely studied in plants that seem to tolerate aneuploid conditions much more readily than do animals. A variety of plant species has provided valuable information concerning the genetic and cytological consequences of aneuploidy, as well as its phenotypic effects. One of the most thoroughly investigated species is the thorn apple or Jimson weed, *Datura stramonium,* a very poisonous plant whose leaves are the source of certain narcotic drugs, among them atropine. Blakeslee undertook work with Datura in 1913, and over a period of years, he and his colleagues were able to isolate a variety of different polysomics. Datura has a haploid chromosome num-

ber of 12, which means that 12 kinds of trisomics are possible, one kind for each of the 12 chromosomes in a haploid set. Thus, any trisomic plant would have 25 chromosomes, two haploid sets plus one extra chromosome. This trisomic can be compared phenotypically with others that carry different extra chromosomes. All 12 possibilities were obtained. Comparisons among them showed the effects of the genic unbalance on such organs as the leaf and the seed capsule. Each of the different extra chromosomes was found to produce characteristic phenotypic departures from the normal (Fig. 10-26).

When the trisomics were examined cytologically, their meiotic cells showed the presence of the extra chromosome. Since a trisomic has three doses of a certain chromosome, meiotic irregularities are to be expected. At synapsis, the three homologues may form an association of three chromosomes, a trivalent. When trivalents (or any multivalents for that matter) are formed, all three chromosomes are never associated throughout their lengths. Instead, the pairing is always two-by-two at certain points, so that a chain of three may be seen (Fig. 10-27). However, the three may not necessarily enter into trivalent formation; a bivalent and an unpaired univalent may result. Because the trisomics in Datura proved to be fertile, they could be crossed in various combinations, either with other trisomics or with normal diploids. This permitted the derivation of other types, such as tetrasomics ($2n + 2$), which could be studied further for the effects of two extra identical chromosomes on genic balance. Crosses between trisomics also enabled Blakeslee and his associates to study the inheritance patterns as well as the modifications of typical Mendelian ratios resulting from the segregation of three homologous loci instead of the normal two. As might be suspected, there is a greater chance in trisomics for the recessive allele to be masked (one dose of wild is often dominant over two recessives). The usual 3:1 ratio is therefore not obtained but is replaced by other ratios in which the homozygous recessive type is less common.

In addition to these polysomics, Blakeslee and Belling obtained other plants whose extra chromosomes had undergone translocations and other kinds of structural alterations. All of this enabled Blakeslee to observe more closely the consequences of extra doses of specific genes on the phenotype. The observations left no doubt that adding extra genes to the diploid resulted in plants whose phenotypes departed in some way from the typical. Increasing the dosage of one

FIG. 10-26. Capsule types in Datura. The capsule of the normal plant with 24 chromosomes differs from the various trisomics. Each trisomic has 25 chromosomes, but the extra one in each represents a different chromosome. The extra chromosome in each case is designated here as A, B, C, and so on. (Redrawn with permission from J. Hered. 25:86–108, 1934.)

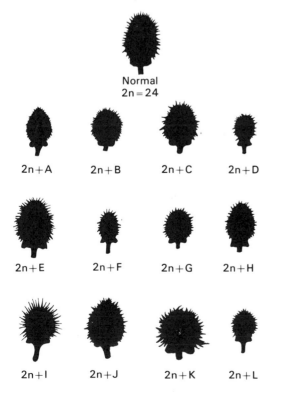

Normal
2n = 24

2n + A 2n + B 2n + C 2n + D

2n + E 2n + F 2n + G 2n + H

2n + I 2n + J 2n + K 2n + L

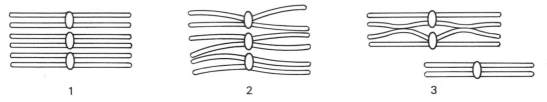

FIG. 10-27. Multivalent associations. When three (or more) homologous chromosomes are present, the meiotic picture is never one in which all three associate throughout their lengths, as shown in *1*. Rather, the pairing is always two by two in a trivalent (*2*). However, not all cells will contain a trivalent, because the three homologues may become distributed as a bivalent and unpaired univalent (*3*).

specific kind of gene increased the phenotypic change in a certain direction. The extensive work with Datura has illustrated the important role of genic balance in the development of the normal characteristics of an organism.

Several plant groups are known in which a range of chromosome numbers exists within a genus or species. One of the most extensive aneuploid series occurs in *Claytonia virginica,* the common spring beauty, a single species in which 20 different diploid chromosome numbers have been reported ranging from 12 to 200. This species is quite unusual in its apparent ability to tolerate fluctuations in chromosome number with no loss of viability or fertility.

Autopolyploidy: its origin and effects

Instead of containing just one or two extra chromosomes, some euploid organisms possess entire extra sets in addition to the typical diploid number. These individuals are called polyploids. Polyploidy is a very characteristic feature of flowering plants and has played an important role in the evolution of many plant groups, including a number of our valuable crop plants, such as wheat and cotton. When we examine a list of the chromosome numbers of species within a plant genus, we often find that the numbers are multiples of a basic one. For example, in different wheats, $2n = 14, 28, 42$ (multiples of 7); in Chrysanthemum, $2n = 18, 36, 54, 72, 90$ (multiples of 9). We can explain these observations on the basis of polyploidy; the groups with the higher numbers have been derived from those with lower numbers through the multiplication of whole chromosome sets.

Before explaining how this multiplication can come about, let us first examine one of the main types of polyploidy. This is autopolyploidy, a condition in which the extra chromosome sets have originated from within the species. A plant species with a diploid number of 10 would have two haploid sets: AA. If one or two

extra sets are present we would have an autotriploid (AAA) and an autotetraploid (AAAA). The origin of the extra sets could have come about through some failure at meiosis. Accidental disruption of the spindle might suppress anaphase movement, giving rise to unreduced gametes (AA). Union of such a gamete with a normal one (A) would produce the autotriploid (AAA; Fig. 10-28). Fusion of two diploid gametes produces an autotetraploid. The accident responsible for autopolyploidy could occur in somatic tissue as well. Nucei of two body cells might fuse; or anaphase of mitosis could fail. The result would be a tetraploid cell (AAAA). If this somatic cell goes on to divide, it can give rise to an entire plant that is tetraploid. Any flowers on a tetraploid branch would produce unreduced gametes (AA). By self-fertilization, tetraploid offspring would result. Although tetraploids arise spontaneously, they may also be induced artificially by a variety of treatments. The most extensively used agent is the drug colchicine. This is a plant alkaloid derived from the European autumn crocus and is used medicinally to alleviate the pain of gout. The primary action of this drug is on the spindle apparatus. Colchicine upsets the organization of the microtubules comprising the spindle and so prevents anaphase movement. All of the chromosomes become incorporated into one nucleus, and a tetraploid cell ($4n$) is formed. Colchicine does not interfere with the duplication of the chromosomes, and so its presence during a series of chromosome replications can lead to high levels of ploidy. Solutions of the drug may be sprayed on buds; seeds or roots may be dipped or soaked for several hours. Because the drug is very toxic, concentration and exposure time must be rigidly controlled.

Polyploids are often produced artificially, because a polyploid at times may encompass some features that make it more desirable than the diploid from which it is derived. Some polyploid cereals, potatoes, and fruits possess desirable commercial qualities. Some

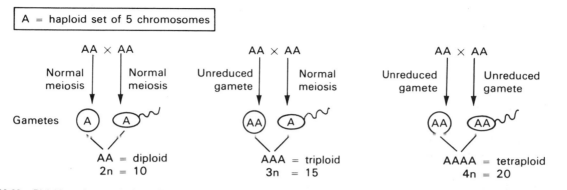

A = haploid set of 5 chromosomes

AA × AA

Normal meiosis | Normal meiosis

Gametes (A) (A)

AA = diploid
2n = 10

AA × AA

Unreduced gamete | Normal meiosis

(AA) (A)

AAA = triploid
3n = 15

AA × AA

Unreduced gamete | Unreduced gamete

(AA) (AA)

AAAA = tetraploid
4n = 20

FIG. 10-28. Diploidy and autopolyploidy. The extra set or sets of chromosomes in an autopolyploid are derived from within the one race or group, so that more than two completely homologous sets are present.

polyploids are larger and more vigorous than the diploid. This has unfortunately given rise to the impression that all polyploids are large, or "gigas," types. It is true that polyploid cells tend to be larger than diploids, but this is not always so. Even the presence of larger cells does not ensure larger plants, as the cell number may be decreased in the polyploid. One generalization that can be made is that the polyploid tends to grow more slowly than the diploid. Consequently, it may flower later and over a longer period of time. Another feature of polyploids is some reduction in fertility compared with the diploid. Some autopolyploids are very sterile, others quite fertile. The safest generalization is that the effects of the ploidy depend on the particular species involved. Genic interaction and the ability to tolerate extra sets of chromosomes may be quite different in different species.

Cytologically, most autopolyploids do carry some meiotic irregularities because of the presence of extra homologous chromosomes. This is well demonstrated by the triploid, which is usually highly sterile. Figure 10-29 shows why this is so. As was the case in the trisomic (see Fig. 10-27), any three homologous chromosomes can enter into a trivalent association, or they can form a bivalent and a univalent. It is even possible that no pairing will be established and that three univalents will result at first meiotic prophase. So in triploid meiotic cells, one may encounter trivalents, bivalents, and univalents. It should be obvious that there is no mechanism to ensure an anaphase separation in which two chromosomes of each kind will move to one pole and one of each kind to the other pole. Such a separation would produce balanced diploid and balanced haploid gametes. However, random segregation of the members of each group of

FIG. 10-29. Irregular segregation in a triploid. Since three of each kind of chromosome are present, trivalents and univalents may occur at meiosis, and segregation will tend to be irregular. Balanced gametes would contain two of each kind of chromosome or one of each kind. Most cells, however, will contain unbalanced combinations. For example, the gamete at the left has one extra long chromosome. The gamete at the right has two extra chromosomes, one of medium size and a small one; if it possessed another long one, it would be a balanced diploid gamete.

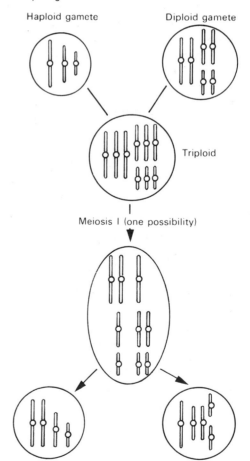

Haploid gamete   Diploid gamete

Triploid

Meiosis I (one possibility)

three homologues usually occurs, so that only a few gametes have a balanced combination of chromosomes. This, in turn, means that the gene dosage is unbalanced in most of the cells after meiosis. Consequently, the pollen of triploids is often very sterile, and seed set is very low. The same sort of meiotic irregularity accounts for the reduced fertility characteristic of polyploids with high but odd chromosome numbers ($5n$, $7n$, etc.). This sterility has commercial value in plants such as the watermelon, where the almost seedless fruit is more desirable than that of the fertile diploid. To maintain sterile triploids, the plant breeder must propagate them asexually by means of leaf or root cuttings; seeds that will grow into triploids can also be obtained by crossing a tetraploid with a diploid.

In the autotetraploid, there is a greater chance than in the triploid for the formation of balanced gametes. Since four chromosomes of each kind are present, multivalents composed of an association of four are seen (tetravalents). Trivalents and univalents are also possible. But there is a higher probability of bivalent formation in the tetraploid (or any polyploid with an even chromosome number). Therefore, anaphase segregation results in a much larger proportion of cells that have balanced gene combinations. In certain species, such as the Jimson weed, Datura, the autotetraploids suffer little decrease in fertility.

Allopolyploidy and plant speciation

Although an autopolyploid may exhibit some phenotypic differences from the diploid, autopolyploidy does not contribute any new genetic information to a species, nor does it bring about any new combinations of alleles. All the alleles of the autopolyploid originate from within the species or the race; any different phenotypic effects are primarily a consequence of the interaction of extra gene doses. The second main type of polyploidy, however, does make possible new combinations of alleles from different sources. This is allopolyploidy, a polyploid condition in which the chromosome sets are derived from different races or species. For an allopolyploid to arise, hybridization must occur between two genetically different groups. One criterion that distinguishes two groups as separate species is their inability to cross and produce fertile offspring. The horse and donkey are certainly separate species. The horse has 64 chromosomes ($n = 32$), whereas the donkey has 62 ($n = 31$). Although they can be readily crossed, their offspring,

the mule ($2n = 63$), is sterile as a result of incompatibility between horse and donkey chromosomes, which fail to pair properly at meiotic prophase. This leads to genic unbalance and defective gametes. That two distinct groups can cross at all is evidence of a relationship between them, as would be expected from an understanding of evolutionary processes. In the following discussion of allopolyploidy, the two groups involved must still share sufficient genetic similarities to permit some gene exchange to take place. Some may be true species in every sense of the word, so that production of offspring between them on the diploid level never or rarely occurs. Others may not be as completely isolated genetically, but may actually be subspecies. In other words, no sharp distinction exists between autopolyploidy and allopolyploidy, because the distinction between species is frequently blurred.

One of the best examples of allopolyploidy is the first to be reported, the experimental production of a hybrid between radish and cabbage. Although both species are members of the mustard family, they are certainly genetically isolated and differ enough to be placed in separate genera by taxonomists. They would rarely, if ever, produce offspring under natural conditions. The radish, Raphanus, and the cabbage, Brassica, both contain a diploid chromosome number of 18. In Fig. 10-30, a haploid set of radish and a haploid set of cabbage are represented by 9R and 9B, respectively. The Russian cytologist Karpechenko succeeded in producing an F_1 between them. This F_1 hybrid was a diploid and contained 9R and 9B chromosomes. It was also quite unfertile because of meiotic irregularities. The two parental species have been reproductively isolated long enough to allow the accumulation of a significant number of changes between their chromosome complements. These changes have produced incompatibilities, which prevent synapsis between a set of R (radish) and a set of B (cabbage) chromosomes. Since no bivalents form, irregular chromosome segregation leads to unbalanced combinations and nonfunctional gametes. Among the abnormal meiotic segregations would be some in which all of the chromosomes would become incorporated into a single cell. These would give rise to diploid gametes. Such occurrences are rare, but nevertheless, two such diploid gametes were produced by the F_1 hybrid, and these combined to form a plant with four sets of chromosomes, 2R and 2B. Since this individual has twice the chromosome number of either original parent, it is a tetraploid. It cannot be called an auto-

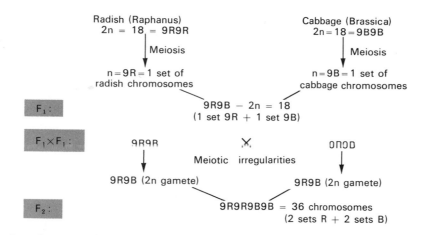

FIG. 10-30. Origin of an allotetraploid. One haploid set of radish chromosomes is represented by 9R and one haploid set of cabbage by 9B. The irregular meiotic behavior in the highly infertile F_1 on rare occasions produces a gamete that is diploid. If two of these unite, an allotetraploid arises that is fertile, because each parental set (9R and 9B) is represented in the diploid condition and can form bivalents.

tetraploid, because the extra chromosome sets have come from different species. And this allotetraploid, unlike the F_1 hybrid that gave rise to it, is very fertile. This is quite understandable, because there are now two sets of radish chromosomes that can form harmonious pairs at meiosis, and similarly two sets of cabbage chromosomes that can also form bivalents. As a result of the normal synapsis and orderly chromosome segregation, the gametes are balanced, each containing one set of radish (R) and one set of cabbage (B) chromosomes.

Unlike the autopolyploid, the allopolyploid contains very new combinations of alleles that would not arise under ordinary circumstances. The allopolyploid in our example was given a new name, Raphanobrassica, because it combined the traits of both parental species. Unfortunately, the combination did not produce a plant with economically valuable attributes. However, some experimentally produced allopoly-

ploids do contain a combination of features that make them more desirable than the diploids, as can be seen in the case of several hybrid wheats. Allopolyploids may be produced experimentally with the aid of colchicine, which is applied to the diploids to obtain diploid gametes. The polyploid at times possesses an advantageous assortment of characteristics that enables it to thrive under conditions that would be unfavorable for either of the diploid parents. The polyploid may actually be more robust than either diploid and may replace them. Allopolyploidy definitely occurs in nature, and the process, more so than autopolyploidy, has operated in the history of the major plant groups. It should be appreciated from a reappraisal of the radish–cabbage cross that the allotetraploid is a species in every sense of the word. It is morphologically distinct from either diploid parent. But more important to the definition of a species is its reproductive isolation from either diploid parent.

FIG. 10-31. Reproductive isolation of the allotetraploid. If an allopolyploid such as Raphanobrassica is backcrossed to either parent species (radish or cabbage), the F_1 is a sterile triploid. The backcross to the radish parent (*left*) leaves the nine cabbage chromosomes unpaired. Similarly, the cross to the cabbage parent leaves nine unpaired radish chromosomes. Irregular segregation of the univalents causes sterility. The allotetraploid can be recognized as a distinct species on the basis of reproductive isolation.

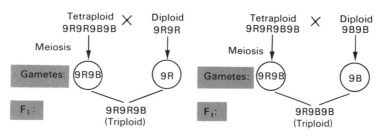

As Fig. 10-31 shows, a cross of the allotetraploid to either one of the parents yields triploid combinations. One set of chromosomes in either kind of triploid would be unpaired at meiosis, with the ensuing formation of unbalanced gametes. Therefore, allopolyploidy provides one mechanism for the evolution of new species. We will return to this subject in Chapter 21 in discussing the topic of speciation.

Somatic cell hybridization and plant hybridization

Besides their application to linkage studies (Chap. 9), techniques that apply cell fusion to plants offer excit-

ing possibilities for the derivation of superior hybrids. Mature, fertile, interspecific hybrids have actually been obtained through cell fusion methods that completely bypass the normal, sexual process. The feat involved two wild tobacco species, *Nicotiana glauca* ($2n = 24$) and *Nicotiana langsdorfii* ($2n = 18$). Leaf cells were subjected to enzyme digestion to free them from their cellulose walls. More than 10^7 protoplasts of each species were plated in a solution that stimulated them to fuse (Fig. 10-32). Approximately 25% of the cells underwent fusion in culture. Some fusions were between two cells of the same species, but others were hybrid fusions. These latter cells were then selected by plating all the cells on a medium that permits

FIG. 10-32. Parasexual hybridization in tobacco. Cells of tobacco leaves are enzyme treated to release free cells. Cells from two different species that differ in chromosome number are cultured together. Cellular fusions occur; some of these are fusions between the same kinds of cells; others are hybrid cells between the species. These are selected by the growth medium. The hybrid cells develop into plantlets lacking roots. The small plants are grafted onto other roots and eventually develop into mature plants bearing flowers and producing seeds. The chromosome numbers as well as all the other characteristics of the plants derived in this way are identical to the interspecific hybrid produced by crossing the two species and doubling the chromosome number to obtain a fertile polyploid.

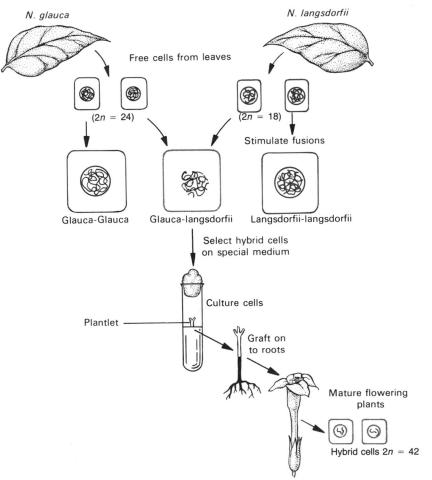

growth only of those containing the genetic information of both parent species. After further culture, the isolates developed rudimentary shoots and leaves but failed to form roots. However, when grafted to other tobacco roots, they developed into mature plants, which were identical in chromosome number ($2n = 42$), and in all other characteristics, to hybrid plants produced by conventional cross-pollination. Moreover, their seeds gave rise to offspring identical to them in all ways, illustrating that these "fusion hybrids" can breed true to type.

As we have just seen, in the typical production of an interspecific polyploid such as this, chromosome doubling is required to produce fertile hybrid offspring from highly infertile F_1 plants. The cell fusion method, called parasexual hybridization because the normal sexual process is circumvented, avoids this difficulty and others involved in crossing species. The parasexual process holds great promise as a means of combining the genetic information from plant species that would otherwise be unable to cross because of reproductive barriers.

Hybrids derived parasexually may very well enable the plant breeder to produce vigorous new species combining the most desirable features of both parents, such as increased disease resistance or increased yield of an edible product. Attempts to produce plant hybrids through cell fusion depend largely on selective techniques that make possible the isolation of only the hybrid cells that would be scattered among the total percentage of cells fused. It seems quite likely that this problem can be solved once differences in growth requirements between the parental species are fully known. The parasexual process may prove to be a very valuable method to increase food production in needy areas by the derivation of hardier and more nutritious plant species.

Polyploidy in animals

Since polyploidy is so characteristic of the plant kingdom, we might expect it to have been important in animal evolution as well. Actually, in humans and other mammals, polyploidy exerts a lethal effect, and the consequences of tetraploidy are apparently more severe than those of triploidy. The frequency of observed spontaneously aborted embryos that are tetraploid is quite low, indicating that the zygotes do not usually develop very far; no such fully developed fetuses have been recovered. However, nearly 4% of all spontaneous human abortions show triploid cells.

These embryos usually develop further than those with tetraploid cells, and may approach the development of a full-term fetus. A very small number of triploids have actually lived for a few hours, but they showed severe bodily defects. In contrast, examples of tetraploid cells in surviving humans have been extremely rare, and these cases involved somatic mosaicism (see the next section for more details).

Polyploid animals have been produced experimentally in some species, such as Drosophila and the silkworm; however, when animal groups are surveyed, it becomes quite evident that the number of species with naturally occurring polyploidy is very low. Good reasons for its scarcity among animals were first presented by H. J. Muller. In the first place, most animals are dioecious—that is, an individual is either male or female. Higher plants, on the other hand, are commonly hermaphroditic. Therefore, if a polyploid flower arises, diploid eggs and sperms are formed on the same individual, and the chances of union between two diploid gametes are good. However, in the animal, in which the sexes are separate, any diploid gamete that arises usually combines with a haploid one. This is so because the probability that a diploid egg and a diploid sperm would arise at the critical time in two separate individuals which chanced to mate is extremely low. The combination of a haploid and diploid gamete would produce a triploid, and we have seen that such individuals are sterile. Thus, autopolyploidy encounters difficulties in just getting started.

But suppose the rare combination of events takes place and a polyploid female and a polyploid male are formed. In Drosophila, the former would have four sets of autosomes and four X chromosomes. AAAAXXXX; a corresponding male would be AAAAXXYY. At spermatogenesis, the Xs of the male would usually pair with each other, as would the Ys (Fig. 10-33). This would produce gametes that are AAXY. These would combine with eggs that are AAXX, resulting in offspring that are AAAAXXXY. This combination of four sets of autosomes and three Xs would be a sterile intersex (Chap. 5), and so the tetraploid races could not be perpetuated and would disappear. Although the situation is not identical for mammals, it is obvious from what has been said about sex determination in humans that the XXXY individual would be a sterile Klinefelter type.

And so we would not expect to find any appreciable amount of polyploidy in those species that have separate sexes. An exception to this is the plant genus

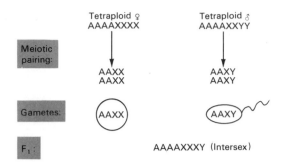

FIG 10-33. Cross of tetraploid fruit flies. A = one set of autosomes. Tetraploid males and female are genetically balanced; the ratio of X chromosomes to autosomes is normal (2:2 in the female and 1:2 in the male). However, the balanced tetraploid condition cannot be perpetuated. Pairing and segregation in the tetraploid male produces a preponderance of gametes that contain one X and one Y. Union of these with a diploid egg results in offspring which are sterile intersexes, because the balance of X to autosomes departs from 2:2 or 1:2.

Melandrium, which has an XY sex-determining mechanism similar to that in animals. Polyploidy occurs in this dioecious species, because the presence of just one Y is sufficient to produce a fertile male, even in the presence of three Xs (XXXY is male, not a sterile intersex). In most animals, however, this is not possible because of the interaction of many sex-determining genes on the X, the Y, and even the autosomes in many species.

It should be added that allopolyploidy, which has been very important in plant evolution, is most unlikely in higher animals, because it requires interspecific hybridization. Mating behavior places a barrier in the way of this, because no animal is likely to respond to the mating signals or devices of a member of another species. In animals, one of the mechanisms for reproductive isolation of species is provided by mating preference. This is obviously not the case in plants.

Chromosome anomalies and the human

The human and many other mammals seem unable to tolerate the unbalance resulting from both aneuploid and polyploid conditions. The significance of this becomes more apparent when we realize that chromosome anomalies are the most common cause of congenital defects in humans. They would cause even more suffering if more than 90% of all abnormal embryos were not spontaneously aborted.

In Chapter 14, we will see that radiations, in addition to their induction of gene mutations, are also potent agents in the production of chromosome anomalies. Exposure of cells to high-energy radiations greatly increases the frequency of all types of structural changes, as well as nondisjunction. These same aberrations may also be induced by certain chemicals. The list of substances that can alter chromosome structure is ever multiplying and is becoming a matter of growing concern in today's world.

In Chapter 5, we mentioned mosaic individuals whose bodies consist of cell populations that differ in their chromosome constitutions. Genetically diverse lines of cells in an organism result from mitotic mishaps during embryological development and may involve either the autosomes or the sex chromosomes. The outcome is an individual composed of a mixture of cell types. The earlier the mitotic error occurs, such as nondisjunction, the greater the chance that a larger number of cells will have the aberration. The mishap may occur in a cell of a specific organ or tissue, so that only a restricted part of the body shows the mixture. Individuals with a variety of types and degrees of mosaicism have been described. The effects of the aberrations also vary accordingly. A number of mosaics for the sex chromosomes have been identified. Klinefelter individuals are known who are mixtures of cells that are XX and others that are XXY (written as XX/XXY). Turner syndrome has been found in persons who are XO/XXX and even XO/XYY.

Although complete tetraploidy has never been reported in a live infant, a 6-year-old mentally retarded child with an assortment of bodily abnormalities was found to be a mosaic of normal cells, tetraploid cells, and those with an extra Chromosome 18. A married man who was fertile and normal was also shown to be a mosaic, a mixture of cell lines. Among the genetically diverse cells, the most common were XXY, but 15% of the fibroblasts were tetraploid. This man had 5 abnormal children out of 10, of whom 2 suffered from Down syndrome. Another appeared to be an XX/XO Turner's mosaic. The case suggests a family tendency for nondisjunction. This example, plus those that follow, illustrate the problem mosaicism presents to the genetic counselor.

Two other phenotypically normal fathers produced children with Down syndrome. After cells were taken from them from various tissues and then grown in culture, both men were found to be mosaics for Chromosome 21. In one man, the extra Chromosome 21 was found in skin and testicular fibroblasts but not in leukocytes. In the other man, mosaicism for the chromosome was found among the leukocytes. In these cases, the Down aberration was undoubtedly coming through the male parent. All of the persons involved were under 30 years of age. Such examples indicate the need to consider both parents in cases of suspected inherited anomalies and to look for mosaicism in different tissues.

Facilities are available today for proper genetic counseling of prospective parents who are concerned about the birth of children defective as a result of chromosome anomalies. Techniques are at hand not only for karyotype analysis of infants and adults but also for the detection of chromosomal and chemical defects in the unborn. These procedures can prevent some of the unfortunate consequences that follow upsets in chromosome structure and number.

REFERENCES

Arrighi, F. E. and T. C. Hsu. Localization of heterochromatin in human chromosomes. Cytogenetics 10: 81–6, 1971.

Astaurov, B. L. Experimental polyploidy in animals. Ann. Rev. Genet. 3: 99, 1969.

Atnip, R. L. and R. L. Summitt, Tetraploidy and 18-trisomy in a six-year-old triple mosaic boy. Cytogenetics 10: 305, 1971.

Beatty, R. A. The origin of human triploidy: an integration of qualitative and quantitative evidence. Ann. Hum. Genet. 41: 299, 1978.

Carr, D. H. Genetic basis of abortion. Ann Rev. Genet. 5: 65, 1971.

Caspersson, T., G. Gahrton, and J. Lindsten, et al. Identification of the Philadelphia chromosome as a number 22 by quinacrine mustard fluorescence. Exp. Cell Res. 63: 238, 1970.

Francke, U. Quinacrine mustard fluorescence of human chromosomes: characterization of unusual translocations. Am. J. Hum. Genet. 24: 189, 1972.

Grouchy, J. De and C. Turleau. Clinical Atlas of Human Chromosomes. Wiley, New York, 1977.

Hamerton, J. L. Human Cytogenetics, vols. I and II. Academic Press, New York, 1971.

Hook, E. B. Behavioral implications of the human XYY genotype. Science 179: 139, 1973.

Hsu, T. C. Human and Mammalian Cytogenetics: A Historical Perspective. Springer-Verlag, New York, 1979.

Knudson, A. G. Jr., A. T. Meadows, S. W. Nichols, and R. Hill. Chromosomal deletion and retinoblastoma. N. Engl. J. Med. 295: 1120, 1976.

Mikkelsen, M., A. Hallberg, and H. Poulsen. Maternal and paternal origin of extra chromosome in trisomy 21. Hum. Genet. 32: 17, 1976.

Niebuhr, E. The cri du chat syndrome. Hum. Genet. 44: 227, 1978.

O'Riordan, M. L., J. A. Robinson, and K. E. Buckton, et al. Distinguishing between the chromosomes involved in Down's syndrome (trisomy 21) and chronic myeloid leukemia Ph' by fluorescence. Nature 230: 167, 1971.

Rothwell, N. V. and J. G. Kump. Chromosome numbers in populations of *Claytonia virginica* from the New York metropolitan area. Am. J. Bot. 52: 403, 1965.

Rowley, J. D. A new consistent chromosomal abnormality in chronic myelogenous leukaemia identified by quinacrine fluorescence and Giemsa staining. Nature (Lond.), 243: 290, 1973.

Stebbins, G. L. Processes of Organic Evolution, 2nd ed. Prentice-Hall, Englewood Cliffs, NJ, 1971.

Stebbins, G. L. Variation and Evolution in Plants, Columbia University Press, New York, 1950.

Wagenbichler, P., W. Killian, A. Rett, and W. Schnedl. Origin of the extra chromosome 21 in Down's syndrome. Hum. Genet. 32: 13, 1976.

Wahrman, J., J. Atidia, and R. Goitein, et al. Pericentric inversion of chromosome 9 in two families. Cytogenetics 11: 132, 1972.

White, M. J. D. Chromosome rearrangements and speciation in animals. Ann. Rev. Genet. 3: 75, 1969.

Wilson, M. G., J. W. Touner, and G. S. Coffin, et al. Inherited pericentric inversion of chromosome no. 4. Am. J. Hum. Genet. 22: 679, 1970.

Yunis, J. J. (ed.). New Chromosomal Syndromes. Academic Press, New York, 1977.

REVIEW QUESTIONS

1. Let the following represent a pair of homologous chromosomes in an inversion heterozygote. (The 0 represents the centromere.)

   ——0 A B C D E F G   ——0 a b f e d c g

   Assume that a crossover takes place between the c and the d loci at meiosis. Give the two products that will result from the crossover event.

2. In the inversion heterozygote in Question 1, what will the two products be assuming a double crossover occurs between the same two strands—the first one between c and d and the second between e and f?

3. Let the following represent a pair of homologous chromosomes in another inversion heterozygote. (The 0 represents the centromere.)

   H I J 0 K L M N   h l k 0 j i m n

   Assume that crossing over takes place between the loci k and l. Give the two products resulting from the crossover event.

4. In the inversion heterozygote in Question 3, what will the two products be, assuming double crossing over between the same two strands, one crossover event between i and j, the other between k and l?

5. Assume that an individual has one chromosome with the gene order as follows:

   ——0 R S T U V W X Y Z

In the homologue, a shift resulted in the insertion of the segment T U V between X and Y. Show how this situation can lead to the origin of a chromosome with a duplication and one with a deficiency.

6. Following are two pairs of homologous chromosomes with the centromeres in a median position. Assume a reciprocal translocation occurs so that whole arms are shifted between two of the nonhomologous chromosomes following the breaks whose positions are indicated:

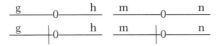

A. What is the composition of a translocation heterozygote carrying the two pairs of chromosomes?
B. If adjacent disjunction occurs, what are the two kinds of gametes possible for each of the two possibilities?
C. What kinds of gametes are possible from alternate disjunction?

7. A certain race of Oenothera carries two complexes of seven chromosomes each. One complex of seven is called alpha; the other set of seven is beta. Following self-fertilization, what will the composition of the zygotes be? Indicate which ones will survive.

8. Assume that a deletion arises in the X chromosome of the mouse and removes a gene required in at least single dose for the survival of the embryo.

A. What would be the result of crossing a female carrying the deletion and an ordinary male?
B. Assume that the deletion of an autosomal gene required in single dose accumulates in a population. What will be the outcome when carriers of the deletion mate with each other?

9. Allow "B" to represent one dose of the Bar locus on the X chromosome. A Bar-eyed female of the constitution B B/B B is mated to a Bar-eyed male of the constitution B B/Y. Assume that "unequal" pairing occurs in some of the meiotic cells of the female. What kinds of offspring can result from the fertilization of both the exceptional kinds of eggs and the eggs in which pairing is exact?

10. Several cases of reduction of eye size to a narrow bar in Drosophila have been related to removal of the Bar locus from its normal location on the X to tiny Chromosome IV, which is rich in heterochromatin. Different doses and locations of the Bar locus are depicted in the following. (B represents one dose of the Bar locus. X and IV represent the X and fourth chromosomes, respectively.) Types 1 and 2 flies have normal eye size. Types 3, 4, and 5 have the same narrow eye. Type 6 has an eye size much narrower than even these. Based on the following picture, explain the eye size differences.

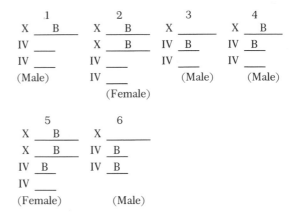

11. Assume that a certain animal species has a haploid chromosome complement that consists of the five chromosomes designated I, II, III, IV, and V. Give the chromosome constitution of a body cell of an individual which happens to be:

A. Trisomic for II.
B. Triploid.
C. Monosomic for IV.
D. Tetrapolid.

12. Species A has a diploid chromosome number of 24; so does another plant, Species B. After many attempts a highly infertile F_1 is produced. From this F_1, a fertile allotetraploid is eventually derived.

A. Give the chromosome number of the F_1 plant.
B. Give the chromosome number of the fertile derivative.

13. Species C has a diploid chromosome number of 30, whereas that for Species D is 20. A highly infertile F_1 arises from a cross between the two species, and eventually a fertile allopolyploid is derived.

A. Give the chromosome number of the F_1 plant.
B. Give the chromosome number of the allopolyploid.

14. The same two species discussed in Question 13 are manipulated in a parasexual hybridization procedure in which cells freed from leaves are used to obtain a hybrid. Briefly outline the procedure used, and give the chromosome number of the hybrid obtained from the two species.

15. In the course of several studies of plant populations, the following cytological pictures were encountered in individual plants. In each case, the somatic chromosome number was 24. Present a likely explanation for each picture:

A. Ten bivalents and a circle of four chromosomes at metaphase I of meiosis.
B. Bridges and fragments in pollen mother cells at anaphase I.

C. Two tetravalents and eight bivalents at metaphase I in pollen mother cells.

D. Five trivalents, three bivalents, and three univalents in pollen mother cells at metaphase I.

E. Twelve bivalents, one of which shows a distinct buckle at late prophase I of meiosis.

F. Two tetravalents, six bivalents, one trivalent, and one univalent at diakinesis.

G. Eleven chromosomes composed of two chromatids in a cell undergoing metaphase II.

H. Twenty-four chromosomes composed of two chromatids in a cell undergoing second meiotic division.

16. Assume plant Species 1 has a diploid chromosome number of 16. Give the following chromosome numbers:

A. An autotriploid derivative of Species 1.

B. A trisomic member of Species 1.

C. An allotetraploid derived from a cross with Species 2, which has a haploid chromosome number of 8.

D. A monosomic member of Species 1.

E. A plant derived from Species 1 and Species 2 using somatic cells and the technique of parasexual hybridization.

17. Give at least one way to account for the origin of humans with the following chromosome constitutions:

A. XXY.

B. XO.

C. XYY.

D. Only one normal metacentric Chromosome 1 but with an unusual chromosome the size of Chromosome 1 but acrocentric.

E. Three complete sets of chromosomes.

F. Only one Chromosome 21 and only one Chromosome 13 plus one unusually large chromosome.

G. XXX.

H. XX/XO mosaic.

I. XXY/XX mosaic.

J. XY/XYY/XO mosaic.

18. Match the disorder in column A with the letter of an appropriate chromosome anomaly in column B (there may be more than one possible choice in a given case and an item in column B may be used more than once):

A	B
—— Turner syndrome	1. polyploidy
—— Klinefelter syndrome	2. aneuploidy
—— Down syndrome	3. monosomy
—— Cri-du-chat syndrome	4. triploidy
—— Philadelphia chromosome	5. inversion
—— XYY male	6. translocation
	7. deletion
	8. nondisjunction

19. Demonstrate or explain how a metacentric chromosome can be derived from two acrocentric chromosomes with centromeres very near the end of a chromosome arm.

20. Occasionally males with Klinefelter syndrome are found to have an XX sex chromosome constitution, with no Y chromosome present! Explain, using information on the sex chromosomes in Chapter 5.

11

CHEMISTRY OF THE GENE

The pneumococcus transformation

By 1940, it had been demonstrated that the chromosomes of eukaryotes contain protein associated with nucleic acids, DNA and ribonucleic acid (RNA). The gene itself was generally considered to be composed of protein, since proteins are the most complex of all the chemical substances in the cell. Only proteins seemed complex enough to afford the diversity needed to build up so many different kinds of genes, each with an important and specific function.

A landmark in biology was reached in 1944 with the report of an investigation that implicated DNA and not protein as the genetic material. The history of the announcement goes back many years earlier (1928) to the English bacteriologist Griffith, who was working with the pneumonia organism *Streptococcus pneumoniae*. This is a spherical cell that typically occurs in pairs surrounded by a sizable common capsule. Such encapsulated cells are virulent and can kill experimental animals. On agar plates, they form colonies that are glistening and mucoid and have been described as "smooth." On occasion, cells from smooth colonies (or S cells) may change and lose the capsule. Such a mutation renders the cells harmless, because they can now be attacked by the white blood cells of the body. Moreover, their colonies on agar appear

granular in contrast to the smooth, encapsulated forms and have been termed "rough," or R.

The S cells from smooth colonies, when injected into an animal (such as a rabbit) can elicit the formation of specific types of antibodies. This is because the capsule, a polysaccharide, acts as an antigen. When different strains of S cells are injected into animals, it is found that the antibodies formed may be of different types. Therefore, there are really different kinds of virulent cells, which can be designated types I, II, III, and so on. Their distinctions are detected by the fact that the antibodies against the several types are specific, so that cells of type I react only with type I antibody, not with that of type II. The specific types are genetic characteristics that are inherited from one cell to another. When a virulent cell of type I mutates to a rough form, the R cells on rare occasions may later revert (back mutate) to the S or virulent form. When they do so, they always mutate back to the S type from which they were derived. For example, descendants of a rough cell that arose from type I would revert only to S type I, never to type III or any other (Fig. 11-1).

A series of experiments by Griffith produced some very unexpected results. He injected virulent cells from smooth colonies into mice. As expected in this case, the animals died, and autopsy revealed the pres-

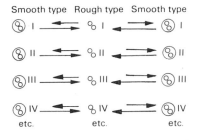

FIG. 11-1. Spontaneous mutation in Pneumococcus. Cells may mutate spontaneously but when they do so, they remain true to their antigenic type, which is a genetic characteristic. The back mutation from rough to smooth occurs at a much lower frequency than the mutation smooth to rough. A smooth or rough of any type never changes spontaneously to the rough or smooth of a different type.

ence of living, smooth cells. When he injected animals with rough cells, they survived (Fig. 11-2). Again, this was expected, because such cells are not virulent. When Griffith injected mice with virulent S cells destroyed by heat treatment, the animals lived because killed bacteria cannot multiply to produce toxin.

A most unusual outcome occurred when the experiment was performed in another way. Griffith injected mice with two kinds of cells: a small amount of living R cells derived from type II and heat-destroyed S cells of type III. The predicted result was that the animals would live, because neither living R cells nor killed S cells alone can cause disease. However, many of the animals died and, when examined,

were found to contain living cells that produced smooth colonies and were of type III. Such cells can kill, but how did they arise? Griffith said that the living R forms had somehow been able to take on the particular capsule type of the killed virulent cells. They had been "transformed," and the phenomenon became known as the pneumococcus transformation. In this process, R cells derived from any type of S could acquire the capsule of any type S cell. Apparently, then, the R cells could somehow be changed genetically through incubation with nonliving S cells. Something was causing the genetic change, and this was termed *TP* for transforming principle.

Others confirmed the work of Griffith and demonstrated that the transformation could be accomplished in the test tube and did not require an animal for incubation. R cells of one type could be transformed in vitro by heat-killed S cells of any type.

Avery, MacLeod, and McCarty of Rockefeller Institute, employing refined chemical procedures, were able to obtain small amounts of highly purified TP extracted from type III S cells. This purified TP was very active in transforming RII cells into SIII. Avery and his associates then subjected the purified TP to an assortment of exacting chemical tests to ascertain its chemical nature. Their extraction procedures removed most of the protein and fat. Tests of the final preparation showed the TP to be rich in DNA and to contain some RNA. The transforming activity of the TP preparation was not affected by treatment with

FIG. 11-2. Pneumococcus transformation. Mice injected with rough cells alone or heat-killed cells alone survive as expected, because the rough are not virulent and dead cells should not kill. However, a combination of living rough and heat-killed smooth cells can cause the death of an animal. Living smooth cells can be recovered after autopsy, and they are always found to be of the same bacterial type as the heat-killed smooth.

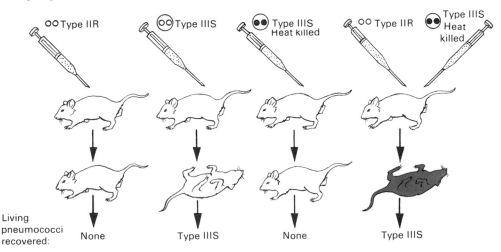

protein-digesting enzymes. Since only traces of protein could be present in the TP after the extractions, it seemed highly unlikely that the transforming ability was due to any protein molecules. When the enzyme RNAase was added to destroy any RNA that was present, the activity of the preparation was still not affected. However, treatment with the enzyme DNAase caused transforming power to be lost. Thus, there was little doubt that the transforming power resided in DNA. This implied that pure DNA by itself could bring about a genetic change. Therefore, the gene must be composed of this substance, at least the gene for capsule type in the pneumonia bacterium. It was not a complex protein as had been believed.

Additional evidence for DNA as the genetic material

Other lines of evidence assembled in the 1940s and early 1950s also implicated DNA as the genetic material. The Feulgen staining reaction, which had been developed in the 1920s, was shown to be highly specific for DNA. When cells were stained by this method, it was found that the chromosomes or nuclear contents stained and not the cytoplasm. Since the genetic material is in the chromosomes, this is exactly where we would expect to detect it by staining procedures. Stains for substances such as protein or carbohydrate showed them to be located throughout the cell. Moreover, the amount of DNA in a single nucleus could be measured quantitatively. This was made possible by the development of a photometric method. Cells can be stained for DNA by the Feulgen reaction and then placed on the stage of a microscope. A beam of light can be directed to a single nucleus. The amount of light passing through depends on how much dye is complexed to the DNA. The more dye that is bound, the more light is absorbed, leaving less light to pass through the microscope. The amount actually passing through can be measured precisely using sensitive photo cells.

Use of this Feulgen photometric procedure on a variety of plants and animals demonstrated that the nuclei of gametes contained exactly half the amount of DNA present in body cells or diploid cells. This is exactly what we would expect of the genetic material, the haploid amount in the reproductive cells, and the diploid amount in somatic lines. Cells known to have three or four sets of chromosomes (triploid and tetraploid cells) have exactly three and four times the amount of the DNA in the haploid gametes. Thus, an exact parallel was found between the amount of DNA

and the level of the chromosome number, haploid, diploid, and polyploid. Cell components that stained for the presence of proteins and other substances did not give these sharp results.

Further observations showed that DNA, unlike protein, fat, or carbohydrate, does not undergo continual breakdown and resynthesis in the cell. It tends to remain the same in the various cells of an organism, whereas other substances are continually metabolized. Again this feature of DNA is what we would expect of the hereditary material. It would certainly seem unlikely that the genetic material would undergo continual change in the cells. If it is needed to supply information, we would expect it to maintain its integrity. Only the DNA appears to do this.

The DNA of the chromosomes is associated with various kinds of proteins, among them the histones. These are relatively simple, basic proteins that are distributed over the entire length of the DNA and are present in about the same amount as DNA on a weight basis. The apparent constant association of the histones with the chromosomes of eukaryotes might suggest that they contain the genetic information. However, there are a few exceptions to the typical association between histones and DNA. The main one is in the sperm of many animals (some birds, fish), where the bulk of associated protein is not histone but protamine, a similar but detectably different type. When the sperm fertilizes the egg, the protamine of the chromosomes from the males is replaced by histones during early divisions of the embryo. If histones compose the gene, we would not expect them to be absent from sperm cells. Nor does it seem logical that genes are usually histones but in certain cells can undergo change into another substance. The DNA, however, is constant and does not change in kind during gamete formation.

A classic experiment performed in 1952 by Hershey and Chase argued strongly against protein as the substance of the gene. Hershey and Chase selected as their objects of study the bacterium *Escherichia coli* and a virus that attacks it. They did so because the T/2 virus or phage has a relatively simple morphology. From electron microscope studies, they knew it was composed of a protein envelope in which a core of DNA resides (Fig. 11-3). This structure is characteristic of this and the other T-even phages that can reproduce only in *E. coli* cells. After infecting the bacteria, the viruses command the cell machinery to manufacture more virus, and consequently the cell's activities become involved with the manufacture of phage particles. As a result, the cell eventually bursts.

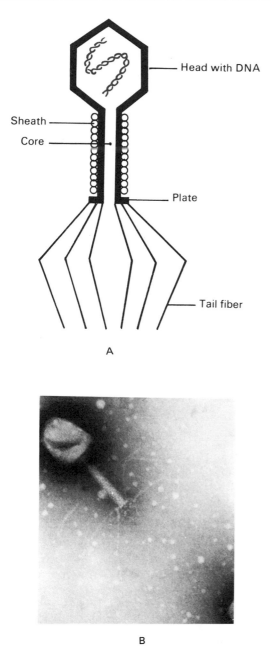

Head with DNA

Sheath

Core

Plate

Tail fiber

A

B

FIG. 11-3. T-even viruses. (*A*) Diagram of some features of the virus. A contractile sheath surrounds the core. Except for DNA in the head, all of the other parts are protein. (*B*) Electron micrograph of phage T4. (Courtesy Dr. James G. Wetmur.)

The viruses released are in every way like the original infecting viruses. Therefore, they must contain genetic information that is continuous from one generation to the next. Somehow, a virus that attacks a bacterial cell must take over the cell and provide it with the information needed to make more virus.

Hershey and Chase set out to determine exactly what is getting into the cell with the information for virus construction. They knew that the phage seemed to attach to the bacterial cell wall, as if it were injecting something into the cell. Was this protein, DNA, or both that was entering the cell with genetic information of the phage? Using a very elegant experimental design, Hershey and Chase obtained striking results (Fig. 11-4). They took a group of bacteria (group A) and raised it on radioactive sulfur, ^{35}S. Sulfur is a component of protein, and so all the bacterial protein would be labeled. Viruses were then allowed to infect these bacteria. The new viruses released from these cells would all contain radioactive protein envelopes, because only labeled protein was available for their coats, which were assembled in the labeled bacterial cells.

A second group of bacteria (group B) was raised on a medium supplied with radioactive phosphorus, ^{32}P. The phosphorus would be incorporated not into the protein but into the nucleic acids of the cell, as phosphorus is one of the critical ingredients of DNA and RNA. The group-B phages were then allowed to infect the labeled cells. The released phage particles would have radioactive phosphorus incorporated into their DNA, which would consequently be tagged or labeled. Hershey and Chase now had two groups of viruses—group A, with protein coats tagged with radioactive sulfur, and group B, with DNA labeled with radioactive phosphorus. The group A phages were then allowed to infect ordinary *E. coli*. A few minutes after infection, the bacterial cells were placed in a blender. The tiny cells cannot be injured by the blades, but any particles adhering to their walls can be torn away. After exposure to the blender, the cells were separated from the fluid medium by centrifugation, to give a cellular fraction and a fluid fraction. Both fractions were then tested for radioactivity of ^{35}S. It was found that almost all of the activity was in the fluid fraction. This meant that only a little of the sulfur was associated with the cells.

Group B phages with tagged DNA were also allowed to infect ordinary *E. coli*. The identical procedure was followed to give a cellular fraction and a fluid fraction. Both were then tested for the labeled DNA. It was found that almost all of the radioactivity was with the cells, in contrast to the results with group A. In other words, most of the DNA of the phages entered the bacteria, whereas most of the protein remained outside in the fluid. It was also shown that the small amount of protein that was found with the cells did not appear in the new viruses, which arose if the cells

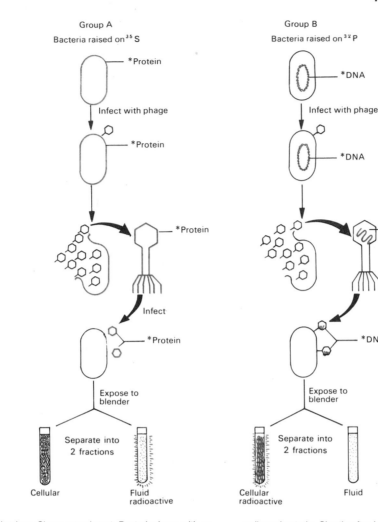

FIG. 11-4. The Hershey–Chase experiment. Bacteriophage with radioactive (*) protein coats were obtained by infecting bacteria whose protein was all radioactive. Phages with radioactive (*) DNA were obtained by infecting another group of bacteria whose DNA was all radioactive. The phages whose protein was tagged (group A) and those whose DNA was tagged (group B) were allowed to infect ordinary bacteria. Shortly after infection, the cultures were separated into a cellular fraction and a fluid one. The radioactivity in Group A was associated with the fluid, meaning that the labeled protein of the infecting phage did not enter the cells. In group B, the cellular fraction was labeled, indicating that the labeled DNA did enter.

were allowed to burst. The protein tended to disappear and not enter into the construction of new viruses from the infected cells. These results are strongly in favor of DNA as the genetic material of the phage that bears instructions for the manufacture of new virus particles.

Chemistry of nucleic acids

Still other lines of evidence implicated DNA as the genetic material, and we will encounter some of these throughout later discussions. It is now necessary to review what was known about nucleic acids in the early 1950s. The first studies go all the way back to the 1860s, those of Miescher in Basel. Studying pus cells and sperm, he demonstrated that nuclei contain a protein–phosphorus compound. His work was followed by other work in Europe and the United States. The best source of nucleic acid from animals proved to be the calf thymus gland. Hydrolysis of the nucleic acid showed that it was composed of (1) phosphoric acid, (H_3PO_4); (2) a five-carbon sugar, later identified as D(−)2-deoxyribose; (3) the purine bases adenine and guanine; and (4) the pyrimidine bases thymine

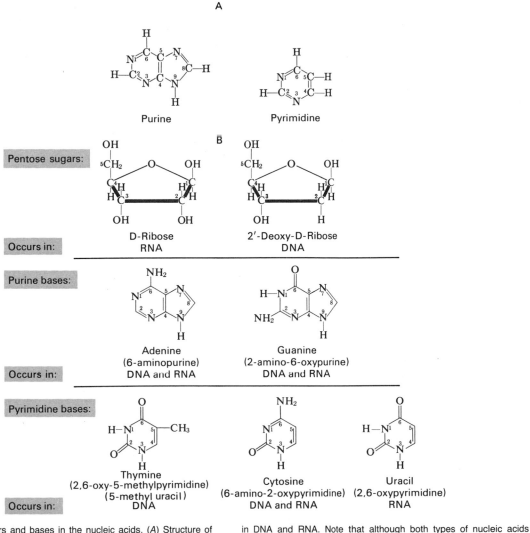

FIG. 11-5. Sugars and bases in the nucleic acids. (A) Structure of purine and pyrimidine, the parent compounds of the nitrogen bases in DNA and RNA. (B) Structure of sugars, purines, and pyrimidines in DNA and RNA. Note that although both types of nucleic acids contain similar building units, an important difference exists in the sugar component and with respect to two pyrimidine bases.

and cytosine (Fig. 11-5A). Both kinds of bases contain significant amounts of nitrogen and are referred to as *nitrogen bases*.

Another rich source of nucleic acid is yeast which is very abundant in RNA. Both DNA and RNA are essential ingredients of all cells. Analysis of the RNA yielded the same kinds of products as the DNA, upon hydrolysis. One big distinction between the DNA and RNA lies in the sugar component; a second difference lies in one of the pyrimidine bases (Fig. 11-5B). The sugar of RNA, D-ribose, is slightly different from the deoxyribose sugar. The main difference is at the carbon 2 position; the deoxyribose has one less oxygen

at carbon 2 than the ribose. In RNA, the pyrimidine base, thymine, does not occur. Instead, thymine is replaced by the pyrimidine uracil, which in turn is not found in DNA. Another distinction between the two kinds of nucleic acid also became evident—their cellular distribution. The DNA is confined almost exclusively to the nucleus, whereas the RNA has a wider distribution, being found in the nucleus and throughout the cytoplasm.

Once DNA was implicated as the genetic material, extra attention was directed toward it; if the gene is composed of DNA, knowledge of how the gene operates on the molecular level depends on an under-

A

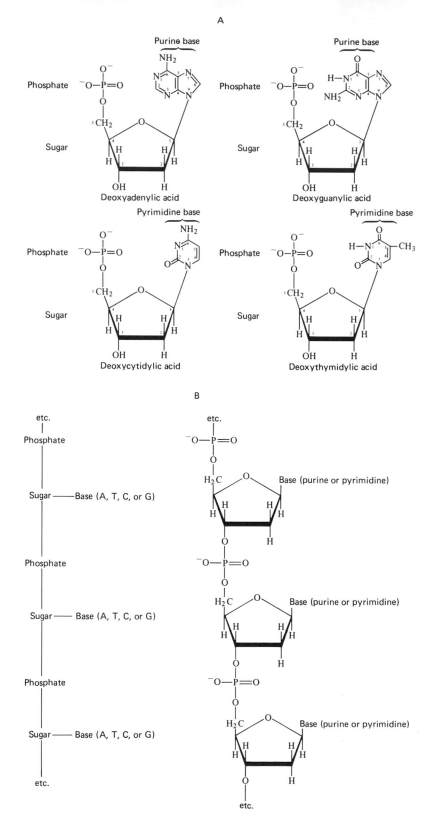

FIG. 11-6. Nucleotides in DNA. (*A*) Four different nucelotides occur in DNA. Each is a combination of deoxyribose sugar joined to a phosphate and to one of the four bases: adenine (A), guanine (G), cytosine (C), or thymine (T). (*B*) DNA is a polynucleotide. The nucleo-

tides are attached to each other through the sugar and phosphate components. Note that the bases are not connected to the phosphate but are linked only to the sugar.

286

standing of how the DNA itself is put together. Chemical analyses showed the DNA (and the RNA also) to be constructed of repeating units called *nucleotides.* Each DNA nucleotide is a combination of a phosphoric acid, a deoxyribose sugar, and one of the nitrogen bases (either a purine or a pyrimidine). Therefore, four different kinds of nucleotides are possible in DNA, depending on the nitrogen base (Fig. 11-6A): adenylic acid, guanylic acid, thymidylic acid, and cytidylic acid. The DNA is thus a polynucleotide. The individual nucleotides are connected through chemical bonds between the sugar and phosphoric acid components (Fig. 11-6B). The nitrogen bases in a strand of DNA are not joined to each other; they are linked to the sugar. As the figure indicates, the bases provide the only distinction among the nucleotides. In any stretch of DNA, one kind of nucleotide need not be followed by any particular nucleotide. As DNA from various sources was analyzed, no restriction seemed to be imposed on the sequence of nucleotides within the DNA molecule.

Note (Figs. 11-5 and 11-6) that the sugar component of the nucleic acids exists in the furanose form, in a ring of five instead of six, as is found in the pyranose form. The sugar may occur in the cell in association with one of the bases without being linked to a phosphoric acid. Such a combination of a pentose sugar and a purine or pyrimidine base is called a nucleoside. Since a nucleoside is equivalent to a nucleotide devoid of the phosphate component, four different kinds correspond to the four nucleotides of DNA: adenosine, guanosine, thymidine, and cytidine (Fig. 11-7A). Uridine is the nucleoside in RNA that corresponds to the RNA nucleotide, uridylic acid (Fig. 11-7B). Chemically, the nucleotides are phosphoric esters of the nucleosides. It should be noted that nucleotide structure is found not only in nucleic acids but also in several biologically important compounds, such as certain coenzymes and the universal energy source, adenosine triphosphate (ATP).

The 1940s yielded a great deal of information on how the nucleotides are assembled in DNA. One theory was that the DNA is composed of repeating tetranucleotides, each tetranucleotide being a sequence of four nucleotides in which the bases adenine (A), guanine (G), cytosine (C), and thymine (T) are always represented. In other words, the four kinds of bases were supposed to occur in equal amounts. Chargaff and his associates showed that this is not the case. The DNA does not contain equal amounts of the four different bases. Nothing restricts the se-

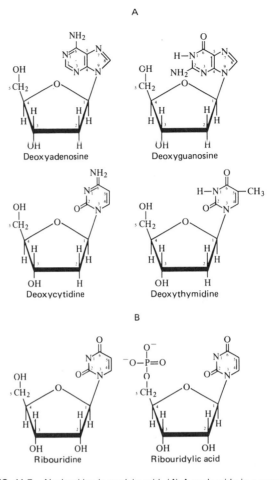

FIG. 11-7. Nucleosides in nucleic acid. (*A*) A nucleoside is a combination of a purine or pyrimidine base with a pentose sugar. The four kinds in DNA are deoxyribonucleosides and correspond to the nucleotides (Fig. 11-6), which contain a phosphate group linked to the sugar. (*B*) Nucleosides and nucleotides of RNA correspond closely to those of DNA. One major difference is in the sugar. The ribose of RNA contains an oxygen at position 2 in the sugar ring which is absent from that position in DNA. Another distinction is that in RNA the nucleoside ribouridine and the nucleotide ribouridylic acid occur in place of thymidine and thymidylic acid.

quence of the nucleotides. Moreover, the proportion of bases varied greatly with the different sources of the DNA. But one very important fact came to light. It was found that the amount of A was always equal to the amount of T. Similarly, the amount of G equaled that of C. Therefore, the A/T ratio is 1 and so is the G/C ratio.

However, the amount of A + T is not necessarily equal to the amount of G + C. Indeed, wide differences were found in these proportions. Some types of DNA have much more A + T than G + C, whereas an abun-

TABLE 11-1 Base composition of DNA from various sources[a]

DNA SOURCE	ADENINE	THYMINE	GUANINE	CYTOSINE
Human (liver)	30.3	30.3	19.5	19.9
Human (sperm)	30.7	31.2	19.3	18.8
Human (thymus)	30.9	29.4	19.9	19.8
Bovine sperm	28.7	27.2	22.2	21.9
Rat bone marrow	28.6	28.5	21.4	21.5
Salmon	29.7	29.1	20.8	20.4
Sea urchin	32.8	32.1	17.7	17.7
Wheat germ	27.3	27.2	22.7	22.8
Yeast	31.7	32.6	18.3	17.4
Escherichia coli	26.0	23.9	24.9	25.2
Streptococcus pneumoniae	30.3	29.5	21.6	18.7
Mycobacterium tuberculosis	15.1	14.6	34.9	35.4
Vaccinia virus	29.5	29.9	20.6	20.3
Bacteriophage T2	32.5	32.5	18.3	16.7[b]
Bacteriophage T7	26.0	26.0	24.0	24.0

[a] Note that although both the A:T and the G:C ratios tend to equal 1, the proportion of A + T/G + C varies from one species to the next. Some have little G + C (sea urchin), whereas others have a great deal (*Mycobacterium tuberculosis*).

[b] Has hydroxymethylcytosine in place of cytosine.

dance of G + C over A + T is found in the DNA from other species (Table 11-1). Thus, the makeup of the DNA was shown to vary from one species to another in its base composition, but the different tissues of any one species showed the same kind of DNA. This, of course, made sense. We would expect the hereditary material to differ among the diverse forms of life but not from one cell type to another within an organism or a species.

The base composition of the DNA is therefore a characteristic of any particular species. Each species has its typical A + T/G + C ratio. However, when the base composition of a species is discussed, rather than talking in terms of the A + T/G + C ratio, reference is usually made to the "G + C content" or the "percent GC content" for short. If it is said that an organism has a G + C content or percent GC content of 40%, this is taken to mean G = 20%, C = 20%, A = 30%, and T = 30%.

A striking finding is that the percent GC content of lower forms (prokaryotes, for example) shows great variability from one species to the next, perhaps 25% in one and 75% in another. With evolutionary progression, the percent GC appears to vary less and less within the divisions of plants and animals and to be a more fixed value. In mammals, the overall average value is about 40%, one species of mammal showing a similar value to the next one. The reason for this stability of G + C value in higher life forms and its evolutionary significance are unknown.

FIG. 11-8. The linkage between nucleotides. Note that carbons 3 and 5 of the sugar are involved in the formation of a linkage with the phosphate component. A phosphoric acid has formed one ester bond with the −OH associated with the carbon in position 3 of the one sugar and another ester bond with the −OH associated with the carbon in position 5 of the adjacent sugar. (Refer also to Fig. 11-6B, which does not show the C atoms in the sugar.)

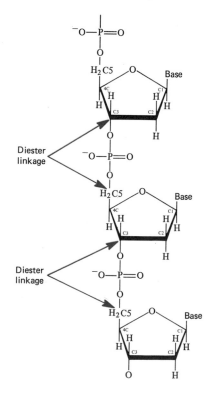

More detailed chemistry showed exactly how the nucleotides are linked to one another through the sugar and phosphate (Fig. 11-8). The number 1 carbon in the sugar is attached to the nitrogen base. The second carbon is the one lacking the oxygen. Carbon 4 is completely in the ring. Carbons 3 and 5 are involved in the internucleotide linkages. The −OH group of carbon 3 in the sugar component of one nucleotide forms an ester linkage with the phosphoric acid. The phosphate in turn forms another ester linkage with the −OH group of carbon 5 in the adjacent nucleotide. Thus, the phosphoric acid engages in the formation of a double ester. It is these diester linkages that hold together the sugar–phosphate backbone. Because positions 3 and 5 of the sugar are involved, we say the nucleotides are joined by 3′–5′ phosphodiester linkages.

Physical structure of DNA and the Watson–Crick model

By the early 1950s a great deal was known about the chemistry of the DNA, but this told little about the architecture of the molecule. Important data on the physical structure of DNA were obtained in the laboratory of Maurice Wilkins and his group at King's College. The information was primarily in the form of X-ray diffraction studies of fibers of DNA. In diffraction studies, the material is subjected to a beam of X-rays. Assuming the material is composed of molecules and atoms arranged in a random or haphazard manner, the rays, in their passage through the material, will not be diverted in any characteristic fashion. If they pass through the material and fall on a photographic plate, they will not describe any definite pattern, and when the plate is processed, little more than a diffuse blackening may be seen. If, however, there is a definite arrangement of the parts within the molecule, the rays will be deflected more in some directions than in others. As the arrangement repeats itself, the photosensitive emulsion may be bombarded over and over again in certain areas. This will heighten the darkening in these areas, with the result that certain dark spots may repeat and form a pattern (Fig. 11-9). Studies of such patterns can therefore tell a great deal about the spatial organization within a molecule and the repetition of the units that compose it. Diffraction studies by Wilkins's group showed that the DNA molecule is not haphazard in arrangement but possesses a definite organization in which the parts repeat themselves. Using the information from

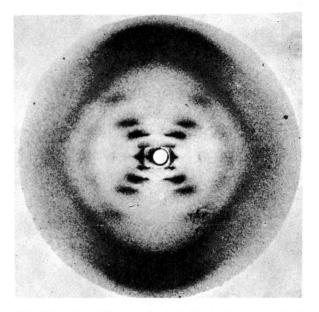

FIG. 11-9. X-ray diffraction. (Copied with permission from J. D. Watson, and F. H. C. Crick, *Cold Spring Harbor Symp. Quant. Biol.* 18: 123–131, 1953.)

chemical analyses along with the diffraction photographs made by Wilkins and his associate, Rosalind Franklin, Watson, and Crick were able to propose a model of the structure of the DNA molecule. The Watson–Crick model is perhaps the greatest landmark in modern biology. It has given rise to the discipline of molecular biology and has provided a deep insight into the very nature of gene action.

The diffraction patterns of Wilkins and Franklin suggested that the DNA molecule is composed not of just a single long strand of nucleotides but rather two strands bonded together in some way. Moreover, the two polynucleotide chains are not simply stretched out but are twisted in a spiral, much as the railings of a spiral staircase. The DNA molecule is in the form of a double helix (Fig. 11-10) having a diameter of about 20 Å. One complete turn of each chain is 34 Å, and this includes 10 nucleotides. Therefore, the distance between one nitrogen base and the next one on a chain is 3.4 Å. The chemical information told Watson and Crick that the nitrogen bases in DNA are in the ratios: A:T = 1 and G:C = 1. Added to what was known about the physical structure of the DNA, this information suggested to them that preferential base pairing occurs in the molecule and that this pairing is responsible for holding the two chains together. The purine adenine would thus pair preferentially

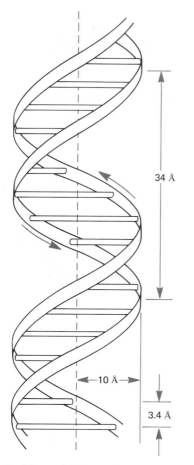

FIG. 11-10. The Watson–Crick double helix. According to the model, the DNA is composed of two polynucleotide chains twisted in a spiral. The sugar and phosphate components provide the backbone of the chains. These in turn are held together by pairing between the nitrogen bases (see Figs. 11-11 and 11-12). The bases would be stacked inside the double helix, much like stairs in a spiral staircase. Note that a large groove and a small one are present.

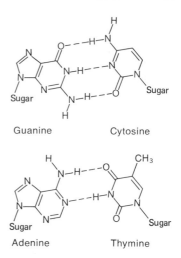

FIG. 11-11. Preferential base pairs in DNA. Adenine on either one of the two chains pairs with thymine; cytosine pairs with guanine. Note that two hydrogen bonds (*dashed lines*) hold the A–T pair together, whereas three are involved in the G–C pair.

with the pyrimidine thymine, whereas the purine guanine would pair with the pyrimidine cytosine (Figs. 11-11 and 11-12). The two bases composing each base pair would be held together by hydrogen bond formation. Hydrogen bonds are much weaker attractive forces than covalent bonds, which entail the sharing of electrons between atoms.

Knowledge of the molecular structures of the four bases indicated that two hydrogen bonds could hold the A-T pair together, whereas three would be formed in the G-C pair. As a consequence of this hydrogen bond formation between the base pairs all along the molecule, the two polynucleotide chains would in turn be held together. The base pairs extend from the

sugar–phosphate backbone and are stacked up inside the double helix resembling the stairs in the analogy of the spiral staircase. The two chains of the molecule run in opposite directions with respect to the attachment between sugar and phosphate groups. Recall that 3′–5′ phosphodiester linkages hold the nucleotide units together in a chain. If we start at the "top" of a DNA molecule, the orientation of one chain would be 3′-5′-3′-5′- and so on. The other chain would be in reverse order and would run 5′-3′-5′-3′- and so on. As a result of this orientation, the molecule looks the same from the top as from the bottom: a 3′ chain and a 5′ chain terminate at each end (Fig. 11-12). Thus, each end of the molecule has one 5′ or phosphate terminating chain and one 3′ or −OH terminating chain. The double-helical DNA is a molecule with a right-handed twist. Note (Fig. 11-10) that it also possesses two grooves—a narrow minor groove and a wider major groove.

The structure of the DNA molecule, as proposed by Watson and Crick, was constructed through careful, painstaking correlation between chemical and crystallographic data. All of the suggestions in the model were in complete agreement with established chemical and physical principles. One critical feature of the model is that the two polynucleotide chains are not identical because of the preferential base pairing. Not only do the chains run in the opposite direction, but most important, they are complementary, the A and C in one strand corresponding to a T and G in

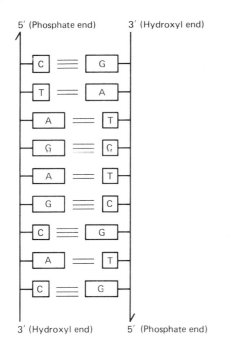

5′ (Phosphate end)   3′ (Hydroxyl end)

3′ (Hydroxyl end)   5′ (Phosphate end)

FIG. 11-12. Diagram of the DNA double helix. Note the difference in orientation of the two chains: one runs in the order 3′–5′ and the other in the order 5′–3′. It can also be seen that the two chains are complementary, not identical.

the other. Since the model imposes no restrictions on the sequence of the bases in any one chain, A could be followed by T, G, C, or another A. However, if one knows the sequence of bases along a length of one of the chains (say, A-T-C-C-A-G), one can be certain of the sequence of bases occurring along the complementary chain (T-A-G-G-T-C).

As research on the DNA molecule continued, it became apparent that DNA can exist in more than one form. The one presented in Fig. 11-10 and described by Watson and Crick is known as the *B form,* the form in which most cellular DNA is believed to exist. Other forms of DNA and their significance are discussed in Chapter 17 in relation to the topic of gene regulation.

At this point, a few remarks are in order concerning the need to designate the polarity of a written sequence of bases in a DNA strand. When just a portion of a single DNA chain is being written, the polarity must be designated in some way, since the two DNA chains of a double helix are antiparallel (Fig. 11-10). Without some convention or understanding, simply writing the sequence, TACG, is insufficient, because it is unknown which base is toward the 5′ end and

which is toward the 3′ end of the chain. A convention is needed to clarify the polarity of the written sequence. One can write the sequence TACG as pTpApCpG with the understanding that the 5′ end of the chain is to the left and that each p between the bases represents a phosphodiester linkage. Another way to represent this same sequence is: p5′-TACG-3′–OH, again with the 5′ end to the left. The writing can now be simplified, however, to TACG, since by convention the 5′ end of a written sequence is taken to be toward the left unless otherwise indicated.

Implications of the Watson–Crick model

A burst of excitement followed the announcement of the Watson–Crick model. One of its many contributions was to provide a key to the interpretation of several familiar genetic phenomena about which little or no insight had been provided. Although everyone knew that the genetic material undergoes self-duplication, there was no well-founded suggestion as to how this might take place. The Watson–Crick model offered a valid basis for this. According to the model, at the time of replication, the hydrogen bonds between the bases of the two complementary chains dissolve (Fig. 11-13). The two chains then unwind and separate. Each strand maintains its integrity in the process and does not break down in any way. Instead, each one acts as a template or pattern for the assembly of another strand, one that is complementary to it. These complementary strands are constructed from their building blocks, the nucleotides present in the cell. As the sugar–phosphate backbone is assembled, each nitrogen base in the original strands attracts the complementary one. The purine adenine (A) in one chain attracts the pyrimidine thymine (T) in the cellular environment. Likewise, T in the other original chain attracts nucleotides of A in the cell. This complementary attraction by all the bases in each of the two original strands, along with the assembly of the sugar–phosphate backbone, results in the construction of two new chains, each complementary to the original "old" ones. The overall effect is two double-helical DNA molecules, each identical to the original.

This fits perfectly with what we would expect if the genetic material remains identical from one cell generation to the next, for the genetic material must contain a store of information. The Watson–Crick model has important implications on this extremely critical point. Since the model assumes no restriction

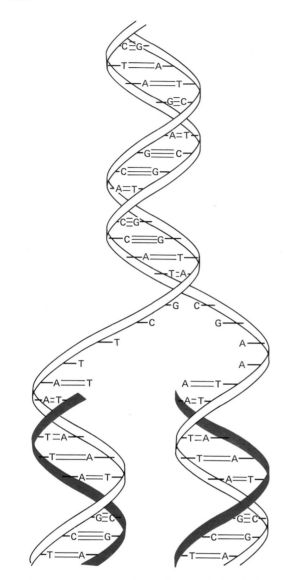

FIG. 11-13. Self-duplication of DNA. During replication, the hydrogen bonds between the strands are dissolved. Each original strand remains intact and acts as a pattern for the assembly of a complementary chain (new strands are shown in *red*).

on the sequence of base pairs, the genetic information could somehow be stored in the form of a code. The code cannot reside in the sugar or phosphate, because these are the same from one nucleotide to the next; only the bases can differ. The sequence of bases could then provide the basis of a code. If the sequence is A-T-T along one stretch of the DNA, this could mean something different from the sequence G-T-C. A gene, a segment of double-stranded DNA, would then contain information coded in the sequence of its base

pairs. And one gene or one allele would differ from the next because of the difference in the sequence. Since a gene is most likely composed of many nucleotides, it would include many base pairs. The unrestricted sequence allows a limitless number of differences and could account for the endless number of different genes found in living species.

But it is known that the genetic material may suddenly change as the result of a spontaneous gene mutation. This implies a change in the coded information. How could this come about? Chemical analysis of the DNA showed that the molecule is tautomeric. This means that the molecule undergoes occasional shifts in the positions of certain protons within it. One form of the molecule would be the most common, but at any one moment, a hydrogen may change position, so that a molecule could be in its rarer form or state (Fig. 11-14A). This rare state might occur at the time the gene is replicating or building more of itself. Since the tautomeric changes can affect hydrogen atoms in the purine and pyrimidine bases, the change could alter the pairing preferences of the bases (Fig. 11-14B).

Let us assume that in a stretch of DNA a shift occurs in the position of a hydrogen in one of the pyrimidine bases, thymine. As a consequence of the shift, it may not be able to pair with its normal partner, adenine, but instead, pairs with guanine (Fig. 11-15). The next time the molecule undergoes replication, there would be no reason to expect a tautomeric shift, because this would be a rare event in any case and would not have to affect the same bases. Normal bonding preferences would therefore be expected at the next replication. However, because G has been inserted in the one strand (the right one in the figure) as a result of the first error, C will now be placed in the corresponding position on the new complementary strand. Instead of the original base pair sequence A-T, T-A, C-G, we now have A-T, C-G, C-G. This T-A pair has been replaced by a C-G. If the base pairs do provide the basis for a code of stored information,

FIG. 11-14. Tautomerism and formation of exceptional base pairs. (*A*) Each of the four bases in DNA typically exists in a common state. However, hydrogen atoms may at times shift position, so that a rare state may exist at a particular time. For example, adenine bears a $-NH_2$ group. At rare times, one of the H atoms shifts so that the $-NH_2$ becomes $-NH$, and the H is in a new position. (*B*) Such tautomerism can change pairing preference. For example, the shift that can occur in adenine now allows it to bond with cytosine. Shifts in the other bases cause similar changes in pairing preference.

A

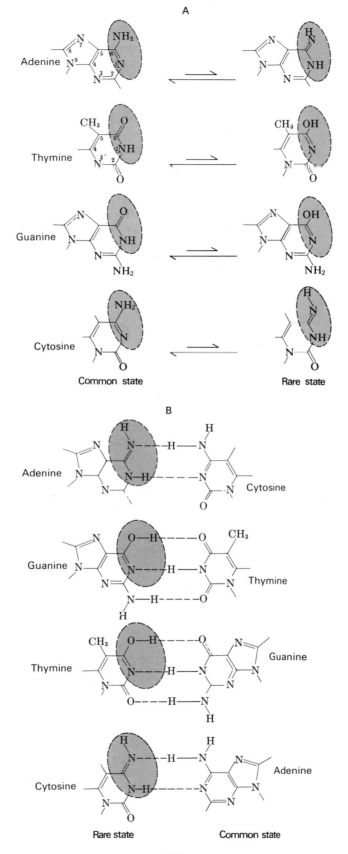

Adenine

Thymine

Guanine

Cytosine

Common state   Rare state

B

Adenine   Cytosine

Guanine   Thymine

Thymine   Guanine

Cytosine   Adenine

Rare state   Common state

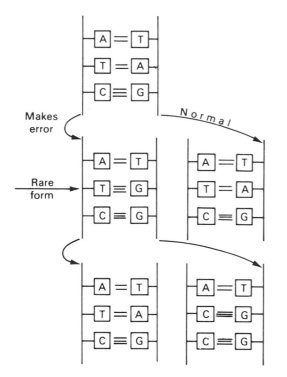

FIG. 11-15. Tautomerism and mutation. A sequence of normal nucleotide pairs (*above*) is typically found in a stretch of DNA. If a base should exist in its rare form (*middle*) in one of the strands at the time of replication, an exceptional base pair may form. At the next replication (*below*), the bases in each strand would usually behave normally and attract their complementary bases. However, because a substitution was made at the previous replication, a new base pair, C:G, now substitutes for the original ,T:A, producing a mutation.

the code would now be changed. The new sequence of base pairs could mean something very different from the original message. This model could also explain the dire effects of a deletion in which a part of the chromosome is lost. If a sequence of base pairs is omitted, then part of the message would be missing. Any change in the structure of the chromosome could alter the base sequence and so upset the genetic message.

If the Watson–Crick model is correct, this means that the coded information in the DNA must somehow be transferred to other parts of the cell. Obviously, a code is of no use if its information cannot be communicated. The model triggered a wealth of investigations on the mechanism of information transfer from nucleus to cytoplasm, a most important topic, which is the subject of Chapter 12.

Experimental testing of the Watson–Crick model

One very admirable quality of the Watson–Crick model is that its formulation permits tests of its validity. Throughout the years, the model has passed the scrutiny of many investigations, the results of which support all the features of the model. In 1958, Meselson and Stahl presented strong support for the concept of replication as pictured by Watson and Crick. According to the model, DNA replication is semiconservative. This means that the entire molecule does not remain intact when new DNA is being formed; instead, the two strands come apart. However, each strand is preserved intact, as shown in Fig. 11-13; it does not dissociate in any way. Besides this semiconservative concept of replication, two other possibilities can be imagined: conservative (the two strands stay together, and a new double-stranded molecule is built up next to them) and dispersive (the two original strands break down and entirely new strands are constructed from these and other precursors in the cell).

To gain information on this critical point, Meselson and Stahl employed a procedure now routine in laboratories—density gradient ultracentrifugation. The technique is based on the knowledge that a solute (sugar, salt, etc.), when spun at a high speed in a centrifuge, becomes distributed from one end of the tube to the other. It eventually reaches an equilibrium, at which time the molecules are most dense at the centrifugal end and gradually decrease in density toward the centripetal end. Such an equilibrium density gradient has practical laboratory application. A mixture of substances that are very similar in density may be separated by this procedure, for when a mixture is spun in the appropriate solution, a density gradient is established in the centrifuge tube. The substances composing the mixture tend to separate out at characteristic levels in the tube. This dissociation occurs because a substance settles at a location where the density of the solute molecules matches its own density. The technique is so sensitive that materials of very similar densities may be separated by using the proper solute. For example, a mixture of DNA molecules of different densities may be separated in a gradient of cesium chloride (Fig. 11-16). Under laboratory conditions, a given kind of DNA can be made heavier than normal by supplying organisms with a medium containing only heavy nitrogen, ^{15}N, an isotope of the common ^{14}N. If "heavy" DNA (heavy because its bases contain ^{15}N) and light DNA are

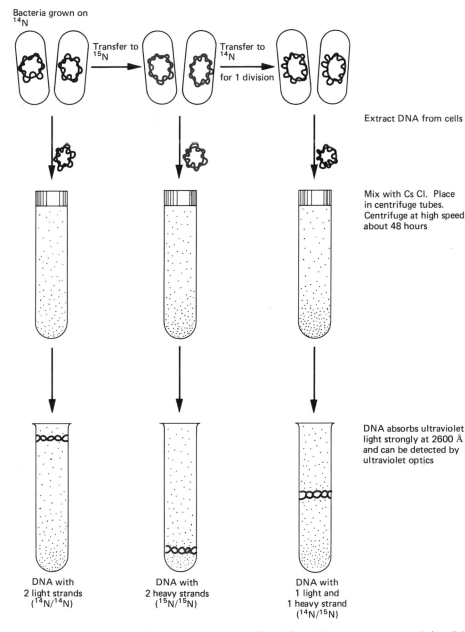

Bacteria grown on ^{14}N

Transfer to ^{15}N

Transfer to ^{14}N for 1 division

Extract DNA from cells

Mix with Cs Cl. Place in centrifuge tubes. Centrifuge at high speed about 48 hours

DNA absorbs ultraviolet light strongly at 2600 Å and can be detected by ultraviolet optics

DNA with 2 light strands ($^{14}N/^{14}N$)

DNA with 2 heavy strands ($^{15}N/^{15}N$)

DNA with 1 light and 1 heavy strand ($^{14}N/^{15}N$)

FIG. 11-16. Density gradient ultracentrifugation. Substances that are very similar in density may be separated by this sensitive method because each will tend to settle where its density matches that of a solute. Different types of DNA can be separated, such as the three kinds indicated here: one type composed of two light chains containing ^{14}N, another composed of two heavy chains with ^{15}N, and an intermediate DNA made of one light and one heavy. (See text for details, and compare with Fig. 11-17.)

mixed, they may be dissociated in a cesium density gradient. This is possible, because they come to equilibrium and form distinct bands at slightly different locations in the centrifuge tube.

In their experiment (follow Fig. 11-17), Meselson and Stahl employed the bacterium *E. coli*. The organism was grown on a source of labeled, heavy ^{15}N for several generations until all of its DNA was heavy. When extracted from the cells, this heavy DNA settled out in the density gradient at a characteristic position.

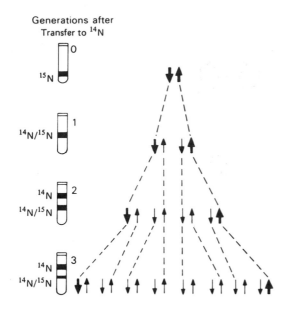

Generations after
Transfer to ^{14}N

FIG. 11-17. Meselson and Stahl experiment. In this highly schematic diagram, a pair of reversed arrows represents a double helix. It can be seen that if the cells are grown on medium with heavy (^{15}N) nitrogen, all of the cellular DNA is heavy if no division is permitted (O generation). As the number of cell divisions increases, following transfer to ^{14}N medium, the number of cells with light strands increases. Note that the strands maintain their integrity. Any strand formed in one generation acts as the template for a new light strand in the next generation. In the third generation, the hybrid (^{14}N/^{15}N) band is narrower than the ^{14}N, because only light strands can be made. Still, the heavy strands continue to act as templates.

Meselson and Stahl removed samples of bacteria from the heavy cultures and placed them on a medium containing ordinary ^{14}N. They tested the DNA of some of these immediately, without allowing any cell division (0 generations). Some samples were allowed to divide once (1 generation), others two and three times (2 and 3 generations), before their DNA was extracted. As Fig. 11-17 shows, all of the DNA from the 0 generation bacteria was heavy, as expected.

It was reasoned that if the Watson–Crick model was correct, each of these heavy strands would separate during the first round of cell division. Each could only direct next to it the construction of a light strand, because only light nitrogen would now be present. The DNA of the first generation cells would thus contain DNA that is hybrid, intermediate in density between double-stranded heavy and double-stranded light molecules. It should therefore settle out at a location in the centrifuge tube distinct from that of the other two.

This is what was found, a single band at a position

characteristic of neither ^{14}N nor ^{15}N. The DNA was then tested from the bacteria that had undergone two divisions. According to the Watson–Crick model, each strand will maintain its integrity. But under the conditions of the experiment, each strand can order the construction of only light strands. The original heavy ones build up light strands, as do the light ones made during the first division. As a result, half of the cells will have DNA with two light strands and half will have DNA that is hybrid (^{14}N/^{15}N). A perfect correlation was found between the prediction and the experimental results, for now *two* bands of *equal* size were present in the density gradient: one at the level of the hybrid DNA and the other at a less dense position, that typical of light DNA. The third generation DNA again gave exactly what would be expected from the Watson–Crick model. Hybrid and light strands are again both present, because the original heavy ones continue to act as templates. However, cells with hybrid DNA are fewer in proportion to cells with completely light DNA, since only light strands can be constructed. Observations clearly showed that the light band in the density gradient was conspicuously thicker than the slightly heavier hybrid one. If replication were conservative or dispersive, such results could not be explained. Replication as proposed by Watson and Crick offers the only answer.

According to the model, as replication takes place, the two chains separate from each other while each base attracts the complementary one. It is conceivable that chain separation is unnecessary for the construction of two new chains. It is also possible that both strands must separate completely along their lengths and exist entirely as single chains at the time of replication. The work of Cairns with *E. coli* DNA provided details on these aspects of the duplication process. Cairns was a pioneer in the study of the "chromosomes" of bacteria. His superb isolations of bacterial DNA showed that the DNA of *E. coli* is in the form of a circle (Fig. 11-18). Bacteria and other prokaryotes do not possess chromosomes in the true structural sense of the word as used for eukaryotes. The genetic material of these simple organisms contains naked strands of DNA that are not complexed with histone protein. Much less DNA is found here than in the true chromosomes of eukaryotes. When extracted from a cell, the prokaryotic DNA can often be followed through its entirety. The overall simplicity of this DNA provides a system for detailed study that is extremely difficult in eukaryotes.

To follow the replication of DNA in *E. coli*, Cairns

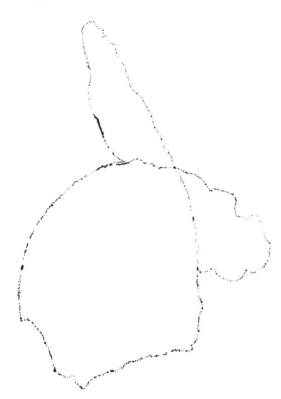

FIG. 11-18. Diagram of an autoradiograph of the *E. coli* chromosome. The chromosome is actually in the act of duplicating. The DNA is from cells that have been grown in the presence of radioactive thymidine for two generations (see the text for details). (Drawn after J. Cairns, *Cold Spring Harbor Symp. Quant. Biol.* 28: 43–46, 1963).

used the technique of autoradiography. When cells of any kind are grown in a medium containing thymidine, this nucleoside is incorporated only into the DNA. The thymidine can be rendered radioactive if it contains tritium (a heavy isotope of hydrogen). As it disintegrates, tritium emits beta particles (electrons). If these fall on a photographic emulsion, the film will contain darkened spots when it is processed. Cairns grew *E. coli* cultures in a medium of tritiated thymidine for different periods of time. He extracted the DNA and covered it with an emulsion that was eventually developed in a fashion similar to that of any photographic film. The dark spots on the resulting autoradiograph indicate the breakdown of radioactive atoms in the DNA. The more such atoms exist in one location, the more spots there are, because of the greater number of beta rays affecting the emulsion. Thus, the number of spots or their density would be proportional to the number of radioactive nucleotide chains in the DNA molecule. If two radioactive chains

are present at a certain location and only one at a second location, the former gives off twice as many emanations as the latter. The darkening is thus twice as dense.

Cairns compared autoradiographs from bacteria that had grown in labeled thymidine for different lengths of time. Particularly significant were autoradiographs from those cultures of cells that had completed one round of DNA replication in the labeled medium and were now undergoing a second replication in the

FIG. 11-19. Point of replication in DNA. (*1*) If one chain of the double helix is labeled with tritiated thymidine, dark spots will appear on the autoradiograph because of the emission of electrons as the label decays. (*2*) If both chains are fully labeled, the density of dots on the autoradiograph will be twice as great as in the case of the half labeled DNA. (*3*) Half labeled DNA that is undergoing a second round of replication in labeled medium will show a region that appears like a fork with one arm having a density of dots twice that of the other. In the denser arm, two radioactive chains are present (the original and the new one being constituted). In the other arm, only the new chain will be radioactive, because it is being formed in a labeled medium on a nonlabeled template.

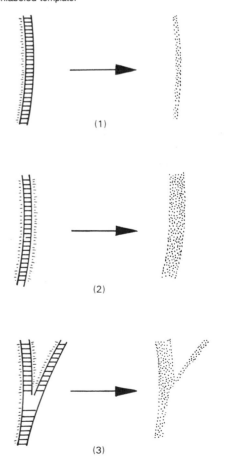

presence of the label. On each of these, Cairns was able to identify a region that appeared as a fork (Fig. 11-19). This single fork appeared at different positions, depending on the length of time the cells had grown in the presence of the radioactive thymidine.

The observations can be readily interpreted according to the Watson–Crick model. The fork marks the point in the DNA where replication is taking place. If one round of replication is completed, the DNA is composed of one labeled and one nonlabeled strand. As a second round starts, the two strands unwind, and each proceeds to direct the formation of a radioactive partner. However, one strand has two labeled chains, the other only one. Consequently, a "Y" figure is produced on the emulsion at this point, but one arm of the Y has a density of dots twice that of the base of the Y and the other arm.

Cairns was able to time the movement of a single replicating site from the initiation point around the circular chromosome by comparing autoradiographs from cultures grown for different lengths of time in the labeled thymidine. A fork was estimated to travel approximately 20 to 30 μ /minute. It was later established that replication of the DNA in the *E. coli* chromosome is initiated at a single site, as indicated by Cairns's original observations, but that two growing points are established at that site. Replication is then bidirectional; these two replicating sites move in opposite directions around the circular bacterial DNA until they eventually converge and fuse. The observations clearly indicate that chain separation goes hand in hand with replication of the DNA strands much as predicted by the Watson–Crick model.

In addition to bacteria, the genomes of many vi-

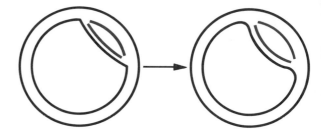

FIG. 11-20. Replicating circular DNA. A replicating circular DNA molecule resembles the Greek letter theta (Θ), since the region in which replication has taken place appears as an eye. Such a molecule is often referred to as a theta structure or theta molecule. The eye becomes progressively larger as the newly synthesized DNA (*red*) continues to lengthen.

ruses, as well as those of organelles (Chap. 19), are known to occur in the form of circular DNA molecules. When the template is circular, as in the case of *E. coli*, the replicating molecule, when viewed with the electron microscope, resembles the Greek letter theta, θ. This is the result of the formation of an eye, which becomes progressively larger as replication proceeds (Fig. 11-20). Another way that circular DNA replicates is discussed in Chapter 15.

The DNA of higher organisms, including the human, has been studied in a similar manner by refined autoradiography and electron microscopy. In eukaryotes, however, there are many initiation points and thus many replication forks—hundreds per molecule—along the length of the DNA. As in *E. coli*, replication proceeds bidirectionally from any one initiation point (Fig. 11-21). The replication forks that originated from different initiation points eventually

FIG. 11-21. Replication forks in eukaryotes, In eukaryotes, replication begins at numerous origin points (O1, O2, O3, etc.) and then proceeds bidirectionally by two replication forks from each point of

origin. Initiation of replication may be staggered; replication may start from one origin point well in advance of another. Eventually, replication forks that arose from different origins meet, and fusion of loops occurs.

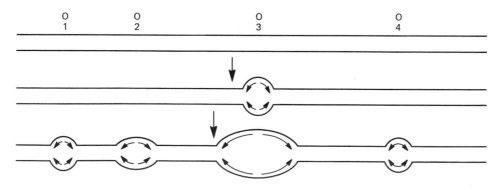

meet. In electron micrographs, replicating regions of eukaryotic DNA molecules show several loops or eyes along the molecule, in contrast with viral and bacterial DNA, which shows only one, since typically only one initiation point exists in the DNA of bacteria and viruses. In eukaryotes, the loops become larger as the smaller loops fuse. Eventually, at the end of replication, two linear, double-helical DNA molecules result.

We see that the entire eukaryotic chromosome includes several units of replication, each with an initiation site or point of origin of DNA replication and two termination sites on opposite sides of the origin. Such a unit of replication has been termed a replicon. On the other hand, a chromosome of a virus or of a prokaryote usually consists of but a single replicon. In later chapters, we will learn that replicons that reside outside the chromosome may exist in a cell.

The work of Cairns and that of Meselson and Stahl are only two of the many experimental approaches that have tested and supported implications of the Watson–Crick model. Some of the other approaches will be encountered as we pursue discussions of gene action at the molecular level.

Other aspects of semiconservative replication

Many important facts have been established since the publication of the Watson–Crick model, and a few of these must be grasped to give a more accurate and complete picture of the replication process.

In the 1950s Kornberg isolated from *E. coli* an enzyme, DNA polymerase I, which was thought to be the enzyme of replication. DNA polymerase I is able to catalyze the joining together of nucleotides and in this way to build up a DNA chain complementary to the template chain. However, the enzyme is only able to recognize the sugar–phosphate portion of the nucleotide. Preformed DNA must be present to act as a template to establish the sequence of the four kinds of nucleotides, as depicted in the Watson–Crick model. Another feature of the DNA polymerase is that it increases the length of a growing chain only in the 5′ to 3′ direction. This means that as a chain grows, the 5′ end of the nucleoside triphosphate being added is joined to the 3′ end (the −OH end) of the nucleotide already in the chain (Fig. 11-22). In other words, a nucleotide just inserted would be found at the 3′ end of the growing chain. The first nucleotide in the growing chain would have a free 5′ end, or phosphate end. Recall that the two complementary chains in DNA show opposite polarity, one running 5′ to 3′ and

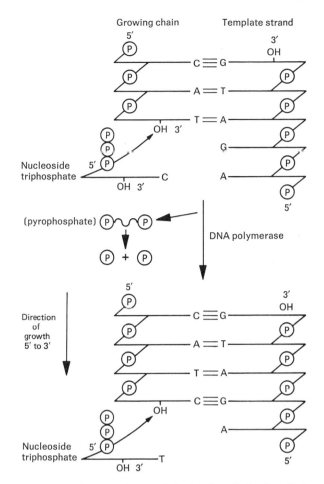

FIG. 11-22. Growth of DNA chain in the 5′ to 3′ direction. Each nucleotide added to a growing DNA chain is in its triphosphate form and is inserted by the union of its 5′ (PO₃) end with the 3′ (OH) end of the nucleotide already in the chain. A pyrophosphate is produced and is hydrolyzed to two phosphates in the elongation process, which is catalyzed by DNA polymerase, working only in the 5′ to 3′ direction.

the other 3′ to 5′ along the length of the molecule (Figs. 11-12, 11-22). Nevertheless, we will see in a moment that as the double helix unwinds, replication in both newly growing strands is 5′ to 3′.

Since DNA polymerase requires a 3′ − OH end, the enzyme by itself cannot initiate the joining of two nucleotides. A primer must be present—a molecular segment that provides the free 3′ − OH for the internucleotide linkage. So both a template and a primer are required for DNA polymerase activity (Fig. 11-23).

It was found that DNA polymerase I had still another catalytic property in addition to its synthetic ability. Besides adding nucleotides in a 5′ to 3′ direc-

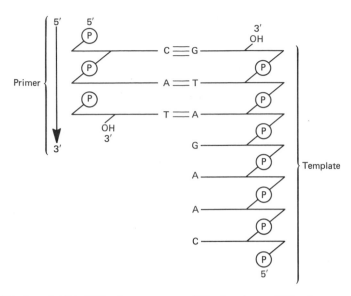

FIG. 11-23. Template and primer in replication. DNA polymerases require a template strand to determine the specific sequence of nucleotides in the growing strand. In addition, they require a primer strand, a short segment of nucleotides to provide a free 3′ —OH end. Without a primer, a DNA polymerase cannot initiate the linking of two nucleotides. The primer strand is increased in length by the addition of nucleotides in the 5′ to 3′ direction. (Compare with Fig. 11-22, which shows growth of the primer.)

tion, it can actually remove nucleotides, one at a time, from the end of a DNA chain. It thus possesses exonuclease activity. It can catalyze the removal of nucleotides, however, from either the 3′ or the 5′ end of the chain. Thus, its exonuclease activity is both 5′ to 3′ and 3′ to 5′.

In addition to exonuclease activity, many important enzymes possess *endonuclease* activity and have the ability to attack internal bonds in the DNA, thus producing nicks in DNA strands. We will encounter endonucleases in many of the following discussions.

Many years after the discovery of DNA polymerase I, a mutant strain of *E. coli* was isolated in which very little of the enzyme was present. Nevertheless, the mutant was still able to synthesize DNA at a normal rate. From this mutant strain, two other DNA polymerases were isolated in 1972—DNA polymerase II and DNA polymerase III. In nonmutant *E. coli* cells, the high activity of DNA polymerase I obscures the activities of the other two polymerases, making them difficult to detect. Like DNA polymerase I, they require a template and can assemble DNA nucleotides only in a 5′ to 3′ direction on the end of a primer. DNA polymerase II differs from DNA polymerase I in that it lacks 5′ to 3′ exonuclease activity, being able to remove DNA nucleotides only in the 3′ to 5′ direction. DNA polymerase III, however, like DNA polymerase I has exonuclease activity in both directions.

Studies of *E. coli* mutants have revealed that DNA polymerase I is not the primary enzyme of DNA replication but it plays an accessory role, which we will describe shortly. DNA polymerase II also apparently plays a secondary role in replication. It is DNA polymerase III that appears to be the primary DNA polymerase of replication in *E. coli,* because mutants that are deficient in DNA polymerase III are unable to link nucleotides together into DNA.

Since all known DNA polymerases can polymerize DNA chains only in a 5′ to 3′ direction, certain problems become apparent. Where does the necessary primer end come from in a replication fork as the double helix is unwinding, so that two new chains can be constructed on the template strands? Moreover, one chain runs in the direction 5′ to 3′ and the other 3′ to 5′. How can replication occur in the same direction on both strands if they have opposite orientations?

Answers to these questions were provided by several discoveries that have filled in some of the details of DNA replication. One very important set of experimental data has shown that DNA synthesis is not entirely continuous but takes place by the discontinuous synthesis of short DNA chains on at least one of the template strands. These fragments, each about 1000 nucleotides long, are called Okazaki fragments after their discoverer. Note (Fig. 11-24) that as the

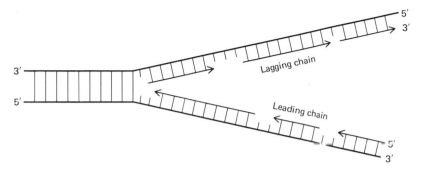

FIG. 11-24. Discontinuous replication. DNA synthesis is not entirely continuous, but takes place, at least on one of the template strands, by the discontinuous synthesis of short DNA chains, Okazaki fragments. The newly growing chain, whose —OH ends are directed toward the fork and whose template runs in a 3' to 5' direction, is the leading chain. The lagging chain has its —OH ends directed away from the fork, and its template runs in the 5' to 3' direction. Therefore, the two growing chains are extended in opposite directions. The diagram shows discontinuous synthesis on both template strands. However, some data indicate that only the lagging chain is extended in short pieces.

double helix unwinds, the short segments, or Okazaki fragments, are synthesized in opposite directions with respect to the replication fork. On the DNA template strand that runs in the 3' to 5' direction, a chain is assembled which is called the leading chain. Synthesis takes place in the direction toward the fork. This is so because any free 3' —OH ends are oriented in this direction. The other chain being assembled is called the lagging chain or the retrograde chain. Its template runs in the 5' to 3' direction, and the direction of synthesis of this lagging chain is away from the fork. Again, this is the result of the orientation of the free 3' —OH ends. Data suggest that discontinuous synthesis may occur at the replication fork only with respect to the lagging strand.

Another important discovery has supplied some answers to how a primer is provided so that a free 3' —OH end will be available to DNA polymerase at a replication fork. It has been shown that RNA is involved in the replication of DNA. As the double helix unwinds and a replication fork arises, short segments of RNA about 10 nucleotides long are assembled. Each of these acts as a primer by supplying a required 3' end. The RNA primers themselves are assembled in a 5' to 3' direction by the activity of a class of RNA polymerases called primases (Fig. 11-25).

Once the RNA primers are assembled, nucleotides of DNA are added to them in the usual 5' to 3' direction. As replication proceeds, it becomes necessary to excise the RNA primers. This removal can be accomplished by the exonuclease activities of the DNA polymerases. There is evidence that DNA polymerase I is the important enzyme in removing the primer through its 5' to 3' exonuclease activity. Following the removal of a primer, the DNA polymerase may exert its synthetic activity on adding nucleotides in the 5' to 3' direction using the parental strand as template. Note (Fig. 11-25) that although DNA polymerase can fill in the gaps left by excision of the primers, it cannot join the 3' end of the very last nucleotide of the "new" segment to the 5' end of the first nucleotide of the following segment. Consequently, a nick will be present at each site where a gap has been filled. These nicks are sealed through the activity of still another enzyme, DNA ligase, which can join the two ends together by forming a phosphodiester linkage. If a mutant cell lacks the ligase, Okazaki fragments accumulate, because they cannot be joined together.

Another important function has been attributed to the DNA polymerases. This is their ability to act as proofreaders. Recall that all of them have 3' to 5' exonuclease activity. This may actually guarantee a high degree of accuracy for DNA replication. Any nucleotide that has been erroneously incorporated into a growing chain does not pair properly with the nucleotide in that position on the template strand. A polymerase will add a nucleotide to a growing chain only if the "last" base is correctly paired. Otherwise, a DNA polymerase can excise the erroneous addition, using its 3' to 5' exonuclease activity. Once the wrong nucleotide is removed, the enzyme again exerts its polymerase activity and can link the correct nucleotide into position (Fig. 11-26).

Although most of the details of DNA replication have been assembled from experiments with prokar-

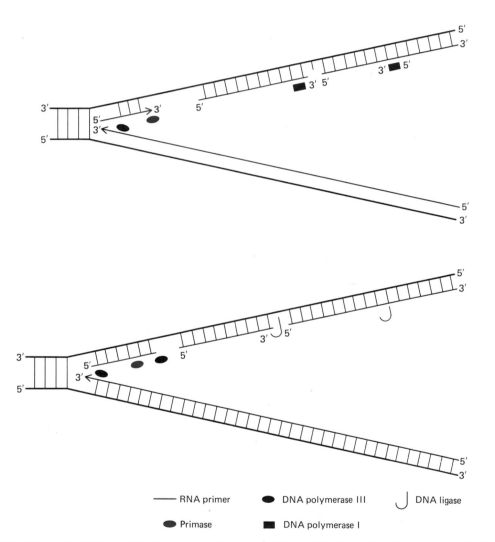

RNA primer	● DNA polymerase III	⊔ DNA ligase
● Primase	■ DNA polymerase I	

FIG. 11-25. RNA primer and replication. In this diagram, discontinuous replication is shown only with respect to the lagging strand. Short RNA segments are assembled in the 5′ to 3′ direction by the activity of primases *(above)*. Each RNA primer provides a 3′ end for DNA polymerase III, which adds DNA nucleotides to the primer segment in the 5′ to 3′ direction. RNA primer is excised by the exonuclease activity of DNA polymerase I *(middle)*, which is believed to fill in the gaps resulting from removal of the primer. DNA polymerases cannot seal the nicks left between segments after gaps are filled. DNA ligase performs this function *(below)*.

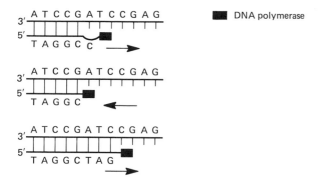

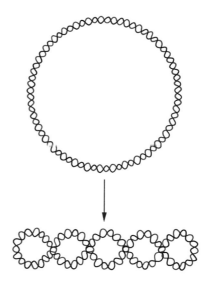

FIG. 11-26. Proofreading activity of DNA polymerases. An incorrect nucleotide is added to a growing DNA chain *(above)*. The 3' end of this chain cannot serve as a primer, because the incorrect nucleotide cannot pair properly with the nucleotide on the template. The DNA polymerase moves in reverse *(middle)* and removes the error through its exonuclease activity. Following excision, the polymerase moves forward again *(bottom)* and continues to add nucleotides to the growing chain.

FIG. 11-27. Supercoiling of DNA. Naturally occurring DNA exists as a circular molecule or as a series of large loops that are held intact, and thus have the form of circles. A circular DNA molecule *(left)* can twist around about itself and assume a supercoiled configuration. Topoisomerases convert circular DNA molecules from one form to another.

yotes, replication in eukaryotes has been found to entail many similarities. Okazaki fragments have been demonstrated in eukaryotes, but these are shorter than those in prokaryotes, consisting of approximately 100 to 200 nucleotides. RNA primers and ligase activity have also been found. Moreover, at least three DNA polymerases are known in eukaryotes, and they have been designated α (alpha), β (beta), and γ (gamma). All three have the ability to increase the length of a DNA chain, but again only in the 5' to 3' direction. The α DNA polymerase is believed to be the primary one in DNA replication in the eukaryotic cell, because only it can add nucleotides to an RNA primer. The other two enzymes may play a role in gap filling or repair. It is known, however, that the γ DNA polymerase is required for the replication of the DNA found in mitochondria (Chap. 19).

Our picture of DNA and its replication would be far from complete if we failed to take notice of still other required factors. Moreover, it must be stressed that naturally occurring DNA is not found as a long extended, linear double helix. Rather the DNA is organized into loops and is twisted around upon itself to form a supercoil (Fig. 11-27).

For replication to take place at all, the double helix must unwind, and the template must be rendered single stranded to form a replication fork. This entails a group of proteins, among them a class called *single-stranded DNA-binding proteins* (SSB proteins). As

their name implies, they have the ability to bind strongly to each single-stranded DNA chain at the site of a fork. The binding proteins do not act enzymatically. They bind only weakly to double-stranded DNA, and their role appears to be to keep the chains in a single-stranded state. In so doing, they may decrease the amount of energy required to unwind the double helix. This unwinding is accomplished mainly by a class of catalysts named *unwinding enzymes*, or *helicases*. These bind at a replication fork and are able to hydrolyze ATP and to use the released energy in the unwinding process.

A third type of protein, the enzyme *DNA gyrase*, also operates at a fork. This enzyme appears to relax the double helix in front of the replication fork. The gyrase does this by nicking a DNA strand. This enables the free end to rotate around the other chain, which performs the role of a swivel. As the DNA in front of the fork becomes relaxed, the gyrase reseals the nick (Fig. 11-28). The gyrase is an example of a group of enzymes known as *topoisomerases*. These enzymes can convert DNA from one topological form to another. During replication, the DNA gyrase relaxes the DNA by removing certain supercoils. We return to DNA gyrase and various topological forms of DNA in Chapter 17.

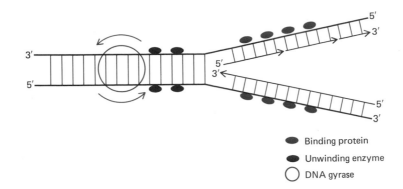

FIG. 11-28.   Proteins at the replication fork. In addition to the DNA and RNA polymerases and the DNA ligase, other proteins function at the replication fork. The binding proteins bind to single-stranded regions and keep the chains in the single-stranded state. The unwinding enzymes help to unwind the double helix. The DNA gyrase relaxes the double helix in front of the fork. It produces a nick in one of the strands, permitting rotation of the nicked strand. The gyrase later reseals to the nick.

The DNA molecule and crossing over at the molecular level

Biologists have long recognized the importance of crossing over in the evolution of sexual species, and speculations on the mechanism whereby it occurs have fascinated cytogeneticists for years (Chap. 8 and 9). With the advent of molecular biology and the wealth of information that has accumulated on the structure of the DNA molecule and its replication, new approaches have been provided to help solve the riddle of the mechanism of crossing over. In the long run, it is events at the molecular level that must be sorted out before any real understanding of crossing over at the level of the chromosome can be gained.

Several models have been proposed to explain crossing over on the molecular level. One that is compatible with various genetic analyses as well as with data from molecular biology is the Holliday model, which has undergone several modifications since its proposal in 1964. Figure 11-29 summarizes some of the major points in the Holliday model. Figure 11-29 (Step 1) shows two single, homologous DNA duplexes. Each duplex represents one duplex DNA of a single chromatid present in a bivalent. These two nonsister duplexes are the ones that will participate in crossing over. Therefore, the duplexes that are sisters to each of these are not shown, because only two of the four chromatids in a bivalent are involved in any one crossover event (Chap. 8). Two pairs of genetic factors are being followed in the given example. One parental duplex carries the linked marker genes A and B; the other duplex carries the corresponding alleles, a and b. Before recombination can

take place, the two homologous duplexes involved in crossing over must recognize each other. They must then arrange themselves in precise alignment. Otherwise a crossover event would result in duplication or deletion of genetic material.

Step 2 in Fig. 11-29 shows two endonuclease-generated nicks that affect strands of the same polarity. Each nicked strand separates from its complementary strand as unwinding takes place at the site of the nick. Each released strand now leaves its complementary partner, invades the homologous duplex, and associates with its complementary strand in the homologous duplex (Fig. 11-29, Step 3). The strands are held together at first by hydrogen bonds, but the nicks in the strands soon become sealed by DNA ligase activity. Each strand that has crossed over continues to migrate, as more regions of pairing occur between the nonsister strands (Fig. 11-29, Step 4). This is called *branch migration,* and its effect is to increase the regions of pairing between the two homologous strands that were originally parts of two separate double helixes. Note that each double helix now has a region at the point of exchange that is *heteroduplex* and that branch migration increases the extent of the heteroduplex regions. The heteroduplex DNA is represented here by the segments in which pairing occurs between complementary single strands with different origins. In the heteroduplex regions, the two strands came from different original (or parental) DNA molecules.

Note also that the two original duplexes are now joined together into one molecule as a result of the strand exchange. They are held together by the two

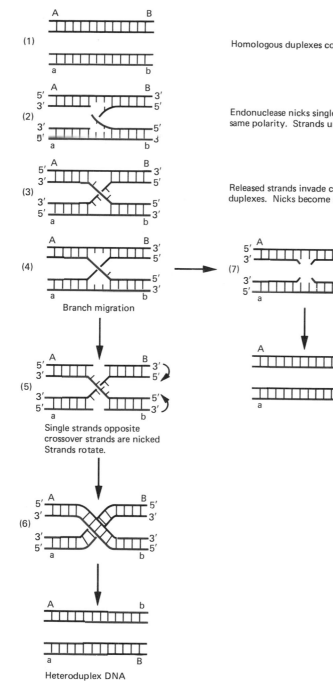

(1) Homologous duplexes come into alignment.

(2) Endonuclease nicks single strands of same polarity. Strands unwind.

(3) Released strands invade complementary duplexes. Nicks become sealed.

(4) Branch migration

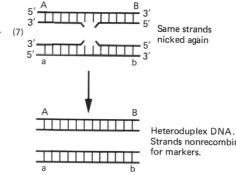

(7) Same strands nicked again

Heteroduplex DNA. Strands nonrecombinant for markers.

(5) Single strands opposite crossover strands are nicked Strands rotate.

(6)

Heteroduplex DNA
Recombinant chromatids

FIG. 11-29. Summary of steps in the Holliday model of crossing over. (From *Cold Spring Harbor Symp. Quanti. Biol. 43.* Cold Spring Harbor Laboratory, New York, 1979.)

single strands that crossed over to pair with their complements in the homologous duplex. This joint molecule must be resolved into two separate duplexes by cutting in some way. One way in which this can take place is shown in Step 5 of Fig. 11-29, where the two strands *opposite* the crossover strands are now nicked. Some digestion of DNA may occur. A rotation of the nicked strands then takes place. Gaps are filled in by DNA polymerase action and nicks are sealed by DNA ligase. The result is two recombinant chromatids, Ab and aB (Fig. 11-29, Step 6).

According to the model, it is also possible for the nicks that resolved the joined duplexes to break the *same* two strands that were nicked at the outset, leaving the other two strands completely unaffected. It can be seen from Step 7 of Fig. 11-29 that if this is the case, the two resulting chromatids will *not* be recombinant with respect to the two marker genes on either side of the exchange. Each retains the parental arrangement, AB and ab. However, it is important to note that, in both cases, the duplexes contain DNA regions that are heteroduplex. The heteroduplex DNA is represented here by those segments in which pairing occurs between complementary strands that had *different origins*. The two paired DNA strands in these regions originated in different parental DNA molecules. As indicated in the depiction of the Holliday model, heteroduplex DNA arises following crossing over, whether all four strands of the two duplexes experience endonuclease-generated nicks or the same two strands are nicked again (Fig. 11-29, Steps 6 and 7).

The Holliday model is compatible with a vast amount of genetic data. Moreover, many of its features are supported by studies of *E. coli* mutants in which genetic recombination is absent or greatly reduced. As a result of studying such mutants, investigators have identified various proteins that appear to be required for recombination. Among these are SSB proteins, which, as noted in the previous section, maintain single strands at the replication fork, and the RecA protein. This RecA protein possesses several remarkable properties. It can promote pairing between two homologous DNA molecules and bring them into alignment, a prerequisite for an exchange of strands between two duplexes. Like the SSB proteins, it too can bind tightly to single DNA strands. It also has the ability to hydorlyze ATP with the release of energy and can apparently use this ability to enable a single DNA strand of one double helix to invade another double helix and to base pair with the com-

plementary strand in the other helix, much as depicted in the Holliday model. The RecA protein can also bring about branch migration, as pictured in the model.

The RecA protein and the SSB proteins must act together to bring about the pairing reaction, since a mutant bacterial cell that lacks just one of the two types of proteins has a greatly reduced amount of recombination. Many details on recombination are yet to be resolved, and further modifications of the Holliday model may yet be made. However, the riddle of crossing over is certain to be solved as refined techniques in molecular biology continue to provide an insight into the details of this process, which is so essential to all sexual species.

REFERENCES

Avery, O. T., C. M. Macleod, and M. McCarty. Studies on the chemical nature of the substance inducing transformation of pneumlococcal types. *J. Exp. Med.* 79: 137, 1944. Reprinted in *The Biological Perspective, Introductory Readings.* W. M. Laetsch (ed.), pp 105–125. Little, Brown, Boston, 1969.

Brutlag, D. and A. Kornberg. Enzymatic synthesis of deoxyribonucleic acid: a proofreading function for the 3′ to 5′ exonuclease activity in DNA polymerases. *J. Biol. Chem.* 247: 241, 1972.

Cairns, J. The bacterial chromosome. *Sci. Am.* (Jan.): 36, 1966.

DePamphilis, M. and P. M. Wassarman. Replication of eukaryotic chromosomes: a closeup of the replication fork. *Ann. Rev. Biochem.* 49: 627, 1980.

Hershey, A. D. and M. Chase. Independent functions of viral protein and nucleic acid in growth of bacteriophage. *J. Gen. Physiol.* 36: 39, 1952. Reprinted in *Papers on Bacterial Viruses,* 2nd ed., G. S. Stent (ed.), pp 87–104. Little, Brown, Boston, 1965.

Holliday, R. A mechanism for gene converstion in fungi. *Genet. Res.* 5: 282, 1964.

Huberman, J. A. and A. D. Riggs. On the mechanism of DNA replication in mammalian chromosomes. *J. Mol. Biol.* 32: 327, 1968.

Kornberg, A. *DNA Replication.* W. H. Freeman, San Francisco, 1980.

Kornberg A. *1982 Supplement to DNA Replication.* W. H. Freeman, San Francisco, 1982.

Kriegstein, H. J. and D. S. Hogness. Mechanism of DNA replication in *Drosophila* chromosomes: structure of replication forks and evidence for bidirectionality. *Proc. Natl. Acad. Sci. U.S.* 71: 135, 1974.

Masters, M. and P. Broda. Evidence for the bidirectional replication of the *Escherichia coli* chromosome. *Nature (New Biol.)* 232: 137, 1971.

Meselson, M. and C. M. Radding. A general mode for genetic recombination. *Proc. Natl. Acad. Sci. U.S.,* 76: 2615, 1978.

Meselson, M. and F. W. Stahl. The replication of DNA in

Escherichia coli. Proc. Natl. Acad. Sci. U.S. 44: 671, 1958. Reprinted in *Biochemical Genetics*, G. L. Zubay (ed.), pp 397–402. Little, Brown, Boston, 1966.

Mirsky, A. E. The discovery of DNA. *Sci. Am.* (June): 78, 1968.

Ogawa, T. and T. Okazaki. Discontinuous DNA replication. *Annu. Rev. Biochem.* 49: 421, 1980.

Radding, C. M. Homologous pairing and strand exchange in genetic recombination. *Ann. Rev. Genet.* 16: 405, 1982.

Wang, J. C. DNA topoisomerases *Sci. Am.* (July): 94, 1982.

Watson, J. D. and F. H. C. Crick. Molecular structure of nucleic acids. A structure for deoxyribose nucleic acid *Nature (Lond.)*, 171: 737, 1953. Reprinted in *Classic Papers in Genetics*, J. A. Peters (ed.), pp 241–243. Prentice-Hall, Englewood Cliffs, NJ, 1959.

Watson, J. D. and F. H. C. Crick. Genetical implications of the structure of deoxyribonucleic acid. *Nature (Lond.)*, 171: 964, 1953. Reprinted in *Papers on Bacterial Genetics*, E. A. Adelberg (ed.), 2nd ed., pp 127–131. Little, Brown, Boston, 1966.

REVIEW QUESTIONS

1. In each of the following, DNA preparations from a donor bacterial strain are incubated with living cells of a recipient. What would be the nature of any transformed cells with respect to each of the specific traits being studied?

    A. Donor: capsulated, type III. Recipient: rough, type V.
    B. Donor: streptomycin sensitive. Recipient: streptomycin resistant.
    C. Donor: penicillin resistant. Recipient: penicillin sensitive.

2. Bacteria were infected with phages labeled with ^{35}S and then whirled in a blender. Cells were then separated from the fluid medium. Cells and fluid were next tested for radioactivity. Where will most of it be found and why?

3. In the Hershey–Chase experiment, suppose labeled carbon, ^{14}C, had been used in place of ^{35}S and ^{32}P. What would have been the results? What conclusions could one have reached? Explain.

4. Of what building units is each of the following composed?

    A. A nucleotide.
    B. A nucleoside.

5. How does DNA differ from RNA and respect to the following:

    A. Its base content.
    B. The chemistry of its sugar component.
    C. Its cellular distribution.

6. Four samples of double-stranded DNA are analyzed, and the following information is obtained:

Sample 1—15% thymine
Sample 2—20% guanine
Sample 3—30% adenine
Sample 4—40% cytosine

For each of the samples, what can you predict as likely percentages for all of the bases? Is it possible that any of these samples represent DNA from the same source?

7. Consider the nucleotide sequence, T-A-G. Assume at the time of DNA replication that A in this strand cannot pair with its normal partner but pairs with C instead. At the replication following this rare event, what will be the resulting sequence in the two new DNA molecules?

8. Using the following segment of a DNA chain, answer the questions that follow:

5′ ACACCCTTTACAAAT 3′

    A. What is the orientation and base sequence in the complementary strand?
    B. What is the $A+T/G+C$ ratio in the intact double helix?
    C. In the intact double helix, what is the $A/T:G/C$ ratio?
    D. In the intact double helix, what is the ratio $A+G/C+T$?

9. Following is the nitrogen-base composition of DNA from three sources. For each, calculate (A) $A+T/G+C$; (B) $A+G/T+C$; (C) A/T; (D) G/C; (E) %$G:C$.

DNA SOURCE	BASE (AND PERCENTAGE OF TOTAL NITROGEN BASES)			
	Adenine	Thymine	Guanine	Cytosine
Human liver	30.3	30.3	19.5	19.9
Drosophila melanogaster	30.7	29.4	19.7	20.2
Yeast	31.3	32.9	18.7	17.1

10. The two strands of a DNA molecule are separated, and one of them is analyzed for its $A+T/G+C$ ratio. This is found to be 0.2.

    A. What is the $A+T/G+C$ ratio in the other strand?
    B. Suppose analysis of a single DNA strand shows the $A+G/T+C$ ratio to be 0.2. What is the $A+G/T+C$ ratio in the complementary strand?
    C. What is the ratio of $A+G/T+C$ in the intact double-stranded DNA in the case of the two DNAs described in A and B?

11. Suppose analysis of a single DNA strand gives an $A+T/G+C$ ratio of 0.5.

    A. What ratio is expected in the complementary strand?
    B. What is this ratio for the whole molecule?

12. If the $A+G/T+C$ ratio in one strand of DNA equals 0.5, what is expected for this ratio in the complementary strand?

13. Consider four separate nucleoside triphosphates—one with the base adenine, one with cytosine, one with guanine, and one with thymine. Each is inserted into a growing DNA chain next to a template through the action of DNA polymerase. The nucleoside triphosphate with thymine becomes the first residue at the beginning of the chain, the one with adenine the second, and that with cytosine the third; the one with guanine is the last residue in the chain. Answer the following:

   A. In the new chain formed, which base will be in the nucleotide that has a free $-OH$ end?
   B. In the new chain, which base will be in a nucleotide with a free phosphate?
   C. Which end of the second nucleotide will be joined to that of the third one that is added to the growing chain?
   D. In the template strand, what will be the base in the nucleotide with the free $-OH$ end.

14. Suppose that DNA did not replicate semiconservatively, as proposed by Watson and Crick. Assume the double helix stays intact and a double helix composed of two new strands is formed at replication. This would be a conservative rather than a semiconservative method of replication. If the DNA replication were conservative, what would have been the observations in the density gradient experiment of Meselson and Stahl?

15. The two chains composing a DNA double helix unwind, and a replication fork is formed. Chain A is presented to the DNA polymerase in the direction $3' \rightarrow 5'$. Its complementary chain, Chain B, is presented in the direction $5' \rightarrow 3'$. Answer the following:

   A. On which of the two chains will DNA synthesis occur in a direction toward the fork?
   B. On which chain will the free $-OH$ ends of replicating DNA be oriented toward the fork?
   C. On which chain will the PO_3 end of the RNA primer be directed toward the fork?
   D. On which chain will the Okazaki fragments have $-OH$ ends directed away from the fork?

16. Suppose that both strands of a DNA molecule are labeled with tritiated thymidine. This DNA is allowed to undergo another round of replication in labeled medium.

   A. What will be the appearance of the Y-like fork on an autoradiograph of this DNA undergoing another round of replication in the labeled medium?
   B. Suppose the DNA that is fully labeled in both strands is allowed to undergo another round of replication in unlabeled medium. What will be the appearance of the Y on an autoradiograph of the DNA undergoing another round of replication in the unlabeled medium?

17. Match the enzymes given on the left with the aspects of DNA listed on the right. More than one item may match some of the enzymes, so an item may be used more than once.

   A. DNA gyrase
   B. DNA ligase
   C. DNA polymerase I
   D. DNA polymerase II
   E. DNA polymerase III
   F. Primase
   G. Helicase
   H. Rec A

   1. Extends DNA chains in a $5'$ to $3'$ direction.
   2. $5'$ to $3'$ exonuclease activity.
   3. The primary enzyme of DNA replication.
   4. Assembles RNA in a $5'$ to $3'$ direction.
   5. Needs a primer and a template to act.
   6. Hydrolyzes ATP to unwind DNA.
   7. Is main enzyme in removing RNA primer.
   8. Is needed to join Okazaki fragments.
   9. Brings about pairing between DNA molecules during a crossover event.
   10. A proofreading enzyme.
   11. Nicks a DNA strand in front of a fork.

12

INFORMATION TRANSFER

Gene control of protein synthesis

Many lines of evidence from an assortment of investigations indicated that the gene exerts control on the organism through its influence on protein formation. Genetic research conducted in the 1940s and 1950s with microorganisms (mainly bacteria and Neurospora) contributed an abundance of information concerning the effect of gene mutation on the chemistry of the cell. Numerous examples accumulated showing that a particular gene mutation could cause the block of a chemical step through its effect on a specific enzyme. As more and more such cases were found, it seemed that one important way in which a gene can influence a cell is through its control over an enzyme. This led to the formation of the "one gene, one enzyme" theory, which states that each enzyme is under the control of a specific gene. Because enzymes are protein in nature, it seemed possible that the gene might control the formation of proteins in general. In support of this idea, gene mutations were found in humans that alter the structure of proteins. The first such case was that of the effect of the recessive allele for sickle cell anemia on the hemoglobin molecule. In 1952, Ingram demonstrated that the adult hemoglobin of victims of sickle cell anemia differs from the normal in a surprisingly small way.

The normal hemoglobin molecule contains two kinds of chains, each composed of more than 140 amino acids (more details in Chap. 13). The defective sickle cell hemoglobin departs from the normal in the replacement of just a single amino acid (glutamic acid in normal hemoglobin) by another (valine in sickle cell hemoglobin) at one position in one of the chains. Comparable alterations at just one position in the hemoglobin also became known for other kinds of hereditary anemias.

Investigations with bacteria and the viruses that attack them (phages) showed that one of the first detectable changes in a bacterium after the entry of the DNA of the virus is the appearance in the cell of proteins that are specific to the virus, *not to* the bacterial host. This indicated that the DNA of the phage exerts it influence on the cell by directing the formation of new kinds of protein. Another argument for a relationship between DNA and protein can be made on the basis of the physical structure of the two types of molecules. Although DNA is a double helix, it is essentially a linear molecule. Each of the two strands is a polynucleotide composed of nucleotide units that follow one another in a linear order along the length of the molecule. Classic data on linkage and crossing over established that genetic maps are also linear. This concept of one gene following another

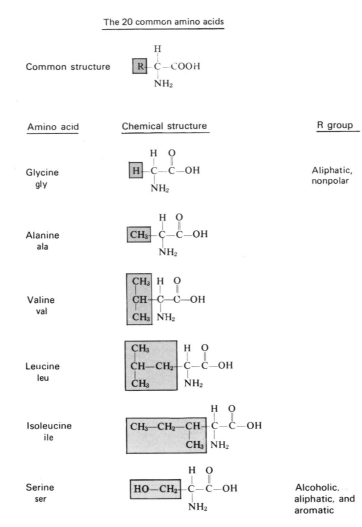

FIG. 12-1. The 20 common amino acids. Amino acids possess a carboxyl group, an amino group, and an R group (outlined in the figure). It is the R group that is the distinguishing feature among the amino acids. Its complexity varies from a single H, as in glycine, to the complex groups seen in the aromatic amino acids. Figure 12-1 continues on following pages.

along the length of the chromosome fits in well with the molecular picture of the gene, a segment of a linear molecule. So the linear genetic map corresponds to the linear sequence of nucleotides composing the DNA. Furthermore, protein molecules are also essentially linear. Protein molecules are composed of amino acids—20 different kinds. The backbone structure of all proteins is provided by linkages between the amino acids, in which one amino acid is joined to another in a linear order.

Although the types of amino acids differ in their complexity, they possess certain common features, with an exception presented by proline (Fig. 12-1). All contain a first carbon, known as the α-carbon, to which characteristic parts are bonded. One of these is a hydrogen atom, a second is a free α-carboxyl group ($-COOH$), and a third is a free α-amino group ($-NH_2$). (In proline, which strictly speaking is an imino acid, the amino group is found in a ring.) It is the fourth position attached to the α-carbon of each amino acid that distinguishes one from the other. This is the R group (R designates "radical"). The R group may represent nothing more than a hydrogen atom, as in glycine, the simplest of the amino acids. It may, however, be a more complex side grouping, as in tyrosine.

When the amino acids unite in the formation of a protein, it is their carboxyl and amino groups that

Amino acid	Chemical structure	R group

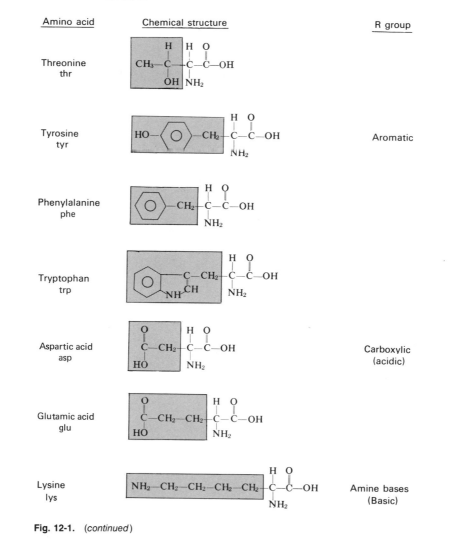

Fig. 12-1. *(continued)*

play the most important role, for the separate amino acids are joined by the reaction of the carboxyl group of one with the amino group of another (Fig. 12-2). A molecule of water is evolved, and a bond is formed known as the *peptide linkage*. As a result of these linkages, short, linear stretches may form, composed of two to several linked amino acids. These are called *peptides,* a dipeptide consisting of two linked amino acids, a tripeptide of three, and so on. A polypeptide is a chain composed of many linked amino acids. Proteins, which may be composed of one or more polypeptide chains, can become very complicated and diverse and are the most complex of all the chemical substances in the cell. Twenty kinds of amino acids arranged in different ways in chains of varying lengths permit countless varieties. In addition, a complex R

grouping on one amino acid may react with the R grouping on another amino acid in the same chain or in an adjacent one. These interactions can produce complex foldings of the separate polypeptide chains. All of this makes possible a diversity of protein molecules that differ not only in their amino acid content but in their three-dimensional architecture as well. But no matter how complex the protein, it is still a linear molecule in the topographical sense; for if we unfold a protein, we see that it is fundamentally a series of amino acids joined together by peptide linkages. We see, therefore, a correspondence from the linear genetic map to the linear DNA molecule to the linear protein molecule.

Based on such considerations, it seemed quite likely that the specific linear order of the amino acids in a

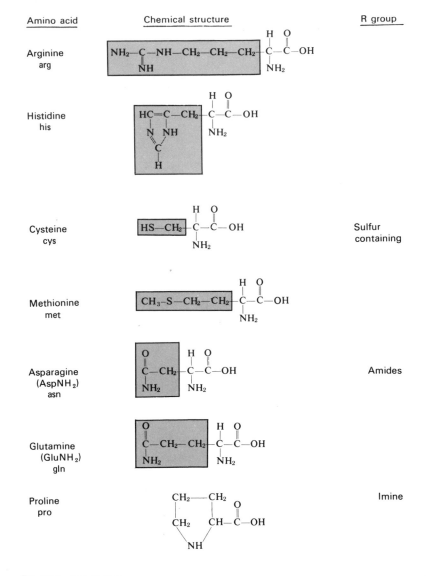

Fig. 12-1. *(continued)*

protein might be determined by the specific linear order of the nucleotides in a gene. A gene mutation (a change in a nucleotide) might bring about a change in a specific amino acid at a specific location in a protein molecule. This could explain such cases as the difference between sickle cell hemoglobin and the normal protein. It could also explain the relationship between the gene and the enzyme, as stated in the one gene, one enzyme theory, for a mutant gene might result in an altered enzyme that is unable to catalyze an essential step in a biochemical pathway.

If the control of the gene on cell activities *is* actually through its control of protein formation, it then be-

comes necessary to discover the mechanism whereby information is transferred. Cell biologists had established that protein formation takes place mainly in the cytoplasm, on the submicroscopic bodies, the ribosomes (refer to Fig. 2-3). If information on protein structure is coded in the DNA, then it becomes necessary to explain how this information is transferred to the sites of protein assembly. It was conceivable that the ribosomes themselves contained some information that determined amino acid sequence in a protein. The composition of the ribosomes was shown to be a complex of protein and RNA. As a matter of fact, most of the RNA of the cell is located in the

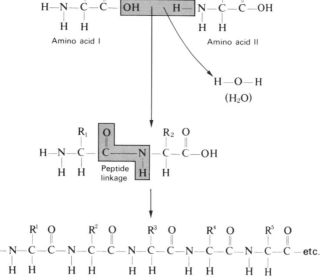

FIG. 12-2. The peptide linkage. The backbone of the protein molecule consists of amino acids joined together through peptide linkages. Reactions may take place between various R groups of the many amino acids, so that the molecule may come to assume a complex architecture. Still, when unfolded, the protein is seen to consist of amino acid units in a linear order joined by peptide linkages *(below)*.

ribosome fraction. It was suspected, therefore, that RNA molecules might somehow play a significant role in the problem of information transfer. Although RNA was known to resemble DNA in many respects, details of its structure were not clear in the 1950s. It was evident that both kinds of nucleic acid, RNA and DNA, are polynucleotides. Both contain the same purine and pyrimidine bases, with one exception. (Review Figs. 11-5 and 11-7.) The pyrimidine base thymine is not found in RNA. In its place, uracil, a different pyrimidine, occurs and this is absent from DNA. Another difference between the two nucleic acids was found in the sugar component. The ribose of RNA and the deoxyribose of DNA are both pentose sugars that are identical except at position 2 in the five-membered ring. The deoxyribose lacks the one oxygen present at the 2 position in ribose.

Transfer RNA and some of its interactions

The close similarities between the two macromolecules suggested that RNA could somehow carry information coded in the DNA. A key to the solution of the problem of information transfer was provided through research conducted at Harvard University by Hoagland and his associates. The important discovery was made that each amino acid, *before* engaging in

protein synthesis, attaches to a special kind of RNA which is *not* the RNA of the ribosome. This special RNA makes up a small portion of the soluble part of the cytoplasm. The ribosomes are not in the soluble fraction; so this soluble kind of RNA was named sRNA to distinguish it from the rRNA of the ribosome. The attachment of amino acids to sRNA was demonstrated not only in the cell but in the test tube as well. For example, if isolated ribosomes are supplied with amino acids and certain other essential substances, they may manufacture polypeptides in vitro, even though intact cells are not present. Test-tube studies demonstrated that in these cell-free systems the amino acids bind to the sRNA independently of the ribosomes. Once each amino acid is attached to an sRNA, it is then transported to the ribosome, where it may then participate in protein formation. Because the sRNA transfers the amino acid to the ribosome, it was also designated *transfer RNA* or *tRNA*. (We will continue to refer to the soluble RNA by the more descriptive term, transfer RNA.)

The tRNA was then shown to be more than a mere vehicle that carries amino acids. Since there are 20 types of amino acids that take part in protein synthesis, there are at least 20 different kinds of tRNA, one for each type of amino acid. The structure of certain tRNA molecules was worked out in the late 1960s. It

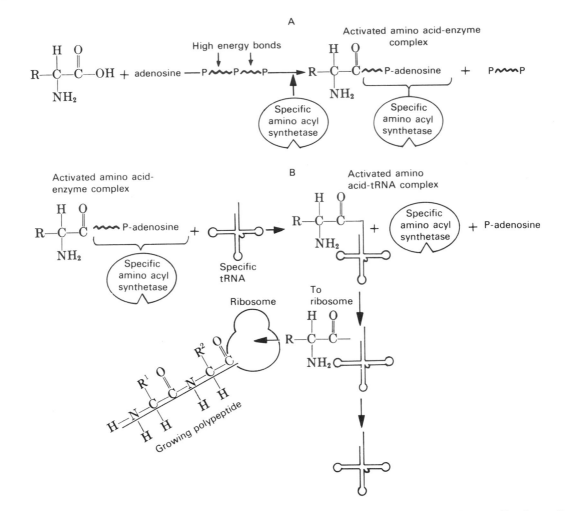

FIG. 12-3. Activation of an amino acid and its attachment to tRNA. *(A)* An amino acid reacts with the high-energy compound ATP. This is brought about by an enzyme specific for the particular amino acid. The enzyme forms a complex with the amino acid in which a high-energy bond is transferred to the amino acid from the ATP. Two phosphates are released. *(B)* The activated amino acid–enzyme complex possesses sufficient energy to react with a tRNA and to transfer the amino acid component to the tRNA. Each type of amino acid has a specific type of tRNA with which it reacts. After the reaction, the enzyme and adenosine are released. The tRNA, carrying its specific amino acid, can now move to the ribosome where the amino acid will be placed in a specific site in a growing polypeptide chain. The tRNA is then free to recycle and to react again with its specific kind of amino acid. Until an amino acid reacts with its kind of tRNA, it has no identity and cannot react with the ribosome.

was shown to be a single-stranded RNA and to have a cloverleaf shape (details in Chap. 13). An amino acid in the cell, once it is joined to its specific tRNA, receives an identity. After attachment to its tRNA, the amino acid is placed in a specific position in a growing polypeptide chain. Without being identified in such a way, the amino acids could not be assembled into any specific sequence. Therefore, the binding of each amino acid to its identifying tRNA is a critical event in protein systhesis, which was shown to entail a series of steps.

Before any free amino acid in the cell can be bound, it must be activated so that it can react with tRNA and also have the energy needed to form peptide linkages with other amino acids (Fig. 12-3A). The energy required to activate an amino acid is supplied by ATP, the universal energy donor of the cell. The transfer of energy from ATP to an amino acid requires a specific enzyme that can recognize both the amino acid and the ATP. Since there are 20 different kinds of amino acids, there are 20 varieties of the enzyme, one for each specific amino acid. These essential enzymes are called *aminoacyl-tRNA synthetases*. Not only are they necessary for the activation of the amino acids, they are also responsible for joining the energized amino acids to their appropriate tRNAs (Fig. 12-3*B*). Each synthetase thus has two critical roles to perform: (1) to make the specific amino acid active by effecting the transfer of energy from an ATP molecule; and (2) to join the activated amino acid to its proper tRNA. Once it is coupled to its tRNA, the amino acid is carried to the ribosome where it can now form peptide linkages with other amino acids that have undergone the same sequence of events. The tRNA drops away and can recycle to pick up another specific amino acid.

Transcription and messenger RNA

Although this information supplied many details on steps in protein synthesis, it also raised many additional questions: How does a specific tRNA recognize the proper amino acid? Is a code involved? If so, where is it located, and what is it? Since the ribosomes are the sites where peptide linkages are formed, do they contain a set of instructions that directs the kind of protein to be made?

A series of brilliant experiments performed by different teams of investigators provided some answers to these important questions. One line of research pursued the synthesis of RNA from its building units, the four different kinds of ribonucleotides. This kind of study was advanced by the isolation of a certain enzyme, RNA polymerase, from the bacterium *Escherichia coli*. It was found that this catalyst can stimulate the synthesis in vitro of RNA. Whole cells were not necessary for synthetic RNA to be made, but certain requirements had to be met (Fig. 12-4). Among these was the need for the ribonucleosides to be present as triphosphates, their activated or energized form. If any RNA is to be made in the test tube, it is also essential for some DNA to be present. The reason for this is that the enzyme RNA polymerase cannot link the nucleotides together on its own. DNA is required as a blueprint or template; RNA formation by the polymerase depends on DNA. The RNA that is synthesized is complementary to the DNA that is present and that serves as the template in the RNA formation. This latter point is clearly seen in the following way. The DNA used as the template in cell-free systems can be derived from different sources. In Chapter 11, we learned that the DNA of each species of organism has its characteristic base com-

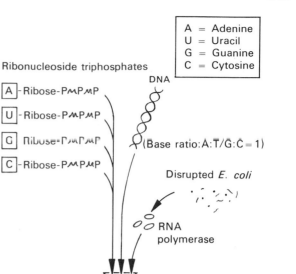

FIG. 12-4. In vitro synthesis of RNA. Cell-free synthesis of RNA can be achieved in the presence of several basic ingredients. Among these are the building units of RNA in their energized triphosphate form, DNA to act as a template in their assembly, and RNA polymerase needed as the catalyst. The RNA formed will possess the same base ratio as the DNA that served as the template.

position, which can be expressed as the A + T/G + C ratio. For example, in *E. coli*, as seen in Table 11-1, this ratio is approximately 1. DNA from calf thymus gives a different value, 1.3, for the A + T/G + C ratio, and so on for any number of organisms. When DNA from *E. coli* was used as the template in the synthesis of RNA, it was found that the base pairs in the new RNA occurred in the same ratio as they did in the template DNA. This means that the ratio of A + U (because uracil replaces thymine in RNA) to G + C is 1. If calf thymus were used, the A + U/G + C ratio would be 1.3

Additional investigations of this type left no doubt that the DNA can direct the manufacture of RNA that closely resembles it. On the basis of the DNA model, it seemed feasible that the double helix can unwind and engage in at least two different, essential processes. The first, DNA replication or the synthesis of more DNA on a DNA template, has already been

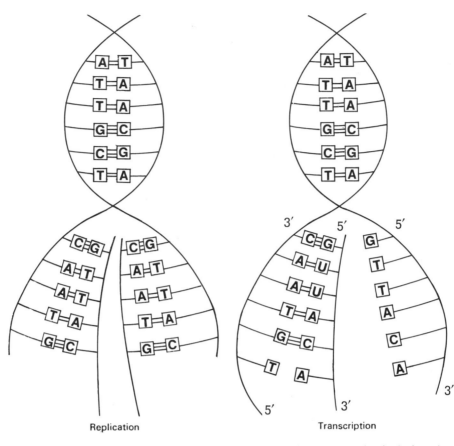

FIG. 12-5. Self-replication and transcription. Self-replication (left) is necessary to ensure continuity of the genetic information from one cell generation to the next. However, this information must be transmitted to the cell if it is to be of any consequence. A critical step in this transfer is transcription (right). A portion of the double helix unwinds and acts as a template for the formation of a stretch of RNA. The RNA resembles the DNA closely, but one major difference is the replacement of T by U in the RNA. Also, the polarity of the RNA is opposite that of the DNA template.

discussed along with the involvement of RNA in this replication process (see Chap. 11). But obviously, if the DNA did no more than direct its own synthesis, there would be no possibility for the origin of anything other than DNA. If evolution is to proceed, genetic material must not only be able to replicate itself precisely (except for rare mutations) but also to transfer stored information. Through both of these processes, replication and information transfer, genetic material can direct cellular activity and make possible the construction of other complex cellular products. At the same time, transmission of the stored information is ensured from one cell generation to the next.

It has now been established that the double helix unwinds, not always to engage in replication, but to permit the formation of RNA which is complementary to it (Fig. 12-5). For example, along one strand the enzyme RNA polymerase, using the DNA as a guide, assembles the RNA precursors (the ribonucleoside triphosphates) into an RNA strand complementary to the DNA strand. The bases along a stretch of DNA (CAATG, for example) pair off with the complementary ones of the ribonucleosides (GUUAC). This process is called *transcription,* the formation of RNA complementary to, and under the direction of, a segment of DNA. The RNA formed after transcription closely resembles the DNA stretch that acted as a template. The sequence of bases in the DNA is reflected in the sequence of the complementary bases in the RNA.

It must be noted here that RNA as well as DNA has a definite orientation, a 5′ and a 3′ end, because of the ester linkages between the phosphate and the sugar. This was discussed in the case of DNA along with the fact that the assembly of a DNA strand occurs in the 5' to 3' direction (see Chap. 11). Similarly, the nucleotides in RNA are joined in the 5' to 3' direction; the 5' end of the nucleotide being added is linked to the 3' end of the nucleotide in the growing chain. Therefore, the polarity of RNA formed in transcription is just the opposite of that of the DNA template on which it was assembled. For example, the DNA sequence reading 3' C A A T G T 5' would form a transcript with the sequence 5' G U U A C A 3' (Fig. 12-5).

If the base order in DNA corresponds to coded information, then the RNA contains information that is stored in the DNA. The RNA formed in this way could therefore be involved in the transfer of information. But exactly what kind of role does this RNA play in the cell? Is this the RNA of the ribosomes, or is it still another kind of RNA that is distinct from the rRNA and tRNA?

Answers came from research using *E. coli* and certain phages that attack it (T/2 and T/4). It had been established that once the DNA of the virus enters the bacterium, the host is directed to make protein that is not normally found in the cell. Clearly, the virus can instruct the host cell to make products specific to the virus. Therefore, it must have some mechanism that enables it to convey the necessary information to the host for the synthesis of the viral necessities. Studies of virus-infected bacteria revealed that immediately after entry of the virus DNA, an RNA appears (Fig. 12-6). This RNA was found to resemble the DNA of the virus, *not* the DNA of the host.

Another very important fact emerged from studies of infected bacterial cells. It was demonstrated that after viral infection *no* new ribosomes are formed in the cell; those present in the cell *before* infection are retained. The new RNA formed under the direction of the virus DNA was shown to associate with the ribosomes of the cell. Newly formed protein, protein specific to the virus, can then be detected in association with the ribosomes. No doubt remains that the bacterial ribosomes can somehow synthesize viral protein by assembling amino acids in the proper order. To do this, the ribosomes must require direction in the form of a set of instructions. This is apparently provided by RNA—an RNA different from the ribosomal and the transfer RNAs. This special variety was named *messenger RNA (mRNA)* because it contains a message that directs the ribosomes to assemble amino acids in an order specific for a given kind of

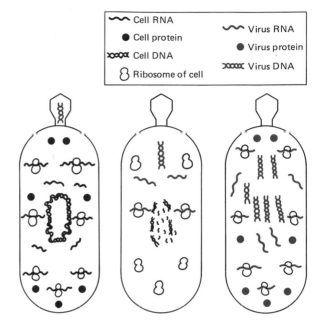

FIG. 12-6. Phage infection of a bacterial cell. (1) Before the phage injects its DNA into the bacterial cell, the DNA of the cell is engaged in transcription, forming RNA that resembles the DNA of the cell. This RNA associates with the ribosomes of the cell to form necessary cell proteins. Shortly after the viral DNA enters the cell (2), a new type of RNA appears. This is a transcript of the viral DNA and it associates with the ribosomes of the cell. The cellular DNA decomposes; RNA and protein synthesis typical of the cell cease to take place. Shortly thereafter (3), more viral DNA and viral RNA appear in the cell. Viral-specific protein also is synthesized. No new ribosomes are made; the viral mRNA associates with the ribosomes that were originally present.

protein. Without the mRNA, the ribosomes can do nothing. The existence of mRNA, as well as tRNA and rRNA, has now been established in all groups of organisms from bacteria to mammals.

Evidence for a triplet code

The process of transcription, the formation of coded information in RNA from a DNA template, is essential for the orderly transfer of genetic instructions. In transcription, passage of information occurs from one kind of nucleic acid to another—from DNA to RNA. To be meaningful, this information must now undergo *translation;* the information coded in the RNA must be read and finally transferred to protein. Before the mechanism involved in translation can be understood, we must decipher the code that resides in the DNA. Any such code must depend on the sequence of the bases in the DNA molecule. If this is so, then the bases must somehow correspond to the amino acid units that compose protein.

It is obvious that one base cannot correspond to one amino acid, because there are only 4 bases to designate 20 amino acids. Perhaps a sequence of two bases, a doublet such as CA, indicates an amino acid. However, if 4 different entities of any kind are involved, such as 4 different bases, and if 2 of these are taken at a time, only 16 different permutations or arrangements of the 4 (4^2) are possible. This is 4 fewer than the needed 20. If the code is based on triplets, three bases taken at a time, then 64 (4^3) distinct arrangements can be realized. This is many more than necessary.

Many ideas were offered on this point; among these was the suggestion that only 20 triplets designate amino acids, 1 triplet for one amino acid. The remaining triplets would be meaningless or nonsense. Another idea was that the code is *degenerate;* most or all of the triplets would represent amino acids. In a degenerate code, 2 or more triplets could code for the same amino acid.

One of the first clear lines of evidence that the code *is* actually based on triplets, adjacent groups of three nucleotides, came from the work of Crick and his colleagues. Crick favored the idea that the simplest explanation might be the correct one. He suggested that the message coded in the gene is based on triplets and that the message is read three nucleotides at a time from a definite starting point. Crick's idea proved to be correct and was supported by a host of later investigations that established the foundations of our knowledge of transcription and translation at the molecular level. We know that only one of the two DNA strands of a gene undergoes transcription to form a messenger RNA transcript coded for the amino acid

sequence of a polypeptide chain. This message is read in triplets, and its beginning is in the form of a specific triplet that indicates the start of the message, and hence the very first amino acid in the polypeptide (Fig. 12-7). The reading of triplets continues up to the endpoint of the message, indicated by another specific triplet. The sequence of triplets in the message constitutes a *reading frame,* which is a reflection of the complementary reading frame in the DNA.

To obtain information on the nature of the genetic code, Crick employed *E. coli* and phage T/4. The wild type bacterial virus can infect *E. coli* and cause cell lysis. On an agar plate, the lysis is evident as plaques, clear areas where cells have been destroyed. The size of the plaque is characteristic of wild T/4. Several types of mutations may affect the ability of the virus to produce lysis so that mutant plaques, or even no plaques at all, form on an agar plate.

Crick studied a number of different mutations in a gene (the B gene of the virus) that affects plaque formation. The mutations were induced by mutagens, and the separate mutation sites within the gene were mapped from recombination data (crossover frequencies).

In the studies, acridine dyes were employed as mutagens. These chemicals, by actually inserting into the DNA, can cause deletions and duplications of one or a few base pairs (Chap. 14). Most other mutagens typically produce their effect by causing base pair substitutions in the DNA. After exposing the T/4 virus to acridines, it was noted that the different mutations within the gene often resulted in complete loss of gene function. The sites of various acridine-induced mutations were mapped. This made it possible to infect bacteria with viruses mutant at two sites. From such a mixed infection, offspring viruses could be

FIG 12-7. Reading of a gene. The messenger RNA *(red)* contains a sequence of nucleotides coded for an amino acid sequence that constitutes a reading frame. There is a definite starting point, a first triplet (AUG), which indicates where the message begins, and the reading proceeds triplet after triplet from that point. The reading will come to a halt when another specific triplet (UGA) indicates the endpoint of the message. The reading frame in the messenger RNA corresponds to the complementary one in the template strand that was transcribed.

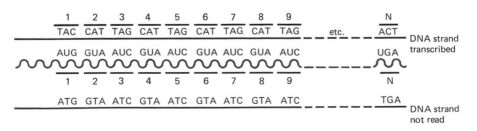

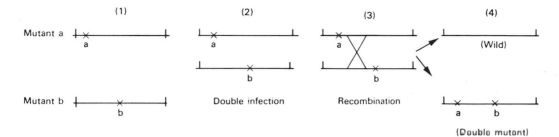

FIG. 12-8. Intragenic recombination. After exposure to a mutagen such as an acridine, mutations may arise in the particular gene under study. (1) The location of each separate mutation can be mapped. A bacterium may be infected with two types of virus (2). A process akin to crossing over between the virus particles can take place (3). Among the progeny viruses, wild types and double mutants can be recognized (4).

obtained. Among these progeny viruses, recombinants could be formed; some would be doubly mutant, others wild (Fig. 12-8). The doubly mutant ones could then be studied for their ability to produce plaques on *E. coli*. It was found that certain double mutants were nonfunctional, whereas others retained some function or behaved as if they were wild. It was possible to classify the single mutants into two categories: + and −. A combination of a + and a − could suppress the mutant effect, whereas two + or two − mutations in one virus DNA would be completely mutant.

Such results can be easily explained according to Crick's hypothesis and fit in perfectly with our present knowledge. As Fig. 12-7 shows, the reading frame starts from a beginning triplet, and the message in the mRNA then proceeds to be read three bases at a time, each triplet designating an amino acid. In the absence of mutant sites within the gene, and hence within the mRNA, a wild-type polypeptide or protein is formed. Allowing + mutations to stand for added bases, we can see that the addition of just one nucleotide will upset the reading of the gene (Fig. 12-9A). After the point of insertion, the reading is out of phase, because there is nothing within the gene to set the triplets apart. The reading of triplets from the point of insertion on is therefore changed. Many of the triplets now designate amino acids that should not be placed in particular positions in the protein chain. The result is an altered protein and a mutant effect. The same sort of reasoning applies to a − or deletion mutation, in which the reading is again out of phase from the point of alteration where the base is deleted (Fig. 12-9B). However, a combination of a + and a − mutation can bring the reading back into phase again (Fig. 12-9C). Therefore, if the stretch between the + and the − is not too long, wild-type

function might be restored. A combination of three + or three − mutations would also result in a return of function if enough of the correct message is in phase. However, two + or two − mutations together would not restore correct reading to the frame, and gene function would be lost (Fig. 12-9D).

Various combinations of mutants were made, and results were all compatible with Crick's hypothesis. They also demonstrated that at least two different classes of gene mutation can be recognized. There are the familiar kinds of mutations in which a single base pair is changed, rather than removed or duplicated (Fig. 12-9E). These can be designated *missense mutations*. Although they may cause a mutant effect, the gene often retains some of its function. Except for one triplet, all the rest of the gene reads correctly. Therefore, only one amino acid is changed in the protein dictated by this mutant gene. Missense mutation can also mutate back to wild on rare occasions.

In contrast with these are the *frame shift mutations*, illustrated by the + and − alterations. In this kind of mutation, a base is added or deleted, with the consequences that the reading of the gene may be shifted over a long stretch of nucleotides. Any protein associated with this type of genetic alteration would thus possess a long stretch of amino acids different from the normal and would very likely be nonfunctional. Spontaneous revertants of the + and − frame shift mutations would not be expected, and this seems to be the case. All the experimental evidence supported Crick's concept of a gene that is read in groups of three nucleotides form a fixed starting point.

According to this model, the portion of the gene that is coded for a polypeptide is "commaless"—it contains within it no punctuation or signals to indicate a stop to translation and then another start. However,

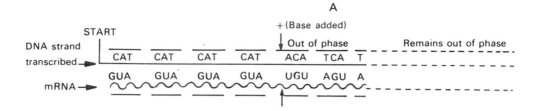

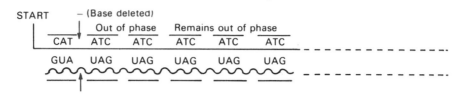

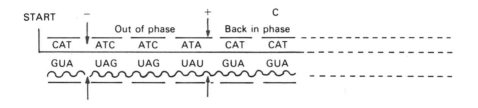

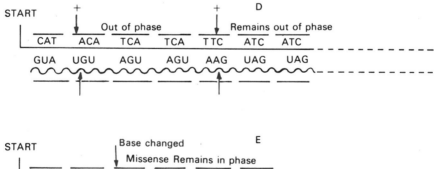

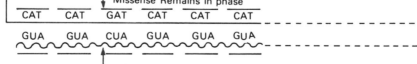

FIG. 12-9. Types of mutations and their effects on the reading of the gene. Assume that a portion of a reading frame in the template strand is CAT, CAT, CAT, and so on. The addition or deletion of a base (A and B) will cause the reading frame to go out of phase, because the bases are read in triplets starting from a fixed point within the gene. Note that the mRNA that is transcribed from the DNA will reflect the change and will also have stretches that are out of phase. A combination of a + and a − mutation (C) can bring the reading frame back into phase. If the stretch that is out of phase is not too extensive, the gene may show some function. A combination of two + mutations (D) cannot bring the frame back into phase. This is also true of a combination of two − mutations. Most spontaneous mutations are probably the result of a base substitution (E), rather than the addition or deletion of a nucleotide. Except for one triplet, which codes for one "incorrect" amino acid, all the bases are in phase and would be read correctly. The gene would retain some function. These missense mutations can revert spontaneously to normal if a later mutation causes the original base to be inserted for the substituted one. A + or a − mutation, (A–D), on the other hand, would not be expected to revert spontaneously with any ease, because a whole nucleotide must be either eliminated or added, rather than just a base substituted.

some kind of punctuation must exist to indicate the actual end of the coded message and to bring translation to a halt. This is necessary because the messenger RNA contains a trailer sequence at its 3' end that lies beyond the end of the message and is thus not coded for the polypeptide and must not be read (see later in this chapter). The work with the T/4 virus gave evidence for the existence of some sort of signal to indicate the end of the message and to bring translation to a halt at the proper point.

A mutation arose that completely eliminated the function of gene B of the virus, even though there was no reason to believe this mutation was a + or − frame shift mutation. It was reasoned that the mutant change had resulted in the formation of a triplet, a nonsense triplet, which did not designate an amino acid but brought the reading of the mRNA to a premature halt. Consequently, an incomplete protein would be formed. If this theory is correct, then it should be possible to eliminate the responsible triplet by combining it with a + and a − frame shift mutation on either side of it (Fig. 12-10). This combination of three mutations actually resulted in a partial restoration of gene function. A frame shift mutation before the nonsense triplet and another one after it shifts the reading frame, eliminates the nonsense triplet, and brings the reading back again into phase.

Further work with the T/4 virus gave evidence for the existence of some DNA encoded signals or punctuation between two adjacent genes. A deletion arose that removed a segment of DNA at the end of viral gene A and the beginning of gene B, which follows A on the chromosome. As a result, the two viral genes became joined into one reading frame and were no longer read as two separate genes. Later in this chapter, we discuss in some detail signals at the end of genes that bring transcription to a halt and thus prevent the reading of one gene into the next, to form an improper message.

FIG. 12-10. Nonsense triplet. (A) Assume that a portion of a normal nucleotide sequence is as shown and results in the formation of a completely functional mRNA that leads to the formation of a normal polypeptide chain or protein. (B) If a mutation that alters just one base in a triplet occurs, it may cause the formation of a nonsense triplet (shown here as ACT) within the gene. Since the triplet codes for no amino acid, the mRNA that forms carries a nonsense triplet (UGA) complementary to the one in the DNA. This will cause premature termination of the protein chain. The gene thus will not be able to perform its function. (C) A + and a − mutation on either side of the nonsense mutation causes a shift in the reading frame. Although the gene is out of phase in the one area, it nevertheless eliminates the nonsense codon. A protein that has a few incorrect amino acids but still has some function may form.

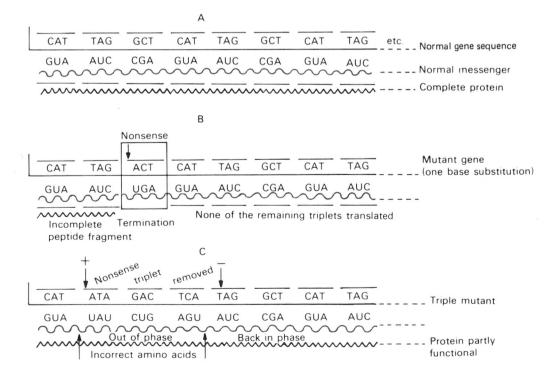

Codons and their assignments

As supporting evidence for the triplet nature of the code was assembled, important investigations were conducted to elucidate the very meaning of the 64 triplets. Which triplets stand for which amino acids, and which of them are nonsense? The pioneer studies in deciphering the genetic code were carried out by Nirenberg and Matthei, who reported a breakthrough in 1961. The logic behind the work was to provide the protein-making machinery of the cell with a *known* genetic message. Any protein made would then reflect the known information that had been supplied by the investigator. Nirenberg and Matthei employed cell-free test-tube systems containing components from ruptured bacteria plus mixtures of amino acids. For any polypeptide to be made in vitro, certain essentials are required. Besides the amino acids, all of the parts of the system needed to assemble them into protein must be present (Fig. 12-11). This includes an energy source, ribosomes, the various tRNAs along with the corresponding synthetases, and a coded mRNA. The ability to manufacture a synthetic messenger RNA with a known sequence of nucleotides provided the critical key to deciphering the genetic code. We noted that the transcription of DNA into mRNA by the enzyme RNA polymerase requires a DNA template. Fortunately, in the mid-1950s Ochoa and Grunberg-

Mangano had isolated from *E. coli* still another enzyme, polynucleotide phosphorylase, which is capable of synthesizing RNA. Although it has another role in the intact cell, it can be used in vitro to assemble RNA nucleotides in a random fashion without following a DNA template. This meant that one could provide the enzyme with certain specific ribonucleotide units and obtain an RNA of known composition. Using the phosphorylase, Nirenberg and Matthei prepared a synthetic messenger that consisted only of repeating units containing uracil (polyuridylic acid). Such RNA could then be used as a messenger and supplied to the protein-making system described earlier (Fig. 12-11).

In the experiments, separate test-tube mixtures of the essential factors were followed (Fig. 12-12A). In each mixture, one of the 20 different kinds of amino acids was labeled with ^{14}C, the remaining 19 being nonradioactive. This made it easy to tell which specific amino acid was being incorporated into any insoluble polypeptide that formed in response to the messenger RNA that had been supplied; if a polypeptide formed and if it was labeled, then it must contain the particular labeled amino acid in that mixture.

When the polyuridylic acid (UUU) was used as mRNA in the cell-free system, a polypeptide was formed. Chemical analysis showed it to be polyphenylalanine, a polypeptide composed of repeating units of the amino acid, phenylalanine (Fig. 12-12B). The messenger contained only repeating units of nucleotides with uracil. Therefore, if the code *is* based on triplets, the RNA triplet UUU must indicate phenylalanine during translation. Since the RNA triplet is a transcript of DNA, the DNA triplet AAA in the template strand must correspond to phenylalanine (Fig. 12-13). A sequence of three nucleotides designating an amino acid or other information was termed a *codon*. Thus, there are DNA codons and RNA codons, such as AAA and UUU, respectively, for phenylalanine.

In the same way, polymers composed of other single kinds of ribonucleotides were prepared and supplied as messengers. Polycytidylic acid (repeating nucleotides containing cytosine) was found to induce the formation of a polypeptide composed of repeating units of the amino acid proline. Thus, the RNA codon CCC and the complementary DNA triplet GGG represent proline. Polymers were then made containing two and three different bases. For example, suppose the enzyme polynucleotide phosphorylase is supplied with equal amounts of two kinds of nucleotides, half of

FIG. 12-11. Cell-free system for polypeptide synthesis. Simple polypeptides can be made in vitro without the presence of intact cells. However, several basic cellular ingredients are required, as shown here.

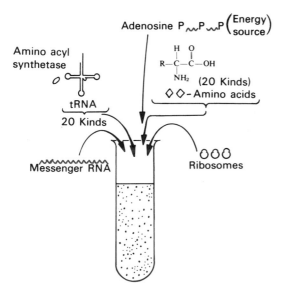

(A)

* = Label

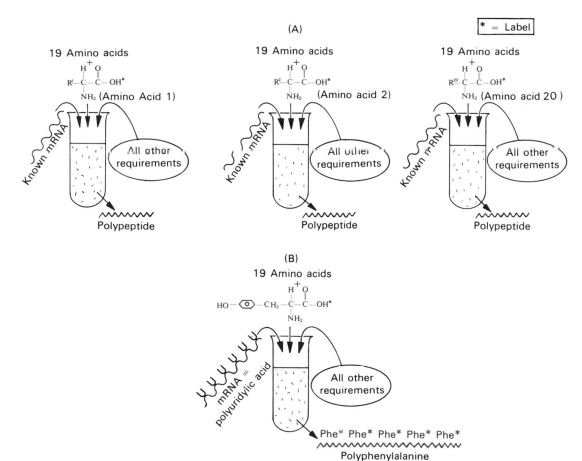

FIG. 12-12. In vitro synthesis of polypeptide using known mRNA. (A) To determine whether or not a specific codon is related to an amino acid, an mRNA of known composition is supplied to cell-free systems. Twenty separate systems must be followed. In each, one of the 20 amino acids is labeled. If the label is recovered in the resulting polypeptide manufactured by the system, it is then known that the messenger used contained a codon for that amino acid. (B) When a synthetic mRNA composed of repeating units of uridylic acid (UUU) is supplied to each of 20 separate systems, only the system containing labeled phenylalanine gives rise to a labeled polyphenylalanine chain. Since none of the other 19 yielded a labeled polypeptide, it could be concluded that UUU is the triplet that codes for phenylalanine.

them containing cytosine and the other half uracil. At random, the enzyme will assemble an RNA containing eight different triplets in equal amounts: UUU, UUC, CUU, UCU, UCC, CCU, CUC, and CCC. When used as messenger, this polyribonucleotide directed the incorporation of phenylalanine and proline, because the codons UUU and CCC are present. But in addition, the polypeptide that forms also contains leucine and serine. This means that these two amino acids must be coded by RNA codons that contain both uracil and cytosine. However, the specific triplets for leucine and serine (or for any amino acid coded by more than one base) cannot be determined by this procedure. This is so because the exact order (UUC, UCU, or CUU, etc.) cannot be established. Later

FIG. 12-13. RNA and DNA codons. If the mRNA codon UUU specifies phenylalanine, then the DNA triplet AAA must designate UUU, because the mRNA is a transcript of DNA. Therefore, each amino acid has a DNA as well as an RNA triplet or codon that is specific for it.

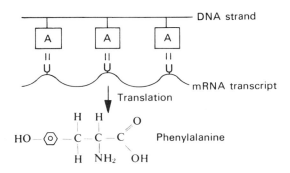

techiques devised by Nirenberg permitted the precise assignments of codons and clarified other features of the genetic code.

In these procedures, simple trinucleotides are synthesized which are composed of just three ribonucleotides in a known sequence. It was found that each simple trinucleotide was able to act as mRNA by binding to both a ribosome and tRNA carrying its specific amino acid. For example (Fig. 12-14), the trinucleotide CCC will form a complex with a ribosome and a transfer RNA charged with the amino acid proline. Because the small trinucleotide is obviously too short to act as a regular messenger, protein synthesis does not take place. However, ribosome-transfer RNA-trinucleotide complexes can be isolated on filters along with free ribosomes. As in the other method, any one mixture contains all 20 kinds of amino acids with only one kind labeled. The ribosome

fraction of the system is isolated on the filters. This also contains the ribosomes that are complexed with the charged tRNA and the trinucleotides. By determining whether or not the fraction is labeled, a specific trinucleotide sequence can be directly related to a specific amino acid.

Features of the genetic code

The procedure just described gives conclusive proof of the concept of a triplet code. But it has also shown that most of the triplets represent amino acids. As we will discuss later in more depth, only 3 of the 64 codons do not correspond to amino acids. This means that the code is "degenerate"; more than one triplet can designate a specific amino acid. This is the case for almost all of the amino acids (Table 12-1). At first,

FIG. 12-14. System for precise codon assignments. A short known sequence of three bases, a trinucleotide, can be synthesized. When added to the in vitro system, it attaches to a ribosome and also binds to a specific tRNA carrying its specific amino acid. In the procedure, only one amino acid of 20 is labeled in any one test-tube system. In this figure, it is proline, and the synthetic trinucleotide is CCC. The complex, composed of the ribosome, trinucleotide, and charged tRNA,

can be isolated on filters. In this case, the ribosome fraction will be labeled because the proline is labeled. The presence of a label in the ribosome fraction of any one test tube indicates the presence of a labeled complex. This, in turn, indicates that the particular labeled amino acid is coded by the known trinucleotide supplied. Such procedures have shown that more than one kind of trinucleotide can code for the same amino acid, meaning that the code is degenerate.

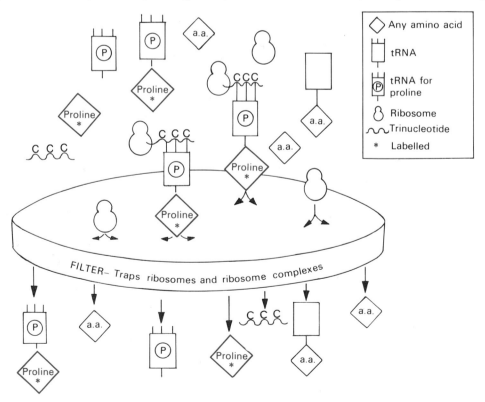

TABLE 12-1   RNA codons and their designations[a]

FIRST BASE (5' end)	SECOND BASE				THIRD BASE (3' end)
	U	C	A	G	
U	phe	ser	tyr	cys	U
	phe	ser	tyr	cys	C
	leu	ser	ter	ter	A
	leu	ser	ter	trp	G
C	leu	pro	his	arg	U
	leu	pro	his	arg	C
	leu	pro	gln	arg	A
	leu	pro	gln	arg	G
A	ile	thr	asn	ser	U
	ile	thr	asn	ser	C
	ile	thr	lys	arg	A
	met	thr	lys	arg	G
G	val	ala	asp	gly	U
	val	ala	asp	gly	C
	val	ala	glu	gly	A
	val	ala	glu	gly	G

[a]See Fig. 12-1 for the amino acid symbols. Ter designates a chain terminating or "nonsense" codon.

it might seem surprising that the code is degenerate. But this in no way means that the code is haphazard; for any one amino acid is designated only by specific codons. In addition, the degeneracy of the code provides an actual advantage to living things. As can be seen by referring to Table 12-1, it is usually the third letter of the codon that differs when two or more triplets signify the same amino acid. Consider the case of arginine, which is specified by six codons. Of these, the first two letters are identical in four of them. In the remaining two codons, the first two letters are the same.

An advantage becomes apparent when we consider the codons more closely. Suppose the CGU sequence that designates arginine is altered by mutation to CGA (Fig. 12-15). This codon also designates arginine. Actually, in four of the codons, any base change at the third position will still code for arginine. A change in the CGA triplet at the first position to A produces AGA, which again corresponds to arginine. We see that the degeneracy of the code actually acts as a buffer against the effects of mutation. If only one triplet coded for one amino acid and the remaining 44 were nonsense, then a mutation would always result in an altered protein, either by causing an amino acid substitution or by bringing about premature termination of the protein chain. Degeneracy, on the other hand, results in a number of "silent mutations"—codon changes that give rise to different codons designating the *same* amino acid. Since no amino acid substitution would follow a silent mutation, the protein product would be normal and the very fact of the mutation would remain unnoticed. Examination of the code also shows that amino acids with similar chemical properties have similar codons that differ by only one letter. This means that if the AGA codon for arginine were changed by mutation to AAA, lysine would be substituted. Inasmuch as both are basic amino acids, the mutation might not cause too great a change in the properties of the protein and could also possibly remain undetected. Natural selection has undoubtedly operated during evolution of the genetic code to reduce the effects of gene mutation as much as possible through the establishment of a genetic code with the greatest buffering power.

Three of the codons, however, do not stand for amino acids. These are the so-called nonsense codons, UAA, UAG, and UGA (also referred to as "ochre," "amber," and "opal," respectively). As suggested by

FIG. 12-15   Silent mutations. Shown here are some of the possible silent mutations that involve codons for the amino acid arginine. If any base substitution occurs at the third position in any one of these four DNA codons shown, the result is still an RNA codon that designates arginine. Therfore, cetain mutations may take place and remain un-detected, because they bring about no amino acid substitutions. The buffering effect of a degenerate code is apparent; if a change in a codon always caused a change to a codon standing for another amino acid or for termination of the protein chain, every mutation would have a detectable effect, most of them harmful.

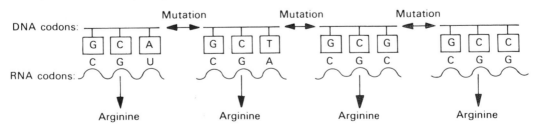

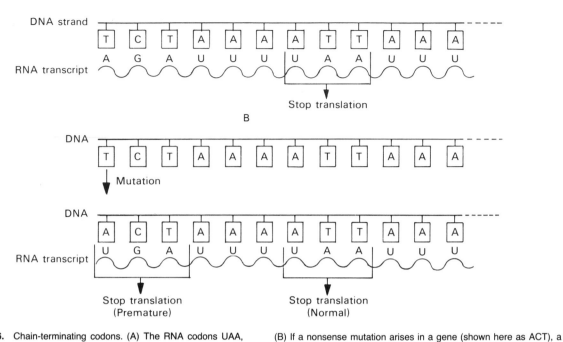

FIG. 12-16. Chain-terminating codons. (A) The RNA codons UAA, UAG, and UGA establish the end of translation of a messenger RNA. The complementary codons on the DNA would be ATT, ATC, and ACT. When the ribosome reads the messenger, the chain-terminating codons indicate the end of a polypeptide and the chain is terminated.

(B) If a nonsense mutation arises in a gene (shown here as ACT), a chain-terminating codon will be transcribed onto the mRNA in a position ahead of normal one. A premature stop signal is thus carried on the messenger and the ribosome does not translate the message past that point. Thus, that particular polypeptide chain is incomplete.

Crick's studies with the T/4 virus, these codons have been shown to provide a type of punctuation. They are better termed *chain-terminating codons* than *nonsense*, because the presence of any one of them on an mRNA indicates the end of the message and hence the end of translation by the ribosome. The existence of these RNA chain-terminating codons also tells us that the complementary DNA codons must be ATT, ATC, and ACT (Fig. 12-16A). The identification of the chain-terminating codons has been made possible through genetic studies with *E. coli* and phage, as well as through work with synthetic RNAs of known nucleotide sequence. For example, presence of the triplet UGA brings about termination of the reading of a synthetic messenger RNA by a ribosome. If it arises in vivo through gene mutation, we say that a nonsense mutation has arisen; translation comes to a premature halt with the production of an incomplete protein or polypeptide fragment (Fig. 12-16B).

We see from all of these studies that on the molecular level, mutations may represent very different alterations of the codons. A *missense mutation* (see Fig. 12-9E), we have seen, is a change from one codon specific for a certain amino acid to another codon specific for a different amino acid. An altered protein results, and the magnitude of the effect depends on which amino acid has substituted for the original. A *nonsense mutation,* on the other hand, occurs when a codon specific for an amino acid is changed to one of the chain-terminating codons and thus brings translation to an end. An incomplete protein or fragment forms. A *silent mutation* (Fig. 12-15) represents a change from a codon specific for an amino acid to another codon specific for the same amino acid. No effect is produced on the protein product because no amino acid substitution has occurred. Figure 12-17 summarizes these classes of mutation. A deleted base or an added one can alter the reading frame and bring about a *frame shift mutation* in which a portion of the coded information in the gene will be out of phase (Fig. 12-9). These can result in very defective protein products with stretches of incorrect amino acids. A frame shift mutation may even cause the origin of a chain-terminating codon with the consequent formation of an incomplete protein.

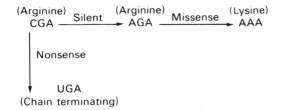

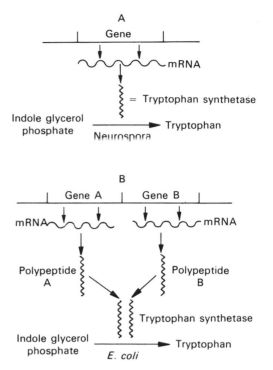

FIG. 12-17. Three types of mutations. If the RNA codon for arginine, CGA, is changed by mutation to AGA, the altered codon still codes for arginine. No change will occur in the polypeptide governed by the gene carrying this mutation. However, if CGA is changed to UGA, the protein will be incomplete. If the codon AGA for arginine is changed to AAA, a missense mutation will have occurred, because another amino acid will be placed in the polypeptide instead of the correct one. The protein may or may not be greatly altered, depending on the particular substitution. In this case, the substituted lysine would probably not cause a pronounced effect, because its properties are similar to arginine.

One gene–one polypeptide chain

The relationship between gene mutation and enzyme activity was very evident long before any insight had been gained into the molecular basis of information transfer. So many examples became known of the effect of a point mutation on an enzymatic step that the concept of "one gene–one enzyme" was widely accepted. We have already noted that the idea implies that each enzyme or protein is controlled by a specific gene. Today we know a great deal more about the structure of enzymes, proteins, and the gene itself than we did more than 30 years ago, and so we must reexamine this theory. In many cases, an enzyme *is* composed of one kind of long polypeptide chain, and it *is* under the control of one gene. An example is the enzyme tryptophan synthetase in the red bread mold, Neurospora. This important catalyst is needed for the final step in the manufacture of the amino acid, tryptophan. A segment of DNA, a gene, controls the formation of this specific protein or long polypeptide chain that makes up the entire enzyme tryptophan synthetase.

We can think of any gene as a functional unit, meaning that it has a specific function to perform. The function of the gene in this example is to guide the formation of a particular kind of polypeptide, tryptophan synthetase (or "t'ase"), which has a specific enzymatic action (Fig. 12-18A). A gene which contains coded information for some structural product such as the amino acid sequence of a polypeptide is called a *structural gene*. A structural gene may also code for an RNA product, as we will see later on.

FIG. 12-18. One gene–one polypeptide. (A) in Neurospora, the enzyme tryptophan synthetase is composed of one kind of polypeptide chain that is under the control of one gene. In this case, the one gene–one enzyme relationship is direct. (B) In E. coli, the enzyme t'ase is composed of two different kinds of polypeptides. Each is controlled by a separate gene. In this case, therefore, the concept of one gene–one enzyme does not strictly hold. However, in both Neurospora (above) and E. coli (below), the function of each gene is to govern the formation of a specific type of polypeptide, and the one gene–one polypeptide relationship is clear.

The one gene–one enzyme concept is very clear-cut in this case; however, this is not always so. In *E. coli,* the enzyme t'ase is under the control of two distinct genes (Fig. 12-18B). Two different kinds of polypeptide chains make up this bacterial enzyme. The function of each of the genes is to transfer information for the formation of a specific polypeptide. The two kinds of polypeptides then assemble to form tryptophan synthetase, essential to the final step in the formation of tryptophan. Since the completed enzyme in this case is composed of two kinds of polypeptides, each specified by a different gene, the relationship of one gene–one enzyme is not clear-cut. Because all proteins really consist of one or more polypeptide chains, however, we can modify the original statement to "one gene–one polypeptide chain." This statement is basically sound, for every kind of polypeptide chain is specified by a particular gene. On the other hand, many proteins are composed of several separate chains that become associated. Indeed, the tryptophan synthetase of *E. coli* is actually composed of four, but only two different kinds are present, each one represented twice. In Chapter 17, we will reexamine the gene in relation to the way its mRNA is processed and learn that certain genes may be associated with more than one similar but nevertheless different polypeptides. Although the one gene–one polypeptide concept holds in most cases, discoveries on the molecular level present exceptions that continue to challenge established ideas.

Choice of DNA strand read in transcription

One very important decision that must be made is the choice of which of the two DNA strands of a gene is to be read at the time of transcription. A moment's thought tells us that the RNA polymerase probably does not read both DNA strands when it forms mRNA. One reason we would not expect this is that the genetic code is degenerate. If both strands were read, then two different kinds of protein would form after the translation of the two kinds of messenger. This is not the case, for only one kind of protein or polypeptide chain is associated with a specific gene, an observation that led to the formulation of the one gene–one enzyme theory. Work with certain microorganisms has clearly shown that only one of the two strands of a gene is read.

How can it be demonstrated experimentally that a certain DNA sequence is responsible for the formation (transcription) of a certain kind of RNA? This can be achieved by the technique of DNA–RNA hybridization, one of the most valuable tools of molecular biology and one to which we will continually refer. In any variation of this procedure, the double-stranded DNA must first be denatured, transformed into a single-stranded state. This can be achieved by performing an alkali treatment or by raising the temperature above the melting point of the DNA, a point at which the two strands of the double helix separate (Fig. 12-19A). If the denatured DNA is quickly cooled, the two complementary strands will not have the opportunity to get together to reform double strands (to reanneal). RNA transcripts that are radioactively labeled may then be added to these single-stranded preparations. The mixture of DNA and RNA is incubated at a temperature that is too low to permit double-

stranded DNA to form but sufficient to allow RNA to combine with any complementary DNA segments. The formation of hybrid duplexes, which can be detected by the radioactive label, indicates specific DNA segments and the RNA transcribed from them.

In one application of this procedure, denatured DNA may be trapped on a filter and then presented with labeled RNA (Fig. 12-19B). The filters are washed and exposed to an RNase that digests away any single-stranded RNA present but does not attack the RNA paired with the DNA. The presence and level of radioactivity can be detected, indicating the presence and amount of the DNA–RNA hybrids.

The technique of DNA–RNA hybridization has provided much information on the choice of DNA strands in transcription. Single DNA strands from certain sources can be separated and isolated from their complementary strands using density gradient ultracentrifugation. This separation and isolation of complementary strands is possible in those cases in which one DNA strand is lighter than its complementary strand. Complementary strands can be recognized as heavy (H) or light (L) when the two differ with respect to the percentage of purines or pyrimidines they contain. This is so because the purines adenine and guanine are heavier than the pyrimidines thymine and cytosine. Thus, a single strand rich in adenine and guanine is heavier than its complement and will settle in a different region in a density gradient. This makes it possible to isolate H and L strands on separate filters. Certain very simple viruses contain a single chromosome composed of a DNA molecule in which one strand is heavy and one is light. This makes possible the separation of the two strands of the DNA molecule and the isolation of the H and L strands on separate filters. Populations of such viruses can be allowed to infect bacterial hosts that are growing in the presence of radioactive phosphorus. The viral mRNA that forms following such an infection is consequently labeled. When this labeled RNA is isolated and presented to the separate H and L filters, it is found that the single-stranded DNA on only *one* filter hybridizes with the labeled mRNA. This tells us that only one of the two strands of the viral DNA has undergone transcription. In the case of these simple viruses, it is usually the H strand.

Experiments with other types of viruses have shown that portions of the H strand and portions of the L strand hybridize with the RNA transcripts. This means that, unlike the situation in the very simple viruses,

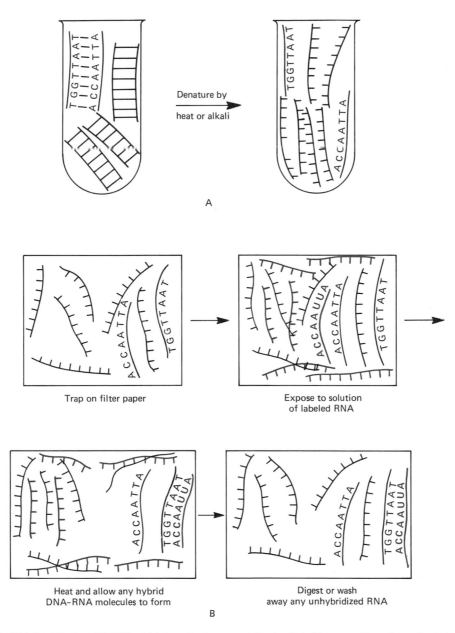

FIG. 12-19. DNA–RNA hybridization. (A) DNA molecules may be rendered single stranded (denatured) by heat or alkali treatment. The single-stranded preparation may then be used to identify any specific DNA sequences coded for specific RNA transcripts. (B) The single strands may be fixed on filter paper and then exposed to labeled RNA molecules that diffuse over them. Any extended complementary sequences on the DNA and RNA will result in the formation of DNA–RNA hybrid molecules. Excess unpaired RNA is washed away. Only the labeled RNA that is part of a hybrid molecule remains. The hybrid duplexes can now be detected as a result of the label.

the H strand and the L strand of the DNA molecule both contain sequences that code for RNA transcripts. Refined analyses have shown that for a given gene it is always the same strand of the double helix that undergoes transcription to form mRNA. This strand is known as the *sense strand*. The noncoding strand of a gene is called the *antisense strand*. The sense strands of all the genes in a genome are not always the same strand of an entire double helix, except in some simple viruses. Thus, in most genomes, the sense strand may be in the H strand for one gene and in the L strand for another. We see, therefore, that, strictly speaking, only one strand of a gene, the sense strand, carries a nucleotide sequence that specifies the amino acid sequence of a polypeptide.

Initiation of transcription

It is not known exactly how the decision is reached as to which of the two DNA strands will be read, but the enzyme RNA polymerase is actively involved. In a living cell, the polymerase must be able to recognize the beginning of a gene. If the enzyme could initiate transcription at any point in a gene, it would indiscriminately copy only portions of either one of the two strands into RNA, and no orderly transfer of the genetic information would occur. A definite starting region for the attachment of the enzyme has been found to exist. This is called the *promoter*, a DNA region containing start signals that are recognized by RNA polymerase. The information coded in the promoter enables the enzyme to recognize the beginning of a gene and the exact point where transcription is to begin.

Four different polypeptides make up the bacterial RNA polymerase: α, β, β', and σ. Since two α chains are present, the complete enzyme, called the *holoenzyme*, can be respresented as $\alpha_2\beta\beta'\sigma$. The sigma chain is quite easily removed from the rest of the enzyme, leaving the core ($\alpha_2\beta\beta'$). Actually, the sigma portion is not involved in catalyzing the formation of RNA. This is accomplished by the core without sigma. However, sigma is essential for the recognition of signals in the promoter region. Without sigma, as can be seen clearly in test-tube systems, transcription may begin at random anywhere along the DNA, with either of the two strands acting as the template.

In eukaryotes, the picture is a bit more complex with respect to RNA polymerase, since several distinct types of the enzyme have been recognized. One of these, RNA polymerase I, occurs in the nucleolus and is concerned with the synthesis of the major classes of ribosomal RNA (Chap. 13). Two other kinds of the polymerase are found in the nuclear sap. RNA polymerase II is specialized for the synthesis of messenger RNA. The other, RNA polymerase III, is reserved for the synthesis of both tRNA and the smallest RNA of the ribosome. The precise structures of these enzymes have not been completely clarified. They are known to be large proteins, composed of 2 large polypeptides and about 10 smaller ones. The interactions of these

chains with the DNA and the specific roles of each in catalytic activity are subjects of investigation.

Eukaryotes also possess at least two additional RNA polymerases, one in the mitochondrion, and the other in the chloroplast (see Chap. 19). These appear to be smaller than the enzymes that occur in the nucleus of the cell.

Promoter regions

When RNA polymerase is bound to any DNA region, it confers on that segment protection from digestion by the DNAases, DNA-digesting enzymes. This fact has facilitated the isolation of promoters. DNA fragments that contain them remain protected when they are complexed with the RNA polymerase and then exposed to a DNAase that digests away any unprotected region (Fig. 12-20). Their isolation has permitted analysis of promoter segments and the determination of their nucleotides sequences. Different promoter regions have been compared to determine

FIG. 12-20. Isolation of promoters by DNAase protection method. One way in which promoters can be isolated takes advantage of the fact that RNA polymerase protects portions of DNA to which it is bound from digestion by DNAases *(left)*. When the holoenzyme is added to any DNA fragments, the sigma chain enables the entire enzyme to bind tightly to any promoter that is present. Exposure to a DNA-digesting enzyme will then degrade all the DNA not covered by enzyme. The protected pieces, including the promoter sequences, may then be retrieved and analyzed *(right)*.

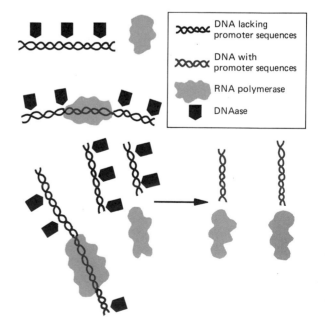

	DNA lacking promoter sequences
	DNA with promoter sequences
	RNA polymerase
	DNAase

whether or not they possess any sequences in common. Promoter mutations are also studied to identify any nucleotide sequences that may be critical to the start of transcription. The isolated promoters may also be used in vitro to identify the first bases that are transcribed from the sense strand of a gene.

Analysis of more than 100 promoters has brought to light many important features of this region and its interaction with RNA polymerase. The enzyme recognizes certain signals in the promoter, however, it does not begin transcription from the site to which it first attaches. The attachment to this recognition segment of the promoter is loose and is a distance from the point where transcription begins (Fig. 12-21A). The terms *upstream* and *downstream* are used to refer to portions of a DNA template in relation to the site of initiation of transcription. The polymerase is pictured as entering the promoter and flowing downstream into the gene. The DNA position at which transcription begins is given the designation "+1." Proceeding downstream, the + values assigned to the positions in the gene increase accordingly (Fig. 12-21B). All positions before the site of initiation are given negative values, starting with −1, next to +1.

When different promoter regions are compared, similarities in nucleotide sequence are looked for. Any sequence that proves to be common to all or most promoters would be considered significant, because a sequence that is essential as a signal for RNA polymerase recognition should be maintained or conserved from one promoter to the next. Any drastic change in such a sequence because of mutation would

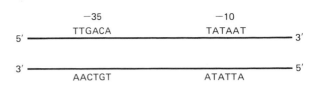

FIG. 12-22. Conserved sequences in the promoter. Two promoter sequences in which little variation exists between different promoters can be recognized. One of these is found at about −35 and is apparently the region to which the RNA polymerase first attaches. Downstream, at approximately −10, is another conserved sequence, the Pribnow box. This is the region where the enzyme attaches firmly and the double helix opens up.

be expected to decrease the recognition of the site by the polymerase, and would thus be detrimental. Comparison shows that a characteristic sequence occurs at approximately position −35, which is the recognition region to which the polymerase first binds. The −35 sequences in the promoters are variants of the sequence T T G A C A (Refer to the antisense strand in Fig. 12-22). Although the sequence given is characteristic, substitutions may exist at a given position, so that some variation is seen when different promoters are compared.

The polymerase then proceeds to move downstream from the −35 sequence to another region that is also conserved from one promoter to the next. The sequence T A T A A T centers approximately at position −10 and is known as the *Pribnow box* (Fig. 12-22). Although variation occurs in this sequence, the TA that begins the sequence, and especially the final T

FIG. 12-21. RNA polymerase and the start of transcription. (A) The enzyme first attaches to the promoter at a site that is a distance before the point where transcription starts. All sequences before the starting site are designated upstream. Those from the site of initiation of transcription on are considered downstream. (B) The position at which transcription is initiated is designated +1, and all positions downstream from it are given + values. The sequences prior to +1 are given negative values, which go in ascending order from position −1, just adjacent to the initiation site.

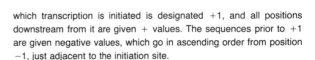

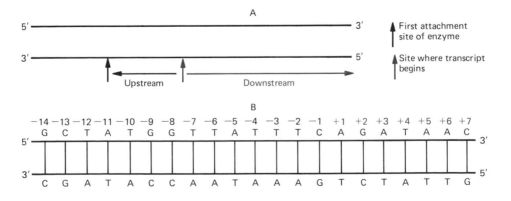

that completes it, are considered to be extremely important to the binding of the polymerase, because they are present in the vast majority of promoters. The distance between the Pribnow box and the − 35 sequence varies from promoter to promoter but is approximately 16 to 18 base pairs in the great majority of cases. It is the region of the Pribnow box where the polymerase binds more firmly, and this is believed to be the location where the double-stranded DNA opens so that the sense strand can interact with the RNA nucleotides to be assembled on it.

It should be noted that the region of the Pribnow box is rich in A:T base pairs, making the region more susceptible to denaturation, because an A:T pair is joined by only two hydrogen bonds as opposed to the three joining a G:C pair. The RNA polymerase is believed to play a main role in causing the region to open up so that single strands are exposed. The open region extends from the Pribnow box to a few positions past position + 1, the starting point of transcription. The first nucleotide in the transcript is typically one with a purine, adenine or guanine. This means that a pyrimidine—thymine or cytosine—marks the position in the sense strand at which transcription begins. As the polymerase moves along the sense strand, ribonucleotides are added in the 5′ to 3′ direction, and the transcript increases in length. The sigma chain of the polymerase detaches from the rest of the enzyme after approximately 10 nucleotides have been added to the growing transcript, so that the core enzyme alone carries out most of the elongation of the RNA. As the enzyme moves along, the open region of DNA extends only a few base pairs in length, because the double helix reforms not far behind the enzyme. The RNA that has formed a hybrid region with the DNA in the opened-up segment is then released from its hydrogen bonding with the DNA as the two DNA strands rejoin.

A few points regarding conventions used in discussing genes must be understood at this point. As we have noted, RNA, like DNA, is assembled in the 5′ to 3′ direction. RNA is written in a way that stresses this fact, with the 5′ end to the left. As Fig. 12-23 shows, the beginning of the DNA sequence of the strand that is to undergo transcription (the sense strand) starts at the 3′ end of the DNA. This is so because of the assembly of the ribonucleotides in the 5′ to 3′ direction as the RNA transcript is synthesized. Therefore, in the case of any DNA template, transcription starts toward the 3′ end of the sense strand. Each growing RNA transcript will thus have a free 3′ end to which the 5′ end of a new nucleotide can be added to increase the length of the chain.

When referring to a gene, however, it is typical to consider the 5′ end as the beginning of the gene and to refer to the antisense strand rather than the sense strand. This is done so that the sequence of nucleotides in the mRNA can be directly compared with the coding sequence of nucleotides in the template. The antisense strand is in the same orientation as the messenger RNA (Fig. 12-24). Moreover, the sequence in the transcript is identical to that in the antisense strand, and the two can be directly compared simply by substituting T's (thymines) for U's (uracils) in the

FIG. 12-23. Relationship of transcript to gene. The first nucleotide of the transcript is typically a purine. Consequently, an A or G will be found at the 5′ end of the transcript, which is assembled in the 5′ to 3′ direction. The beginning of the sequence that undergoes transcription in the sense strand is toward the 3′ end of the strand. Consequently, the leading RNA *(red)* nucleotide will have a free 5′ end and a free 3′ end to which the next RNA nucleotide can join. The final nucleotide in the transcript will have a 3′ end.

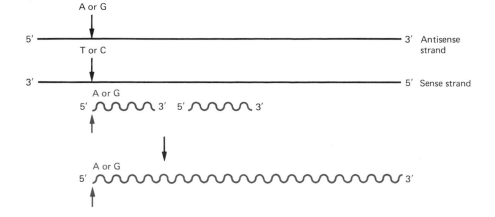

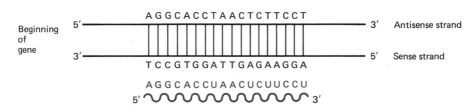

FIG. 12-24. Orientation of gene and transcript. When reference is made to a gene, the 5' end is typically considered its beginning, and the sequence of the antisense strand is presented when the nucleotides are named going from upstream *(left)* to downstream *(right)*. This convention is followed, because the 5' to 3' orientation of the messenger RNA *(red)* is preserved and can be related to a 5' to 3' orientation in the DNA. A comparison of the transcript with the antisense strand shows that the two correspond exactly in nucleotide sequence, substituting Us in the transcript for Ts in the DNA strand.

transcript. In our discussion of the Pribnow box and the −35 sequence, reference was made to sequences on the antisense strand (Fig. 12-22).

Promoters and effector molecules

In *E. coli,* before efficient transcription of certain genes can take place, the presence of specific regulatory molecules is required. For full transcription of these genes to occur, two factors are necessary. One is a simple molecule, cyclic adenosine monophosphate (cyclic AMP) formed from ATP. The small cyclic AMP molecule has been designated the *second messenger,* because of its interactions with hormones in their control of cell activity. But cyclic AMP (cAMP) has also been shown to stimulate the transcription of certain genes in *E. coli.* However, it does not do this directly. It must activate a specific cell protein, catabolite activator protein (CAP). When CAP is activated

FIG. 12-25. Promoters and effector molecules. Certain promoters require specific effectors for efficient transcription to occur. Promoters of several genes in *E. coli* have a CAP-binding site in addition to, and upstream from, the site bound by RNA polymerase. In the absence of binding at the CAP site *(above),* the double helix downstream from the polymerase tends to remain stably paired. The presence of cyclic AMP *(below)* activates the CAP. A CAP–cAMP complex can now bind to the DNA. This opens up the region downstream from the polymerase, which can now proceed to initiate transcription.

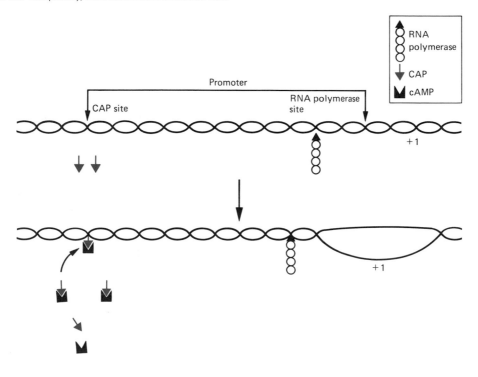

by cAMP, the CAP–cAMP complex binds directly to the DNA (Fig. 12-25). Although the CAP binding site is upstream from the RNA polymerase binding site in all promoters, its exact location varies from one specific gene promoter to another. Those promoter regions in *E. coli* that require CAP and cAMP for transcription at a maximum rate are associated with DNA sequences that code for enzymes needed in the use of various sugars such as galactose, arabinose, and lactose. (The lactose region in *E. coli* is discussed in some detail in Chap. 17). In the absence of the CAP–cAMP complex, the RNA polymerase becomes bound to the DNA in the region of the Pribnow box, but apparently the DNA in the area does not tend to open up. The action of the CAP–cAMP complex is not completely known, but one possibility is that the complex may destabilize the DNA in the Pribnow box region, causing it to denature. In those promoters requiring CAP–cAMP, the polymerase by itself is unable to do this and cannot proceed with transcription. With low levels of CAP–cAMP, the efficiency of such promoters may fall greatly, so that only a small percentage of the potential amount of transcription occurs.

Termination of transcription

The transcript proceeds to elongate as the polymerase continues downstream, adding ribonucleotides according to the information provided by the sense strand. Obviously transcription must be brought to a halt at some point to prevent reading into the next gene. Signals to halt transcription are built into the DNA in the form of sequences called *terminators*. These have been well studied in bacteria and their viruses, but not much detail is known about them in eukaryotes. Studies of terminators reveal a few features they share. Very significant is the presence of inverted repeats. As Fig. 12-26A shows, reading of an inverted repeat in the same direction (5′ to 3′, for example) on either of the two DNA strands yields the same sequence of nucleotides. You will note that on either of the two strands, the inverted repeat provides a region where base pairing can take place between segments of the same strand if the two strands are separated (Fig. 12-26B). The importance of this in termination will be evident in a moment. An inverted repeat of a terminator is typically associated with a region rich in G:C.

FIG. 12-26. The terminator region. *(A)* A terminator typically contains an inverted repeat *(Inv. Repeat)*. Note that when opposite strands are read in the same direction (such as 5′ to 3′), the nucleotide sequence is identical: CGGACG. Reading will also be identical when the inverted repeat is read in the direction 3′ to 5′. The inverted repeat is rich in G:C pairs. A region rich in A:T pairs is often found at the end of a terminator. *(B)* If the two DNA strands are separated, each single strand, as a consequence of the inverted repeat, has a region that makes possible intrastrand base pairing. As a result of this, a loop or hairpin configuration forms.

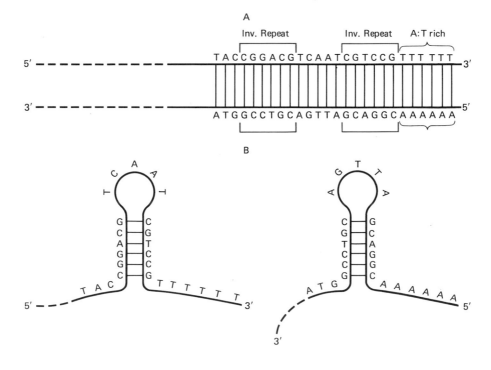

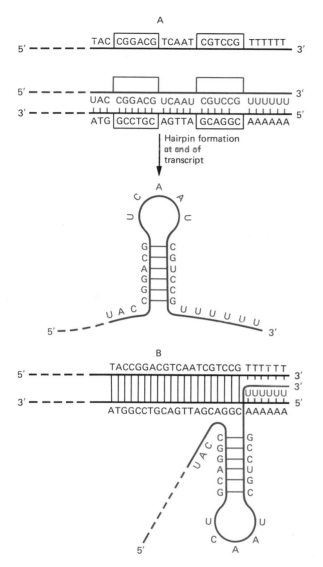

FIG. 12-27. Events at the terminator. *(A)* When transcription proceeds into the A:T-rich region, the final transcript will contain at its 3′ end a copy of the inverted repeat *(brackets)* and a terminal string of Us because of the string of Ts in the sense strand. As a result of the inverted repeat, a hairpin can form at the 3′ end of the transcript because of intrastrand pairing. *(B)* According to one model, the formation of the hairpin at the 3′ end of the transcript favors the disengagement of the transcript from the sense strand, allowing the two DNA strands to reassociate and reform a double helix in the region.

Another characteristic of a terminator is the presence of an A:T-rich region following the inverted repeat and directly preceding the point at which transcription comes to a halt. Transcripts consequently end with a sequence of six to eight uracils as a result of transcription of the series of As on the sense strand (Fig. 12-27A). The exact manner in which termina-

tion takes place is still uncertain, but research at the molecular level is consistent with the following idea. The evidence indicates that both the inverted repeat and the stretch of A:Ts provide signals to halt transcription. The RNA polymerase has been shown to pause as it approaches the region of the inverted repeat. This slowdown may start a series of reactions that bring about changes in the enzyme, leading to termination. Note from Fig. 12-27B that when the polymerase proceeds past both the inverted repeat and the A:T-rich region, the end of the transcript will contain a segment in which base pairing can take place. As a result, a hairpinlike configuration results. The formation of this hairpin is believed to favor the disengagement of the transcript from the sense strand. Once the transcript is dislodged, reassociation of the sense and antisense strands in the region is favored. More details remain to be worked out, however. The string of A:Ts with the formation of the string of Us appears to be quite important in the dissociation of the transcript from the template. Evidence for this is in the fact that deletions that cause a reduction in the number of Us in the transcript interfere with the termination of transcription.

The story of termination includes even additional factors at the molecular level, since certain terminators are known that require a special protein factor called *rho* for termination to occur. This group of terminators is known as rho-dependent terminators, as opposed to the simple or rho-independent type of terminator just discussed. Like the rho-independent terminators, those terminators that require rho protein also contain an inverted repeat that results in the formation of a hairpin in the transcript. However, there is less G:C in the region of the inverted repeat and no segment that is rich in A:T. Consequently, there is no string of Us at the end of the transcript.

Much remains to be learned about the action of rho, which apparently binds weakly to DNA. However, it can bind tightly to RNA, and according to one theory, the rho protein attaches to the 5′ end of an elongating transcript and moves along, using its ability to degrade ATP and obtain energy. As the polymerase pauses in the terminator region, the rho factor interacts with the RNA polymerase, with the result that the transcript, enzyme, and rho dissociate from the template strand. Different terminators vary in their responses to the presence of rho protein; some are able to effect termination at low rho concentrations, and others require very high concentrations. Although two classes of terminators are recognized— simple and rho-dependent—much remains to be

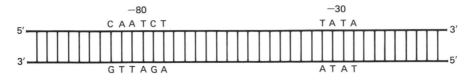

FIG. 12-28. The promoter region of eukaryotic genes. Two regions have been recognized in the promoters of eukaryotic genes that seem to be required for the proper start of transcription. They occur 80 and 30 base pairs upstream from the site of initiation of transcription.

learned about the interplay of rho on the molecular level with RNA polymerase, template, and transcript.

Eukaryotic promoters

Much less is known about eukaryotic promoters in comparison to the information on prokaryotes; however, several important features of eukaryotic promoter regions have been identified. In the case of promoters that interact with RNA polymerase II, two conserved sequences have been recognized. These have been found in organisms as diverse as mammals, birds, and insects. As in bacteria, these sequences appear to be required for the proper initiation of transcription. One sequence, the Hogness box located located about 30 nucleotides upstream, is also known as the TATA box because of the characteristic nucleotide sequence: T A T A A A. Although some nucleotide substitutions are found in different promoters, the majority do not vary greatly in the typical TATA sequence. This sequence apparently determines the exact site of the start of transcription, but evidently it does not work alone. It is apparently influenced by another nucleotide sequence found about 80 nucleotides upstream, with the characteristic pattern C A A T C T, referred to as the CAAT box (Fig. 12-28). There is evidence that sequences even farther upstream exert an influence on the initiation of transcription.

Promoters for RNA polymerase III also appear to contain TATA and CAAT boxes. However, the very surprising finding was made that the promoter sequences are *not* located upstream from the site of initiation of transcription but actually occur downstream, *past* the initiation site, starting at positions such as +8 and +50. This means that the promoter is actually within the transcription sequence itself (Fig. 12-29). Certain transfer RNA genes have been found to contain promoters that are split into two parts, so that two separate promoter segments exist, separated by gene regions coded for the tRNA.

RNA polymerase I acts exclusively on a single kind of transcription unit, the gene for the major classes of ribosomal RNA (Chap. 13). One might expect the promoter sequences for ribosomal RNA genes to have many common regions when different species are compared, since only one kind of transcript is involved. However, although certain homologous segments can be found when members from closely related groups are compared, no conserved promoter sequences are evident in comparisons of rRNA genes from less closely related species, such as yeast, Drosophila, Xenopus, and the mouse. The data suggest that nucleotide sequences located immediately around the initiation site play an important role. There is also evidence that species-specific transcription factors are involved, since successful transcription in vitro of a DNA template by RNA polymerase I depends on components that can be supplied from cell extracts. Ribosomal genes isolated from the mouse, the human, and protozoans are transcribed accurately only if all the components supplied are from the same species. Extracts from the mouse will not support the in vitro transcription of human ribosomal genes, or vice versa.

FIG. 12-29. Promoters for RNA polymerase III. Promoter sequences *(red)* that are recognized by RNA polymerase III are located within the sequences transcribed. When the enzyme binds to the promoter, it somehow starts transcribing bases upstream from the promoter, as opposed to other promoters that start the transcription of bases downstream.

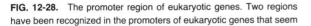

Another unexpected finding in eukaryotes is that some promoters are greatly influenced by other DNA sequences named *enhancers*. Enhancer elements have been identified in certain specific cell types where they are responsible for stimulating transcription from promoters of specific genes. For example, an enhancer has been recognized that greatly stimulates the promoter associated with the insulin gene, but *only in* insulin producing cells. Enhancers and their stimulation of transcription from specific promoters are currently the focus of research in several laboratories. They are especially relevant to the problem of regulation of gene expression and will be discussed at greater length in Chapter 17.

Much more is known about eukaryotic promoters than is known for eukaryotic termination signals; however, investigations can be expected to clarify the picture somewhat within the next few years. In Chapter 13, we will learn that the immediate RNA transcript of a eukaryotic gene undergoes processing, a feature that presently clouds the series of events that bring transcription of eukaryotic mRNA to a halt.

The DNA template and its messenger RNA transcript

The information presented in the preceding sections tells us that the transfer of coded information from the DNA template to its messenger RNA entails the interaction of many factors. Moreover, not all of the nucleotide sequences in the DNA template are reflected in the transcript. A promoter region is present that contains signals for the binding of certain factors, such as the sigma chain of RNA polymerase. The actual mRNA transcript, coded for a specific polypeptide product, begins past the promoter region. Note (Fig. 12-30) that the mRNA contains a segment at its 5′ end (the leading end) called the *leader*. The length of this leader varies from approximately 20 to hundreds of nucleotides.

As mentioned earlier, a purine nucleotide occurs at the 5′ end of the mRNA strand. The complementary DNA segment from which the leader is transcribed is referred to as the *antileader*. It must be emphasized that the leader sequence of the mRNA does not contain coded information that designates the amino acid sequence in a polypeptide. The leader is not translated by the ribosome. Its role involves the binding of mRNA to the ribosome, so that translation of the nucleotide sequences that *do* designate amino acids can take place properly. The codon AUG usually designates the point where translation is to begin (Chap. 13). The sequence of the transcript that codes for the amino acid sequence of a polypeptide is found in that part of the mRNA between the leader and the *trailer*. This trailer sequence, like the leader, is not translated. A chain-terminating codon (UAG, UGA, or UAA) signals the end of translation. The trailer is thus any sequence at the 3′ end of the mRNA that follows a chain-terminating codon. The terminator region of the template strand contains sequences complementary to the trailer. Trailer length can vary from one gene transcript to another, because transcription does not appear to terminate at any one specific site. As noted in the previous section, the messenger transcript of eukaryotes undergoes modification or processing after the first, or primary, transcript is formed.

That part of the mRNA containing coded informa-

FIG. 12-30. General relationship between a DNA template and its transcript. Only some of the nucleotide sequences of the template strand are reflected in the transcript. Moreover, only the part of the transcript that bears the information encoded in the gene proper is translated (see the text for details).

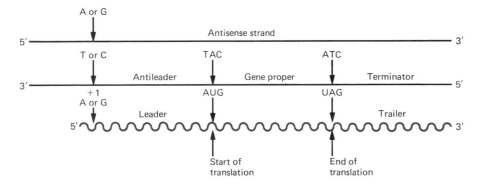

tion for the amino acid sequence of a polypeptide lies between the two stretches that are *not* translated, the leader at the 5′ end and the trailer at the 3′ end. If we consider a gene to be a sequence of nucleotides in a DNA molecule that contains coded information for some functional product such as a polypeptide, then it is just that portion of the sense strand with this coded information that is the *gene proper*.

Still other interactions involving mRNA, tRNA, and the ribosome in translation are required to achieve the final transfer of information from the DNA that is needed to complete the formation of a specific polypeptide. These form the basis of discussion in the following chapter.

REFERENCES

Adhya, S. and M. Gottesman. Control of transcription and termination. *Annu. Rev. Biochem.* 47: 967, 1978.

Avers, C. J. *Molecular Cell Biology.* Benjamin-Cummings, Menlo Park, CA, 1986.

Brenner, S., A. O. W. Stretton, and S. Kaplan. Genetic code: the "nonsense" triplets for chain termination and their suppression. *Nature* 206: 994, 1965.

Burgess, R. R. Separation and characterization of the sub-units of ribonucleic acid polymerase. *J. Biol. Chem.* 244: 6168, 1969.

Cold Spring Harbor Symposium on Quantitative Biology. *The Genetic Code,* vol. 31. Cold Spring Harbor Laboratory, New York, 1966.

Cold Spring Harbor Symposium on Quantitative Biology. *Transcription of Genetic Material,* vol. 35. Cold Spring Harbor Laboratory, New York, 1970.

Crick, F. H. C. The genetic code. III. *Sci. Am.* (Oct.): 55, 1966.

Farnham, P. J. and T. Platt. Rho-independent terminator: dyad symmetry in DNA causes RNA polymerase to pause during transcription in vitro. *Nuc. Acids Res.* 9: 563: 1981.

Freifelder, D. *Molecular Biology,* 2d ed. Jones and Bartlett, Boston, 1987.

Goldstein, L. and D. M. Prescott (eds.). *Gene Expression: The Production of RNA's. Cell Biology, A Comprehensive Treatise,* vol. 3. Academic Press, New York, 1980.

Lake, J. A. Amino-acyl-tRNA binding at the recognition site is the first step of the elongation cycle of protein synthesis. *Proc. Natl. Acad. Sci.,* 74: 1903, 1977.

Nirenberg, M. W. and J. H. Matthaei. The dependence of cell-free protein synthesis in *E. coli* upon naturally occurring or synthetic polyribonucleotides. *Proc. Natl. Acad. Sci. U.S.,* 47: 1588, 1961.

Pastan, I. Cyclic AMP. *Sci. Am.* (Aug.): 97, 1972

Roberts, J. W. Termination factor for RNA synthesis. *Nature* 224: 1168, 1969.

Rosenberg, M. and D. Court. Regulatory sequences involved in the promotion and termination of RNA transcription. *Annu. Rev. Genet.* 13: 319, 1979.

Roth, J. R. Frameshift mutations. *Ann. Rev. Genet.* 8: 317, 1974.

Sarabhai, A. S., A. O. W. Stretton, and S. Brenner. Colinearity of the gene with the polypeptide chain. *Nature* 201: 13, 1964.

Yanofsky, C., R. Drapeau, and J. R. Guest, et al. The complete amino acid sequence of the tryptophan synthetase A protein (a subunit) and its colinear relationship with the genetic map of the A gene. *Proc. Natl. Acad. Sci.* 57: 296, 1967.

Zubay, G., D. Schwartz, and J. Beckwith. Mechanism of activation of catabolite-sensitive genes: a positive control system. *Proc. Natl. Acad. Sci.* 66: 104, 1970.

REVIEW QUESTIONS

1. Give three features that the various kinds of amino acids have in common.

2. In addition to the 20 kinds of amino acids, the appropriate buffers, and inorganic substances, give five major essentials that must be supplied in a cell-free in vitro system for polypeptide synthesis to occur.

3. Assume a segment of a DNA strand has the following nucleotide sequence:

   3′   TACAAATCTCATTGTATAGGA   5′

   A. Give the base sequence and polarity in the complementary DNA strand.
   B. Give the base sequence and polarity of an mRNA strand formed by transcription from the first-mentioned 3′ strand.
   C. How many amino acids would this segment represent?

4. A. Assume that exposure to a mutagen brings about a base substitution so that the first nucleotide in the third codon from the 3′ end of the DNA strand is changed to A. What would the ultimate effect be? (Consult the coding dictionary, Table 12-1).
   B. What do we call this class of mutations?

5. A. Assume exposure to a mutagen brings about a base substitution, so that the third nucleotide of the second codon from the 3′ end is changed to G. What would the ultimate effect be?
   B. Name this class of mutations.
   C. Assume the second nucleotide of the second codon from the 3′ end was changed to C. What would the effect be?
   D. Name this class of mutations.

6. A. Suppose exposure to an acridine causes elimination of the third nucleotide in codon 1 and elimination of the first nucleotide in codon 3. What will this do to the reading of the gene, and what will be a likely outcome?

B. Suppose the second nucleotide in codon 4 were removed in addition to the two already mentioned. Now what would the effect be?

C. Suppose the first elimination, removal of the third nucleotide in codon 1, is followed by the addition of a nucleotide containing G between the first and the second nucleotides of codon 3. What would happen to the reading of the gene, and what would be a likely outcome?

D. What do we call the class of mutations described here?

7. Of the two amino acids, tryptophan and arginine, which is more likely to be substituted in a polypeptide chain by another amino acid as a result of a mutation? Explain why after consulting a coding dictionary.

8. Suppose the base ratio is determined for each of the two strands of a certain structural gene. The transcript of the gene is isolated and the base ratio is then determined for it. Using the following results, answer these questions:

A. Which is the H Strand?
B. Which is the sense strand, and why?

	A	T	G	C	U
DNA Strand A	31.1	26.3	29.1	13.5	–
DNA Strand B	26.1	30.9	13.7	29.3	–
Transcript	26.2	–	13.3	29.3	31.2

9. Synthetic trinucleotides are added to cell-free test tube systems that contain all the ingredients for polypeptide synthesis, except mRNA. In each tube, one of the 20 kinds of amino acids present is labeled, and the labeled amino acid type varies from tube to tube. Ribosomes and any complexes formed with them can be trapped on filters. Features of five such systems are given here. Answer the pertinent questions that follow.

Tube	Trinucleotide added	Labeled amino acid present
1	AAU	Leucine
2	AAC	Proline
3	AGU	Serine
4	AGC	Leucine
5	UGA	Serine

A. After consulting a list of codon assignments, tell what amino acid, if any, will be bound in a complex in each tube.

B. Tell whether or not the filter will be labeled in each case as a result of the trapping of labeled complexes.

C. If we had not known the codon assignments at the outset, what would the results have indicated?

10. Slow heating of a DNA solution brings about separation of the DNA strands. The DNA is said to undergo melting or denaturation. The melting temperature is denoted by T_m, the temperature at which half of the double-stranded molecules are denatured. Of the three samples of duplex DNA represented here, which would you expect to have the highest melting temperature (T_m) and which the lowest? Explain.

Sample 1—AGAAGATTATCACAT
TCTTCTAATAGTGTA

Sample 2—AGCCATCGACAGGTG
TCGGTAGCTGTCCAC

Sample 3—AGTTGTCATTAGGAC
TCAACAGTAATCCTG

11. Of the following two double-stranded DNA molecules, which would require a higher temperature for complete strand separation, and why?

Sample 1—AGAAGTCACAAACTTTGATTC
TCTTCAGTGTTTGAAACTAAG

Sample 2—AATTAGCCGGCAAACATTATA
TTAATCGGCCGTTTGTAATAT

12. Following is a portion of an antisense strand with an upstream position noted. Give the transcript that would be associated with this portion of the DNA along with its polarity.

$$5'\ \overset{-6}{T}ATAACGAACATAAATTT$$

13. Below is a portion of a sense strand with a downstream position noted. Give the transcript expected along with its polarity.

$$+6$$
$$\downarrow$$
5' CCCCATAAATTATAT 3'

14. A transcript is represented here. Give the sense and antisense strands that are related to it, along with their polarities.

5' AGACCCUAUGCAGAG 3'

15. Following is a section of duplex DNA containing inverted repeats. Identify the repeats and show the hairpins that can form when both strands are separated. Also give the polarities of the strands.

5' ATCAAATTATGAGTTACAACTCATAACTAGGG 3'
3' TAGTTTAATACTCAATGTTGAGTATTGATCCC 5'

16. In Question 15, assume that the lower strand is the sense strand. Show the hairpin that can form from the transcript along with its polarity.

17. Assume that the following sequence represents a sense strand for five amino acids.

TACGAGCCCTCAAAG

A. What must be the polarity of the strand? Explain.

B. What amino acids would the strand designate reading the transcript in the direction 5′ to 3′?

18. Following are two DNA strands composing a region of a gene that codes for nine amino acids. Determine which of the two is the antisense strand and which the sense strand. Deduce the polarity for each strand. Explain.

CCCATTATGATCACTGTGTATTTAGAT

GGGTAATACTAGTGACACATAAATCTA

19. Following is a portion of an antisense strand. The entire complementary sequence in the sense strand undergoes transcription. Keeping in mind that the antisense strand shown contains a segment that corresponds to a portion of the leader in the transcript, give the amino acid sequence for which this portion is coded.

5′ AATTTTATGTTCTCTATCGGG 3′

20. Following is a portion of a sense strand of DNA. Give the amino acid sequence for which this strand is coded,

numbering the amino acids 1, 2, and so on. Keep in mind that the transcript is translated in the direction 5′ to 3′.

3′ TACTATTGCGGA 5′

21. In some viruses, the genetic material is DNA, whereas in others it is RNA. On the basis of the composition shown here for four viruses determine whether each virus is a DNA or an RNA virus and whether the nucleic acid is single or double stranded.

NUCLEIC ACID SOURCE	BASE (AND PERCENTAGE OF TOTAL NITROGEN BASES)				
	Adenine	Guanine	Cytosine	Thymine	Uracil
Virus 1	25	24	18	33	
Virus 2	28	22	22		28
Virus 3	31	19	19	31	
Virus 4	22	19	26		33

13

MOLECULAR INTERACTIONS IN TRANSCRIPTION AND TRANSLATION

The ribosome and some of its interactions

The burst of genetic research in the late 1950s and early 1960s contributed valuable insights into the mechanism of translation. The process had been shown to encompass interactions among the ribosomes, the mRNA, and the tRNAs, each kind charged with its specific amino acid. Although ribosomes themselves cannot assemble amino acids into protein, it soon became evident that they are much more than passive entities that somehow simply form peptide linkages. Ribosomes from different kinds of cells and creatures are surprisingly uniform in their physical and chemical characteristics. Their size and weight may be measured by the rate at which they sediment or settle out in a centrifugal field. The rate of sedimentation is expressed in S (Svedberg) units. On this basis, two size classes of ribosomes can be recognized, those from prokaryotes such as *E. coli*, with an S value of 70, and those from the cytoplasm of eukaryotes, with an S value of 80. Ribosomes are also found in the chloroplasts and mitochondria of eukaryotic cells, and these are similar to the 70S ribosomes of prokaryotes. The significance of these is discussed more fully in Chapter 20.

A ribosome is composed of two subunits that can be separated by lowering the magnesium concentration in the environment (Fig. 13-1). The smaller of the two subunits sits like a cap on the larger. The subunits also have characteristic S values, 50S and 30S, respectively, for the larger and smaller ones of prokaryotes and 60S and 40S for the corresponding subunits in eukaryotes. The subunits may reassociate to form a complete ribosome if the magnesium concentration is again raised. Each of the subunits is composed of RNA and protein. The RNA found in each ribosomal subunit has a characteristic S value also. In prokaryotes, the smaller subunit contains a 16S RNA molecule. The 50S subunit contains two kinds of RNA molecules—a larger one with a sedimentation value of 23S and a smaller one with a value of 5S. The smaller (40S) subunit of eukaryotes contains an RNA molecule with a value of 18S. The 60S subunit has one large RNA component whose sedimentation value differs with the species grouping (Table 13-1). For multicellular animals, the S value of the large RNA molecule is 28S; for protozoans, fungi, and plants, the S value is 25 to 26. Two smaller kinds of RNA molecules are also present in the 60S subunit of all eukaryotes. One is a 5S molecule, and the other is a different type with an S value of 5.8. This latter species of RNA molecule, not found in

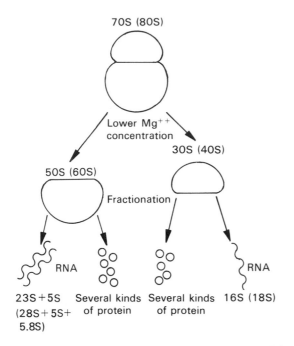

FIG. 13-1. Ribosomal components of prokaryotes and multicellular animals. A ribosome consists of two subunits, which may dissociate and associate depending on the Mg^{++} concentration. Each subunit may be fractionated into its RNA and protein components. The sedimentation value is characteristic for the ribosomes and their parts. In the figure, the S values are given for prokaryotes with the values for multicellular animals in parentheses.

TABLE 13-1   Characteristic S values of cytoplasmic ribosomes and their components

SOURCE	INTACT RIBOSOME	RIBOSOME SUBUNITS	rRNA IN SUBUNIT
Prokaryotes	70S	30S	16S
		50S	23S, 5S
Eukaryotes	80S	40S	
		60S	
Protozoa		40S	18S
		60S	25–26S, 5S, 5.8S
Fungi		40S	18S
		60S	25–26S, 5S, 5.8S
Plants		40S	18S
		60S	25–26S, 5S, 5.8S
Animals		40S	18S
		60S	28S, 5S, 5.8S

prokaryotes, is very closely associated in the larger subunit with the larger RNA molecule.

The proteins of the subunits are very complex. In eukaryotes, approximately 30 different kinds are found in the smaller one and about 50 in the larger. Some of them have been shown to be essential for forming peptide bonds and binding tRNA molecules. The complexity of the ribosome is undoubtedly related to the active role it plays in protein synthesis. The larger subunit must be able to recognize mRNA and tRNA; it must also join amino acids by peptide linkages and move the mRNA along so that the coded information in the message can be completely translated. Important roles of the smaller subunit include binding to the mRNA, recognizing where translation is to begin, and ensuring that the mRNA retains a single-stranded configuration so that all the codons are exposed for reading. If the mRNA looped on itself, certain codons might not be available for translation.

Also binding to the smaller subunit is the tRNA bearing the amino acid that will be the first one in the protein that is to be constructed (more details follow). Evidence indicates that the two subunits of the ribosome actually exist in the cell as separate parts and do not associate until they engage in protein synthesis (Fig. 13-2). Once a ribosome reads a messenger RNA strand, the two subunits again dissociate and are free to recycle. This means that the "top" of a ribosome may associate with the "bottom" of another if they engage again in translation. As Fig. 13-2 also shows, single ribosomes do not read a messenger strand. Instead, groups of ribosomes associate with a single mRNA, an association termed a *polyribosome* or *polysome*. It must be emphasized that in the polyribosome, each single ribosome forms a separate protein molecule independently. One ribosome does not cooperate with another to form a protein. Therefore, at any moment as a message is being read, protein molecules of the same kind in different stages of completion will be associated with different ribosomes along the length of the mRNA. Once a ribosome reaches the end of the messenger, the completed protein molecule is released. The ribosome leaves the mRNA and dissociates into subunits that may participate again in translating this same or a different message. The messenger RNA is read in the 5' to 3' direction, the same direction in which it is assembled on the template strand.

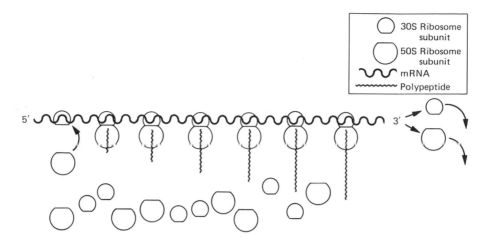

FIG. 13-2. Polysome and ribosome. The ribosomal subunits exist as separate entities in the cell. A 30S and a 50S subunit associate toward the beginning of the mRNA, the 5′ end. Each mRNA strand contains a group of ribosomes, each in the process of forming a polypeptide. The entire association is termed a *polysome.* As the ribosome moves toward the 3′ end of the mRNA, its polypeptide grows in length. Near the end of the messenger, the finished polypeptide is released and the two subunits dissociate to recycle and enter the pool of subunits in the cell. (The diagram does not show tRNA and other details.)

Molecular interactions in translation

For the genetic information to be read, a mRNA must be singled stranded. The messenger strand is composed of ribonucleotide units, and these are complementary to the DNA of the sense strand from which they were transcribed. Since these DNA segments represent specific genes and therefore contain different kinds of genetic information, the lengths of genes must vary. Consequently, mRNAs in a cell would be different in size. Since the polypeptide chains of proteins occur in an assortment of sizes, so do the corresponding mRNAs. Unlike the RNA of the ribosomes, the mRNA of a cell, as expected, is far from uniform. Nor is the mRNA as simple to characterize as the tRNA. The tRNA molecule is a relatively small one composed of only about 80 nucleotides and with an S value of only 7.

Since each kind of tRNA gives identity to a specific amino acid, transfer RNA molecules cannot all be identical. They must contain different kinds of coded information. The structure and exact nucleotide sequence is known for several tRNA molecules, and this prediction has been upheld. The tRNAs have been found to be single stranded, but X-ray diffraction studies indicate a double-helical structure in certain portions of the molecule. The reason for this apparent contradiction is that the single strand of the tRNA loops to form folds, somewhat in the shape of a clo-

verleaf (Fig. 13-3). Four loops can be recognized: a large one near the 3′ end followed by a small one that varies in size, and two other large loops moving toward the 5′ end. The molecule is held in this shape because of hydrogen bonds between complementary base pairs that lie opposite each other within a fold.

Chemical analysis of the various tRNAs has shown that the 5′ end usually begins with a nucleotide containing guanine. The 3′ end always terminates in the sequence of bases: cytosine, cytosine, and adenine. It is this adenine-terminating end (the 3′ end) that joins to the activated amino acid. It will be recalled that this takes place when the tRNA and amino acid are linked by a specific enzyme, an aminoacyl-tRNA synthetase. In addition to the four bases typically found in RNA (A, U, G, and C), the transfer RNAs have been found to contain exceptional bases, those that differ slightly from the usual ones. These unusual bases are derived from the more familiar ones once the latter have been inserted into the RNA strand. Since they do not engage in base pairing, they may prevent pairing in the loop region and thus permit these regions to interact with certain specific molecules and cell components. One of these odd bases, pseudouradine (ψ) always occurs in the sequence G C ψ in the loop near the 3′ end (Fig. 13-3). This sequence is thought to be involved in attaching the tRNA to the ribosome. Note the presence in this transfer RNA loop of ribothymidine, an exception to

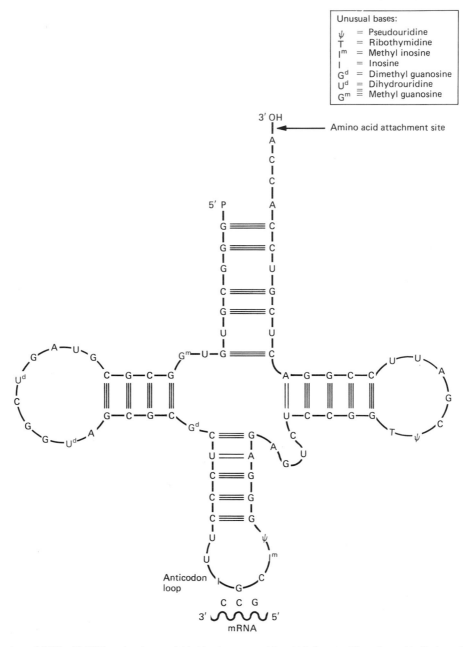

FIG. 13-3. Structure of tRNA. All tRNA molecules are folded back on themselves, as in the case of a yeast alanine tRNA, depicted here. The clover leaf shape is maintained as a result of complementary base pairing, G:C and A:U. All tRNAs terminate at the 3' end with an adenine nucleotide followed by two cytosine nucleotides. It is this 3' end to which the specific amino acid attaches. All tRNA molecules terminate at the 5' end with a guanine nucleotide, and all contain unusual bases along with the usual ones: A, U, G, and C. The lower loop, the anticodon loop, contains a triplet that can base pair with a codon in the mRNA.

the rule that thymine is not found in RNA. The large loop near the 5' end contains the unusual base dihydrouridine, and this loop is thought to be the portion of the molecule that binds to an aminoacyl synthetase.

One important difference among the kinds of tRNA is found in a sequence of three nucleotides in the lower loop of the molecule (Fig. 13-3). This triplet is critical in placing the amino acid in the correct position in the growing protein. It accomplishes this by recognizing the complementary triplet in the messenger. Because it is complementary to the codon in the mRNA, it is called the *anticodon*. The anticodon in the transfer RNA can base pair with the codon of the messenger.

It is important to note here some points relating to the polarity of codons and anticodons. Remember that a DNA strand and its RNA transcript possess opposite polarities (refer to Fig. 12-5). The RNA codons given in Table 12-1 are written as they would occur in mRNA reading in a 5′ to 3′ direction. The corresponding DNA codons are written in the reverse order, 3′ to 5′. Thus, the RNA codon 5′ AUG 3′ (for methionine) corresponds to the DNA codon 3′ TAC 5′. The anticodon in the tRNA, though complementary to the codon in the mRNA, possesses the opposite polarity, and in this case would be 3′ UAC 5′. The ability of the anticodon to base pair with a codon in the mRNA enables the tRNA to position its amino acid properly in a growing polypeptide chain according to the scheme in Fig. 13-4.

The smaller unit of the ribosome attaches to the messenger RNA and to the first transfer RNA. This first tRNA contains an anticodon that pairs with the first codon in the messenger that denotes an amino acid. It is the portion of the messenger strand toward

the 5′ end that becomes attached to the ribosome and this first tRNA. Recall (see Fig. 12-30) that the mRNA contains a leader segment that is not translated. Analyses of the leader segments of several prokaryotes show it includes a portion that is complementary to a stretch of nucleotides in the 16 S RNA of the smaller subunit. This may facilitate the initial binding and orientation of mRNA before translation commences. The first codon to designate an amino acid is located immediately after the leader segment. The mRNA, as noted previously, is read (translated) in the 5′ to 3′ direction, the same direction in which it was synthesized.

After the complex is formed, composed of the smaller ribosomal subunit, the mRNA, and the tRNA, the larger subunit of the ribosome associates with the smaller to form a complete ribosome. The larger subunit contains two adjacent sites that can bind tRNA molecules. One of these sites, the *amino site* (or A site), is the one that accommodates all the tRNA molecules arriving *after* the first one. For example, the second tRNA, carrying its specific amino acid, moves into the amino site of the larger subunit. The other site, which is bound to the first charged tRNA, is the *peptide* or P site. After entry of the second

FIG. 13-4. Beginning of translation. (1) tRNA, carrying the first amino acid coded by the first codon in the messenger, attaches to the smaller subunit. (2) The larger subunit then associates with the smaller one. A second tRNA with its amino acid moves toward the A site of the larger subunit. (3) The first tRNA is now at the P site and the second one is at the A site. A peptide bond forms between the NH₂ group of amino acid 2 and the COOH of amino acid 1. (4) The first tRNA is

ejected from the ribosome and the second one, now carrying both amino acids, moves to the P site. Meanwhile, a third charged tRNA has moved to the A site. The process repeats until a chain-terminating codon is reached. Note that it is the 5′ end of the mRNA that is read first by the ribosome. The ribosome translates the messenger from the 5′ to the 3′ end.

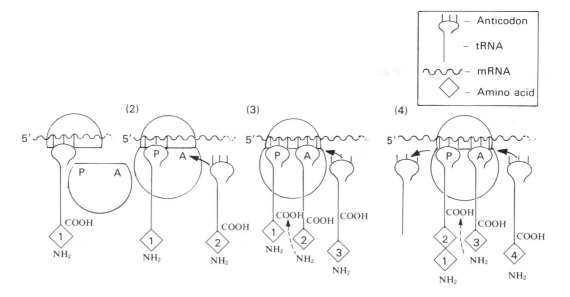

transfer RNA molecule to the A site, a peptide bond is formed between the carboxyl group of amino acid 1 at the P site and the amino group of amino acid 2 at the A site.

This bond formation requires a specific enzyme, peptidyl transferase, found in the larger subunit of the ribosome. Transfer RNA 1 is now ejected from the ribosome. It can now recycle and pick up another specific amino acid. The second tRNA moves from the A to the P site, carrying the two linked amino acids. As this takes place, the mRNA moves so that the next codon (the third in our example) is exposed in the proper position. A third charged tRNA now moves to the A site. Again this involves the interaction of the codon (the third one here) in the mRNA with the complementary anticodon of the tRNA. The amino group of the third amino acid, now in the A position, forms a peptide bond with the carboxyl group of amino acid 2 at the P site. The second tRNA is now ejected; tRNA 3, holding a tripeptide, moves to the P site, while the mRNA moves again to expose the next codon. A fourth charged tRNA enters, and the process repeats over and over until the polypeptide chain is completed and finally released from the ribosome.

It is evident that the first amino acid in the polypeptide chain has a free amino group and thus an "amino end." Each amino acid that is added to the growing polypeptide chain is linked by its amino group to the carboxyl group of the preceding one. The last amino acid will therefore have a free carboxyl end. Once the linear polypeptide or protein chain is completed, it can assume other configurations because of the interaction of amino acid residues along its length. The complex foldings typical of many proteins result from such interactions of amino acid units along one polypeptide chain or between two polypeptide chains in those cases in which the protein is many stranded. The folding, therefore, takes place after polypeptide chains are formed; the ribosome does not guide the three-demensional architecture of a protein. The messenger RNA in the cell may or may not be used over and over again, depending on the cell type and its stage in development. If no longer needed, it may be degraded by enzymatic action, probably by polynucleotide phosphorylase.

Initiation of translation

Once mRNA is formed, we see that it is translated by forming a complex with the ribosome and the charged tRNA. As a transcript of the entire gene, a messenger RNA strand includes transcriptions of many DNA codons, each designating an amino acid. In addition, there are leader and trailer sequences. We will see later in this chapter that still other regions occur in the RNA transcripts of eukaryotes. As we will learn in Chapter 17, one messenger RNA may even carry the transcriptions of two or more genes that control steps in the metabolism of the same substance. Therefore, a messenger strand must contain some kind of signal to indicate where translation into a polypeptide by the ribosome is to begin. This is achieved by the RNA codon AUG, which occurs following the leader sequence. The coding dictionary (see Table 12-1) shows that this triplet designates the amino acid methionine, and that this is the only RNA codon assigned to methionine. Chemical analyses of proteins from microorganisms demonstrate that this amino acid is the one most frequently found at the beginning (the amino end) of the polypeptide chain. Moreover, it has been demonstrated that the polypeptide chain formed in vitro begins with n-formyl methionine, a form of methionine in which the α-NH_2 group is bonded to a formyl group (Fig. 13-5A). In the same polypeptide made in vivo in the bacterial cell, the n-formyl methionine is not present. This special form of methionine is normally placed in the first position in a polypeptide chain in prokaryotes. Reference to its structure shows that it has a blocked amino group. Therefore, it cannot join to the carboxyl group of any amino acid. However, it has a carboxyl group of its own with which any amino acid can form a peptide linkage.

This feature of the blocked amino group makes n-formyl methionine suitable as the first unit starting a polypeptide chain. It is not found in normal protein that is formed in the living cell, but is later changed by enzyme action to a typical methionine residue containing an amino group and lacking the formyl grouping. It may also be cleaved from the chain, so that the second amino acid residue in the chain later becomes the leading one with the free-NH_2 end (Fig. 13-5B).

Analyses on the molecular level have revealed a special kind of transfer RNA in the prokaryotic cell. It is called "initiating tRNA," or tRNA$_f^{met}$, because it is complexed with n-formyl methionine. Actually, the formyl group is added by enzyme action *after* ordinary methionine has become attached to the tRNA$_f^{met}$. This special transfer RNA thus makes the enzyme reaction possible, whereas the ordinary methionine tRNA (tRNA^{met}) cannot, and so any methionine attached to

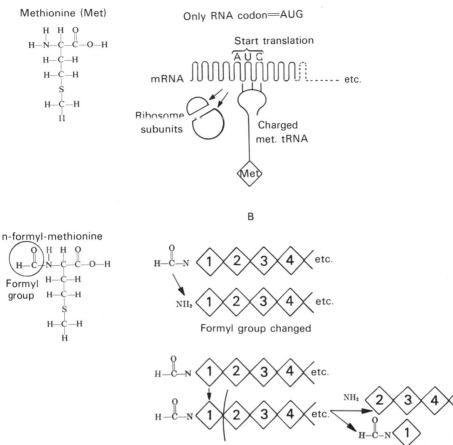

FIG. 13-5. Start of translation. (A) The amino acid methionine is represented by only one RNA codon. Methionine is the most frequently occurring amino acid at the beginning of a polypeptide chain, as shown by analyses of protein from cells of microorganisms. It appears that the RNA codon, AUG, signals the start of translation at that point in the messenger. The specific tRNA carrying methionine would thus be the first to complex with the messenger and the ribosome. (B) In the test tube, however, n-formyl methionine is frequently found at the first position. Evidently, this form of methionine is placed first in the chain. Its formyl group is then later changed to an amino group. It may also be cleaved from the polypeptide chain so that the second amino acid becomes the leading one.

it remains unchanged. The tRNA $_f^{met}$ can also recognize the triple AUG when it is toward the 5′ end, following the leader segment. If AUG is found elsewhere along the messenger strand, ordinary methionine is inserted by tRNA^{met} (Fig. 13-6). Thus, we see that a codon may mean different things, depending on its position in the mRNA. The first AUG toward the beginning of the coding sequence indicates that the polypeptide is to be started from that point, and it is recognized by the special tRNA charged with n-formyl methionine. Located elsewhere, the triplet AUG is recognized by the usual tRNA (tRNA^{met}) carrying methionine.

Although AUG is the major initiating codon, it is known that the codon GUG may in some cases serve as the initiating codon. Ordinarily, GUG is recognized as a valine codon, but when found at the beginning of a coding sequence on the mRNA, it is recognized by tRNA$_f^{met}$ as a methionine codon.

Both tRNA^{met} and tRNA$_f^{met}$ have been found to have the same anticodon, 3′ UAC 5′, and therefore are coded by the same triplet, AUG. One of the main distinctions between them appears to depend on their ability to react with certain other protein factors in the cell. Only the initiating tRNA carrying n-formyl methionine can bind to certain initiating factors re-

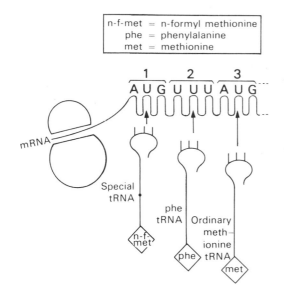

FIG. 13-6. Initiating transfer RNA. A special kind of tRNA carries *n*-formyl methionine and recognizes AUG when this triplet is the first on the messenger to designate an amino acid. The triplet AUG, when located at any other position on the messenger, designates ordinary methionine that is identified by the ordinary methionine tRNA. This diagram shows AUG at the first and the third positions and the codon for phenylalanine at the second. Phenylalanine will thus be placed at position 2 and ordinary methionine at position 3 in the polypeptide chain as the ribosome reads the messenger.

quired for binding of the tRNA to the 30S subunit of the ribosome and to the mRNA marking the beginning of translation (Fig. 13-4). On the other hand, it cannot bind to another kind of factor called an elongation factor. Ordinary tRNA^{met} as well as the other kinds of tRNAs can react with this factor and enter the A site of the ribosome as a polypeptide chain grows during translation.

We will see in Chapter 17 that in prokaryotes two or more genes may be transcribed on the same mRNA strand. There must, therefore, be stop signals along the messenger to indicate the end of translation. Otherwise, two or more normally separate proteins would become joined as one large abnormal unit. Moreover, a trailer sequence is found at the 3′ end of the mRNA, following the segment coded for the polypeptide (refer to Fig. 12-30). A stop signal is needed at the end of the coded segment to prevent translation of the trailer and the formation of a polypeptide containing additional amino acids. Such signals are just as important as the stop signals on the DNA and the rho factor, which indicate when transcription is to terminate. We

have already encountered the three chain-terminating codons, UAA, UAG, and UGA. These codons can be recognized by certain specific proteins, the "release factors." Somehow a release factor acts to remove the "last" tRNA from the ribosome and bring translation to a halt.

In the human, certain mutations are known that apparently have altered chain-terminating codons so that a normal termination signal is erased. Polypeptide chain variants of human hemoglobin have been identified that are called chain-elongation variants because of the presence of additional amino acids at the carboxyl end of the polypeptide chain. The codon UAG is now known to be the signal for termination of one of the polypeptide chains of hemoglobin, the alpha (α) chain. If the codon is altered to CAG as a result of a mutation, it is then read as glutamine by the ribosome, which continues to translate the transcript into the trailer region until it eventually encounters another signal to cease transcription. Several chain-elongation variants are known, and these affect both the α and β (beta) hemoglobin chains (see later in this chapter). Some of the variants are believed to have resulted from frame-shift mutations that have shifted the reading of the transcript near the 3′ end of the coding sequence and have eliminated the normal chain-terminating signal.

The normal length of the α polypeptide chain is 141 amino acid residues. As many as 172 amino acids compose some of the variants as a result of translation of sequences in the trailer, which are normally not read and are not coded for the normal hemoglobin chain. We see that the three chain-terminating codons are essential for normal translation and make up one part of the complex of interacting factors necessary for orderly information transfer.

The anticodon and the wobble hypothesis

It has been well established that the genetic code is degenerate and that 61 of the 64 codons designate amino acids. It seems logical to expect to find 61 different kinds of tRNA molecules, one kind with a specific anticodon for one kind of codon. And indeed, more than one kind of tRNA is known for certain specific amino acids. However, the 1 to 1 correspondence does not seem to hold. Moreover, it has been demonstrated that one given kind of tRNA (and hence one given anticodon) can recognize more than one RNA codon! In addition, some anticodons contain the unusual base inosine (see Fig. 13-3). This base is derived by enzymatic action from adenine, which is originally in that position in the tRNA loop.

Crick has proposed the wobble hypothesis, which helps explain these several points. Remember that the RNA codon in the mRNA is written in a 5′ to 3′ direction and that the anticodon that base pairs with it has the opposite polarity, 3′ to 5′. According to the wobble hypothesis, when the codon and anticodon interact, the first base pairing occurs between the bases at the 5′ end of the codon and the one at the 3′ end of the anticodon. The middle bases pair next, and finally the two at the third position, the base at the 3′ end of the codon and the 5′ end of the antico- don. The first two pairings are specific. The base at the 5′ end of the anticodon, however, can wobble in the sense that it is freer to move about spatially than the other two and, in some cases, may form hydrogen bonds with more than one kind of base at the 3′ end of the RNA codon. Consider an anticodon for arginine: 3′ UCU 5′. The base uracil is at the 5′ or wobble position. At this position, uracil can pair with either the base adenine or guanine in a codon, hence with either the codons 5′ AGA 3′ or 5′ AGG 3′. Both of these RNA codons are for arginine.

The unusual base inosine (I on Fig. 13-3) at the wobble (5′) position may pair with A, U, or C at the 3′ position in an RNA codon. G at the wobble position may pair with U or C in the codon. U as the 5′ base in the anticodon can pair with A or G. There are restrictions, however; C in the wobble position may pair only with G (as in the case of the codon–antico- don interaction for methionine). Similarly, A at the 5′ position can base pair only with U. No single kind of tRNA can recognize four different codons.

The wobble hypothesis explains why fewer than 61 different tRNAs may exist and also why one purified kind of tRNA has been shown to interact with more than one kind of RNA codon. In addition, it can account for the presence of inosine in certain anti- codons. Moreover, the hypothesis has been strength- ened by the fact that certain predictions made from it have been verified. It correctly predicted, for ex- ample, the minimum number of tRNAs required for the amino acids serine and leucine. According to the wobble hypothesis, 32 tRNAs are sufficient to read all 61 codons.

In the process of translation, we see the active role of the ribosome in reading the message, forming the peptide linkages, and moving along the mRNA. We also see that tRNA is not just a passive carrier of an amino acid. Through its anticodon, it actively places its amino acid in the proper position, following the instructions coded in the messenger strand. The in- teractions of the mRNA, tRNA, and ribosome in trans- lation achieve the final transfer of information from the DNA codons within the double helix—the for- mation of a specific polypeptide.

Ribosomal genes and the processing of ribosomes

Cells that are actively growing and synthesizing pro- tein at a high level are very rich in ribosomal content. In a cell of *Escherichia coli*, it may account for as much as 30% of the cell mass! A decrease in growth rate is followed by a corresponding drop in the number of ribosomes. Ribosome production is very closely coordinated with other parts of the cell in the many aspects of normal metabolism and entails a variety of molecular interactions.

In prokaryotes, the ribonucleic acid of the ribo- somes (rRNA) is coded by specific regions of the deoxyribonucleic acid, the ribosomal genes, or ribo- somal DNA (rDNA). The ribosomal genes are also structural genes. The ribosomal RNA that is tran- scribed from this DNA is released into the cell, where it is clothed in protein. In *E coli*, analyses of hybridi- zation between rRNA and DNA indicate that about 10 genes exist per genetic complement for ribosomal RNA (the 16S, 23S, and 5S RNAs).

It is known that a precursor rRNA transcript arises from each of these genes (each called an rrn). How- ever, as transcription occurs, this precursor RNA is cleaved into smaller units to produce 16S, 23S, and 5S rRNA. In addition to these, a small 4S RNA is also produced, and this turns out to be a tRNA! Figure 13-7 shows one way in which the transcript process- ing can take place. Actually, because no set pattern exists, it occurs in more than one way. The position of the tRNA in the 30S precursor molecule with respect to the locations of the 23S and 5S rRNA is not set. Instead, its location varies among the tran- scripts from different rrn genes. In prokaryotes, it appears that genes for tRNA may occur within the rDNA regions. In *E. coli*, the genes for rRNA are continually undergoing transcription. As one RNA polymerase moves down the promoter region of an rrn, another one attaches to the recognition site in the promoter. Most of the RNA in an *E. coli* cell is

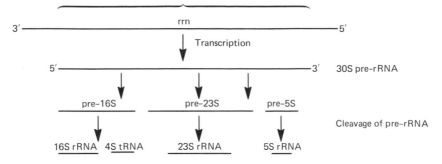

FIG. 13-7. Processing of ribosomal RNA in *E. coli*. Each rRNA gene (an rrn) undergoes transcription to produce a 30S pre-rRNA. This precursor molecule is cleaved, producing smaller pre-rRNAs as transcription takes place. Further processing finally results in the production of mature 16S, 23S, and 5S rRNAs plus a 4S tRNA. The pattern shown here is just one of several that can exist (see the text for details).

rRNA. This continual transcription of the ribosomal genes can account for the large supply of ribosomes found in the bacterial cell.

In eukaryotic cells, the story is more complex. Cytochemical work has shown that the cytoplasm of a metabolically active cell stains deeply with basic dyes because of the presence of a large number of ribosomes. Such a cell typically contains a nucleolus that is larger than that in a less active cell (Fig. 13-8). Conclusive evidence that the nucleolus is intimately involved with ribosome formation was presented in the mid-1960s in investigations with the African clawed toad, *Xenopus laevis*. In this organism, the normal diploid cell contains two nucleoli. A mutation (actually a deletion of the nucleolar organizer) was discovered that can affect the formation of the nucleoli. An individual heterozygous for this change has cells that

contain only one nucleolus. When heterozygotes are mated, about 25% of their offspring die in the larval stage. The cells of these homozygotes lack nucleoli entirely. It was shown in the following way that the cells of these anucleolate mutants are incapable of producing normal amounts of ribosomal RNA. When exposed to labeled CO_2, these larvae fail to incorporate the label into 28S and 18S RNA, in contrast to the cells of animals that contain nucleoli. The cells of the anucleolate mutants are unable to form these types of RNA on their own. Therefore, no label is found in the 28S and 18S RNA present in their cells. The unlabeled 18S and 28S rRNA that does occur is actually derived from the egg. The embryo reaches the tail-bud stage of development only because of the ribosomes of maternal origin. Once these are depleted, the embryo dies.

Additional work with Xenopus using density gradient ultracentrifugation gave more information on the genesis of the ribosomes. When DNA is isolated from eukaryotic cells and subjected to density gradient ultracentrifugation in cesium chloride, any fragment of the DNA will move to a position in the tube at which its density matches that of the solution (review Fig. 11-16). Most of the DNA fragments settle in a characteristic position as a major band. This is so because most of the fragments (the bulk DNA) contain a G + C content of 40%. If fragments are denser than fragments in the bulk DNA (i.e., they contain a higher G + C content) or if these are less dense (having a lower G + C content), they will separate from the bulk DNA as smaller satellite bands (Fig. 13-9). One of the attributes of rDNA is its unusual base composition that helps to distinguish it from the rest of the nuclear DNA. The ribosomal

FIG. 13-8. Epidermal cells of a grass. A cell forming a root hair is much more active metabolically than a cell that fails to produce one. Note the large nucleolus in the nucleus of the hair (above) in contrast to the smaller nucleolus in the nucleus of the hairless cell (below).

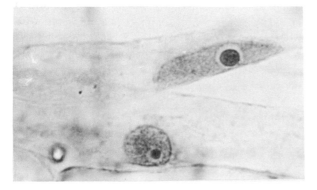

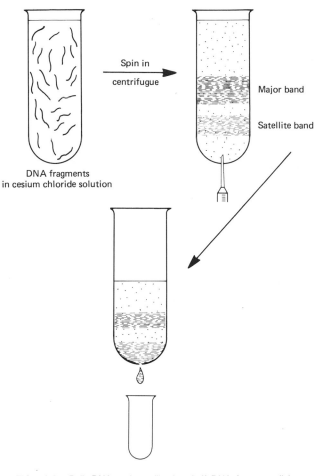

FIG. 13-9. Bulk DNA and satellite band. If DNA from a cell has regions that contain more (or less) G + C than the major portion of the DNA, the two kinds of DNA can be identified and separated, using density gradient ultracentrifugation. When extracted from the nuclei of cells, the DNA fragments at random during the course of the procedure. In the density gradient, those fragments with a higher G + C content will separate from those having a lesser G + C content because of the density difference between them. Most of the DNA has a G + C content of 40% and settles out as a major band. Any DNA fragments with either a higher or lower density will separate from it as one or more satellites. The separation permits isolation of the bands, as seen here for the band, which is denser because of its higher G + C content. The tube may be punctured and samples collected as the tube drains off drop by drop.

DNA is especially rich in its G + C content. The DNA in a given band may be isolated from the DNAs in bands of different densities and then tested with RNA from a known source to identify it, using the technique of DNA–RNA hybridization (see Fig. 12-19B). Since ribosomal RNA is abundant in the cells, it is relatively easy to determine whether or not a given sample of DNA represents ribosomal genes, rDNA. This is so because only the rDNA stands will hybridize with rRNA, the transcripts of the rDNA. Ribosomal RNA of wild-type Xenopus was made radioactive by labeling with tritium. When such labeled RNA was used to form DNA–RNA hybrid molecules, it was found that only the DNA of a satellite band, not that of the major band, could form hybrid molecules with labeled 18S and 28S rRNA.

The genes for the 5S RNA (the 5S DNA) separate as a band at a different position in the density gradient, but these can also be isolated and identified as described for the other rDNA. Since the rDNA for the 18S RNA and the 28S RNA plus the 5.8S RNA band together in the density gradient, the DNA sequences for these ribosomal RNAs must be in close proximity and must be at a distance from the 5S rDNA that has a different density and forms a distinct band of its own. The location of the 18S and 28S DNA in Xenopus (the DNA coding for the 18S and 28S RNA of the ribosomes) was clearly shown to be confined to the region of the nucleolar organizer, one of which is found per haploid set of Xenopus chromosomes. A clear-cut demonstration of this has been made possible by a technique known as in situ hybridization, a variation of the DNA–RNA hybridization procedure in which the denatured DNA is left in place in the chromosomes instead of being extracted. Cells are fixed on a slide and subjected to a mild alkali treatment to denature the DNA. The single strands can then be presented with single-stranded radioactive fragments of DNA or RNA from some known source. To identify the rDNA regions of the chromosomes, radioactive rRNA (18S and 28S rRNA) is used. Any DNA sequences complementary to these species of RNA will bind to the rRNA and will hence pick up the radioactivity. Slides are washed to rid them of any unbound RNA and coated with a photographic emulsion so that an autoradiograph can be prepared. Silver grains will be developed wherever there is a radioactive emanation from any bound, labeled rRNA. It has been shown that the region of the nucleolar organizer exclusively binds the 18S and 28S rRNA, as shown by the fact that silver grains develop only over the nucleolar organizing region of the chromosome (Fig. 13-10). However, the 5S DNA is not found in the region of the nucleolar organizer but is located at the tips of all the chromosomes.

It was not surprising to find that the 5S DNA is located at sites other than the nucleolar organizer, because anucleolate toads that cannot form 28S and

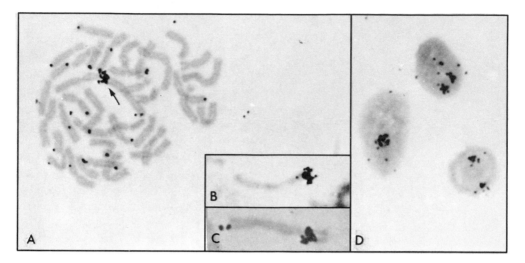

FIG. 13-10. Autoradiographs of cytological preparations from *Xenopus mulleri* and *Xenopus laevis* hybridized in situ with labeled cRNA transcribed in vitro from DNA sequences coding for 18S and 28S ribosomal RNA. (A) Metaphase plate from *Xenopus mulleri* showing hybridization of the labeled RNA over the terminal secondary constrictions on a pair of the largest submetacentric chromosomes (arrow). (B) The nucleolus organizer chromosome from *Xenopus mulleri*, showing localization of the sequences coding for 18S and 28S ribosomal RNA at the extreme end of the shorter arm. (C) The nucleolus organizer chromosome from a *Xenopus laevis* cultured cell line, showing localization of the sequences coding for the 18S and 28S ribosomal RNA over the secondary constriction. (D) Interphase nuclei from *Xenopus mulleri*, showing hybridization over nucleoli. (From Cold Spring Harbor Symp. Quant. Biol 38: 475, 1973. Courtesy of Dr. Mary Lou Pardue.)

18S rRNA (as shown by the failure of their cells to incorporate label into these RNA types) are able to produce 5S rRNA. The rDNA for the 5S rRNA has now been shown in almost all eukaryotes to be located at a distance from the 18S and 28S rDNA that is associated with the nucleolar organizing region. This has also been found to be the case in the human whose 5S genes are located in Chromosome 1 and several other chromosomes, in contrast to the rDNA for the other species of rRNA. The rDNA for the latter is localized at the short arms of Chromosomes 13, 14, 15, 21, and 22, all of which possess satellites and have a nucleolar organizing function.

The availability of the ribosomal RNA, the product of the ribosomal genes, has permitted an estimate to be made of the number of ribosomal genes present in a set of chromosomes of different eukaryotes. Again, this was made possible by DNA–RNA hybridization. A sample of a known amount of single-stranded cellular DNA is allowed to hybridize with a known quantity of pure ribosomal RNA. The larger the number of ribosomal genes in the DNA sample, the greater the amount of the rRNA that will hybridize with it. In the process (similar to that shown in Fig. 12-19*B*), the radioactive rRNA is added to single-stranded DNA. The DNA–RNA hybrid molecules are trapped on filter paper, the remaining unbound RNA being washed

away. The amount of bound RNA in the form of hybrid molecules can be measured, since the RNA is radioactive. The hybridization procedure thus acts as an assay and reveals that about 450 copies of the genes for each 18S, 28S, and 5.8S RNA are present in a haploid set of Xenopus chromosomes!

Isolation of the ribosomal genes of various species has made it possible to subject them to very detailed analyses using an assortment of chemical procedures, as well as electron microscopy. As a result, we now have a very precise picture of the genes for ribosomal RNA in several species. A ribosomal DNA region that governs the synthesis of the 18S, 28S, and 5.8S RNAs was found to consist of repeating units of a given length, each unit containing certain recognizable sequences (Fig. 13-11). The unit includes a DNA sequence for 18S RNA, one for 28S RNA, and one for 5.8S RNA. Each unit is separated from the next by an untranscribed spacer of variable length. This spacer DNA has never been found to hybridize with any cellular RNA. The units and spacers are repeated hundreds of times. These repeated rRNA genes give an excellent example of a *simple multigene family,* a set of repeated genes that, in this case, are clustered together on a chromosome and repeated next to each other in a linear array.

When a unit of rDNA undergoes transcription in

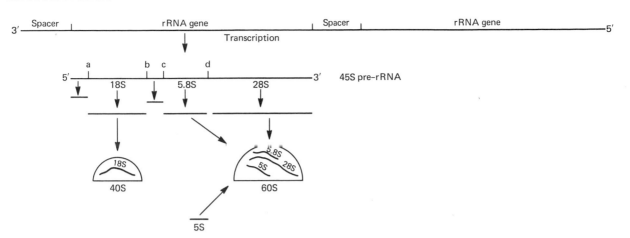

FIG. 13-11. Processing of ribosomal RNA in eukaryotes. The rDNA for 5.8S, 18S, and 28S is found in the nucleolar organizer. The rRNA genes occur as repeating units, each gene found in many copies and separated from the next rRNA gene by a spacer region that does not undergo transcription. When an rRNA gene is transcribed, a 45S precursor rRNA is produced. Processing of this transcript then takes place as four cuts are made. One cut *(a)* removes a leader sequence from the 5' end. The second *(b)* results in the formation of an 18S RNA, which becomes part of the 40S subunit of the ribosome. A third cut *(c)* removes a small transcribed spacer. The final cut *(d)* yields the 5.8S and 28S RNAs of the larger subunit of the ribosome. The 5S rRNA of the larger subunit is a transcript of DNA found outside the nucleolar organizing region.

mammals, a 45S preribosomal RNA transcript is produced that contains the rRNA for the 18S, 5.8S, and 28S RNAs of the ribosome. RNA polymerase I, which is found in the nucleolous, is responsible for the formation of this transcript. (This enzyme is actually responsible for more than 50% of the RNA synthesis in the cell.) The 45S RNA transcript is then processed. The transcript is cleaved in four places (Fig. 13-11). One cut removes a leader sequence from the 5' end of the transcript. The second cut produces the 18S rRNA for the smaller subunit of the ribosome. A 36S RNA fragment remains. The next cut removes from it a small transcribed spacer region, leaving a 32S fragment. A final cut produces the 5.8S and 28S RNAs of the larger ribosomal subunit.

Transcription from the DNA of the nucleolar organizing region can actually be observed by studying photographs made with the electron microscope (Fig. 13-12). The untranscribed spacers are seen as regions between stretches of ribosomal DNA to which RNA transcripts are attached. The function of the untranscribed spacers is unknown.

Analyses of the ribosomal DNA molecules for the 5S RNA have shown that the 5S DNA also contains units that repeat, making up a simple multigene family. The portion that is transcribed into 5S RNA makes up only one seventh of a unit and is separated from the next unit by an untranscribed spacer of variable length. RNA polymerase II, which is responsible for

FIG. 13-12. Visualization of ribosomal DNA from oocyte nucleoli of the spotted newt, *Triturus viridescens*. The linearly repeated ribosomal genes (rDNA) are separated by untranscribed spacer segments. Each ribosomal gene actively transcribes rRNA molecules. The shorter transcripts are the more recently synthesized and are closer to the beginning of each gene. The difference in their length accounts for the feathery, arrowhead appearance. (Courtesy O. L. Miller, Jr., from Miller, O. L., Jr., and B. Beatty. Science 164:955–957, Fig. 2, 1969. Copyright © 1969 by the American Association for the Advancement of Science.

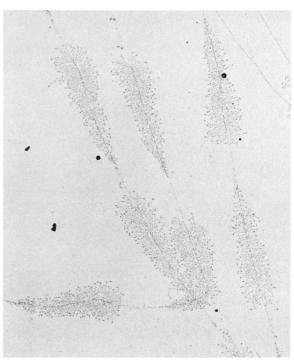

the formation of mRNA, also produces the 5S RNA, the template of which is *not* in the nucleolus.

The RNA of the ribosomes finally becomes complexed with protein. This ribosomal protein is formed on the ribosomes of the cytoplasm. In eukaryotes, some of the protein moves into the nucleolus, where it unites with RNA to complete the larger ribosomal subunit. It appears that the RNA of the smaller subunit moves out to the cytoplasm, where it is combined with protein. Somehow the 5S rRNA, found on a completely different molecule from the other ribosomal RNAs, is packaged into the larger ribosomal subunit along with the 28S rRNA, a process that must entail complex molecular interactions.

Transfer RNA, its genes and its processing

The formation of ribosomal RNA tells us something about information transfer: DNA may be coded for a product other than a polypeptide. Such a gene is also a structural gene. The mature rRNAs contain no information for the amino acid sequences of polypeptides and are not translated. Transfer RNA is an example of still another kind of RNA that does not undergo translation. Moreover, the production of functional tRNA has many similarities to the formation of mature rRNA.

The genetic information for a minimum of 40 different kinds of tRNA in *E. coli* is known to occur at various positions on the single chromosome. Some of this information for tRNA is present in more than one copy. In some cases, these identical tRNA genes, structural genes, are closely linked together in a cluster. In other cases, the identical genes may be separate from each other in different chromosome regions. Still other portions of the chromosome include clusters that contain genes for two or more different tRNA species. We have also seen (in Fig. 13-7) that in *E. coli* a DNA sequence that codes for a tRNA lies between the DNA sequences coding for 16S and 23S RNA. Various parts of the prokaryotic genome thus appear to contain information for the various kinds of tRNA. In eukaryotes, transfer RNA genes appear to occur in adjacent clusters. Hybridization between the DNA and tRNA from a given eukaryote has shown that genes for tRNA occur in more than one copy. A haploid set of Drosophila chromosomes contains about 8 for each kind of tRNA. In Xenopus, the figure is about 200.

As in the case of the rRNA transcript, the immediate tRNA transcript is a precursor molecule which undergoes processing. Several events must occur before the production of any mature tRNA, such as that in Fig. 13-3. Most of the information on tRNA processing comes from bacteria, although the processing of tRNA precursor molecules in eukaryotes is believed to entail similar steps. The larger tRNA precursor molecules may contain sequences for one or more tRNA molecules. Surplus RNA segments in any precursor are eliminated as the result of RNase activity. If a precursor contains information for just a single tRNA, a leader segment that may contain 40 nucleotides is cleaved from the 5′ end, and a shorter segment is often cleaved from the 3′ end. If a tRNA precursor contains sequences for more than one tRNA, similar sequences at the 5′ and 3′ ends are also removed. In addition to these surplus segments, spacers are located in these cases between the individual RNA sequences, and these must also be removed. Still other smaller sequences are known to occur within a remaining tRNA sequence, and these too are finally excised, and the free ends of the remaining portions joined. Note from Fig. 13-3 that the sequence -CCA occurs at the 3′ end of the mature tRNA. This sequence, essential for joining to the amino acid, may be added to the tRNA during the processing of the tRNA precursor. In other cases, the sequence -CCA is already present in the precursor, having been transcribed from the DNA for these genes.

We noted earlier that unusual bases occur within the tRNA (see Fig. 13-3) and that these arise as a result of modification of more familiar bases. These modifications, which may be very complex, are also a part of tRNA processing. Some modifications take place before the removal of the surplus segments from the precursor, others after the segments have been eliminated.

Colinearity

We have noted that both DNA and protein are essentially linear molecules. And we have seen that a codon (a sequence of three nucleotides in a DNA segment and in the complementary RNA triplet transcribed from it) is typically specific for a certain amino acid type and determines the position of that kind of amino acid in a polypeptide chain. It is tempting to speculate that perhaps the linear sequence of the nucleotides in the DNA and RNA corresponds to the linear sequence of the amino acids along a stretch of a protein molecule. Actually, this has been demonstrated quite clearly in microorganisms. For example, Yanofsky and

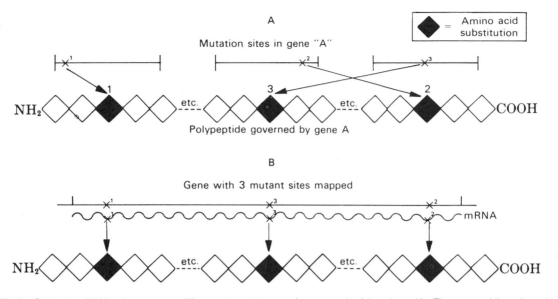

FIG. 13-13. Colinearity. (A) Mutations occur at different sites within gene A of E. coli, which determines the amino acid sequence of one of the polypeptide chains of tryptophan synthetase. The relative positions of the separate mutations can be mapped. This schematic representation shows that if a mutation maps at the beginning of a gene, an amino acid substitution is made at the beginning of the polypeptide chain. A mutation at the end of the gene (mutation 2) is reflected by an amino acid substitution near the other end of the protein, the carboxyl end. Mutations, such as mutation 3, which map between ends of the gene are associated with amino acid substitutions between ends of the polypeptide. The gene and the polypeptide chain it governs are thus colinear. (B) Messenger RNA, carrying the information for amino acid sequence, is a transcript of the DNA. Therefore, alterations in the nucleotides of the DNA are reflected at corresponding positions in the mRNA. These in turn are responsible for the amino acid substitutions at positions in the protein. These amino acid changes, therefore, correspond to altered nucleotide sites in the mRNA and in the DNA. The gene, its RNA transcript, and the polypeptide are thus colinear.

his colleagues studied in detail one of the two genes whose function is to guide the formation of the bacterial enzyme, tryptophan synthetase. We discussed (Chap. 12) that the *E. coli* enzyme contains two different kinds of polypeptide or protein, each controlled by a separate gene. Several separate mutations in one of these two genes, gene A, were mapped very precisely (Fig. 13-13A). The defective protein associated with each of the mutations was analyzed in detail. Each defective protein was found to have a single amino acid substitution. The substitution was at a different position in the protein chain from one mutation to the other. When the position of the amino acid substitution was related to the site of mutation in the gene, it became evident that a mutation mapping at one end of a gene produced an amino acid change at one end of the protein—let us say the amino end. A second mutation might map at the other end of the gene. The corresponding protein was shown to have an amino acid alteration at the other end of the protein, the carboxyl end. A third mutation might map between the first two. The amino acid substitu-

tion associated with it would then fall at a position in the chain between the other two amino acid changes. In other words, the position of a mutation within a gene corresponds to the position of an amino acid substitution in a protein chain. The relative position of a triplet in the DNA and in the complementary RNA transcribed from it is related directly to the position of a specific amino acid in a protein. The nucleic acid molecules and the protein are *colinear* (Fig. 13-13B). A change at a site in the DNA is reflected at a corresponding site in the messenger RNA. This, in turn, is reflected at a corresponding position in the polypeptide after translation of the messenger.

Some differences between prokaryotes and eukaryotes

Genetic research with microorganisms has built a foundation for our understanding of replication, transcription, and translation and has given us insight into the gene and its interactions at the molecular level. The information from microorganisms has also provided a point of departure for investigations into related phenomena in higher cell types. Since most of the pioneer work has involved prokaryotes, we must not reach premature conclusions about eukaryotes and assume that the situations are identical. We have already noted some differences between prokaryotes and eukaryotes in several of our discussions. It is appropriate at this point to review some of these and to call attention to still other ways in which eukaryotic cells have now been found to differ from prokaryotes.

Since prokaryotic cells lack a membrane-bound nucleus, and hence a nucleolus, processing of the ribosomes, as we have seen in detail in this chapter, is different, and the S values of the ribosomes in the two groups are distinct (see Fig. 13-1). The enzyme RNA polymerase has been described for prokaryotes (see Chap. 12). You will recall that eukaryotes possess three kinds of RNA polymerase, in contrast with bacteria.

An interesting fact relating to translation is that, although the eukaryote possesses two types of tRNA for the amino acid methionine, and one of these acts as initiating tRNA, as does $tRNA_f^{met}$ in prokaryotes, ordinary methionine is always the first amino acid placed in the polypeptide chain. No formylation of methionine occurs in the cytosol of the eukaryotic cell. (In Chap. 19, we will learn that the mitochondria and the chloroplasts of eukaryotic cells possess an enzyme that can convert methionine to n-formyl methionine as well as an initiating tRNA, $tRNA_f^{met}$.) It must be pointed out that although methionine is the first amino acid positioned in the growing polypeptide chain, as in the case of prokaryotes it may be removed

during processing of the polypeptide, so that any amino acid may be at the NH_2 end in a finished polypeptide of a eukaryote.

Several very important distinctions between prokaryotes and eukaryotes on the molecular level center on messenger RNA. In prokaryotes, the ribosomes begin translation of the mRNA as it is still being formed by transcription of the sense strand of the DNA. This means that the template and its transcript are in close proximity when translation begins. This is definitely not the case in eukaryotes, because the immediate transcript (the primary transcript, or the pre-mRNA) cannot act as functional mRNA. As in the case of the preribosomal RNA, the pre-mRNA must first undergo several modifications known as RNA processing. The processing of the pre-mRNA alters it and converts it to the final messenger which then travels from the nucleus into the cytoplasm, where translation takes place. Actually, the first modification of the primary transcript occurs about 1 second after the initiation of transcription, while the RNA strand is fewer than 30 nucleotides long. This involves the addition of an unusual base, 7-methyl guanosine, to the transcript's 5′ end. This cap becomes linked by a triphosphate connection to what was the first base of the transcript. The linkage involved is an unusual 5′ to 5′ linkage, which results from an interaction between the 5′ triphosphate end of the pre-mRNA and the triphosphate of the guanosine (Fig. 13-14). A further modification of the cap takes place in all eukaryotes, except for one-celled organisms. This is the addition of another methyl group to the 2′ position of the sugar of the first nucleotide of the primary transcript. Such a cap is

FIG. 13-14. Capping of primary transcript. The base 7-methyl guanosine is joined by a 5′ to 5′ linkage to the 5′ nucleotide (nucleotide 1) of the transcript. Such a cap is called a cap O. Methylation of the sugar of the first nucleotide can occur to produce a cap 1. Further methylation at the position of the second sugar produces a cap 2.

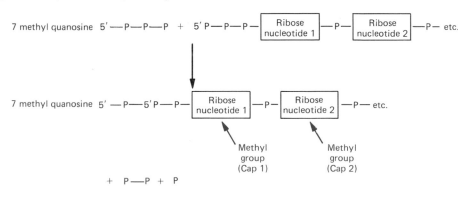

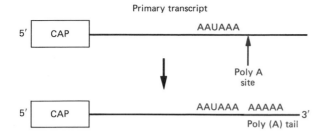

FIG. 13-15. Processing at the 3' end of the primary transcript. A nucleotide sequence, AAUAAA, occurs a short distance upstream from the poly(A) site, where the transcript will be cut. An enzyme then adds to the 3' end a poly(A) tail that may include up to 300 nucleotides.

called a *cap 1* type. Lacking this methyl group, the cap is called *cap 0*, typical of one-celled eukaryotes. In a few eukaryotes, still another methyl group may be added to the sugar of the second nucleotide, also at its 2' position. This is a *type 2* cap. The mRNAs of all eukaryotes and of most of the viruses that infect them have been found to have caps. The role of the cap appears to involve binding of the mRNA by the ribosome and the initiation of translation. In the test tube, mRNAs with altered caps show a slowdown in the start of translation and are not translated efficiently.

Another change in the pre-mRNA of eukaryotes occurs at its 3' end or trailer end. This modification takes place after the primary transcript has been formed. The primary transcript contains a site known as a poly(A) site. The pre-mRNA is cut at this spot, and an enzyme then proceeds to add to the 3' end a stretch of up to 200 nucleotides, each containing the base adenine (Fig. 13-15). The result is the production of a poly(A) tail at the 3' end of a processed mRNA. Approximately 20 nucleotides upstream from the poly(A) site, the following nucleotide sequence occurs on the transcript: AAUAAA. This sequence appears to be involved in the recognition of the poly(A) site downstream from it. The actual cutting is believed to involve a *small ribonucleoprotein particle (snRNP or snurp)*. Each snurp includes several proteins associated with one RNA molecule between 100 and 300 nucleotides in length. Snurps, which are confined to the nucleus, vary in their RNA types. Those RNAs that are rich in uracil are designated U-RNAs. The snurp believed to play a role in cutting the primary transcript contains a uracil-rich RNA referred to as U_1RNA. We will return to these interesting ribonucleoprotein particles in the next section, where we discuss further pre-mRNA processing. A few eukaryotic mRNAs (those for histones, for example) lack the poly(A) tail, the function of which remains unknown. It has been suggested that the tail may help to bind the mRNA to the ER of the cytoplasm, thus facilitating translation, and that it may also act to prevent degradation by the many ribonucleases present in the cytoplasm. The poly(A) tail is also typical of viruses that infect eukaryotic cells.

Figure 13-16 gives a general summary of some of the aspects of eukaryotic mRNA following posttran-

FIG. 13-16. Posttranscriptional modification of eukaryotic messenger RNA. Pre-mRNA, the immediate transcript of the sense strand undergoes modification before the final mRNA is produced and leaves the nucleus. The 5' end of the final messenger RNA is capped and the 3' end bears a long stretch of nucleotides containing adenine. Prokaryotic messenger RNAs also contain a trailer and a leader but lack the 5' cap. Very short poly(A) tails appear to be characteristic of the 3' ends of prokaryotic mRNA. (Compare with Fig. 12-30, and see the text for details.)

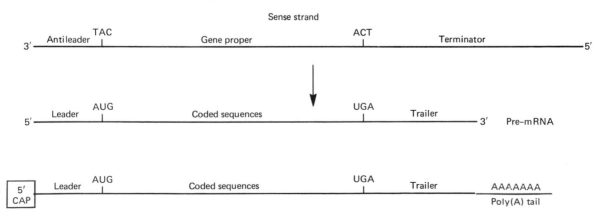

scriptional modification. It can be seen that the 5′ end is capped and that an untranslated leader segment, which ends with the initiating codon AUG, follows the cap. From the initiating codon to one of the chain-terminating codons, nucleotide sequences that code for the polypeptide are found. Following the chain-terminating codon is an untranslated trailer to which is attached the poly(A) tail at the 3′ end. Before the mRNA leaves the nucleus, it becomes linked to at least two types of protein units. Complexed with the protein, it moves into the cytoplasm. However, this is by no means the entire story, as we shall now see in the following section.

Fragmented genes

One of the biggest surprises in molecular biology was the finding that the genes of eukaryotes are not strictly colinear with their mRNA transcripts. It has been firmly established by the technique of DNA sequencing (Chap. 18) that most eukaryotic genes and also those of animal viruses are fragmented. The sequences that are coded for a functional product are interrupted by noncoding intervening sequences designated *introns*. The coding portions of the gene are named *exons*. This means that the typical eukaryotic gene is in pieces and contains a much longer nucleotide length than is required to code for the amino acid sequence of its polypeptide (Fig. 13-17A). Although genes for histone proteins and for actin lack introns entirely, most of the genes studied are spread out into 20 or more pieces. The gene for alpha-collagen con-

tains 52 pieces! The length of an intron may be appreciable. In the mouse, for example, the gene for beta-hemoglobin contains two introns whose total length is 762 base pairs, whereas the sum of the coding sequences is only 432 base pairs. A gene that may require a coding sequence of 1000 nucleotide bases to code for a polypeptide could be spread out over a length 10 times as great.

Before the transcript and its poly(A) tail can serve as mRNA, it must undergo further processing to remove the introns. This must be followed by a linking of the exons, the segments coded for the amino acid sequence of the gene's polypeptide (Fig. 13-17B). The snipping out of the introns and the joining together of the exons to form the final, functional mRNA is called *RNA splicing*. Clues to the events occurring during RNA splicing have come from experiments in which the excising of introns and joining of exons have been achieved in the test tube. These experiments have implicated snRNPs (snurps) in splicing as well as in the cutting of the pre-mRNA at the poly(A) site, as discussed in the previous section. Successful in vitro splicing depends on certain cell extracts, which apparently contain factors essential for the splicing process. In vitro splicing will not take place without the addition of the proper extracts and hence the splicing factors. Antibodies against the protein found in the snurps can be prepared. If these antibodies are added to extracts that have the potential to bring about splicing, the extracts then lose their splicing ability. Moreover, these same extracts are ineffective in causing poly(A) tails to be added to the

FIG. 13-17. The fragmented gene of the eukaryote. *(A)* The region of the gene containing information coded for a polypeptide is fragmented in the sense that sequences coded for amino acid sequences (the exons) are interrupted by noncoding sequences (introns). *(B)* Both the introns and exons are transcribed and are thus represented in the immediate transcript, the pre-mRNA. The introns are snipped out enzymatically and the exons joined together during the processing of the pre-mRNA to produce the final mRNA, which contains an uninterrupted region coding for the amino acid sequence of a polypeptide.

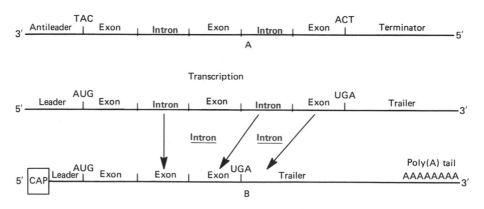

transcript's 3' end. Such observations indicate that snurps are needed both for cutting the transcript at the poly(A) site and also for recognizing introns, excising them, and joining the exons together.

It now appears almost certain that splicing involves a snurp that contains U₁RNA, as was also the case in the recognition of the poly(A) site. When splicing takes place, it is essential that the factors that excise the introns recognize the junctions between the exons and the introns. DNA sequencing of intron–exon junctions has revealed that all introns in transcripts begin with GU at the 5' end and terminate with AG at the 3' end. The GU and AG in turn are part of *consensus sequences*, which occur at the exon–intron junctions (Fig. 13-18). (A consensus sequence is the most prevalent sequence of nucleotides found in a particular genetic region or composing a genetic factor. There may be slight departures in a given case when a sequence of interest is compared with a consensus sequence, the sequence showing the nucleotides that are most often present in that genetic segment.) Removing an intron must be a very precise event, because an error that involves just a single nucleotide can upset the entire reading frame of the transcript, resulting in a defective polypeptide or an incompleted one. A snurp appears to participate in cutting at the exon–intron junction. The U₁ snRNA has at its 5' end a sequence complementary to the consensus sequence at the 5' exon–intron junction. The U₁RNA apparently binds to the complementary sequence at the beginning of the intron. The binding seems to depend on the entire snRNP, since U₁RNA by itself cannot accomplish this. Once the transcript is bound to the particle, a cut is made at the left intron–exon junction (Fig. 13-19). The intron is then folded in such a way that the 5' end of the intron is joined to a nucleotide at the right end of the intron. As a result, the intron assumes a lariat configuration.

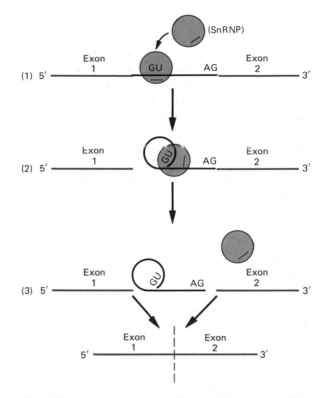

FIG. 13-19. Splicing of transcript. *(1)* An snRNP carries an RNA molecule *(red)* that is complementary in part to the consensus sequence at the 5' end of the intron. A particle binds to the transcript at the junction. *(2)* A cut is made at the 5' end of the intron, which is then folded to produce a lariat. *(3)* Another is then made at the 3' end of the intron, which is removed. *(4)* The exons are joined together.

Cutting then takes place at the right junction, and the intron is removed. Exons are finally joined together by the formation of 5' to 3' linkages. Many details are yet to be filled in, but the transcript, with the 5' cap, the 3' tail, and excised introns, is now ready to be transported through the nuclear envelope to the cytoplasm, where it can act as mRNA in translation. Processing and splicing of primary transcripts hold significant implications for the problem of gene regulation and gene expression, and we will discuss them in this context in Chapter 17.

It should be noted at this point that there is a difference between introns and the spacers, which were discussed in reference to the processing of rRNA and tRNA. Introns, as we have just seen, are found *within* a gene, whereas spacers typically are not and do not interrupt a sequence coded for a specific product. The transcribed spacer encountered in the case of the rRNA genes (see Fig. 13-11) does not interrupt the sequences coded for the 18S and 5.8S rRNAs. It

FIG. 13-18. Exon–intron junctions of transcript. *(Above)* A portion of a primary transcript is depicted bearing exons separated by an intron. *(Below)* An intron begins with the sequence GU at the 5' end *(left arrow)* and ends with AG *(right arrow)* at the 3' end. These sequences are part of two consensus sequences *(brackets)*. The most frequently occurring nucleotides of the two are shown.

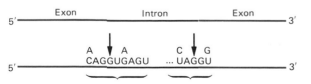

has been found that in some cases introns *do* occur within certain genes coded for tRNA and rRNA. In a few lower eukaryotic species (among the protozoans and slime molds) a single intron interrupts the sequence coded for the larger rRNA. In yeast, an intron is found in some, but not all, of the genes coded for tRNA. When present, this single intron is always located near the beginning of the sequence for the anticodon loop. We see that introns *do* occur elsewhere in certain genomes, but they are by far more characteristic of the eukaryotic genes coded for mRNA transcripts.

Significance of introns

The discovery of the introns that interrupt the coding sequences of genes has helped to clarify several puzzling observations about eukaryotic genes and their transcripts. First of all, eukaryotes, in contrast with prokaryotes, appeared to contain DNA far in excess of the amount needed to code for all their proteins, even considering the DNA, which might not have a coding function but may be essential for chromosome structure and gene control. In the human, it has been calculated that about 90% of the DNA does not contain coding sequences for polypeptides. Equally puzzling was the observation that not only is RNA also present in excess, but large RNA strands occur that do not seem to leave the nucleus. These strands were called *heterogeneous nuclear RNA (hnRNA)* to reflect their varying sizes. Their significance was unknown, and only much shorter RNA strands were found to leave the nucleus and enter the cytoplasm. It now appears that some of the excess DNA of eukaryotes is found in noncoding sequences or introns and that the hnRNA represents primary transcripts containing introns and exons, as well as transcripts in various stages of processing and splicing.

Although the discovery of noncoding sequences answers certain questions, it raises others, the most obvious being, "Why are eukaryotic genes typically fragmented and spread out over a nucleotide length many times longer than that needed to code for their products?" One suggestion is that such an arrangement has value for evolutionary progress. The presence of noncoding intervening sequences stretch out the coding sequences of the genes. Each exon, it is argued, carries information for a part of the polypeptide that plays a distinct functional role in the completed polypeptide chain. This has actually been demonstrated for many genes. For example, each exon of

the genes encoded for antibodies, hemoglobin, and certain enzymes is coded for a portion of the protein that has a distinct function to perform. Being spread out in pieces, it is easier for the coded sequences to undergo recombination through crossing over than if they were closer to one another in one coded segment. By spreading out, the exons, each coded for an important part of the polypeptide, can more easily be brought together in new combinations. This can result in polypeptides with new combinations of important functional regions. Some of these may entail an advantage over the old arrangement and be selected by the forces of evolution.

Many proponents of this idea have considered introns to be additions to the gene that were introduced into the chromosomes during the evolution of the eukaryotic cell. There is reason to question this belief, because splicing of mRNA has now been observed in certain bacteria, although it is nowhere as extensive as in eukaryotes, and the intron number is very small. Moreover, studies of the chemistry of RNA support the concept that the very first genes were composed of RNA. Chains of RNA can form spontaneously under conditions of a high concentration of salt and RNA molecules. One can argue that before the origin of cells, the first genes were short chains of RNA coded for very simple proteins. The useful information coded into these simple genes very possibly would have been interrupted by stretches that contained no useful information. These stretches would be equivalent to introns and similar to the picture noted earlier regarding the genes for antibodies and hemoglobin in which segments coded for functional parts of the protein (exons) are separated by noncoding introns. And so, according to this theory, introns were present in genes from the start and were not introduced later at random into the gene as eukaryotes evolved.

At the earliest time, RNA may have achieved the excising of introns by self-splicing, since it could have had catalytic activity. Examples of the catalytic activity of RNA are now known (see the next section). Eventually the gene became DNA instead of RNA. How this transfer could have taken place is highly speculative, but there is some reason to believe that the enzyme *reverse transcriptase* (Chap. 16) may have been present very early in cellular evolution. Through the action of this enzyme, the genetic information found in RNA would have been converted to DNA. Being a more stable molecule than RNA, DNA would have been selected as the genetic material. This DNA would have included both the exons and the introns that were present in the RNA from which it was derived. As bacteria evolved, introns would have been almost completely eliminated, in contrast with the eukaryotes that retained them. Several other theories concerning the significance of introns have been proposed, including the ones regarding their role in gene regulation. However, final conclusions on the role of introns in the economy of the cell as well as on their origin still await further information.

RNA as a catalyst

Proteins, with their highly diversified structures, have been considered to be the only molecules with sufficient variation and complexity to act as enzymes in biological systems. Two cases in which RNA has been shown to behave as a catalyst with true enzyme activity force us to reevaluate this concept.

In *E. coli*, the unusual enzyme ribonuclease P contains both an RNA and a protein portion, each encoded by its own separate gene. This enzyme is responsible for generating the 5' end of all tRNA molecules in *E. coli* by making appropriate cuts in tRNA precursor molecules. It is possible to dissociate the enzyme into its protein and RNA components. Under in vitro conditions, it has been demonstrated that the RNA part, by itself (which makes up the greater part of the enzyme), can make accurate cuts in a tRNA precursor. In vivo, however, both the RNA and protein portions are required for the enzyme activity, as can be seen by the fact that a mutation in either the gene for the protein or that for the RNA part of the enzyme can abolish ribonuclease P activity. But it is actually the RNA component of the intact enzyme, not the protein, that provides the catalytic activity.

A second example of RNA as a catalyst is found in the protozoan Tetrahymena. In this organism, a 26S ribosomal RNA is formed from a 35S rRNA precursor following removal of a single intron. Excision of the intron and splicing of the RNA to form the 26S rRNA has been shown to result from an intrinsic splicing ability of the RNA itself. Furthermore, it has been demonstrated that the intron, when it is removed, is shortened during a series of reactions in which 19 nucleotides are eliminated. This shortened form of the intron, when present in vitro with certain short polynucleotides, is able to cleave and to rejoin them! The short RNA acts as a true enzyme with RNA polymerase activity. At a higher pH, it has been shown to possess ribonuclease activity. Not only does this

example hold implications for the versatility of RNA and its roles in the cell, it also adds to our speculations on the true significance of introns, a topic to which we return in Chapter 20.

Heterogeneous nature of eukaryotic DNA

We have just seen that much more DNA exists in a eukaryotic than a prokaryotic cell. It was also noted in Chapter 11 that the percent GC content of the DNA in lower forms, especially in prokaryotes, shows great variability from one species to the next, whereas in eukaryotes, the value is much more stabilized—about 40% in mammals.

Another distinction between prokaryotic and eukaryotic DNA is seen when DNA is isolated from cells, fragmented into segments of comparable size, and then subjected to density gradient ultacentrifugation. Handled in this way, the prokaryotic DNA yields one band following centrifugation, indicating that the separate fragments of comparable size are approximately the same in density and hence in their base composition. We have seen in this chapter that when eukaryotic DNA is processed in this way, centrifugation reveals that the DNA is heterogeneous and consists of a main band (including the bulk of the DNA) and one or more satellite bands. This means that the G:C and A:T pairs are not uniformly distributed on the fragments but may be clustered in certain regions, yielding DNA fragments of quite different densities. Some of the satellite DNA represents the DNA of the nucleolar organizer. Moreover, certain satellite bands represent the DNA found outside the nucleus in organelles and chloroplasts. However, the term *satellite band* is generally taken to refer only to those minor bands that are nuclear in origin.

A most unusual feature of eukaryotic DNA is its inclusion of base sequences of varying lengths, which are repeated throughout the bulk of the DNA. This difference between prokaryotes and eukaryotes can be demonstrated as follows. DNA extracted from a prokaryotic cell may be fragmented into pieces, all approximately the same size, perhaps 500 nucleotide base pairs long. The DNA is then subjected to a high temperature so that the DNA strands separate (denature), and single-stranded DNA fragments arise. The material is then cooled, permitting the single strands to rejoin and form double-helical DNA segments (renaturation). If only one copy of a stretch of a nucleotide sequence is present, then renaturation can occur only when a single-stranded DNA fragment

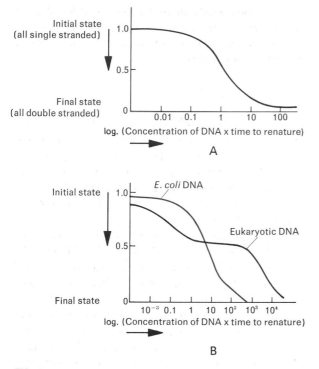

FIG. 13-20. DNA renaturation curves. (A) Generalized renaturation curve expected for DNA in which stretches of nucleotide sequences occur only in single copies. Curves for prokaryotes, such as E. coli, approach this idealized curve. (B) DNA renaturation curves of E. coli and a typical eukaryote compared.

encounters its original partner or the same complementary stretch from another cell. This means that renaturation tends to occur slowly. The course of the renaturation of the DNA is expressed in the form of a Cot curve in which the fraction of DNA molecules that have reassociated is plotted against the log of Cot. A Cot value is equal to Co×t, where Co is the starting concentration of single-stranded DNA in moles per liter and t is the time, in seconds, that the incubation has proceded. In E. coli (Fig. 13-20A), renaturation occurs slowly, progressing at a constant rate for the first 10 minutes and then slowing down even more. This is so because as strands slowly find their complements, fewer and fewer single strands are present, and the chance that a complementary single-strand encounter will occur is reduced.

When this same procedure is followed with DNA from eukaryotic cells, a different picture emerges. At first a very fast renaturation takes place, as indicated by the very low Cot values (Fig. 13-20B). This tends to level off and the appearance of the curve then comes to resemble that of E. coli, except that the DNA

reassociations occur at higher Cot values. Renaturation studies have revealed that the DNA of most eukaryotes is heterogeneous in the sense that it includes very many copies of certain nucleotide sequences and just single copies of others. Human DNA, for example, contains approximately 2.5×10^9 base pairs, and about 20 to 30% of this consists of multiple copies of certain nucleotide sequences. The DNA of eukaryotes can be classified into three general categories on the basis of the kinetics of DNA renaturation. The DNA that renatures slowly (at a high Cot value) is nonrepetitive, composed of single-copy sequences. When the eukaryotic DNA is fragmented, a given stretch of nonrepetitive DNA, represented as just a single copy on a given chromosome, can unite only with its original complementary partner or with that same complementary stretch from another cell. This DNA renatures slowly, since most chance encounters do not bring together complementary strands.

A renaturation experiment can be monitored to permit the isolation of fragments that have reassociated to produce duplexes at a particular Cot value. This is accomplished by passing the DNA preparation over hydroxyapatite columns at a given incubation time. The fragments that are double stranded bind to the hydroxyapatite and can then be isolated from those that have not renatured and used in further analyses.

The slow renaturing, single-copy DNA has been shown to include the bulk of the structural genes, those sequences coding for polypeptides. Not all the single-copy sequences in this class consist of such genes, however. Many other single-copy sequences occur that are not associated with any specific product.

A second class that can be recognized by the kinetics of reassociation is the moderately repetitive DNA. In this group of fragments, the nucleotide sequences are not all identical by any means. The fragments include various stretches that are sufficiently alike to permit a single strand to pair not just with its exact complement but also with other strands that bear similar, though not completely identical, sequences. Within this kinetic class, one can recognize families. The repeated sequences recognized as a family, though clearly similar, are not necessarily identical because of some evolutionary change. The most frequently occurring base sequence is the one that is taken to represent a given family. Within a family, the related single-stranded fragments are able to pair during a renaturation experiment and to form stable duplex DNA that can bind to the hydroxyapatite. The conditions of the experiment can be so altered (by changing the temperature, for example) that fragments that are less related in nucleotide sequence can hybridize. The duplex DNA formed under these conditions would thus contain several regions that are matched but some regions that remain unpaired. Depending on the experimental conditions, these may be sufficiently stable to permit their binding to the hydroxyapatite and thus to fall into the middle repetitive class. This DNA class, therefore, includes an assortment of fragments from different families, and duplexes may form that are not completely complementary through their length.

A third class is the highly repetitive DNA. Sequences occur in this group, which may be repeated a million or more times! The fragments bearing these highly repeated sequences therefore reassociate at very low Cot values. The term *satellite DNA* is often used to refer to this group of sequences, since this highly repetitive DNA frequently forms a satellite band in a cesium gradient. This is because of the differences in its $A + T / G + C$ ratio compared with the bulk of the cellular DNA.

The distinction between highly repetitive, moderately repetitive, and single-copy DNA is not clear-cut but is somewhat arbitrary. Other classes or subclasses can even be recognized for certain eukaryotic DNAs. One additional type of DNA has now been identified and can be considered a fourth kinetic DNA class. This is referred to as *fold back, snap back,* or *palindromic* DNA. The naming and the behavior of this DNA are based on the fact that a fragment in this class contains an inverted repeat (refer back to Fig. 12-26). When a DNA fragment with an inverted repeat is rendered single stranded, each strand can fold back on itself very rapidly during a renaturation experiment, since a collision with another strand is unnecessary. The fold-back DNA can thus bind almost immediately to the hydroxyapatite just a few seconds after the reassociation experiment has begun.

What is the significance of these different kinetic classes that are so typical of eukaryotic DNA? What are the roles of these different classes, and how are they arranged along the intact. unfragmented DNA of the chromosome? Many investigations have been performed to answer these questions. DNA representing a given kinetic class that has reassociated at a certain Cot value can be prepared for in situ hybridization studies to determine its chromosome distribution. The unique or single-copy DNA has been shown to be rather evenly distributed along the chromosome, not concentrated in any particular locations. In the moderately repetitive fraction, which represents approximately 20% of the total cellular DNA, some families have been shown to be dispersed, as is true of the unique DNA, whereas other families are clustered in tandem repeats at certain chromosomal regions such as the centromeres and telomeres. In this moderately repetitive fraction, some genes are found whose functions are well known. These include genes for the histone proteins, tRNA, ribosomal RNA, β-globins, and others. However, these sequences with known functions represent only a small fraction of the moderately repetitive DNA and constitute less than 1% of the total chromosomal complement. Most of this kinetic class of DNA, as is true of the highly repetitive DNA, is made up of sequences that have no known function. These moderately repetitive sequences are dispersed all over the genome and may be repeated up to 10^5 times.

Some of the highly repetitive DNA sequences are dispersed, but most of them occur in tandem arrays concentrated at the centromeres and telomeres. The amount of this highly repetitive or satellite DNA can vary greatly in amount when different eukaryotes are compared. In mammals the satellite DNA constitutes approximately only one third of the total repetitive DNA. In contrast to the moderately repetitive DNA, this highly repetitive class does not contain as many different families, and most of it is not transcribed. When two closely related higher eukaryotic species are compared, similarities are seen in the sequences of their highly repetitive DNA. At the same time, however, the sequences of the one species are always distinct from those of the other. These decided differences evident in the nucleotide sequences of satellite DNAs taken from related species indicate that, once a change occurs in a repeat family, it spreads very rapidly throughout the genome. Within any one species, however, the sequence in a family of satellite DNA is exceptionally uniform. This indicates that some device must keep these repeated sequences the same.

Patterns of repetitive DNA in eukaryotes

Unlike the highly repetitive DNA sequences, which tend to occur as clusters following each other in tandem, the moderately repetitive DNA contains many sequences that are dispersed throughout the chromosome in hundreds of thousands of copies, each

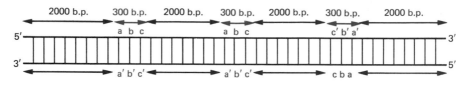

FIG. 13-21. Pattern of dispersed repeats in eukaryotic DNA. A portion of a double-stranded DNA is depicted. Along the length of the DNA, in eukaryotes, dispersed repeats *(red)* approximately 100 to 300 base pairs in length are typically interspersed with unique sequences approximately 2000 base pairs long. A repeated sequence may occur as a direct or as an inverted repeat *(right)*.

about 300 nucleotide base pairs long. These dispersed sequences have not changed as rapidly during the evolution of mammalian species as have the highly repetitive DNA discussed previously. When genomes from very different eukaryotic species are compared, a characteristic arrangement of the dispersed sequences and the unique ones is seen. The dispersed repeats, measuring about 100 to 300 base pairs in length, are typically interspersed with single-copy sequences approximately 1000 to 2000 base pairs long (Fig. 13-21). Note that a repeat may be direct or inverted which could be detected as snap-back DNA in a renaturation experiment.

This pattern of short, interspersed repeated sequences is common in eukaryotes and is found in species as diverse as sea urchins, amphibians and mammals. In the sea urchin, however, the short repeats belong to many unrelated families; no one kind of repeat is the most numerous. On the other hand, the DNAs of rodents and of primates contain a family of dispersed repeats that is much more abundant by far than any other in the genome. This is the *Alu family,* which in human DNA consists of 300 base pair long sequences repeated between 3×10^5 and 5×10^5 times and accounts for about 3% of the total human DNA! Approximately 300,000 copies or more are found in the haploid chromosome complement. The name *Alu* derives from the fact that sequences occur in these fragments that can be cut by a certain restriction endonuclease, AluI (see Chap. 16 for restriction enzymes). These Alu sequences undergo transcription to give repetitious RNA. (In contrast with the short, interspersed repeats such as the Alu sequences, the highly repetitive DNA sequences do not appear to be efficiently transcribed.) The Alu sequences make up a large population of the total interspersed 300 base pair repeats in human DNA. In human DNA, the distance between Alu repeats varies when different regions of the chromosomes are compared. Members of this abundant, widely dispersed

Alu family are not necessarily identical, although they are related in sequence. However, although evolutionary divergence has occurred in members of the Alu family, there has at the same time been a great deal of conservation in nucleotide sequence over the length of the 300 base pairs in the human. The characteristic stretch of nucleotides composing an Alu sequence in the human has been identified. The DNAs of other primates and of rodents contain a prominent Alu-equivalent family of interspersed repeats, but unlike human Alu, the length is only about 130 base pairs.

Although the Alu family is indeed a prominent one, several other very abundant repeats are now being described in mammalian DNA. It is estimated that Alu repeats, along with two other family members, make up about half of the repeat families in human DNA. It is also now known that, in addition to the short, interspersed repeats in which the repeat is less than 500 base pairs long, there are long interspersed repeats that measure more than 5000 base pairs in length and are repeated about 10^5 times in the genome.

The recognition of the highly repetitive nature of eukaryotic DNA raises the question of the significance of the repeated sequences. The fact that highly repetitive DNA does not seem to undergo transcription to any significant degree, if at all, suggests that any role it may have does not concern the coding of polypeptide or of any product. Rather, it has been suggested that its role is involved with the structural organization of the chromosome or with chromosome pairing. Still, the highly repetitive DNA remains largely a mystery. In contrast with the satellite DNA, the short, dispersed sequences *do* undergo transcription. Some members of the Alu family appear to be transcribed by RNA polymerase II. However, the transcripts seem to be part of the hnRNA and to be removed before transport of mature mRNA to the cytoplasm. It has been suggested that the Alu sequences may have some role in handling and maturation of the hnRNA. Some Alu family members do *not* appear to occur within units of RNA transcription but may be located *between* pairs of genes that seem to be closely related, such as between the human beta-globin-like genes. Members of the Alu family also may be transcribed as short RNA molecules by RNA polymerase III. Very interesting is the observation that only a small proportion of any interspersed repeat family ever undergoes transcription, suggesting that most of it is inactive. There have also been suggestions that the interspersed repeats may have roles in DNA replication by serving as origin points and that they may serve as regulatory sequences in gene transcription. As in the case of the highly repetitive satellite DNA, however, their true function, if any, remains in question.

There is also the school of thought that the multiple-copy DNA may actually have no function at all. Indeed, it is highly probable that most of the DNA of higher forms does not code for polypeptides. For many years, several perplexing facts about eukaryotic DNA have been apparent. When the amount of DNA in a haploid chromosome set (the *C value*) of one eukaryote is compared to that in others, we can see that the range is extreme. Quite amazingly the degree of complexity of an organism does not seem to be related directly to its C value. Although the amount of DNA in the human is 800 times that in *E. coli*, certain plant species and amphibians have 30 times as much DNA as humans! Very perplexing is the fact that the DNA content varies greatly among certain closely related species. In amphibians, DNA content among species may vary by a factor of 100. In buttercup species of similar karyotype, the content between some types may differ by a factor as great as 80! How can we explain the fact that the karyotypes of the plants are similar, as are the morphological resemblances, and yet their C values are quite different? In some cases, species differing in C value can be crossed to produce offspring. Bivalent formation in the hybrid looks quite normal, implying that the extra DNA is fairly evenly distributed over the chromosomes.

Various estimates suggest that in mammals, not more than 1% of the genome may be involved in regulation or in coding for proteins. The mammalian genome has enough DNA to code for about 3 million proteins, but it does not seem likely that it is coded for more than 30,000 of them. While some of the noncoding DNA may indeed be involved in the structure and organization of the chromosome, there still

appears to be too much DNA. According to the "selfish DNA" concept, many of the repeated DNA sequences have no function whatsoever and are not essential for cell survival. Their persistence in the genome results from the fact that the sequences have been selected for their ability to spread and duplicate efficiently in the genome without harming the host organism. We will return to this subject in Chapter 17, where the ability of the dispersed sequences to spread throughout the genome will be discussed in terms of the possibility that they may represent mobile DNA elements. Perhaps the repetitive DNA sequences may eventually be shown to perform a function, but at the moment, suggestions remain no more than speculations.

Fingerprinting and mutant hemoglobins

Although most of our early knowledge of molecular interactions came from elegant experiments with microorganisms, it was actually a human disorder, sickle cell anemia, that provided the first experimental demonstration that the amino acid sequence in a polypeptide is under the control of a gene. The unfortunate effects of the disease had been shown to stem from some sort of change in the hemoglobin molecule, but it was not known whether the change was an alteration in the amino acid content or in the folding of the molecule.

An answer to the question demands the detailed analysis of the amino acid sequence in hemoglobin. We can see this task is immense when we realize that each hemoglobin molecule in an adult is not just a single polypeptide chain, but consists of four chains—two of one kind and two of another. Each kind of chain is controlled by a different gene at a different genetic locus. In the adult, most of the hemoglobin in a red blood cell is hemoglobin A, composed of two alpha chains combined with two beta chains. About 2% of the hemoglobin present is a type called hemoglobin A_2 in which two alpha chains are combined with two delta chains. Each alpha chain contains 141 amino acids and is quite distinct from the beta and the delta chains. However, each beta and each delta chain is composed of 146 amino acids, and the two differ from each other in only 10 of these. The beta and delta loci, which control the formation of these two chains, are adjacent to each other on the chromosome and are believed to have arisen as the result of a duplication.

In sickle cell anemia, normal hemoglobin A is not present in the red blood cell, since a change has occurred in the beta chain. Inasmuch as hemoglobin A_2 is not involved, it will not be considered further in this discussion.

The analysis of hemoglobin was greatly facilitated by a shortcut procedure known as *fingerprinting,* which was used by Ingram in his pioneer determination of amino acid sequence in the protein. In this method, a protein that is to be analyzed is broken up into short fragments by cleaving it with a protein-digesting enzyme such as trypsin (Fig. 13-22). The resulting fragments, the peptides, then must be separated from one another. This is accomplished by using the technique of paper chromatography coupled with that of electrophoresis. The products of the enzyme action are placed near one edge of a piece of filter paper. The paper is then exposed to a specific solvent, which flows across it in one direction. The peptide fragments travel along in the direction of the solvent, but because they do not all move at the same rate in a particular solvent they can be partially separated. To obtain a more complete separation of the fragments, the paper is then turned 90 degrees and exposed to an electric field. Because of differences in their net electric charges, the fragments move at different rates to the + or − pole. The overall result is a good separation of peptide fragments, which then may be stained by methods that color proteins.

The technique is called *fingerprinting,* because a particular protein gives a characteristic picture. When the fingerprints of normal and sickle cell hemoglobin were compared, they were found to be identical except for one fragment which was out of place. This indicated the presence of some difference that alters the rate of migration of the specific fragment in the solvent and electric field. Because the peptide fragments are small (approximately eight amino acids each), their exact amino acid composition can be determined with relative ease.

Ingram was able to show that the peptide fragment difference between normal and sickle cell hemoglobin was a result of a difference in just a single amino acid (Fig. 13-23). Position 6 in the beta chain of normal adult hemoglobin (numbering from the amino acid with the free amino group) was shown to be occupied by glutamic acid, whereas this position was occupied by valine in sickle cell hemoglobin. After Ingram's attack on the problem, the different peptide fragments were eventually pieced together to show the exact sequence of amino acids in hemoglobin. It was finally shown this single amino acid substitution demonstrated by Ingram is the only difference between normal adult hemoglobin (hemoglobin A) and sickle cell hemoglobin (hemoglobin S).

Shortly after this was established, still another abnormal hemoglobin was analyzed—hemoglobin C, which also causes an anemia. It results from a recessive mutation, but the cells do not sickle, and the effects are not as severe as those of sickle cell anemia. However, hemoglobin C was shown to differ from the normal hemoglobin A at exactly the same position as hemoglobin S (see Fig. 13-23). In hemoglobin C, position 6 is occupied by lysine. The two amino acids (valine and lysine) substituting at this position for the glutamic acid in normal hemoglobin have electric

FIG. 13-22. Fingerprinting technique. The figure shows a fingerprint of the enzyme ribonuclease from the sheep pancreas. The enzyme was exposed to trypsin. A portion of the digested material was then applied to the small spot at the left. It was next subjected to paper chromatography, as indicated by the arrow below. After allowing the solvent to evaporate, the paper was moistened with buffer solution and subjected to electrophoresis. The sheet was sprayed with ninhydrin solution, which stains the areas containing peptides. These areas can be cut out and the peptides washed from the paper. The amino acid composition of each peptide can then be determined by further analysis. (Reprinted with permission from C. B. Anfinsen, The Molecular Basis of Evolution, p. 145. Wiley, New York, 1959.)

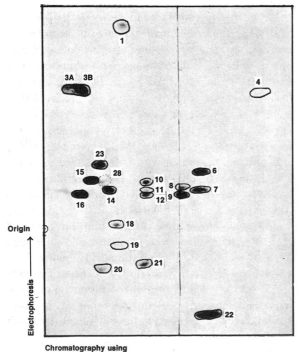

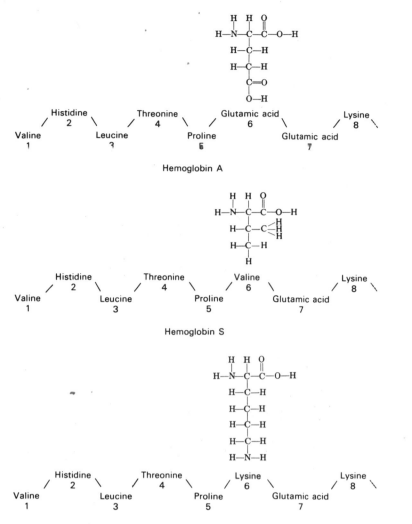

FIG. 13-23. The difference between normal hemoglobin and two mutant types. The precise amino acid sequence was determined for normal hemoglobin (hemoglobin A) and for sickle cell hemoglobin (hemoglobin S). A difference between them was found at one position. When the exact amino acid sequence was established for the entire beta chain of the hemoglobin, this was shown to be the only difference in the entire chain. At position 6 (near the NH_2 end), the glutamic acid in hemoglobin A is replaced by valine in hemoglobin S. Another mutant hemoglobin, hemoglobin C, is also altered at position 6. In this case, lysine is substituted.

properties that differ from each other and from those of glutamic acid as well. These differences may cause changes in the shape of the beta polypeptide chain. An alteration in the shape of this chain may, in turn, affect the architecture of the entire hemoglobin molecule and so cause a series of metabolic disturbances.

A normal beta (β) chain for normal hemoglobin A depends on an allele that can be represented as Hg_{β}^{A}. The alleles for β chains, which result in sickle cell hemoglobin and hemoglobin C, can be represented respectively as Hb_{β}^{S} and Hb_{β}^{C}.

Knowledge of the specific amino acid alteration involved in sickle cell anemia made it possible to relate the change to certain codons that are specific for the amino acids involved. Reference to the dictionary of codons (see Table 12-1) shows that two RNA codons designate glutamic acid (GAA and GAG), whereas four represent valine (GUG, GUU, GUC, and GUA). Advances on the molecular level have shown that the mRNA for the beta chain of hemoglobin A contains the codon GAG to designate position 6 in the beta chain, whereas in the mRNA associated with sickle cell hemoglobin, the codon GUG is substituted in its place.

Details on the molecular basis of sickle cell anemia and the techniques enabling us to pinpoint the specific alterations associated with the disorder are presented in Chapter 19. We can nevertheless appreciate the progress that has been made in our understanding of the molecular change that accompanies a genetic disorder. The whole syndrome of serious effects associated with sickle cell anemia can now be related to just a single nucleotide change. Dozens of other mutations are known which alter the polypeptide chains of hemoglobin. Compared with normal chains, each atypical chain is found to contain one specific amino acid substitution, which may involve either the alpha or the beta chain.

All the evidence with the mutant hemoglobins indicates that two separate functional units (genes) govern the production of hemoglobin A. The function of one of them is the formation of an alpha chain and that of the other a beta chain. Two alpha and two beta chains then associate to form normal adult hemoglobin. A point mutation (a change in a single nucleotide) at any site in one of the genes can result in an abnormal alpha chain. Similarly, a mutation at a site within the other gene can cause an abnormal beta chain to be produced. The hemoglobin story presents a good example of two units of function associated with a complex protein product made of more than one polypeptide chain.

As discussed in Chapter 19, the molecular information assembled in the past few years has deepened our understanding of the molecular nature of various genetic disorders. We are now in a position that enables us to detect the presence of certain defective alleles in unsuspected carriers and to undertake approaches to the treatment of some genetic afflictions.

Allelism and the functional unit at the molecular level

Our knowledge of the gene and its interactions at the molecular level has provided us with a sound physical basis for some established genetic concepts. This is well demonstrated by a return to the subject of allelism.

In Chapter 4 (Fig. 4-17), we discussed the multiple allelic series which included vg$^+$ (wild), vg (vestigial wing), and vg^a (antlered wing). Since vg and vg^a represent separate recessive mutations in the same gene at the vg locus, a cross of vestigial (vg vg) and antlered (vg^a vg^a) flies results in mutant offspring (vg vg^a). The antlered and vestigial mutations cannot complement each other because a fly of either type lacks a wild-type allele at the vg locus.

Now let us consider another recessive mutation, dumpy (dp), which also alters the shape of the wing, making it inadequate for flying. The dp locus is located on Chromosome II, which is also the case for the vg locus. When dumpy flies are crossed to vestigial (or antlered flies), the F$_1$ is found to consist of all wild-type flies. We can readily understand this, because the dp and vg loci, though they both affect the wing and are on Chromosome II, are nevertheless separate and distinct. Genes residing at the two loci are different genes. In other words, dp and vg are *not* allelic to each other. When two wing mutants are crossed (Fig. 13-24A), each parent contributes a normal form of the "wing" gene.

Here again we see that two separate genes at different loci can affect the same characteristic. Since the loci, dp and vg, are far apart, we can expect crossing over to occur between them. As this does indeed take place, the dihybrid with the *trans* arrangement can produce chromosomes with both mutant alleles on the same chromosome and both wild alleles on the other (Fig. 13-24B). As a result, dihybrid flies that have the *cis* arrangement can arise.

Although these points are simply a review of basic principles, it is important to realize here that when we are dealing with two recessive mutations at different loci, the F$_1$ heterozygote in *cis* or *trans* will be wild. This is so because both heterozygotes contain a normal dominant allele for each of the defective ones. We say that mutations such as dp and vg complement each other. This means that when two mutant forms are crossed, the recessive defect carried by either mutant parent can be "covered up" by the corresponding normal allele contributed by the other mutant parent. The F$_1$ resulting from a cross between dp and vg is normal, since the dp parent contributes a functional vg$^+$ allele and the vg parent a normal dp$^+$ allele. Two completely normal wing alleles are present, and so complementation takes place. This is in contrast with the cross between vestigial and antlered, in which the two mutations do not complement each other (Fig. 4-17).

From the information presented here, we see that the detection of a multiple allelic series is not difficult. It is relatively simple to tell if two recessive mutations are allelic or not. For if we cross two organisms showing mutant phenotypes and obtain mutant in-

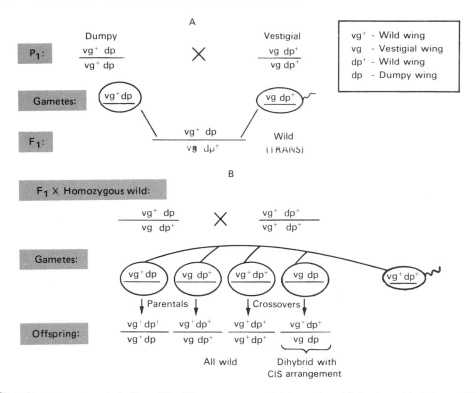

FIG. 13-24. Cis and trans arrangements in Drosophila. (A) a cross of two mutants, homozygous for recessives at two different loci produces an F_1 that is wild and carries alleles in the trans arrangement. The F_1 is wild in phenotype, even if the parents showed a mutant trait for the same characteristic. This is so in this example because the dumpy parent contributes the wild allele of the vestigial, and the vestigial parent contributes the wild of dumpy. We say that such alleles complement each other. (B) Since vestigial and dumpy are separate loci, crossing over may occur between them. This can give rise to offspring that are dihybrid and have the cis arrangment. These dihybrids are wild, just as those with the trans arrangement.

stead of wild offspring, we then know that the two mutant genes are not complementary. They must be allelic to each other. However, if they are mutations at separate and distinct sites within the same gene, we should be able to detect crossing over between them. We know today that *intragenic* crossing over is frequently observed in microorganisms, because it can be more readily detected in these forms than in higher organisms.

The reason crossing over within a gene is easy to detect in microorganisms should be evident from what we have learned about linkage and chromosome mapping. Genes that are closely linked on the DNA of the chromosome tend to stay together in the parental combination. There is less crossing over between them than between two genes that are widely separated. And obviously, different sites within the same gene are very closely linked, usually more so than two sites in two different genes. Therefore, to detect crossing over within a gene, we need to raise a very large number of offspring, because the new combinations (the recombinants) will be very few. To detect intragenic crossing over, even in a species as prolific as the fruit fly, requires the breeding and examination of thousands of offspring. With microorganisms (viruses, bacteria, and certain molds), it is possible to raise millions of progeny in a very short time. And techniques are available for the detection and scoring of recombinants. This extremely rapid rate of multiplication is one of the main reasons that crossing over within genes was fully analyzed in microorganisms. Intragenic mapping is discussed in more detail in Chapter 16.

Cases of crossing over within the gene came to light slowly in Drosophila, and their true nature had to await clarification from studies with microorganisms. Since mapping sites within a gene entails vast numbers of offspring, intragenic maps in higher organisms are few compared with those in microorganisms. In the fruit fly, several genes have been mapped for a few mutant sites, and progress has been made in mapping certain genes in mice that influence coat color (the yellow gene) and the development of the embryo (the tailless gene). Failure to demonstrate recombination between two alleles does not necessarily mean that crossing over cannot occur between them. It may mean only that we have as yet failed to detect it because the mutant sites are very close together. Crossing over between mutant sites in the white gene (for eye color) in Drosophila, such as between the allele "white" and the allele "apricot," was not noted at first. When crossovers between the two were finally observed, the interpretations were not always correct and were often made to conform to the concept that the gene is indivisible and that crossing over cannot take place within it. Work with microorganisms changed this idea and clearly showed that intragenic crossing over is a fact. There is little wonder that many genes have been mapped in bacteria and viruses in contrast with higher organisms, with their longer generation times and smaller number of offspring.

Let us now see how allelism can be explained in light of our knowledge of molecular interactions. We have considered the gene in relationship to its function at the molecular level, the transfer of information to direct the formation of a specific polypeptide chain. Any gene, because it has a certain function to perform, therefore, may be considered as a *unit of function*. Each gene, or unit of function, contains smaller units within it, the nucleotides. Mutation may occur at any one of these sites within a gene, and as we will see in more detail in Chapter 16, crossing over can take place between nucleotides within the gene. Therefore, the gene, the unit of function, is composed of smaller units of mutation and crossing over, the nucleotides. Each gene is a stretch of DNA with a specific function to perform, and each includes separate sites of mutation and crossing over. The term *cistron* (Chap. 16) has been coined to refer to the gene as a unit of function. The expression should not confuse us if we understand what has been said about the physical basis of the gene and its molecular interactions. The words *gene* and *cistron* mean the same thing insofar as both designate a genetic unit with a specific function. The term cistron implies that the functional unit, the ordinary gene, is divisible. It represents a genetic region within which separate mutations can occur and crossing over can take place. It is equally correct to say "one gene–one polypeptide" or "one cistron–one polypeptide."

Remember that a cross between two recessive types produces normal progeny if the mutations being followed are in separate genes. However, if two separate mutations represent defects at different sites within the *same* gene, mutant offspring are produced. In the latter case, we say that the two are therefore allelic. If no mutant offspring had resulted, we would know that the mutations *do* complement each other. The mutations would then be defects in separate genes and would *not* be allelic.

Let us now examine these ideas in terms of gene function. Suppose two separate recessive mutations arise that cause the blocking of a metabolic step controlled by a certain enzyme (Fig. 13-25A). The blocking is shown to result from the production of a defective enzyme by the mutant types. The genetic defects in each mutant are mapped and are shown to lie at closely linked, but nevertheless separate, sites on the chromosome. The question now arises, "Are these defects separate mutations within the same gene or are they in different genes?" A cross is made between the two mutant types, each carrying a mutation that maps at a different site. When the offspring are produced, they are also found to be mutant, defective for this same enzyme.

What does this mean in terms of the unit of function (Fig. 13-25B)? Since the mutations are not complementary, no normal enzyme can be produced by the F_1 offspring. This is so for the following reasons. One parent contributed a genetic factor with a defect at a certain site, and this results in the production of a defective polypeptide, one incapable of catalyzing the needed metabolic step. The other parent also contributed a defective genetic factor—a form of the same gene that is defective in the first parent—and so the same polypeptide is affected. The only difference between the two mutants is that they carry defects at different sites within the same gene. The two factors are allelic to each other. Therefore, the *trans* arrangement produces mutant types, because no functional polypeptide governed by that one gene is being produced. No intact enzyme can be formed.

Crossing over in the F_1 offspring with the *trans* arrangement may yield a double defective allele and one that is normal (Fig. 13-25C). Any individual carrying both a normal allele and a gene form with two defects within it would produce normal enzyme. This is so because the individual with the *cis* arrangement possesses one completely normal allele (B in Fig. 13-25C). Contrast this with the F_1 heterozygote with the *trans* arrangement. The latter carries no completely

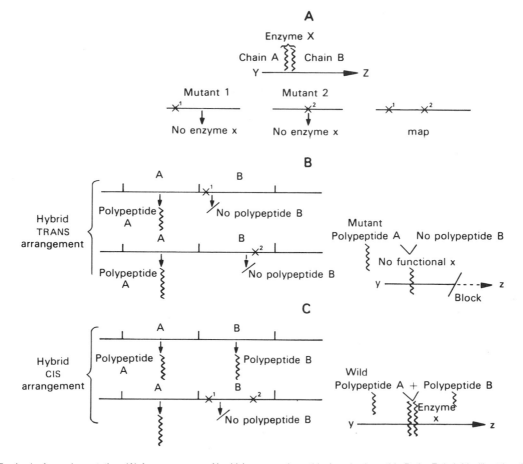

FIG. 13-25. Lack of complementation. (A) Assume enzyme X, which is composed of two kinds of polypeptide chains, is needed to catalyze the step in which Y is changed to Z (above). Two separate mutations occur that block formation of enzyme X, so that the step does not proceed (below). When the sites of the mutations are mapped, they are found to occur at separate, closely linked positions. (B) When the two mutants are crossed in this example, the combination of the two separate mutations gives no enzyme. This is to be expected if the two defects are in the same gene. In this diagram, both are represented in gene B. If the normal enzyme structure depends on both polypeptide A and polypeptide B, the F_1 hybrid will not be able to form normal enzyme, because no normal polypeptide B can arise. This follows from the fact that the F_1 received a defective B allele from each parent. The two separate defects or mutations do not complement each other. (C) An individual with the mutations in the cis arrangement will be able to produce normal enzyme. This is so because at least one completely normal B allele is present to form the required polypeptide. Both defects are within the same stretch of DNA, giving a doubly defective allele, but the completely normal one can carry out the essential function.

normal unit of function for the formation of the necessary polypeptide and is consequently mutant. Therefore, when crossed, if two recessive mutant types produce an F_1 that is defective for the same characteristic (here a specific enzyme), the two separate mutations are allelic; they affect the same unit of function. The *trans* arrangement gives a mutant phenotype, whereas the *cis* yields a wild type. This is a demonstration of the *cis–trans* effect.

Now let us apply exactly the same kind of reasoning to another case of two separate recessive mutations (Fig. 13-26A). In this example, however, a cross of two mutants defective for the same enzyme gives an F_1 that is wild, one which can produce normal enzyme and carry out the essential step. This means the two mutations can complement each other in the *trans* arrangement. The defects must be in separate units of function, separate cistrons. They complement each other because gene A controls a polypeptide chain that is distinct from the polypeptide chain controlled by gene B. For the enzyme to be functional, there must be at least one good chain A and one good chain B. Mutant A contributes a defective A but a good B. Mutant B contributes a good A but a defective B. As

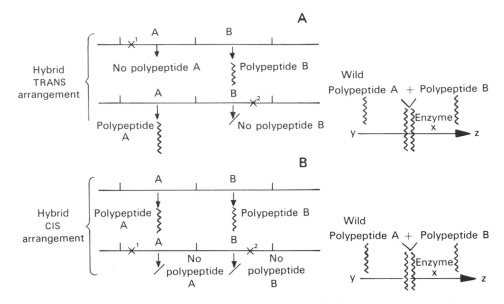

FIG. 13-26. Cis–trans arrangements. Complementation. (A) In this example, an F_1 hybird is formed between two mutants (A and B), each with a defect in a separate gene. Consequently, one functional A allele and one functional B allele are present to allow formation of the enzyme. (B) This cis arrangement is wild, because one chromosome contains both the functional A and the functional B alleles of two different genes.

a result, the F_1 has some good A and some good B, and can threfore form functional enzyme. The hybrid with the *cis* arrangement will also form functional enzyme (Fig. 13-26*B*). No *cis–trans* effect is apparent.

The *cis–trans* test is a fundamental one used to determine allelism. The different results in the *cis* and *trans* arrangements are easily understood on the basis of gene function and molecular interactions.

REFERENCES

Britten, R. J. and D. E. Kohne. Repeated DNA *Sci. Am.* (April): 36, 1970,

Cech, T. R. RNA as an enzyme. *Sci. Am.* (Nov): 64, 1986.

Chabot, B., D. L. Black, D. M. Le Master, and J. A. Steitz. The 3′ splice site of pre-messenger RNA is recognized by a small ribonucleoprotein. *Science* 230: 1344, 1985.

Chambon, P. Split genes. *Sci. Am.* (May): 60, 1981.

Clark, B. F. C. and K. A. Marcker. How proteins start. *Sci. Am* (Jan): 36, 1968.

Clegg, J. B., D. J. Weatherall, and P. F. Milner. Haemoglobin Constant Spring—a chain termination mutant? *Nature* 234: 337, 1971.

Cold Spring Harbor Symposium on Quantitative Biology. *The Mechanism of Protein Synthesis,* vol. 34. Cold Spring Harbor Laboratory, New York, 1969.

Crick, F. H. C. Codon-anticodon pairing: the wobble hypothesis. *J. Mol. Biol.* 19: 548, 1966.

Crick, F. H. C. Split genes and RNA splicing. *Science* 204: 264, 1979.

Darnell, J. E. Jr. The processing of RNA. *Sci. Am.* (Oct): 90, 1983.

Darnell, J. E., Jr. RNA *Sci. Am.* (Oct.): 68, 1985.

De Jong, W. W., P. Meera, and L. F. Bernini. Hemoglobin Koya Dora: a high frequency of a chain termination mutant. *Am. J. Hum. Genet.* 27: 81, 1975.

Federoff, N. V. On spacers, *Cell* 16: 697, 1979.

Fuchs, F. Genetic amniocentesis. *Sci. Am.* (June): 47, 1980.

Gall, J. G. Chromosome structure and the C-value paradox. *J. Cell Biol.* 91: 3s, 1981.

Gall, J. G. and M. L. Pardue. Formation and detection of RNA-DNA hybrid molecules in cytological preparations. *Proc. Natl. Acad. Sci.* 63: 378, 1969.

Gilbert, W. Genes in pieces revisited. *Science* 228: 823, 1985.

Jelinek, W. R. and C. W. Schmid. Repetitive sequences in eukaryotic DNA and their expression. *Annu. Rev. Biochem.* 51: 813, 1982.

Konkel, D. A., S. M. Tilghman, and P. Leder. The sequence of the chromosomal mouse β-globin major gene: homologies in capping, splicing, and poly(A) sites. *Cell* 15: 1125, 1978.

Kozak, M. How do eucaryotic ribosomes select initiation regions in messenger RNA? *Cell* 15: 1109, 1978.

Lake, J. A. The ribosome. *Sci. Am.* (Aug.): 84, 1981.

Lerner, M. R. and J. A. Steitz. Snurps and scyrps. *Cell* 25: 298, 1981.

Lewin, R. Repeated DNA still in search of a function. *Science* 271: 621, 1982.

Miller, O. L. The nucleolus, chromosomes, and visualization of genetic activity. *J. Cell Biol.* 91: 15s, 1981.

Nomura, M. (ed.). *Ribosomes*. Cold Spring Harbor, New York, 1973.

Orgel, L. E. and F. H. C. Crick. Selfish DNA: the ultimate parasite. *Nature* 284: 604, 1980.

Padgett, R. A., S. M. Mount, J. A. Steitz, and P. A. Sharp. Splicing of messenger RNA precursors is inhibited by antisera to small nuclear ribonucleoprotein. *Cell* 351: 101, 1983.

Perry, R. P. RNA processing comes of age. *J. Cell Biol.* 91: 28S, 1981.

Proudfoot, N. J. and G. G. Brownlee. 3′Non-coding region sequences in eukaryotic messenger RNA. *Nature* 263: 211, 1976.

Rich, A. R. and S. H. Kim. The three dimensional structure of transfer RNA. *Sci. Am.* (Jan.): 52, 1978.

Ruskin, B. and M. R. Green. An RNA processing activity that debranches RNA lariats. *Science* 229: 135, 1985.

Schmid, C. W. and W. R. Jelinek. The Alu family of dispersed repetitive sequences. *Science* 216: 1065, 1982.

Sharp, P. A. Splicing of messenger RNA precursors. Science 235: 766, 1987.

Sibley, C. G. and J. E. Ahlquist. Reconstructing bird phylogeny by comparing DNA's. *Sci. Am.* (Feb.): 82, 1986.

Spiegelman, S. Hybrid nucleic acids. *Sci. Am.* (May): 48, 1964.

Zaug, A. J. and T. R. Cech. The intervening sequence RNA of *Tetrahymena* is an enzyme. *Science* 231: 470, 1986.

REVIEW QUESTIONS

These questions are based on information in Chapters 11, 12, and 13.

1. Match the number of the term on the right with the appropriate description on the left. A number may be used more than once or not at all.

A. Region of gene that contains recognition signals.	1. promoter
B. Needed for normal termination of transcription of some genes.	2. sigma factor
	3. cyclic AMP
C. Signal for the start of translation.	4. AUG
	5. n-formyl methionine
D. Is formed only on initiating tRNA.	6. rho factor
E. Actively recognizes the signals in the gene for the start of transcription.	7. methionine
	8. valine
	9. glutamic acid
F. Needed to link an amino acid to tRNA.	10. UUU
	11. UGA
G. Signal to end a polypeptide.	12. CTT
H. Found at position 6 in the beta chain of hemoglobin A.	13. CAP
I. Protein needed to stimulate transcription of certain genes before the binding of RNA polymerase.	14. peptidyl transferase
	15. aminoacyl-tRNA synthetase
J. Needed to form linkages between carboxyl and amino groups of amino acids.	

2. When pure DNA of the nucleolar organizer is hybridized with 18S and 28S RNA from the same cell type, only about 25% of the DNA present hybridizes with the RNA. Explain.

3. Answer the following questions in relation to this DNA strand:

    3′ TACAAATCTCATTGTATAGGA 5′

    A. How many individual tRNA molecules are required to translate the transcript of this segment?
    B. What are possible anticodons that could be used for translation of the transcript of this segment?

4. For each of the following anticodons, give the possible corresponding RNA codons and the DNA triplets in the sense strand. Include the polarities.

    A. 3′ UUA 5′.
    B. 3′ UCU 5′.
    C. 3′ AAU 5′.
    D. 3′ CCI 5′.

5. Consult the genetic dictionary (Table 12-1), and give the amino acid designation in each of the four cases in Question 4.

6. From the genetic dictionary, note the RNA codons for the amino acid, serine. What would be the *minimum* number of tRNAs needed to insert serine into a growing polypeptide chain during translation?

7. Answer Question 6 for the amino acids leucine and arginine.

8. Consider an mRNA carrying ribonculeotides numbered 1, 2, 3, and so on, from the beginning of the message for a specific polypeptide. The normal polypeptide has 300 amino acids.

    A. Ribonucleotide 14 undergoes a change resulting in a missense mutation. At which position from the NH_2 end in the polypeptide will an amino acid substitution occur?
    B. Assume Ribonucleotide 23 changes and a "nonsense" mutation arises. How many amino acids would you expect in the peptide fragment?

9. A certain polypeptide chain is composed of 100 amino acids. During its synthesis, a methionine molecule was picked up by a specific tRNA and placed in position 1 in the growing polypeptude. An asparagine molecule was placed in the second position, a leucine in the third, a histidine in the fourth, and so on, until a tryptophan is placed in the last position. Identify these five particular molecules in the completed polypeptide for the following.

    A. The tRNA carrying this amino acid was the first to move to the A site of the ribosome.
    B. This amino acid will have a free carboxyl group.

C. The tRNA carrying this amino acid did not associate with the A site of the ribosome.

D. The carboxyl group of this amino acid formed a peptide linkage with the amino group of leucine.

E. This amino acid has a free amino group.

10. Do the following concerning ribosomes:

A. Give the characteristic S value of the intact eukaryotic ribosome.

B. Give the S value of the subunit in prokaryotes that attaches to the mRNA and to the first tRNA.

C. Give the S value of the subunit of *E. coli* that contains A and P sites.

D. Give the S value of the eukaryotic subunit that contains the enzyme required for the formation of peptide linkages.

E. Name the ion required for the cohesion of the two subunits.

F. Give the S value of the eukaryotic rRNA whose genes are not in the nucleolar organizer.

11. The following represents a tripeptide: NH_2 met-glu-trp COOH. Give the double-stranded sequences that could code for it, giving the polarity as well as the sense and the antisense strands.

12. Assume that a particular DNA template normally undergoes transcription, and that the transcript is in turn translated to produce a polypeptide. A series of alterations arises that affects the template. For each of the following, give that part of the template that has most likely been altered.

A. Transcript produced but is unable to bind to ribosome. No polypeptide is produced.

B. No transcript formed because RNA polymerase cannot start transcription. Therefore, no polypeptide is produced.

C. Polypeptide of normal length is produced but is nonfunctional because of an amino acid substitution.

D. Defective polypeptide of abnormal length is produced in which additional amino acids occur at the carboxyl end.

E. Only a nonfunctional peptide fragment is produced, consisting of the first four amino acids of the normal polypeptide.

13. Indicate the polarity of each of the following as either 5' or 3'.

A. Translation of mRNA begins at this end.

B. The beginning of the gene proper is toward this end of the template.

C. The base at this end of the anticodon is in the wobble position.

D. The first nucleotide in the mRNA has this end free.

E. The amino acid with the carboxyl end was designated by a codon toward this end of the transcript.

14. The DNA associated with five different structural genes can be represented as shown here, where H and L designate the heavy and light strands:

H5' _ 3'	H5' _ 3'	H5' _ 3'	H5' _ 3'	H5' _ 3'
L3' _ 5'	L3' _ 5'	L3' _ 5'	L3' _ 5'	L3' _ 5'
(1)	(2)	(3)	(4)	(5)

Indicate for each structural gene whether transcription will occur in the direction left to right ($\rightarrow$) or right left ($\leftarrow$) after considering the following for each DNA region.

A. The core enzyme transcribes nucleotides of the gene proper in the L strand.

B. The antileader is toward the 3' end of the H strand.

C. The terminator ending in a series of A-containing nucleotides is in the H strand.

D. The first 5'ATG3' sequence is in the H strand.

E. The first 5'TAG3' sequence is in the H strand.

15. In a certain human blood disorder, one of the β chains of hemoglobin is absent. However, transcripts have been isolated from the cells of afficted persons. These transcripts form hybrid molecules with the DNA known to code for the β chain. Offer explanations for the nature of the genetic disorder.

16. DNA analyses of three different mammalian species shows the following base percentages:

	A	T	G	C
Species A	29.6	29.6	20.4	20.4
Species B	29.5	29.5	20.5	20.5
Species C	29.8	29.8	20.2	20.2

Denaturation–renaturation experiments are performed with two species at a time. In such an experiment, the separate DNA chains of one species are left intact; in the other, they are fragmented. It is found that almost 100% of the DNA from species A and C hybridize. Only about 5% of the DNA of species B hybridizes with A or C. Offer an explanation.

17. Next to each of the following statements, place a "P" if it is more applicable to prokaryotes, an "E" if it is more applicable to eukaryotes, and "P+E" if about equally applicable to both.

A. ——Little variation G:C value.

B. ——5' to 3' growth of DNA chains.

C. ——Three types of RNA polymerase.

D. ——*n*-formyl methionine in cytosol.

E. ——*m*RNAs terminating in long stretches of A-containing nucelotides.

F. ——Initiating type of tRNA.

G. ——Satellite DNA.

H. ——Very fast renaturing DNA fraction.

I. ——hn RNA.

J. ——Trailer and leader.

K. ——AUG-initiating codon.

L. ——tRNA genes within rDNA.

M. ——7-methyl guanosine.

N. ——Degenerate code.

O. ——Translation accompanies formation of transcript.

18. Answer the following:

A. The anticodon ACG will pair with what codon or codons for what amino acid?

B. The anticodon ACA will pair with what codon or codons for what amino acid

C. The codon UGG for tryptophan will pair with what anticodon, and what is its polarity?

19. Answer the following:

A. Considering the fact that two specific anticodons, as noted in Question 18, are used for cysteine, why is there no anticodon at all with the sequence ACI, since this, too, could pair with the two codons for cysteine?

B. There is no ACU anticodon; yet it could pair with the codon for tryptophan. Explain.

20. In sickle cell anemia, the mRNA coded for the beta chain of hemoglobin contains the codon GUG at the sixth amino acid coding position. In the case of normal hemoglobin, this position in the transcript is occupied by the codon GAG. Show this position in duplex DNA, indicating the sense and antisense strands along with their orientation. Indicate the specific change at the site in the DNA that is responsible for sickle cell hemoglobin.

21. In a renaturation experiment, rank the following four fractions from lowest to highest expected Cot values: (1) satellite DNA; (2) sequences encoded for polypeptides; (3) snap-back DNA; (4) moderately repetitive sequences encoded for histones, tRNA, and rRNA.

22. The term $Cot_{1/2}$ refers to that Cot value required for the reassociation of half of all the single DNA strands in a renaturation study. $Cot_{1/2}$ is related directly to the amount of DNA present. In a certain renaturation study, the same concentration of bacterial DNA and DNA from a higher eukaryote was used. When the renaturation curves are compared, would you expect the eukaryotic $Cot_{1/2}$ to be higher, lower, or the same as that for the bacterial $Cot_{1/2}$?

23. Does the Cot curve shown here represent a Cot curve from a prokaryotic or eukaryotic source? Explain.

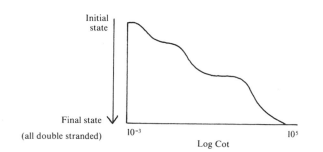

24. A gene in the chicken that codes for a specific egg protein was isolated. In one set of experiments, the duplex DNA was denatured and allowed to hybridize with mature, processed messenger RNA, which was isolated from cells that secrete the protein. The electron microscope revealed heteroduplex molecules formed between the mRNA and single DNA strands. Such a heteroduplex is shown here. Answer the following questions regarding this heteroduplex:

A. Is the DNA represented by the blue or the red strand? Explain.

B. How many introns appear to be preseent?

C. The longer strand shows an unpaired segment at each end. What do these represent?

D. The shorter strand shows an unpaired segment at one end. What does this represent, and is it at the 3' or 5' end of the strand?

25. In a certain fungus, several recessive mutations occurred independently, each of which prevents the manufacture of adenine. Normal wild-type fungi of this species have the ability to make adenine. The several adenineless stocks have been named ad_1, ad_2, ad_3, and so on to distinguish them according to their order of occurrence. Various crosses among some of the stocks were made with the following results:

1. $ad_1 \times ad_3$ (wild)
2. $ad_1 \times ad_8$ (wild)
3. $ad_1 \times ad_{16}$ (wild)
4. $ad_8 \times ad_{16}$ (adenineless)
5. $ad_3 \times ad_8$ (wild)
6. $ad_3 \times ad_{16}$ (wild)

How many loci are being followed in these crosses, and which of the adenine mutations seem to be allelic to each other?

26. Assume that a, b, and c are recessive mutations in a certain mold and that the sites of the mutations are very closely linked. The following combinations give the results indicated for each:

$$\frac{a^+ \ b^+}{a \ b} = wild \qquad \frac{a^+ \ b}{a \ b^+} = mutant$$

$$\frac{b^+ \ c}{b \ c^+} = wild \qquad \frac{b^+ \ c^+}{b \ c} = wild$$

Do any of the mutations complement each other? Are any of them allelic?

27. Assume that in a certain biological step substance A is converted to substance B by enzymatic action. The step can be blocked if an individual is homozygous for the mutations a^1, a^2, a^3, or a^4. These separate mutations map at closely linked but different sites that are separable by crossing over. A series of crosses gives the following results. Explain on the concept of the cistron, and diagram the crosses.

    A. $a^1 \times a^1$—step blocked.
    B. $a^1 \times a^4$—step blocked.
    C. $a^1 \times a^2$—step proceeds.
    D. $a^1 \times a^3$—step proceeds.
    E. $a^2 \times a^3$—step blocked.
    F. a^1a^4 double mutant $\times$ normal—step proceeds.
    G. a^1a^2 double mutant $\times$ normal—step proceeds.
    H. a^2a^3 double mutant $\times$ normal—step proceeds.

28. Two married persons suffer from an anemic condition that follows an autosomal recessive pattern of inheritance in each family line. Fingerprinting of their hemoglobin A shows that the female has one amino acid substitution in the alpha chain and the male has one amino acid substitution in the beta chain. Their children, however, do not suffer from the anemia. Explain.

14

GENE MUTATION

Somatic and germinal point mutations

Without knowing their physical basis, De Vries called attention to the importance of sudden inheritable changes as the ultimate source of variation in the process of organic evolution. Any advantageous changes that may benefit the species must result from variations in the hereditary substance of the germ cells. Although most mutations might be harmful, evolutionary progress necessarily depends on the ability of the gene to mutate at times to another form (an allele) that conveys an advantage. Indeed, all hereditary variation seen in living things, good as well as bad, has ultimately stemmed from mutations in the hereditary material.

The pioneer geneticists, such as T. H. Morgan, realized the importance of mutation in genetic research. If no variation at all occurred in a character, nothing could be learned about its mode of inheritance. It was possible to work out the inheritance of eye pigments in the fruit fly only because mutations to white and various shades were uncovered in the laboratory. Crossing only pure breeding red-eyed flies is meaningless and contributes no information on the genetic loci that interact and influence color of the eye. Although the importance of spontaneous mutation was appreciated, the physical basis of the hered-

itary alteration was unknown and had to await the discovery of the chemical nature of the gene. Let us first examine certain significant facts about point mutation that were assembled in earlier studies. We will then review them in the light of more recent information acquired through research at the molecular level.

Since point mutations are changes in the DNA (Chaps. 11 and 12) we can expect them to occur in any kind of cell, somatic as well as germ cells. In diploid organisms, a somatic mutation at a specific locus tends to remain undetected. This is so because the wild form of the gene at that locus on the homologous chromosome is also present in the cell, and its effect is probably dominant and would offset that of the mutant form (Fig. 14-1A). However, somatic mutations in some plants have given rise to plant varieties of commercial value, such as the navel orange. In each case, a mutation arose in a single somatic cell. By repeated mitotic divisions, cells derived from the original mutant cell produced a branch composed of genetically identical cells, each with the original mutant allele. Since the desirable fruits of these mutant forms are often sterile (no seeds in the navel orange), the varieties must be propagated asexually by cuttings or grafts. Somatic mutations are also known that result in mosaic individuals, those with phenotypically

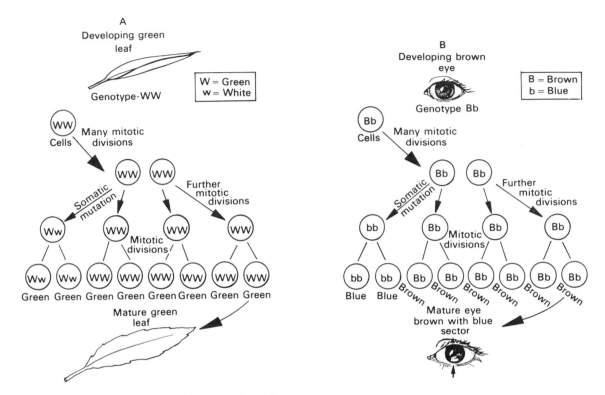

FIG. 14-1. Effects of somatic mutations. (A) Many somatic mutations remain undetected. The wild allele is usually present in the cell and is typically dominant to the mutant form. If a recessive mutation arises as a result of a mutation during development, the dominant is still expressed in the heterozygous cells. All the cells, heterozygous and homozygous, show the same phenotype. (B) If an individual is heterozygous, a somatic mutation may affect the dominant allele and thus generate a cell that is homozygous recessive. Further cell divisions will give rise to two populations of cells, which differ phenotypically as well as genotypically.

different patches of tissue. Variegated flowers with segments of different color, as seen in certain varieties of roses, are examples. In humans, cases of a brown-eyed person with a blue segment in one iris (or vice versa) can be explained on the basis of somatic mutation (Fig. 14-1*B*).

Although somatic mutations may be of consequence to the individual, the germinal mutations are of greater significance to the population, as these can be transmitted from one generation to the next. Because the consequences of gene mutation are so great for all species of living things, it is important to appreciate the kinds or types of mutations that can occur. Those mutations that arise for no apparent reason are called *spontaneous*. A spontaneous mutation may occur at any site on a chromosome. If we could follow every locus within a species, we would undoubtedly find that at some time or other a mutation would arise at each locus. However, the frequency of mutation at the different loci would vary greatly.

In one of the first precise studies of spontaneous mutation rates, Stadler followed eight loci that affect the phenotype of the kernels on an ear of corn. Since each kernel represents one individual, many hundreds of offspring of one plant can be observed by examining several ears. Because spontaneous mutation of any one gene is such a rare event, it is essential to examine very large numbers of individuals to estimate the mutation rate. During the course of his study, Stadler noted (Table 14-1) that mutations occur much more frequently at some loci than at others. Indeed, during the course of the investigation, no mutations at all were detected at the waxy locus (Wx) in contrast to the high number that arose at R and I. Mutation analysis in microorganisms has shown that certain sites on the DNA are more likely to undergo mutation than others. Controlled studies of mutation rate are difficult to conduct in higher forms. However, variation in mutation frequency among different loci has also been demonstrated in several species, including Drosophila, and is even indicated in humans.

TABLE 14-1 Spontaneous mutation rates of eight genes in corn

GENE	NO. OF MUTATIONS PER MILLION GAMETES
R	492
I	106
Pr	11
Su	2.4
Y	2.2
Sh	1.2
Wx	0
C	2.3

Some consequences of spontaneous mutation

When the effects of gene mutations on the individual are examined, it is found that the overwhelming majority is harmful in some way. This is to be expected on the basis of the interrelated pathways in metabolism. An unplanned change in a complex system is certainly more likely to produce a harmful effect than one that shows foresight or planning. And spontaneous mutation *is* a random, unplanned event that leads to a genetic alteration without regard for the consequences. We have already considered this fact as the reason for the high incidence of lethal alleles resulting from gene mutations (Chap. 4). A complete lethal has an effect so drastic that it eliminates the individual before it reaches reproductive age. Lethals far outnumber those mutant alleles that produce some visible effect on a characteristic. Indeed, most mutant alleles that bring about a visible change in the phenotype carry some sort of detriment with them.

By far the most common mutations, however, are those that produce just a slight effect. This should not seem surprising on the basis of our knowledge of gene interactions (Chap. 4). We have learned that many genes influence the expression of a character and that many biochemical pathways are interrelated in cellular metabolism. Some genes control chemical steps that are more important or critical than other steps. Although many genes influence a character, the majority of them influence it only slightly (consider the many modifying genes and genes with quantitative effects). Therefore, any mutation, being random, will more likely occur in a gene with a slight effect than in one that can alter the character drastically.

Well-controlled studies with the fruit fly have shown that gene mutations giving rise to alleles with a visible effect on a character are in the minority and that the more pronounced the visible effect is, the greater the accompanying detriment will be. But more common than these mutations are the complete lethals that have no visible effect but can completely remove the individual before the age of sexual maturity. Most frequent of all, however, are those mutant alleles that have no visibly detectable effect but contribute a slight degree of harm. We may call these *detrimentals*. H. J. Muller called attention to detrimentals, maintaining that no matter how slight their impairment to an individual they were nevertheless of consequence. In the case of a 100% lethal, the one individual is removed at once, because the entire burden falls on the individual organism. Muller referred to any burden imparted by mutant genes as the *genetic load*. He argued that in the case of detrimentals, the load, instead of falling on one individual, is distributed in smaller amounts and usually passes through more individuals. The load is thus shared by all of them, perhaps with just a small effect but each individual's chances of survival are decreased by a real amount. And in the long run, a genetic death results. It occurs at once for the full lethal, but for the detrimental, though the allele passes through more individuals, the outcome is the same—genetic death. Genetic death does not necessarily mean that the individual must die. Any allele causes genetic death if it prevents the individual from reproducing. In terms of genetic consequences, loss of reproductive capacity because of deleterious genetic factors is equivalent to an outright killing. In both cases, the individual fails to pass genetic information down to the next generation.

Usually genetic load is expressed in terms of *lethal equivalents*. One such equivalent can be viewed as a deleterious allele that results in genetic death when it is present in the homozygous condition. There is also one lethal equivalent in the case of two alleles, each of which causes genetic death half the time when it is homozygous. Suppose a population is composed of individuals, each of whom carries eight deleterious recessives. If each of these recessives, when homozygous, has a 50% chance of causing genetic death, the population would have a store of four lethal equivalents.

Several studies in the human indicate an average of one to three lethal equivalents per genome. Considering one lethal equivalent to be the average, this means that each human gamete carries one allele that can lead to genetic death if it unites with another gamete carrying the same deleterious allele. Muller and other geneticists have stressed the need to prevent an increase in the number of detrimentals in the population by exposure to agents known to increase the mutation rate.

A mutation that changes a gene from its wild form to a mutant form is called a *forward mutation*. Once a gene has mutated, the new allele continues to duplicate itself, unless, of course, it is eliminated at once. The new gene form, as well as the original one, may mutate further, resulting in still additional allelic forms, as seen in cases of multiple allelic series. A mutant allele at times may revert to the original or wild form of the gene. This is called *reverse mutation* or *back mutation*. Although the frequency of back mutation is characteristic for a specific gene, back mutation is rarer than forward mutation, the change from wild to mutant. We will see why this is so in the following section.

When a gene mutation arises, it may result in the origin of either a dominant or recessive mutant allele; but by far the most numerous mutant forms of a gene have recessive effects. This is quite understandable when we consider that natural selection has operated over long periods of time, so that today the wild alleles in any species are the products of millions of years of screening. Those alleles that are most conducive to the survival of the species will have been established by natural selection. Any wild allele would thus be one that is beneficial to the individual organism living in its specific environment. Part of the benefit of any wild allele is its ability to express itself in the presence of mutant forms. In other words, wild alleles have actually been selected for their dominant effects.

We can appreciate the value of this when we consider that most spontaneous alterations in a gene are harmful. If the wild form can "cover up" the deleterious effect of the mutant, the heterozygote will not be harmed. More individuals will survive than if the deleterious mutant allele were dominant. If a rare recessive mutation that carries an advantage superior to that of the wild does occur, natural selection favors the mutant. It also favors those forms of the mutant gene that later arise to make it more dominant. Eventually the mutant allele may supplant the original wild, and thus become the new wild type.

Reversions

The mutant effect of a gene mutation, as noted earlier, may be reversed by the effects of a second mutation. A *true reversion* or *true back mutation* entails an exact reversal of the change that produced the for-

ward mutation. For example, suppose that a forward mutation results in the substitution of the base pair G:C for the pair A:T and produces a defective allele. In a true reversion, the same position in the gene would be involved, and the G:C pair would be replaced by the original A:T, restoring the wild gene form. Since a true reversion requires a second specific change at one specific site, it is a much rarer event than a forward mutation, which can involve any base pair change at a site within the gene. Other kinds of reversions can occur, however. These are more common than the true reversions and are due to suppressor mutations, mutations at sites other than the original, which can restore the wild phenotype. We have already been introduced to certain mutations of this type. One example is the frameshift mutation (Chap. 11). A deletion or addition of a base in a gene can shift the reading frame, producing a mutant allele that can no longer code for the normal polypeptide. A second frameshift mutation nearby in the same gene, however, can restore function, such as a + mutation following a − one. This is an example of intragenic suppression or *second-site reversion,* a suppressor mutation that occurred at a different site in the same gene as the first mutation and restores function.

Second-site reversions can occur in other ways besides frameshift mutations. For example, as a result of a missense mutation, an amino acid that now substitutes for the original one may alter the folding of a polypeptide. This can render the polypeptide defective by disrupting the position of some active site such as one involved in the catalytic action of an enzyme. A second missense mutation at a different position in the gene may cause the substitution of still another amino acid at a second site in the polypeptide. By itself, this second mutation could very well be harmful, because it, too, can alter the folding of the polypeptide chain. However, the two missense mutations in the same cell may have the effect of substituting two amino acids that can now interact to restore normal or near-normal folding of the polypeptide chain. Consequently, normal activity may be restored to the enzyme by bringing the active catalytic site back into proper position.

Another way in which a reversion may take place and restore normal function to a gene stems from a mutation that, instead of being intragenic, occurs at a site in another gene. Such *intergenic mutations,* which restore normal function and thus suppress the deleterious effects of a first mutation, are usually called *intergenic suppressors* (Chap. 4). Let us see

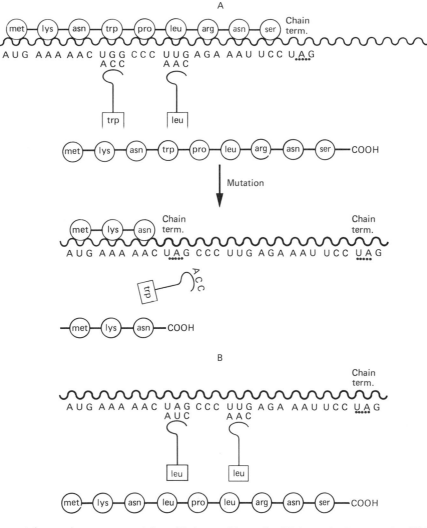

FIG. 14-2. Nonsense mutations and suppressor mutation. *(A)* A portion of an mRNA strand is depicted with the corresponding amino acids above the codons. One chain-terminating codon, UAG, is shown bound to release factors *(red dots)*. Two charged tRNAs, tRNA^trp and tRNA^leu, are shown with their anticodons. (The other tRNAs, as well as ribosomes, are omitted for clarity.) The amino acids will be joined together until the chain-terminating codon brings translation to a halt. A nonsense mutation changes the codon UGG for tryptophan to UAG, a chain-terminating codon. The tRNA^trp can no longer recognize that site, which is now recognized instead by release factors, bringing translation to a premature halt. An incomplete, nonfunctional polypep-

tide results. *(B)* A mutation in a gene for tRNA^leu changes the anticodon in the tRNA from ACC to AUC. This is complementary to the chain-terminating codon UAG. The mutant tRNA can now position leucine at the site normally occupied by tryptophan. There is more than one gene for tRNA^leu, and so the cell will contain tRNA^leu molecules with the normal anticodon. These will continue to position leucine at its normal sites. The resulting polypeptide, though normal in length, will carry a substituted amino acid. Nevertheless, some function may now be restored because of the mutation in the tRNA^leu gene, which has acted as an intergenic suppressor.

how these suppressors may be able to cause reversion from mutant to wild. In the virus SV40 (Chap. 16), a certain mutation causes a base pair substitution near the site of initiation of DNA replication of the viral genome. This alteration prevents the effective binding of a specific protein to the DNA area around the

initiation site. The binding of the protein to the DNA is apparently needed to trigger DNA replication. Hence, the rate of replication of the mutant virus is greatly reduced, because the protein can no longer bind efficiently to the required site. However, a mutation in another gene, the one coded for the protein, can cause suppression of the mutant effect. The altered mutant protein, unlike the normal one, *is able* to recognize the altered initiation site very effectively and can bind well to the DNA, restoring the normal replication rate.

Suppressor mutations that occur in genes coded for tRNA have been very well studied. You will recall that a point mutation can change a codon that designates a specific amino acid to a chain-terminating codon: UAG, UAA, and UGA (Chap. 12). Such a nonsense mutation will bring translation to a premature halt, resulting in an incomplete protein. Let us suppose that the RNA codon 5′-UGG-3′ for the amino acid tryptophan is converted to 5′-UAG-3′ as a result of a mutation in the template strand of a structural gene. The mutation converts the DNA triplet ACC to ATC. This means that the normal tryptophan tRNA in the cell no longer recognizes this codon (Fig. 14-2A). Instead, it is recognized only by release factors that bring translation to a halt. Later, a second mutation may arise, this time in a gene coded for a tRNA—let us say one for leucine, tRNA^leu. This tRNA normally picks up leucine and recognizes one of the mRNA codons for leucine, 5′-UUG-3′. This means that its anticodon would be 3′-AAC-5′. The mutation may bring about a change in the anticodon to 3′-AUC-5′. The mutation does not alter the part of the tRNA that enables it to pick up leucine. However, the mutant tRNA can now only recognize the nonsense codon, UAG, and can add leucine at that site (Fig. 14-2B). Although leucine is substituting for tryptophan at a specific site in the amino acid chain, and the completed polypeptide may not be as efficient in its activity as the wild type, some activity has been restored. The very damaging effects of the nonsense mutation are averted, and the second mutation has acted as a suppressor mutation. Certain intergenic mutations, however, cannot prevent the damaging effects of a first mutation if the substituted amino acid is one that disrupts a crucial site in the polypeptide.

It is fortunate that multiple copies of most tRNA genes are found in cells. Otherwise a suppressor mutation in a transfer RNA would probably have lethal effects. This is so because a change in the anticodon, from ACC to AUC in our example, means that the one altered tRNA^leu can no longer deposit leucine at the correct locations in the mRNA. It can now deposit leucine only at sites of UAG, a chain-terminating codon. Every protein containing the amino acid leucine would be incomplete, and the cell would

not survive. But does this mean that the normal termination of translation will now be altered, since the mutant tRNA can deposit leucine at a normally positioned UAG codon? This does not happen, because the binding between release factors and UAG is stronger than the binding of the mutant tRNA^leu to UAG. This means that UAG is not always read by the altered tRNA. Therefore, polypeptides with UAG as a stop signal to translation will be terminated properly most of the time.

The site of mutation in any tRNA gene can change any one of the three bases in an anticodon. And so, there are several mutations which cause a change in the tRNAs of amino acids other than the one for leucine and which also enable these altered tRNAs to recognize the UAG codon and to deposit their specific amino acids at the UAG site. While our discussion here has centered around the UAG codon, a similar story holds for the other two chain terminating codons, UAA and UGA.

Mutation rate and some factors that influence it

Although spontaneous mutation rates for individual loci are very low (a probability, on the average, of 1–10 per million gametes), the frequency of spontaneous mutation should not be underestimated, as any higher organism contains thousands of genetic loci. Because so many genes are present, the chance that *one* of them will undergo a change is quite significant. Studies on the total mutation rate in Drosophila indicate that in one generation the probability is that 5% of the gametes will contain a mutation that arose in that generation time. Calculations of total mutation rate in other organisms, including the human, give a comparable figure.

It is important to understand what this 5% value means in the life cycles of different species. Note that the value for the overall mutation frequency is expressed, *not* in relation to definite time units such as months or years, but in relation to generation time. Different species obviously have different generation spans, approximately 2 weeks in the fruit fly in contrast to approximately 30 years in the human. But when different species with their different generation times are compared (excluding microorganisms), the overall mutation rates expressed in *generation time* are comparable. Obviously, in terms of weeks or months, this would mean that a species with a short life cycle of just a few weeks would have a mutation rate hundreds of times higher than humans or any other species with a much longer life cycle. The very fact that the figure is similar when expressed in generation time implies that natural selection has somehow geared the spontaneous mutation rate to the life cycle. Evolutionary progress depends on the ability of the gene to change. However, a very high or uncontrolled amount of gene mutation would prove fatal to a species, because most mutations are harmful.

Thus, it seems logical to expect that mutation rate itself is a genetic feature, a characteristic adjusted to the life cycle of the species. If this is so, there must be genes that can influence the mutation frequency of other genes. This brings us to a discussion of factors that can affect the typical mutation frequency. The genetic composition itself has indeed been shown to play an important role. In the fruit fly, spontaneous mutation rates may vary as much as tenfold among different stocks. Such marked increases in mutation frequency have been attributed to definite genes, *mutator genes,* which can increase the mutation rate of other genes. Normally, we would not be aware of the presence in the genotype of genes that keep the mutation frequency adjusted to the life cycle. However, if a mutation occurs in such a gene, the adjustment is interrupted. An increase in mutation frequency follows, and a mutator gene can be recognized.

The first factor found to alter mutation frequency was temperature. Morgan and his students noted that fruit flies raised at 27°C had two to three times the mutation rate per generation than those raised at 17°C. Later studies showed that not only high temperature but also abnormally low ones elevated the mutation frequency. Increased mutation rate also followed sudden temperature shifts back and forth from high to low. This increase was noted even if the shift was from a high to a low temperature within the normal range for the species. So it would appear that departures from temperature conditions to which the organism is normally adapted may increase the mutation rate.

Observations made on some plant and animal groups indicate that aging can increase the mutation rate. An increase is seen after the aging of seed and pollen in plants and the aging of sperm in Drosophila. There is even evidence that in humans a similar effect pertains. As stated before, a mutation has occurred in humans if a condition, known to be inherited as a dominant defect, suddenly appears in an offspring of two unaffected parents. Certain dominant disorders, such as achondroplastic dwarfism, have been found to appear with a greater frequency when the male parent is considerably older than the average.

The importance of malnutrition as a causative agent in mutation is indicated by the doubling of the mutation frequency in plants, such as the snapdragon, when they are deprived of certain mineral elements. Such effects as those of malnutrition, aging, and other environmental influences on mutation rate may be related to the failure of repair mechanisms in the cell, a topic discussed more fully later in this chapter.

Mismatch repair

You will recall that DNA polymerase possesses proofreading ability (see Fig. 11-26). At the time of replication, it can correct an error resulting from the incorporation of an incorrect base into the DNA strand that is being assembled on the template. At times, this editing fails, and a "wrong base" is not removed. As a result, a site is created within a double helix where mismatching occurs, such as between a thymine and a guanine. A mismatch can arise not only as a failure in proofreading at the time of replication but also following recombination when heteroduplex DNA regions form (see Fig. 11-29). In these regions, where the two strands of the DNA had their origins in different parental duplexes, the strands may not be perfectly complementary, and therefore certain base pairs may be mismatched.

Still another way in which duplexes may come to have mismatched base pairs results from tautomeric shifts. Suppose that thymine exists in its "rare" state at the time of replication, so that it cannot pair with adenine but pairs instead with guanine, as depicted in Figs. 11-14 and 11-15. The unusual base pair, T:G, results and is not recognized by the proofreading mechanism, because hydrogen bonding will be satisfied at the time and pairing relations will appear to be correct. Since the tautomeric shift is a rare condition, however, the base—thymine in this example—returns to its normal state. A mismatched site is now present in which the hydrogen bonding is not correct.

What is the outcome of the formation of the mismatch? At least two possibilities are in store. At the next replication, two duplexes can form (see Fig. 11-15), but the two will differ from each other at the site that had been mismatched. The base pair change actually represents a mutation. Another possible outcome, however, is removal of the mismatch and a small area surrounding it by a mechanism known as *mismatch repair* (Fig. 14-3). In this repair process, the mismatched region is recognized and removed by nuclease activities and then filled in by DNA polymerase. The repair removes the incorrectly incorporated base (guanine, in this example) and replaces it with the correct one, restoring a duplex with the correct base pair.

A moment's consideration, however, tells us that the mechanism has an equal chance of removing the "correct base" on the template strand as removing the "incorrect one" on the new strand. This would produce a mutation. However, the cell has another mechanism that guards against this outcome and makes it more likely that the erroneous base will be the DNA region removed in mismatch repair. The template strand contains certain adenine and cytosine nucleotides to which methyl groups have been attached. (Methylation is discussed in some detail in Chapter 17.) Methylation takes place at the time of DNA replication, but it does not occur immediately when a base is incorporated at the replication fork. Therefore, for a short time there are more methyl groups in the template DNA segment than in the DNA, which is being newly assembled. This difference in the degree of methylation can be detected by the mismatch repair system, which can therefore distinguish between the

parental strand and the new one. Consequently, the chance is greater that the incorrect base will be excised following mismatch repair, than the correct one in the template strand.

We see that the cell's repair system reduces the frequency of spontaneous mutations by backing up the proofreading capacity of DNA polymerase that acts at the time of replication. If mismatch repair does not occur shortly after the incorrect base has been added, however, there may be sufficient time for the new strand to become methylated in the region of the mismatch. In such an event, discrimination between parental and daughter strands is not possible, and the correct segment on the parental strand may be removed, generating a mutation. In some bacterial strains the ability to methylate bases has been lost or the enzyme mechanism involved in excision by mismatch repair is defective; such strains show a mutation frequency far above that of other strains.

Gene conversion

In Chapter 9, the importance of Neurospora and certain other fungi in genetic studies was related to the topic of tetrad analysis (see Figs. 9-11 and 9-12). We learned that ascospore segregation patterns could be used to determine the distance in map units between two linked genes as well as between a gene and its centromere. Many years ago, genticists became aware that at times segregations arose in an ascus that produced patterns other than the expected. Figure 9-11 shows that if two linked genes, a and b, are being followed in the cross $a^+b^+ \times a\,b$, tetrads of three types can arise. Most numerous are the parental ditypes (PD); least numerous the nonparental ditypes (NPD). The tetratypes (TT) occur with intermediate frequency. It is important to note that, regardless of the tetrad type, the products are reciprocal. Each allelic form is equally represented ($4a^+:4a$; $4b^+:4b$). Let us suppose, however, that a tetrad arises in such a cross in which the ascorpores are as follows: $2a^+b^+:2a^+b^+:2a^+b:2a\,b$. It can be seen that although $4b^+:4b$ are present, the a alleles occur in the proportion 6:2 or 3:1. It is evident that the products are not reciprocal. Such a tetrad is an exception to the 4:4 segregation of alleles and cannot be explained on the basis of classic events. The one marker, a^+, is represented six times and its alternative form only twice, as if the one gene form were being converted to the other. Thus, the term *gene conversion* was applied to the process giving rise to such unexpected tetrads.

FIG. 14-3. Mismatch repair. A mismatched base pair, T:G in this case *(1)*, may arise in several ways. One way a mismatch can be corrected is by mismatch repair. A small segment of one of the strands containing a mismatched base is removed by nuclease activities *(2)*. The gap is then filled in by DNA polymerase, using the complementary strand as a template *(3)*. Correct base pairing is then restored.

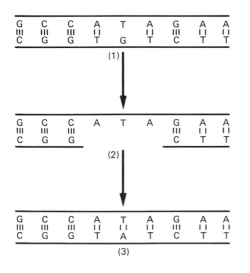

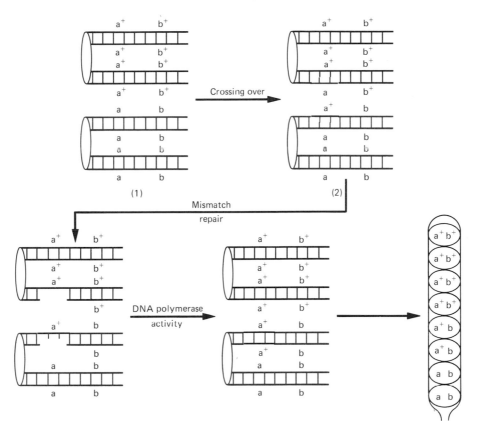

FIG. 14-4. Gene conversion by mismatch repair. *(1)* The duplex DNA in four chromatids of a bivalent are depicted. Two chromatids are a^+b^+ and the other two are a b with respect to two very closely linked markers. Following crossing over *(2)*, heteroduplex DNA is generated. The two chromatids that participated in the crossover event now have heteroduplex segments in the a region, where a mismatch is now found. Mismatch repair mechanisms may excise a portion of one of the strands of a heteroduplex *(3)* Excision in this case is shown to take place in both chromatids. Filling in of the gaps by DNA polymerase *(4)* brings about a gene conversion. Meiosis is completed and a mitotic division follows. The ascus that forms in this case contains products that are not reciprocal *(5)*. Although the ratio for the b marker is 4 b^+: 4b (2:2), the proportion of the a marker is 6a^+: 2a (3:1).

Gene conversion was found to occur when the markers are very closely linked, so close that they may even reside in the same gene. The number of asci showing conversion involving such very closely linked markers at times exceeded that arising with the expected or typical pattern. These unexpected conversions can be explained on the basis of mismatch repair accompanying genetic recombination. A review of Fig. 11-29 shows that heteroduplex DNA is generated if crossing over occurs according to the widely accepted Holliday model. As explained in the last section, in a heteroduplex region the two strands may not be perfectly complementary, and a mismatch may be present. In a dihybrid cross in which the alternative genetic forms are found at very closely linked sites, even within the same gene, a crossover event can generate heteroduplex segments containing mismatched pairs in the region where the markers reside (follow Fig. 14-4). This heteroduplex is found in two of the four chromatids, the two that participated in the crossover event. A mismatch repair may follow that excises a portion of the DNA in one of the strands of the heteroduplex. In Fig. 14-4, excision takes place in both strands where heteroduplexes are found. (However, this need not be the case. Neither strand, or only one of them, may be repaired. In such cases the outcome would be different from those depicted in the figure.) DNA synthesis fills in gaps resulting from excision. The final result is a departure from the expected pattern. In this case, instead of the expected 2a^+:2a, we find 3a^+:1a. As noted previously, this is not the only possible outcome. For example, if the segments excised had been the two carrying a^+ instead of a, the outcome would have been 3a:1a^+.

High-energy radiation and its general effects on matter

Some of the features of spontaneous mutation that were noted in early studies indicated something about the physical basis of the mutation process. H. J. Muller, one of Morgan's students, was impressed by the fact that the gene, a very stable entity that goes through countless accurate duplications, can change suddenly to another form, which in turn can duplicate itself accurately. These rare sudden changes in a gene were seen to be random; a particular environment did not favor a particular kind of mutation over another. And when a rare gene change occurred in a cell, it was always *just one* change in one gene. Even in a diploid cell, if a gene mutates, the same gene on the homologous chromosome is not affected.

These observations all indicated to Muller that mutation was some kind of chance disturbance on the molecular level. Although environmental factors might conceivably increase the frequency of such disturbances, they would not favor one kind over another. *Any* point in the hereditary material might be affected at a given moment. A disturbance affecting a particular gene did not have to influence the homologous position in the same cell, because it was a random event. Muller reasoned that if a mutation was actually the result of a chance pointwise excitation, then high-energy radiations should be able to increase the frequency of mutation. This would be expected, because radiations were known to cause chance disturbances as they pass through matter. Any agent that could cause such changes might be considered a potential instrument for raising the mutation rate. In his classic experiment performed in 1927, Muller obtained almost immediate results showing conclusively that X-rays increase the frequency of both gene mutations and structural chromosome changes.

To appreciate this classic work and its relevance today, let us review a few basic facts concerning radiation. The radiations pertinent to the topic of mutation are those that can travel through space that is empty of matter. They do not, therefore, include sound waves. Of the two major types of radiation, the electromagnetic ones are probably the most familiar (Table 14-2). In one sense, they may be considered to be waves, whose lengths can be measured from the long radio waves to the extremely short cosmic rays. Note that they form a continuous gradation or spectrum from the longest extreme to the shortest without any sharp boundaries between them. Visible

TABLE 14-2   Spectrum of electromagnetic radiations

WAVE TYPE	WAVELENGTH	$\text{Å} = 1/10,000 \; \mu$
Radio	10^7–0.04 cm	Insufficient energy to excite or ionize atomic electrons
Infrared	20,000 Å–7,800 Å	
Visible light	7,800 Å–3,800 Å	
Ultraviolet	3,800 Å–150 Å	Energy to excite
X-rays	150 Å–0.15 Å	Sufficient energy to excite and ionize
γ Rays	0.15 Å–0.005 Å	
Cosmic rays	0.005 Å–0.00008 Å	

light includes certain wavelengths in part of this continuum. It is, however, those electromagnetic waves whose lengths are shorter than those of light that are effective as mutagenic agents—factors which can increase the rate of mutation.

To appreciate part of the reason for this, we must consider the energy associated with waves of different lengths. Although this type of radiation has many features of waves, in certain other respects it resembles particles. For when absorbed by matter, it is absorbed in definite units. Similarly, when given off by a source, such as an X-ray tube or ultraviolet lamp, this type of radiation is emitted in definite units. These units are called *photons,* and a photon of a particular wavelength has a certain characteristic amount of energy associated with it, called the *quantum.* As the wavelength decreases, the quantum energy associated with a unit (a photon) of the electromagnetic radiation increases. Thus, *more* energy is associated with photons of *shorter* wavelength.

This energy value is important to the material through which the radiation passes. If sufficient quantum energy is associated with the photons, the energy may be enough to influence the internal organization of the molecules composing a substance. Visible light and the longer wavelengths do not possess enough energy to disorganize subatomic particles.. A photon of X-radiation, however, has sufficient energy to be absorbed by submolecular elements, usually electrons.

Normally, these electrons are found at a characteristic distance from the nucleus, as they are attracted and held by the nuclear binding forces. The inner orbital electrons are held more strongly by the attractive forces of the nucleus than those in the outer orbits. If a single electron absorbs excess energy, it may move farther away from the nucleus than it normally would. We say that such an atom is in an "excited state," meaning that a single particle in the atom now possesses excess kinetic energy. This means that the electron concerned is bound less firmly by the nuclear attractive forces. If sufficient energy is absorbed, the electron may break away from the nucleus to which it belongs. In such a case of an outright separation of electric charges, we say that ionization has occurred. The electrons in outer orbits are most likely to be affected, because the inner ones are more firmly held.

Once an electron has been knocked out of an atom, it can in turn knock out electrons in other atoms. Eventually, the energy is dissipated. The free electrons attach to other atoms, resulting in positively and negatively charged particles. The overall effect of excitations and ionizations is to make the material more reactive. So much energy is associated with gamma and cosmic radiations that inner electrons and atomic nuclei themselves may be influenced. Ionizing radiations cause both excitations and outright ionizations. Ultraviolet light, however, with its relatively long wavelength, does not cause ionizations. Its effects are discussed more fully later in this chapter.

The other major class of radiations includes corpuscular radiations (Table 14-3). Unlike electromagnetic radiation, these possess a definite mass. They are discrete bodies that may behave as streams of particles. They affect matter in the same way as described for the other class. We may think of a stream of protons, neutrons, or electrons actually hitting a subatomic particle and imparting excess kinetic energy to it which is sufficient to cause ionization. As mentioned earlier, once an electron is separated from an atom, it may in turn collide with other electrons

and ionize a second atom. Similarly, a proton or neutron may pass through matter, knocking out electrons, which can then affect still other electrons in other atoms. From the behavior of both classes of radiation, we can thus appreciate what is meant by a chance, pointwise disturbance at the level of the atom and why Muller thought it likely that high-energy radiations could induce mutations.

Radiation and mutation rate

To obtain quantitative data on any possible effects of X-rays on the mutation rate, Muller designed a method that is ingenious for its simplicity and lack of ambiguity. Instead of selecting for study mutations that produce visible effects, Muller chose sex-linked lethals that occur with a higher frequency than the visibles. Moreover, any sex-linked recessives will express themselves sooner than autosomal recessives. For his investigation, Muller derived stocks of flies in which the females were ideally suited for the detection of new mutations on the X chromosome (Fig. 14-5). In these females, one X is an ordinary wild or standard one. The other, however, possesses two mutant sites: the Bar mutation (B), which behaves as a dominant, and a recessive lethal (1). This same X also contains an inverted segment (C), which acts as a suppressor of crossing over in that part of the X between B and 1. Such females are called ClB females and are phenotypically Bar eyed.

In the mutation study, two series of crosses were made with these females. In one series, the ClB flies were crossed with males that had received known doses of radiation. The other group, the controls, were mated with ordinary, untreated males. It can be seen from Fig. 14-5 that the X chromosome of the treated P_1 males will go to all their daughters, both the wild and the ClB daughters. The latter can be recognized by their Bar phenotype. Half of the F_1 male offspring will die because they receive an X chromosome that carries a lethal (the ClB) from their mothers. The remaining males, as well as the wild daughters, are rejected. Only the F_1 Bar-eyed daughters are saved for further mating. The reason for this is that we know these Bar-eyed flies must contain the original lethal, because the Bar locus is linked to it, and crossing over is suppressed. We are now interested in the "treated X" received from the P_1 males. Did the treatment with the X-rays induce any "new" lethal mutations?

To answer this, each F_1 Bar female is crossed with

TABLE 14-3   Some of the corpuscular radiations

PARTICLE	WEIGHT (ATOMIC MASS UNITS)	CHARGE
Electron (beta particle)	0.00055	Negative
Positron (positive beta particle)	0.00055	Positive
Proton (nucleus of common isotope of hydrogen)	1.007	Positive
Deuteron (nucleus of heavy isotope of hydrogen)	2.013	Positive
Alpha (nucleus of helium)	4.002	Positive
Neutron	1.009	Neutral

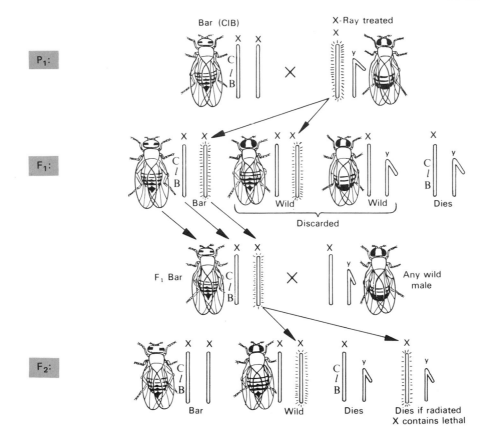

FIG.14-5. The ClB method. When a ClB female is mated, half of the normal number of male offspring will develop, because half of them receive the ClB chromosome, which carries a lethal. If the ClB female is mated to a male treated with radiation, the treated X chromosome is transmitted only to the F$_1$ daughters. Only the Bar-eyed daughters are selected, because it is known that they carry at least one lethal, that linked to the Bar locus. When these F$_1$ Bar females are mated to any male, it is known that at least half of the males will not develop, again because of the lethal linked to Bar. The other half receive the treated X, which traces back to the original treated P$_1$ male. If it, too, carries a lethal because of the treatment, no male offspring at all appear in the F$_2$. The appropriate controls must be run to consider spontaneous mutation (see the text for details).

any untreated male. It is known in advance that in the F$_2$, half of the male offspring will die for the same reason that half of the F$_1$ males died; they receive the ClB chromosome with the lethal. The other X chromosome in these F$_1$ Bar females is derived from the P$_1$ males which were subjected to radiation. If the radiation produced a new lethal on the other X carried by the ClB females, the remaining male offspring will also die, and the females will produce no sons at all. The reason is simply that the female would have a different lethal on both X chromosomes; the males, being hemizygous, cannot survive. The F$_1$ females can survive, even with two lethals, one on each X, because any new lethal arising by chance would almost always be at a locus other than that of the original lethal on the ClB chromosome. If by improb-

able chance it is at the same locus, then the F$_1$ Bar female will not develop, because she will be homozygous (*ll*) for the recessive lethal.

Each F$_1$ ClB female was bred separately. If any male offspring were seen in the breeding bottle, it was immediately known that no new mutation had been produced. If, however, a ClB female failed to produce sons, then a new lethal mutation must have arisen on the other X. Since spontaneous mutations would also be occurring, the results on the X-ray–treated series had to be compared with the nonexposed controls. The method also permitted the detection of any visibles produced on the X by the radiation. The procedure gives very clear-cut results; when sex-linked lethals are followed, one class, the male sex, is either present or absent. There is no source of error

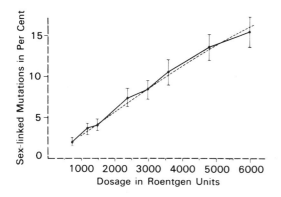

FIG. 14-6. Sex-linked mutations and X-ray dosage in Drosophila. The number of point mutations is directly proportional to the dosage of X-rays as measured in roentgens. At higher doses, the relationship falls off as the radiation decreases the viability of the flies. (Reprinted with permission from E. W. Sinnott, L. C. Dunn, and T. Dobzhansky, Principles of Genetics, 5th ed. McGraw-Hill, New York, 1958.)

through difficulty in recognizing a particular phenotype.

Several thousand F_1 flies were followed in both the treated and untreated groups. The data were conclusive from the start. Similar results were later obtained by other investigators (Fig. 14-6). It was clear that the number of sex-linked lethals that arose was directly proportional to X-ray dosage, as measured in roentgen units [1 roentgen (R) is equal to an amount of radiation dosage sufficient to produce 2×10^9 pairs of ions in 1 cc of air]. A dose of approximately 60 R was found to be enough to double the spontaneous production of sex-linked lethals. A given amount of radiation would result in a certain percentage of lethals. Doubling or tripling the dosage would cause a corresponding increase in the number of lethals obtained. Of course, there is a limit to the dosage that can be applied. At higher levels, the viability of the flies is affected, so that a strict linear relationship begins to fall off. Moreover, at higher doses the probability that more than one lethal will be induced in the same X chromosome increases. Because two or more lethals would be counted as one, the direct relationship tends to fall off for this reason as well. The direct relationship that does exist between radiation dosage and genetic effect has been found to apply generally to the production of other kinds of mutations besides sex-linked lethals. It also holds for wavelengths other than those of X-rays.

The corpuscular radiations, neutrons and protons, cause ionizations, and these were shown to be effective in producing mutations. However, their ionization distribution is much more condensed than that produced by electromagnetic radiations, so that certain differences are apparent in the relationship of the dosage applied to the number of mutations produced. Nevertheless, all the ionizing radiations can induce genetic change.

It must be kept in mind, however, that treatment with any mutagenic radiation is not specific for a particular gene and does not ensure that one certain kind of gene mutation will be produced in preference to another. No more visibles of a given kind are produced by this or that agent. What *does* happen is that the entire spectrum of mutation is increased in frequency. Nothing new is produced. The mutation rate is elevated for all the loci. But with the increase, the loci still mutate with the same *relative* frequency. A locus with a spontaneous rate double that of another will still show this same relationship to the second one, even when the frequencies of both have been raised.

Suggested mechanisms of radiation action on the genetic material

The observation that the production of point mutations is generally proportional to the dose applied led to the idea that induced mutations result from ionizations produced within the gene itself. According to this concept, the *target theory,* the gene itself must be struck for a mutation to occur, just as a target being hit by bullets. With twice the number of bullets, there would be twice the chance of hitting the target (the gene) and of causing ionizations within it. Also, it should not matter how the dose is distributed. If small doses were given over an extended period of time, the outcome should be the same as if one large dose had been applied. This would be similar to using a few bullets over a longer time period or firing them all in one burst. The chance of hitting something would be related to dose, not to distribution of the dose. The target theory seemed to apply to the production of point mutations in Drosophila.

Besides their ability to induce gene mutations, ionizing radiations were also found to be very effective in the production of structural chromosome changes. The frequency of all types, deletions, inversions, translocations, and so on was increased by application of radiations. The rule of direct proportion between mutation and dose of radiation, however, did not apply for the gross structural changes. This makes sense, when we consider the fact that two breaks are nec-

Radiation dose (arbitrary units)	Chance of one break	Chance of two breaks together
1	1/10	1/100
2	2/10	4/100
3	3/10	9/100
4	4/10	16/100

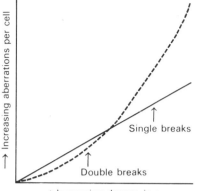

FIG. 14-7. Radiation dosage and production of chromosome breaks. Single breaks in a chromosome show a direct proportion to dosage applied, as is true for the production of point mutations. The chance of two breaks occurring in the same nucleus is equal to the product of two single breaks. Consequently, large structural chromosome alterations increase in proportion to the square of the dose. This means, as can be seen on the right, that at lower doses more single breaks are produced, but with increased dosage, the chance is greater for two breaks, and hence structural aberrations, to take place.

essary for a structural change to take place. A given dosage (Fig. 14-7) would correspond to a certain probability for a break to occur in a chromosome. The chance of two breaks occurring together would be equal to the product of the chances of the single breaks. Therefore, the production of gross structural chromosome aberrations would increase *not* in a linear relationship to dose, but in proportion to the square of the dose. The higher the dose, the greater the chance of producing chromosome aberrations. Very small deletions and inversions might be caused by just one hit; these would be proportional to the dose, as seems to be the case. And the intensity of the dose seems to play a role in the production of structural changes. More of them are obtained if a large dose is given at one time rather than distributed in small packages. This can be explained by assuming that, when a dose is weak, the probability that two breaks will be produced in the same nucleus is less. The break tends to be repaired before another one was formed. Thus, if an intense dose is applied at one time, there will be more single breaks overall at one time that can engage in new arrangements.

Later work with mice has shown that these animals react differently from Drosophila in response to the intensity of the radiation. The intensity exerts an effect in the production of point mutations similar to that in the production of structural aberrations. Such differences between species in response to radiation are one of several observations that suggest the mech-anism of inducing gene mutation is not as direct as the target theory states.

Interaction of radiation and other factors in the production of mutations

For years, several factors have been known which can cause the number of mutations associated with a given dose of radiation to vary, even in the same organism. In Drosophila, a given dose produces more point mutations if mature sperm are radiated rather than earlier stages. The chromosomes of mature sperm are also much more sensitive to chromosome break-age than are those in immature germ cells. Generally, cells with chromosomes in a condensed state, such as at nuclear division, are also much more susceptible to chromosome damage than those in the nondividing nucleus. This accounts in part for the greater suscep-tibility of cancer cells to radiation damage than the normal cells around them, which would not be divid-ing as actively. Moreover, the number of mutations that arise after a given radiation dose may vary de-pending on the stock or strain. In fruit flies, just as some stocks have a higher spontaneous mutation frequency, so some stocks vary in their sensitivity to the radiation. The influence of environmental condi-tions on the yield of induced mutations has also been definitely demonstrated by the effect of oxygen. Fewer mutations for a given dose are found at lower oxygen tensions. Among microorganisms, those forms that can exist either aerobically or anaerobically undergo a higher rate of both spontaneous and induced mutation under aerobic conditions.

The temperature of the surroundings at the time of applying the radiation has an influence that is probably related to the oxygen effect. Colder conditions strengthen the induction of chromosome alterations, as well as gene mutations; this is most likely associated with the presence of more oxygen at lower temperatures. Also connected with the amount of oxygen is the action of certain reducing agents present in the environment at the time of irradiation. The number of chromosome breaks is reduced in the presence of —SH compounds, alcohols, and various strong reducers. These compounds may afford a certain amount of protection by removing oxygen dissolved in tissues.

Ultraviolet and the induction of mutations

We noted earlier in this chapter that the amount of energy associated with ultraviolet wavelengths is too small to produce ionizations. The relatively long waves are not highly penetrating and so do not usually reach those cells that give rise to gametes. Nonetheless, it has been demonstrated in several higher organisms that ultraviolet has a definite mutagenic effect once it does reach the DNA. Microorganisms can be readily exposed to an ultraviolet source and have provided a wealth of information on its mutagenic action. Many years ago it was observed that the mutagenic activity seems to result from the hereditary material itself actually absorbing the ultraviolet. This was shown by comparing the effectiveness of the different wavelengths in their ability to induce mutations. Some wavelengths were found to be more effective than others, those of 2600 Å being the most mutagenic. When a curve is plotted showing these differences, it is found to agree with a curve plotting the ability of DNA to absorb wavelengths of ultraviolet. In other words, the wavelengths that cause more mutations are those that are more strongly absorbed by the genetic material. Little activity is found by those wavelengths only weakly picked up by the DNA.

Later work with isolated DNA demonstrated changes that ultraviolet can produce in nucleic acid. It is the bases in the DNA that are primarily responsible for the absorption of the ultraviolet. The pyrimidine bases thymine and cytosine are particularly sensitive. A major consequence of the absorption is the formation of stable bonds between two thymine bases or between two units of cytosine. Such associations between two identical molecules or units of a molecule are called *dimers*. Production of thymine dimers seems to be a very important means by which ultraviolet alters the DNA (Fig. 14-8).

Ultraviolet absorption may also cause stable bond formation between two different primidine bases, giving a "mixed" dimer (e.g., cytosine–thymine). When dimer formation occurs, it is two pyrimidine bases adjacent to each other on a DNA strand that become stably bonded. At very high doses, ultraviolet has been shown to be capable of breaking the sugar–phosphate backbone of isolated DNA. Whether or not it operates this way in the intact cell, ultraviolet can cause chromosome breakage in the fruit fly and some other species. Its ability to induce breakage, however, is much less than that of the high-energy ionizing radiations.

From our knowledge of the Watson–Crick model of DNA, we can surmise how ultraviolet may cause some of its mutagenic effects. Duplication of the DNA depends on clean separation of the two chains, each of which acts as a template in the construction of new, complementary chains. Obviously, any process that interferes with this, such as the formation of stable bonds by dimerization, could very possibly result in DNA with altered activity or DNA unable to transmit the information stored within it.

FIG. 14-8. Dimerization by thymine. Ultraviolet light may cause stable bonds to form between two identical molecules of thymine or cytosine. The result is a dimer. If dimers form between two bases located along the same strand of a stretch of DNA, replication of the DNA molecule could become impaired.

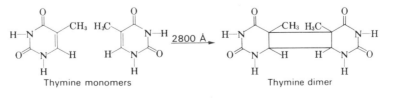

Thymine monomers   Thymine dimer

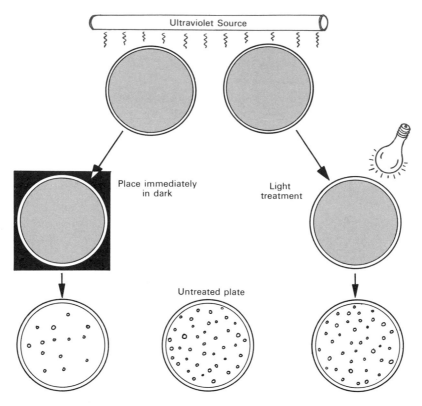

FIG. 14-9. Photoreactivation. Two plates spread with the same number of bacteria are subjected to a dose of ultraviolet radiation. One is placed immediately in the dark. The other is kept in the light for at least 1/2 hour. After incubation of the plates, the light-treated one is found to have a much larger number of bacterial colonies than the plate placed immediately in the dark. The number may approach that found on a plate not exposed to the ultraviolet at all. This untreated plate can be placed in the dark or given a light treatment, but the number of colonies on it will not be altered. The photoreactivation operates on the damage induced by the ultraviolet.

One unusual aspect of ultraviolet is that its mutagenic effect may be reduced by white light. A simple way to demonstrate this (Fig. 14-9) is to spread each of two agar plates with the same number of bacteria and subject each to a dose of ultraviolet known to have killing activity. The plates are then incubated, one completely in the dark but the other after a ½ hour exposure to a strong source of white light. Observation of the plates the next day will show a much better growth on the plate treated with the white light. The killing effect of the ultraviolet is reduced so that only about 40% of the original killing activity remains. Therefore, if a certain dosage can kill half the bacteria on a plate, exposure to white light can reduce this lethal action from 50 to 20% (0.50×0.40), so that 80% of the bacteria will survive. This reversal of the lethal effect of ultraviolet light by white light, called *photoreactivation*, has also been demonstrated in Drosophila and higher plants. The mechanism of photoreactivation remained a mystery until the dis-

covery of "repair enzymes," which can undo damage to the DNA (more about this later in this chapter).

Evidence exists that, in addition to its direct effect on the DNA, some of the mutagenic activity of the ultraviolet may be through its action on other parts of the cell, which in turn exert a mutagenic effect. This is clearly indicated by experiments in which culture medium alone is exposed to ultraviolet light and then later inoculated with microorganisms. A definite increase in mutation is noted, even though the microorganisms were plated without direct contact with the ultraviolet light. The mutagenic effect has been linked to the formation of hydrogen peroxide (H_2O_2) in the medium by the radiation. Treatment of amino acids with peroxides was shown to make them capable of inducing mutation. It would thus seem that the ability of radiation to induce mutations (both point and chromosomal) is not dependent solely on ionization directly within the gene. Indeed, a great many of the observations we have just discussed suggest that mutation frequency might be altered by chemical conditions surrounding the hereditary material.

Chemical mutagens

It was not surprising when Auerbach and Robson discovered the first chemical mutagen in 1941. These investigators turned their attention to mustard gas and related compounds as possible mutagenic agents because their burning effects on somatic tissue were similar to those of high-energy radiation. And it was found that, like high-energy radiation, the mustard compounds could produce a high incidence of both gene mutations and structural aberrations. The same types of gene mutations result as with the radiation (detrimentals, lethals, etc.). The mutagenic action was demonstrated on a variety of plant and animal species. The mustard compounds can also inflict great damage on rapidly growing tissue by producing chromosome changes, and they have been used as well as radiation in the treatment of malignancies.

The mutagenic action of mustard gas, however, differs from that of radiation in a very important respect. The effect of treatment with mustard gas is frequently delayed; a mutation is not necessarily induced in the cell that was exposed but may arise in a descendant of that cell, as if the treated DNA had somehow become unstable. The mutation may be delayed for two generations, and when it does occur, it may arise in a body cell as well as in a germ cell. A mutation occuring during the development of a fruit fly, for example, could take place in just one body cell. This mutant cell could then give rise, through cell division, to a patch of cells surrounded by nonmutant cells. The result is a mosaic individual. This delayed effect makes mustard gas a more serious mutagenic agent in many ways than one whose action is expressed directly.

After the discovery of the first mutagenic chemicals, many other substances were also found to possess mutagenic activity, and the list is still increasing. Few have the strong effect of the mustard compounds, and the mechanism by which many of them operate to increase the frequency of mutations is not known. The chemical structures of the many mutagens have nothing in common. Furthermore, a substance found to be mutagenic in one organism may be ineffective in another. Caffeine, for example, is highly mutagenic in some organisms, but is apparently broken down in others before it can exert any action. Formaldehyde markedly increases the mutation rate in the fruit fly,

but only if applied to the food of the larval male. It is ineffective at other stages of the life cycle and is entirely nonmutagenic in the female! Most chemical mutagens have also been found to be capable of inducing structural chromosome changes as well as point mutations.

The interesting discovery was made that some substances have an antimutagenic action—that is, they tend to suppress the mutagenic activity of other agents. Antimutagens were first detected by Novick, working with bacteria. The strong mutagenic action of purines, such as caffeine and adenine, was suppressed by the presence in the medium of purine ribonucleosides, such as adenosine and guanosine. The antimutagens were ineffective toward the mutagenic action of ultraviolet and gamma radiations. Even the spontaneous mutation rate was shown to be reduced when the antimutagens were present. It is significant that some purine nucleosides were found to have a high antimutagenic activity and that many of the nontoxic chemical mutagens were purines or their derivatives. Indeed, the mutation rate seemed to depend in part on the physiological state of the organism. These various observations suggest that the study of mutation might be approached on a biochemical basis, which leads us to suspect that the spontaneous mutation rate, which is characteristic for a species and geared to the generation time, results from an interaction between mutagenic and antimutagenic substances arising during the course of the metabolic activities of the cell.

Repair of damaged DNA

We are now well aware of an assortment of enzymes in cells that can repair damage and eliminate distortions in the DNA molecule. These distortions, resulting from the formation of dimers, cause problems at the time of replication and transcription by preventing a DNA strand containing dimers from acting as a proper template. As a result, a strand newly synthesized on such a template will contain gaps that reflect the positions of the dimer-containing regions. Mismatched bases, discussed earlier in this chapter, that are found in heteroduplex DNA segments will *not* generate gaps, because a mismatch does not prevent a strand from acting as a template in replication or transcription.

The enzyme involved in photoreactivation is an example of a repair enzyme that can recognize a distortion in the DNA caused by dimer formation

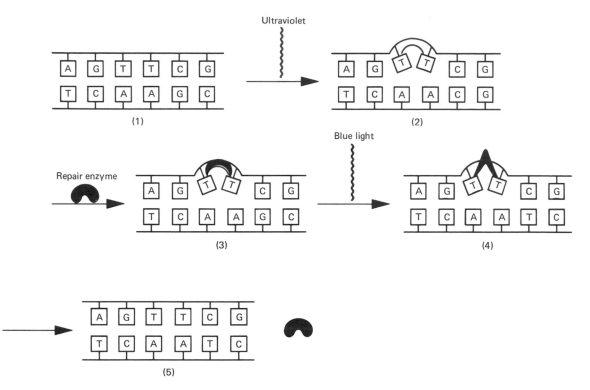

FIG. 14-10. Photoreactivation. *(1)* A segment of DNA is subjected to ultraviolet light. *(2)* A dimer forms between two thymines. *(3)* This is recognized by a repair enzyme that can bind to it in the dark. *(4)* Blue wavelengths activate the enzyme, which then cleaves the bond between the bases. *(5)* The segment of DNA is restored, and the enzyme is released.

following exposure to ultraviolet light (Fig. 14-10). Although the enzyme can bind to the altered site in the dark, it requires energy in visible light, actually those wavelengths in the blue portion of the spectrum, to become activated. It is then able to cleave the bond between the two bases, thus restoring the DNA to its original state.

In addition to photoreactivation, repair systems that do not depend on light energy for their action are known to exist in prokaryotic cells. One of these is the excision repair system, which entails the interaction of a battery of enzymes, the actual removal of the damaged segment, and its replacement by a new portion of DNA. Some of the enzymes involved have nuclease activity enabling them to attack either the 3′ or the 5′ end of a phosphodiester linkage. *Endonucleases* attack only those bonds that occur in the interior of a polynucleotide chain, producing an internal nick in a strand of a double helix. The class of nucleases designated *exonucleases* also attack 3′, 5′ phosphodiester bonds, but they do so only from the free end of a phosphodiester chain, either the 3′ or the 5′ end.

A very essential part of excision repair is DNA ligase, which catalyzes phosphodiester linkages and can restore a double helix to its intact condition after an internal nick is produced by an endonuclease. DNA polymerase is also needed in DNA repair to catalyze the assembly and linkage of nucleotides on a DNA template. (DNA ligase and the DNA polymerases, as well as the nucleases, were discussed in Chapter 11 in relation to DNA replication.)

Excision repair takes place in the following way. The initial step entails an endonuclease that can recognize the damaged region, probably a result of the change in configuration of the DNA brought about by the presence of the dimer (Fig. 14-11). The endonculease produces a nick in the region of the damage on the 5′ side of the distortion, causing the strand to break. The free end that arises is then recognized by an exonuclease, which digests away part of the strand including the dimer. Once the damaged portion is removed, the enzyme DNA polymerase synthesizes the missing segment in a 5′ to 3′ direction using as a template the complementary region that lies exposed on the other strand. Finally, DNA ligase catalyzes a phosphodiester linkage between the 3′ end of the newly formed segment and the 5′ end of

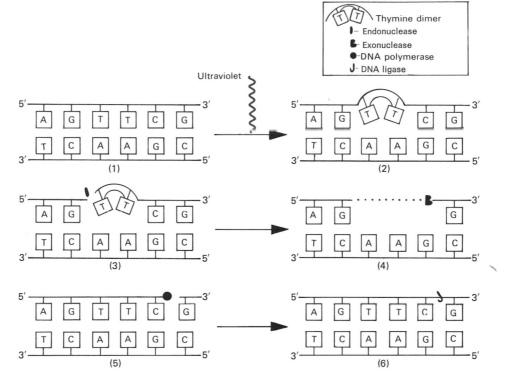

FIG. 14-11. Excision repair. A stretch of DNA *(1)* is exposed to ultraviolet light, and a thymine dimer forms, distorting the DNA *(2)*. An endonuclease recognizes this region and produces a nick in the vicinity of the damage *(3)*. A second enzyme, an exonuclease, recognizes the free end and digests away a segment of the DNA strand that includes the damage *(4)*. DNA polymerase in a 5′ to 3′ direction then synthesizes the portion that has been eliminated, making a linkage to the 3′ end of the original strand *(5)*. Finally, DNA ligase links the newly formed segment to the original strand, and the DNA is restored to normal *(6)*.

the original portion of the strand from which the damage was removed.

This excision repair pathway just described is a general scheme. In *E. coli*, many enzymes are involved, not just four. A gene mutation affecting any one of several genes can cause a deficiency in the excision repair system.

A third type of repair mechanism that has been studied largely in *E. coli* is known as *postreplication repair*. This system, which can work in more than one way, depends on the interactions of many kinds of proteins. The RecA protein involved in recombination plays a vital role in postreplication repair, so named because the repair takes place after replication of the DNA containing the dimers. In one kind of postreplication repair, a recombination event may take place that generates an undamaged duplex DNA. According to one model (Fig. 14-12), when the DNA, altered by the presence of dimers, undergoes replication, it encounters difficulty because of the distorted regions that cannot act as templates to form new strands. DNA polymerase is blocked at these positions, but it does manage, after a lag period, to resume replication at a site beyond the distorted area. As a result, new DNA strands are formed, but these contain gaps—postreplication gaps. Excision repair systems cannot take care of these postreplication gaps; however, a recombination event may take place between the two sister duplexes. An intact, undamaged duplex can arise if segments of sister duplex strands are switched during the recombination event. As seen in Fig. 14-12, damaged DNA, which may not survive, is still present. The repair has nevertheless produced one intact, undamaged duplex. Postreplication repair mechanisms must play a very important part in the repair of DNA in *E. coli*, because cells deficient in recombination repair are exceedingly sensitive to ultraviolet light, more so than mutant cells that are defective in the other repair mechanisms.

Other types of postreplication repair systems are also known to exist. One of these requires DNA polymerase III and is known as an "error-prone repair

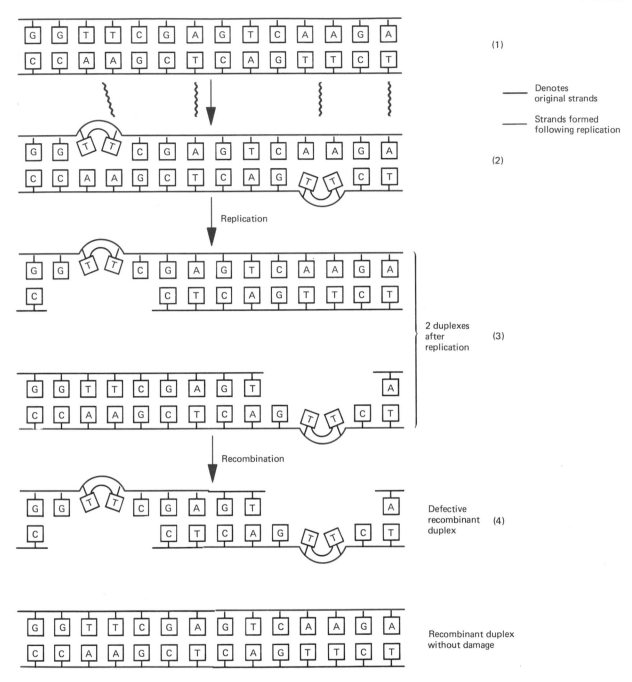

FIG. 14-12. Postreplication repair. *(1)* A segment of DNA is subjected to ultraviolet light. *(2)* Dimers are produced in the irradiated strands. *(3)* When the damaged duplex replicates, complementary strands are produced, but these contain postreplication gaps. Therefore, the two new duplexes are doubly defective. *(4)* If recombination takes place, the four strands of the two duplexes may rearrange to form a defective duplex as well as one with no damage.

system," so called because the repair of the distorted DNA regions is accompanied by the generation of mutations. In *E. coli*, an error prone response may be called into play following various treatments that damage DNA or interfere with its replication. The re-

sponse to ultraviolet exposure has been studied in greater detail than has the response to other treatments, such as exposure to alkylating agents. Remember that DNA polymerase has the ability to proofread or edit the DNA and to eliminate errors when "wrong" nucleotides are incorporated at the time of replication. In error-prone repair pathways, also known as *SOS repair,* this editing capacity of the polymerase is relaxed. As the DNA polymerase assembles nucleotides on the template strand during replication and encounters a region distorted by thymine dimers, it nevertheless continues to insert bases, even if these are not the correct ones! This insertion of incorrect bases constitutes the origin of gene mutations, and so the mutation frequency increases as a result of the loss of the proofreading function of DNA polymerase during SOS repair.

Error-prone repair systems involve the interactions of several gene products, some of which act as regulators of gene expression. Again, a pivotal role is played by the RecA protein, which acts as a switch to turn on the SOS repair mechanism in response to damage produced by ultraviolet light. If the ultraviolet has produced more dimers than the excision repair system can accommodate, the RecA protein starts the chain of events leading to repair by the SOS mechanism. Since the SOS mechanism is error prone, mutations can accumulate. It would therefore be detrimental to the cell for SOS repair to go on indiscriminately in response to every DNA damage that arises. Hence, various genes in the system are coded for products that repress the expression of other genes in the SOS system and thus prevent the system from activating when excision repair can handle the situation. Only when DNA damage caused by ultraviolet light signals an emergency is the SOS repair system, with its complex series of events, called to action.

The significance of repair enzymes to an organism cannot be overestimated. Repair enzymes are undoubtedly needed in cells for protection against the mutagenic action of factors found in the average environment. There is no doubt that cells, both eukaryotic and prokaryotic, have various ways to restore damaged DNA to its original state in the face of mutagens within and outside the cell. The development of such mechanisms was critical to the evolution of life on land, where cells are exposed to doses of radiation higher than those reaching cells in an aquatic environment.

This is demonstrated by observations of persons expressing a serious inherited skin disorder, *xeroderma pigmentosum.* These persons are extremely sensitive to sunlight and develop skin cancer at an early age. It has been shown that their cells are deficient in the DNA excision repair process that excises ultraviolet-damaged DNA. Apparently, it is the first step in this process that is defective. Its failure results in the accumulation of ultraviolet-damaged DNA and the development of malignancies. In mammals, excision repair appears to involve a large number of different genes and enzymes. There is also evidence that recombination repair mechanisms exist in mammals. Repair systems appear to be important in the maintenance of normal cell activities. There is even reason to suspect that failure of repair mechanisms may contribute to the process of aging, during which damage to the DNA may accumulate, leading to eventual cell death (Chap. 17).

Another critical activity of enzymes that participate in repair processes relates to the role they may play in crossing over. It is becoming increasingly evident that crossing over entails enzymes that may also participate in the repair of damaged DNA. This is seen in the case of certain mutant strains of *E. coli* in which the capacity for genetic recombination has been lost along with the ability to repair ultraviolet damage to the DNA. Further support for the involvement of repair enzymes in recombination is the demonstration that, when crossing over is believed to occur in eukaryotes, DNA synthesis takes place at pachynema with the production of short DNA segments. This has been clearly demonstrated in meiotic cells in the lily anther. Moreover, at the outset of meiotic prophase I, enzymes such as DNA polymerase, DNA ligase, and endonuclease start to appear, increasing in level until pachynema. The RecA protein appears to be vital to the process of crossing over (Chap. 11) as well as to the SOS repair system. The versatility of the repair enzymes may thus extend from preserving the cell to generating new combinations of the genetic material, essential to the process of evolution.

Effect of chemical mutagens at the molecular level

As a result of tautomeric shifts the usual pairing relationships among the bases may be altered. Instead of adenine pairing with its complementary base, thymine, the rare form of adenine may pair with cytosine to give an A:C pair instead of the normal A:T. Conversely, the occurrence of a rare tautomeric form of cytosine may result in a C:A pair instead of C:G. Likewise (Fig. 14-13), instead of T:A or G:C, the unusual pairs C:A and A:C may form. The ultimate effect of these pairing errors is shown in Fig. 14-13. It can be seen that the A:C error results in the formation of a G:C nucleotide pair. The original po-

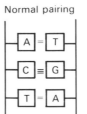

Normal pairing

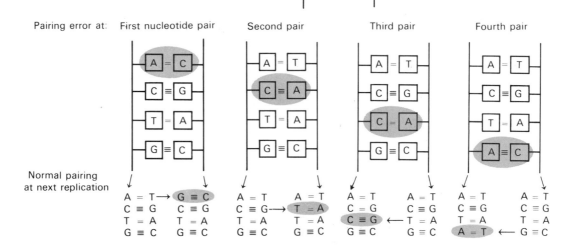

FIG. 14-13. Transitions. A stretch of four nucleotides with normal pairing is shown above. Abnormal pairs may form at the time of replication because of the occurrence of a rare tautomeric state of one of the bases. If adenine and cytosine form a pair, instead of the normal A:T, the result is a transition, a substitution of one purine–pyrimidine pair by another or the substitution of one pyrimidine–purine pair by another. In this example, we see (1) G:C substitutes for A:T, (2) T:A for C:G; (3) C:G for T:A, and (4) A:T for C:G.

sition of the purine adenine on one of the two strands of the double helix is now occupied by the purine guanine. In other words, one purine:pyrimidine pair is now represented by a different purine:pyrimidine pair. It can be seen from the figure that, in each case, the purine:pyrimidine relationship is maintained— one purine:pyrimidine pair substitutes for another or one pyrimidine:purine pair substitutes for another. Base pair substitutions of this type have been called *transitions*.

In contrast to this type of change is the *transversion,* the substitution of a purine:pyrimidine pair by a pyrimidine:purine pair (or vice versa; Fig. 14-14). In the transversion, there is a new orientation of purine and pyrimidines. A purine on one strand is replaced by a pyrimidine; a pyrimidine on the complementary strand is replaced by a purine. The significant fact is that, in the case of either a transition or a transversion, a different sequence arises at a site on the DNA. This alteration thus modifies the genetic information at that point and can cause a mutant effect as a consequence.

It was reasoned that if spontaneous mutation did actually result from such transitions and tranversions, then it should be possible to induce mutations with chemicals whose activities are known. An example of such a chemical is nitrous acid, which has the ability to remove free —NH$_2$ groups from the bases found in nucleic acids (Fig. 14-15A). After deamination by nitrous acid, adenine is converted to another base, hypoxanthine. In hypoxanthine, the arrangement of atoms involved in base pairing is different from the

FIG. 14-14. Summary of types of transversions. In an alteration of this kind, the end result is a substitution of a purine–pyrimidine pair by a pyrimidine–purine pair (or vice versa). For example, in 1, the purine–pyrimidine pair A:T is replaced by the pyrimidine–purine pair T:A (or vice versa), and so on for the other substitutions.

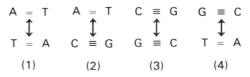

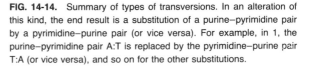

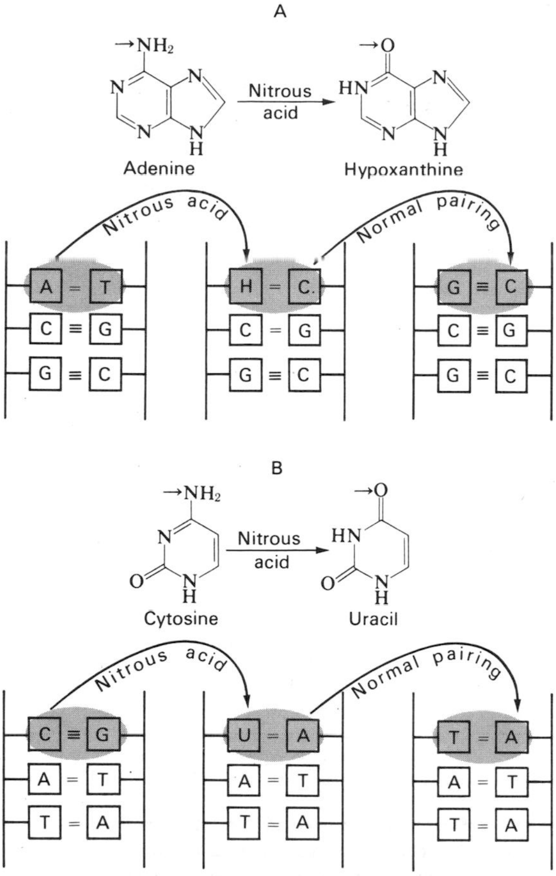

FIG. 14-15. Mutation induction by nitrous acid. Nitrous acid is able to remove free amino groups from bases in nucleic acid, as can be seen above, in the conversion of adenine to hypoxanthine. Hypoxanthine has pairing properties similar to those of guanine, and so it tends to pair with cytosine (A). The end result is a transition, the substitution here of G:C for the original A:T. (B). Cytosine is converted to uracil, which behaves like thymine and so pairs with adenine. The result again is a transition, in this case a change from C:G to T:A. Nitrous acid can also cause deamination of guanine and its conversion to xanthine. This product, however, behaves like guanine in its pairing preference for cytosine.

adenine (A), but similar to guanine (G). This means that if adenine is changed to hypoxanthine (H) at a point on the double helix, a transition from A:T to G:C can occur. By deamination, nitrous acid converts cytosine (C) to uracil (U, Fig. 14-15B), whose pairing characteristics are like those of thymine (T). Again a transition can result, this time from C:G to T:A. Since nitrous acid can cause such changes in nucleic acid, it is therefore not surprising that it is a very powerful mutagenic chemical. Several other chemical agents are known that can cause alterations in the bases of nucleic acids, and these too show mutagenic activity. Most of these investigations have been performed on microorganisms. Data from phages strongly support the concept of induced mutation arising from changes in base pairing at the time of gene replication.

Another class of mutagenic substances is known as the *base analogues*. They are so called because they resemble the normally occurring nucleic acid bases both in structure and activity. For example, thymine has a base analogue, 5-bromouracil (5-BU), so similar to thymine that it can substitute for it. This means it can be attracted to adenine and form hydrogen bonds with adenine in much the same manner as thymine.

One important difference between thymine and 5-BU, however, is that the latter undergoes tautomeric shifts more often than the thymine. In its rarer form, 5-BU behaves like cytosine and tends to pair with guanine. Therefore, the 5-BU could be in its "'cytosinelike" state when it is being taken into the DNA, as shown in Fig. 14-16. Or it may be in this form later at the time of replication, *after* it has been

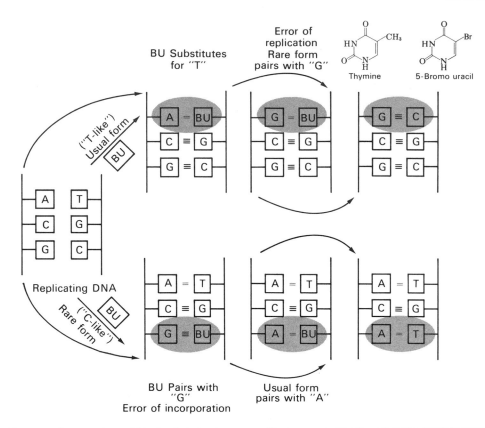

FIG. 14-16. 5-bromouracil as a mutagen. This chemical closely resembles thymine in structure (above), and it can substitute for it at the time of DNA replication. The upper part of the figure shows this. At a following replication, it may be in its rarer form and behave like cytosine. At the next replication, the incorporated guanine pairs with cytosine. The result is a substitution of a G:C pair for the original A:T.

The lower part of the figure indicates that 5-bromouracil may be in its rarer C-like state before it enters a DNA chain. It can thus pair with guanine. At the next replication, it would behave typically (like thymine) and attract adenine. At the next replication the adenine attracts thymine. The result is a substitution of an A:T pair for the original G:C.

incorporated into a strand of the nucleic acid. The latter is called a mistake in replication; the former is a mistake in incorporation. Again the end result is a transition, G:C to A:T (from the error of incorporation) or A:T to G:C (from the error of replication). Highly mutagenic base analogues such as 5-BU undoubtedly exert their mutagenic activity in this manner.

The alkylating agents, which include the nitrogen mustards and ethyl sulfonate, constitute another class of mutagenic agents. In contrast to substances such as nitrous acid, which are effective mainly in prokaryotes, these are very efficient mutagens in eukaryotes. The alkylating agents can impair the normal hydrogen bonding of guanine and thymine by adding an alkyl group to the hydrogen-bonding oxygen of these two bases. As a consequence, mispairing of G with T takes place, leading to transitions (A:T to G:C and G:C to A:T). Alkylation can also bring about the loss of gua-

nine from the DNA by breaking the bond that joins the base to the sugar. Mutation can result from this depurination (Fig. 14-17), but more likely excision repair mechanisms will restore the correct nucleotide. An error-prone repair system may come into play, however, to repair positions in the DNA that have been alkylated. An alkylated nucleotide is removed, but errors of incorporation are often made, frequently bringing about transversions.

Acridine dyes are mutagenic substances frequently used in mutation analysis. Acridines can insert themselves into the double helix between two adjacent nucleotides. Following exposure to acridines, there is an increase in frame shift mutations in which there is either the addition of an extra base, a (+) mutation, or the deletion of a base, a (−) mutation. Frame shift mutations, which cause very serious consequences by upsetting the proper amino acid sequence of poly-

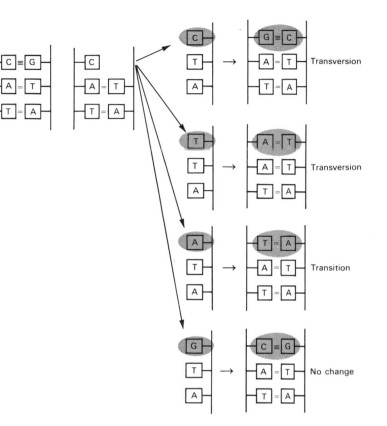

FIG. 14-17. Effect of mustard compounds. Sulfur and nitrogen mustards can cause the base guanine to be eliminated from a strand of DNA. The missing base may later be replaced by any one of the four bases. When this strand replicates, a transition or a transversion may occur, depending on which base substitutes for the original guanine.

peptides, are also known to occur spontaneously as well as following exposure to ultraviolet light. We have seen that mutations of this type were very important in deciphering the genetic code (Chap. 12). No matter what the exact mode of action, all known chemical mutagens bring about the same result—a change in the coded information through some type of alteration in nucleotide sequence.

Although we know that certain chemicals can cause characteristic molecular changes, this information must not be misconstrued to mean that the investigator can bring about a specific kind of gene mutation by using a specific chemical, for these agents raise the entire mutation rate, and the four bases—adenine, thymine, guanine, and cytosine—are found throughout the DNA. It is known from work with microorganisms that one mutagen may be more likely than another to induce a certain *class* of mutation. For example, Novick found that ultraviolet treatment of *E. coli* caused a higher mutational change to resistance to a kind of virus, phage T/6. In contrast to this, certain purine mutagens caused a greater probability of mutation to resistance against a different virus, T/5. Comparison of some chemical mutagens has shown that some genes may be more susceptible to mutation by one chemical than to another. The chemical 2-amino purine favors a certain kind of back mutation (from mutant to wild) at a site in a gene controlling the enzyme tryptophan synthetase in *E. coli*. However, no mutagen specific for just one kind of gene is known. Certain techniques now being used with viruses have an important bearing on this point and will be discussed in Chapter 18.

Bacteria and carcinogens

Bacterial genetic systems have valuable applications in procedures designed to yield information on the potential dangers that certain environmental factors present because they are carcinogens, or cancer-inducing agents. Although the steps triggering the transformation of a normal cell to the malignant state are far from clear, there is little doubt that alteration of some kind in the cellular DNA is a key factor. (Recall the story of repair enzymes and xeroderma pigmentosum.) When a cell is transformed to the malignant state, it has somehow been released from the normal genetic controls that regulate its DNA replication and its interactions with neighboring cells. Once the cell is changed, its malignant state is transmitted to all those cells that trace back to it.

Today we are surrounded by an ever-increasing number of chemical compounds in our daily lives in the food we eat, in substances we handle, in the air we breathe, and in many other ways. As chemicals continue to pour into the environment, it is essential to have a quick, inexpensive way to alert us to any that may be carcinogenic. The use of mammalian test systems, however, to screen only a fraction of the chemicals we contact is not only very expensive but very time-consuming. Fortunately, tests with microorganisms have been devised that reveal a chemical's mutagenic potential. The most widely used of these was devised by Bruce N. Ames of the University of California at Berkeley. One might wonder whether the response of bacteria to any test substance can be equated to that of a cell from a mammal. However, the results have shown that the bacterial tests distinguish between *known* carcinogens and *known* noncarcinogens with an accuracy over 90%. Moreover, they have singled out certain previously unsuspected substances as potential carcinogens. When tested later in animals, these chemicals were indeed shown to have cancer-inducing ability.

The values of the Ames test are manyfold. It is much less expensive than those conducted with animals. The results are available much more quickly— in days versus years! It can also detect substances that may be only weak carcinogens. Once a chemical is shown to be suspicious, on the basis of the Ames procedure, it can then be followed further in mammalian systems. The bacterial test thus makes it possible to check a large number and variety of compounds that otherwise could not be screened by the animal tests. Once we are alerted to the potential danger of a substance, it can be studied further.

The basis of the Ames test is the demonstration that a gene mutation has occurred in response to a specific substance. This is then taken as evidence of DNA damage, and hence the ability of that substance

to effect a malignant transformation. Although it is true that not all mutagens have been found to be carcinogens, the correlation is extremely high. If a substance is shown to be mutagenic, the probability is about 90% that it is also a carcinogen.

The Ames test employs strains of a particular colon bacillus, *Salmonella typhimurium*. The bacteria used in the test carry a mutation (his⁻) that makes them nutritionally deficient because they cannot synthesize one of the enzymes needed to manufacture histidine. Some of the his⁻ strains used in the Ames test carry base pair substitutions (transitions or transversions); others are defective because of frame-shift mutations. Cells that are his⁻ can, of course, undergo back mutation. However, the reversion to his⁺ occurs spontaneously at a very low rate. If some agent can increase the rate of mutation in the cells, it will consequently increase the chance that his⁻ will revert to his⁺. Since different his⁻ tester strains are used (those with base pair substitutions and those with frame-shift mutations), a chemical can be scrutinized for its ability to achieve one or more kinds of genetic alterations. The ability to cause reversion from his⁻ to his⁺ is thus taken as a measure of an agent's mutagenic, and hence carcinogenic, potential.

To use the his⁻ strains, however, it was essential to circumvent the fact that many substances are unable to pass through bacterial cell walls. The strains used therefore carry a mutation that makes the wall more penetrable to chemicals. The strains also carry another mutation that causes a defect in the DNA repair mechanism which excises defective DNA (see Fig. 14-11). Such a mutation permits easier detection of various kinds of alterations in the DNA, since a defect would not be removed. Very important consideration was also given to the need to create a system that has certain mammalian characteristics. This was necessary, because it is known that chemical carcinogens are not necessarily active in their original forms. Metabolic processes may take place in the mammal that would change the chemical somehow and render it carcinogenic. Conversely, cells may convert a potentially harmful compound to an innocuous one. Ames therefore added an extract of rat liver to the bacterial system. In this mixture, a test chemical is thus exposed to mammalian metabolic processes.

In the Ames test, the test substance is added to the his⁻ bacteria–rat liver mixture. The mixture is then plated on solid, minimal medium, which contains only enough histidine to support a few cycles of replication, but not enough to permit colonies to form. After about 2 days of incubation, the plates are examined. Those cells that have experienced a his⁺ reversion will give rise to colonies on the minimal medium. The number of these can be related to controls and to the amount of the test substance used, to give a quantitative result. The value of the Ames test has been recognized by government and private industry and is currently in use in thousands of laboratories throughout the world. The speed, accuracy, and inexpensive features have already resulted in the screening of thousands of chemicals, a figure that would not have been approached by the more cumbersome animal tests.

REFERENCES

Ames, B. N. Identifying environmental chemicals causing mutations and cancer. *Science* 204: 587, 1979.

Devoret, R. Bacterial tests for potential carcinogens. *Sci. Am.* (Aug.): 40, 1979.

Drake, J. W. Comparative rates of spontaneous mutation. *Nature* 221: 1132, 1969.

Haseltine, W. A. Ultraviolet light repair and mutagenesis revisited. *Cell* 33: 13, 1983.

Hollaender, A. and F. J. de Serres (eds.). *Chemical mutagens. Vol. 5: Principles and Methods for Their Detection.* Plenum, New York, 1978.

Howard-Flanders, P. Inducible repair of DNA. *Sci. Am.* (Nov.): 72, 1981.

Kenyon, C. J. and G. C. Walker. DNA-damaging agents stimulate gene expression at specific loci in *Escherichia coli. Proc. Natl. Acad. Sci.* 77: 2819, 1980.

Lewin, B. *Genes.* Wiley, New York, 1983.

Lindahl, T. *DNA repair enzymes. Annu. Rev. Biochem.* 51: 61, 1982.

Little, J. W. and D. W. Mount. The SOS regulatory system of *E. coli. Cell* 29: 11, 1982.

Marx, J. L. DNA repair: new clues to carcinogenesis. *Science* 200: 518, 1978.

Muller, H. J. Artificial transmutation of the gene. *Science* 66: 84, 1927.

Novick, A. and L. Szilard. Experiments on spontaneous and chemically induced mutations of bacteria growing in a chemostat. *Cold Spring Harbor Symp. Quant. Biol.* 16: 337, 1951.

Setlow, R. B. Repair deficient human disorders and cancer. *Nature* 271: 713, 1978.

Singer, B. and J. T. Kusmierek. Chemical mutagenesis. *Annu. Rev. Biochem.* 51: 655, 1982.

Stadler, L. J. Mutations in barley induced by X-rays and radium. *Science* 68: 186, 1928.

Vogel, F. and R. Rathenberg. Spontaneous mutations in man. *Adv. Hum. Genet.* 5: 223, 1975.

West, S. C., E. Cassuto, and P. Howard-Flanders. Postreplication repair in *E. coli:* strand-exchange reactions of gapped DNA by Rec A protein. *Mol. Genet.* 187: 209, 1982.

REVIEW QUESTIONS

1. The allele W for chlorophyll development is dominant over the recessive w for albino in a certain plant species. As a result of spontaneous mutation, which of the following plants would be the most likely to exhibit mosaicism if mutation arises in cells of a developing leaf? Which would be the least likely?

   A. WW.
   B. Ww.
   C. ww.
   D. WWWw, a tetraploid.

2. Suppose a certain human disorder which is inherited as a dominant trait is associated with the allele B. The normal allele is the recessive b. It is found that in a given period of time five abnormal children are born among a total of 100,000 births to normal parents with no family histories of the disorder. From these figures, what appears to be the mutation rate per germ cell per generation from b to B?

3. In corn, the genetic factor R has a spontaneous mutation rate of 492 per million gametes, whereas the factor Pr has a rate of 2.4 per million gametes. What is the chance that a plant of genotype RR PrPr will produce a gamete of the constitution r pr?

4. A mutation converts the DNA triplet ATA in the template strand of a structural gene to the triplet ATT. As a result of the mutation, translation comes to a premature halt with severe consequences for the cell. A second mutation arises in a gene encoded for a glutamine transfer RNA and brings about a change in the anticodon. Some activity is now restored as a result of the second mutation. Explain. What was the nature of the first change? What anticodon change must have been involved in the suppressor mutation? (Consult the codon dictionary—Table 12-1.)

5. A portion of a double helix has a segment with the following nucleotide sequence:

   GGATCC
   CCTAGG

   Suppose that at the time of replication a tautomeric shift causes A in the upper strand to pair with C. For the duplex DNA with the mismatch, answer the following:

   A. What is the outcome at the next replication if no mismatch repair occurs?
   B. What is the more likely outcome if mismatch repair does occur shortly after the error is made?
   C. What would be the outcome in a mutant cell that has lost the ability to methylate bases?

6. In *E. coli,* assume that the amino acid tyrosine occurs at amino acid position 70 in a certain polypeptide that is 110 amino acids long. A mutant strain arises sponta-

neously in which the normal polypeptide is no longer formed. Instead, a short polypeptide 69 amino acids long occurs. The mutant strain is subjected to mutagenic chemicals. Among the cells derived from the treated cells are those that are wild and produce the normal polypeptide of 110 amino acids.

    A. Explain the nature of the spontaneous mutation, indicating the codon or codons involved.

    B. Would you expect all of the wild-type cells that arose following chemical treatment to be true revertants? Explain.

7. Suppose a wild-type male fruit fly is treated with X-rays. It is then crossed to an ordinary wild female carrying no known mutant alleles. If a recessive, X-linked lethal is induced in any of the radiated X chromosomes, what would be the effect on the sex ratio:

    A. In the first generation?

    B. In the second generation?

8. Answer the following questions pertaining to Muller's classic ClB method:

    A. If no inversion were present, what effect would this have in the P_1 female that would influence the results?

    B. Answer Question A regarding the ClB F_1 females.

9. In the ClB experiment:

    A. What would be the phenotype of the males in the F_1 and the F_2 generations with regard to eye shape?

    B. Answer Question A, assuming no inversion in the P_1 female and in an F_1 ClB Bar female.

10. A certain dose of radiation applied to a culture of actively growing cells is found to induce single chromosome breaks at random with a frequency of 1/50.

    A. How many single breaks would be expected in the cell culture by increasing the dosage three times and then four times?

    B. How many structural chromosome changes (inversions, reciprocal translocations, etc.) would be expected by increasing the dosage three times and then four times?

11. In Drosophila, the percentage of sex-linked lethals is directly proportional to the X-ray dosage as measured in roentgens. Suppose that, among the offspring of males treated with 600r the frequency of sex-linked lethals was 3%. Among males irradiated with 1600r the frequency of sex-linked lethals found among the offspring was 8%. What is the percentage of sex-linked lethals expected among offspring of males exposed to a dosage of 2500r?

12. A. Suppose male fruit flies were irradiated with 1000r and that 72 sex-linked mutations were found among 720 offspring. In another group of males irradiated with 1700r, 136 sex-linked mutations were produced among 800 offspring. How many mutants would you predict among 780 offspring of males irradiated with 2250r?

    B. Suppose a group of male fruit flies were given a dose of 1250r in one experiment and a dose of 1000r in another. How many mutations would you expect among 780 offspring of these males?

13. In Neurospora, a single gene mutaation can block a step in a biochemical pathway by causing a defect in an enzyme that governs that specific step. A mutant mold with such a defect can grow on a simple, basic growth medium only if the medium is supplemented with one or another growth requirement: substance A, B, C, D, etc. Four nutritonally deficient strains were studied in relation to the substances that, when added to the basic medium can satisfy their growth requirements. In the results given in the following table + = growth and 0 = no growth. From this information determine the sequence of reaction steps in the pathway and the specific step at which each mutant is blocked because of an enzyme defect.

STRAIN	SUBSTANCE ADDED			
	A	B	C	D
1	+	+	0	0
2	0	+	0	0
3	+	+	+	0
4	+	+	+	+

14. Suppose a fifth mutant Neurospora strain is found which can grow on none of the substances mentioned in Question 13 except substance B. However, it can also grow if supplied with another substance, substance E, which the other strains do not require for growth on the basic medium. Where is the block in the case of this mutation?

15. H. J. Muller estimated that the total background radiation from natural sources is equivalent to about 0.045r for a 4 week period. Would you expect this background radiation to be more significant in the production of spontaneous mutations for a species like the human or one like the fruit fly with a life span of about 4 weeks? Why?

16. Assume that a bacterial suspension is diluted so that a given volume contains approximately 200 cells. This same amount of suspension is placed on several plates divided into three groups: A, B, and C. Nothing is done to group A. Groups B and C are exposed to a dosage of ultraviolet light sufficient to kill 30% of the cells. Group C plates are given a half-hour exposure to strong white light. The three groups are then placed in an incubator. Since about 60% of the killing effect of ultraviolet can be neutralized by photoreactivation, on the average about

how many colonies would you expect to see the next day on plates of groups A, B, and C?

17. Suggest why exposure to excess radiation might cause severe anemia as a result of blood cell destruction as well as lower male fertility.

18. Certain human cells are found to accumulate DNA containing thymine dimers, particularly after exposure to doses of ultraviolet light. When samples of the cells are exposed to ionizing radiation, which can induce chromosome breakage, the chromosomes show an ability to undergo repair. Explain.

19. Cells of a certain bacterial strain are exposed to ultraviolet light. A sample is exposed immediately to white light. A second sample is placed in the dark. The cells in the first sample show a very high rate of survival; only a small percentage dies. In the case of the second sample, a very high mortality rate results, which is far in excess of that shown by other strains that received the same ultraviolet exposure and dark treatment. Explain.

20. Consider the following sequence of nucleotide pairs along a molecule of DNA:

$$A\ G\ T$$
$$T\ C\ A$$

Exposure to nitrogen mustard causes G to become less firmly linked to the sugar–phosphate backbone, and it is eventually eliminated. Give the possible base sequences after the missing base is replaced and the strand replicates. Classify each change as a transition or a transversion.

21. Assume that the polynucleotide strand with the sequence A G T is replicating but that it has been stretched between G and T because of an acridine dye. On the growing complementary strand a nucleotide with thymine is inserted in the stretched region. What nucleotide pairs will result when this complementary strand has undergone a replication producing its own complementary strand?

22. A polynucleotide strand with the segment

$$A\ G\ T$$
$$T\ C\ A$$

is exposed to nitrous acid, and the base cytosine is changed to uracil.

A. What strand will be formed that is complementary to this one in which the change has occurred?
B. What will the base pairs at the next replication be, and will this outcome be a transition or a transversion?

23. The amino acid sequence of a portion of a certain polypeptide in *E. coli* is as follows: NH_2-met-pro-ser-ser-asp-lys-tyr-arg-leu-. . . . Cells were exposed to an acridine, and mutants arose in which this same segment of the polypeptide had the following amino acid sequence: NH_2-met-pro-ser-gln-ile-asn-thr-asp-leu-. . . . The mutant cells were again exposed to an acridine, and offspring cells arose in which this same polypeptide segment was found to be as follows: NH_2-met-pro-ser-gln-ile-lys-tyr-arg-leu-. . . . Determine the following:

A. The most likely kind of mutation that has arisen. Using the dictionary of codons, determine the nucleotide sequence in the original mRNA, and show the change produced as a result of the first treatment with the mutagen.
B. The second change produced by the acridine in the mRNA from the first mutant cells.

24. Artificial mRNA can be synthesized and then used in an in vitro translation system. Four different kinds of artificial messenger RNA can be made that contain only a single kind of ribonucleotide. These are poly (U), poly (A), poly (C), and poly (G). As indicated in the following table, each homopolymer can be translated into a polypeptide containing only one kind of amino acid. As the table also shows, when streptomycin is present in the system, each kind of homopolymer is translated into a polypeptide containing three or four kinds of amino acids. From the information in the table, and with the aid of the dictionary of codons, answer the following:

A. How does streptomycin appear to influence translation and the reading of codons?
B. Why is it that some polypeptides that are synthesized in the presence of the drug contain only three kinds of amino acids whereas other polypeptides contain four?

| | STREPTOMYCIN | |
HOMOPOLYMER	Absent	Present
poly (U)	phe	phe, leu, ile, val
poly (A)	lys	lys, gln, glu
poly (C)	pro	pro, ser, thr, ala
poly (G)	gly	gly, trp, arg

25. A certain food additive is supplied to bacteria in the same test situation as that in the Ames test except for the elimination of the rat liver extract. It is found that a statistically significant number of his$^+$ cells arise in relation to controls. When the chemical is tested again in the presence of the liver extract, there is no significant increase in the number of his$^+$ cells over the controls. Explain.

15

BACTERIAL GENETIC SYSTEMS

Variations in microorganisms

Early work with microorganisms showed that these lower forms are very diverse. For example, even within a given kind of bacterium, specific strains may be recognized by colony type, host specificity, pigment formation, drug susceptibility, nutritional requirements, and other distinctions. Although investigators before 1940 had demonstrated many such variations in microorganisms, the results of their studies were equivocal and did not permit a definite answer to the question, "Are the variations in microorganisms hereditary?" It was not even certain that true genetic mutations occur in bacteria as they do in higher organisms.

As more and more drugs were studied for their bacterial killing properties, a variety of responses by microorganisms was noted. A typical picture to emerge was as follows: a particular drug dosage might destroy most of a bacterial culture but would leave a few surviving members. Upon isolation, these resistant cells would be able to form the basis for a strain completely resistant to that first drug dosage. Treating cultures from this derived strain with a still higher dosage could result in the isolation of survivors able to withstand the higher level. Repeating the procedure could lead to the establishment of strains resis-

tant to a drug concentration that would have killed cells outright in the original culture.

The correct interpretation of such observations remained in doubt as long as work failed to distinguish between genetic and environmental factors. Two explanations were equally feasible. According to the one, the reaction of the cells to the drug stems directly from contact of the cells with the drug. In other words, an external agent can *cause* cells to adapt or change in direct response to its presence in the environment. A contrasting idea, incompatible with the first, is that spontaneous mutations take place in the bacterial population. Since these arise at random, some of the genetic changes result in a small number of cells that possess an increased drug tolerance. These changes in the cellular ability to withstand higher drug levels occur *independently* of the presence of the drug. The drug would be acting as a screening agent, selecting out those cells *already* immune to a certain drug level even *before* it was present in the environment.

The two conflicting viewpoints are reminiscent of the controversy between the Lamarckian idea of acquired characters and the neo-Darwinian concept. According to the former, the environment directly alters individuals and brings about hereditary change. Adherents of the second school of thought would say that development of a new hereditary strain depends on random gene mutation followed by selection of those forms best suited to the immediate environment.

Spontaneous mutation in microorganisms and the fluctuation test

Since it was not even clear that true spontaneous mutations arise in microorganisms, no well-founded arguments were available to resolve the role of the environment in the origin of new strains. It was not until 1943 that the matter was settled through the elegant experimental design of the *fluctuation test* developed by Luria and Delbruck. These pioneers in microbial genetics presented a methodology to answer the question once and for all. The publication of their work laid the foundation for all later research in the discipline of the genetics of microorganisms.

The procedure was designed to distinguish between changes in population structure through (1) direct modification by an environmental agent or (2) spontaneous mutation and subsequent selection by an environmental agent. The specific bacterium used was *Escherichia coli* strain B, which is sensitive to the bacteriophage, T/1. If the bacteria are exposed to the T/1 virus on solid media, most of the cells are destroyed, but a few colonies manage to develop. In broth, the turbid culture clears and later becomes cloudy. These observations indicate that lysis is rampant at first, killing a majority of the cells. But a few survivors, resistant to the virus, remain and grow. When these in turn are exposed, they show resistance to the virus. This is the same situation as that resulting from exposure of bacterial cells to a drug, and it raises the same questions: Did the agent (the virus here) cause some of the cells to change and become resistant by direct contact? Or were some cells already resistant to the virus before it was applied, so that only these cells were able to survive and form the basis of a resistant culture?

It was known that in a culture of 10^8 *E. coli* cells, an average of approximately 100 could be expected to show phage resistance. To answer the question of the origin of these resistants, the experiment was set up in two different ways. In the first (Fig. 15-1A), a culture containing a large number of cells was divided into 10 parts. For the sake of simplicity, let us say that a culture of 10^8 was divided into 10 parts of 10^7 cells each. Therefore, in this part of the investigation, a culture of cells was simply divided into equal parts and then tested for any virus-resistant cells that might

be present. No further growth of the samples from the original large culture was allowed up to this point. The divisions were plated on phage-impregnated media and then incubated to permit growth of any resistant cells. The results showed that a small portion, approximately 100 cells per 10^8 (or 10 cells on the average in each subdivision), were phage resistant. How can this be explained?

When the results were subjected to statistical analysis, it was found that the *mean* number of resistants in the subdivisions was of the same order of magnitude as the variance. The variance, it will be recalled from Chapter 7, is equal to the square of the standard deviation. Statistical analysis has demonstrated that, in any random distribution, the value of the variance is close to that of the mean. Therefore, the results here are to be expected and can be explained by either of the two contrasting ideas. One could argue that in the culture of 10^8 cells, spontaneous mutations occur from time to time, and some of these genetic changes result in phage resistance. Some resistant cells have recently arisen in the culture, whereas others have descended from mutant cells that arose earlier. When the culture is divided and tested for mutants, one would obtain a random sampling of the culture and therefore a distribution of mutants with a variance equal to the mean. Alternatively, one could argue that no mutations existed in the culture as it grew to the original total of 10^8 cells. When the culture of 10^8 was divided and plated out in contact with the phage, most of the cells lysed. A few, however, adapted or were induced to mutate by the phage. This occurred at random in the samples, and so the variance should be close to the mean.

Therefore, this setup does not permit a distinction between the two hypotheses. However, the second part of the experimental design was crucial and allowed a clear choice between the two arguments. A small culture of bacteria was taken and divided (Fig. 15-1B). Each of these contained 50 to 500 cells and was allowed to grow until the total number of cells was equal to the total number in the first setup, in our example 10^8 cells. Note the difference between the two setups. In the first, a culture of 10^8 cells was subdivided and allowed no further growth before being brought into contact with the phage. In the second approach, each subdivision grew up independently from a small sample until a total of 10^8 cells was reached. According to the adaptation hypothesis, there is little difference between the two setups. In both cases, we have 10^8 cells. It should not matter that in

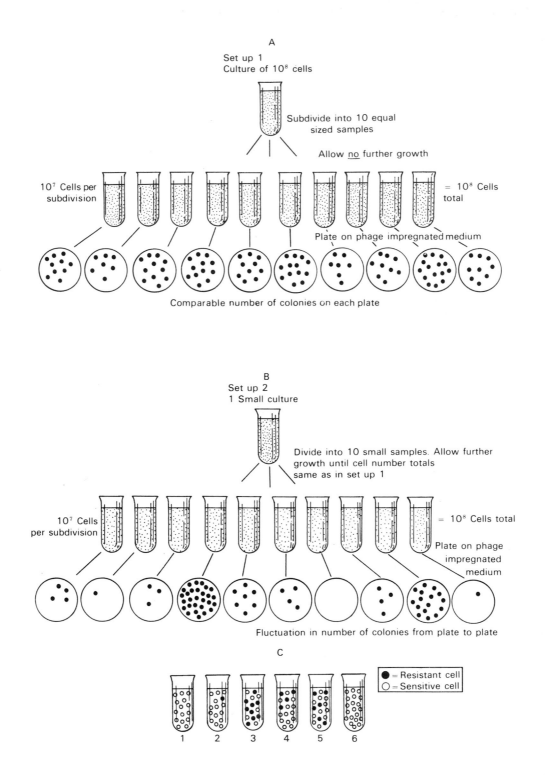

A
Set up 1
Culture of 10⁸ cells

Subdivide into 10 equal sized samples

Allow no further growth

10⁷ Cells per subdivision

= 10⁸ Cells total

Plate on phage impregnated medium

Comparable number of colonies on each plate

B
Set up 2
1 Small culture

Divide into 10 small samples. Allow further growth until cell number totals same as in set up 1

10⁷ Cells per subdivision

= 10⁸ Cells total

Plate on phage impregnated medium

Fluctuation in number of colonies from plate to plate

C

1 2 3 4 5 6

● = Resistant cell
○ = Sensitive cell

the one case a big culture of 10^8 was subdivided into 10 smaller ones and in the other case 10 smaller cultures were grown separately until the same total of 10^8 was reached. There would be no relevant difference between them, because the adaptation (or induced mutation) hypothesis assumes that the resistants appear *only* when exposed to the phage. Thus, when the samples are tested, the variance should reflect the randomness and should be about the same as the mean.

According to the spontaneous mutation concept, however, we would expect a wide distribution in the number of mutants in the separately grown cultures (Fig. 15-1C). This is so because the separately grown cultures will differ in the number of mutations that arose within each during growth, and they will also differ in the stages of culture growth at which the mutations occurred. If a single mutation arises at the end of the growth of the culture, there will be fewer mutant cells than if the mutation occurs early. In the latter case, many descendants of the mutant cell would be present in the culture, because the bacteria multiply exponentially. Therefore, even in cultures in which the same number of mutations took place, the

number of resistants recovered would be very different. This is *not* the same as taking random samples from one big culture of cells, and one would not expect the variance to be equal to the mean.

The results of many separate experiments of this type were comparable and gave an order of variance very far from the mean. Since the number of resistants in the second setup fluctuates from one culture to the next, the procedure has been named the *fluctuation test*. Its application to investigations of variations in microorganisms leaves no doubt that spontaneous mutations do occur in these lower forms and that inherited variations can arise independently of an external agent.

Replica plating technique

After the fluctuation test, other procedures were developed that also established that spontaneous mutations arise in microorganisms. One that is now routinely used in laboratory investigations is the technique of replica plating, devised by Lederberg and Lederberg. Figure 15-2 demonstrates the procedure in reference to streptomycin resistance in bacteria. Many strains of *E. coli* cells are sensitive to streptomycin, and colonies developing from them will not grow on a solid medium that contains the drug. However, cells at times arise that produce resistant colonies.

It can be shown by the replica plating that this change to drug resistance is completely independent of the drug itself. A piece of velvet is secured over a block. The velvet is gently applied to the surface of the colonies growing on the plate. The nap of the velvet picks up sufficient cells to transfer to other plates containing various media. Therefore, imprints or replicas of the original colonies can be easily made onto a plate containing streptomycin. This second plate must be oriented in the same way as the original one, so that any colonies that develop on it can be related back to colonies on the first plate. A colony may arise on the plate with the drug. Since the two plates are oriented, the colony on the first plate that produced this drug-resistant colony can be identified. Cells taken from the colony on the plate without the drug can then be exposed to a tube of broth containing streptomycin. These cells will grow in the presence of the drug, whereas other cells from neighboring colonies will not. Because no streptomycin was present on the first plate, the cells had never come into contact with the drug until they were placed in the broth. We would not have known that streptomycin-

FIG. 15-1. The fluctuation test (A). In setup 1, a large culture of cells, 10^8 in this example, is not permitted further growth. It is subdivided into 10 smaller samples of 10^7 each. Each of these samples is then plated on medium containing phage. Colonies that are phage resistant develop, and the number from one plate to the next is comparable. In one experiment, the mean number of resistant colonies was 10.7 and the variance 15, a figure close to the mean. Repetitions of this set up give comparable results. The observations can be explained on either the adaptation hypothesis or the spontaneous mutation theory. (B) In setup 2, a small culture of cells is subdivided, and each is allowed to grow further in the separate tubes until the number of cells is the same as in the first setup. Thus, in both parts of the experiment, there are 10 tubes of 10^7 cells giving a total of 10^8. However, in setup 2, the results on the phage-impregnated plates are quite different. Certain plates may have no resistant colonies; some few; others many. There is a great deal of variation from one to the next, and the variance is far from the mean number. (C) This fluctuation from one plate to the next is to be expected only on the basis of the spontaneous mutation theory. According to this concept, mutation can occur at any time in any one of the separately grown cultures. Certain tubes (1 and 6) may have no mutant cells at all. Others may contain a mutant cell that arose later in the growth of the culture (tube 2) and did not have the chance to divide further. Other mutant cells may arise earlier and at different times, giving different numbers of mutant cells at the end of the growth period. It is possible for two or more mutant cells to arise in one culture. When plated on phage-impregnated medium, the number of resistant colonies will vary from one plate to the other, reflecting the spontaneous nature of the mutations to phage resistance.

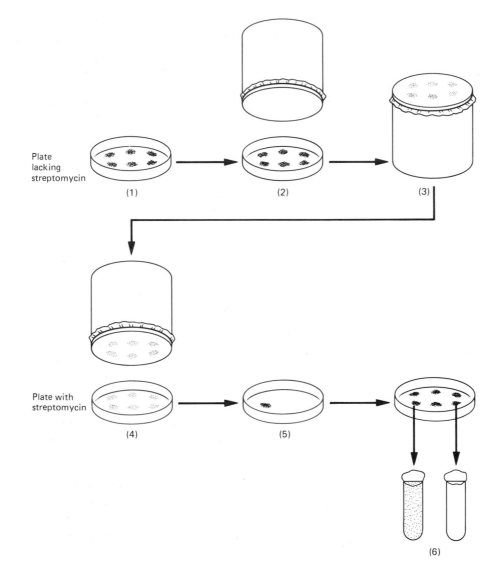

Plate
lacking
streptomycin

(1)   (2)   (3)

Plate with
streptomycin

(4)   (5)

(6)

FIG. 15-2. The replica plating technique. *(1)* Separate bacterial colonies, each originating from a single cell, are allowed to grow on a nutrient plate lacking streptomycin. *(2)* A block covered with a piece of velvet is brought into contact with the colonies on the plate. *(3)* The velvet picks up cells from the colonies. The cells will be in the same orientation as they were on the plate. *(4)* A plate containing streptomycin is oriented with the first plate, and the velvet on the block is allowed to contact the surface of the second plate. Cells are trans-ferred from colonies on the first plate. *(5)* In this example, one colony grows on the plate with the antibiotic. The two plates are compared, and this one colony is related to a colony on the plate without the streptomycin. *(6)* Cells are taken from this colony on the plate lacking streptomycin and placed in broth containing the antibiotic. The cells grow in the broth. A sample taken from another colony on the same plate proves to be sensitive to the streptomycin and cannot grow in the broth.

resistant cells had arisen and were present unless we had identified the colonies by replicating onto the second plate that has the drug. The procedure leaves no doubt that the drug-resistant cells were there on the first plate before any contact had been made with the streptomycin.

Detection of sexual recombination in bacteria

Once mutation in microorganisms was firmly established by such demonstrations, it became possible to use bacteria and viruses as tools in genetic analysis. Before the mid-1940s, bacteria were thought to lack any mechanism akin to a sexual process that could give rise to genetic recombination. The only way for new combinations of genes to arise in such organisms was thought to be mutation. These lower forms would therefore be totally asexual and would be at a disadvantage because beneficial gene combinations would arise more slowly than in sexual organisms.

Until the mid-1940s there was no conclusive proof of a sexual mechanism in bacteria. The first unambiguous studies were those of Lederberg and Tatum who worked with multiply mutant strains of *E. coli*. These strains had been obtained through the mutagenic action of ultraviolet light and were biochemically deficient in one or more ways. Any wild type or normal bacterial cell can grow on a basic or minimal medium that contains just a few simple substances from which the cell can manufacture all the complex organic molecules needed for its metabolism. Such

cells, which thrive on a minimal medium, are called *prototrophs*. In contrast, other cells may be nutritional deficients in that they lack the ability to make one or more essential growth factors from the minimal source. Such deficient cells are designated *auxotrophs* and must be provided with the growth factors they cannot synthesize. These nutritional mutants can grow only on a supplemental medium, a minimal medium to which one or more defined factors have been added.

Lederberg and Tatum mixed together different auxotrophic strains. For example, strain 1 might require biotin and methionine. Such an auxotroph can be represented as B^-M^-. Strain 2, requiring threonine and proline to supplement the minimal medium, can be designated T^-P^-. After the two strains have been mixed together, samples are incubated for different time periods and then placed on minimal medium

FIG. 15-3. Prototrophs from auxotrophs. (A) An auxtrophic strain, such as strain 1 or 2, cannot grow on minimal medium. Essential substances must be added. When the strains are mixed and samples of the mixture plated on minimal medium, some plates develop colonies. The cells in these must be prototrophs, because no essential metabolites have been added to the medium. (B) The mixture can be depicted as a cross of bacterial strains in which the genes are all linked. The prototrophs could arise through a crossover event. Cells with the reciprocal crossover product cannot develop, because they need to be supplemented with four essential metabolites. Other crossovers are possible, but these cannot develop because the cells will require one or more essential substances. Only the prototrophs $(B^+M^+T^+P^+)$ can grow on the minimal medium.

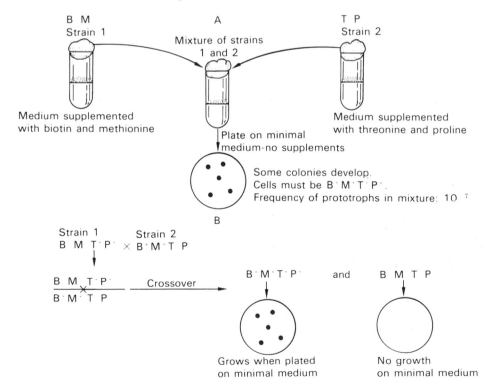

(Fig. 15-3*A*). The logic is that only $B^+M^+T^+P^+$ cells will grow. Colonies growing on minimal medium were actually obtained after mixture and incubation of the strains. How did these arise? One suggestion is through sexual recombination. Strain 1, although deficient for biotin and methionine, carries the genetic information to manufacture threonine and proline ($B^-M^-T^+P^+$). Likewise, strain 2, although unable to manufacture threonine and proline, has the wild genes for biotin and methionine ($B^+M^+T^-P^-$). The mixture can be represented as a cross and the appearance of the wild colonies explained on the basis of cell fusion followed by recombination (Fig. 15-3*B*).

If we assume the loci are linked, we can imagine the operation of some process similar to crossing over that is able to produce the prototrophic recombinants, $B^+M^+T^+P^+$. The parental cells from strains 1 and 2 cannot grow on the minimal medium, nor can any recombinants that lack the ability to carry out all the biochemical steps required for nutrition. The results, however, can be interpreted on the basis of mechanisms other than sexuality. Of the cells in the mixture, only a small number, approximately 1 in 10^7, is prototrophic. Perhaps these are not recombinants at all but revertants—wild cells that have arisen through back mutations: $B^-M^-T^+P^+{\rightarrow}B^+M^+T^+P^+$ (strain 1) and $B^+M^+T^-P^-{\rightarrow}B^+M^+T^+P^+$ (strain 2).

To answer this objection to the sexuality hypothesis, a second mixture was studied, this time employing two strains that were mutant for three factors. When triply deficient parental strains were mixed together, prototrophs arose with the same frequency as when doubly deficient auxotrophic strains were used. If back mutation were responsible for the origin of the prototrophs, the frequency of their occurrence when triply deficient strains were used should have been much lower than when doubly deficient strains were used, because the former require three instead of two separate back mutations to produce a prototrophic cell. These observations provide a strong argument against the origin of the prototrophs through spontaneous mutation and favor the sexuality concept. Additional evidence showed that cell contact was essential for the origin of prototrophs after mixing multiply mutant strains. Experimental work eliminated the possibility of transformation (see more on this later in this chapter), symbiotic association of two different cell types, and other phenomena as explanations for the origin of recombinants. Only the sexuality hypothesis was upheld.

The fertility factor

More refined analyses of the *E. coli* sexual process indicated that the various markers (mutant alleles for auxotrophy as well as those that did not affect the nutritional requirements) were linked into just one group, in contrast to the two or more linkage groups typical of higher forms. Evidence for linkage and gene order was based on reasoning similar to that used in classic genetic analysis. Those genes closer together would undergo less recombination than those farther apart on the chromosome. Double crossovers would be less common than singles, and triples even less frequent than the doubles. Therefore, by crossing various multiply mutant strains, counting the resultant recombinant colonies (in the procedure each colony is derived from one cell), and scoring their phenotypic features, a single linkage map emerged.

During the course of the linkage analysis, however, it became apparent that the *E. coli* sexual system possessed several features that departed from the picture typical of sexual recombination in higher forms. One unexpected observation indicated that the production of recombinants depended on the continued survival of just *one* of the parents. For example, suppose strain 1 is $B^-M^-T^+P^+$ and is also streptomycin sensitive. Strain 2 is $B^+M^+T^-P^-$ but is resistant to the drug. The two strains are mixed and finally prototrophs are isolated on minimal medium containing streptomycin. Now suppose that exactly the same procedure and the same markers are followed in a second cross but now strain 1 is streptomycin resistant whereas strain 2 is streptomycin sensitive (Fig. 15-4).

In such a cross, absolutely no recombinants form. Crosses of this type, in which the survival of only one of the parental strains was guaranteed, repeatedly demonstrated that only one of the parents was giving rise to the recombinants. In our example, they are coming from strain 2, for if it is killed, no recombinants are formed. Strain 1 is transferring genetic material to strain 2, but once the transfer has been achieved, it is no longer needed for the formation of recombinants. Further studies revealed that the appearance of recombinants definitely depends on two types of cells, a donor and a recipient. Genetic material is transferred from the donor cell to the recipient, which then acts as a "zygote" whose survival is necessary for the segregation of recombinant cells. It was also discovered that at times cells could arise in a

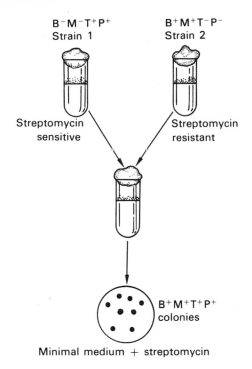 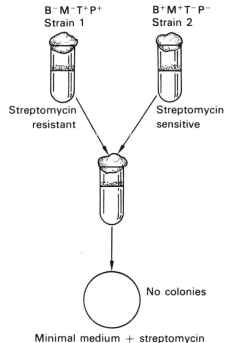

FIG. 15-4. Survival of one parent. To obtain any recombinants, the survival of one of the parental strains must be ensured. One way to demonstrate this is by performing two types of crosses between two auxotrophic strains. The only difference between the crosses is that a parental strain is sensitive to streptomycin in one cross and resistant to it in the other. As the diagram shows, one of the two strains must not be killed if recombinants are to be obtained (strain 2 must survive here). This would appear to be the recipient strain. Strain 1 would be a donor that transfers genetic material to a recipient.

culture of donor types that no longer had the ability to act as donors. They lacked the capacity for transferring genetic material to the recipient. Some factor appeared to be required for a cell to act as a donor type. If it was lost, a cell originally from a donor culture could be used as a recipient. This fertility factor needed to confer donor capacity was designated F. Donor cells were thus called F^+ (with fertility factor) and recipient cells F^- (lacking fertility factor).

Observations then showed that the fertility factor (F) could be transferred with relative ease to F^- cells, thus converting them to the donor state. Simply bringing donor and recipient cells into contact for 1 hour or so would change the F^- cells to the F^+ state. The following summarizes what was now evident about the sexual system in *E. coli:*

1. Sexual recombination depends on the presence of two cell types, a donor or male (F^+) and a recipient or female (F^-). $F^- \times F^-$ crosses are ineffective. Mixing together F^+ cultures with other F^+ cultures yields some recombinants because of the presence of F^- cells that arise spontaneously, but the amount of recombination is much less than in crosses of $F^+ \times F^-$.

2. The F^+ or donor state results from the presence of F, the fertility factor, which can be transferred by cellular contact to F^- cells, which are then converted to F^+ donors.

3. The appearance of recombinants depends on the continued viability of the F^- parent.

It must be realized at this point that, although $F^+ \times F^-$ crosses yield recombinants, the actual number of the recombinant cells is low, in the vicinity of 10^{-6}. In other words, although the F factor is transferred with a high frequency or efficiency, the chromosome itself is transferred only in a small minority of the cells. Hence, the chromosomal genes undergo recombination infrequently.

Partial genetic transfer

Another unorthodox characteristic of the *E. coli* sexual system was revealed. It was noted that when

Cross 1

Strain 1 F⁺ (auxotrophic-requires histidine)   Strain 2 F⁻ (auxotrophic-requires arginine)

$$A^+B^+C^+D^+E^+G^+H^- \times A^-B^-C^-D^-E^-G^-H^+$$

Place on minimal medium

$$A^+B^-C^-D^-E^-G^-H^{+-} \quad \text{Prototrophs}$$
Genes mainly from F⁻ strain 2

Cross 2

Strain 1 F⁻ (auxotrophic-requires histidine)   Strain 2 F⁺ (auxotrophic-requires arginine)

$$A^+B^+C^+D^+E^+G^+H^- \times A^-B^-C^-D^-E^-G^-H^+$$

Place on minimal medium

$$A^+B^+C^+D^+E^+G^+H^{+-} \quad \text{Prototrophs}$$
Genes mainly from F⁻ strain 1

A⁺ makes arginine
A⁻ requires arginine
H⁺ makes histidine
H⁻ requires histidine
C,D,E,G,- other genes not influencing basic nutrition

Fig. 15-5. Partial transfer. F⁺ and F⁻ types were found within strains. This makes possible reciprocal crosses. In this example, strains 1 and 2 are both auxotrophs, because each lacks the ability to manufacture a specific amino acid (histidine in 1 and arginine in 2). The other loci indicated are for traits that do not influence the ability to grow on minimal medium. When the strains are crossed and the cells placed on minimal medium, only the prototrophs (A⁺H⁺) will grow. These prototrophs can be tested for the other markers. It is found that the reciprocal crosses differ. Although various combinations are possible, most of the alleles in the A⁺H⁺ prototrophs will have been derived from the F⁻ parent. The transfer is only partial from the F⁺ to the F⁻.

genetic material was transferred from F⁺ to F⁻, the transfer was usually partial; only a portion of the genetic material of the F⁺ seemed to enter the F⁻. Certain genes were not transferred at all from donor to recipient. Evidence for this was obtained from reciprocal crosses between strains. For example, in *both* strain 1 and strain 2, F⁺ and F⁻ types were isolated. This made possible reciprocal crosses, and the results of these proved to be quite different (Fig. 15-5). Only some of the genetic material of the F⁺ parent entered the F⁻ cell to contribute to the offspring. The F⁻ parent, on the other hand, contributes all of its genetic material. Recombinants, therefore, will contain a greater percentage of their genes from the F⁻ cells than from the F⁺. This is quite a different situation from that in higher forms in which both parents make equal donations of the genetic material to the zygote.

High-frequency types and the transfer of genetic markers

Some clarification of the unusual features of recombination in *E. coli* was provided through the discovery of still another donor or male type. These donors behave as "supermales." When F⁻ cells are crossed with them instead of with ordinary F⁺ males, the percentage of recombination is increased approximately 1000 times. Therefore, the supermales were termed high frequency (Hfr), denoting their ability to bring about genetic transfer.

Various Hfr males arose independently in cultures of ordinary F⁺ donors. When different strains of Hfr were examined, it was noted that they could become F⁺ again. This suggested that the sex factor (F) was present in the Hfr cell. But some puzzling facts emerged when different Hfr × F⁻ crosses were com-

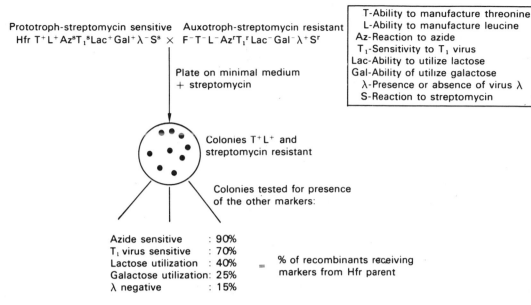

Prototroph-streptomycin sensitive   Auxotroph-streptomycin resistant
Hfr T⁺L⁺AzˢT₁ˢLac⁺Gal⁺λ⁻Sˢ × F⁻T⁻L⁻AzʳT₁ʳLac⁻Gal⁻λ⁺Sʳ

| T-Ability to manufacture threonine |
| L-Ability to manufacture leucine |
| Az-Reaction to azide |
| T₁-Sensitivity to T₁ virus |
| Lac-Ability to utilize lactose |
| Gal-Ability of utilize galactose |
| λ-Presence or absence of virus λ |
| S-Reaction to streptomycin |

Plate on minimal medium
+ streptomycin

Colonies T⁺L⁺ and
streptomycin resistant

Colonies tested for presence
of the other markers:

Azide sensitive   : 90%
T₁ virus sensitive   : 70%
Lactose utilization : 40%   =   % of recombinants receiving
Galactose utilization: 25%   markers from Hfr parent
λ negative   : 15%

FIG. 15-6. Cross of Hfr and F⁻. Prototrophs are selected by plating on minimal medium. These cells will all possess the T⁺L⁺ markers, which come from the prototrophic Hfr parent. The Hfr parent, however, is destroyed by the presence of the streptomycin in the minimal medium; so all the prototrophic colonies are truly recombinant. The F⁻ parent, being an auxotroph, cannot grow on the minimal medium. These recombinant offspring are then tested for the presence of the other markers from the F⁺ parent. It can be seen that the markers have very different chances of being transferred to the F⁻ cells to enter into the formation of recombinants.

pared with one another and also with ordinary F⁺ × F⁻ crosses. In the latter type of cross, it will be recalled, the F factor itself is transferred quite readily to the F⁻ cells, converting them into donor types. And although a low frequency of recombinants is obtained in F⁺ × F⁻ crosses, *any* portion of the genetic material can be transferred from F⁺ to F⁻. In contrast to this, when any specific Hfr strain is used as the donor parent, only a restricted portion of its linkage group is readily transferred. Some genes rarely go from the Hfr to the F⁻ parent. Moreover, the F factor that must be present in the Hfr cell usually is *not* transferred. Consequently, in a cross of Hfr × F⁻, the offspring are almost always F⁻. For example, consider the cross in Fig. 15-6. After the strains are mixed, samples are plated out on a medium lacking threonine and leucine but containing streptomycin. By doing this, we are selecting prototrophs that are resistant to streptomycin. Therefore, only T⁺L⁺ colonies will grow. These cells must be recombinant, because the drug will have killed the Hfr parent; the F⁻ parent, being an auxotroph (T⁻L⁻), cannot grow on minimal medium. The recombinant colonies on the plates will thus all be

T⁺L⁺. They can subsequently be tested for any other markers transferred by the Hfr parent.

In one strain of Hfr, the streptomycin marker is rarely transmitted, whereas the segment of genes T through lambda is transmitted with a high frequency. Figure 15-6 shows that very curious ratios are found among the recombinants. Of the T⁺L⁺ offspring, 90% are azide sensitive (meaning that 90% received the Az marker from the Hfr parent), 70% received the T₁ marker, and so on for the others. It would seem that not all the marker alleles have the same chance of entering the F⁻ cells from the Hfr. Moreover, the recombinants are usually F⁻, not F⁺ or Hfr as would be expected if the F factor had a good chance of entering the recipient cell.

An insight into the basis of this odd mating system was provided in an ingenious procedure employed by Wollman and Jacob. Using strains similar to those just discussed, they mixed together Hfr and F⁻ and then interrupted the mating at various time intervals. They did this by subjecting the mixture to a blender, which can pull apart pairs of cells without damaging them. They then plated out samples and tested for

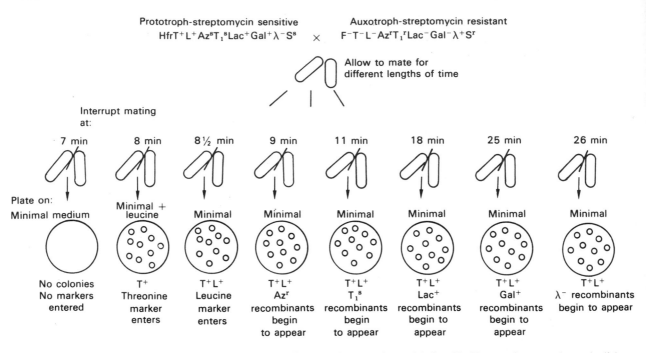

FIG. 15-7. Interrupted mating. The same two strains shown in Fig. 15-6 may be allowed to mate for different periods of time. Samples of a mating mixture are removed at specific times and subjected to a Waring blender, which pulls the cells apart. The sample is then tested for recombinants. It is found in this cross that no markers enter if the cells are separated before 8 minutes. The markers being followed in the cross are then found to enter at different characteristic times.

the presence of Hfr markers in the recombinant cells. They found (Fig. 15-7) that no prototrophs were formed before an 8-minute period. Then the threonine marker was transferred and shortly thereafter the leucine marker.

Although only 9 minutes of contact between cells was needed for entry of the Az marker, 25 and 26 minutes were necessary for Gal and lambda. If mating was interrupted at 18 minutes, for example, some cells would have received Lac, but none would be Gal⁺. This suggested to Wollman and Jacob that the genetic material or chromosome of the Hfr parent entered the F⁻ cell at a definite rate and in an established order (Fig. 15-8A), in this case, the order T through lambda. If the mating was interrupted and the chromosome broken at a certain time, some genes would not have yet been able to penetrate into the F⁻ cell.

The breaking of the chromosome does not prevent the integration of the donor genes into the recipient by recombination. The peculiar recombination ratios that occur under normal conditions of mating can be explained by the spontaneous breakage of the donor chromosome as it is being transferred. The closer a

gene is to the origin—the chromosome end that penetrates the recipient first—the greater its chance of being transferred. The spontaneous breakage point would vary from one mating pair to the next. The result would be different percentages of recombination for the different genetic markers carried by the male parent. The bacterial mating system, therefore, has unusual features that permit us to map genes in two ways: by recombination percentages and by time of entry into the female parent (Fig. 15-8B). We will see a bit later that the donor cell does not lose genetic material, since DNA replication takes place as genetic material is transferred.

Relationship among the *E. coli* mating types

Still unsettled was the fundamental difference between Hfr and ordinary F⁺. Why did the latter show a high frequency of transfer of the F factor but a low frequency of transfer of the marker alleles? Since Hfr originated from F⁺ cultures, it was considered possible that the Hfr were mutants and *only* these mutants of F⁺ were actually able to transmit the chromosome. The F⁺ cells would really be ineffective in bringing

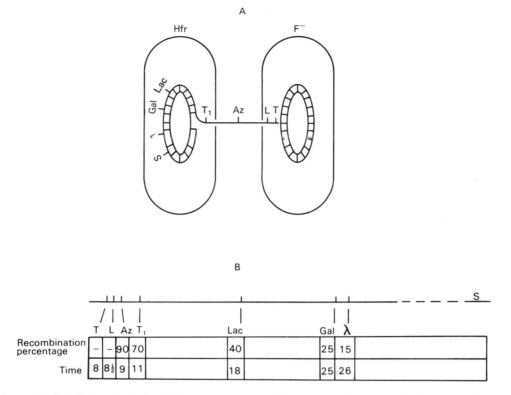

FIG. 15-8. Entry of Hfr DNA. The results depicted in Figs. 15-6 and 15-7 can be explained in part as shown here. The chromosome of E. coli is known to be circular. At the time of mating, the circle in the Hfr cell generates a linear chromosome strand that penetrates into the F⁻ cell at a specific rate and in a definite order. The donor cell loses no genetic material, as the material transferred is replaced by replication of DNA. Only portions of the donor chromosome material enter the F⁻ cell to engage in recombination, since mating cells usually break apart before the transfer of an entire chromosome strand. The closer a locus is to the beginning of the Hfr chromosome, the greater the chance that it will be transferred. The T and L loci, as shown here, have a greater chance of entry than those following them. Some loci, such as the one for streptomycin tolerance, are so far toward the end that breakage usually occurs in front of them, and they rarely enter. (B) According to this concept, the chromosome may be mapped in either of two ways, by recombination percentages or by times of entry. In this example, all of the cells would be T⁺L⁺, because only prototrophs were selected in the cross. Different percentages of these prototrophs carry the other markers. We see that 90% of them will receive the azide marker, only 25% the Gal marker. The difference is explained by the different times of entry, 9 versus 25 minutes. The sooner a marker enters, the closer it is to the beginning and the greater the number of cells that will receive it. The whole chromosome may be mapped in time units. At 37°C, approximately 90 minutes are required for any cells to receive the entire Hfr chromosome.

about transfer of genes from male to female cells. Wollman and Jacob applied the fluctuation test and showed that the origin of Hfr through mutation of F⁺ was a correct concept. They also showed this through application of the replica plating technique, which in addition enabled them to isolate different strains of Hfr in pure culture. Recall that replica plating enables us to obtain copies of an original plate and thus permits study of the same bacterial cells or colonies under a variety of conditions.

Using the replica plating method, Wollman and Jacob started with a population of F⁺ cells that were T⁻L⁻ (Fig. 15-9). They grew discrete colonies on plates supplemented with threonine and leucine. A second set of plates contained minimal medium, and these were spread with F⁻ cells that were M⁻B⁻. Imprints of the F⁺ colonies were then made onto the second set of plates. The original plate was oriented in exactly the same way as the replica plates. After cells were transferred in this way from the F⁺ plate to the plates containing the F⁻, it was found that some colonies formed on the latter. These must be recombinant prototrophs, because only they can grow on minimal medium. The positions of the colonies on the second set of plates were then related back to the positions of certain F⁺ colonies on the first plate.

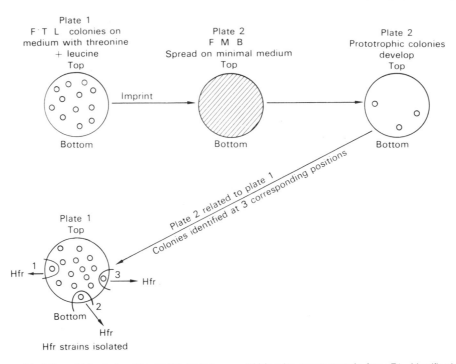

FIG. 15-9. Origin and isolation of Hfr strains. The replica plating procedure demonstrated that only a few cells in an F⁺ population are capable of transferring genetic material to the F⁻. These are Hfr cells, which arise spontaneously from F⁺. Identification of Hfr colonies growing among F⁺ colonies on the same plate permitted the isolation of pure Hfr cultures.

Cells from these F⁺ colonies, which had been picked up by the velvet, must have contacted the F⁻ cells on the second plate and transferred genetic material to them. After genetic recombination, prototrophs arose from the F⁻ cells and were selected on the minimal medium, where they formed colonies.

The replica plating technique thus permitted the identification of just those few F⁺ colonies that contained cells capable of acting as donor or male cells and could transfer chromosome markers to F⁻. Each of the F⁺ colonies had arisen on the plate from a single cell and, therefore, each donor colony represented a cell that was different from the other cells on the plate that did not behave as donors. Each chromosome donor in the original F⁺ population is in the minority and must have arisen independently by spontaneous mutation.

After the donor colonies were identified in this way, they were isolated to establish different, pure Hfr strains. Crosses were made between these several Hfr types and F⁻ cells. Further analysis employing the interrupted mating method contributed information of fundamental importance to bacterial genetics. The results clearly showed that different Hfr types had a different gene at the anterior end of the chromosome (Fig. 15-10A). Moreover, some blocks of genes that were *not* transferred with a high frequency in one strain were transferred readily in another. (The segment M through S is transferred in strain 1 but not in strain H; the segment Az through lambda is transferred with high frequency in strain H but not in strain 1.)

When the different strains are compared in this way, an unexpected picture emerges. All of the observations indicate only one chromosome or linkage group, and this linkage group appears to have no beginning or end. It can be interpreted only in the form of a circle (Fig. 15-10B). Comparison of the Hfr strains shows that the arrangement of their genes is the same; however, the order in which the genes are transferred by the several strains to the F⁻ parent differs. In addition, for each Hfr type, there is a characteristic segment of the circular chromosome that is usually not transmitted at all.

When the circular chromosome map was first proposed, the actual physical shape of the E. coli chromosome was unknown. As discussed in Chapter 11, the refined procedures of Cairns clearly demonstrated

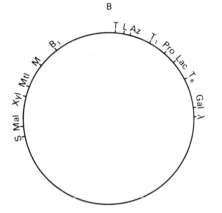

Hfr Type	A Leading End
H	T L Az T₁ Pro Lac T₆ Gal λ
1	L T B₁ M Mtol Xyl Mal S
2	T₁ Az L T B₁ M Mtol Xyl Mal S
3	T₆ Lac Pro T₁ Az L T B₁ M Mtol Xyl Mal S
4	B₁ M Mtol Xyl Mal S λ Gal
5	M B₁ T L Az T₁ Pro Lac T₆ Gal λ

FIG. 15-10. Hfr types and partial linkage map. (A) Several Hfr strains were isolated (only six are shown here). When mated to F⁻, the strains were found to differ in relation to the genes that were transferred with a high frequency and to the order in which the markers enter the F⁻ cell. (B) The observations on the six Hfr types can be explained on the basis of a circular linkage group. When genetic material is transferred to the F⁻ cell, a linear chromosome is derived from the circle. The anterior end that is established for the linear chromosome would differ among the Hfr strains, so that the first loci to enter the F⁻ would vary from one Hfr to the other. The arrangement of the loci, however, remains the same; only the order of entry is different. Those genes near the leading end are transferred with a high frequency. Those farther back go over to the F⁻ with a low frequency or not at all.

the actual circular nature of the *E. coli* chromosome. This represents another superb correlation between genetic analysis and cytology.

Most of the peculiarities of the *E. coli* mating system have now been clarified and make perfect sense in light of the findings. In the F⁻ cell, the circular chromosome remains as such. No DNA replication occurs to generate the transfer of genetic material. This is generally the case in the F⁺ cell, which in addition to the circular chromosome contains an independently existing F element, an entity that is also circular and composed of double-stranded DNA. When

an F⁺ cell contacts an F⁻ cell, the free existing DNA of the F element is replicated and transferred to the F⁻ cell, which then becomes F⁺.

In an occasional F⁺ cell, the F factor may insert itself into the chromosome of the bacterial cell, thus converting it to an Hfr type. The inserted element is generally replicated along with the bacterial chromosome as if it were a natural part of it. However, contact with an F⁻ cell sets off a cycle of DNA replication in the Hfr cell that results in transfer of chromosome material to the F⁻ cell. To appreciate the events taking place in the transfer of genetic material from both F⁺ and Hfr cells to F⁻, we must become acquainted with the rolling circle method of DNA replication described in the following section.

Rolling circle method of replication

In Chapter 11, the replication of the circular *E. coli* chromosome was discussed. Another method by which a circular DNA molecule can replicate is the rolling circle mechanism. Rolling circle replication (summarized in Fig. 15-11) is initiated by a nick in only one strand of the circular DNA molecule by an endonuclease. This cut produces a free 5′ end and a free 3′ end. The intact circle then proceeds to act as a template strand as nucleotides are added to the free 3′ end of the nicked strand. Growth of the new DNA (always in the 5′ to 3′ direction) causes the 5′ end to become displaced and to move away from the circle. We can imagine that the growing strand rolls or revolves around the circular template as new nucleotides continue to be added to the 3′ end. As new growth continues at the 3′ end and proceeds around the circle, the 5′ end of the strand rolls out and appears as a tail, which in turn continues to grow in length. After the growth has completed one turn around the circular template, the tail length represents the whole length of the original nicked strand that was displaced.

The tail, even before it reaches full length, may serve as a template for the assembly of DNA and be converted to the double-stranded state. It is the equivalent of a lagging strand, so that the 5′ to 3′ replication would be discontinuous (recall Fig. 11-24). If the replication continues and two revolutions of the circle take place, then two whole, connected tail lengths would be generated, as actually occurs in the replication of some viruses (Chap. 16). On the other hand, if replication ceases before one revolution is com-

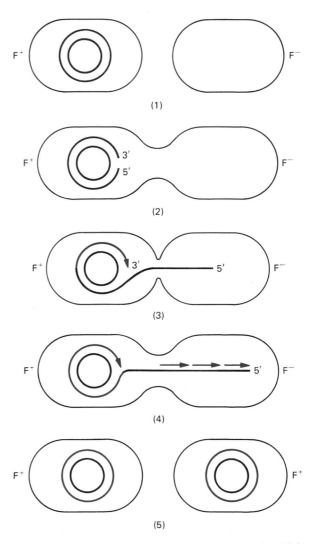

pleted, then the tail length that is displaced will be less than the full length of the original strand.

Rolling circle and the *E. coli* sexual system

The transfer of sex factor (F) from F^+ to F^- entails rolling circle replication, as outlined in Fig. 15-12.

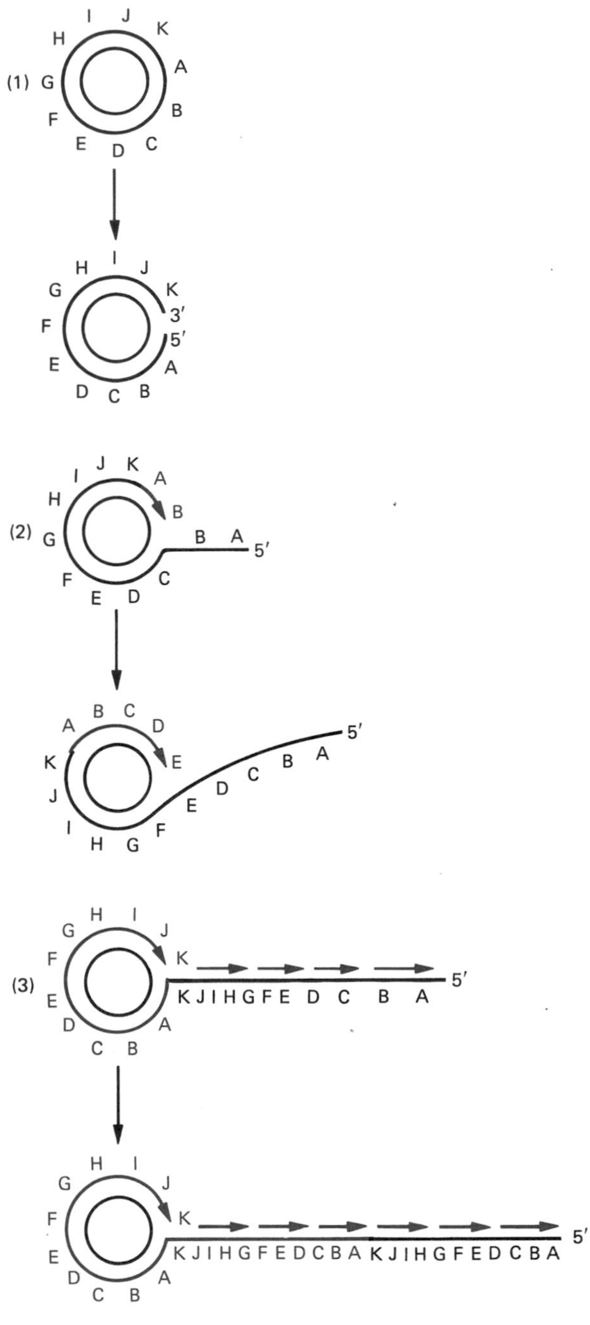

FIG. 15-11. Rolling circle replication. *(1)* A circular DNA molecule is nicked in one of the strands by an endonuclease. This produces a free 3' end and a free 5' end. *(2)* Nucleotides are added *(red)* to the 3' end, bringing about growth of the strand *(left)*. This addition of nucleotides causes displacement of the strand at its 5' end. As growth continues at the 3' end *(right)*, the tail at the 5' end increases in length. *(3)* After one revolution *(left)*, the tail length represents the length of the original strand that was displaced. Discontinuous replication may take place on the tail to form double-stranded DNA *(arrows)*. Two revolutions *(right)* produce two tail lengths, whereas less than one revolution produces a tail shorter than the original strand *(2)*.

FIG. 15-12. Transfer of sex factor by rolling circle replication. *(1)* An F^+ cell contains an F element that the F^- cell lacks. The F factor exists free of the chromosome (not shown here for clarity). *(2)* Contact with an F^- cell brings about a nick in one of the strands of F, and free 3' and 5' ends are produced. *(3)* A round of replication by the rolling circle mechanism is initiated, and the displaced 5' end enters the F^- cell. *(4)* As the 5' end moves, DNA replication takes place in the F^- cell, and the transferred strand becomes double stranded. Transfer ceases after one round of replication. *(5)* Synthesis of the complementary strand is completed. The two cells separate, and circularization of the linear strand occurs. DNA ligase seals all nicks. Both cells are now F^+.

FIG. 15-13. Transfer of chromosome from Hfr to F⁻. *(1)* A Hfr cell contains an F element *(1, 2, 3, 4)* integrated into its circular chromosome. An F⁻ cell lacks the F element. (Chromosomal genes are designated A through F.) *(2)* Contact between an Hfr and an F⁻ cell triggers a round of DNA replication. (The chromosome is not shown in the F⁻ cell, for clarity.) An endonuclease nicks one strand of the F element, establishing free 3′ and 5′ ends. The nick splits the F factor. *(3)* The 5′ end enters the F⁻ cell through the conjugation tube. The leading 5′ end carries a part of the genetic material of the F factor *(2, 1)*. Chromosomal genes enter behind the portion of the sex factor. *(4)* As more of the choromosome strand is transferred, it is replaced in the Hfr by rolling circle replication. DNA replication also occurs in the F⁻ cell *(red arrows)* with the transferred segment acting as a template. *(5)* Breakage of the strand usually occurs before the entire chromosome strand is transferred. The transferred segment (now double stranded) can participate in recombination with a homologous chromosome region of the F⁻ cell. The F⁻ cell remains F⁻, because it carries only a portion of the F factor *(2, 1)*. The Hfr cell has lost no genetic material, because rolling circle replication replaces any DNA that is transferred (segment carrying 2, 1, A, B, C).

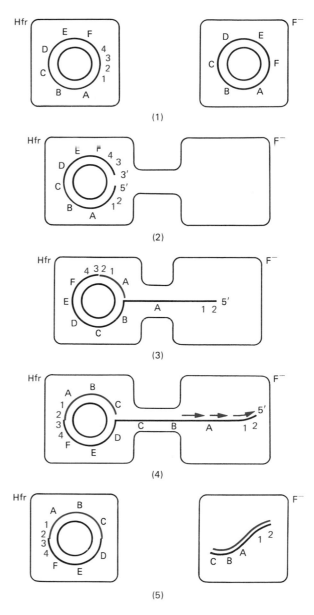

Conjugation triggers a nick in one of the strands of F, producing free 5′ and free 3′ ends. A cycle of replication is initiated. Growth at the 3′ end displaces the 5′ end, which moves through the conjugation tube from the F⁺ to the F⁻ cell. Replication at the 5′ end takes place in the F⁻ cell.

After DNA growth has completed one revolution, both the transfer and the synthesis of DNA ceases. Transfer of the strand takes about 2 minutes. The strand transferred to the F⁻ cell is completely replicated and becomes circularized. The original F⁻ cell has now been converted to F⁺.

When an Hfr cell contacts an F⁻ cell, the F factor that is inserted into the bacterial chromosome becomes activated, and a round of DNA replication is initiated which takes place by means of the rolling circle mechanism. When Hfr transfers chromosome material to F⁻, the Hfr parent remains viable; it does not lose genetic material because rolling circle replication takes place. An endonuclease-induced nick occurs within the sex factor itself (Fig. 15-13) and affects just one of the two strands of the inserted F element. Note that the nick splits the F factor. Consequently, the 5′ end that enters the F⁻ cell contains only some of the genetic material of F. The remaining portion of F is at the 3′ end and can be transferred to the recipient cell only if replication by the rolling circle mechanism completes one revolution. However, breakage usually occurs before the entire amount of DNA is transferred. Once inside the F⁻ cell, the transferred strand acts as a template for DNA synthesis, and the portion becomes double stranded. It is

this double helical DNA, representing a portion of the Hfr genetic material, that can then engage in crossing over with the F⁻ chromosome.

It should now be apparent why the F⁻ cell usually remains F⁻, since the breakage of the transferred strand before one revolution of replication is completed results in transfer of only a part of the F factor to the F⁻ cell. Those Hfr genes in the region away from the leading 5′ end (i.e., near the 3′ end) will have little chance of entering the recipient. Note (Fig. 15-14) that the different Hfr strains simply represent cells with the sex factor integrated at different posi-

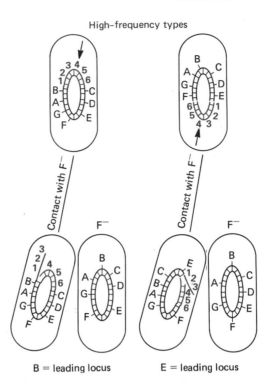
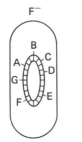

B = leading locus   E = leading locus

FIG. 15-14. Differences in mating type. The F⁻ cell contains a circular chromosome, as do the F⁺ and Hfr types. In the F⁺, however, the fertility factor, F, is present as a free entity. The Hfr type contains F in an integrated state on the circular chromosome. The various Hfr types differ in the position at which F is located *(arrows)* and its orientation. Contact with an F⁻ cell triggers DNA replication in the Hfr, and a break is produced within F. This establishes a leading point and an endpoint. Because F can be located at different positions and in different orientations on the chromosome, the order of entry of the genes varies from one Hfr strain to the next.

tions in the circular chromosome. When replication is triggered by conjugation and the F factor is nicked in one strand, a leading end is established, and this will obviously differ from one Hfr strain to the next, depending on the site of insertion of F and its orientation. We can now understand why the various Hfr strains differ with respect to which chromosomal segments they transfer efficiently and which they rarely transfer at all to F⁻ cells.

Chromosome transfer and recombination

Once a portion of the Hfr chromosome has penetrated the F⁻ parent, it may undergo recombination with the genetic material in the F⁻ cell. If any recombinant bacterial cells are to arise, the introduced Hfr segment must pair with the homologous chromosome region of the F⁻ cell. Genes are then reciprocally exchanged between the recipient chromosome and the introduced segment (Fig. 15-15). Note that all the chromosomal genes contributed by the Hfr parent do not

necessarily enter into the formation of the recombinants (e.g., A in Fig. 15-15). The free fragment that results is apparently discarded from the recombinant cell along with the portion of the sex factor associated with the introduced segment.

Rarely, a recombinant cell is found that has received a gene that is typically *not* transmitted by a particular Hfr strain (e.g., gene F in Fig. 15-13). When this does occur, the recombinant may then prove to be Hfr instead of F⁻. We now understand how this can come about, because the entry of a chromosomal gene at the extreme end means the entire chromosome strand must have been transferred. Therefore, there is a good chance that the other end of the split F factor may have entered the F⁻ cell as well; thus, all of the genetic material of the F factor is now present. In such an unusual case, in which premature strand breakage does not take place, an intact sex factor can integrate into the chromosome, converting the F⁻ to Hfr (details on insertion are discussed in the following section). The Hfr re-

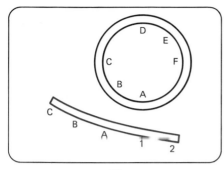

(1)

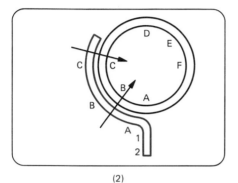

(2)

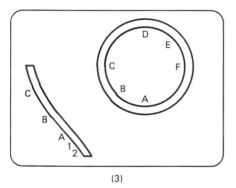

(3)

FIG. 15-15. Formation of a recombinant F⁻ cell. *(1)* Chromosomal segment transferred from Hfr to F⁻ includes chromosomal genes (A, B, C) and a portion of the F factor (2, 1). *(2)* Pairing takes place between segment and homologous region of the chromosome. The F⁻ cell carries no region homologous to the portion of the F factor. Pairing is followed by crossing over *(arrows)*. *(3)* The reciprocal crossing over results here in exchange of portion CB and the formation of a recombinant chromosome in the F⁻ cell and a recombinant fragment. The latter is discarded along with the portion of the F factor.

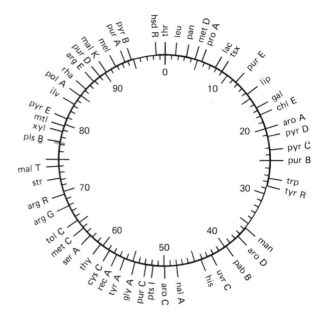

FIG. 15-16. The 100-minute linkage map of *E. coli,* showing some important genes. (From B. J. Bachmann, K. B. Low, and A. L. Taylor. *Bact. Rev.* 40: 116, 1976.)

required. As noted earlier in this chapter, genes may be mapped according to time of entry from Hfr to F⁻ cell (see Fig. 15-8). The *E. coli* map is therefore scaled on the basis of 100 minutes (Fig. 15-16). Note that the time, in minutes, starts arbitrarily at 0 with threonine. All the other loci are related to the threonine locus on the basis of their entry times, as revealed by interrupted mating experiments.

An ordinary population of F⁺ cells contains occasional Hfr cells that have arisen independently by insertion of the F factor at different positions. So when F⁺ and F⁻ populations are mixed, only the occasional Hfr types in the population are able to transfer the chromosome. But because the point of insertion of the F factor varies from one Hfr type to another, the linear chromosomes that arise in the several Hfr types at the time of conjugation have different anterior and different terminal ends. Consequently, different segments are transferred from one mating pair to another. The F⁺ population is thus a mixed one containing an assortment of Hfr types, in contrast to an isolated population pure for one Hfr strain. Thus, when F⁺ and F⁻ are mixed, *all* genes seem to be transferred at a low frequency. The F particle itself goes over at a high frequency because it exists free in most of the cells in the F⁺ population.

combinants derived in this way may also revert to the F⁺ state, indicating that the F factor can exist in a free state or attached to the chromosome.

For the entire chromosome strand to be transferred from Hfr to recipient cell, about 100 minutes are

FIG. 15-17. Electron micrograph of conjugating *E. coli* cells. (Reprinted with permission from T. F. Anderson, E. L. Wollman, and F. Jacob, *Sur les processus de conjugaison et de recombinaison genetique chez E. coli. III. Aspects morphologiques en microscope electronique.* Annales Institut Pasteur, 1957.)

F factor as an episome

Jacob and Wollman coined the term *episome* to designate any factor with a behavior such as that of the F factor. An episome is a genetic entity that may exist free in a cell or be integrated into the chromosome. The episome is also a self-replicating unit, a replicon (see Chap. 11). Different techniques have demonstrated that the F factor of *E. coli* is a circular DNA element 94,500 base pairs long, approximately 2% the size of the bacterial chromosome. As we have seen in the case of the F factor, an episome may be present or absent in a given kind of cell. The F factor also illustrates that an episome may have several remarkable attributes. The F factor of *E. coli* confers an assortment of properties on the bacterial cell. Not only does the episome convert the cell from a nondonor to a potential donor cell (sometimes called a male cell), its presence has also been shown to bring about the production of particular types of appendages. Termed *sex pili* or *sex fimbriae*, these appendages are necessary for conjugation. Electron micrographs of mating cells reveal cytoplasmic connections or bridges between them. Each such bridge may represent one of the sex pili required to establish a connection for the transfer of the chromosome between mating cells (Fig. 15-17).

Presence of the F factor renders a cell sensitive to certain single-stranded DNA and RNA phages, the so-called male-specific phages. It also causes the cell to become resistant to certain other phage types that can infect F⁻ cells. Presence of a single F factor also renders the *E. coli* cell immune to the entry of or the maintenance of any additional F factors, making the relationship 1:1 between F factor and circular chromosome. Since the F factor confers several traits on the cell that harbors it and since it is a replicon of appreciable size, it is not surprising that several genes are now recognized that occur on the episome. Both mutant and nonmutant F elements have been isolated and analyzed with respect to the genetic regions carried on the F factor itself. A cluster of 13 genes (*tra* genes) is known to be needed to establish sex pili and transfer DNA during conjugation. One genetic sequence is known to be required for the origin and initiation of transfer. Other known genetic regions of

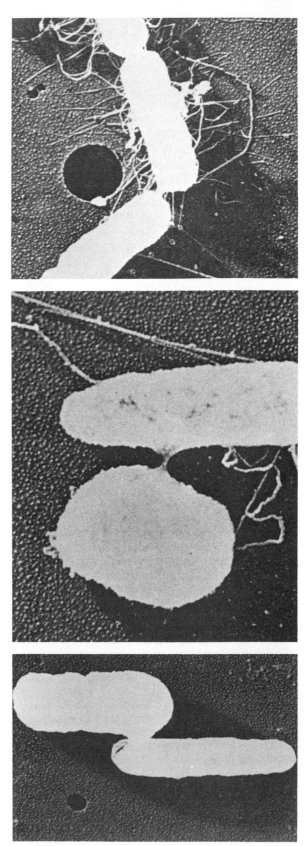

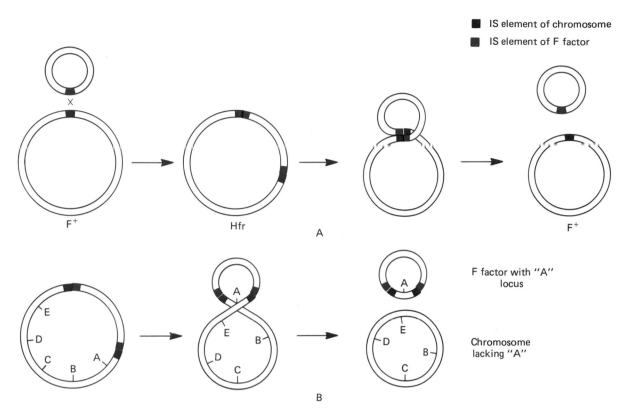

FIG. 15-18. Insertion and removal of F factor. (A) The factor possesses certain regions, IS elements, which are homologous to segments of the chromosome. This enables the fertility factor to pair with the chromosome at different points. A recombination event inserts F into the chromosome of an F⁺ cell, thereby converting it to Hfr. At times, the F factor may remove itself from its integrated state and become free, converting the cell to an F⁺. Normally this would occur by a reversal of the integration process. Homologous regions would again pair, and a recombination event would take place. (B) At rare times, the F factor may carry away a portion of the chromosome. This is believed to come about when small homologous portions of the chromosome on either side of the inserted F factor pair and a crossover event takes place. (Only one IS element is shown in the F⁺ chromosome, for simplicity.)

the F factor are responsible for the reaction of the host cell to specific phages and for its resistance to the establishment of additional F elements in the cell. Other genes are also present that are required to replicate the F element itself.

Very important is the presence on the F factor of certain short sequences called *insertion sequences* or *IS elements*. These range from 800 to 1400 base pairs in length, and four of them are known to exist on the F factor. These IS elements are evidently responsible for the integration of the episome into the *E. coli* chromosome. The *E. coli* chromosome also contains IS elements, some identical to those in the episome. It is the positions of the IS elements in the different F factors and in the circular chromosome of the different strains of *E. coli* that are evidently responsible for determining the different sites of integration of the episome as the cell is converted from F⁺ to Hfr. The IS elements are able to bring about the

recombination between the episome and the chromosome (Fig. 15-18A), an event that results in insertion of the episome into the host chromosome. By a reversal of the process, it can again become free, converting the cell back to an F⁺. The IS elements have great bearing on other phenomena that have come to light; these will be discussed later in this chapter.

When the F factor removes itself from the chromosome, it may, at rare times, become associated with a portion of the host chromosome it carries away with it (Fig. 15-18B). As a result, an F factor can originate that contains a marker allele, such as Gal⁺. The piece of the chromosome now replicates along with the genetic material of the episome and is part of a replicon containing genetic material of both F factor and the chromosome.

Harboring pieces of genetic information in this way, F factors can enter F⁻ cells. This means that genes

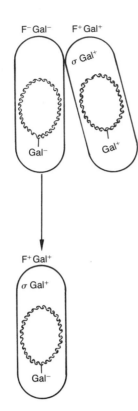

F⁻ Gal⁻   F⁺ Gal⁺

σ Gal⁺

Gal⁻   Gal⁺

F⁺ Gal⁺

σ Gal⁺

Gal⁻

FIG. 15-19. Sexduction. This diagram shows an F⁻ cell that cannot ferment galactose and an F⁺ cell that carries the allele for galactose use on its F factor and also on its chromosome. After the two cells make contact, the fertility factor may be transferred to the F⁻, which is therefore converted to an F⁺ with the ability to use galactose. It carries the allele for this ability on its F factor, whereas its chromosome carries the allele that does not enable the cell to use the sugar.

may be transferred from one cell to another by an episome, which can thus confer a new genetic property on a recipient cell. Suppose that an F⁻ cell is Gal⁻ and thus cannot ferment galactose (Fig. 15-19). This cell may contact an F⁺ cell that contains an F factor carrying a Gal⁺ allele. The F⁻, Gal⁻ cell in turn can be converted to an F⁺, Gal⁺ type by receiving the fertility factor, which not only makes it a potential donor cell but also gives it the capacity to ferment galactose. This transfer of genetic material from one cell to another by way of a sex factor has been called *sexduction*. The term *F-genote* was coined to designate a sex factor that has incorporated a portion of the chromosomal DNA. Those bacterial strains that carry F-genotes are known as F prime (F′) strains. A bacterial cell therefore can acquire the ability to perform a function as a result of an allele contained in an episome it harbors and not as a result of an allele

on its chromosome. Such a cell is haploid except for a small chromosome portion. In this short region, it can possess a pair of alleles and would therefore be partially diploid as well as a partial heterozygote. A *merozygote* is any partially diploid bacterial cell resulting from conjugation or a process such as sexduction. The term *heterogenote* describes a cell heterozygous for just a segment of the genetic material, as opposed to *heterozygote*, which is reserved for diploid cells.

Plasmids and drug resistance

The discovery of the many remarkable properties of the F factor focused attention on the possible occurrence of similar particles in various kinds of cells. Of great importance to humans are certain replicons that have been recognized in bacteria and can confer drug resistance on the cells harboring them. Like episomes, these nonchromosomal genetic entities may be absent from a cell but when present they can replicate autonomously. However, they do not integrate with the chromosome of the cell as do F factors. The more general term *plasmid* is used to define any replicon that may exist in a cell independently of the chromosome. Episomes are plasmids that *do* have the ability to integrate with the chromosome. Most plasmids are not usually essential to the survival of the cell in which they occur.

Most strains of bacteria are known to harbor some kind of plasmid. Plasmids have been isolated from cells by density gradient procedures and then subjected to detailed analysis. They have also been visualized with the electron microscope. All plasmids are naked, circular, double-stranded DNA molecules. They vary greatly in the number of genes they carry, from just a few to hundreds. Plasmids also vary in the number of copies in which they exist in bacterial cells. Certain plasmids occur in only a single copy, others are found in a few copies, whereas some exist in large numbers, even up to 100. The replication of the plasmid depends on many of the enzymes of replication coded by the host cell. However, the plasmid itself is coded for sequences that regulate the initiation of replication and the number of copies of the type of plasmid that may occur in a given cell. Regulation of copy number may involve repressor protein substances that can bind to DNA sequences and prevent the initiation of DNA replication (see Chap. 17 for details of repressor action). In the case of a plasmid like F, with low copy number, the level of repressor in the cell is sufficient to inhibit replication of the plasmid, but as the cell grows, repressor becomes less concentrated, and the plasmid can replicate. A plasmid type that has high copy number requires more of its specific repressor protein to inhibit its replication. Therefore, if the plasmid number is low in this case, so is repressor concentration. Plasmid replication therefore continues until sufficient plasmids of that type, with their repressor genes, are present to raise the repressor concentration to a level that effectively inhibits further plasmid replication.

The concept of repressor involvement in control of copy number explains why certain plasmids cannot coexist in the same cell. Assume that two plasmids, A and B, are identical or very much alike. They would then be coded for the same kind of repressor. The level of repressor would consequently double when they are present in the same cell. The repressor from one can bind to the DNA of the other and inhibit replication. Two plasmids that cannot coexist in the same cell are said to be *incompatible*. Presence of one F factor in a cell precludes the maintenance of an-

other F factor in the same cell. If, however, F and some unrelated plasmid are present together, each plasmid would code for its own kind of repressor, and replication of one plasmid would not be regulated by the other's repressor. Two such plasmids that can coexist in the same cell are called *compatible*. In addition to repressor proteins, various other mechanisms are believed to be involved in the control of plasmid copy number and compatibility among plasmids.

One way in which plasmids are classified relates to the kind of genetic information coded in their DNA. The fertility factor, F, and the F' factors derived from it constitute one class of plasmid. Bacteria may also carry plasmids that enable them to produce proteins or toxins that kill sensitive bacterial cells. Of extreme importance is a class of plasmid-carrying genes that confer resistance to one or more antibiotics. Let us explore this topic more closely.

Along with the use of antibiotics, the number of bacterial strains that have become resistant to one or more drugs has increased with amazing rapidity. Several loci on the bacterial chromosome are known to

FIG. 15-20. R plasmids. A bacterial cell may carry genetic determinants that make it resistant to certain drugs *(1)*. Some of these determinants may be on the chromosome, but others may be on plasmids. R plasmids can exist in a cell independently of the F factor. A conjugal R plasmid can promote conjugation with a cell *(2)*. In this way, a previously drug-sensitive cell can be converted to one that is multiply drug resistant if it receives a plasmid that carries resistance to two or more drugs *(3)*. It is possible for a cell to carry a resistance determinant to a specific drug, say streptomycin, on either its chromosome or on an R plasmid (see the text for more details).

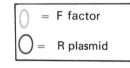

$\bigcirc$	= F factor
$\bigcirc$	= R plasmid

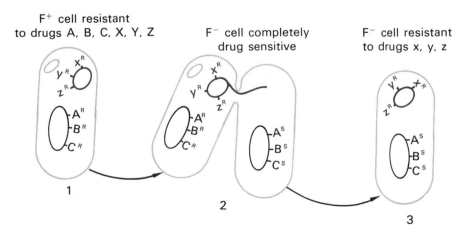

control drug resistance. Antibiotics such as chloramphenicol commonly exert their lethal effect on the bacterial cell by associating with the ribosomes, thus interfering with translation. Chromosomal gene mutations that confer drug resistance generally bring about changes in the ribosome itself, so that the antibiotic can no longer combine with it; this permits normal translation and protein synthesis in the presence of the drug. Another well-known class of drug-resistant determinants does not represent genetic information native to the chromosome. Instead, these genes have originated outside the chromosome, as part of a plasmid known as an *R plasmid* (previously called a resistance-transfer factor, or RTF). Some of the R plasmids are conjugal, meaning that, like the F factor, they can bring about conjugation between two cells and in this way promote their own transfer to cells which lack them. This transfer is independent of F, which can coexist in a cell with a conjugal R plasmid (Fig. 15-20). Since the plasmid may carry several genes for drug resistance, conjugation can

confer multiple drug resistance on a previously sensitive cell.

Conjugal plasmids have been shown to cause a cell to produce sex fimbriae resembling those under the control of the F factor. The cells with the R plasmid–induced fimbriae also become sensitive to phages that can lyse male cells. Both the R plasmid and the F factor may be removed from cells after treatment with acridines. These are intercalating agents that inhibit the replication of F and other plasmids but do not affect the replication of the chromosome. Loss of F factor and R plasmid is accompanied by the inability of the cells to produce sex fimbriae and to adsorb the male-specific phages. We see from this that these two types of replicons have many features in common. The fertility factor, however, serves mainly to bring about transfer of the chromosome once it becomes integrated, whereas the R plasmid is responsible for its own transfer and for that of the resistance genes it may be carrying.

While nonchromosomal genes for drug resistance

FIG. 15-21. Conjugal and nonconjugal plasmids. Certain determinants to drug resistance do not reside on conjugal plasmids but are found instead on other R plasmids that are nonconjugal and that can replicate independently. Nonconjugal plasmids cannot bring about their own transfer to another cell *(above).* A conjugal plasmid carrying one or more resistance determinants can bring about its own transfer. When this takes place *(below),* nonconjugal plasmids, if present in the same cell, may also be transferred.

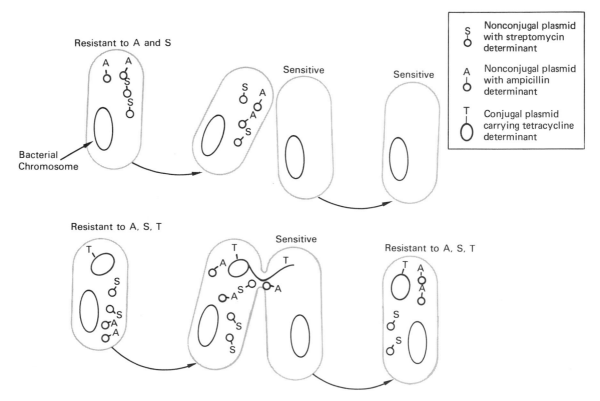

may exist as part of a conjugal R plasmid, they may also occur on R plasmids that are not conjugal. For example, in *E. coli,* a gene that confers tetracycline resistance may be integrated with a conjugal R plasmid that regularly transmits itself and the genetic factor for tetracycline resistance. However, other genes for resistance, such as those for resistance to streptomycin and ampicillin, may exist on other replicons that are nonconjugal plasmids. By themselves, non conjugal plasmids cannot be transferred from one cell to another; their transfer requires the presence of a conjugal plasmid. If a conjugal R plasmid is also present in a cell with nonconjugal plasmids, one or more of the latter may be transferred (Fig. 15-21). A variety of R plasmids is known, and these carry different combinations of resistance-determining genes.

The existence of such an assortment of R plasmids, both conjugal and nonconjugal, raises questions concerning their origin. Unlike the chromosomal genes, which can confer drug resistance by altering the structure of the ribosome, the nonchromosomal determinants exert their effect by directing the synthesis of products that act directly on the drug and inactivate it. The ribosome remains unaltered. This different method of conferring drug resistance suggests that the genes of the plasmid are distinct from their chromosomal counterparts. As we will learn in the following section, R plasmids have been found to possess several very remarkable properties.

Plasmids and transposons

A great deal of information on plasmids has come to light throughout the course of investigations with microorganisms. Unexpected discoveries have been made that are pertinent to eukaryotic cells as well, and have significant bearing on such important matters as gene regulation and cellular differentiation. Here we will discuss only those findings that relate to certain facets of plasmid behavior in microorganisms. Further discussion of plasmids and transposons appears in later chapters.

A class of mutations discovered in *E. coli* produces most unusual properties. When such a mutation arose, its effect extended beyond the site of the gene that had mutated. It was found that the mutated DNA had become longer than the normal DNA, indicating that the mutation resulted from a piece of DNA that had become inserted into the gene that had mutated (see Fig. 17-14)! Such DNA segments were found to be able to insert into many different sites on the bacterial

chromosome. When they did so, the activity of a gene at that position was abolished. These segments were called *insertion sequences* or *IS elements*—the same IS elements discussed in reference to insertion of the F factor into the *E. coli* chromosome.

A curious feature of the IS elements became apparent. Evidently they could insert at a large number of chromosomal sites. This indicated that nonhomologous recombination was involved, because an IS element could not be expected to contain DNA sequences homologous to so many different sites on the chromosome. Other unusual observations were being made at this time by those studying antibiotic resistance in bacteria. It was found that certain antibiotic resistance genes (genes that code for proteins that can confer resistance to certain drugs) can move about in the cell from one site to another. For example, the gene for chloramphenicol resistance can move from one replicon to another in the cell, from an R plasmid to the chromosome to a virus. A relationship to the IS elements became evident when it was found that transfer from one replicon to another, such as from one plasmid to another, was always accompanied by an increase in the length of the DNA of the recipient replicon. Apparently the gene for antiobiotic resistance was being carried by a DNA sequence that could move from one molecule to another. The name *transposon* was given to a unit that could move about in this way.

It was found that transfer by a transposon could occur in bacterial cells that were mutant for a particular gene, rec A, whose protein product was known to be required for recombination in the cell. This, too, implied that a kind of nonhomologous recombination event was taking place, enabling the transposon to insert at nonhomologous positions in different molecules. Although transposons are apparently able to insert by nonhomologous recombination, the process is not completely random, because some transposons are known to have certain preferential sites of insertion on the DNA molecule. Some transposons seem to prefer one or a small number of specific sites; others seem to be less discriminate and insert at a large number of sites, whereas most are intermediate. Certain parts of the *E. coli* genome, known as "hot spots," seem to be selected for insertion over and over again, whereas other regions are excluded by transposons for residence.

IS elements are small insertion sequences that contain no known genes related to any properties other than insertion; however, transposons that carry genes

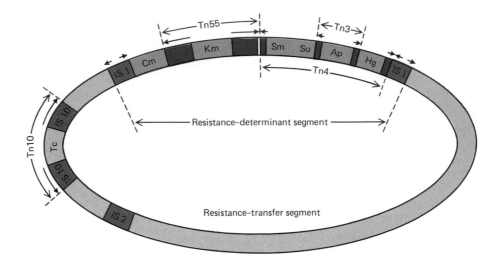

FIG. 15-22. Transposons and conjugal R plasmids. IS elements contain only properties related to insertion. Other transposons may carry genes for drug resistance. Some of these transposons may be compound and include other transposons. A conjugal R plasmid is constructed from smaller units. A segment is present, the resistance transfer segment, which carries genes required for the conjugal properties of the plasmid. A conjugal R plasmid also contains a segment bearing genes for drug resistance. This resistance determinant segment may contain several transposons. Note that transposon 3 (Tn3) is within Tn4. Each transposon can be transferred independently to another molecule. The small arrows pointing outward indicate inverted repeats at the end of each transposon. Note the presence of IS elements at the junctions of the resistance transfer segment and the resistance determinant segment. Note that Tn10, a transposon encoding resistance to tetracycline (Tc) occurs here on the resistance transfer segment and that it includes IS elements (see the text for further details). (From S. Cohen and J. Shapiro, *Transposable Genetic Elements.* Copyright © 1980 by Scientific American, Inc. All rights reserved.)

FIG. 15-23. Inverted repeats. A transposon carries at each end nucleotide sequences that are identical but in the reverse order. When DNA bearing a transposon is denatured and each single strand is allowed to anneal with itself, the complementary bases in the inverted repeat form pairs. This results in single strands that bear a stem and a loop. The loop contains the genes carried by the transposon. The stem–loop configuration can be detected with the electron microscope and indicates the presence of a transposon.

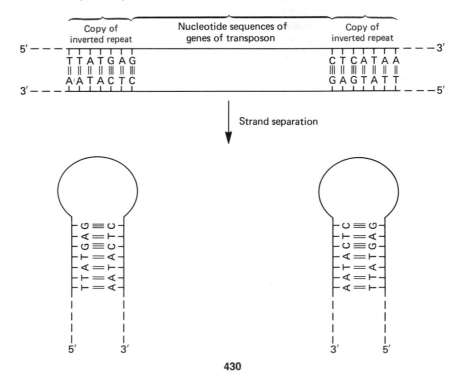

430

for resistance are larger and more complex. They may even include IS elements. The IS elements were the first class of transposons to be discovered. Now by convention, all transposons, regardless of the number of genes they contain, are designated by the symbol *Tn* plus an identifying number, such as Tn10 or Tn55 (Fig. 15-22). A distinguishing feature of all transposons is that the DNA of the transposable unit has inverted repeats at its ends. This means that a copy of the very same sequence is present at each end of the double-stranded DNA, but in reverse order (Fig. 15-23). (We encountered inverted repeats in our discussion of the terminator region in Chap. 12.) As a result of the inverted repeat, each *single* strand carries

nucleotide sequences that are complementary but in reverse order. When the two strands of the transposon are separated by denaturation, each separate strand can reanneal with itself. A loop forms, which can be visualized with the electron microscope.

The investigations with transposons have given us a very clear insight into how R plasmids are constructed. A conjugal R plasmid is a compound entity, built up from discrete units. A conjugal plasmid contains a segment that carries those genes necessary for the conjugal or transfer properties of the plasmid. These are genes similar to the *tra* genes of the *E. coli* F element. This segment of the conjugal plasmid is called the RTF (resistance transfer factor) segment.

FIG. 15-24. Reversible dissociation of a conjugal R plasmid. A conjugal R plasmid is a compound entity that can dissociate reversibly at sites where IS elements reside. Dissociation may result in a conjugal and a nonconjugal plasmid *(above)*. A large, compound plasmid in turn can be built up from smaller ones by the reverse process *(middle)*, and this can dissociate *(bottom)*.

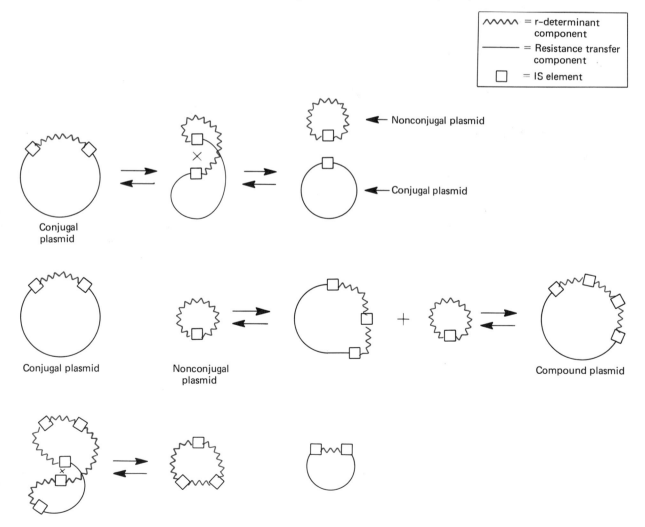

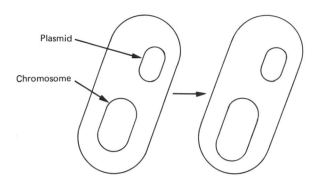

FIG. 15-25. Transposition. Transposition from one replicon to another does not involve loss of the transposon from its original location. A copy of the transposon *(red),* originally on a plasmid, may later occur on the chromosome as well.

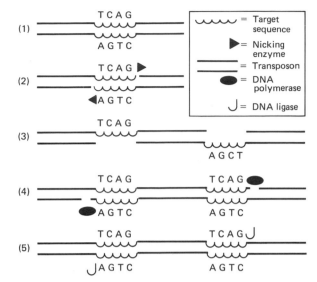

FIG. 15-27. Generation of direct repeats. *(1)* A target sequence is present in one copy in the recipient. *(2)* An enzyme of the transposon produces two single-stranded, staggered breaks. *(3)* The transposon links to staggered ends of cut strands, producing two gaps. *(4)* The gaps are filled in by DNA polymerase, using target sequence on strands as template. *(5)* DNA ligase seals the nicks. The target sequence is now present in two direct copies.

The RTF segment also carries genes that regulate copy number and DNA replication. The RTF is flanked at its two ends by IS elements. The R plasmid contains, in addition to the RTF, a segment bearing the genes for drug resistance—a segment known as the R determinant, or resistance-determinant, segment. Note from Fig. 15-22 that an R segment may contain several genes for drug resistance and several transposons. Two or more R determinant segments may be present in a complex R plasmid. Conjugal plasmids reveal their compound nature by their ability to dissociate reversibly at various sites, yielding a conjugal and a nonconjugal plasmid. At these sites, IS elements are found and evidently account for the dissociation. The process, as seen in Fig. 15-24, is reversible, as a larger plasmid may be built up from smaller ones by the reverse process.

The RTF segments of different plasmids have been compared by DNA–DNA hybridization procedures. These show that RTFs from different sources are quite similar. The R determinant segments appear to be collections of various transposons carrying genes for resistance to antibiotics. All the evidence indicates that large R plasmids have been constructed from separately derived R determinant segments that become bound to a basic RTF segment. The separately

evolving R determinants are joined to the RTF in different combinations, generating a great variety of R plasmids, which has important medical implications. Also of medical concern is the fact that identical transposons are found not only in related species of bacteria but can move from one unrelated species to another.

Transposition

The exact way in which transposons move from one replicon to another is still unknown, but a number of findings provide some clues to the mechanism of the process. First, it should be stressed that when we say a transposition has occurred or a transposon has moved from one replicon to the next, the transposable element is not necessarily lost from the original location.

FIG. 15-26. Features of transposon insertion. Insertion of a transposon *(red)* is accompanied by the direct repeat of a short sequence of nucleotides in the recipient DNA. At the ends of the transposon are nucleotide sequences that are inverted repeats of each other.

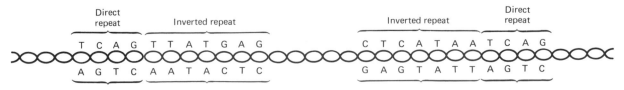

The transposition is fundamentally a process entailing duplication of the transposon, so that one copy remains in the original location and another occurs at a different site (Fig. 15-25).

The transposition process requires proteins that are coded by genes of the transposon. A characteristic feature of transposon insertion relates to the site at which the transposon takes up residence. As noted previously, some sites may be preferred over others by some transposons. Nevertheless, insertion of a transposon always involves the duplication of a short nucleotide sequence of the recipient DNA. These duplications range in size from about 3 to 12 base pairs and flank the inverted repeats on either side of the transposon (Fig. 15-26). The sequence in the recipient that is duplicated is known as the *target sequence*, and it is believed to be repeated in the

manner shown in Fig. 15-27. An enzyme of the transposon brings about two single-stranded breaks that are staggered, one in each strand, at the ends of the selected target sequence. The transposon DNA links to the ends of the cut strands in such a way that two gaps are produced, one on each strand. A DNA polymerase fills in these gaps, and DNA ligase seals the nicks. The result is duplication of the target sequence, the length of which is characteristic for a given transposon. This length, in turn, is determined by the degree to which the particular enzyme involved staggers the single-stranded nicks. Duplication of a target sequence is so characteristic of transposons that duplication of the direct repeat type flanking certain DNA sequences in a genome is taken as a clue that a transposon has been inserted in that region (Chap. 17).

FIG. 15-28. Cointegrate formation and transposition. *(A)* Two replicons in a cell are represented, plasmid x with the transposon and plasmid Y, which lacks it. *(B)* Replicon fusion produces a cointegrate, a composite of the two plasmids. In the process, a copy of the transposon has been generated. Each copy is at the junction uniting the two replicons, and each of the copies is a direct repeat with the same orientation. *(C)* A recombination event that depends on an enzyme encoded by a gene of the transposon resolves the cointegrate into two replicons, each with a copy of the transposon.

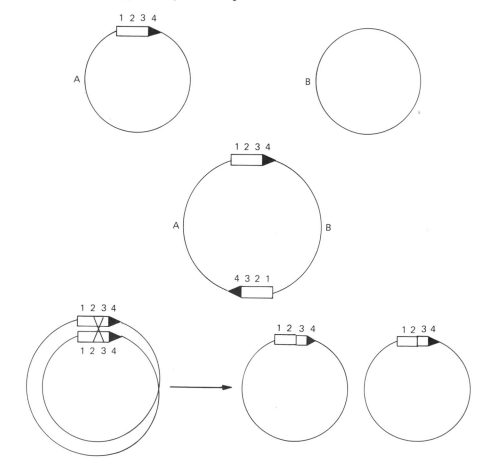

Another insight into transposition comes from observations of what are known as *cointegrates*. A cointegrate is a composite replicon formed by the fusion of the replicon containing the transposon with a replicon lacking one. Note from Fig. 15-28A and B that the cointegrate contains two copies of the transposon, indicating that replication of the transposon DNA has accompanied the fusion process. The transposon copies are located at the sites where the two replicons are joined, and each is in the same orientation. The formation of a cointegrate depends on a *transposase*, a large protein coded by one of the genes of the transposon. After many generations, cells containing cointegrates may give rise to other cells that no longer contain the cointegrate but now contain two plasmids, each with a copy of the transposon. The two plasmids are believed to arise as a result of a crossover event between the two transposons in the cointegrate, a step that depends on another enzyme coded by a gene of the transposon (Fig. 15-28C). This enzyme, named *resolvase*, brings about the reduction of the cointe-

grate with the formation of two plasmids, each with one copy of the transposon.

A very well-known transposon, Tn3, found in bacteria is about 5000 nucleotides long and contains three genes. Two of these are coded for the enzymes required for transposition—the transposase needed for cointegrate formation and the resolvase needed to reduce the cointegrate (Fig. 15-29A). The third one is coded for an enzyme, β-lactamase, which inactivates ampicillin and thus confers resistance to that antibiotic on cells harboring Tn3. The expression of both the transposase and the resolvase genes is regulated by the resolvase enzyme itself. Besides its role in transposition, the resolvase has repressor activity and can bind to a region between the transposase and resolvase genes, thereby preventing their expression and thus preventing a high rate of transposition (Fig. 15-29B). Note from Fig. 15-29 that the transposase and resolvase genes are transcribed in opposite directions, because their sense strands are opposite. Therefore, when the repressor binds to the binding site

FIG. 15-29. Transposon Tn3 and regulation of transposition. *(A)* Tn3 contains three genes between its inverted repeats. The two involved in transposition are transcribed in different directions. The transposase and resolvase result from translation of the corresponding mRNAs. *(B)* A high frequency of transposition is apparently prevented by the

ability of the resolvase to act as a repressor when its concentration reaches a certain level. It can bind to a region between the two genes that contains the sites for the start of transcription of both transposition genes. The repressor does not affect transcription of the gene for ampicillin resistance.

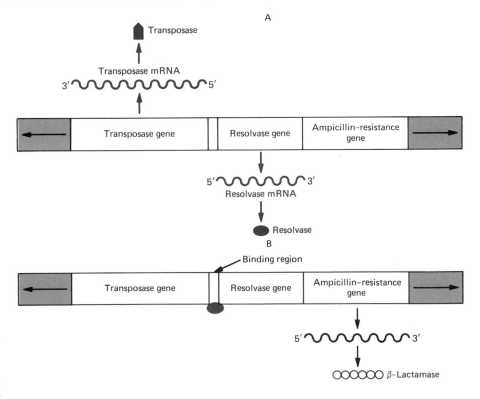

between the two genes, transcription from initiation sites in both genes is blocked.

Note that at no time during the process of transposition did the transposon move about freely by itself in the cell from one replicon to another. The transposition is seen as a process requiring DNA replication and a recombination event in which the transposon remains integrated with replicons within the cell. Cointegration formation is believed to be one route by which many transposons bring about transposition. Certain transposons, however, appear to transpose in a manner that does not entail cointegrate formation and is yet to be clarified.

Transformation in relationship to recombination

Before leaving this discussion of bacteria, let us reexamine one method of genetic change in bacteria that may have implications for higher forms of life. In Chapter 11, the phenomenon of transformation in Pneumococcus was discussed in relation to the identification of the chemical nature of the gene. It will be recalled that, in transformation, DNA preparations from a donor strain may be applied to a recipient. After a period of incubation, the latter may incorporate some of the donor DNA and acquire a new trait. When completed, this transformation is stable and represents a true genetic alteration.

At first, transformation was considered more or less a curiosity in the pneumonia organism, but it now seems to be rather widespread. Transformation has been recognized in at least 17 bacterial species, among them such well-known groups as Streptococcus, Bacillus, and Hemophilus. Many characteristics of these microorganisms are known to be susceptible to transformation—drug resistance and ability to synthesize specific enzymes, among others. It seems that any genetic locus may be transferred from donor to recipient in transformation. We must emphasize that the DNA donated to the recipient does not require cell contact; pure DNA preparations extracted from donor cells are effective. Nor does the transfer depend on any vector, such as an episome or virus. Bacterial cells appear to have an affinity for uptake of DNA from the environment during a certain phase of the growth cycle of the bacterial population, near the end of the log phase. This receptive stage is of brief duration in the recipient cells and entails physiological changes that are not completely understood. A cell in the receptive stage that can take up extraneous DNA and be transformed by it is said to be *competent*.

Competence may also refer to the portion of the growth cycle of the bacterial population when most of the cells are physiologically capable of transformation.

Several very ingenious experiments have revealed many facts about transformation that are of general biological significance. In the early 1960s, studies were conducted in which labeled donor DNA was followed from the moment of contact with recipient cells until the latter were transformed. It was found that immediately after entering the recipient cells, some of the donor DNA is converted to a single-stranded state, while the rest is degraded. This labeled, single-stranded donor DNA is then inserted into the recipient DNA (Fig. 15-30). Once inserted, it replaces the homologous segment of the recipient

FIG. 15-30. Steps in transformation. To bring about transformation, the donor DNA must be double stranded; otherwise it will not penetrate the recipient cell. Once inside the recipient, some of the donor DNA is degraded and some transformed to the single-stranded condition (1). This single-stranded form may associate with the region on the chromosome with which it is homologous. It may then replace a single-stranded portion of the recipient DNA (2). This recipient segment is eventually lost. The cell, now carrying the hybrid DNA, is transformed. After it divides (3), it gives rise to nontransformed and to transformed cells. In the latter, the region involved in the transformation has both strands identical to the original donor DNA. This results because the inserted donor strand forms a complementary copy of itself at the next replication.

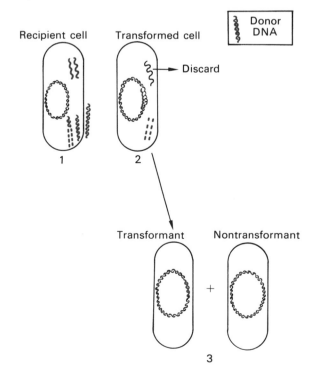

DNA, which is then discarded. It is this newly inserted DNA that is responsible for the transformation. It is able to effect a genetic change and become a part of the genetic material of the recipient cell. This cell, in turn, may be used as a donor in a subsequent transformation. Thus, the labeled DNA introduced in the first transformation can be followed further to see how it behaves in later transformations or in later cell divisions.

Extremely valuable information has come from following DNA throughout all phases of transformation in a single cell and throughout two or more divisions of transformed cells. One of the most important discoveries, with great implications for genetic theory, was the demonstration that the introduced donor DNA is physically integrated into the recipient DNA. This observation supports the concept of crossing over as a result of synapsis followed by breakage and reunion, much as Janssens pictured it in 1909, a matter that had been controversial for decades. The studies with the labeled donor DNA clearly showed that the introduced transforming DNA penetrates the cells and becomes associated with the homologous DNA segment of the recipient. It is then incorporated into the recipient chromosome by an actual physical insertion, which is accompanied by the genetic transformation of the cell.

As noted, only one strand of the donor DNA participates, and the evidence indicates that *either* strand of the donor double helix can bring about the transformation. It is interesting that the DNA that is picked up by the recipient cell in the first place *must* be double stranded. It must also be above a certain minimal molecular weight (5×10^5). Single-stranded DNA is completely ineffective in transformation. But once inside the recipient cell, the double-stranded donor DNA is altered, and it is only a single strand that engages in the transformation process. This means that when this single donor strand replaces one of the two strands of the recipient, heteroduplex DNA results. One strand of the double helix in the involved region would be a strand of the recipient, the other the strand of the donor. And this has been found to be true. Therefore, just one strand, either of the two of the original double helix, can cause a transformation. At the cell divisions that follow the transformation, the DNA of the transformed cells will come to have both strands of the *original* donor DNA in the region of the transformation (Fig. 15-30).

Although most studies of transformation employ DNA preparations that are extracted from donor cells by an investigator, it is now known that transformation may take place spontaneously. Aging cultures of some bacterial strains contain cells that are undergoing self-digestion and release DNA into the medium. This DNA can cause transformations of other cells in the population. Pneumococcus transformations may occur spontaneously when two different bacterial strains are injected into mice. Such observations raise the possibility of transformation in higher species. Actually, this has been reported in the fruit fly. Eggs of Drosophila containing very young embryos have been treated with DNA from other stocks carrying certain marker alleles. The treated embryos have given rise to stocks of flies that are true genetic transformants, whose cells have incorporated some of the genetic material applied to them. However, in the flies, the donor DNA does not replace the resident DNA of the recipient chromosome, as it does in bacteria, but manages somehow to persist along with it in the cell. Moreover, the donor DNA does not always undergo transcription. Its expression is sporadic. Whether the information contained in the DNA of the donor or in that of the host will be transcribed seems to vary from cell to cell, so that the flies are always mosaics. The mosaic flies pass down to their offspring the information introduced from the donor stock, but these offspring are in turn mosaics. Transformation, a genetic phenomenon discovered in bacteria, may prove to have many implications for various higher life forms.

REFERENCES

Adelberg, E. A. and S. N. Burns. Genetic variation in the sex factor of *Escherichia coli. J. Bacteriol.* 79: 321, 1960.

Cold Spring Harbor Symposium on Quantitative Biology. *Movable Genetic Elements*, vol. 45. Cold Spring Harbor, NY, 1981.

Cohen, S. N. and J. A. Shapiro. Transposable genetic elements. *Sci. Am.* (Feb.): 40, 1980.

Fox, A. S., S. B. Yoon, and W. M. Gelbart. DNA-induced transformation in *Drosophila*: genetic analysis of transformed stocks. *Proc. Natl. Acad. Sci.* 68: 342, 1971.

Gurney, T., Jr., and M. S. Fox. Physical and genetic hybrids in bacterial transformation. *J. Mol. Biol.* 32: 83, 1968.

Hayes, W. *The Genetics of Bacteria and Their Viruses,* 2nd ed. Wiley, New York, 1968.

Kingsman, A. and N. Willets. The requirement for conjugal DNA synthesis in the donor strand during F'Lac transfer. *J. Mol. Biol.* 122: 287, 1978.

Lacks, S. Molecular fate of DNA in genetic transformation of *Pneumococcus. J. Mol. Biol.* 5: 119, 1962.

Lederberg, J., and E. L. Tatum. Gene recombination in *Escherichia coli. Nature* 158: 558, 1946.

Luria, S. E. and M. Delbruck. Mutations of bacteria from

virus sensitivity to virus resistance. *Genetics* 28: 491, 1943.

Novick, R. P. Plasmids. *Sci. Am.* (Dec.): 102, 1980.

Shapiro, J. A. Molecular model for the transposition and replication of bacteriophage mu and other transposable elements. *Proc. Natl. Acad. Sci.* 76: 1933, 1979.

Taylor, A. L. and C. D. Trotter. Linkage map of *E. coli* K12. *Bact. Rev.* 36: 504, 1972.

Vapnek, D. and W. D. Rupp. Identification of individual sex factor DNA strands and their replication during conjugation in thermosensitive DNA mutants of *Escherichia coli*. *J. Mol. Biol.* 60: 413, 1971.

Willets, N. and R. Skurray. The conjugation system of F-like plasmids. *Annu. Rev. Genet.* 14: 41, 1980.

REVIEW QUESTIONS

1. In one part of an experiment with a strain of penicillin-sensitive cells, a sample of bacteria is taken and allowed no further growth. The sample is divided into several equal parts and then each is plated immediately on a medium containing penicillin. In another part of the experiment, a small sample of the sensitive culture is taken and divided further into several subdivisions. Each subdivision is then allowed to grow until it contains the same number of cells as each division in the other part of the experiment. From the following data, tell which part of the experiment is represented by the data in column A and which by that in column B. Explain.

	A	B
Total number of cells	10^9	10^9
Mean number of resistant cells	22	15
Variance	4480	17

2. Colonies derived from eight different cells are grown on a plate supplemented with all the necessary growth substances. Colonies 2, 5, 6, and 8 are prototrophs. The others are auxotrophs. Colonies 3 and 7 require leucine; 1 and 4 require arginine. Answer the following:

   A. What colonies will grow following incubation after a replica is made onto a plate with minimal medium?

   B. What colonies will grow if a replica is made onto a plate with minimal medium supplemented with leucine?

   C. What colonies will grow on a replica plate containing minimal medium plus arginine?

3. Let A^+, B^+, C^+, D^+, E^+ represent markers enabling an *E. coli* cell to manufacture amino acids required for growth on a minimal medium. *E. coli* Strain 1 is $A^+B^+C^-D^-E^-$. Strain 2 is $A^-B^-C^+D^+E^+$. The two strains are mixed and incubated together. After plating on minimal medium, a small percentage of prototrophs is selected.

   A. What would the genotype of the prototrophs be with respect to these markers?

   B. Two other *E. coli* strains, Strains 3 and 4, can manufacture all the amino acids required for growth on minimal medium. Strain 3 is Lac$^+$ and can consequently use lactose, a sugar that can be used as an accessory energy source. Strain 4 is Lac$^-$ and cannot use lactose when this sugar is provided. Which of these strains is auxotrophic? Explain.

4. In *E. coli*, five different single-point mutations resulted in the origin of five different auxotrophic strains, each with a genetic block in the biochemical pathway leading to the synthesis of an essential substance. An auxotrophic mutant can grow if the minimal medium is supplemented with one or more substances: A, B, C, D, E, and F. From the following data, determine the sequence of steps in the biochemical pathway and indicate the step at which each auxotrophic strain is blocked (+ = growth; 0 = no growth).

STRAIN	MEDIUM SUPPLEMENT					
	A	B	C	D	E	F
1	+	+	0	0	+	0
2	0	+	0	0	0	0
3	0	+	+	0	+	0
4	0	+	0	0	+	0
5	0	+	+	0	+	+

5. Two reciprocal crosses are made:

Strain 1	Strain 2
1. $A^+B^+C^-D^+E^-$	$A^-B^-C^+D^-E^+$
streptomycin resistant	streptomycin sensitive
2. $A^+B^+C^-D^+E^-$	$A^-B^-C^+D^-E^+$
streptomycin sensitive	streptomycin resistant

After incubation, plating is made on minimal medium containing streptomycin. From cross 1 colonies develop, but none arises from cross 2. Explain.

6. Explain the differences between the results of the following two crosses in *E. coli*:

   A. F^+ cell having chromosome markers $T^+L^+Pro^+Lac^+ \times F^- \rightarrow$ . . . F^+ cells with no chromosome markers transferred.

   B. Hfr cell having chromosome markers $T^+L^+Pro^+Lac^+ \times F^- \rightarrow$ . . . F^- cells with chromosome markers transferred.

7. Let A^+ and B^+ represent markers for essential growth substances needed for survival on minimal medium. The other markers, L^+, M^+, and so on, are for traits that do not affect the minimal nutritional requirements. From the cross and the information given here, arrange the marker alleles in the proper order of entry.

Hfr $A^+B^+L^+M^+N^+O^+Q^+ \times F^-$ $A^-B^-L^-M^-N^-O^-Q^-$
strep. sensitive   strep. resistant

Colonies on minimal medium with streptomycin: A^+—100%; B^+—100%; L^+—25%; M^+—88%; N^+—none; O^+—73%; Q^+—35%

8. Markers are transferred from five Hfr strains to F^- cells in the following order:

Hfr strain	Order of entry
1	B K A R M
2	D L Q E O C
3	O E Q L D N
4	M C O E Q L D N
5	R A K B N

   A. Draw a map showing the sequence of these markers on the chromosome.
   B. For each strain, indicate on the map the site of insertion of the fertility factor by placing an arrowhead so that the first gene to be transferred by a strain is behind the arrowhead.

9. Assume that Hfr recombinants are desired and that crosses are made with the Hfr strains in Question 8. For each strain, indicate which marker should be selected to derive the highest number or recombinants that will also be Hfr. Explain.

10. Assume a DNA of an Hfr *E. coli* has the following arrangements of nucleotides:

$$5' \ A\,G\,C\,T\,A\,T \ 3'$$
$$3' \ T\,C\,G\,A\,T\,A \ 5'$$

Assume the Hfr cells have been growing on medium supplied with ^{14}C and ^{15}N so that all its DNA is heavy. Mating of the Hfr with F^- cells is performed on ordinary medium. In the mating, the lower DNA strand is transferred from its 5' end.

   A. Show this segment of the double helix in the donor cell after DNA replication, indicating light (L) and heavy (H) strands.
   B. Do the same for the recipient, assuming this whole strand is incorporated.

11. A certain Hfr strain normally transmits the lac^+ marker as the last one during conjugation. In a cross of this strain with an F^-lac^-, some recombinant lac^+ cells received the lac marker too early in the mating process. When these lac^+ cells are mixed with F^-lac^- cells, the majority of the latter are converted to F^+lac^+, but other markers are not transferred. Explain. What can you say about the genotypes of the cells from the F^- strain that became lac^+?

12. Suppose Strain A of Pneumococcus is streptomycin resistant and Strain B is streptomycin sensitive. In a transformation experiment, competent B cells are incubated with DNA from Strain A. DNA of Strain A carrying the resistant markers enters some B cells and associates with the B DNA. The latter cells become transformed. Answer the following:

   A. Will the DNA of the transformed recipient have A or B DNA for the marker region?
   B. When any transformed cell divides, what will be the nature of its two cell products with regard to streptomycin resistance, and what will the DNA of these cells be like regarding the A and the B DNA?

13. In a transformation experiment, donor DNA is randomly broken up into fragments in the process of extraction. Therefore, genes closely linked will be included on the same fragment more often than they will become separated onto different fragments. Genes that are far apart from each other will tend to become separated onto different fragments during extraction of the donor DNA.

   DNA from a donor strain of a species of Streptococcus was used to transform recipient cells. The donor cells are a^+b^+ in relation to two growth requirement substances, whereas the recipients are doubly defective nutritional deficients, a^-b^-. Three types of transformants arose in the following frequences:

Type of transformed cell	Number of transformed cells
a^+b^+	650
a^-b^+	155
a^+b^-	95
	900

   A. Based on this information, would you say that the two genes are closely linked or far apart? Explain.
   B. What appears to be the distance in map units between the two genes?
   C. Why do the recombinant classes occur in unequal numbers?

14. Suppose the following sequence is a target sequence in a recipient cell:

$$5' \ \underline{A\,T\,C\,G\,T} \ 3'$$
$$3' \ T\,A\,G\,C\,A \ 5'$$

Let the following represent sequences in a transposon that recognizes the particular target sequence (the transposon inserts in the sequence X Y Z, as shown):

$$5' \ \underline{X\,Y\,Z} \ 3'$$
$$3' \ X'Y'Z' \ 5'$$

Show the arrangement of the DNA after the transposon is inserted.

15. Allow A and B to represent determinants for drug resistance that are found on a conjugal plasmid in certain *E. coli* strains. C and D represent drug-resistance determinants that occur on two different nonconjugal plasmids. E and F are genetic drug-resistance determinants

found on the chromosome. For each of the situations given here, tell the likely maximum number of resistance determinants transferred to any cell that completely lacks them:

A. F^+ ABCDEF mixed with F^-.
B. F^- ABCDEF mixed with F^-.
C. F^- CDEF mixed with F^-.
D. Hfr ABCDEF mixed with F^-.

10. A strain of bacterial cells carries genetic factors for streptomycin and tetracycline resistance. Both resistance factors are regularly transferred together, on contact, to sensitive cells. Cells are isolated from the original strain that are still resistant to both antibiotics but do not always transmit both determinants. Some recipient cells become resistant only to streptomycin and in turn transmit that drug resistance determinant by itself to other sensitive cells. Other cells isolated from the strain are resistant only to tetracycline, and these no longer transmit resistance of any kind to other cells. Explain these observations.

17. In a certain F^+ *E. coli* strain, a factor for streptomycin resistance is transferred along with the factor for chloramphenicol resistance on a compound R plasmid. Hfr cells arise in this F^+ population, and these are isolated and found to be resistant to both antibiotics. However, it is found that some of the Hfr cells contain an R plasmid that transfers only the determinant for chloramphenicol resistance. The factor for streptomycin resistance is transferred only when chromosomal markers are transferred to an F^- cell. Explain these findings.

16

VIRUSES AND THEIR GENETIC SYSTEMS

Bacteria and the viruses that infect them (bacterio-phages) have proved to be extremely valuable tools for the molecular biologist. Because of their rapid reproductive rate, microorganisms enable the geneticist to examine in a very short period of time many generations consisting of extremely large numbers of individuals. This has made possible a refinement of genetic analysis that would be impossible with higher forms. In addition, studies of viruses, phages in particular, have disclosed mechanisms of genetic recombination and control that were unsuspected by the classic geneticist. Research with viruses has led to the development of methodologies that have been extended to permit probing into the genetic systems of prokaryotes and eukaryotes. Moreover, the interactions between viruses and their host cells have contributed to an understanding of aspects of cellular differentiation and cell pathology.

Events following infection by phages

It has been noted throughout several chapters that phages can destroy the bacterial hosts they infect. These are *virulent* phages that undergo a lytic cycle resulting in death and lysis of the host cell. Phages T2 and T4 of *E. coli* are examples of virulent phages, all of which must encounter a supply of host cells to continue their reproduction. We have seen that it is the nucleic acid of the phage that enters the host cell; the protein envelope is left behind (refer to Fig. 11-4). Once inside the cell, the phage particle takes control of the cellular machinery. One of the first "new" substances to appear after infection is an mRNA that is complementary to the DNA of the virus, *not* to that of the host. This is followed by the appearance of proteins that represent certain enzymes needed for the activities of the virus. Meanwhile, the normal activities of the host cell cease. The bacterial DNA becomes degraded and enters a pool in the cell. Substances from the cell medium are added to this pool, from which virus-specific substances will be made. In short, after entering the host cell, the genetic material of the virus subverts that of the bacterium (refer to Fig. 12-6). By the ability of its nucleic acid to undergo transcription in the host cell, virus-specific mRNA is formed. When this is translated by the ribosomes of the host, specific viral enzymes are constructed that are needed to build new viruses. Following the appearance of the enzymes, products foreign to the cell then form in the host cell. All of this is accomplished by means of the ability of the virus DNA to be transcribed and its mRNA to be translated in the host cell.

In the first few minutes after entry, no mature virus

particles can be detected in the bacterial cell because the preliminary activities needed for virus construction are going on. After the degradation of the nucleic acid of the host, construction of the phage DNA begins. This occurs approximately 6 minutes after infection. After the appearance of the protein, which has an enzymatic function, another class of protein begins to appear. This is the protein that will form the envelope of the completed viruses. Finished phage particles may appear in the cell approximately 12 minutes after the infection, and their number will increase until lysis of the cell occurs. The exact time of this cell burst depends on the particular bacterial host and the strain of the virus. Approximately 1/2 hour after infection by a phage, the cell bursts and releases 100 or so mature phage particles.

The lysis results not just from pressure of the virus particles in the cell, but apparently from the activity of an enzyme (lysozyme) in the tail of the virus, which can break down the cell wall. This same enzyme also enables the phage to inject its DNA into the cell. After lysis, the released phage particles are free to enter other susceptible cells and repeat the process. It is evident that the virus is much more than just a simple association of nucleic acid and protein. Indeed, its ability to use cell machinery for the complex reactions necessary for its own reproduction implies a high degree of specialization. This must have involved a long history in which the forces of evolution acted on both the virus and the host. Figure 11-3 shows the virulent virus, phage T4, that infects E. coli. Although not cellular in structure, it is more complex than was first suspected. The DNA is stuffed into the protein of the headpiece during maturation of the virus in the host cell. The tailpiece, which is needed to attach the infecting phage to the cell, is associated with a plate equipped with prongs. The contraction of the tailpiece causes the viral DNA to pass through the hollow core into the bacterial cell.

Not all phages, however, are virulent like phage T4. Many can infect cells and take up residence in them without causing lysis or cell destruction. Such a nonvirulent virus residing in a bacterium is called a *symbiotic* or *temperate phage*. The bacterial cell remains unharmed by the presence of the phage and may actually receive some benefits from the particle it contains. If we artificially burst open a bacterial cell that harbors a temperate phage, we cannot detect any mature phage particles. The virus seems to have disappeared. Yet its presence in the cell can still be demonstrated. Assume we have a strain of bacterial

cells that contains a temperate virus. Although we find no mature virus in the cells, the fluid in which the cells are growing will always contain some mature phage particles. The reason for this is that in a few rare cells of the bacterial strain, the virus becomes virulent, enters a lytic cycle, and causes lysis of its host. We can detect these released particles if we bring the culture fluid into contact with an indicator strain of bacteria. The latter would be a strain that is susceptible to the phage. If a drop of the culture fluid is added to a cloudy tube containing the susceptible cells in broth, the broth clears because of the destruction of the sensitive cells by the virus. Many bacteria normally contain unsuspected virus of some kind whose presence is revealed by contact with a sensitive strain.

Phage lambda and temperate phages

Phage lambda (λ) of *E. coli* has been extensively studied, and a great deal of information has been assembled on events following its infection of a bacterial cell. After entry, the viral DNA may undergo a lytic cycle in the cell or it may behave as a temperate phage. Which pathway it follows depends on a balance of several regulatory proteins, a topic discussed in some detail in Chapter 17.

The viral DNA is replicated at first as a circle during a lytic pathway. Intermediate theta-shaped DNA molecules arise, as when the *E. coli* chromosome replicates (see Fig. 11-20). When rapid replication of the viral DNA ensues later in the lytic cycle, replication takes place through the rolling circle method (see Figs. 15-11 and 15-12). The lambda DNA that is to be packaged into the empty protein envelopes occurs in the form of a concatamer (Fig. 16-1A). This is a chain or linear series of genomes linked together. Figure 15-11 shows how more than one genome's worth of DNA can arise using the rolling circle method. An enzyme recognizes certain nucleotide sequences of the λ DNA that define the ends of each genome in the concatamer. Snipping of the enzyme may result in slightly more than one genome's worth of DNA being inserted into a protein envelope (Fig. 16-1B). Concatamers occur in the replication and maturation of certain other phages, such as phages T1, T7, and T4. They may arise through methods other than the rolling circle method of replication.

The DNA that is packaged into a lambda envelope is linear, and it enters the *E. coli* cell in this form. However, it soon assumes the form of a circle. This

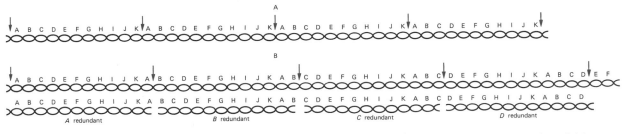

FIG. 16-1. Concatamers and their processing. *(A)* The concatamer consists of a series of genomes (each A–K) linked together. Snipping *(arrows)* by an enzyme results in the formation of separate genomes. Only one genome's worth of DNA (A–K) will be packaged in phage lambda into a protein envelope. *(B)* The snipping of the concatamer in phage T4 can start at a random point and produce slightly more than one genome's worth of DNA. There is thus terminal redundancy, a repetition of genes at each end. As a result, certain genes are represented twice in the T4 viruses, and the duplicated genes differ from one mature virus to the next.

circularization can occur because the linear DNA bears 12-nucleotide-long, single-stranded projections at its 5′ ends (Fig. 16-2). We say the ends are "sticky" or "cohesive," because they are complementary and base pairing can take place between them. DNA ligase activity seals the nicks, and the circular form of the virus is stabilized. The sticky ends are generated as a result of staggered, single-stranded cuts made by the enzyme involved in snipping.

In its circular form inside the *E. coli* cell, the lambda DNA is replicated if a lytic pathway is followed (Fig. 16-3). If phage lambda behaves as a temperate virus after entering the cell, however, the circular form of the DNA does not undergo replication. Instead, it inserts into the chromosome by a recombination process, in much the same way that the F factor of *E. coli* becomes inserted into the Hfr strain, although IS sequences are not involved (see Fig. 15-18). Instead,

FIG. 16-2. Linear and circular forms of phage lambda. Phage lambda enters the *E. coli* cell as a linear molecule with single-stranded extensions *(red)* at the 5′ ends *(above)*. The linear molecule can circularize *(below)* as a result of the complementarity of the ends. DNA ligase seals at the positions of nicks *(arrows)* to form an intact circle.

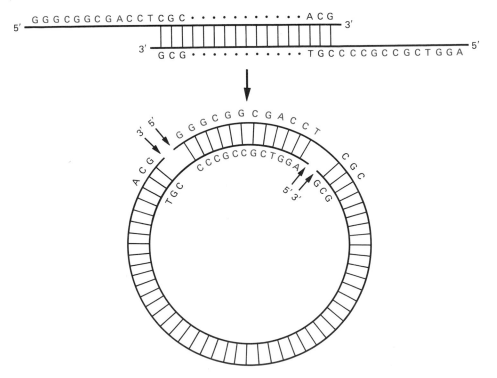

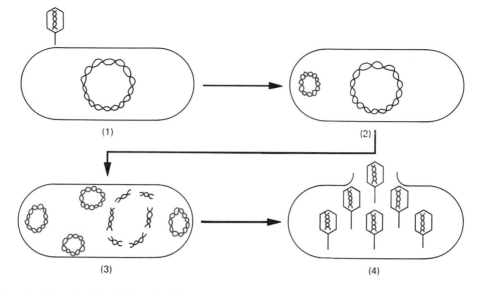

FIG. 16-3. Lytic cycle of phage lambda. *(1)* Phage lambda occurs in a linear form in the infective virus. The injected DNA circularizes after entry into the cell *(2)*. The virus replicates as circles *(3),* while the host DNA is degraded. The infective phages released from a lysed cell contain the DNA in its linear form *(4)*.

specific attachment sites play a role. One of these is located on the bacterial chromosome and is designated att B. This site is located next to the bacterial gal locus, a region concerned with the ability of the cell to produce an enzyme needed for using galactose. The corresponding attachment site, att P, occurs on the phage DNA (Fig. 16-4). Each attachment site—att B on the bacterial DNA and att P on the phage DNA—contains a homologous segment, and it is within this homologous portion of the DNA that recombination takes place, resulting in the insertion of the phage into the chromosome as a linear sequence. An enzyme coded by a phage gene can recognize the attachment site. This phage protein, along with a protein factor coded by the host DNA, is able to bring about the recombination event that results in the integration of phage λ. We see that a bacterial cell harboring an integrated virus gives no visible evidence of the virus. In its reduced state, the temperate phage is called *prophage.* The cell contains only one intact prophage of a given kind, such as λ, because the presence of an integrated virus renders the cell immune to further infection by the same kind of virus.

In its integrated state, the temperate virus behaves in every way as if it were part of the bacterial replicon. In an occasional cell, the virus may come out of the chromosome by a reversal of the process of insertion (Fig. 16-4). This excision process depends on the two proteins that are also required for insertion, as well as a third protein that is coded by a phage gene. Excision from the chromosome may come about spontaneously, or it can be stimulated to do so by certain agents, such as ultraviolet light. Most of the time, the temperate, integrated virus is prevented from leaving the chromosome and thus from becoming detrimental to the cell by a repressor substance, coded by the virus itself (Chap. 17). If this repressor is somehow prevented from acting, as from the effects of ultraviolet radiation, the genes of the virus that are concerned with viral replication will be expressed. The result is replication of the virus, cell lysis, and the entry of mature phage particles into the medium. These may then take up residence in other appropriate host cells, or they may destroy cells that are sensitive to them.

Since the bacterial cell with the temperate phage contains information to make virus that can cause lysis of sensitive cells, it is called *lysogenic.* The establishment of an integrated state between viral and bacterial DNAs is known as *lysogeny.* A lysogenic strain, therefore, has the potential to make virus and to bring about lysis of cells from an indicator strain. On the level of microorganisms, we find many different strains of cells that vary in their viral susceptibility as well as different strains of viruses that vary in their virulence. A spontaneous mutation may arise that

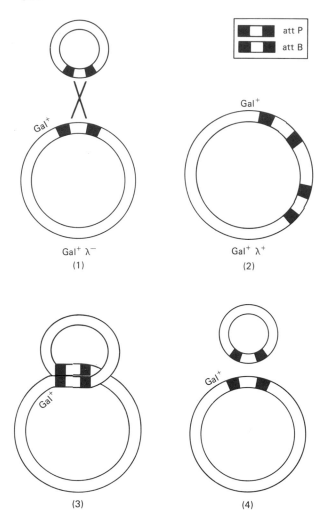

FIG. 16-4. Integration and excision of phage lambda. Phage lambda DNA *(red)* and the host cell DNA *(black)* contain corresponding attachment sites, att P and att B, respectively. Both sites contain a core region that is a DNA segment homologous to both sites. It is within this region that recombination takes place *(1)*. The attachment sites are close to the gal locus of the cell. Integration converts a λ^- cell to one that is λ^+ *(2)*. The integrated virus (a prophage) on rare occasions undergoes a reversal of the process of integration *(3)*. Pairing between the two attachment sites, followed by recombination, produces a chromosome free of the virus plus an intact phage *(4)*. The latter can now replicate and enter the lytic pathway, ending with destruction of the host cell.

changes a sensitive bacterial cell to one resistant to a specific virus. In turn, mutation in a phage gene can make the virus still more virulent to the bacterial strain. A mutation in a phage gene may even change the virus to a less virulent or even temperate one. The establishment of a symbiotic relationship between a bacterium and its temperate phage is the product of

mutation operating along with the force of natural selection.

The closeness of the phage–bacterium relationship becomes apparent when we realize that a given virus such as λ, in its prophage state, resides not at just any chromosome location, but at a very specific site, unlike the F factor, which can insert at any of several sites. As noted in Figs. 15-8*B* and 16-4, the attachment site for phage λ is near the bacterial marker Gal. Phage λ may or may not be present in a cell at this site, depending on the bacterial strain. A Hfr lysogenic strain harboring λ at the attachment site near Gal is designated *E. coli* (λ). A Hfr cell that is λ^- can transmit the absence of lambda to a λ^+ cell, just as if the lambda state (presence or absence of lambda phage) were inherited as a specific bacterial gene (Fig. 16-5*A*). If a Hfr λ^+ cell is mated to an $F^-\lambda^-$, the chromosome of the former may penetrate into the F^- cell and transfer phage. However, the prophage is immediately activated when transmitted in this way to an $F^-\lambda^-$ cell (Fig. 16-5*B*). The virus removes itself from the chromosome and causes destruction of the F^- cell. Thus, some phages can be induced to become virulent as a result of their transfer during conjugation to cells lacking them. The activation of the virus to a virulent state by transfer during conjugation is called *zygotic induction*. Not all phages, however, may be induced. A host cell may contain one or more different kinds of viruses in their prophage state, each inserted at a characteristic location on the chromosome and each behaving as if it were a bacterial locus.

Phages and transduction

When the virus in its prophage state removes itself from the chromosome, it may carry with it an adjacent piece of the host chromosome. Phage lambda may on rare occasions (1 in 10^6 cells) pick up a Gal^+ marker from a cell (follow Fig. 16-6 in this discussion). As it does this, it leaves behind a portion of itself that remains inserted in the host chromosome. The virus, now containing a bacterial gene, may be released into the medium, where it is free to enter an appropriate cell, containing no lambda prophage. The virus may insert itself as prophage in the second cell. As a result, it may confer a new property on the host. If the latter, for example, is Gal^- on the chromosome, it may acquire the ability to use galactose because of the presence of the Gal^+ allele associated with the prophage that it is now harboring. This transfer of genetic

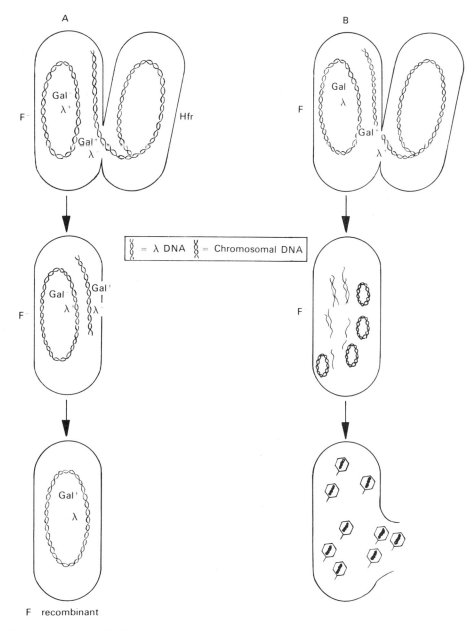

FIG. 16-5. Transfer of λ prophage. (A) A prophage occupies a definite location on the bacterial chromosome and may behave like an ordinary bacterial gene. The phage λ occupies a position adjacent to the marker, Gal. In conjugaton, the absence of λ (λ⁻) state may be transferred to the F⁻ cell, just as any other marker. Recombinants can arise which are λ⁻. (B) The prophage cannot be transferred from the Hfr in the λ⁺ condition without causing lysis of the F⁻ cell. This is because of zygotic induction, in which the phage, on entering the F⁻ cell, becomes activated to manufacture more λ virus. Lysis of the cell results with release of mature λ phage.

material from one cell to another by way of a virus is called transduction. The phenomenon of transduction was discovered in 1952 by Zinder and Lederberg working with another species of bacterium, *Salmonella typhimurium* and another phage, phage P_1.

It is now well known that variations occur in trans-

duction, depending on the host and the phage under consideration. Phage (λ), which has been extensively studied in *E. coli*, illustrates specialized transduction. The behavior of this virus brings to light several interactions that can take place between the host cell and the virus. When transduction takes place through

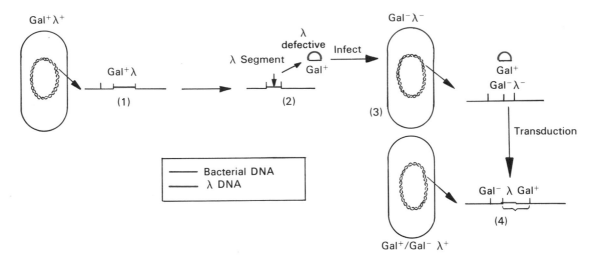

FIG. 16-6. Prophage and transduction. Phage λ resides adjacent to the Gal locus (1) in a λ$^+$ cell. When it comes out of the prophage state and deintegrates from the chromosome, it may carry the Gal marker with it (2), but in so doing, it leaves a portion of itself behind. The phage is now defective, having lost some of its abilities. If the detective virus, carrying a Gal$^+$ marker infects a Gal$^-$ cell that does not contain λ (3), it may insert itself in the Gal region. The cell is transduced (4) and contains the defective λ plus the Gal locus in the diploid condition. The cell is a heterogenote. The defective nature of λ is seen in the fact that, by itself, it cannot remove itself from the chromosome and reproduce. This function was lost when the phage left a segment of itself behind (2) and incorporated the Gal locus.

a virus such as lambda which has existed as a pro-phage in a previous host cell, the second host becomes diploid for a very small segment of the genetic material, the Gal region in our example. The cell is a partial diploid, heterozygous for just a small portion of the genome. We again use the term *heterogenote* for such a condition.

Since the virus left a part of itself inserted in the chromosome of the first host as it picked up the Gal marker, we might expect the virus to show some alteration in its behavior. And indeed, the virus is now changed. This is seen in its inability to become infective again; it cannot cause cell lysis or be released into the cellular environment. The defective prophage can be liberated again, however, if the cell also contains an intact λ. A transduced bacterial cell (one that has received a marker by means of a virus) may contain both the defective transducing phage with the marker allele and an intact phage containing all of its own genetic information (Fig. 16-7). The latter is able to supply the information missing in the former and can bring about the lysis of an occasional cell. Such transducing cells are immune to further infection by λ particles. Bacterial cells from a strain that has been transduced may thus undergo lysis and contribute to the culture medium both kinds of phages, the defective ones carrying a bacterial gene and the intact ones with all the needed virus information to effect lysis. Both types of particles, in turn, would be able to enter other cells of the same host strain and again take up residence.

It should be apparent that transduction by a virus clearly resembles sexduction—transfer of a bacterial gene by a sex factor. Indeed, the two processes are so similar that they may be considered aspects of the same phenomenon. The F factor, as well as the phage, may also leave a piece of itself behind when it picks up a bacterial gene. We may consider the phage to be an episome, because it satisfies the definition in the same way as the F factor.

Another kind of transduction is generalized transduction, which can be effected by phage P_1 in *E. coli* and by phage P_{22} in Salmonella. As the name implies, any region of the host chromosome may be involved and thus be carried over to another bacterial cell by the phage particle. This is quite different from specialized transduction in which a specific phage particle such as phage λ is able to carry away to another cell only a specific bacterial marker gene, such as Gal.

Another difference relates to the manner in which the bacterial DNA segment is incorporated into a phage particle. Recall that in specialized transduction, the phage carries away a portion of the chromosome during the process by which it emerges from the chromosome. In the case of generalized transduction,

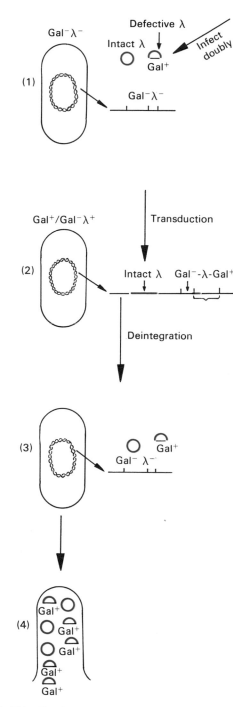

FIG. 16-7. Double infection. A bacterial cell may become infected simultaneously with an intact virus and a defective one carrying a marker (1). Both viruses take up residence on the chromosome, and the cell is transduced and becomes a heterogenote (2). In a rare cell, the two viruses may deintegrate (3). This is possible only because the normal intact virus is present to supply the function missing in the defective phage. Both types of phages are able to reproduce, and the cell lyses (4), releasing the two types of particles.

the fragment of host chromosome becomes incorporated into the headpiece of the virus during the lytic cycle of the phage. Remember that when a virulent virus is completing its lytic cycle the host DNA is broken down and enters a DNA pool. A piece of bacterial DNA that has not been completely degraded to its constituent nucleotides may be packaged accidentally into the headpiece. It appears that P_1 particles capable of generalized transduction carry very little, if any, phage DNA. As a matter of fact, experimental evidence indicates that at times only bacterial DNA may become packaged into the headpiece of certain kinds of phage particles.

When such a phage particle encounters another bacterial cell of the proper strain, the DNA fragment of the first bacterial host is injected. This fragment would contain a region homologous to a region on the second host's chromosome. If a crossover event takes place between the two regions, the second host may become a recombinant cell if the donor DNA carries an allelic form different from that present on the homologous region of the recipient.

Unlike specialized transduction with phage λ, the P_1 transducing particles do not contain defective viral DNA integrated with bacterial genes. These particles do not undergo multiplication in the presence of any helper phages, and the transduced cells, the transducants, are not immune to infection by other P_1 particles, as in specialized transduction.

It has been demonstrated that the very presence of a phage in a bacterium may impart a new characteristic to the cell, one that is not due to a gene introduced from another bacterium. For example, certain nontoxic strains of *Corynebacterium diphtheriae* gain the ability to produce the virulent diphtheria toxin when they are infected with particular phages. The production of the toxin by the cell does not result from the presence of any bacterial genes that have been carried by the virus; rather, it depends on both the genetic machinery of the cell and that of the phage. If the phage is lost from the cell, so is the ability to form the toxin. This is an example of conversion, the acquisition of an inherited cell property as the result of the interaction of host and virus genes.

Recombination and the genetic system of phages

What kinds of characters can be followed when the genetic system of a virus is being investigated? Since mutations occur in viral genes just as they do in higher forms, the study of the inheritance of a mutant

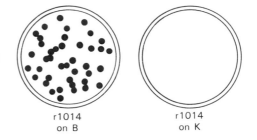

| Wild | r1017 | r1014 | r1014 |
| on *E. coli* B and K | on B and K | on B | on K |

FIG. 16-8. Plaque formation. Viruses may differ in the type of plaque they form and also in the range of hosts in which they can reproduce. Phage T4 produces the small, wild-type plaques on *E. coli* strains B and K. Certain mutants (r mutants) produce large plaques on B and K. Still others, the rII mutants, produce the large mutant plaques on B but no plaques at all on K.

allele and its standard form permits genetic analysis of the virus and the mapping of its genes. We have noted that viruses grow only in certain host strains. Phages thus have characteristic host ranges. A mutation, however, may widen or narrow that range. For example, the *T-even* phages can infect and lyse *E. coli*. A variant form of these viruses can infect and lyse strain B of *E. coli* but cannot cause lysis of strain K. Host range is thus a genetic characteristic that can be recognized in viruses. Another common one is the type of plaque produced by a virus when it infects a lawn of bacteria spread on an agar dish (Fig. 16-8). Each plaque, a clear zone on the bacterial layer, represents an area where a single infecting virulent phage originally entered a bacterial cell and was reproduced, followed by cell lysis and the release of new phage particles. These, in turn, infected other bacterial cells in the area and continued the process, resulting in the formation of a visible plaque or hole in the bacterial lawn. The wild-type plaque is quite distinct for a given type of virus. It may be small with smooth edges, whereas a mutant form may be large. Some plaques are cloudy; others are clear. Another trait that is easily followed is the rapidity with which the virus bursts the host cells.

Studies conducted in the late 1940s by Hershey and Doermann and several others brought out the features typical of phage crosses. Suppose host range and speed of lysis are being followed. The wild allele (h^+) enables the virus to grow on E. coli strain B but not on E. coli B/2. The mutant allele (h) increases the host range to include the B/2 strain. The wild allele (r^+) determines small plaques, whereas its allele (r) results in large plaques as a consequence of very rapid lysis. To determine whether or not recombination can take place between the loci h and r, we may

proceed as follows (Fig. 16-9A). The two bacterial strains, B and B/2, are mixed and then spread on agar plates. We know that if a virus is h^+r, it can grow only on strain B and will produce large plaques. We can easily identify plaques produced on a plate containing both B and B/2 bacteria. The virus will lyse only the B cells, but the large plaques will be turbid or cloudy because they will contain unlysed cells of strain B/2. On the other hand, a virus that is hr^+ will produce small clear plaques because r^+ determines slower (wild) lysis and the h allele enables the phage to lyse both B and B/2 cells. The resulting plaques will contain no intact cells and therefore will be clear.

A mixture of the two parental virus types (h^+r and hr^+) is allowed to infect a liquid culture of strain B host cells (Fig. 16-9B). Both parental viruses can infect and lyse the B cells. This B culture is then diluted by liquid broth and allowed to undergo lysis by the phages. The viruses that are released after lysis of the B cells are then plated on a mixture of fresh host cells, this time both B and B/2. By plating on this mixture, we can detect virus offspring with normal host range (turbid plaques) and mutant (clear plaques). The plaques that form are then examined for the size and clearness. The two types caused by the two parental phages are seen (large, turbid and small, clear types). But we also find some that are large and clear and others that are small and cloudy. These latter two types must result from the presence of viruses with the respective genotypes, hr and h^+r^+. This tells us that recombination has occurred. We can express the percentage of recombination by using the same reasoning as was used to determine crossover frequency in higher organisms. The proportion of the number of recombinant offspring, hr and h^+r^+ (as measured by features of the plaques), to the num-

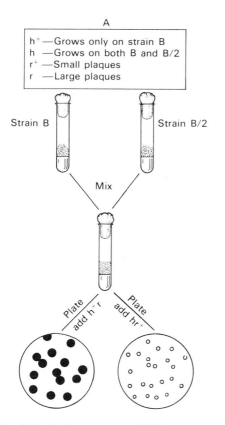

FIG. 16-9. Recombination in phages. (A) Mixture of strains B and B/2 can be used to identify the kinds of phages present. The h⁺r type produces large, turbid plaques, because it produced rapid lysis on strain B but cannot grow in B/2. Therefore, the plaques are cloudy because of the unlysed B/2 cells. The hr⁺ produce wild-type (small) plaques, but they are clear, because both B and B/2 cells are lysed.

(B) One can determine the amount of recombination in this procedure by looking at the type of plaques. Again a mixture of B and B/2 is used as an indicator, and the four types of plaques are easily identified. The amount of crossing over is determined by counting the total numbere of plaques and expressing the number of recombinants as a percentage.

ber of total progeny (the total of all the plaques) is obtained. The percentage value tells the recombination frequency, just as it would if we were studying eye color and body color in fruit flies.

The classic genetic principles used in making three-point testcrosses (Chap. 8) may also be applied to phages. Suppose that three loci are being mapped: m, r, and tu (Table 16-1). Again, plaque characteristics are examined. From the data, it can be seen that the most numerous type of plaque and the least numerous represent the parental types and the double crossovers, respectively. The locus r is established as the one in the middle, because it appears to have switched position. Establishment of regions 1 and 2 allows us to determine the corresponding crossovers in the two regions. Again we must add the frequency of doubles to each class (3.3 here). And so the map becomes

m   12.9   r   20.8   tu

By applying detailed linkage analysis to phage T/2 of E. coli, it was found that all of the genes fall into one linkage group and that the map is circular. The circularity of the map, however, does not mean the DNA of the virus "chromosome" is necessarily in the form of a circle as in E. coli.

Actually, the T/2 and T/4 chromosomes are rod-shaped in the mature virus particle, even though the map of the chromosome is circular. The reason for this is that the DNA that is to be packaged, as in the case of phage lambda, occurs in the form of a concatamer. As a result of random snipping of the concatamer, however, terminal redundancy occurs at each end of the chromosome (see Fig. 16-1A and B). The

TABLE 16-1   Progeny from three-factor linked cross in phase T/4[a]

CATEGORY	CLASS	GENO-TYPE	TOTAL PLAQUES	PER-CENT-AGE
Parental	A	m r tu	3467	33.5
Parental	A	$m^+ r^+ tu^+$	3729	36.1
Single crossover	B	$m r^+ tu^+$	520	5.0
Single crossover	B	$m^+ r tu$	474	4.6
Single crossover	C	$m r tu^+$	853	8.2
Single crossover	C	$m^+ r^+ tu$	965	9.3
Double crossover	D	$m r^+ tu$	162	1.6
Double crossover	D	$m^+ r tu^+$	172	1.7

[a]A, noncrossovers; B, crossover between m and r; C, crossover between r and tu; D, double crossover; m, minute; r, rapid lysis; tu, turbid.

Source: Reprinted with permission from A. H. Doermann, Cold Spring Harbor Symp. Quant. Biol. XVIII: 4–11, 1953.

part of the chromosome that is repeated varies from one phage particle to the next. Since mapping the chromosome entails a large population of phage particles, it will always seem as if any gene in the sequence is next to the following one. No ends will be detected as a consequence of the terminal redundancy.

Repeated mating in phages

Although we may build up genetic maps employing the familiar logic applied to higher organisms, close inspection reveals several unusual features peculiar to phage systems. When the crosses are performed in the manner described earlier, a tremendous number of bacterial cells and phage particles are obviously involved. Techniques are available, however, that enable us to make critical dilutions. From these, we may study the offspring phages coming from the burst of a single bacterial cell that had been infected by two types of phage particles, let us say h^+ r and hr^+. When such dilutions are made, it is found that the recombinant classes, hr and h^+r^+, are not present in equal amounts. On the basis of our understanding of crossing over, we would expect the recombinant classes to be reciprocal and hence equal in number. In the case of phages, these are usually found to be very

unequal. Moreover, if a cell is infected with three different kinds of virus particles, such as m r^+ t^+, m^+ r t^+, and m^+r^+ t, we might find offspring that are m r t. For such a combination to arise, three parents must be involved! How can such an observation be explained?

Some clarification is provided by bursting the infected cells artificially several minutes before they would lyse naturally. When this is done, it can be demonstrated that recombinant viruses can exist in the cell many minutes before the natural cell burst. As time increases toward natural burst, the number of recombinants also increases. In other words, the acts involved in recombination start well in advance of bursting of the host cell. This means that before lysis, more than one round of mating and recombination can take place. According to the theory of repeated mating, the phages enter the cell and then multiply. The phage units in the cell can now mate at random repeatedly in pairs, so that several rounds of mating and multiplication take place. Exchange of genetic material by crossing over takes place and produces recombinants. If reciprocal recombinants are formed early in the mating process, they may be changed by participating in later mating. Therefore, the phage offspring coming from a lysed cell represents several generations of multiplication. We can now understand how three parents may be involved in the production of a phage genotype (Fig. 16-10). A mating of m^+r t^+ in one round of multiplication can give $m^+r^+t^+$ and m r t^+. If the latter mates with m^+r^+t in the next round of mating, the genotype m r t can arise.

When we study the products of many cell bursts after cell infection by two different phage types (h^+r and h r^+), we are unaware that multiple rounds of mating have taken place. Equal numbers of reciprocal recombinants are found among the collection of offspring from the bursting of the entire population cells as a result of randomness. Since an extremely large number of phage units is involved, the recombinant types will be produced in comparable amounts.

Phage systems and the recognition of the cistron

One of the greatest contributions of phage analysis is the revised concept it has given us on the nature of the gene. Previous to the 1950s, classic ideas held the gene to be an indivisible unit. It performed a certain function. As a unit, the gene could mutate and engage

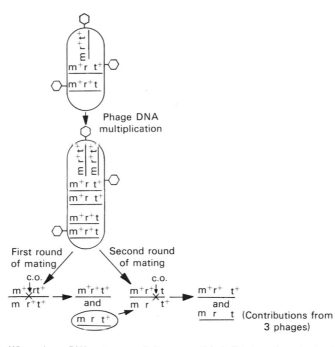

FIG. 16-10. Repeated mating. When phage DNA enters a cell, it multiplies. The resulting phage units may undergo several rounds of random mating during which recombination can take place. If a cell is infected with phages of three different genotypes, as shown here, offspring may arise which carry contributions from three separate phage units (in this case, m, r, and t were originally in three separate particles). This is easily understood when it is realized that more than one mating takes place in the infected cell before mature viruses are released. In this example, one of the products of the first round of mating (m r t⁺) participates in a second mating. The genotype mrt can thus be found in a mature phage released upon lysis of the cell.

in crossing over. When the gene mutated, all of it was somehow changed. Mutation at different sites within the gene was not considered possible. Nor was crossing over within a gene recognized. Because it was an intact unit, breakage and crossing over could occur on either side of it but not within it. Contradictory cases arose in the fruit fly as well as in other well-studied organisms, but these apparent exceptions were explained by coining new terms and adapting them to fit the classic, accepted idea of the gene.

The work of Benzer, undertaken in the 1950s, revolutionized the picture of the gene as the indivisible unit of function, mutation, and recombination. Benzer worked with the T-even viruses that infect *E. coli*. He was one of the first investigators to try to relate genetic units, such as units of crossing over and units of mutation, to the actual physical basis of DNA. This procedure requires the construction of very detailed chromosome maps. In turn, constructing such maps relies on the examination of extremely large numbers of offspring. We have seen that microorganisms are ideal for such purposes, especially phages, in which millions of progeny can be examined in a short time.

Benzer studied in detail the genetics of plaque formation in phage T4 of *E. coli*. The wild-type plaque, as seen on an agar plate, is a small clear area with rough edges. These wild-type plaques are produced when phage T4 is plated on either strain B or strain K of E. coli. Certain mutants arising in phage T4 are called rII mutants. When plated on strain B, they form a so-called r plaque, one that is larger than the wild type (r⁺) and has smooth edges. These rII mutants, however, produce no plaques on strain K. (Fig. 16-8 summarizes these points.) The rII mutants infect K cells and start the synthesis of phage DNA and other viral products, but they cannot produce infective offspring. As a result of this defect, they cannot cause cells of strain K to lyse.

This growth defect of the rII viruses on strain K provided Benzer with a very critical tool. One of his goals was to take different rII mutants that had arisen independently and cross them in various combinations. The point was to determine whether the nu-

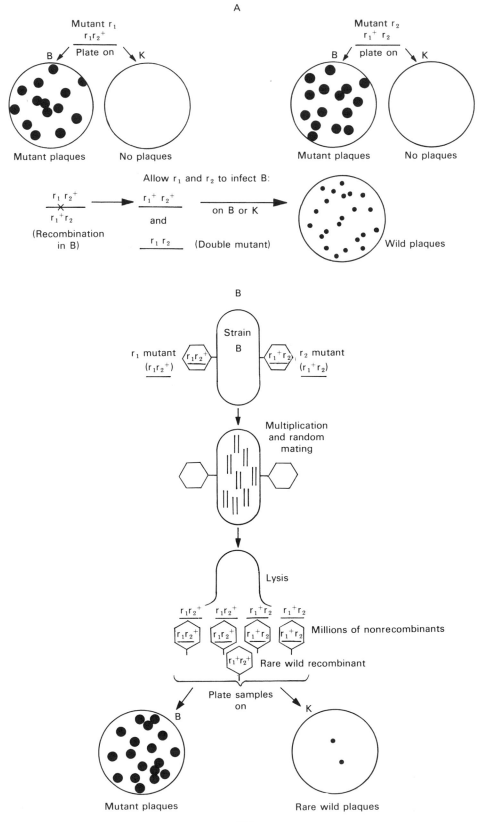

A

Mutant r_1
$r_1 r_2^+$
Plate on

B → K

Mutant plaques   No plaques

Mutant r_2
$r_1^+ r_2$
plate on

B → K

Mutant plaques   No plaques

Allow r_1 and r_2 to infect B:

$$\frac{r_1\,r_2^+}{r_1^+ r_2} \times$$

(Recombination in B)

→

$$\frac{r_1^+\,r_2^+}{}$$ and

$$\frac{r_1\,r_2}{}$$ (Double mutant)

on B or K →

Wild plaques

B

r_1 mutant
$(r_1 r_2^+)$

$\langle r_1 r_2^+ \rangle$

Strain
B

$\langle r_1^+ r_2 \rangle$

r_2 mutant
$(r_1^+ r_2)$

Multiplication
and random
mating

Lysis

$$\frac{r_1 r_2^+}{\langle r_1 r_2^+ \rangle} \quad \frac{r_1 r_2^+}{\langle r_1 r_2^+ \rangle} \quad \frac{r_1^+ r_2}{\langle r_1^+ r_2 \rangle} \quad \frac{r_1^+ r_2}{\langle r_1^+ r_2 \rangle}$$ Millions of nonrecombinants

$\langle r_1^+ r_2^+ \rangle$ Rare wild recombinant

Plate samples
on

B → K

Mutant plaques   Rare wild plaques

452

merous independent mutants represented separately occurring mutations at the same or at different sites on the genetic material. For example (Fig. 16-11A), assume that two r mutations, r_1 and r_2, are actually alterations at two different positions within the DNA. Being rII mutants, the phages produce a mutant plaque on strain B and none on strain K. However, a strain B cell, infected with the two types, could give rise by recombination to wild, $r_1{}^+r_2{}^+$. This is possible, because the r_1 mutant is carrying a normal r_2 region and would have the genotype $r_1r_2{}^+$. Similar reasoning tells us that the r_2 mutant would be $r_1{}^+r_2$. Recombinant wild-type phages, $r_1{}^+r_2{}^+$, could arise by crossing over. The resulting wild phage would be able to produce wild plaques on both strains B and K. But if the two mutant sites, r_1 and r_2, are sites within the same gene and are therefore extremely closely linked, the wild recombinants would be exceedingly rare. This means that the investigator would be forced to search through perhaps millions of mutant plaques on strain B to find the one wild plaque. This is a feat that would be truly impossible. Fortunately, the growth defect of the rII virus on strain K can avoid the need to do this. It can act as a screen to uncover the rare recombinant among the millions of others.

In the procedure, rII mutants are therefore recognized by the r plaques they produce on strain B and their inability to produce any plaques on K. Strain B is then infected with two rII mutants that arose independently (Fig. 16-11B). The infected E. coli B are allowed to lyse. Samples of the resulting phages are then plated on agar dishes containing strain K. Only

FIG. 16-11. Crossing of rII mutants. (A) Recognition of phage types. rII mutants are recognized by the fact that they produce a mutant plaque (large) on E. coli strain B and no plaques on strain K. If strain B is infected with two rII mutants that have alterations at different sites within the gene, a very rare recombinant may be formed, $r_1{}^+r_2{}^+$. This can produce wild (small) plaques on either B or K. On strain B, both the rII mutants and the wild type can grow. The rare wild plaque would be lost among the mutant ones. Strain K, however, acts as a screen, because only the wild type can grow on it. (B) Recombination frequency in the rII region. The different rII mutants are recognized as shown in A. Two types of rII mutants are allowed to infect sensitive strain B. When B cells lyse, they will release phages that are primarily mutant, because recombinants within the short rII region under study are rare. However, by plating on strain K and observing for wild plaques, the number of these can be established. These can then be compared with the number of mutant plaques produced on strain B. A comparison of the number of mutant plaques on B with the number of wild ones on K allows one to calculate the frequency of crossing over between the two mutant sites. (The figure does not attempt to reflect actual numbers of plaques.)

the wild recombinants will form plaques. Their numbers will be few, because the recombinants are rare. But this number of wild plaques on K can now be compared to the number of r plaques produced by samples plated on strain B. From these counts, the frequency of recombination is easily determined; the logic is again the same as for the estimation of crossover frequency in higher organisms.

Benzer's method was so sensitive that he could have detected one recombinant among 10^8. The wild plaque that this recombinant can form would be lost among the r plaques if only strain B were available. The rare recombinant would never be found if only strain B were used, therefore, no maps could be constructed. With his method, Benzer found that the rII mutations cluster in a small portion of the T4 map. Within that small region, the positions of different r II mutations can be established. Actually, 300 separate sites have been recognized within the small map segment (Fig 16-12). Well over 2000 separate mutations were studied, and some of these were found to be mutations at the very same site. As a matter of fact, Benzer recognized "hot spots," sites where large numbers of the mutations map in contrast with other locations where only one or a few mutations have arisen. The smallest recombination frequency was approximately 0.02%, meaning that two different rII mutant sites may be only 0.02 map units apart. Benzer translated this into terms of nucleotides along a stretch of DNA. Making certain assumptions, the 0.02% value indicated that a distance of not more than six nucleotide pairs would separate two mutant sites along a strand of DNA. From the wealth of genetic analysis of microorganisms that followed Benzer's studies, we now know that two mutant sites may be adjacent to each other, representing two nucleotide pairs side by side.

Does the fact that two r mutant sites may be so close together mean there are different mutant sites within the gene and they can undergo recombination by crossing over within the gene? Consequently, is the unit of function different from the unit of mutation and recombination? We have already discussed these points in relation to the gene and molecular interactions (Chap. 13), but it was Benzer's work that reformed thought on the nature of the units of heredity. Using strain K of E. coli, Benzer demonstrated that two genes of function are present in the rII region of phage T4. He discovered that a mixed infection of strain K with a wild-type phage plus any rII mutant results in the lysis of the cell and the reappearance

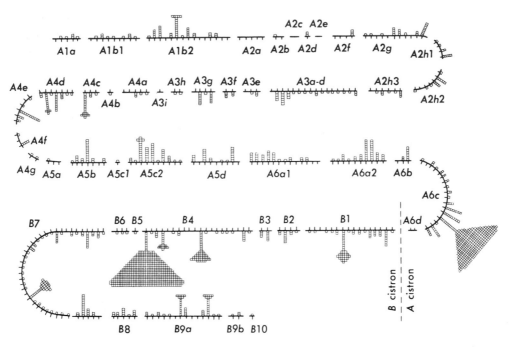

FIG. 16-12. Map of rII region. A large number of different mutations that originated independently have been assigned to the rII region. The exact order of the mutations in a small segment is not known in most cases. Each square represents an independent mutation whose site has been mapped. It can be seen that more mutations occur at certain sites than at others. (Reprinted with permission from S. Benzer, *Proc. Natl. Acad. Sci., U.S.A.* 47: 410, 1961.)

of both wild-type and mutant viruses (Fig. 16-13). This means that the presence of the wild phage makes possible the reproduction of the mutant type. In a sense, it is as if the wild ($+$) form is dominant over a recessive r allele.

Benzer then infected strain K with different pairs of rII mutants (follow Fig. 16-13). From the mapping he had done, he knew how far apart any two rII mutants would be. He now wanted to see if two rII mutants could interact in some way to cause lysis of strain K with the production of offspring phages; this is an effect which no rII mutant can bring about on its own. Benzer found that certain pairs of rII mutants could interact to effect lysis on strain K, whereas others could not. The small rII region was shown to be divisible into two subregions, A and B. A combination of any two mutants in the same region—both in A or both in B—cannot cause lysis. However, a combination of two mutant types—one with a mutation anywhere in A and the other with one anywhere in B—can cause lysis.

This observation told Benzer that all of those sites within subregion A had more in common with one another than with any site in B. The same relationship would hold for the B sites. The common feature that distinguishes all of the sites within one of the subregions is evidently a specific function to which they are all related. In this case, the function is one that is necessary for cell lysis and the production of phage. Thus, in the rII region we have two subregions, A and B, which are units of function. For lysis to occur, each unit must be carrying out its specific function. When a wild phage is crossed with any rII mutant, the wild particle contributes two normal functional units, A and B, and so lysis can occur. Similarly, a cross of any A mutant with any B produces lysis, because a mutant defective in A contributes an intact B, and vice versa. If a cell of strain K is infected with two B mutants, however, there is no normal, functional B unit in the cell (and likewise for any two A mutants). Because one of the two functions is lacking, lysis cannot occur. The work with rII phage clearly demonstrates that the unit of function is larger than the unit of mutation and crossing over. Benzer coined the term *cistron* to designate the unit of function. And the concept of the cistron, as we have seen, is applicable to higher species as well as to microorganisms.

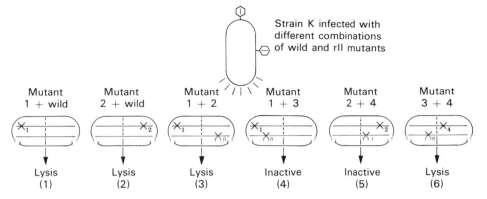

Strain K infected with different combinations of wild and rII mutants

| Mutant 1 + wild | Mutant 2 + wild | Mutant 1 + 2 | Mutant 1 + 3 | Mutant 2 + 4 | Mutant 3 + 4 |

Lysis (1) | Lysis (2) | Lysis (3) | Inactive (4) | Inactive (5) | Lysis (6)

Fig. 16-13. Demonstration of the functional unit. Strain K can be infected with wild-type phages plus one of the rII mutant types. Any combination of a wild and a mutant results in lysis of strain K (1 and 2), indicating that the wild-type phage is able to carry out the function that is deficient in the mutant (inability to grow on strain K by itself). Certain combinations of two different rII mutants on strain K result in lysis (3 and 6), whereas other combinations do not (4 and 5). It is found that the rII region can be divided into two segments (indicated by the *dotted line*). Any combination of mutants whose mutant sites are within the same region fails to cause lysis. This indicates that two such mutants carry defects in the same functional region, even though the defects are not at exactly the same site. When the defects are in separate regions, the one mutant contains a normal, functional unit that the other lacks (3 and 6). Lysis occurs because the two functional units needed for growth in K are present in the cell.

In Chapter 13, the cistron, or gene of function, was discussed in relation to its control of a polypeptide chain. In light of the knowledge assembled by molecular biologists since Benzer's original work, the concept of the gene as a unit of function (a stretch of DNA) composed of smaller units (nucleotide pairs that may mutate independently and engage in crossing over) agrees perfectly with the known physical basis of the gene itself. As was stressed in Chapter 13, no confusion need exist about use of the terms *cistron* and *gene*. They are equivalent if we picture the "familiar gene" as a stretch of DNA controlling a specific function (formation of a specific polypeptide chain) and composed of smaller sites that may mutate independently and engage in crossing over (the nucleotide pairs). The expressions *muton* and *recon* were also coined by Benzer to refer to these sites of mutation and crossing over within a cistron. However, they are now less frequently used, because they are known to represent the same thing in physical terms, one nucleotide pair.

A cistron, or gene of function, can be recognized in any group of organisms by the performance of a cis-trans test (Chap. 13). When any two mutants are crossed, the mutant sites in the F_1 offspring are found in the trans arrangement. In our discussion of allelism in higher forms, we saw that a cross of two recessive mutants produces mutant offspring if the two parents are carrying mutations within the same gene. Benzer was actually performing cis-trans tests in his work with strain K and the rII mutants. For a cross of any rII mutants gives a trans arrangement (see Fig. 16-13). If the resulting phenotype is wild (here lysis and plaque formation), the mutations complement each other and hence belong to separate genes or functional units. If a mutant phenotype (no lysis) follows from a mixed infection by two rII mutants, then the mutations do not complement each other but are in the same gene or cistron.

Phage with single-stranded DNA

The variation displayed by viruses has led to the clarification of many critical points related to the nature of gene action. One virus that has been used as a valuable tool is ϕX 174. After its discovery in 1958, chemical analysis of ϕX 174 revealed an unexpected feature. Although its DNA contains the familiar building blocks of phosphate, sugar, and bases, the proportion of adenine to thymine is not 1:1, nor is that of guanine to cytosine! At first, this seemed to contradict the rule of preferential base pairing. The ϕX 174 DNA was subjected to a variety of tests, including ultracentrifugation and exposure to enzymes. The data clearly showed that the DNA of this virus is single stranded rather than in the form of a double helix. Moreover, it exists in the form of a closed circle.

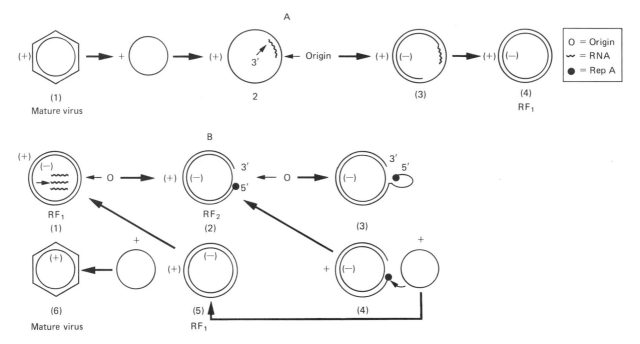

FIG. 16-14. Stages in replication of φX 174 DNA. *(A)* An infective (+) strand of a φX 174 virus enters a cell *(1)*. Using enzymes of the host, an RNA primer *(red)* is assembled to a 3′ end of the origin *(2)*, and a negative DNA strand complementary to the (+) strand is started *(3)*, and then finally completed *(4)*. The primer is removed, the gap left is filled, and the new strand is circularized. A replicative form, RF1, is now completed. *(B)* The (−) strand of the replicative form acts as the sense strand for all the genes of φX 174. Transcripts of the genes *(1)* are coded for assembly of a protein, Rep A, which can nick the (+) strand at its origin and bind to it at the 5′ end *(2)*. This form is referred to as RF2, and from it, rolling circle replication may ensue *(3)*. Nucleotides in a new round of replication *(red lines)* are added to the 3′ end of the nicked (+) strand as the intact (−) strand acts as a template. The (+) strand is displaced from the circle. The (+) strand then is cleaved from the new DNA to which it is linked by the Rep A protein, which also circularizes it *(4)*. The Rep A protein then attaches to the 5′ end of the newly assembled (+) strand. The (+) strand may act as a template for assembly of a new (−) strand *(5)*, as depicted in *A*. This new RF1 can then be converted to RF2 and the cycle is repeated. Eventually (+) strands will be captured by phage envelopes and packaged *(6)* to form mature virus. The (−) strand with the newly assembled (+) strand, seen in step *4*, can serve as an RF2, which then starts another cycle of replication, forming another (+) strand. The cycle can keep repeating.

The φX 174 is just one of several DNA viruses now known to exist primarily in a single-stranded form. One question that arose concerned the manner in which such a virus could replicate. It was found that the single-stranded form is infective, capable of invading its *E. coli* host. Once inside the cell, however, the single-stranded DNA (the + strand) can direct the formation of a complementary strand (the − strand), producing a double-stranded form of the virus, the so-called replicative form. The latter is thus composed of the infective strand (+) and its complementary strand (−) made inside the cell (Fig. 16-14*A*). The formation of this first double-stranded replicative form (designated RF1) depends entirely on enzymes present in the host cell. A short primer RNA is synthesized, DNA polymerase III adds nucleotides to its 3′ end and continues to lengthen the strand until the template (+) strand is followed completely. The RNA primer is removed, the gap is filled in, and

ligase seals the nick to produce a circular (−) strand, and consequently a replicative form (RF1).

The (−) strand that is now present acts as the sense strand for the few genes of φX 174. Actually, this virus was among the first forms in which it was demonstrated that only one strand of a gene undergoes transcription (see Chap. 12). Transcription of a specific gene from the (−) strand of RF1 results in the formation of a protein, Rep A. This protein brings about a nick in the (+) strand, always at the replication origin (Fig. 16-14*B*). The protein remains attached to the 5′ end of the nicked strand. The viral form with the intact (−) and nicked (+) strands is referred to as RF2. Rolling circle replication then ensues. Nucleotides are added to the free 3′ end of the (+) strand as it is displaced. After one replication cycle is completed, the Rep A protein cuts the (+) strand off from the tail and then circularizes it. Depending completely on host cell enzymes, it can now also act as a template for the synthesis of a new ($-$) strand to form another RF1, which in turn can be converted to RF2 and undergo the cycle just described.

After it cuts and circularizes the ($+$) strand, the Rep A protein then links to the 5′ end of the new ($+$) strand made in the first round of replication. Another cycle of replication can then ensue with the Rep A protein again cleaving and sealing the second new ($+$) strand. The cycle can keep continuing in this fashion as the intact ($-$) strand continues to act as a template for ($+$) strands that, as noted, may serve to generate more RF1 forms. However, some ($+$) strands start to be packaged as protein phage envelopes begin to appear in the cell. More and more of the ($+$) strands are packaged as phage heads arise, until no more RF1 forms are produced. New single-stranded ϕX 174 phages, each with a ($+$) strand, are ready for release from the cell and eventual infection of other *E. coli*.

Overlapping genes

Later investigations with the phage ϕX 174 have yielded some unexpected results that demand a new look at the nature of the gene and its reading. The single DNA strand of the phage has been subjected to intense investigations, many involving the cleavage of the DNA with enzymes and then separation of the resulting DNA fragments. The length of the ϕX 174 DNA has been found to be 5375 nucleotides, and their exact sequence has been completely determined. This is quite a feat in itself, the sequencing of the entire hereditary material of a genetic element. A dilemma, however, presented itself regarding the length of the DNA of the phage and the number of proteins for which the DNA is apparently coded. The virus can bring about the production of nine different proteins whose molecular weights are known. The amount of DNA required to code for these is far in excess of the amount of DNA that seems to be present in the phage.

The answer to the problem was quite unanticipated. Research teams in the laboratory of F. Sanger in Cambridge, England, studied in detail several genetic regions of ϕX 174, one of which contains two genes, genes D and E, which code for proteins D and E, respectively. The D protein, whose exact amino acid sequence was known, is produced in great amounts in the host cell and is required for the production of single-stranded viral DNA, although its exact role in the process is unclear. The E protein is required for the lysis of the host cell when hundreds of new viruses

have been produced and are to be released. The two genes, D and E, on all counts seem like any other separate and distinct genes. Nonsense mutations are known to occur in each one, without apparently affecting the function of the other.

The exciting find was made that a certain stretch of the phage DNA contains the coded information for both the D and the E proteins and that the D and E genes overlap! The start of the E gene is in the middle of the D gene. Both genes end close to the same spot. However, the E gene is not just a segment of the D gene or just the end part of the D gene. The two proteins, D and E, are very different in amino acid sequence. The DNA, however, in the D and E genetic region is read in different frames. Recall frame-shift mutations (see Chap. 12), which were recognized in Crick's pioneer studies on the triplet code. Reading of a gene begins at a fixed starting point, three nucleotides at a time. A $-$ or a $+$ mutation may cause the entire reading of a gene to shift. In the case of the D and E genes, the starting sequence for the E gene was found to be within the D gene. The D gene and the E gene within it, however, are read in different frames (Fig. 16-15).

Another interesting finding is that the termination codon of the D gene overlaps by one nucleotide the start codon of the next gene, gene J. Moreover, overlapping of reading frames does not seem to be confined to just this one region of ϕX 174 DNA. Gene A, which contains coded information for another protein needed for viral construction, overlaps with gene B, coded for a protein that produces a nick in the DNA when new copies of the DNA are needed. Note in Fig. 16-15 that it is the antisense strand that is shown. The triplets indicated have the same sense as the mRNA. Usually it is the antisense strand that is given when a DNA nucleotide sequence is presented (review Fig. 12-24). Overlapping genes are also known to occur in simian virus 40 (SV40), another virus that has been studied in great detail and is discussed later in this chapter. The discovery of overlapping genes is quite an exception to the general picture of the gene. The genetic code was found to be nonoverlapping, each single nucleotide being part of only one code word or codon. However, in ϕX 174 and SV40, a single nucleotide may be part of two code words, depending on the reading of the frame. The finding of the overlapping genes helps solve the problem of fitting the nine genes of ϕX 174 into a supposedly less than adequate length of DNA. The overlapping may have evolved in viruses as a device to accommodate a larger number of genes than would other-

```
                    G-A-G-T-C-C-G-A-T-G-C-T-G-T-T ·C-A-A-C-C-A-C-T-A-A-T-A-G-G ·T-A-A-G-A-A-A-T-C· │A-T-G·│
                                    ↑                                                                    ↑
                                mRNA start                                                          D start
D        Ser - Gln - Val - Thr - Glu - Gln - Ser - Val - Arg - Phe - Gln - Thr - Ala - Leu - Ala - Ser - Ile - Lys - Leu - Ile -
         A-G-T-C-A-A-G-T-T-A-C-T-G-A-A ·C-A-A-T-C-C-G-T-A-C-G-T-T-T ·C-C-A-G-A-C-C-G-C-T-T-T-G-G ·C-C-T-C-T-A-T-T-A-A-G-C-T-C·A-T-T-
                   10            20            30                                                                    60

D        Gln - Ala - Ser ·- Ala - Val - Leu - Asp - Leu - Thr - Glu - Asp - Asp - Phe - Asp - Phe - Leu - Thr - Ser - Asn - Lys -
         C-A-G-C-C-T-T-C-T-G-C-C-C-T-T ·T-T-G-G-A-T-T-T-A-A-C-C-G-A ·A-G-A-T-G-A-T-T-T-C-G-A-T-T ·T-T-C-T-G-A-C-G-A-G-T-A-A-C·A-A-A-
                   70            80                                                          110           120

D        Val - Trp - Ile - Ala - Thr - Asp - Arg - Ser - Arg - Ala - Arg - Arg - Cys - Val - Glu - Ala - Cys - Val - Tyr   - Gly -
         G-T-T-T-G-G-A-T-T-G-C-T-A-C-T ·G-A-C-C-G-C-T-C-T-C-G-T-G-C-T ·C-G-T-C-G-C-T-G-C-G-T-T-G ·A-G-G-C-T-T-G-C-G-T-T-T· │A-T-G·│G-T-
                  130           140           150           160           170                                          E start
                                                                                                                        ↑
                                                                                                                        Leu

E        Val - Arg - Ser - Thr - Leu - Trp - Asp - Thr - Leu - Ala - Phe - Leu - Leu - Leu - Leu - Ser - Leu - Leu - Leu - Pro - Ser -
D        Thr - Leu - Asp - Phe - Val - Gly - Tyr - Pro - Arg - Phe - Pro - Ala - Pro - Val - Glu - Phe - Ile - Ala - Ala - Val -
         A-C-G-C-T-C-G-A-C-T-T-T-G-T-G ·G-G-A-T-A-C-C-C-T-C-G-C-T-T ·T-C-C-T-G-C-T-C-C-T-G-T-T-G ·A-G-T-T-T-A-T-T-G-C-T-G-C-C·G-T-C-
                                          210           220           230

E        Leu - Leu - Ile - Met - Phe   - Ile - Pro - Ser - Thr - Phe - Lys -   Arg - Pro - Val - Ser - Ser - Trp   - Lys - Ala - Leu -
D        Ile - Ala - Tyr - Tyr - Val -   His - Pro - Val - Asn - Ile - Gln -   Thr - Ala - Cys - Leu - Ile - Met - Glu - Gly - Ala -
         A-T-T-G-C-T-T-A-T-T-A-T-G-T-T ·C-A-T-C-C-C-G-T-C-A-A-C-A-T ·T-C-A-A-A-C-G-G-C-C-T-G-T-C ·T-C-A-T-C-A-T-G-G-A-A-G-G-C·G-C-T-
                  250                                                                            270

E        Asn - Leu - Arg - Lys - Thr - Leu - Leu - Met - Ala - Ser - Ser -   Val - Arg - Leu - Lys - Pro - Leu - Asn - Cys - Ser -
D        Glu - Phe - Thr - Glu - Asn   - Ile - Ile - Asn - Gly - Val - Glu - Arg - Pro - Val - Lys   - Ala - Ala - Glu - Leu - Phe -
         G-A-A-T-T-T-A-C-G-G-A-A-A-A-C ·A-T-T-A-T-T-A-A-T-G-G-C-G-T-C ·G-A-G-C-G-T-C-C-G-G-T-T-A ·A-A-A-G-C-C-G-C-T-G-A-A-T-T-G·T-T-C-
                  310           320                                          340           350           360

E        Arg - Leu - Pro - Cys - Val - Tyr - Ala - Gln - Glu - Thr - Leu -   Thr - Phe - Leu - Leu - Leu - Thr - Gln   - Lys - Lys - Thr -
D        Ala - Phe - Thr - Leu - Arg - Val - Arg - Ala - Gly - Asn - Thr -   Asp - Val - Leu - Thr - Asp - Ala - Glu - Glu - Asn -
         G-C-G-T-T-T-A-C-C-T-T-G-C-G-T ·G-T-A-C-G-C-G-C-A-G-G-A-A-A ·C-A-C-T-G-A-C-G-T-T-C-T-T-A-C-T-G-A-C-G-C-A·G-A-A-G-A-A-A-A-C-
                  370                   390           400           410           420

E        Cys - Val - Lys - Asn - Tyr - Val - Arg - Lys - Glu                                                                    ‖
D        Val - Arg - Gln - Lys - Leu - Arg - Ala - Glu - Gly - Val - Met        J      Ser - Lys -  Gly - Lys - Lys - Arg - Ser -
         G-T-G-C-G-T-C-A-A-A-A-A-T-T-A ·C-G-T-G-C-G-G-A-A-G-G-A· G-│T-G-A│·T-G·│T-A·A│·T-G ·T-C-T ·A-A-A-G-G-T-A-A-A-A-A-A-C-G·T-T-C-T-G
                  430           440                       ↑          ↑       ↑                470           480
                                                      E stop    D stop   J start
```

FIG. 16-15. Overlapping genes in φX 174. The starting sequence and the end sequence of the E gene are within the D gene. In the overlapping genes, codons are read in different frames. The amino acid sequences of the D and E proteins, as well as that of the J protein (product of the gene following gene D) are aligned with the DNA sequences. Note the overlapping between the terminating codon of gene D and the start codon of gene J. The DNA triplets shown here have the same sense as the RNA triplets of the messenger RNA, because the strand shown here is not the one that undergoes transcription, but is the antisense strand. Since it has the same sense as the mRNA, the mRNA sequence can be immediately deduced by changing all the Ts to Us.

wise be possible in so small an amount of DNA. The discovery of the phenomenon focuses attention on such matters as genetic controls and expression and also raises the possibility that overlapping genes may yet be found in other organisms.

Use of viruses in the isolation of a gene

One group of investigators made use of transducing viruses to accomplish a very exacting feat, the actual isolation of a specific gene found in *E. coli*! The particular gene that was isolated was *lac*, which determines the ability of the cell to utilize the sugar lactose. At the beginning of this chapter, we discussed the ability of certain viruses to pick up pieces of genetic material from bacterial chromosomes and to integrate these pieces with their own DNA. The phage may then carry the integrated gene to another cell (transduction). In isolating the lac gene, Beckwith and his group employed two different kinds of viruses (lambda plac 5 and φ80 plac 1), each of which had incorporated the lac marker of *E. coli*.

The team took advantage of certain special features of these transducing phages when they selected them for their work. In each phage, the base composition of the DNA is such that one of the two strands is

FIG. 16-16. Isolation of the lac gene. (A) The two phages selected for the isolation possess certain special features. In each, the DNA is composed on one strand that is heavy and one that is light. This enables them to be easily separated under laboratory conditions. Both phages had incorporated the bacterial gene, lac, but the gene is inserted in the opposite direction in each case. Consequently, the sense strand of the lac gene is associated with the heavy strand in one phage and with the light strand in the other. The same is obviously true for the antisense strand. (B) The two strands of each virus can be separated by centrifugation after melting of the double-stranded DNA. The heavy strands of each are then brought together. The DNA strands of the lac gene, incorporated into the phage DNA, form a double helix, because they are complementary. The tails of the DNA of the two phages remain attached. Enzyme digestion removes the phage DNA that has remained single stranded, because the strands are from different phages and are not complementary. This leaves isolated the complete lac gene, composed of a sense (+) and an antisense (−) strand.

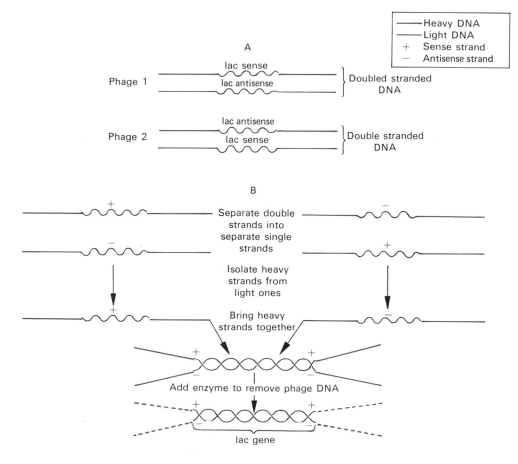

heavier than the other. Thus, each phage normally possesses DNA composed of a heavy strand and a light one. Moreover, the bacterial DNA carried by one phage was inserted in a direction opposite to that in the other phage. Now recall that of the two DNA strands, one is the sense strand and one is the antisense strand. Because of the difference in orientation (Fig. 16-16A) of the bacterial DNA that is integrated into the two phages, the sense strand is attached to a heavy strand in one phage, whereas the antisense (or complementary) strand is attached to the heavy strand in the other phage. This means that one strand of the lac gene is associated with a heavy strand of one of the phages, and its complimentary strand is associated with the heavy strand of the other phage. If the two strands can be brought together and freed from the

FIG. 16-17. Electron micrograph of lac gene isolated by procedure summarized in Fig. 16-16. (Reprinted with permission from J. Shapiro, L. MacHattie, L. Eron et al., *Nature* 224: 768–774, 1969.)

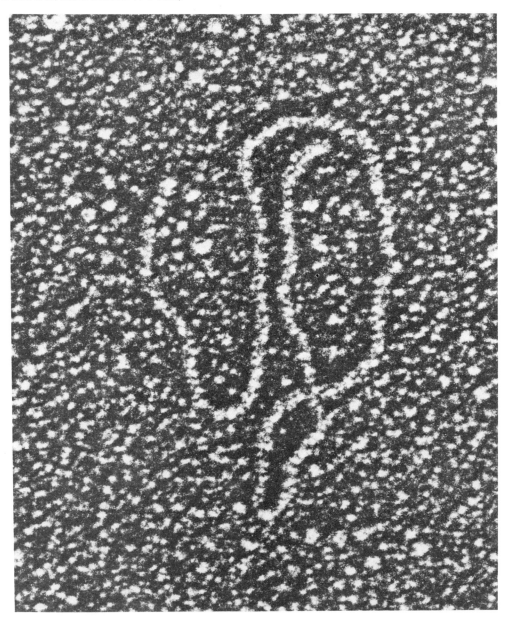

associated viral DNA, then the complete lac gene, composed of a sense and an antisense strand, will be isolated by itself.

In the actual procedure (Fig. 16-16B), the DNA of each virus was separated into two separate strands. The heavy strands could be easily isolated from the light ones in each case. The lac "sense" is on the heavy strand of one phage and the lac "antisense" is on the heavy strand of the other. When the two strands are brought together, they form double-helical DNA, because they are complementary. However, they now have "tails" associated with them. These, of course, represent the DNA of the phages that had incorporated the bacterial DNA. The DNA strands of the phages are not complementary; they are completely different DNA segments from two different viruses. Therefore, the DNA strands of the two viruses do not associate to form a double helix. These single-stranded tails were removed by treatment with an enzyme that is specific for single-stranded DNA but does not harm double-stranded DNA. Removal of the tails by the enzyme leaves double helices, and these represent pure lac genes. The isolated gene has been visualized by the electron microscope (Fig. 16-17) and has been shown to be 1.4 mμ in length. This first isolation of a gene opened up new pathways for the study of mechanisms involved in the control of gene expression. Progress in the area of genetic manipulation has since proceeded at an incredible pace and has led us into a whole new world of genetic investigation, the major topic of Chapters 18 and 19.

Cellular defenses against foreign DNA

When the foreign DNA of a phage enters a bacterial cell, a cellular defense mechanism comes into operation to degrade the DNA of the virus, thus protecting the cell against destruction. This *restriction activity* is accomplished by a certain class of enzymes, the *restriction endonucleases*. Restriction endonucleases fall into two main classes. Let us consider first those designated type II enzymes, which are the more common type. These are the enzymes at the very core of genetic manipulation that have made possible the incredible advances in molecular genetics (see Chap. 18).

You will recall that an endonuclease is capable of breaking an internal phosphodiester bond. Unlike other endonucleases that are nonspecific as to where they cleave a DNA strand, these restriction endonucleases produce internal double-stranded nicks only within

certain nucleotides sequences—target sites they are able to recognize. Protective enzymes such as these have been isolated from various species and strains of bacteria. The enzymes in this group are characteristically dimers, each composed of two identical subunits. When these endonucleases recognize a target site on the foreign DNA, they produce a double-stranded, highly specific cut at the site of recognition. The DNA sequences recognized by most of these enzymes contain a point or axis around which a symmetrical arrangement of base pairs occurs (Fig. 16-18A). Note that in such a stretch of DNA the reading in the 5' to 3' direction on either of the two paired strands produces the same base sequence. Such a symmetrical arrangement is known as a *palindrome*. A palindrome is equivalent to inverted repeats that are side by side. Some restriction endonucleases produce breaks in the two DNA strands that are both at the axis of symmetry. This results in the formation of DNA segments with blunt ends (Fig. 16-18B). Other restriction enzymes produce breaks in the complementary DNA strands at points actually several nucleotides apart in each strand. As a result of this staggered cleavage, DNA regions are generated that have single-stranded projections.

When the foreign DNA of a virus enters a bacterial cell, it may be broken up into fragments as a result of cleavage by restriction enzymes at specific sites they recognize on the viral DNA. The fragments may then be degraded by exonucleases of the cell. Because any one of these restriction enzymes cleaves only at a specific target site, the number of fragments resulting from attack by that enzyme depends on the number of specific target sites found on the viral DNA that the enzyme can recognize. This means that such an enzyme can cleave a given kind of DNA into a specific number of fragments of characteristic size. As we will see in Chapter 18, this specific property of these restriction enzymes has provided the molecular biologist with a tool making it possible to probe into the very nucleotide sequence of a given gene.

Since cells of a bacterial strain possess enzymes that can cleave foreign DNA, thus restricting the growth of certain phages within them, it is necessary for the cells to protect their own DNA from the cleaving potential of their own restriction enzymes. If a given site occurs on the bacterial DNA as well as on the viral DNA, a particular restriction endonuclease of the cell would be able to attack targets on both kinds of DNA. The cell, however, has a device to prevent this, because it possesses methylase activity.

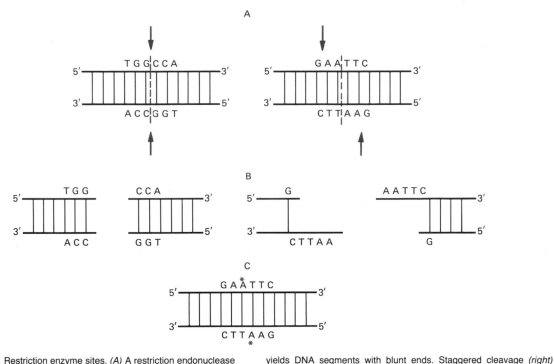

FIG. 16-18. Restriction enzyme sites. *(A)* A restriction endonuclease can recognize a specific nucleotide sequence that occurs in a symmetrical arrangement or palindrome. Two such sequences are shown. (The axis of symmetry is indicated by the *broken red line*.) A restriction enzyme produces very specific cuts at sites in each DNA strand *(arrows)*. *(B)* Cleavage in both strands at the axis of symmetry *(left)* yields DNA segments with blunt ends. Staggered cleavage *(right)* results in DNA segments that bear complementary single-stranded projections. *(C)* Modification methylases protect the DNA of the cell from restriction enzymes by methylating certain bases *(asterisks)* in a recognition sequence.

This gives the cell the ability to modify a target site that can be recognized by an endonuclease. In the case of the type II restriction enzymes, a corresponding *modification enzyme* is found that can recognize the same nucleotide sequence as a given restriction endonuclease. It can modify the particular DNA sequence recognized by both, and thus protect it from cleavage. A modification enzyme achieves protection of the cell's DNA by methylating certain bases at the site of recognition (Fig. 16-18C). The modification methylase transfers methyl groups to particular bases after these bases have been incorporated into a DNA strand. Following modification by the methylase, the site in the cell's DNA that can be recognized by the corresponding restriction endonuclease is now immune to cleavage.

The entering phage DNA, however, is not necessarily modified, and therefore is vulnerable to the restriction enzyme; it thus has a reduced chance of establishing itself in the cell. Usually, entering phages become restricted in the presence of restriction endonucleases. At times, however, sequences at recog-

nition sites in the viral DNA do manage to become methylated by the cell's modification enzymes before the cell's restriction enzymes can cleave the viral DNA. In such a case, the viral DNA becomes resistant to the cell's restriction endonucleases and can therefore multiply in the cell and effect lysis. We see here an excellent example of the interplay between a host cell and its virus.

The second class of restriction endonucleases is composed of enzymes, each of which possesses *both* exonuclease and methylase activities. Two subclasses are recognized. In one of these (type I enzymes), each enzyme is multimeric, composed of three types of subunits—one that is needed to recognize the target sequence, another type capable of restriction, and a third type that is responsible for methylation. Since such an enzyme is capable of both restriction and modification, what determines whether a target site will be cleaved or methylated? The decision is apparently made in response to the condition of the target sequence. If the nucleotide sequence is completely methylated, the enzyme may bind to the target, but is released without acting further. If only one of the strands of the duplex DNA at the site is methylated, the enzyme will then proceed to bring about complete methylation of the target site. On the other hand, if the site is not methylated at all, cleavage of the DNA takes place following binding.

Unlike the type II enzymes, these type I enzymes, when they cleave the DNA, do so at sites that are more than 1000 base pairs away from the sites of recognition! Moreover, the site that is cut is not a specific sequence. Certain sites, however, appear to be preferred over others by the enzymes in this group. The second subclass of this group (type III enzymes) consists of two kinds of subunits, one for restriction and one that accomplishes both recognition of a target and methylation. Only a few enzymes in this subclass have been studied in detail.

Tobacco mosaic virus—an RNA virus

The study of viruses has also shown that the hereditary material is not universally DNA. Long before the foundations of molecular genetics were established, it was known that tobacco mosaic virus (TMV) is composed of a protein coat enclosing a core of (RNA). The RNA was later shown conclusively to be the hereditary material of this plant virus, which can also attack members of the sunflower and nightshade families. The protein coat of TMV is necessary to protect the RNA; without it, the ability of the virus to enter the plant cells decreases. The protein coat and the RNA can be separated, and it has been clearly demonstrated that the coat alone cannot infect a cell and bring about the production of either new virus protein or virus RNA. On the other hand, TMV RNA by itself can do this. Experiments have been performed using different strains of the virus. For example, in addition to the standard strain (S), there is a more virulent strain (H). After separating the protein coats from the RNA of viruses of each strain, hybrid virus particles can be reconstituted, those with H-RNA and S coat and those with S-RNA and H-coat protein. When these are allowed to infect plant cells, it is found that the progeny viruses that form are always typical of the strain that provided the RNA. As seen in Fig. 16-19, H-RNA with S coat governs the formation of typical H viruses that have both H-RNA and H protein.

This shows not only that the RNA is the genetic material of these viruses but that it governs the formation of more RNA and also of virus-specific protein,

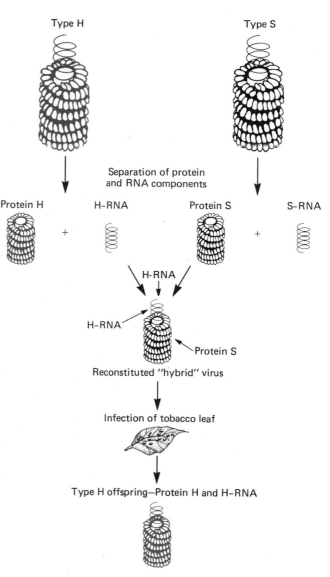

FIG. 16-19. Reconstitution of tobacco mosaic virus. The RNA of one strain of virus can be combined with the protein coat of another strain. When such hybrids infect the tobacco plant, the viruses that are later isolated are typical of the one from the strain that supplied the nucleic acid. The RNA of the virus must therefore direct the production of more specific RNA and also of the specific protein surrounding it.

much as DNA governs the formation of more DNA and protein. The coat of the TMV is composed of many protein subunits. Each of these however, is identical and is in turn made up of 158 amino acids. Since the genetic code is a triplet one, 474 nucleotides are needed to translate the protein of the TMV. Surprisingly, analysis of the TMV RNA shows it to be much longer, 6400 nucleotides long and apparently in linear rather than circular form. Thus a great deal more genetic information is present than is needed for the production of the single type of protein found in the TMV coat. When the RNA of the virus enters a host cell, it acts as a template for the synthesis of more viral RNA, and it also governs the formation of the coat protein, which then combines with the RNA. These processes undoubtedly involve several enzymes, proteins that would not be present in the completed coat. This accounts for some of the other messages that are evidently coded in the viral RNA.

RNA viruses and the direction of flow of genetic information

RNA viruses are not peculiar to plant hosts. A number of them is known to attack bacterial and animal cells, among them those responsible for polio, influenza, and the common cold. The RNA of many of the animal viruses, as well as the RNA of many bacteriophages, infects the cell as a single RNA strand surrounded by a protein coat. After entry, the single RNA strand of a phage, the + strand, can act as a template for the synthesis of a complementary strand, the − strand. In turn, the − strand may act as a template for the formation of new + strands. Any + strand can also act as messenger RNA and attach to ribosomes (Fig. 16-20A). Indeed, after entry of the + strands into a cell following infection by virus particles, + strands at first act as mRNA. This is essential for the synthesis of the enzyme RNA replicase, required to form RNA on an RNA template. This enzyme is not produced in an uninfected cell. An uninfected cell never contains

RNA molecules capable of serving as templates for forming other RNA. The + strands are coded for the formation of this enzyme needed for the replication of the virus. They also contain the information for the coat proteins of the infective virus. These coat proteins eventually surround + strands that were formed in the cell on the − strand templates. Finally, an infected cell bursts, releasing infective viruses, each with a single + strand.

This process is similar to the manner in which the DNA transfers information (Fig. 16-20A). The flow of genetic information pictured in Fig. 16-20B was first proposed by Crick. It seemed to be the universal process for transferring coded information from DNA, and so the concept became established as the central dogma of molecular biology. The central dogma was taken by many biologists to mean that the flow of genetic information from DNA to protein could occur only in the one direction, as indicated in the figure. Protein could not impart information to other protein or to RNA. RNA could not impart information to DNA. Any such reverse flow of information, as from RNA to DNA, would be a violation of the central dogma.

Therefore, it was with a great deal of excitement that certain findings were reported from studies with a group of RNA viruses; these appeared to violate the central dogma. The particular assortment of viruses includes some that are capable of inducing tumors in several animal species: rats, mice, and chickens, for example. The Rous sarcoma virus has been one of the most widely studied members. After infecting a chicken cell, the virus can transform it into a cancer cell, which can in turn divide and produce new virus

FIG. 16-20. Flow of genetic information. (A) In many RNA viruses, the nucleic acid can bring about the formation of more RNA. Some of the RNA can act as mRNA that is translated into protein. (B) DNA contains information that can guide the formation of more DNA or of RNA. The latter contains information imparted to it by the DNA, and this is translated into protein. The concept of the flow of genetic information in only one direction, from DNA to RNA to protein, is known as the *central dogma*.

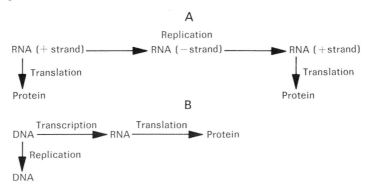

particles. The first clue to the unexpected way in which this virus replicates came through the investigations of Temin, who worked with the antibiotic, actinomycin D, as part of his analytical procedure. This substance had been shown to prevent transcription of DNA into RNA by blocking the DNA as a template. Therefore, it has become a commonly used tool for determining DNA-dependent RNA synthesis. If applied to a cell infected with an RNA virus, RNA would still be produced (Fig. 16-21A). This is so because the RNA template of the virus is not blocked. Therefore, RNA-dependent RNA synthesis would still go on, because only DNA-dependent RNA synthesis is shut off by the antibiotic.

When Temin applied the drug to cell cultures infected with the Rous sarcoma virus, he found that all RNA production ceased. This was perplexing and suggested that DNA might somehow be involved in

FIG. 16-21. Effect of actinomycin D. (A) The drug blocks DNA as a template, so that no RNA transcript can be formed when it is applied to a cell (above). When it is applied to a cell harboring an RNA virus (below), more viral RNA is made, because the drug cannot block an RNA template. (B) Reproduction of the RNA of the Rous sarcoma virus entails DNA that evidently depends on the RNA template. This DNA in turn gives rise to more viral RNA. If actinomycin D is applied, production of the viral RNA ceases, because the DNA template, which is needed to guide the formation of the RNA of the virus, is blocked.

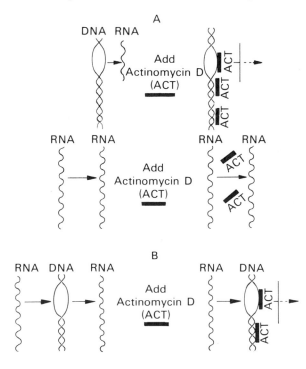

the production of the RNA virus. A variety of experiments was then performed by Temin's group and others, and the data strongly supported the idea that new DNA is produced in a cell infected with the virus. This DNA is actually viral DNA and is formed on an RNA template, the RNA of the infective virus (Fig. 16-21B). If actinomycin D is applied to a cell containing the Rous virus, the viral DNA formed from the viral RNA template cannot undergo transcription into RNA, and so no new viral RNA is made by the cell.

Temin's group succeeded in demonstrating the presence of a DNA polymerase in the Rous sarcoma virus particles as they occur outside the cell. This DNA polymerase can form DNA from an RNA template as well as from a DNA template. It has been designated *RNA-directed DNA polymerase*. In popular usage, this polymerase is called *reverse transcriptase*. In the Rous sarcoma virus, this DNA polymerase preferentially uses RNA as a template for the assembly of DNA, whereas most DNA polymerases use a DNA template. It is apparent that a strict interpretation of the central dogma must be modified to include the reverse flow of information from DNA. The findings, as we will now see, also demand a new outlook on the interactions between the genetic material of a virus and the host cell.

Retrovirus replication

Since the discovery of the Rous sarcoma virus, other RNA viruses are now known that also use reverse transcriptase to generate a duplex DNA form of the virus which can then integrate into the host DNA. The term *retrovirus* designates this unusual class, which includes several viruses of birds and mammals, among them the well-publicized AIDS virus. Most retroviruses are capable of evoking tumor formation and for this reason are said to be *oncogenic*—having the ability to transform the cells they infect to a state in which they grow uncontrollably. The first retrovirus to be isolated from the human and the first virus to be linked to a human cancer is HTLV-I, human T-cell leukemia virus. Although retroviruses were known to cause a variety of cancers and other diseases in animals, none had ever been found in the human until the isolation of HTLV-I. This virus is capable of infecting T lymphocytes (T4 cells) and transforming them to the malignant state. The AIDS virus, originally designated HTLV-III but now referred to as HIV (human immunodeficiency virus), shares a number of properties with HTLV-I, such as the ability to infect T4 cells. Despite certain similarities, HIV differs from HTLV-I in many biological aspects. For example, infection by HIV does not transform the lymphocytes to a cancerous state but results in cell death.

Figure 16-22A summarizes the replication cycle of a retrovirus such as the Rous sarcoma virus. The mature viral particle, the *virion,* contains two identical single-stranded RNA molecules and carries the enzyme reverse transcriptase. Inside the cell, the RNA is released from the protein coat, and the reverse transcriptase then uses a single-stranded RNA as a template to produce a complementary DNA strand in the cytoplasm of the cell. The enzyme possesses DNA polymerase activity and then catalyzes the production of a complementary DNA with the formation of linear duplex DNA. Since the reverse transcriptase also has RNAase activity, it can digest away the RNA template that formed a DNA–RNA hybrid with its complementary DNA strand. The duplex DNA enters the nucleus where it apparently assumes a circular form and then integrates into the host DNA. In this integrated state, it is known as a *provirus.* Actually several viral DNAs, as many as 10 copies, may integrate into a host cell's DNA, apparently at random sites.

In its linear DNA form, the virus has at each end *long terminal repeats,* a left and a right one. These are long, directly repeated sequences that are identical at each end of the virus (Fig. 16-22B). A portion of each long terminal repeat (LTR) functions as a promoter. One promoter, the left one, is responsible for transcription of the provirus. The right promoter in the other LTR can start the transcription of any host genes lying directly next to it. When it integrates, the

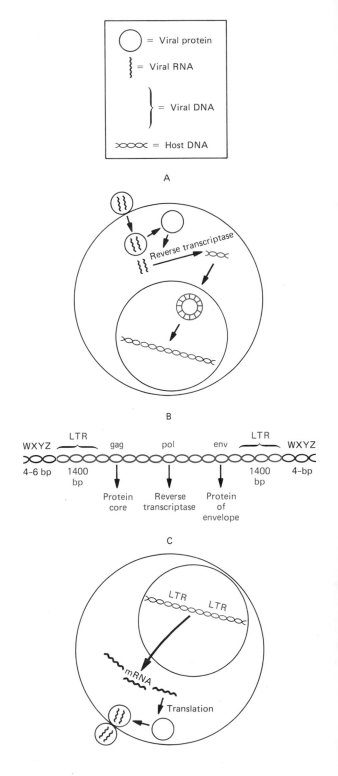

FIG. 16-22. The retrovirus and its replication in the host cell. *(A)* The viral particle enters a cell and releases from the viral envelope two identical RNA strands as well as reverse transcriptase. As a result of reverse transcriptase activity, each strand can now serve as a template for the formation of a complementary DNA strand. A duplex linear DNA is produced, which enters the nucleus where a circular form of the DNA arises and can now integrate with the host DNA. *(B)* The integration of the viral DNA brings about duplication of host DNA (WXYZ) flanking the provirus. The provirus has long terminal repeats (LTR), a left and a right one, at each end. Each LTR contains a promoter. The provirus possesses genes (gag, pol, env) encoded for proteins. (The DNA segments depicted are not drawn to scale.) *(C)* The provirus undergoes transcription, and some of the transcripts act as mRNA that undergoes translation with the formation of viral proteins. These may associate with other transcripts to form mature virions that are released from the cell.

virus brings about a short direct repeat of the host DNA. Therefore, at either end of the provirus, 4 to 6 base pairs of the host DNA will be directly repeated. This is characteristic of the behavior of transposable genetic elements, a major topic of Chapter 17. Although the RNA form of the virus disappears from the cell, all the information needed to make viral RNA is present and is carried along in the provirus with the host's own genetic material.

The provirus typically carries at least three genes. One of these, gag, codes for a protein that is cleaved into four smaller proteins, which become part of the protein core of the mature virus particle. Another gene, pol, is coded for reverse transcriptase, and a third, env, for the protein of the envelope of the virion. The Rous sarcoma virus has still another gene coded for a protein involved in malignant transformation. This gene, src, is coded for a protein required for *oncogenesis* and is classified as an *oncogene,* a genetic sequence capable of inducing tumor formation. The three genes essential for retrovirus replication are thus gag, pol, and env. In addition to these three genes, the genome of HIV, the AIDS virus, contains at least four additional genes. Two of these, tat and art, are critical for replication of the virus, the latter gene regulating the proteins encoded by genes env and gag. The two other genes, sor and 3'-orf, code for proteins, the functions of which remain unknown.

When the proviral DNA is transcribed, some transcripts may serve as mRNA and therefore be translated. The resulting viral proteins coded by the genes may then associate with other viral transcripts to form new virus particles (Fig. 16-22C), which are released from the cell. The cell may not be destroyed in the process, as in the case of Rous sarcoma virus, and the provirus continues to be transmitted to newly formed cells in its integrated state as it is replicated along with the cell's DNA.

Oncogenes and tumor viruses

Tumor-inducing viruses occur among several groups of DNA viruses, but of the RNA viruses, oncogenesis is found only among the retroviruses. Since the isolation of the first retrovirus from a chicken tumor, the Rous sarcoma virus, retroviruses have now been isolated from other chicken tumors and from those of cats, mice, humans, and other primates. Two groups of retroviruses can be recognized. Those in the first group cause tumors very efficiently after just short

periods of incubation. They also transform cells in culture, causing them to proliferate wildly with the production of disoriented piles of cells on the culture plate. However, the viruses are defective and cannot replicate. The Rous sarcoma virus is the only exception in this respect to other members of this group. The second category includes viruses that can cause tumors, but with a low efficiency and then only after a long period of incubation. Moreover, although they can enter cells growing in culture, they are unable to transform them. They can replicate in the cells, however, so that cells or animals will shed viral particles that are able to infect other cells.

An explanation of these differences is now at hand. It is known from the well-studied Rous sarcoma virus that its ability to cause tumors is due to the gene, src, the single oncogene it carries. A virus that has a mutant defect in the src gene can infect a chicken cell and replicate but cannot cause tumors. On the other hand, any virus with a mutation in one of the three genes—gag, pol, or env—cannot replicate unless a helper virus is present to perform the missing function. The typical Rous sarcoma virus, unlike other members of the first group, is nonmutant with respect to these three genes. Its oncogene has been added to them without replacing any portion of them. However, in other members of the viral group, the oncogene has caused a deletion by replacing one of three genes or a portion of one of them. Such a virus nevertheless has a nonmutant oncogene that enables it to transform cells. The oncogene appears to vary among the different viruses, because the proteins associated with the oncogenes of the various viruses differ from one another and from the protein encoded by src, the oncogene of the Rous sarcoma virus. The oncogene encodes a protein that has been purified and whose amino acid sequence is known. It acts as an enzyme, a protein kinase, capable of adding phosphate groups to tyrosine. Oncogenes are pursued further in Chapter 17 in relation to cell regulation and differentiation.

DNA tumor viruses

Tumor-inducing DNA viruses are found among the adenoviruses, the polyoma, and the papilloma viruses. A very well-studied polyoma virus is simian vacuolating virus (SV40). The DNA of this virus is double stranded, and the exact sequence of its 5226 base pairs has been determined as well as the locations of various genes (see Chap. 18). SV40 can follow either of two cycles, depending on the host cell. After it infects monkey cells growing in culture, a lytic cycle is initiated that leads to cell death. The monkey cells become infected by a linear viral DNA, which circularizes inside the cell. Both an early and a late phase of the viral lytic cycle can be recognized. Shortly after infection, the virus migrates to the cell nucleus, where viral replication and transcription take place. Two of the proteins encoded by the virus are found during the early stage; these are known as the T and t antigens or proteins. These two proteins are encoded by the same stretch of DNA and are identical in their amino acid sequence at the N-terminal end. The difference between them at the C-terminal end results from two different ways in which the transcript of the region is spliced and introns are removed during the processing of the mRNA. (Differential processing is discussed in Chapter 17 in relation to regulation of gene expression.) The T and t proteins are both required for replication of the viral DNA, which starts about 12 hours or more after a cell is infected. The replication of the viral DNA in turn appears to be necessary for stimulation of the late genes to undergo transcription. The late proteins finally assemble with the viral DNA to form thousands of mature virions that are released 2 days or so after infection, resulting in death of the monkey cells.

If SV40 is allowed to infect mouse cells in culture, a different scenario is followed. Approximately 24 hours after they are infected, the mouse cells begin to show some of the characteristics of malignant cells as the T antigen is produced in the cell nuclei. Late genes of the virus are not expressed, nor do mature virions appear in these cells. Most of the infected cells revert back to the normal state, but a number of them become permanently transformed and can grow into fatal tumors when injected into mice.

During the conversion of normal to cancer cells by SV40 and various other DNA tumor viruses, the viral DNA is integrated into a host chromosome. This can be demonstrated by turning to the technique of DNA–DNA hybridization. If we demonstrate that viral DNA hybridizes with the cellular DNA, we can conclude that viral DNA sequences are present along with certain cellular DNAs. To determine whether or not this is the case, we extract the DNA from the nuclei of transformed cells and fragment it with the aid of a specific restriction endonuclease (Fig. 16-23). The resulting fragments are then placed on an agarose gel and subjected to electrophoresis, a procedure that separates the mixture of fragments on the basis of size, smaller fragments traveling faster down the gel. Fragments of a given specific size can be visualized in the gel as a band. A band at a given location reflects a fragment of a specific size. (Chapter 18 contains details on procedures used in genetic manipulation.) Since our goal is to detect any hybridization between viral and cellular DNA, we must render both DNAs single stranded. The two DNAs must then be incubated together to permit hybridization to take place. Because, hybridization cannot take place when one of the DNAs is in a gel, it is necessary to transfer this DNA from the gel to a surface, where it can come into contact with DNA from the other source. Moreover, transfer from the gel must be done in such a way that the order of the fragments in the gel is preserved.

A technique developed by E. Southern, known widely as *Southern blotting,* enables us to do just this. The gel containing the separated bands of fragments is placed in a solution that denatures the DNA so that all of the fragments consist of single strands. The gel is then laid on a large piece of nitrocellulose paper so that the gel and the filter paper are in direct contact. A buffer is allowed to flow through the gel toward the filter, at right angles to the direction in which the fragments were separated by the electrophoresis. The denatured DNA is carried by the buffer out of the gel onto the filter to which the single-stranded DNA binds tightly. The original positions of the fragments in the gel will correspond almost exactly with their positions on the filter. This routinely used Southern blotting procedure, which enables us to transfer denatured fragments on a gel to a filter, is invaluable in molecular biology.

The denatured fragments are then exposed to denatured labeled viral DNA. Pairing between the viral DNA and any DNA on a chromosome fragment can be detected by autoradiography and indicates that strands complementary to the viral DNA are integrated with the host DNA. The integrated viral DNA thus behaves as inherited elements in transformed cells. When this procedure (Fig. 16-23) is repeated using DNA from different clones of infected cells, the results continue to show that the viral DNA is integrated, but the DNA is not necessarily associated with the same fragments from one clone to the next. This indicates that SV40 virus, unlike phage λ, does not have a single preferential site of integration. It is possible to find clones in which two viruses are inte-

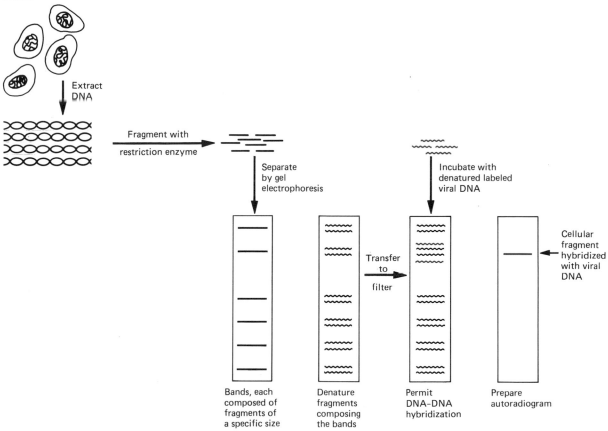

FIG. 16-23. Demonstration of SV40 integration. DNA is extracted from cells composing a population that traces back to one cell that was transformed by SV40. The DNA is exposed to a restriction enzyme that fragments the DNA into pieces of specific sizes. These fragments can be separated into bands by gel electrophoresis, the larger fragments toward the top of the gel. Each band represents a fragment of a specific size. The DNA is rendered single stranded and then transferred to a filter, where it can be used in a DNA–DNA hybridization procedure. After exposure to labeled, single-stranded SV40 DNA, a certain fragment hybridizes with the viral DNA, indicating that a viral DNA sequence is present. The band, representing the specific fragment, can be detected when an autoradiograph is prepared, because of the labeled DNA of the virus. If this same procedure is repeated with DNA from cells of another clone of transformed cells, the fragment that pairs with the viral DNA may be different from the one shown here.

grated close to each other, so that occasionally more than one fragment on the filter may hybridize with the viral DNA.

The T antigens of SV40 and other DNA tumor viruses are now known to be able to transform normal cells to the malignant state. As in the case of SV40, some DNA viruses may be able to transform the cells of some species in cell culture but not the cells of other species. Moreover, although some of these viruses can transform the cells of certain species, the transformed cells, when injected into healthy animals, do not cause tumors to form. This is true even though the cells express T antigen and have malignant characteristics when growing in cell cultures. If these same cells are injected into animals with suppressed immune systems, however, tumors result. This sort of observation indicates that the ability of cells that were transformed in culture to produce tumors depends on interaction between the virally transformed cells and the immune system of the host animal. A low resistance of the transformed cells to the immune system of the host results in their rejection; a high resistance results in the growth of fatal tumors. There is good evidence that the early appearing T antigens of the tumor viruses, by interacting with the host DNA, regulate the level of resistance that transformed cells show to the host immune system and are responsible for the pathway followed. In Chapter 18, we present more details on the T antigen and genes of SV40.

REFERENCES

Alwine, J. C., D. J. Kemp, and G. R. Stark. Method for detection of specific RNAs in agarose gels by transfer to diazobenzylomethyl-paper and hybridization with DNA probes. *Proc. Natl. Acad. Sci.* 74: 5350, 1977.

Arya, S. K., C. Guo, S. F. Josephs, and F. Wong-Stall. Transactivator gene of human T-lymphotropic virus type III (HTLV-III). *Science* 229: 69, 1985.

Benzer, S. On the topology of the genetic fine structure. *Proc. Natl. Acad. Sci.* 47:403, 1961.

Bishop, J. M. Oncogenes. *Sci. Am.* (March): 80, 1982.

Campbell, A. M. How viruses insert their DNA into the DNA of the host cell. *Sci. Am.* (Dec.): 1976.

Cold Spring Harbor Symposium on Quantitative Biology. *Replication of DNA in Microorganisms,* vol. 33. Cold Spring Harbor, NY, 1968.

Contreras, R., R. Rogiers, A. Van de Voorde, and W. Fiers. Overlapping of the VP$_2$-VP$_3$ gene and the VP$_1$ gene in the SV40 genome. *Cell* 12: 529, 1977.

Fiddes, J. C. The nucleotide sequence of a viral DNA. *Sci. Am.* (Dec.): 54, 1977.

Fraenkel-Conrat, H. and B. Singer. Virus reconstitution II. Combination of protein and nucleic acid from different strains. *Biochim. Biophys. Acta* 24: 540, 1957.

Gajdusek, D. C. Unconventional viruses and the origin and disappearance of huru. *Science* 197: 943, 1977.

Gallo, R. C. The first human retrovirus. *Sci. Am.* (Dec.): 88, 1986.

Gallo, R. C. The AIDS virus. *Sci. Am.* (Jan.): 40, 1987.

Holland, J. J. Viruses, slow, inapparent, and recurrent. *Sci. Am.* (Feb.): 32, 1974.

Landy, A. and W. Ross. Viral integration and excision: structure of the lambda att sites. *Science* 197: 1147, 1977.

Luria, S. E. and J. E. Darnell. *General Virology,* 2nd ed. Wiley, New York, 1968.

Nash, H. A. Integration and excision of bacteriophage lambda: the mechanism of conservative site specific recombination. *Annu. Rev. Genet.* 15: 143, 1981.

Nathans, D. Restriction endonucleases, Simian virus 40, and the new genetics. *Science* 206: 903, 1979.

Pratt, D. Single stranded DNA bacteriophages. *Annu. Rev. Genet.* 3:343, 1969.

Ptashne, M., A. D. Johnson, and C. O. Pabo. A genetic switch in a bacterial virus. *Sci. Am.* (Nov.): 128, 1982.

Razin, A. and A. D. Riggs. DNA methylation and gene function. *Science* 210: 604, 1980.

Sanger, F., G. M. Air, B. G. Barrell, and N. L. Brown et al. Nucleotide sequence of bacteriophage φX174 DNA. *Nature* 265: 687, 1977.

Simons, K., H. Garoff, and A. Helenius. How an animal virus gets into and out of its host cell. *Sci. Am.* (Feb.): 58, 1982.

Smith, H. O. Nucleotide sequence specificity of restriction endonucleases. *Science* 205: 455, 1979.

Smith, M., N. L. Brown, G. M. Air, and B. G. Barrell et al. DNA sequences at the C termini of the overlapping genes A and B in bacteriophage φX174. *Nature* 265: 702, 1977.

Southern, E. M. Detection of specific sequences among DNA fragments separated by gel electrophoresis. *J. Mol. Biol.* 98: 503, 1975.

Stahl, F. W. Genetic recombination. *Sci. Am.* (Feb.): 90, 1987.

Stent, G. S. and R. Calendar. *Molecular Genetics. An Introductory Narrative.* Freeman, San Francisco, 1978.

Temin, H. M. RNA-directed DNA synthesis. *Sci. Am.* (Jan.): 24, 1972.

Temin, H. M. Function of the retrovirus long terminal repeat. *Cell* 28: 3, 1982.

Tjian, R. T antigen binding and the control of SV40 gene expression. *Cell* 26: 1, 1981.

Varmus, H. E. Form and function of retroviral proviruses. *Science* 216: 812, 1982.

Wong-Staal, F. and R. C. Gallo. Human T-lymphotropic retroviruses. *Nature* 317: 395, 1985.

REVIEW QUESTIONS

1. Show how the following can be converted into a duplex DNA with extended 5′ sticky ends that would allow the molecule to form a circle.

5′   TTACCTTCATAAAGCGCACATTACCTTCAT   3′

3′   AATGGAAGTATTTCGCGTGTAATGGAAGTA   5′

2. Let A, B, and C represent three strains of *E. coli.* Drops of culture fluid are interchanged, as shown in the following. The recipient tubes are examined in each case for clearing (+) or no change (−). Explain the results.

Drop derived from:	Recipient	Clearing (+)
A	B	+
A	C	+
B	A	+
B	C	+
C	A	−
C	B	−

3. The order of entry of markers in a certain Hfr strain is

T   L   Lac   λ   S

←————————

The following three crosses are made. Answer the questions that follow, and offer an explanation.

(1). Hfr T$^+$ L$^+$ Lac$^+$ Gal$^+$ λ $^+$ S^s x F$^-$ T$^-$ L$^-$ Lac$^-$ Gal$^-$ λ$^-$ S^r

(2). Hfr T$^+$ L$^+$ Lac$^+$ Gal$^+$ λ$^-$ S^s x F$^-$ T$^-$ L$^-$ Lac$^-$ Gal$^-$λ$^+$ S^r

(3). Hfr T$^+$ L$^+$ Lac$^+$ Gal$^+$ λ^+ S^s x F$^-$ T$^-$ L$^-$ Lac$^-$ Gal$^-\lambda^+$ S^r

A. From which of the crosses would it be possible to obtain recombinants that are Lac$^+$ Gal$^+$ λ^-

B. From which of the crosses would it be possible to obtain recombinants that are Gal$^+$ λ^+?

C. From which of the crosses would it be possible to obtain rare recombinants that are S^s

4. A. In what way does transformation differ from specialized transduction?

B. How does conversion resemble and differ from transduction?

5. A culture of E. coli cells that are λ^- Gal$^-$ is infected with lambda phages, which have been residing in cells that are λ^+ Gal$^+$. The concentration of virus used to infect the E. coli is very low. A small percentage of Gal$^+$ cells is isolated. However, none of these transduced cells is found to be lysogenic, and none can be made to liberate virus. Explain.

6. E. coli cells which are λ^- Gal$^-$ may become transduced by phage lambda to Gal$^+$ cells. These cells can give rise to colonies. It is then found that 1 to 10% of the cells in most of the colonies cannot use galactose. Explain.

7. Circle any answer that is correct. In generalized transduction:

A. A helper phage enables the transducing particle to replicate in the host cell.

B. The transduced cell becomes lysogenic.

C. The transduced cell is vulnerable to further infection by the same virus.

D. The transducing virus can be activated in the bacterial cell by zygotic induction.

E. The viral particle can integrate at any one of various sites on the host chromosome.

8. Formation of a phage particle capable of generalized transduction is a very rare event. Consequently, if particles capable of transduction frequently bring about cotransduction of two genes, this is taken to mean that the two genes are closely linked. Suppose a donor strain of bacteria is c$^+$d$^+$ and the recipient is c$^-$d$^-$. Phages capable of generalized transduction are allowed to infect donor cells. Phage particles from this infection are then allowed to infect recipient cells, and recombinants are formed among the recipient cells. The following kinds and numbers of transductants arise. Give the recombination frequency and map unit distance between the genes.

$$
\begin{array}{rr}
c^+d^- - & 150 \\
c^-d^+ - & 300 \\
c^+d^+ - & \underline{650} \\
& 1100
\end{array}
$$

9. Suppose you are studying three genes, r, s, and t, in relation to cotransduction in generalized transduction

experiments. It is found that r and s are cotransduced fairly often. The same is true for r and t. However, s and t are cotransduced with a much lower frequency. What appears to be the gene order?

10. Bacterial Strain 1 cannot grow on a minimal medium, because it is unable to form the essential growth substances m and n. However, it can form the essential metabolite, o. The strain can be represented as m n o$^+$. This strain is infected with a phage capable of generalized transduction. The offspring phages are used to transduce Strain 2. Strain 2 is also auxotrophic, since it is unable to form substance o, and can be represented as m$^+$n$^+$o$^-$. After the second infection, Strain 2 cells are plated on minimal medium. A reciprocal experiment is performed in which Strain 1 is the recipient and is transduced with phage particles derived from an infection of Strain 2. From the results given here, tell the order of the three genes, and explain.

Experiment 1—Strain 1 donor; Strain 2 recipient	Experiment 2—Strain 2 donor; Strain 1 recipient
200 wild type cells per 10^8	20 wild type cells per 10^8

11. One type of T/2 virus is h$^+$r$^+$ and another is h r. The two types are allowed to infect Strain B of E. coli. Plating of viruses from the lysate is made on a mixture of Strains B and B/2 of E. coli. Plaques are then scored for size and cloudiness. Based on the following information, give the percentage of recombination and the viral genotypes represented by the plaques.

Small, clear	42
Small, cloudy	2195
Large, clear	2230
Large, cloudy	33

12. When an experiment is performed such as that described in Question 11, the virus particles emerging from a single cell may be followed further. It is often found that one recombinant class, h$^+$r or h r$^+$, may not be found at all. Explain.

13. Assume that a, b, and c are recessive mutations and are very closely linked. Suppose the following double heterozygotes have been constructed and the resulting phenotypes are as indicated here:

$$\frac{a\ b^+}{a^+\ b} = mutant \quad \frac{a\ b}{a^+\ b^+} = wild \quad \frac{b\ c^+}{b^+\ c} = wild$$

$$\frac{b\ c}{b^+\ c^+} = wild \quad \frac{a\ c^+}{a^+\ c} = wild \quad \frac{a\ c}{a^+\ c^+} = wild$$

Which of the following would appear to be true? Circle any answer that is correct.

A. b and c are engaged in different functions.

B. All three are engaged in the same function.

C. a and b are complementary to each other.

D. a and b are alleles.

E. The three sites are all part of the same cistron.

F. b and c are alleles.

14. Assume that a DNA segment of the infective form of the virus ϕX 174 has the nucleotide sequence: A A G T T A C C A. The single-stranded DNA infects an *E. coli* and mRNA is formed by the replicative form RF₁. Give the sequence of bases in the mRNA that would relate to the preceding segment.

15. The sequence in the following gene has the same sense as the mRNA. Consulting a genetic dictionary, if needed, determine where it is possible in the sequence for two genes to overlap, assuming a difference in reading frames.

    AAATTAGCTCCCGGAGCGTGATGTCTAAAGGT

16. Suppose the genome of a DNA virus contains overlapping genes. The same base sequence can be read in three different frames, depending on the starting point of the messages. The DNA segment depicted here is from the antisense strand of the virus. What would you predict as the amino acid sequence in the translation products if the codons are read three at a time beginning with (1) the first base; (2) the second base; (3) the third base?

    5′ G C T G C T G C T G C T G C T G C T G C T   3′

17. Assume that the "+" strand of an RNA virus enters a bacterial cell. A stretch of the RNA in the "+" strand is as follows:

    AACAGGACGCAG

    A. What will be the nucleotide sequence in the nucleic acid that attaches to the ribosomes of the host?
    B. What RNA sequence is required for replication of the preceding sequence?
    C. Name the enzyme required for replication of the viral RNA.

18. The following is part of the base sequence of a particular viral mRNA:

    5′ A C G C G U U A A U C A A A 3′

    A. An in vitro system is provided with reverse transcriptase, and a DNA strand is synthesized. Give the base sequence in this DNA reading in the direction 5′ to 3′.
    B. What will the results be if actinomycin D is also added to the system?

19. A. In each of the following nucleotide sequences pick out the palindrome and indicate the axis of symmetry.
    1. 5′ CTTGGAAGCTTAATAC 3′
       3′ GAACCTTCGAATTATG 5′
    2. 5′ AACGTTAACTTCCTA 3′
       3′ TTGCAATTGAAGGAT 5′
    B. The palindrome in sequence 1 is cleaved by a restriction enzyme at staggered sites. In the upper strand, the cut is two nucleotides to the left of the axis of symmetry. In the lower strand, it is two

nucleotides to the right. Show the fragments that will be generated after endonuclease action.
    C. The palindrome in sequence 2 is cleaved exactly at the axis of symmetry in both strands. Show the fragments that will be generated by endonuclease action.

20. A particular type II restriction enzyme can attack DNA with the palindrome found in the following DNA sequence. However, a particular methylating enzyme can methylate all the cytosines in the palindrome. Indicate by asterisks in this sequence all the cytosines that would become methylated.

    5′ CCACCGGTCCTCATACCGGTC 3′
    3′ GGTGGCCAGGAGTATGGCCAG 5′

21. A particular oncogenic virus, virus A, is allowed to infect mouse cells growing in culture, and a number of infected cells is transformed to the malignant state. From single, transformed cells, clones are established: 1, 2, 3, and so on. DNA is extracted from cells constituting a clone and is then subjected to a specific restriction enzyme. The resulting fragments are placed on a gel, subjected to electrophoresis, and then denatured. The denatured fragments are transferred to a filter by Southern blotting and then exposed to denatured labeled DNA from the oncogenic virus A. DNA is also extracted from nontransformed cells and handled in exactly the same way as described. Autoradiographs are then prepared for each DNA source. When the autoradiographs are compared, the following picture is seen. Explain the results.

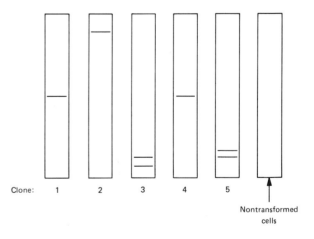

Clone: 1   2   3   4   5

Nontransformed cells

22. A. In the procedure discussed in Question 21, why was it necessary to transfer the denatured fragments from the gel?
    B. Suppose oncogenic DNA virus B is allowed to infect other mouse cells in culture. Exactly the same procedure is followed as was discussed in Question 21. Explain the following results of a comparison of autoradiographs.

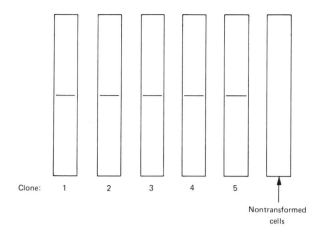

Clone: 1   2   3   4   5

Nontransformed
cells

23. The MS$_2$ virus of *E. coli* is a single-stranded RNA phage in which the RNA serves as both mRNA and the genetic material of the virus itself. The RNA codes for a few genes, one of which governs the formation of the coat protein. This protein consists of 129 amino acids, and the exact amino acid sequence has been established. Work was undertaken to determine the precise nucleotide sequence of the gene encoded for the amino acid sequence of the coat protein. In the procedure followed, the RNA of the virus was fragmented and the fragments isolated on gels. The nucleotide sequence was then determined for each fragment. Here we show a sequence of 15 amino acids in the coat protein along with the nucleotide sequences of four fragments.

NH$_2$–pro-phe-tyr-ala-thr-ala-asn-ser-
  1   2   3   4   5   6   7   8
gly-ile-tyr-arg-gly-gly-val –COOH
 9 10 11 12 13 14 15

Fragment 1:   U U U C U C U G U A A A A A G
                              A U G
Fragment 2:   G C A A A C U C C G G U A U C
                              U A C C G U
Fragment 3:   U G U U G G C A U U U U G A U
                              C A C C U C C A U
Fragment 4:   U U U U A U G C A A C U G C A
                   A A C U C C G G U A U C U A C

With the aid of the genetic dictionary, determine which fragments relate to this portion of the gene for coat protein, and establish as much of the nucleotide sequence as possible for the gene. (Assume each fragment starts at the 5′ end of a complete codon.)

17

CONTROL MECHANISMS AND DIFFERENTIATION

A moment's thought tells us that the genetic information is the same in all the cells of the body of a multicellular organism. From the zyogote onward, a human arises through repeated nuclear divisions; except for cells in the germ line, all of these are mitotic. The first few mitotic divisions of the embryo produce cells that are all essentially the same, but then something happens to cause them to diverge and give rise to different cell types. If specialization or differentiation had never occurred in living things, only masses of identical cells could form, and life as we know it would not exist.

Not only is differentiation required for the orderly development of a normal organism, it must not be upset once it is established. A pathological disorder, such as a malignancy, reflects some type of change in the cell that causes it to depart from its normally defined role. Differentiation and its maintenance depend mainly on cell mechanisms that control the formation of specific protein products. Obviously, only a part of the total genetic information is expressed in any one cell. A liver cell will not translate information pertinent only to a cell of the eye, even though all of the information is there in the liver cell nucleus. How do diverse cell types arise from a population of identical cells? What devices permit only certain genes to function in a given kind of cell? These questions are

still far from completely answered, but more information is being assembled on possible control mechanisms in eukaryotes. Pioneer genetic studies with microorganisms have contributed important information that has formed a point of departure for investigations into differentiation and cell change in higher species.

Genetic control of enzyme induction in bacteria

Several critical investigations have depended on the isolation from bacteria of certain mutants in which control mechanisms have gone awry. For example, a normal cell of *E. coli* can use the sugar lactose, when it is supplied in place of glucose. The ability to split the lactose, a disaccharide, into the monosaccharides glucose and galactose depends on a specific enzyme, β-galactosidase, which the normal cell can manufacture. However, in the absence of lactose, this enzyme is not needed, and it is not produced. Indeed, it would be a waste of energy for the cell to produce any protein it does not require. Normally, therefore, the enzyme is made only when lactose, the substrate of the enzyme, is supplied. Such a cellular reaction is called induction, the formation of a specific enzyme in response to the presence of the substrate. However, in populations of bacteria, certain mutant cells arise in

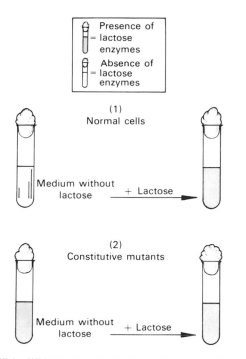

FIG. 17-1. Wild-type cultures and constitutive mutants. Normal *E. coli* cells do not waste energy by producing the enzyme beta-galactosidase in the absence of the enzyme's substrate (1). A culture of the wild *E. coli* cells can be induced to produce the enzyme when lactose is supplied as the sugar source. In contrast, constitutive mutants produce the enzyme in both the absence and presence of the lactose (2). These mutants have some defect in a control mechanism.

which enzyme is formed regardless of the presence of the substrate. These are called constitutive mutants, and they produce large amounts of the enzyme in both the absence and the presence of substrate (Fig. 17-1).

Associated with β-galactosidase are two other enzymes, β-galactoside permease and galactoside transacetylase. The former is involved in the transport of lactose across the cell membrane into the cell, where it is cleaved by the galactosidase into glucose and galactose. The function of the transacetylase is still unclear. All three enzymes are formed by the constitutive mutants in the absence of the substrate lactose. Chemical analysis of these three protein products in the constitutive mutants shows them to be normal. Therefore, nothing has gone wrong to alter the composition of the protein. Instead, a mutation has interfered with the cell's ability to produce the three enzymes only when they are needed. The constitutive mutant cell cannot control the level of enzyme pro-

duction; it produces normal enzyme indiscriminately, a process that wastes cell energy.

The lactose system is just one of several in which mutations are known that can upset normal controls, but it has been studied in the greatest detail. The information gained from it seems to apply to the other systems as well and has permitted Jacob and Monod to formulate a model of genetic control. This model is based on the results of crossing various kinds of mutants in different combinations. As in the case of any other proteins, there are also mutations that can alter the structure of the three enzymes of the lactose system. From crossing cells containing mutations of this sort, it has been possible to map the sites of the genes that code for the formation of the three enzymes. Mapping has shown (Fig. 17-2A) that the three loci are clustered together in a sequence on the chromosome (Z, Y, and A for the galactosidase, the permease, and the transacetylase, respectively). A mutation in any one of these can alter the structure of its polypeptide product. Since these three genes direct the amino acid sequences in polypeptides, they are structural genes.

FIG. 17-2. The lactose operon. (A) Three enzymes are involved in the metabolism of lactose. Each is controlled by a structural gene, and the three genes map together in a portion of the *E. coli* chromosome. These genes undergo transcription to form mRNA when the cell is induced by the presence of lactose. A mutation within any one of the genes can alter the structure of the protein it governs. (B) Other genes, the operator and the regulatory gene, can affect the lactose system. However, a mutation within either of these does not alter the structure of any one of the three enzymes. Instead, the mutation can result in a failure of the enzymes to be produced at the appropriate time. The operator is located adjacent to the structural genes. The promoter region to which RNA polymerase attaches to begin transcription is next to the operator. The structural genes for the enzymes plus the operator and promoter compose an operon. All the genes and sites within an operon are transcribed on the same mRNA. The regulatory gene (I) maps on the other side of the promoter.

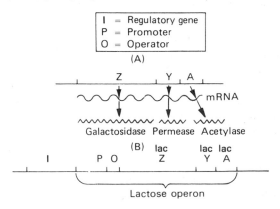

It is important to understand that mutations which affect the regulation of enzymes and produce constitutive mutants do not occur in their structural genes. Instead, they map in other locations. One kind of constitutive mutation maps in a portion of the chromosome that is a distance away from the three clustered structural genes. However, these mutations result in the unregulated production of the three enzymes of the lactose system. Such mutations enabled Jacob and Monod to identify a regulatory gene, a gene that is not concerned with the formation of an enzyme but rather with a regulator substance (Fig. 17-2B). But the regulatory gene is not the only genetic factor involved in control of the three genes of the lactose system. A genetic region called the operator has also been recognized as a result of mutations that have arisen within it. It has been mapped and shown to be located adjacent to structural gene Z. Like the regulatory gene, it is concerned with control and not with the amino acid sequence of an enzyme. (It was gene Z, along with the promoter and operator, that was isolated using the transducing phages, as described in Chapter 16.)

Jacob and Monod were able to show that the regulatory gene is responsible for the production of a repressor substance that is produced by the gene in the absence of the substrate lactose (Fig. 17-3A). When lactose is not present in the cell, the repressor can combine with the operator. As a consequence, RNA polymerase that attaches to the promoter site cannot transcribe the information of the three genes into messenger RNA, even in the presence of cylic AMP and CAP (see Chap. 12). Transcription is thus effectively blocked by the interaction of the repressor with the operator site.

Now let us suppose that lactose is added to the medium and enters the cells (Fig. 17-3B). The substrate lactose has the ability to act as an effector, which binds the repressor molecules. The latter would normally be present only in a small amount. As a result of the interaction between repressor and effector (lactose), the operator is freed. The RNA polymerase bound at the promoter encounters no barrier and can proceed with transcription. However, if a constitutive mutation arises in the regulatory gene, the latter may be unable to produce a normal repressor. This altered repressor may be unable to combine with the operator. Consequently, the RNA polymerase is free to transcribe the structural genes at any time. Transcription can thus take place even in the absence of the substrate when the three enzymes are not needed (Fig. 17-3C).

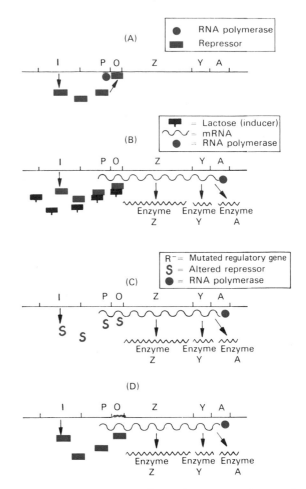

FIG. 17-3. The repressor. (A) The regulatory gene produces a substance that can combine with the operator. In the absence of the substrate, lactose in this example, combination of the repressor and the operator prevents the RNA polymerase from moving along the DNA to bring about transcription. (B) When an inducer is added (the lactose molecules), it binds to the repressor. The operator site is freed. The polymerase can now move along the DNA to transcribe it into mRNA. (C) If a mutation arises in the regulatory gene, the repressor may be altered in structure so that it cannot combine with the operator. Therefore, even in the absence of the lactose, the RNA polymerase is free to transcribe genes in the operon. The mRNA is then translated, and the three enzymes are formed in the cell, even when they are not needed. (D) A mutation can affect the operator in such a way that it no longer can combine with the normal repressor. The result, again, is freedom of the RNA polymerase to transcribe genes in the operon. The three enzymes are formed constitutively, without regard to presence of the substrate.

If a constitutive mutation occurs in the operator gene, again the enzymes are produced without regard to the presence of substrate. This is so because the mutation in the operator renders it incapable of binding with the normal repressor (Fig. 17-3D). Consequently, the RNA polymerase is again free to transcribe the structural genes. One should note that the three structural genes are undergoing transcription together on the same stretch of messenger RNA. Thus, the promoter, the operator, and the adjacent genes define a unit—a unit of transcription called the operon. An operon is composed of an operator, a promoter, and associated structural genes (or cis-trons), which are all transcribed on the same piece of mRNA. A piece of mRNA that contains the coded information for two or more genes is polycistronic; it carries the information that can direct the formation of more than one protein. In transcription of the operon, the promoter region is apparently not included in the transcript, but the operator or a portion of it may be. We can now appreciate more fully the need for punctuation codons in the mRNA to distinguish between two or more polypeptide chains controlled by two or more different genes in an operon. The RNA codons—UAA, UAG, and UGA—along the messenger are needed to prevent a runon protein. Without them, the ribosome would be unable to complete one protein and start the next. The three enzymes in the lactose operon would all be hooked together and would not exist as separate functional entities.

The model of the operon explains beautifully the control of inducible enzymes. Jacob and Monod did not identify the repressor, but they had sufficient evidence to propose its existence. For example, it was possible through sexduction (see Chap. 15) to obtain *E. coli* cells that were diploid in the lactose region of the chromosomes (Fig. 17-4). Partial heterozygotes were derived that possessed a normal regulatory gene and a normal operator on the sexduced fragment; the chromosome contained a defective regulatory gene. Such a cell behaves normally; it is not constitutive but produces enzyme only when needed. Figure 17-4 shows clearly why this is so. The defective repressor cannot bind with the operator. If the cell were haploid, lacking the F factor with the chromosome segment, enzyme would be produced indiscriminately (Fig. 17-3C). However, the presence of a normal regulatory gene anywhere in the cell, even on the episome, gives rise to the production of normal repressor. This is a substance that can diffuse throughout the cell and bind with both normal operators, the one on the chromosome and the one carried by the sex factor. The structural genes on both DNAs are therefore not transcribed unless substrate can tie up the normal repressor. In effect, the constitutive regulatory gene mutations are behaving as recessives to the normal gene form.

In contrast to this, the constitutive operator mutations were shown to be dominant. A study of Fig. 17-5 shows why this would be expected. In the diploid state, one of the DNA segments has a normal operator

FIG. 17-4. A heterogenote for the regulatory gene. A bacterial cell may possess a fertility factor that carries a lac operon and a regulatory gene. The regulatory gene on the chromosome may be defective and produce ineffective repressor. However, if a normal regulatory gene is present in the sex factor, it will cause normal repressor to be formed. This combines with the operator on the chromosome and also with the operator associated with the lac genes on the F factor itself. The cell consequently behaves normally and does not produce enzyme in the absence of the substrate. The regulatory gene mutation is thus behaving as a recessive to the normal.

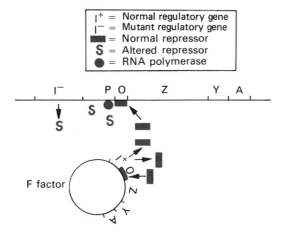

FIG. 17-5. A heterogenote for the operator. If a mutation occurs in an operator gene, it may not bind with the normal repressor. This means that those structural genes adjacent to it (in the cis arrangement) will be freely transcribed. The figure shows a normal operator in the sex factor combining with the repressor, which is being produced by both regulatory genes, one on the chromosome and one on the episome. The operator on the chromosome does not combine with the repressor, so that the three enzymes are produced in the presence or absence of substrate. The operator mutation therefore behaves as a dominant.

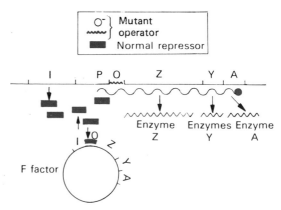

(the F factor in the figure). The repressor substance produced by either normal regulatory gene can bind to it. But this is not so with the mutant operator (the one on the chromosome). Since it cannot bind with the repressor, the structural genes adjacent to it (in the cis arrangement) are free to be transcribed in both the absence and the presence of the substrate.

Repressors and their interactions

Results such as these from various arrangements of mutant and normal alleles in operator, regulatory gene and structural genes allowed Jacob and Monod to propose the model of the operon in 1961. However, it was not until later in the decade that genetic repressors were actually isolated by the team of Gilbert and Muller-Hill, working with the lactose operon, and by Ptashne studying a repressor of phage in *E. coli*. It was expected that the repressors would contain an RNA component to allow them to recognize complementary DNA sequences in specific operators. Surprisingly, the repressor turned out to lack an RNA component and to be a rather ordinary acidic protein. The lactose repressor is an aggregate, consisting of four identical polypeptide subunits. Each is composed of 147 amino acids whose exact sequence is known. Since the regulatory gene is coded for the amino acid sequence of a polypeptide, it too is a structural gene, even though the polypeptide for which it is coded is regulatory rather than enzymatic.

In one of the experiments, isolated repressor was made radioactive and mixed with the appropriate DNA. For example, lac repressor was placed in a centrifuge tube along with DNA known to contain the lactose genes. When spun in a density gradient, the repressor bound tightly to the DNA and sedimented with it. In contrast, when the repressor was spun in the centrifuge tube with DNA from cells containing lactose-constitutive mutations, the repressor did not bind. These results provided strong support for the model of the operon and the concept of a repressor that physically blocks genetic transcription.

Extensive investigations have been made on the lactose operon over the years. When the Z gene was isolated using transducing phages (see Fig. 16-16), the adjacent operator and promoter regions were obtained along with it. The exact nucleotide sequence of the operator region is now known. This has been accomplished by subjecting *E. coli* DNA to nuclease treatment. Lactose repressor protein, when added to the DNA before exposure to the nuclease, protects from digestion the operator region to which it binds.

Analysis of the protected portion has shown it to consist of about 26 base pairs and to contain two regions where there is a degree of symmetry:

$$5' \quad \overrightarrow{\text{A A T T G T}}_____\text{A C A A T T} \quad 3'$$
$$3' \quad \text{T T A A C A}_____\underleftarrow{\text{T G T T A A}} \quad 5'$$

(You will recall that such inverted repeats were discussed in relation to terminator sequences in Chapter 12 and to restriction enzymes in Chapter 16. We will continue to encounter such DNA regions throughout the following discussions. Inverted repeats and palindromes apparently act as signals in various kinds of molecular interactions.) The particular portion in the case of the lac operon can apparently be recognized by the repressor, two subunits of which bind to the regions of symmetry. Alteration within the 26-base-pair portion as a result of mutation can result in defective binding of the repressor. Not more than 20 copies of the lac repressor are normally found within a cell.

The 26 base pairs of the operator region are now known to contain the starting point for synthesis of lac messenger RNA. The length of the promoter region for the lactose operon is 80 base pairs long and contains a CAP site as well as an RNA polymerase binding site. It has been shown that a bit of overlapping of nucleotides exists between the end of the lac promoter region and the beginning of the lac operator. We can now picture how the bound repressor prevents transcription in this operon. It does so by preventing RNA polymerase from binding to the appropriate signals in the promoter region where the overlap occurs (Fig. 17-6). The operator and promoter are referred to as controlling elements, because they respond to positive or negative environmental signals, thus controlling the transcription of the gene or genes with which they are associated.

It appears, however, that even when the lac repressor is removed, the rate of transcription of genes in the operon is still low. The operon requires an activator, a regulator molecule that acts in a positive fashion in controlling transcription, as opposed to a repressor. CAP is the activator, and it must bind to the promoter of the lac operon if RNA polymerase is to bind at a high rate. As a positive regulator molecule, CAP is required for maximum transcription of genes coding for enzymes needed to use sugars such as lactose, arabinose, and galactose. If glucose is present in the environment, certain microorganisms such as *E. coli* will use it preferentially, even if other sugars are present. The presence of glucose in the medium

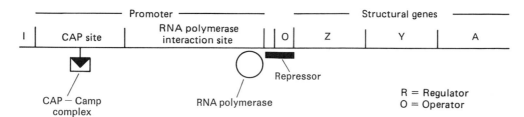

FIG. 17-6. Blockage of transcription by repressor in the lac operon. The lac promoter has been found to contain a CAP site to which the CAP–cAMP complex binds before transcription can proceed. The promoter also contains a region with signals for binding RNA poly-merase. At the end of this region is the site at which RNA polymerase starts transcription. This site overlaps with the beginning of the operator. If repressor is bound to the operator, this site is blocked to RNA polymerase, and transcription cannot proceed.

brings about inhibition of the synthesis of enzymes required for metabolism of the other sugars, a phenomenon known as *the glucose effect.* A metabolite of glucose is somehow able to reduce the level of cAMP, which CAP requires to stimulate the promoter, and hence transcription of an operon such as the lac operon. Consequently, transcription of the operon is reduced as cAMP levels fall and glucose levels increase. Conversely, when glucose amounts decrease, cAMP level increases, and so does the positive effect of CAP as a stimulus of transcription. We see here clearly another aspect of control, positive control as opposed to negative control that is effected by repressors. In various microorganisms, the maximum amount of energy is provided by glucose as the carbon source. The lactose operon and others that show the glucose effect illustrate very well the interactions between positive and negative controlling factors for the highest degree of cell efficiency.

The structural gene coded for the amino acid sequence of CAP and the one coded for the amino acid sequence of the enzyme required for cAMP formation have also been identified. A mutation in either of these structural genes greatly reduces the amount of transcription from those operons such as lac, which show the glucose effect. Moreover, operons are known in bacteria (the L-arabinose and L-maltose operons, e.g.) in which a specific regulatory gene associated with a specific operon is coded for a substance required for the activation of the structural genes in the particular operon. Both kinds of control, positive and negative, are undoubtedly involved in regulating most structural genes.

Repression of enzyme synthesis

The amino acid tryptophan is synthesized in five steps. The completion of these steps requires five different polypeptides, encoded by five structural genes that lie adjacent to each other on the *E. coli* chromosome (Fig. 17-7A). Three enzymes, essential to the five steps leading to tryptophan synthesis, are derived from these polypeptides. The last step in forming tryptophan (the union of indole and serine) requires the enzyme tryptophan synthetase. This enzyme is composed of two of the polypeptides (β and α) encoded by the structural genes B and A, respectively. The five contiguous structural genes, associated with an operator and promoter, constitute the tryptophan operon. The regulatory gene, trp R, associated with the trp operon, in contrast with the picture in the lac operon, lies an appreciable distance away from the operator with which it interacts. Trp R is responsible for the production of the trp repressor which, like the lac repressor, functions as a tetramer composed of four identical subunits.

Unlike the lac operon, the trp operon in *E. coli* is repressible and illustrates the repression of enzyme synthesis. The bacteria need tryptophan for growth and usually must synthesize it. Therefore, they manufacture tryptophan synthetase and the other catalysts involved in the synthesis of tryptophan. However, if tryptophan is supplied to the cells, production of the five polypeptides ceases as the five structural genes are repressed. This is economical, because it would be wasteful for the cell to make all five polypeptides and the end product, tryptophan, when a ready supply of the required substance exists. The operon model explains this repression in a manner similar to that used for the lactose operon (Fig. 17-7A and B). The main difference between the tryptophan and lactose control systems is that the operator in the former does *not* react with the repressor in the absence of the effector, in this case the tryptophan. But when tryptophan is added to the cell, it acts as a *corepressor* and alters the repressor. It is this changed

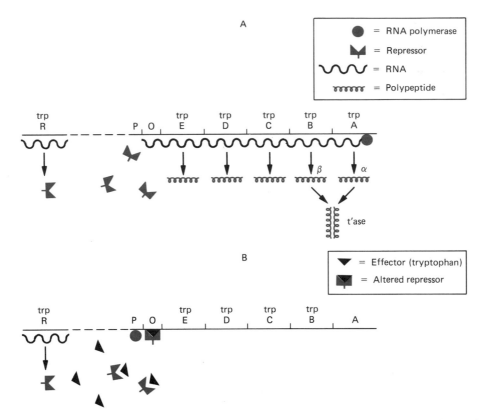

FIG. 17-7. The tryptophan operon and repression of enzyme synthesis. *(A)* The trp operon includes five structural genes lying next to each other, each coded for a polypeptide needed as a catalyst to form tryoptophan. Two of these, β and α, encoded by tryp B and trp A, associate to form tryptophan synthetase (t'ase). The regulatory gene, trp R, is responsible for producing a repressor that cannot by itself combine with the operator (O). Consequently, transcription is allowed from the promoter (P). The five polypeptides are formed, β and α combining to form t'ase. *(B)* If tryptophan is added to the cell medium, t'ase and the other proteins needed to catalyze the synthesis of tryptophan are no longer necessary. The tryptophan acts as an effector or corepressor, which alters the repressor. This altered repressor can now combine with the operator and block transcription.

repressor that can bind to the operator and block transcription. In a repressible system, as seen in the example of tryptophan, therefore, the effector (the end product in a reaction) alters the repressor so that it *can* combine with the operator and block transcription. In inducible systems, on the other hand, as in the lactose operon, the effector (substrate of an enzyme) alters the repressor so that it *cannot* combine with the operator, thus allowing transcription to take place (See Fig. 17-3B).

Choice between lysis and lysogeny in phage lambda

An excellent example of gene control that entails more than one operon is provided by phage λ. Whether λ establishes itself as prophage after entry into the cell or enters a lytic cycle depends on the interaction and balance among six regulatory proteins that form in the host cell and are coded by the DNA of viral genes. Establishment of the virus as prophage depends ultimately on the presence of the main lambda repressor protein, which is formed following transcription of one of the viral genes. As noted in Chapter 16, phage λ is linear in the phage particle, but it becomes circularized shortly after the *E. coli* cell is infected. It is in this form before it becomes integrated to take up residence as prophage.

When a genetic map of lambda is presented, the letters R and L (for right and left) are used to designate the direction of synthesis of two mRNAs, transcripts of genes N and cro, which are produced very early after the entry of lambda (Fig. 17-8). The designation also indicates that mRNA synthesis is taking

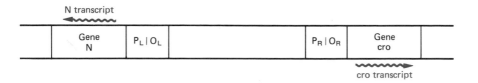

FIG. 17-8. Direction of transcription from operators in phage λ. Not all of the genes of phage λ have the same sense strand. When a map is presented, the symbols R and L are used to distinguish two different promoter/operator regions. Hence, a left promoter/operator place on opposite strands of the λ DNA. When a map of λ is observed (Fig. 17-9), it can be seen that genes with similar or related functions tend to cluster together. Gene clustering often occurs with related genes in organisms from bacteria to diverse eukaryotic species, including the human.

(P_L/O_L) and right one (P_R/O_R) can be recognized for genes that are expressed very early after phage infection (genes N and cro). The transcripts of these genes are seen to be derived from opposite strands of the DNA.

Let us first follow the chain of events that take place when a lytic cycle results following lambda infection. In the lytic cycle, three sets of genes and three stages are recognized to indicate the timing of events in the cycle. These are *immediate early, delayed early,* and *late.* Right after the entry of λ, the RNA polymerase of the *E. coli* cell recognizes two promoter–operator regions, P_L/O_L and P_R/O_R. Transcription from these two promoters occurs in opposite directions, as noted earlier and as the names imply. Transcription from P_R leads to the formation of mRNA from gene cro. (Follow Fig. 17-10 throughout the discussion.) Transcription of this gene gives rise to cro protein, which can interact with the two operators. The importance of this will become clear a bit later. Transcription from P_L leads to formation of an mRNA from gene N and the formation of N protein. This protein is necessary for the next set of genes, the delayed early genes, to be read. The N protein combines with the host RNA polymerase and modifies it so that it ignores termination signals at the end of both the N and the cro genes. (Refer to Fig. 17-9 for positions of genes.) Lacking N protein, transcription would halt at these points. N product is thus an activator that enables transcription to continue and to yield transcripts of two delayed early genes, cII and cIII. These transcripts are not involved in the lytic cycle, and so we will return to them a bit later.

N protein also enables the delayed early genes O, P, and Q to undergo transcription. Genes O and P are required for lysis, because their products are needed for replication of λ DNA. The fourth delayed early gene, Q, is coded for Q protein, which, like N protein, behaves as an activator or antiterminator. It is required for the late genes to undergo transcription. These transcripts of the late genes are coded for the protein of phage heads and tails and also for the enzyme, a lysozyme, which is needed for the bursting of the host cell. We see here in phage λ a neat regulatory system that enables critical events in the lytic cycle to follow an orderly time sequence.

Let us now examine the relationship of these events to the lysogenic cycle of λ. We noted that two delayed early genes, cII and cIII, were activated by N protein (Fig. 17-10). Both of these genes encode products that also have activator properties. The two products can bring about transcription from a promoter, P_{RE},

FIG. 17-9. Partial map of phage λ showing the relative positions of some genes involved in the lytic and lysogenic cycles. Note that genes that have similar functions tend to occur in clusters. Two promoter/operator regions, P_L/O_L and P_R/O_R *(red),* are found on either side of gene cI. The red arrows indicate the different directions in which transcription occurs from these two regions. A third promoter, P_{INT}, is associated with genetic sequences for integration of phage into the chromosome of the host.

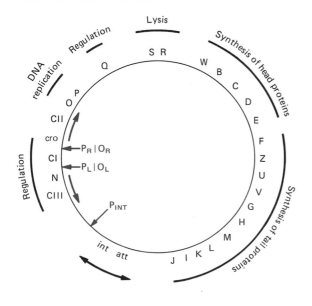

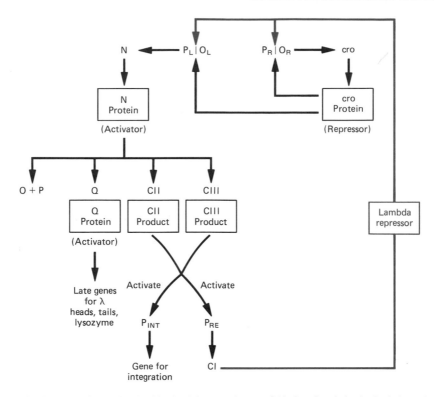

FIG. 17-10. A scheme showing some factors involved in the lytic and lysogenic cycles of phage λ. Whether the virus establishes itself as prophage or follows a lytic cycle depends on a balance of several factors. Critical to the choice is the balance between cro protein, a repressor, and the main λ repressor *(red)*, encoded by gene cI.

the promoter for the establishment of the main repressor protein. When formed, this strong repressor has an affinity for the two operators, O_L and O_R. These operators overlap the corresponding promoters, P_L and P_R. When the λ repressor is bound to the operators, transcription can no longer occur from P_L and P_R. Thus, N protein is no longer produced. Consequently, the genes involved in the delayed early and late parts of the lytic cycle are not activated. Note from Fig. 17-10 that the cII and cIII products also activate another promoter, P_{INT}, the promoter for the integration of the phage (also see Fig. 17-9). Remember that phage integration requires a crossover event between phage and host-cell DNAs. The crossing over requires two proteins, one coded by a viral gene and one by a gene of the cell (see Chap. 16). In this case, the viral gene coded for the crossing over is under the control of P_{INT}, which in turn requires activation by cII and cIII products.

But what determines whether lysis or lysogeny occurs, since we see that N protein activates not only genes O, P, and Q of the lytic cycle but also cII and cIII, which are required for lysogeny? This takes us back to the very beginning of the story (Fig. 17-10). Recall that right after infection, the immediate early gene cro is read. This gene encodes cro protein, which also has repressor properties. Like the λ repressor protein, it too can combine with operators O_R and O_L. It is not as strong a repressor, however, as the main λ repressor, does not bind as well, and is less stable. So although it slows down transcription from O_R and O_L, it does not completely prevent it. Therefore, some N protein is produced, even though the cro repressor is bound. Just how much cro binds is central to the decision whether lysis or lysogeny will occur. Above a certain level, cro causes a decrease in the amount of N protein, which causes lysis to be favored. This is so because at the low level less activation of cII and cIII occurs. Their products are required for activation of the promoter for the repressor protein, P_{RE}. A low level of the cII and cIII products is insufficient to activate P_{RE}, and hence strong repressor is not formed to any extent. However, this lower level of N *can* sufficiently stimulate genes O, P, and Q, which are

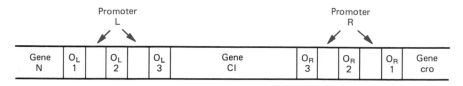

FIG. 17-11. Operator regions of phage λ. Each operator is composed of three binding sites. Mutations in these alter the binding of λ repressor. Promoter mutations map between these operator sites, indicating that the promoters and operators in each of the two operons overlap.

required for the lytic cycle. Even though a lesser amount of N and even Q products is produced, this is still sufficient for completion of the lytic cycle. Therefore, gene cro is the critical gene in the choice between a lytic or a lysogenic pathway. Above a certain level of cro protein, a lytic cycle will be followed; below this level, a lysogenic one will occur. The amount of cro protein can be influenced by such factors as temperature, the genetic constitution of both λ and the host, and the metabolic state of the host cell.

A moment's thought tells us that once the lysogenic state is established, no more N protein will be produced. Therefore, lysis does not occur, but neither does stimulation of cII and cIII, whose products are needed to stimulate the promoter of the repressor gene. A small amount of repressor must be synthesized to replace any that breaks down and thus to maintain lysogeny. This small amount results from transcription from a promoter other than P_{RE}—one called P_{RM}, for promoter for repressor maintenance. The control of this promoter depends on the concentration of lambda repressor itself. When lambda repressor falls to a low level, the repressor somehow stimulates its own synthesis, whereas it inhibits transcription when it is present in higher amounts.

The monomer of the lambda repressor, isolated by Ptashne, has a molecular weight of 26,000, and it may bind to O_R and O_L as a dimer or a tetramer. Ptashne has succeeded in isolating both operators and has found that each is really composed of three regions separated by spacers (Fig. 17-11). Mutation analysis indicates that these spacer regions are promoter sites. Recall that the operator and promoter regions overlap in the lac operon. The arrangement is more striking in the case of phage λ and accounts for the fact that the two regions are referred to as P_R/O_R and P_L/O_L.

There are mutations that affect lambda which are comparable to the constitutive operator mutations in the lac operon. Certain mutant phages cannot establish themselves as prophage because of the inability of the mutant operator to bind to the repressor efficiently. As a result, N and cro genes remain active, and a lytic cycle is favored. Mutations are also known in the promoter regions that overlap with the operators.

In Chapter 16, the phenomenon of zygotic induction was discussed briefly. The activation of the prophage after entry into an $F^-\lambda^-$ cell results from the lack of repressor in the λ^- cell. When the chromosome segment of the Hfr cell bearing λ enters the recipient, transcription can start from both operators, to begin a lytic cycle. It should also be clear why a lysogenic cell is immune to another λ particle. If another λ enters such a cell, the repressor being made by the integrated prophage will immediately bind to O_L and O_R of the entering phage and prevent it from starting the lytic cycle.

Treating lysogenic *E. coli* with ultraviolet light, nitrogen mustard, or most carcinogens will activate the prophage with destruction of the host cells. These agents apparently can damage the host DNA and thus reduce the efficiency of binding of the repressor to the host chromosome.

Transposons and gene expression

Transposable genetic elements (Chap. 15) are very relevant to the subject of control of gene expression. It is now well established that certain transposons may act as biological switches. For example, the insertion of a particular IS element (IS3) into a particular plasmid can turn off the expression of a nearby gene, which can confer tetracycline resistance. Excision of the IS element from the region of the gene restores its expression for drug resistance.

IS elements are also known that can insert at multiple sites on bacterial and phage chromosomes and produce polar mutations. In these cases, insertion of the IS element not only switches off the function of the gene into or near which the element inserts but also affects the expression of other nearby genes. For

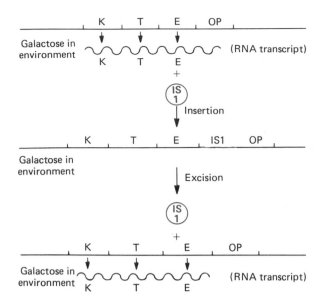

example, the element IS1 can insert into the gal operon of *E. coli* (Fig. 17-12). The gal operon includes three structural genes coded for enzymes involved in galactose metabolism. When the IS1 element inserts next to the operator–promoter region, expression of genes distal to the insertion site is turned off. When the IS element leaves the site, transcription is restored and the three structural genes can be expressed upon induction. IS elements have also been shown to produce a similar effect in the lac operon.

Moreover, it is known that in the case of some IS elements the orientation or direction in which the element inserts is significant. Inserted in one direction, the IS element turns off genes nearby; inserted in the opposite direction, a nearby gene that is unexpressed may be turned on! A particular IS sequence (IS2) can insert in either of two orientations in the operator–promoter region of the gal operon (Fig. 17-13). When it is inserted in a direction opposite to the direction in which the gal structural genes are transcribed, the expression of the genes is decreased. This means that the three gal enzymes are produced in lower than normal amounts upon induction of the enzymes.

If, however, the element is inserted in the same direction in which transcription of the structural genes takes place, it then acts like a positive control element. The structural genes are turned on, and the enzyme production becomes constitutive! The element IS1

FIG. 17-12. Insertion element and polar mutation. The three structural genes (K, T, and E) in the normal inducible galactose operon are coded for three enzymes required in galactose metabolism. The operon undergoes transcription when galactose is in the environment. If the insertion element IS1 inserts in the operator/promoter region (OP) of the operon, the three genes in the operon fail to undergo transcription when galactose is present and the three enzymes are required. When the IS element leaves the OP region, normal, and inducible activity is restored to the operon.

FIG. 17-13. Effect of orientation of IS element. The three structural genes of the galactose operon normally undergo transcription only when induced by galactose. If the insertion element IS2 inserts into the operon in a direction opposite to that of transcription of the structural genes *(left),* transcription is greatly reduced when galactose is present and the three enzymes are required. If IS2 inserts so that it is oriented in the same direction as that of transcription of the operon *(right),* transcription then occurs indiscriminately, and enzyme synthesis is thus constitutive.

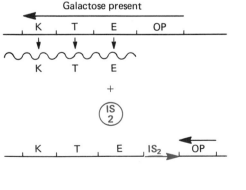

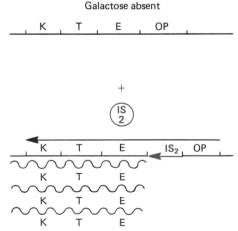

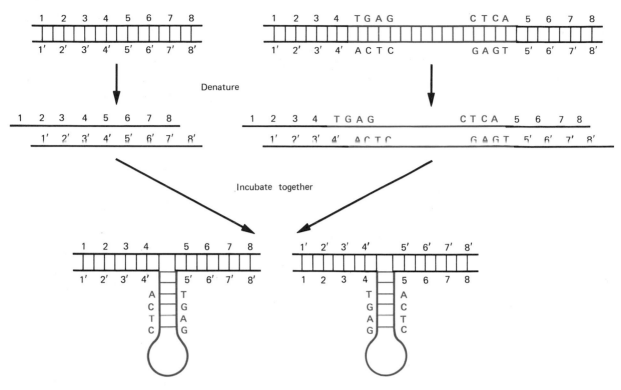

FIG. 17-14. Evidence for an inserted transposon. DNA from phages carrying a specific genetic sequence is denatured, as is the DNA from other phages that carry this same sequence and that show a polar mutant effect. When single-stranded DNA from the two sources is brought together and allowed to anneal, DNA that contains loops can be distinguished. A loop represents an inserted sequence with the base of the loop paired as a consequence of the inverted repeats at the ends of the transposon.

mentioned previously does not show any difference in its effect related to its orientation when it is inserted in the gal operator–promoter region. Transcription of the genes distal to its site of insertion is abolished regardless of its orientation.

It can be shown that the mutated DNA associated with these polar effects has increased in length compared with the corresponding nonmutant DNA. One way to illustrate this entails the use of transducing phages, one group carrying a specific unmutated DNA sequence, the other carrying the same sequence but showing a mutant polar effect. DNA is extracted from the phages and rendered single stranded ((Fig. 17-14). When denatured single strands lacking the mutation are incubated with strands carrying the polar mutation, a loop can be seen with the electron microscope. This loop represents an additional DNA sequence that has been inserted into the normal sequence. The loop reflects the base pairing between the inverted repeats flanking the transposon. The length of the loop is a direct measure of the length of the inserted DNA.

There is no question that transposable elements such as IS sequences may alter gene expression. We have already seen (Chap. 15) that they are actually needed in *E. coli* for the conversion of an F^+ cell into a Hfr type, and hence for the transfer of genes from one cell to another. Perhaps they actually play some basic role in gene regulation. We can speculate that reversal of orientation of certain IS elements such as IS2 can determine whether or not certain genes are expressed. Transposons, or "jumping genes" as they are often called, are the focus of much current research at the molecular level and may prove to have very important implications for normal gene control and expression.

Further unusual aspects of some transposons are seen in a very unusual virus discovered in 1963. This virus has the ability to exist as prophage within a bacterial cell, but, in addition, it possesses properties of a transposable element. The virus can insert itself at several sites in the host chromosome, and when it does so, it causes various kinds of mutations in the cell. Because of this property, the virus was named mu, for mutator.

This peculiar virus, which is both an infectious particle and a transposon, is able to bring about a variety of chromosomal rearrangements. As is the case of other transposons, virus mu has inverted repeat sequences at each of its two ends. When the particle

FIG. 17-15. The virus mu. (1) If phage mu is inserted into a bacterial chromosome, it can bring about the fusion of the chromosome with a phage λ, which enters the cell. Phage λ is then found between two copies of phage mu. (2) Inserted into the chromosome, phage mu can cause transposition of a bacterial gene (his$^+$) to a plasmid in the cell. The transposed gene is then found between two copies of mu. (3) Phage mu can bring about inversion of a segment of the chromosome. A copy of mu is then found at each end of the inserted segment.

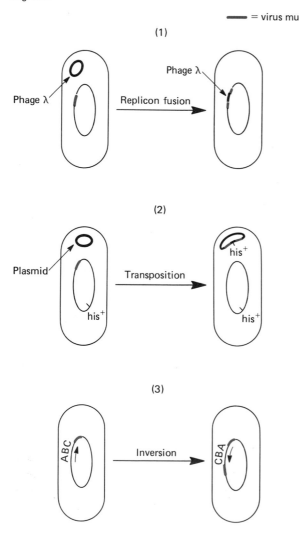

leaves a bacterial chromosome, it picks up pieces of the bacterial DNA on either side of its own DNA. Upon entering another host cell, it sheds the old DNA and can insert at any site in the host chromosome. In the host cell, while it is established as prophage, mu can bring about the fusion of two different replicons. For example, it can cause a lambda virus that has entered the cell to integrate with the cell's chromosome. It can cause portions of the host chromosome to be transposed to a plasmid. Moreover, it can cause deletions and inversions of portions of the host chromosome (Fig 17-15). Following such changes, the particular rearranged DNA segment is found to have a copy of prophage mu at each of its ends. A certain gene of the mu virus must be expressed if the virus is to bring about these various rearrangements of DNA. This same gene is also required for replication of the virus and for its transposition. For mu to replicate itself, it must be inserted into the bacterial chromosome.

Other transposable elements are now know which have properties similar to mu and can catalyze rearrangements of cellular DNA. The recognition of such elements as mu, which have properties of both a virus and a transposon that can alter gene expression, is of great significance in studies of cellular control mechanisms.

Transposable elements in corn

The very existence of transposable genetic elements was first deduced in the 1940s in a brilliant series of genetic studies by Barbara McClintock, who was studying the inheritance of color and pigment distribution in kernels of the corn plant. Certain kernels were mottled or variegated in appearance, because they contained different sized patches of pigment. In these plants, certain genes were being turned off at abnormal times. These plants had also experienced chromosome breakage. McClintock followed many generations of such plants throughout a complex genetic analysis. She concluded that transposable genetic elements (which she called controlling elements) were responsible for the mottled pattern and that these could move about the genome from one corn chromosome to another. As they did so, they acted as switches. Genes could be turned on or off by particular transposable elements. More than one gene could be affected at the same time. Examination of corn chromosomes with the microscope showed that chromosomes with transposable elements could in-

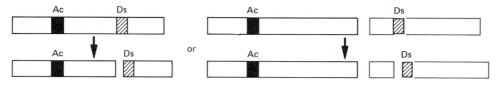

FIG. 17-16. The Ac and Ds elements in corn. The Ds element may bring about chromosome breakage at the site it occupies, but it can do so only if the Ac element is present in the same cell. Ac and Ds can occur on the same chromosome *(left)* or on different chromosomes *(right)*. In both cases, breakage is at the Ds site.

deed undergo breakage and rejoining with the production of alterations in chromosome structure.

McClintock's report in 1947 that a genetic element could move about the genome was considered a peculiarity probably confined to a few very special cases. However, well over 20 years later, with the description of transposable elements in viruses and bacteria, McClintock's discovery was recognized as one of general importance, not just an isolated phenomenon. For her discovery of transposition, now considered a major discovery of the twentieth century, McClintock was awarded the Nobel Prize in 1983.

The two most thoroughly studied transposable elements in corn studied by McClintock are Ds and Ac. The Ds element was the one McClintock first found to be capable of transposition. Associated with the movement of Ds was a breakage of the chromosome at the site or locus occupied with Ds. Hence, McClintock named the locus Dissociator (Ds). However, it became clear to her that although the Ds site is the point where the chromosome breaks, Ds by itself cannot bring about the breakage. Only if another element at a different locus is present will the breakage occur at the Ds position. This second locus was named Activator, or Ac, to denote its ability to trigger chromosome breakage at the Ds site. Ac may occur on the same chromosome as Ds or on a different one. Nevertheless, in either case, Ac can activate breakage at the Ds site (Fig. 17-16).

McClintock soon realized that in a small percentage of the offspring, after certain corn plants were crossed, Ac could move to a new location (Fig. 17-17). This movement could be to the same or to a different chromosome. Moreover, Ac could even disappear. She showed that Ds could also move about, but just as Ds depends on Ac for chromosome breakage, it also depends on Ac for its movement from one site to another (Fig. 17-18). Since movement of Ac or Ds entails loss of the element from the original site and its transfer to a new site, the picture is different from that of transposition in bacteria, in which the transposon is not lost from the original site. Because Ac can move about by itself, it is classified as an autonomous element in contrast with nonautonomous elements such as Ds, which depends on the presence of Ac for its movement. As we will see later in this discussion, Ac contains genes coded for proteins that apparently act as enzymes of transposition. Lacking these genes, the nonautonomous Ds depends on Ac for the proteins required for movement from one chromosome site to the next.

Early in this century, before McClintock's studies, it had been noted that a certain mutation in corn could affect the synthesis of reddish pigment in the layer of cells surrounding and protecting each individual kernel. The unusual feature of this mutation was that, although it prevented formation of pigment, it was very unstable and appeared to revert back to the normal pigment allele very often as each corn kernel was developing. In those cells in which the mutation had reverted, the pigment could be produced. The overall affect was mottling or variegation, a kernel with regions of pigmented and unpigmented cells.

Studies by M. M. Rhoades in the 1930s had shown

FIG. 17-17. Transposition of Ac. Ac can move to a new location, either to one in the same chromosome *(above)* or to one in another chromosome *(below)*. In either case, Ac is lost from its original site. Ac may even be lost from the cell entirely.

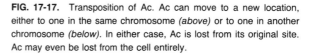

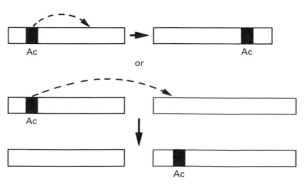

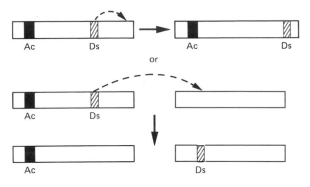

FIG. 17-18. Transposition of Ds. Ds can move to another site on the same chromosome *(above)* or to a site on another chromosome *(below)*. However, its movement depends on the presence of Ac in the same cell. In the absence of Ac, Ds will not undergo transposition.

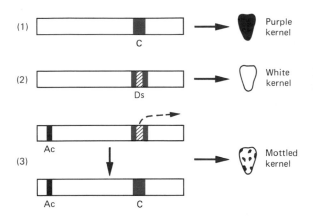

FIG. 17-19. Mutation at the C locus in corn. *(1)* The allele C is necessary for the production of purple pigment in cells of the kernel. *(2)* If the Ds element inserts into C, the allele is no longer functional in pigment production, and cells of the kernel are white. To move, Ds requires the presence of Ac in the genome. If Ac is lost, Ds will no longer move, and all cells of the kernel will be colorless. *(3)* If Ac remains present in the genome, Ds will undergo transposition in some cells of the kernel as it develops. The mutation for colorless will thus revert to the gene form for pigment production in such cells. The kernel will appear mottled, showing patches of purple (pigmented cells in which Ds has moved out of C) against a white background of cells (those in which Ds is still inserted in C).

that certain apparently stable gene mutations in corn can become unstable when a particular gene is present. This results in many reversions from the mutant allele back to the normal. The mutation in this case was one that affected the synthesis of purple pigment in the outermost layer of cells in the endosperm of the kernel, the layer just under the protective covering of the kernel. As a result of the mutation, no synthesis of purple pigment occurs, and a kernel may be white. However, if a certain gene is present in cells carrying the mutation, the mutation reverts frequently, so that the kernel has patches of purple amid the white background of the kernel. Experiments of McClintock with the Ac–Ds system clarified the story of the unstable mutations. McClintock found a case in the course of her studies in which Ds moved from its original site to the site of a gene designated "C." The allele C is required for the synthesis of purple pigment in the outermost layer of the endosperm of the kernel. A mutation at the locus occupied by C can result in lack of pigment. McClintock was able to show that a certain new mutation at C giving colorless endosperm was the result of the insertion of the Ds element into C. The kernels in such a case are white, but only if Ac is absent from the rest of the chromosome complement (Fig. 17-19). If Ac remains present in the genome along with Ds, then chromosome breakage occurs at the C locus, just as it did at the site of Ds before it moved into C. Consequently, the mutation to colorless then reverts back to pigment production in some cells when Ac is present. The result is a kernel with splotches of pigment against a colorless background.

If Ds becomes lost from the C locus in germ cells

of a plant, then offspring plants would have completely pigmented kernels, having reverted completely. The C gene functions perfectly well again after Ds removes itself. If Ac is present in the genome after Ds is gone, no further breakage occurs at the site of C. These unstable mutations studied by McClintock were similar to those studied earlier by Rhoades and other investigators and cast them in a new light.

As McClintock's studies continued, she and others identified still additional transposable elements in corn. Now the techniques of recombinant DNA analysis (Chap. 18) are at last making it possible to analyze the transposable elements in corn on the molecular level and have resulted in the isolation of Ac and Ds. Ac proved to be a DNA element 4500 nucleotides in length. It is very similar to the Tn3 transposon described earlier (in Chap. 15) and is also flanked by terminal inverted repeats. However, it lacks the gene for ampicillin resistance (refer to Fig. 15-22). Two genes were recognized that code for proteins, and these are similar in size to the transposase and resolvase of Tn3. Moreover, the transcription of the two genes occurs in opposite directions, as is the case for Tn3. Ds elements have also been isolated and compared with Ac. The Ds element resembles Ac but is shorter because of a small deletion that occurs in the gene believed to be encoded for the transposase. It appears that a Ds element is a defective Ac because of a missing nucleotide sequence in the gene that may code for the transposase. However, further investigations indicate that some Ds elements may vary in structure from Ac, and these may not all be defective Ac elements. However, all Ds elements do carry the same inverted terminal repeats as Ac. There seems to be little doubt that these inverted repeats are essential for recognition by the enzymes of transposition supplied by Ac, and hence for movement of Ds.

The techniques of molecular biology have made it possible to scan the corn genome to determine the frequency of elements such as Ac and Ds. The results indicate that corn may contain as many as 100 Ac or Ac-like sequences scattered throughout the chromosomes. Most of these scattered sequences, however, do not appear to be complete Ac elements, and probably depend on the presence of complete Ac sequences to supply the enzymes required for them to move. The Ac–Ds system represents only one category of transposable elements in corn. Other families containing autonomous and nonautonomous elements exist. Within each, the elements appear to depend on family-specific enzymes and recognition signals for their movements. It seems very likely that the corn genome contains quite a variety of different types of transposable elements in addition to the best studied family, the Ac–Ds system.

Although it has not yet been demonstrated in eukaryotes, transposable elements similar to those in corn may prove to be involved in the normal control of gene expression during development. As more transposable elements in eukaryotes are identified, their possible role in gene regulation will continue to be a focus of study. Extremely intriguing is the significance of such elements in organic evolution. The ability of some transposable elements to break and rearrange chromosomes makes them good candidates as important agents in changing the genome and providing another source of genetic variability.

Evidence for other mobile elements

In Chapter 13, we learned that eukaryotic DNA is characterized by short interspersed repeats. There are several reasons for suspecting that these sequences are mobile DNA elements, even though no family of repeats in mammals, amphibians, or the sea urchin

has been directly shown to be transposable. For example, analysis of the DNA sequences of Alu family members, typical of mammals, shows that each sequence is flanked on either side by direct repeats, which range in size from 7 to 20 base pairs. Moreover, the repeat on each side of an Alu sequence is unique to each particular Alu sequence. In other words, when one Alu sequence is compared with others, the flanking sequences are found to differ. This is, again, reminiscent of the direct repeats that flank transposons and IS elements in bacteria. A good argument can be made that the direct repeats on either side of an Alu sequence have resulted from duplication of the chromosomal DNA sequence at the target site where the mobile Alu sequence inserted (refer to Figs. 15-26 and 15-27). The RNAs transcribed from these inserted sequences would then function in moving the sequences to other locations throughout the genome. The process involves formation of duplex DNA from these RNA transcripts, using reverse transcriptase, and then the insertion of the DNA sequences into the chromosome at target sites. In this way, the sequences are dispersed throughout the chromosomes. If this proves to be the actual case for Alu and certain other dispersed repeats in higher eukaryotes, they could very possibly play important roles in gene expression, depending on their sites of insertion.

There is also some similarity between insertion elements and the RNA viruses that can use their RNA as a template for the synthesis of a DNA copy (Chap. 16). When these retroviruses synthesize DNA using reverse transcriptase, the final duplex DNA copy may integrate into the host chromosome. This DNA then causes a small segment of the host chromosome to be duplicated. No similarity exists between the nucleotide sequence of the virus and that of the chromosome. Temin has proposed that retroviruses may have evolved from moveable DNA sequences in cells.

Transposable elements in Drosophila and yeast

In 1976, a certain class of DNA was recognized in the fruit fly *Drosophila melanogaster,* which undoubtedly represents an assortment of transposable elements. This DNA consists of repeated sequences dispersed throughout the chromosomes. Some 20 to 30 families of dispersed repeated sequences are known in the fruit fly, and they may constitute as much as 5% of the DNA of the fly. The best studied of the

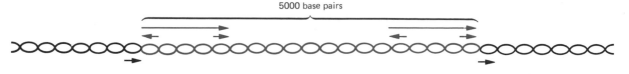

FIG. 17-20. The copia element in Drosophila. A copia element *(red)* consists of 5000 base pairs. The element carries at each of its ends direct repeats, DNA sequences in the same orientation *(large red arrows)*. Each of these repeats includes, at each of its ends, inverted repeats, sequences that have opposite orientations *(small red arrows)*. On either side of the copia element is a small direct repeat of a chromosomal sequence *(dark arrows)*, apparently representing a target sequence.

families has been named *copia* to stress the presence of very large amounts of RNA transcripts of the repeated sequences of this family in the mRNA fraction of the cell. The behavior and structure of copia can be used to illustrate the general characteristics of several other families, even though the nucleotide sequences differ from one family to the next. Copia is present in about 50 dispersed copies per cell. Chromosome mapping has shown that the repeats vary in their locations in different strains, even within individuals of the same strain. The picture appears to be one in which the dispersed repeats wander from site to site within the genome.

Each copia sequence is 500 base pairs in length, although smaller and larger sequences occur in other families. A copia element is flanked on each side by direct repeats, each 276 base pairs long. Each repeat is itself flanked by inverted repeats, each 17 base pairs long (Fig. 17-20). On either side of the copia element, a short (5 base pairs), direct repeat is found that seems to indicate the insertion of the copia element at a target site. The characteristics of copia in its structure and locations all imply that these are transposable elements. Moreover, fruit fly cells grown in tissue culture show that the number of copia repeats per cell increases and new copies of copia appear at new chromosome sites.

Yeast is also known to harbor a family of dispersed DNA sequences that have many of the characteristics of transposable elements. There is evidence that moveable elements in yeast and the fruit fly are the cause of certain mutations. These mutations are unstable and can revert to wild type when an element removes itself from a chromosome site. Moreover, deletions are associated with some of these mutations. The presence of transposons in various eukaryotes seems to be well established. The extent of their occurrence and their exact roles continue to be subjects of discussion among molecular biologists.

Hybrid dysgenesis

An interesting phenomenon associated with transposable elements in Drosophila is known as *hybrid dysgenesis*. This term refers to a collection of correlated genetic abnormalities that arise spontaneously in hybrid offspring after certain strains of flies are crossed. The defects do not occur in somatic cells but are seen in the germ line, which shows a high increase in the frequency of chromosome aberrations, visible gene mutations, abnormal chromosome disjunction, and sterility. One of the factors responsible for hybrid dysgenesis appears to be a transposable element called P. The strains with the element are known as *P strains*, and those lacking it are *M strains*. Strains of flies that have been established for years in the laboratory are M strains, as if the P element has been lost from them. On the other hand, flies in the wild harbor P elements, scattered in large numbers all over the chromosomes. When P strains are compared, however, the chromosome sites occupied by the P elements differ. The various P elements have been shown to contain the same nucleotide sequences, but although they are homologous, deletions apparently occur so that certain DNA segments may be missing. This results in P factors that vary in size. All of the P factors have been shown to possess direct terminal repeats flanked by an 8-base-pair duplication of the chromosome, representing duplication of the target site. As many as 50 P elements may be present in the genome of a given P strain.

Very striking differences are seen when flies of the two strains are crossed. The cross of an M female with a P male yields flies that appear phenotypically normal but prove to be sterile. The reciprocal cross, P female with M male, gives rise to completely normal, fertile offspring. The other crosses, P with P and M with M, are normal. An explanation for these results can be given on the supposition that the P factor is

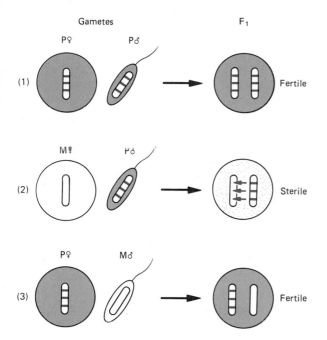

FIG. 17-21. Hybrid dysgenesis in Drosophila. *(1)* Crossing flies from a strain that harbors P elements *(red)* yields fertile offspring, because repressor present in the cytoplasm *(pink)* prevents transposition of P elements, and hence disruption of the genome in the F_1. *(2)* The cross of M females with P males results in sterile offspring, because the cytoplasm of the egg *(colorless)* lacks repressor. In the hybrids, insufficient repressor is present. P elements in the chromosomes from the male can transpose to new locations, disrupting the genome. *(3)* The cytoplasm of the egg contains large amounts of repressor in P females. The P elements contributed by the female cannot undergo transposition in the F_1, because sufficient repressor is present in the F_1 cytoplasm. Consequently, the hybrids are fertile.

coded for a transposase that is required for its movement and insertion into a new chromosome site (Fig. 17-21). It is also coded for a repressor that prevents expression of the translocase gene and hence transposition of the P element. In P strains, the cytoplasm contains the repressor, lacking in M cytoplasm, which inhibits the expression of the P element gene coded for the transposase. Therefore, crosses of P with P are normal, because P elements in these cases cannot move about and insert at new locations. Dysgenesis does not take place because each of the scattered factors is prevented from expressing its translocase gene. On the other hand, when a P male is crossed to an M female, the cytoplasm of the egg lacks repressor. The transposase gene is expressed, and the P element is able to move about and insert at various new locations, causing the hybrid dysgenesis. In the

reciprocal cross, M male with P female, the egg contains sufficient repressor to prevent transposition of the P elements contributed by the female, whereas the M male, of course, contributes no P elements at all.

Besides being of interest as an example of a transposable element in a eukaryote, the P element and related factors are significant from an evolutionary point of view. Hybrid dysgenesis renders sterile offspring when certain strains are crossed. Inability to produce fertile offspring is a primary feature of species divergence. The phenomenon of hybrid dysgenesis suggests that two separated populations of a given species might come to possess different families of elements like the P element. When they are brought together, only sterile offspring would be produced because of hybrid dysgenesis. The two subpopulations would be well along the path to the origin of two species. The recognition of transposable elements such as these are sure to raise many provocative ideas concerning the evolution of species.

Proteins of chromatin

Any approach to the problem of the control of gene expression in eukaryotes must take into consideration the assortment of proteins associated with the DNA of the chromosomes. These proteins fall into two major classes, the histone and the nonhistone proteins. Histones are relatively small proteins (MW 10,000–20,000) and have a very basic nature because of the inclusion of large amounts of the basic amino acids lysine and arginine. This high proportion of positively charged amino acids enables them to bind tightly to the DNA, which carries negatively charged phosphate groups. The total mass of the histones is about equal to that of the DNA to which it is bound.

Five different classes of histone proteins occur in chromatin. Four of these—H2A, H2B, H3, and H4—are exceptionally constant in their composition from one cell type to another and from one species to the next. Of the 102 amino acids composing H4, 100 are identical in chromatin from the cow and in that from the pea plant! Very little change in the amino acid sequence of these four histone classes has taken place over the course of evolution of eukaryotes. Such a high degree of conservation implies that the histones are perfected for an important role they must play in the structure of the chromosome, a function so crucial that almost no amino acid substitution can be tolerated.

Except for one central region, the amino acid sequence found in the fifth histone class, H1, shows much less conservation. Variation in H1 is found when chromatin from different organisms is compared. Moreover, several closely related varieties of H1 exist in each cell of an organism.

Enormous quantities of histone proteins exist in a cell, about 60 million copies of each type in a cell! A tremendous amount of histone must be synthesized during the S period of the cell cycle, so that the histones can bind to the DNA as it is replicated. The demand for such large amounts of histone during S may be met as a consequence of the repetition of the genes for the five histone types. As is true of the ribosomal RNA genes (the rDNA), the histone genes also occur in clusters and are found in the moderately repetitive DNA fraction, as revealed by renaturation kinetics (see Chap. 13). Each gene in a cluster (or gene family) is transcribed separately on its own mRNA. Unlike most eukaryotic genes, sequences coded for the five histone proteins lack introns.

The size of the spacers that separate the histone genes, as well as their order in a cluster, varies from one species to the next, as can be seen in Fig. 17-22. For example, in the sea urchin, the order is H1, H4, H2B, H3, H2A, whereas in Drosophila, it is H1, H3, H4, H2A, H2B. In Drosophila, two of the genes (H4 and H2B) are transcribed left to right on the strand opposite the sense strand for the other three genes. In contrast, all transcription comes from the same strand in the sea urchin.

A great deal of variation among species is also seen in the number of clusters found per genome. Several

hundred copies of each cluster occur in sea urchins. Drosophila possesses about 100 repeats of a histone cluster, whereas in the mammal, only about 20 repeats are found. In the chicken, the number of repeats is only 10. The gene clusters in a given species are located at specific chromosomal regions. In the human, in situ hybridization shows them at a certain region in the long arm of Chromosome 7.

Although the histone proteins make up a very well-defined conservative class of proteins, the nonhistone proteins of chromatin compose an assortment of thousands of different proteins with many roles. Included are various acidic proteins and the functional proteins such as the RNA and DNA polymerases. Most of the nonhistone proteins are present in very small amounts, making it difficult to isolate each kind for study. In contrast with the histones that are very highly conserved, the nonhistones are variable in kind. Evidence shows variation from one species to the next, as well as from tissue to tissue within a species. The nonhistones also vary greatly in relation to the mass of the DNA in chromatin, from less than 50% of the DNA in chromatin to more than 100% of the DNA amount. We will return later to a discussion of certain kinds of nonhistone proteins, following a look at the organization of the chromatin fiber.

The chromatin fiber and the nucleosome

The unreplicated chromosome in an interphase nucleus consists of a single continuous chromatin fiber composed of DNA associated with histones and nonhistone proteins. Various procedures have shown that only one double-stranded DNA molecule runs the length of the continuous fiber. Following replication

FIG. 17-22. Histone gene clusters. The genes for the five classes of histone proteins occur in clusters repeated over and over. Variation is seen in the order in which the specific genes occur in a cluster and in the size of the spacers separating them. Each gene is transcribed separately on its own mRNA. Transcription may be all from one strand, as in the sea urchin, or from both strands, as in Drosophila, in which three genes are transcribed from right to left and two from left to right.

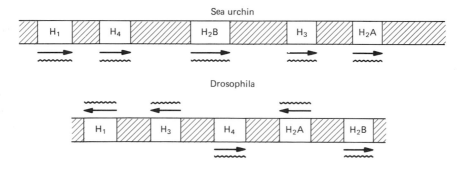

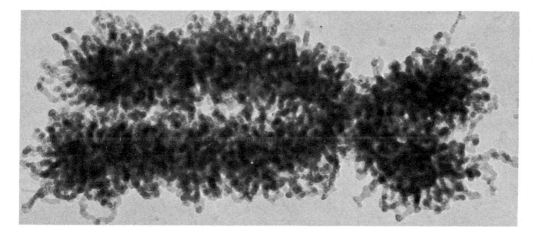

FIG. 17-23. An electron micrograph of a whole mount of human Chromosome 12 at metaphase. Each of the two chromatids is composed of a compactly folded 30-nm fiber, which in turn arises from a narrower 10-nm fiber. (Courtesy E. J. DuPraw.)

at S of interphase, one duplex is present per chromatid. The chromatin fiber of each chromatid folds back on itself repeatedly to form the compact chromosome of metaphase (Fig. 17-23).

In the 1970s, the chromatin fiber was analyzed in detail to gain an insight into its organization. Roger Kornberg pioneered the biochemical studies and Olins and Olins produced superb electron micrographs of chromatin fibers. The investigations have resulted in the formulation of the *nucleosome model* of chromatin fiber organization. After its extraction from cells, chromatin may be treated to remove its nonhistone proteins. When such chromatin is then exposed to different salt concentrations, its organization is seen to vary. At very dilute salt concentrations, which do not exist in the living cell, the chromatin fiber is seen at its thinnest and appears to have a diameter of about 10 nm. This continuous fiber has a repeated, beaded appearance (Fig 17-24). Many details on fiber organization have come from biochemical studies that make use of an endonuclease obtained from micrococci. The fiber, composed of the DNA and histone, is subjected to different concentrations of the digesting enzyme for various periods of time. A brief exposure to the enzyme yields particles that are rather uniform in length and shows that the continuous fiber consists of repeating units. These were called *nucleosomes*. The nucleosome includes an average of 200 nucleotide base pairs, 146 of which are wrapped in two turns around an octamer of histone proteins and form a bead (Fig. 17-25A). The octamer consists of two molecules each of histones H2A, H2B, H3, and

FIG. 17-24. Electron micrograph of 10-nm chromatin fiber. Note the repeated, beaded appearance. (Courtesy Dr. Barbara Hamkalo.)

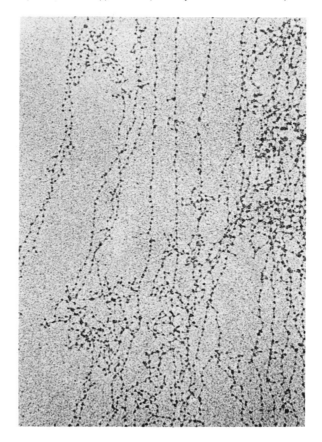

H4. Each nucleosome also includes one molecule of histone H1 located at a position where the DNA leads into and leaves the bead.

Further exposure to the nuclease yields the nucleosome core particles. Each core particle consists of 146 nucleotide base pairs wrapped around the octamer of histones. H1 protein is no longer present (Fig. 17-25B). These results come from the fact that a brief enzyme treatment attacks very sensitive sites on what

is called the *linker DNA*, the DNA that joins together the nucleosome core particles. Units consisting of about 200 nucleotide base pairs, one H1 molecule, and the octamer of histones are recovered. Once the linker is cut by the enzyme, it is digested away rather rapidly. The H1 protein is lost along with it in the process. What is left is the core particle, 146 base pairs wrapped around the histone octamer. The DNA of the core receives some protection from digestion as

FIG. 17-25. The nucleosome model of chromatin fiber organization. *(A)* According to the model, the chromatin fiber is continuous and consists of repeating units, the nucleosomes. Each nucleosome includes a bead containing an octamer of histone proteins around which DNA is wrapped in two turns. Also included is an Hl protein found outside the bead. *(B)* Detail of a portion of a chromatin fiber. A nucleosome *(left)* arising after a brief enzyme treatment can be exposed to further enzyme digestion. The linker DNA and H1 protein are lost, yielding the nucleosome core particle or bead.

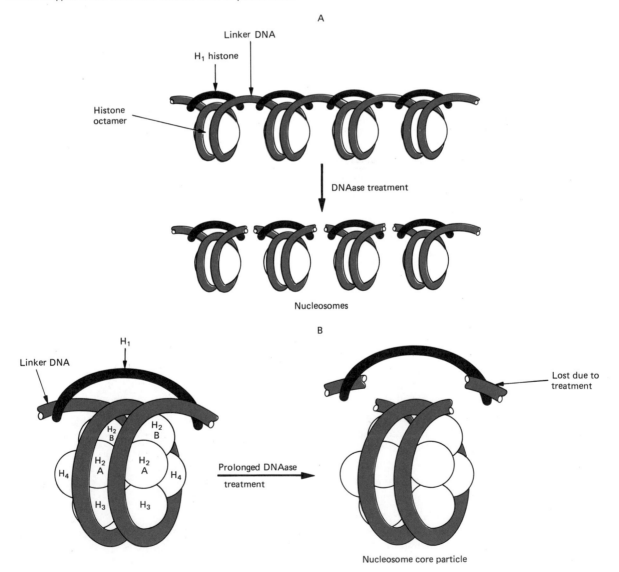

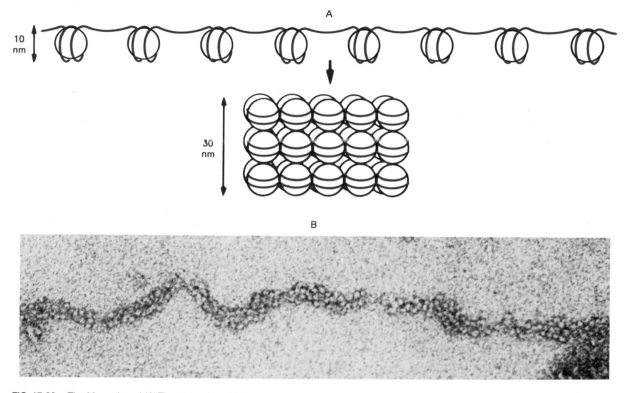

FIG. 17-26. The 30-nm thread *(A)* The 10-nm thread folds to produce a thicker, 30-nm thread, the chromatin thread of interphase. Histone HI is responsible for the condensation. The 30-nm thread is twisted further into a supercoil and is also called a solenoid. *(B)* Electron micrograph of 30-nm thread. (Courtesy Dr. Barbara Hamkalo.)

a result of its association with the eight histone molecules. While the core particle, also called the *nucleosome bead,* always includes 146 base pairs, the length of the linker DNA can show considerable variation, from approximately 10 base pairs per nucleosome to more than 100. This variation becomes apparent when different species are compared, and even when nucleosomes from different cell types within an organism are compared.

The very high degree of conservation of the amino acid sequence of the nucleosome core proteins fits in well with the fact that nucleosome organization of chromatin is found in almost all eukaryotes, the only exception being found among the fire algae or dinoflagellates. This indicates a very crucial role for nucleosome organization in the biological world. The consensus is that nucleosome structure represents the first level of packaging of the DNA. The fundamental need to pack DNA to form the compact chromosome at nuclear divisions becomes obvious when we realize that the DNA in each human chromosome would be about 5 cm long if it were stretched out!

Packing such a long, flexible duplex into a compact body that permits distribution at nuclear divisions is a critical process.

The function of the H1 protein appears to be to generate further condensation of the chromatin. Kornberg and his associates have shown that the H1 protein brings about folding of the 10-nm chromatin fiber to form a fiber 30 nm wide twisted further into a supercoil (Fig. 17-26). They have named this 30-nm structure a *solenoid,* a helix with about 6 nucleosomes per turn. Under experimental conditions, the 10-nm and 30-nm fibers can be interconverted, but if H1 is removed, only irregular chromatin clumps result. It has been proposed that H1 brings about coiling of the 10-nm fiber into the 30-nm one through contact of adjacent H1 proteins with one another. As mentioned previously, the 10-nm fiber is seen when chromatin is exposed to very dilute salt concentrations. At the concentration found in the living cell, the 10-nm fiber is converted to the 30-nm fiber. It is this 30-nm fiber that is considered to be the chromatin fiber of interphase and the fiber of the metaphase chromosome (see Fig. 17-23). The chromatin is thought to adopt the extended "beads-on-a-string" form only rarely. When cells are lysed following a very gentle treatment, the chromatin is seen in electron micrographs as a fiber 30 nm in diameter, not as a 10-nm fiber.

Histone proteins and gene regulation

Much attention has been focused on histones as potential regulators of gene expression. Histones can undoubtedly influence the regulation of genes, since DNA associated with the core of the nucleosome would not be as exposed as linker DNA to regulatory molecules in the cell. This brings us to the question of *nucleosome phasing.* If nucleosomes are phased in a given region, this means that the very same DNA sequences would be in the same position in all cells of the same type. Therefore, each specific DNA sequence would always be found in a given position in relation to the nucleosome. The same sequences would be available for interaction with regulatory molecules. The linker DNA sequences would all be the same. If, on the other hand, the nucleosome beads are randomly placed along the DNA, then the linker sequences would differ from one cell to the next. Different DNA sequences would be associated with the bead from one cell to another, thus making different sequences available for interaction with regulatory molecules (Fig. 17-27). There is evidence for nucleosome phasing in at least some parts of the chromatin. One idea is that, in different cell types, the nucleosomes are arranged in different phases. This means that a given gene could be bound to a bead in one cell type, whereas in a cell type with different phasing, the same gene would be unbound to the histone octamer of a bead and thus more accessible to regulatory substances.

Whatever the case, it seems quite clear that DNA remains associated with its histones, even during transcription and replication. It appears that the histone octamer rarely, if ever, leaves the DNA region to which it binds. It remains somewhat of a puzzle to explain how the DNA bound up in a nucleosome can lend itself to transcription by RNA polymerase. Some temporary change in the nucleosome would seem necessary. It is even harder to envision how transcription can occur without some packing alteration in the 30-nm fiber, where the chromatin is even more condensed than in the 10-nm fiber.

Evidence exists that the H1 protein, whose role

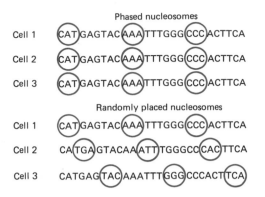

FIG. 17-27. Phased and randomly placed nucleosomes. If nucleosomes are phased *(above)* in a given region of the chromatin, specific DNA sequences would always be found in the same position when different cells are compared. The linker regions would be the same from one cell to the next, and the same sequences would be associated with the nucleosome bead. If nucleosome beads are placed at random, then different sequences are found in the linker DNA from cell to cell when this chromatin region is compared *(below).*

appears to involve packing the nucleosomes together, is less tightly bound to chromatin that is actively undergoing transcription, although the histones of the bead appear unchanged. However, certain modifications of the histones of the core particle are known to occur. A certain enzyme can add acetyl groups to some of the lysines. Highly active chromatin appears to have highly acetylated histone octamers. Acetylation of the histones has been shown in vitro to decrease the tendency for the nucleosomes to pack together. Both the increased acetylation and reduced binding of H1 may be factors in altering the conformation of the 30-nm fiber, making certain DNA regions more accessible to substances in the nuclear sap. The pattern of acetylation constantly changes in the chromatin. As acetyl groups are added, others may be removed because of the activity of another enzyme.

Besides acetylation, the histones of the bead can be modified in at least two other ways. Serines may become phosphorylated and dephosphorylated as a result of the activities of still other enzymes. The other kind of modification involves certain lysines in the H2A histone. These become complexed with a protein called *ubiquitin,* which consists of 74 amino acids. The H2A–ubiquitin complexes seem to occur only in the interphase nucleus, and there is some evidence that the ubiquitin-modified histones are associated with actively transcribing genes. Ubiquitin may exert its effect by causing the chromatin to unpack, thus favoring gene transcription in that region.

Nonhistone proteins and gene regulation

It will be recalled that a large assortment of nonhistone proteins is also associated with chromatin and that, unlike the histones, they are quite variable. Since most of the nonhistone proteins are usually present in only small amounts, isolating a given kind for study is difficult. Only those nonhistone types present in very large quantities are available for detailed investigations. These nonhistone proteins are the HMG proteins. They are relatively small, charged molecules that move rapidly when electrophoresis is performed. This characteristic gives them their name—HMG for *high-motility-group* proteins. Two HMG proteins have been implicated with nucleosome beads that are associated with active genes (see the next section). Most of the nonhistone proteins, because of their small amounts, are yet to be characterized and studied for their possible interactions with chromatin. It is considered very possible that a number of these nonhistones may act as regulatory proteins by binding to specific DNA sequences, thus determining which genes undergo transcription.

Some nonhistone proteins may even interact with hormones. Hormones are chemical messengers that can stimulate the synthesis of certain proteins in target cells—those cells that are receptive to the specific hormones. When given to animals in vivo, steroid hormones increase the rate of RNA synthesis. It is known that the action of most steroids depends on their binding to very specific protein receptors in the cell cytoplasm. After the arrival of a steroid hormone, a receptor–hormone complex is formed, and this complex moves to the nucleus and eventually binds to the chromatin. Transcription is then somehow stimulated. Perhaps certain nonhistone proteins may act as acceptors in the chromatin for the binding of the steroid–receptor complex. As a result of the binding, the chromatin would become altered and the capacity of the chromatin for transcription could then be changed. Many observations show that when chromatin is activated for RNA synthesis (e.g., the stimulation of gene activity in the prostate by testosterone), an associated increase in the rate of phosphorylation of nonhistone proteins occurs.

The bulk of the nonhistone proteins are yet to be studied for their effects on chromatin. They may produce changes in gene activity in a number of ways. Possibly they may sometimes interact with the histones; at other times they may interact directly with enzymes, such as the polymerases, which are also components of the chromatin.

Chromatin structure and enzyme sensitivity

Very good evidence indicates that when eukaryotic genes undergo transcription, the activation is accompanied by changes in the chromatin structure. This is clearly seen by differences in sensitivity of genes in response to various digestive enzymes. For example, compared with inactive genes, active ones are much more sensitive to DNAase I, S1 nuclease, and restriction endonucleases. DNAase hypersensitive sites are located toward the 5' end of the gene, anywhere from a short distance to hundreds of base pairs upstream from the starting site of transcription. Hypersensitive sites have also been found in or near DNA regions that serve as binding sites for proteins. These hypersensitive sites have been shown, in some cases, to be present at the time the gene becomes active and, in other cases, before transcription begins. One suggestion is that certain critical controlling regions of the DNA that interact with regulatory molecules are exposed to the nuclear environment and are not bound up in nucleosome beads. Hence, in cell differentiation, parts of the DNA may become unassociated with a nucleosome bead, thus becoming exposed to various enzymes and factors that can initiate transcription. In an inactive gene, on the other hand, such regions would be bound to the nucleosome core and not exposed to controlling molecules.

It has been shown that hypersensitive sites may be passed down from one cell generation to the next, and the suggestion has been made that such sites could possibly act as signals in development. It is known that certain hypersensitive sites become associated with two types of nonhistone proteins (two HMG proteins) that are necessary for maintaining the sensitivity of the genetic region. The information indicates that, in gene activation, a region undergoes a change that enables it to bind to a nonhistone protein necessary to keep the site sensitive and hence available for interactions with other molecules. Sites bound to the nonhistone protein could thus serve during embryonic development as signals indicating which genetic regions are available for interaction with the enzymes of transcription and possibly various controlling molecules as well.

Enhancers

We mentioned in Chapter 12 that several promoters in eukaryotes are known to be greatly influenced by other DNA sequences named *enhancers*. Presence of an enhancer can increase the rate of transcription from some promoters by as much as 200 times. Actually, these genetic elements were discovered first in several animal viruses, such as the monkey tumor virus, SV40 (Chap. 16). In SV40, transcription starts from two promoters that lie next to each other on the circular DNA. Transcription of the early genes from the early promoter proceeds in the direction opposite that from the late promoter (see Fig. 18-5). It was demonstrated that removal of the enhancer sequence from the early promoter region decreases early gene transcription to less than $\frac{1}{100}$th its normal amount!

Following their discovery in viruses, enhancer sequences were recognized as activators of several different eukaryotic genes. Investigations of enhancers have revealed that they possess several distinguishing features that set them apart from other controlling elements known to influence the rate of transcription. For one thing, experimental manipulation shows that the enhancer's stimulating effect can be felt at different distances and locations in relation to the gene. It may be located as far as 5 to 10 kilobases upstream or downstream from the gene it influences. In some cases, the enhancer is located *within* the segment that undergoes transcription! In the latter case, it would thus lie downstream from the promoter it stimulates. Experiments in which enhancers are moved about also reveal that an enhancer will increase the rate of transcription of any promoter that is placed nearby, showing that a given enhancer is not limited in its effect to just one promoter of a given gene. In addition, an enhancer can work efficiently even if it becomes inverted and thus oriented in a direction opposite to its usual one.

Enhancer sequences have now been discovered that greatly increase the transcription of specific promoters in specific cell types. They are known, for example, to activate the insulin gene in insulin-producing cells, genes that are coded for the chemotrypsin enzyme, and immunoglobulin genes, among others. A given enhancer—let us say the one for immunoglobulin stimulation—is active only in those

cells in which the immunoglobulin genes are being expressed.

It is not yet clear how enhancers exert their stimulating effect, but there is evidence from work with the beta-globin and immunoglobulin genes for the involvement of regulatory proteins. For example, in chicken erythrocytes, a specific protein appears only in the globin-producing cells, only when the globin genes are active, and only when the globin genes become hypersensitive to nucleases. The observations support the idea that in a specific tissue a specific protein may bind to the enhancer of a specific gene, such as an immunoglobulin gene. As a result of the binding to the enhancer, a regional change takes place altering the conformation of the DNA. The change could make the region more accessible to RNA polymerase and other molecules involved in transcription. If specific regulatory proteins do become implicated in the stimulation of enhancers in a given cell type, another problem arises. It will then be necessary to explain the tissue-specific device that stimulates the gene coded for the enhancer-specific protein. Attention continues to be focused on these curious enhancer sequences as they relate to the problem of gene regulation.

Methylation and gene activity

The chromatin structure associated with active genes has also been shown to be related to the degree of methylation. It was noted in the late 1940s that certain cytosines in the DNA were converted to 5-methylcytosine as a result of enzyme action. The modification involves the covalent linkage of a methyl group to carbon 5 in cytosine. This alteration will not interfere with replication or transcription, because it does not affect the hydrogen bonding involved in these processes. The 5-methylcytosine acts as a fifth base, since it is chemically distinct from the cytosine.

When methylation of cytosine takes place, it is always at sites where the cytosine is followed by guanine in the nucleotide sequence. Thus, it can be seen from Fig. 17-28 that both DNA strands in the region are symmetrically methylated. Once methylation takes place, this state is inherited from one cell generation to the next. At replication, as a new strand forms on the template, methylation of the guanine–cytosine sequence occurs immediately. Those guanine–cytosine sequences that do not become methylated also tend to be stably inherited in that way.

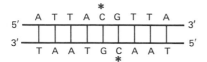

FIG. 17-28. Methylation pattern. A cytosine that is methylated *(asterisks)* on a DNA strand is one that is followed by a guanine. Both strands are symmetrically methylated.

It was not until the mid-1970s that a possible association was made between DNA methylation and cell differentiation. It was realized that a very large amount of the DNA in a mammalian cell, approximately 90%, is permanently inactive during development. Variation in which specific genetic regions are inactive occurs from one cell type to another. But the inactive regions are inherited, as such, for any one kind of cell. For example, a cell specialized in the pancreas for a digestive enzyme would not express genes for beta-globin formation. The latter must be permanently turned off and the former left free as cell divisions occur.

The facts that about 80% of mammalian DNA is methylated and cellular methylation patterns are inherited prompted investigators to observe cells for any connection between gene activation and methylation. And indeed it was found that in many cases, hypermethylation of cytosine within or near a coding sequence in a structural gene toward the 5' end is associated with gene suppression. Those genes in a cell known as "housekeeping genes," on the other hand, are permanently active in all cells and lack methyl groups. The pattern that emerged indicated that genes are active only if hypomethylated. However, contradictions arose. In about 20% of the genes studied, no correlation was evident between gene activity and the degree of methylation. In those vertebrates lower than mammals, only about 20% of the DNA is methylated, and in the invertebrate Drosophila, no methylation at all is found. Despite such contradictions, many investigators believe that methylation does play a role in gene regulation and development. Its role in determining which genes should be switched on or off, however, is not a primary one. Mammalian cells must have many different ways by which they accomplish differential gene activity.

There is good experimental evidence to show that altering methylation patterns in the DNA can switch genes on or off. Manipulation of the methylation pattern has been carried on with the inactive X chromosome of the female mammal (Chap. 5). The X that appears to have most of its genetic information turned off has been treated with the drug, 5-azacytidine. DNA that has incorporated the drug cannot be methylated because a nitrogen blocks the carbon atom that normally accepts the methyl group. Otherwise, the DNA behaves like ordinary DNA in most respects. Genes on an inactive X chromosome that is *not* exposed to the drug remain inactive. When DNA fragments containing these genes are transferred to other cells (see Chap. 19 for procedure), the genes remain inactive. After drug treatment of a deactivated X, however, genes on it become active, as shown by their activity when they are introduced into other cells. Additional examples could be presented to show that demethylation of a gene with the drug is associated with gene activation, although some investigators suspect the drug may be producing unknown effects. However, experiments with promoter regions of certain viruses have shown that the promoter does not function when it is methylated. Absence of promoter methylation, in contrast, is associated with promoter activity.

Clarification of the exact role methylation plays in gene regulation faces many problems. The enzyme that originally adds methyl groups to those genes premanently turned off during development has not yet been found. However, "maintenance enzymes" have been recognized. These are responsible for keeping those already methylated sequences in that state. One proposal on the role of methylation is that it keeps certain nucleosomes in a set position on the DNA. Those regions of the gene that are involved in control would not be wound into nucleosome beads. Since DNA in the bead is much more highly methylated than regions between the beads, the idea is that the methyl groups stabilize the nucleosome beads, preventing them from associating with other DNA regions. If methyl groups are removed, then a bead is free to shift and associate with another DNA region. This in turn influences gene expression in the cell.

It has been suggested that certain proteins may interact with the methylation. These proteins, determinator proteins, would first fix the nucleosome beads in a certain pattern. Then methylation of the DNA would keep them in that pattern permanently. According to this theory, methylation is a secondary device in gene regulation rather than a primary one. An organism such as the fruit fly, having few cell divisions in its life cycle compared with mammals, would not require methylation. It would make the proteins to fix the nucleosome pattern every time cell divisions occur. Since there are so few cell cycles, methylation would not be required in such a species. Methylation may, therefore, be just one of many controls involved with gene regulation in higher eukaryotes. It may be a device superimposed on others, to ensure that certain basic patterns of gene expression are preserved.

Gene regulation and plasticity of DNA

As noted in Chapter 11, more than one form of the DNA molecule can exist. The B form discovered by Watson and Crick is the one we have depicted in our discussions. However, it was apparent even at the time of Watson and Crick's announcement of DNA structure that another form, known as the A form, exists.

Today, improved techniques permit the synthesis of short DNA molecules composed of any nucleotide sequences. Pure preparations of a single kind of DNA can now be analyzed with precision using X-ray diffraction methods. A most significant outcome of these studies is the realization that the DNA molecule is far from a rigid, inflexible molecule, as might be assumed from the depiction of its familiar B form. The evidence indicates that the DNA can undergo changes in conformation and thus assume a form other than the familiar one in certain localized areas. This flexibility undoubtedly enables the DNA to interact in different ways with the assortment of cellular molecules that regulate gene expression. An acquaintance with some of the DNA configurations is therefore in order.

Three basic types of duplex DNA have been defined by the improved methodology. The familiar B form is stable under the conditions of high relative humidity (92%) that occur in cells and is undoubtedly the most typically occurring form in vivo. The A form mentioned earlier can be assumed by the DNA when the humidity falls to about 75%. Changes in ion concentration also play a role in changes from the B to A form. The A form of the molecule is shorter and fatter than the B form, and its bases are tilted differently in relation to the axis of the double helix (Fig. 17-29A). It is very likely that the A form is assumed when DNA–RNA hybridization takes place in regions undergoing transcription. Both the A and B forms of DNA are molecules with a right-handed twist.

Studies of short synthetic DNA molecules composed of dinucleotides of alternating purines and pyrimidines (G:C, C:G) led to the discovery of an unexpected DNA form, the Z form (Fig. 17-29B). Unlike the A and B forms, the Z form does not have a right-handed twist. The sugar–phosphate backbone follows a zig-zag course (hence Z DNA) and is left handed. In the Z form, a single, deep minor groove replaces the major and minor grooves found in B and A DNAs. The Z DNA is thin and elongated and has 12 base pairs in each turn of the helix—more than the other forms.

The question of whether or not the Z form exists in the living cell is still unsettled. For the Z form to arise, the alternating purine:pyrimidine sequence appears to be essential. It is considered possible that certain stresses imposed on the molecule may cause the B form to unwind and favor the formation of the Z form. This Z configuration could provide recognition signals for certain important regulatory molecules.

The information from in vitro studies has enabled us to look at the DNA molecule in a new light. Far from being inflexible, the commonly occurring B form, in response to certain environmental changes may undergo transitions to the A or Z forms in certain limited regions, whereas the B configurations of the molecule on either side of the altered site are preserved.

You will recall that inverted repeats occur throughout the DNA. When the two strands of the double helix separate, either strand may form a hairpin as the result of intrastrand pairing (see Fig. 12-26). It is considered possible that factors favoring the unwinding of specific regions of the DNA may result in the formation of these hairpins and, as in conversion to the Z form, may serve as recognition sites for regulatory molecules.

The ability of the DNA to assume localized changes

FIG. 17-29. Forms of the DNA molecule. A. In the A form of DNA (left), the base pairs (represented here as planks) are tilted and pulled away from the axis of the double helix. In B DNA (right), the base pairs are perpendicular to the axis with one base of a pair on each side of it. The sugar phosphate backbones are represented as ribbons. Both forms are molecules with a right-handed twist. (From R. E. Dickerson, The DNA helix and how it is read. Copyright © 1983 by Scientific American, Inc. All rights reserved). B. The B form of DNA (left) is clearly seen to have a right-handed twist as well as a major and a minor groove, neither of which extends into the helix axis. The smooth heavy line traces the backbone of the molecule, going from phosphate to phosphate. The Z form of the DNA molecule (right) lacks a major groove but possesses a deep minor groove which extends to the axis of the left handed helix. The backbone, as seen by the heavy line, is irregular and follows a zig-zag course.

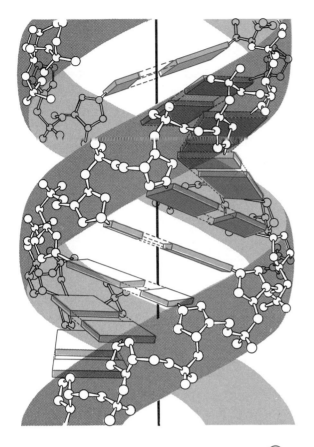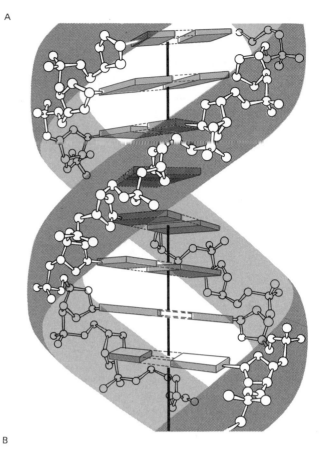

A

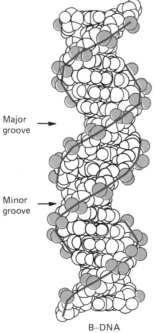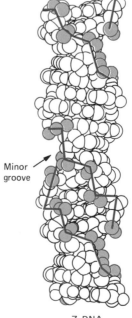

Major
groove

Minor
groove

Minor
groove

B–DNA   Z–DNA

B

in its form is now being considered as an important factor in the interactions between the DNA and regulatory molecules. It is known that the lac repressor recognizes nucleotide sequences in the operator in preference to other sequences in the bacterial DNA. Hydrogen bonds link certain amino acid side chains of the repressor to bases in the nucleotides of the operator. It is the formation of these specific hydrogen bonds that is largely responsible for the specificity of the repressor for the operator. Electrostatic forces involving positively charged amino acids of the repressor and negative charges in the sugar–phosphate backbone contribute to the strength of the binding of the repressor.

The cro repressor protein that was discussed in this chapter, and that plays a central role in the lysogenic cycle of *E. coli,* fits into the major groove, where it forms hydrogen bonds with certain bases in the DNA. Again, it is the hydrogen bonding between specific amino acids of the repressor and bases of the DNA that provides the specificity for the protein–DNA interaction. The ability of the DNA molecule to alter its shape is probably a factor in its interactions with regulatory proteins such as cro and lac. There is good evidence that deformation of the DNA *does* take place in some interactions. This is seen in the case of the interaction of CAP protein with the lac promoter. When free of the protein, the lac DNA is straight, but when the CAP–cAMP complex binds to it, the DNA bends. This can be detected by comparing differences in the way the free and the compressed DNAs travel in a gel electrophoresis experiment.

An important fact that has emerged about DNA is

that all naturally occurring DNA (as mentioned in Chap. 11) exists in a supercoiled conformation rather than as an extended, linear duplex. Remember that the duplex DNA is right handed. The twisting that produces the supercoil, however, runs in the direction opposite to that of the turns of the duplex DNA. The DNA is said to be negatively supercoiled (review Fig. 11-27). To gain an insight into the generation of such a conformation, consider the following (see Fig. 17-30). Let us suppose that we take a circular duplex DNA molecule and cut both strands at a particular site. We then take the ends of the two strands and rotate them about each other in the direction *opposite* to the direction of the twist of the helix. This results in an unwinding of the helix. If the cut ends are then joined to reform a closed circle, the DNA will assume the form of a twisted circle. It assumes the negatively supercoiled configuration to relieve the strain which was imposed on it as a consequence of the unwinding.

In both bacteria and eukaryotes, the DNA is organized into loops, each one held together at its base by an RNA–protein complex. A loop behaves as if it were a closed circle and is found in the supercoiled state. In bacterial cells, the enzyme DNA gyrase, a topoisomerase, is responsible for bringing about the formation of supercoils and maintaining them. When DNA is replicating, the unwinding of the DNA strands can generate positive supercoils in the part of the molecule that has not yet undergone replication. These twists would become very tight and interfere with replication. The DNA gyrase prevents this by introducing negative supercoils.

The supercoiling is believed to exert an important

FIG. 17-30. Generation of supercoiled DNA. *(1)* The two strands of a circular duplex DNA are cut at a particular location *(dotted line)*. *(2)* The strands are then rotated about each other in the direction opposite to that in which the helix is twisted, and the helix begins to unwind. *(3)* The number of times one strand goes around the other is reduced as a result of the unwinding. The cut ends are then rejoined *(arrows)*. *(4)* The DNA now assumes the form of a twisted circle, relieving the strain imposed on it by the unwinding. (From J. C. Wang, DNA topoisomerases. Copyright © 1982 by Scientific American. All rights reserved.)

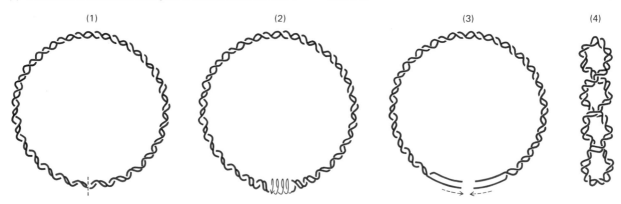

(1)   (2)   (3)   (4)

influence on transcription. The negative supercoiling favors strand separation, which is a necessary preamble to transcription. An increase in negative supercoiling in a region would favor the transcription of certain genes. Negative supercoiling also stabilizes Z DNA, as well as regions with hairpins formed by intrastrand pairing. By favoring changes in the conformation of DNA, supercoiling may play a significant part in the interactions between the DNA and regulatory molecules. We see that the DNA molecule is one of unexpected flexibility, which enables it to interact in various ways with diverse cellular components as it transfers its information to the rest of the cell.

Gene expression and differential processing

In Chapter 13, we learned that many cellular and viral genes are fragmented and that their transcripts undergo processing and splicing (see Figs. 13-15 and 13-19). These events are very pertinent to the subject of the control of gene expression. We can envision the primary transcript of a gene being processed and spliced in two or more different ways so that, in effect, more than one kind of mRNA would result. This means that one gene could carry information for more than one kind of mRNA and hence for more than one kind of product. How the transcript becomes processed in a given cell type could be the basis for how

that gene is expressed. We now know that several such genes do exist among eukaryotes and viruses. A gene of this kind is known as a *complex transcription unit*, and it can give rise to different species of mRNA, depending on the cell type and several other factors.

As Fig. 17-31 shows, a primary transcript could possibly contain more than one poly(A) site as well as several exons. Depending on where the cut is made at the 3' end and how the splicing takes place, more than one kind of final mRNA could result. Transcripts from complex transcription units are known in mammals, and these appear to be processed in different ways, depending on the cell type. For example, the primary transcript of a particular gene in one type of muscle cell may be processed and spliced in a certain way. In another type of muscle cell, the same gene's transcript may be processed in another way. Since the transcripts are different, their translation will result in similar but nevertheless different proteins. Even though splicing can occur in more than one way, as in this example, it is still essential that a transcript, to be an active mRNA, retain its 5' cap and be fitted with a poly(A) tail.

A particular complex transcription unit in the rat has been shown to be associated with at least two different mRNAs—one typical of the pituitary and one found in the thyroid. However, the same primary transcript is found in both thyroid and pituitary. The difference between the mRNAs is explained as a

FIG. 17-31. A complex transcription unit. Assuming a transcript of a specific gene contains more than one poly(A) site *(arrows 1, 2, and 3)*, the transcript could be cut at site 1 in one type of cell, site 2 in a second type, and site 3 in a third. The 5' cap must be retained, and a poly(A) tail must be added. In the processing, certain exons *(red)* may be discarded along with the introns in a given cell type. In cell type 1, exons 4, 5, and 6 are discarded along with the introns. In cell type 2, exons 3, 4, and 6 are discarded, whereas in type 3, only exon 5 is discarded. The one gene has thus given rise to three different mRNAs as a result of differential processing from one cell type to another.

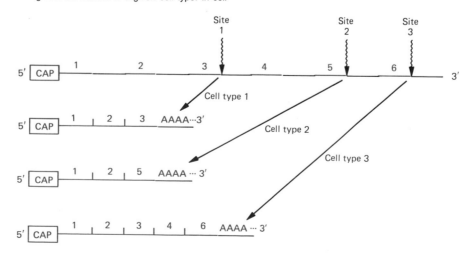

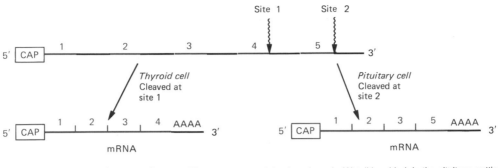

FIG. 17-32. Differential processing in two cell types. The same primary transcript *(above)* of a given gene occurs in both the thyroid and the pituitary of the rat. However, this same transcript is processed differently *(below)*. In the thyroid, cutting of the transcript occurs at site 1. Introns *(black)* are discarded, the first four exons *(red)* are joined, and a poly (A) tail is added. In the pituitary, cutting takes place at site 2. One of the exons kept in the thyroid (exon 4) is discarded, but the one at the 3' end is included. Two different mRNAs have resulted from the same transcript and hence from the same gene.

result of differential processing of the primary transcript, which contain two poly(A) sites (Fig. 17-32). In thyroid cells, this primary transcript is cleaved at the first site, and the first four exons are then spliced together to give a prehormone protein. In the pituitary, cleavage occurs at the second poly(A) site. Splicing then results in elimination of the introns along with one of the exons that is found in thyroid mRNA. However, the coded region at the end of the primary transcript is then joined to the first three exons. The outcome is production of an mRNA for a different prehormone in the pituitary.

These complex transcription units cause us to pause in our definition of a gene, because in these cases a given gene is associated with more than one kind of final product. Although differential processing of the primary transcript can determine which particular products may be produced in different cell types from transcripts of a specific gene, the primary way in which genes are regulated is still at the level of transcription. This is seen in that a given gene, though its transcript may be handled differently depending on the cell types, is not actually expressed at all in most cells of the organism. Housekeeping genes are indeed expressed in all cells; however, only a small percentage of the genome is being expressed in any given cell type. We now know conclusively that genes that are *not* expressed in a cell are *not* transcribed. Any expression of a gene depends on its undergoing transcription. Although the transcript may be handled differently in different cells, as in the case of complex transcription units, it is still the factors that switch transcription on and off that are dominant in gene control.

Giant chromosomes and gene activity

Some excellent studies of differentiation have made use of giant polytene chromosomes, described in Chapter 2. Such giant chromosomes can be seen in the salivary gland nuclei of the larval Drosophila (see Fig. 2-18). They are also found in other fly genera and can be seen in cells of other organs as well—the midgut, rectum, and excretory organs. These unusual structures provide a unique opportunity to study gene action. Each giant chromosome is a linear body composed of about 2^{10} single-unit chromatin fibers. As a result of the juxtaposition of so many individual chromosome threads, the polytene structure reflects features of the single chromosome that would otherwise go unnoticed. For example, the bands or chromomeres along the salivary chromosomes are regions where DNA appears to be concentrated because of tight packing of the individual chromatin fibers; the interbands, on the other hand, are more extended. We can see the band and interband arrangement, because more than 1000 separate fibers are closely associated next to each other.

The nuclei in which these giant chromosomes occur are in a permanent interphase state. They are not preparing for cell division, but are concerned only with the synthesis of products needed in the nondividing nucleus. They are thus in a more active, stretched out condition than they would be if they were preparing to divide. Structural chromosome changes (see Chap. 10) have enabled us to associate some of the bands with specific genes. An inversion, duplication, or deletion is often related to a comparable change in a portion of one of the giant chromosomes. Each band appears to represent a site occupied by one gene. Besides DNA, the chromosomes contain all of the other molecules typically associated with chromatin, such as RNA and protein.

In the early 1950s, Beerman, working with Chironomus, noted that some bands of the polytene chromosomes appear in a swollen or puffed conditions. (These puffs are also found associated with giant chromosomes of several fly species, including Drosophila.) Comparison of the giant chromosomes in different tissues of the same individual also showed puffing, but significantly, the pattern of puffing differed from one kind of cell to another. Moreover, the pattern seen in a tissue is characteristic of a certain stage of development; the pattern changes at a later stage (Fig. 17-33).

Cells with polytene chromosomes were supplied with labeled amino acids and labeled RNA precursors. The results clearly showed that swollen bands are actively engaged in RNA synthesis but are not incorporating amino acids at an increased rate. All of the evidence taken together strongly suggests that the bands are sites of genes and their puffed state indicates that a gene or genes have become active. The gene activity is reflected by both the puffing and the increased rate of RNA formation. Observations with the electron microscope indicate that the puffing itself results from an unwinding of chromosome fibers in the more condensed regions of the bands. This unpacking exposes a much greater length of the individual chromosome fibers, and this makes them available as templates for RNA transcription (Fig. 17-34). Beerman and his colleagues have actually been able

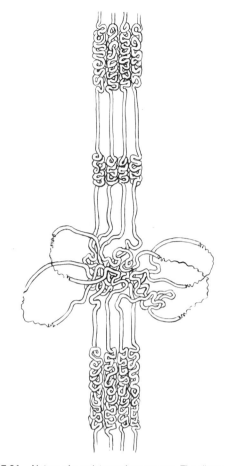

FIG. 17-34. Nature of a polytene chromosome. The diagram shows that the giant chromosome consists of many unit chromosome threads or chromatin fibers side by side. Tight folding in all the threads at a specific site produces a band. A puff arises when the threads at a site become unpacked. The threads in their extended state can then engage in RNA synthesis. RNA isolated from puffs at different sites has been shown to be different in its composition, indicating that the RNA is mRNA. Different bands along a chromosome would thus include different genes that form different mRNAs during the puffed state. (Reprinted with permission from E. J. DuPraw, and P. M. M. Rae, *Nature (Lond.* 212: 598, 1966.)

FIG. 17-33 Chromosome puffs in *Rhynchosciara angelae.* The chromosome shows puffs that were not present at earlier stages of larval development but appear at a characteristic time. (Reprinted with permission from W. V. Brown, Textbook of Cytogenetics, p. 27. C. V. Mosby Co., St. Louis, 1972.)

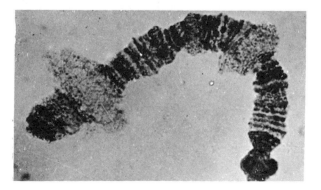

to isolate RNA associated with different puffs. If the RNA formed at separate puffs represents different kinds of mRNA coded by different genes, we would expect the base composition of the isolated RNA samples to vary. And this is what has been found. Not only have different types of RNA been demonstrated, but the various puffs show different rates of RNA synthesis. There is little doubt that the polytene chromosomes, with their puffing patterns and associated species of RNA, afford an exceptional opportunity to study differential gene activity at various stages of development.

If the puffs represent genes that have been switched on at specific times by an unpacking of the DNA, an important question arises concerning the nature of the mechanism that accomplishes this. A clue was provided by the discovery that a specific hormone, the insect molting hormone ecdysone, can induce the formation of very specific puffs. When ecdysone is released by the larva at a certain stage in development, the puffs that were already present decrease in size, and new ones quickly appear, eventually more than 100 of them, until the pupal stage is reached. Each puff has a characteristic time of appearance and duration.

The hormone ecdysone is a steroid, and it is believed that the first puffs appearing after its release are caused directly by the hormone's ability to promote transcription. Various steroid hormones, as noted earlier in this discussion, are believed to exert their effects by directly promoting the transcription of specific genes. Ecdysone will cause the earlier puffs to form even if protein inhibitors are applied. This indicates that synthesis of new proteins is not essential for the formation of these early puffs but that the hormone is acting directly on the chromatin. Later puffs, however, do seem to depend on proteins made as a result of the earlier puffing, because they will not arise if protein synthesis is blocked. The hormone level also seems to be important, because several puffs may appear at ecdysone concentrations that bring about no changes in other bands along the length of the chromosome. Certain puffs may actually decrease in size, whereas others increase in response to the hormone when it is given experimentally.

Observations indicate that the various genetic regions along the length of the chromosome may react quite differently to a given hormone level. Although the changes in puffing patterns are taking place, the histone content remains constant, again indicating that these proteins are not removed from the DNA when it becomes active in transcription. Any modifications that may take place in histone or nonhistone components of the chromatin of the polytenes and permit transcription at a specific time await clarification.

Cytoplasm and external environment in cellular differentiation

The importance of the cytoplasm must not be overlooked in considering cellular differentiation. Very striking experiments that show the involvement of the cytoplasm are those that entail the transplantation of nuclei in amphibians. In such an experiment, the nucleus is removed from an egg cell. Another nucleus is then excised from a different kind of cell and placed into the egg cytoplasm (Fig. 17-35). Shortly after this is done, synthesis of RNA and DNA takes place. This occurs even when the nucleus introduced into the egg is from a highly differentiated cell type. A brain cell nucleus, for example, does not normally divide again; yet when inserted into the egg, it can reacquire this ability. Nuclei taken from the gut not only can divide but some may give rise to normal tadpoles! Such results demonstrate that the genetic material in differentiated as well as nondifferentiated cells possesses all of the information needed to direct the development of a complete individual. Cytoplasmic substances undoubtedly play a regulatory role by signaling the expression of specific genes at the proper time.

Superb experiments with plants have shown both the ability of a single differentiated cell to develop into a new individual and the importance of the external environment in cellular control. Most cells in the root of a carrot or of the underground stem (tuber) of the potato no longer divide when the plant reaches a mature state. It is common knowledge, however, that such quiescent cells express their capacity for cell division when the organ is wounded by cutting and the cells become exposed. Once the wound is healed, division again comes to a halt. Evidently, the information for cell division and growth is present in these cells. There must be some kind of signal that indicates when these processes are to start and stop. Based on observations with healing, it seemed likely that some signal could come from outside the cell itself. Steward and his associates, after exposing fragments of carrot root to a variety of substances, found that coconut milk acts as a potent growth stimulator. Fragments exposed to the milk increased in weight about 80-fold in approximately 3 weeks. The coconut milk must contain something that triggers the mechanism for cell division in these typically quiescent cells.

A very graphic illustration of the potential of a mature cell to express all of its coded genetic information was shown when individual root cells were freed completely from their neighbors. Apparently, surrounding cells in the intact root can restrict the growth of other cells in the vicinity. Separate root cells cultured in the coconut milk may develop into complete, normal plants! This is clear-cut evidence

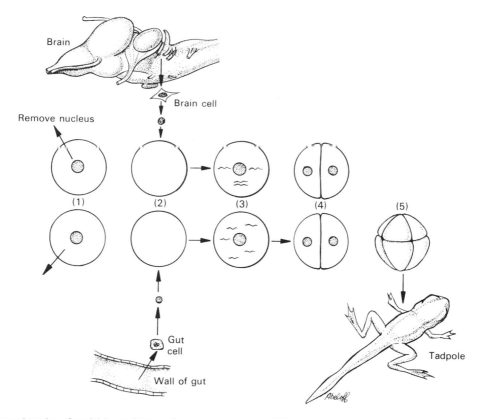

FIG. 17-35. Transplantation of nuclei in amphibians. The nucleus can be removed from an unfertilized egg (1). Another nucleus may be transplanted to such an egg from a cell that is differentiated, such as one from the brain or from the gut (2). After the transplant, DNA and RNA synthesis can be detected in the cell with the transplanted nucleus (3). The cell may divide several times (4). The nuclei taken from some cells trigger more divisions than the others (5). Some result in the formation of normal tadpoles.

that all of the genetic information for directing the formation of another individual resides in the nucleus of the mature root cell. In the intact root tissue, however, the ability to express all of this information is suppressed by surrounding cells. When freed from this inhibitory effect and subjected to substances that can stimulate growth, the full potential of the mature nucleus is realized.

Another striking demonstration of this was seen when separate cells were obtained from an intact carrot embryo and then were cultured in coconut milk. Thousands of such individual embryo cells developed into thousands of other embryos, just as if each separate cell had regained the power of the fertilized egg (Fig 17-36). Moreover, the stages of development through which these secondary embryos passed were similar to those of embryos that develop in the usual way in the intact plant.

We see that although an adult, differentiated cell contains all the coded information of the fertilized egg, it is normally restricted from expressing the entire set of instructions. Apparently, this suppression takes place during normal development by the inhibitory effect of certain environmental substances, as well as by the influence of surrounding cells. Other substances may trigger the information for division and growth at the proper time or "turn on" the transcription of the appropriate genes when their products are needed. By a series of coordinated inhibitory and stimulatory signals, the genetic information is so controlled that essential gene products are present at the needed time. When not required, superfluous gene products could be detrimental and would waste valuable cell energy, as in the case of the constitutive bacterial mutants.

It is, of course, imperative that cells do *not* express all of their genetic information; the result would be masses of identical cells. Normal differentiation from the fertilized egg on requires that only a fraction of the total coded information be translated. Expression

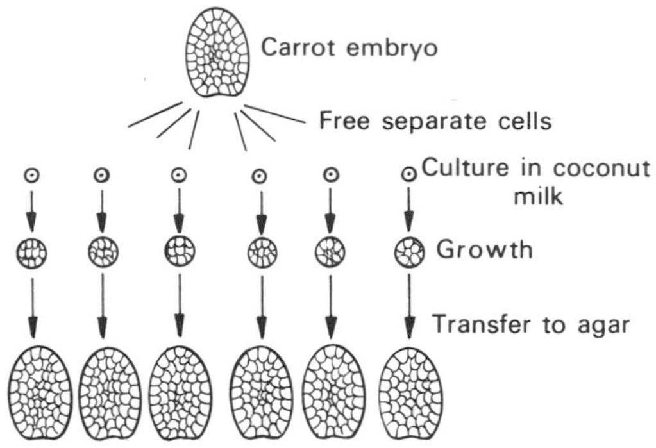

Carrot embryo

Free separate cells

Culture in coconut milk

Growth

Transfer to agar

FIG. 17-36. Embryos from embroys. Separate cells can be freed from a growing carrot embryo. When these separate cells are cultured in coconut milk, they can multiply to form a small group of cells. Each of these in turn can form a carrot embryo when raised on agar medium.

of extra information in a growing cell or a mature cell may cause chaotic or abnormal growth.

Control mechanisms and the cancer cell

The problem of cancer is related directly to the subject of cell regulation and differentiation. A normal differentiated cell that was once performing certain functions in cooperation with its neighbors can somehow undergo a fundamental change that converts it to a malignant state. It is released from various restraints that hold cell division in check. When it is transformed to a cancerous state, a cell that was noncycling (see Chap. 2) reenters the cell cycle and proliferates wildly to produce billions of altered cells, which constitute the tumor.

Uncontrolled growth is just one feature a cell acquires when it is transformed to the malignant state. The cancer cell also gains the property of immortality in the sense that it can multiply indefinitely in cell culture, unlike normal cells that lose the ability to divide after about 50 cell generations and then die (see the next section). Cancer cells also become less responsive to feedback mechanisms that control the growth of normal cells both in culture and in vivo. For example, normal cells divide and form a neat, single-cell layer when grown on a culture dish. Cell division shuts down when the entire surface of the dish is covered. Malignant cells, on the other hand, are not confined to an orderly single layer but grow on top of one another in disordered piles.

Cell division has not shut down in response to the presence of other cells. The malignant cell has lost the property of *contact inhibition,* a term that may be

somewhat misleading. Experiments suggest that it is not the actual contact among normal cells that is the basis for shutting down cell division but rather the degree to which cells spread out. A cell that is able to spread out on a surface has a shorter growth cycle. The rate of protein synthesis decreases in the more rounded-up cell, even in the absence of contact with other cells. Apparently, when a normal cell contacts other cells, it cannot spread out to its fullest extent. This reduces the rate at which it makes proteins, one or more of which may be needed for cell division. Cancer cells, even when they cannot flatten out on a culture dish, still appear unaffected and continue to proliferate. This may be related to the fact that cancer cells require fewer protein factors than normal cells for their growth and survival in cell culture. Moreover, the cancer cell usually differs in shape from its normal counterpart. Unlike normal cells, cancer cells engage in anaerobic respiration to a large extent. In addition, the cell membrane of a cancer cell displays differences from that of the normal cell, among these the possession of special antigens that can be detected by immunological procedures. Cancer cells also frequently have chromosomes in excess of the normal diploid number.

These are just a few of the features that distinguish a cancer cell from the normal type from which it arose (Fig. 17-37). Such a list of changes accompanying the malignant transformation indicates that the cellular control mechanisms have been profoundly altered in the transition from the normal to the cancerous state. There is reason to suspect that a genetic alteration of some kind is the basis of the malignant state, because carcinogenic agents such as high-energy radiation and various chemicals are known to have the potential to damage DNA and cause gene mutations. However, for many years no evidence existed in the cancer cell of any specific genes that had undergone change, which could be considered responsible for the transformation and maintenance of the malignant state.

The first clue that cells of higher eukaryotes might contain specific genes that can cause cancer came from studies with viruses. In Chapter 16, we learned that some retroviruses carry, along with other genes in their genome, a single gene, an oncogene, that can evoke tumor formation. The oncogene of Rous sarcoma virus was the first to be identified, and it was called *src* ("sarcoma"). Other oncogenes associated with other retroviruses were then identified, and these too were designated by three letters (e.g., *fes* for

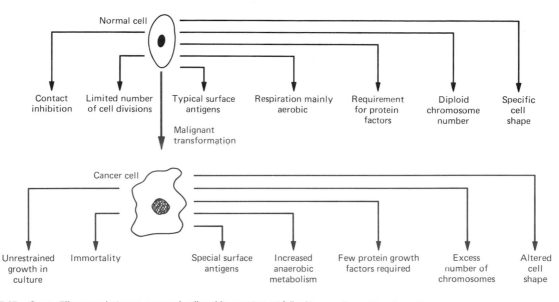

FIG. 17-37. Some differences between a normal cell and its counterpart following a malignant transformation.

"feline sarcoma virus"). Those retroviruses carrying oncogenes are the fastest acting carcinogens known and can efficiently convert susceptible cells to the malignant state very shortly after infection. They have been isolated only from animals with tumors. The retroviral oncogenes are the only genes known that can initiate and maintain the cancerous state.

As noted in Chapter 16, three genes are essential for retroviral reproduction, *gag, pol,* and *env.* The nucleotide sequence of the oncogene typically replaces one of the three essential genes, so that the virus is defective and depends on a helper virus to replicate. Rous sarcoma virus is an exception in that its oncogene, src, is added to the other three and does not replace any part of them.

In the mid-1970s, the surprising discovery was made that the src viral oncogene is not truly viral but is an almost exact copy of a gene found in all chicken cells! Approximately 20 genes have now been isolated from retroviruses that are oncogenes in various birds and also in rats, mice, the cat, and the monkey. Each of these has been shown to be very closely related to a normal gene in the chromosome complement of the animal. The viral oncogene encodes for an oncogenic protein, a protein similar to a normal one encoded by a cellular counterpart. Each normal gene that corresponds to a viral oncogene is designate a *protooncogene.* A retrovirus can apparently pick up a protooncogene from a cell and incorporate it into its genome.

Once in the genome of the retrovirus, the protooncogene is activated to become an oncogene, which can then transform a susceptible animal cell to the cancerous state.

Does this mean that the normal cell carries genes that can potentially trigger a malignant transformation? Evidence for this comes from gene transfer experiments. In one type of experiment, DNA is extracted from cancer cells, such as mouse fibroblasts. The DNA is next precipitated with calcium phosphate and then applied to normal mouse fibroblasts in culture. (Procedures used in gene transfer are described in Chapter 18.) Human DNA from tumors of the lung, colon, and bladder, as well as from leukemic cells, was also applied to normal mouse fibroblasts. It was demonstrated in all cases that the diverse donor DNA from the various cells could transform normal mouse cells in culture. When DNA from normal cells is used as the donor and is applied to normal fibroblasts, no transformation takes place. Clearly, the donor malignant cells of the human or mouse must have some common elements not present in normal cells capable of transforming the fibroblasts (Fig. 17-38). It does not matter that the donor DNA may come from a different species or tissue. The donor DNA, evidently, must contain some genetic information capable of evoking the transformation in normal cells, even those of a different species or tissue type.

Other evidence indicated that the DNA sequences

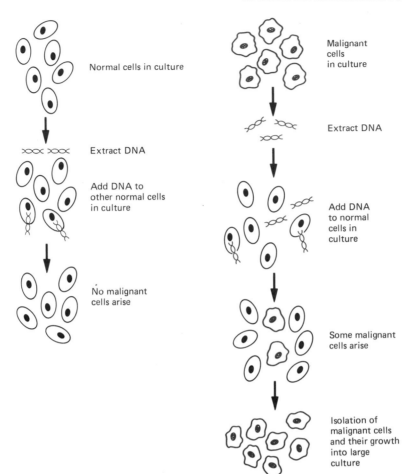

FIG. 17-38. Demonstration of genetic factors in malignant transformation. *(Left)* DNA taken from normal cells can be added to cultures of other normal cells. The extracted DNA has been treated to enable it to enter cells more readily than untreated DNA. No malignant cells arise. The same procedure can be followed using DNA extracted from malignant cells growing in culture *(right)*. When this DNA is applied to normal cells, some of the latter are transformed to cancer cells. These transformed cells can then be isolated from the culture and used to produce a large culture of cancer cells. DNA can, in turn, be isolated from these cells, and the entire process repeated again and again. After several repetitions, the results are the same as shown. Some normal cells are still transformed. This indicates that the factor or factors responsible for the transformation occur on a small segment of DNA, since repeating the process over and over causes loss of more and more of the original transforming DNA.

responsible for the transformation must be present in just a small DNA region. This becomes apparent, because the DNA from the first transformed cells (Fig. 17-38) can be extracted and used to transform additional normal cells. The procedure can be repeated through several cycles of DNA extraction and infection of normal cells. Transformation still occurs throughout the course of several transfers. Only short pieces of the transforming DNA from the original malignant cells used at the beginning can be expected to persist throughout several repetitions of the procedure. It becomes apparent that the genetic se-

quences responsible for the malignant change must be present on a very small DNA segment. This eliminates the possibility that many genetic elements must work together to produce a cancerous change, because very few small segments, at most, could survive the several transfers and enter the normal cells.

Attention was next focused on identification of the sequences responsible for the oncogenesis. Techniques of recombinant DNA technology involving cloned DNA made this possible (see Chap. 18 for details). In each case, several teams of investigators were able to recognize a single gene from the malignant donor DNA that was responsible for transforming the normal cells. Oncogenes or cancer genes were thus demonstrated in tumors from the mouse and the human. The next question to be answered concerned the origin of the oncogenes in mammalian cells. Again using the techniques of recombinant DNA analysis, several research teams were able to characterize specific oncogenes isolated from malignant cells. For example, the oncogene associated with a human bladder tumor was isolated from malignant cells and was shown to be almost identical to a normal gene present in human cells. The oncogene can bring about malignant transformation, but its normal counterpart, the protooncogene, cannot do so when transferred to normal cells. DNA sequencing (see Chapter 18) revealed that the protooncogene is converted into the oncogene as a result of a point mutation that changes a codon for glycine (GGC) to a codon for valine (GTC). This is the only change in the 5000-base-pair-long gene associated with the bladder tumor (Fig. 17-39).

What brings about the change converting a protooncogene to an oncogene? A spontaneous mutation or a carcinogenic agent might be responsible for the point mutation. (Recall that most carcinogens are also mutagenic.) As a result, any product encoded by the protooncogene would be altered. As we have seen, a protooncogene may also be activated as a result of its integration into a retrovirus. Most viral oncogenes are hybrids of coding regions of protooncogenes linked to coding regions of genes of the retrovirus. A few viral oncogenes are composed of coding regions from a protooncogene linked to regulatory regions (promoter, enhancer) of the retrovirus. It appears that the same protooncogene can be activated in two ways: by mutation in a human cell and by integration with the retroviral DNA.

There appear to be still other ways in which a protooncogene can be activated to bring about a cellular transformation. A protooncogene can become amplified in some way, so that a large number of copies are present in the cell instead of the normal diploid amount. As a result, an unusually high level of the product is found to be associated with the gene. This amplification of a protooncogene seems to be quite common in tumor cells, where the gene may have about 30 copies. A normal protein that is necessary for proper cell regulation and is associated with the gene might very well cause a malignant transformation when it is present in excessive amounts.

Structural chromosome changes may also be responsible for the activation of protooncogenes. A re-

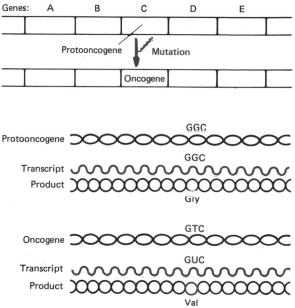

FIG. 17-39. Oncogene and protooncogene. Recombinant DNA methodology has permitted oncogenes to be characterized. An oncogene has been found to have a counterpart, a protooncogene, in a normal cell. The protooncogene (gene C, *above*) can experience a point mutation and give rise to an oncogene. In one known mutation, the only difference between oncogene and protooncogene is a change in one codon that substitutes valine for glycine in the polypeptide product.

ciprocal translocation between human Chromosomes 8 and 14 has definitely been shown to be associated with a malignancy of B cells in the human immune system, Burkitt's lymphoma (Fig. 17-40). The translocation moves the protooncogene, *myc*, whose oncogenic counterpart has been found in human and mouse tumors, from its normal location on Chromosome 8 to Chromosome 14. In its new location, it is situated next to genes responsible for the synthesis of antibodies. As a result, the protooncogene is released from controls that normally keep its expression at a low level in antibody cells. The translocation apparently has an activating effect only in cells that produce antibodies and not in other kinds of cells. Enhancer sequences have been shown to occur within the regions that are coded for antibodies, and it has been suggested that the translocation of myc from Chromosome 8 to 14 places the protooncogene next to an enhancer. Released from normal controls, it would then be expressed along with the antibody genes as if it were part of the activity of these specialized cells. The abnormally high level of the myc

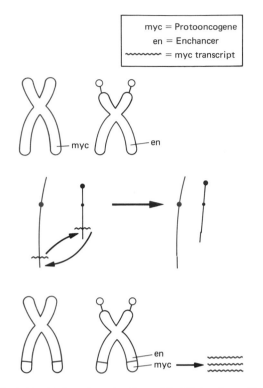

myc = Protooncogene
en = Enchancer
~~~~~~~ = myc transcript

FIG. 17-40. Reciprocal translocation and Burkitt's lymphoma. Normal Chromosome 8 *(above)* carries a protooncogene, myc, near the tip of its longer arm. Chromosome 14, which bears satellites, contains an enhancer region associated with antibody production in its longer arm. A reciprocal translocation *(middle)* between the two chromosomes places segments of the longer arms in new locations. The result of the translocation *(below)* is the production of mature chromosomes with the rearrangement in both chromatids. The myc gene from Chromosome 8 is now placed near the enhancer for antibody production. This can release the protooncogene from controls imposed on it in its normal location, enabling it to undergo transcription at a highly increased rate.

product would somehow upset regulatory mechanisms, producing a malignancy.

Products of oncogenes

What is the role of the protooncogenes in a cell and the function of their products? Although the precise role of protooncogenes in the normal cell is unknown, we can reason that it must be extremely fundamental to normal cellular activities. This is so because protooncogenes have remained practically unchanged during the course of evolution. Protooncogenes related to those in the human occur not only in other mammals but in all vertebrates, and even in Drosophila! They must be indispensable to the cells of major eukaryotic groups. It has been estimated, however, that not more than 100 protooncogenes occur in the human genome out of a total of about 30,000 genes.

The protein products of seven viral oncogenes have been found to have protein kinase activity. Protein kinases transfer energy-rich phosphate from ATP to a particular protein. Most protein kinases add the phosphate to serine or threonine residues in the protein. Unlike these, the protein kinases of the seven oncogenes phosphorylate tyrosine. Normal cells have been found to possess protein kinases that are tyrosine specific. Apparently, a protein encoded by a retroviral oncogene usually has a function similar to that of the corresponding protein in the normal cell. The similarity between the two strongly suggests that a viral oncogene product may produce its effect by acting in place of the normal product. As a consequence, an upset in cell activity is triggered.

The protein products of several other oncogenes are also believed to possess tyrosine-specific protein kinase activity. In a normal cell, phosphorylation of tyrosine occurs very infrequently. Almost all of the transferred phosphate is linked to serine or threonine. In cells transformed by the viruses with the protein kinase activity, however, the amount of phosphotyrosine increases 10-fold. How could phosphorylation of tyrosine residues in cellular proteins cause all the cell changes that accompany malignant transformation (see Fig. 17-37)? Although the answer has not been established, we can suggest that such phosphorylation alters proteins that play key roles in various regulatory pathways that control cell division, the shape of cells, and so on. Since tyrosine protein kinase activity in a normal cell is low, a critical balance between phosphorylated and nonphosphorylated proteins involved in regulation could be disturbed. The kinases of oncogenes, being similar to those of the corresponding normal cellular kinases, could well mimic the effect of the latter. Moreover, various human cancers show evidence of multiple copies of the protooncogene myc. The normal product of this gene, present at much higher levels than normal, might also be associated with malignant transformation resulting from too much protein kinase activity. In such cases, no mutation has altered the protooncogene coded for the product, but somehow the normal gene has been amplified, a situation that seems to occur frequently in tumor cells.

The discovery of oncogenes has provided many new approaches to the cancer problem. Malignant transformation can now be studied more closely on the molecular level using many of the procedures of molecular biology. It must be noted that, although the discovery of oncogenes has provided new insight into the cancer problem, this is not the entire picture. For example, activated protooncogenes are not always associated with tumors. Moreover, in some cases the activation appears to occur *after* the tumor has started, suggesting that the activation is a consequence rather than a cause of the malignant change. In the case of Burkitt's lymphoma, the protooncogene myc is *not* always transferred to a new location, as shown in Fig. 17-40, but sometimes remains at its normal site.

Finally, it must be realized that the malignant transformation is not considered to be a one-step process. Introduction of a single oncogene into cultured cells taken directly from an animal, as opposed to introduction into cells established in culture for a longer time, does not effect a malignant change. The cooperation of certain pairs of oncogenes appears to be required for the transformation to occur. It is clear from various lines of experimental evidence that cancer induction in mammals entails not one but many steps.

Armed with the knowledge that cells carry genes that can cause cancer, research can now focus on the effects of the interactions of such genes and their products on various metabolic pathways. There is now the hope that the tools available for molecular approaches to the cancer problem will eventually solve the mystery of malignant transformation, an alteration that disturbs critical control mechanism of the normal cell.

Differentiation in relation to time and aging

We must remember that differentiation in a many-celled organism begins with the development of the fertilized egg. And it does not cease upon completion of the individual but continues throughout adulthood until death terminates the process. Perhaps the very earliest sign of differentiation occurs *even before* zygote formation, with the appearance in the oocyte of DNA polymerase. No active form of this enzyme can be found in the oocyte until the latter is ready for fertilization. Once the oocyte becomes receptive, the polymerase appears in the cytoplasm and moves to the nucleus, where it can activate the synthesis of DNA.

In the zygote, most of the structural genes are repressed. During ontogeny, only certain genes will be activated at certain times by signals from cyto-

plasmic substances and from the outside environment. There is a correct time for specific genes to be called to expression. Such genes as those for development of limbs and digits would be turned on at the appropriate embryonic stage. After birth, the DNA segments coding for mature eye pigments undergo transcription. At puberty, the information for beard development in the male or breast development in the female is switched on. Throughout the normal life span, the whole array of genes may have come to expression in the appropriate cells at the correct time. Indeed, aging and the life span itself may be a genetic characteristic of all living things. We all realize that, as one gets older, the chances of dying increase. A person of 50 has a greater probability of death than a 30-year-old. The 20-year-old has a lower probability of dying than the other two. Certainly more people reach an older age today than was true years ago. But has the true limit to the span of life really been increased? Or have our modern medical practices and technology allowed more people to reach the upper age limit, which cannot be exceeded? An intriguing series of experiments by Hayflick suggests that the latter case may be correct and that life span and aging are genetic characteristics, just as much as eye color or blood groups.

Hayflick has employed cell cultures of human fibroblasts, cells that produce collagen and fibrin and continue to divide in the body of the adult. If there is no restriction on the capacity of these cells to divide, then one might expect them to continue on forever in tissue culture from one cell generation to the next. However, immortality does not seem to be a property of normal human cells. Fibroblasts taken from 4-month-old embryos divided over and over again for approximately 50 cell generations. Then the population died. That the number of doublings is somehow genetically controlled was suggested by the amazing fact that the *total number* of doublings was not changed, even if cell division was interrupted for different periods of time. For example, cells were allowed to undergo 30 population doublings and were then suspended in cold storage for years. Upon thawing, they went on to divide an average of 20 times more!

Fibroblasts were taken from persons in different age groups. These divided fewer times than the average of 50, which is typical of embryos. Moreover, cells from older donors underwent fewer doublings than those from younger persons. A limit to length of life span through genetic restriction on cell division

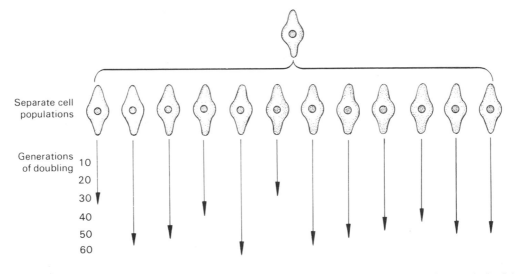

FIG. 17-41. Clonal aging. Groups of cells that trace back to the same ancestral cells are genetically identical and compose a clone. When populations of identical cells are followed through cell divisions, it is found that the separate populations cease division at different times but that more and more of them lose the capacity for division as they approach the fiftieth doubling generation. Eventually all the cell populations cease to divide further and die.

was strongly suggested by observations made on cells from species with short life spans. Fibroblasts taken from rat embryos undergo no more than 15 doublings in cell culture, compared with the 50 divisions for the human. And those from the adult rat undergo even fewer.

Single human embryonic cells were isolated and allowed to divide. Any cells arising from a common cell would all be genetically identical. From these, groups of genetically identical cells, *clones,* can be established (Fig 17-41). The ability of these genetically identical populations to divide was then followed. It was shown that a gradual decline in the capacity for cell division took place. More and more of the clones lost the ability to divide as the fiftieth doubling generation was approached. Finally, division ceased in all the clones. The clonal studies strongly supported the idea that aging was determined largely by factors internal to the cell, rather than by substances in the outside environment.

We know that certain kinds of body cells in vertebrates cease active division as a normal part of their development or maturation. Such a restriction on division is seen in the pronephros or metanephros during embryology. We also know that at old age, many organs may weigh less and contain fewer cells. Perhaps this is a reflection of a limit to the number of cell doublings somehow coded in the genetic material. If so, the DNA would even determine the limit to the life span of normal cells.

It is also possible that aging involves chromosome aberrations as well as mistakes during DNA replication or even in transcription or translation of the genetic message. An accumulation of such errors could increase the number of critical metabolic pathways that become blocked and could cause a deterioration of chemical activities necessary to maintain normal cell functions. If such factors are someday clearly shown to be involved in the aging process, perhaps techniques will be devised to correct them. We now know that prokaryotic and eukaryotic cells both contain enzymes that can repair damage to the DNA (see Chap. 14). Those species with longer life spans may have evolved more efficient repair mechanisms than those with shorter life spans, as Hayflick suggests. If these repair systems become precisely understood, perhaps science can supplement the normal mechanism by injecting the active enzymes into cells. We may even learn how to correct faulty messages that arise as copy errors at transcription or at the time of gene replication. The accumulation of genetic blocks could thus be prevented or at least decreased in number.

REFERENCES

Amara, S. G., V. Jonas, M. G. Rosenfeld, E. S. Ong, and R. M. Evans. Alternative RNA processing in calcitonin gene expression generates mRNAs encoding different polypeptide products. *Nature* 298:240, 1982.

Ashburner, M., C. Chihara, P. Meltzer, and G. Richards. Temporal control of puffing activity in polytene chromosomes. *Cold Spring Harb. Symp. Quant. Biol.* 38: 655, 1974.

Bauer, W. R., R. H. C. Crick, and J. H. White. Supercoiled DNA. *Sci. Am.* (July): 94, 1982.

Beerman, W. and U. Clever. Chromosome puffs. *Sci. Am.* (April): 210, 1964.

Bingham, P. M., M. G. Kidwell, and G. M. Rubin. The molecular basis of P-M hybrid dysgenesis: the role of the P element, a P-strain-specific transposon family. *Cell* 29: 995, 1982.

Bishop, J. M. The molecular genetics of cancer. *Science* 235: 305, 1987.

Brown, D. D. Gene expression in eucaryotes. *Science* 211: 667, 1981.

Bukhari, A. I., J. Shapiro, and S. L. Adhya. *DNA: Insertion Elements, Plasmids, and Episomes.* Cold Spring Harbor, 1977.

Cohen, S. N. and J. A. Shapiro. Transposable genetic elements. *Sci. Am* (Feb.): 40, 1980.

Croce, C. M. and G. Klein. Chromosome translocations and human cancer. *Sci. Am.* (March): 54, 1985.

Darnell, J. E., Jr. Variety in the level of gene control in eukaryotic cells. *Nature* 297: 365, 1982.

De Robertis, E. M. and J. B. Gurdon. Gene transplantation and the analysis of development. *Sci. Am.* (Dec.): 74, 1979.

Dickerson, R. E. The DNA helix and how it is read. *Sci. Am.* (Dec.): 94, 1983.

Dickerson, R. E., H. R. Drew, B. M. Conner, R. M. Wing, A. V. Fratini, and M. L. Hopka. The anatomy of A-, B-, and Z-DNA. *Science* 216: 475, 1982.

Dickson, R. C., J. Abelson, W. M. Barnes, and W. S. Reznikoff. The lac control region. *Science* 187: 27, 1975.

Doerfler, W. DNA methylation and gene activity. *Annu. Rev. Biochem.* 52: 93, 1983.

Dunsmuir, P., W. J. Brorein, Jr., M. A. Simon, and G. M. Rubin. Insertion of the *Drosophila* transposable element *copia* generated a 5 base pair duplication. *Cell* 21: 575, 1980.

Elgin, S. C. R. DNase I- hypersensitive sites of chromatin. *Cell* 27: 413, 1981.

Emerson, B. M., C. D. Lewis, and G. Felsenfeld. Interaction of specific nuclear factors with the nuclease-hypersensitive region of the chicken adult β-globin gene: nature of the linking domain. *Cell* 41: 21, 1985.

Federoff, N. V. Transposable genetic elements in maize. *Sci. Am.* (June): 84, 1984.

Federoff, N., S. Wessler, and S. Shure. Isolation of the transposable maize controlling elements Ac and Ds. *Cell* 35: 235, 1983.

Felsenfeld, G. Chromatin. *Nature* 271: 115, 1978.

Felsenfeld, G. DNA. *Sci. Am.* (Oct.): 58, 1985.

Felsenfeld, G. and J. McGhee. Methylation and gene control. *Nature* 296: 602, 1982.

Gilbert, W., N. Maizels, and A. Maxam. Sequences of controlling regions of the lactose operon. *Cold Spring Harb. Symp. Quant. Biol.* 38: 845, 1974.

Gurdon, J. B. Transplanted nuclei and cell differentiation. *Sci. Am.* (Dec.): 24, 1968.

Hayflick, L. The cell biology of human aging. *Sci. Am.* (Jan.): 32, 1980.

Herskowitz, I. and D. Hagen. The lysis-lysogeny decision of phage lambda: explicit programming and responsiveness. *Annu. Rev. Genet.* 14: 399, 1980.

Hunter, T. The proteins of oncogenes. *Sci. Am.* (Aug.): 70, 1984.

Jacob, F. and J. Monod. Genetic regulator mechanisms in the synthesis of a protein. *J. Mol. Biol.* 3: 318, 1961.

Jagadeeswaran, P., B. G. Forget, and S. M. Weissman. Short interspersed DNA elements in eucaryotes: transposable DNA elements generated by reverse transcription of RNA P1 III transcripts? *Cell* 26: 141, 1981.

Jamrick, M., A. L. Greenleaf, and E. K. F. Bautz. Localization of RNA polymerase in polytene chrmosomes of *Drosophila melanogaster.* *Proc. Natl. Acad. Sci.* 74: 2079, 1977.

Khoury, G. and P. Gruss. Enhancer elements. *Cell* 33: 313, 1983.

Klein, G. Specific chromosomal translocations and the genesis of B-cell derived tumours in mice and man. *Cell* 32: 311, 1983.

Kolata, G. Fitting methylation into development. *Science* 228: 1183, 1985.

Kolata, G. Is tyrosine the key to growth control? *Science* 219: 377, 1983.

Kolata, G. New clues to gene regulation. *Science* 224: 588, 1984.

Korge, G. Direct correlation between a chromosome puff and the synthesis of a larval saliva protein in *Drosophila melanogaster.* *Chromosoma* 62: 155, 1977.

Kornberg, R. D. and A. Klug. The nucleosome. *Sci. Am.* (Feb.): 52, 1981.

Lewis, A. M., Jr., and J. L. Cook. A new role for DNA virus early proteins in viral carcinogenesis. *Science* 227: 15, 1985.

Maniatis, T. and M. Ptashne. A DNA operator-repressor system. *Sci. Am.* (Jan.): 64, 1976.

Marx, J. L. The case of the misplaced gene. *Science* 218: 983, 1982.

Marx, J. L. Change in cancer gene pinpointed. *Science* 218: 667, 1982.

Marx, J. L. Z-DNA: Still searching for a function. *Science* 230: 794, 1985.

McClintock, B. The control of gene action in maize. *Brookhaven Symp. in Biol.* 18: 162, 1965.

McGhee, J. D., J. M. Nichol, G. Felsenfeld, and D. C. Rau. Higher order structure of chromatin: orientation of nucleosomes within the 30 nm chromatin solenoid is independent of species and spacer length. *Cell* 33: 831, 1983.

Olins, D. E. and A. L. Olins. Nucleosomes: the structural quantum in chromosomes. *Amer Sci.* 66: 704, 1978.

O'Malley, B. W. and W. T. Schrader. The receptors of steroid hormones. *Sci. Am.* (Feb.) 32: 1976.

Palo, C. O. et al. The lambda repressor contains two domains. *Proc. Natl. Acad. Sci.* 76: 1608, 1979.

Ptashne, M. and W. Gilbert. Genetic repressors. *Sci. Am.* (June): 36, 1970.

Richmond, T. J., J. T. Finch, B. Rushton, D. Rhodes, and A. Klug. Structure of the nucleosome core particle at 7 Å resolution. *Nature* 311, 532, 1984.

Sachs, L. Growth, differentiation and the reversal of malignancy. *Sci. Am.* (Jan.): 40, 1986.

Smith, G. R. DNA supercoiling: another level for regulating gene expression. *Cell* 24: 599, 1981.

Steward, F. C. The control of growth in plant cells. *Sci. Am.* (April): 104, 1963.

Wang, J. C. DNA topoisomerases. *Sci. Am.* (July): 94, 1982.

Weinberg, R. A. The action of oncogenes in the cytoplasm and nucleus. *Science* 230: 770, 1985.

Weinberg, R. A. A molecular basis of cancer. *Sci. Am.* (Nov.): 126, 1983.

Weisbrod, S. Active chromatin. *Nature* 297: 289, 1982.

Yunis, J. J. The chromosomal basis of human neoplasia. *Science* 221: 227, 1983.

Yunis, J. J. and A. L. Soreng. Constitutive fragile sites and cancer. *Science* 226: 1199, 1984.

REVIEW QUESTIONS

For Questions 1 through 4, allow R^+ and O^+ to stand for regulatory gene and operator and R^- and O^- to represent their deficient alleles. Z^+, Y^+, and A^+ represent normal alleles of structural genes for enzyme production and Z^-, Y^-, A^- their alleles for absence of enzyme. In normal cell types, the enzymes are inducible unless otherwise stated.

1. In cells of the following genotypes, indicate whether enzyme production will be inducible or constitutive:

   A. $R^+ O^+ Z^+ Y^+ A^+$

   B. $R^+ O^- Z^+ Y^+ A^+$

   C. $\dfrac{R^+ O^+ Z^+ Y^+ A^+}{R^+ O^- Z^+ Y^+ A^+}$

   D. $\dfrac{R^+ O^- Z^+ Y^+ A^+}{R^+ O^- Z^+ Y^+ A^+}$

   E. $\dfrac{R^+ O^+ Z^+ Y^+ A^+}{R^- O^+ Z^+ Y^+ A^+}$

   F. $\dfrac{R^- O^+ Z^+ Y^+ A^+}{R^- O^+ Z^+ Y^+ A^+}$

2. For each of the following, tell which enzymes will be produced constitutively and which by induction:

   A. $\dfrac{R^+ O^- Z^+ Y^- A^-}{R^- O^+ Z^- Y^+ A^+}$

   B. $\dfrac{R^+ O^+ Z^- Y^+ A^+}{R^- O^+ Z^+ Y^- A^-}$

3. Let O^0 represent an operator mutation causing the operator to bind irreversibly with the normal repressor. Let R^s represent a regulatory gene mutation that causes the formation of an altered repressor, which cannot react with the inducer even though it can bind with the normal operator. For each of the following, tell which enzymes will be produced constitutively and which by induction:

   A. $R^s O^+ Z^+ Y^+ A^-$

   B. $R^+ O^0 Z^+ Y^+ A^+$

   C. $\dfrac{R^+ O^0 Z^+ Y^+ A^-}{R^- O^+ Z^- Y^- A^+}$

   D. $\dfrac{R^- O^+ Z^+ Y^+ A^-}{R^s O^- Z^- Y^- A^+}$

4. Assume that A^+ and B^+ govern enzyme formation in an operon that is a repressible system, such as in the case of tryptophan synthetase. In such a system, the effector (end product) must combine with the regulatory gene product for the operator to block transcription. For each of the following, tell what enzymes would be ex-

pected from the cells in the absence of effector and in the presence of effector:

   A. $R^+ O^+ A^+ B^+$

   B. $R^+ O^- A^+ B^+$

   C. $R^- O^+ A^+ B^+$

   D. $\dfrac{R^+ O^+ A^+ B^-}{R^+ O^- A^- B^+}$

5. Merozygotic *E. coli* may be partially diploid for genes in the lac region. From the results given in the table (1) determine the dominance relationships among the I alleles, and (2) indicate evidence showing that the *lac* repressor is a diffusible substance ($+$ = enzyme synthesized; 0 = no enzyme synthesis).

		LACTOSE	
STRAIN	GENOTYPE (Chromosome/F factor)	Present	Absent
1	$I^+O^+Z^+/I^sO^+Z^-$	0	0
2	$I^sO^+Z^+/I^+O^+Z^-$	0	0
3	$I^sO^+Z^+/I^-O^+Z^-$	0	0
4	$I^-O^+Z^+/I^sO^+Z^-$	0	0
5	$I^+O^+Z^+/I^-O^+Z^-$	+	0
6	$I^-O^+Z^+/I^+O^+Z^-$	+	0

6. Suppose the lac repressor were not diffusible. Show the results that would then be expected for the six strains shown in Question 5 when in the presence and absence of lactose.

7. The operator-repressor system is considered to be a negative control, whereas the cAMP-CAP mechanism is considered to be positive. Why is this so?

8. Some bacterial cells harbor a plasmid known as R6-5. The plasmid does not confer drug resistance. From a population of such cells, tetracycline-resistant cells have been isolated, and the resistance determinant has been found to be associated with the plasmid. Analysis of the plasmid from the resistant cells shows that it is about 1.4 kilobases shorter than R6-5. Explain these observations.

9. A certain inducible operon includes three structural genes coded for enzymes A, B, and C, respectively. Cells have been isolated that produce the three enzymes constitutively. Other cells have been isolated that cannot produce the enzymes at all and are thus completely uninducible. DNA analysis of both kinds of cells shows that the region of the operon is about 2.5 kilobases longer than the same region in cells of the wild type. How can this be explained?

10. In some variant *E. coli* cells, all the structural genes of the gal operon are permanently turned off, in contrast with those in the wild-type operon, which is inducible. Viruses capable of transducing the operon were isolated and the DNA was extracted. DNA was also isolated from transducing viruses carrying the mutant operon. The

two kinds of DNA were denatured, and the single strands were separated. Single strands from the normal and the mutant cells were then allowed to hybridize. The electron microscope reveals DNA that is duplex but contains an unpaired segment. This segment appears as a stalk or stem with a loop at the top of the stem. Account for all these observations.

11. A particular IS element bears the following sequence at the end of one of its DNA strands: 3′ GGTGATGTA. This is followed by nonduplicated sequences that can be represented as XYZ. The sequence at the 3′ end exists as an inverted repeat at the 5′ end.

    A. Give the double helical structure of this IS element, using XYZ and X′Y′Z′ to represent the regions not involved in the inverted repeat.
    B. If the strands are separated, show what should be expected as a result of base pairing in each of the isolated strands.

12. Following are nine strains of phage lambda, each with the specific mutant defect noted. In which of these strains will lysogeny be favored over lysis?

    A. Gene O deleted.
    B. Gene cII deleted.
    C. P_{RE} deleted.
    D. Gene cro deleted.
    E. P_{RM} deleted.
    F. Overexpression of cro.
    G. Underexpression of N.
    H. O_R with strong affinity for lambda repressor.
    I. P_{INT} deleted.

13. Red kernel color in corn depends on the dominant allele, R, whereas its recessive counterpart, r, is associated with white kernels because of lack of pigment production. Assume that certain plants of genotype RR are mated to plants with white kernels, genotype rr. Plants of genotype RR produce gametes of the following five types:

    1. A gamete with the Ds element located on a chromosome other than the one with the pigment locus. No Ac element is present.
    2. A gamete with Ds inserted into the pigment locus. Ac is on the same chromosome.
    3. A gamete with Ds inserted into the pigment locus. Ac is on another chromosome.
    4. Gamete with Ds inserted into pigment locus. Ac is lost.
    5. Gamete with Ac near pigment locus. Ds is lost.

    How would the kernels be expected to appear in the next generation in each case, when the gamete described unites with a gamete carrying the recessive r?

14. The transposable element "P" is found in P strains of the fruit fly and is associated with hybrid dysgenesis in certain crosses. M strains lack the element.

    A. Suppose highly fertile flies with normal chromosome behavior arise from a cross involving a P and an M strain. From which parental strains were the females and males in this cross?
    B. With respect to M and P strains, what other crosses will yield fertile offspring with normal chromosome behavior?

15. Assume that a certain nucleotide sequence is known to occur in the DNA of the rat. Chromatin is extracted from cells of different tissues and then exposed to an endonuclease treatment that digests away linker DNA, leaving the nucleosome beads. When the beads derived from chromatin of pancreatic cells are subjected to further analysis, it is found that the particular nucleotide sequence can always be recovered. However, when chromatin taken from cells of the skin, liver, and other sites is treated in the same way, this same DNA sequence is found to be either lost, recovered only in part, or recovered at times in its entirety. Explain these findings.

16. A particular gene is known to be highly active in most tissues of the rat. Would you expect the promoter region at its 5′ end to be bound to the histone octamer? Explain. How could you determine whether or not the sequence is tightly bound?

17. In a particular DNA virus, a certain early gene becomes very active shortly after infection of the cell. Its transcripts are needed to turn on genes that become active later in the viral cycle. A strain of the virus is isolated in which transcription from the early gene has decreased by well over three fourths, greatly interfering with replication of the virus. Analysis of the viral DNA shows the entire early gene to be perfectly intact along with its promoter. However, a deletion of a DNA segment is found more than a kilobase away from the gene. Explain.

18. Would you expect a gene associated with each of the following to be active or inactive: (1) associated tightly with histone octamer; (2) associated with acetylated histones; (3) associated with a high number of methyl groups; (4) associated with ubiquitin; (5) associated with receptor-steroid hormone complex; (6) associated with increase in negative supercoiling?

19. Indicate by an asterisk any site at which methylation may take place in the following DNA sequence:

    5′ C A C G T C A C G T A C C 3′
    3′ G T G C A G T G C A T G G

20. Assume in a certain animal species that the pancreas, pituitary, and thyroid each secretes a particular prehormone. Certain missense mutations that cause amino acid substitutions toward the amino end of the three different prehormones are known. When such a missense mutation alters the thyroid prehormone, the other two prehormones remain unaffected. However, a missense mutation that brings about an amino acid substitution toward the amino end of the pancreatic protein

always has the same effect on the pituitary protein though never altering the thyroid prehormone in any way. Molecular analysis reveals that the pancreatic and pituitary prehormones are identical for about the first 50% of their amino acids, but the latter portion toward the carboxyl end is very different. The amino acid sequence of the thyroid prehormone does not resemble those of the other two. How many genes appear to be involved in the case of these three prehormones? Explain the observations.

21. A Drosophila larva is fitted with a bond in such a way that portions of the salivary glands become tied off from the rest of the glands. The effect is such that one portion continues to be exposed to ecdysone, whereas the rest does not receive the hormone. The polytene chromosomes in cells receiving hormone continue to undergo puffing, and the cells continue to develop. The chromosomes in cells cut off from the hormone show little puffing, and tissues do not differentiate.

   A. Using radioactively labeled precursor molecules, how might you show that chromosome puffs undergo transcription whereas unpuffed regions do not?
   B. What can you suggest as to the role of ecdysone, based on information regarding steroid hormone–chromatin interactions?

22. In normal mouse cells the protooncogene (p-onc) resides at a certain chromosome site. The DNA in the region of the protooncogene can be broken into specific fragments by a restriction enzyme, as indicated here by arrows:

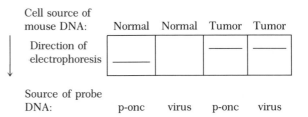

p-onc

   In the course of an experiment, normal mouse cells growing in culture are transformed into tumor cells following infection with a retrovirus. You wish to determine if the malignant transformation is the result of viral integration near p-onc. DNA is extracted from both normal and transformed cells. The DNA from both sources is exposed to a restriction enzyme, which fragments the DNA and attacks the sites indicated previously by arrows. Following gel electrophoresis, DNA denaturation, and Southern blotting, the fragments are exposed to (1) denatured labeled p-onc DNA and (2) DNA of the specific virus. (The DNA from these two sources is being used as probes to detect the presence of that specific DNA.) Autoradiographs are then finally prepared. Interpret the comparison of the four autoradiographs shown here:

Cell source of mouse DNA:	Normal	Normal	Tumor	Tumor
Direction of electrophoresis	——		——	——

Source of probe DNA: p-onc   virus   p-onc   virus

23. Suppose in the case of another virus, the following results were obtained in a situation exactly like that described in Question 22. What would you conclude?

Mouse DNA:	Normal	Normal	Tumor	Tumor
	——		——	

Probe DNA: p-onc   virus   p-onc   virus

18

THE IMPACT OF GENETIC MANIPULATION

A completely artificial gene

In 1977, the announcement was made that Khorana and his associates at MIT had succeeded in synthesizing the first truly artificial gene, one that can function in a living cell. Khorana's group synthesized a gene by chemical procedures alone, without relying on a natural DNA template. The gene that was synthesized is one found in phage lambda, which can infect *E. coli* cells. The naturally occurring viral gene codes for a new kind of tRNA that appears in the bacterial cells after viral entry, a tRNA specific for the amino acid tyrosine. (Certain viruses carry genes that code for new types of tRNA. These new tRNA species may be necessary for viral replication in the cell, because some viruses, after entry, are known to alter the ribosomes of the host. Perhaps the new types of tRNA are needed to work efficiently with the altered ribosomes.)

The sequence of nucleotides in the gene proper (the part that undergoes transcription into RNA) includes 126 nucleotide pairs. This sequence had been deduced by English investigators from analyses of the RNA transcript. For reasons still unknown, the transcript is cleaved by enzyme action after its formation. Forty nucleotides are removed in the host cell, leaving a tRNA composed of 86 nucleotides. In addition to

the gene proper, which is transcribed into RNA, the gene has a promoter region of 56 nucleotide pairs and, at its other end, a stop signal that is 25 nucleotide pairs long. Khorana and his group determined the nucleotide sequences of these other gene parts. They also artificially synthesized short gene fragments, each containing a dozen or so nucleotide pairs. In the procedure, each segment that was used bore a short piece of single-stranded DNA extending from each end (Fig. 18-1). These extensions act as joints or splints, which enable them to complex more readily as a result of the complementarity of bases in the extensions. The segments are then joined together by DNA ligase.

By joining the short segments, the entire gene (gene proper, promoter, and terminator regions) composed of 207 nucleotide pairs was constructed without the use of any natural DNA product. Khorana then tested the ability of the synthetic gene to function. A certain mutant strain of phage lambda cannot produce the new tyrosine tRNA in the host cell because of a defect in the gene. The defect appears to be in the terminator region. As a result of this defect, and hence the defective tRNA, the mutant strain produces short, nonfunctional proteins in the cell and cannot multiply in the host. The artificial gene was incorporated into mutant viruses, and these were then allowed to infect

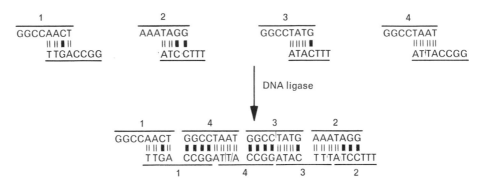

FIG. 18-1. Joining of DNA fragments. Each fragment is synthesized bearing single-stranded projections to act as splints so that stretches of complementary bases on the extensions are exposed and can complex, bringing the segments together. After the segments are properly positioned, DNA ligase activity *(arrows)* joins them together by forming phosphodiester linkages. (The sequence shown here is purely illustrative.)

E. coli cells. The originally defective viruses could now multiply in the host cells as well as normal viruses. This conclusively showed that the artificial gene was functioning. The new and necessary tyrosine tRNA was proven to be present in the host, because normal translation took place with the formation of normal viral proteins.

The construction of totally artificial genes opens up many new pathways in genetic research. One extremely important one can lead to a better understanding of a gene, how it is regulated, and how it is expressed. The investigator can produce alterations in the gene at will and observe any ensuing effect. Information of this type is of profound significance in studies of cell differentiation and pathological cell changes. Although the accomplishments of Khorana and others were with microorganisms, equally incredible strides are now being made in the manipulation of genes of higher forms, as we will see later in this chapter.

Cleavage maps

Restriction endonucleases (Chap. 16) have made it possible to obtain what are known as cleavage maps, which have been extremely important in deciphering the nucleotide sequences of DNA fragments and eventually of entire genomes. Recall that a given restriction endonuclease of type II recognizes certain specific regions on DNA and breaks up any given DNA into a characteristic set of short pieces, from a few hundred to a few thousand base pairs long. The DNA of simian virus 40 (SV40; see Chap. 16) has

been analyzed in detail with the aid of restriction endonucleases.

When the DNA of virus SV40 is subjected to a particular restriction enzyme isolated from the bacterium Hemophilus, 11 fragments are produced. This enzyme was called Hind but is now known actually to be a combination of two enzymes, HindII and HindIII. The 11 fragments obtained as a result of cleavage by Hind can be separated from one another by the process of gel electrophoresis, a technique used routinely in molecular biology and discussed in Chapter 16. In this process, a sample of the material is placed on top of a narrow column of a polyacrylamide or an agarose gel, which is then subjected to an electric field. The concentration of the polyacrylamide can be varied to govern the size of the molecules that can move into the gel. The more concentrated the gel, the greater the restriction on entry of larger molecules into it. Under the influence of the electric field, fragments of proper size may enter the gel, the smaller ones traveling faster than the larger ones in a given time.

The procedure may be modified for the preparation of autoradiographs. This requires the labeling of DNA with ^{32}P and the preparation of the gel as extremely thin sheets. After electrophoresis, the paper-thin gel is placed in contact with X-ray film. This is finally developed to reveal black lines that represent the different positions of the labeled DNA fragments. Figure 18-2 shows such an autoradiograph of ^{32}P-labeled SV40 DNA that has been digested with Hind. The largest of the 11 fragments, A, is near the top of the gel; the smallest, K, is near the bottom. The sizes

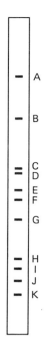

FIG. 18-2. Tracing of an autoradiograph of ^{32}P-labeled SV40 DNA prepared following digestion with restriction endonuclease Hind and separation of fragments by gel electrophoresis. The shortest fragment is K, the largest A. (Used with permission from D. Nathans.)

of the fragments can be checked by electron micrograph measurements and compared with the sizes calculated on the basis of mobility in the gel. The next step is to identify these fragments precisely and determine their exact order in the intact circular virus chromosome.

This can be achieved by using the same procedure over again; subjecting the DNA of the virus to a battery of different restriction endonucleases used alone or in combination. For example, the viral DNA can be treated with Hind and the resulting fragments exposed to a restriction enzyme from *E. coli*, EcoRI. The fragments produced by the activities of both enzymes are then separated by gel electrophoresis. It is now found that fragment F is not present in its typical position. Instead, two smaller fragments are generated, and these together equal the size of fragment F. From this we can deduce that the F fragment carries a site that is recognized and cleaved by EcoRI. Only one such site must be present in the entire genome, because all the other fragments are characteristically present, as if they had been treated by Hind alone. One can then turn to still other restriction

enzymes and study their effects on the viral genome. Treatment of the SV40 DNA with the enzyme HpaI, for example, produces three large fragments, Hpa-A, B, and C. Each of these can be isolated by itself from the gel, exposed to other restriction enzymes, and finally subjected to electrophoresis to establish the order of digest products in a single fragment. When isolated Hpa-C is exposed to Hind, two bands are produced (Fig. 18-3). These are identical in position to fragments Hind-B and Hind-I. This tells us that Hind fragments B and I must be found right next to each other. Together they form the fragment Hpa-C, produced following treatment of the chromosome with HpaI.

Using an assortment of restriction endonucleases, D. Nathans and his associates proceeded in this way, isolating separately from gels the partial digest products of one enzyme, such as the product Hpa-C. Each such partial digest was then subjected to further digestion with another enzyme. The electrophoretic patterns of the products of the second digestion were studied and compared with known standards. The result has been the production of cleavage maps of the SV40 chromosome. The initial map was based on cleavage sites generated by three enzyme treatments, EcoRI, HindII plus HindIII, and HpaI plus HpaII.

Let us make sure we understand this simple map (Fig. 18-4). The outer circle shows the sites of cleavage of the Hpa endonucleases. The inner circle shows the 11 Hind fragments, A through K, which were mentioned earlier in this section. The true positions of these fragments were established by the reasoning we have just presented. Note that the inner circle also shows the site of cleavage of restriction enzyme EcoRI. This single cleavage site, indicated by the arrow, was designated coordinate O. The other map units (0.1, 0.2, etc.) represent the fraction of the DNA length of SV40 from that EcoRI site in an arbitrary direction around the map.

$$\text{Map units} = \frac{\text{distance from 0 coordinate}}{\text{length of SV40 DNA}}$$

More detailed maps have been produced using the initial one as a reference. To obtain more detailed cleavage maps, still other restriction endonucleases were used. Since the SV40 genome can be cleaved by these enzymes at any of several different sites, one can generate large or small fragments from any part of the genome. In the more detailed maps, the sites

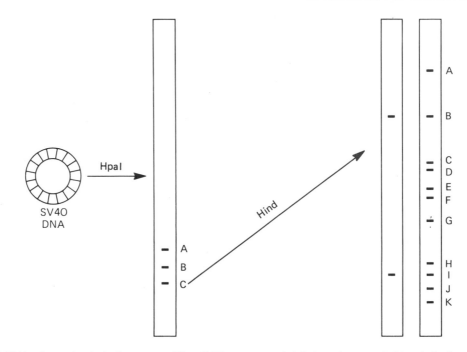

FIG. 18-3. Establishing fragment order in chromosome. When SV40 DNA is subjected to the restriction enzyme HpaI, only three large fragments are produced. Each one of these can be isolated from the gel and exposed further to other restriction enzymes. When fragment C is isolated from the gel and exposed to Hind, two fragments are generated that can be separated by electrophoresis. The resulting pattern can then be compared with that obtained when SV40 is treated with Hind alone *(right)*. Comparison shows that the two fragments are identical in position to fragments Hind B and Hind I. We now know that segments B and I must be adjacent in the intact chromosome.

FIG. 18-4. Initial cleavage map of SV40 chromosome (see the text for details). (Reprinted with permission from K. J. Danna, G. H. Sack Jr., and D. Nathans, J. Mol. Biol. 78: 363, 1973. Copyright by Academic Press Inc. (London) Ltd.)

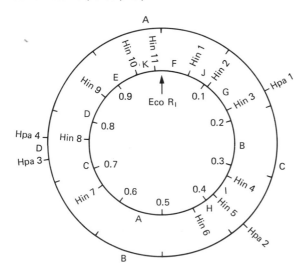

of cleavage of all the various additional restriction enzymes are localized and indicated.

Localization of genes in SV40

The cleavage map and the sequencing of the entire SV40 genome (see the next section) have permitted identification of the positions of structural genes, as well as the site of origin of viral DNA replication. DNA replication in SV40 is known to begin at a definite site, about map coordinate 0.67 in segment C (Fig. 18-4), and to proceed bidirectionally until termination of replication occurs in segment G. RNA transcripts that are present early in an infected cell before the start of viral DNA replication were isolated. These were then exposed to denatured restriction fragments of the viral DNA that had been transferred to a filter following Southern blotting (see Chap. 16). DNA–RNA hybrids were formed between the early RNA and approximately half the viral genome between 0.17 and 0.67 (Fig. 18-5). The transcription was found to follow a counterclockwise direction from an early promoter. The late mRNA, originating from a late pro-

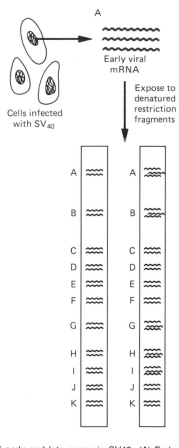

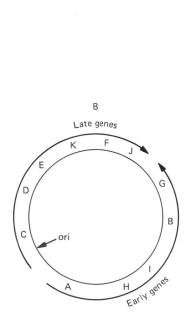

FIG. 18-5. Localization of early and late genes in SV40. (A) Early appearing viral mRNA is extracted from cells and allowed to hybridize with specific restriction enzyme fragments (A–K). Results with the early mRNA show that only certain fragments, composing about half the genome, contain sequences that were transcribed into the early RNA. The same procedure followed with late mRNA shows hybridization with the other fragments. (B) Hybridization of the early and late transcripts to the specific fragments enables the genes from which they were transcribed, and which were coded for them, to be related to the SV40 cleavage map. The early genes, composing about half the genome, are transcribed in a counterclockwise direction. The late genes are transcribed from the opposite strand in a clockwise direction. The site of viral DNA replication is found in fragment Hind C.

moter adjacent to the early one, was shown to be coded by the other half of the genome and to be synthesized in a clockwise direction. (Recall that the early promoter is known to be activated by an enhancer.)

Nathans and his collegues were able to locate the positions of certain structural genes. For example, they studied temperature-sensitive viral mutants (ts mutants), viruses that cannot replicate at certain temperatures that permit replication of wild type. One way in which the mutant genes were located on the viral DNA made use of heteroduplex DNA and mismatch correction. The circular DNA from a mutant strain was denatured to single strands. Wild-type DNA was then exposed to a restriction enzyme to yield specific fragments, which were then also denatured. The denatured DNA fragments were allowed to incubate with the mutant, single-stranded circles. If a fragment is used that corresponds to the region containing the site of the mutation causing temperature sensitivity, a partial heteroduplex will form (Fig. 18-6A). Two possibilities now follow. Replication of the DNA can result in the formation of two daughter homoduplexes after separation of the two imperfectly matched strands. The formation of wild virus (non-ts) indicates that the fragment *did* cover the mutant site. If it did not, then only mutant viruses would arise after two rounds of replication (Fig. 18-6B).

Another possibility entails mismatch repair (Fig. 18-6C). If a heteroduplex forms, digestion of part of one of the strands takes place. The intact strand is then copied. Again, wild-type viruses can arise. If the strand did *not* cover the mutant site, no mismatch repair will take place and hence no wild-type viruses will be generated (Fig. 18-6D). Therefore, use of the various restriction fragments to determine which spe-

cific ones cover up specific mutations enables one to identify the location of mutant sites. This is possible because wild-type viruses arise as a result of replication or mismatch repair if the specific restriction fragment corresponds to (and hence can cover up) a region carrying the mutation.

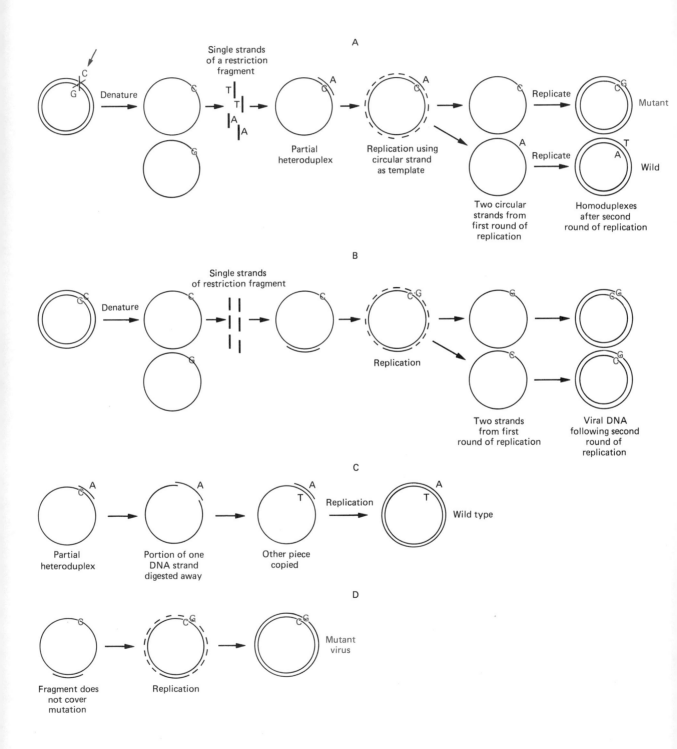

In this way, it was shown that all the ts mutants that were defective in the initiation of DNA replication mapped in the early region of the viral genome. From various mutation analyses, it became clear that the early region of the SV40 genome codes for T antigen and that this region is sufficient for viral DNA replication and the transformation of the host cell.

Mutation at preselected sites in SV40

As was stressed in Chapter 14, the application of a mutagen raises the overall mutation rate. It does not enable the investigator to predetermine which specific gene or genetic region will be altered. The possibility of bringing about a mutation at a site selected in advance has now been realized in SV40 through a series of brilliant studies by Nathans and his associates. Some of the mutations produced were deletions; others were base pair substitutions. Production of the latter is summarized here.

First, the circular viral DNA is nicked in one of its two strands at a specific site, with the aid of a specific restriction endonuclease—let us say at the short sequence CCGG. With the aid of an exonuclease, a small gap is then produced that causes a nucleotide

sequence to be exposed on the intact, complementary strand (Fig. 18-7). A mutagen that can react with single-stranded DNA is then applied. Sodium bisulfite is a mutagenic chemical that can cause deamination of cytosine to uracil. If a mutation is produced in the exposed region, the sequence GGCC on the complementary strand could be altered to GGUC. Next, DNA polymerase and ligase are added to repair the gap. As a result of these steps, a double-stranded circular DNA with a sequence that was originally

$$
\begin{array}{c}
CCGG \\
GGCC
\end{array}
$$

is now changed to

$$
\begin{array}{c}
CCAG \\
GGUC
\end{array}
$$

Realize that if the original sequence occurs at only one site in the viral DNA, then only this one spot will be attacked by the restriction enzyme and any alteration will occur only at this spot. Also, the location of this region on the cleavage map is known at the outset. Therefore, if a change has been produced as a result of the procedure, its location is known.

Any changes that are induced would be expected to occur only in a certain percentage of the treated viral genomes. Therefore, circular molecules that still have the original sequence would also be present in the preparation. These molecules are unaffected by the mutagen and are easily eliminated by exposing the preparation to the restriction enzyme employed at the outset. The enzyme attacks only the sequence CCGG and destroys all those molecules with the original sequence. Only the double-stranded circular molecules carrying the mutation will remain. These can now be allowed to infect cells for study of changes in viral characteristics. Nucleotide sequence analysis can also be done on restriction fragments of the mutant viral DNA, to pinpoint precisely the site of the base substitution in the area.

Such local mutagenesis procedures have been used to construct mutant viruses that carry base pair substitutions associated with the point of origin of DNA replication—a region that is palindromic and that, you will recall, maps to a coordinate of 0.67. Specific restriction enzymes were used to nick the SV40 DNA in the region of this palindrome (Fig. 18-8). Local mutagenesis was carried out, and base pair substitutions were produced. The mutant viruses with these substitutions were then studied for any alterations in their ability to replicate. A base pair change from G:C

FIG. 18-6. Locating positions of mutant genes in SV40. *(A)* Viral DNA containing a mutant site *(arrow)* is extracted from a temperature-sensitive mutant and rendered single stranded. (Only one of the two strands is followed here, for simplicity.) Single circular strands are exposed to a specific, denatured restriction fragment from wild-type virus. Assuming the fragment from the wild-type virus is homologous to the region in which the mutant site is found, it will pair with that region following incubation to form a partial heteroduplex. In the heteroduplex, there is no correspondence at the position of the base substitution (C versus. A). A second round of replication can result in formation of homoduplexes. Wild, double-stranded viral DNA can arise. *(B)* If a fragment is used that does not cover the mutant site, no wild-type virus can arise, because completion of replication of the fragment will copy the mutant site in the circular DNA. (Only one of the two strands is followed here.) After a second round of replication, two mutant viral DNA types will arise from the orginal circular DNA. *(C)* If the fragment covers the mutant site, as it also did in *A*, mismatch repair can take place. A stretch of DNA in the vicinity of the mutation is digested away. Digestion can remove either a piece of the fragment or a piece of the circular DNA. (Only the latter is shown here.) Completion of replication of the fragment, using the rest of the circular DNA as a template, produces wild-type viruses. Even if the nonmutant sequence is removed at times by mismatch repair, wild-type viruses will also be produced, in addition to mutant ones, because the picture depicted here will also be operating. *(D)* If the fragment does not cover the mutant site, no wild-type virus can possibly be formed by mismatch correction, since as in *B*, completion of replication of the fragment always results in copying the mutant site.

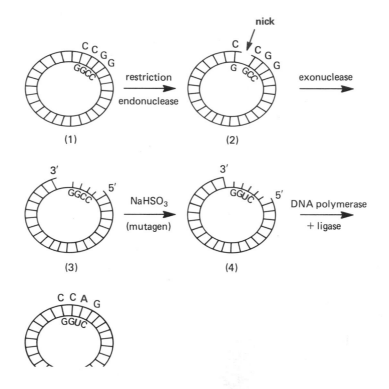

FIG. 18-7. Summary of steps in producing a mutation at a preselected site. *(1)* A restriction enzyme is selected that will attack a sequence of nucleotides occurring at one site in the viral DNA. The position of this site is known as a result of the construction of a cleavage map and DNA sequencing. *(2)* The particular restriction enzyme nicks one strand of the duplex, circular DNA. *(3)* An exonuclease attacks the free ends and digests away DNA, producing a gap.

(4) A mutagenic chemical is applied that can cause a base change in the exposed area of the intact circular strand. *(5)* DNA polymerase is added and makes a complementary copy of the missing region using the intact strand as a template. It thus inserts a nucleotide that is the complement of the one changed. Ligase seals the DNA, resulting in a double-stranded, circular molecule carrying a base pair substitution at the preselected site.

to A:T at the point of axial symmetry was found to have no effect on the rate of DNA replication. However, other changes seen in Fig. 18-8 led to the production of either mutants with definite decreases in DNA replication or ts mutants. The evidence indicates that the T antigen plays a critical role in the initiation of viral DNA replication and that it must first bind to a portion of the SV40 genome that includes this palindromic segment associated with the point of origin. A single base pair change in the region may alter the ability of the T protein either to bind in the proper amount or to interact properly with the origin site.

The ability to select within the genome a region for mutations is a monumental step in refining genetic procedures that permit the manipulation of DNA. Through studies of DNA regions involved in the control of transcription or replication, it may eventually

be possible to establish the precise location of sites critical to normal gene expression, growth, and development in higher forms.

Sequencing DNA

The production of specific fragments of a given size from a specific DNA source, following exposure to a restriction endonuclease, was a major factor permitting the analysis of stretches of DNA for their exact nucleotide sequence. This, in turn, made it possible to determine the nucleotide sequences for larger and larger DNA stretches, and eventually for an entire genome! Techniques developed by Sanger and Coulson and Maxam and Gilbert allowed rapid sequencing of DNA fragments. Let us follow the method of the latter team of investigators to analyze the sequence of nucleotides in a DNA fragment.

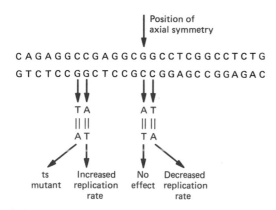

FIG. 18-8. Mutagenesis of palindromic sequence at the site of origin of SV40 DNA. Following local mutagenesis, specific base-pair substitutions were made in the palindrome. A G:C pair occurs at the position of axial symmetry in the palindrome. Transition to an A:T pair does not affect replication of the viral DNA. However, other base-pair substitutions were shown to have effects. A similar change to an A:T pair at the next position to the right results in a decrease in replication rate. The change from C:G to T:A on the left leads to a temperature-sensitive mutant. The change to A:T at the position next to it leads to an increased rate of viral DNA replication.

First of all, we must start with a preparation of pure DNA segments, let us say 50 base pairs long, generated by a given restriction endonuclease. The fragments are then labeled by another enzyme that attaches ^{32}P to the 5' end of each of the two strands composing the segment (Fig. 18-9). The DNA must then be denatured to yield pure preparations of the single strands, each strand now labeled at its 5' end. This can be accomplished by taking advantage of the fact that one of the two strands may be heavier than the other because it contains more purine nucleotides and the lighter one has more pyrimidines. Gel electrophoresis enables us to separate the two kinds of strands. We can then proceed to analyze either of these, the heavy or the light strand. Assuming that we select a pure preparation of the labeled light strand, we then divide it among four test tubes (Fig. 18-9). Each of the tubes contains a different chemical that *preferentially* destroys one or two of the bases present in the DNA. As a result, the strand is cleaved in two at the location of the particular base that is attacked.

FIG. 18-9. Some steps in nucleotide sequencing of DNA. Pure DNA segments of a given length are labeled at the 5' end of each of the two strands. The two strands are separated. Many such single H and L strands would be present. Strands of one kind (H or L) are then divided among four tubes, each one containing a different chemical that can break a single-stranded DNA preferentially at some site along its length.

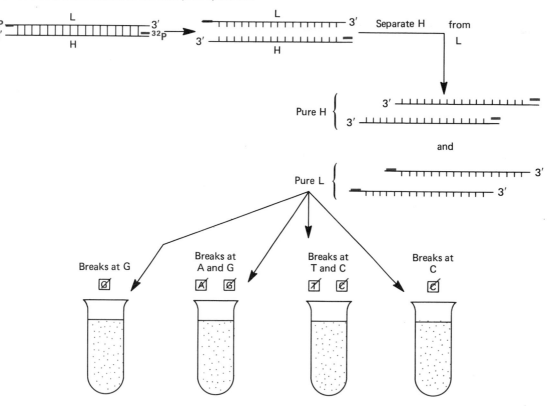

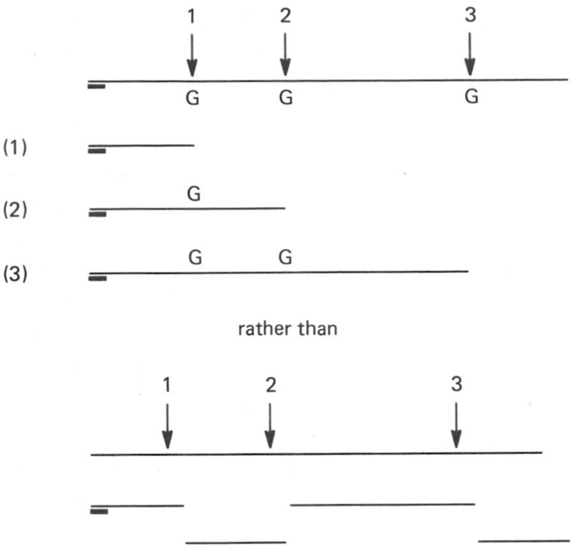

FIG. 18-10. Breaking strands in nucleotide sequencing. The reaction must be controlled so that a single strand is broken only occasionally at a site rather than at every possible breakage site every time. Assume that a chemical can break a strand at sites where G occurs (above) and that there are three Gs on a particular strand. Breaking some of the strands at site 1 only, some at site 2 only, and others only at site 3 yields three labeled fragments of different lengths. Because of the label, these strands can be detected. We now know that a destroyed G must have been present at the end of each of the strands of different size. If cleavage occurred at all possible sites (below), pieces would be generated, but only one of them of a given length could ever be detected, because this is the only one that would carry the label. All we would know is that a G was present at the end of that segment; the distribution of the other Gs would remain unknown, because the other fragments, being unlabeled, would be lost to us.

For example, dimethyl sulfate can cause methylation of guanine or adenine and thus break a strand at the position of a G or an A.

The chemical interacts with G five times faster than it does with A, however, and thus will break the strand at sites where G occurs much more often than where A is found. When the strand is broken, fragments result. Some of these are radioactive; others are not. It is the labeled ones that are critical to the procedure, for these are the only pieces we are able to follow. A radioactive strand of a given length tells us that the base that the chemical alters must have been at the broken end of that strand. It is therefore very important that only some of the strands are broken at the positions of a given base along the DNA. Figure 18-10 shows why this is so. If every possible break always occurs at the sites where the base is found, pieces

would always be generated which were not labeled, and hence would never be detected. We must control the procedure so that strands are occasionally broken only at site 1, other times at site 2, and so on. The chemical reactions in the four different sets of tubes preferentially break the DNA at different sites as follows: set 1 at G alone, set 2 at Gs and As, set 3 at Ts and Cs, and set 4 at Cs alone (Fig. 18-9).

Following the chemical treatments, the four samples are placed on a gel which is then subjected to electrophoresis. The fragments in each sample are allowed to migrate in separate, adjacent lanes in the gel. The pieces of DNA in each lane thus become separated by size, the smaller pieces travelling farther in the gel than the larger ones. The separation of the pieces in the four lanes is then visualized by preparing an autoradiograph which reveals the distance travelled by the pieces in each lane (Fig. 18-11). We are enabled, therefore, to distinguish side by side the pieces resulting from the four kinds of treatments:

FIG. 18-11. Sequencing a DNA strand. The fragments generated following four different chemical treatments (Fig. 18-9) are separated side by side in four lanes, each lane representing a given treatment. An autoradiograph detects the label at the 5′ end of a piece of DNA of any given size. Shorter fragments in each lane are toward the lower end of the gel, since they travel faster. By comparing the four lanes, we can see what labeled fragments of a given size are present. Starting from bottom to top, the sequence of bases can be read off directly (see text for details).

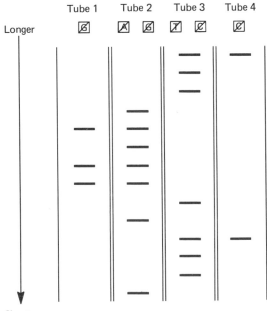

one in which breaks occur at G alone; a second at A or G; a third at T or C, and a fourth at C alone. The sequence of bases is easily read off by comparing the dark lines in the separate lanes. The lines indicate the positions of the radioactive fragments in each lane. Since the 5′ end is responsible for the labeling, the shortest fragment at the bottom (Fig. 18-11) indicates that the 5′ end of a DNA segment which was broken at the site of a particular base is present at that position. In our example, the base nearest the 5′ end must be A. We can see that this must be so, because in tube 1, no radioactive fragment was generated to produce a very small fragment at that position. This means the DNA was not broken at *that* position in the case of the chemical treatment. In tube 2, where a DNA strand could have been broken at the site of A or G, we *do* have a small fragment. It must have been generated at the site of an A, since tube 1 tells us that no G is present there. Obviously, no T or C occurs at the first position, because there are no fragments at that size position in tubes 3 and 4. Continuing from bottom to top, we read the nucleotide sequence of the fragment as 5′ATTCATGGAGATTC3′. From this sequence, we can obviously deduce the sequence of the complementary strand as 3′TAAGTACCTCTAAG5′. In this way, the sequence of 200 to 300 nucleotide base pairs can be worked out in a single experiment.

The determination of the exact nucleotide sequences of specific restriction fragments has made it possible to determine the exact nucleotide sequences of many genes and even of certain entire genomes. From the data of several teams of investigators, an exact nucleotide sequence map of the entire SV40 genome has been established. This was accomplished by piecing together the sequences determined for fragments that were produced by using different restriction enzymes. For example, suppose enzyme 1 produces a series of small fragments and these are sequenced. The same is then done for fragments generated by treatment with enzyme 2. We then compare the sequenced fragments and look for sequences that overlap. Suppose the following sequence has been worked out for a particular fragment generated by enzyme 1: 5′TGGAGGACCCGGAT3′. A fragment generated by enzyme 2 is found to have the following sequence: 5′AGAATTTAGTGGAGGACC3′. When the two are compared, an overlap is evident, as indicated by the areas underlined. From this, it can be deduced that the two fragments have been generated from the same region of the chromosome. A larger sequence can now be established: 5′ AGAATTTAGTGGAGGACCCGGAT3′.

Proceeding to look for overlapping sequences among fragments in such a way, investigators have completed the entire genome of 5375 nucleotides for ϕX 174 and the 5226 nucleotide pairs of SV40. The exact positions of attack by each of the restriction enzymes is known for SV40. This knowledge of a complete nucleotide map has made it possible to proceed further to localize exactly the structural genes and signals involved in control.

Progress in DNA sequencing is proceeding at such a rapid rate that within a few years a complete sequence should be available for the entire genome of *E. coli*, which consists of about 10,000 kilobases! At about that time, approximately one tenth of the DNA sequence of a human chromosome should have been made! DNA sequencing has revealed many unexpected features of DNAs from diverse sources. It revealed the amazing fact that overlapping genes occur in such forms as ϕX 174 and SV40 (Chap. 16). It established the fact that the genes of eukaryotes and their viruses are typically in pieces, consisting of introns and exons. Very significantly, it is making possible the recognition of nucleotide sequences in prokaryotes and eukaryotes that are involved in genetic control or act as signals (promoters, operators, terminators enhancers). It has even brought to light some very unexpected features of mitochondrial DNA (Chap. 19).

Especially exciting are the advances being made in the development of automated DNA sequencing machines. A DNA sequenator has been designed which is able to sequence nucleotides at a rate at least ten times higher than that which can be achieved by more costly and time consuming manual methods. Now that automatic DNA sequencing technology permits almost 8000 nucleotides a day to be read off, molecular biologists are able to turn their attention to certain projects which would not be feasible by manual methods. Among these is the sequencing of the entire human genome, a feat which automation can accomplish within a few years! DNA sequencing is a vital part of the new genetic methodology and will continue to play a pivotal role in research at the molecular level.

Genetic manipulation

The strides made in genetic manipulation, in which genetic materials from different sources are combined to produce a recombinant DNA, are at the forefront of the most dramatic scientific research in modern times. So rapid has been the breakthrough in recombinant DNA methodology that the widest bridges between species have been spanned. Let us familiarize ourselves with some of the basic aspects of recombi-

nant DNA technology, and then examine some of its refinements and applications.

You will recall that in his synthesis of the first truly artificial gene, Khorana used single-stranded projections as splints to join together separate DNA fragments (see Fig. 18-1). Recombinant DNA procedures

FIG. 18-12. Steps in genetic manipulation. (A) A palindrome may occur in a stretch of DNA and be recognized by a specific restriction endonuclease. The enzyme EcoR1 cleaves at staggered sites four nucleotides apart on each complementary strand. Consequently, the resulting fragments possess single-stranded projections. (B) If the circular DNA of a plasmid contains only one such palindrome that can be recognized by EcoR1, it is broken at a specific site in each strand *(arrows)* and becomes linear with single-stranded ends projecting. (C) If the same palindrome exists more than once in DNA, the DNA will be cleaved at a specific site in each region *(arrows)* with the production of many fragments, each with projecting, single-stranded ends. Note that these extensions in the fragments are complementary to those in the plasmid (B). (D) A fragment of DNA from some source, and with the proper complementary base sequences on the ends *(red)*, can pair with exposed bases in the plasmid. DNA ligase can then join the ends together, resulting in the production of a recombinant DNA molecule.

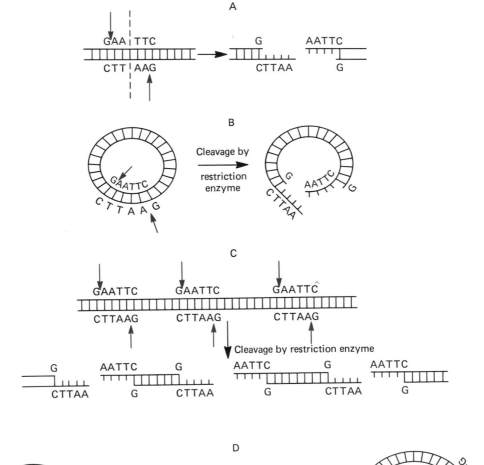

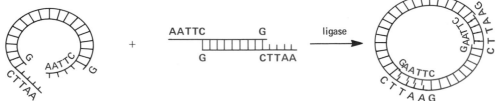

commonly use such single-stranded "sticky" ends. These single-stranded projections are generated by many restriction enzymes which produce staggered nicks within DNA sequences that are palindromes and contain an axis around which a symmetrical arrangement of base pairs occurs. Figure 18-12*A* shows the site of cleavage of the widely used restriction enzyme EcoR1. This enzyme produces breaks in complementary DNA strands at points that are actually several nucleotides apart. As a result of this property, DNA regions are generated with single-stranded projections that will facilitate their joining to strands bearing the complementary nucleotide sequences.

Remember that fragments or segments of DNA do not usually possess the ability to replicate unless they are part of an intact replicon. Therefore, if any DNA is to be introduced into a cell, it must be part of a suitable replicon. The first vehicle selected to carry foreign DNA into a cell, and a vector still commonly used, was an R plasmid, which may contain factors providing antibiotic resistance (see Chap. 15). R plasmids can be isolated from cells by density gradient ultracentrifugation. It was discovered that when bacteria are treated with calcium chloride, their cell membranes become permeable to extraneous DNA. When salt-treated cells are exposed to R plasmids, about one in a million becomes transformed by taking up an R plasmid.

This information and the available techniques set the stage for combining DNA from different sources to produce recombinant DNA. One kind of genetic engineering combines plasmid DNA from two distinct species of bacteria. For example, *E. coli* and *Staphylococcus aureus* are distinct species and cannot exchange genetic material. Either one may carry its own kind of R plasmid. Assume that the *S. aureus* plasmid carries a factor for penicillin resistance and that the *E. coli* plasmid carries one for resistance to tetracycline. From cultures of cells carrying known R plasmids, the DNA may be extracted and subjected to density gradient ultracentrifugation. Certain chemical treatments are used to impart a greater density to the plasmids, which are then easily separated from the chromosomal DNA in the gradient.

Once isolated, the two kinds of plasmids are mixed and subjected to a restriction endonuclease such as EcoRI, which can recognize a specific palindrome and cleave the DNA in the region of axial symmetry at staggered sites. The *E. coli* plasmid used as a carrier in the experiments had only one such site, so that the circular plasmid becomes linear but is not broken up into short fragments (Fig. 18-12*B*). Fragmentation

occurs in a plasmid, or a stretch of DNA if the palindrome is repeated in various regions, since the enzyme will attack each of them. The same sequence may occur in DNAs from different sources. The particular *E. coli* plasmid used as a carrier has the same palindrome as the *S. aureus* R plasmid. In the latter, however, the palindrome may occur more than once, so that fragments are produced from the *S. aureus* R plasmid as a result of the enzyme activity (Fig. 18-12*C*). Keep in mind that the action of enzymes such as EcoRI also causes broken DNA to have projecting single-stranded ends and that these ends are complementary. Complementary projections on any DNA pieces, no matter what the source, permit base pairing. If this pairing occurs between DNA pieces from different sources, exposure to DNA ligase can join them by forming phosphodiester links, and thus give rise to a recombinant DNA (Fig. 18-12*D*).

Plasmids treated in this way are then mixed with *E. coli* cells, which have been treated with calcium chloride to render them permeable to DNA. Since the majority of salt-treated cells will not be transformed, and since some of those that have been transformed will contain two plasmids of the same kind joined together, it is necessary to screen out cells that carry a composite plasmid, composed of DNA from the two bacterial species. It is possible to do this by growing cells in the presence of the antibiotics for which the two types of plasmids carry resistance factors—in our example penicillin and tetracycline. Only doubly resistant cells can grow and therefore must carry antibiotic resistance factors from both species.

Following these steps, such cells carrying recombinant DNA have been isolated. In these cells, the integrated foreign DNA combined with the native DNA continues to replicate as if it were native to the cell. The process bridges the natural barrier to crossing or genetic exchange between two species. In the example given, most of the genetic information expressed is typical of *E. coli,* but the cells also carry DNA derived from a genetically separate species, and this DNA is also expressed in *E. coli!*

Further applications of genetic manipulation

The term *cloning* is applied to those recombinant DNA procedures in which a foreign stretch of DNA is replicated in cells, using a carrier such as a plasmid or virus. The cells continue to grow, proliferating the specific recombinant DNA. A pure culture or population of identical cells, a clone, is thus established. The introduced DNA is multiplied thousands of times in the clone. Plasmids carrying the foreign DNA can easily be isolated from bacterial cells and then subjected to further study. The DNA carried on a plasmid is thus always readily available for various lines of investigation.

One goal of genetic manipulation is to introduce into bacteria those genes known to control the formation of some rare but vital substance, such as insulin. If an introduced gene can function and produce the valuable product in the cells, vast amounts of costly and rare substances might become readily available to help alleviate human suffering. However, recombinant DNA procedures are often rather hit or miss, in the sense that the DNA from some given source, such as the human, is broken enzymatically into a large number of fragments, from a few hundred to a few thousand nucleotide pairs. Each piece is then inserted into bacteria to establish clones. The term *shotgun experiment* is often used to denote the collection of a large sample of cloned DNA fragments representing the DNA of a given species. The collection can serve as a reserve of cloned DNA, and from it one can later select clones of interest for further studies. An important aim of genetic manipulation is to be able to place into a plasmid for cloning *only* the DNA that carries some desired genetic sequence, such as the DNA coded for the amino acid sequence of insulin.

How can one insert only the gene or sequences one specifically wants for cloning and thus avoid the need to examine many thousands of clones to find the desired one? One approach is applicable in those cases in which pure RNA transcripts of the gene in question are available at the outset. The mRNAs for various human globin chains are examples of specific transcripts that can be isolated in pure form. The availability of the globin transcripts has been responsible for the wealth of information we now possess on the arrangement of sequences that code for these polypeptides.

We can make good use of the fact that the information for the amino acid sequence of a polypeptide is carried in the messenger RNA and this coded information is complementary to the DNA strand that underwent transcription to produce it. One can convert the RNA information into DNA information by the use of the enzyme reverse transcriptase (Fig. 18-13A). DNA that is formed on an RNA template by reverse transcriptase is cDNA (copy DNA). Once a cDNA is obtained, the mRNA is destroyed. The single-stranded cDNA can then be rendered double stranded with DNA polymerase. The two DNA strands, how-

ever, are joined at the end by strong covalent linking. Another enzyme (S_1) solves this problem by breaking the linkage to give a normal double helix.

In a commonly used procedure, this double helix is treated further to provide it with projecting ends. Returning to Fig. 18-12, you will recall that both the plasmid DNA and the DNA segments to be inserted bore single-stranded projecting ends and that those of the plasmid were complementary to those of the fragments. Projecting complementary ends are not always present, as is the case here where the double-helical cDNA bears no projections at all. The properties of still another enzyme—terminal transferase—can be called on to provide sticky ends. This enzyme has the ability to add to the 3' ends of DNA strands a series of identical nucleotides (Fig. 18-13B). The cDNA is exposed to the enzyme in the presence of, let us say, molecules of guanosine triphosphate. This results in the production of an extension at each 3' end of the cDNA composed of a sequence of repeating bases, all with G. The plasmid that is to be the carrier is processed separately. It is first rendered linear by treatment with a restriction enzyme. In our example, it would then be subjected to terminal transferase in the presence of molecules of cytidine triphosphate. Complementary ends of repeating sequences of C are thus added to the plasmid to produce projecting ends.

The plasmid and the cDNA are then mixed together (Fig. 18-13B) to permit complementary base pairing. Any gaps are filled in by adding DNA polymerase and extra nucleotides. DNA ligase finally seals breaks between the inserted DNA and the plasmid to produce a recombinant DNA. The plasmid selected as a carrier contains at least one genetic factor that confers on a bacterial cell resistance to a specific antibiotic. As

FIG. 18-13 Summary of major steps in recombinant DNA production starting with transcripts of a specific gene. (A) The mRNA is used as a template in the presence of reverse transcriptase to form a complementary DNA strand. The RNA is then destroyed, leaving the single-stranded cDNA. This is then used as a template to form a complementary DNA strand. An enzyme is required to separate the two strands that are linked. A cDNA duplex results which is now available for further processing. (B) The cDNA and the plasmid are handled separately to add single-stranded projecting ends that are complementary. This is accomplished by terminal transferase, an enzyme that can add strings of identical nucleotides to 3' ends. The plasmid must first be opened up before the application of the transferase. This is accomplished by a restriction enzyme specific for a palindrome in the plasmid. Finally, the cDNA and plasmid are brought together. Extensions on the cDNA may pair with their complements on extensions of the plasmid. Filling in of gaps and sealing yield a circular duplex recombinant DNA carrying a copy of the specific gene.

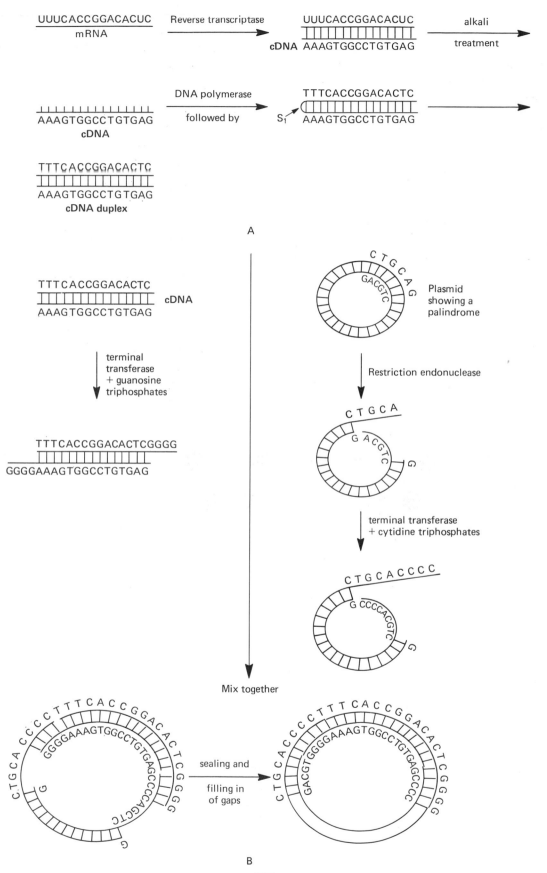

A

B

described earlier, this enables the investigator to screen out any transformed cells that have arisen following exposure of the bacterial cells to calcium chloride and then to the R plasmid.

It is now necessary to determine which of the transformed clones contain, not just fragments, but all the DNA of the desired gene. Since purified RNA transcripts of the gene are at hand, the task proves to be rather simple. Excellent use is made of DNA–RNA hybridization (Fig. 18-14). First of all, separate bacterial colonies, each representing a clone, are maintained on nutrient agar. Copies of these colonies are transferred and grown on nitrocellulose filters using the replica plating technique. The bacterial cells are lysed, and their DNA is released onto the filter paper and then denatured. Pure radioactive transcripts of the gene being searched for are presented to the denatured DNA. Any single-stranded DNA present on the filters that has appreciable nucleotide sequences complementary to the RNA will hybridize with it. These DNA–RNA molecules are radioactive and are easily detected by autoradiography after the filters are washed. Since replicas of the clones have been maintained, the investigator can return to the original plate and identify the clone on the nutrient agar that corresponds to that on the filter. In this way, using the mRNA as a probe, one can identify those clones that have successfully incorporated the entire desired gene sequence into the plasmid. Such clones are now available for further study.

FIG. 18-14. Identifying the proper clone. Clones are maintained on master plates *(1)*. To determine which, if any, carry all the DNA of the desired gene, a copy of the master is made by replica plating *(2)*. The DNA on this plate is then treated to denature it. Labeled transcripts of the gene in question are added *(3)*. DNA from one or more clones may form hybrid molecules with the added RNA if that RNA carries sufficient nucleotide sequences complementary to the DNA. Hybrid formation indicates that most or all of the desired gene is present in that clone *(4)*. To detect any DNA–RNA hybrid molecules, the plate is covered with a photographic emulsion to produce an autoradiograph *(5)*. The clone with the desired gene (C) is identified by relating it back to the clone held in reserve on the master plate.

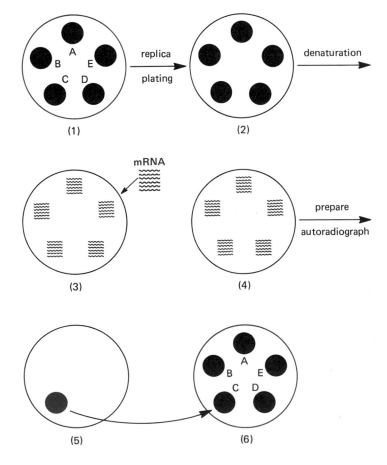

The importance of probes in cloning and genetic manipulation cannot be overestimated, and indeed, advances in these and related areas depend on the availability of the proper probes. Probes are specific, labeled stretches of RNA or DNA that can hybridize with the complementary nucleic acid stretches. A probe for a specific nucleic acid sequence may be a pure mRNA species, cDNA that has been prepared from the mRNA, or a stretch of DNA. Having a probe for a specific gene or a specific DNA segment makes it possible to screen out the gene or segment that otherwise might remain undetected or unidentified among countless other pieces of DNA.

Problems in searching for the desired clone

Pure RNA transcripts are not available for most of the genes one wishes to clone. Therefore, the task of producing and locating a clone that contains a desired gene becomes even more laborious than the procedure already described, although in general the same steps are followed. First, one must turn to the tissue that is specialized to produce the protein whose gene is being sought for cloning.

If insulin is the product of the desired gene, one would turn to those pancreatic cells specialized for production of the hormone. The reason for this is that these cells produce mRNA transcripts for the protein, whereas other cells, such as those of the kidney, do not, since genetic information for producing insulin is repressed in the kidney. Of course, the insulin-producing cells contain other mRNAs in addition to that for insulin, because all cells need to produce products for basic cell survival. This means that the insulin mRNA is mixed in with significant amounts of other kinds of RNA.

The next steps in the procedure are the same as those described previously and in Fig. 18-13. However, one now has cDNAs of various kinds, not just those desired. Different clones are thus produced that carry everything from fragments of various kinds to the structural information for various genes, including those with complete information for the gene desired, in this case the one for insulin. Without a specific pure RNA probe to locate the desired clone or clones, additional steps must be taken.

The clones, whose complete genetic contents are unknown, are numbered and a DNA sample is taken from each. This DNA is denatured in a test tube to which is added that RNA taken from the specialized cells at the start of the procedure. This is the very RNA that was used to make the cDNA that was inserted into the bacteria. This RNA is, of course, unpurified, a mixture of various species of RNA. Any of this added RNA with nucleotide sequences complementary to a DNA sequence present in a clone will hybridize with that DNA. This fact of hybridization can now be used to identify the desired clone or clones. This stems from the fact that RNA that has formed a hybrid molecule with DNA is unavailable for translation. Therefore, no polypeptide can be formed from hybridized RNA when hybrid DNA–RNA is placed under conditions favoring translation. To find out whether a specific hybrid DNA–RNA is present, all the requirements for polypeptide formation are added to the test tube (Fig. 18-15; review Fig. 12-11).

The amino acids that are added are labeled with radioactivity. Polypeptide is formed as a result of translation of any *unhybridized* mRNA. To tell whether or not the system is making the desired polypeptide, one adds antibodies to that polypeptide, the polypeptide specified by the gene one wishes to clone. The antibodies will react specifically to the polypeptide and will bind to it, causing it to precipitate out of the solution if it is present. If the desired gene is present in the system, the mRNA for that gene will be bound up in hybrid DNA–RNA and will not be available for translation. The antibody, being specific, does not react with any other proteins formed by translation of free, unhybridized mRNAs, and so no precipitate is formed.

We are now in a position to identify those clones that contain the sought-after DNA. In a clone in which the precipitate forms, the DNA of the desired gene *cannot* be present. If it were, it would have bound the mRNA complementary to it. Since the gene is not present, the transcript is available for translation, and the protein (insulin in our example) precipitates out, because it is recognized by the antibody. Other proteins are also formed, since much of the added mRNA would not be hybridized. However, this protein is *not* precipitated; only the insulin is. We then know that this clone does *not* contain the structural gene for insulin.

In a clone that *does* contain the structural gene for insulin, the DNA–RNA hybrids are formed between the DNA of that gene and its messenger. Consequently, translation is blocked. In that system, no precipitate forms, because no insulin could be synthesized to permit a reaction with the specific antibody against it. Again, other proteins are formed here,

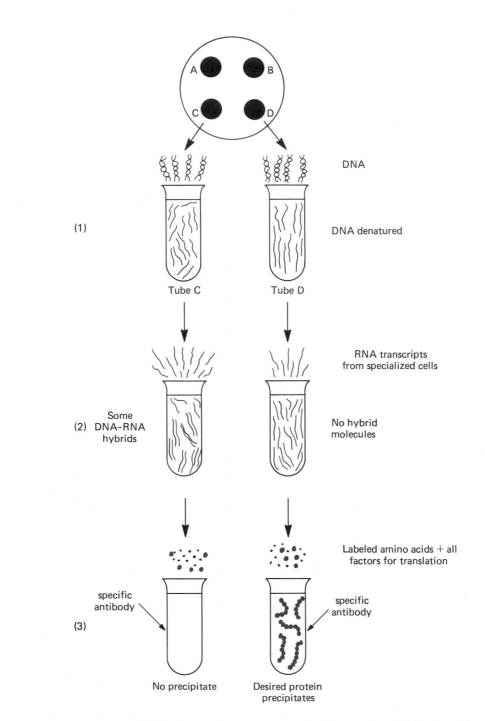

FIG. 18-15. Identifying the proper clone in the absence of a pure RNA sample. *(1)* From clones held in reserve (C and D here), DNA samples are taken and denatured. *(2)* Messenger RNA that has been isolated from cells known to produce transcripts of the desired gene is then added. These transcripts will be mixed with other transcripts from the cells. However, if the desired gene sequences are present in any of the denatured DNA, hybrid molecules will form (tube C). This hybridization will tie up the transcripts of the gene, making them unavailable for translation. *(3)* In a translation system, the protein product of the desired gene can form only if the mRNA coded for its amino acid sequence is present. This is possible in tube D but not in tube C. Antibody precipitates the protein out if it is present. The absence of the protein in tube C tells us the transcript is tied up and the coded information for that protein is present. Hence, clone C, held in reserve *(above),* carries the desired gene.

but they are not precipitated out. In this way, a desired clone can be recognized.

A further check can be made on the identity of a clone. The DNA–RNA hybrid molecules can be separated from the unbound RNAs present in the system. The RNA bound in the hybrid molecules can be released from the DNA. This RNA is then used to direct the synthesis of protein in another translation system and identify the protein.

Once a clone has been identified as a population of cells that contains a specific structural gene, further genetic manipulation is generally required if that gene is to express itself in the bacteria. We know that the normal expression of a genetic message depends on regulatory signals. If a foreign gene, lacking appropriate signals, is inserted into a plasmid, there is no reason to expect expression of the gene and formation of its associated product. The gene could simply be replicated along with the plasmid DNA, without transcripts being formed from it. If transcripts are formed which lack the proper "start" and "stop" signals, they could fail to undergo proper translation.

The task of providing a gene with the proper regulatory signals to permit its transcription, followed by translation of the transcript, is being aproached in various ways. Using restriction enzymes and ligase, the investigator may insert a foreign gene *into* a bacterial gene residing on the plasmid. The introduced gene is thus provided with the regulatory signals of a resident bacterial gene. The result of such a manipulation is the production of a hybrid genetic region that starts off as a bacterial gene, changes to the introduced gene, and ends as the bacterial gene. The hybrid region has, however, the signals required for the beginning and ending of transcription.

If all the necessary steps are followed, the cells of the clone may secrete into the medium a polypeptide that begins as a bacterial protein, then changes to the desired one, and ends as a bacterial protein again. The desired portion may then be retrieved using protein-digesting enzymes. Relying on the various procedures for clone production, identification of the proper clone, and production of a desired protein, several teams of investigators have demonstrated that recombinant DNA technology can indeed be used to obtain from bacteria biologically active proteins typically secreted by specialized eukaryotic cells. Such valuable products as human insulin and growth hormone have been produced using such methods and may be made available in ever larger amounts as cloning techniques become

even more refined. Yields of exceptionally scarce substances in increased quantities, such as the potential anticancer protein interferon, may enable the design of critical experiments to evaluate their biological properties and their value in medicine.

The procedure described here and in previous sections to form recombinant DNA with the use of projecting, sticky ends is one that is commonly used. It is not, however, the only method available today. One can now use *blunt end ligation*. This procedure makes use of the ability of the DNA ligase from phage T4 to join together two DNA molecules that do not have projecting sticky ends but instead have blunt ends. (As was noted in Fig. 16-18, some restriction enzymes do not produce staggered cuts and consequently yield fragments with blunt ends.) Each procedure—the one using sticky ends and the other joining together blunt ends—has advantages and disadvantages, as well as variations. Which specific procedure is used will depend on the specific aim of a particular experiment, in addition to the circumstances under which it is conducted.

Gene cloning and chromosome mapping

In addition to its impact on the various areas of molecular biology and genetics discussed in the previous sections, gene cloning is greatly accelerating the rate of human chromosome mapping. In Chapter 9 we discussed advances in the mapping of human chromosomes made possible through somatic cell genetics. This procedure, although responsible for the assignment of well over 100 genes to specific chromosomes, entails a definite limitation. You will recall that the cell hybridization procedure depends on the ability of the specific gene being mapped to express itself in the hybrid cell. Failure of a gene to produce some detectable product or effect in the hybrid cell makes it impossible to assign it to a chromosome using the procedure of somatic cell genetics by itself. Let us see how gene cloning, coupled with somatic cell procedures, has been able to overcome this severe limitation and to make it possible to detect the presence in a hybrid cell of a gene that is not being expressed.

The ability to detect the presence of a specific gene or specific DNA sequence using gene cloning ultimately depends on the availability of a specific probe for the sequence or gene under consideration. Earlier in this chapter we discussed the importance of probes.

At the very outset of our discussion of gene cloning, we also noted that the abundance of RNA transcripts of globin genes facilitated the cloning of these genes. Not only could the pure transcripts be used to obtain cDNA, but labeled copies of the transcripts could be used as probes to detect only those clones, among a large number of different clones, that contain the specific gene. Actually, once any specific gene has been successfully cloned, it becomes available for a host of studies and uses. Among these is the use as a probe of the cloned gene, such as a globin gene or the insulin gene. Let us suppose that we can use as a probe a labeled cDNA of the gene for beta (β) globin. We are now in a position to locate the β gene sequence on a specific chromosome. To accomplish this, we first perform somatic cell hybridization. Human and mouse cells are fused, as described in Chapter 9, and stable clones containing just a few human chromosomes are established. Because the β gene is not expressed in the cell hybrid, we were previously prevented from screening the clones, since no detectable evidence of the globin gene in the form of a product would be present in any hybrid cell containing the gene. Now, however, we can proceed further. DNA is extracted from human cells and then subjected to a particular restriction enzyme. Mouse DNA is similarly treated, and so is the DNA taken from the hybrid cell containing, let us say, human Chromosomes 5, 8, and 11. The three DNA samples that have now been cut by the specific restriction enzyme are placed on an agarose gel and subjected to electrophoresis, which separates the fragments on the basis of their sizes. Our ultimate goal is to hybridize fragments with a probe. To do this, it is necessary to transfer and bind the fragments to a nitrocellulose filter. This can be accomplished using the Southern blotting procedure (see Chap. 16). The filter with the denatured fragments is now available for hybridization with our probe.

The filter contains many separated DNA fragments—mouse in one lane, human fragments in another, and mouse DNA plus DNA for human Chromosomes 5, 8, and 11 in a third lane. Our labeled probe of denatured cDNA of the globin gene is then brought into contact with the filter. Any single-stranded DNA fragments on the filter that contain sequences complementary to sequences on the probe will form hybrid duplex DNA with it. Such complementary sequences can exist in rodent as well as human DNA fragments because the rodent also carries genes coded for globin proteins. Since our probe is labeled, any

fragments on the filter that hybridize can be detected by autoradiography. As Fig. 18-16A shows, signals from the radioactive probe are now evident. We see that the lane with human DNA alone contains two fragments that hybridized with the probe, indicating that those fragments contain DNA sequences complementary to sequences in the probe, and hence contain sequences of the beta-globin gene. The mouse lane gives a different pattern. One fragment contains the sequences for the gene. The DNA from the hybrid cell shows a pattern that combines both the human alone and the mouse alone. We now know that human beta-globin DNA sequences are present on either Chromosome 5, 8, or 11.

To ascertain which human chromosome actually carries the beta-globin gene, we must screen clones containing other combinations of human chromosomes. It is found that only when Chromosome 11 is present do we find the pattern shown in Fig. 18-16A. DNA from any hybrid cell lacking Chromosome 11 shows only the mouse pattern (Fig. 18-16B). We can thus deduce that the gene for β-globin is on Chromosome 11.

This procedure enables us to assign a gene, even though that gene is not expressed in the hybrid cell. What we are detecting are specific fragments, generated by restriction enzymes, which contain stretches of the gene. Our probe enables us to detect these fragments. Coupling this procedure with the procedure of somatic cell genetics, we can now relate the fragments to a specific chromosome. We simply look at our filter to see which chromosome, when absent from a hybrid cell, is associated only with the mouse pattern and when present is always associated with the human plus mouse pattern. The reasoning is the same as that used in Chapter 9 when screening panels for a gene associated with a detectable product. Using the procedure made possible by strides in molecular biology, human chromosome mapping is advancing at a rapid rate. Not only does it enable us to assign to a chromosome a gene that is not expressed in a hybrid cell, it also allows us to assign such a gene to a specific chromosome band. The basic reasoning is again the same as was described in Chapter 9 and is illustrated in the following case of the insulin gene.

Earlier in this chapter, we discussed procedures permitting the identification of clones containing insulin cDNA. Once clones with cDNA of any gene have been identified, they provide a source for further investigation of a specific gene. Using insulin cDNA as a probe, investigators were able to assign the in-

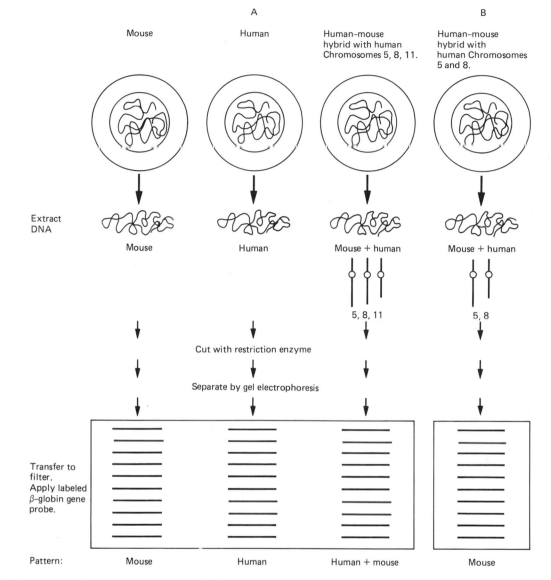

FIG.18-16. Assigning the gene for beta globin. *(A)*. Total mouse DNA, when subjected to a particular restriction enzyme, yields one band *(red)* that hybridizes with the labeled probe, indicating that sequences for the globin gene are present on that fragment. Similarly treated total human DNA yields two bands that hybridize with the probe, hence carry beta-globin gene sequences. When DNA from human–mouse cell hybrids is similarly treated, the mouse plus human pattern is found when Chromosomes 5, 8, and 11 are present. *(B)*. In the absence of Chromosome 11, only the mouse pattern is found. The gene for human beta-globin must reside on Chromosome 11.

sulin gene to Chromosome 11. It was then assigned to a specific region of Chromosome 11. This was made possible through the use of a human cell culture containing a specific reciprocal translocation. Such structural rearrangements occur spontaneously in humans, as we noted in Chapter 10, and they account for an appreciable number of spontaneous abortions and congenital defects. Cells from individuals carry-

ing specific translocations have been used to establish hundreds of different lines, each representing a specific translocation. Each specific reciprocal translocation can be identified, because it alters the banding pattern of the chromosome. Held in reserve, such cell lines are available for further use. In the case of the insulin gene that had been assigned to Chromosome 11, it was possible to pinpoint its location on the

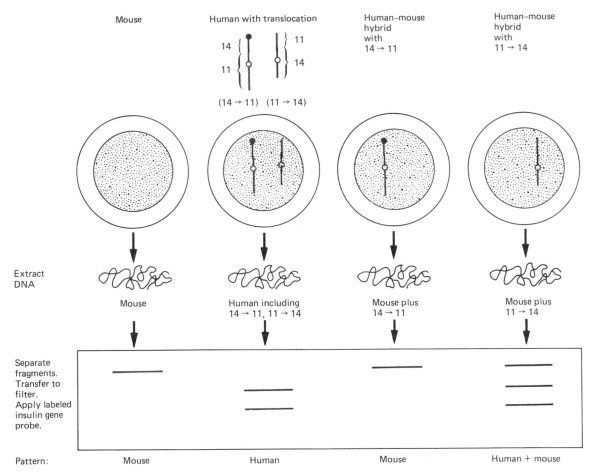

FIG. 18-17. Localization of gene for insulin. The gene for insulin had been assigned to Chromosome 11 in a manner similar to that shown in Fig. 18-16. Using a reciprocal translocation involving Chromosomes 11 and 14, we can show, following somatic cell hybridization, that the end portion of Chromosome 11 carries the insulin gene. Only when the end of human Chromosome 11 is present in a cell do we find hybridization with the insulin gene probe, giving the mouse plus human pattern.

chromosome. Figure 18-17 shows that only when the very end portion of the short arm of Chromosome 11 is present does the hybridization give the human plus mouse pattern. In the absence of that specific chromosome segment from Chromosome 11, only the mouse pattern is revealed following hybridization with the human cDNA probe, indicating that the probe is hybridizing only with complementary sequences on the mouse DNA.

Use of structural aberrations such as translocations and deletions along with the advances in molecular procedures is rapidly increasing our knowledge of the precise chromosome positions of genes that previously would not have been assigned to a chromosome at all because of their failure to express themselves in the human–rodent hybrid cell.

Establishing gene libraries

Restriction enzymes have been responsible for many highly significant advances in molecular genetics and cell biology. Among these is the progress made in the molecular analysis of the DNA of entire genomes of species such as the human and the mouse. For example, total DNA is extracted from human fibroblasts and then exposed to a given restriction enzyme such as HindIII, which cleaves it into a certain number of specific fragments. These fragments are now ready for insertion into a suitable plasmid containing only one site vulnerable to the restriction enzyme. The general procedure is identical to that described in Fig. 18-12.

A very popular plasmid used as a vector in cloning

is one designated *pBR322*. The reason for its wide use is easily understood. The plasmid carries two genes that confer antibiotic resistance, one to ampicillin (Amp^r), the other to tetracycline (Tet^r). Fortunately, each gene contains sites that can be recognized by several restriction enzymes. Certain of these endonucleases recognize sites only in the Tet^r gene, others only in the Amp^r gene. The value of this fact is that the plasmid may be cut selectively in either of these two genes. The cuts are staggered so that projecting, sticky ends are generated. If a fragment of foreign DNA, such as from the human, is inserted into either resistance gene, the activity of that gene is lost. For example, cleavage by endonuclease HindIII produces staggered cuts in gene Tet^r at one site, rendering the plasmid linear. (The restriction enzyme EcoRI can also cut the plasmid, but its specific site is found outside both of the resistance genes.) Human DNA cleaved by HindIII will produce many fragments, each with sticky ends complementary to those in the plasmid. Endonuclease-treated plasmids and the endonuclease-generated human fragments are incubated together. Following DNA ligase treatment, the material is mixed with calcium chloride–treated bacterial cells and then screened to detect those cells that are resistant to ampicillin only. Loss of resistance to tetracycline because of insertion of the human DNA enables us to screen out those cells that have been transformed and that contain plasmids into which a fragment of human DNA has been inserted (Fig. 18-18). In this way, pieces of human DNA may be inserted into bacterial cells, which in turn produce a number of clones. Each clone that is established harbors a specific piece of human DNA.

You will recall that the steps in the cloning of human DNA began with the extraction from cells of the total DNA content. This total DNA was then fragmented by the endonuclease and exposed to the plasmid. It is obvious that the procedure is a shotgun approach, hit or miss as to which fragment enters a particular plasmid. It is possible to follow the procedure using a more purified type of DNA. For example, a cesium gradient may supply us with a small amount of a particular band of DNA. If this DNA contains a region vulnerable to the proper restriction enzyme, then it may be handled in the manner just described. The result is the establishment of clones covering a smaller but much more specific portion of the genome.

By joining fragmented DNA from the human or some other species to plasmids and then establishing

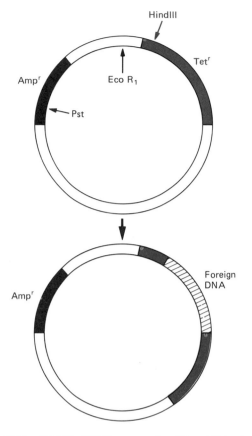

FIG. 18-18. Plasmid pBR322 as a vector. *(Above)* This plasmid contains a gene for resistance to ampicillin (amp^r) and one for resistance to tetracycline (tet^r). Various restriction enzymes recognize sequences in both of these genes. Others, such as Pst, recognize sites found only in the amp^r gene. Still others, such as HindIII, recognize sites found only in the tet^r gene. The EcoRI site is found outside the resistance genes. Using recombinant DNA procedures, we can insert a fragment of foreign DNA into one of the two resistance genes, making use of a restriction enzyme that will cleave within one of the genes but not the other. In this example *(below)*, the foreign DNA has been inserted into the tet^r gene after the plasmid was cut by HindIII in the tet^r gene. Carrying the foreign DNA, the tet^r gene no longer confers resistance to tetracycline. Loss of resistance to this antibiotic, but not to ampicillin, makes it possible to screen out those cells that have been transformed and also carry an inserted foreign DNA.

clones, we can construct a gene library. The library consists of all the different clones, as if each one were a volume. At this point, however, we do not know what is in any of the clones or volumes, because our procedure has not enabled us to take a specific gene and place it in a given plasmid. Somehow, we must now try to determine what gene or sequences are located in a given clone. If we can achieve this, then we can turn to the clone at any time to study further a particular gene that is of interest to us. Obviously, our problem can be solved if the proper probes are available, as discussed in the following section.

Probes and the screening of gene libraries

Knowing that we have a cDNA probe for a specific gene, such as the beta-globin gene, we can turn to our library (all the various clones containing different specific fragments of the entire genome) and test each clone to determine whether or not it will hybridize with the specific probe. The procedure is essentially the same as that explained earlier in this chapter (see Fig. 18-14). Unfortunately, cDNAs are not as readily obtained for specific genes as they are for the globin genes. The problem is similar to that discussed for the insulin gene. Lacking a pure RNA sample or cDNA derived from it, we were able, in a rather roundabout way, to find the clone we desired (see Fig. 18-15). The RNA bound to the DNA in the hybrid molecule can be released and retrieved. Knowing that it is an mRNA transcript of the insulin gene in this case, we can use it as a probe in further investigations

or use it to derive cDNA probes for insulin DNA sequences.

Another way to obtain specific probes is to turn to a specific differentiated cell type, as we did in the case of the pancreatic cells in our search for the insulin mRNA. Mixed mRNAs are isolated from the specialized cells and used to form a mixture of cDNAs after a treatment with reverse transcriptase. Again, the procedure is similar to that shown in Fig. 18-13, except that we would be starting with many different transcripts of unknown identity and would generate a mixture of cDNAs, also of unknown identity. These different cDNAs are inserted into a suitable plasmid type, such as pBR322. We now have a mixture of plasmids carrying different cDNAs. Bacterial cells are then exposed to the plasmids. Any transformed cells are identified, and clones are finally established from single bacterial cells. In essence, we are constructing a cDNA library (Fig. 18-19). Each bacterial cell carries a certain unknown cDNA, which is being multiplied in each clone.

We must now try to identify the cDNA in a given clone. To do this, the cDNA is extracted from cells in a clone, rendered single stranded, and then attached

FIG. 18-19. Establishing a cDNA library. Transcripts are isolated from a differentiated cell type. These mRNAs are mixed and represent transcripts of many different genes. They are used as templates in the formation of cDNA in a fashion similar to that shown in more detail in Fig. 18-13. The mixture of cDNAs that are derived represent many different genes. These are then inserted into a suitable plasmid type such as pBR322, which in turn is used to transform bacterial cells. Those cells that are transformed and that contain inserted cDNA can be selected and allowed to form clones. A library is thus constructed. Each clone will harbor a cDNA representing a genetic sequence whose identity remains unknown until a suitable probe is found.

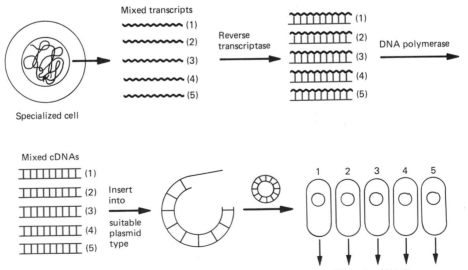

to a nitrocellulose filter. The bound cDNA is next exposed to mixtures of mRNA. If any mRNA in a mixture hybridizes with this cDNA, it can be located on the filter and then isolated by itself. Although this RNA represents a single kind of transcript, we still do not know what information it carries. To ascertain this, we may proceed to place the mRNA in a cell-free system, as explained in Chap. 12 (Fig. 12-11). Any protein that is made would reflect information carried in the mRNA used, and hence would indicate the specific gene that gave rise to the transcript.

The particular protein must now be characterized using one or more procedures. Once its identity is established, we have also identified a probe for the gene that specifies this protein. The probe is the

cDNA that was able to bind to the mRNA coded for the protein and that has been multiplied in the bacterial cells. Armed with the cDNA probe, we can now return to the volumes in our DNA library to locate the clone that carries genetic sequences corresponding to the probe and hence identifying the presence of a given gene.

These are by no means the only methods used to obtain probes for specific genes or even to obtain gene libraries, as we learn in the following section. Good use can be made of somatic cell hybrids. For example, mouse–human hybrid cells were formed in which the only human chromosome present was Chromosome 11. A DNA library was constructed using the DNA from the hybrid cells. Most of the DNA fragments in

FIG. 18-20. Obtaining a probe for human Chromosome 11. DNA is extracted from mouse–human hybrid cells that carry in addition to the mouse DNA, only human Chromosome 11. A library is constructed from the DNA of the hybrid cells. Most of the clones of the transformed bacterial cells contain plasmids with mouse DNA only. The occasional

clone with a plasmid containing an inserted piece of Chromosome 11 can be recognized by performing DNA–DNA hybridization procedures between DNA from the clones and labeled denatured fragments from total human DNA. Only when a clone is harboring a piece of Chromosome 11 does hybridization occur between the DNAs.

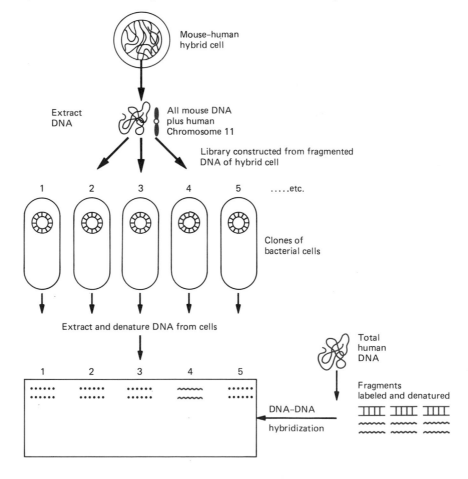

this library would represent mouse DNA, but a small number of the clones would contain fragments of human Chromosome 11. These can be identified by taking *total* human DNA that is fragmented and labeled to screen the mouse–human library for human fragments (Fig. 18-20). In this way we can identify DNA fragments that represent sequences of a specific chromosome. These fragments are then available for further investigations, such as DNA sequencing and may even be used as probes themselves for screening other DNA samples for corresponding homologous sequences now known to be on Chromosome 11.

Cloning in bacteriophages

In addition to plasmids, human DNA may be inserted into a suitable bacteriophage and then cloned. Phage lambda (λ) has been used extensively to establish libraries of human DNA. The procedure is similar to that in which foreign DNA is inserted into a plasmid. In this case, certain strains of phage λ may be subjected to the restriction enzyme EcoRl, cutting it into pieces (Fig. 18-21). Luckily, two large end pieces are generated, and these end pieces are the only portions of viral DNA needed for replication of λ in its *E. coli*

host. The end pieces can be separated from the other fragments and then mixed with DNA from another species, such as the human, which has been cut by EcoRl. A phage particle with a piece of inserted foreign DNA can then replicate in *E. coli*, resulting in the multiplication of the recombinant DNA.

Any one phage carrying a particular foreign DNA segment can thus give rise to a clone. A plate covered with *E. coli* can be infected with a large number of different phage particles carrying different human DNA fragments. Each plaque produced has been derived from a single infecting phage and represents a clone of phages (Fig. 18-22). The problem now is to locate clones carrying specific DNA fragments. The DNA from a clone of phages can be denatured and exposed to a specific probe to determine if the cloned human DNA carries sequences homologous to those in the probe. In this way, an entire phage library can be screened by a probe to determine the location of a given gene.

Searching for an oncogene

Phage λ libraries were used to search for a genetic sequence that can bring about a malignant transfor-

FIG. 18-21. Inserting foreign DNA in phage λ. *(1)* The DNA of certain λ strains, when exposed to EcoRl, is cut into two large fragments representing the left (L) and right (R) ends of the λ DNA plus several small fragments. The DNA of the latter is not required for the replication of λ in *E. coli*. The two large end fragments, which bear sticky ends, are separated from the small ones *(2)*. Foreign DNA *(red)* that has also been cut by EcoRl, will also bear sticky ends that are complementary to those of the L and R fragments of λ. Foreign DNA can thus be inserted between the two end fragments, yielding recombinant DNA *(3)*. A particle bearing foreign DNA can now be packaged in the protein coat of the virus *(4)*. Such a virus can replicate in *E. coli*.

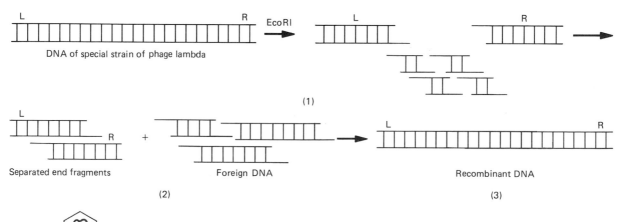

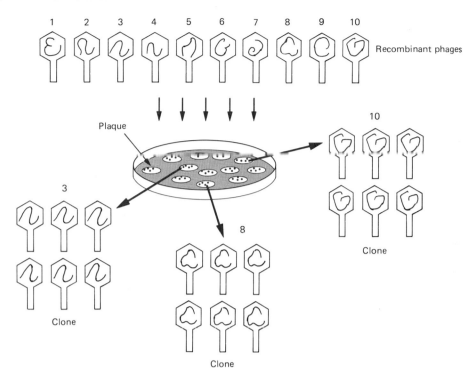

FIG. 18-22. Establishing a phage library. Recombinant virus particles bearing different portions of a foreign genome (1, 2, 3, etc.) are allowed to infect a lawn of *E. coli* growing on a plate. The phages infect the bacterial cells, multiply in them, and lyse them, resulting in the formation of plaques on the lawn. Each plaque represents a clone of phages, each one derived from a single infecting recombinant phage. Consequently, each plaque is equivalent to a clone of recombinant DNA. An assortment of clones such as these covering the entire genome of a species constitutes a genomic library.

mation (see Chap. 17). In one approach, DNA was extracted from cells of a human bladder tumor. This DNA was then used to transform mouse cells to a malignant state. DNA from these malignant cells was extracted, in turn, and used to transform a second set of mouse cells. From these cells in the second cycle of transformation, a phage library was constructed.

The following reasoning was applied. The amount of human DNA in the second set of transformed mouse cells must be very small, because during the course of the gene transfer, starting with the original amount of human DNA through the second mouse transfer, much human DNA would have been lost. *Only* those cells that had been transformed were being selected on the plate (Fig. 18-23). In effect, only that DNA which had transforming ability was being selected for. Much of the DNA *not* involved in the malignant transformation would be discarded along with the cells that were not transformed. It was also reasoned that any human DNA fragment had a high probability of being associated with *Alu* sequences.

You will recall from Chapter 13 that these are widely scattered throughout the human chromosomes. A probe for the Alu sequence was available. The DNA in the phage library from the second round of infection could now be screened using the Alu probe.

The phage library would contain only a small number of phage particles carrying the transforming gene. This would be so because the DNA incorporated into the phages was extracted from mouse cells that had been transformed by a small amount of human DNA. The library was plated on *E. coli* cells (Fig. 18-24). The resulting plaques represented DNA clones, and these were replicated onto filter paper. A labeled Alu probe was then allowed to contact the plaques. Autoradiography revealed the location of any plaques containing DNA with Alu sequences. DNA from such a plaque was then used to infect mouse cells. It was found that this DNA could indeed transform cells to the malignant state. The DNA containing the genetic sequence for the human bladder cancer oncogene had thus been isolated.

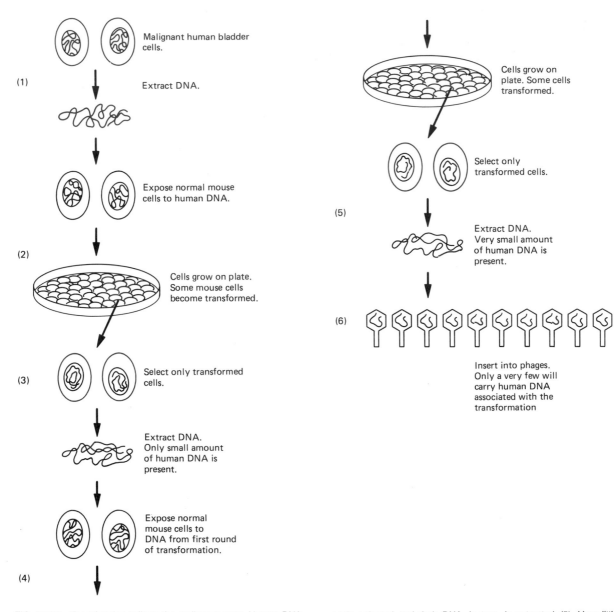

FIG. 18-23. Transforming cells to the malignant state. Human DNA *(red)* is extracted from cells of a malignant tumor *(1)*. This DNA is applied to normal mouse cells that take up the DNA. Some of them undergo a malignant transformation *(2)*. These cells are selected and their DNA is extracted *(3)*. Less human than mouse DNA is present, because only malignant cells are selected. These must contain the human sequences for the malignant transformation. Any other human DNA in the untransformed cells is ignored. This extracted DNA is used to transform more normal mouse cells *(4)*. Malignant cells are again selected and their DNA, in turn, is extracted *(5)*. Very little human DNA is now present, and again it must contain the sequences for the transformation. Any human DNA in the untransformed cells is again ignored. The DNA from the second round of transformation is used to establish a phage library *(6)*. Most of the inserted pieces will be mouse DNA. Only a very few phages will carry the small amount of human DNA. This DNA is associated with transforming ability, because most of the rest of the human DNA was not selected during the transfers.

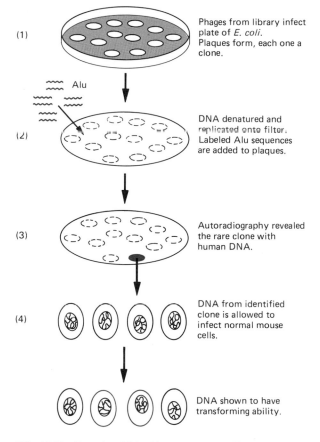

(1) Phages from library infect plate of *E. coli*. Plaques form, each one a clone.

Alu

(2) DNA denatured and replicated onto filter. Labeled Alu sequences are added to plaques.

(3) Autoradiography revealed the rare clone with human DNA.

(4) DNA from identified clone is allowed to infect normal mouse cells.

DNA shown to have transforming ability.

FIG. 18-24. Detecting DNA with an oncogene. Phages from the library (Fig. 18-23) are allowed to infect plates of *E. coli* cells *(1)*. Plaques form, each one a DNA clone. DNA is denatured and transferred to a filter *(2)*. Each DNA clone is exposed to a labeled probe for the repetitive human Alu sequence. The rare clone with the human DNA can then be identified by autoradiography *(3)*. DNA from this clone, when added to normal mouse cells, has transforming ability. The oncogene for the human bladder cancer must be in the DNA of this clone. Other clones lack transforming ability.

Gene transfer and DNA uptake

Especially exciting is the progress being made in gene transfer experiments. Several methods are available to introduce foreign DNA (donor DNA) into nuclei of recipient cells, frequently those of another species. One of the goals of such work is to obtain more information on the mechanism of gene control and expression. Another goal is to develop efficient methods for gene therapy, to correct the genetic defect in individuals handicapped by certain inherited disorders. The amazing advances in recombinant DNA technology are greatly accelerating research in gene transfer and are making possible more efficient and precise ways of placing specific genes or DNA segments into recipient mammallian cells.

Three main methods are in use. One of these entails incubation of a monolayer of recipient cells with donor DNA that has been precipitated with calcium phosphate. Although the reasons are not clear, it was noted back in 1973 that DNA treated with calcium phosphate had a much greater chance than untreated DNA fragments of being taken up by the recipient cells. Although the calcium phosphate treatment enhances DNA uptake, the number of cells that acquire the donor DNA is nevertheless low. One problem is to locate those few cells that have incorporated the donor DNA. One way around this is to use donor DNA, which carries a gene enabling only those cells that have picked up the DNA to survive under certain conditions. For example, recipient cells that cannot produce thymidine kinase (tk^-) cannot grow on HAT medium (refer to Fig. 9-2). In a pioneer experiment, calcium phosphate–precipitated DNA carrying the tk^+ allele from the herpes simplex virus was incubated with tk^- mouse cells on HAT medium. Only transfected cells (those that took up the donor DNA carrying the tk^+ allele) could grow, permitting them to be isolated. If proper selection methods are available, a variety of marker genes in addition to the tk gene can be used to transfect cells.

It was soon discovered that *cotransformation* occurs in mammalian cells. Most cells, for example, that take up DNA with the tk^+ allele can also be shown to take up other genes at the same time—genes that were *not* being selected for by the medium. This was shown in the following way. If gene transfer is to be effective at all, a certain minimum amount of DNA must be used to treat donor cells. The efficiency of the DNA uptake is increased by using more DNA precipitate to layer over the recipient cells. Unexpectedly, it was found that tk^+ cells that had been selected on HAT medium also contained the carrier DNA. Obviously the selector marker (tk^+ in this case) and the unselected marker do not have to be linked on the same DNA fragment to be taken up together. In one experiment, two unlinked genes, tk and the rabbit gene for β-globin, were tied together after entering the cell. Once inside, the joined genes could integrate into a cell's chromosome.

It was found that certain cell lines are unstable and tend to lose the foreign genes they have taken up. These unstable lines lose the joined DNAs before

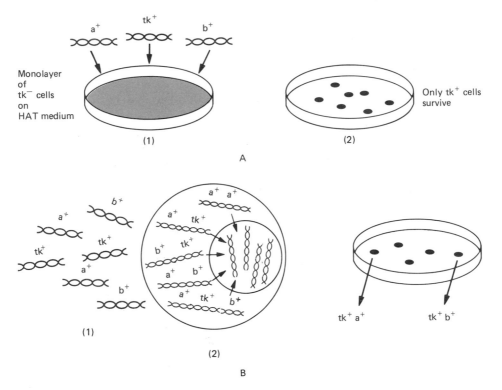

FIG. 18-25. DNA uptake and cotransformation. *(A)* (1) A monolayer of tk⁻ cells may be incubated with calcium phosphate–precipitated DNA, which carries the tk⁺ allele. However, a certain minimum amount of DNA must be present in the precipitate if DNA uptake is to be efficient. The DNA of the carrier DNA *(red)* used to increase efficiency may also contain certain alleles: a⁺, b⁺, for ex. (2) The only cells that survive on HAT medium are those that have taken up the DNA carrying the tk⁺ allele. Any carrier DNA that entered along with the tk⁺ allele is not selected for by the medium, and so it is not now apparent if any genes associated with the carrier DNA (a⁺, b⁺) are present in the tk⁺ cells. *(B)* (1) Actually, DNA from several sources (the DNA with the selected marker as well as any other DNA in the precipitate) may enter a cell. (2) The DNA from different sources are joined together after entering the cell to form a larger molecule. This means that carrier DNA can be joined to DNA with the marker. (Only some arrangements of the alleles in the large molecule are shown here). Once formed, the larger molecule may become integrated into one of the chromosomes of the host cell. (3) Those tk⁺ cells that survive on the HAT medium can then be treated further. Most tk⁺ cells can be shown to contain the a⁺, b⁺, or some allele that was not selected for on the HAT medium.

integration takes place. Other lines are stable and retain the new DNA. In these lines, it was shown that the donor DNA has become integrated into a chromosome, a different chromosome in each of the cell lines. It was demonstrated that two originally separate genes are joined together *before* they are integrated. This is demonstrated by the fact that two added DNAs are always found next to each other on the chromosome. If each had integrated with the chromosome separately, each one would usually be found by itself, bounded on each side by the cellular genes of the recipient, instead of next to each other. Moreover, it is now clear that the introduced DNA becomes part of a large molecule, a concatamer (see Chap. 16). Apparently any DNA in the precipitate (DNA with a marker, part of the carrier DNA, as well as specific

genes from still other sources) can be taken up by cells, then joined together as a large molecule that may then be integrated into the chromosome (Fig. 18-25). We thus see that cells can be readily cotransformed. Evidently, any gene can be introduced into a mammalian cell growing in culture, along with a suitable selectable marker gene, although a given nonselectable gene is not necessarily expressed.

Gene transfer and micromanipulation

A second method used to introduce foreign genes into a cell entails direct injection of donor DNA into the nucleus of a recipient cell by means of a micropipette drawn out to a fine tip with a diameter of approximately 0.5 μm. Although this procedure requires special equipment and expertise in manipulation, its efficiency is much grater than that of direct uptake of calcium phosphate–precipitated DNA. Moreover, it permits insertion into the recipient of any DNA of choice. The recipient cell may be a body cell, or even a zygote. After being injected with foreign DNA, a mouse zygote, for example, may be inserted into the reproductive tract of a foster mother to complete its development.

A major goal of such experiments is to study gene expression during development. Recipient cells are typically injected with a solution containing plasmids, into which cloned donor DNA has been inserted. Injecting a recipient cell with plasmids rather than with pure copies of the linear gene favors multiplication or amplification of the transferred donor sequence in the recipient cell. Some of these amplified sequences may be retained as the embryo develops, and some may become integrated into the host genome, and even transmitted through the germ line. Although foreign DNA can be introduced into cells of experimental animals, getting the proper expression of the transferred gene is difficult. Therefore, an attempt is made to introduce foreign DNA that has a chance of being not only expressed in the recipient but in the proper tissues in a regulated way.

The dramatic story of the "giant mice'" illustrates many aspects of microinjection techniques. The investigators selected the rat growth hormone gene for introduction into mouse zygotes. Since regulated gene expression was desired in the recipient mouse, the rat growth hormone (GH) gene was modified. It was realized that achieving regulated expression would depend on the proper regulatory sequences associated with the introduced gene. Since the structural gene for the GH was associated with rat regulatory sequences, it might not respond properly to mouse controls in the recipient mouse cells. Therefore, using recombinant DNA procedures, the rat gene was altered by genetic engineering. The regulatory sequences at the 5′ end (the end of the gene where transcription begins) was replaced with a corresponding region from a mouse gene, the metallothionein (MT) gene.

The MT story is of great interest by itself. Actually two linked structual genes, MTI and MTII, are found at the MT locus. These genes are coded for two distinct proteins known as *metalloproteins*. These low-molecular-weight proteins are found in species varying from blue-green algae to humans, a fact that underscores the importance of their role in the cell.

These proteins have a strong affinity for various trace metal ions (copper, zinc, cadmium) and play key parts in regulating the level of essential trace elements in the cell. By binding to cadmium, metalloproteins prevent it from accumulating in a toxic form. Each MT gene is associated with a strong promoter that acts as an active switch or control region to be turned on or off, depending on the concentration of certain metallic ions in the environment. For example, an appropriate concentration of Zn^{++} induces metalloprotein synthesis.

After the switch that regulates the MI gene of the mouse was located, it was joined to several other unrelated genes by recombinant DNA methods and then used in various gene transfer experiments. It was demonstrated that the expression of the recombinant genes could be induced by exposing cells to proper doses of Zn^{++} and Cd^{++}. Because it can respond to metals and certain other inducers, transcriptional regulation of the recombinant structural genes is possible in these systems. The MT promoter or switch proved to be an excellent means of manipulating the expression of other genes joined to it.

Armed with the modified rat GH gene, which was now combined with the mouse MT switch (Fig. 18-26), the investigators injected about 600 copies of the fusion gene into each of almost 200 recently fertilized eggs. These were then implanted into foster mothers, and 21 mice developed, 7 of which had one or more copies of the modified GH gene. At a young age, the mice were placed on a diet containing zinc to stimulate the MT switch combined with the GH gene. Six of the 7 mice grew significantly faster than the others in the litter. Greater than normal levels of GH were produced in these mice, resulting in gigantism. The high level of GH in the animals corresponded to a high level of GH mRNA in the liver and increased levels of GH in the serum. All the data indicate that the altered mouse phenotype results from the integration and expression of the modified GH gene. The gene was transmitted and inherited by 10 of the 19 offspring of one mouse and has been followed to at least the seventh generation.

Although the transfer and expression of the recombinant gene have been successful, we must note that the normal location of expression of the GH gene is the pituitary, whereas here it is the liver. Moreover, certain controls are apparently not operating properly, because higher than normal hormone levels were produced, resulting in gigantism, which indicates the failure of some feedback mechanism in liver cells.

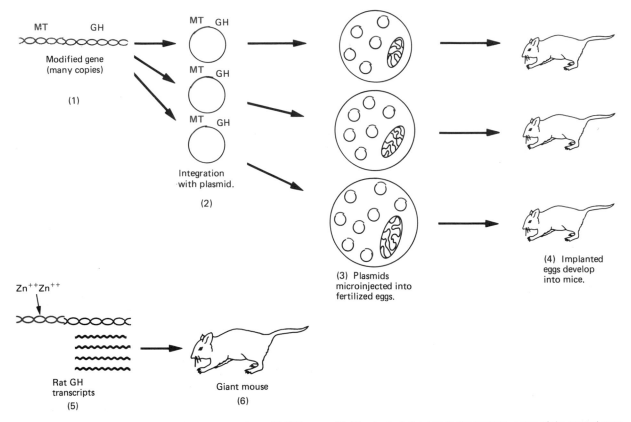

FIG.18-26. Transfer of a gene and its expression in host. *(1)* With the use of recombinant DNA procedures, the rat growth hormone gene (GH) can be combined with the mouse MT switch, a strong promoter that responds to Zn^{++}. *(2)* The modified gene is then integrated with a plasmid. *(3)* Plasmids containing the modified gene are then microinjected into newly fertilized eggs. Integration of the donor DNA into a plasmid favors its amplification in the recipient cells.

(4) After implantation into foster mothers, some of the treated eggs develop into mice that contain copies of the modified gene. *(5)* Zn^{++} in the diet of the mice with the altered gene causes stimulation of the MT promoter. A high level of transcripts of the rat GH is produced which results in a high level of rat growth hormone in some mice, *(6)* which are conspicuously larger than those lacking high levels of the hormone.

Nevertheless, such an exciting accomplishment has many implications besides those for studies of development and gene expression. The approach may lead to a means of correcting genetic defects, a way of accelerating animal growth, and a way of injecting desirable genes into farm animals to produce new breeds. Controlled, tissue-specific expression of genes in transgenic animals still remains a central problem in this approach. However, some degree of success has been reported in several experiments of this type. In one case, a rat gene for a pancreatic enzyme was transferred by microinjection into mice. In four out of five of the transgenic animals, the enzyme was expressed selectively in the pancreas at levels characteristic of the differentiated state of the gland.

Viral vectors and gene transfer

A third method employed in gene transfer involves vectors, and great success is being reported, especially in the use of certain viral vectors to transfer foreign DNA into animal cells. There is good reason to hope that, in a few years, it will be possible with this method to treat patients with certain rare genetic disorders safely. The structure and cycle of many DNA and RNA viruses have been known for years, and we understand a great deal about how a given virus enters and leaves a cell (see Chap. 16).

It is also a simple task to infect cells with many viral particles. If the virus has been altered and is carrying a chosen foreign DNA, this makes it likely

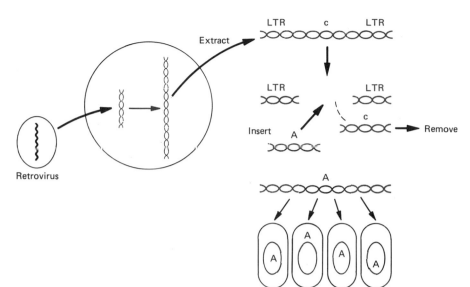

FIG. 18-27. Cloning a desired gene in a retrovirus. A retrovirus infects a mouse cell, and its RNA acts as a template for synthesis of a DNA duplex copy of the viral genome *(red)*, which inserts into the chromosome of the cell. Viral DNA can be extracted from infected cells and then processed using recombinant DNA procedures. A number of viral genes can be snipped out, including gene c, required for the formation of the protein coat of the virus. The virus has long terminal repeats (LTR) at each end, which contain promoters. A desired gene, A, is inserted between the segments with the promoter regions. The engineered recombinant virus with the desired gene is cloned in bacteria for use as needed. The virus, however, is now defective. Because it lacks gene c, it cannot make coat protein, and therefore cannot produce infectious viral particles.

that many copies of the desired DNA will be introduced into recipient cells along with the virus. Retroviruses are especially attractive as vectors. In their double-stranded DNA form, they integrate into the host's chromosome and behave as a part of it. Moreover, integrated retroviruses have very strong promoters located in the long terminal repeats at either end of the virus. (We have just seen, in the preceding section, the value of a strong promoter in the case of the MT gene.) If the promoter of the virus is activated to bring about transcription, then any foreign gene spliced into the virus will also be transcribed.

Retroviruses that have been integrated into cellular DNA can be isolated. A genetic region may be cut out using the proper enzymes and a cDNA of a specific foreign gene inserted in its place. This recombinant virus can then be cloned in bacteria using standard cloning procedures (Fig. 18-27). The goal is to establish a cell line that harbors a recombinant virus carrying a desirable gene and that can be used as a ready source of the virus to infect other cells, such as bone marrow cells.

But a problem with the recombinant virus becomes evident. Although it can perform all the necessary retroviral functions, it remains noninfective. It cannot produce protein envelopes that will enable it to leave the cell and enter another cell. Somehow we must enable noninfective engineered viruses to leave cells from an established stock and to enter other cells (e.g., bone marrow cells), insert the desired gene, and *do no more*. A helper virus (see Chap. 16) can be called on to enable us to establish a cell line carrying such an engineered retrovirus. A helper virus in the same cell with the defective recombinant virus will supply the coat needed by the defective virus to replicate and to enter other cells. However, another problem becomes obvious. The helper virus supplies coats for itself and the defective virus. Both of these are then free to leave and enter other cells indiscriminately. This is not at all desirable; it means that the vector can enter human cells, leave, and transfer itself and the inserted gene into any tissue, perhaps with fatal consequences. Therefore, the vector must be able to enter a cell but not reproduce.

Amazingly, this problem has been solved with the discovery that a specific segment of the retroviral genetic material is required for the virus to package or wrap itself in its protein envelope. Missing the

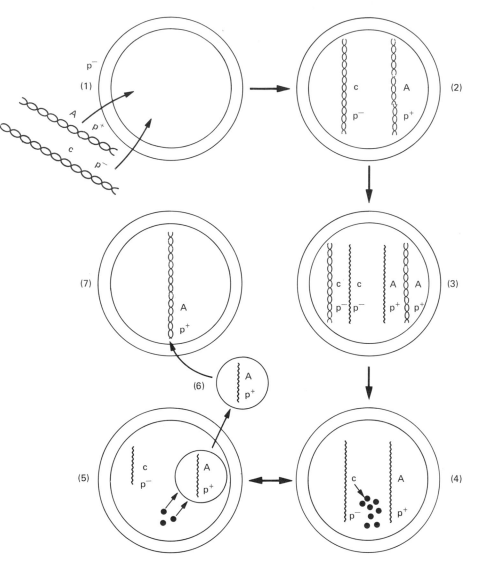

FIG. 18-28. Establishment of cell lines to supply noninfective retrovirus free of helper. *(1)* Cells are incubated with many copies of the engineered gene and also with DNA copies of a helper virus containing gene c, to make coat protein, but lacking a gene (p−)necessary for wrapping the protein around the virus. The DNA of the engineered virus carries this gene (p+), even though it lacks the coat protein gene, c. *(2)* Some DNA of both viruses is taken up by cells and inserted into the chromosomes (not shown). *(3)* Both the helper virus and the one with the desired gene (A) undergo transcription, making RNA copies *(single dark lines)*. *(4)* RNA copies of the helper are translated and coat protein is produced *(dark dots)*. *(5)* The helper cannot package the protein, and so remains noninfective. The viral RNA with the desired gene can wrap itself in the coat protein, produce infective particles, and leave the cell. *(6)* Such a virus from this cell line can then readily enter other cells, carrying the desired gene along with it. *(7)* Once inside another cell, it again makes a duplex DNA copy that inserts into a chromosome. However, the virus can no longer become infective and leave the cell, because it cannot bring about the formation of coat protein. Free of helper virus, the cell contains the desired gene (A), which will be replicated along with the rest of the chromosomal DNA.

genetic segment with the packaging gene, such a retrovirus, although it has the gene to produce protein for the envelope, cannot be wrapped in its coat and cannot leave the cell. Employing molecular manipulation, a team at MIT was able to construct a retrovirus-packaging mutant that can be used to establish a cell line yielding a source of helper-free defective retroviruses (follow Fig. 18-28). The helper supplies the protein coat for the recombinant, defective virus, but cannot itself be wrapped in the coat, because it lacks the packaging gene. Therefore, it cannot leave the cell or enter others. The recombinant virus can get into another cell and insert into the chromosome, but it is unable to leave, because it lacks the genes required to make the protein coat, even though it *does* have the packaging gene. But now there is no helper virus to supply the protein.

Experiments are being conducted with mice to use defective viruses from established cell lines to infect cells of the bone marrow. If precursor cells in the marrow pick up foreign DNA containing a desirable gene, then that gene may become integrated into the cellular DNA and continue to be produced indefinitely in new cells. Theoretically, there is promise that, in the future, human marrow could be removed from a person with a genetic blood disorder and treated with a retrovirus that can insert a desirable gene, and then the marrow could be injected back into the patient. Genetic disorders involving blood cell production may lend themselves to this type of gene therapy.

Although there is reason to be otpimistic that gene transfer mediated by retroviruses will offer a way to correct certain genetic disorders, many problems still await solution. Foremost among these is the fact that, as yet, there is no way to insert the transferred DNA at a preferred site. Moreover, it has not been possible to dictate the cell type in which an introduced gene will be expressed. And even if a gene is expressed in the proper tissue, its expression must be regulated. For example, enabling a cell to produce normal β-globin in place of sickle cell chains is not the entire solution. The inserted gene must express itself *only* in red blood cells, but it must also express itself in relation to other globin genes, such as α and δ. Excess β-globin chains in relation to α-globin would present a host of other medical problems.

Thus, although barriers to the use of this type of gene therapy prevent its use in humans at the moment, we realize that many exceedingly difficult problems have already been solved and that the present hurdles may be overcome. Use of retroviruses to cure a variety of human disorders may someday be widespread.

REFERENCES

Botstein, D. and D. Shortle. Strategies and applications of *in vitro* mutagenesis. *Science* 229: 1193, 1985.

Cohen, S. N. The manipulation of genes. *Sci. Am.* (July): 24, 1975.

Danna, K. J., G. H. Sack, and D. Nathans. Studies of simian virus 40 DNA. VII. A cleavage map of the SV40 genome. *J. Mol. Biol.* 78: 363, 1973.

D'Eustachio, P. Gene mapping and oncogenes. *Amer. Sci.* 72: 32, 1984.

D'Eustachio, P. and F. H. Ruddle. Somatic cell genetics and gene families. *Science* 220: 919, 1983.

Gilbert, W. and L. Villa-Komaroff. Useful proteins from recombinant bacteria. *Sci. Am.* (April): 74, 1980.

Hopwood, D. A. The genetic programming of industrial microorganisms. *Sci. Am.* (Sept.); 126, 1981.

Khorana, H. G. Total synthesis of a gene. *Science* 203: 614, 1979.

Kornberg, A. The synthesis of DNA. *Sci. Am.* (Oct.): 64, 1968.

Kucherlapati, R. S. Introduction of purified genes into mammalian cells. *ASM News* 50: 49, 1984.

Maniatis, T., E. F. Fritsch, and J. Sambrook. *Molecular Cloning: A Laboratory Manual.* Cold Spring Harbor Laboratory, Cold Spring Harbor, NY, 1982.

Maniatis, T., R. C. Hardison, E. Lacy, J. Lauer, C. O'Connell, D. Quon, G. K. Sim, and A. Efstratiadis. The isolation of structural genes from libraries of eucaryotic DNA. *Cell* 15: 687, 1978.

Mann, R., R. C. Mulligan, and D. Baltimore, Construction of a retrovirus packaging mutant and its use to produce helper-free defective retrovirus. *Cell* 33: 153, 1983.

Marx, J. L. Gene transfer moves ahead. *Science* 210: 1334, 1980.

Maxam, A. M. and W. Gilbert. A new method for sequencing DNA. *Proc. Natl. Acad. Sci.* 74: 560, 1977.

Myers, R. M., T. Tilly, and T. Maniatis. Fine structure analysis of a β-globin promoter. *Science* 232: 613, 1986.

Nathans, D. Restriction endonucleases. Simian virus 40, and the new genetics. *Science* 206: 903, 1979.

Robins, D. M., S. Ripley, A. S. Henderson, and R. Axel. Transforming DNA integrates into the host chromosome. *Cell* 23: 29, 1981.

Ruddle, F. H. A new era in mammalian gene mapping: somatic cell genetics and recombinant DNA methodologies. *Nature* 294: 115, 1981.

Sanger, F. and A. R. Coulson. A rapid method for determining sequences in DNA by primed synthesis with DNA polymerase. *J. Mol. Biol.* 94: 441, 1975.

Smith, M. The first complete nucleotide sequencing of an organism's DNA. *Amer. Sci.* 67: 57, 1979.

Swift, G. H., R. E. Hammer, R. J. MacDonald, and R. L. Brinster. Tissue-specific expression of the rat pancreatic elastase gene in transgenic mice. *Cell* 38: 639, 1984.

Watson, J. D., J. Tooze, and D. T. Kurtz. *Recombinant DNA: A Short Course.* W. H. Freeman, New York, 1983.

REVIEW QUESTIONS

1. Khorana's artificial gene is a DNA unit 207 nucleotide pairs long. Suppose this DNA were hybridized with the tRNA for which it is coded. About what percentage of the DNA would you expect to form hybrids with the RNA? Why?

2. A viral DNA is subjected to the restriction enzyme Alu. Five fragments are produced and separated by gel electrophoresis. These are designated Alu A, B, C. D, and E, respectively, in order of decreasing size. When the same viral DNA is subjected to the restriction enzyme Hpa, only two fragments are produced, Hpa A and Hpa B. Hpa A is isolated and subjected to Alu. Two fragments are found on the gels that correspond in position to Alu B and Alu D. When the viral DNA is subjected to Alu and then to the restriction enzyme Hae, the gels show fragments that correspond in position to Alu A, B, C, and D. Two additional very small fragments are found. What information do we now have about fragments Alu A, B, C, D, and E?

3. A DNA sequence 10 kbp (kilobase pairs) long contains one site at which it may be cut by the restriction enzyme PstI. This same 10-kbp-long sequence also contains one site which can be cut by the restriction enzyme BglII. Three separate fractions of this DNA were treated separately as follows: (1) digestion by PstI;(2) digestion by BglII;(3) digestion by both enzymes. The three treated fractions were then subjected to gel electrophoresis. The following table shows the lengths of the restriction fragments associated with each treated fraction. From the results, draw a restriction map that shows the PstI and the BglII cutting sites in the 10-kbp DNA sequence.

ENZYME IN MIXTURE	FRAGMENT LENGTHS (kbp)
PstI	6.4, 3.6
BglII	8.8, 1.2
PstI + BglII	5.2, 3.6, 1.2

4. A particular virus has a duplex DNA genome in the form of a circle. This circular molecule can be attacked by the restriction enzymes HpaI, HindIII, and EcoRI. Each enzyme attacks the circular DNA at only one site, which is specific for that enzyme. The circular genome is cut first with HpaI, yielding a linear duplex. This linear molecule was then subjected to further enzyme treatment with the production of restriction fragments. The results are shown in the following table:

ENZYME TREATMENT	FRAGMENT LENGTHS (KBP)
HpaI	5.6
HpaI + HindIII	3.0; 2.6
HpaI + EcoRI	4.8; 0.8
HpaI + EcoRI + HindIII	2.6; 2.2; 0.8

From these results do the following:

A. Draw a restriction map of the genome after it has been linearized, showing the HindIII and the EcoRI restriction sites relative to the HpaI site.
B. Draw the restriction map as a circle, indicating all three restriction enzyme cutting sites.

5. Let us suppose that a ts (temperature sensitive) mutant strain arises from the virus described in Question 4. How could you establish where the mutant site is located on the circular restriction map derived? (See answer 4B.)

6. In a DNA sequencing analysis, an autoradiograph is prepared following chemical treatment of pure DNA segments of a given length that were labeled at the 5′ end. Give the sequence of nucleotides and the polarity of the double-stranded DNA involved using the following information on one of the strands.

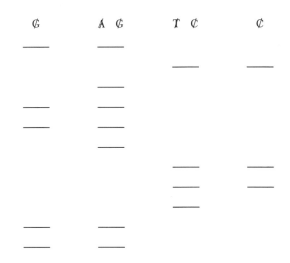

7. Suppose the strand complementary to the one sequenced in Question 6 had been the strand that was analyzed. What pattern would result when the autoradiograph is prepared?

8. The following is an RNA transcript of a segment of DNA you wish to clone:

5′ CCAUUUCGAAUGGAGCAA 3′

Double-stranded DNA can be derived from this transcript using reverse transcriptase and DNA polymerase.

A. Show the derived DNA and how it can be provided with projecting ends employing the appropriate enzymes and thymidine triphosphate.
B. Show how you would then process the plasmid selected to carry this DNA.

9. Following is a segment of DNA containing a palindrome.

C T A C G A A T T C G T C T
G A T G C T T A A G C A G A

A restriction enzyme that recognizes this produces breaks four nucleotides apart from the axis of symmetry in each strand. The break is to the left in the top strand and to the right in the lower one. Show the fragments that can be generated with complementary single-stranded projections.

10. Following is a segment of a circular plasmid.

GACTATTACGTAATAGCA
─────────────────
CTGATAATGCATTATCGT

A restriction enzyme specific for the palindrome present produces breaks four nucleotides apart from the axis of symmetry, to the left in the inner or upper strand, and to the right in the outer, or lower one.

A. Show the ends that can be generated.
B. Could the fragments generated in Question 9 be inserted? Explain.

11. Assume that you have a number of bacterial clones, each of which contains human DNA inserted into a plasmid. The goal is to locate any clones that contain the entire gene for the production of human enzyme Z. DNA is taken from clones 1 through 5 and is rendered single stranded. It is then added to systems containing all the requirements for polypeptide formation, including labeled amino acids. The RNA transcripts present were taken from cells of the tissue in which the enzyme is produced. Following are the observations made after addition of anti-Z antibody to each of the five systems:

(+ = precipitate; − = no precipitate)

Clone or System

1	2	3	4	5
−	−	+	+	+

A. What clones contain the gene coded for enzyme Z? Explain your reasoning.
B. Are any polypeptides formed in systems 1 and 2? Explain.

12. The restriction enzyme BamI recognizes the following sequence in which the arrows indicate cut sites:

```
          ↓
5'-G-G-A-T-C-C-3'
3'-C-C-T-A-G-G-5'
              ↑
```

The enzyme was allowed to attack a cloned DNA sequence from a mouse genome. Two fractions were obtained following BamI digestion of this genomic sequence. Each of these fractions contained a different duplex restriction fragment. The duplex fragment digest in each fraction was melted to single strands, and only one strand from each fraction was sequenced by the Maxam–Gilbert method. The gel patterns of the two fragment digests were as follows:

Digest 1

G	A G	T C	C

Digest 2

G	A G	T C	C

Determine the following:

A. The DNA sequence from digest 1.
B. The DNA sequence from digest 2.
C. The nucleotide sequence of the cloned duplex DNA before it was cut by the enzyme. (Show the 5'–3' orientation).
D. The nucleotide sequences of the two duplex fragments after the cut. (Show the orientation.)

13. A type of plasmid used frequently as a vector in recombinant DNA procedures carries gene markers for resistance to tetracycline (tet-r) and to ampicillin (amp-r). In one experiment, such plasmids were exposed to a restriction enzyme that cuts within the amp-r gene but leaves the tet-r gene intact. Plasmids exposed to the enzyme were next mixed with genomic restriction fragments from a cloned library of mouse DNA. The aim was to produce and detect recombinant DNA among

transformed drug-sensitive *E. coli* cells. Following their exposure to the treated plasmids, *E. coli* cells were plated on nutrient media and allowed to form colonies. Replica plates of these colonies were made on selective drug-containing media to find bacterial colonies that carry recombinant DNA. The results are as shown:

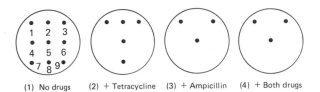

(1) No drugs   (2) + Tetracycline   (3) + Ampicillin   (4) + Both drugs

Answer the following questions, giving a brief explanation:

A. Which colonies have neither plasmid nor recombinant DNA?
B. Which colonies carry intact plasmid DNA but no recombinant DNA?
C. Which colonies carry recombinant DNA?

14. Suppose you have a plate on which several transformed *E. coli* colonies are growing. You wish to identify any colony that carries the mouse gene that codes for thymidine kinase. Briefly outline steps that would enable you to locate such a colony.

15. Let us suppose that one wishes to assign to a particular human chromosome the gene coded for a certain critical enzyme. DNA is extracted from human cells, mouse cells, and hybrid human–mouse cells that carry different combinations of human chromosomes. The DNA from the various sources is cut by a specific restriction enzyme, and the fragments are separated by gel electrophoresis on the basis of size. Following Southern blotting and denaturation of the fragments on a filter, the fragments are exposed to a labeled cDNA probe of the gene. Autoradiography is performed with the following results. (The numbers represent the human chromosomes present in the hybrid cells.)

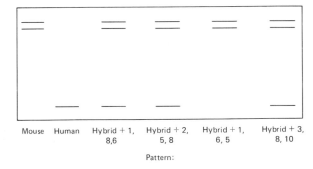

Mouse   Human   Hybrid + 1,   Hybrid + 2,   Hybrid + 1,   Hybrid + 3,
                              8,6   5, 8   6, 5   8, 10

Pattern:

On which human chromosome does the gene for the human enzyme reside?

19

APPLICATIONS
OF MOLECULAR BIOLOGY

Molecular biology and prenatal diagnosis

Recombinant DNA procedures are accelerating our understanding of many human genetic defects. The methodology now permits the prenatal detection of certain mutant alleles that previously could not be identified by the techniques available. Sickle cell anemia illustrates well the role the new approaches are playing in prenatal screening.

To obtain fetal blood for prenatal detection of a deleterious allele entails a risk almost 10 times that associated with amniocentesis. Consequently, use of fetal blood to check for the presence of the sickle cell allele has been limited. Molecular biology now offers a way to overcome this problem by precluding the need for blood to detect the defective allele. In normal β-globin, position 6 in the polypeptide chain is occupied by glutamic acid, whereas in sickle cell hemoglobin, this position is occupied by valine (see Fig. 13-23). Human β-globin mRNA has been isolated and used to obtain double-stranded cDNA following treatment of the transcript with reverse transcriptase. Sequencing the cDNA has indicated that the mRNA codon corresponding to the glutamic acid residue normally found at position 6 in the β-globin polypeptide is GAG. This RNA codon, if changed to GUG as the result of point mutation would now code for valine.

As Fig. 19-1 shows, the sickle cell mutation at this position is accounted for by a transversion in which a base pair alteration in the β-globin gene has changed an A:T pair to a T:A pair.

This single base pair change happens to affect a region in the gene that can be recognized by a certain restriction enzyme, Dde. The enzyme recognizes the sequence CTNAG (in which N represents any nucleotide). As a result of the mutation, the enzyme can no longer identify the site and will therefore not cut the DNA at that location. Armed with this knowledge, we can now screen DNA from fetal cells obtained by amniocentesis to determine whether or not the DNA has been altered at that restriction enzyme site. If it has, we know the fetus is carrying the sickle cell allele.

DNA is extracted from human cells known to contain the normal β-globin allele on each of the two homologous chromosomes (Chromosome 11). The DNA is treated with the restriction enzyme, which recognizes the site in the normal gene and consequently cuts the DNA at that location. DNA extracted from cells of a fetus at risk for sickle cell anemia is also exposed to the enzyme. Let us assume that the DNA from the fetal cells carries the A:T→T:A transversion and that the enzyme will therefore not cut the DNA at that site. Both DNA samples, the one

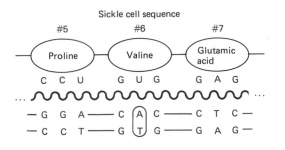

FIG. 19-1. The sickle cell mutation. On the left, the normal DNA sequence is shown, as well as the specific sites of cutting by enzyme Dde (*red arrows*). This restriction enzyme recognizes the sequence CTNAG. A transversion (*red circling*) that results in the sickle cell allele eliminates the palindrome that the enzyme recognizes. The cut, therefore, cannot take place.

with the normal allele and that with the sickle cell gene form, are subjected to gel electrophoresis and Southern blotting. Many fragments are generated, because the enzyme will cut the DNA wherever the recognition sequence occurs in the two total DNA samples. To find and visualize the β-globin DNA fragments among the many others that have been separated on the gel, we must turn to a suitable probe. One such probe is a labeled cDNA specific for the 5′ end of the β-globin gene. This probe enables us to locate the fragments on the filter containing those

DNA sequences that code for the beginning of the β-globin polypeptide.

Comparison of the fragments formed from the DNA of the normal β-globin gene with that carrying the sickle cell allele shows a difference. It is seen that a larger fragment of the sickle cell DNA hybridized with the probe than is the case for the normal DNA (Fig. 19-2). The reason, of course, is that elimination of the restriction site in the sickle cell DNA eliminated a cut, allowing a larger than normal fragment to be generated. In this case, the normal hemoglobin will

FIG. 19-2. Restriction fragments formed from DNA of normal β-globin gene and the sickle cell allele. DNA from cells with the normal allele and cells with the sickle cell allele is subjected to the restriction enzyme Dde. Many fragments are produced, because the sequence CTNAG occurs at many sites along the DNA in addition to the site within the start of the β-globin gene. To detect fragments containing the sequence at the beginning of the gene (*red arrows*), Southern blotting and a suitable probe are required. The labeled probe used can visualize the 201-, 175-, and 376-base-pair fragments when an autoradiograph is prepared. The red arrows indicate the specific fragments on the gel that are undetected until they are exposed to the probe (*right*).

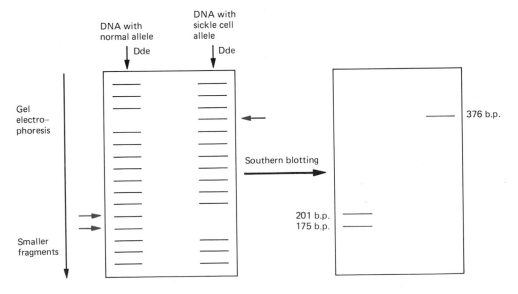

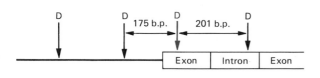

D–Dde sites
D–Dde site affected by
  sickle cell allele

FIG. 19-3. Variation in restriction fragment length because of gene mutation. The reason for the results seen in Fig. 19-2 is the elimination of the Dde restriction site occurring in the first exon of the gene *(red D)* by the sickle cell mutation. Therefore, the 201-base-pair fragment cannot form. Along with the 175-base-pair fragment, it becomes part of a 376-base-pair fragment. Other fragments are generated, because other Dde sites occur along the DNA and are not associated with the 5′ end of the β-globin gene.

produce two bands that hybridize with the probe, a 175-base-pair band and a 201-base-pair band. On the other hand, sickle cell DNA produces only a 376-base-pair band. The 175-base-pair fragment cannot form but becomes part of the 376-base-pair fragment (Fig. 19-3).

Other enzymes besides Dde can be used to resolve fragments and show a difference between normal and sickle cell DNAs. Different probes are available that vary in the length of the DNA sequences they cover. Therefore, the fragment size that can be detected varies, but the procedures are nevertheless the same as that outlined here and can resolve differences between normal hemoglobin A and hemoglobin S. The need for taking fetal blood and the dangers it entails are avoided.

Molecular biology and detection of persons at risk

We have encountered, throughout our discussions, many examples of genes whose protein products are known. Knowing the product that is encoded by a particular gene enables us to associate a defect in the protein with a specific defective allele of that gene. It may also permit detection of carriers of a defective allele (Tay-Sachs, sickle cell anemia). Knowing the protein product also makes it possible to locate the gene and to clone it (globin genes, insulin gene). Unfortunately, in the case of most genetic disorders,

the protein product associated with the gene in question is unknown. One way around this dilemma is to evaluate the inheritance pattern of the defective allele and look for markers to which it may be linked. We saw the feasibility of this in the case of myotonic dystrophy (see Chap. 9). The dystrophy gene, whose product is unknown, is linked to a gene associated with the ability to secrete blood antigens into body fluids. Depending on the pedigree, this enables us to calculate the risk, let us say, a person who is a secretor has of carrying the defective allele (see Fig. 9-5).

For many genetic disorders, however, the defective allele has not been associated with any other gene that can serve as a marker. Consequently, the location as well as the linkage relationships of such a defective gene remain unknown. Now the methodology from molecular biology and recombinant DNA procedures is enabling us to overcome this barrier. In our discussion of sickle cell anemia in the previous section, we saw that a gene mutation may alter a specific restriction site within a gene. Such a mutation, which alters a restriction site, can occur anywhere along the DNA. If two individuals vary in a particular DNA sequence at a given site, a specific restriction enzyme will produce fragments of different sizes when it cuts the DNA from the two sources (Fig. 19-2 and 19-3). In other words, a mutation, by affecting the cutting site of a restriction enzyme, will produce variations in the DNA pattern. Pairs of variations in the DNA are inherited as traits, just as if they were variant forms of a given gene (Fig. 19-4).

In the same way that we can detect differences in a specific protein because of genetic variation, we can now detect differences in DNA sequence by observing fragments of different sizes generated by cutting with a given restriction enzyme. Whereas traditionally we used protein variations (protein polymorphisms) as markers to indicate the presence of a particular allele, we can now use DNA polymorphisms (restriction fragment polymorphisms) as genetic markers. Since the DNA polymorphisms can affect any type of DNA sequence, this means that an alteration producing a restriction fragment polymorphism can occur within a coding sequence of a gene (as was the case with the β-globin gene), noncoding sequences (introns), sequences between genes, and even DNA with no known function, such as repetitive DNA. Because the DNA polymorphisms can affect any part of the genome and they need not be expressed as a protein product, they are extremely valuable markers.

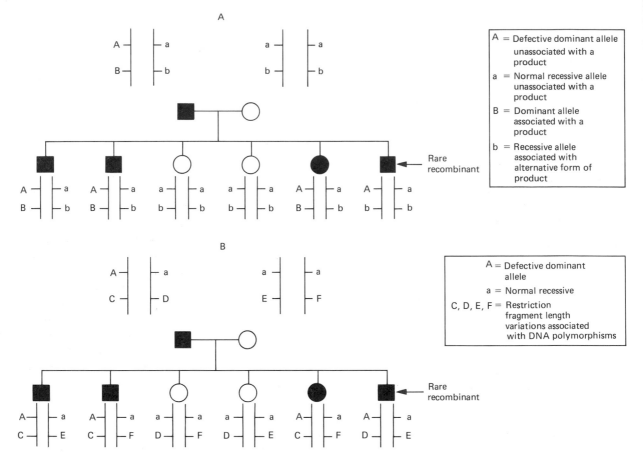

FIG. 19-4. Inheritance of a defective allele in relation to markers. *(A)* The A and a alleles are unassociated with any protein product, but their locus, a, is linked closely to the b locus occupied by alleles B and b, which *are* associated with products. Even though the dominant allele, A may not be expressed until later in life, a person carrying it can be detected because of the close linkage between the two loci. The dominant defect in this pedigree is traveling with allele B. The presence of B product indicates that allele A is probably present. Therefore, the b locus acts as a marker for alleles at the a locus. A recombinant resulting from crossing over may arise with a frequency depending on the distance between the a and b loci. If the two loci are 2 map units apart, then there is a 2% chance that a recombinant will arise at any birth. *(B)* This pedigree is exactly the same as the pedigree shown in *A,* only here the marker is a DNA region on the chromosome that is variable in its response to a certain restriction enzyme because of DNA sequence variations. Consequently, restriction fragments of different lengths are generated, and these can be identified as different patterns following hybridization with an appropriate probe. Of the four patterns, pattern C is associated with the defect in this pedigree. Again, a rare recombinant can arise.

Restriction fragment polymorphisms and Huntington's disease

A prime example of the use of restriction fragment polymorphisms to detect the presence of a gene whose primary protein defect remains unknown is Huntington's disease. This devastating disorder is inherited as a dominant, but its chromosome location remained unknown until the advent of recombinant DNA techniques. Huntington's disease is a neurological disorder in which a person with an apparently normal central nervous system experiences premature death of nerve cells. Its frequency in the Caucasian population is about 1 in 10,000. The afflicted individual, whose disease may progress over a 15- to 20-year period, becomes completely disabled. Since the primary biochemical defect in the disorder remains a mystery, no successful approach to treatment has been found to prevent the deterioration and ultimate death of the victim of Huntington's disease.

Moreover, ignorance of the gene product has not permitted assignment of the Huntington's gene to a specific chromosome location or the development of a test to identify persons at risk. This is particularly significant in the case of this disorder, because the

Huntington's allele appears to be 100% penetrant. Although the disorder can strike a victim at almost any time in the life span, it typically manifests itself between the ages of 35 to 45. This means that when the symptoms start, the individual may be a parent. Since it behaves as an autosomal dominant, the occurrence of Huntington's disease in one parent gives each child a 50% chance of inheriting the defective allele and then developing the disorder, usually well after the age of sexual maturity. It is evident that the disorder is an especially insidious one that can be passed down by an unsuspecting person to one or more offspring.

For years, investigators attempted to find a linkage relationship between the Huntington's disease (HD) locus and some other genetic locus that is associated with a detectable product such as an enzyme, as was the case with myotonic dystrophy. This would make it possible to calculate the risk that a person will inherit the defective allele and would also lead to its eventual assignment to a chromosome.

After years of failing to achieve these goals, researchers in HD turned their attention to the techniques provided by molecular biology. They were also able to study a large number of families in the United States in which HD occurs. In addition, they were alerted to a large number of Venezuelan families that were interrelated. Many members of these families were at appreciable risk for HD, because they included 100 HD victims as well as 1100 children. Tissue samples were collected from the U.S. and South American families. DNA was then extracted in a search for restriction fragment length polymorphisms (RFLPs) that might be associated with the HD locus. DNA samples from the families were subjected to various restriction enzymes, and RFLPs were detected. Fragments of different length generated by the enzymes were then inserted and cloned in phage vectors. These cloned fragments representing different portions of the genome were now available as potential probes to search later for any DNA fragment polymorphisms that might be associated with the HD locus. Not long after the search was started, a particular RFLP was identified that was linked to the HD locus. A fragment of this marker region, cloned in a phage vector, was available to act as a probe.

As Fig. 19-5 shows, four different DNA patterns can be generated from a DNA sample representing genetic material from this chromosome region (now known to be on Chromosome 4). Each pattern (A, B, C, and D) had been designated a *haplotype*. The

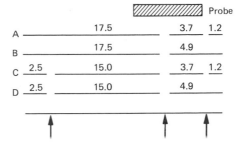

FIG. 19-5. HindIII restriction sites in a region of human Chromosome 4 linked to the HD locus. This region of Chromosome 4 contains three sites *(arrows)* that can be recognized by restriction enzyme HindIII. Two of these sites are variable *(red arrows)*. Consequently, when DNAs from different persons are analyzed, fragments of different lengths can be generated, depending on whether or not the enzyme cuts at both of these variable sites, one of them, or neither. The probe used to detect the variability identifies fragments of 17.5, 3.7, 4.9, and 15.0 kbp in length. The probe does not overlap the two smallest fragments (2.5 and 1.2 kbp) and therefore cannot detect them. Nevertheless, four different haplotypes or patterns can be distinguished: A (17.5 + 3.7 kbp fragments), B (17.5 + 4.9 kbp fragments), C (15 + 3.7 kbp fragments), and D (15 + 4.9 kbp fragments). Each haplotype is inherited as if it were a combination of tightly linked alleles on a chromosome. Each person carries two chromosomes 4, and thus can have any combination of the haplotypes.

haplotype depends on whether or not the variable sites (shown by the red arrows in Fig. 19-5) allow recognition and cutting by the enzyme HindIII at that region of the DNA. If a change in base sequence is found at a variable site, the enzyme will fail to cut at that location. The probe can detect fragments of 17.5, 3.7, 4.9, and 15 kilobase pairs. Thus, 10 different genotypes are possible because one could have any combination of haplotypes, having inherited one haplotype from one parent and the same or another haplotype from the other parent. Figure 19-6 represents autoradiographs of these different genotypes. The autoradiographs are produced following DNA digestion, gel electrophoresis, Southern blotting, and hybridization with the radioactive probe.

When the DNA of 51 persons from unrelated families was analyzed, it was found that more than half were heterozygotes in relation to the DNA markers (the RFLPs) in this chromosomal region. When the U.S. and Venezuelan families known to be at risk for HD were studied, it was found that the HD locus *was* indeed associated with the RFLP in this region. For example, in a U.S. family, the HD gene was found to be associated with a chromosome carrying the A haplotype. In a Venezuelan family, it was found to be associated with the C haplotype. In other words, we

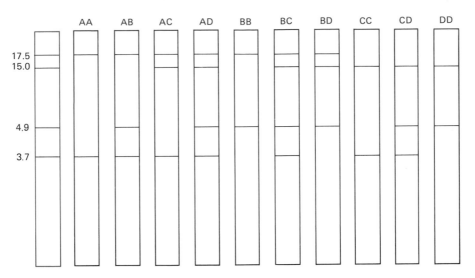

FIG. 19-6. Depiction of autoradiograph patterns of the 10 genotypes possible from the different combinations of the A, B, C, and D haplotypes. The pattern on the left is the one seen when the genotype permits detection of the four fragments of different kilobase pair lengths. Reference to Fig. 19-5 shows that these patterns are to be expected because of the different combinations of the haplotypes and because the radioactive probe will identify on the gel fragments of four different kilobase pair lengths (17.5, 3.7, 4.9, and 15 kbp).

now have a marker (the DNA region that generates the RFLPs) known to be linked to the HD locus. We can use the RFLPs to assess the risk that a person may have received the dominant HD allele from a parent. If it is shown, for example, that the allele is traveling through family members and that any member with the disease always has haplotype C, we then know that the disorder is traveling with haplotype C in that family line (Fig. 19-7).

Suppose a female parent with HD carries haplotypes A and C and the C haplotype of that parent is known from pedigree studies to be linked to the HD allele. The male parent is homozygous for haplotype B (BB). There are two offspring. Analysis of DNA from one child shows the AB pattern, whereas the DNA of the other child show the BC pattern (see Fig. 19-6). We now know the former is probably free of the allele (barring recombination, estimated to be about 2%), whereas the latter probably carries the genetic defect and will later become afflicted with HD.

Human–mouse hybrid cells carrying different combinations of human chromosomes have now been screened. It is finally possible to assign the polymorphic DNA marker, and hence the haplotypes, to Chromosome 4. Hybrid cells containing only Chromosome 4 or only a portion of the short arm of Chromosome 4 yield DNA that reveals the presence of the marker. Since the marker is associated with the HD locus, the HD locus must also reside on Chromsome 4.

The result of this new information on HD makes it possible to inform certain persons in whose families HD occurs if they are likely to have inherited the allele. When a routine test does become available, it will not, of course, be able to answer the question for everyone. A person would need to have cell samples (blood, skin) not only from himself but also from his parents and even grandparents. Unfortunately, such a test will raise many additional problems. How will an otherwise healthy young person cope with the knowledge that he or she is probably carrying the HD allele and will most likely suffer from the devastating disease? Should a person at risk be required to take such a test in order to prevent transmitting the disorder to an offspring? Should a spouse have the right to insist on having such knowledge?

Now that the HD locus is known to reside on Chromosome 4, work continues with the hope of zeroing in on the gene, to finally isolate and clone it. Once this is accomplished, the basic defect in the HD allele can be described and the primary product associated with the gene may be discovered. The ultimate goal of the molecular procedures is to rid families of the suffering and fear this dreadful disorder inflicts. Perhaps it will someday be possible to prevent the

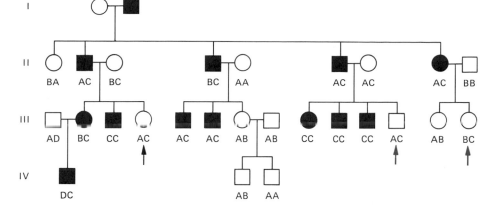

FIG. 19-7. Pedigree showing inheritance of HD. This pedigree shows that, wherever it was possible to determine the haplotypes, each HD victim *(shaded)* also has haplotype C. Those persons with haplotype C who marry into the family do not have this rare allele (persons II-3 and II-7). An offspring in the pedigree who has received haplotype C from such a parent will remain free of the disorder, barring a rare crossover event (Person III-4, indicated by *arrow*). Person III-12, *(first red arrow)* is at risk because the C haplotype may be derived from the afflicted or the nonafflicted parent. Person III-14 *(second red arrow)* is at extremely high risk, because the C haplotype has been transmitted by the afflicted female parent, and the recombination frequency is only 2%.

onset of the disease in those known to carry the HD allele and to prevent its transmission through counseling of those known to be at high risk.

Interactions of cells of the immune system

The cells that manufacture antibodies (immunoglobulins) are *B lymphocytes*, small white blood cells that are formed and mature in the bone marrow. *T cells* constitute another class of lymphocytes, also found in bone marrow, which complete their maturation in the thymus gland. Although both types of cells appear very much alike, they perform differently in the immune response. A T cell participates in a direct attack, killing any cell bearing foreign antigens. T cells also act against viruses. The T cell recognizes the discarded protein of the virus that remains attached to an infected cell's membrane. The cytotoxic killer T cell then destroys the virally infected cell. It is the killer T cell that is involved in directly attacking cells following a tissue transplant. There are also *T helper cells* that, after recognizing an antigen, stimulate other T cells on guard against the same antigen. Moreover, the helper T cells appear to be necessary for the alerted B lymphoctyes to undergo divisions to complete their differentiation. Another class of T cells, the T suppressor cells, are required to cut down on the activity of active T killer cells, and hence keep the response within the proper limits. Acting together, the T helper and T suppressor cells act in a regulatory capacity.

When a B lymphocyte, which has recognized a specific antigen, is activated by a T helper cell, it goes on to proliferate. Some of the resulting B cells then proceed to act as *memory cells*. These B lymphocytes continue to circulate around the body, alert to the specific antigen. They are able to bring about a swift response following an infection by the same antigen and account for the speed with which the immune system reacts to a second infection by a given antigen. Those B lymphoctyes that do not become memory cells after stimulation by the T helper no longer divide but undergo a terminal differentiation to become *plasma cells*—B cells that live only a few days but secrete very large quantities of antibodies. A given B cell is committed to produce a single kind of specific antibody when stimulated by the antigen for which it is on the alert. Although the antibodies secreted by plasma cells interact with their specific antigens as they circulate in the blood system, they do not destroy them directly. Rather, they mark them so that the antigens are recognized by other components of the immune system, such as *macrophages*, which do the actual destruction.

Antibody diversity

The different kinds of foreign antigens that can infect an individual number in the millions. The body must therefore have the ability to produce a corresponding number of different antibodies to interact with them. It would seem that the genome of a mammal must include many millions of genes set aside for the coding of millions of different antibodies. However, estimates of gene number in mammals indicate that the grand total of genes actually found in the genome can certainly not be more than a million. Only a small percentage of these are coded for specific antibodies. Yet this fraction of the genetic complement nevertheless specifies countless kinds of antibodies to counteract the effects of countless types of antigens. Laborious research by geneticists and immunologists over the years has sought an explanation for this puzzling situation. Although certain details are yet to be clarified, the basis for antibody diversity has now been determined. The fruits of the research have also demonstrated the unexpected versatility of the mammalian genome and have forced us to reexamine some established concepts, such as the one gene–one polypeptide theory.

An antibody is a Y-shaped molecule composed of four polypeptide chains held together by disulfide bridges (Fig. 19-8). Two of the chains are called *light chains*. These are identical, each composed of about 200 amino acids. The other two chains—the *heavy chains*—are longer, composed of more than 300 or 400 amino acids; these are also identical in any one antibody molecule. Antibodies fall into five different types or classes. Each type of antibody or immunoglobulin (Ig) is distinguished by the kind of heavy

chain it possesses: M, D, G, E, or A. All five types may be produced by lymphocytes responding to the very same antigen. Consequently, the five different types of antibodies—IgM, IgD, IgG, IgE, and IgA—are specific for that one antigen. Just how the particular antibody operates against the antigen, however, differs from one type to another. For example, the IgG antibody circulates in the blood; IgD antibody remains attached to its cell's surface.

When we compare the heavy and light chains from antibodies specific for different antigens, we find that certain differences exist in the amino acid sequence of the chains. These differences are found only in one portion of each chain. This is the so-called *variable region*, which makes up about half of the light chain and one fourth of the heavy chain (Fig. 19-8). The remainder of the chains are called *constant regions*, because they have these same amino acid sequences in a given type of antibody, *regardless* of their differences in specificity for different antigens. For example, all IgG$_3$ antibodies have the same constant heavy chain region, even though two IgG$_3$ molecules may be specific for different antigens. Note from Table 19-1 that there are four kinds of constant gamma chains, which makes four subclasses of IgG possible. It is the constant region of the heavy chain—mu (μ), delta (δ), gamma-3 (γ_3), and so on—that determines the type of the antibody and the way it will carry out its job.

Although there are five kinds of heavy chains (remembering that the gamma type actually divides into

FIG.19-8. Representation of antibody molecule. Each antibody molecule is composed of two identical light chains *(black)* and two identical heavy chains *(red)* held together by disulfide links (S–S). A portion near the NH$_2$ ends of each chain, the variable region *(shaded)*, is composed of amino acid sequences that vary when antibodies of the same type are compared. The constant region *(unshaded)* is the same for all antibodies of the same type. Within each variable region are three hypervariable regions.

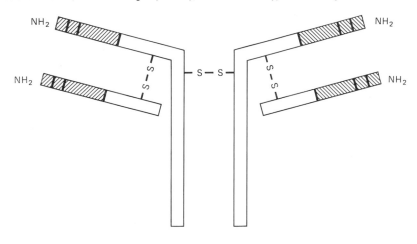

TABLE 19-1   Classes of antibodies and their chains[a]

CLASS	HEAVY CHAIN	LIGHT CHAIN
Ig M	mu (μ)	kappa (K) or lambda (λ)
Ig D	delta (δ)	kappa (K) or lambda (λ)
Ig G	gamma (γ)	
Subclasses:	gamma-3 (γ_3)	kappa (K) or lambda (λ)
	gamma-1 (γ_1)	kappa (K) or lambda (λ)
	gamma-2b (γ_{2b})	kappa (K) or lambda (λ)
	gamma-2a (γ_{2a})	kappa (K) or lambda (λ)
Ig E	epsilon (ϵ)	kappa (K) or lambda (λ)
Ig A	alpha (α)	kappa (K) or lambda (λ)

[a] The heavy chain defines the antibody class. The gamma class is divided into four subclasses of heavy chain. Any one kind of heavy chain can be combined with either a kappa or a lambda light chain to form the antibody molecule.

four subclasses of its own), there are two kinds of light chains, kappa and lambda (Table 19-1). Either two kappas or two lambdas will be present along with two identical heavy chains of a given type in every antibody. As in the case of the heavy chain, the constant region of any kappa is identical to that of all other kappa light chains in all other antibodies. The same is true for the constant region of the lambda chain. The variable regions are those parts of an antibody molecule that can recognize a specific antigen and interact with it. The constant regions do not participate by reacting directly with the antigen, although, as we have noted, the constant heavy region determines how that type of antibody carries out its job against the antigen.

The variable regions of the heavy and light chains are found at that part of the polypeptide near the beginning of the NH$_2$ end of the chain. Each variable region of a heavy and a light chain contains within it three *hypervariable regions*, segments where the amino acid sequences are especially different when antibodies of different specificities are compared.

Theories of antibody diversity

The presence of a variable region in each kind of antibody chain makes possible a large number of different kinds of light and of heavy chains—at least 1000 of each kind. According to one theory, each haploid genome contains 1000 genes coded for the 1000 different kinds of light chains and 1000 genes coded for the different heavy chains. Besides the large number of genes required by this theory to code for

the vast assortment of specific antibodies, another difficulty accompanies this concept. Let us say that there are 1000 genes for light chains. What then is the device that keeps all 1000 constant regions identical in each of the 1000 genes? Some mechanism would be required to prevent the effects of gene mutation, and thus to conserve perfectly the constant region in each of the chains. And yet the variable region of each of these genes would be permitted to mutate and change rapidly. To explain how such a system could be maintained requires the surveillance of some unknown kind of biological device and argues against the one gene–one specific antibody chain idea.

A different theory proposed in the mid-1960s suggested that the genome contains many genes that code for all the variable regions of a chain of a given type but only one gene for a constant region of a given type. For example, in the case of heavy chains, there would be one gene for the constant region of type μ chain, one for the constant region of a δ chain, and so on. There would, however, be many genes for the different variable portions. In the case of the light chain—kappa let us say—one gene would exist for the constant portion and perhaps hundreds for the variable. According to this theory, as an antibody-producing cell matures, the DNA is shuffled, and a constant gene is combined at random with one of the variable genes to produce an active antibody gene. In another cell, the same constant region gene might combine with a different variable gene to give another active antibody gene. The proposal is a radical one, because it implies that two separated genes or DNA segments can be brought together to form one segment or gene, which is then coded for a polypeptide chain (Fig. 19-9). Nevertheless, the basic concept of this unusual proposal has now been shown to be essentially correct. According to this scheme, as a lymphocyte matures, one constant region gene can be brought into combination with any one of many variable genes that lie a distance from it. This recombination then yields an active gene with a variable region and a constant region, which is the same for every polypeptide chain of a given specific type.

Evidence supporting this DNA rearrangement theory has been provided by experiments using recombinant DNA procedures. Mouse DNA coding for the variable (V) and constant (C) regions of the light chain, lambda, were cloned in phage λ. DNA was obtained from two cell types—embryonic and mature, antibody-secreting cells. Restriction enzymes were

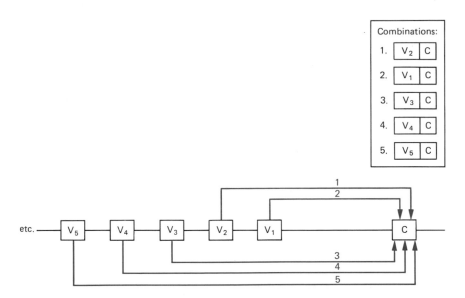

FIG. 19-9. Theory of gene shuffling. According to one proposal, diversity of antibody genes can be accounted for by assuming that many genes or segments code for variable regions (V1, V2, etc.). Only one or a few segments would code for the constant region (one C segment is depicted here). In one cell, an active gene can be formed by combining the C segment or gene with the segment coded for V1, which lies a distance away from it. In other cells, different combinations of C and V segments occur to produce many different kinds of active antobody genes, which are then able to form transcripts.

FIG. 19-10. Detection of DNA shuffling. Suppose *(above)* that a C gene in an embryonic cell or a nonantibody-producing cell is located between two DNA sites *(arrows)* that are sensitive to a particular restriction enzyme and are 5000 nucleotides apart. Located a distance away is a V gene with a sensitive site close to it. When the cell is exposed to the enzyme, C can be located with a probe and will be found on a fragment that is 5000 nucleotides long. If shuffling of DNA occurs, the restriction site to the left of the gene may be eliminated, and V may be brought close to C. Cleavage with the restriction enzyme, along with probing for gene C, now shows that C is associated with a much smaller fragment.

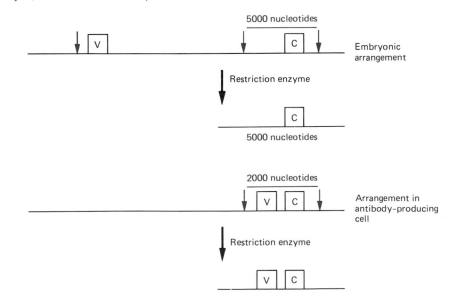

then used to cut the DNA obtained from the two sources. If some sort of shuffling or recombination of the DNA does take place as an antibody-secreting cell matures, a restriction enzyme site could be altered. This would mean that when DNA containing V and C genes from an embryo is compared with the corresponding DNA in a mature cell, different-restriction fragments would be obtained. And this proved to be the case. In the embryonic cells (or any cells that are not antibody secreting), the DNA sequences for the V and C regions of the light chain are much farther apart than they are in a cell that is secreting antibodies. This was clearly demonstrated by the size of the restriction fragment bearing the C segment (Fig. 19-10). Such experiments leave no doubt that maturation of a B lymphocyte into an antibody-producing cell is accompanied by rearrangement of the DNA. A rearrangement of this type, which takes place during the course of differentiation, is called *somatic recombination*.

Before proceeding with details of the arrangement of antibody genes, it must be pointed out that in any antibody-producing cell, the antibody genes on *only one* of the two chromosomes undergo somatic recombination. It is these recombined segments that constitute the DNA that will undergo transcription and then be transcribed (Fig. 19-11). The DNA corresponding to these antibody genes on the homologous

chromosome does not undergo somatic recombination but remains in the original arrangement found in embryonic or in nonantibody-producing cells. This DNA will not undergo transcription, and therefore does not produce an mRNA for translation. This phenomenon is known as *allelic exclusion*. Allelic exclusion is the rule, and only rarely do both corresponding segments in a lymphocyte undergo shuffling.

Recombination and the generation of antibody genes

DNA sequencing, combined with other methods from molecular biology, has filled in many details on the arrangement of the DNA sequences coded for antibodies. It was demonstrated that even when the variable and constant regions are brought close together during differentiation, they still remain separated by a distance of about 1500 nucleotides. This is because of a joining segment known as *J*. As is shown for the human kappa light chain in Fig. 19-12*A*, there is a single C gene for the constant region plus five J segments and hundreds of V genes. Both the J and the V sequences are encoded for the variable sequences of the chain. Each V sequence is separated from a leader by a short intervening sequence. A long noncoding sequence separates the four J sequences from the cluster of V genes.

During differentiation of a lymphocyte to a plasma cell, somatic recombination can shuffle a V gene with its leader into a combination with one of the J segments, which in turn combines with the single C gene. This recombination brings about formation of a gene that can then act as a kappa gene and undergo transcription (Fig. 19-12*B*). A primary transcript forms containing all the sequences of the recombined gene. This pre-mRNA undergoes processing during which intervening sequences are removed and segments are spliced together. A poly(A) tail is added, and the transcript is ready for translation into a kappa light chain. The leader sequence of the transcript results in the assembly of a leader portion of the polypeptide chain composed of about 20 amino acids. This leader portion is part of the original polypeptide formed and is thought to be important for the transport of the newly formed antibody chain through the membrane of the cell. As the chain passes through the membrane, the leader is snipped off.

In the case of the human lambda chain, the situation is similar to that for the kappa chain. However, there are six constant genes, not just one, for this

FIG. 19-11. Allelic exclusion. In an antibody-producing cell, the DNA of only one of the two homologous chromosomes undergoes somatic recombination to produce an active antibody gene *(red)*. Only the rearranged DNA on the one chromosome will form an active gene that will produce transcripts coded for a specific antibody. The other chromosome retains the gene sequence of the immunoglobulin family in the original (embryonic) condition.

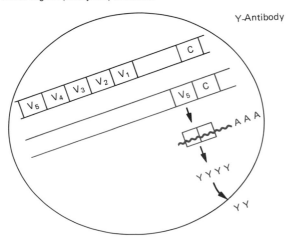

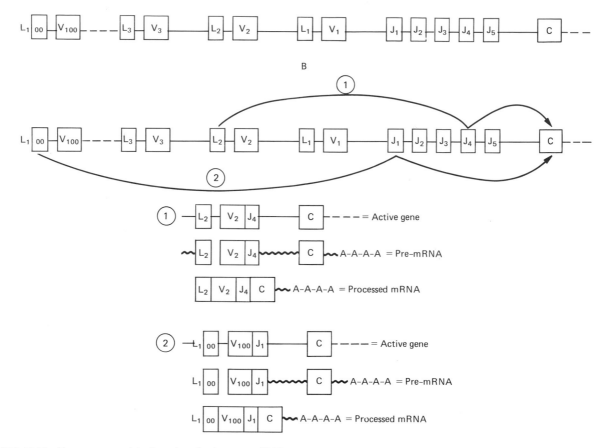

FIG. 19-12. Kappa genes and the formation of active genes. *(A)* The kappa light chain is encoded by a single C gene, five J sequences, and hundreds of V genes. Each V gene has a leader segment separated from it by an intron. *(B)* In forming an active kapa gene in a lymphocyte, somatic recombination can combine any V gene with a J segment. These then join the single C gene. In cell 1, the combination joins V2, J4, and C. In cell 2, the combination is V100, J1, and C. Many combinations are possible from one lymphocyte to another. The active kappa gene undergoes transcription to form a pre-mRNA, which is then processed. Introns and intervening segments are removed, and the V, J, and C segments are spliced together to form a functional mRNA.

light chain, and each of the six is closely associated with its own J segment, forming a pair. About 300 V genes occur, and any one of these can be brought into combination with any pair of J and C genes (Fig. 19-13). The genes for the lambda and kappa chains have been assigned to Chromosomes 2 and 22 respectively. In each case, the cluster of V genes linked to the C region constitutes a gene family.

A picture similar to that for the light chains holds for the genes coded for the variable and constant regions of a heavy chain. An active heavy-chain gene results from the shuffling of four sets of sequences, V (with the leader), C, J, and D. Thus, much greater diversity is possible by shuffling the heavy-chain sequences than is the case for the light chain. These sequences are located on Chromosome 14, where they constitute another immunoglobulin gene family. The D (for diversity) genes or segments, each composed of about 13 nucleotides, are responsible for the greater diversity of the different kinds of heavy chains. A D segment in the finished heavy chain gene is located in a position corresponding to that of one of the hypervariable regions of the heavy chain.

In the embryonic or "nonshuffled" state, the human DNA containing antibody genes includes about 100 V segments, approximately 10 D segments, and 4 J segments. In the assembly of a functional heavy-chain gene, somatic recombination brings one of the V genes (with its leader) together with a D segment and a J segment, which also code for the variable region of the heavy chain. The segments for the constant sequences are grouped together at a distance from

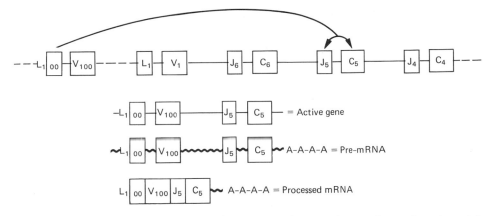

FIG. 19-13. Lambda genes and the formation of active genes. Six genes are coded for the constant region of the λ chain (only three are shown here). Each is paired with a J segment. About 300 variable genes occur. Any one of them can combine with any J–C pair to form an active gene that can then undergo transcription. The primary transcript is processed. Introns and intervening sequences are removed, and a completed mRNA is produced.

the V genes (Fig. 19-14). Unlike the situation with light chains, several different separate C genes exist, each designating the type of heavy chain (mu, delta, gamma, etc.). Since there are four different classes of gamma chains, eight different C sequences exist. These occur in the order mu, delta, gamma, types epsilon and alpha. Actually, the situation is even more complex, because each of the constant genes is separated by introns into three to six separate coded segments.

In the development of a mature B lymphocyte, a stage is recognized in which the cell is called a *pre-B lymphocyte.* Such a cell produces a heavy chain of the mu type. The heavy chain synthesis occurs before light-chain synthesis. Following the formation of light chains (kappa or lambda in a given cell), the pre-B lymphocyte produces a delta chain, and the cell produces completed IgM and IgD molecules. In other words, all maturing B lymphocytes first produce heavy chains of the mu class. Then delta chains are produced. This comes about as a result of both DNA shuffling and RNA processing (Fig. 19-15). In the pre-B lymphocyte, the somatic recombination joins a V, a D, and a J segment close to the constant region. This produces an active gene, which forms an RNA transcript carrying the information for a mu constant sequence. Processing of this transcript removes introns within the mu gene as well as other intervening sequences, to give a finished IgM heavy-chain transcript. A primary transcript can also include information from a delta gene. Splicing may then remove the mu gene sequences and connect the segment carrying the delta information to the V-D-J segment. Therefore, differential splicing, as well as DNA shuffling, is involved in the production of IgM and IgD heavy chains in the pre-B lymphocyte. Both immunoglobulins, IgM and IgD, at first appear on the surface of the prelymphocyte cell. This simultaneous production of IgM and IgD is the only exception to the statement that a mature lymphocyte synthesizes *only one* type of immunoglobulin.

Further development of the B lymphocyte depends on its interacting with some specific antigen. When a specific antigen becomes bound to the cell, the

FIG. 19-14. Unshuffled DNA sequences coded for the heavy antibody chain. In the human, the sequences coded for the heavy antibody chain include about 100 variable genes, 10 D sequences, 4 J sequences, and 8 constant genes, each of which designates a specific class of heavy chain. There are five main classes, but the gamma class includes four subclasses.

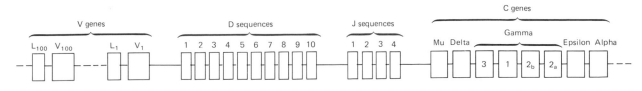

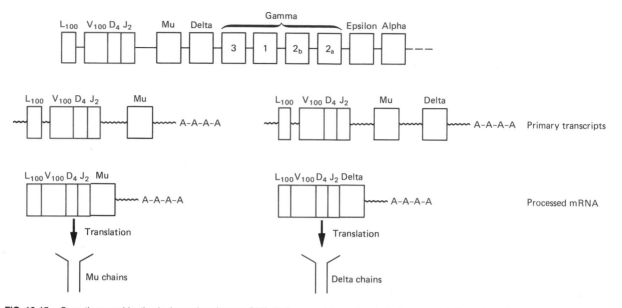

FIG. 19-15. Somatic recombination in the prelymphocyte. DNA shuffling brings together a combination of a V gene and a D segment and a J segment (in this example, V100, D4, and J2). This combination produces an active gene. Primary transcripts arise *(left)* coded for the mu constant gene. Processing results in the formation of mRNA, which can then be translated. Some primary transcripts *(right)* include information for both mu and delta chains. The processing eliminates the information in the transcript for the mu chain. An mRNA forms, which can be translated for a delta chain. In this cell, therefore, the same V region will be associated with two different constant regions, and the cell will produce both M and D chains.

lymphocyte completes its development. It goes on to divide to produce a clone of mature antibody-producing B lymphocytes. During this transition, both IgM and IgD disappear from the cell surface, and the cell then secretes one of the immunoglobulins: IgM, IgG, IgE, or IgA. This means that the *same V region* can now become associated with a heavy-chain constant region *other than* the constant mu region, a process called *heavy-chain class switching*. This switching involves additional DNA shuffling and differential processing of the mRNA (Fig. 19-16). Between the constant region genes are nucleotide sequences that seem to act as signals in switching. These signal sequences have some similarity to a sequence located between the mu gene and the variable genes. These signal sequences may be involved in bringing about a recombination event that joins the V-D-J variable sequences to one of the other constant regions. After this next DNA recombination occurs, the same variable region is joined to a different C region so that the variable region is directly adjacent to one of the C genes. After the DNA has undergone this second reshuffling, it is transcribed. The mRNA is processed to form a messenger that codes for a gamma, an epsilon, or an alpha constant region, whichever one is adjacent to the variable sequence. Signal sequences similar to

those associated with the DNA coded for the heavy chains are also known to occur in the DNA coded for the light chains. In the case of the kappa chain, one of these is found near each V gene and one near the J segment.

It should be noted that the IgM that can be secreted by a mature lymphocyte is a different form of the IgM that appears bound to the surface of the prelymphocyte along with IgD. The bound form of mu contains an amino acid sequence at its end that is responsible for binding it to the membrane. The mu that is secreted lacks this amino acid chain. The two different forms of mu arise from translation of two slightly different kinds of mRNA transcripts. The primary transcript for the bound form of mu arises as a result of the complete reading of six separate DNA segments of the mu gene. The last two segments are coded for the tail portion, which anchors mu to the membrane. The primary transcript for the secreted form results from the termination of transcription after the first four segments are read. Hence, the final mRNA, after processing of the primary transcript, lacks the two segments coded for the tail. Consequently, a secreted form of mu arises after translation of the shorter messenger.

The amount of diversity possible from the somatic

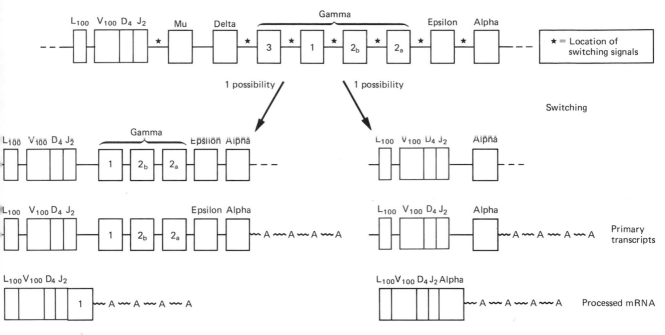

FIG. 19-16. Heavy chain class switching. Further DNA shuffling as the lymphocyte matures can place the same V-D-J sequence (refer to Fig. 19-15) directly next to still a different C gene. A gamma 1 gene, for example, could be placed next to V-D-J *(left)* with the deletion of the genes to the left of it. Any one of the other C genes could be placed adjacent to V-D-J, for example, alpha *(right)*. All constant genes still present along with the one directly next to V-D-J will undergo transcription with it, and a primary transcript is produced. We see here on the left that the primary transcript includes information from gamma 1 through alpha. Differential processing, however, produce an mRNA that carries information only for gamma 1, the C gene placed directly next to the V-D-J segment following heavy chain class switching.

recombination that joins different constant and variable genes for both the light and heavy chains is astronomical. In the mouse, more than 10,000 possible combinations for the heavy chain have been calculated. Each kappa or lambda light chain, in turn, can occur in a large number of different variations, since each combines different V with different C regions. Therefore, combining light chains with heavy chains yields more than 10 million possibilities! Added to the possible diversity resulting from these different combinations is the role played by somatic mutation. For reasons that are not clear, the genes in antibody-producing cells appear to be unstable and to undergo mutation at a rate far in excess of that in other kinds of cells. Some of these mutated genes in lymphocytes have been shown to differ from the corresponding variable gene before it is shuffled by just a single nucleotide.

It has been estimated that one change in the V region occurs in every 3 to 30 cell divisions. The B cell may contain some factors for inducing such a high mutation rate. Supplementing mutation and adding to the potential diversity is the fact that the joining of V and J segments can occur in more than one way. The two segments do not necessarily always recombine at the same site. The point of recombination can vary over several nucleotides. As Fig. 19-17 shows if V and J segments always combine at site 1, then sequences coded for proline and tryptophan would always result at the site of recombination. If they sometimes recombine at site 2, however, the sequence would be for proline and arginine. This flexibility at the V-J combining sites and also at V-J-D, makes even more variation possible in sequences coded for variable regions of the antibody.

Taking all the possibilities for antibody genes into consideration, the somatic recombination along with mutation and different ways of joining segments, molecular biologists have estimated the possible number of different antibodies as high as 18 billion. Such a fantastic number of diverse molecules is encoded by only a few hundred separate genes or segments, a far cry from the 18 billion that would be called for if each antibody type were encoded by a separate gene!

Many answers must still be found for events taking place during the maturation of B cells and the for-

FIG. 19-17. Joining of V and J segments. When recombination takes place to join the V and J segments, it does not always occur at exactly the same site. If joining or recombination between a given V and a given J always occurred at point 1 *(arrow 1)*, the two triplets, as depicted in the figure, would always designate proline–tryptophan in the recombinant DNA at the position where the recombination took place. If, however, there is flexibility allowing recombination to take place at another point *(arrow 2)*, a new sequence, proline–arginine, can be generated.

mation of antibodies. The unexpected finding of gene shuffling or somatic recombination requires machinery to bring together the different V, D, C, and J segments. Perhaps the signal sequences mentioned earlier, which are associated with genes for the heavy and light chains, are involved with enzymes that must be able to cut the DNA during the shuffling process. The complex story of antibody diversity makes us look at the gene in another light and reconsider how many genes may be involved in forming a polypeptide gene. We still need many explanations regarding the mechanism that controls the recombination of the coded segments or genes to produce an active gene, one that can form a transcript to carry information for the assembly of an antibody chain. The knowledge that gene shuffling occurs in the formation of active antibody genes raises the possibility that similar reshuffling may occur in cases of other DNA regions to generate diversity of genes and hence proteins involved in certain critical developmental pathways.

Hemoglobin chains and their genes

Recombinant DNA procedures have provided us with an excellent picture of the arrangement of the various genes coded for globin chains of hemoglobin as well as revealing some unexpected findings. To appreciate the accomplishments in this area, we must examine a few details on human hemoglobin.

Most of the hemoglobin of the adult, hemoglobin A, is composed of four polypeptide chains, two alpha (α) combined with two beta (β). However, more than one type of hemoglobin appears during the human life cycle. Nevertheless, all hemoglobin molecules consist of a tetramer composed of four globin chains. Two of these polypeptides are alphalike and two are betalike, a distinction made on the basis of their structure. Each kind of polypeptide chain is encoded by its own specific gene, and at least seven genes occur for the globin polypeptides of the human. Throughout the life span, different globin genes are called to expression, and some of these become repressed later on.

As Table 19-2 shows, the first hemoglobin to arise in the human is composed of two epsilon (ϵ) chains, which are betalike, and two zeta (ζ) chains, which are alphalike. This hemoglobin $\zeta_2\epsilon_2$, is known as Gower I. As the embryo reaches the eighth week of development, the genes encoded for these chains become repressed as other globin genes are called to expres-

TABLE 19-2   Human globin chains and hemoglobins

STAGE	GLOBIN CHAINS		HEMOGLOBIN CLASS
	α-like	β-like	
Embryonic (up to 8 weeks)	ζ_2	ϵ_2	$\zeta_2 \epsilon_2$ Gower I
Fetal	α_2	$^G\gamma_2$ or $^A\gamma_2$	$\alpha_2 \gamma_2$ (general formula) Hg F
Adult (after birth)	α_2	β_2	$\alpha_2 \beta_2$ Hg A
Adult (after birth)	α_2	δ_2	$\alpha_2 \delta_2$ HgA$_2$

sion. These are genes for the α chains and for two different types of gamma (γ) chains, which are beta-like. These chains, $^G\gamma$ and $^A\gamma$, are encoded by two separate genes and are identical except for amino acid position 136. A glycine is present at that site in $^G\gamma$, whereas it is an alanine in $^A\gamma$.

A combination of two α chains and two of the γ chains ($\alpha_2{}^G\gamma_2$ or $\alpha_2{}^A\gamma_2$) forms hemoglobin F (general formula $\alpha_2\gamma_2$), which gradually replaces the embryonic hemoglobin and becomes the main hemoglobin throughout the life of the fetus. (During the transition from embryonic to fetal hemoglobin, one can detect other hemoglobin types such as $\alpha_2\epsilon_2$ and $\zeta_2\gamma_2$.) The two γ genes, in turn, start to become repressed shortly before birth as the globin chains characteristic of adult hemoglobin appear. Genes coded for β globin and delta (δ) globin polypeptide are now called to expression. At 6 months of age, the child has a hemoglobin profile typical of the adult. About 97% of the hemoglobin is hemoglobin A ($\alpha_2\beta_2$), the remainder consisting mainly of hemoglobin A_2 ($\alpha_2\delta_2$). Approximately 1% fetal hemoglobin ($\alpha_2\gamma_2$) continues to be produced. The α-globin chain is composed of 141 amino acids. The β, δ, and γ chains all contain 146 amino acids.

The nucleotide sequences of the different globin genes have been worked out using recombinant DNA techniques. All the α-like and β-like globin genes have been shown to be split, each containing two introns. Moreover, all the globin genes have been mapped. The α-like genes compose a family that occurs in a cluster on Chromosome 16. Two α-globin genes, α_2 and α_1, typically occur in a cluster (Fig. 19-18A). The β-like genes constitute another family clustered on Chromosome 11 (Fig. 19-18B).

One reason that such rapid advances have been made in our understanding of globin genes is the ready availability of their mRNAs, as we noted previously. In relation to other kinds of mRNA, the immature red blood cells manufacture an abundance of transcripts for the globin chains. Transcripts isolated from the immature red blood cells can then be used in various kinds of investigations. For example, isolated human transcripts have been placed in cell-free translation systems composed of ingredients obtained from nonhuman sources—wheat cells in some cases. Even though the tRNA, ribosomes, and other factors are from nonhuman cells, the human globin transcripts are nevertheless translated into human globin chains. This is an important bit of information; it tells us that whatever factors regulate the transcription of the globin genes must not operate, to any large extent, at the level of translation. Instead, they must exert their major influences at the level of transcription or of processing the RNA.

The globin gene transcripts have been used to make cDNA, which is then inserted into bacterial plasmids. All the human globin genes have been cloned and are now available for use in various kinds of experimental approaches. One obvious use of the cDNA is as a probe to locate globin DNA sequences in any kind of cell or from among any group of DNA fragments. Use of probes led to the discovery of the α and β gene clusters and the sequence of the genes in each cluster. Several very interesting observations have been made on these α and β clusters. In each of them, the sense strand of the separate genes proved to be the same DNA strand. With the 5' end of each gene written on the left, the reading of each gene is from left to right (Fig. 19-18). For example, the strand used as the template to produce γ mRNA is the same as the one used for the δ and β mRNAs.

As Fig. 19-18 also shows, the arrangement of the genes in each cluster corresponds to the order in which they are expressed (ϵ before γ before β and δ

FIG. 19-18. The globin gene clusters. Sequence of genes and pseudogenes *(red)* in the alphalike cluster on Chromosome 16 *(A)* and the betalike cluster *(B)* on Chromosome 11. The 5' end of each gene is toward the left. The direction of transcription for each gene *(arrows)* is from left to right.

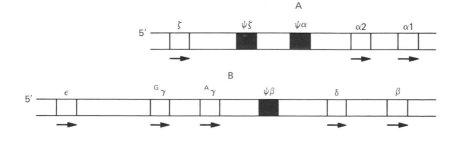

in the β group). In addition, the figure shows that both the α and the β clusters contain nucleotide sequences known as *pseudogenes*. These sequences have been detected using labeled DNA from globin genes as probes. Neither the β-like nor the α-like pseudogene sequences have ever been shown to be associated with any known globin polypeptide. Apparently, they are not expressed, even though they contain large stretches of DNA homologous to the functional genes in each cluster (about 75% homology). All the globin gene clusters studied so far have pseudogenes between the embryonic or fetal genes and the adult genes (Fig 19-18). Pseudogenes are considered to have arisen from functional genes as the result of gene duplications.

Duplications and the globin genes

Duplications are considered to be one class of structural chromosome changes that have played a highly significant role in the evolution of the genome (see Chap. 10). Once a gene duplication arises, each gene copy becomes subject to its own history of point mutation and structural chromosome change. For a while, both genes may retain the same function. One of the duplicates, however, may accumulate mutations that render it incapable of forming a functional product. Pseudogenes are believed to represent cases

of this. The detrimental mutations can accumulate because the gene involved originated as a duplication, and the other copy, lacking detrimental mutations, is able to perform the function associated with the gene. The genetic alterations in the pseudogenes of mammals include deletions that shift the reading frame, point mutations that alter intron–exon boundaries, and missense mutations that would substitute amino acids at critical positions in any associated polypeptide chain. Taken together, the alterations in the pseudogenes would be detrimental to both normal transcription and translation, even if a transcript should arise. Pseudogenes are believed to be common in vertebrates and possibly in other organisms as well. Many gene families are now known to include them other than the globin cluster—for example, the sequences coded for the immunoglobulins and those for the histocompatibility antigens.

The two adjacent genes for the α-globulin chains on Chromosome 16 are clear-cut examples of very recent duplications, as are the two almost identical γ-globulin genes. In both cases, the members of the pair have clearly maintained the same function. As evolution proceeds, however, one of the two genes in a pair of duplicates may maintain the original function while the other evolves to perform a somewhat different but related function. For many years, the globin genes have been considered a primary example of a

FIG. 19-19. Equal and unequal crossing over in the delta–beta region. If normal alignment of the homologous chromosomes takes place *(above),* the two delta genes pair with each other, as do the two beta genes. A crossover event *(red arrows)* anywhere along the length of the paired region will produce equal crossover products with intact delta and beta genes. However, if misalignment occurs *(below)* and crossing over takes place between a delta and a beta gene, unequal crossover products are generated. A Lepore gene is formed,

composed of the left portion of the delta gene and the right portion of the beta gene. Depending on the point of crossing over, Lepore genes will differ in the proportion of delta and beta sequences they contain. The other product of the unequal crossing over is a chromosome that contains an anti-Lepore sequence, the left part of the beta gene, and the right part of the delta gene. A normal delta and a normal beta gene are also present.

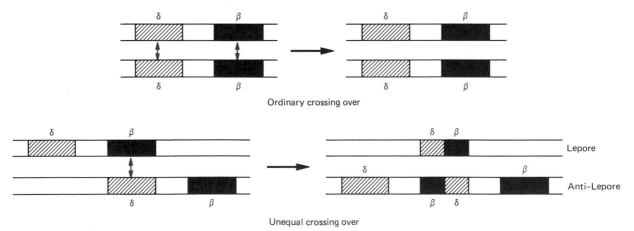

Ordinary crossing over

Unequal crossing over

group of related genes, originating from the duplication of ancestral genes. Before recombinant DNA procedures were developed, excellent evidence for this concept was provided by the striking similarity of the amino acid sequences of the several globin chains that are encoded by several separate genes. The β and δ chains differ by only 10 amino acids; the two γ chains differ from each other by only a single amino acid. The β, δ, and γ chains are all the same length. The β chain differs from the γ chain by 30 amino acids, but it differs from the α chain by 84.

The β and δ genes are now known to be identical over 90% of their nucleotide sequences. Close homology between the β and δ genes has been apparent to investigators for many years, since crossing over can occur between them. In Chapter 10, we saw that, as a result of tandem duplications in the Bar region of Drosophila, crossing over can take place in such a way as to yield unequal crossover products. A similar situation pertains to the β and δ genes. Since the two genes may have stretches of homology, a certain amount of mispairing can occur, giving rise to unequal crossing over. As Fig. 19-19 shows, pairing between β and δ genes, followed by a crossover event, leads to the formation of two dissimilar or unequal products. A variant gene is generated that starts off

as a δ gene and ends as a β gene. Such a gene, known as a Lepore gene, is responsible for the production of a variant globin chain that starts at the amino end as a δ chain and ends at the carboxyl end as a β chain. Since crossing over between the mispaired genes may take place at more than one site, more than one type of Lepore gene and hence Lepore globin is expected, and this is the case. Note from Fig. 19-19 that the reciprocal product yields a chromosome region that contains a δ gene, an anti-Lepore gene (which begins as a β gene and ends as a δ gene), and finally a complete β gene. Anti-Lepore hemoglobins are known. In such cases, the person has a full set of normal β and δ genes plus the β–δ fusion gene.

These and other observations lend support to the theory that the genes for the four globin chains have been derived from a single gene, one that was also ancestral to the gene encoded for myoglobin, an oxygen-binding muscle protein. The myoglobin gene, like the genes for the globin units, is also interrupted by two introns. Myoglobin, though showing many differences from the amino acid sequences of the globin polypeptides, is comparable to them in length, consisting of 153 amino acids. The α chain is composed of 141 amino acids, and the β, γ, and δ chains are made up of 146 of them. According to the proposed relation-

FIG. 19-20. Duplications and the globin genes. An ancestral gene is believed to have duplicated. One copy evolved into the present-day gene for myoglobin. The other copy underwent another duplication. One of these evolved down a pathway leading to an alphalike gene. Several later duplications eventually gave rise to the alphalike gene cluster. The other gene copy evolved in the betalike direction, and further duplications also took place in this pathway leading to the origin of the betalike cluster. Two introns (red) are present in all the genes involved.

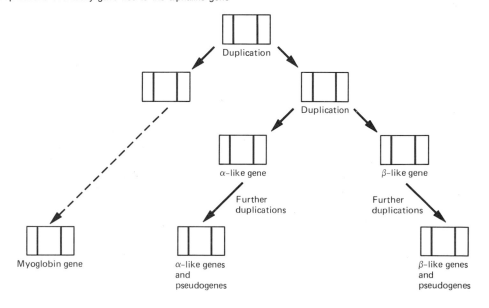

ships among the globin genes and that for myoglobin, an ancestral gene duplicated. One copy proceeded down the pathway leading to the present-day myoglobin gene. The other copy underwent another duplication. One of these evolved into a gene that was ancestral to the α cluster; the other was ancestral to the β cluster (Fig. 19-20). Further duplications in the more recent past set the stage for the origins of the various α-like and β-like genes, as well as the pseudogenes associated with them.

All the gene members of the β group are more similar to one another than any one of them is to the α sequence. Very interesting is the fact that the two introns that split the five β-like genes occur at the same location in each gene in relation to the reading frame. For example, in all five genes, the first intron occurs between codons 30 and 31, the second one between codons 164 and 165. The two introns that interrupt the genes in the α-like cluster also split the reading at exactly the same location, the first between codons 31 and 32, the second between codons 99 and 100. Such observations lend strong support to the concept that gene members of the β cluster resemble one another more than they do members of the α gene cluster, since they stem from duplications that occurred in the more recent past. Not enough time has elapsed for major differences in nucleotide sequence to have arisen among them. Nor have they yet been separated onto different chromosomes by structural rearrangements. In contrast, the α sequence stems from a more ancient duplication of an ancestral gene and has diverged appreciably from those in the β group, to which it is no longer linked but is now on a separate chromosome.

Molecular basis of globin chain disorders

Recombinant DNA procedures have supplied a wealth of information on the molecular level concerning several human disorders related to variant hemoglobin molecules. Two main classes of human blood disorders can be recognized. In the first type, a globin chain is altered, usually because of an amino acid substitution resulting from a point mutation. Sickle cell anemia is a typical example of this class. In the second type of disorder, the globin chain is not altered in any qualitative way. Instead, there is a quantitative change. The globin chains that are produced have the normal amino acid sequence, but they are present in reduced amounts. The thalassemias are disorders that fall into this class. Both α and β thalassemias occur in which production of either α or β chains is reduced. Let us first concentrate on β thalassemias.

An individual who is heterozygous for β thalassemia, and thus has one normal β allele and one defective one, has no symptoms, because the normal allele compensates for the defect, and sufficient β chains are produced. Persons homozygous for β thalassemia suffer from a very severe anemia and from iron toxicity, a condition commonly known as Cooley's anemia. β thalassemia, in turn, falls into two main types, β^+ and β^0. In β^+ thalassemia, there is a decrease in the amount of β-chain mRNA, which causes a decrease in the amount of normal β-globin polypeptide chains. Individuals with β^+ thalassemia vary in the amount of β-globin synthesis that does take place, indicating that not one but several kinds of molecular defects are associated with the β^+ disorder. These may possibly be associated with regulation of transcription or with mRNA processing.

In individuals homozygous for β^0 thalassemia, no β-globin chains at all are produced. Studies at the molecular level show that the absence of the chains may result from an assortment of defects here as well as in the β^+ condition. Some β^0 individuals have no β-globin mRNA whatsoever. This could also be attributed to some defect at the level of transcription or to some abnormality associated with mRNA processing. A base substitution or a deletion could also be responsible. In some β^0 thalassemia cases, β mRNA is present in almost normal quantities, whereas in others the amount is reduced. This strongly suggests that for some reason the mRNA cannot be translated. In one case, a premature chain-terminating codon was found to be present in the β-globin mRNA.

There are two other rarer conditions related to the β^+ and β^0 thalassemias. These are the hereditary persistence of fetal hemoglobin (HPFH) and delta–beta thalassemia. In persons with these disorders, no δ or β-chain synthesis occurs at all. However, fetal hemoglobin (Hg F) is produced in increased amounts. As a result, delta–beta thalassemia is a mild condition, and in HPFH, anemia is absent. In the delta–beta condition, the fetal hemoglobin compensation is not quite sufficient to prevent a mild anemia. However, in HPFH, the production of Hg F is greatly increased and eliminates anemia. How can this be explained? The level of Hg F is also increased in the β^+ and β^0 thalassemias. However, Hg F production is not raised sufficiently as in the delta–beta and HPFH condi-

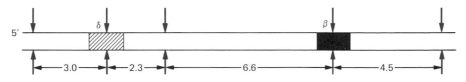

FIG. 19-21. Organization of the delta–beta region. Molecular details on the arrangement of the delta and beta genes and the regions surrounding them were obtained using restriction enzymes and labeled cDNA probes. When human DNA from persons with normal hemoglobin chains is exposed to EcoR1, the restriction enzyme attacks the DNA at several sites *(arrows)* in the delta-beta region. As a result, four fragments are generated. The four fragments differ in size, expressed in kilobases, and were distinguished from all other DNA fragments in the enzyme-exposed DNA by the labeled cDNA probes. The sequence of the four fragments was worked out, and the 3.0 kb fragment was found to contain the 5′ end of the delta gene whereas the 5′ end of the beta gene is in the 6.6 kb fragment. The 2.0 and 4.5-kb fragments contain, respectively, the 3′ ends of the delta and the beta genes. The distance between the delta and beta genes is approximately 7 kb.

tions. Consequently, serious anemia results, because compensation for the decreased or absent hemoglobin A is inadequate.

In a series of investigations, cDNA probes were used to detect globin genes in DNA obtained from persons with no blood disorders and from persons with HPFH, delta–beta, β^+, and β^0 thalassemias. The organization of the δ and β-globin gene sequences and the region surrounding them was determined. With the use of restriction enzyme EcoR1, four fragments were identified that hybridize with the appropriate cDNA probes, and the sequence of the fragments was worked out for the normal condition (Fig. 19-21).

DNA from persons with normal hemoglobin was subjected to a series of different restriction enzymes. Appropriate cDNA probes were then used to identify fragments containing DNA sequences found in the delta–beta region. The same was done with the DNA from persons with variant hemoglobins. The results indicated that in persons homozygous for delta–beta thalassemia, the 5′ end of the δ gene plus sequences immediately surrounding it are present, but the rest of the delta–beta sequences are absent. In persons homozygous for HPFH, all the delta–beta fragments normally found in this chromosome region are entirely absent (Fig. 19-22)!

In the case of the β^+ and β^0 thalassemias, there is no evidence for any extensive deletions, and it appears that the defect in these serious conditions results mainly from single nucleotide changes. (As noted, in one β^0 thalassemia case, a single change produced a chain-terminating codon, resulting in premature termination of translation.) In some β^+ thalassemias, there is reason to suspect that the problem lies at the level of RNA processing, as a result of nucleotide changes affecting exon–intron junctions. Thus, it seems likely that the β^+ and B^0 thalassemias may stem from different kinds of defects at the molecular level.

It is very significant that in the delta–beta thalassemias, where symptoms are not too severe, and in HPFH, where there is no anemia at all, extensive deletions exist in the delta–beta region. The production of Hg F compensates for lack of intact genes for the δ and β chains; the compensation is greater in HPFH, in which the entire delta–beta region has been deleted! The increase in Hg F production above the normal in β^+ and β^0 thalassemias (in which there is no evidence for extensive deletions) is not at all sufficient to prevent severe anemia. One interpretation of these observations is that the delta–beta region of the chromosome contains sequences that regulate the expression of the gamma genes and decrease the level of their expression before birth. In the case of HPFH, these sequences are lost because of deletion, and γ chains can consequently be produced at high levels.

FIG. 19-22. Deletions in the delta-beta region. Labeled cDNA probes show that persons homozygous for delta–beta thalassemia have a deletion in the delta–beta region. Only the sequences toward the 5′ end of the delta gene and immediately to the left of it are present. The rest of the delta–beta region is missing. In HPFH, the entire delta–beta region has been deleted.

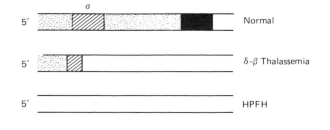

FIG. 19-23. Deletion of an alpha gene from one chromosome. Unequal crossing over in the alpha-gene cluster can result in the deletion of one of the alpha genes from a chromosome and the origin of a chromosome with three alpha genes.

In delta–beta thalassemia, the deletion does not remove all of this controlling region, so that some reduction in expression of the γ genes occurs. In the β^+ and β^0 thalassemias, in which little if any deletion exists, the controlling region is present and tends to suppress the γ genes. As a result, Hg F levels are kept too low to compensate for a decrease in the normal amount of Hg A.

Although many details are yet to be uncovered, the information now available on the β-gene cluster and the changes that can alter it at the molecular level has provided us with a deeper understanding of the nature of the β thalassemias and offers hope that ways to lessen the severity of the β^+ and β^0 conditions may be found.

Alpha thalassemias

As noted in Fig. 19-18*A*, two alpha-globin genes may occur on Chromosome 16. In most human populations this is the case. Since alpha genes exist in duplicate copies, this again makes possible the occurrence of unequal crossing over, as discussed for the δ and β genes. As the result of an unequal crossover event, a deletion of an α gene can occur (Fig. 19-23). α thalassemias are known that may have resulted from different deletions of α-globin genes and may have been generated by unequal crossing over. Such deletions result in a decrease in the number of α-globin chains. These α gene deletions have been detected using α-globin cDNA probes. Four different possibilities exist regarding the number of α genes. Deletion of just one gene gives an individual three α-globin genes, two on one chromosome and one on the other. This results in a decrease in α-globin synthesis, but no anemic symptoms are present. When two α-globin genes are deleted, there is a slight anemic effect. If three of them are deleted, the β chains accumulate, because an insufficient number of α chains are being produced to combine with them.

The result is the formation of unstable tetramers of four β chains, which break down, resulting in a mild anemia associated with hemoglobin H disease. If all four α chains are deleted, the fetus dies as a result of a very severe anemia, *hydrops fetalis*. (Figure 19-24 summarizes some of the possibilities in the number of α-globin genes.) Human populations differ in the incidence of hydrops fetalis. Knowing what we now know about the α- and β-gene clusters, and armed with rapidly advancing techniques, it may be possible in the near future to transform human marrow cells with isolated, nonvariant globulin genes, thus preventing the severe effects associated with aberrant or reduced globin chains.

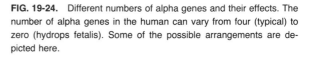

FIG. 19-24. Different numbers of alpha genes and their effects. The number of alpha genes in the human can vary from four (typical) to zero (hydrops fetalis). Some of the possible arrangements are depicted here.

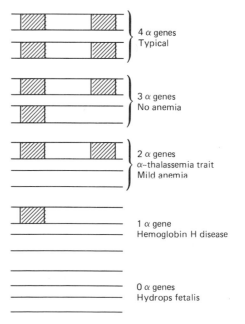

REFERENCES

Anderson, W. F. and E. G. Diacoumakos. Genetic engineering in mammalian cells. *Sci. Am.* (July): 106, 1981.

Anderson, W. F. and J. C. Fletcher. Gene therapy in human beings: when is it ethical to begin? *N. Engl. J. Med.* 303: 1293, 1980.

Bank, A., J. G. Mears, and F. Ramirez. Disorders of human hemoglobin. *Science* 207: 486, 1980.

Drayna, D. and R. White. The genetic linkage map of the human X chromosome. *Science* 230: 753, 1985.

Efstratiadis, A. et al. The structure and evolution of the human β-globin gene family. *Cell* 21: 653, 1980.

Geever, R. F., L. B. Wilson, F. S. Nallaseth, P. F. Milner, M. Bittner, and J. T. Wilson. Direct identification of sickle cell anemia by blot hybridization. *Proc. Natl. Acad. Sci.* 78: 5081, 1981.

Gusella, J. F., R. E. Tanzi, M. A. Anderson, W. Hobbs, K. Gibbons, R. Raschtchain, T. C. Gilliam, M. R. Wallace, N. S. Wexler, and P. M. Conneally. DNA markers for nervous system diseases. *Science* 225: 1320, 1984.

Hozumi, N. and S. Tonegawa. Evidence for somatic rearrangement of immunoglobulin genes coding for variable and constant regions. *Proc. Natl. Acad. Sci.* 73: 3628, 1976.

Kolata, G. Gene therapy method shows promise. *Science* 223: 1376, 1984.

Kolata, G. Huntington's disease gene located. *Science* 222: 913, 1983.

Laurence, J. The immune system in AIDS. *Sci. Am.* (Dec.): 84, 1986.

Leder, P. The genetics of antibody diversity. *Sci. Am.* (May): 102, 1982.

Lewin, R. DNA fingerprints in health and disease. *Science* 233: 521, 1986.

Marrack, P. and J. Kappler. The *T* cell and its receptor. *Sci. Am.* (Feb.): 36, 1986.

Proudfoot, N. Pseudogenes. *Nature* 286: 840, 1980.

Sakano, H., K. Hüppi, G. Heinrich, and S. Tonegawa. Sequences at the somatic recombination sites of immunoglobulin light-chain genes. *Nature* 280: 288, 1979.

Tonegawa, S. The molecules of the immune system. *Sci. Am.* (Oct.): 122, 1985.

Tonegawa, S. Somatic generation of antibody diversity. *Nature* 302: 575, 1983.

Weatherall, D. J. and J. B. Clegg. Thalassemia revisited. *Cell* 29: 7, 1982.

REVIEW QUESTIONS

1. Suppose a rare and very serious blood disorder is associated with an abnormality of the alpha-globin chain. DNA is extracted from persons with normal hemoglobin and from those suffering from the disorder. DNA from both sources is subjected to a certain restriction enzyme. This is followed by gel electrophoresis and Southern blot hybridization. Two probes are available. One of these is a labeled cDNA specific for the 5′ portion of the alpha-globin gene and the other a labeled cDNA specific for the 3′ portion of the gene. Two autoradiographs are prepared, and the results are given here. Explain the differences between the autoradiographs and the molecular basis of the disorder (A = DNA from nonafflicted persons; B = DNA from those with the disorder).

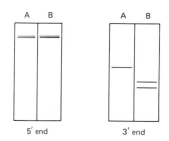

2. A 28-kbp segment is known which is vulnerable to a certain restriction enzyme. The enzyme can cut the segment at three sites to yield fragments of different kilobase pair lengths as seen in the following, with arrows indicating cutting sites. However, the site at the left arrow is variable. A probe, as seen here, can detect restriction fragments resulting from the action of the enzyme.

Based on the information presented, answer the following questions:

A. What fragment sizes can the probe detect?
B. What haplotypes are possible? (Use the A, B naming system to designate haplotypes.)
C. What are the possible genotypes in reference to the haplotypes?
D. What fragments would one observe on autoradiographs of the possible genotypes?

3. In reference to Question 2, suppose the sites at both the left and right arrows are variable in relation to cutting by the enzyme. Answer the following.

A. How many haplotypes are possible, and what are they in regard to fragment sizes? (Use the A, B system to name them.)
B. How many genotypes are possible, and what would they be?

4. In a particular pedigree, it is found that Huntington's disease (HD) is associated with haplotype D in respect to HindIII restriction fragments (see Fig. 19-5). Two related families are studied in which one parent is afflicted and the unafflicted parent is not related by blood

to other members in the pedigree. The genotypes of the parents in each family are as follows, in reference to the haplotypes:

Family 1—Afflicted parent: AD; unafflicted parent CD
Family 2—Afflicted parent: CD; unafflicted parent CD

DNA is extracted from offspring in each family, treated with HindIII, and autoradiographs are finally prepared. The results are presented here with the pattern at left showing all possible HindIII fragments. Deduce the genotypes of the five persons involved in the two families. Give the relative degree of risk each has of developing HD, along with a brief explanation.

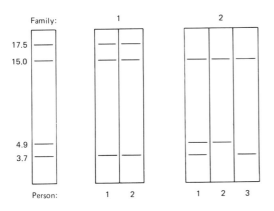

5. If T-helper cells are destroyed by a virus such as the one responsible for AIDS, the immune system breaks down. Why is this so, since the T-helper cells do not directly destroy foreign cells or produce antibodies?

6. DNA was extracted from fibroblast cells and also from antibody-producing cells. The DNA from both sources was subjected to a particular restriction enzyme, and the fragments were separated on a gel. A radioactively labeled available cDNA probe can detect a C gene for the constant region of a light chain. Following Southern blotting, the fragments were exposed to the probe, and autoradiographs were prepared. The results of a comparison of the autoradiographs are shown here, with the numbers indicating the fragment lengths in nucleotide base pairs. Explain the differences seen (F = fibroblasts; P = plasma cells).

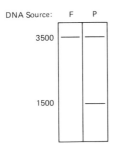

7. For each of the following, indicate whether the item is relevant to heavy or light chains, or to both types: (1) kappa chain; (2) C gene; (3) mu chain; (4) constant region; (5) gamma chain; (6) allelic exclusion; (7) lambda chain; (8) J sequence; (9) V gene; (10) D sequence; (11) alpha chain; (12) somatic recombination; (13) chain class switching; (14) RNA processing.

8. As diagrammed here, with just two V (variable) genes, three J sequences, and one C (constant) gene, give the combinations possible for an active, light chain gene following DNA shuffling.

$$\underline{V_{18} \quad V_6 \quad J_1 \ J_2 \ J_3 \quad C_1}$$

9. As diagrammed here, with just one V gene, two D sequences, two J sequences, and two C genes, give the combinations possible for an active heavy chain following DNA shuffling.

$$\underline{V_2 \quad D_1 \quad D_2 \quad J_1 \quad J_2 \quad C_1 \ C_2}$$

10. During the development of immature B lymphocytes, IgM and IgD are produced and bind to the cell membrane. After antigenic stimulation, the prelymphocytes mature into antibody-secreting plasma cells, and IgM and IgD disappear from the cell surface. A mature cell may begin to secrete IgG, IgE, or IgA. Explain the basis for regulation of differential gene expression in mature plasma cells.

11. Following is the arrangement of heavy chain sequences in a pre-B lymphocyte. The symbol (S) designates a signal sequence involved in bringing about a shuffling of the DNA. (Only one gamma sequence is indicated for the sake of simplicity.)

$$\underline{V_{50} \quad D_3 \quad J_2 \quad (S) \quad \mu \quad \delta \ (S) \ \gamma \ (S) \ \epsilon \ (S) \ \alpha}$$

Answer the following:

A. Give the arrangements that can form as mature antibody-producing B lymphocytes arise following antigenic stimulation and heavy-chain class switching.
B. For each arrangement, give the sequences that can be found in the pre-mRNA transcript.
C. Give the sequences for each that will be found in each processed mRNA and the class of chain for which the mRNA is encoded.

12. Human DNA can be obtained from the spleen or from white blood cells. When the DNA is subjected to restriction enzyme EcoR1, fragments are generated. With the use of cDNA probes, four fragments can be identified as DNA in the region containing the genes for the delta and the beta chains of hemoglobin. These four fragments have the following lengths in kiloases: 2.3, 3.0, 4.5, and 6.6.
DNA is taken from patients with hemoglobin Lepore

in which the "beginning" portion of the delta gene is fused with the "end" portion of the beta gene. The DNA from these persons yields only the 3.0 and the 4.5 kb fragments.

DNA is then taken from persons with a very rare disorder affecting the B gene. A mutation has altered a site in the B gene that cannot now be recognized by EcoR1. The DNA from these persons gives three fragments: 11.1, 2.3, and 3.0 kb.

Give the arrangement of the four fragments generated by EcoR1 from DNA taken from persons with normal hemoglobin, starting with the beginning of the delta gene. Explain the logic used in deducing the arrangement.

20

NONCHROMOSOMAL GENETIC INFORMATION

Life cycle of Paramecium

Investigations performed by Sonneborn and his colleagues with the protozoan *Paramecium aurelia* were among the first to emphasize the importance of the cytoplasm in inheritance and to focus attention on the existence of nonchromosomal genetic determinants. An acquaintance with a few features of the life cycle of Paramecium is essential for an understanding of the genetic implications of certain inheritance patterns in this organism. The nondividing Paramecium contains a large macronucleus and two tiny micronuclei (Fig. 20-1A). Each of these divides when the cell divides; however, the nuclear membranes remain intact. Each micronucleus contains the diploid number of chromosomes, and these can be seen at mitosis. The macronucleus, on the other hand, never shows chromosomes or a spindle; only granules are evident within it. However, a macronucleus is essential for the life of the cell. Therefore, the essential information it contains is somehow equally distributed at cell division. The micronuclei are dispensable to a cell but are necessary for sexual reproduction; it is these that demand our attention in this discussion.

As Figure 20-1B shows, when two cells of proper mating type conjugate, the animals touch on the mouth side. Each micronucleus undergoes meiosis so that eight haploid micronuclei arise in a cell. Seven of these eight micronuclei degenerate, leaving only one haploid nucleus in each conjugating cell. The remaining nucleus then undergoes *mitosis*, producing two identical haploid nuclei per cell. Of these two, one is capable of migrating to the other cell by way of a small connection that forms between the mating cells. The result is that each member receives a haploid nucleus from the other. The migratory nucleus then unites with the stationary nucleus of the cell, restoring the diploid condition. This fertilization nucleus in each conjugating cell contains equal amounts of genetic material from each of the two parents. The two conjugants thus become identical with respect to their chromosomal information.

While these events are taking place, the macronucleus in each cell breaks down. The fertilization nucleus then divides to produce two diploid nuclei, each of which divides in turn to give four per cell. Two of these four begin to enlarge. When the ex-conjugant cell divides mitotically after having mated, the two daughter cells receive two micronuclei and one of the two enlarging ones. This latter will become the new macronucleus.

The important point is that after conjugation and fertilization, each parent cell contains new genetic information that it has received from its mate, and

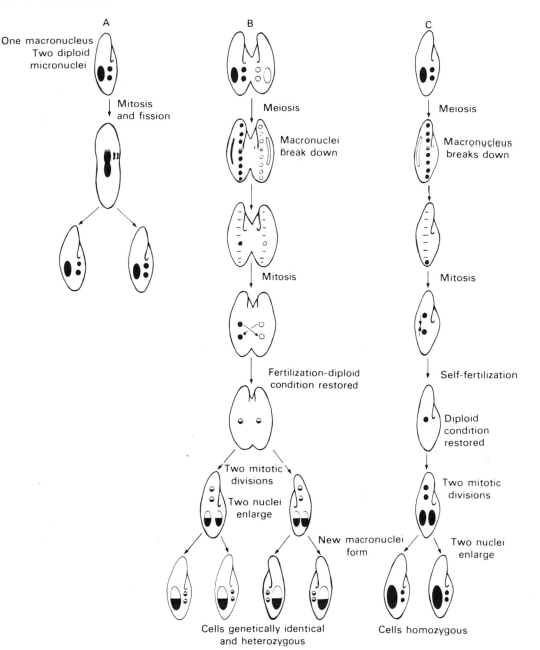

FIG. 20-1. Life cycle of *Paramecium aurelia* (A) Reproduction by cell division and mitosis. (B) Conjugation and exchange of micronuclei. Note that the outcome is the production of cells that are genetically identical because the mating cells exchanged a haploid nucleus. (C) Autogamy or self-fertilization. Note that this process results in homozygosity. (Follow the text for details).

the two parent cells are identical in their chromosomal complements. The diploid condition is present again in the micronuclei.

If a paramecium is prevented from engaging in conjugation, it may undergo a self-fertilization process known as autogamy (Fig. 20-1C). The sequence of

steps in this process is identical to those just described for conjugation, except that no nuclear exchange takes place between cells. The two haploid nuclei present after meiosis simply unite with each other. This means that after autogamy, a cell will be homozygous for *all* of its chromosomal genes, because the two haploid nuclei that unite in the self-fertilization process are identical.

The genetics of the killer trait

In the 1930s, Sonneborn observed that at times, when two stocks of Paramecium are mixed together, some of the animals become abnormal and die. It was easy to mark one of the stocks by feeding the animals colored food and to demonstrate that one of the stocks in each case was responsible for the killing effect. Such stocks were designated "killers," and those affected by them were called "sensitives." Studies were undertaken to clarify the nature of the killing action. Examination of the fluid in which killers had lived demonstrated that the cell-free filtrate contained killing action. Evidently, something was being liberated from the bodies of the killers that could destroy sensitive cells swimming in the fluid. Many observations suggested that the killing agent is very different from other known antibiotic agents and that it exists in the form of discrete, lethal particles. A certain number of killers will liberate a certain number of particles in a given period of time, each of which can kill a sensitive.

This thesis was clearly demonstrated in various ways, such as the following. Immediately after fission, a single killer may be placed in a given amount of fluid (Fig. 20-2A). At the end of 1 hour, it is transferred to a fresh quantity in another container. This process is repeated so that the cell remains for 1 hour in each of several separate equal volumes. The outcome is that at the end of 5 hours, which is about the length of a fission cycle, one of the volumes will contain enough killing agent to kill one sensitive cell. If two killers had been used (Fig. 20-2B), then two killing particles would be liberated into one or the other of the five volumes. This clearly illustrates that a discrete particle is suddenly liberated from the body of a killer into the medium.

Investigations were undertaken to clarify the genetic basis for the formation of this unusual killing agent. Fortunately, sensitive stocks are immune to the killing action during conjugation, so that matings may be performed between killers and sensitives. Let us examine such a cross, one between killer stock 51 and sensitive stock 47 (Fig. 20-3). After conjugation, each of the parent cells can be isolated and followed through subsequent divisions. It is found that the ex-

FIG. 20-2. Particulate nature of killing action. (A) One killer cell may be transferred from one to the other of several containers after a stay of 1 hour in each. The amount of fluid is fixed. When a given number of sensitives is added to each, it is found that one cell will be killed in one of the first five containers. (B) If the same procedure is followed with two killers, then two sensitives will be killed, in either the same or different containers.

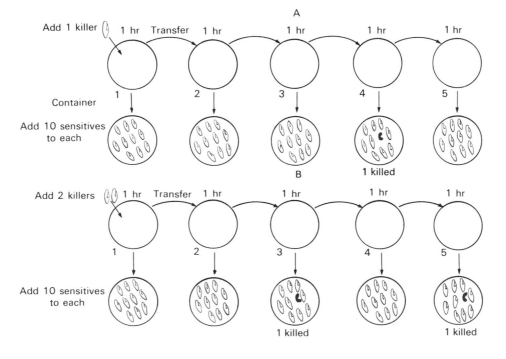

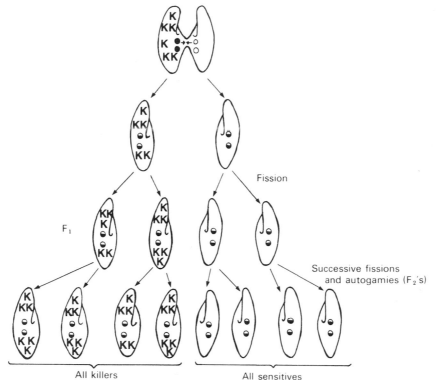

FIG. 20-3. A cross of killer stock 51 with sensitive stock 47. After conjugation, the killer cell remains a killer and the sensitive cell remains sensitive. The killer and sensitive traits persist throughout further cell divisions and autogamies. Since all the cells are identical for nuclear genes, the genetic basis for the killing must entail a cytoplasmic element.

conjugant cell that had been the killer cell gives rise only to killers by fission, and the killer trait persists throughout further self-fertilizations (autogamies). The sensitive ex-conjugant produces only sensitive cells when it is followed in the same way. This immediately tells us that something other than Mendelian inheritance is involved in the transmission of the killer trait. Both ex-conjugant cells must have identical chromosomal genes, because they exchanged micronuclei. They and the cells they produce by fission are F_1s resulting from a cross of two pure-breeding stocks. Any difference between them must be due to something other than the expression of genetic information in their nuclei. Apparently, the killer trait is following the cytoplasm in this cross. If a parent cell is sensitive, all of the cells arising from it after conjugation with a killer will also be sensitive, even though they must contain the same chromosomal information as the killer.

The importance of the cytoplasm becomes even more apparent when conjugating cells exchange an appreciable amount of cytoplasm during mating. This exchange takes place at times when the mating period is prolonged. A sizable bridge that can be seen with the aid of a dissecting microscope may form between the mating cells. When conjugation is completed and the ex-conjugants are followed, the results of a cross between killer stock 51 and sensitive stock 47 are now different from those of an ordinary mating (Fig. 20-4). We see that *both* the original killer and the original sensitive give rise to killers through subsequent fissions and autogamies.

Obviously, something has passed from the cytoplasm of the killer through the cytoplasmic bridge to the body of the sensitive. This factor, which can convert a sensitive to a killer and resides in the cytoplasm, was named *kappa*. Additional studies showed that if a potential killer cell divides very rapidly, it may lose kappa. A cell can regain it and become a killer only by obtaining kappa from another cell.

Further crosses between killer and sensitive stocks showed that more than cytoplasm is involved in the

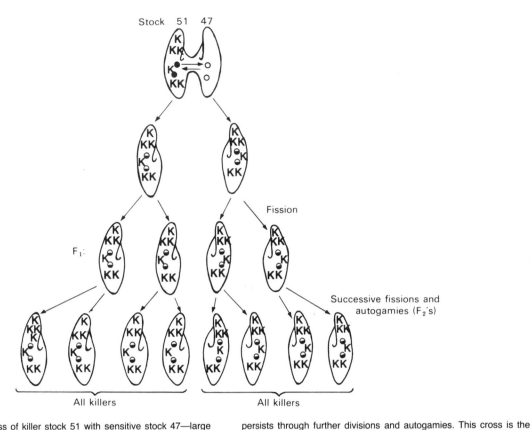

FIG. 20-4. A cross of killer stock 51 with sensitive stock 47—large bridge. When a large cytoplasmic connection forms between the two conjugating animals, the sensitive cell is then converted to a killer. Both ex-conjugant cells now give rise to killers, and the killing trait persists through further divisions and autogamies. This cross is the same as that in Fig. 20-3, except for the amount of cytoplasm exchanged.

inheritance of kappa. Let us next examine a cross between our killer stock 51 and another sensitive, stock 29. In an ordinary mating in which no large cytoplasmic bridge forms between the cells, it is found that the two ex-conjugants retain their original phenotypes, the killer giving rise to killers and the sensitive to sensitives by fission (Fig. 20-5). This is similar to the F_1 results shown in Fig. 20-3. However, a significant difference is seen when the F_2 results are compared. In this cross of 51×29, the sensitive F_1s continue to produce sensitives after self-fertilization, but the killer F_1s can give rise to either killers or to sensitives. A population of F_1 killers such as these will produce killer and sensitives in a ratio of 1:1. This 1:1 segregation of killer and sensitives is also seen when a large amount of cytoplasm is exchanged between two parents (Fig. 20-6). Only now, the original sensitive parent is converted to a killer and produces killer F_1s by fission; it also gives rise to an F_2 with a 1:1 segregation of killer and sensitive.

How do we explain the different results from the cross of this same killer parent (stock 51) with the two different sensitive stocks, 47 (the first example) and 29 (the second cross)? The 1:1 ratio produced in the second cross tells us that a pair of alleles is segregating. Apparently, the cytoplasmic factor, kappa, depends for its maintenance on a dominant (K); its recessive allele (k) cannot support kappa. Killer stock 51 must be homozygous (KK) and sensitive stock 29 homozygous recessive (kk) (Fig. 20-7A; compare with Figs. 20-5 and 20-6). When the two are crossed, the F_1s are heterozygotes that can support kappa, if it is present in the cytoplasm. At autogamy, however, meiosis occurs, and the alleles segregate. A haploid nucleus may receive either K or k. This means that half of the F_1 cells in a population will become KK after self-fertilization and will be able to support kappa; the other half become kk and cannot maintain kappa.

Sensitive stock 29 differs in its genic constitution from sensitive stock 47 in the first cross. Stock 47,

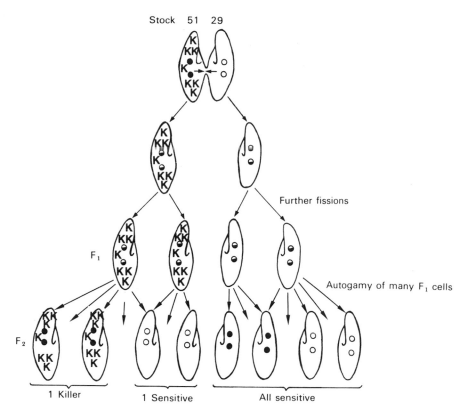

FIG. 20-5. A cross of killer stock 51 with a sensitive stock 29—small bridge. The results of the F₁ are the same as in the cross of 51 × 47; the killer cell gives rise to killers and the sensitive to sensitives. When the F₁ cells are allowed to undergo autogamy, however, half of the killer F₁s give rise to sensitives.

like killer stock 51, is homozygous KK (Fig. 20-7B). It is sensitive only because it lacks kappa. Once it acquires kappa through a cytoplasmic bridge, it can maintain it and is converted into a killer. We see from these crosses that an inherited characteristic may have a basis that is both chromosomal and cytoplasmic. To be a killer, a cell must have kappa in its cytoplasm. It also requires an allele K to support kappa, but this same allele does not enable the cell to make kappa. Kappa must be introduced into the cytoplasm. When kappa is present, it somehow converts a cell of the proper genotype to a killer by enabling it to liberate killing particles into the environment.

The nature of kappa and related particles

Various studies were undertaken to elucidate the nature of kappa. Preer found that X-rays could inactivate kappa when they were applied to killer cells. The dosage required to do this indicated that kappa should be of a size that can be resolved by the light micro-scope. Killer and sensitive cells were stained by various procedures that revealed the presence of large numbers of particles in the cytoplasm of killer stocks (Fig. 20-8A) These particles were never seen in sensitive stocks, and all the observations indicated that the cytoplasmic bodies were indeed kappa. Detailed examination later showed that the kappa particles in a killer are not all the same but two kinds exist—"brights" and "nonbrights" (Fig. 20-8B). The two are very similar except that the former type contains a refractile body that is evident with bright-phase contrast. Both the brights and nonbrights are bounded by a double membrane and contain appreciable amounts of DNA distributed throughout.

Killer cells have been followed cytologically throughout all phases of their life cycle. The observations have shown that the maintenance of the killing activity depends on both kinds of kappa particles. The brights seem to develop from the nonbrights. The nonbrights alone have the ability to divide; if they are lost, neither nonbrights nor brights can reappear. It

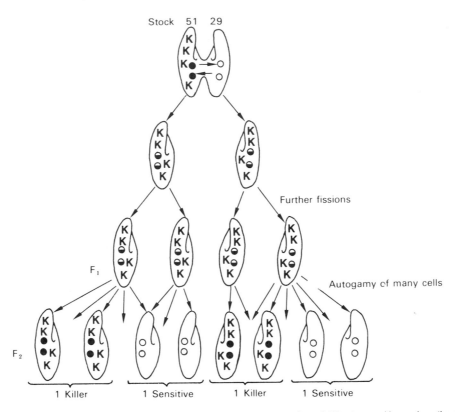

FIG. 20-6. A cross of killer stock 51 with sensitive stock 29—large bridge. When a large amount of cytoplasm is exchanged, the sensitive is converted to a killer, as is true when stocks 51 and 47 exchange a lot of cytoplasm. In this case, however, at autogamy, there is now a 1:1 segregation of killer to sensitives when the F_1s from both the original killer and the original sensitive are allowed to undergo the self-fertilization process.

is the bright, however, that is evidently responsible for the killing activity.

In some cells mutations have taken place in the kappa particles. One kind of mutant kappa is called *pi*. Cells carrying pi possess no killing activity and are also sensitive. Cytological examination shows that the pi particles all resemble the nonbrights; no brights are present. It seems that in these cells, brights cannot develop from the nonbrights. In ordinary killer cells, the disappearance of brights is followed by their reappearance. Many observations of this kind indicate that the bright particle is actually liberated from the body of the killer and that this particle is responsible for the killing action on a sensitive cell.

We can now understand some of the unusual features of the killing action that seemed so puzzling when it was first analyzed. The killing event is a result of the liberation of a discrete entity, a bright. This develops from a nonbright in the cytoplasm of a killer. But the killer must have a nuclear genetic factor

(K) to support the cytoplasmic kappa. These unusual features have raised many questions concerning the origin of kappa. It is certainly a genetic entity containing DNA. However, kappa is not essential to the life of the cells and is absent from most paramecia.

This suggests that is not native to the species but has arisen from another source, an idea supported by the fact that it cannot be made by a cell but must be introduced. As is shown by the electron microscope, the structure of kappa resembles bacterial cells in many respects. Moreover, cytochemical tests have demonstrated that kappa is a Gram-negative body and contains cytochromes that resemble those of bacteria and not those of any eukaryote, including Paramecium itself. Most authorities agree that kappa is an infective agent with definite relationships to bacteria.

Further aspects of the story raise additional questions about the significance of such an extra genetic element in the cytoplasm of a eukaryotic cell. For kappa has been found to be but one of numerous

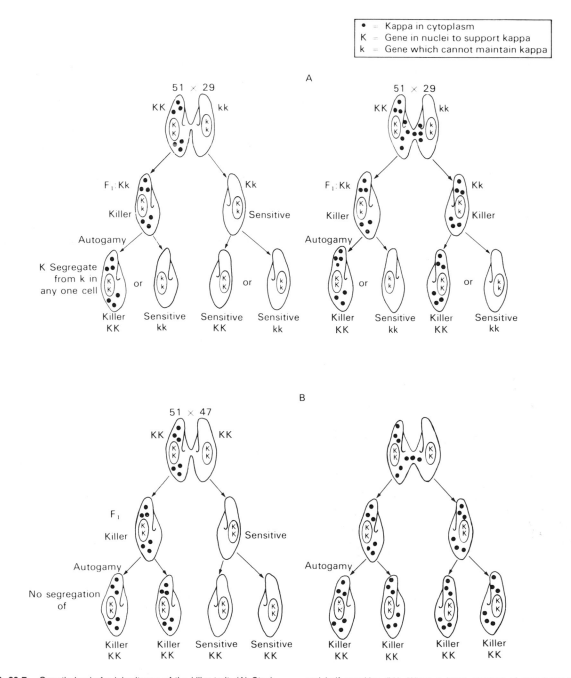

FIG. 20-7. Genetic basis for inheritance of the killer trait. (A) Stock 51 × 29—small bridge (left). Stock 51 is a killer because it possesses kappa in the cytoplasm and carries in its nuclear material allele K needed to support it. Stock 29 is sensitive, carrying the allele k, which cannot maintain kappa. A cross between the two makes all of the F_1 cells heterozyotes (Kk). However, if no kappa goes over to the sensitive mate, it will remain sensitive even though it has the K to support kappa. The F_1 cells from the killer cells produce a ratio of 1 killer to 1 sensitive at autogamy, because in any one cell that undergoes the self-fertilization, there is a 50:50 chance that the two haploid nuclei that unite will carry K or k. Therefore, half of the F_2 become killer (KK) and half sensitive (kk). When a large amount of cytoplasm is exchanged (right), kappa converts the sensitive to a killer. Kappa is supported, because the original sensitive cell is now genetically heterozygous (Kk). At autogamy, there will be a 1:1 segregation of killer to sensitive from this cell line, as well as for the other. The reason is the same as was just given. (B) Stock 51 × 47. Stock 47 is genetically identical to stock 51. It is sensitive only because it lacks kappa. Once kappa is received through a large bridge, all the cells remain killers. There is no recessive k present to give rise to sensitives if kappa is also present in the cytoplasm.

589

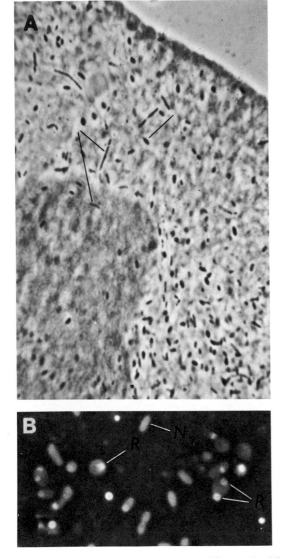

FIG. 20-8. Kappa particles. (A) Kappa in a stained Paramecium killer cell. The nucleus is in the lower left of picture and appears slightly darker than than the cytoplasm. The kappa particles are very numerous in the cytoplasm. Those in the nuclear area are actually on the surface of the nucleus. (B) Brights and nonbrights. These particles are from a freshly broken animal which has not been stained. The phase contrast microscopy reveals the refractile body (R) in the bright particle that is absent from the nonbright (N). (Reprinted with permission from J Cell. Sci. 5:65–91, 1969. Courtesy Dr. J. R. Preer, Jr.)

kinds of entities that may infect Paramecium. We have already mentioned pi, which appears to be a mutant form of kappa. Indeed, any Paramecium may be infected by some kind of extraneous entity whose maintenance depends on the genotype of the host cell. This fact reflects the close association that has

evolved in the relationship between the cell and the infective agent. It suggests that some sort of advantage may be conferred on the host cell by the presence of kappa and related particles, all of which may be thought of as symbionts.

If the presence of the symbiont does confer an advantage, any genetic changes in the host to support the infective particles would be selected by the forces of evolution. However, no advantage from these particles to Paramecium has yet been demonstrated other than the fact that cells with a symbiont are immune to any toxic effects it may produce. Most cells, however, would not seem vulnerable to the lethal effects, because a killing particle or a liberated toxin would find little chance in the natural environment of contacting a sensitive cell swimming by. Whatever the reason is for such a variety of kappalike entities in the cells of Paramecium, it raises interesting questions relating to the establishment of host–symbiont relationships and the benefits that can be derived from the presence of chromosomal and infective DNA in the same cell.

Extranuclear DNA in eukaryotes

Two organelles of eukaryotic cells, the mitochondrion and the chloroplast, both of which are essential to all aerobic life on earth, carry genetic information. All the eukaryotes that have been examined show DNA in mitochondria. Moreover, the chloroplasts from algae through flowering plants contain even larger amounts of DNA than are found in mitochondria. The electron microscope reveals that both the mitochondrial DNA (the mt DNA) and the DNA of the chloroplast (the ct DNA) are typically circular molecules. In some protozoans—Paramecium, for example—the mt DNA is linear. Both the mt DNA and ct DNA are double stranded and occur in their respective organelles unassociated with proteins and unbounded by any membrane.

Organelle DNA usually differs sufficiently in its G + C content from nuclear DNA that it forms a distinct satellite band when cellular DNA is extracted and subjected to density gradient ultracentrifugation. This makes it possible to isolate organelle from nuclear DNA for use in further studies. Organelle DNA undergoes semiconservative replication in the typical 5′ to 3′ manner, although the process may be modified for mt DNA. Both DNA strands do not necessarily begin replication at the same time. One strand, the

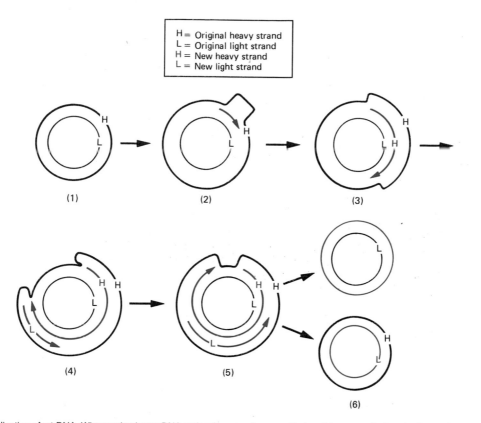

H = Original heavy strand
L = Original light strand
H = New heavy strand
L = New light strand

FIG. 20-9. Replication of mt DNA. When a circular mt DNA molecule *(1)* composed of H (heavy) and a L (light) strand undergoes replication, displacement of the H strand may occur, and a loop forms *(2)*. Synthesis of a new H strand then begins using the original L strand as a template. Displacement of the original H strand continues *(3)* as the new H strand increases in length. Synthesis of a new L strand then begins, using the original H strand as the template *(4)*. As synthesis on both original strands proceeds, H loop displacement continues *(5)*. Completion of DNA synthesis *(6)* results in two mt DNA molecules, one with a new H strand, the other with a new L strand.

heavy or H strand, becomes displaced and forms a displacement or D loop (Fig. 20-9). Replication occurs on the L (light) strand before replication on the H strand. Consequently, the replication is staggered, as the D loop becomes more pronounced. Finally, two circular duplexes are formed, each with an old and a new strand.

In mammals, the DNA composing a genome of mt DNA consists of approximately 16,000 to 17,000 base pairs, a tiny fraction of that found in the nucleus. The human mitochondrial genome has been completely sequenced and consists of 16,569 base pairs, a figure that falls within the size range for animals in general (13,500–18,000 base pairs). In higher plants, the mt genome is much larger, and great variation is found from one species to another, ranging approximately from 250,000 to more than 2 million base pairs! When algae and higher plant species are compared, the size of the chloroplast genome, the ct DNA, is seen to be

much less variable than that of the mt DNA, ranging approximately from 128,000 to 180,000 base pairs.

More than one copy of mt DNA is found in a mitochondrion, where it occurs in the matrix of the organelle (2–50 copies in yeast; 5–10 in mouse and rat cells). Even with the multiple copies, the amount of mt DNA in a mammalian cell is less than 1% of the total amount of the cellular DNA. In plants, however, with their larger mitochondrial genomes and in which 20 to 40 copies may exist, an appreciable percentage of the cellular DNA (15%) may be accounted for by the organelle.

Mitochondria and chloroplasts always arise by means of growth and division of the preexisting organelles. Replication of the ct and mt DNAs takes place throughout the cell cycle and is not confined to the S phase, when the nuclear DNA replicates. Moreover, in plants, the mt DNA and the ct DNA replicate throughout the cell cycle, completely independently of each other. In all eukaryotic cells, however, there must be some mechanism that keeps replication of the organelle DNA in check, since the *total* amount of organelle DNA just doubles during a cell cycle, resulting in a constant amount of organelle DNA from one cell generation to the next.

Extranuclear machinery for information transfer

In addition to their DNA, mitochondria and chloroplasts also possess components required for the synthesis of protein. Isolated mitochondria and chloroplasts can be supplied in the test tube with labeled amino acids. When this is done, the organelles incorporate the amino acid units, which can later be identified in certain protein fractions. Both kinds of organelles contain ribosomes, and these have been isolated and characterized. Ribosomes of mitochondria are typically composed of two subunits of unequal size. The intact ribosomes vary in sedimentation values from 55S (typical of multicellular animals) to 80S for the protozoan Tetrahymena (an exception whose ribosomes dissociate into two subunits of equal size). The ribosomes of chloroplasts, composed of two unequal subunits, have been found to be much less variable than those of the mitochondria. An S value of 70 has been obtained for chloroplast ribosomes from many sources. Their smaller and larger subunits have S values of 30 and 50, respectively. The chloroplast ribosomes closely resemble bacterial cell ribosomes.

The ribosomes of both kinds of organelles contain RNA, a smaller molecule in the small subunit and a larger one in the larger subunit. These ribosomal RNAs of chloroplasts are very similar to those of bacterial cells (16S and 23S). A 5S RNA is also present in the large subunit. The small and large rRNAs of the mitochondrial ribosomes, however, are much more variable in size and consequently cannot be said to be like those of bacteria. A 5S RNA has not been reported in the large subunit. The ribosomal units of both kinds of organelles contain an assortment of proteins, approximately 30 and 25 for the large and small subunits, respectively. These proteins are quite distinct from those that are constituents of the ribosomes found in the cytosol (the fluid portion of the cytoplasm, exclusive of the organelles) of the same cell.

In addition to DNA and ribosomes, the mitochondria and chloroplasts contain other parts of the machinery associated with replication and information

transfer. However, these are quite distinct from their counterparts found in the cytosol. These include DNA polymerase, tRNAs for the different amino acids, aminoacyl synthetases, and RNA polymerase. Messenger RNA has also been identified in mitochondria and chloroplasts. Both organelles possess a formylase, an enzyme that can convert methionine to *n*-formyl methionine, as well as an initiating tRNA, $tRNA_f^{met}$.

The electron microscope and biochemical procedures have demonstrated the presence of polysomes in mitochondria and chloroplasts. The ribosomes of the polysomes are held together by a thin strand that has been identified as RNA as a result of digestion by RNase. Polypeptides in various stages of growth have also been isolated from the polysome fraction of mitochondria. When the mitochondria from cells as diverse as those of Drosophila and human HeLa cells have been burst open and their DNA isolated, it has been found that polysomes are associated with the DNA. This indicates that translation is taking place while transcription is occurring from the DNA template. Putting the entire picture together, we can see many striking resemblances between the organelle systems and the protein synthesizing system of prokaryotes, such as the occurrence together of transcription and translation and the presence of formylase. Moreover, components of the organelle system respond similarly to those agents that inhibit nucleic acid and polypeptide synthesis in prokaryotes, whereas the counterparts in the cytosol are not affected. Very interesting is the fact that components of the organelle and prokaryotic protein-synthesizing systems (tRNAs, various enzymes) are interchangeable in in vitro investigations and can substitute for each other in translation. No such substitutions can be made between the organelle components and the equivalent parts in the cytosol. All the foregoing observations suggest that a certain degree of autonomy may reside in the mitochondria and chloroplasts of eukaryotic cells. Let us become familiar with the results of some investigations that bear on this point and have called attention over the years to the existence of extranuclear genetic determinants.

The petite mutations in yeast

Excellent genetic evidence accumulating over the years points to the mitochondrial DNA as the site of mutations responsible for certain traits that are inherited in a non-Mendelian fashion. The most thoroughly studied of these are the mutations in yeast

(Saccharomyces). Cells carrying a petite mutation grow slowly and form tiny colonies on agar, in contrast with the large ones of the wild phenotype. These petites have been shown to possess enzyme defects and to be deficient in aerobic respiration. They require a substrate containing a fermentable product such as glucose, which they use in the presence of oxygen as if they were growing anaerobically.

To understand the inheritance of the petite trait, we must be familiar with the very simple life cycle of yeast (Fig. 20-10). Like Neurospora, yeast is an ascomycete, or sac fungus. It is a unicellular organism, and the haploid cells can be classified into either of two mating types, + or −. The diploid zygote formed from fusion of a + and a − cell may grow by budding to produce a diploid colony. The diploid cells can also be stimulated to undergo meiosis. The cell then enlarges and forms four haploid nuclei, each of which becomes the nucleus of a spore. The meiotic cells behave like an ascus or sac, and so the four spores are considered to be ascospores.

We can see in Fig. 20-10 that two of the ascospores will be mating type + and two will be −. This 2:2 segregation indicates that the genetic determinants of mating type are nuclear and exist as one pair of alleles that segregate at the meiosis preceding ascospore formation. Similarly, many other characteristics known in yeast behave in the same way, such as the ability to produce adenine or some other essential metabolite. In all of these cases, 2:2 segregations are found when a single pair of alleles is followed. Linkage as well as independent assortment are easily detected in yeast and have permitted the construction of chromosome maps using the same reasoning followed for other organisms.

Mutations to the petite phenotype have arisen in both mating types, + and −. Some of these petites behave in the expected Mendelian fashion: a cross of a petite with a wild produces wild diploid cells; the ascospores yield a 2:2 segregation of wild type to petite. Petites such as these are called *nuclear petites*. In contrast to these are others that are decidedly non-Mendelian in their pattern of inheritance. One kind of nonchromosomal petite is known as the *neutral* or *recessive* petite. A cross of a neutral petite with a wild (Fig. 20-11) produces diploid cells that are normal in phenotype. When sporulation is induced, the ascus yields spores that produce only wild-type cells. The segregation is thus 4:0. The petite phenotype has disappeared and does not reappear when these wild cells are followed further in similar crosses. Clearly,

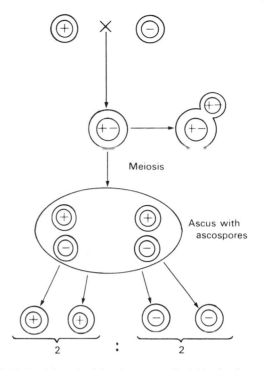

FIG. 20-10. Life cycle of Saccharomyces. Haploid cells of opposite mating types (+ and −) fuse to form a diploid zygote. This can give rise to a colony of diploid cells by budding. Diploid cells may also be stimulated to undergo meiosis. The cell enlarges, and four haploid nuclei result, each the nucleus of an ascospore. The mating type as well as other nuclear alleles segregate 2:2 from the ascus.

the neutral petite is not behaving as a Mendelian trait, as indicated in the departure from the 2:2 segregation in the ascus.

Experiments were performed in which yeast was treated with acriflavine. (It has been found that acridine dyes can eliminate F factors from *Escherichia coli* cells.) The results showed that almost a whole population of normal cells could be transformed to petite after exposure. No known mutagen can affect nuclear genes to such an extent that every exposed cell contains an induced mutation. Other observations of this type strongly indicated that the determinants for the petite phenotype reside in the cytoplasm.

Another class of petites was discovered in which the trait also behaved in a non-Mendelian fashion, but it differed in its pattern from the neutral or recessive petite. This is the *suppressive petite*. When it is crossed with a wild type, the results depend on when the zygotes are induced to sporulate (Fig. 20-12). If ascospore formation takes place very soon after the zygote forms, it is found that most of the asci will give a segregation of 0:4; that is, all the spores will give rise to petites.

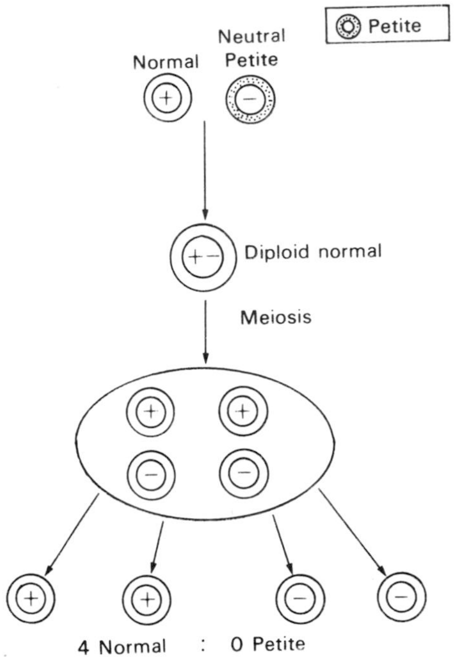

The zygotes, if immediately plated out on agar after the mating, form diploid colonies that are also petite. In contrast to the neutral petite, it is as if the wild type were tending to disappear. However, different results can be obtained from the same cross. Instead of being induced to sporulate immediately, the zygotes may be subcultured in liquid medium for a period of time. If the diploid colonies are then plated out, they are almost all wild type! Similarly, if these zygotes are induced to sporulate *after* being subcultured, the ascus gives a 4:0 segregation, four wild to no petite.

Thus, depending on the treatment after mating, two very different pictures may emerge. The appearance of the wild type after subculture in the liquid can be explained by the fact that the liquid culture favors the survival of wild-type cells. Any cells of normal respiratory phenotype that are present immediately after mating would be selected in the liquid medium, because it is not conducive to the growth of the petites. Therefore, wild-type cells would increase and take over the culture, giving normal diploids upon plating and 4:0 segregation after sporulation.

FIG. 20-11. Wild-type yeast cells and neutral petite. In a cross between the two types, diploids arise, and these are normal in phenotype. After meiosis, the mating types segregate 2:2 as expected, but all the resulting ascospores produce wild-type cells. The petite trait seems to have disappeared and is not behaving in a Mendelian fashion.

FIG. 20-12. Wild-type yeast × suppressive petite. The suppressive petite, unlike the neutral one, tends to express itself over the wild. The diploids are petite, and if diploid cells are induced to sporulate immediately, all the spores give rise to petite colonies. However, if diploids are subcultured and later sporulated, all the spores give rise to wild colonies.

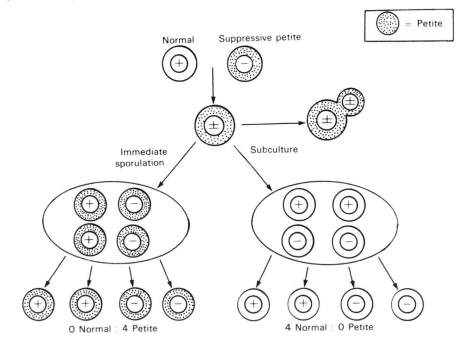

There are many other features of the petite story, but it should be evident from this short discussion that the inheritance of the neutral petite and of the suppressive petite is decidedly non-Mendelian. For many years it had been considered likely that the petite mutation is a result of some change in the mitochondrion, because the respiratory capacity of the petite cell is deficient. It has now been demonstrated by density gradient ultracentrifugation that the mt DNA of cytoplasmic petites has undergone some sort of alteration. The mt DNA of suppressive petites differs in its buoyant density from the mt DNA of wild-type yeast. There is a shift to a lower G + C value. These changes generally arise as deletions of portions of the mt DNA. The deletions differ when the mt DNAs from various strains are compared. Although usually about as much DNA exists in the suppressive petites as in the wild-type yeast cell, most neutral petites have been shown to lack mt DNA altogether! We can well understand why the cytoplasmic petites do not undergo spontaneous reversion to wild type as is the case for point mutations.

Inheritance of drug resistance in yeast

Another class of non-Mendelian mutations has been recognized in yeast. These bring about resistance to various antibiotics, such as erythromycin and chloramphenicol. After a cross of erythromycin-sensitive (wild type) to erythromycin-resistant (mutant) cells, the zygotes are allowed to undergo mitotic divisions. Repeated divisions eventually yield cells that are parental types, either erythromycin sensitive or resistant. When one of these diploid cells is allowed to sporulate, the four ascospores will be identical, either all sensitive or all resistant. Such results are decidedly non-Mendelian (Fig. 20-13). We can easily understand why this occurs if we consider the fact the yeast cell produces a bud when it divides mitotically. Originally, the zygote contains a mixture of both mutant (resistant) and wild (sensitive) mitochondria. However, when a bud is produced, only a few mitochondria enter it. Continued mitotic divisions of the diploid cells eventually lead to buds that contain either all resistant or all sensitive mitochondria. As a result of this mitotic segregation, when the diploid cell sporulates, the four spores produced will contain the same kind of mitochondria, either all resistant or all sensitive. Resistance to chloramphenicol follows the same pattern.

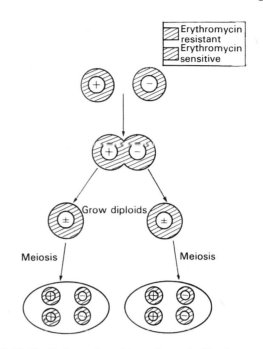

FIG. 20-13. Erythromycin resistance in yeast. After the cross of resistant × sensitive, diploid cells can be established from growth of the zygotes. Upon sporulation, the diploids give rise either to all resistant cells or to all sensitive ones. (The two kinds of diploid cells do not necessarily occur in a 1:1 ratio.)

If a cross is made between two strains, one resistant to erythromycin and the other resistant to chloramphenicol, it can be seen that recombination can give asci containing all doubly resistant spores and asci with all doubly sensitive spores (Fig. 20-14). This tells us that recombination of nonchromosomal genetic determinants can occur as well as recombination of nuclear alleles. Such a process adds even more diversity to the combinations of inherited material possible after sexual reproduction, and its full significance in all species is yet to be assessed. There is good evidence that the determinants of drug resistance in yeast also reside in the mitochondrial DNA. For example, when mutations to petite occur and produce a physical alteration in the mitochondrial DNA, markers for drug resistance may be lost. This implies that the physical change in the organelle DNA, the elimination of a segment, has brought about the petite phenotype and the simultaneous loss of the resistance determinants. The inheritance in yeast of the petite trait and of the resistance to certain drugs offers excellent evidence for the localization of specific nonchromosomal genes in the mitochondrion.

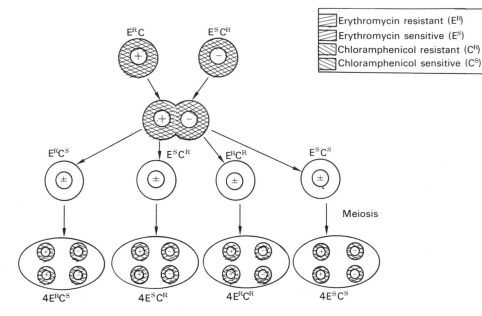

FIG. 20-14. Two-factor cross involving drug resistance. If chloramphenicol resistance is followed alone, it shows the type of non-Mendelian inheritance illustrated in Fig. 20-13 for erythromycin. If chloramphenicol and erythromycin resistances are followed together in a two-factor cross, it can be demonstrated that recombination of the non-Mendelian factors can take place. After the mating, diploid cells are cultured. From these, four types can be isolated (not necessarily in a 1:1:1:1 ratio). Each gives rise to ascospores, which are either all parental or all recombinant. The recombinants here are the E^RC^R and E^SC^S. (Modified slightly with permission from R. Sager, *Cytoplasmic Genes and Organelles*, p. 132. Academic Press, New York, 1972.

Nonochromosomal genes in Chlamydomonas

Very good evidence for the storage of genetic information in the chloroplast has been provided by studies of Sager with the unicellular green alga Chlamydomonas. The life cycle of this species is very simple and is comparable to that of yeast. Again two mating types, + and −, are recognized. Mating between cells of opposite mating type produces a zygote (Fig. 20-15). When the zygote matures, meiosis occurs, and four zoospores (the motile products of the meiotic division) arise. Half of these are mating type + and half are type −. The zoospores may undergo mitotic division to form clones of cells. Again, the mating type is seen to depend on a pair of nuclear alleles inherited in typical Mendelian fashion.

Many characteristics in Chlamydomonas are known to depend on nuclear genes: spore color and requirements for certain metabolites, for example. Both linkage and independent assortment have been demonstrated for these genes associated with Mendelian traits. There are, however, some genetic determinants in Chlamydomonas that are definitely nonchromosomal. Sager has worked extensively with certain strep-

tomycin-resistant strains that exhibit non-Mendelian inheritance. As Figure 20-16 shows, the results of a cross between streptomycin-resistant and streptomycin-sensitive strains depends on the phenotype of the mating type + parent.

Although mating type itself segregates among the zoospores in the Mendelian 2:2 fashion, the zoospores always have the streptomycin trait shown by the mating type + parent. This suggests maternal inheritance, but here, the + and the − cells are identical. Neither class of gamete contributes more cytoplasm and hence more cytoplasmic genes than the other. Something is operating to prevent the transmission of the nonchromosomal genes of the mating type − parent from the zygote to the meiotic products. Since the discovery of the nonchromosomal streptomycin determinants, many others have been found. Some of these were induced by mutagens, particularly by streptomycin itself, which in Chlamydomonas is able to induce mutations in nonchromosomal genes but not in those of the nucleus.

It was later found that rare exceptions occur to this mating type + pattern of inheritance. In fewer than 1% of the zygotes, exceptions arise in which both sets

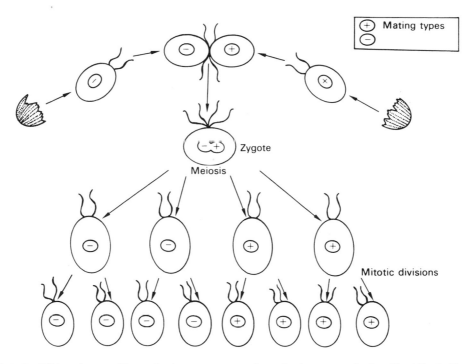

FIG. 20-15. Life cycle of Chlamydomonas. Two mating types occur. After fusion of two cells of opposite type, meiosis takes place, and the mating type segregates in a Mendelian fashion. The motile cells can divide mitotically to form clones of other identical cells.

FIG. 20-16. Inheritance of streptomycin resistance in Chlamydomonas. It can be seen that inheritance of sensitivity or resistance to steptomycin depends on the phenotype of the parent that was mating type +. Although mating type is segregating as expected, the cells are always like those of the + parent in relation to tolerance of streptomycin.

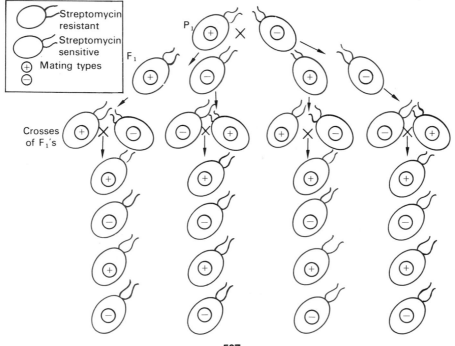

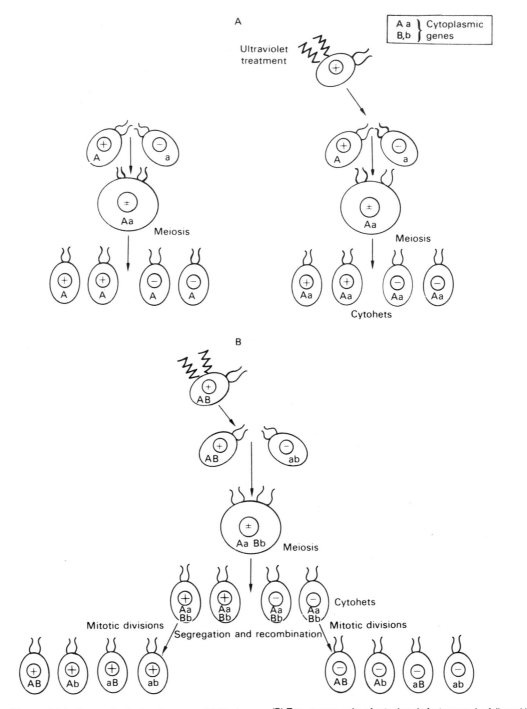

FIG. 20-17. Biparental inheritance of cytoplasmic genes. (A) Typically, the cytoplasmic genes of the mating type − parent are not transmitted from the zygote, so that the haploid products of meiosis, the zoospores, show the phenotype of the + parent (left). If, however, the mating type + parent is treated with ultraviolet light and then mated, the cytoplasmic genes from both parents may be transmitted to the zoospores (right). Although the nuclear factors (e.g., + and −) segregate at meiosis, the cytoplasmic ones do not. The F₁ cells thus have two complete sets of cytoplasmic genes and are called *cytohets*.

(B) Two or more pairs of cytoplasmic factors can be followed in cases of biparental inheritance. The haploid zoospores are cytohets. If each zoospore is then allowed to divide further to form a clone, segregation of the alleles takes place, as does recombination among the members of the different allelic pairs. Thus, nuclear alleles in Chlamydomonas segregate and undergo recombination at meiosis, whereas these processes do not take place for the cytoplasmic genes until the mitotic divisions of the haploid zoospores, after meiosis has occurred.

of nonchromosomal genes, the set from the mating type − as well as the one from the mating type + parent, are transmitted. Sager also found that if the mating type + parent is treated with a dose of ultraviolet light before mating, 50% of the zygotes exhibit biparental inheritance. All of the offspring arising from such a zygote contain a complete set of nonchromosomal genes from each parent. Such cells are called *cytohets*, indicating that they are heterozygous for cytoplasmic genetic determinants (Fig. 20-17A and B).

It is possible to follow more than one pair of cytoplasmic markers in biparental inheritance. The cytoplasmic markers do *not* segregate when meiosis takes place. The haploid zoospores are therefore heterozygous for the cytoplasmic genes. When a heterozygous zoospore divides mitotically and forms a clone, the alternative forms of the cytoplasmic genes segregate and undergo recombination. This segregation and recombination continues with each additional mitotic division of the zoospores until there are no more heterozygous markers to detect. The recombination frequency of these non-Mendelian genes has been measured. This has permitted the construction of a map and the assignment of several nonchromosomal genes to definite positions. It appears that the cytoplasmic genes are all part of a single circular linkage group.

Sager has presented several lines of evidence that support the hypothesis that the physical basis of the nonchromosomal genes (and hence the linkage group to which they belong) is the DNA of the chloroplast. For example, most of the cytoplasmic determinants that have been mapped were caused to mutate by streptomycin, a drug known to exert its influence almost exclusively on the development of the chloroplast, with no direct effect on other parts of the cell. Other work with Chlamydomonas continues to provide greater insight into the overall significance of nonchromosomal genes, as well as into the mechanism of uniparental inheritance.

Plastid autonomy in higher plants

Examples from higher plants have been known for years which show that the plastids possess a certain degree of autonomy. Rhoades studies the inheritance of a gene in corn known as *iojap*. When a plant is a homozygous recessive (ij ij), there is a tendency for plastids to become so altered that chlorophyll is not produced. Not all plastids necessarily undergo this change, but the tendency is sufficient to produce streaks of white tissue, giving the plant a striped or variegated appearance. Some seedlings appear colorless and do not survive because the majority of their plastids have been altered and lack chlorophyll. The results of a cross between a striped iojap plant and a normal green one differ, depending on the way the cross is made.

Figure 20-18A shows that the inheritance is strictly maternal and seems to depend on what the female gamete contributed. If the female parent is normal green (Ij Ij) and the male parent is striped (ij ij), all of the F_1 plants will be normal in appearance. On the other hand, if the female parent in a cross is striped (ij ij) and the male parent is green (Ij Ij), three phenotypes occur among the F_1: normal green, striped, and colorless. It thus seems that the genotype of the male parent is ineffective in the expression of the iojap trait. The inheritance is maternal when one of the parents is of the genotype ij ij.

Particularly significant is the result of backcrossing (Fig. 20-18B). In a backcross, if a striped F_1 is used as the female parent and is crossed with a homozygous wild (Ij Ij), three classes of progeny again arise: green, striped, and colorless. This means that plants heterozygous or homozygous for the wild-type allele (Ij ij or Ij Ij) may still show the mutant effect. Some plastids have been altered, but they are obviously reproduced and carried along in this changed condition. The original mutation (ij) occurred in a nuclear gene, but once the alteration in the plastid has arisen, it becomes established and persists; it no longer depends for its perpetuation on the recessive allele.

There is obviously some degree of plastid autonomy. This is clearly indicated by the striped and colorless phenotypes that persist in homozygous wild plants (Ij Ij). The genus Oenothera also has given evidence for a certain amount of plastid autonomy, although nuclear genes also exert an effect on plastid development. Let us now examine some of the information assembled from studies on the molecular level that have cast light on the interactions between organelles and the rest of the cell.

Organelle genetic systems

As discussed earlier in this chapter, organelles have the ability to replicate their DNA and they possess the machinery for protein synthesis. We may now wonder about the precise kinds of information encoded that may be in the organelle DNA and the

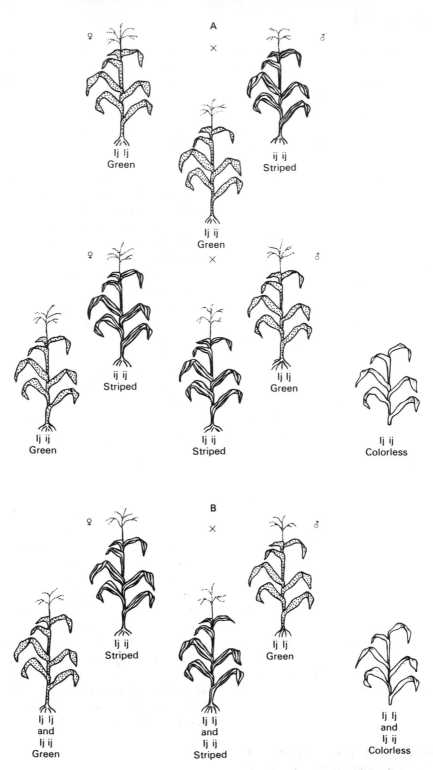

FIG. 20-18. lojap mutation and plastid inheritance. (A) Plants that are homozygous for the recessive lojap will carry altered plastids and appear variegated or striped with white streaks. If an lojap plant is used as a male parent *(above)*, the offspring are green, but the reciprocal cross produces three classes of offspring (below). The altered plastid condition can evidently be transmitted by the female gamete but not by the male. (B) When a striped heterozygous plant is used as the female parent and crossed with a homozygous normal, offspring arise that have altered plastids. This means that individuals homozygous for the wild allele (lj lj) are expressing the mutant effect and may be either striped or colorless. The original plastid alteration resulted from the presence of the mutant nuclear allele in homozygous condition (ij ij), but the plastid, once changed, can still reproduce itself even though the wild genotype is present. The plastid thus has a certain amount of independence.

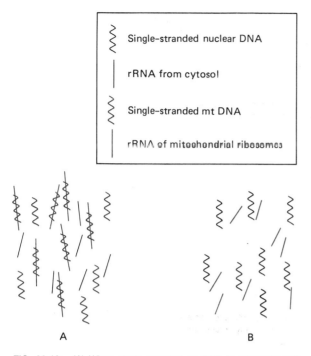

Single-stranded nuclear DNA

rRNA from cytosol

Single-stranded mt DNA

rRNA of mitochondrial ribosomes

A   B

FIG. 20-19. *(A)* When single-stranded mt DNA is presented with RNA derived from the cytosol along with mitochondrial rRNA, the only DNA–RNA hybrid molecules that form are those between the mt DNA and the rRNA of the mitochondrial ribosomes. Here, for example, the rRNA of the ribosomes from the cytosol does not compete with the rRNA of the mitochondria for sites on the mt DNA. *(B)* No DNA–RNA hybrid molecules form when the nuclear DNA is presented with rRNA from the mitochondrial ribosomes of the same cell.

kinds of interactions that may take place between the organelle system and the nucleocytoplasmic component of the cell.

Organelle ribosomes can be isolated and then separated into their RNA and protein components. The rRNA from mitochondria or chloroplasts may then be presented with the DNA from the same kind of organelle in the presence of RNA from some other source, such as rRNA from the cytoplasmic ribosomes. When this is done (Fig. 20-19), it is found that all the rRNA from mitochondria forms hybrid DNA–RNA molecules only with DNA from mitochondria. Other kinds of RNA from outside the organelle do not compete with it, which demonstrates that the "foreign" RNAs do not contain a significant number of nucleotide sequences in common with the mt DNA. The same is true for the rRNA and the DNA of the chloroplast. The rRNA from an organelle (mitochondrion or chloroplast) does not form hybrid molecules with the DNA taken from the nucleus of the same cell (Fig. 20-19*B*). Results from such DNA–RNA

hybridization studies leave no doubt that the chloroplast and the mitochondrion each contains information in its DNA coded for the specific rRNA of the organelle.

Other DNA–RNA hybridization studies that involve tRNA from yeast have indicated that all the mitochondrial tRNAs of yeast are transcripts of genes found in the mt DNA. These are tRNAs that represent all 20 amino acids. In contrast, most of the genes in yeast that code for the proteins of the ribosome and for the aminoacyl synthetases have been found to exist in the cell nuclei. These proteins are synthesized on the ribosomes found in the cytosol and then move into the organelle. The ribosomal proteins finally associate with the rRNA made in the mitochondrion to form the complete functional ribosomes of the organelle. The same holds for the chloroplast.

It is not unexpected to find that information for many of the parts of the mitochondria or chloroplasts is coded in genes of the cell nucleus. This is so because many different nuclear gene mutations are known to affect the structure of the mitochondria or the chloroplasts. Moreover, both organelles contain hundreds of different kinds of proteins (ribosomal proteins, membrane proteins, enzymes), and the amount of organelle DNA is simply not sufficient to code for all of these.

In addition to the genes of the mt DNA in yeast that code for tRNA and rRNA, a second important class has been recognized. These genetic determinants code for polypeptides that function in electron transport and oxidative phosphorylation. Their existence was established by using mutant cells in which the mt DNA had been altered by deletions. Obviously, since neutral petites contain no mt DNA, the absence of any genes in such strains would indicate that their genetic information resides in the mt DNA. Since the amount of DNA lost in the different suppressive petites is variable, the various strains of suppressive petites can supply valuable information on the proximity of two or more genes as well. For example, if two genes are lost together when a given deletion occurs in the mt DNA, then they must have been positioned close to each other on the mt DNA (Fig. 20-20).

Recognition of extranuclear point mutations in yeast (e.g., those affecting drug resistance and the cytochrome oxidase complex) was extremely valuable in pinpointing genetic sites on the mt DNA. Unlike the petites, which cannot revert back to wild type because of deleted DNA, these mutations can revert. The point mutations exist in different allelic forms, and they

Type of cell	Status of mt DNA	Status of mitochondrial genetic factors

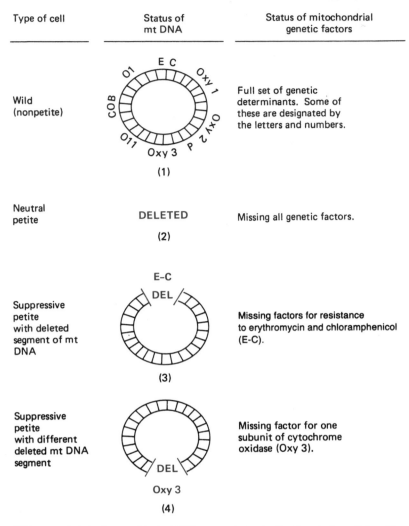

Wild (nonpetite)	(1)	Full set of genetic determinants. Some of these are designated by the letters and numbers.
Neutral petite	DELETED (2)	Missing all genetic factors.
Suppressive petite with deleted segment of mt DNA	(3)	Missing factors for resistance to erythromycin and chloramphenicol (E-C).
Suppressive petite with different deleted mt DNA segment	(4)	Missing factor for one subunit of cytochrome oxidase (Oxy 3).

FIG. 20-20. Nonpetite, wild-type yeast strains have an intact mt DNA and carry a complete set of genetic determinants. *(1)*. These genetic factors are absent in neutral petites, which also lack all the mt DNA, indicating that these missing genetic elements reside on the mt DNA *(2)*. Two different suppressive petites *(3, 4)* carry deletions in the mt DNA that differ in size and location. The genetic factors that are missing in the two strains will be different. Those factors closer together on the mt DNA will have a greater chance of being lost together when a deletion of the mt DNA occurs. By studying different deletion mutants in this way, the positions of the mt genes can be estimated.

have made it possible to recognize nonchromosomal loci that undergo recombination and can be analyzed genetically, as described earlier in this chapter for drug resistance in yeast. Moreover, acridine dyes can induce the origin of petites, all of which have defects of aerobic respiration. However, petites do differ in their extent of DNA loss. Yeast strains with different point mutation markers can be treated with an acridine mutagen to convert them to petites differing in the deletion of their DNA. As mentioned, the higher the frequency of simultaneous loss of two markers (or, conversely, their retention) in the different petite strains, the closer together the two genes must be. Mutant forms also known in Paramecium and the mold Neurospora are believed to be the result of alterations in mt DNA.

Translation in organelles

Although genetic information is present in organelles, not enough of it is there to code for most organelle characteristics. Much information coding for features

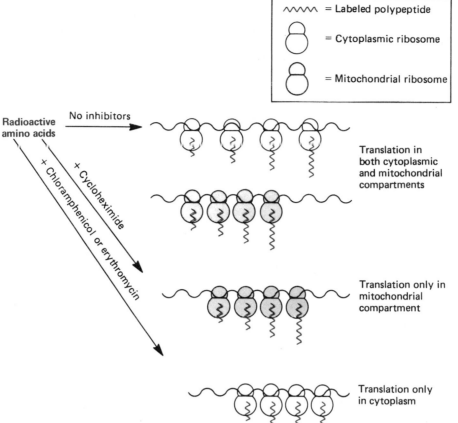

∿∿∿ = Labeled polypeptide

= Cytoplasmic ribosome

= Mitochondrial ribosome

Radioactive amino acids

No inhibitors

+ Cycloheximide

+ Chloramphenicol or erythromycin

Translation in both cytoplasmic and mitochondrial compartments

Translation only in mitochondrial compartment

Translation only in cytoplasm

FIG. 20-21. (Top) When labeled amino acids are supplied to cells, labeled protein appears in all parts of the cell. Both the cytosol and mitochondrial translation systems are operative, and the polypeptides produced in each compartment are labeled. (Middle) When cycloheximide is added, translation is shut off in the cytosol but remains unaffected in the organelles. Label can appear only in those polypeptides made in the organelle. (Bottom) When the antibiotics are added, the organelle translation system is shut off, but the cytosol apparatus is intact. Label appears only in those polypeptides synthesized in the cytosol.

of the organelles must reside in the cell nucleus. It has been possible to sort out some of the interactions that take place between the organelle and nucleocytosol systems. When labeled amino acids are supplied to cells, the newly made proteins are radioactive and are found in all parts of the cell, including the mitochondria. How can we determine whether a protein found in the mitochondria was made in the organelle or in the cytosol and then transported into the organelle? Since the 1960s investigators have been distinguishing between these two alternatives through the use of drugs that selectively shut off either the protein-synthesizing apparatus in mitochondria or the protein-synthesizing machinery of the cytosol. For example, cycloheximide can inhibit protein synthesis by ribosomes in the cytosol. It does not affect the ribosomes of the organelle. Therefore, if one finds

labeled protein that has arisen in cells supplied with labeled amino acids and grown in the presence of cyclohexamide, it is evident that the protein was assembled on mitochondrial ribosomes. Conversely, chloramphenicol and erythromycin can affect the mitochondrial ribosomes and shut off any translation that may result from their activities. Any labeled polypeptides formed in this setup must have been made on ribosomes in the cytosol that remain unaffected by those antibiotics (Fig. 20-21).

From such selective inhibition studies, significant information has been obtained on the site of origin of various polypeptides found in the mitochondria. The proteins of the mitochondrial ribosomes must be made by the ribosomes of the cytosol in yeast. No labeled amino acids are found in the ribosomes of yeast mitochondria following cycloheximide treatment, which

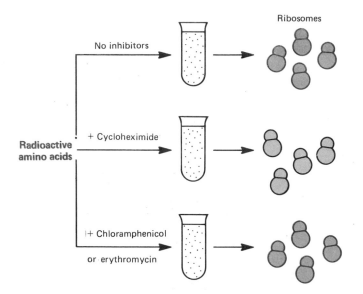

FIG. 20-22. (Top) When labeled amino acids are supplied to cells in the absence of inhibitors, the isolated mitochondrial ribosomes contain labeled protein, whereas (middle) no label is found in these ribosomes when cycloheximide is present. However (bottom), the mitochondrial ribosomal proteins are labeled in such an experiment when one of the antibiotics is present. Because cycloheximide shuts off the translation system in the cytosol, the results indicate the protein of the mitochondrial ribosomes is formed in the cytosol, since label appears in ribosomal protein only when the cytosol compartment is operative (compare with Fig. 20-21).

shuts off the ribosomal activities in the cytosol (Fig. 20-22). Conversely, when yeast cells are treated with chloramphenicol or erythromycin, labeled amino acids appear in all the proteins of the mitochondrial ribosomes, indicating that they must have been synthesized on the ribosomes of the cytosol and then shipped into the organelle. Once there, they assemble with rRNA made in the organelle to form an essential part of the functional mitochondrial ribosomes.

Such procedures have been employed to study the complex enzyme ATPase, which is bound to the inner membrane of the mitochondria. The ATPase of the mitochondria is composed of 10 polypeptide subunits and functions in oxidative phosphorylation coupled to electron transport. Some of the ATPase subunits are made in the cytosol, others in the organelle. In yeast, four are mitochondrial in origin, coded by the mt DNA; in Xenopus three are mitochondrial, and in Neurospora only two. It has also been shown that in the chloroplast some of the subunits of this enzyme are made in the organelle.

Another complex enzyme of the inner mitochondrial membrane, cytochrome oxidase, consists of seven polypeptide chains. Selective inhibition studies have shown that in yeast the three largest ones are made in the mitochondria and the four smaller ones in the cytosol (Fig. 20-23). Again we see cooperation be-

tween the cytosol and the organelle systems in the formation of a complex protein. When mt DNA is altered by a deletion, so that the genes coding for the three subunits of cytochrome oxidase are lost, as in petites, transcripts of the missing genes are consequently absent, and the three subunits cannot be formed in the mitochondria. The four smaller subunits *are* formed, however, because nuclear DNA is intact, but no functional enzyme is assembled inside the organelle (Fig. 20-24). All the results clearly show that the three larger subunits of cytochrome oxidase are coded by mt DNA and that the transcripts for them are translated in the organelle.

One of the seven or eight subunits of the complex coenzyme QH2-cytochrome c reductase is translated in the mitochondria of yeast and Neurospora. All the others are synthesized in the cytosol. We see another good example of cooperation between the translation systems of the mitochondria and cytosol in forming a complex mitochondrial component. What is the reason for the separate origins in the cell of certain essential mitochondrial components? It has been noted that those polypeptides made in the mitochondria, such as the subunits of the complex enzymes, contain large numbers of nonpolar amino acids, which make them hydrophobic and hence less able to pass from the cytoplasm through the membranes of the orga-

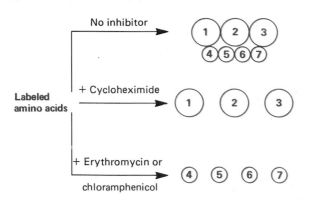

FIG. 20-23. The enzyme, cytochrome oxidase of yeast, is composed of seven polypeptide subunits, three large and four small. When cells are supplied with labeled amino acids in the absence of inhibitor (top), cytochrome oxidase is formed, and all seven subunits are labeled. In the presence of cycloheximide (middle,) no complete enzyme is made. Only the three large subunits are synthesized and are thus labeled. In the presence of one of the antibiotics (bottom), again no complete enzyme is made, but now the four smaller subunits are formed and are labeled. No large ones are present. Since the translation machinery of the cytosol is shut off with cycloheximide, the three subunits that appear in its presence must have been made in the mitochondria (middle). The four that appear in absence of cycloheximide but when the mitochondrial machinery is turned off (bottom) must have been made in the cytosol.

FIG. 20-24. When wild-type yeast cells are supplied with labeled amino acids in the absence of any inhibitors, the complete cytochrome oxidase enzyme is formed, composed of its seven polypeptide subunits. In petites with deleted mt DNA, no intact enzyme is formed. The three large subunits are missing. These are also missing when wild yeast is subjected to chloramphenicol or erthromycin, as seen in Fig. 20-23. The results tell us that the genetic information for the three large subunits (always absent in the petites) is coded in the mt DNA and that they are synthesized in the organelle. The four small subunits, always present in petites (and when the machinery of the cytosol is operative) are coded by the nuclear DNA and synthesized on the ribosomes of the cytosol.

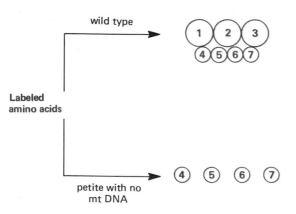

nelle. Such hydrophobic molecules can more easily become associated with the inner mitochondrial membrane if they are made on the ribosomes within the mitochondrion itself. Those polypeptides made in the cytosol are more hydrophilic and can readily pass through the outer mitochondrial membrane to associate with the inner membrane. This explanation may not be completely satisfactory, however, because certain hydrophobic polypeptides of some mitochondria are now known to be synthesized on ribosomes of the cytosol. For example, a hydrophobic subunit of ATPase in Neurospora is made in the cytoplasm and evidently passes easily into the organelle to form the complex enzyme. This same subunit in yeast is synthesized in the mitochondria.

Mitochondrial genomes of the human and yeast

In yeast, mutants with deletions in the mt DNA were shown to be defective in their cytochrome oxidase, ATPase, and coenzyme QH2-cytochrome c reductase. Genetic analysis of the circular yeast mt DNA, as noted earlier in this chapter, has entailed the simultaneous loss and retention of point mutations in different strains of petites as well as their recombination values in two- and three-point test crosses. The efforts made possible the construction of a genetic map. Mitochondrial genes have also been positioned on a physical map with the aid of restriction enzymes, in a manner similar to that described in Chapter 18. Such analyses led the way to identification and mapping of mitochondrial genes that are encoded for specific products. The development of methods permitting sequencing of the mt DNA has reinforced the earlier approaches and led to a more precise mapping of both the yeast and human mitochondrial genomes (Fig. 20-25).

The 16,569 base pairs of the human mt DNA includes genes for two ribosomal RNAs, 22 tRNAs, and 5 membrane proteins. In addition, eight regions appear to be reading frames encoded for polypeptides, but no products have yet been associated with these sequences. On the human mt DNA, the genes are packed so closely together that there appears to be little room left for noncoding regions between them. Moreover, not only do the genes themselves lack introns, but in some regions genes actually overlap, the same nucleotides being read in different frames in two different genes. There also appears to be a scarcity of nucleotide sequences to serve as promoters.

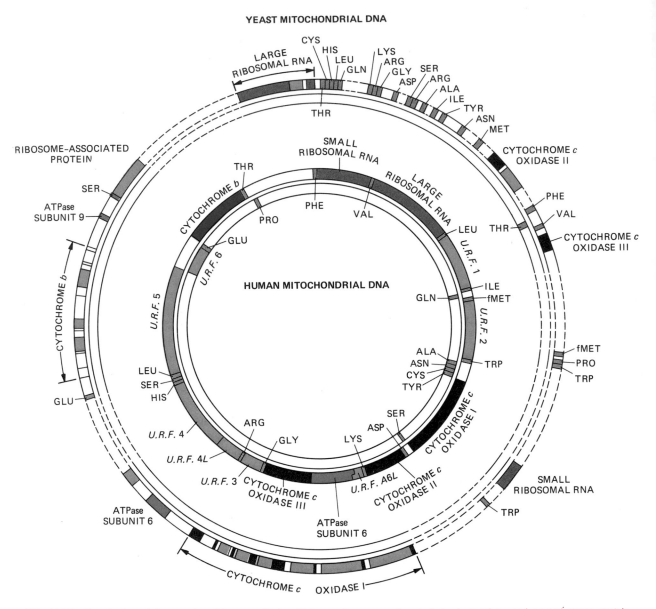

FIG. 20-25. Organization of the yeast and human mitochondrial genomes. The mitochondrial DNAs of yeast and the human are shown as concentric double circles. The two single circles of each double circle represent the two strands of the double helix. All darkened segments on the circles indicate coding regions, such as those for specific proteins, rDNAs, and tRNAs. Each abbreviation for an amino acid indicates the location of the gene for that specific tRNA. The designation U. R. R. indicates a reading frame that is as yet unassociated with any known product. In yeast, the mt genome contains long, noncoding *(colorless)* stretches, and several genes contain introns, some of which in turn contain unassigned reading frames. The human mt genome contains little noncoding DNA between genes, and the entire nucleotide sequence of the human mitochondrial genome is known. In the case of the yeast mitochondrial genome, the broken lines indicate regions that have not yet been sequenced. In actuality, the yeast mitochondrial DNA molecule is five times the length of that of the human. (From L. A. Grivell; *Mitochondrial DNA.* Copyright © 1983 by Scientific American, Inc. All rights reserved.)

In contrast, not only do the eukaryotic nuclear genes contain introns, but the transcription of each gene is controlled by its individual promoter, which interacts with the enzymes of transcription. On the human mt DNA only one major promoter region exists on each of the two strands of the circular DNA. This means that when transcription occurs, the full length of a strand is transcribed. Each long transcript is then processed to yield the tRNAs, rRNAs, and mRNAs. One of the transcripts (that made on the H strand) gives rise to 2 rRNAs, most of the tRNAs, and about 10 mRNAs. Each mRNA carries a poly(A) tail about 55 nucleotides in length. This tail is added after transcription has taken place through the action of a mitochondrial poly(A) polymerase. The messenger RNAs, which lack the 5′ cap typical of eukaryotic mRNA in the cytosol, are then ready for translation. Upon processing, the other long mitochondrial transcript (that made on the L strand) yields eight tRNAs and only one small RNA with a poly(A) tail. Apparently, 90% of the transcript is devoid of information and becomes degraded.

The cleavage of a large mitochondrial transcript into separate smaller ones occurs while the RNA is being formed on the mt DNA template. An interesting observation is that sequences coded for tRNAs exist on each side of a longer gene and that cleavage of the long transcript takes place exactly at the beginning and end of each tRNA sequence. Since there is little noncoding DNA in the human mitochondrial genome, these tRNA sequences would seem to be acting as recognition sites for the enzymes that process each long mitochondrial transcript. The tRNA sequences may accomplish this by folding into a shape that the processing enzymes recognize.

Unlike nuclear transcripts, the mitochondrial mRNAs lack the leader sequences to which ribosomes bind at the start of translation. The ribosomes of human mitochondria are able to recognize certain codons for methionine (more on this later) as start signals and manage to bind to the transcripts without the aid of leaders. Moreover, sequencing both the human mt DNA and its transcripts has shown that most of the genes have incomplete chain-terminating codons and also lack sequences coded for the trailer. For example, a sense strand may have the sequence -AT at its 5′

end (Fig. 20-26). Its transcript therefore terminates in -UA, an incompletion of the chain-terminating codon UAA. The problem is solved by the posttranscriptional addition of the poly(A) tail mentioned earlier. Therefore, the transcript is extended to include the sequence UAA, a chain-terminating codon. In the same way, posttranscriptional extension enables a transcript ending in a single -U to include the UAA chain-terminating sequence.

Yeast mitochondrial DNA, which is about five times the length of human mt DNA, has been mapped for genes encoded for 2 rRNAs, about 30 tRNAs, 6 membrane proteins, and several other proteins yet unidentified. The yeast mitochondrial genome contrasts sharply with the human in many respects. Unlike human mitochondrial genes, those of yeast are scattered, separated by long noncoding regions, 95% of which are composed of A:T base pairs. The mt genes of yeast *do* code for chain-terminating codons in the transcript as well as noncoding trailer sequences. The yeast transcripts do *not* have poly(A) tails added posttranscriptionally. Some mitochondrial mRNAs have been found to have leader sequences that are much longer than the sequences that code for the product. Moreover, some mitochondrial genes in yeast are split by long introns. The genes for cytochrome b and for subunit I of cytochrome oxidase contain almost 10 times the amount of DNA needed to code for the protein product. (Yeast strains vary in the number and size of their mitochondrial introns.) We can certainly conclude that much of the yeast mt DNA, unlike that of its human counterpart, is noncoding.

A very interesting discovery was made relating to one of the introns found in the yeast mitochondrial gene for cytochrome b. Mutations that arise in this intron influence the splicing of the RNA transcript of

FIG. 20-26. Incomplete chain-terminating codons of human mitochondrial transcripts. Mitochondrial genes in the human are typically very closely packed, the end of a gene next to a sequence coded for tRNA, as depicted above for the gene on the left and one on the right. Most of the genes do not contain complete chain-terminating codons.

As a result, the transcript formed on the sense strand ends in UA *(left)* or simply U *(right)*. Following transcription, an enzyme adds a long poly (A) tail (red). This extension of the transcript brings about completion of the chain-terminating codon, UAA.

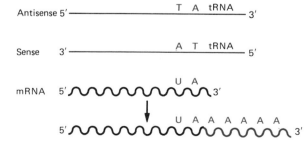

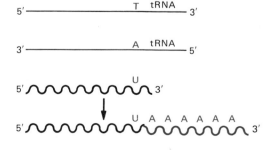

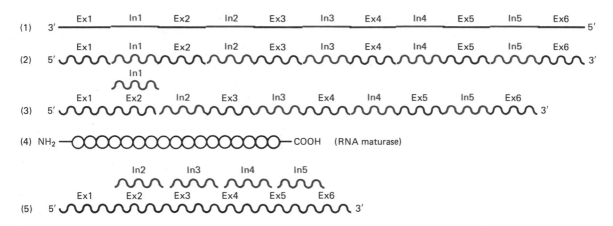

FIG. 20-27. The cytochrome b gene of yeast mt DNA. The gene includes five introns and six exons. When the sense strand *(1)* undergoes transcription, a transcript is formed *(2)* containing the introns and exons. Intron 1 is then removed *(3)*, joining together exons 1 and 2 and putting their coding sequences in phase with those of a reading frame in intron 2. Translation of this portion of the transcript produces a short protein, RNA maturase *(4)*, which aids in the removal of intron 2. Final excision and splicing removes the remaining introns, joins together the six exons *(5)*, and brings about the formation of the transcript for cytochrome b.

FIG. 20-28. (A) A segment of a DNA strand may be sequenced and the order of its nucleotides established. A DNA sequence can then be related to some polypeptide whose amino acid sequence is known. For example, if the DNA strand is part of the sense strand for the polypeptide, it is possible to recognize a region where the sequence of DNA triplets (and hence the RNA condons complementary to them) corresponds to the sequence of amino acids in the polypeptide. If such a correspondence does exist between the two, the DNA would appear to be coded for that particular amino acid sequence. (B) Apparent exceptions to the reading of RNA codons in translation have been uncovered. This is seen where a particular DNA triplet (and hence the RNA codon complementary to it) corresponds in position to an unexpected amino acid. The RNA codon, UGA, is a chain-terminating codon, but in the case of certain mitochondrial systems, UGA appears to be read as tryptophan. Cases have also been found in which AUA (for isoleucine) appears to be read as methionine (usually associated only with the codon AUG). The sequences shown here are purely illustrative and do not depict any actual sequences, in DNA or a polypeptide.

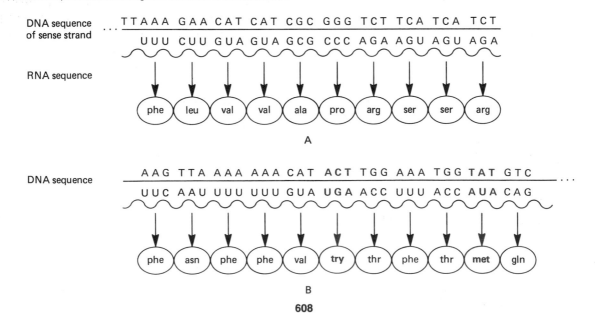

that gene. Research on the cytochrome b gene revealed the following picture (Fig. 20-27). The gene includes five introns and six exons. Transcription gives rise to a long transcript containing the introns and exons. Intron 1 is soon removed through the activity of a splicing enzyme encoded by a nuclear gene. As a result of the removal, exons 1 and 2 become linked together and to intron 2. This intron has been found to include a portion that contains a reading frame. This reading frame is now in phase with the reading sequences of the first two exons. These joined sequences (exons 1 and 2 and the reading frame of intron 2) act as an mRNA, which is translated into a protein. The first 143 amino acids of this protein, encoded by exons 1 and 2, are those of cytochrome b. The carboxyl end of this protein is composed of amino acids encoded by the reading frame of intron 2. This odd protein has been called an *RNA maturase*. It has the ability to aid in the catalytic removal of intron 2. By so doing, it limits its own concentration in the organelle because it destroys its own mRNA!

The discovery of such an unusual intron is significant, because the introns of chromosomal genes have never been found to code for anything. Introns 3 and 4 of the cytochrome b gene also contain reading frames, as do four of the introns of a gene that codes for a unit of cytochrome c oxidase. There is reason to suspect that these introns may also be partly encoded for intron-specific maturases. Such unexpected findings again raise the question of the true purpose of introns.

Mitochondrial translation and the genetic dictionary

The reading of RNA codons in the messenger RNA during translation has been thought to be invariable, a particular triplet always designating the same amino acid. The reading is precisely the same in viruses, bacteria, lower eukaryotes, plants, and animals. Consequently, the genetic code has been considered to be universal. Violations of this concept have now come to light from studies of mitochondrial DNA sequencing. When a nucleotide sequence for a segment of DNA has been determined, it can be compared with the amino acid sequence of the polypeptide for which it is encoded. The two sequences can be related, a triplet of nucleotides for an amino acid (Fig. 20-28A). Such comparisons between the nucleotide sequences

TABLE 20-1   Comparison of codon recognition by mammalian and yeast mitochondrial translation systems

CODON	UNIVERSAL CODE	CODE IN MITOCHONDRIAL TRANSLATION SYSTEM	
		Mammals	Yeast
UGA	Chain terminating	trp	trp
AUA	ile	met	met
AGA	arg	Chain terminating	arg
AGG	arg	Chain terminating	arg
CUU	leu	leu	thr
CUC	leu	leu	thr
CUA	leu	leu	thr
CUG	leu	leu	thr

of mitochondrial genes and the amino acid sequences of polypeptides for which they are encoded has brought to light several exceptions to the reading of the genetic code (Fig. 20-28B; Table 20-1).

In the mitochondria of yeast and mammals, the codon UGA is not recognized as a chain-terminating signal, as is the case in the cytosol. Instead, it is read as tryptophan in addition to the usual UGG codon for tryptophan. These mitochondria also recognize AUA as methionine instead of isoleucine. However, not only may codon reading in the mitochondria depart from the "universal" genetic dictionary, the reading of codons can actually vary among different organisms. In corn, for example, CGG is read as an extra codon for tryptophan instead of UGA. Moreover, while the mammalian mitochondria read AGA and AGG as chain-terminating codons, yeast mitochondria translate these as arginine, following the rules of the universal code. Human mitochondria, however, follow the genetic dictionary by translating the four codons beginning with CU as leucine, but yeast mitochondria recognize these as threonine (Table 20-1).

In the mammalian mitochondria, only 22 tRNAs are found, whereas no fewer than 32 should occur according to the wobble hypothesis (Chap. 13). It appears that although certain mitochondrial tRNAs *do* follow the wobble rules, others do not. No one kind of tRNA should be able to recognize four different codons, but that is exactly what seems to be the case for certain tRNAs in human mitochondria. These particular tRNAs all have U in the wobble position. Somehow, this U may manage to pair with all four possible bases at the 3' position in the codon. It is also possible that the anticodons of these mitochondrial tRNAs ignore the base on the codon at the 3' position and recognize only the first two. It is not known what mechanism enables some mitochondrial tRNAs such as these with U at the wobble position to read all four 3′ positions in the codons (U, C, A, and G), whereas other mitochondrial tRNAs with U at this position follow the wobble rules and pair only with codons ending in G and A. Since many mitochondrial tRNAs have shapes that differ from the typical cloverleaf configuration, it is possible that the structure of the tRNA somehow influences the specificity of the anticodon.

The unusual mitochondrion, with its various exceptions to the nucleocytosol "rules," as seen in this and previous sections, causes us to reexamine many established concepts. The organelle may eventually provide us with deeper insight into the nature of the mechanisms controlling both transcription and translation.

The chloroplast genome and its interactions

Algae and plants contain two kinds of genetic systems in their cytoplasm, that of the mitochondrion and that of the chloroplast. Moreover, the ct DNA, as noted earlier, is much larger than the mt DNA of yeast and animals. As a consequence, information on the organization of the chloroplast genome on the molecular level has been more difficult to obtain than on that of the mitochondrial genome. In plants, for example, it is often very difficult to determine whether a cytoplasmic mutant genetic factor resides in the mt DNA or the ct DNA.

As noted in our discussion of Chlamydomonas, Sager has presented evidence that several nonchromosomal genes reside in the ct DNA and has placed them on a linkage map based on her recombination analyses. The single circular linkage group that has emerged

FIG. 20-29. Restriction endonuclease map of the chloroplast DNA of *Chlamydomonas reinhardii*. The three circles from the inside to the outside represent, respectively, restriction maps using the enzymes BglII BamHI, and EcoRI. (Reprinted with permission from J. D. Rochaix, *J. Mol. Biol.,* 126: 596, 1978. Copyright by Academic Press Inc. (London) Ltd.)

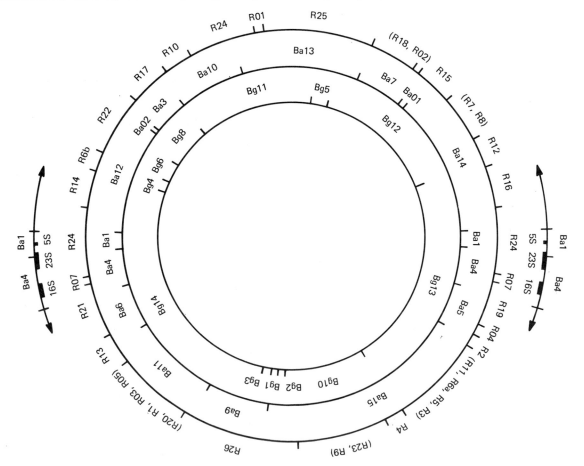

from these studies is in agreement with the actual physical circularity of the ct DNA. The ct DNA of Chlamydomonas has been subjected to restriction endonucleases, and a physical map has been constructed (Fig. 20-29). Restriction fragments have been hybridized with chloroplast rRNAs and tRNAs. In this way, the locations of the coded information for these transcripts have been determined.

In the case of the iojap mutation (see Fig. 20-18), we saw that plastids have a certain degree of autonomy. In this example, as is also true in most higher plants, chloroplast inheritance is strictly maternal, because the plastids in the pollen do not enter the zygote. If chloroplast DNA from the female parent is absent or contains some sort of mutant defect, the plastids do not develop properly and may cause a lethal condition in the plant. In those plants in which plastids enter the zygote through both the pollen and egg parents, a plastid defect in one of the parents can lead to the production of offspring that are variegated, containing a mixture of tissue with both normal and defective plastids, seen as streaks of green and white patches.

As is true for the genome of the mitochondrion, the ct DNA contains genes that code for subunits of several complex enzymes. For example, the very im-

portant enzyme critical to the dark reaction of photosynthesis is composed of a small subunit that enters the chloroplast from the cytosol and combines with the large subunit made on the chloroplast ribosomes to form the active enzyme, ribulose 1,5-diphosphate carboxylase.

The DNA polymerase that replicates the DNA of the plastid does not arise through translation of mRNA in the plastid. Actually, the polymerase forms in the cytosol as a result of translation of mRNA by the ribosomes found there. This can be shown in the following way. Cells may be treated with the drug rifampicin, which prevents the activity of the RNA polymerase of the plastid. This interference actually causes a loss of the ribosomes of the plastid, because no rRNA can be transcribed from the DNA of the plastid. The drug does not inhibit the RNA polymerase of the nucleus, and so ribosomes of the cytosol are not affected by it. When ribosomes are lost from the plastids after treatment with rifampicin, the plastids still manage to replicate their DNA. Therefore, the DNA polymerase needed for this replication must enter the plastid, because it is a protein and cannot be made there in the plastid if no ribosomes are present (Fig. 20-30). Unlike mitochondria, which import most of their lipids from the cytosol, the chloro-

FIG. 20-30. Interaction of chloroplast and cytosol. (1) The normal chloroplast contains, among other things, its own DNA, RNA, and ribosomes. Some of the chloroplast DNA may undergo transcription to form the rRNA of the ribosomes of the plastid. (2) The drug rifampicin blocks transcription of the plastid DNA. It does not block transcription of the nuclear DNA. (3) Plastids arise that lack ribosomes, because no ribosomal RNA is formed in presence of the drug. (4) In the plastids lacking the ribosomes, the DNA manages to replicate. This requires the enzyme DNA polymerase. The polymerase could not be formed in the plastids because they lack ribosomes, and these are needed for protein formation. The polymerase must have entered the plastid from the cytosol. The cytosol contains ribosomes because the drug does not prevent RNA formation from the nuclear DNA.

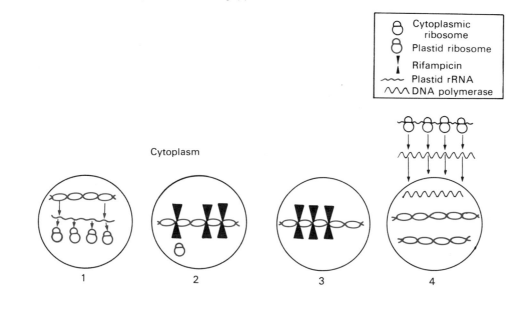

Cytoplasm

⊖ Cytoplasmic ribosome
⊖ Plastid ribosome
▼ Rifampicin
⌇ Plastid rRNA
⌇⌇ DNA polymerase

1   2   3   4

plasts themselves apparently synthesize most of the lipids needed to construct the organelle.

Although both ct DNA and mt DNA contain genes coded for parts of the structure of the organelles, it is still the nucleus that exerts the greatest degree of control over the biogenesis of the chloroplast and the mitochondrion. Most of the proteins of both organelles are imported from the cytosol, where translation of nuclear transcripts occurs. The nucleus apparently exerts control over the amount of protein the ribosomes of the organelles make. However, evidence also supports involvement of the organelles in the mechanisms that regulate the balance between the products synthesized in the organelles and those made in the cytosol for import into the organelles. Much is yet to be learned about the processes that enable the nucleus and organelle genetic systems to coordinate their activities in order to ensure efficient biosynthesis of mitochondria and chloroplasts and their orderly distribution from one cell generation to the next.

Biological implications of genetic information in organelles

The presence of genetic information in the mitochondria and chloroplasts and the interactions between the organelles and the rest of the cell have led to the formulation of interesting hypotheses concerning the origin of mitochondria and chloroplasts. We have already noted several ways organelles and prokaryotic organisms resemble each other. The DNA of the mitochondria, the chloroplasts, and bacteria are not associated with histones or bounded by membranes. The ribosomes of the organelles and of prokaryotes are inhibited by drugs such as streptomycin and chloramphenicol, antibiotics that do not interfere with the activity of the 80S ribosomes in the cytoplasm of the eukaryotic cell. The RNA polymerases of bacteria, mitochondria, and plastids are inhibited by rifampicin. Although this drug can bind to these enzymes, it does not affect the DNA-dependent RNA polymerase of the nucleus. Moreover, in bacteria the enzymes of aerobic respiration and photosynthesis form a part of the cell membrane. This is also true of the internal membrane of the plastid and the mitochondrion.

In contrast, the plasma membrane of the eukaryotic cell does not contain these enzymes, which are confined to the specific organelles. Recall that n-formyl methionine is the initiating unit in both bacterial and mitochondrial translation. Also the bacterial and organelle components of polypeptide synthesis can sub-

stitute for one another in translation. This cannot be done with the parallel components of the nucleocytoplasmic compartment of the cell. Translation and transcription also appear to be simultaneous in both prokaryotes and organelles, but these events do not take place in the cell nucleus simultaneously. These and other observations suggest a similarity between organelles and prokaryotes which has led some biologists to propose that the mitochondria and plastids in eukaryotic cells represent the descendants of ancient symbiotic organisms that took up residence in ancient prokaryotic host cells.

The mitochondrial ancestor would have been a primitive type of aerobic bacterium and the ancestral chloroplast a primitive type of blue-green alga. According to the thesis, their residence gave a decided advantage to the primitive prokaryotic host. Those with the mitochondrial ancestor were provided with an efficient energy source. The host cytoplasm contained the machinery for anaerobic respiration, whereas the symbiont contained the aerobic mechanism. As a result of the benefits conferred to the host, other features of the eukaryotic cell became established: the nuclear membrane, mitotic apparatus, and endoplasmic reticulum, and so on. Those cells that also carried the primitive blue-green algae evolved into eukaryotic plant cells, which had acquired the ability to trap the energy of light.

Thus, according to this concept, which actually dates back to the late nineteenth century, all eukaryotic cells have arisen from a host–symbiont relationship, and mitochondria and plastids are the descendants of the original symbionts we see today in the cells of every eukaryote. During the course of evolution, natural selection would favor any mutations in the genetic material of the host cell that helps to preserve the relationship or make it still more efficient. Thus, the host cell eventually took over some of the functions of the symbiont as the relationship became more and more advantageous. This picture is reminiscent of Paramecium with kappa and the related particles that probably *do* represent symbionts. We saw that the host cells carried chromosomal genes that affect the maintenance and development of the particles. Symbiotic relationships between green algae and cells of Paramecium and Hydra are also well known.

It is therefore conceivable that the eukaryotic cell organelles arose in a similar fashion and that today most of the properties of the symbionts are controlled by nuclear genes of the host. Most of the proteins and the developmental stages of the mitochondria and plastids are certainly under the control of nuclear genes. And most of the essential structural components of the organelles are made outside of them in the cytosol of the cell. The amount of information in the mitochondrion is so small that not very many proteins could be coded for. Thus, one can argue that during the course of evolution, most of the original mitochondrial functions were assumed by the cell nucleus because mutations in this direction were favored by natural selection. But the annoying question still remains as to why any genetic information at all continues to reside in the mitochondrion or plastid. If the symbiont hypothesis is correct, the relationship was established countless ages ago, long before the origin of the simplest eukaryotic cell. If it were beneficial for the nucleus to assume control over the organelles, why has it not been complete? The force of evolution usually weeds out any superfluous features of an organism.

According to alternative hypotheses, mitochondria evolved gradually, not in a large step the endosymbiont theory proposes. Supposedly, an aerobic prokaryote was ancestral to the eukaryotic cell, and it contained the ingredients for aerobic respiration in its plasma membrane. Gradually, over a very long period of time, changes occurred in which the original plasma membrane was altered and formed mitochondrial membrane. This occurred as a result of selective pressure for a larger surface area needed for respiration as the organism became more and more complex. The membrane-bound organelle would have arisen as a result of infoldings of the plasma membrane of the cell. (Infoldings of the plasma membrane actually occur in bacteria, as seen in bacterial mesosomes.) The present-day machinery for protein synthesis would thus trace back to that which was present in the ancestral prokaryote. As the organism became more complex with time, the mitochondrial system and the cytosol diverged, the one in the organelle remaining more like the original bacterial ancestor.

Some alternative hypotheses propose that the DNA of the mitochondria resulted from the introduction of a plasmid into the organelle. Once this happened, the DNA, ribosomes, and other parts of the organelle evolved into the mitochondrion of today. Another suggestion is that a piece of the genetic material of the ancestral cell became separated and enclosed in the membranes that contained the ribosomes, producing the mitochondrion.

When evaluating the two main theories on the origin of organelles (endosymbiotic and gradual), we must keep in mind that mitochondria could have arisen in one way and chloroplasts in another. The chloroplasts show many similarities to blue-green algae. The amount of DNA contained in a chloroplast is appreciable compared with the very small amount in a mitochondrion. In Chlamydomonas, for example, the plastid DNA content is almost equal to that of a prokaryotic cell. Mitochondria differ from prokaryotes much more than plastids do. Recall the variation seen in mitochondrial ribosomes, discussed earlier in this chapter. The argument for the origin of mitochondria by endosymbiosis is not as strong as that for the origin of plastids.

Definitive explanations of the origin of organelles cannot yet be given, but adherents of one or the other of the two main hypotheses continue to offer evidence for their points of view. Clarification may someday be provided through investigations of nonchromosomal genetic determinants in a variety of species.

REFERENCES

Anderson, S., A. T. Bankier, B. G. Barrell, M. H. L. deBruijn, A. R. Coulson, J. Drouin, I. C. Epron, D. P. Nierlich, B. A. Roe, F. Sanger, P. H. Shreier, A. J. Smith, R. Studen, and I. G. Young. Sequence and organization of human mitochondrial genome. *Nature* 290: 457, 1981.

Attardi, G. Organization and expression of the mammalian mitochondrial genome: a lesson in economy. *Trends Biochem. Sci.* 6: 86, 1981.

Birsky, C. W., Jr. Transmission genetics of mitochondria and chloroplasts. *Annu. Rev. Genet.* 12: 471, 1978.

Borst, P. and L. A. Givell. Small is beautiful-portrait of a mitochondrial genome. *Nature* 290: 443, 1981.

Butow, R. A., P. S. Perlman, and L. I. Grossman. The unusual *var* I gene of yeast mitochondrial DNA. *Science* 228: 1496, 1985.

Dujardin, G., C. Jacq, and P. P. Slonismki. Single base substitution in an intron of oxidase gene compensates splicing defects of the cytochrome b gene. *Nature* 298: 628, 1982.

Gillham, N. W. *Organelle Heredity.* Raven, New York, 1978.

Goddard, J. M. and D. R. Wolstenholme. Origin and direction of replication in mitochondrial DNA molecules from *Drosophila melanogaster.* *Proc. Natl. Acad. Sci.* 75: 3886, 1978.

Grivell, L. A. Mitochondrial DNA. *Sci. Am.* (March): 78, 1983.

Leaver, C. J. and M. W. Gray. Mitochondrial genome organization and expression in higher plants. *Ann. Rev. Plant. Physiol.* 33: 373, 1982.

Ledoigt, G., B. J. Stevens, J. J. Curgy, and J. Andre. Analysis of chloroplast ribosomes by polyacrylamide gel electrophoresis and electron microscopy. *Exp. Cell. Res.* 119: 221, 1979.

Margulis, L. *Symbiosis and evolution*. W. H. Freeman, San Francisco, 1981.

Ojala, D., J. Montoya, and G. Attardi. tRNA punctuation model of RNA processing in human mitochondria. *Nature* 290: 470, 1981.

Preer, L. B. and J. R., Preer, Jr. Inheritance of infectious elements. In *Cell Biology, A Comprehensive Treatise*, Vol. 1, L. Goldstein, and D. M. Prescott (eds.), p 319. Academic Press, New York, 1978.

Rochaix, J. D. Restriction endonuclease map of the chloroplast DNA of *Chlamydomonas reinhardii*. *J. Mol. Biol.* 126: 597, 1978.

Sager, R. *Cytoplasmic Genes and Organelles*. Academic Press, New York, 1972.

Schwartz, R. D. and M. O. Dayhoff. Origins of prokaryotes, eukaryotes, mitochondria, and chloroplasts. *Science* 199: 395, 1978.

Shumway, L. K. and T. E. Weier. The chloroplast structure of *iojap* maize. *Am J. Bot.* 54: 773, 1967.

Sonneborn, T. M. Kappa and related particles in *Paramecium*. *Adv. Virus Res.* 6:229, 1959.

Strathern, J. N., E. W. Jones, and J. R. Broach (eds.). *The Molecular Biology of the Yeast Saccharomyces: Life Cycle and Inheritance*. Cold Spring Harbor Laboratory, Cold Spring Harbor, NY, 1981.

Strathern, J. N., E. W. Jones, and J. R. Broach (eds.). *The Molecular Biology of the Yeast Saccharomyces: Metabolism and Gene Expression*. Cold Spring Harbor Laboratory, Cold Spring Harbor, NY, 1982.

Tzagoloff, A., G. Macino, and W. Sebals. Mitochondrial genes and translation products. *Annu. Rev. Biochem.* 48: 419, 1979.

Wurtz, E. A., J. E. Boynton, and N. W. Gillham. Perturbation of chloroplast DNA deoxyuridine. *Proc. Natl. Acad. Sci.* 74: 4552, 1977.

REVIEW QUESTIONS

1. Assume that each of the following kinds of Paramecium cells undergoes autogamy. Give the genotypes of the two cells that will arise. Indicate whether they are killer or sensitive.

   A. KK with kappa.
   B. KK without kappa.
   C. Kk with kappa.
   D. Kk without kappa.

2. In each of the following, give the genotypes of the two stocks or strains of Paramecium that, when crossed, produce the results described:

   A. A killer stock and a sensitive stock whose F_2 are all killers following cytoplasmic exchange.
   B. A killer stock and a sensitive stock whose F_2 are killers and sensitives in a 1:1 ratio following cytoplasmic exchange.
   C. Two killer strains whose cells, when mated, given an F_1 producing killers and sensitives in a 3:1 ratio with or without cytoplasmic exchange.

3. A. In Question 2A, assume there is no cytoplasmic exchange. What would the genotype and the phenotype of the F_1 cells arising after conjugation from the P_1 sensitive be? What would be the result from the P_1 killer?

   B. In Question 2B, assume there is no cytoplasmic exchange. What would the genotype and the phenotype of the F_1 cells arising after conjugation from the P_1 sensitive be? What would be the result from the P_1 killer?

4. A. In yeast, would you expect the wild or the petite condition in the zygotes and from sporulation of the zygotes after crossing two neutral petites?

   B. Answer the same question for the cross of a neutral petite and a nuclear petite.

   C. How would you expect mating type to segregate in each of the preceding crosses?

5. A genetic determinant for resistance to oligomycin has been found to reside on yeast mt DNA. A cross is made between oligomycin-sensitive and oligomycin-resistant strains. Zygotes arise that are allowed to undergo mitotic divisions. Single zygotes are eventually allowed to sporulate. What is to be expected regarding sensitivity to the drug among the resulting ascospores?

6. In Chlamydomonas, assume the nuclear alleles C, c and D, d are being followed along with the cytoplasmic determinants P, p and Q, q. Give the genotype expected in the zygote and in the products of the zygote in each of the following crosses:

   A. Mating type "+"; CD; Pq × mating type "−"; cd; pQ.
   B. The same cross as in A, assuming that ultraviolet light was applied to the "+" parent.

7. In the mold Neurospora, there are two mating types A and a. Spores may be pigmented or nonpigmented. Some strains of the mold are slow growing (poky) in contrast to the normal, fast-growing habit. Nuclei of the mold are haploid, except for the fertilization nucleus. Meiosis occurs immediately in the ascus giving four haploid nuclei that undergo one mitotic division. The eight haploid spores from a single ascus can be retrieved and tested.

   Two crosses are given here. From the results of testing the spores, what can you say about the transmission of mating type, spore color, and growth habit?

   Cross 1. Female, mating type A, fast growing, colored spores; male, mating type a, poky, nonpigmented spores.

   Cross 2. Female, mating type a, poky, nonpigmented spores; male, mating type A, fast growing, colored spores.

   Results of Cross 1. All fast growing; 50% A, 50% a; 50% pigmented spores, 50% nonpigmented spores.

   Results of Cross 2. All poky; 50% A, 50% a; 50% pigmented spores, 50% nonpigmented spores.

8. Three different petite strains are shown here. The presence (+) or absence (−) of certain marker genes is noted for each strain. What do the data suggest concerning the location of the individual markers?

	MARKER					
	A	B	C	D	E	F
Suppressive petite Strain 1	+	+	−	+	−	+
Suppressive petite Strain 2	+	−	+	+	+	−
Neutral petite	−	−	−	+	−	−

9. After consulting the map of yeast mt DNA, give a reason why you would expect most alterations of the mt DNA to result in the origin of cells that are respiratory deficients.

10. A certain complex coenzyme consists of seven different polypeptide chains, four large (I, II, III, IV) and three small (V, VI, VII). Suppose that labeled amino acids are supplied to cell populations of a certain protozoan and of a certain mold under the conditions indicated in the following. What would the results suggest about the site of synthesis of the various chains of the enzyme in the two organisms?

	LABELED CHAINS FORMED	
	Protozoan	Mold
Erythromycin added	I, IV, VI	V, VI, VII
Cycloheximide added	II, III, V, VII	I, II, III, IV
Control	All 7	All 7

11. Suppose the following sequence occurs at the end of a human mitochondrial gene that codes for a unit of a respiratory enzyme. The end of the coding sequence for the enzyme is next to a sequence that codes for a tRNA. When the transcript associated with the enzyme sequence is eventually read by the ribosomes, how can chain termination be established?

5′   A A T T T G C C G T tRNA sequence   3′
      lys   leu   pro

12. Suppose a random point mutation brings about a base substitution in human mt DNA. Another point mutation alters yeast mt DNA. Which mutation has the greater chance of affecting some important product—the mutation in the human or the yeast mt DNA? Explain.

13. Following is a nucleotide sequence of a segment of a strand of mt DNA. This segment is part of a region of mt DNA that codes for the amino acid sequence of a certain polypeptide. The amino acid sequence shown is related to the given nucleotide sequence. Consult a dictionary of codons, and determine whether or not any exceptions to the reading of the genetic code exist.

5′CCTTTTCCTCATCCATGAGTTATACTA 3′
pro-phe-pro-his-pro-try-val-met-thr

For each of the following questions, select the correct answer or answers, if any:

14. Which of the following is (are) true about kappa and the killing action?

A. Nonbrights are required to maintain brights and the killing action.
B. Nonbrights are eliminated from the cell and cause the killing action.
C. DNA is found in both the brights and the nonbrights.
D. The number of sensitive cells killed is directly proportional to the number of killer cells.
E. Kappa is essential for the survival of the cell.
F. Kappa cannot mutate independently of the nucleus.

15. Which of the following is (are) true about the petite story in yeast?

A. Following the cross of a nuclear petite with a normal, the segregation of ascospores from the zygote is 2 normal:2 petite.
B. Following the cross of a neutral petite with a normal, the diploid cells are petite and give rise to all petites following ascospore formation.
C. Following the cross of a suppressive petite with a normal, the diploids will be petite if they are plated on agar soon after mating.
D. Following the cross of normal with suppressive petite, wild-type cells are favored if subculturing precedes the plating on agar.
E. Petites cannot be induced by any known mutagen.

16. In yeast:

A. Following the cross of an erythromycin-sensitive cell with a resistant one, the sensitive trait disappears, and all cells become resistant.
B. Following the cross of an erythromycin-resistant, chloramphenicol-sensitive cell (E^RC^S) with an erythromycin-sensitive, chloramphenicol-resistant cell (E^SC^R) asci are produced that contain spores only of the parental types.
C. The cross $E^RC^S \times E^SC^R$ produces four kinds of asci: E^RC^S, E^SC^R, E^RC^R, and E^SC^S.
D. The cross $E^RC^S \times E^SC^R$ produces asci, each of which

produces four different kinds of spores, the two parentals and the two recombinant types.

E. The determinants for drug resistance may be associated with a physical alteration in mitochondrial DNA.

17. In higher plants:

A. Nuclear genes do not seem to influence the development of the plastid from the proplastid.

B. There is more DNA in a chloroplast than in a mitochondrion.

C. All of the enzymes involved in photosynthesis are apparently formed from transcripts of the plastid DNA.

D. The RNA of the ribosomes of the plastid arises from transcription of the plastid DNA.

E. Replication of the plastid DNA depends on the formation of plastid DNA polymerase by the ribosomes of the plastid.

18. In corn:

A. The iojap mutation is due to an alteration in the DNA of the chloroplast.

B. Altered plastids due to iojap become normal if they are transmitted to a plant carrying the dominant allele Ij.

C. An individual homozygous for the wild allele (Ij Ij) may have mutant plastids.

D. If a normal, fully green plant (Ij Ij) is used as a female, the offspring will be all green, even though the male parent is striped and is genotype ij ij.

E. If a striped plant used as female parent (ij ij) is crossed to a fully green one (Ij Ij), the F_1 may consist of three types of plants: green, striped, and nongreen.

21

GENES, POPULATIONS, AND EVOLUTIONARY CHANGE

Populations as units of evolutionary change

Genetic discussions often concentrate on inheritance patterns found in families. However, it must be kept in mind that the various allelic forms of genes within a species are ultimately of consequence not to just a few individuals, but to the entire population. In turn, mutations and changes in frequencies of alleles within populations influence the very course of evolution of the species. The life span of the individual runs but a short course, during which time the genotype remains constant. In contrast, the population, which persists over generations, has a genetic constitution that continues to vary. The population, not the individual, is the main unit of evolution. Evolution can be well defined as change in the diversity and adaptation of populations of organisms. The concept of Darwin focuses on the central role in evolution of random inheritable changes in the individuals of a population. Those inheritable modifications that prove to be advantageous in the immediate environment become established, whereas those that entail a disadvantage are eliminated through natural selection.

The individuals within a population have become adapted to a set of environmental conditions under the direction of natural selection, the prime force of evolution. Thus, members of a population tend to exist in a defined location, somewhat different from that of any other genetic group. One characteristic of a population is that its members tend to mate with one another more frequently than with individuals from other populations. Therefore, population members share more gene forms with individuals belonging to their group than with those in related populations.

Most plant and animal species are composed of more than one population, and the number of individuals varies from one population to another. The different populations that constitute a species may all be very similar, exhibiting only minor variations. On the other hand, they may be so different that one can recognize varieties, races, or subspecies within the species (Fig. 21-1A). Nevertheless, an important feature of all the populations comprising a species is that they have the potential to interbreed with one another when brought together. Although population members usually mate within their own group, they are still sufficiently similar in their genetic composition to cross with members from other populations within the species.

We are certainly aware of this in the human species, which is composed of many populations. Groups of human populations may be so different that we call them races, but still, any human from any population can exchange genetic material freely with any other

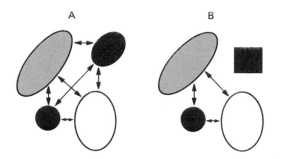

FIG. 21-1. Populations and species. (A) One species. A species may be composed of populations of different sizes. These populations may be quite distinct in appearance, so that we can recognize races or subspecies. Nevertheless (as shown by arrows), all members of a species can freely exchange genes, so that when members of the different subspecific population groups contact, they can mate and produce viable offspring. (B) Two species. If members of a certain population cannot exchange genes with individuals in certain other populations, more than one species can be recognized. In this example, we see two species, one of which is composed of three subspecies.

when the two are brought together, no matter how different they may appear phenotypically. On the other hand, the inability of members from a certain group or population to cross with those in another would suggest that the two populations belong to two different species (Fig. 21-1B). Once barriers to crossing arise between two groups, often after a long period of isolation between them, they are considered members of two distinct species. They can no longer be recognized as subunits of a larger group, the species, for within a species, members can exchange genetic material, even if they are separated into phenotypically distinct populations.

Within any one of the populations that constitute a species, the individuals form a collection of information coded in DNA. This genetic information occurs as genes distributed among all the members. For any one pair of alleles (A and a), the information may be present in the homozygous (AA and aa) or the heterozygous (Aa) condition. All the pairs of alleles within a breeding population form a gene pool, which is the total of all the genes in the breeding population. It is from this gene pool that individuals in subsequent generations derive their genes. Breeding members of one generation generally pass their genetic information back into the gene pool of their own population.

Although independent assortment and crossing over produce a great deal of variation within a population, these processes by themselves do not contribute new genetic information to the population gene pool. New information, in the form of different kinds of alleles, may be added through matings of individuals with members of a different population. Also, genetic changes can be added to a population without contact with other populations, through the origin of spontaneous point mutations or structural aberrations. As a matter of fact, it is an accumulation of these changes that tends to make one population distinct from another. Any population, at a given time, is the result of natural selection operating toward the highest level of adaptability to the specific environment. Frequencies of alleles may change in response to selection. Those individuals better adapted at one time pass their favorable alleles to their offspring, and the frequencies of these alleles tend to become higher. Conversely, individuals with less suitable alleles or those that are deleterious under a given set of environmental conditions, on the average, leave fewer offspring, and the frequencies of these alleles fall. A population, then, is a dynamic group. Even without contact with another population, levels of alleles change from one generation to the next as certain allelic combinations impart more adaptive value than others at a given time. Much of this chapter deals with factors that influence the frequency of an allele in a population and factors that change allele frequencies from one generation to the next.

Equilibrium frequency of genes in populations

To appreciate more fully the interaction of these factors, it is essential to grasp a concept that is fundamental to an understanding of population genetics. Remember that a dominant gene form is not necessarily superior to a recessive one, nor does the fact of dominance by itself ensure that an allele will be more common than the recessive. An allele may increase in frequency if it imparts a reproductive advantage to those who carry it. But this is not directly a consequence of its being dominant or recessive. If the allele, dominant or recessive, is very beneficial, those who carry it will leave more progeny, and it will spread through the population. A deleterious genetic factor will be kept at a low frequency because it restricts in some way the number of offspring left by its carriers. Eventually, an allele will approach an equilibrium frequency in a population when the force of natural selection, for or against it, is balanced by the force of mutation that adds it. A wild-type allele and its mutant form thus reach an equilibrium that tends to be maintained from one generation to the next.

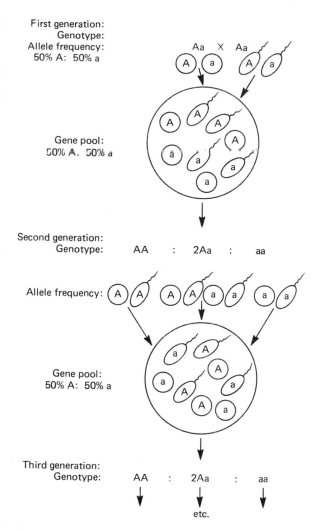

First generation:
Genotype:
Allele frequency:
50% A: 50% a

Aa X Aa

Gene pool:
50% A. 50% a

Second generation:
Genotype: AA   :   2Aa   :   aa

Allele frequency:

Gene pool:
50% A: 50% a

Third generation:
Genotype: AA   :   2Aa   :   aa

etc.

FIG. 21-2. Equilibrium frequency and population gene pool. All members of a population share genes from the same pool. Starting with all individuals of genotype Aa, the frequency of the two alleles is 50:50 in the population. Gametes carrying either allele A or a are contributed in that same proportion to the gene pool that gives rise to the next generation. As a result of random combinations of alleles, the next generation will occur in a genotypic frequency of 25:50:25. However, the frequency off the alleles A versus a in the population has not changed. It is still 50:50, and this same amount will be contributed to the gene pool for the next generation. The population has reached equilibrium with respect to the pair of alleles, A and a. The frequencies of genotypes and phenotypes for this pair of alleles will not change as long as random mating persists.

To illustrate this idea in another way, consider a pair of alleles (the dominant, A, and its recessive form, a). Suppose a group of heterozygotes becomes isolated and forms the basis for the origin of a new population. Individuals can mate only with others of the same genotype, Aa. The mating in the population (Fig. 21-

2) is similar to that between any two heterozygous individuals. However, in this case, the alleles A and a represent genetic information in the gene pool of the population, not solely that in just one pair of parents. We see from the illustration that the original allele frequencies are 50% A and 50% a. The genotypic frequency is 100% Aa, and the phenotypic is 100% A, because A is dominant to a. After one generation of mating, notice that the allele frequency is still the same: 50% A and 50% a. However, the genotypic frequency has now changed to 1AA: 2Aa: 1aa and the phenotypic to 3A: 1a.

Let us assume that these alleles, A and a, are neutral in their effect on the reproductive capacity of the individual. Also assume that persons of phenotype A or a are just as likely to mate with individuals of the opposite phenotype as they are with each other— that is, mating is random, A does not prefer either A or a, over the other, and likewise for a. From Fig. 21-2 and Table 21-1, we can see that the random mating will again result in a genotypic ratio of 1:2:1 and a phenotypic ratio of 3:1, exactly as in the preceding generation. The frequencies of A and a are still 50% and 50%. Nothing about the gene pool has changed with respect to these factors. It is as if mating were still Aa × Aa. We see that genetic equilibrium has been reached for this pair of alleles. And so it will continue generation after generation, as long as mating stays random, the population remains isolated, and no selective advantage or disadvantage is suddenly im-

TABLE 21-1   Random matings[a]

MATING	GENOTYPES			PHENO-TYPES	
AA × AA	AA			A	
AA × Aa	AA	Aa		2A	
Aa × AA	AA	Aa		2A	
AA × aa		Aa		A	
aa × AA		Aa		A	
aa × Aa		Aa	aa	A	a
Aa × aa		Aa	aa	A	a
Aa × Aa	AA	2Aa	aa	3A	a
aa × aa			aa		a
Total	4AA	8Aa	4aa	12A	4a
	1AA	2Aa	1aa	3A	1a

[a] If the three genotypes Aa, Aa, and aa show no mating preference, various combinations can occur. All the possibilities are shown here. The outcome is that the genotypic and phenotypic frequencies will not change from generation to generation because the proportion of A versus a alleles in the gene pool remains the same.

parted to any of the three genotypes. It does not matter whether or not A is dominant to a. Mendelian principles dictate that the frequency of one allele will not increase over that of another.

The Hardy–Weinberg equilibrium

This concept of genetic equilibrium was grasped independently by G. H. Hardy, a British mathematician, and W. Weinberg, a German physician, each of whom published on idealized populations in 1908. The population model that is derived from their conclusions is known as the Hardy–Weinberg law, the Hardy–Weinberg formulation, or the Hardy–Weinberg equilibrium, and is the cornerstone of population genetics and theoretic evolutionary genetics.

The Hardy–Weinberg law can be represented by the simple algebraic expression $p^2 + 2pq + q^2$. Although this formulation is simply an expansion of the binomial $(p + q)^2$, its implications are great for the genetic analyses of populations. With this simple expression, we can estimate the frequency of an allele in an idealized population, as well as the frequencies of genotypes. At equilibrium, the Hardy–Weinberg formula tells us that for any given pair of alleles the genotypes in a population at equilibrium will be distributed as $p^2 + 2pq + q^2$. We see here (Fig. 21-3) that $AA = p^2$ (25%), $Aa = 2pq$ (50%), and $aa = q^2$ (25%), since the frequency of allele $A = 0.5$ and the frequency of allele $a = 0.5$.

The Hardy–Weinberg formulation describes attain-

FIG. 21-3. Distribution of genotypes in a population. The relative frequencies of the alleles (A versus a) in a population gene pool are responsible for the relative frequencies of the three genotypes in a population. At equilibrium, these will occur with frequencies of p^2, $2pq$, q^2.

$$p = \text{freq. A} = .5$$
$$q = \text{freq. a} = .5$$

	A(p) .5	a(q) .5
A(p) .5	AA(p^2) .25	Aa(pq) .25
a(q) .5	Aa(pq) .25	aa(q^2) .25

= 25% AA + 50% Aa + 25% aa
(p^2 + 2pq + q^2)

ment of equilibrium for population gene pools under ideal situations. It is essential at the outset to consider the conditions assumed by this formula, for no real population can be expected to conform to all the conditions of a Hardy–Weinberg population. One requirement of such an idealized population is that the potential parents in the population mate at random with respect to those genetic factors that are being followed. Such random mating is in contrast to *assortive mating,* in which case individuals prefer to select as mates individuals of a specific type. For example, AA individuals might select only other AA types and avoid Aa and aa individuals.

Another requirement is that no forces operate to alter allelic frequencies within the population. This, in turn, imposes several restrictions. To maintain such a static condition, there must be no gene mutation (e.g., from A to a) to increase one allelic form at the expense of another or to add any new genetic information to the gene pool. There must be no selective process that favors in any way one genetic type over another. The three genotypes, AA, Aa, and aa, must have equal chances of leaving offspring. Furthermore, to maintain a Hardy–Weinberg population, there must be no migration into the population of individuals from other populations, and no emigration of individuals from the population. Either of these factors would upset the allelic frequency by adding more AA perhaps, losing more aa, or introducing other allelic forms of the gene—a^1, for example.

A further requirement for the idealized population is that it must be an infinitely large one; otherwise, chance may play a role in altering allelic frequencies. For example, in a small population more matings of the type $AA \times AA$ could occur by chance at any time. We examine later in this chapter the effects of population size as a factor in the alteration of allelic frequencies. Chance exerts a less pronounced effect as population size increases. In very large populations, the effect of chance is small, although it still operates at times, even though all the other requirements for an idealized population are satisfied.

Probably no real population ever conforms to all the restrictions of an idealized Hardy–Weinberg population. We may ask why we should pursue further a formulation that relates to a situation that rarely, if ever, occurs in nature. The answer to this will become apparent as we see in the following sections that the Hardy–Weinberg law enables us to describe and measure real populations and their dynamics against the idealized, or Mendelian, population.

Application of the Hardy–Weinberg law to gene frequencies in populations

In the example given earlier in our discussion (Fig. 21-3), we assumed that the frequencies of A and a were both 50%, or ½. We can summarize this by the expression: $p + q = 1$. It is also evident that the frequencies of the three genotypes, AA (p^2), Aa ($2pq$), and aa (q^2), must total 1 (25% + 50% + 25%). Therefore, $p^2 + 2pq + q^2 = 1$. By using just these simple facts, we can deduce a great deal of valuable information on populations.

Obviously, not every pair of alleles occurs with a frequency of 50:50. More often, one gene form has a higher frequency in a population than its allele. But this does not upset anything we have discussed so far (Fig. 21-4). Let us first look at this situation in a very simple way. For example, suppose that in a certain Indian tribe, whose members share a common gene pool, the frequencies of the M and N blood group alleles (L^M and L^N) are 70% and 30%, respectively. With the Hardy–Weinberg expression we can calculate the expected frequency of the three genotypes in the population. Representing L^M and L^N by p and q, we know that at equilibrium the three genotypes will occur with a frequency distribution of $p^2(L^M L^M)$: $2pq(L^M L^N)$: $q^2(L^N L^N)$. Simple arithmetic gives us 49% $L^M L^M (0.70 \times 0.70)$: 42% $L^M L^N (2 \times 0.70 \times 0.30)$:9% $L^N L^N (0.30 \times 0.30)$, and we can then see that $L^M L^M + L^M L^N + L^N L^N = 1$ (0.49 + 0.42 + 0.09 = 1). Exactly the same principles

FIG. 21-4. Distribution of genotypes at equilibrium for any allelic frequencies. Usually a pair of alleles in a population occurs with a frequency other than 0.5:0.5. Nevertheless, the same principles hold as in Fig. 21-3, and at equilibrium the three genotypes occur in a distribution of p^2, $2pq$, q^2.

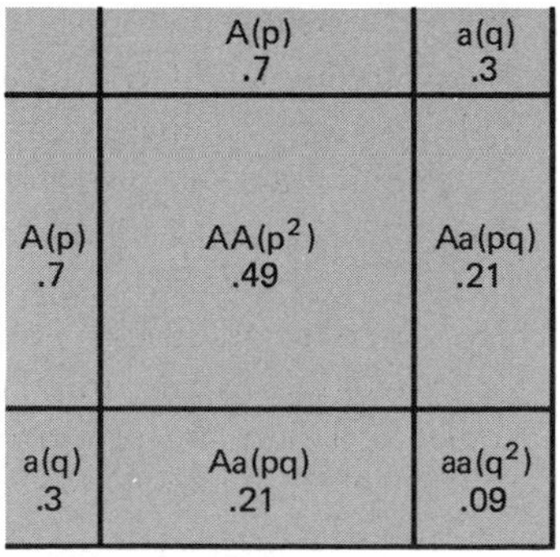

= 49% AA + 42% Aa + 9% aa
($p^2 + 2pq + q^2$)

TABLE 21-2   Estimation of allele frequencies in cases of codominance[a]

BLOOD GROUP	GENOTYPE			FREQUENCY IN POPULATION
M	$L^M L^M$			49% (p^2)
MN	$L^M L^N$			42% ($2pq$)
N	$L^N L^N$			9% (q^2)
	49% (p^2)	42% ($2pq$)	9% (q^2)	
	$L^M L^M$	$L^M L^N$	$L^N L^N$	

Frequency allele L^M = 49% + ½(42%)   = 70%
Frequency allele L^N =   ½(42%) + 9% = 30%

[a]If there is lack of dominance, the frequencies of the two alleles in the population can be easily estimated because the three phenotypes and hence the three genotypes, can be counted directly. The allele frequency p^2, $2pq$, q^2 is attained after one generation of random mating. Therefore, the frequency of allele L^M here must equal the number of M individuals plus half the number of MN. Similarly, the frequency of allele L^N must equal the number of N persons plus half the number of MN.

hold for any pair of alleles, no matter what their frequencies.

In this example, the frequencies of L^M and L^N were given at the outset. Such information is, of course, not automatically known. But in the case of alleles that are incompletely dominant or codominant such as these, the allelic frequencies can be directly calculated. The two homozygous classes can be distinguished from the heterozygotes. We can use the proportion of genotypes ($L^M L^M$:$L^M L^N$:$L^N L^N$) to calculate the frequencies of the alleles L^M and L^N. Finding that the frequencies of the M, MN, and N blood groups are, respectively, 49:42:9, we can estimate the frequencies of the L^M and L^N alleles. For we know (Table 21-2) that 49% represents $L^M \times L^M (p^2)$; 42% indicates $2 \times L^M \times L^N (2pq)$; and 9% is the product of $L^N \times L^N$. The frequency of allele L^M must be 49% + ½ 42%, or 70%. This is so because *only* allele L^M is present in blood type M, whereas only half of the alleles in **MN** types are allele L^M. By using the same reasoning, **the** frequency of allele L^N is 9% + ½ 42%, or 30%.

Testing for equilibrium

Let us explore in a bit more detail the calculation of allelic frequencies from population data when **the** genotypic classes can be directly identified. In actuality, the MN blood grouping happens to provide **an**

TABLE 21-3   The frequencies of MN blood groups in different populations

POPULATION		MM	MN	NN	TOTAL	FREQ. (M)	FREQ. (N)
U.S. whites	Observed number (ON)	1787	3039	1303	6129		
	Expected proportion (EP)	0.2916	0.4968	0.2116		0.540	0.460
	Expected number (EN)	1787.2	3044.9	1296.9			
U.S. negroes	ON	79	138	61	278		
	EP	0.2835	0.4989	0.2186		0.532	0.468
	EN	78.8	138.7	60.8			
U.S. (red) Indians	ON	123	72	10	205		
	EP	0.6015	0.3481	0.0504		0.776	0.224
	EN	123.3	71.4	10.3			
East Greenland eskimos	ON	475	89	5	569		
	EP	0.8335	0.1589	0.0076		0.913	0.087
	EN	474.3	90.4	4.3			
Ainus (Japan)	ON	90	253	161	504		
	EP	0.1845	0.4901	0.3234		0.430	0.570
	EN	93.0	247.0	163.0			
Australian aborigines	ON	22	216	492	730		
	EP	0.0317	0.2926	0.6757		0.178	0.882
	EN	23.1	213.6	493.3			

U.S. whites: Freq. of $M = (1787 + \frac{1}{2} \ 3039/6129)$
  Freq. of $N = (1303 + \frac{1}{2} \ 3039/6129) = 0.540 = p = 0.460 = q$
Expected proportion of $MM = p^2 = (0.540)^2 = 0.2916$
Expected proportion of $MN = 2pq = 2(0.540)(0.460) = 0.4968$
Expected proportion of $NN = q^2 = (0.460)^2 = 0.2116$
These are obtained from the Hardy–Weinberg law.
Expected number of $MM = p^2T = 0.2916 \times 6129 = 1787.2$
Expected number of $MN = 2pq \ T = 0.4968 \times 6129 = 3044.9$
Expected number of $NN = q^2 \ T = 0.2116 \times 6129 = 1296.9$

Source: Data from Stern, C. (1960). *Principles of Human Genetics*, p. 158, 2nd ed. Freeman, London.

excellent example, because the alleles L^M and L^N, being codominants, permit the ready recognition of the heterozygotes and homozygotes through serological testing. Most persons are unaware of their MN blood grouping, and so mating can be expected to be random with regard to the alleles L^M and L^N. Furthermore, there is no evidence to indicate any selection for any one of the three genotypes relating to longevity or increased fertility. Therefore, we can assume that some of the requirements essential for an idealized Hardy–Weinberg population are satisfied for this characteristic, the MN blood grouping

Let us see how we can calculate allelic frequencies for a population by directly counting specific phenotypes which, in turn, reflect the frequencies of specific genotypes. We can then proceed to determine whether or not the population departs from a Hardy–Weinberg equilibrium with respect to the distribution of genotypes.

From Table 21-3 we see that actual counts for a

white population reveal 1787 persons of genotype L^ML^M, 3039 L^ML^N, and 1303 L^NL^N. (Note that the table uses only M and N to designate alleles L^M and L^N.) From these figures, the frequencies of alleles L^M and L^N can be calculated as follows. The frequency of $L^M = 1787 + \frac{1}{2} \ 3039$ (since there is only one dose of L^M in genotype L^ML^N) divided by 6129 (the total number of persons in the sample). This gives a frequency of 0.540 for L^M which we can symbolize as p. Since $p + q = 1$, the frequency of L^N (or q) must be 0.460. We could, however, estimate it directly from the counts, as we did for L^M, because the frequency of L^N (q) must be $1303 + \frac{1}{2} \ 3039$ divided by $6129 = 0.460$.

Next we take these frequencies ($p = 0.540$ and $q = 0.460$) and substitute them in the Hardy–Weinberg formula to obtain the proportions of the three genotypes (L^ML^M, L^ML^N, L^NL^N) expected for a population at equilibrium. The expected proportion of $L^ML^M = p^2 = (0.540)^2 = 0.2916$. The expected proportion of $L^M L^N = 2pq = 2(0.540)\ (0.460) = 0.4968$. The expected proportion of $L^N L^N = q^2 = (0.460)^2 = 0.2116$.

We are now in a position to calculate the expected *numbers* of individuals of the three genotypes in the total of 6129 individuals (T). To do this, we multiply the total number of individuals (T) times the expected proportion in each class (p^2, $2pq$, or $q)^2$. The expected number of individuals of genotype $L^M L^M$ then equals $p^2 T = 0.2916 \times 6129 = 1787.2$. Those of genotype $L^M L^N$ are expected at equilibrium to be equal to $2pqT = 0.4968 \times 6129 = 3044.9$. The expected number of $L^N L^N = q^2 T = 0.2116 \times 6129 = 1296.9$. These expected numbers agree extremely well with those numbers actually counted in the sample. Such agreement between the numbers counted in the sample of 6129 and the numbers expected in a Mendelian population does not tell us with certainty, however, that the population is in equilibrium with respect to this system. The allelic frequencies may change in a subsequent generation, for we do not know whether all the requirements for an idealized population actually pertain. The agreement does tell us that the frequencies of the alleles L^M and L^N, respectively, at this time are 0.54 and 0.46 in the gene pool of the population. We also know that we can expect to find the same genotypic frequencies in samples taken in later generations *if* the population is in equilibrium.

Such agreement is not always the case. A departure from Hardy–Weinberg equilibrium would indicate that one or more of the requirements for an idealized population do not pertain and that some factor is in effect, such as assortive mating, selection favoring one of the genotypes, or chance. When our expected and observed figures do not agree as closely as in the example just presented, we must determine if the departure of our observed data from the expected figures could result from chance or if the deviation is indeed significant. To estimate the probability that the departure could be due to chance factors, we turn to the chi-square test (see Chap. 6). Recall that $\chi^2 = \Sigma d^2/e$. In Table 21-4, a χ^2 value is derived substituting the figures from the example given here.

Once we have the χ^2 value, we must turn to a probability table (see Table 6-5). However, before proceeding to a determination of the probability value, an important point must be grasped relating to the application of χ^2 to this kind of situation. This has a bearing on the number of degrees of freedom. From what we learned in Chapter 6, we might assume 2 degrees of freedom in this case, since we have a total of three classes. However, the case under discussion

TABLE 21-4   Calculations of chi-square (χ^2) value

CLASS	o	e[a]	d	d^2	d^2/e
$L^M L^M$	1787	1787.2	−0.2	0.04	0.00002
$L^M L^N$	3039	3044.6	−5.6	31.36	0.01030
$L^N L^N$	1303	1296.9	+6.1	37.21	0.02869
					$d^2/e = 0.03901$

[a]Total number of individuals in sample times expected Hardy–Weinberg frequency.

here is not quite the same as that depicted in Tables 6-4 and 6-6. In that example, we did *not* take the observed values and then use them to calculate the expected. Instead, we took the total and estimated what the figures would be *if* they fell into a certain ratio (3:1 e.g.). In our present case of the L^M and L^N alleles, we took the *observed figures* to obtain a proportion, which was in turn used to derive the expected figures. This is a very different situation. In cases like this, the number of degrees of freedom is equal to the number of phenotypic classes minus the number of alleles. This gives us here 3 (phenotypes M, MN, and N) − 2 (alleles L^M and L^N) = 1 degree of freedom. A χ^2 value of 0.03 for 1 degree of freedom gives a P value greater than 0.80, indicating that the observed deviation could be the result of chance 80% of the time. Thus, our hypothesis that the sample figures are compatible with an equilibrium frequency rests on very firm ground.

In those cases in which a χ^2 test shows that the observed deviations depart significantly from the figures predicted by the Hardy–Weinberg law, we can conclude that the formulation cannot be applied to the population for the particular genetic factors under consideration. One or more of the conditions required for an idealized Mendelian population are not being met. We might then direct our efforts toward identifying the specific agents that are responsible for producing a change in the frequencies of the alleles.

The Hardy–Weinberg law and dominance

The estimations of allelic frequencies that we have been discussing up to this point apply only to those cases in which heterozygotes can be distinguished from the homozygotes. The situation is a bit different when dominance is operating, for in these cases we cannot distinguish between the heterozygotes (Aa) and the homozygotes (AA). However, the Hardy–Weinberg formula can be applied to estimate allelic frequencies in these instances *if* the population is assumed to be in a Hardy–Weinberg equilibrium.

Suppose a geneticist wishes to estimate the frequency of the recessive a for albinism in a certain island population. The geneticist is particularly concerned about the number of heterozygous carriers (Aa) who appear normally pigmented and cannot be distinguished from the homozygotes (AA). To solve the problem, the investigator focuses attention on the recessives (aa), who *can* be recognized by their albino phenotype. Suppose that counts of such persons show they occur with a frequency of 1/10,000. Simple arithmetic then tells us the approximate frequencies of the alleles A and a, as well as the frequencies of the three genotypes, AA, Aa, and aa. For the value 1/10,000 represents q^2 (if we allow p to represent A and q to represent a). The figure is nothing more than the product of two separate probabilities (Table 21-5A). Two gametes came together with a frequency of 1/10,000. This is the product of the chance of a male gamete and a female gamete occurring separately in the gene pool of the population. Therefore, we take the square root of 1/10,000, which is 1/100. We now

know the frequency of allele a. Since $p + q = 1$, we also know the frequency of A (99/100). The frequencies of the three genotypes AA, Aa, and aa are, respectively, 9801/10,000, 198/10,000, and 1/10,000 (refer to Table 21-5A). And their sum is again 100%, which in this example is 10,000. We can, of course, work just as well with percentages. Expressed in percentage, the wild-type allele is found with a frequency of 99%.

It is important to note (Table 21-5B) that although the homozygous recessive individuals constitute only 0.01% of the population, the heterozygous carriers are found with a frequency of almost 2%. Applying the Hardy–Weinberg formula in this way tells us that appreciable numbers of mutant alleles persist in the population, *mainly* in heterozygotes. In effect, there is a pool of *mutant* alleles in any population, and this mutant gene pool constitutes an important portion of the total gene pool for the population. We see that the number of heterozygous individuals in a population at equilibrium will always be greater than the homozygous recessives in all cases of rare recessive alleles.

It must be understood that in this case the fre-

TABLE 21-5   Estimation of gene frequency when one allele is dominant[a]

A. Calculation of frequency of allele from observed frequency of homozygous recessives[b]

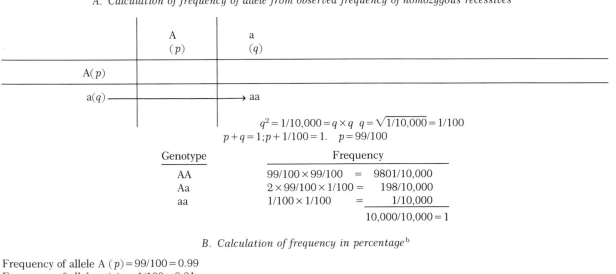

	A (p)	a (q)
A(p)		
a(q) ———→		aa

$$q^2 = 1/10,000 = q \times q \quad q = \sqrt{1/10,000} = 1/100$$
$$p + q = 1; p + 1/100 = 1. \quad p = 99/100$$

Genotype	Frequency		
AA	$99/100 \times 99/100$	=	9801/10,000
Aa	$2 \times 99/100 \times 1/100$	=	198/10,000
aa	$1/100 \times 1/100$	=	1/10,000
			10,000/10,000 = 1

B. Calculation of frequency in percentage[b]

Frequency of allele A (p) = 99/100 = 0.99
Frequency of allele a (q) = 1/100 = 0.01

Frequency of genotype:	AA (p^2)	Aa $(2pq)$	aa (q^2)
	$0.99 \times 0.99 =$	$2 \times 0.99 \times 0.01 =$	$0.01 \times 0.01 =$
	0.9801	+ 0.0198	+ 0.0001 = 1

[a] Normal = A = p; albino = a = q.
[b] Note that $p + q$ always equals 1, as does the sum of AA + 2Aa + aa.

quencies of the alleles, A and a, have been computed assuming that the population *is* in a Hardy–Weinberg equilibrium. Consequently, the figures cannot be used to test for equilibrium, as was done for the codominant alleles L^M and L^N. In that case, counts were made directly, and equilibrium was not assumed at the outset. We can, however, use the standard error of the mean to express the chance that the allelic frequencies we have estimated from the homozygous recessive phenotype reflect the real frequency in the population. The standard error in this case is

$$\sqrt{\frac{1-q^2}{2n}}$$

Assuming that the 1/10,000 value for q^2 was based on 100,000 individuals

$$SE = \sqrt{\frac{1-0.0001}{200,000}} = 0.0022$$

Sex-linked alleles and the Hardy–Weinberg law

We can employ the same reasoning to estimate the frequencies of sex-linked alleles. Because in mammals, males are the heterogametic sex, they carry only one of any two possible X-linked alleles. The number of males expressing the wild phenotype versus those expressing a sex-linked recessive trait gives the frequency of the alleles in the population for *both* males and females directly.

For example, suppose in a certain human population, the incidence of deutan color blindness is found to occur among 5% of the males (Table 21-6). This means that 5% of the males carry the defective allele, d, and 95% carry the normal factor, D. If the entire population is assumed to be at equilibrium, then the frequency of these alleles among the females must be the same as for the males, 95% and 5% for D and d, respectively. Treating the females as a separate population, the frequencies of the three genotypes (DD, Dd, and dd) among them must be as follows: $p^2 = 0.95 \times 0.95 = 90.25\%$; $2pq = 2 \times 0.95 \times 0.05 = 9.50\%$; and $q^2 = 0.05 \times 0.05 = 0.25\%$. Males, of course, are either DY (95%) or dY (5%), which are the values of p and q. So although five different genotypic frequencies will exist in a population for X-linked alleles, there is no contradiction to the Hardy-Weinberg formula. The same frequency of alleles is found among both sexes at equilibrium. Only because the male is hemizygous will the genotypic and phenotypic classes show a difference in relation to sex.

TABLE 21-6 Estimation of frequency of a sex-linked recessive allele[a]

GENOTYPE AND FREQUENCY[b]		ALLELE AND FREQUENCY	
Males			
Normal	DY = 95%	D = p =	95%
Color blind	dY = 5%	d = q =	5%
	100%		100% = p + q

GENOTYPE	FREQUENCY		
Females			
DD	$p^2 = 0.95 \times 0.95$	=	90.25%
Dd	$2pq = 2 \times 0.95 \times 0.05 =$		9.50%
dd	$q^2 = 0.05 \times 0.05$	=	0.25%
			100.00%

[a] In a population at equilibrium, the frequency of the trait in the heterogametic sex indicates directly the frequency of the allele in the population as a whole. The frequency of the genotypes in the homogametic sex is calculated by applying the Hardy–Weinberg formula and treating it as a separate population.
[b] D = normal; d = color blind.

The force of natural selection

If Hardy–Weinberg equilibrium persists for all the genetic factors in the gene pool of a population, evolution will reach a standstill, because evolution is genetic change. Natural selection is the major force that upsets this equilibrium, and its effect is to raise the population to a higher level of adaptiveness in a given environment. Natural selection brings about evolutionary change, because it favors the differential reproduction of certain allelic gene forms. Differential reproduction brings about changes in the frequencies of alleles from one generation to the next, encouraging those alleles that provide the highest level of adaptation between the population and the environment.

To measure the effect of natural selection on a population, we must become familiar with certain other basic concepts employed in population genetics. One of the terms used to measure the intensity of selection is denoted by W to designate adaptive value or relative fitness. It measures the proportion of individuals of a given genotype which survive and reproduce relative to the fittest genotype, the one that has the greatest reproductive rate. We can calculate fitness values for different genotypes (AA, Aa, aa) if we know the frequencies of these genotypes both before and after selection has operated. Let us suppose we begin with equal numbers of young annual plants of the genotypes AA, Aa, and aa. We observe these plants later after a period of drought to determine how many have matured and produced seeds. We find that AA and Aa are equally fit and that for every 100 of each of these that survived, only 70 of genotype aa survived. If we allow the fitness of plants of genotypes AA and Aa to equal 1, then the fitness of plants of genotype aa equals 0.7, since only 70 of these remain to produce seeds for the next generation compared with 100 each for AA and Aa.

A second measurement of the intensity of selection is the selection coefficient, s, which is closely related to W and is very easily determined as follows: $s = 1 - W$. The two terms, W and s, are actually different ways of looking at the same thing, the intensity of selection (Table 21-7). It should be obvious that s measures the proportionate *reduction* in the contribution of a given genotype to following generations, whereas W is a measure of the proportionate or relative success of a given genotype in contributing to later generations. If we are given an s value of 0.37 for a certain genotype, we know that $W = 0.63$. This means that for every 100 fertile offspring produced by the fittest ($s = 0$ and $W = 1$), only 63 are produced by that genotype against which selection is operating. Suppose a certain genotype entails complete sterility or is lethal before reproductive age. This means that $s = 1$ and $W = 0$ and that no offspring at all are left by organisms of that genotype.

Before proceeding with applications of W and s values to specific problems, we should bear in mind several important aspects of fitness and selection. We have just seen that the selection coefficient, s, is expressed negatively. It refers to the relative intensity of selection *against* a particular genotype and thus measures a force that acts in a negative way. Selection eliminates certain genotypes. By weeding these out, it brings about a proportionate increase of those genotypes that are less severely acted on by selection. The s values for a set of genotypes are relative. In other words, selection does not directly create any-

thing new. When we say that something is "selected for," we imply that this is an indirect effect. An s value of 0 ($W = 1$) does not mean that selection fails to act against that given genotype, let us say genotype AA. What is meant is that selection acts with less intensity against AA than it does against aa, which has an s value of 0.75 (see Table 21-7).

Natural selection may be considered any factor in the environment that brings about a relative change in the reproduction of certain alleles. We must also keep in mind that any fitness values are ones that reflect those genotypes' relative advantages or disadvantages in the specific environment at a given time. They are immediate values for fitness under those environmental conditions in which they were derived. The values calculated under controlled laboratory conditions could prove to be very different from those operating in nature, where environment is dynamic and the force of selection may change significantly for any given genotype throughout one or more generations.

A very important point to remember is that models of selection, such as those discussed so far, concentrate on just one gene locus and the allelic forms of the gene that resides there. We have stressed many times throughout our discussions that no one gene acts completely independently, and no one allele at a locus is selected for or against without reference to the rest of the genetic background. Actually, thousands of selective processes are being summed up as if they were just one just one event. Remember, therefore, it is the entire organism that is subject to the force of selection, not alleles at one locus. Any locus under consideration is interacting with countless other loci in complex ways. Therefore, the effects of an allele at one locus are modified by the effects of those at other loci. Although we may concentrate on a given locus and determine the s value for a genotype, we are really referring to the fitness of that genotype averaged with the effects of the rest of the genetic background.

Application of fitness values

Let us turn to a specific example that illustrates the effects of selection against a completely recessive allele in a population. Consider the three genotypes in Table 21-8 along with their fitness values. We see that the s value for both AA and Aa equals 0, whereas for genotype aa it equals 0.6. In problems of this type, the proportionate contribution for each genotype after

TABLE 21-7   Relationship between relative fitness values and selection coefficients

| | GENOTYPES | | |
	AA	Aa	aa
Relative fitness (W)	1.00	0.68	0.25
Selection coefficient (s)	0	0.32	0.75

TABLE 21-8   Application of fitness values

| | GENOTYPES | | | |
	AA	Aa	aa	TOTAL
Initial frequency	$p^2 = 0.25$	$2pq = 0.50$	$q^2 = 0.25$	1.00
Relative fitness (W or $1-s$)	1 or $1-0$	1 or $1-0$	0.4 or $1-0.6$	
Proportionate contribution after selection	0.25	0.50	0.10	$\overline{W} = 0.85 = 1 - sq^2$

selection (bottom row) is simply obtained by multiplying the initial frequency and the fitness value indicated by W (or $1-s$). It is important to see that the *total* value after selection for the frequency of genotypes (the total proportionate contribution) is less than 1.00, in this case 0.85. This is because the force of selection acted against genotype aa and reduced the contribution of this genotype to the gene pool. The total proportionate contribution (symbolized by $\overline{W}$, W bar) equals $1 - sq^2$, the *original* total minus those lost through selection. In this case, $\overline{W} = 1 - 0.6 \times 0.25 = 0.85$.

We must not conclude that as a result of selection the size of the population has necessarily decreased. Remember that the values reflected are relative ones. The population may have actually produced an excess of individuals, and those which were lost may have been a part of the number lost as a consequence of overproduction. We must, however, correct for the apparent decrease. This is accomplished simply by dividing each proportionate contribution after one generation of selection by the new total $\overline{W}$. Keep in mind that $\overline{W}$ equals $1 - sq^2$ or 1 minus the selection coefficient of aa times its original frequency. Also remember that, in problems of this type, genotype aa is the group being selected *against*. In the example given in Table 21-8, the corrected proportionate contributions of the genotypes to the gene pool after one generation of selection become as follows:

AA	Aa	aa	Total
$\dfrac{0.25}{0.85} = 0.294$	$\dfrac{0.50}{0.85} = 0.588$	$\dfrac{0.10}{0.85} = 0.118$	$\dfrac{0.85}{0.85} = 1$

Now we can calculate the change in the frequencies of the alleles A and a following one generation of selection. Before selection, the population consisted of 25AA: 50Aa: 25aa, giving $p = A = 0.5$ and $q = a = 0.5$ After selection, we have the following proportionate

contributions: 0.294AA: 0.588Aa: 0.118aa. The allelic frequencies are now as follows:

$$p = A = 0.294 + \frac{0.588}{2} = 0.588 \qquad q = a = 0.118 + \frac{0.588}{2} = 0.412$$

The change in the frequency of a (Δq)in one generation is simply the difference between the frequencies after and before selection, where q_1 represents the frequency in the first generation after selection and q_0 designates the frequency of q in the original population. Thus, $\Delta q = q_1 - q_0 = 0.412 - 0.5 = -0.088$. It is from these new allelic frequencies (0.588 and 0.412) that the genotypic frequencies will be derived in the next generation. These new frequencies then become as follows: AA $= (0.588)^2 = 0.345$; Aa $= (2 \times 0.588 \times 0.412) = 0.485$; aa $= (0.412)^2 = 0.170$. (The frequencies before selection were: AA $= 0.25$; Aa $= 0.50$; aa $= 0.25$.)

A very valuable formula can now be derived to express the change in the frequency of q (Δq, the change in the frequency of the recessive being selected against) in *any* one generation:

$$\Delta q = q_1 - q_0 = \frac{q_0(1 - sq_0)}{1 - sq_0^2} - q_0 = \frac{-sq_0^2(1 - q_0)}{1 - sq_0^2}$$

This formula can be simplified even further and used to demonstrate a very important point. If the value of s, the selection coefficient, is small, in the vicinity of 10^{-2}, the denominator becomes close to unity and can be disregarded. This gives us a general formula for Δq instead of just the change in q in one generation. Now let us see what this illustrates. A plot of this equation (Fig. 21-5) for all the values of q enables us to inspect the rate of change of q (a recessive allele) when selection continues to operate against it over a period of generations. However, the change in frequency of the deleterious allele q (Δq) is by no means a constant value from one generation to the next. The most rapid change in frequency of q takes place when the q frequency in the population

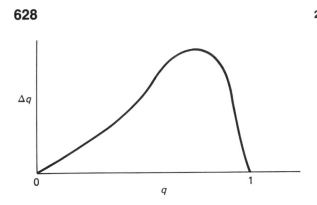

FIG. 21-5. Rate of change in frequency of deleterious recessive, q, in relation to frequency of q in population.

is approximately ⅔. It may be surprising to realize that at either allelic frequency extreme, 1 or 0, the frequency change of q takes place more slowly, even with the same s value, the same intensity of selection acting against it. As the frequency of q falls below ⅔, selection becomes less and less effective. Why should this be so? Remember that if q is a recessive, then the heterozygote (Aa) is *not* being selected against, and the deleterious recessive becomes sheltered in the heterozygote. In real populations, deleterious alleles usually occur at a low frequency. We can now see one reason why selection is ineffective in completely eliminating such undesirable alleles. Even if the force of selection against them should increase many times, the rate of change in their frequency would not be greatly affected.

The equation also tells us that when the recessive is very common (when q approaches 1), the rate of elimination again takes place very slowly. Another way of looking at this is with respect to a desirable

recessive allele that arises in a population. Obviously, its increase in frequency in the population will be slow at first. When the frequency of the allele is near 0 or 1, there is less genetic variability in the population. Without much genetic variability, natural selection works much less effectively than when considerable genetic variability is present—when q is about ⅔, for example.

Selection against a lethal allele

One might question what has been said about the elimination of deleterious alleles when it comes to those alleles that act as complete lethals. Here the selection coefficient equals 1, and the homozygous recessive (aa) does not reproduce at all. We can see (Table 21-9A) that after selection, the total population can be represented as p^2 (AA) $+ 2pq$ (Aa). Obviously, all of the recessive lethal alleles must be contributed by the heterozygotes, Aa. From Table 21-9B, we can calculate the genotypic frequencies that will be derived in the next generation as follows:

$$AA(p^2) + Aa(2pq) + aa(q^2) = (0.665)^2 + (2 \times 0.665 \times 0.335) + (0.335)^2 = 0.442\ AA + 0.446Aa + 0.112aa$$

From one generation to the next, the process continues, the maximum intensity of selection (1) operating against the lethal, q (a). Let us see how this affects the rate of change of the lethal allele and its elimination from the population. We return to the formula given in the last section for the rate of change of a deleterious recessive allele:

$$\Delta q = \frac{-sq^2(1-q)}{1-sq^2}$$

TABLE 21-9   Effect of selection against a recessive lethal allele

(A)	AA	Aa	aa	
Frequency before selection	$p^2 = 0.25$	$2pq = 0.50$	$q^2 = 0.25$	
Fitness (W)	1	1	0	
Proportionate contribution after selection	0.25	0.50	0	
(B)	AA	Aa	aa	Total
Contribution to gene pool after selection	$\frac{0.25}{0.75} = 0.33$	$\frac{0.50}{0.75} = 0.67$	$\frac{0}{0.75} = 0$	$\frac{0.75}{0.75} = 1$
Frequency of A(p) after selection	$\left(0.33 + \frac{0.67}{2}\right) = 0.665$			
Frequency of a (q) after selection	$\left(0 + \frac{0.67}{2}\right) = 0.335$			

We now substitute 1 for s which gives us

$$\Delta q = \frac{-q^2(1-q)}{1-q^2} = \frac{-q^2(1-q)}{(1+q)(1-q)}$$

Thus

$$\Delta q = \frac{-q^2}{1+q}$$

Notice that when the value of q is very small (substituting $0.01q$), Δq is approximately -0.0001. This tells us that even with a constant maximum intensity of selection against the homozygous recessive, the effect of selection decreases (Fig. 21-6). When the selection coefficient is less severe than for a full lethal (less than 1, as for a deleterious allele), the rate of decrease is even slower. An asymptote is approached, so that even in the case of a full lethal, selection against the homozygote does not completely eliminate the undesirable allele from the population. At this point it reflects the rate at which the recessive, a, is being added to the gene pool as a result of mutation from A. Although they steadily decrease, recessive deleterious alleles are not readily eliminated entirely from the population through selection. This does not mean that their frequency cannot be greatly reduced, but this requires many generations.

Suppose we wish to calculate the number of generations (n) required for a completely lethal allele to be reduced from some original value (q_0) to a later lower value, q_n. This can be estimated from the following formula:

$$n = \frac{1}{q_n} - \frac{1}{q_0}$$

The number of generations required to reduce the frequency of a lethal from 1/1000 to 1/10,000 = $1/0.0001 - 1/0.001 = 10,000 - 1000 = 9000!$ If we are considering a species with a long generation time, such as the human, this would mean that about 180,000 years would be required, a figure that extends well before the advent of the modern human!

We can now well understand why a harmful lethal recessive persists in the population even though it

FIG. 21-7. Equilibrium frequency of a fully lethal allele. A genetic factor that kills the individual before reproductive age still persists in the population at a certain frequency. Equilibrium is reached when the number of lethal alleles eliminated by the death of homozygotes is balanced by the number added to the gene pool through gene mutation (above). If the lethal were not replenished by recurrent mutation (below), it would eventually disappear, because the homozygote cannot reproduce, and the alleles are continually being lost from the population gene pool. This, however, does not happen because of the force of mutation pressure (above).

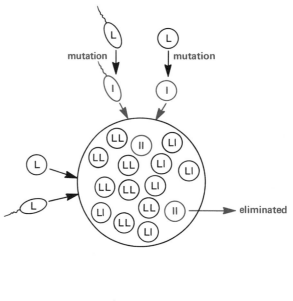

FIG. 21-6. Change in frequency of a lethal allele over a period of generations with selection against recessive aa.

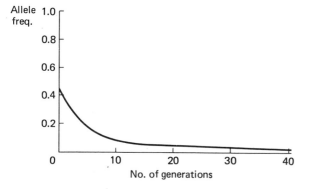

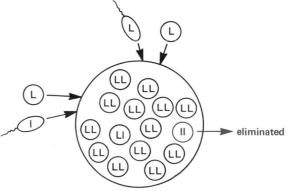

kills before reproductive age. It is sheltered in heterozygous individuals, and selection becomes less and less effective against it as its frequency falls. Eventually an equilibrium is reached between the force of selection against it and the force of mutation that adds it to the population. The number of deaths from the genetic disorder will be balanced by the number of births (Fig. 21-7). Let us say that a certain lethal allele causes the death of individuals with a frequency of 1/100,000. The defective allele is thus being eliminated with this frequency. Yet it does not disappear because 1/100,000 victims are being born. The number of newly arisen mutant alleles balances the number lost. Therefore, for a rare, fully lethal allele such as this in a population at equilibrium, we can approximate the mutation rate, if we know the number of deaths it causes (or the number of individuals born with the defect). In this case, the mutation rate would be in the vicinity of 1/100,000. This reasoning also assumes that there is no advantage conferred by the lethal allele on the heterozygotes, as occurs in the case of certain alleles that are lethal to the homozygote (see later in this chapter).

The Hardy–Weinberg formula tells us that the number of heterozygotes in the population will be still more frequent than the number of victims: the square root of $1/100,000 = 1/316 =$ the frequency of the lethal, l; thus $315/316 =$ the frequency of the wild-type allele, L. Consequently, the number of heterozygotes must be $2 \times 1/316 \times 315/316 = 1/158$. Again, we see that the number of heterozygotes, even for a rare recessive lethal, far outnumbers the homozygous recessives in the population.

Selection for dominant alleles

From our discussion up to this point, it is apparent that selection *for* some desirable recessive allele would proceed slowly when its frequency is low, just as the elimination rate of a deleterious allele proceeds slowly when *its* frequency is low. The effect of selection, however, is much greater in the case of dominant or semidominant alleles. This is well illustrated in nature by the rapid spread of darkened coloration in moths found in industrial regions of England.

In 1850, a new form of the common peppered moth, *Biston betularia,* was caught near Manchester, England. This new form with its darkened body and wings was given the variety name *carbonaria.* During the following years, this darkened or melanic form became quite common, and similar dark forms appeared even in other species of moths. All moth species in which melanism has developed have in common the fact that they rest on tree trunks when not in flight. The color pattern of the lighter form renders it almost invisible when it rests on trunks covered with lichens (Fig. 21-8). In the industrial regions of the country, however, the trunks are bare and darkened because of the quantity of soot that has destroyed the lichens. Consequently, the lighter colored moth stands out. The melanic form, which is conspicuous on the lichen-covered trunks, however, is difficult to detect on the blackened trees.

It was suspected that the spread of melanism is related to the concealment it provides. This was not established until the completion of painstaking work by H. Kettlewell, who demonstrated that the peppered moth is the prey of various species of birds. On the soot-laden trees, a much greater number of lighter than darker moths is destroyed. Conversely, the melanic form is destroyed in greater numbers on the cleaner, lichen-covered trunks. Clearly, predation is very great, and concealment is the selective agent favoring melanism.

The development of melanism has proceeded very rapidly. This is a result of the high intensity of selection for concealment and also of the nature of the responsible genetic factors. The spread of melanism in the peppered moth has been brought about mainly through selection for dominant allelic forms. The reason for this is very easy to understand. Assume first that we have a recessive allele, c, which can produce the darkened coloration when present in double dose, cc. First, in order to bring the desired darkened phenotype to expression, the recessive allele would have to occur in the gene pool with considerable frequency so that homozygous individuals could arise. Any rare homozygotes that do appear would be more likely to mate with the standard form, CC, than with other homozygotes, cc. As a result, they would leave no offspring with the desired darkened coloration. The valuable trait possessed by an individual would fail to be expressed in that individual's offspring. The effectiveness of selection would be extremely low.

Next consider another allele, B, which behaves as a dominant or semidominant to the recessive gene form, b, and also brings about darkened coloration. Now moths of genotypes BB and Bb will be dark and those of genotype bb light. Selection is now considerably simplified. Letting allele B equal q and assuming the population to be a randomly breeding one, the frequency of the darkened individuals in the population is equal to $2pq + q^2$. The frequency of the allele B itself is equal to q. We see that the proportion of

FIG. 21-8. The peppered moth *Biston betularia*. (Left) Three color forms on tree trunk covered with lichens. Note the inconspicuous light form at the top in contrast with the more conspicuous darker forms below. (Right) Three forms on a bare, darkened tree trunk. Note the conspicuous light form above and the more inconspicuous darker forms below. (Courtesy D. R. Lees and E. R. Creed).

individuals with the desired trait is greater than that of the allele that brings it about! Each darkened moth, no matter how rare the trait in the population, will always leave offspring of which at least half also show the darkened trait. For example, a heterozygote with a dark phenotype, Bb, crossed with a light form, bb, will produce dark (Bb) and light (bb) offspring with equal frequency. There is a high degree of correspondence between the phenotypes of parent and offspring, much higher than in the case of a rare recessive allele. Selection, therefore, is highly effective for a desirable dominant gene form.

Alleles at various loci other than those at the single B locus seem to be involved in the development of melanin in the peppered moth. Evidence for this comes from the fact that the melanic variety today is far darker than the original one. Heterozygous individuals (Bb) have grown progressively darker throughout the years, the heterozygote of today being scarcely distinguishable from the homozygous me-

lanic (BB). The dominance of the allele for pigmentation is not due to the single allele alone. A host of modifying genes present in the genome, each modifier having but a slight effect on darkening, interacted with the main allele for melanism to intensify its phenotypic effect, rendering it more dominant. We see once more the interaction of several genes in the production of a phenotype. In the peppered moth, dominant or semidominant alleles interacting with modifiers have made possible the most rapid spread of melanism under intense selection pressure.

The sickle cell allele and heterozygote superiority

The familiar hereditary disorder, sickle cell anemia, offers a very good illustration of selection operating in favor of the heterozygote. In this affliction, the presence of the allele Hb_B^S in place of the normal one, Hb_B^A, results in the substitution of just one amino acid at a critical position in the hemoglobin molecule (see Chap. 13). This change is sufficient to produce an assortment of physical disorders lethal to the homozygote. Because the effects of this mutant allele are so harmful, we would expect natural selection to operate strongly against it, keeping it at a very low level in the population. We would expect it to be maintained only by mutation pressure, if at all, because the relative fitness of the homozygote, Hb_B^S Hb_B^S, is approximately 0 $(s \simeq 1)$. Indeed, the allele Hb_B^S is rare or apparently absent in most human populations.

Surprisingly, however, its frequency is high on the west coast of Africa, where it may reach a prevalence of 20% in some populations. Such a high value for a

very deleterious genetic factor was quite puzzling until it was discovered that the frequency of the allele is correlated with the distribution of malaria. Where the incidence of malaria is low or absent, a corresponding drop is seen in the occurrence of the mutant allele and, consequently, of sickle cell anemia. The high level of the mutant allele in some groups indicated that somehow natural selection was operating to maintain it in certain environments. Obviously, those homozygous for the mutant allele did not usually live long enough to reproduce; so the defective allele was not being returned to the gene pool through the victims of the disorder. Any selection in favor of the defective allele must therefore be on the heterozygote, who carries both a normal and a mutant

FIG. 21-9. Effect of selection on the heterozygote for sickle cell anemia. Although the allele Hb_B^S causes the death of the homozygote, it imparts an advantage to the heterozygote in regions of high malaria incidence. Individuals homozygous for the normal allele, Hb_B^A leave

relatively fewer offspring than the superior heterozygotes, who keep adding the sickle cell allele to the gene pool. The frequency of Hb_B^S therefore remains relatively high in the population.

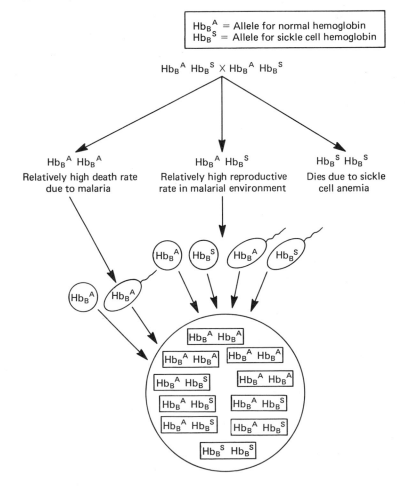

gene form. These heterozygotes were shown to possess a certain number of abnormal, sickled red blood cells. The mutant allele is actually a codominant, because its effect in the heterozygote can be detected.

It was finally demonstrated that the heterozygote possesses an advantage in childhood in his or her natural environment over persons homozygous for the normal allele because of an increased resistance to the malarial parasite, which apparently cannot complete its life cycle in the blood cells of the heterozygote. The red blood cells of these people contain both normal hemoglobin (Hb A) and sickle cell hemoglobin (Hb S). Protected in this way, the heterozygote can survive a malarial attack as a young child and build up antibodies that provide resistance against any later attack. The heterozygote has a sufficient amount of Hb A in the red blood cells to prevent severe anemia. At most, the person will experience only a mild form, and this may not even become evident.

The unfortunate result of the story is the maintenance of the deleterious allele at a high level in the population, because the heterozygotes will have a higher survival rate than those homozygous for the normal allele in areas of high malaria incidence and as a group will leave more offspring (Fig. 21-9). The heterozygotes are thus superior in regard to this characteristic. A certain balance is eventually reached between the frequency of the wild-type allele and its mutant form in different environments.

Let us see how selection in favor of the heterozygote can be studied using the same selection models we have been applying to other kinds of populations. Actually, the mathematical analysis is practically the same as that used to study selection against recessive alleles. We can allow $p^2 + 2pq + q^2$ to represent the proportions before selection of the three genotypes $Hb_B^A Hb_B^A$, $Hb_B^A Hb_B^S$, and $Hb_B^S Hb_B^S$. Since the fitness of genotype $Hb_B^A Hb_B^S$ is superior to those of the other two, it is given a fitness value of 1 ($s = 0$). We now have two other fitness values to consider. Let the adaptive value of the genotype $Hb_B^A Hb_B^A$

$(p)^2$ be equal to $1 - s$ and that of genotype $Hb_B^S Hb_B^S$ $(q)^2$ be equal to $1 - t$. In other words, s and t are selection coefficients, respectively, for the genotypes homozygous for the standard allele and for the mutant allele. Following the usual procedure, the proportionate contributions after selection for the genotypes can be determined as shown in Table 21-10. Notice that after selection, the total proportionate contribution for the population

$$\overline{W} = p^2(1 - s) + 2pq + q^2(1 - t) = 1 - sp^2 - tq^2.$$

With a selection process of this type, it can be shown mathematically that the allelic frequencies p and q (for HB_B^A and Hb_B^S, respectively) will not change from one extreme to the other but will come to assume an intermediate value. In other words, the Δq equation discussed in a previous section of this chapter becomes $\Delta q = 0$. Therefore, an equilibrium value for the frequencies of p and q is reached, and this value is determined by the relative values of s and t. The equilibrium values of q and p ($\hat{q}$ and $\hat{p}$) are

$$\hat{q} = \frac{s}{s + t} \qquad \hat{p} = \frac{t}{s + t}$$

Assuming $t = 1$ and $s = 0.25$, then $\hat{q} = 0.2$. Then assuming $t = 1$ and $s = 0.11$, $\hat{q} = 0.1$. These equilibrium values of 0.2 and 0.1 for q fit in well with actual measurements. The frequency value of Hb_B^S in various localities in Africa has been estimated to fall within the range of 0.1 and 0.2. Therefore, a range of 0.11 to 0.25 appears to be a good estimate for the value of s.

Other studies have been conducted in certain regions of Africa to estimate the adaptive values for all three genotypes. This has been done by recording the frequencies of the genotypes among newborn infants and comparing these with the same genotypic frequencies among adults. The allelic frequencies Hb_B^A and Hb_B^S found in the adult population were applied to the Hardy–Weinberg formula and were found to agree closely with the genotypic frequencies among newborns. In adults, however, there is an excess of

TABLE 21-10 Selection in favor of the heterozygote carrying the sickle cell allele

	GENOTYPES			
	$Hb_B^A Hb_B^A$	$Hb_B^A Hb_B^S$	$Hb_B^S Hb_B^S$	TOTAL
Frequency before selection	p^2	$2pq$	q^2	1
Relative fitness	$1 - s$	1	$1 - t$	
Proportionate contribution	$p^2(1 - s)$	$2pq$	$q^2(1 - t)$	$\overline{W}$

heterozygotes. As expected, genotype Hb_B^S Hb_B^S is heavily selected against and genotype Hb_B^A Hb_B^A has a slight disadvantage in relation to Hb_B^A Hb_B^S. The adaptive values were estimated to be 0.860, 1.00, and 0.169, respectively, for Hb_B^A Hb_B^A, Hb_B^A Hb_B^S, and Hb_B^S Hb_B^S. An adaptive value of 0.86 would give an s value of 0.14, which fits in well with the value for s estimated previously.

In those areas where malaria is no longer a problem, the heterozygotes have no advantage over those persons homozygous for the normal allele. Consequently, the frequency of the mutant allele Hb_B^S can be expected to fall, and this seems to be the case in regions of Africa where the malarial parasite is being eliminated. Those heterozygotes who are descended from African populations and who are now living in marlaria-free regions obviously possess no reproductive advantage. Calculations indicate that the frequency of the sickle cell allele among American blacks has actually decreased over the generations.

Gene mutations as an evolutionary force

The ultimate source of new genetic information is gene mutation. By itself, spontaneous gene mutation takes place continuously in a population. This mutation pressure is a force in evolution, because it brings about a change in the frequency of alleles. As we will see, however, since it is such a rare event, mutation by itself would change the genetic constitution of a population so slowly the change would be almost negligible. We learned in Chapter 14 that point mutation recurs continually and that the spontaneous mutation rate varies from one locus to another. An average rate of 10^{-5} (1/100,000) per generation has been estimated. How intense is spontaneous mutation as a driving force that brings about rapid evolutionary change in a population? This can be readily answered by considering certain simple facts about spontaneous mutation and population equilibrium.

Assume that a single mutant allele is produced: A→a. If the mutation occurs only once, its effect on the population would be exceedingly small, regardless of whether it produces an advantage or deleterious effect. If the population consists of 50 individuals (100 A alleles), the allelic frequencies after the mutation are: A=0.99 and a=0.01. In the next generation, the frequencies of the three genotypes are then $(0.99)^2 : 2 \times 0.99 \times 0.01 : (0.01)^2 = 0.9801$ AA:0.0198 Aa:0.0001 aa. In the absence of any selective force, this equilibrium would remain the same generation after generation. We see here the unimportance of a single gene mutation by itself.

However, a mutation for a given gene recurs each generation at a characteristic rate. Suppose we have a mutation rate for a given gene of 492 per million (the spontaneous per-generation rate for R, a gene for kernel color in corn; refer to Table 14-1). Assuming that we start with a population in which all individuals are homozygotes RR (pp), gene frequency will change from one generation to the next as follows: $\Delta p = up_0$, where Δp = the rate of change in the allele, u = the mutation rate from p to q (R to r here), and p_0 = the frequency of p in the original population. In one generation, the change in the allelic frequency is as follows: $\Delta p = (0.492 \times 10^{-3}) \times 1 = 0.000492$. The frequency of p in the next generation (p_1) is now easily calculated: $p_1 = p_0 - \Delta p = 1 - 0.000492 = 0.999508$. Since $p + q = 1$, $q = 0.000492$.

To obtain the allelic frequencies in subsequent generations, we proceed as follows. Since $\Delta p = up_0$, we can substitute in the p_1 equation to give $p_1 = p_0 - up_0 = p_0(1 - u)$. We then substitute p_2 for p_1 and p_1 for p_0, giving $p_2 = p_1(1 - u)$. We can, in turn, take this equation for the allelic frequency in the second generation and proceed further. We have just seen that $p_1 = p_0(1 - u)$. Therefore, $p_2 = p_0(1 - u)(1 - u) = p_0(1 - u)^2$. If we continue in this way generation after generation, we will find that after t generations, the allele frequency will be: $p_t = p_0(1 - u)^t$. Should the process continue unopposed, we see that p will eventually disappear and only q will be represented. In other words, starting out with 100% allele R in a population, eventually only r would be found.

The number of generations required for any significant change to take place in the allelic frequency is very high. Using 10^{-5} as an average rate of mutation, the speed at which allele frequency changes as a result of mutation pressure by itself is exceedingly slow. The value for u in the last equation is very small in relation to 1.0. Therefore, $(1 - u)^t$ can be considered to be approximately e^{-ut}, where e is the base of the natural logarithm 2.718. The equation then becomes: $p_t = p_0 e^{-ut}$. Let us use this to estimate the number of generations (t) required to reduce the frequency of allele A by half in a population.

We substitute $\frac{1}{2}p_0$ for p_t in the equation and then take the natural logarithm of both sides. If $u = 10^{-5}$, then $t = 0.69/u$ or 69,000 generations! Considering the human generation time to be 20 years, the number of years required to reduce allele A from a frequency of 1.0 to 0.5 would be 1,380,000! We can certainly conclude from such a vast figure that gene mutation by itself cannot be responsible for rapid evolutionary change that occurs in populations in nature.

Back mutation and mutation pressure

We know that gene mutation is not unopposed in one direction, A→a. Actually, reverse mutation takes place, so that the A allele is being replenished by mutation of a to A, a→A. We learned in Chapter 14 that these two rates are not equal, the forward rate being about 10 times higher than the reverse rate. How does the fact of reverse mutation affect what has been said so far about the effect of gene mutation on alteration of allelic frequency?

As a consequence of back mutation, A would not disappear from the population gene pool. If we allow the frequency of A to equal p and that of a to equal q, we then need to consider two mutation rates: u, the rate of mutation from p to q and v, the rate of back mutation. This can be diagrammed as

$$p \underset{v}{\overset{u}{\rightleftarrows}} q$$

In the absence of any other evolutionary forces, an equilibrium frequency will become established, at which time Δp and Δq will each equal O. We can see this from the following consideration: Δq equals the gain in forward mutation (up) minus the loss of q as a result of back mutation. Thus, $\Delta q = up - vq$. Using similar reasoning, $\Delta p = vq - up$, the gain resulting from back mutation minus the loss from forward mutation. At equilibrium, $up = vq$. Since $p = 1 - q$, we can substitute for p in the equilibrium formula just given:

$$u(1-q) = vq = u - uq = vq$$

We can then continue to solve for q at equilibrium $(\hat{q})$:

$$u - uq - vq = 0; \quad uq + vq = u; \quad q(u+v) = u; \quad \hat{q} = \frac{u}{u+v}$$

Since $q = 1 - p$, we can substitute for q in the equilibrium formula, $up = vq$ and derive the equilibrium value for p $(\hat{p})$: $\hat{p} = \frac{v}{u+v}$. Suppose that $u = 0.0006$ and $v = 0.00002$, then the equilibrium frequency of p equals

$$\hat{p} = \frac{v}{u+v} = \frac{0.00002}{0.00008} = 0.25$$

The equilibrium value for $q(\hat{q}) = 0.75$. Mutation pressure by itself, both forward and reverse, leads to an equilibrium in the absence of any other evolutionary forces. Therefore, if the mutation rates, u and v, remain the same, the population arrives at a Hardy–Weinberg equilibrium in which evolution no longer occurs and $\hat{p}$ and $\hat{q}$ are the familiar p and q of the Hardy–Weinberg formulation.

Moreover, since the forward mutation rate (u) is usually greater than the rate of back mutation (v), the mutant allele would almost always become more common in a population than the standard one. We know that this is not the case and that evolution does occur in populations. We must therefore conclude that mutation pressure alone is not responsible for the frequencies of alleles in natural populations. In nature, gene mutations provide different allelic forms of a gene, and these are spread throughout the population by sexual reproduction, which entails independent assortment and recombination through crossing over. This makes possible many different combinations of newly arisen mutant gene forms with each other and with those already established in the gene pool; and as a result, the effect of gene mutation is amplified.

These processes operate in the population to produce variation, and variation in the form of new combinations of the genetic information is the raw material acted on by natural selection, the primary evolutionary force that changes the frequency of alleles in populations. Natural selection changes the gene pool by giving a reproductive advantage to those individuals with favored combinations of alleles, in certain environments. Some genetic factors increase in the gene pool over their allelic forms. The effect of natural selection is to steer the population toward an increased level of adaptability and efficiency within the environment.

Migration pressure

The frequencies of alleles in populations may be changed as a result of migration or gene flow, which introduces foreign gametes from other populations. A very simple example illustrates how allelic frequencies may be altered by the addition of alleles from one population to another. In this example, we will assume that the migration is exclusively from population 2 to population 1. Thus, population 2 can be considered the migrant population and population 1 the original or resident population. We will again consider alleles A and a, represented respectively by p and q in the following equations.

Assume that in the resident population the original allelic frequencies are A = 0.6 and a = 0.4. The original frequency of q can be represented symbolically as q_0. Thus, $q_0 = 0.4$ (follow Fig. 21-10). The frequency of q in the migrant population $(q_m) = 0.7$. Let us suppose that in any given generation there is a certain proportion of new immigrants into the resident population. We can represent this proportion by m. The rest

of the population, the original residents, is now $1 - m$. Assuming that the proportion of immigrants = 0.2, then $m = 0.2$ and $1 - m = 0.8$ (the proportion of natives in the original population after migration). Thus, the frequency of q (representing allele a) in the generation after migration can be calculated in this way: q_1 (frequency of q in generation 1 after migration) = q_0 $(1 - m) + q_m m$. In our example, this becomes: $q_1 = 0.4(0.8) + 0.7 \times 0.2 = 0.32 + 0.14 = 0.46$.

The change in gene frequency in one generation in the new population is thus seen to depend on the size of the proportion of migrants, m, and the difference between q_m and q_0: $\Delta q = q_1 - q_0$, or otherwise expressed: $\Delta q = m(q_m - q_0)$. From either of these equations, we can see that the change in frequency of allele a in our example equals 0.06.

This example also shows that migration may significantly alter the allelic frequency of a resident population in just a single generation when the two populations differ before migration with respect to a specific allele. Migration and gene flow provide a source of genetic variation for a population. When the difference in the allelic frequency is appreciable between the populations (when $q_m - q_0$ is large), the frequencies of the alleles are significantly altered. This may have the effect of imparting much more variability to a population that originally had a very low frequency of a given allele and was thus much more uniform with respect to genotypes. Moreover, migration and constant gene flow are factors that retard the divergence of separate related populations and thus retard the process of speciation (more later).

FIG. 21-10. Effect of migration on allele frequency in a resident population (see the text for details).

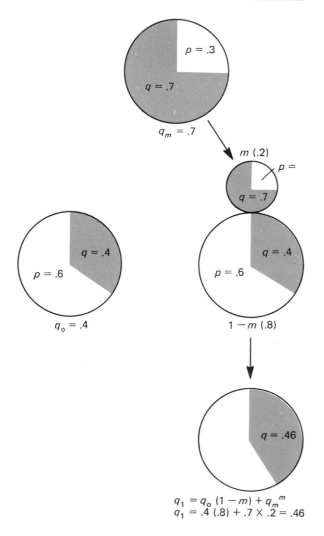

q_0 = Original frequency of q in resident population
q_m = Frequency of q in migrant population
m = Proportion of immigrants
$1 - m$ = Proportion of natives

$q_1 = q_0 (1 - m) + q_m m$
$q_1 = .4 (.8) + .7 \times .2 = .46$

Role of chance in evolution

In addition to mutation, natural selection, and migration pressure, another force operates on populations, and its effect may actually counter that of natural selection. This is the action of *genetic drift*, fluctuations of allelic frequencies within a population through random processes. Genetic drift can be appreciated in light of what we learned about probability. When we are dealing with small numbers, chance fluctuations may arise, such as a run of six heads in tosses of a coin or the birth of four boys in a row in a family. Ideal genetic ratios may not be realized when a small sample is involved. In a similar way, it is possible in a small population for certain alleles to become established or to become lost by chance without any relation to their adaptive value. A beneficial allele may be lost or a deleterious one may even become established as a consequence of genetic drift. Some details can give us insight into the operation of this evolutionary force.

A given generation of a population has a certain number of individuals, represented by size N. N has resulted from a sampling of the gametes of the previous generation, a gamete total represented as $2N$. If N were an infinite number, then the gametes that united to form any generation would reflect accurately the frequency of the alleles in the gene pool of the parental generation. Since N is finite, the gametes that unite to form any given generation represent only a sampling of the gene pool of the previous generation.

Hence, if we were to take different samples of gametes from the gene pool, we would expect variation from one sample to the next and occurrence of sampling error. The smaller the population size, the smaller the sample of gametes and the greater the chance for departure from the mean in the sample and for sampling error. The outcome is random fluctuation in allele frequency from one generation to the next because of the sampling errors. The frequency of allele a can change from one generation to the next, decreasing or increasing in value from the generation before. It is impossible to say with any degree of certainty in which direction the allele will go or by how much. Let us see, however, how it is possible to ascertain the probability that it will lie within a certain range of values.

The *variance* is a value that measures the expected variation from the mean in the frequency of an allele in a sample. Allowing p and q to designate alleles A and a, respectively, the variance of a in one generation is given by $p \times q/2N$. Since the variance is the square of the standard deviation, the standard deviation for a (σ_q) becomes $\sigma_q = \sqrt{pq/2N}$. The variance or the standard deviation will be larger as the population size (N) becomes smaller and the chance becomes greater that allele frequency will vary as the result of random sampling. To illustrate this, consider three populations, each with $p = 0.5$ and $q = 0.5$. Population 1 consists of 50 breeding individuals, population 2 of 100, and population 3 of 5000. The σ values for $q(a)$ can then be determined for each, respectively.

$$\sigma_q = \sqrt{\frac{0.5 \times 0.5}{2 \times 50}} = 0.05$$

$$\sigma_q = \sqrt{\frac{0.5 \times 0.5}{2 \times 100}} = 0.04$$

$$\sigma_q = \sqrt{\frac{0.5 \times 0.5}{2 \times 5000}} = 0.005$$

What do these figures tell us? (Recall from Chapter 7 the use of standard deviation.) In these three populations, the mean of the possible values of q in the next generation $= 0.5$. The symbol q_1 is used to represent a *possible* value of q in the next generation: σ_q measures the expected variation of q_1 from the mean. For population 1 with a σ_q value of 0.05, the chance is 68% that q_1 will have a value of 0.5 ± 0.05 or 0.55 to 0.45. The chance is 32% that it will be greater than 0.55 or less than 0.45. We can see that as the population size increases, σ decreases. This tells us that although a large population can give rise to *more* possible different values in a given sample (or a given new generation), more of these values will be found within a narrower range around the mean than in a small population. In the latter, fewer different allele frequency values can be produced. However, there is a greater chance that any *one* will depart significantly from the mean. The smaller population is thus less predictable than the larger one. Therefore, in a small population, unpredictable random variations in allele frequency arise from one generation to the next and appear to drift aimlessly.

The eventual outcome of this is that q may become lost entirely $(q = 0)$ or that p may be lost and q may become fixed $(q = 1)$. When this happens, no further change is possible, because at that locus all individuals in the population would be homozygous (ignoring other facts such as gene mutation and immigration, etc.). Remember that it is chance that determines the loss or the fixation of an allele in the case of genetic drift. As a result of the operation of chance, very small populations suffer a loss of variability through a decrease in heterozygosity.

When references are made to the size of any population, it is essential to recognize that in sexual populations, not every member is breeding or contributing to the gene pool in a given generation. We are concerned only with the size of the population that is actually breeding and contributing to the formation of individuals in the next generation. These breeding members compose the "effective population size," generally represented as N_e. The effective population size may not be apparent. A population that now has a very large N_e may have once had a small effective breeding population over several generations. The effects of genetic drift at that time could account for certain allele frequencies in the larger population. In populations, the probability of loss of variability is estimated to be $\frac{1}{2}N_e$ per generation. This is also a reflection of the rate at which the population is becoming homozygous at different loci. In a population with 200 breeding members, we can calculate that 1 locus in 400 can be expected to become homozygous in the following generation. Later we will discuss the serious effects of reduction in effective population size on natural populations of the cheetah.

The founder effect

A special case of genetic drift is well illustrated by a type of sampling error called the *founder effect* by Ernst Mayr. Suppose the frequency of allele a in a given sample (q_s) is very different from the frequency of a in the parent population from which the sample was taken (q). The few gametes in the sample thus contain a frequency of alleles that is not representative of the larger, parental population. Now suppose that these gametes represent the genetic contribution of a few individuals acting as founders for the establishment of a new population. The gene pool of this new population will thus be very different from that of the original population (Fig. 21-11).

The founder effect can result from some drastic reduction in the size of a large population, which would produce a small effective breeding population for the following generation. That generation would carry an allelic frequency very different from that in the original population. The founder effect may also result from the migration of just a few individuals from a parent population. These then become the founders of a new population in a different geographic region.

FIG. 21-11. The founder effect. A deleterious recessive allele may be present in a few individuals in a population. Since the frequency of the allele is very low, the deleterious allele is usually present only in heterozygotes. Few homozygotes expressing the recessive trait are seen. If, however, a few carriers become isolated and form the basis of an isolated population, the frequency of the allele in the new population will be greater than in the original, and the recessive will then come to expression much more often.

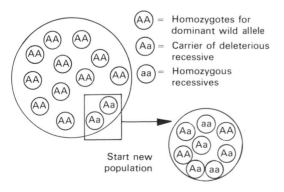

The founder effect is believed to have operated in various plant and animal groups. There is also strong evidence that this special kind of genetic drift has also acted in certain human populations, in which the founders of certain orders or sects just happened to possess allelic frequencies that were not representative of the parent populations from which they emigrated. The Amish people in some localities show a high incidence of a particular recessive trait, a type of polydactyly. This has apparently been derived from the gene pool of the founders, which happened to contain a a high frequency of the recessive allele—a frequency quite different from that in the parental gene pool.

An especially well known case concerns the Dunker religious community of Pennsylvania. During the early part of the eighteenth century, about 50 families migrated from western Germany. At the present time, the community consists of about 300 individuals. However, the size of the group has been under 100 persons for several past generations. For religious reasons, he Dunkers tend to marry only within the group. They have been closely studied by Bentley Glass, who selected several traits for comparison between the Dunkers, western Germans, and neighboring Americans. The traits that were selected for study were chosen because they are considered to be relatively neutral ones with regard to natural selection.

The data show that the Dunkers resemble neither the present-day western German population nor their American neighbors who share a similar environment. In the case of blood groups, for example, the frequency of type A blood is 45% in West Germans, 40% in the American group, but 59% among Dunkers. In the German and the American populations, the total of blood groups AB plus B equals about 15%, but among Dunkers, it is only 5%. Blood type M is about 30% in both of the major groups but is about 45% in Dunkers. Among Dunkers, type N is about 14% and MN 42%, whereas in the other two populations, the values are 30% and 50%, respectively.

The differences between the Dunkers and the other populations are statistically significant. The Dunkers resemble no other populations with regard to the traits studied. It would appear that the founding fathers of the Dunker community did not represent a random sample of the German population from which they emigrated. There seems to be little doubt that this kind of genetic drift, the founder effect, has played a large role in establishing allele frequencies in this and certain other small populations.

Variation in natural populations

Throughout our discussions, we have stressed the importance of variation in the process of evolution. With only a small amount of variation in a population, natural selection would have little raw material to act on, and the rate of evolution would proceed very slowly as a consequence. Indeed, if no variation at all exists at a given locus (all population members are AA or aa, e.g.), then no evolution can take place at this site. In contrast, if two or more alleles exist at a locus, one allele may increase at the expense of another through the operation of natural selection. Since the environment is not static, the selective value of any allele is not fixed throughout time. One allele at a locus may be favored now, but a different form of the gene may have a greater selective value at a later time. The variability within the gene pool may actually preadapt a population to cope with some environmental change in a later generation.

Darwin recognized random inheritable modification of individuals in a population as the basis of evolution. According to Darwinian concepts, advantageous modifications become established through the force of natural selection, which raises the population to a greater adaptive level in its environment. The variation present within a population, however, was not thought to be very widespread. Except for some uncommon variants carrying mutant alleles, population members were regarded as a somewhat standard type, homozygous at most loci for wild-type alleles. This picture of somewhat limited variability is now rapidly changing and does not appear to be the correct one in nature, as we shall now see.

Experiments using artificial selection have revealed that a large amount of genetic variation is stored in natural populations. In such studies, the investigator selects for breeding only those population members expressing some desirable trait to the greatest extent. The selective process is continued over several generations. If the population eventually changes in the desired direction (say taller plants) by artificial selection, the implication is that the original population contained a significant store of variation for that particular trait. When populations are subjected to artificial selection for a wide range of traits, the desired result is almost always obtained. It would seem that natural populations contain a large store of variation for just about every characteristic of the particular species under study.

A knowledge of the extent of genetic variation in a population is valuable for the population geneticist who wishes to know how flexible a species is in its ability to adapt or survive in a changing environment. Controlled experiments with fruit flies have clearly demonstrated that the speed of evolution is related to the amount of genetic variation in the population. A fly population with twice the genetic variation of another was found to adapt much more quickly to laboratory conditions in competition for food and space. Such experiments clearly underscore the importance of variation.

The determination of the amount of genetic variation in natural populations by genetic procedures is an impossible task, because one cannot study *every* locus of an individual to calculate the fraction that is heterozygous. Moreover, it is not possible to obtain a random sample of *all* the loci, since classic genetic procedures detect only those loci where a gene occurs in two or more allelic forms. One loses sight of those loci where only one gene form exists. Therefore, any sample would be biased. Fortunately, the problem has been overcome through the application of molecular biology and the technique of gel electrophoresis to the study of protein variation. We certainly know that structural genes code for the amino acid sequences of proteins. Using protein variation in individuals as the guide, one can estimate the amount of variation in the genetic material, since the variation in a given kind of protein indicates variation in the genetic information that codes for it. If a certain protein shows no variation among the individuals of a population, then the genetic material that coded for it is most likely invariant as well.

One can thus select a number of proteins to represent a random sample of genes in an individual. The proteins are then subjected to gel electrophoresis, a technique so sensitive that it can detect one amino acid difference out of 100 between two proteins. In the procedure, protein samples from several population members can be run side by side on a gel, thus allowing comparison to be made of various specific proteins among individuals.

Let us suppose that blood samples are taken from different population members and that soluble proteins are to be studied to estimate the degree of genetic variability. The blood is treated to separate cellular components from the serum that contains the proteins. Samples containing the soluble proteins are then placed on a gel and subjected to an electric current. Exposure to the electric field causes a protein to migrate through the gel in a direction and at a rate that depends on its net electric charge and molecular size. Each type of protein will come to assume a characteristic position in the gel. After completion of the run, the gel is treated with a chemical solution that will stain a specific protein and reveal its position (Fig. 21-12).

Let us next suppose that a given polypeptide chain is encoded by a specific gene and that two of the chains unite to form an enzyme. Different allelic forms of the gene that specify slightly different versions of the polypeptide chain may exist. Such chains may therefore combine to produce forms of the enzyme that differ in molecular structure and charge, and consequently have different mobilities in the electric field. An individual homozygous for one allelic form of the gene (A1 A1) will produce two identical polypeptides and only one form of the enzyme, which will be visualized as one band. An individual homozygous for another allelic form of the gene (A2 A2) will also produce one enzyme form, but the resulting band may assume a different position in the gel, because the A2 polypeptide is slightly different from the A1. An individual heterozygous for these alleles can produce three different kinds of polypeptides, and thus three different enzyme forms and three bands (A1 A1, A1 A2, A2 A2). Some genetic loci may be associated with genes that exist in more than two allelic forms. In such a case, more than two kinds of homozygotes can be identified as well as a greater number of heterozygotes. This procedure and its modifications can be used in this way to study blood and other tissue samples taken from a number of individ-

uals in a population, and thus to estimate the variation at a given genetic locus.

Gel electrophoresis has revealed unsuspected amounts of genetic variation in natural populations. Studies of more than 250 species have shown that between 10 and 60% of the genes in each species are polymorphic and that, on the average, between 1 and 36% of the genetic loci of an individual are heterozygous. We can conclude that natural populations contain a significant amount of *genetic polymorphism*. This term is used when two or more alleles occur at the same locus in the population with a minimum frequency of at least 1%. A population is considered to be polymorphic when two or more forms exist within the group in such proportions that the rarest form occurs at a frequency too high to be maintained solely through recurrent mutation. A good example is the existence of both the melanic and light forms of the peppered moth, discussed earlier in this chapter.

The genetic polymorphism in populations is much greater than anticipated according to classic Darwinian theory. Members of a population are not homozygous for certain standard alleles at most loci, but instead are heterozygous at a significant proportion of them. The implications of this finding for the generation of new combinations of alleles are staggering. Heterozygosity in the human has been estimated to occur at about 6.7% of the genetic loci. Assuming that there are 100,000 human loci, this indicates that a person is heterozygous at approximately 6700 loci and could potentially produce 2^{6700} kinds of gametes, a figure so vast it cannot be comprehended! Most of

FIG. 21-12. Detection of genetic variability in a specific protein. Soluble protein samples *(black)* from five different individuals are subjected to an electric field that will separate the mixture of proteins in each sample. After the separation, a specific strain is applied to reveal the position in the gel of a specific protein *(red)*, such as an enzyme. If the gene for the enzyme exists in two allelic forms (A1 and A2) in a population, and if the enzyme is composed of two polypeptide chains encoded by the gene, some individuals may be genotype A1 A1 (individuals 1 and 4), some A2 A2 (individual 2), and others A1 A2 (individuals 3 and 5).

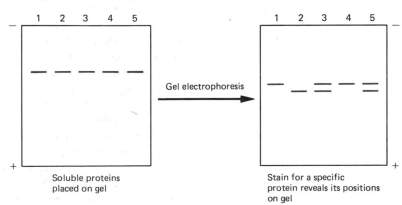

the genetic variation present in a population stems not from the occurrence of new mutations in each generation, as believed by classic Darwinists; rather, it is generated through recombination. The reshuffling of stored alleles reveals the amount of variation that lies hidden in a population over generations. Gene mutation is, of course, the ultimate source of genetic variation, but it is a rare event, adding only a minute amount at each generation, as discussed earlier in this chapter. A given generation does not depend on recurrent mutation to provide it with the needed variation.

Types of polymorphism

Some of the genetic polymorphism in a population is constantly changing, characterized by a continual increase in frequency of one allele, say allele a, and the decrease of another, allele A. This *transient polymorphism* does not attain a steady state. It is governed solely by natural selection and leads toward the fixation of one allele and the elimination of the other. During the period of evolutionary change, over the course of generations, two or more forms will be present in the population. The peppered moth with its melanic and light forms illustrates such transient phenotypic polymorphism. Over a short period of just 50 years a dramatic increase in the frequency of melanic forms took place in some localities, with the corresponding decrease in the lighter forms. The strong selection pressure operating on these populations is the responsible factor. With eradication of pollution from the areas, the process would be expected to reverse itself: the light forms would then increase over the melanic ones, which could eventually disappear if a changed environment continued to select strongly against them.

In contrast to transient polymorphism is *balanced polymorphism,* in which the alleles at a locus, A and a, reach a frequency that tends to remain constant. We might well wonder how such a steady state can be maintained, since it implies that an allele may be maintained in the population even though it does not impart an advantage. Several explanations have been offered.

A mechanism called *frequency-dependent selection* has been suggested to explain balanced polymorphism in some populations. In such a situation, the advantage imparted by an allele at a given locus is determined by its frequency in the population. The advantage it imparts would thus fluctuate with the frequency. In other words, some alleles may have a selective advantage when they are found at a low level in the population. A good example is provided by the self-sterility alleles in various plant species (review Fig. 4-19).

Consider a population in which at first only three sterility alleles are found in equal frequency (S_1, S_2, S_3). All the plants in the population are heterozygotes: S_1S_2, S_1S_3, S_2S_3. Pollen grains carrying any one of the three alleles is compatible with only one third of the plants in the population. Pollen with allele S_1, for example, can grow only on plants of genotype S_2S_3. Now, if a new allele, S_4, arises as the result of a point mutation, it has a high selective advantage at first, because pollen with S_4 can grow on the stigma of any plant in the population. At first allele S_4 will be rare. As S_4 pollen grows on plants of the three prevailing genotypes, the number of individuals carrying S_4 increases as plants of genotypes S_1S_4, S_2S_4, and S_3S_4, incompatible with S_4 pollen, rise in frequency in the population. As the proportion of such plants increases, and hence the frequency of allele S_4, the initial advantage of S_4 decreases. A point of equilibrium is eventually reached at which all the alleles come to exist in a balanced state.

Another theory to explain balanced genetic polymorphism is one with which we are already familiar. This is heterozygote superiority, as seen in the case of the sickle cell allele. In situations like this, the heterozygote, Aa, is favored over either homozygote AA or aa, and thus has a reproductive advantage. Consequently, neither allele can be entirely eliminated.

Consequences of the lack of population variation

The endangered plight of the cheetah in the wild exemplifies the serious consequences of lack of genetic variability in a population, probably resulting from extreme reduction in effective population size (N_e; see earlier this chapter) sometime in the past. To a striking degree, the cheetah exhibits effects that are associated with extreme inbreeding (Chap. 7). As recently as 1900, the cheetah was widely distributed throughout Africa, the Middle East, and India, but today the species is confined to a few small areas in Africa, where its numbers are not thought to exceed 25,000. The animals show reduction in breeding ability and a high mortality rate of cubs, and may be on the road to extinction.

Examinations into the reasons for the cheetah's

decline reveal strong evidence that the gene pool of cheetahs in the southern African population contains very little genetic variability. Analyses of semen disclosed a high incidence of abnormal sperm. Sperm morphology is under genetic control, and an increase in the amount of defective sperm is known to be associated closely with inbreeding in livestock and in laboratory animals. Electrophoretic studies of blood samples in which 52 proteins were extracted revealed that no genetic variation in these proteins occurs from one cheetah to the next. For each protein, all the animals were monomorphic, having only one form of the protein. This is a most unusual situation for a species in the wild, and contrasts with other species in nature where significant polymorphism is found among individuals in respect to a given protein (see Fig. 21-13).

Even more unexpected were the results of studies of the major histocompatibility (MHC) locus in the cheetah. This MHC locus, which is found in mammals (and which is known as the HLA complex in the human), is a complex locus and actually consists of several genes (Chap. 4). Moreover, it is the most polymorphic genetic region found in mammals. A review of Fig. 4-24 will show that many allelic forms are possible at each of three tightly linked histocompatibility genes in the complex. The chance that any two unrelated persons will possess the same combination of alleles is very remote—less than 1 in 10,000. Other mammalian species also show extreme polymorphism at the MHC locus. Variation at this genetic region is largely responsible for the graft rejection seen when tissue is transplanted from one unrelated individual to another. Members of a highly inbred strain of laboratory animals readily accept grafts from each other, as is true for exchanges between identical twins. The reason for this is the homozygosity of the genes at the MHC locus.

Amazingly, all skin grafts between any two cheetahs were readily tolerated. Any animal accepted skin from another just as if it were having a skin transplant from one part of its own body to another. However, a cheetah will not tolerate a skin graft from another species, such as the domestic cat, showing that the cheetah's immune system is intact. This leads to the conclusion that the cheetah's ability to exchange grafts freely is due to homozygosity at the MHC locus. Such a lack of genetic variation in a natural population is unprecedented and poses a great threat to the cheetah's survival.

What can account for such genetic uniformity in the southern African cheetah population? The most likely explanation is that some crisis occurred in the history of the cheetah that greatly reduced the size of the breeding population. We know that earlier than 12,000 years ago, the cheetah had a worldwide distribution and included many distinct species. Then a great extinction, affecting many mammalian species, occurred. The cheetah was reduced to one species, and its range became more and more restricted. Perhaps more than one crisis may have occurred in the cheetah's past—some fairly recent—to reduce the effective population size perhaps to just a very few individuals. Earlier in this chapter, we examined the loss of variability in small populations and the effects of reduction in effective population size (N_e). These effects are dramatically demonstrated in the cheetah, whose lack of genetic variability poses a serious threat to the species' ability to adapt to a changing environment.

Biochemical studies on the cheetahs from eastern Africa are still to be done. It is hoped that these animals, which also appear to be inbred, will nevertheless differ in the composition of their gene pool from the southern African population. If this proves to be the case, programs could be instituted to bring about breeding between animals from the southern and eastern populations. The results could be an increase in the variability in the cheetah's gene pool, which may counteract the dire effects of the popula-

FIG. 21-13. Polymorphism and monomorphism in a protein. Variation is seen in some of the proteins when different population members of a species are compared. A specific protein may prove to be very polymorphic *(left)* when it is identified by a specific stain *(red)*. In contrast, all the blood proteins studied in the cheetah were monomorphic. A given protein *(right)* shows no variation from one individual to the next, even though that same protein may be polymorphic in another species, as seen on the left.

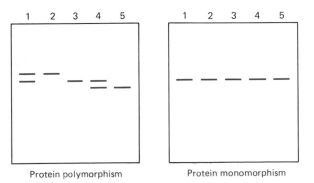

Protein polymorphism   Protein monomorphism

tion bottleneck and divert the species from the path to extinction.

Role of genetic isolation in speciation

Earlier in this chapter we saw that the forces of evolution operate over the course of generations within a population and produce genetic change. In addition to the genetic changes that accumulate in a given population over the course of time, another very important aspect of evolution is the origin of two reproductively isolated groups from one original population. For two distinct species to arise from one ancestral group, two segments of the original population must be separated, preventing genetic exchange between them. Operating under two different sets of environmental circumstances, natural selection may favor certain alleles and genetic combinations in one population segment; these alleles may be selected against in the other segment, where a different assemblage of genetic factors is favored. Moreover, different spontaneous gene mutations may occur in the two separated groups. In effect, there are now two separate populations, each one undergoing changes in its own gene pool. If the two gene pools eventually become sufficiently different, individuals from the two populations may become so genetically dissimilar that they cannot cross when brought together. Two different species will be produced from one original species.

When we observe the distribution of species in nature, we find that certain ones, considered to be related on the basis of morphology, chemical resemblances, karyotype, and so on, are geographically or spatially isolated from one another. Such populations are said to be *allopatric*. No gene flow can take place between them because of the spatial barrier caused by their distribution. Allopatric populations may later come to overlap and to share a portion of each other's range if the barrier between them is eliminated. For example, some environmental change at one time may have fragmented one large population into two and created between them a region inhospitable for survival. The forces of evolution, operating on the two isolated groups over a sufficiently long period, could bring about sufficient genetic differences to cause reproductive isolation if they should come in contact again. A subsequent change in environment, for example, might render the zone between them hospitable again. Perhaps one or both of the populations will expand and the two ranges may come to overlap.

Species whose ranges overlap or occupy the same general area are said to be *sympatric*. Closely related sympatric species maintain their reproductive isolation because of the physiological expression of genetic differences established when they were allopatric (Fig. 21-14A). The two species may now exploit different ecological niches within the same region. They may have different flowering times in the case of plants, or different mating habits in the case of animals. If any crossing took place between them, the offspring would be sterile. The expression of genetic differences established during the period of spatial isolation will keep them reproductively isolated and enable them to diverge even further.

On the other hand, two allopatric populations may become sympatric before the accumulation of sufficient genetic differences renders them reproductively isolated, so that they may still be able to cross with relative ease and produce a high proportion of fertile offspring. Natural selection would operate to eliminate those genetic differences that arose when they were separate populations and that now cause some reproductive incompatibility and sterility. Eventually, the two groups could become part of one new interbreeding population (Fig. 21-14B).

Sympatric speciation

The extent of speciation without the intervention of spatial isolation has been a subject of debate. Special conditions must be involved for sympatric speciation to occur. For example, suppose a population covers a range that is very diversified and encompasses an assortment of local conditions that provide different ecological niches. A given genotype may have an advantage in one of these, whereas a different genotype will do better in a second one. The population, as a whole, may profit if all the local environments within the range are occupied and used to the fullest. For such a set of circumstances to effect sympatric speciation, assortive mating must become established, so that members occupying a given niche mate preferentially with others in the same niche. If divergence leading to reproductive isolation is to become established, genetic exchange between members of the various local areas must not take place.

At one time, the concept of sympatric speciation was discarded entirely, but it is now recognized as a powerful influence in plant speciation. In Chapter 10, we discussed at length the increase in chromosome

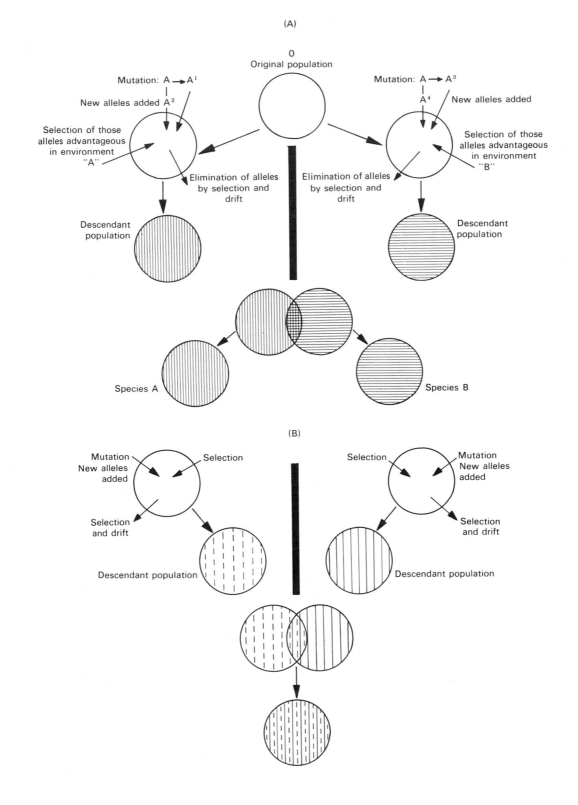

(A)

O
Original population

Mutation: A → A¹

New alleles added A²

Selection of those alleles advantageous in environment "A"

Elimination of alleles by selection and drift

Descendant population

Mutation: A → A³

A⁴

New alleles added

Selection of those alleles advantageous in environment "B"

Elimination of alleles by selection and drift

Descendant population

Species A

Species B

(B)

Mutation New alleles added

Selection

Selection and drift

Descendant population

Selection

Mutation New alleles added

Selection and drift

Descendant population

number by the multiplication of whole chromosome sets through autopolyploidy and allopolyploidy. It should be appreciated from a review of the radish–cabbage cross (see Fig. 10-30) that the derived allopolyploid (the tetraploid Raphanobrassica) is a true species on the basis of several criteria.

There is very strong evidence that the process of allopolyploidy does operate in nature and has produced tetraploid derivatives from diploids that still exist in nature today. This has been demonstrated experimentally by the actual synthesis of certain natural species as a result of crossing known diploids that also occur in nature.

An excellent, early demonstration was made by Muntzing, working with European herbs of the genus Galeopsis. Muntzing crossed two different species, *Galeopsis pubescens* and *Galeopsis speciosa* (Fig. 21-15). Each has a haploid chromosome number of 8. He managed to obtain an F_1 hybrid, but this proved to be quite sterile. However, from this F_1, he finally succeeded in obtaining an F_2 offspring. This plant proved to have 24 chromosomes and was actually a triploid. Muntzing then backcrossed this plant to *G. pubescens* and obtained one seed. This in turn grew into a perfectly fertile plant that was exactly the same as the naturally occurring species, *Galeopsis tetrahit*. When he crossed the experimentally synthesized plant to a natural *G. tetrahit,* he obtained perfectly fertile *G. tetrahit* offspring. This proved that the natural plant and the synthesized one were genetically identical and that natural *G. tetrahit* was an allopolyploid that undoubtedly arose in nature through hybridization of *G. pubescens* and *G. speciosa.*

Many similar syntheses have now been performed with other groups and have established the allopoly-

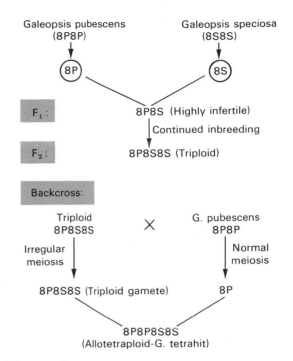

FIG. 21-15. Synthesis of a natural allopolyploid. Muntzing crossed two species and obtained a highly sterile F_1, but continued inbreeding of this F_1 yielded one offspring. This was a triploid, containing two sets of speciosa but only one set of pubescens chromosomes. Backcrossing to the pubescens parent finally produced a fertile plant. This arose as a result of a combination on an unreduced gamete from the triploid parent and a normal haploid pubescens gamete. The plant is fertile because two of each parental set are present. It was identical to a naturally occuring species, *Galeopsis tetrahit,* and crossed easily with the natural species. The latter undoubtedly arose in nature as an allotetraploid between *Galeopsis pubescens* and *Galeopsis speciosa.*

FIG. 21-14. Isolation and the evolution of populations (A) Population O (original) may become separated into two segments existing under different sets of environmental conditions. The two segments do not exchange genes because of an isolating barrier. Some of the spontaneous mutations arising in the two populations may represent different allelic forms of the original gene. Very different allele combinations may become established in two separated groups as a result of the forces of natural selection and drift. The eventual descendant population may become so different as a result of these processes that members of the two populations cannot exchange genetic material when the isolating barrier is removed. Two separate species can be recognized. (B) The barrier may be removed before the process has resulted in the establishment of very different allele combinations in the two groups. When brought together, genetic isolation may not be complete. If interbreeding continues, eventually all may become part of one new population.

ploid origin of several natural species, among them many commercially valuable wheats and cereal grasses. Indeed, cytotaxonomic studies leave no doubt that some of our most economically important plants (tobacco, cotton, potatoes, and others) arose through allopolyploidy. In somewhat of a reverse process to the kind of speciation requiring spatial isolation, allopolyploidy occurs when two genetically different groups become associated in the same locality and produce offspring. Moreover, it occurs abruptly, not as the result of the accumulation of point mutations over a long period of time.

Neutral theory of evolution

Another concept that offers an explanation for the protein variation in natural populations maintains that most of the mutant alleles detected through the application of chemical techniques are neutral, not advantageous or detrimental. A mutant allele giving rise to a protein variant would be neither more nor less advantageous than the gene form from which it arose.

According to the neutral theory of evolution (popularly called *non-Darwinian evolution*), most evolutionary changes that take place at the molecular level are brought about by mutation pressure plus random drift of alleles that are equal in adaptive value. Proponents of the neutral theory consider the Darwinian force of selection to be operative mainly on phenotypes shaped by the activities of many genes. The phenotypic level is considered to be the level of form and function produced through the operation of many genes.

Adherents of the neutral theory recognize the role of the environment in the selection of particular phenotypes over others and point out that selective forces are not concerned with *how* genotypes determine these phenotypes. They envision the forces of molecular evolution to be quite different from those operating at the level of the phenotype of the individual. They believe that at the level of the molecular structure of the genetic material, much evolutionary change is driven by random genetic drift.

Because the neutral theory of evolution has spurred a great deal of controversy, as well as a great deal of further research in population genetics, we should be familiar with some of its details and the evidence for and against them. Until the arrival of special techniques of molecular genetics, studies of evolution dealt mainly with the phenotypic level, the level of form and function. There was no way to connect the observations made at this level to the concept of allelic frequencies in populations, the main approach used in population genetics. Techniques of molecular genetics made it possible to compare individual RNA molecules and proteins among related species. The RNA and proteins represent, respectively, the direct products and ultimate products of genes. It became possible to study the molecular genetic variability within a species and, by comparing this to related species, the rate at which one allele substitutes for another in evolution.

To the proponents of the neutral theory, the information obtained from the molecular studies are not compatible with what is expected according to the ideas of selectionism or neo-Darwinism. For example, when different species of vertebrates are compared with respect to a specific protein (e.g., the α chain of hemoglobin), it is found that amino acid substitutions occur at about the same rate in many of the now divergent genetic lines. Rather than following a pattern, the substitutions found seem to be random. The rate does not appear to depend on such factors as generation time, living conditions, or population size. In the α hemoglobin chain of 141 amino acids, the substitution rate is about 1 amino acid per 7 million years or 1 substitution per billion years for any specific amino acid site.

As discussed earlier in this chapter, the procedure for detecting small differences among proteins revealed a great deal of genetic variability, in turn revealing that a significant proportion of the genes in different groups of creatures is polymorphic, that is, such genes occur in various allelic forms. Many of the protein variants seem to have no visible phenotypic effect and appear to have no correlation with environmental conditions. To leading proponents of the neutral theory, such as M. Kimura, this suggests that most of the nucleotide substitutions in the population gene pool during the course of evolution do not stem from the force of Darwinian selection. Rather, they result from the random fixation of essentially neutral mutants. These neutral mutant alleles are as effective as the original gene forms in promoting survival and reproduction of the individual. Moreover, the observed protein polymorphisms would be neutral with respect to selection. They would be maintained in the population as a result of mutational input and random elimination. Thus, a neutralist maintains that some mutants can spread through a population by themselves, even though they may impart no selective advantage to the population.

Considering the two alleles of a gene, if the mutant form gives the same advantage as the original gene form, its maintenance in the population is left to chance. Over time, its frequency fluctuates as it increases and decreases by chance. This effect stems from the fact that, of the gametes produced in a given generation, only a few are sampled at random in reproduction. As such neutral mutant alleles drift randomly, most are lost by chance, but a minority becomes fixed and thus reaches a frequency of 100% in the population. This means that the genetic constitution of a population can continue to change if the drifting continues over millions of generations. The neutral theory makes various mathematical predictions about the time and rate of fixation of a neutral gene form in a population. The rate of fixation should simply be equal to the rate of mutation per gamete, independent of population size.

The observed rates of molecular evolution, according to the neutralists, show remarkable constancy and agree well with this prediction. This relationship would apply only to neutral alleles. On the other hand, the rate of evolution for those alleles carrying a selective advantage depends on that advantage, the rate of mutation, and population size. Neutralists also point out that certain DNA regions between genes and even within genes do not participate in protein formation. These should therefore be less subject to natural selection. There should be a lack of selective constraint in these cases, unlike the selective pressures for other regions of the DNA. There is evidence of a high incidence of nucleotide substitution in these regions. Similarly, some proteins and certain parts of other proteins appear to have relatively small or unimportant functions. These have been found to have a higher rate of amino acid substitution than those proteins or protein regions that are directly involved in metabolic activities.

Observations such as these support the neutralist theory, which predicts that as selection relaxes, the rate of evolution (as reflected in nucleotide and amino acid substitution) proceeds to a maximum value established by the mutation rate. If selection is not relaxed for a protein or for part of a protein, the amount of polymorphism would be reduced because the protein or protein part under consideration would have an important functional role and would therefore be under functional constraint.

Most of these ideas are not in accord with those of selectionists. In their view, a rare mutant allele must have some selective value to spread through a population and to become fixed. Occasionally, a neutral allele may persist for a while if its locus is closely linked to the locus of an allele that is selected for. According to the selectionist idea, the molecules or parts of molecules that appear to be evolving rapidly would indeed have some important function that is not yet apparent. The rapid evolution would indicate the accumulation of beneficial changes leading to a higher degree of adaptation.

Earlier in this chapter, we discussed mechanisms that can be offered to explain the maintenance in a population of genetic variability in the form of protein polymorphism. According to the selectionists, balanced polymorphism is maintained mainly through heterozygote advantage and frequency-dependent selection. The neutralists propose that most of the protein polymorphisms are selectively neutral and are maintained, as we have just noted, through a balance

between mutation and random elimination. A number of neutral alleles is added per generation. With time, they may become fixed in the population or lost to it. As this is occurring over the generations, these alleles add variability to the population as reflected in protein polymorphism. The neutralists and selectionists have presented experimental evidence to support their opposing theories. For example, selectionists have offered evidence for differences in fitness among variant forms of certain enzymes and also for frequency-dependent selection. The neutralists have countered by interpreting these findings in a manner that does not contradict their basic theories.

From the selectionist point of view, it is the environmental conditions that mainly determine protein polymorphism. Therefore, the more variable the environment, the greater the genetic variability and polymorphism (and vice versa). Selectionists predicted that deep sea organisms would display little genetic variability, because their environment is a rather unchanging one. By contrast, intertidal organisms, living in a changeable environment, should exhibit much genetic variability. The prediction, however, was not supported by the findings, which revealed the bottom-dwelling population to have a high degree of genetic variability and the intertidal organisms very little! This supports the view of the neutralists, who consider molecular structure and function rather than environmental conditions to be the major factors determining protein polymorphisms.

Calculations derived by proponents of the neutralist theory predict that the amount of heterozygosity within a species is determined by N_e (effective population size) and the mutation rate per gene per generation. The larger either the N_e or the mutation rate per generation, the closer the amount of heterozygosity or variability to 100%; that is, the more likely that 100% of the population will be heterozygous at each genetic locus. The results of F. Ayala with a population of *Drosophila willistoni* are not in agreement with these predictions. Studying what he believed to be a very large effective breeding population, Ayala arrived at a figure of 18% for the amount of heterozygosity, a figure that should be closer to 100% according to the neutralist theory. Neutralists answer that the N_e may not always have been as large as Ayala maintains, but that it was greatly reduced several times in the past. Mathematically, they show that it would take millions of generations for the amount of variability to rise significantly from its reduced state and reach levels compatible with those expected from a population maintained constantly at large numbers. Moreover, arguments surround the possibility that certain supposedly neutral alleles may not be neutral and theoretical predictions may be upset as a consequence.

In summary, we are left with two opposing ideas. The Darwinian selectionists maintain that selection operates at all levels and is primarily responsible for the frequency of an allele. The environment is a decisive factor that determines what is to be selected. At the level of the structure of the DNA itself, however, the neutralists contend that much evolutionary change is driven by random genetic drift and not by environmental change. Although the matter is not yet settled, the controversy between the neutralists and the selectionists has propelled evolutionists and population geneticists to reexamine ideas. The controversy has resulted in the design of better experiments and in more reasoned models of the forces that shaped the evolutionary past.

Gene regulation and evolution

Besides the importance of genetic regulatory mechanisms in differentiation and normal cell activity, there is good reason to believe that mutation in regulatory mechanisms may be the basis for anatomical evolution in higher life forms. For example, the human and the chimpanzee differ considerably in anatomical aspects. However, when their body proteins (e.g., blood proteins) are examined, it is found that very little difference exists. The protein differences between the two do not seem to account for the anatomical distinctions. Mutation in regulatory systems appears to be a more likely explanation. Different regulatory activities in two groups that possess similar proteins could affect the timing of gene expression and the pattern of development of the organisms. Creatures with very similar proteins would be unable to hybridize if they were members of separate groups in which the regulatory mechanisms operating during development are very different.

Mutation in regulatory genes may also account for the very rapid evolution placental mammals have undergone in their anatomy in only 75 million years. The rate of protein evolution has not proceeded at a comparable pace. The proteins in this diverse group show fewer distinctions than would be expected on the basis of the anatomical differences that are found. Lower vertebrates, such as the thousands of frog species, show significant protein differences. Yet the various frog species are very similar anatomically. Some biologists recognize two types of evolution that proceed independently and can have different rates, protein evolution and anatomical evolution, the latter due largely to changes in regulatory mechanisms and responsible for the great diversification seen in the anatomy of higher vertebrates.

As investigators continue to scrutinize ever more closely the genetic material of eukaryotes, more and more evidence accumulates to indicate that the genome is far from static and unchanging. Instead, the emerging picture shows the eukaryotic genome as a dynamic system, one capable of undergoing changes in a time period much shorter than previously suspected. Evidence for this is seen in the ability of some transposable elements to break and rearrange chromosomes, as occurs in strains of corn. There is a strong case for the existence of movable elements in eukaryotic groups as diverse as Drosophila, with its copia family, and the human species with its Alu family of sequences.

The observations lend credence to the possibility that such elements, by moving about in the genome, can insert at different target sites in the chromosome and alter the expression of genes. By influencing regulatory sequences, these elements could exert a profound influence on the course and rate of evolution in a population. The phenomenon of hybrid dysgenesis in the fruit fly points to one way in which such elements could operate on two separated populations of a species to lead them down separate pathways, resulting in their divergence into two distinct species in a relatively short time (Chap. 17). Armed with new technologies from molecular biology, the population geneticist may uncover additional mechanisms of evolution that provide population variability and interact with and complement those recognized by classic Darwinian concepts.

REFERENCES

Allison, A. C. Polymorphism and natural selection in human populations. *Cold Spring Harb. Symp. Quant. Biol.* 29: 137, 1964.

Ayala, F. J. The mechanism of evolution. *Sci. Am.* (Sept.): 56, 1978.

Bishop, J. A. and L. M. Cook. Moths, melanism, and clean air. *Sci. Am.* (Jan.): 90, 1975.

Bush, G. L., S. M. Case, A. C. Wilson, and J. L. Patton. Rapid speciation and chromosomal evolution in mammals. *Proc. Natl. Acad. Sci.* 74: 3942, 1977.

Cavalli-Sforza, L. L. The genetics of human populations. *Sci. Am.* (Sept.): 80, 1974.

Crow, J. F. Genes that violate Mendel's rules. *Sci. Am.* (Feb.): 134, 1979.

de Grouchy, J., C. Turleau, and C. Finaz. Chromosomal phylogeny of the primates. *Annu. Rev. Genet.* 12: 289, 1978.

Dobzhansky, T. *Genetics of the Evolutionary Process.* Columbia University Press, New York, 1968.

Dobzhansky, T. F., F. J. Ayala, G. L. Stebbins, and J. W. Valentine. *Evolution.* W. H. Freeman, San Francisco, 1977.

Futuyma, D. J. *Evolutionary Biology,* 2d ed. Sinauer Assoc. Inc., Sunderland, MA, 1986.

Goodman, M., G. W. Moore, and G. Matsuda. Darwinian evolution in the geneology of hemoglobin. *Nature* 253: 603, 1975.

Kimura, M. The neutral theory of molecular evolution. *Sci. Am.* (Nov.): 98, 1979.

Lewontin, R. C. Adaptation. *Sci. Am.* (Sept.): 212, 1978.

Lewontin, R. C. *The Genetic Basis of Evolutionary Change.* Columbia University Press, New York, 1974.

Li, C. C. *First Course in Population Genetics.* Boxwood Press, Pacific Grove, CA, 1977.

Mayr, E. Evolution. *Sci. Am.* (Sept.): 46, 1978.

Mayr, E. *Populations, Species, and Evolution.* Harvard University Press, Cambridge, 1970.

McDowell, R. E. and S. Prakash. Allelic heterogeneity within allozymes separated by electrophoresis in *Drosophila pseudoobscura. Proc. Natl. Acad. Sci.* 73: 4150, 1976.

Miller, D. A. Evolution of primate chromosomes. *Science* 198: 1116, 1977.

O'Brien, S. J., D. E. Wildt, and M. Bush. The cheetah in genetic peril. *Sci. Am.* (May): 84, 1986.

Stebbins, G. L. and F. J. Ayala. The evolution of Darwinism. *Sci. Am.* (July): 72, 1985.

Thoday, J. M. Non-Darwinian evolution and biological processes. *Nature* 255: 675, 1975.

White, M. J. D. *Animal cytology and evolution,* 3rd ed. Cambridge University Press, Cambridge, 1973.

Wilson, A. C. The molecular basis of evolution. *Sci. Am.* (Oct.): 164, 1985.

REVIEW QUESTIONS

1. In a certain Indian tribe, the distribution of the MN blood types (which are based on the codominant alleles, L^M and L^N) is found to be Type M, 64%; Type MN, 32%; Type N, 4%. What are the frequencies of the L^M and of the L^N alleles in this population?

   In Questions 2 through 6, assume that equilibrium exists for the alleles involved in each case.

2. A farmer finds that among his flock of about 8100 birds, one is affected with an undesirable trait that affects the feathers. This trait is known to be inherited as an autosomal recessive. How many birds in the flock of 8100 could be expected to be carriers of the undesirable allele?

3. Approximately 70% of Americans can taste the bitter chemical PTC. This ability depends on the dominant, T, whereas inability to taste it results from homozygosity for the recessive allele, t. The pair of alleles has been found to exert no other detectable effect.

   A. Is the frequency of the dominant allele much higher in the population than the recessive?
   B. Can the dominant allele be expected to increase further?

4. Suppose that on a certain island the frequency of persons homozygous for the autosomal recessive (a) for albinism is 1 in 2500.

   A. How many persons in the population can be expected to be heterozygotes, and how many homozygotes for the wild type allele?
   B. What is the probability that a normally pigmented person with no history of albinism in the family is a heterozygote?

5. Assume that in a certain population 5 males in 100 are found to be color blind because of the X-linked recessive, d.

   A. What is the frequency of the alleles D and d in the population?
   B. What is the frequency of carrier women in the population?
   C. What is the frequency of color-blind women in the population?

6. Assume that a certain autosomal recessive, a, causes severe anemia and death before the age of 5 years. In the population, 1 infant in 200,000 is born with the disorder.

   A. What appears to be the mutation rate at the a locus?
   B. What is the frequency of the recessive allele in the population?
   C. What is the fraction of persons in the population heterozygous for this recessive?

7. Cystic fibrosis occurs in about 1 birth in 2500 among members of certain European populations. What would be the chance that two unrelated, healthy persons might produce a child with the disorder?

8. The alleles R and r affect pigment formation in a certain plant species. Genotypes RR and rr result in red and white flowers, respectively, whereas the heterozygote is pink.

   An examination of 4040 plants in a population gives the following: 920 red, 2280 pink, and 840 white.

   A. Calculate the frequencies of alleles R and r in the population.
   B. What would be the expected number of individuals of each type if the population is at equilibrium with respect to alleles R and r?
   C. Determine a chi-square value, and give the probability that any departure from the expected is due to chance factors.

9. Following are the initial frequencies of three genotypes in a population before selection along with the respective selection coefficients:

	AA	Aa	aa
Initial frequency	0.64	0.32	0.04
s	0	0	0.8

   A. Calculate the corrected proportionate contributions of the three genotypes to the gene pool after selection.

   B. Calculate the frequencies of alleles A and a after one generation of selection.

   C. What is the value of Δq?

10. Following are the initial frequencies of three genotypes in a population before selection, along with the adaptive values:

   A. Calculate the corrected proportionate contributions of the three genotypes to the gene pool after selection.

   B. Calculate the frequencies of alleles A and a after one generation of selection.

	AA	Aa	aa
Initial frequency	0.49	0.42	0.09
W	0.8	1	0.2

11. Following are the initial frequencies of three genotypes in a population before selection, along with the respective selection coefficients and adaptive values:

	AA	Aa	aa
Initial frequency	0.36	0.48	0.16
s	0.4	0.2	0
W	0.6	0.8	1

What is the change in the frequencies of alleles A and a after one generation of selection?

12. Suppose the frequencies of three genotypes in a population are as follows with respect to a certain pair of alleles: LL = 0.64; L1 = 0.32; ll = 0.04. Suppose that "l" acts as a full lethal while the selection coefficient equals zero for the other two genotypes.

   A. What would be the frequency of the lethal allele after one generation of selection?

   B. Suppose that an adaptive value of zero persists for the genotype ll generation after generation. Will the lethal allele l become quickly eliminated? Explain.

13. The frequency of a certain fully lethal recessive allele is 1/50,000. How many generations will be required to reduce this frequency to a value 10 times less?

14. In a certain population, 1 baby out of 200,000 is born homozygous for a fully recessive lethal allele. What would be the approximate value of individuals heterozygous for this deleterious factor, assuming the lethal has reached equilibrium in the population?

15. The selection coefficients for the genotypes AA, Aa, and aa are, respectively, 0.3, 0, and 0.9. What are the expected equilibrium values for alleles A and a?

16. A certain dominant allele A mutates to the recessive form a at a rate designated u. Back mutation from a to A occurs at a rate designated v. Assume that no other evolutionary forces are operating and that $u = 0.00004$ and $v = 0.00001$.

   A. What are the expected equilibrium values of p ($\hat{p}$) and q ($\hat{q}$)?

   B. At an equilibrium value reached in the absence of any other evolutionary force except that of mutation pressure, which allele would almost always have the higher value—the original (A) or the mutant (a)? Explain.

17. In a certain population of butterflies, the frequency of the recessive allele (o) for light orange coloration = 0.1. The frequency of the dominant for deep orange (O) = 0.9. In a second population of butterflies of the same species as the first, the frequency of o = 0.4 and O = 0.6.

   A large migration takes place from population 2 to population 1. The proportion of immigrants added to population 1 is 0.3. What would be the frequency of allele o in the new population one generation after the migration, and what would be the value for Δq?

18. Population 1 consists of 750 breeding individuals. Population 2 has an effective breeding size of 250. The original frequencies of A and a in both populations are respectively 0.7 and 0.3.

   A. For each of the two populations, calculate σ_q, the standard deviation for a.

   B. In which of the two populations will there be a greater chance that the frequency of a in the next generation will vary from 0.3 as a result of random sampling? Give the chances in both populations for a possible value of a in the next generation.

19. Population A has an effective breeding size of 400, whereas that of population B is 4000.

   A. What is the probability of loss of variability per generation in populations A and B?

   B. How many loci can one expect to become homozygous in the next generation in each of the two populations?

20. Assume that two human populations, one in the United States and the other in Australia, both originated from migrants who left the same European population several generations ago. In the American population, there is a high incidence of a recessive disorder that causes deformed skeletons. The disorder is unknown in the Aus-

tralian population but is found in a very low frequency in the European population. Explain.

21. Assume that three separate populations of plants, A, B, and C have many similar phenotypic features. However, many combinations of traits make each population distinct from the other. When plants from population A are crossed with those of population B, the F_1s are intermediate in phenotype, but they are completely sterile. When plants from population B are crossed with members of population C, very fertile intermediates arise. Plants from population A crossed with those from C give F_1s that are intermediate between A and C, but they are sterile. Explain these observations.

22. Two related plant species (species A and B) both have a diploid chromosome number of 14. The ranges of the two species overlap in a certain region, and an occasional hybrid results from crossing between the two species. The hybrid is highly infertile because of failure of chromosome pairing at meiosis.

   A. Explain how a fertile tetraploid could be derived, giving the chromosome constitution of the sterile F_1 and the tetraploid.
   B. Can the tetraploid be considered a separate species? Explain, giving the chromosome constitutions of any offspring between the tetraploid and the parental species.
   C. Suppose two unreduced gametes from the tetraploid unite and that an offspring arises. What would the

chromosome constitution of such a plant be? How many bivalents could it form? How many quadrivalents (associations of four chromosomes) could it form?

23. Suppose that each of three closely related plant species—A, B, and C—has a diploid chromosome number of 12. How could a fertile allohexaploid be derived from the three of them, assuming that A and B cross first to produce an allotetraploid?

24. Suppose soluble protein samples extracted from the blood of animals from population A are placed on a gel and exposed to electrophoresis. The gel is then exposed to a chemical solution, which contains a stain specific for one particular enzyme. The different patterns detected are shown here in A.

   When this same procedure is performed on protein samples from animals of population B, the results shown in B are always obtained. Explain.

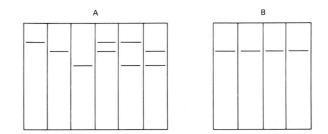

GLOSSARY

acentric. An abnormal chromosome or chromosome fragment lacking a centromere.

acidic amino acid. An amino acid that bears a net negative charge at neutral pH.

acquired characteristics. Those features that an organism takes on during its lifetime through the effect of the environment on somatic tissue and which are not transferred to the next generation.

acrocentric. A chromosome with its centromere very close to one end, giving it one very short arm. The centromere located near one end of a chromosome.

activator. A regulatory substance that binds to a controlling element and acts in a positive fashion, stimulating transcription of a structural gene or genes.

adaptive value (W). A measure of the proportion of individuals of a given genotype that survive and reproduce relative to the genotype that has the greatest reproductive rate. A measure of the relative success of a given genotype in contributing to the next generation.

adenine. A purine base found in DNA and RNA.

adenosine triphosphate (ATP). A nucleoside triphosphate that provides a source of high energy for metabolic processes requiring energy transfers. The main energy-storing molecule in the cell.

allele. A given form of a gene that occupies a specific position (locus) on a specific chromosome. Alternative forms of a gene occur and can thus occupy that specific locus. These variant forms are said to be alleles or to be allelic to one another.

allelic complementation. The production of wild or near wild-type phenotype in an organism that carries two different mutant alleles in the trans arrangement.

allelic exclusion. The expression in plasma cells of the antibody genes found on only one member of a pair of homologous chromosomes.

allopatric. Geographically or spatially isolated.

allopolyploidy. A polyploid condition in which the extra chromosome sets are derived from different groups or species following hybridization.

Alu family. The most common set of dispersed and related DNA sequences in the human genome; each is about 300 base pairs long and repeated approximately half a million times.

amber codon. The chain-terminating codon UAG in mRNA.

amino acid. Any one of the 20 units that are joined by peptide linkages to form a polypeptide.

amino group. The $-NH_2$ chemical group, basic in nature, and characteristic of amino acids.

amino terminal end.   The end of a polypeptide that has a free amino group.

aminoacyl–tRNA synthetase.   One of a group of enzymes required to activate an amino acid and link it to its specific tRNA carrier.

amniocentesis.   A procedure in which a sample of amniotic fluid is removed from a pregnant woman, so that the fluid and fetal cells present in it can be subjected to various analyses.

anaphase.   The stage of nuclear division at which the chromosomes move to opposite poles.

aneuploidy.   A chromosome anomaly in which the number of chromosomes in a cell or individual departs from the normal diploid number by less than a whole haploid set.

Angstrom (Å).   A unit of measurement equal to 10^{-8} cm.

anneal.   Form a double-stranded molecule from two single-stranded polynucleotide chains as a result of hydrogen bond formation between complementary nucleotides. The single strands that anneal may be two complementary DNA chains, two complementary RNA chains, or a DNA chain with an RNA chain complementary to it, thus forming a DNA–RNA hybrid molecule. (Also see *reanneal*.)

antibody.   A protein molecule, also known as an immunoglobulin, that is formed in response to a specific antigen and can recognize and react with it.

anticodon.   A sequence of three nucleotides in a tRNA molecule that pairs with a specific codon in the mRNA and thus enables the tRNA to position its amino acid properly in a growing polypeptide chain.

antigen.   Any substance or large molecule that can stimulate the production of antibodies following entrance into the tissues of a vertebrate.

antisense strand.   That strand of a gene which is not coded for a product and does not undergo transcription.

ascus.   An enlarged cell in which the sexual spores are produced; typical of one major class of fungi, the Ascomycetes.

assortive mating.   Mating that is not at random but entails either the selection of mates with a certain trait or traits or, conversely, the rejection of mates possessing a certain trait or traits.

aster.   The region that marks the poles of dividing cells and includes rays of microtubules surrounding a clear area within which two centrioles are located.

att sites.   Loci occurring on both a phage chromosome and that of its bacterial host, where recombination brings about integration of the phage into, or its excision from, the host chromosome.

autopolyploidy.   A polyploid condition in which the additional chromosome set or sets have been derived from within the same group or species.

autoradiography.   A technique that permits the localizing of sites occupied by a radioactive substance in biological material by covering slide preparations of the material with a photographic emulsion that is later developed to reveal the darkened points at which radioactive emanations have caused decay in the emulsion.

autosomes.   All the chromosomes in the complement other than the sex chromosomes.

auxotroph.   A nutritional deficient that must be supplied with one or more nutrient factors which it cannot manufacture from a minimal medium.

axial symmetry.   See palindrome.

backcross.   A cross of a member of the F_1 generation to a member of one of the parental lines or to an individual that is a parental type.

back mutation.   A mutation in a mutant allele that results in a change back to the wild or standard form of the gene.

bacteriophage.   See phage.

Barr body.   See sex chromatin.

base pair.   A pair of nitrogen bases (one purine and one pyrimidine) held together by hydrogen bonds.

basic amino acid.   An amino acid with a net positive charge at neutral pH.

bidirectional replication.   DNA replication in which two replication forks move in opposite directions away from the same origin.

bivalent.   An intimate association of two homologous chromosomes seen at first meiotic division.

C bands.   Deep-staining regions that are produced following certain procedures for the visualization of the chromosome complement, and which indicate sites of heterochromatin concentration.

C value.   The amount of DNA found in the haploid genome of a species.

cAMP.   See cyclic AMP.

CAP (catabolite gene activator protein).   A positive control element that complexes with cyclic AMP and must bind to a region of the promoter before transcription of certain genes can begin.

carboxyl group. The $-COOH$ chemical grouping, acidic in nature, and found in all amino acids.

carcinogen. A chemical substance or physical agent that can effect a malignant transformation in a cell. A substance or agent capable of producing cancer in an individual.

cDNA. Copied DNA; a DNA strand complementary to an RNA strand from which it was synthesized in vitro using reverse transcriptase.

centriole. A microtubular organelle marking the poles of the spindle in certain types of dividing cells.

centromere. That part of the chromosome responsible for chromosome movement. The region of the chromosome known as the primary constriction, where the sister chromatids of a prophase and metaphase chromosome are held together. (The structural part of the centromere, the kintochore, becomes attached to the spindle fibers.)

centrosome. A clear region of the cytoplasm that contains the centrioles.

chain-terminating codon (nonsense codon). One of three nucleotide sequences (UAG, UGA, UAA) in a mRNA transcript which indicates that translation is to be halted, and thus the assembly of amino acids into a growing polypeptide chain is to be terminated.

characteristic. A general attribute of an organism. In some characteristics, variation is recognizable as distinct, contrasting traits. For others, the variation is continuous, showing a gradual transition from one extreme to the other.

charged tRNA. A transfer RNA molecule attached to its specific amino acid; aminoacylated tRNA.

chi-square. A value that indicates the probability that the figures in a set of numerical data are compatible with those ideally expected for a certain ratio.

chiasma. A crosslike figure seen at first meiotic division and associated with the exchange between two homologous chromatids.

chromatid. A strand of a chromosome that has replicated. A chromosome is composed of two chromatids joined at the centromere region during mitotic prophase and metaphase.

chromatin. The chromosome material composed of DNA and associated proteins.

chromomere. A region of a chromosome thread that appears as a beadlike structure.

chromonema. The chromosome when it is seen as an extremely thin thread.

chromosome. A structure in the cell nucleus composed of chromatin that stores and transmits genetic information.

chromosome aberration (chromosome anomaly). Any modification that alters the morphology or number of chromosomes in the complement.

cis arrangement. That arrangement in a doubly heterozygous individual in which two linked wild alleles occur together on one homologue and the two recessive mutant alleles occur on the other.

cis–trans effect. A position effect in which two individuals, dihybrid with respect to the same genetic sites, show a difference in phenotype as a consequence of the arrangement of the mutant sites on the chromosome, the cis arrangement producing a more normal phenotype than the trans.

cistron. The gene considered as a unit of function; a segment of DNA coded for one polypeptide.

class switching. See heavy chain class switching.

clone. A population of genetically identical cells.

cloned DNA. A collection of identical DNA molecules or sequences produced as the result of replication of a single molecule in a suitable cellular or viral system.

cloned library. See gene library.

cloning. The formation of clones or exact genetic replicas.

codominance. The expression in the heterozygote of both alleles. Codominance is generally used to refer to those cases in which two products can be detected in the heterozygote, each associated with one of the allelic forms. Also used interchangeably with incomplete dominance, as the distinction between the two is often not clear.

codon. A sequence of three adjacent nucleotides (an RNA or a DNA triplet) that designates a specific amino acid or indicates that translation is to be terminated.

coefficient of coincidence. A figure that takes into consideration interference and is used to indicate the chance that a double crossover will occur within a certain map distance.

cohesive ends. Complementary single DNA strands that extend from opposite ends of a duplex molecule or from the ends of different duplexes; can be produced as a result of staggered cutting by a restriction enzyme; also called sticky ends.

cointegrate. A circular molecule formed as the result of fusion of two replicons, one originally possessing a transposon and the other lacking one. The cointegrate has two copies of the transposon occurring as direct repeats, one at each of the two junctions of the replicons.

colinearity. The correlation between the sequence of codons in the DNA of a gene (cistron) and the sequence of amino acids in the polypeptide it specifies.

complementation. See allelic complementation.

complex locus. A cluster of very closely linked, functionally related genes.

complex transcription unit. A gene whose primary transcript is processed differently in different cell types, yielding different mRNA species coded for different products.

concatamer. A chain or linear series of genomes joined together.

consensus sequence. An idealized sequence that presents the nucleotides most often present at each position in a given DNA segment of interest.

constitutive. Unchanging or produced at a constant rate, such as a constitutive enzyme that continues to be produced in abundance, regardless of the cell's requirements.

constitutive heterochromatin. That chromatin which remains constantly condensed from one cell to the next, such as the heterochromatin associated with the centromere.

continuous variation. That phenotypic variation shown by some characteristics in which there is a gradation from one extreme to the other without any delineated categories or distinct traits.

controlling element. A genetic region, such as the promoter or operator, which can respond to an environmental signal and determine whether or not its associated gene will undergo transcription.

conversion. See gene conversion.

copia. A family of closely related DNA sequences in Drosophila that code for very large amounts of RNA and are capable of transposition.

copy DNA. See cDNA.

core particle. The product resulting from nucleosome digestion that consists of a histone octamer around which is wrapped 146 base pairs of DNA; also called a nucleosome bead.

corepressor. A small molecule needed to interact with a specific repressor before the latter can combine with an operator to block transcription.

$C_o t$ plot. A graph representing the reassociation kinetics for a given sample of denatured DNA in which the fraction of molecules remaining single-stranded is plotted against the product of DNA concentration and time ($C_o t$ value).

covalent bond. A strong interaction between atoms that share electrons.

crossing over. The event that entails a reciprocal exchange of segments between two homologous chromosomes and results in the recombination of linked alleles.

crossover. The recombinant product of a crossover event that has a new combination of linked alleles.

ct DNA. The DNA of the chloroplast.

cyclic AMP (3′,5′-cyclic adenylic acid). A small molecule derived from ATP that interacts with specific hormones in cells and must complex with CAP before transcription of certain genes can take place.

cytogenetics. The science that combines the methods and findings of cytology (microscopy) and genetics (breeding or molecular analysis).

cytohet. A cytoplasmically heterozygous eukaryotic cell; a eukaryotic cell containing two genetically different kinds of a specific organelle.

cytokinesis. Division of the cytoplasm producing two daughter cells from a parent cell.

cytoplasm. All the living parts of the cell outside the nucleus.

cytosine. A pyrimidine base found in DNA and RNA.

cytosol. The fluid portion of the cytoplasm exclusive of organelles.

degenerate code. The genetic code that is characterized by the fact that two or more codons may designate the same amino acid.

deletion. A chromosome alteration in which a portion of a chromosome has been lost; loss of a part of a DNA molecule from the genetic complement.

denaturation (of DNA). The separation of the two strands of the double helix as a result of hydrogen bond disruption following exposure to high temperature or chemical treatment.

density gradient centrifugation. A technique in which a mixture of substances is spun at high speed in an appropriate solution until each component of the mixture reaches an equilibrium position, where its density matches the density of the solute molecules at that position in the centrifuge tube.

deoxyribonucleic acid. See DNA.

detrimental allele. Any gene form that imparts a disadvantage to the individual carrying it and decreases the carrier's chances of reproduction or survival compared with that of an individual not possessing the allele.

diakinesis. Last stage of first meiotic prophase during which the bivalents are extremely condensed and well separated from one another.

dicentric. An abnormal chromosome having two centromeres.

dihybrid (cross). A cross in which two pairs of alleles are being followed; also, any individual carrying two pairs of alleles and thus heterozygous at two loci under consideration.

dimer. A compound formed between two identical molecules and thus having the same percentage composition but twice the molecular weight of one of the original molecules.

diploid. Having two complete haploid sets of chromosomes typical for the species or group.

diplonema (diplotene stage). Stage of first meiotic prophase, during which homologous chromosome threads repel and chiasmata appear.

direct repeats. Two or more identical or very closely related nucleotide sequences present in the same orientation in the same DNA molecule.

discontinuous replication. The synthesis of short DNA chains on a template strand. The short chains later become linked to form a long chain.

discontinuous variation. That phenotypic variation shown by some characteristics that can be recognized by clear-cut differences or distinct traits.

disjunction. The separation of chromosomes at anaphase of a mitotic or meiotic division.

DNA (deoxyribonucleic acid). The molecular nature of the genetic material that is composed of the four nitrogen bases—adenine, guanine, thymine, and cytosine—covalently bonded to a repeating chain of deoxyribose and phosphate residues and that occurs in cellular forms as a double helix or duplex molecule.

DNAase (DNase). Any enzyme that digests DNA to fragments composed of only a few nucleotides.

DNA–DNA hybridization. A procedure in which single-stranded DNA from one source is presented with single-stranded DNA from a second source under conditions that allow the formation of duplex DNA molecules, thus permitting an estimate of the degree of similarity between the two kinds of DNA.

DNA ligase. An enzyme that can catalyze phophodiester linkages and can thus restore intact a polynucleotide chain containing one or more nicks.

DNA polymerase. An enzyme that catalyzes the synthesis of DNA on a DNA template from deoxyribonucleotide triphosphate precursors.

DNA–RNA hybridization. A procedure in which single-stranded DNA is presented with RNA under conditions that permit any complementary DNA–RNA sequences to combine.

dominant. Any gene form that expresses itself in some way in the presence of its allele in the heterozygote. Also, any trait that is phenotypically expressed when the responsible gene form is present singly with its allele in the heterozygote.

downstream. The direction on a DNA template from the 3′ end toward the 5′ end and toward the site of initiation of transcription.

drift. Random or chance fluctuation in the frequencies of alleles in a population independent of natural selection and mutation.

duplex. Composed of two strands. A molecule composed of two strands, such as double-stranded DNA.

duplication. An extra copy of a gene or genetic region in the complement. A duplication refers to any situation in which one or more additional copies of a genetic region are present.

electrophoresis. The movement of charged molecules in an electrical field.

endonuclease. An enzyme that can cut phosphodiester bonds that occur internally in a DNA chain.

endoplasmic reticulum (ER). System of membranes distributed within the cytoplasm of eukaryotes which provides sites of protein synthesis and intracellular channels of transport.

enhancers. DNA sequences that can greatly increase the rate of transcription of genes and whose influence can be felt far upstream or far downstream from the promoters they stimulate.

episome. A nonessential hereditary factor that may exist either free within the cell or in a state in which it is integrated with a chromosome.

epistasis. A type of genic interaction in which a gene form at a given locus masks the expression of another genetic factor located at a different locus, and therefore not its allele.

euchromatin. Those chromosome regions that are noncondensed and active in the interphase nucleus.

eukaryote. A cell or organism with a distinct membrane-bound nucleus as well as one or more membranous subcellular components (e.g., mitochondria, Golgi) that are specialized for the performance of certain specific cellular functions. (See also prokaryote.)

exons. Those nucleotide sequences in the gene and its transcript that are coded for a functional polypeptide.

exonuclease. An enzyme that attacks either a free 3′ or 5′ end of a polynucleotide chain and cuts phosphodiester bonds.

expressivity. The variation in the phenotypic expression of a specific allele or genotype when that genetic constitution is penetrant.

F factor (fertility factor). An episome that determines whether a bacterial cell will act as a donor (F^+ and Hfr) or a recipient (F^-) of DNA transferred by the donor cell during conjugation.

F_1. See first filial generation.

facultative heterochromatin. Chromatin that may behave as heterochromatin in some cells and as euchromatin in others, such as one of the X chromosomes of a mammalian female.

F-genote. A fertility factor that carries a portion of DNA that is typically found in the chromosome of a bacterium.

F′ (F prime factor). A bacterial fertility factor, F, which carries a portion of the bacterial chromosome.

fingerprinting. A technique in which a protein is subjected to enzyme digestion and the resulting peptide fragments separated by a combination of paper chromatography and electrophoresis.

first filial generation (F_1). The generation of individuals produced by the first parental generation (P_1) or the first parents being considered.

first parental generation (P_1). The first set of parents considered in any pedigree or mating.

5′ end. That end of a nucleic acid chain terminating in a free phosphate group.

5′ to 3′ growth. The synthesis of a nucleic acid chain by joining the 5′ (PO_3) end of a nucleotide to the 3′ (OH) end of the last nucleotide already in the uncompleted chain.

forward mutation. A mutation that results in the formation of a mutant gene form from the wild or standard allele.

founder effect. A special case of genetic drift in which an isolated population diverges in frequency of alleles from the parent population as a result of sampling error, which produces two different gene pools, the parental and the one derived from it.

frame shift mutation. A deletion or duplication of a base in the DNA that causes a portion of the coded information in the gene to go out of phase.

G bands. Deeply stained bands along the length of a chromosome which alternate with slightly stained regions and are revealed following a procedure using the Giemsa stain.

gel electrophoresis. A technique in which molecules are separated, because of their different shapes, sizes, and electric charges, as they migrate through a gel in an electric field.

gene. Classically a unit of inheritance occupying a specific site (locus) on a chromosome that has one or more specific effects on an organism and can both recombine with other such genetic units and mutate independently to other allelic forms. Many genes contain coded information for some functional product such as a polypeptide, rRNA or tRNA, whereas other genes serve as recognition sites for molecules involved in such processes as regulation of transcription and DNA replication.

gene conversion. A situation attributed to mismatch repair of DNA in which a pair of alleles in a diploid cell of genotype Aa segregates at meiosis in some ratio other than that expected, such as a 3A:1a segregation of ascospres instead of 2A:2a.

gene flow. The exchange of new allelic forms by migrants between two different populations of the same species.

gene library. A collection of cloned DNA fragments generated by the activity of restriction enzymes that includes all or part of the genome of a species.

gene pool. The total of all the genes carried by all the breeding individuals in a population at a given time.

gene proper. That portion of the sense strand containing coded information for a functional product.

generalized transduction. The one-way transfer of genes from one bacterial cell to another by way of a generalized transducing phage, which packages bacterial genetic material instead of its own during its replication cycle in the host cell. In this process, the phage can transfer any gene from the host to the recipient cell.

genetic code. The DNA and RNA triplets (codons) that carry the genetic information specifying the 20 amino acids for polypeptide synthesis as well as the start and stop signals for translation by the ribosome.

genetic drift. See drift.

genetic engineering. Manipulation in which the genetic material is purposefully altered in some way by combining hereditary materials from different sources or by removing a native segment and inserting one from another source.

genetic load. The sum of the mutant alleles in a population that have a detrimental effect and accumulate largely in the heterozygotes.

genetic marker. Any gene whose presence can be readily detected by its phenotypic expression and which is used to identify a cell, chromosome, or individual carrying it.

genome. The genetic content of a single set of chromosomes, or a single set of bacterial or viral genes.

genotype. The genetic constitution of a cell or an individual.

guanine. A purine base found in DNA and RNA.

gynandromorph. A sex mosaic; an individual with certain body segments that can be recognized as phenotypically male and others that are female.

gyrase. A topoisomerase that can introduce negative supercoils in DNA through its nicking and closing activities.

hairpin. A double helical region formed as a result of complementary base pairing between adjacent inverted complementary sequences in a single DNA or RNA strand.

haploid. Having one complete set of chromosomes typical for the species.

haplotype. A combination of very closely linked alleles or markers that tend to be transmitted as a unit to the next generation; also refers to a pattern of DNA restriction fragments for a chromosome region that can be recognized in an autoradiograph.

Hardy–Weinberg law. A mathematical expression showing that the frequency of alleles and the resulting genotypes will remain constant in a population from one generation to the next and will not change as a result of dominance or how the alleles are transmitted.

heavy chain class switching. During differentiation of a B lymphocyte, the association of the same V region with a heavy chain constant region other than mu as a result of DNA shuffling and differential processing of mRNA.

HeLa cell. A widely studied, established cell line originally obtained from a human cervical carcinoma.

helicase. An enzyme that unwinds a double helix at the time of DNA replication.

helix. A spiral of constant diameter and pitch.

helper virus. A virus that, when present in a cell also infected with a defective virus, can provide a function missing in the latter, thus enabling it to multiply in the cell.

hemizygous. Having certain loci present in a single rather than a double dose, as the X-linked genes in a mammalian male.

heritability. The percentage of the variation seen in the expression of a character or trait that can be attributed to genetic factors.

heterochromatin. The chromatin or chromosome region that tends to remain condensed in the interphase nucleus.

heteroduplex DNA. A DNA region composed of two strands, each of which is derived from different duplex DNA molecules.

heterogametic sex. The sex in a given species that produces two kinds of gametes in respect to the type of sex chromosome. The mammalian male is the heterogametic sex because half his sperm carry an X chromosome and the other half a Y.

heterogeneous nuclear RNA (hn RNA). Those RNA strands of varying lengths that do not leave the nucleus of a eukaryotic cell.

heterogenote. A partially diploid bacterial cell heterozygous for just a segment of the genetic material.

heterokaryon. A cell that contains two or more genetically diverse nuclei.

heteropycnosis. The differential reactivity to staining displayed by a chromosome or a part of a chromosome in relation to the rest of the complement.

heterosis. Hybrid vigor; the more robust nature of a heterozygote compared with the more inbred parental lines.

heterozygous. Having alternative forms of a gene (alleles) at a given locus, one allelic form on each of the two homologues.

Hfr strain. A mating strain of *E. coli* that carries the fertility factor, F, integrated into the chromosome and which brings about a high frequency of recombination following conjugation with an F^- cell.

histocompatibility antigens. Antigens in the cell membrane that typically vary from one individual to the next and can be recognized by the immune system of a host, thus determining acceptance or rejection of a transplant.

histones. Basic proteins of low molecular weight that are complexed with the DNA of eukaryotes.

hnRNA. See heterogeneous nuclear RNA.

Hogness (TATA) box. A DNA segment found in those eukaryotic promoters that interact with RNA polymerase II, which contains the characteristic sequence TATAAAA and plays a role in determining the exact site at which transcription will start.

holandric. Any allele whose locus is on the Y chromosome; any trait associated only with the Y chromosome.

homogametic sex. The sex in a given species that produces a single kind of gamete with respect to the type of sex chromosome. The mammalian female is the homogametic sex, because every normal egg carries an X chromosome.

homologous. Having corresponding genetic loci.

homozygous. Having the same allele (gene form) present at a given locus on both homologous chromosomes.

housekeeping genes. Those genes that are expressed in all cells because they are coded for essential functions basic to all cell types.

H-Y antigen. The transplantation antigen present on the tissue of the heterogametic sex of mammals and believed to be required for testis formation.

hybrid. A heterozygote; an individual resulting from a cross between genetically different parents.

hydrogen bond. A weak chemical interaction between an electronegative atom and a hydrogen atom that is covalently linked to another atom.

hydroxyapatite. A form of calcium phosphate that binds to double-stranded DNA.

in situ hybridization. A variation of the DNA–RNA hybridization procedure in which the denatured DNA is left in place in the chromosome and challenged with RNA or DNA chains from another source without being extracted.

in vitro. Occurring outside the living organism, for example, in a test tube or other artifically designed environment.

in vivo. Occurring within the living organism.

incomplete dominance. The expression in the heterozygote of both allelic forms; generally used to refer to those cases where the phenotypic effect of one gene form in the heterozygote appears more pronounced than the other. Incomplete dominance and codominance are frequently used interchangeably, because the distinction between them is often not clearly defined.

incomplete sex linkage. The situation in which a specific locus is found on both the X and the Y chromosomes in that portion of the sex chromosomes that have homologous regions.

independent assortment. The segregation at meiosis of a pair of alleles independently of other allelic pairs whose loci are found on different chromosomes.

induction (of enzyme synthesis). The stimulation of specific enzyme synthesis in a cell when the specific substrate of the enzyme is supplied.

initiating tRNA. A special form of tRNA that can recognize the triplet AUG at the site in the mRNA which marks the start of translation.

interference. The effect exerted by a crossover that decreases the probability of a second crossover occurring within a certain map distance in the chromosome.

interphase. The metabolic nucleus of a cell between nuclear divisions. The condition of a nucleus that has not entered the mitotic or meiotic cycle.

introns. Those nucleotide sequences of a structural gene and its transcript that are noncoding and interrupt those sequences containing information for the assembly of a polypeptide.

inversion. A chromosome aberration entailing two breaks in a chromosome followed by a reversal of the segment and consequently of the gene sequence in the segment. Pericentric inversions include the centromere in the inverted segment, whereas paracentric inversions do not. In an overlapping inversion a point of breakage in a particular inversion occurs in a segment that was reversed in a previous inversion.

inverted repeats. Two identical DNA sequences oriented in opposite directions on the same molecule. Adjacent inverted repeats constitute a palindrome.

ionizing radiation. Any radiation possessing sufficient energy to cause an outright separation of electric charges in the material it strikes.

IS element (insertion sequence). One of a group of small (less than 2000 base pairs long) transposons found in bacteria that are able to move about and insert at a number of sites on the chromosome and are coded for no known properties other than those of insertion.

isoenzymes. Two or more enzymes capable of catalyzing the same reaction but differing slightly in their structures and consequently in their efficiencies under various environmental conditions.

kappa. A DNA-containing cytoplasmic particle found in certain strains of Paramecium that is capable of self-reproduction and can exert a killing effect on sensitive cells when it is liberated in the culture medium.

karyotype. The chromosome constitution of a cell or individual.

kinetochore. The structural part of the centromere into which spindle fibers insert.

labeled compound. A compound containing a radioactive atom that enables the compound or its breakdown products to be followed through a series of reactions or steps by detecting its radioactivity.

lagging chain. The DNA strand that is assembled at the time of replication in the direction away from the replication fork.

leader. That portion of the mRNA toward the 5′ end that is not translated but binds to the ribosome for the proper initiation of translation.

leading chain. That DNA strand that is assembled at the time of replication in the direction toward the replication fork.

leptonema (leptotene stage). The first stage of first meiotic prophase before synapsis, during which each chromosome is in an extremely extended state.

lethal allele. Any allele that can cause the death of the individual who possesses it. A complete lethal removes the individual sometime before the age of reproduction.

library. See gene library.

ligase. See DNA ligase.

linkage. The association of genes or genetic loci on the same chromosome. Linked genes tend to be transmitted together.

linker DNA. That DNA of a nucleosome in excess of the 146-base-pair core DNA to which histone H1 is bound and that connects adjacent core particles.

locus. The specific position occupied by a given gene on a chromosome. At a particular locus, any one of the variant forms of a gene may be present.

long terminal repeats (LTRs). Sequences several hundred base pairs long that are directly repeated at both ends of the DNA of a retrovirus.

lysis. The bursting of a cell following dissolution of the cell membrane.

lysogenic bacteria. A strain of bacteria-harboring prophage that causes the lysis of a strain sensitive to that virus.

lysogeny. A state in which the genetic material of a virus and its bacterial host are integrated.

lytic virus. A virus that causes lysis following its multiplication in the host cell.

map unit. A measure of genetic distance between two linked genes that corresponds to a recombination frequency of 1%.

marker. See genetic marker.

mean. The numerical average; the sum of all the values of a group of figures divided by the number of individual measurements in the group.

megaspore. A haploid plant cell derived from meiotic division of a megaspore mother cell that may give rise to a female gamete following mitotic divisions.

meiosis. The nuclear process that includes two divisions and results in the reduction of chromosome number from diploid to haploid.

melting. Denaturation of double-stranded DNA to single strands.

Mendelian trait. Any trait that is transmitted in accordance with Mendel's laws.

merozygote. A partially diploid bacterial cell.

messenger RNA (mRNA). The complementary RNA copy of DNA formed on the DNA template during transcription and carrying coded information for the amino acid sequence of a polypeptide.

metacentric. A chromosome with its centromere in a median position, thus having two arms of equal length; the median location of the centromere in the chromosome.

metaphase. That stage of nuclear divisions when the chromosomes are arranged at the equatorial plane of the spindle.

microspore. A haploid plant cell derived from the meiotic division of a microspore mother cell that may give rise to male gametes following mitotic divisions.

minimal medium. A medium that provides only those substances essential for the growth and reproduction of wild-type cells or organisms.

mismatch repair. Enzymatic correction of nucleotide mispairing that entails the removal of defective, single-stranded segments followed by the synthesis of new segments.

missense mutation. A point mutation in which a codon is changed to a different codon designating a different amino acid.

mitochondrion. A double-membrane cytoplasmic organelle found in eukaryotes and the site of aerobic respiration.

mitosis. The nuclear division that results in the accurate distribution of the genetic material and produces two nuclei exactly like the parent nucleus in chromosome content.

modification enzyme. An enzyme that recognizes the same nucleotide sequence as a corresponding restriction endonuclease and methylates certain bases within the sequence, thus conferring protection from the endonuclease activity.

modifier (modifying gene). A gene whose expression affects the expression of a gene or genes at other loci, which are thus not allelic to it. A modifier often has no other known effect.

molecular biology. A modern branch of biology that investigates biological phenomena on the molecular level, often using techniques of physical chemistry to pursue genetic and other biological problems.

monohybrid. An individual heterozygous at a locus under consideration, thus having a pair of alleles at that locus. Also, a cross in which a pair of alternative gene forms (alleles) is being followed.

monosomy. A chromosome aberration in which only one of a given kind of chromosome is present instead of the normal two characteristic of a typically diploid organism.

mosaicism. A situation in which the body of an individual is composed of two or more genetically distinct cell types.

mt DNA. The DNA of a mitochondrion.

multigene family. A set of genes derived by duplication of some ancestral gene followed by a history of independent mutations. Genes composing a family may be either clustered together on the same chromosome or dispersed on different chromosomes.

multiple alleles. Three or more forms of a gene, any one of which can occur at a given locus on a specific chromosome.

multiple-factor inheritance. Inheritance that entails many nonallelic genes and the complex interaction of environmental factors, many of which are unknown (see quantitative inheritance).

mutagen. Any agent that can cause an increase in the rate of mutation in an organism.

mutation. A sudden inheritable change that includes gene (point) mutation and chromosome aberrations in its broadest sense.

mutation pressure. Spontaneous gene mutation occurring continuously in a population.

natural selection. The differential reproduction of alleles that occurs in a population from one generation to the next and results in an increase in the frequency of certain alleles and a decrease in others.

negative control. This type of genetic regulation in which the product of the regulatory gene acts (alone or with a corepressor) to repress transcription of specific structural genes.

nondisjunction. The failure of homologous chromosomes or sister chromatids to separate at mitotic or meiotic division, resulting in aneuploid cells or individuals.

nonhistone protein. An assortment of proteins, mostly acidic in nature, that vary in kind with the species and cell type and that appear to function in controlling transcription in eukaryotes.

nonsense codon. A chain-terminating codon.

nonsense mutation. A point mutation in which a codon specific for an amino acid is changed to a chain-terminating codon.

nuclease. Any enzyme that can break the phosphodiester bonds joining nucleotides together in a polynucleotide chain.

nucleoid. A region within a prokaryote that contains the DNA.

nucleolar organizer. That region of a specific chromosome in the complement (the nucleolar organizing chromosome) that contains genes for the RNA of the ribosome and is associated with the formation of the nucleolus.

nucleolus. A dense body within the nucleus produced by the nucleolar organizer and associated with the processing of the ribosomes.

nucleoside. A molecule composed of a 5-carbon sugar linked to a nitrogen base.

nucleosome. The basic structural subunit of chromatin, consisting of an average of 200 base pairs of DNA plus an octamer of histone proteins.

nucleosome bead (core particle). An octamer of histones plus 146 base pairs of DNA wrapped around it in two turns.

nucleotide. A nucleic acid unit composed of a 5-carbon sugar joined to a phisphate group and a nitrogen base.

ochre codon. The chain-terminating codon UAA.

Okazaki fragments. Short DNA chains that result from discontinuous replication on at least one of the template strands of a duplex DNA.

oncogene. A gene that can initiate and maintain a tumorous state in an organism and that arises from a protooncogene of a normal cell.

oogenesis. The entire series of events in a female in which a gamete is produced from the maturation of an immature germ cell.

oogonium. An immature germ cell in the female that can give rise to a primary oocyte.

opal codon. The chain-terminating codon UGA.

operator. That segment of DNA associated with one or more structural genes, which interacts with the product of a regulatory gene in the control of transcription.

operon. A unit of transcription composed of one or more structural genes associated with an operator and a promoter.

organelle. A subcellular component containing particular enzymes and specialized for a certain cell role.

overlapping genes. Different genes that share a common base sequence but are read in different frames, so that they specify different products.

overlapping inversion. See inversion.

P$_1$. See first parental generation.

pachynema (pachytene stage). Stage of meiotic prophase I after the completion of synapsis and characterized by a detectable thickening of the chromosome threads. The stage of meiosis when crossing over takes place.

palindrome. In DNA, a region in which a symmetrical arrangement of bases occurs around a point in the molecule, with the result that the base sequence is the same on either side of it, reading either of the two paired strands in the 5′ to 3′ direction; adjacent inverted repeats.

paracentric inversion. See inversion.

penetrance. The ability of a specific allele or genotype to express itself in any way at all when present in an organism.

peptide. A short linear stretch of two or more chemically linked amino acids.

peptide bond. A covalent linkage between two amino acids formed when the amino group of one becomes bonded to the carboxyl group of the other and water is eliminated.

peptidyl transferase. The enzyme found in the larger subunit of the ribosome that catalyzes the formation of peptide linkages between amino acids at the P and A sites.

pericentric inversion. See inversion.

phage. A virus with a bacterial cell as its host.

phenocopy. A nongenetic, environmentally induced imitation of the effects of a specific genotype.

phenotype. Any detectable feature of a living organism. The phenotype is a product of the interaction between the genetic material and the environment.

photoreactivation. The reversal of the deleterious effect of ultraviolet light on DNA by white light.

pilus. A filamentous, hollow appendage extending from the surface of a bacterial cell.

plaque. A clear area on an otherwise cloudy growth of bacteria or cells growing in tissue culture where cells have been lysed by a virulent virus.

plasmid. Any replicon that may exist in a cell independently of the chromosomes.

pleiotropy. The multiple phenotypic effects of an allele.

point mutation. A mutation in which one nucleotide substitutes for another.

polar body. A tiny cell that receives little cytoplasm, produced by the division of a primary or a secondary oocyte.

polar mutation. A genetic alteration that affects the expression not only of the gene in which it occurs but also of other genes in the vicinity.

poly (A) tail. The stretch of adenine nucleotides at the 3′ end of eukaryotic mRNA that is added to the pre-mRNA as it is processed, before its transport from the nucleus to the cytoplasm.

polygenic inheritance. Inheritance involving alleles at many genetic loci, which interact with environmental factors. *Polygenic* refers to the many genetic factors as opposed to the environmental component. (Also see quantitative inheritance.)

polymorphism. The existence within a population of two or more forms in a proportion that cannot be attributed to recurrent mutation alone.

polypeptide. A single long chain composed of amino acid units joined by peptide linkages.

polyploidy. A condition in a cell or individual in which one or more whole haploid sets of chromosomes are present, in addition to the normal diploid number typical for the species or group.

polysome. An assembly of ribosomes active in translation and connected by the same mRNA strand.

polytene chromosome. A chromosome composed of many intimately associated strands or unit chromatids formed as a result of endoreduplication.

population. An interbreeding group of organisms of the same species.

position effect. A phenotypic change resulting solely from the placement of a gene or genetic region in a new location in the chromosome complement, without any change within the newly located genetic material itself.

positive control. The type of genetic regulation in which a substance produced in the cell is required to activate transcription of specific structural genes.

posttranscriptional modification. Any alteration made to pre-mRNA before it leaves the nucleus as mature mRNA.

pre-messenger RNA (pre-mRNA). The immediate transcript of a sense strand that undergoes modification in the eukaryotic nucleus and then yields the mRNA that leaves the nucleus to undergo translation.

Pribnow box. The promoter sequence TATAAT centered about 10 base pairs before the point at which transcription of bacterial genes starts, which is believed to play a very important role in the binding of RNA polymerase and initiation of transcription.

primary oocyte. A cell in the female derived from an oogonium and which undergoes first meiotic division.

primary spermatocyte. A cell in the testis derived from a spermatogonium and which undergoes first meiotic division.

primase. An enzyme required for the assembly of an RNA primer at the time of DNA replication.

primer. A short nucleic acid segment that provides a free 3'-OH end for internucleotide linkage in the 5' to 3' direction.

probe. Any biochemical that is usually labeled or tagged in some way so that it can be used to identify or isolate a gene, gene product, or protein.

processing. The posttranscriptional modification of pre-mRNA.

prokaryote. Organisms or cells that lack a nucleus bounded by a membrane as well as specialized membrane bound organelles. (Also see eukaryote.)

promoter. The segment of DNA containing start signals that can be recognized by RNA polymerase for the start of transcription of a gene.

prophage. A temperate virus devoid of its protein components, which is integrated with the DNA of the host and behaves like a genetic factor of the infected cell.

prophase. That phase of nuclear divisions following DNA replication, during which the chromosomes become evident as they gradually shorten and thicken.

protooncogene. A normal cellular gene that can be converted to an oncogene as the result of somatic mutation or recombination with a viral gene.

prototroph. A microorganism capable of supplying its own nutrient requirements from a minimal medium.

provirus. The DNA form of an RNA virus that can integrate into the chromosome of a cell.

pseudogene. A gene that closely resembles a known functional gene at another locus but has become nonfunctional because of an accumulation of defects in its structure that prevent normal transcription or translation.

purine. A nitrogen-containing compound with a double ring structure, and the parent compound of adenine and guanine.

pyrimidine. A nitrogen-containing compound with a single ring structure, and the parent compound of cytosine, thymine, and uracil.

Q bands. Those regions along the length of the chromosome that fluoresce after staining with a fluorescent dye.

quantitative inheritance (variation). That type of hereditary transmission or variation that involves many alleles at different genetic loci, each allele producing some measurable effect (often used synonomously with polygenic and multiple factor inheritance).

R plasmid. An extrachromosomal replicon carrying genetic determinants for drug resistance.

rDNA. See ribosomal DNA.

reading frame. A sequence of codons that can be translated, which contains a codon indicating where translation is to begin and a chain-terminating codon indicating where it is to end.

reanneal. To form duplex molecules from single-stranded complementary nucleotide chains. (*Reannealing* implies that the single-stranded chains came from the same source, whereas *annealing* implies they originated from different sources.)

reassociation. Reannealing or renaturation.

reassociation kinetics. See *renaturation kinetics*.

recessive. Any gene form that is not expressed in the presence of its allele in the heterozygote. Also, any trait that is expressed phenotypically only when the responsible gene form is present in double dose in a homozygote.

recombinant DNA. DNA molecules resulting from the union of DNA derived from different sources.

recombinant DNA technology. The techniques employed for splicing DNA molecules derived from more than one source and also for amplifying the recombinant DNA in a suitable host, to obtain large amounts of the DNA or of its products.

recombination. A new combination of alleles resulting from rearrangement following crossing over or independent assortment. (Some restrict usage of the term to those new combinations produced by crossing over.)

regulatory gene. A DNA sequence that functions primarily to control the expression of other genes by modulating the synthesis of their products. (A regulatory gene coded for a specific product is also a structural gene.)

renaturation (of DNA). The reconstitution of double-helical DNA by the reassociation of single strands derived from it by melting. (See also *reanneal.*)

renaturation kinetics. A technique that measures the rate of reassociation or reannealing of complementary single DNA strands derived from a single source that is used to indicate genome size and complexity. (Also see C_ot *plot.*)

repetitive DNA. Repeated nucleotide sequences that may occur in hundreds, thousands, or more copies in the chromosome complement of a eukaryote.

replication. The synthesis of a macromolecule identical to and under the guidance of a parent or template macromolecule.

replication fork. The Y-shaped region within a double-stranded DNA molecule that is undergoing replication and marks the site at which complementary strands are being synthesized at that time.

replicon. A unit of DNA replication with a specific point of origin at which replication begins.

repression (of enzyme synthesis). The cessation of production of a specific enzyme by a cell when the end product of the enzyme reaction is supplied.

repressor. The protein product of a regulatory gene that can combine with a specific operator and block transcription of the structural genes in an operon.

resistance transfer factor (RTF). That segment of a conjugal R plasmid carrying the genetic factors that confer on the plasmid those properties needed for transfer of the plasmid from one cell to the next.

restriction enzyme (restriction endonuclease). A type of endonuclease that recognizes and produces internal cuts only within a certain specific nucleotide sequence.

restriction fragment length polymorphisms (**RFLPs**). Variations found within a species in the length of the DNA fragments generated from a particular DNA region by a specific restriction enzyme.

restriction map. A physical map or depiction of a gene or genome derived from the ordering of restriction fragments produced by restriction enzymes that indicates the sites at which one or more restriction endonucleases cleave the molecule.

retrovirus. An RNA virus that uses reverse transcriptase to assemble a DNA copy of its RNA genome using its RNA as the template.

reverse mutation (reversion). A change in a mutant allele that restores it to the wild type.

reverse transcriptase. A polymerase that preferentially assembles a DNA strand using an RNA strand as the template; RNA-dependent DNA polymerase.

rho factor. A protein required to halt transcription of certain genes.

ribonuclease. An enzyme that hydrolyzes RNA; RNAase.

ribosomal DNA (rDNA). DNA whose transcripts are processed into the RNA components of the ribosomes.

ribosomal RNA (rRNA). That class of RNA molecules present in both the small and large subunits of prokaryotic and eukaryotic ribosomes.

ribosome. A complex cellular component, composed of RNA and protein, that interacts with mRNA and tRNA to join together amino acids into a polypeptide chain. Found in the cytosol, mitochondria, and chloroplasts.

RNA (ribonucleic acid). A category of polynucleotides in which the component sugar is ribose and uracil takes the place of thymine. (RNA composes the genome of certain viruses and falls into three main classes in prokaryotes and eukaryotes: tRNA, mRNA, and rRNA.)

RNAase. See *ribonuclease.*

RNA polymerase. An enzyme that catalyzes the synthesis of RNA from ribonucleoside triphosphate precursors using a template DNA strand.

RNA replicase. An enzyme required for the assembly of RNA strands on an RNA template.

rolling circle. A model for the manner of replication of certain DNA molecules, in which a replication fork proceeds around a circular template for an indefinite number of revolutions and a newly synthesized DNA strand displaces the strand synthesized in the previous revolution.

rRNA. See *ribosomal RNA.*

S phase. The portion of interphase during which DNA replication takes place.

S value. A sedimentation coefficient or constant expressed in Svedberg units.

satellite. A terminal end of a chromosome arm produced by a secondary constriction, connected to the rest of the chromosome by a very narrow region, and typically associated with nucleolus formation.

satellite DNA. The DNA of a eukaryotic cell that has a different density from the bulk of the cellular DNA and equilibrates at a different position from the main band following density gradient centrifugation.

second site reversion. A suppressor mutation that restores function to a gene and occurs in the same gene as the first mutation that produced the aberrant effect.

secondary oocyte. The larger of the two cells produced by the division of a primary oocyte.

secondary spermatocyte. One of the two cells derived from the division of a primary spermatocyte that has completed first meiotic division.

selection coefficient(s). A measure of the proportionate reduction in the contribution of a given genotype to the following generation.

selection pressure. The operation of natural selection on the allele frequency in a population resulting in an increase in the frequency of certain alleles and a decrease in the frequency of others.

semiconservative replication. The manner in which double-stranded DNA is synthesized and which produces two double-stranded molecules from the original DNA, each containing one parental strand and one strand that is newly formed.

sense strand. The one of the two DNA strands composing a gene that undergoes transcription.

sex bivalent. An association composed of the paired X and Y chromosomes in a primary spermatocyte.

sex chromatin. A condensed or inactivated X chromosome (or part of an X) in the interphase nucleus seen as a densely straining body in cells containing more than one X chromosome.

sex chromosomes. Those chromosomes involved in sex determination (e.g., the mammalian X and Y) that show a difference in morphology or number between the sexes.

sex-influenced allele. Any allele that is expressed as a dominant in one sex and a recessive in the other.

sex-limited allele. Any allele that expresses itself in only one of the sexes.

sex-linked (X-linked) allele. Any gene form with its locus on the X chromosome. Also any trait associated with such a genetic factor.

sex pilus. An appendage on a bacterial cell produced by the presence of a fertility factor and necessary for conjugation

sexduction. The transfer of a chromosomal gene from one cell to another by way of a sex factor, such as F, the fertility factor.

shift. A chromosome aberration in which a chromosome segment is transposed to a new location in the same or a different chromosome. The shift is a simple type of translocation.

shotgun experiment. The cloning of an entire genome in the form of randomly generated fragments to create a "gene library" for the species that is available for later studies.

sigma chain. One of the polypeptide chains composing RNA polymerase, which is essential for the recognition of signals in the DNA for the start of transcription.

silent mutation. A point mutation in which a codon specific for a given amino acid is changed to a different codon that designates the same amino acid.

small ribonucleoprotein particle (sn RNP). A unit confined to the nucleus and composed of one RNA molecule and several proteins; a snurp. (Snurps differ in RNA type, the type rich in U_1 RNA playing a role in processing pre-mRNA.)

somatic cell(s). The cells of the body, in contrast to gametes and the cells from which they were derived.

somatic cell hybridization. the process of fusion between somatic cells to produce somatic cell hybrids for genetic analysis.

SOS response. An error-prone mechanism to repair damaged DNA in *E. coli* that involves the coordinate induction of many enzymes in response to irradiation or other damage to DNA.

Southern blotting. A procedure for transferring denatured DNA from an agarose gel to a nitrocellulose filter, where it can be hybridized with complementary single-stranded nucleic acid.

specialized transduction. One-way gene transfer in bacteria in which a phage that had been integrated as prophage carries some of its own genome and some of the host's into a recipient cell. Only genes adjacent to the site of integration of the virus in its prophage form can be transferred.

spermatocyte. See *primary spermatocyte.*

spermatid. One of two cells derived from the division of a secondary spermatocyte that has completed second meiotic division.

spermatogenesis. The series of events starting with the origin of a primary spermatocyte and culminating with the production of sperms.

spermatogonium. A type of cell found in the wall of the testis that can give rise to a primary spermatocyte.

spermiogenesis. The portion of spermatogenesis during which a spermatid is transformed into a mature sperm.

spindle. An aggregation of microtubules essential for the positioning and distribution of the chromosomes at nuclear divisions.

spindle fiber. A microtubule composed of an assemblage of protein units.

splicing. Joining together separated component parts. RNA splicing in eukaryotes entails the removal of introns and the joining of exons of pre-mRNA to produce mature mRNA.

standard deviation. A statistic that describes the amount of variation on either side of the mean in a sample and which can be used to calculate the standard error of the mean.

standard error of the mean. A statistic that indicates the probability that the mean calculated from a sample is close to the true mean for the entire population.

statistics. Measurements or values obtained from samples rather than those measurements made on an entire population, the parameters.

steroid. One of an assortment of complex lipids, composed of four interlocking rings of carbon atoms, which are often biologically important compounds, such as vertebrate male and female sex hormones.

sticky ends. See *cohesive ends.*

structural gene. A segment of DNA coded for the amino acid sequence of a polypeptide chain or for a tRNA or rRNA molecule. (Regulatory genes are structural genes whose products control the expression of other genes.)

structural heterozygote. An individual with a normal chromosome and some kind of change in the homologue that affects its morphology or structure.

supercoil. A coil superimposed on a coiled duplex; a conformation resulting from the coiling of a circular duplex

DNA molecule upon itself, so that it crosses its own axis.

suppressor mutation. A mutation at a genetic locus separate from the one in which a first mutation occurred that reverses the effect of the first mutation and produces a wild-type phenotype in the double mutant.

symbiosis. An intimate association of two organisms of different species living together, usually for mutual benefits.

sympatric. Overlapping or occupying the same general area.

synapsis. The intimate pairing, locus for locus, of homologous chromosomes at first meiotic prophase.

synaptonemal complex. A complex structure that forms during first meiotic prophase and facilitates the intimate pairing between homologous chromosomes.

syndrome. The collection or group of symptoms or features associated with a specific disease or abnormality.

synteny. The association of genes or genetic loci on the same chromosome. Often used synonomously with linkage. Some restrict the term *linkage* to those genes that clearly do not undergo independent assortment, as shown by genetic analysis. *Syntenic* is used to indicate that procedures such as somatic-cell hybridization have shown two genes are on the same chromosome, even if genetic analysis has not yet demonstrated this.

T cell (T lymphocyte). A white blood cell that becomes differentiated in the thymus gland and can recognize foreign antigens on cell surfaces and bring about the removal of the foreign cells.

TATA box. See *Hogness box.*

tautomerism. The reversible shifting of the location of a proton in a molecule that changes certain chemical properties of the molecule.

telocentric. A chromosome with its centromere appearing to be in a terminal location; thus having only one evident arm, the centromere located at one end of the chromosome.

telomere. A region of a chromosome marking the extreme end of a chromosome arm. Every normal chromosome possesses two telomeres.

telophase. The stage of nuclear division during which the nuclear membrane reforms and the chromosomes gradually become less and less evident.

temperate (symbiotic) phage. A bacterial virus that tends to take up residence in the host cell without destroying it.

template. A macromolecule that serves as a blueprint or mold for the synthesis of another molecule. A preformed nucleic acid chain that is required to establish the nucleotide sequence in a complementary chain.

terminator sequence. A stretch of DNA that contains a signal for RNA polymerase to bring transcription to a halt.

testcross. The cross of an individual to one that expresses a specific recessive trait or traits under consideration.

tetrad. The four nuclei or four cells that are the immediate results of meiosis in a parent cell (sometimes also used to designate a bivalent).

tetraploid. A cell or organism with four complete haploid sets of chromosomes instead of the typical two.

tetravalent (quadrivalent). An association at first meiotic division of four chromosomes that are completely or partially homologous.

3′ end. That end of a nucleic acid chain terminating in a free $-OH$ group.

thymine. A pyrimidine base found in DNA but not in RNA.

topoisomerase. An enzyme that can convert DNA from one topological form to another.

trailer. That portion of an mRNA toward the 3′ end that follows a chain-terminating codon and does not undergo translation.

trait. A distinct alternative form of a characteristic.

trans arrangement. That linkage arrangement in a double heterozygote in which a wild allele and a recessive mutant gene form occur together on one chromosome, and the corresponding recessive and wild alleles occur together on the homologue.

transcript. The single-stranded RNA chain that is assembled on a DNA template.

transcription. The assembly of a complementary single-stranded molecule of RNA on a DNA template.

transductant. A cell that has been transduced.

transduction. The transfer of genetic material from one cell to another by a viral vector. (Also see *generalized* and *specialized transduction.*)

transfer RNA (tRNA). A small RNA molecule that recognizes a specific amino acid, transports it to a specific codon in the mRNA, and positions it properly in a growing polypeptide chain.

transformation. A genetic change effected in a cell as the result of the incorporation of DNA from a virus or some genetically different cell type; also refers to the taking up of extraneous genetic material by salt-treated bacterial cells in genetic manipulation.

transgressive variation. The quantitative variation in a characteristic of the offspring which exceeds that shown by the same characteristic in either parent.

transition. A point mutation in which a purine is replaced by a different purine or a pyrimidine by a different pyrimidine.

translation. The assembly of a polypeptide chain from the coded information in the mRNA that directs the amino acid sequence of the chain.

translocation (reciprocal). A chromosome alteration in which a chromosome segment or arm is transposed to a new location. A reciprocal translocation involves a mutual exchange of chromosome segments or arms between two nonhomologous chromosomes.

transposon. A genetic unit capable of moving from one chromosome site to another or from one replicon to another.

transversion. A point mutation in which a purine is replaced by a pyrimidine or a pyrimidine is replaced by a purine.

triploid. A cell or organism with three complete haploid sets of chromosomes instead of the normal two.

tRNA. See *transfer RNA.*

trisomy. A condition in a cell or individual of a typically diploid species in which a particular chromosome is present in three doses instead of the normal two.

trivalent. An association at first meiotic division of three chromosomes that are completely or partially homologous.

unique DNA. A class of DNA representing sequences that are present only once in a genome.

univalent. A chromosome that remains unpaired at the first division of meiosis.

upstream. The portion of the DNA template that lies away from the initiation site of transcription toward the 3′ end.

uracil. A pyrimidine base found in RNA but not in DNA.

variance. A statistic that measures the amount of variability in a sample and can be broken up into separate components to ascertain the role of responsible genetic and environmental factors in the variation; the square of the standard deviation.

vector. An agent that transfers material from one host to another.

virion.   A complete virus particle consisting of its nucleic acid core and protein coat.

virulent phage.   A bacterial virus that typically destroys the host cell it infects.

wild type.   The form of a gene or an individual that is considered the standard type, typically found in nature.

wobble hypothesis.   The concept that the base at the 5′ end of the anticodon is somewhat free to move about spatially and can thus form hydrogen bonds with more than one kind of base at the 3′ end of a codon in the mRNA.

X chromosome.   In mammals and various other groups, the sex chromosome that is found in two copies in the homogametic sex and in one copy in the heterogametic sex.

X linked.   See *sex linked*.

Z DNA.   A form of DNA existing as a left-handed double helix whose sugar–phosphate backbone follows a zigzag course and may possibly play a role in gene regulation.

zygonema (zygotene stage).   The stage of first meiotic prophase during which homologous chromosomes pair.

zygotic induction.   The activation of a virus to a virulent state following its transfer as prophage from a Hfr bacterial cell to an F⁻ cell that lacks the virus.

ANSWERS TO REVIEW QUESTIONS

CHAPTER 1

1. A. OO
   B. oo
   C. Oo
   D. oo

2. A. 1 OO:1 Oo. All round
   B. 1 OO: 2 Oo: 1 oo. 3 round: 1 oblong
   C. oo. All oblong
   D. 1Oo: 1 oo. 1 round: 1 oblong

3. Short hair is the dominant trait. Genotypes: female 1 is Ss; female 2 is SS. The male is ss.

4. A. Horned and hornless in a ratio of 1:1
   B. One fourth
   C. Three fourths

5. A. Red (RR), roan (Rr), white (rr) in a ratio of 1:2:1
   B. One roan (Rr); one white (rr)

6. A. hhrr
   B. H—Rr
   C. HhRR

7. A. All with black fur and short hair
   B. One out of four; BBLL—black short; BBll—black long; bbLL—albino short; bbll—albino long

8. A. All blue and mildly frizzled
   B. 1 blue and mildly frizzled: 1 white, mildly frizzled

C. 1 blue, mildly frizzled: 1 blue frizzled: 1 white, mildly frizzled: 1 white frizzled
D. 1 black frizzled: 2 blue frizzled: 1 white frizzled: 2 black, mildly frizzled: 4 blue, mildly frizzled: 2 white, mildly frizzled: 1 black straight: 2 blue straight: 1 white straight

9. A. Aamm
   B. A—Mm
   C. aamm

10. A. Parental genotype: AaMm. Gametes: AM, Am, aM, am
    Parental genotype: Aamm. Gametes: Am, am.
    B. Offspring: nonalbino with migraine, nonalbino without migraine, albino with migraine, albino without migraine in a ratio of 3:3:1:1

11. A. No chance
    B. One fourth
    C. Two thirds

12. All the children will be genotype ss, homozygous for the defective allele, just as the parents are. The children would require treatment, because the environment has not altered the genetic material that is transmitted through the gametes. A Lamarckian would claim that the treatment has produced the normal condition and this trait can now be transmitted, so that the children would be normal eventually regardless of treatment.

13. A. (1) ffSs (2) FfSs
    B. Free, no sickling; free, some sickling; free, with anemia; attached, no sickling; attached, some sickling; attached with anemia. Ratio 1:2:1:1:2:1

14. A. An allele does not increase in frequency simply because it is dominant. The frequency is the result of the interaction of many factors. Dominance by itself does not make an allele increase. Dominance means that a gene form will express itself in the presence of its allele.
    B. The person has a 50:50 chance of developing the disorder. The afflicted parent is most likely genotype Hh, since the allele is so rare.

15. A. 4
    B. 16
    C. 16

16. A. 16
    B. 8
    C. 32

17. A. 81
    B. 27
    C. 243

18. (1) TG, tG
    (2) TGW, TGw, tGW, tGw
    (3) TGW, TgW, tGW, tgW, TGw, Tgw, tGw, tgw

19. (1) 1 tall yellow: 1 dwarf yellow
    (2) 1 tall yellow round: 1 tall yellow wrinkled: 1 dwarf yellow round: 1 dwarf yellow wrinkled
    (3) 1 tall yellow round: 1 tall green round: 1 dwarf yellow round: 1 dwarf green round: 1 tall yellow wrinkled: 1 tall green wrinkled: 1 dwarf yellow wrinkled: 1 dwarf green wrinkled

20. 9 tall yellow round: 3 dwarf yellow round: 3 tall yellow wrinkled: 1 dwarf yellow wrinkled

CHAPTER 2

1. A. Prophase
   B. Telophase
   C. Metaphase
   D. Prophase
   E. Telophase
   F. Anaphase

2. A. (1) 92
      (2) 92
   B. (1) 46
      (2) 46
      (3) 46

3. A. 22 pairs
   B. 22 pairs
   C. 2

D. 2
E. 7

4. A. AAXX
   B. AAXY

5. 1. B
   2. E
   3. F
   4. A
   5. E
   6. C
   7. I
   8. J
   9. G
   10. H

6. Prokaryotes lack mitochondria, a nuclear envelope, chloroplasts, endoplasmic reticulum, centrioles, mitotic spindles (additional distinctions are made in later chapters).

7. A. Locus
   B. Homologous
   C. Homozygote or homozygous
   D. Alleles
   E. DNA
   F. Metacentric
   G. Polytene
   H. Karyotype
   I. Satellite
   J. Acrocentric
   K. Colchicine
   L. S of interphase
   M. Q bands
   N. C bands
   O. q arm

8. A. +
   B. +
   C. −
   D. −
   E. −
   F. +
   G. −
   H. +
   I. −

9. a. La La
   b. La⁺ La⁺ or + +
   c. La⁺ La or + La

CHAPTER 3

1. A. Diplonema
   B. Zygonema
   C. Diakinesis
   D. S of premeiotic interphase
   E. Leptonema

F. Pachynema
G. Metaphase II

2.

	Number of DNA Molecules	Chromosome Number	Number of Bivalents
A.	92	46	23
B.	92	46	23
C.	92	46	23
D.	46	23	0
E.	46	23	0
F.	23	23	0

3. A. At mitotic metaphase single chromosomes are oriented at the equator, whereas bivalents are at the equator of first meiotic metaphase.
   B. In mitosis, each chromosome moving poleward is composed of a single chromatid; at meiotic first anaphase, each is composed of two chromatids.
   C. No bivalents would be present at mitotic prophase. At the earliest meiotic prophase, the chromosome threads are much more extended than at mitosis and show obvious chromomeres.

4. A. Only AaBb
   B. AB, Ab, aB, ab

5. A. 46
   B. 46
   C. 23
   D. 23
   E. 23
   F. 23

6. A. AX
   B. AX and AY
   C. AAXY
   D. AAXX

7. Only in the primary oocyte.

8. A. (1) 4000, (2) 2000, (3) 1000, (4) 8000
   B. (1) 1000, (2) 1000, (3) 1000, (4) 250

9.
   A. 10
   B. 10
   C. 11
   D. 6
   E. 5

10. A. 10
    B. 20
    C. 20
    D. 20
    E. 10
    F. 10

11. (1) Bo, (2) Bo, (3) Me, (4) Bo, (5) Me, (6) Me, (7) Bo, (8) Bo, (9) Me, (10) Me, (11) Me, (12) Bo

CHAPTER 4

1. The dominant E shows variable expressivity and reduced penetrance. The man with the extra toe on each foot definitely carries allele E and is most likely genotype Ee, since the allele is rare. The wife is probably genotype ee, because the allele is rare, and she does not come from her husband's family line. Both their children are genotype Ee, even the one of normal phenotype. The normal offspring's wife is almost certainly ee, again because the allele is rare, but the offspring here is again Ee, having received the dominant from the phenotypically normal parent who carried it but did not express it.

2. 60%

3. Pleiotropy

4. A. CcPp × Ccpp
   B. White and colored in a ratio of 5:3

5.
   A. Red
   B. White
   C. Yellow
   D. White

6. A. Brown × white. Offspring: 2 white: 1 black: 1 brown
   B. White × black. Offspring: 2 white: 1 black: 1 brown
   C. White × white. Offspring: 3 white: 1 black

7. A. Deaf × deaf. No chance of a deaf child
   B. Normal × normal. No chance of a deaf child
   C. Normal × normal. One chance out of four
   D. Normal × normal. One chance out of four

8. A. Pea × rose. Offspring: all walnut
   B. Walnut × walnut. Offspring: 9 walnut: 3 rose: 3 pea: 1 single
   C. Walnut × single. Offspring: 1 walnut: 1 rose: 1 pea: 1 single

9. A. RrPp × rrpp
   B. RRpp × rrPp
   C. Rrpp × RrPp

10. A. RRppBB × rrPPbb. F_1 birds are walnut, blue: RrPpBb
    B. RrPpBb × rrppBB. F_2: Black birds of the four different types of combs and blue birds of the four different types of combs

11. A. BBCCoo × BBccOO. F_1: BBCcOo
    B. BBCcOo × BBCcOo. F_1: 9 BBC—O—; 3 BBC—oo; 3 BBccO—; 1 BBccoo. The ratio is 9:7.

C. BbCCOO × BbCCOO. F$_1$: BBCCOO; BbCCOO; bbCCOO. The ratio is approximately 1:2:1.

12. The F$_2$ plants fall into a ratio of 9:7, a modified dihybrid ratio. Two pairs of alleles are involved, say Aa and Bb. High cyanide content requires both dominants, A and B. All those plants lacking either A or B will have a low cyanide content.

13. The ratio among the F$_2$s is approximately 15:1, indicating a modified dihybrid ratio. Assuming the allelic pairs Aa and Bb, the F$_1$ parents must have been dihybrids AaBb. The P$_1$s would have been AAbb and aaBB. Only double recessives, aabb, have unfeathered legs. Duplicate dominant epistasis is involved, so that the presence of either dominant, A or B, will give feathers on the legs.

14. Mexican hairless dogs are all heterozygotes, carrying an allele that can be represented as H. This acts as a dominant with respect to the hairless condition, but it also has a recessive lethal effect. Other breeds of dogs are homozygous for the recessive allele h, which permits hair growth. They thus lack the lethal. Genotypes are Mexican hairless, Hh; dogs with hair, hh; dead pups, HH.

15. A. Hhww
    B. hhww
    C. hhWW

16. A. 1 hairless: 1 wire-haired
    B. 4 hairless: 3 wire-haired: 1 straight-haired

17. When two plants are crossed and produce albino offspring, it is known they are heterozygotes. These plants of known genotype Ww can be crossed to those whose genotype is still unknown. If the cross of a known heterozygote and an unknown produces a large number of offspring (say, more than 20) without the appearance of any albinos, the unknown would most likely be genotype WW. Recognizing WW plants in this way and then using them exclusively in further crosses would eliminate the lethal.

18. A. 1 black: 1 albino.
    B. Same results as in A because ovaries are aa.
    C. All black offspring.

19. Pp (no PKU: no mental retardation) and pp (PKU but no mental retardation if placed on special diet).

20. Pp and pp both suffering mental retardation as a result of damage from toxic substances that crossed the placenta.

21. A. S$_1$S$_4$; S$_1$S$_5$; S$_2$S$_4$; S$_2$S$_5$
    B. S$_3$S$_5$; S$_4$S$_5$
    C. No offspring can result, because all the pollen will abort.
    D. S$_3$S$_4$; S$_3$S$_5$

22. A. c^{ch}c^h (light gray) and c^{ch}c (light gray)
    B. c$^+$c (wild) and c^{ch}c (light gray)
    C. c^{ch}c^{ch} (chinchilla); c^{ch}c^h (light gray); c^{ch}c (light gray); c^{h}c (Himalayan)
    D. c$^+$c^h (wild); c$^+$c (wild); c^{h}c (Himalayan); cc (albino)

23. A. 3 Himalayan: 1 albino
    B. 3 fully pigmented: 1 albino

24. A. I^{A}i and ii
    B. I^{A}I^B and I^{A}i
    C. I^{A}I^A and I^{B}I^B

25. A. I^{A1}I^{A2} (Type A$_1$); I^{A1}i (Type A$_1$); I^{A2}i (Type A$_2$); ii (Type O)
    B. I^{A1}I^{A2} (Type A$_1$); I^{A2}I^B (Type A$_2$B); I^{A1}i (Type A$_1$); I^{B}i (Type B)
    C. I^{A1}I^{A2} (Type A$_1$); I^{A1}I^B (Type A$_1$B); I^{A2}I^B (Type A$_2$B); I^{B}I^B (Type B)
    D. I^{A1}i (Type A$_1$); I^{B}i (Type B)

26. Neither. A type O person such as man 1 cannot have AB offspring. The second man, being blood type A$_2$, cannot contribute an I^{A1} allele required for the child's I^{A1}I^B type.

27. A. I^{A}i Se se (will secrete A antigen) and I^{B}i Se se (will secrete B antigen).
    B. I^{A}i Se Se; I^{A}i Se se (both types secrete A antigen); I^{B}i Se Se; I^{B}i Se se (both types secrete B antigen); I^{A}i se se; I^{B}i se se (both types are nonsecretors).
    C. I^{A}I^B Se Se (secretes antigens A and B); I^{A}I^B Se se (secretes antigens A and B); I^{B}i Se Se (secretes B); I^{B}i Se se (secretes B); I^{A}I^B se se (nonsecretor); I^{B}i se se (nonsecretor); I^{A}i Se Se (secretes A); I^{A}i Se se (secretes A); I^{A}i se se (nonsecretor); ii Se Se; ii Se se; ii se se (no A or B antigen can be produced with O type, even in presence of Se).

28. A. 3 type A; 3 type B; 6 type AB; 4 type O
    B. 3 type A; 5 type O
    C. 5 type O; 1 A; 1 B; 1 AB

29. A. All type O; 3 chances out of 4 for any one to secrete H antigen; 1 chance out of 4 to secrete no H antigen.
    B. 50% chance for an offspring to be type O who will secrete no A, B, or H antigens; ⅛ chance for an O who will secrete H antigen; ⅛ chance for type AB (secretor of A, B, and H antigens); ⅛ chance for type A (secretor of A and H antigens); ⅛ for type B (secretor of B and H antigens).
    C. All type O; 9 chances out of 16 for H antigen to be secreted; 7 chances out of 16 for no H antigen.
    D. ⅜ chance for type AB who will secrete antigens A, B, and H; ⅜ chance for type AB who will secrete none of the antigens; ¼ type O who will secrete none of the antigens.

30. Man 1 cannot be the father because the father must contribute the L^N allele. Man 2 could be the father, but

one cannot say that he is simply because he has the appropriate genotype.

31. A. O Rh$^+$ and O Rh$^-$
    B. AB Rh$^+$; AB Rh$^-$; A Rh$^+$; A Rh$^-$; B Rh$^+$; B Rh$^-$; O Rh$^+$; O Rh$^-$
    C. A Rh$^+$; A Rh$^-$; B Rh$^+$; B Rh$^-$

32. Both men could be.

33. A. Type O; secretes antigens H, M, and N.
    B. Type A; secretes none of the antigens.
    C. Type O; secretes antigens M and N.
    D. Type AB; secretes none of the antigens.
    E. Type O; secretes antigens D and N.

34. There would be a risk in matings B and C, because the female is Rh$^-$ (dd) and the male is Rh$^+$, possessing allele D. The risk is greater in mating B, because the male is homozygous for allele D.

35. A. $\dfrac{B_2C_1A_2}{B_4C_3A_4}, \dfrac{B_2C_1A_2}{B_3C_4A_3}, \dfrac{B_1C_2A_1}{B_4C_3A_4}, \dfrac{B_1C_2A_1}{B_3C_4A_3}$
    B. They can neither donate to nor accept from the parents.
    C. The boy will be able to accept from either parent or his siblings, because he is deficient in the immune response, but he will not be able to donate tissue to his parents or to sibs with different haplotypes, because his tissue carries antigens that can trigger an immune response.

36. A. Mother: $\dfrac{B_9A_2}{B_8A_1}$ Father: $\dfrac{B_6A_3}{B_8A_3}$

    Child 1: $\dfrac{B_9A_2}{B_8A_3}$

    Child 2: $\dfrac{B_8A_1}{B_6A_3}$ Child 3: $\dfrac{B_9A_2}{B_6A_3}$

    B. $\dfrac{B_8A_2}{B_8A_3}$ (crossover from mother)

37. A. All are normally pigmented; genotype AaBb.
    B. Chance for an albino is 7 out of 16.

CHAPTER 5

1. A. AaPp
   B. AapY
   C. AaPY

2. A. aP, ap
   B. Ap, AY, ap, aY
   C. AP, Ap, aP, ap

3. A. All children with normal pigment distribution. Genotypes: AaOo, AaOY
   B. Daughters with normal pigmentation; sons lacking eye pigment. Genotypes: AaOo, AaoY

4. A. Daughters all carriers with normal iris (Ii); sons all with normal iris (IY)
   B. Daughters all carriers with normal iris (Ii); sons all with cleft iris (iY)
   C. Daughters: half carriers with normal iris (Ii) and half with cleft iris (ii); sons: half with normal iris (IY) and half with cleft iris (iy)

5. Girl: Mmii; mother: mmIi; father: M—iY

6. Migraine, normal iris; migraine, cleft iris; no migraine, normal iris; no migraine, cleft iris in a ratio of 1:1:1:1 among both sons and daughters

7. A. ppTT × PYtt. F$_1$ PpTt (daughters with normal vision); pYTt (sons with red-green color blindness)
   B. Normal vision, red-green color-blind, and totally color-blind in a ratio of 3:3:2 among both sons and daughters

8. A. All daughters with rickets (Rr); all sons normal (rY)
   B. Half of the sons and half of the daughters with rickets (RY;Rr) and half of the sons and half of the daughters normal (rY; rr)

9. A. Females: bY (non-Bar); males: Bb (Bar)
   B. Females: BY (Bar) and bY (non-Bar); males: Bb (Bar) and bb (non-Bar)

10. A. The females carrying the lethal would produce only half the expected number of male offspring, since no male could be born with the allele. The 1:1 sex ratio would become distorted to 2:1 in favor of females.
    B. The males carrying the lethal would produce only half the expected number of female offspring. No hen could carry the allele. The sex ratio of 1:1 would become distorted in favor of males.

11. To have the affliction, a female must receive an X chromosome bearing the recessive from both parents. Since males with this affliction do not live to reproduce, offspring receive the recessive only from a carrier mother. Because a male has only one X chromosome, a male receiving the allele will express it.

12. The butterfly must be female, because the allele for whiteness can be expressed only in the female of this species. The genotype could be WW or Ww. The tortoise kitten must be female, because the genotype for tortoise must be Bb, and this requires two X chromosomes. The sex of the yellow kitten is in doubt. It could be a male (bY) or a female (bb).

13. Non-color-blind female with three X chromosomes (PPp); non-color-blind Turner female (PO); non-color-blind Klinefelter male (PpY). The YO zygote would not survive.

14. Non-color-blind Turner female (PO); non-color-blind Klinefelter male (PPY); color-blind Turner female (pO); non-color-blind Klinefelter male (PpY).

15. (A) 100, (B) 100, (C) 50, (4) none, (5) none, (6) 50, (7) 100, (8) 100

16. A. Apricot males and intermediate apricot females
    B. Apricot males and females intermediate between cherry and apricot
    C. Cherry males and red-eyed females
    D. Males: 1 wild: 1 cherry; females: 1 red: 1 intermediate between cherry and apricot

17. This male must be carrying certain alleles that cause reduction in milk yield. These would be sex-limited alleles that are not expressed in the male. However, in a female these alleles can come to expression whether received from a male or a female parent. In this example, the bull is carrying them and transmitting them to female offspring, where they are expressed, causing a reduction in milk yield.

18. A. Half of the sons nonbald (bb) and half of them bald (Bb); daughters all nonbald (Bb and bb)
    B. Sons all bald (Bb); daughters all nonbald (Bb)
    C. Same as A

19. Daughters: all nonbald, but half will be color blind. Sons: equal chances for bald and normal vision; bald and color blind; nonbald and normal vision; nonbald and color blind

20. (1) ll, (2) LL, (3) Ll (long fingers), (4) Ll (short fingers)

21. No early hair loss among the daughters. Early hair loss in three out of four sons.

22. (A) +, (B) −, (C) −, (D) +, (E) +, (F) −, (G) −, (H) +, (I) +

23. (1) Older sister or sister who is fraternal twin; (2) younger brother (sisters preferred in respect to HY antigen); (3) mother; (4) aunt; (5) unrelated girl

24. D, E

CHAPTER 6

Answers—Pedigree I:

2. There is no chance, because the allele is dominant. If neither parent shows the trait, they are not carrying the allele and cannot pass it down.

3. The answer is the same as in 2. As far as this trait is concerned, the fact that they are cousins does not matter because neither shows it and so neither carries the responsible allele.

Pedigree II:

5. The chance is 1/3 that V-1 is a heterozygote. The chance is 2/3 that her mother is a heterozygote. If she is, there is a chance of 1/2 that she will pass the recessive to an offspring. Thus: $2/3 \times 1/2 = 1/3$.

6. The answer is 1/6. The chance is 1/3 that she is a carrier. If a carrier marries an affected individual (Aa × aa), the chance is 1/2 for an affected child. The chance for the events to happen together is $1/3 \times 1/2$ or 1/6.

Pedigree III:

7. The answer is 1/2 because it will be passed to all the males.

8. No chance, because person V-5 is not carrying it on his Y chromosome.

9. The chance is 100%, or 1, because all of the male children will be affected.

Pedigree IV:

11. The answer is 1/8. The chance that V-3 is a carrier is 1/2. Her mother must be Aa, because person V-1, her brother, shows the sex-linked trait. There is a chance of 1/2 that the mother passed it to V-3. If she is a heterozygote, the cross is Aa × aY and 1/2 of the children will be affected. This means the chance of an affected child is $1/2 \times 1/2$ or 1/4. Because there is a chance of 1/2 for a boy, the probability for an affected son is 1/8.

12. There is no chance. Since V-6's father does not have the rare sex-linked trait, she has not received it from him. Her mother is not a blood relative of the father and would not be carrying it. Although person IV-10 has the trait, the cross would be AA × aY, and no children will show it.

13. $(1/2)^6$

14. 1/2

15. (A) 1/64, (B) 1/64, (C) 1/32, (D) 5/16

16. A. 1/8
    B. 3/8
    C. 3/8 + 1/8 = 1/2

17. A. $3/8 \times 1/8 \times 1/8 \times 3/8 \times 3/8 = 27/32,768$
    B. $10 (3/4)^3 \times (1/4)^2 = 270/1024 = 135/512$
    C. The chances are three out of four that the child will be normally pigmented and one out of four that the child will be albino.

18. The answer is 1/3. The reasoning is as follows: In a family of two children, the probability is 1/4 for two girls, 1/2 for a boy and a girl, and 1/4 for two boys. You can eliminate the two boys in this case, because you know one of the children is a girl. This leaves the following possibilities: a girl, then a girl; a girl, then a boy; or a boy, then a girl. Since two of the remaining possibilities are boy–girl combinations, one of the three is girl–girl.

19. A. $(1/3)^3 = 1/27$
    B. $(2/3)^3 = 8/27$

20. $2/3 \times 1/2 = 1/3$

21. A. 3/64
    B. $4 \times (1/4)^3 \times 3/4 = 12/256 = 3/64$

22. A. 1/2
    B. 1/4
    C. 1/4
    D. 1/2

23. A. $3/16 \times 9/16 \times 3/16 = 81/4096$
    B. $1/4 \times 1/2 \times /1/4 = 1/32$

24. $\dfrac{4!}{2!1!1!} \times (3/16)^2(3/16)(1/16) = 324/65{,}536$

25. Calculation of χ^2:

o	e	d	d^2	d^2/e
33	37	4	16	0.43
74	74	0	0	0
41	37	4	16	0.43
				$0.86 = \chi^2$

P is between 0.50 and 0.80 (see Table 6-5). χ^2 does not tell you that it is a certain ratio. In this case, the figures are compatible with the hypothesis of a ratio of 1:2:1, which has become distorted by chance factors.

26. If genes are assorting independently, a ratio of 1:1:1:1 is expected in a dihybrid testcross. The χ^2 value based on this hypothesis turns out to be 53.06 in this case, as is shown by the following calculations. Such a high χ^2 gives a P value much less than 0.01 for 3 degrees of freedom. Therefore, from these data one cannot say that the genes are assorting independently, because the data are incompatible with a ratio of 1:1:1:1. The probability that chance factors have been operating to distort the ratio is extremely low. Calculation of χ^2:

o	e	d	d^2	d^2/e
539	600	61	3721	6.20
659	600	59	3481	5.80
712	600	112	12544	20.90
490	600	110	12100	20.16
				$53.06 = \chi^2$

CHAPTER 7

1. A. AABb or AaBB
   B. AaBb
   C. aabb

2. The white parent is aabb. The intermediate is probably either AAbb or aaBB. This is so because intermediates with genotype AaBb would make it possible for children of various shades to arise.

3. Intermediate: AaBb
   Light: Aabb and aaBb
   White: aabb

4. The two parents in the first case are probably not heterozygous at both loci. Both could be AAbb; both could be aaBB; or one could be AAbb and the other aaBB. The second two parents are probably both AaBb.

5. A. 7 (including white)
   B. 1/4096 red and 1/4096 white

6. A. Mean = 10 in.
   B. Variance = 1.68
   C. Standard deviation = 1.30
   D. Standard error = 0.29

7. A. 69 in. to 72 in.
   B. 67.5 in. to 73.5 in.

8. A. 0.15
   B. The chance is 68% that the true mean for the whole population lies within the values 70.50 ± 0.15 (70.35 and 70.65 in.).

9. A. Mean = 107. Calculated as follows:

v	f	fv
90	1	90
93	2	186
94	2	188
97	1	97
101	4	404
108	7	756
111	1	111
115	1	115
119	3	357
125	3	375
25		2679	Mean = 2679/25 = 107

B. Variance = 114.75. Calculated as follows:

d	d^2	f	fd^2
17	289	1	289
14	196	2	392
13	169	2	338
10	100	1	100
6	36	4	144
1	1	7	7
4	16	1	16
8	64	1	64
12	144	3	432
18	324	3	972
			$2754 = \Sigma fd^2$

Variance = 2754/24 = 114.75

C. Standard deviation = $\sqrt{114.75}$ = 10.72

10. A. Mean = 110. Calculated as follows:

v	f	fv
99	6	594
103	2	206
110	3	330
111	4	444
112	3	336
118	1	118
119	2	238
120	2	240
126	2	252
	25	2758

mean = 2758/25 = 110

B. Variance = 74.08. Calculated as follows:

d	d^2	f	fd^2
11	121	6	726
7	49	2	98
0	0	3	0
1	1	4	4
2	4	3	12
8	64	1	64
9	81	2	162
10	100	2	200
16	256	2	512

$1778 = \Sigma fd^2$

Variance = 1778/24 = 74.08
C. Standard deviation = $\sqrt{74.08} = 8.60$

11. Four pairs of alleles are involved. Each effective allele contributes 1/4 in. to ear length.

12. Three pairs of alleles are involved. Each effective allele contributes 1 g.

13. Four pairs of alleles are involved. Each effective allele contributes 1 mm.

14. The two parents are obviously not extreme types. Several genotypes are possible. For example, each could be AaBbCcDd. One other possibility is that one parent is genotype AABbCcdd and the other parent is genotype AABbccDd.

15. A. (Nonhygienic queen) UURR × ur (hygienic drone). UuRr—both workers and queen; workers' behavior is nonhygienic
    B. UR drones
    C. UR, Ur, uR, ur
    D. UURR, UURr, UuRR, UuRr—all nonhygienic

16. A, B, and C

17. A, B, C, and D

18. Any cheetah would reject grafts from the cat because the two species differ genetically at the equivalent of the HLA complex. They thus possess different tissue antigens, and the cat skin is rejected as a consequence of the immune response of the cheetah. Since a cheetah tolerates a skin graft from any other cheetah, there must be a great deal of homozygosity at the genetic region governing tissue antigens. The cheetahs show the consequence of a high degree of inbreeding and have apparently lost genetic variability at the loci that govern the formation of tissue antigens.

CHAPTER 8

1. A. $\underline{r^+ \quad s}$ and $\underline{r \quad s^+}$
   B. $\underline{r^+ \quad s^+}$ and $\underline{r \quad s}$

2. A. $\underline{t^+ \quad u}$ (35%); $\underline{t \quad u^+}$ (35%); $\underline{t^+ \quad u^+}$ (15%);
      $\underline{t \quad u}$ (15%)
   B. $\underline{t^+ \quad u^+}$ (35%); $\underline{t \quad u}$ (35%); $\underline{t^+ \quad u}$ (15%);
      $\underline{t \quad u^+}$ (15%)

3. $\underline{t^+ \quad u^+}$ and $\underline{t\, u}$, the original combinations from the dihybrid

4. A. $\dfrac{pr \quad vg^+}{pr^+ \quad vg} \times \dfrac{pr \quad vg}{pr \quad vg}$ B. 15%

5. Purple flies and vestigial flies in a ratio of 1:1.

6. A. $\dfrac{o^+ \quad s}{o \quad s^+} \times \dfrac{o \quad s}{o \quad s}$ B. 20%

7. A. $\underline{o^+ \quad s^+}$ (40%); $\underline{o^+ \quad s}$ (10%); $\underline{o \quad s^+}$ (10%);
      $\underline{o \quad s}$ (40%)

   B. Round, simple (66%); round, branched (9%); long, simple (9%); long, branched (16%)
   C. $1000 \times 9\% = 90$

8. P_1: $\dfrac{I^+ \quad F}{I^+ \quad F} \times \dfrac{I \quad F^+}{I \quad F^+}$; F_1: $\dfrac{I^+ \quad F}{I \quad F^+}$

9. A. $\dfrac{I^+ \quad F}{I \quad F^+} \times \dfrac{I^+ \quad F^+}{I^+ \quad F^+}$
   B. $\dfrac{I \quad F^+}{I^+ \quad F^+}$ (white, normal);
      $\dfrac{I^+ \quad F}{I^+ \quad F^+}$ (colored, mildly brittle);
      $\dfrac{I^+ \quad F^+}{I^+ \quad F^+}$ (colored, normal);
      $\dfrac{I \quad F}{I^+ \quad F^+}$ (white, mildly brittle)
   C. 20%

10. P_1: $\dfrac{w \quad y^+}{w \quad y^+} \times \dfrac{w^+ \quad y}{Y}$
    F_1: $\dfrac{w \quad y^+}{w^+ \quad y}$ (red-eyed, gray-bodied females);

$\dfrac{w \ - y^+}{Y}$ (white-eyed, gray-bodied males)

11. $15/685 =$ approximately 2%. Consider only the male offspring.

12. A. $\dfrac{d \ p^+}{d \ p^+} \times \dfrac{d^+ \ p}{Y}$

Offspring: $\dfrac{d \quad p^+}{d^{\,\prime} \ p}$ (non-color-blind daughters);

$\dfrac{d \quad p^+}{Y}$ (color-blind sons)

B. $\dfrac{d^+ \ p}{d \ p^+} \times \dfrac{d^+ \ p^+}{Y}$;

Offspring:

$\dfrac{d^+ \ p}{d^+ \ p^+}$ and $\dfrac{d \quad p^+}{d^+ \ p^+}$ (non-color-blind daughters)

$\dfrac{d^+ \ p}{Y}$ and $\dfrac{d \quad p^+}{Y}$ (sons color blind)

C. Daughters all with normal vision:

$\dfrac{d^+ \ p}{d^+ \ p^+}$; $\dfrac{d \quad p^+}{d^+ \ p^+}$;

$\dfrac{d^+ \ p^+}{d^+ \ p^+}$; $\dfrac{d \quad p}{d^+ \ p^+}$

Sons: $\dfrac{d^+ \ p}{Y}$; $\dfrac{d \ p^+}{Y}$; $\dfrac{d \ p}{Y}$ (color blind)

$\dfrac{d^+ \ p^+}{Y}$ (normal vision)

13. A. $\dfrac{o \ d^+}{o \ d^+} \times \dfrac{o^+ \ d}{Y}$

Offspring: $\dfrac{o \quad d^+}{o^+ \ d}$ (daughters normal)

$\dfrac{o \quad d^+}{Y}$ (sons with ocular albinism)

B. $\dfrac{o^+ \ d^+}{o \ d} \times \dfrac{o^+ \ d^+}{Y}$

Offspring:

$\dfrac{o^+ \ d^+}{o^+ \ d^+}$ and $\dfrac{o^+ \ d^+}{o \ d}$ (daughters all normal)

$\dfrac{o^+ \ d^+}{Y}$ and $\dfrac{o \ d}{Y}$ (sons normal and sons with both ocular albinism and color blindness)

14. $\underline{e \ 4 \ r \ 7 \ s}$ or $\underline{s \ 7 \ r \ 4 \ e}$
(both maps are the same)

15. A. $0.18 \times 0.14 = 0.025$ (expected);

coincidence $= \dfrac{0.015}{0.025} = 0.6$

B. The locus cannot be placed with certainty on the basis of the information given. It could be located 13 units to the right of p or 13 units to the left. Information is needed on crosses involving genes at the s locus and those at m or d.

16. A. $\dfrac{v^+ \ 1 \ b^+}{v \ 1^+ \ b}$ B. $\underline{v \ \ 17.4 \ \ 1 \ \ 26.7 \ \ b}$

C. $0.174 \times 0.267 = 5\%$ (approx.)

coincidence $= \dfrac{40 \text{ (obtained)}}{50 \text{ (expected)}} = 0.8$

17. $\underline{1 \ \ 11.8 \ \ r \ \ 33.7 \ \ g}$

18. A. P_1 females: $\dfrac{x^+ \ z^+ \ y}{x \ \ z \ \ y^+}$; males: $\dfrac{x^+ \ z^+ \ y^+}{Y}$

B. $\underline{x \ \ 6.4 \ \ z \ \ 7.3 \ \ y}$

C. Little, if any, interference, because five doubles were obtained, and this is about what should be expected, assuming no interference.

19. A. Would expect half of the flies to be females. Therefore, approximately 2000 would be wild females, $r^+ s^+ \ t^+$.

B. Would expect 2000 males. Approximately 6 should be r^+ s t. This is a double crossover type. The total number of doubles expected on the basis of the map is $2000 \times 1.2\%$ or 24. However, the coincidence is 0.5, and so a total of 12 doubles is expected. Of these, 6 would be r^+ st and 6 would be $r \ s^+ \ t^+$.

C. These are parental types. They are determined by subtracting the sum of the doubles (12) and the singles in region I (228) and the singles in region II (188). The singles in region I, for example, are determined in the following way: $12\% \times 2000 = 240$ expected from the map information. However, the doubles (12) must be subtracted, because they were included in the construction of the map. This gives 228. The same reasoning gives 188 for region II. The parental types among the male flies, $r^+ s^+ t$ and $r \ s \ t^+$, would total about 1572 ($2000 - 428$, the sum of the doubles and the singles in I and II).

20. A. 2 wild (red eyes, straight wings): 1 sepia-eyed, straight wings: 1 red-eyed, curled wings

B. $\dfrac{se \ e^+}{se \ e^+} \times \dfrac{se^+ \ e}{se^+ \ e}$; F_1: $\dfrac{se \ \ e^+}{se^+ \ e} \times \dfrac{se \ \ e^+}{se^+ \ e}$

	$se \ e^+$	$se^+ \ e$	$se^+ \ e^+$	$se \ e$
$se \ e^+$	$\dfrac{se \ e^+}{se \ e^+}$ (sepia)	$\dfrac{se^+ \ e}{se \ e^+}$ (wild)	$\dfrac{se^+ \ e^+}{se \ e^+}$ (wild)	$\dfrac{se \ e}{se \ e^+}$ (sepia)
$se^+ \ e$	$\dfrac{se \ \ e^+}{se^+ \ e}$ (wild)	$\dfrac{se^+ \ e}{se^+ \ e}$ (ebony)	$\dfrac{se^+ \ e^+}{se^+ \ e}$ (wild)	$\dfrac{se \ \ e}{se^+ \ e}$ (ebony)

The amount of recombination cannot be estimated from a cross in Drosophila in which the linked alleles are in the trans arrangement in the F_1s. This is so because there is no crossing over in the male fruit fly. Consequently, both the crossover and non-crossover offspring will arise in the F_2 in a ratio of 2 wild: 1 mutant: 1 mutant.

21. A. 25% crossovers and 75% noncrossovers. (The percentage of noncrossovers equals all the gametes formed from those meiotic cells in which no crossing over took place plus 1/2 of the gametes from meiotic cells in which crossing over did occur, in this case $50\% + 25\%$).

B. 50%.

CHAPTER 9

1. 2%

2. 90%

3. $\dfrac{gd^+ \quad P}{gd \quad p} \times \dfrac{gd^+ \quad P}{Y}$

   Offspring:
   Females—all enzyme producers with normal vision
   Males—enzyme producer, normal vision, 47%; no enzyme, color blind, 47%; enzyme producer, color blind, 3%; no enzyme, normal vision, 3%

4. A. $\dfrac{pro^+ \quad leu^+}{pro \quad leu^+}$ (2); $\dfrac{pro^+ \quad leu}{pro \quad leu}$ (2);
   $\underline{pro \quad leu^+}$ (2); $\underline{pro \quad leu}$ (2)
   B. $\underline{pro^+ \quad leu^+}$ (4); $\underline{pro \quad leu}$ (4)
   C. $\underline{pro^+ \quad leu}$ (4); $\underline{pro \quad leu^+}$ (4)
   D. $\underline{pro^+ \quad leu^+}$ (2); $\underline{pro \quad leu}$ (2);
   $\underline{pro^+ \quad leu^+}$ (2); $\underline{pro \quad leu}$ (2)

5. A. $\underline{pro^+ \quad ser^+ \quad leu^+}$ (4); $\underline{pro \quad ser \quad leu}$ (4)
   B. $\underline{pro^+ \quad ser^+ \quad leu^+}$ (2); $\underline{pro^+ \quad ser \quad leu}$ (2);
   $\underline{pro \quad ser^+ \quad leu^+}$ (2); $\underline{pro \quad ser \quad leu}$ (2)
   C. $\underline{pro^+ \quad ser^+ \quad leu^+}$ (2); $\underline{pro^+ \quad ser^+ \quad leu}$ (2);
   $\underline{pro \quad ser \quad leu^+}$ (2); $\underline{pro \quad ser \quad leu}$ (2)
   D. $\underline{pro^+ \quad ser^+ \quad leu^+}$ (2); $\underline{pro^+ \quad ser \quad leu^+}$ (2);
   $\underline{pro \quad ser^+ \quad leu}$ (2); $\underline{pro \quad ser \quad leu}$ (2)

6. The crossover event will produce two chromosomes, in each of which the two chromatids are different genetically:

   $\dfrac{sn \qquad y^+}{sn^+ \qquad y}$ and $\dfrac{sn \qquad y^+}{sn^+ \qquad y}$

   An arrangement on the spindle, as shown here in some of the cells at mitosis, will give rise to cells that are homozygous for y and cells that are homozygous for sn. These cells, in turn, will give rise to yellow and singed patches among the cells that show the gray and the normal bristle traits.

7. The h locus must be on Chromosome III, linked to sepia. The reason is that no double recessives can be found among the F_2 offspring of a cross in Drosophila when two mutant stocks are crossed and the recessives being followed are linked. The F_1 dihybrid offspring would have alleles in the trans arrangement:

   $$\dfrac{se \quad h^+}{se^+ \quad h}$$

   Since there is no crossing over in the male fruit fly, only parental combinations can be passed to the gametes: se h^+ and se^+ h. When these fertilize the four kinds of gametes formed by the female, a 2:1:1 ratio results. No double recessive can be produced.

8. Enzyme X—Chromosome 7; enzyme Y—Chromosome 5; enzyme Z—Chromosome 1.

9. A chromosome alteration such as a shift or translocation should be suspected. A karyotype analysis should be performed using such techniques as fluorescent staining to bring out the chromosome banding pattern. This person's karyotype should then be compared with that of typical persons. If a band normally present in Chromosome 1 is absent from that chromosome of the atypical person, and if a similar band is now present in Chromosome 5, where it is not typically seen, then good evidence can be offered for a shift. The locus governing the enzyme's production can now be more precisely assigned to a specific region of Chromosome 1.

10. B, C

11. 10 map units (20% of the asci show second division segregation patterns. Thus, crossing over occurred in 20% of the meiotic cells. Only half of these are recombinants.)

12. 50 map units

13. Cross I—Type 1 asci are NPD; type 2 are PD; Types 3, 4, and 5 are TT. The two genes are not linked because the numbers of NPD and PD asci are about equal.

    Cross II—Type 1 asci are PD, as are Type 2. Type 3 are NPD and Type 4 are TT. The two genes are linked because the NPD asci are significantly fewer than the PD asci.

14. A. 10% ($80/400 \times 100 = 20\% \times \frac{1}{2} = 10\%$)
    B. 40% ($80 + 240 = 320$; $320/400 \times 100 = 80\% \times \frac{1}{2} = 40\%$)
    C. 31 ($240/400 \times 100 = 60\% \times \frac{1}{2} = 30\% + 4/400 \times 100 = 31\%$) = 31 map units

15. In addition to the second division segregation seen in asci of Type 2, the TT (Type 4) asci clearly show that the recombinant types of spores and the parental types can occur in the same ascus. This would not be possible if crossing over occurred in the two-strand stage. If this were the case, only recombinant-type spores could be formed in those asci in which crossing over had taken place. The parentals could not occur along with them.

CHAPTER 10

1. The dicentric $\dfrac{\text{0}\;\; A\,B\,C\,d\,e\,f\,b\,a}{\;\; g\,c\,D\,E\,F\,G}\,\text{0}$
   and the acentric

2. $\dfrac{\text{0}\;\; A\,B\,C\,d\,e\,F\,G}{\text{0}\;\; a\,b\,f\,E\,D\,c\,g}$ and

3. $\dfrac{H\,I\,J\;\text{0}\;K\,l\,h}{n\,m\,i\,j\;\text{0}\;k\,L\,M\,N}$ and

4. $\dfrac{H\,I\,j\;\text{0}\;k\,L\,M\,N}{h\,l\,K\;\text{0}\;J\,i\,m\,n}$

5. At meiosis, each chromosome will form a loop, which includes the genes T U and V. Crossing over anywhere between the loops, before W or before X, will lead to

$$\underbrace{}_{\text{(a duplication)}}0\ \underline{\text{R S T U V W X T U V Y Z}}$$

and

$$\underbrace{}_{0}\ \underline{\text{R S W X Y Z}}\atop{\text{(a deficiency)}}$$

6. A.
$$\begin{array}{cccc}\text{g} & \quad\text{h} & \quad\text{m} & \quad\text{n}\\ \hline\quad 0 & & & 0\\ \text{g} & \quad\text{n} & \quad\text{m} & \quad\text{h}\\ \hline\quad 0 & & & 0\end{array}$$

B. g-h, m-h and m-n, g-n (one possibility); m-h, m-n and g-n, g-h (the other possibility)

C. g-h, m-n and g-n, m-h

7. Alpha-alpha (dies); alpha-beta (survives); beta-beta (dies)

8. A. Half of the male embryos would die, so that the sex ratio would be 2 females: 1 male. Half of the females carry the deletion.

B. One fourth of the offspring from two carrier parents will fail to survive, just as if a recessive lethal gene were present.

9. From eggs with exact pairing: $\dfrac{\text{B \ B}}{\text{B \ B}}$ (Bar females) and $\dfrac{\text{B \ B}}{\text{Y}}$ (Bar males)

From exceptional eggs: $\dfrac{\text{B \ B \ B}}{\text{B \ B}}$ (very narrow Bar females); $\dfrac{\text{B \ B}}{\text{B}}$ (wide-Bar females); $\dfrac{\text{B \ B \ B}}{\text{Y}}$ (ultra-Bar males); $\dfrac{\text{B}}{\text{Y}}$ (normal-eyed males)

10. When Bar is taken out of its normal location on the X chromosome and placed next to heterochromatin of Chromosome IV, a narrowing of the eye occurs. The absolute number of Bar regions in a fly is not the sole determinant of eye width. The position of the Bar locus is important. In its normal location in the X, Bar does not reduce eye size, but when Bar is placed on Chromosome IV, eye size decreases, and the decrease is proportional to the number of abnormally placed Bars on Chromosome IV.

11. A. I I II II II III III IV IV V V

B. I I I II II II III III III IV IV IV V V V

C. I I II II III III IV V V

D. I I I I II II II II III III III III IV IV IV IV V V V V

12. A. 24
B. 48

13. A. 25
B. 50

14. (1) Leaf cells fused following removal of cell walls; (2) selection of hybrid cells on medium, permitting growth of hybrid cells only; (3) grafting of plantlets, if necessary, to roots of compatible species such as a parental type; (4) development of mature plants with chromosome number of 50.

15. A. A reciprocal translocation involving two pairs of non-homologous chromosomes is present.

B. A paracentric inversion is present. The bridge and fragment have resulted from crossing over within the inverted region.

C. The plant would appear to be derived as a result of autopolyploidy and would be a tetraploid, as evidenced by four chromosomes undergoing synapsis in each tetravalent.

D. The plant would appear to be an autotriploid, as suggested by the trivalents. The three univalents represent chromosomes that did not succeed in entering into trivalent formation. This also accounts for the bivalents.

E. The buckle suggests that a deletion is present in the chromosome lacking the buckle. It is also possible that the buckle reflects a duplication present in the chromosome showing the buckle.

F. Autotetraploidy is indicated by the tetravalents. The trivalent and univalent represent a potential tetravalent that failed to form.

G. Nondisjunction at first meiotic division would produce two cells—one with 13 chromosomes, the other with 11. These may complete meiotic division.

H. Complete failure of chromosome separation at first meiotic division would yield a cell unreduced in chromosome number. This may complete second meiotic division and eventually give rise to unreduced gametes.

16. A. 24
B. 17
C. 32
D. 15
E. 32

17. A. Nondisjunction of X chromosomes at first anaphase of oogenesis giving an XX egg, which is fertilized by a Y-bearing sperm. Also nondisjunction of the X and Y chromosomes at first anaphase of spermatogenesis, giving an XY sperm that fertilizes an egg.

B. Nondisjunction of sex chromosomes at spermatogenesis or oogenesis, producing gametes with no sex chromosomes. Also loss of one X chromosome at first mitotic anaphase of an XX zygote.

C. Nondisjunction of Y chromosome at second anaphase of spermatogenesis.

D. A pericentric inversion.

E. Complete failure of anaphase movement at meiosis, producing an unreduced sperm or egg that unites with a haploid gamete.

F. A translocation.
G. Nondisjunction of X chromosomes at oogenesis producing an XX egg that is fertilized by an X-bearing sperm.
H. Loss of an X chromosome during mitotic divisions of an XX embryo.
I. Loss of a Y chromosome during mitotic divisions of an XXY embryo.
J. Nondisjunction of Y chromosome during mitotic divisions of an XY embryo.

18. Turner syndrome: 2, 3, 8
Klinefelter syndrome: 2, 8
Down syndrome: 2, 6, 8
Cri-du-chat syndrome: 7
Philadelphia chromosome: 6 (previously considered a deletion)
XYY male: 2, 8

19. A metacentric chromosome may be formed following two breaks, one in each of two acrocentrics. In one of the acrocentrics, the break is just before the centromere, in the long arm of the chromosome. In the other, the break is just behind the centromere, in the short arm. The latter long piece with the centromere joins the other long arm piece, which lacks a centromere. The tiny pieces left will generally consist largely of heterochromatin. These may become lost or may unite to form a tiny chromosome that is later lost.

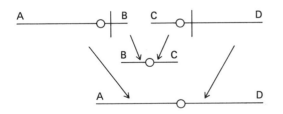

20. Testis determination seems to depend on very few (possibly just one) Y-linked loci. This small region could have been inserted into an X chromosome at the time of chromosome pairing at meiotic prophase. The rest of the Y chromosome could have been lost following this at meiotic divisions. The outcome could be a gamete lacking a visible Y but containing the tiny, testis-determining region.

CHAPTER 11

1. A. Capsulated, type III
   B. Streptomycin sensitive
   C. Penicillin resistant

2. Most will be in the cell-free fluid, because the radioactive sulfur was in the protein coats which did not enter the host cells.

3. Radioactivity would have been found in appreciable amounts in both the fluid and the cellular fractions.

Carbon is an element common to all organic molecules, and therefore it would be incorporated into both protein and DNA. Consequently, the results would have been inconclusive.

4. A. Phosphoric acid, purine or pyrimidine base, sugar component
   B. Purine or pyrimidine base, sugar component

5. DNA contains thymine, never uracil. RNA contains uracil in place of thymine.
   B. The deoxyribose of DNA has one fewer oxygen at the No. 2 position in the sugar ring than does the ribose of RNA.
   C. DNA is confined mainly to the chromosomes, whereas RNA is much more widely distributed in the nucleus and the cytoplasm.

6. Sample 1—$A = 15\%$; $G = 35\%$; $C = 35\%$; $T = 15\%$
   Sample 2—$A = 30\%$; $G = 20\%$; $C = 20\%$; $T = 30\%$
   Sample 3—$A = 30\%$; $G = 20\%$; $C = 20\%$; $T = 30\%$
   Sample 4—$A = 10\%$; $G = 40\%$; $C = 40\%$; $T = 10\%$
   Samples 2 and 3 could be from the same source.

7. T-A-G paired with A-T-C (the original sequence of base pairs) and T-G-G paired with A-C-C (A G-C pair now replaces the original A-T.)

8. A. 3′ TGTGGGAAATGTTTA 5′
   B. 2 or 66.6%: 33.3%
   C. 1:1
   D. 1:1

9.

	Human	Drosophila	Yeast
A.	1.54	1.51	1.79
B.	0.99	1.02	1.00
C.	1.00	1.04	0.95
D.	0.98	0.98	1.09
E.	39.4	39.9	35.8

10. A. 0.2, since the amount of $A + T$ in the complementary strand must be the same as in its partner. The same is true for the amount of $G + C$.
    B. $1/0.2 = 5$
    C. 1 in both. The sum of the purine bases must equal the sum of the pyrimidines as a consequence of the specific base pairing.

11. A. 0.5, since the sum of $A + T$ in one strand must be the same as the sum of $A + T$ in the other strand.
    B. 0.5, since $A + T = 0.5$ in both strands.

12. $1/0.5 = 2$

13. A. The one with guanine
    B. The one with thymine
    C. The 3′ end or —OH end
    D. Adenine

14. No hybrid DNA band would form. At 0 generations, one heavy band. At one generation, two bands, one heavy (^{15}N/^{15}N) and one light (^{14}N/^{14}N). At two generations, two bands, one heavy and one light, but the latter about three times as wide as the former. At three generations, almost all light.

15. A. −A
    B. −A
    C. −B
    D. −B

16. A. The stem and both arms of the Y will be equal.
    B. The stem will be twice as heavily labeled as the two arms of the Y.

17. A. k
    B. h
    C. a, b, e, g, j
    D. a, e, j
    E. a, b, c, e, j
    F. d
    G. f
    H. i

CHAPTER 12

1. A carboxyl group, an amino group, and an R group

2. ATP, aminoacyl synthetases, tRNA, mRNA, ribosomes

3. A. 5′ A T G T T T A G A G T A A C A T A T C C T 3′
   B. 5′ A U G U U U A G A G U A A C A U A U C C U 3′
   C. 7

4. A. Translation would halt prematurely.
   B. Nonsense.

5. A. No effect on the protein, because both code for phenylalanine.
   B. Silent.
   C. Cysteine would be substituted for phenylalanine.
   D. Missense.

6. A. The codons from the point of the first deletion would go out of phase and will remain out. Likely outcome is a completely nonfunctional protein because of amino acid substitutions and also possibly one that is incomplete because of the formation of a nonsense codon.
   B. Reading will come back into phase with codon 5. Polypeptide may function if the first four amino acid substitutions are not too detrimental.
   C. Reading will come back into phase with codon 4. Polypeptide may function.
   D. Frame shift mutations.

7. Tryptophan, because it is designated by only one codon, whereas arginine is represented by six.

8. A. Strand A
   B. Strand A, because it is complementary to the transcript

9. A. (1) Aspartic acid; (2) aspartic acid; (3) serine; (4) serine; (5) none
   B. Only the contents of tube 3 will produce a labeled filter.
   C. We would know that AGU codes for serine. We would also know that the other codons do not designate the particular amino acids that are labeled in the tubes to which they were added.

10. Since G:C pairs form three hydrogen bonds, whereas A:T pairs form only two, the melting temperature increases along with increase in G + C content. Therefore, sample 2 would have the highest melting point, whereas sample 1 would have the lowest.

11. Although the G + C content is the same for both samples, sample 2 would require a higher temperature for complete strand separation because it contains a stretch of G:C pairs. This G + C region would separate after the other regions surrounding it.

12. 5′ GAACAUAAAUUU 3′

13. 3′ GGGGUAUUUAA 5′

14. 5′ AGACCCTATGCAGAG 3′—antisense strand;
    3′ TCTGGGATACGTCTC 5′—sense strand

15. Reading in the 5′ to 3′ direction on each of the two strands, the inverted repeat is TTATGAG. Upon separation, the two strands can form the following hairpins:

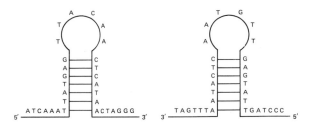

16.

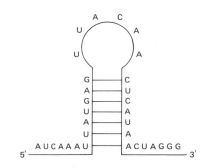

17. A. It must be written in the 3' to 5' direction because reading in the reverse direction would produce the triplet ACT, which would yield UGA, a chain-terminating codon; thus, five amino acids could not be designated.
    B. met, leu, gly, ser, phe

18. The upper strand is the antisense strand and is written in the 5' to 3' direction. The lower strand, the sense strand, is written 3' to 5'. The upper strand cannot be the sense strand. Reading from left to right reveals three codons—ATT, ATC, and ACT—which would be complementary to UAA, UAG, and UGA, chain-terminating codons. Reading the strand from right to left reveals ATT as the second codon, which would yield UAA upon transcription. The lower strand will not give rise to chain-terminating codons if read from left to right. If read right to left, the first codon, ATC, would be complementary to the chain-terminating codon UAG. Only in the orientation noted previously can the double-stranded DNA specify nine amino acids.

19. met, phe, ser, ile, gly

20. met, ile, thr, pro

21. Virus 1—single-stranded DNA; Virus 2—double-stranded RNA; Virus 3—double-stranded DNA; Virus 4—single-stranded RNA

CHAPTER 13

1. A. 1
   B. 6
   C. 4
   D. 5
   E. 2
   F. 15
   G. 11
   H. 9
   I. 13
   J. 14

2. Not all the DNA of the nucleolar organizer is complementary to the 18S and 28S RNA, because a considerable amount is spacer, which is not transcribed at all. The transcribed spacer would not form part of the final rRNA, and this would also account for some of the DNA that does not hybridize with the RNA of the ribosomes.

3. A. 7
   B. U A C A A A U C U C A U U G U A U A G G A

4. A. RNA codons: 5' AAU 3'
      DNA codons: 3' TTA 5'
   B. RNA codons: 5' AGA 3' and 5' AGG 3'
      DNA codons: 3' TCT 5' and 3' TCC 5'
   C. RNA codons: 5' UUA 3' and 5' UUG 3'
      DNA codons: 3' AAT 5' and 3' AAC 5'
   D. RNA codons: 5' GGA 3'; 5' GGU 3'; 5' GGC 3'
      DNA codons: 3' CCT 5'; 3' CCA 5'; 3' CCG 5'

5. A. Asparagine
   B. Arginine
   C. Leucine
   D. Glycine

6. A minimum of three would be needed. Four of the degenerate codons have the same bases in the first two positions; however, no single anticodon can recognize four different codons. The base G at the wobble position can recognize U or C at the 3' position in the codon. Similarly, U at the wobble position can recognize A or G. Thus, two tRNAs can suffice for these four. The other two codons for serine can be recognized by one transfer RNA for the same reasons.

7. Three in each case

8. A. The fifth position
   B. 7

9. A. Asparagine
   B. Tryptophan
   C. Methionine
   D. Asparagine
   E. Methionine

10. A. 80S
    B. 30S
    C. 50S
    D. 60S
    E. Mg^{++}
    F. 5S

11. 5' ATGGAGTGG 3'(antisense)
    3' TACCTCACC 5' (sense)
    or
    5' ATGGAATGG 3'(antisense)
    3' TACCTTACC 5' (sense)

12. A. Antileader
    B. Promoter
    C. Gene proper
    D. Alteration in chain-terminating codon at end of gene proper
    E. Origin of chain-terminating codon at beginning of gene proper

13. A. 5'
    B. 3'
    C. 5'
    D. 5'
    E. 3'

14. A. ⟶ (left to right)
    B. ⟵ (right to left)
    C. ⟵ (right to left)
    D. ⟶ (left to right)
    E. ⟶ (left to right)

15. It would appear that the transcript is untranslatable for some reason. The defect could be one that permits DNA–RNA hybrid molecules to form. Some possibilities

are origin of a chain-terminating codon near the 5' end of the sequence coding for the polypeptide; defect in leader sequence; improper processing of pre-mRNA, so that introns are not properly excised.

16. Simply because the percentage values of the bases in two DNAs are the same, it does not necessarily follow that the distribution of the bases in the two DNAs is the same. Nucleotide sequences would be very different in species that are not closely related. Genes from more closely related species would be expected to be more similar and to be coded for related protein products. In this case, B apparently has many nucleotide sequence differences (and hence gene differences) from A and C. The latter two species would appear to be closely related, because most DNA stretches are complementary.

17. A. E
    B. P+E
    C. E
    D. P
    E. E
    F. P+E
    G. E
    H. E
    I. E
    J. P+E
    K. P+E
    L. P
    M. E
    N. P+E
    O. P

18. A. UGU and UGC for cysteine
    B. UGU for cysteine
    C. 3'ACC5'

19. A. An anticodon with the sequence ACI would also be able to pair with UGA, a chain-terminating codon. There are no anticodons for chain-terminating codons.
    B. An ACU anticodon would be able to pair with the chain-terminating codon UGA. Hence, no such anticodon is found.

20. Normal: 5'GAG3'   Sickle cell: 5'GTG3'
           3'CTC5'   3'CAC5'
    The sense strands are the lower ones in both cases. The change involves the substitution of a T:A pair for an A:T pair at the second position.

21. From lowest to highest: snap-back DNA; satellite DNA; moderately repetitive sequences; sequences coded for polypeptides.

22. Higher, because the bacterial genome, being smaller than that of a higher eukaryote, will provide a sample in which many more copies of each specific DNA sequence will be present than in the eukaryote. Eukaryotic DNA is larger and more complex than prokaryotic DNA, so that in a DNA sample equal in amount to that of a prokaryote, relatively fewer copies of each specific eu-

karyotic sequence are present. Therefore, because **fewer** copies of any given sequence are present, the **time** required for reassociation will be greater, giving a **higher** Cot½ for the eukaryote.

23. The Cot curve is typical for a eukaryotic genome **and** clearly shows distinct steps indicating fractions with **fast** annealing sequences (left) to very slow (right).

24. A. The DNA is represented by the dark strand. **The** mRNA represented by the red strand has undergone processing, and introns have been excised from **it**. These sequences are present in the DNA and **form** loops in the hybrid molecule because no complementary sequences are available in the mRNA **to** permit pairing.
    B. Seven, because each loop formed by the DNA represents an intron that has been excised from **the** mRNA.
    C. At one end is the promoter region that is not **transcribed**. At the other end is a segment containing sequences in the terminator that are not found **in** the mRNA.
    D. This represents the poly (A) tail at the 3' end of **the** strand. It is added after transcription is **completed**, and there is no complement to it in the template strand.

25. Three loci are being followed: adenine 1, adenine 3, **and** the locus at which the adenine 8 and adenine 16 **mutant** sites have occurred. The ad^8 and ad^{16} mutations do **not** complement each other, and so the ad^8 and ad^{16} **sites** are allelic.

26. Complementation is seen between b and c. However, a and b are alleles. Therefore, a and c will most **likely** complement each other in a cross.

27. The enzyme involved is composed of two different polypeptide chains, each controlled by a separate cistron. Mutant sites a^1 and a^4 are in the same cistron that is a unit of function separate from the cistron containing both a^2 and a^3. The crosses are

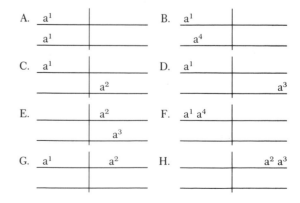

28. The two different chains of hemoglobin A are coded by two different genes. The female parent may be represented as genotype aaBB (defect in the gene for the

alpha chain), and the male parent as AAbb (defect in the gene for the beta chain). The offspring would be genotype AaBb, having received a functional alpha gene from one parent and a functional beta gene from the other. Sufficient normal hemoglobin A may form. Complementation has taken place.

CHAPTER 14

1. Plants of genotype Ww will have the greatest probability, because mutation is rare and is usually to an allele that is recessive to normal or standard. The tetraploid is the least likely to show the effect, because two dominant alleles would still be present if one of them mutates in this case.

2. 1/40,000, because 5 children out of $100,000 = 1/20,000$ afflicted individuals. Each individual represents two sex cells, and the new mutation could have arisen in either the sperm or the egg.

3. $(492 \times 10^{-6})(2.4 \times 10^{-6}) = 1180.8 \times 10^{-12}$

4. The first change converted the RNA codon UAU for tyrosine to UAA, a chain-terminating codon. The second mutation must have altered the anticodon GUU, which is complementary to the codon CAA for glutamine. A change at the 3′ position in the anticodon must have resulted in the anticodon AUU, which would recognize the chain-terminating codon UAA. Consequently, glutamine would be deposited at the site of the chain-terminating codon.

5. A. $\begin{array}{c} \text{GGATCC} \\ \text{CCTAGG} \end{array}$ and $\begin{array}{c} \text{GGGTCC} \\ \text{CCCAGG} \end{array}$ (a transition)
   B. A more likely outcome is the restoration of the correct base pairing, since the mismatch repair system will recognize the region with the incorrect base pair, remove a portion of the new strand that includes the incorrect base, and then fill in the region by DNA polymerase activity. Chances are that the system will be able to discriminate between the template and the new strand shortly after the error is made.
   C. Since the cell cannot methylate bases, there will be no way for the repair system to discriminate between the template strand and the newly synthesized one with the incorrect base. Therefore, it is equally possible that the correct base will be restored as described earlier, or that the correct base will be removed from the template strand. In this case a G:C pair will substitute for the original A:T, and a transition will result.

6. A. Two codons designate tyrosine: UAU and UAC. A single-step mutation in either of these can give rise to UAA, a chain-terminating codon.
   B. Not all the phenotypically wild-type cells would be expected to be true revertants. Assuming, first, that UAU is the codon specifying tyrosine in the wild-

type cell, a true revertant would be one in which UAA was changed back to UAU. However, if UAA changes to UAC, the polypeptide would still have tyrosine at position 70 and would be phenotypically wild, although the cell would not be a true revertant because a codon different from the original one has been formed. The same reasoning applies if UAC was the codon in the original wild.

7. A. Since the P_1 females are carrying no X-linked recessives, the sex ratio should be 1:1, even if any treated male X chromosomes carry a recessive, because no F_1 male would receive an X from a P_1 male.
   B. If a sex-linked lethal has been induced, any F_1 female carrying the recessive and crossed to an F_1 male will produce female:male in a ratio of 2:1.

8. A. The dominant B serves as a marker to indicate that any Bar-eyed female carries one recessive lethal on the X with Bar. Without the inversion, crossing over could separate the Bar marker from the lethal and place the B marker on the X chromosome without the lethal. These F_1 flies would be assumed, however, to carry the lethal on the X along with Bar. Breeding them would produce a sex ratio of 1:1 if no lethal were produced in the P_1 male and a sex ratio of 2 females: 1 male, even if a lethal were produced. Seeing any males at all would be counted as "no induced lethals," even though there was one.
   B. If crossing over occurs in an F_1 ClB female, it can produce some males, even if a lethal is on the treated X. If a lethal has been induced, crossing over could thus produce an X that carries two lethals—the original one and the induced. The other X would then lack the lethals and could give rise to males.

9. A. Must be non-Bar, because the Bar trait is carried on the X with the lethal.
   B. Crossing over in the P_1 female could separate B and l and would give rise to F_1 males which are Bar-eyed. Crossing over in the F_1 ClB would give rise to some Bar-eyed males, whether or not the treated X carried a lethal.

10. A. 3/50 and 4/50
    B. 9/2500 and 16/2500

11. Since the percentage of sex-linked mutations is directly proportional to the dosage applied, there must be a constant value in each experiment for percentage of mutants/dosage in r. The constant here is seen to be 0.005 (3/600 and 8/1600). The constant is then multiplied by 2500 to find the expected mutations at that dosage in r, giving 12.5%.

12. A. Since 10 and 17% mutations were produced by dosages of 1000r and 1700r, respectively, the constant is found to be 0.01. Therefore, at 2250r, we would expect 22.5% mutations.
    B. The answer is 22.5%, because the total exposure is the same as in Part A. It does not matter in Droso-

phila whether the total dosage is given at one time or broken up into smaller doses in a series of experiments. The effect of ionizing radiations is cumulative over the life of an organism.

13. $\xrightarrow{4} D \xrightarrow{3} C \xrightarrow{1} A \xrightarrow{2} B$

The table shows that substance B permits growth of all four strains. Therefore, it must be the final product in the pathway. Any block at an earlier step will be satisfied if the final substance (B in this case) is supplied. Strain 4 is the only strain that can grow with substance D. Moreover, strain 4 can grow with any of the substances. These observations indicate that substance D is the earliest of the products in the pathway, and strain 4 must be blocked in the step leading to D. Strain 2 is blocked in the last step leading to substance B, since that is the only substance it can use for growth. Strain 1 can use substance A or B, and strain 3 can use A, B, or C. This tells us that substance C comes before A and that A comes before B in the pathway.

14. Because the fifth strain can use only substance E or B, this tells us that E must be produced in an independent pathway parallel to the one described in Question 13. Since strain 5 can use the same end product—B, as all the others—it must be able to use E to make B. Substances A and E may combine to make B in the last step for which strain 2 is deficient. This is indicated as follows:

$$C \longrightarrow A$$
$$Precursor \xrightarrow{5} E \quad \Big\rangle \xrightarrow{2} B$$

15. More significant to the human with his longer life span, because the effects of radiation are cumulative. The average fruit fly would be getting very little natural radiation. A human with a longer life span would accumulate about 15r in 30 years.

16. About 200 on plates of group A, about 140 on B, and about 176 on C. (About 40% of the killing action remains; therefore, 12% killing action is left following photoreactivation.)

17. Cells in the bone marrow and in the testes are actively dividing. In such cells, the chromosomes are in a more condensed state than in nondividing cells. With their condensed chromosomes, these cells are much more susceptible to destruction by mutagens, because the chromosomes can be more readily damaged.

18. It would appear that the cells are deficient in the first step of excision repair, which requires an endonuclease to nick the DNA so that the other enzymes involved may come into action. When the DNA is broken by the radiation so that breaks or minute deletions are produced, the presence of the other enzymes can repair some of the damage.

19. The strain is normal with respect to the genetic information required to code for the enzyme of photoreactivation. However, it is deficient in one or more of the genes coding for the other enzymes of repair such as endonuclease, exonuclease, and DNA ligase.

20. A G T (Base replaced with guanine, no change)
    T C A
    A C T (Base replaced with cytosine, a transversion)
    T G A
    A T T (Base replaced with thymine, a transversion)
    T A A
    A A T (Base replaced with adenine, a transition)
    T T A

21. A G A T
    T C T A

22. A. A A T
    B. A A T (a transition)
       T T A

23. A. The most likely explanation is that a frame shift mutation took place following the first exposure, since too many amino acids have been changed. Missense mutations cannot reasonably account for this. We can deduce the following sequence and change for the mRNA:

$\downarrow$ base deleted

5′ AUG CCA AGU UCA GAU AAA UAC CGA CUU A
   met pro ser ser asp lys tyr arg leu

5′ AUG CCA AGU CAG AUA AAU ACC GAC UU G
   met pro ser gln ile asn thr asp leu

B. The mutant cells exposed to further treatment have a polypeptide that is similar to the normal one except for the fourth and fifth amino acids. The reading frame was probably restored as a result of the second treatment, which brought about addition of a base in the sequence, as indicated here:

$\downarrow$ base added

5′ AUG CCA AGU CAG AUA AAA UAC CGA CUU
   met pro ser gln ile lys tyr arg leu

24. A. Streptomycin causes a misreading of the message in the mRNA. In addition to the expected amino acid in each case, others are incorporated into the polypeptide. In each case, the first or 5′ base in the codon is the only one misread. For example, UUU (phe) can be misread as AUU, GUU, and CUU so that ile, val, and leu, respectively, can be incorporated as well as phe.
    B. Misreading of poly(A) can specify UAA, a chain-terminating codon that does not specify an amino acid and would cause shorter polypeptides to be made. Misreading of poly (G) produces the codons AGG and CGG, both of which designate arginine.

25. The substance is obviously mutagenic in bacteria, but the mammalian cell can evidently convert the substance into a product or products that lack this property.

CHAPTER 15

1. The fluctuation from the mean is very great in A, as shown by the large difference between the mean and the variance. The A data must refer to the second part of the experiment, since spontaneous mutations will result in fluctuation from one sample to the next because their occurrence is random.

2. A. Colonies 2, 5, 6, 8
   B. Colonies 2, 3, 5, 6, 7, 8
   C. Colonies 1, 2, 4, 5, 6, 8

3. A. $A^+B^+C^+D^+E^+$
   B. Neither. Both are prototrophs, because both can grow on minimal medium.

4.

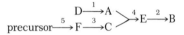

5. Strain 2 is the donor strain. The recipient strain is strain 1, and it must survive because the recombinants are derived from the F^- cells.

6. In cross 1, the F factor is not integrated into the chromosome but exists as a free entity that can readily transfer itself to an F^- cell and convert it to F^+, but it cannot effect chromosome transfer. In cross 2, the F factor is integrated into the chromosome and can bring about transfer of chromosomal markers. However, it is split before transfer of a strand of the chromosome. In most cases, the recipient cell will remain F^-, because only an occasional cell receives the entire chromosome strand with all portions of the F factor.

7. M-O-Q-L-N. Because the procedure selects only A^+ and B^+ cells, the order of these cannot be established here.

8. A and B:

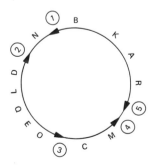

9. The last known marker to enter should be selected in each case. **Select for M in the case of strain 1, C in the**

case of strain 2, and so on. Since the site of the fertility factor will be the last to enter, chances are greater that selection for a marker near the terminus of the transferred donor chromosome will screen out cells that have received the entire donor strand along with the fertility factor.

10. A. (H) 5′ A G C T A T 3′
       (L) 3′ T C G A T A 5′
    B. (L) 5′ A G C T A T 3′
       (H) 3′ T C G A T A 5′

11. The cells that first received lac$^+$ too early received the sex factor that had removed itself from its site of insertion on the chromosome along with the adjacent lac$^+$ marker. These cells then acted as donor cells, transferring F^+lac$^+$ to F^-lac$^-$ with a high frequency. This is an example of sexduction. The derived F^+lac$^+$ cells are heterogenotes. They carry lac$^-$ on the chromosome and lac$^+$ on the sex factor that is existing free in the cell.

12. The DNA will be hybrid: one strand A, one strand B.
    B. One cell will be streptomycin resistant and the DNA will be composed of two A-type strands in the marker region. The other cell is sensitive and will have two B-type strands.

13. A. The genes a and b are closely linked, because the number of double transformants (a^+b^+) is much higher than the number of singles. If the two genes were far apart, they would tend to occur on separate fragments, and two independent events would be required to produce double transformants. The probability of doubles would be much lower than in this case and would equal the product of the separate chances for each single recombinant (155/900 × 95/900).
    B. 155 + 95/ 900 × 100 = 27.8% = 27.8 map units. When strength of linkage is estimated between two genes, the only transformant classes considered in the calculations to be recombinants are those that arose as the result of a crossover event between the two genes. In the case of the double transformants, no crossover event took place between the two genes.
    C. The recombinant classes a^-b^+ and a^+b^- are not equal because they did not result as reciprocal products from the same crossover event. Each recombinant class arose as the result of separate crossover events in which separate donor DNA fragments became incorporated in the recipient chromosome.

14. 5′ A T C G T X Y Z A T C G T 3′
    3′ T A G C A X.Y′Z′ T A G C A 5′

15. A. ABCD, because the conjugal plasmid and the nonintegrated F factor cannot transfer the chromosome.
    B. ABCD
    C. No determinants will be transferred.
    D. EF, because integrated F factor mainly brings about chromosome transfer.

16. The original strain carried a compound conjugal R plasmid with an R determinant component containing genetic factors for resistance to both drugs. This conjugal plasmid dissociated into a conjugal and a nonconjugal plasmid, the latter carrying the factor for tetracycline resistance. At times only the conjugal plasmid with the streptomycin-resistance factor is tranferred. Other cells of the original strain appear to have lost the conjugal plasmid, so that the factor for tetracycline resistance is not transferred.

17. The factor for streptomycin resistance is found on a transposon that moved to another replicon, the F factor or the chromosome. It is now part of the replicon that is composed of the chromosome and the inserted F factor, and is thus transferred along with the chromosomal markers.

CHAPTER 16

1. The first and the last 10 nucleotide pairs represent a repetition or terminal redundancy. Nicking of each strand and removal of the last 10 bases toward the 3' end yields

5' T T A C C T T C A T A A A G C G C A C A 3'
      3' T T T C G C G T G T A A T G G A A G T A 5'

2. A and B are both lysogenic. Each harbors a different prophage, so each can be lysed by the other. C harbors no known prophage, and is thus susceptible to both A and B but cannot lyse either of them.

3. A. From either 1 or 2, since phage lambda would not be transferred from the Hfr, and zygotic induction would not occur.
   B. From either 2 or 3. In 2 the prophage is not transferred. In 3, both the Hfr and F⁻ carry the prophage, so induction will not occur.
   C. From either 2 or 3. Impossible in 1, because site of lambda is well before the S locus and therefore lambda would be transferred before S, and lysis would follow.

4. A. Transformation requires no vector. Transformation can occur by uptake of pure donor DNA alone. Transduction requires a vector; the DNA is introduced into the cell by the vector and does not enter by itself. Transformation requires that the cell be in a state of "competence." This is not so in transduction. Transduction depends on the ability of a cell to harbor appropriate prophage. This is not required in transformation.
   B. Both depend on introduction of a virus into the bacterial cell, but in conversion the introduced virus does not necessarily carry any bacterial genes. The new features of the bacterial cell are not a result of any introduced genes from another bacterial strain but of the interaction of the genetic material of the virus and of the cell itself.

5. If the level of infection is very low, only one virus particle would be likely to infect a cell. Those that are transducing the cells are defective. Because no intact helper phage is present, the transducing phage cannot effect lysis.

6. The transduced cells are heterogenotes. Some of them may lose the Gal⁺ marker and revert to the Gal⁻ trait.

7. C

8. 40%; 40 map units

9. s r t or t r s

10. Gene o is in the middle, and this can be deduced in the following way. The minimal medium permits growth only of prototrophs, $m^+n^+o^+$. In experiment 1, prototrophs will arise when o^+ alone is transferred to strain 2. In experiment 2, prototrophs will arise if m^+ and n^+ but not o^- are transferred to strain 1. If o is the gene in the middle, more $m^+n^+o^+$ transductants are expected in experiment 1 than in experiment 2. This is so because the first case then requires a double crossover to get $m^+n^+o^+$. However, a much rarer quadruple crossover is required to get prototrophs in the second case. This is seen here:

1. donor: m⁻ o⁺ n⁻   2. donor: m⁺ o⁻ n⁺
recipient: m⁺   o⁻   n⁺   recipient: m   o⁺   n⁻

Other arrangements of the genes require double crossovers only to derive the prototrophs in the reciprocal experiments.

11. 1.7% recombination; small clear hr^+; small cloudy, h^+r^+; large clear, hr; large cloudy; h^+r

12. Repeated mating occurs. One recombinant class formed early in the cycle would probably engage in later matings, and thus not be represented at the time of burst.

13. A and D

14. AAGUUACCA

15. Taken three nucleotides at a time from the left, the seventh triplet is TGA, which corresponds to UGA, a chain-terminating codon in the messenger. Starting a new frame with the A in this triplet produces ATG, which corresponds to the start signal AUG in mRNA. It is thus at this point that an overlapping could occur.

16. (1) A string of alanines; (2) a string of leucines; (3) a string of cysteines

17. A. The same as the + strand, since the + strand acts as mRNA.
   B. UUGUCCUGCGUC
   C. RNA replicase

18. A. 5′ T T T G A T T A A C G C G T 3′
    B. The same results because actinomycin D blocks DNA-dependent RNA synthesis, not RNA-directed DNA synthesis.

19. A. 5′ CTTGGAAG|CTTAATAC 3′
      3′ GAACCTTC|GAATTATG 5′

      (1)

      5′ AACGTT|AACTTCCTA 3′
      3′ TTGCAA|TTGAAGGAT 5′

      (2)

    B. 5′ CTTGGA   AGCTTAATAC 3′
      3′ GAACCTTCGA   ATTATG 5′

    C. 5′ AACGTT AACTTCCTA 3′
      3′ TTGCAA TTGAAGGAT 5′

20. 5′ CCACCGGTCCTCATACCGGTC 3′
    3′ GGTGGCCAGGAGTATGGCCAG 5′

21. In nontransformed cells, the viral DNA has not integrated with the DNA of the cell, as indicated by lack of a signal from labeled DNA. In the transformed cells, the virus has integrated with the cellular DNA, and it has no preferential site of integration, as can be seen by the location of signals at different positions, corresponding to the fragments of different sizes (smallest at the bottom). More than one virus may integrate with the cellular DNA, as seen in clones 3 and 5.

22. A. Hybridization between single-stranded nucleic acids cannot be performed on a gel. The transfer is necessary to achieve this essential step.
    B. Virus B, unlike virus A, has a preferential site of integration with the cellular DNA.

23. Fragments 1 and 3 do not correspond to this amino acid sequence of the coat protein. Fragment 2 corresponds to amino acids 6 through 12 in the sequence. Fragment 4 overlaps fragment 3 so that we can establish the sequence of nucleotides in the coat protein gene that corresponds to amino acid sequence 2 through 12 in this portion of the protein. The nucleotide sequence is

U U U U A U G C A A C U G C A A A C U C C G G
U A U C U A C C G U

CHAPTER 17

1. A. Inducible
   B. Constitutive
   C. Constitutive
   D. Constitutive
   E. Inducible
   F. Constitutive

2. A. Z constitutively; Y and A by induction
   B. A, Y, and Z only by induction

3. A. No enzymes produced constitutively or by induction
   B. Same as A
   C. No enzymes produced constitutively; only A produced by induction
   D. A produced constitutively; nothing else produced

4. A. A and B produced only in absence of effector
   B. A and B produced in absence and in presence of effector.
   C. Same as B
   D. A and B produced in absence of effector; B produced in presence of effector

5. (1) I^s is dominant to both I^+ and I^-; I^+ is dominant to I^-. (2) The altered lac repressor acts in both the *cis* and *trans* arrangement with O^+Z^+. This is seen in strains 1, 2, 3, and 4, where all are repressed. Allele I^s is expressed in both arrangements despite the presence of the + type allele in strains 1 and 2 and the I^- allele in 3 and 4. Strains 5 and 6 are both inducible. The I^+ allele is acting in both *cis* (5) and *trans* (6) in relation to O^+Z^+. In strains 1, 4, and 6, the repressor must be diffusing to the DNA in the *trans* arrangement in order to influence the alleles on the other molecule. If the repressor acted only at its point of formation or on the DNA on the same molecule, then it would not be diffusible, and the results presented here would be different.

6.

| | Lactose ||
Strain	Present	Absent
1	+	0
2	0	0
3	0	0
4	+	+
5	+	0
6	+	+

7. In the case of a positive control system, a substance must be added to stimulate the process. A negative control mechanism requires removal of the substance to stimulate the process. When the cAMP–CAP complex is added, it binds to the promoter and stimulates transcription by influencing the binding of RNA polymerase. This is in contrast to the operator–repressor system in which the repressor must be removed from the site of the operator for transcription to proceed.

8. An IS element is present in R6-5, and its presence switches off the determinant for tetracycline resistance. Excision of the element results in a shorter plasmid but restores the activity of the determinant for tetracycline resistance.

9. The additional length suggests that an IS element has been inserted in the region of the operon in the case of

both variants. The difference in expression can be explained as a result of differences in the direction of insertion of the element. Insertion in one direction causes transcription of the structural genes, even in the absence of inducer. Insertion in the opposite direction results in complete suppression of transcription from the structural genes.

10. An IS element appears to have been inserted in the region of the operon, resulting in switching off of the structural genes in the operon. The electron microscope reveals the region where the IS element is located, since the strand from the wild type does not bear the element. The "stem" results from the pairing of the inverted repeats at the ends of the single strand of the element. The "loop" contains nonduplicated sequences between the inverted repeats. No pairing of bases occurs in the loop.

11. A. 3′ GGTGATGTA . . . X Y Z . . .
        TACATCACC 5′
        5′ CCACTACAT . . . X′ Y′ Z′ . . .
        ATGTAGTGG 3′

    B.

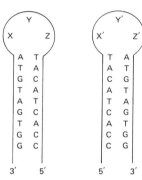

12. Lysogeny will be favored in strains A, D, and H.

13. (1) Red, (2) mottled, (3) mottled, (4) white, (5) red

14. A. The females must have been from the P strain and the males from M, since P cytoplasm has repressor activity that prevents transposition of P. Moreover, M males carry no P elements. If the females were from strain M in the cross, insufficient repressor would be present to prevent transposition of P coming from the males.
    B. M × M and P × P

15. In cells of the pancreas, this sequence does not occur as part of the linker DNA. Nucleosome phasing appears to be the case here. In other types of cells, nucleosome beads appear to be randomly placed, so that the sequence may be associated entirely with the linker. It may also sometimes be associated in part or sometimes in its entirety with the octamer. Only that part of the DNA associated with the bead will be protected from digestion.

16. The promoter region of this gene would not be expected to be tightly bound to the octamer, because an active promoter undoubtedly must interract with various regulatory molecules governing transcription. If it were bound to the bead, such interractions would be curtailed. To determine whether or not it is tightly bound, expose the chromatin from cells in which the gene is active to a nuclease. If the region is not closely associated with the bead, it should be highly sensitive to nicking with an endonuclease. One might also attempt to detect association of the site with HMG.

17. Although the entire gene is intact, an enhancer sequence appears to have been deleted. Without the stimulating effect of the enhancer, transcription from the promoter of the gene is greatly reduced.

18. (1) Inactive, (2) active, (3) inactive, (4) active, (5) active, (6) active

19. 5′ C A C G T C A C G T A C C 3′
    3′ G T G C A G T G C A T G G 5′

20. Two genes are involved. One is coded for the thyroid prehormone. The other gene is a complex transcription unit. Differential processing of the primary transcript occurs in pancreatic and pituitary cells. Coding sequences toward the first half of the transcript from its 5′ end are preserved in both kinds of cells, so that the amino acids toward the amino end of the two polypeptides are the same. However, the exons that are retained toward the 3′ end of the transcript must be different in the two cell types. Coding sequences discarded during processing of the pancreatic mRNA would be retained in processing of the pituitary mRNA, and vice versa. The result of translation of the two mRNAs resulting from differential processing of the primary transcript is two different proteins, identical in the first 50% of their amino acids but different in the latter portion.

21. A. Incorporate labeled uridine into the food. Newly synthesized RNA will be labeled, and this can be visualized by preparation of autoradiographs made of the larval salivary gland chromosomes. If RNA synthesis and puffing are associated, dots will appear only over the puffs and not over unpuffed regions.
    B. Like most steroids, ecdysone binds to protein receptors. The hormone–receptor complex binds to the chromatin and turns on transcription at the site of binding. The hormone effects cell differentiation by turning on certain genes and not others, causing different gene products to be made at different times in development. Depending on these products, different kinds of cells may originate.

22. We see that the protooncogene is present in normal cells but that the viral DNA is not. Both viral and p-onc DNA are present in the tumor cells. The viral DNA must have integrated near p-onc in the same restriction fragment, since the fragment containing p-onc is longer in the tumor cells than in the normal ones.

23. DNA of the protooncogenes is present in both normal and transformed cells, but there is no evidence that the viral DNA has integrated with the cellular DNA in the tumor cells.

CHAPTER 18

1. Somewhat under 20%. This is so because only half of the DNA would be involved in transcription. Of this amount, not all the DNA would be transcribed, such as the promoter and the terminator. Of that part that is transcribed, a portion is cleaved from the final RNA product. Only 86 nucleotides are present in the final product.

2. Alu B and Alu D are adjacent to each other. One site is present that is recognized by Hae, and it is in fragment E.

3. As indicated in the following PstI could cut toward either the right or left of the sequence to produce fragments 6.4 and 3.6 kbp in length.

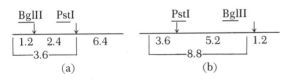

We must now locate the BglII cutting site in relation to the PstI site. Placing the BglII site nearer the left end of the sequence (a) would result in a double enzyme digest of fragments 1.2, 2.4, and 6.4 kbp long. Placing the BglII site nearer the right end, the double digest will yield fragments that are 3.6, 5.2, and 1.2 kbp long, as shown in b. This is what was obtained, and therefore the map shown in b is consistent with the results.

4.

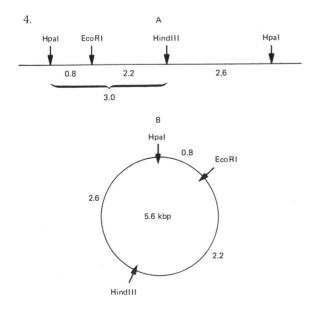

5. Denature the double-stranded circular DNA of the mutant virus to yield two single-stranded circles. Expose DNA from wild-type virus to HpaI, which will open the circle, and then to both HindIII and EcoRI. This yields the 2.6, 2.2, and 0.8-kbp-long fragments that can be separated from one another by gel electrophoresis. A fragment of a given length, let us say the 2.6-kbp fragment, can then be denatured to the single-stranded state. These single strands are incubated with the single-stranded circles from the mutant under conditions permitting DNA replication. If the resulting viruses all prove to be mutant, we know that the fragment did not cover the mutant site, which must be located elsewhere. Conversely, origin of wild-type viruses following DNA replication would indicate that the fragment does cover the mutant site, because wild-type viruses can now arise by replication that may entail mismatch repair. The mutant site would thus be localized to the 2.6-kbp region.

6. 5′ GGTCCAGGACG 3′
   3′ CCAGGTCCTGC 5′

7. G   A G   T C   C

   (gel electrophoresis band diagram)

8. A. The derived DNA is

   3′ GGTAAAGCTTACCTCGTT 5′
   5′ CCATTTCGAATGGAGCAA 3′

   Terminal transferase can be used to add projecting ends of T at the 3′ ends to the DNA piece:

   3′ TTTTTTGGT...GTT 5′
              5′CCA...CAATTTTTT 3′

   B. Render the plasmid linear by subjecting it to a restriction endonuclease that cleaves it at only one site. Then supply the plasmid with projecting ends of A at its 3′ ends using terminal transferase and adenosine triphosphate.

9. C T A
   G A T G C T T A A G C
   and

                              C G A A T T C G T C T
                                        A G A

10. A. $\underline{G\ A\ C\ T\ A}$
$\underline{C\ T\ G\ A\ T\ A\ A\ T\ G\ C\ A\ T\ T}$
and
$\underline{T\ T\ A\ C\ G\ T\ A\ A\ T\ A\ G\ C\ A}$
$\underline{A\ T\ C\ G\ T}$

B. No. Different palindromes are involved, and the ends generated in the plasmid are not complementary to those in the DNA segments of Question 9.

11. A. Clones 1 and 2 contain the desired gene. The absence of precipitate in these two systems indicates that the transcript for the enzyme is unavailable for translation. Therefore, no polypeptide is present to react with the antibodies. The mRNA for the specific enzyme Z has complexed with its template DNA, the gene for enzyme Z. Presence of precipitate in the other three systems means the mRNA for enzyme Z is available for translation. Its template DNA, the desired gene, must not be present in systems 3, 4, and 5 to remove it by pairing with it.

B. It is most likely, because an assortment of transcripts was added. Most of these are free to undergo translation, because their template DNA is not present. No precipitate forms in systems 1 and 2, since the antibody is specific only for Enzyme Z which is not being formed. Other protein can be present that would not react with the antibodies.

12. A. 5′ GATCCTTCGACGTT 3′
B. 5′ GATCCTGATA 3′
C. 5′ TATCAGGATCCTTCGACGTT 3′
3′ ATAGTCCTAGGAAGCTGCAA 5′
D. 5′ TATCAG 3′
3′ ATAGTCCTAG 5′ and
5′ GATCCTTCGACGTT 3′
3′ GAAGCTGCAA 5′

13. A. Colonies 4, 6, 7, and 9, since these are sensitive to both drugs. If plasmid DNA with two intact resistance genes were present, the host cells would be resistant to both drugs. If recombinant DNA is present, the host cells would be resistant to tetracycline but sensitive to ampicillin.

B. Colonies 1, 3, and 5, since these are resistant to both drugs.

C. Colonies 2 and 8, since they are resistant only to tetracycline, indicating that they have the plasmid with the intact tet-r marker and that mouse DNA may have been inserted into the cut amp-r gene, rendering it incapable of conferring resistance to ampicillin.

14. (1) Imprint colonies from the master plate onto filter paper. (2) Break open cells to extract the DNA. (3) Denature the DNA on the paper, retaining the original orientation of the colonies. (4) Add a radioactively labeled probe of thymidine kinase mRNA or of radioactively labeled single strands of the cloned thymidine kinase gene, and permit molecular hybridization to take place. (5) Wash away any unhybridized probe molecules. (6) Prepare an autoradiograph of the materials on the filter paper. (7) Any colony with radioactivity is then revealed as one with DNA sequences complementary to the labeled probe that was added (the known thymidine kinase sequences).

15. On Chromosome 8. Only when 8 is present do we see the human plus mouse pattern. In the absence of 8, only the mouse pattern is seen.

CHAPTER 19

1. The cDNA probe specific for the 5′ portion reveals no difference between the DNA from the two sources. However, the probe for the 3′ end detects one fragment in unafflicted persons and two smaller ones in DNA from those with the disorder. The affliction appears to be associated with a gene mutation that has created a restriction enzyme site in the 3′ end of the gene. This indicates a substitution of one nucleotide base pair for another and an amino acid substitution in the alpha-globin chain toward the carboxyl end.

2. A. 20, 15, 6, and 2 kbps.
B. Two haplotypes are possible. We can call one A and the other B, or vice versa. For example, A—20 + 6 + 2 kbp fragments; B—15 + 6 + 2 kpb fragments.
C. Three genotypes are possible: AA, AB, and BB.
D. An autoradiograph of genotype AA would reveal fragments of 20, 6, and 2 kbps. One of genotype BB would show 15 + 6 + 2 kbps, and one of AB would reveal 20, 15, 6, and 2 kpb fragments.

3. A. Four haplotypes are possible, which can be designated as follows: A—20 + 8 kbps; B—20 + 6 + 2 kbps; C—15 + 8 kbps; D—15 + 6 + 2 kbps.
B. Ten are possible. AA, AB, AC AD, BB, BC, BD CC, CD, DD.

4. Four genotypes are possible among the offspring in family 1: AC, AD, CD, and DD. If haplotype D is present, fragments of both 15.0 and 4.9 kbp must also be present. Therefore, person 2 cannot have haplotype D and must be genotype AC, and is consequently at low risk because the recombination frequency is low for the region. Since person 1 shows all four possible fragments, this indicates the genotype is AD. However, the person must have received haplotype D from the unafflicted parent and A from the afflicted one. Again, the person would be at little risk.

In family 2, the possible genotypes are CC, CD, and DD. Person 2 is at very high risk, because the genotype must be DD, and the person must have received one haplotype D from the afflicted parent. Person 1 is at risk, because the genotype must be CD, and there is a 50% chance that the D haplotype was contributed by the afflicted parent. Person 3 is at little risk, because the genotype must be CC.

5. T-helper cells play a pivotal role in coordinating the interactions among various cell types involved in the immune response. Helpers are needed to stimulate other T cells against a particular antigen. It is the T-helper cells that cause certain B cells to divide and complete their differentiation to circulating memory cells, which are on guard against a specific antigen. Other B lymphocytes differentiate to become the antibody-secreting plasma cells after stimulation by T helpers. T cells are also critical to the interaction between T suppressor and active T-killer cells. Lacking T helpers, for example, the suppressors may continue to cut down the activity of the killer cells. Hence, without T-helper cells, activities of B lymphocytes and other T cells go awry, resulting in a lack of both sufficient antibodies and killer-T cells to combat foreign antigens.

6. C gene DNA is seen to be associated with two fragments in the plasma cells, one the same size as the fragment in the nonantibody-producing (fibroblast) cells. However, C gene DNA in plasma cells is also associated with a much smaller fragment, indicating that rearrangement or shuffling of some of the DNA has taken place in the development of these cells. The shuffling may have altered a restriction site on one of the chromosomes of a plasma cell, with the result that C gene DNA becomes part of a smaller fragment. Because of allelic exclusion, the C gene DNA on the other chromosome in each plasma cell will remain in the same (original) arrangement as in nonantibody-producing cells, and will thus be associated with the larger fragment.

7. (1) Light, (2) both, (3) heavy, (4) both, (5) heavy, (6) both, (7) light, (8) both, (9) both, (10) heavy, (11) heavy, (12) both, (13) heavy, (14) both

8. V_{18} J_1 C_1; V_{18} J_2C_1; V_{18} J_3 C_1; V_6 J_1 C_1; V_6 J_2 C_1; V_6 J_3 C_1

9. V_2 D_1 J_1 C_1; V_2 D_1 J_2 C_1; V_2 D_2 J_1 C_1; V_2 D_2 J_2 C_1; V_2 D_1 J_1 C_2; V_2 D_1 J_2 C_2; V_2 D_2 J_1 C_2; V_2 D_2 J_2 C_2

10. In the maturation of plasma cells, additional gene shuffling occurs in which heavy chain class switching takes place. Before chain class switching, a specific VDJ combination has been brought adjacent to the cluster of C (constant) region genes. In the cluster, the C genes occur in the following order: mu, delta, gamma 3, gamma 1, gamma 2b, gamma 2a, epsilon, alpha. Mu and delta are immediately adjacent to VDJ, and these two classes of chains can be produced. If further rearrangement takes place, deletion may remove one or more C genes. The class of chain produced depends on the C genes that are deleted. If mu, delta, and gamma 3 are deleted, then gamma 1 will be directly next to the specific VDJ combination. Chains of class gamma 1 will then be produced following transcription of the entire contiguous region and differential processing of the transcript. If the region mu through gamma 2a is deleted in the class switching, then the epsilon gene will be placed next to the VDJ combination, and the cell will secrete

epsilon class chains following formation of pre-mRNA and its differential processing to form mature mRNA. Chains of class alpha can be produced when all the genes except C alpha are deleted.

11. A. (1) V_{50} D_3 J_2 (S) gamma (S) epsilon (S) alpha
     (2) V_{50} D_3 J_2 (S) epsilon (S) alpha
     (3) V_{50} D_3 J_2 (S) alpha
   B. Answers are the same as in Part A, because the entire C cluster adjoining the VDJ segment can be transcribed.
   C. (1) V_{50} D_3 J_2 gamma—codes for IgG
      (2) V_{50} D_3 J_2 epsilon—codes for IgE
      (3) V_{50} D_3 J_2 alpha—codes for IgA

12. The order must be 3.0, 2.3, 6.6, 4.5. This can be deduced because the "beginning" of the delta gene and the "end" of the beta gene are present in persons with hemoglobin Lepore. The other persons are known to have an alteration that eliminates an EcoR1 cutting site in the beta gene only. The 3.0 fragment is present in the DNA from these individuals. This must be the start of the delta gene. The 4.5 fragment that includes the end of the beta gene must be part of the 11.1 fragment. The 6.6 fragment must make up the rest of this large fragment, and it must include the beginning of the beta gene. The 2.3 fragment is present in these persons, and it must include the end of the delta gene.

CHAPTER 20

1. A. KK, killers
   B. KK, sensitives
   C. Two KK killers or two kk sensitives
   D. Two KK sensitives or two kk sensitives

2. A. KK and KK
   B. KK and kk
   C. Kk and Kk

3. A. KK sensitives from the sensitive and KK killers from the killer
   B. Kk sensitives from the sensitive and Kk killers from the killer

4. A. Zygotes and ascospores petite
   B. Zygotes wild. Ascospores 2 wild: 2 petite
   C. $2^+: 2^-$

5. Two classes of zygotes are to be expected: those that, upon sporulation, will yield spores that are all drug resistant and those that will yield spores that are all sensitive.

6. A. Zygote: CcDdPpQq. Products of zygote: (1) CDPq, (2) CdPq, (3) cDPq, (4) cdPq. Each cell type will produce cells of the same type.
   B. Zygote: CcDdPpQq. Products of zygote: (1) CDPpQq, (2) CdPpQq, (3) cDPpQq, (4) cdPpQq. Type 1 will give rise to CDPQ, CDPq; CdPQ; CdPq. Type 2 will

give rise to CdPQ; CdPq; CdpQ; Cdpq. Type 3 will give rise to cDPQ; cDPq; cDpQ; cDpq. Type 4 will give rise to cdPQ; cdPq; cdpQ; cdpq.

7. The transmission of growth habit is non-Mendelian. The reciprocal crosses give different results. In each case, growth habit is like that of the female parent, whereas the other traits show Mendelian inheritance.

8. D would appear to be in the nuclear DNA, because it is present in the neutral petite, which lacks mt DNA. The other markers would appear to be in the mt DNA. C and E are lost together in one strain and appear to be adjacent. The same is true for B and F in the other strain.

9. Because the genetic sequences coded in the mt DNA for the polypeptide chains of various respiratory enzymes are not localized in one part of the mt DNA. The map indicates that they are in various locations around the circular DNA.

10. In the protozoan, chains I, IV, and VI must be made in the cytosol, since erythromycin inhibits the ribosomes of the organelles. The remaining four chains are made in the mitochondria, since cycloheximide inhibits translation by the ribosomes of the cytosol. In the mold, the three small chains are made in the cytosol and the four large ones in the organelles.

11. A large transcript will be formed as a result of transcription of the entire DNA strand. Processing of the transcript will result in cleavage of the enzyme sequence from the tRNA sequence. The enzyme transcript will then terminate in a U. The transcript will next be extended by posttranscriptional addition of a poly (A) tail. The solitary U can finally be recognized as part of a chain-terminating codon, UAA.

12. The mutation affecting the human mt has the greater chance of altering some product, because human mitochondrial genes are packed closely together with little noncoding DNA present. This contrasts with yeast mt DNA, which contains an appreciable amount of DNA that is apparently not coded for specific products.

13. Three violations are seen: TGA should indicate a chain-terminating codon (UGA), ATA (AUA) isoleucine, and CTA (CUA) leucine.

14. A, C, and D

15. A, C, and D

16. C and E

17. B and D

18. C, D, and E

CHAPTER 21

1. $M = 0.8$; $N = 0.2$

2. 178 out of 8100

3. A. The frequency of the recessive allele must be equal to $\sqrt{0.30}$, or about 54.8%; that of the dominant must therefore be less, about 45.2%.
   B. The allele will not increase simply because it is dominant. Unless some selective value is imparted to T or to t, the frequencies of the alleles will remain the same except for chance fluctuations.

4. 98/2500 heterozygotes and 2401/2500 homozygotes for the wild allele
   B. 98/2500 divided by $98/2500 + 2401/2500 = 2/51$

5. A. $d = 1/20$; $D = 19/20$
   B. 38/400
   C. 1/400

6. A. 1/200,000
   B. 1/450
   C. $2 \times 1/450 \times 449/450 =$ about 1/220

7. $q^2 = 1/2500$; $q = 1/50 = 0.02$. $p = 0.98$. Frequency of heterozygotes $= 2 \times 0.02 \times 0.98 = 0.0392$. Probability of two heterozygotes mating $= (0.0392)^2 = 0.0015$. Probability of recessive trait appearing $= 1/4$. $1/4 \times 0.0015 = 0.000375 = 375$ per 1,000,000

8. A. Frequency of $R = 920 + 1140 = 2060$.
   $2060/4040 = 0.51$
   Frequency of $r = 840 + 1140 = 1980$.
   $1980/4040 = 0.49$
   B. Red: $(.51)^2 \times 4040 = 1050$
   Pink: $(2 \times 0.51 \times 0.49) \times 4040 = 2019$
   White: $(.49)^2 \times 4040 = 970$

Expected	o	(d)	d^2	d^2/e
Red 1050	920	−130	16,900	16.1
Pink 2019	2280	261	68,121	33.7
White 970	840	−130	16,900	17.4

$$\Sigma d^2 e = 67.2$$

Probability value is less than 1% and indicates that this population is not at equilibrium for this pair of alleles. One or more conditions required for an idealized population are not being met.

9. A. $AA = 0.64/.968 = 0.661$; $Aa = 0.32/0.968 = 0.331$; $aa = 0.008/0.968 = 0.008$. (The figure $0.968 = \overline{W}$, the sum of the proportionate contributions of each genotype after each initial genotypic frequency has been multiplied by W or $1 - s$.)
   B. Frequency of $A = 0.661 + 0.166 = 0.83$. Frequency of $a = 0.008 + 0.166 = 0.17$.

C. $\Delta q = q_1 - q_0$. The frequency of a before selection $(q_0) = 0.2$. Therefore $\Delta q = 0.17 - 0.20 = -0.03$.

10. A. $AA = 0.392/0.83 = 0.472$; $Aa = 0.42/0.83 = 0.506$; $aa = 0.018/0.83 = 0.022$
    B. Frequency of $A = 0.472 + 0.253 = 0.725$. Frequency of $a = 0.022 + 0.253 = 0.275$

11. Before selection, frequency of $A = 0.6$ and of $a = 0.4$. After selection, frequency of $A = 0.28 + \frac{1}{2}(0.51) = 0.535$, and frequency of $a = 0.21 + \frac{1}{2}(0.51) = 0.465$. Change in frequency of $A = 0.535 - 0.6 = -0.065$. Change in frequency of $a = 0.465 - 0.4 = 0.065$.

12. A. Frequency of lethal recessive $= 0 + .165 = 0.165$.
    B. No. The effect of selection decreases when the value of q (l in this case) becomes very small. Full elimination will not take place, as l will be added back to the gene pool through mutation.

13. $n = 1/q_n - 1/q_o = 1/0.000002 - 1/0.00002 = 500,000 - 50,000 = 450,000$

14. $q^2 = 1/200,000$. $q = 1/447$, $p = 446/447$. Heterozygotes $= 2 \times 1/447 \times 446/447 = 892/199809 = 1/224$

15. Letting $A = p$ and $a = q$, then $\hat{p} = t/s + t = 0.9/1.2 = 0.75 =$ equilibrium value for A. $\hat{q} = 0.3/1.2 = 0.25 =$ equilibrium value for a.

16. A. $\hat{p} = v/u + v = 0.00001/0.00005 = 0.2$, $\hat{q} = u/u + v = 0.00004/0.00005 = 0.8$
    B. The mutant allele, since forward mutation usually has a higher value than back mutation. Thus, in absence of selective forces, for example, the mutant allele would assume a high value.

17. $q_0 = 0.1$; $p_m = 0.6$; $q_m = 0.4$; $m = 0.3$; $1 - m = 0.7$. Thus, $q_1 = q_0(1 - m) + q_m m = 0.1\ (0.7) + 0.4\ (0.3) = 0.07 + 0.12 = 0.19$. $\Delta q = 0.09$

18. A. $\sigma_q = \sqrt{\dfrac{pq}{2N}}$ population $1 = \sqrt{\dfrac{0.7 \times 0.3}{1500}} = 0.012$

    population $2 = \sqrt{\dfrac{0.7 \times 0.3}{500}} = 0.02$

    B. It is greater in population 2, where the chance is 68% that a possible value of q will be 0.3 ± 0.02. In population 1, the chance is 68% that a possible value of q will be 0.3 ± 0.01.

19. A. Population A: $\dfrac{1}{2N_e} = 1/800 = 0.00125$

    Population B: $\dfrac{1}{2N_e} = 1/8000 = 0.000125$

    B. For A, 1/800; for B, 1/8000

20. This can be explained on the basis of the founder effect. The undesirable allele, though low in the ancestral European population, was undoubtedly present in some of the few members who founded the American population. This would have been a chance event. Those who founded the Australian population happened to be free of the allele.

21. Two species can be recognized. Populations B and C, though phenotypically distinct, have not accumulated sufficient genetic differences to limit free exchange of genetic material. The two populations may be considered varieties of the same species (let us say, species B). Population A is genetically isolated from members of the other two populations (members of species B), as seen by the lack of fertile hybrids when A members are crossed to B and C members. Population A is thus composed of individuals of a different species (let us say, species A).

22. A. The F_1 would have 14 chromosomes (7A 7B). A tetraploid could be derived by the union of unreduced gametes (7A 7B). The tetraploid would have 28 chromosomes (7A 7A 7B 7B). Fourteen bivalents can now form.
    B. Yes. The tetraploid is fertile and when crossed to either parental species will produce offspring with 21 chromosomes (7A 7A 7B or 7A 7B 7B). These are sterile triploids with one set of chromosomes unpaired. The tetraploid would thus be reproductively isolated from either parental species.
    C. This would result in an autoallooctoploid: 7A 7A 7A 7A 7B 7B 7B 7B. This can form as many as 28 bivalents or as many as 14 quadrivalents.

23. A very infertile F_1 with 12 chromosomes (6A 6B) could be produced that gives rise to a tetraploid with 24 chromosomes as a result of the union of unreduced gametes. A cross between this tetraploid (6A 6A 6B 6B) and species C could produce a sterile triploid (6A 6B 6C). From gametes of this plant containing all 18 chromosomes, an allohexaploid could result with 36 chromosomes: 6A 6A 6B 6B 6C 6C. At meiosis 18 bivalents can form.

24. Genetic polymorphism exists at the enzyme locus in population A, as indicated by the variation seen in the enzyme from one individual to the next. The enzyme appears to be composed of two polypeptide chains encoded by a specific gene. The gene exists in three allelic forms, let us say A1, A2, and A3. This makes possible six different genotypes. Population B, on the other hand, is monomorphic with respect to the gene that codes for the enzyme, because only one enzyme form has been detected.

INDEX